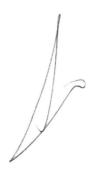

Quadrature Amplitude Modulation

Quadrature Amplitude Modulation

From Basics to Adaptive Trellis-Coded,

Turbo-Equalised and Space-Time Coded

OFDM, CDMA and MC-CDMA Systems

Second Edition

L. Hanzo
University of Southampton, UK

S.X. Ng
University of Southampton, UK

T. Keller
UbiNetics, UK

W. Webb
Ofcom, UK

IEEE PRESS

IEEE Communications Society, Sponsor

John Wiley & Sons, Ltd

Other Wiley Editorial Offices

John Wiley & Sons Inc., 111 River Street, Hoboken, NJ 07030, USA

Jossey-Bass, 989 Market Street, San Francisco, CA 94103-1741, USA

Wiley-VCH Verlag GmbH, Boschstr. 12, D-69469 Weinheim, Germany

John Wiley & Sons Australia Ltd, 33 Park Road, Milton, Queensland 4064, Australia

John Wiley & Sons (Asia) Pte Ltd, 2 Clementi Loop #02-01, Jin Xing Distripark, Singapore 129809

John Wiley & Sons Canada Ltd, 22 Worcester Road, Etobicoke, Ontario, Canada M9W 1L1

Wiley also publishes its books in a variety of electronic formats. Some content that appears in print may not be
available in electronic books.

IEEE Communications Society
Sponsor COMMS-S Liaison to IEEE Press, Mostafa Hashem Sherif

British Library Cataloguing in Publication Data

A catalogue record for this book is available from the British Library

ISBN 0470 09468 0

Typeset by the author using LaTex software
Printed and bound in Great Britain by Antony Rowe Ltd, Chippenham, Wiltshire.
This book is printed on acid-free paper responsibly manufactured from sustainable forestry
in which at least two trees are planted for each one used for paper production.

Contents

About the Authors

**Lajos Hanzo**, Fellow of the Royal Academy of Engineering (FREng), received his Master degree in electronics in 1976 and his doctorate in 1983. In 2004 he was awarded the Doctor of Sciences (DSc) degree by the University Southampton, UK. During his 28-year career in telecommunications he has held various research and academic posts in Hungary, Germany and the UK. Since 1986 he has been with the Department of Electronics and Computer Science, University of Southampton, UK, where he holds the chair in telecommunications. He has co-authored 10 John Wiley/IEEE Press books totalling about 8000 pages on mobile radio communications, published in excess of 500 research papers, organised and chaired conference sessions, presented overview lectures and been awarded a number of distinctions. Currently he is managing an academic research team, working on a range of research projects in the field of wireless multimedia communications sponsored by industry, the Engineering and Physical Sciences Research Council (EPSRC) UK, the European IST Programme and the Mobile Virtual Centre of Excellence (VCE), UK. He is an enthusiastic supporter of industrial and academic liaison and he offers a range of industrial courses. Lajos is also an IEEE Distinguished Lecturer of both the Communications Society and the Vehicular Technology Society as well as a Fellow of both the IEEE and IEE. For further information on research in progress and associated publications, please refer to http://www-mobile.ecs.soton.ac.uk

**Dr Soon Xin Ng** received a first-class B.Eng. degree in Electronics Engineering and the Ph.D. degree in mobile communications from the University of Southampton, U.K, in July 1999 and December 2002, respectively. Currently, he is continuing his research as a postdoctoral research fellow at the University of Southampton, U.K. His research interests are mainly in adaptive coded modulation, channel coding, turbo coding, space-time coding and joint source and channel coding. He published numerous papers in this field.

Thomas Keller studied Electrical Engineering at the University of Karlsruhe, the Ecole Superieure d'Ingenieurs en Electronique et Electrotechnique, Paris, and the University of Southampton. He graduated with a Dipl.-Ing. degree in 1995. Between 1995 and 1999 he was with the Wireless Multimedia Communications Group at the University of Southampton, where he completed his PhD in mobile communications. His areas of interest include adaptive OFDM transmission, wideband channel estimation, CDMA and error correction coding. He recently joined Ubinetics, Cambridge, UK, where he is involved in the research and development of third-generation wireless systems. Dr. Keller has co-authored two monographs and about 30 various research papers.

Professor William Webb Head of R&D, Ofcom. William joined Ofcom as Head of Research and Development and Senior Technologist in 2003. Here he manages a team of 35 engineers providing technical advice and managing research across all areas of Ofcom's regulatory remit. He also leads some of the major reviews conducted by Ofcom. Previously, William worked for a range of communications consultancies in the UK in the fields of hardware design, computer simulation, propagation modelling, spectrum management and strategy development. William also spent three years providing strategic management across Motorola's entire communications portfolio, based in Chicago.

William has published seven books, sixty papers, and four patents. He is a Visiting Professor at Surrey University. He is a Fellow of the IEE where he sits on the Publications Board, chairs the Advisory Committee for the Communications Magazine and is a Series Editor for the IEE Telecommunications Books. His biography is included in multiple "Who's Who" publications around the world. He sits on the judging panels of the Wall Street Journal "European Innovation Awards" and the GSM Association annual awards.

William has a first class honours degree in electronics, a PhD and an MBA. He can be contacted at william.webb@ofcom.org.uk.

Related Wiley and IEEE Press Books [1]

- R. Steele and L. Hanzo (ed): *Mobile Radio Communications: Second and Third Generation Cellular and WATM Systems*, John Wiley and IEEE Press, 2nd edition, 1999, 1064 pages

- L. Hanzo, F.C.A. Somerville and J.P. Woodard: *Voice Compression and Communications: Principles and Applications for Fixed and Wireless Channels*; IEEE Press and John Wiley, 2001, 642 pages

- L. Hanzo, P. Cherriman and J. Streit: *Wireless Video Communications: Second to Third Generation and Beyond*, IEEE Press and John Wiley, 2001, 1093 pages

- L. Hanzo, T.H. Liew and B.L. Yeap: *Turbo Coding, Turbo Equalisation and Space-Time Coding*, John Wiley and IEEE Press, 2002, 751 pages

- J.S. Blogh and L. Hanzo: *Third-Generation Systems and Intelligent Wireless Networking: Smart Antennas and Adaptive Modulation*, John Wiley and IEEE Press, 2002, 408 pages

- L. Hanzo, C.H. Wong and M.S. Yee: *Adaptive Wireless Transceivers: Turbo-Coded, Turbo-Equalised and Space-Time Coded TDMA, CDMA and OFDM systems*, John Wiley and IEEE Press, 2002, 737 pages

- L. Hanzo, M. Münster, B.J. Choi and T. Keller: *OFDM and MC-CDMA for Broadband Multi-user Communications, WLANs and Broadcasting*, John Wiley and IEEE Press, May 2003, 980 pages

- L. Hanzo, L-L. Yang, E-L. Kuan and K. Yen: *Single- and Multi-Carrier CDMA: Multi-User Detection, Space-Time Spreading, Synchronisation, Standards and Networking*, John Wiley and IEEE Press, June 2003, 1060 pages

---

[1] For detailed contents and sample chapters, please refer to http://www-mobile.ecs.soton.ac.uk

Preface

Since its discovery in the early 1960s, quadrature amplitude modulation (QAM) has continued to gain interest and practical application. Particularly in recent years many new ideas and techniques have been proposed, allowing its employment over fading mobile channels. This book attempts to provide an overview of most major QAM techniques, commencing with simple QAM schemes for the uninitiated, while endeavouring to pave the way towards complex, rapidly evolving areas, such as trellis-coded pilot-symbol and transparent-tone-in-band assisted schemes, or arrangements designed for wide-band mobile channels.

The rationale of the book is that at the time of writing virtually all wireless communications systems, including the family of digital audio and video broadcast systems, wireless local area networks and even wide-area mobile communications systems, such as the High Speed Data Packet Access (HSDPA) mode of the third-generation mobile radio systems as well as the next-generation proposals employ QAM schemes for the sake of increasing the achievable effective throughput. Part I constitutes a light-hearted yet rigorous textbook, while Parts II-IV are targeted at the more advanced reader, providing a research-oriented outlook using a variety of novel QAM-based single- and multi-carrier arrangements in the spirit of the turbo-detection era.

More explicitly, Part I - constituted by Chapters 1-8 - is a rudimentary introduction for those requiring a background in the field of modulation, radio wave propagation, clock- and carrier recovery, trained as well as blind channel equalisers and classic QAM modems designed for telephone lines and other Gaussian channels. Readers familiar with the fundamentals of QAM and the characteristics of propagation channels, as well as with Nyquist pulse shaping techniques may decide to skip Chapters 1-5. Commencing with Chapter 6, each chapter describes individual aspects of QAM. Readers wishing to familiarize themselves with a particular subsystem, including clock and carrier recovery, equalisation, standardised trellis coded telephone-line modem features, etc. can turn directly to the relevant chapters, whereas those who desire a more complete treatment might like to read all the introductory chapters.

Part II, including Chapters 9-14, are concerned with single-carrier QAM-based transmissions over fading mobile radio channels. These chapters are dedicated to the more advanced reader. Specifically, Chapter 9 concentrates mainly on coherent QAM schemes, including reference-aided transparent-tone-in-band and pilot-symbol assisted modulation arrangements. By contrast, Chapter 10 focuses on low-complexity differentially encoded QAM schemes and on their performance achieved both with and without forward error correction coding and trellis coded modulation. Chapter 11 details various timing recovery schemes, while Chapter 12 is dedicated to high-rate wide-band transmissions over dispersive channels and proposes a novel equaliser arrangement. Various so-called Q^2AM orthogonal signaling techniques are proposed in Chapter 13, while the spectral efficiency of QAM in co-channel interference-limited cellular frequency re-use structures is characterised in Chapter 14, which concludes Part II of the book.

Part III of the book concentrates on multi-carrier modulation. Specifically, following a rudimentary introduction to Orthogonal Frequency Division Multiplexing (OFDM) in Chapter 15, Chapters 16-22 detail a range of implementational and performance aspects of powerful OFDM schemes designed primarily for wideband fading channels. Chapter 16 is concerned with transmission over Gaussian channels, Chapter 17 with wideband fading channels, while the focus of Chapter 18 is frequency- and time-domain synchronisation. Our discussions evolve further in Chapter 19 to sophisticated turbo-coded single- and multi-user adaptive OFDM schemes. The subject of Chapters 20 and 21 is the employment of both channel coded as well as space-time coded adaptive OFDM schemes, with the aim of achieving the highest possible effective throughput. Part III of the monograph is concluded with a detailed study on the design of optimum thresholds for adaptive modulated systems and with the comparison of space-time coded adaptive OFDM and frequency-domain-spread MC-CDMA using multiuser detection.

Finally, Part IV of the book first appeared in this new edition and contains about 250 pages new research results, which have not been published elsewhere in the literature. This part commences in Chapter 23 with a theoretical discussion on the capacity of wireless channels. The discussions continue by contriving sophisticated iterative coded modulation systems, such as TCM, TTCM, BICM, BICM-ID designed for turbo-detected QAM-based space-time coded OFDM and CDMA systems operating over wireless channels, which are characterised in Chapters 24-29. Chapters 28 and 29 considered TCM, TTCM, BICM, BICM-ID aided QAM CDMA systems, in the latter one additionally invoking also space-time coding. Finally, Chapter 24 concentrates on the performance aspects of various standard-compliant and enhanced OFDM-based Digital Video Broadcasting (DVB) systems designed for transmission to mobile receivers.

Whilst the book aims to portray a rapidly evolving area, where research results are promptly translated into products, it is our hope that you, the reader, will find this new edition comprehensive, technically challenging and above all, an enjoyable as well as informative read.

Lajos Hanzo
Soon-Xin Ng
Thomas Keller
William Webb

Acknowledgements

The authors would like to express their warmest thanks to Prof. Raymond Steele. Without his shrewd long-term vision the research on single-carrier QAM would not have been performed, and without his earnest exhortations a book on the subject would not have been written. Furthermore, Professor Steele has edited some of the chapters and given advice on the contents and style of this book.

Contributions by Dr. P.M. Fortune, Dr. K.H.H. Wong, Dr. R.A. Salami, D. Greenwood, R. Stedman, R Lucas and Dr. J.C.S. Cheung who were formerly with Southampton University are thankfully acknowledged. We thank Multiple Access Communications Ltd. for supporting the work on QAM, particularly in the framework of the DTI LINK Personal Communications Programme, dealing with high data rate QAM transmission over microcellular mobile channels. Special thanks goes to Peter Gould and Philip Evans for the major part they played in the construction of the star QAM test-bed. We are grateful to John Williams of Multiple Access Communications Ltd. for the production of many of the figures involving simulated waveforms. Much of the QAM work at Multiple Access Communications Ltd. derives from the support of BT Laboratories, Martlesham Heath, the DTI and the Radio Communications Agency. Specifically, we thank the latter for their support of the research on spectral efficiency which facilitated Chapter 14.

Many of the results in Chapters 17-20 are based on our work conducted as a sub-contractor of Motorola ECID, Swindon, UK as part of our involvement in a collaborative Pan-European Wireless Asynchronous Transfer Mode (WATM) project known as Median, which was generously supported by the European Commission (EC), Brussels, Belgium. We would like to acknowledge all our valued friends and colleagues - too numerous to mention individually - who at some stage were associated with the Median consortium and with whom we have enjoyed a stimulating collaboration under the sterling management of IMST, Germany. Our gratitude is due to Andy Wilton and to Paul Crichton of Motorola, who have whole-heartedly sponsored our research. Further thanks are also due to Dr. Joao Da Silva, Bartolome Aroyo, Bernard Barani, Dr. Jorge Pereira, Demosthenes Ikonomou and to the other equally supportive members of the EC's programme management team in Brussels for their enthusiastic support. Furthermore, we enjoyed the valuable support of EPSRC, Swindon UK, and the Mobile VCE, for which we are equally grateful.

Finally, we express our gratitude for the creative atmosphere to our colleagues Derek Appleby, Steve Braithwaite, Sheng Chen, David Stewart, Jeff Reeve, Lie-Liang Yang as well as Stephan Weiss at Southampton University, UK and gratefully acknowledge the stimulating embryonic discussions with Prof. G. Gordos (Technical University of Budapest, Hungary), Prof. H.W. Schüssler (University of Erlangen-Nürnberg, Germany) and Dr.Ing. H.J. Kolb as well as the numerous thought-provoking contributions by many established authorities in the field, who appear also in the Author Index Section of the book.

A number of colleagues have influenced our views concerning various aspects of wireless communications and we thank them for the enlightenment gained from our collaborations on various projects, papers and books. We are grateful to J. Brecht, Jon Blogh, Marco Breiling, M. del Buono, Clare Brooks, Peter Cherriman, Stanley Chia, Byoung Jo Choi, Joseph Cheung, Peter Fortune, Lim Dongmin, D. Didascalou, S. Ernst, Eddie Green, David Greenwood, Hee Thong How, W.H. Lam, C.C. Lee, M.A. Nofal, Xiao Lin, Chee Siong Lee, Tong-Hooi Liew, Matthias Muenster, V. Roger-Marchart, Redwan Salami, David Stewart, Juergen Streit, Jeff Torrance, Spyros Vlahoyiannatos, John Williams, Jason Woodard, Choong Hin Wong, Henry Wong, James Wong, Lie-Liang Yang, Bee-Leong Yeap, Mong-Suan Yee, Kai Yen, Andy Yuen and many others with whom we enjoyed an association. Special thanks are due to Dr. Lie-Liang Yang for his insightful contributions on the theory of the coding schemes used in Chapter 20, to Tong-Hooi Liew for his kind assistance in the preparation of a co-authored paper, which was the basis of Chapter 20, which resulted in a joint journal submission. Similarly, the contributions of Matthias Muenster in Sections 19.3 and 19.7 are thankfully acknowledged along with those of Chee-Siong Lee and Spyros Vlahoyiannatos to the papers, which constituted the basis of Chapter 30. Lorenzo Piazzo's permission to expand the material of his Electronics Letter on optimum power- and bit-allocation of OFDM is thankfully acknowledged. We are also grateful to our editors, Mark Hammond and Sarah Hinton and Daniel Gill at Wiley. Finally, the authors warmly thank Rita Hanzo, Denise Harvey and Dr. Peter Cherriman for their dedicated and skilful assistance in typesetting the manuscript in Latex as well as in amalgamating the new material of the second edition with the first edition.

Lajos Hanzo
Soon-Xin Ng
Thomas Keller
William Webb

Part I

QAM Basics

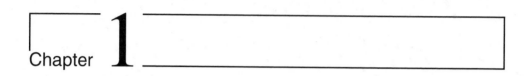

Chapter **1**

Introduction and Background

This book is concerned with the issues of transmitting digital signals via multilevel modulation. We will be concerned with digital signals originating from a range of sources such as from speech or video encoding, or data from computers. A typical digital information transmission system is shown in Figure 1.1. The source encoder may be used to remove some of the redundancy which occurs in many sources such as speech, typically reducing the transmission rate of the source. The forward error correction (FEC) block then paradoxically inserts redundancy, but in a carefully controlled manner. This redundancy is in the form of parity check bits, and allows the FEC decoder to remove transmission errors caused by channel impairments, but at the cost of an increase in transmission rate. The interleaver systematically rearranges the bits to be transmitted, which has the effect of dispersing a short burst of errors at the receiver, allowing the FEC to work more effectively. Finally, the modulator generates bandlimited waveforms which can be transmitted over the bandwidth-limited channel to the receiver, where the reverse functions are performed. While we will discuss all aspects of Figure 1.1, it is the generation of waveforms in the modulator in a manner which reduces errors and increases the transmission rate within a given bandwidth, and the subsequent decoding in the demodulator, which will be our main concern in this book.

It is assumed that the majority of readers will be familiar with binary modulation schemes such as binary phase shift keying (BPSK), frequency shift keying (FSK), etc. Those readers who possess this knowledge might like to jump to Section 1.2. For those who are not familiar with modulation schemes, we give a short non-mathematical explanation of modulation and constellation diagrams before detailing the history of QAM.

1.1 Modulation Methods

Suppose the data we wish to transmit is digitally encoded speech having a bit rate of 16 kbit/s, and after FEC coding the data rate becomes 32 kb/s. If the radio channel to be used is centred around 1 GHz, then in order to transmit the 32 kb/s we must arrange for some feature of a 1 GHz carrier to change at the data rate of 32 kb/s. If the phase of the carrier is switched at the rate of 32 kb/s, being at 0° and 180° for bits having logical 0 or logical 1, respectively, then

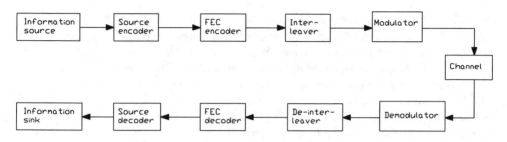

Figure 1.1: Typical digital information transmission system

there are $1 \times 10^9 / 32 \times 10^3 = 31,250$ radio frequency (RF) oscillations per bit transmitted. Figure 1.2(a) and (b) show the waveforms at the output of the modulator when the data is a logical 0 and a logical 1, respectively. On the left is the phasor diagram for a logical 0

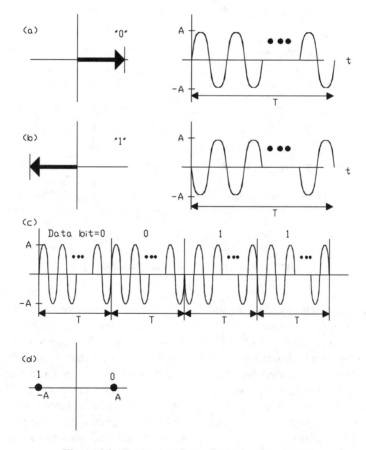

Figure 1.2: Carrier waveforms for binary input data

and a logical 1 where the logical 0 is represented by a phasor along the positive x-axis, and the logical 1 by a phasor along the negative x-axis. This negative phasor represents a phase shift of the carrier by 180°. Figure 1.2(c) shows the modulator output for a sequence of data bits. Note that no filtering is used in this introductory example. Here the waveform can be seen to change abruptly at the boundary between some of the data symbols. We will see later that such abrupt changes can be problematic since they theoretically require an infinite bandwidth, and ways are sought to avoid them. Figure 1.2(d) is called a constellation diagram of phasor points, and as we are transmitting binary data there are only two points. As these two points are at equal distance from the origin we would expect them to represent equal magnitude carriers, and that the magnitude is indeed constant can be seen in Figure 1.2(c). The bandwidth of the modulated signal in this example will be in excess of the signalling

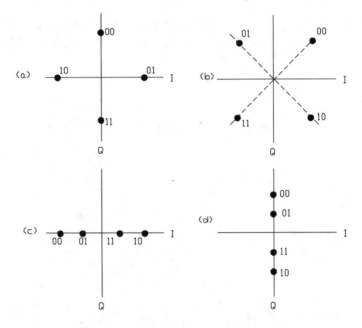

Figure 1.3: Examples of four-level constellations

rate of 32 kb/s due to the sudden transition between phase states. Later we will consider the bandwidth of modulated signals in-depth. Suffice to say here that if we decreased the signalling rate to 16 kb/s the bandwidth of the modulated signal will decrease. If the data rate is 32 kb/s, and the signalling rate becomes 16 kb/s, then every symbol transmitted must carry two bits of information. This means that we must have four points on the constellation, and clearly this can be done in many ways. Figure 1.3 shows some four-point constellations. The two bits of information associated with every constellation point are marked on the figure. In Figure 1.3(a) and (b) so-called quadrature modulation has been used as the points can only be uniquely described using two orthogonal coordinate axes, each passing through the origin. The orthogonal coordinate axes have a phase rotation of 90° with respect to each other, and hence they have a so-called quadrature relationship. The pair of coordinate axes can be associated with a pair of quadrature carriers, normally a sine and a cosine waveform, which can be independently modulated and then transmitted within the same frequency band. Due to their orthogonality they can be separated by the receiver. This implies that whatever symbol is chosen on one axis (say the x-axis in the case of Figure 1.3(a)) it will have no effect on the data demodulated on the y-axis. Data can therefore be independently transmitted via these two quadrature or orthogonal channels without any increase in error rate, although some increase may result in practice and this is considered in later chapters. We have used I and Q to signify the in-phase and quadrature components, respectively, where in-phase normally represents the x-axis and quadrature the y-axis. Figure 1.3(c) and (d) show constellations where the four points are only on one line. These are not quadrature constellations but actually represent multilevel amplitude and phase modulation where both the carrier amplitude and phase can take two discrete values.

For the constellations in Figure 1.3(a) and (b) we have a constant amplitude signal, but the carrier phase values at the beginning of each symbol period in Figure 1.3(b) would be either $45 \deg, 135 \deg, 225 \deg \, or \, 315^{o}$. There are two magnitude values and two phase values for the constellations in Figure 1.3(c) and (d).

In order to reduce the bandwidth of the modulated signal while maintaining the same information transmission rate we can further decrease the symbol rate by adding more points in the constellation. Such a reduction in the bandwidth requirement will allow us to transmit more information in the spectrum we have been allocated. Such capability is normally considered advantageous. If we combine the constellations of Figure 1.3(c) and (d) we obtain the square QAM constellation having four bits per constellation point as displayed in Figure 1.4. We will spend much time dealing with this constellation in this book. In general, grouping n bits into one signalling symbol yields 2^n constellation points, which are often referred to as phasors, or complex vectors. The phasors associated with these points may have different amplitude and/or phase values, and this type of modulation is therefore referred to as multilevel modulation, where the number of levels is equal to the number of constellation points. After transmission through the channel the receiver must identify the phasor transmitted, and in doing so can determine the constellation point and hence the bits associated with this point. In this way the data is recovered. There are many problems with attempting to recover data transmitted over both fixed channels such as telephone lines and radio channels and many of these problems are given a whole chapter in this book. These problems are generally exacerbated by changing from binary to multilevel modulation, and this is why binary modulation is often preferred, despite its lower capacity. In order to introduce these problems, and to provide a historical perspective to quadrature amplitude modulation (QAM), a brief history of the development of QAM is presented.

1.2 History of Quadrature Amplitude Modulation

1.2.1 Determining the Optimum Constellation

Towards the end of the 1950s there was a considerable amount of interest in digital phase modulation transmission schemes [4] as an alternative to digital amplitude modulation. Digital phase modulation schemes are those whereby the amplitude of the transmitted carrier is held constant but the phase changes in response to the modulating signal. Such schemes have constellation diagrams of the form shown in Figure 1.3(a). It was a natural extension of this trend to consider the simultaneous use of both amplitude and phase modulation. The first paper to suggest this idea was by C.R. Cahn in 1960, who described a combined phase and

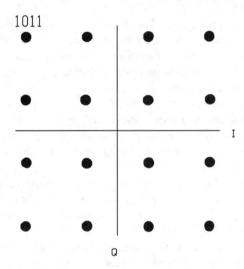

Figure 1.4: Square QAM constellation

amplitude modulation system [5]. He simply extended phase modulation to the multilevel case by allowing there to be more than one transmitted amplitude at any allowed phase. This had the effect of duplicating the original phase modulation or phase shift keying (PSK) constellation which essentially formed a circle. Such duplication led to a number of concentric circles depending on the number of amplitude levels selected. Each circle had the same number of phase points on each of its rings. Only Gaussian channels characteristic of telephone lines impaired by thermal noise were considered. Using a series of approximations and a wholly theoretical approach, he came to the conclusion that these amplitude and phase modulation (AM-PM) systems allowed an increased throughput compared to phase modulation systems when 16 or more states were used and suggested that such a system was practical to construct.

1.2.1.1   Coherent and Non-Coherent Reception

The fundamental problem with PSK is that of determining the phase of the transmitted signal and hence decoding the transmitted information. This problem is also known as carrier recovery as an attempt is made to recover the phase of the carrier. When a phase point at, say, $90°$ is selected to reflect the information being transmitted, the phase of the transmitted carrier is set to $90°$. However, the phase of the carrier is often changed by the transmission channel with the result that the receiver measures a different phase. This means that unless the receiver knew what the phase change imposed by the channel was it would be unable to determine the encoded information.

   This problem can be overcome in one of two ways. The first is to measure the phase change imposed by the channel by a variety of means. The receiver can then determine the transmitted phase. This is known as coherent detection. The second is to transmit differences in phase, rather than absolute phase. The receiver then merely compares the previous phase with the current phase and the phase change of the channel is removed. This assumes that any phase change within the channel is relatively slow. This differential system is known as non-coherent transmission. In his paper, Cahn considered both coherent and non-coherent transmission, although for coherent transmission he assumed a hypothetical and unrealisable perfect carrier recovery device. The process of carrier recovery is considered in Chapter 6, and the details of differential transmission are explained in Chapter 4.

1.2.1.2 Clock Recovery

Alongside carrier recovery runs the problem of clock recovery. The recovered clock signal is used to ensure appropriate sampling of the received signal. In Figure 1.2(c) a carrier signal was shown which had vertical lines indicating each bit or symbol period. It was the phase at the start of this period which was indicative of the encoded information. Unfortunately, the receiver has no knowledge of when these periods occur although it might know their approximate duration. It is determining these *symbol periods* which is the task of clock recovery. So carrier recovery estimates the phase of the transmitted carrier and clock recovery estimates the instances at which the data changes from one symbol to another. While the need for carrier recovery can be removed through differential or non-coherent detection, there is no way to remove the requirement for clock recovery.

Clock recovery schemes tend to seek certain periodicities in the received signal and use these to estimate the start of a symbol (actually they often attempt to select the centre of a symbol for reasons which will be explained in later chapters). Clock recovery is often a complex procedure, and poor clock recovery can substantially increase the bit error rate (BER). The issue of clock recovery is considered in Chapter 6. In his work, Cahn overcame the problem of clock recovery by assuming that he had some device capable of perfect clock recovery. Such devices do not exist, so Cahn acknowledged that the error rate experienced in practice would be worse than the value he had calculated, but as he was unable to compute the errors introduced by a practical clock recovery system, this was the only course open to him.

1.2.1.3 The Type I, II and III Constellations

A few months later a paper was published by Hancock and Lucky [6] in which they expanded upon the work of Cahn. In this paper they realised that the performance of the circular type constellation could be improved by having more points on the outer ring than on the inner ring. The rationale for this was that errors were caused when noise introduced into the signal moved the received phasor from the transmitted constellation point to a different one. The further apart constellation points could be placed, the less likely this was to happen. In Cahn's constellation, points on the inner ring were closest together in distance terms and so most vulnerable to errors. They conceded that a system with unequal numbers of points on each amplitude ring would be more complicated to implement, particularly in the case of non-coherent detection. They called the constellation proposed by Cahn a Type I system, and theirs a Type II system. Again using a mathematical approach they derived results similar to Cahn's for Type I systems and a 3 dB improvement for the Type II over the Type I system.

The next major publication was some 18 months later, in 1962, by Campopiano and Glazer [7]. They further developed the work of the previous papers but also introduced a new constellation - the square QAM system, which they termed a Type III system. They described

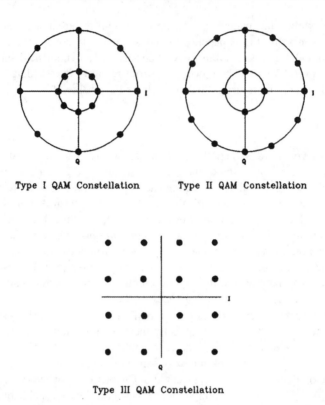

Type I QAM Constellation   Type II QAM Constellation

Type III QAM Constellation

Figure 1.5: Examples of types I, II and III QAM constellations

this system as "essentially the amplitude modulation and demodulation of two carriers that have the same frequency but are in quadrature with each other" - the first time that combined amplitude and phase modulation had been thought of as amplitude modulation on quadrature carriers, although the acronym QAM was not suggested. They realised that the problem with their Type III system was that it had to be used in a phase coherent mode, that is non-coherent detection was not possible and so carrier recovery was necessary. Again, a theoretical analysis was performed for Gaussian noise channels and the authors came to the conclusion that the Type III system offered a very small improvement in performance over the Type II system, but thought that the implementation of the Type III system would be considerably simpler than that of Types I and II. Examples of the different types of constellation are shown in Figure 1.5.

Three months later another paper was published by Hancock and Lucky [8] in which they were probably unaware of the work done by Campopiano and Glazer. They attempted to improve on their previous work on the Type II system by carrying out a theoretical analysis, supposedly leading to the optimal constellation for Gaussian channels. In this paper they decided that the optimum 16-level constellation had two amplitude rings with eight equispaced points on each ring but with the rings shifted by 22.5^o from each other. This constellation is shown in Figure 1.6.

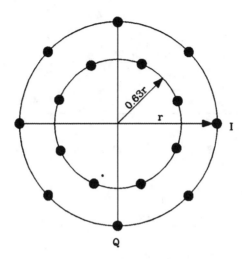

Figure 1.6: Optimum 16-level constellation according to Lucky [8] ©IRE, 1962

Again, they concluded that 16 was the minimum number of levels for AM-PM modulation and that a SNR of at least 11 dB was required for efficient operation with a low probability of bit error.

After this paper there was a gap of nine years before any further significant advances were published. This was probably due to the difficulties in implementing QAM systems with the technology available and also because the need for increased data throughput was not yet pressing. During this period the work discussed in the above papers was consolidated into a number of books, particularly that by Lucky, Salz and Weldon [9]. Here they clearly distinguished between quadrature amplitude modulation (QAM) schemes using square constellations and combined amplitude and phase modulation schemes using circular constellations. It was around this period that the acronym QAM started appearing in common usage along with AM-PM to describe the different constellations.

One of the earliest reports of the actual construction of a QAM system came from Salz, Sheenhan and Paris [10] of Bell Labs in 1971. They implemented circular constellations with 4 and 8 phase positions and 2 and 4 amplitude levels using coherent and non-coherent demodulation. Neither carrier nor clock recovery was attempted. Their results showed reasonable agreement with the theoretical results derived up to that time. This work was accompanied by that of Ho and Yeh [11] who improved the theory of circular AM-PM systems with algorithms that could be solved on digital computers which were by that time becoming increasingly available.

Interest in QAM remained relatively low, however, until 1974. In that year there was a number of significant papers published, considerably extending knowledge about quadrature amplitude modulation schemes. At this time, interest into optimum constellations was revived with two papers, one by Foschini, Gitlin and Weinstein [12] and the other by Thomas, Weidner and Durrani [13]. Foschini *et al.* attempted a theoretical derivation of the ideal constellation using a gradient calculation approach. They came to the conclusion that the ideal

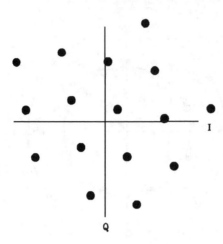

Figure 1.7: Optimum 16-level constellation according to Foschini [12] ©IEEE, 1974

constellation was based on an equilateral triangle construction leading to the unusual 16-level constellation shown in Figure 1.7. This constellation has not found favour in practical applications, since the complexities involved in its employment outweigh the associated gains that were claimed for it.

Their conclusions were that this constellation, when limited in terms of power and operated over Gaussian channels, offered a performance improvement of 0.5 dB over square QAM constellations. Meanwhile Thomas *et al.*, working at COMSAT, empirically generated 29 constellations and compared their error probabilities.

1.2.2   Satellite Links

In the paper by Thomas *et al.* [13] they also mentioned the first application of QAM, for use in satellite links. Satellite links have a particular problem in that satellites only have a limited power available to them. Efficient amplifiers are necessary to use this power carefully, and these had tended to employ a device known as a travelling wave tube (TWT). Such a device introduces significant distortion in the transmitted signal and Thomas *et al.* considered the effects of this distortion on the received waveform. They came to the interesting conclusion that the Type II constellation (in this case 3 points on the inner ring and 5 on the outer) was inferior to the Type I (4 points on inner and outer rings) due to the increased demand in peak-to-average power ratio of the Type II constellation. Their overall conclusion was that circular Type I constellations are superior in all cases. When they considered TWT distortion they discovered that AM-PM schemes were inferior to PSK schemes because of the need to significantly back off the amplifier to avoid severe amplitude distortion, and concluded that better linear amplifiers would be required before AM-PM techniques could be successfully used for satellite communications. They also considered the difficulties of implementing various carrier recovery techniques, advising that decision directed carrier recovery would be most appropriate, although few details were given as to how this was to be implemented. Decision directed carrier recovery is a process whereby the decoded signal is compared with the closest constellation point and the phase difference between them is used to estimate the error in the recovered carrier. This is discussed in more detail in Chapter 6.

Commensurate with the increasing interest as to possible applications for QAM were the two papers published in 1974 by Simon and Smith which concentrated on carrier recovery and detection techniques. In the first of these [14] they noted the interest in QAM that was then appearing for bandlimited systems, and addressed the problems of carrier recovery. They considered only the 16-level square constellation and noted that the generation of a highly accurate reconstructed carrier was essential for adequate performance. Their solution was to demodulate the signal, quantise it, and then establish the polarity of error from the nearest constellation point, and use it to update the voltage controlled oscillator (VCO) used in the carrier generation section. They provided a theoretical analysis and concluded that their carrier recovery technique worked well in the case of high signal-to-noise (SNR) ratio Gaussian noise, although they noted that gain control was required and would considerably complicate the implementation. They extended their work in [15] where they considered offset QAM or O-QAM. In this modulation scheme the signal to one of the quadrature arms was delayed by half a symbol period in an attempt to prevent dramatic fluctuations of the signal envelope, which was particularly useful in satellite communications. They noted similar results for their decision-directed carrier recovery scheme as when non-offset modulation was used.

1.2.2.1 Odd-Bit Constellations

Despite all the work on optimum constellations, by 1975 interest had centred on the square QAM constellation. The shape of this was evident for even numbers of bits per symbol, but if there was a requirement for an odd number of bits per symbol to be transmitted, the ideal shape of the constellation was not obvious, with rectangular constellations having been tentatively suggested. Early in 1975 J.G. Smith, also working on satellite applications, published a paper addressing this problem [16]. He noted that for even numbers of bits per symbol "the square constellation was the only viable choice." In this paper he showed that what he termed "symmetric" constellations offered about a 1 dB improvement over rectangular constellations and he considered both constellations to be of the same implementational complexity. Figure 1.8 shows an example of his symmetric constellation when there are 5 bits per symbol,

1.2.3 QAM Modem Implementations

About this time, the Japanese started to show interest in QAM schemes as they considered they might have application in both satellite and microwave radio links. In 1976 Miyauchi, Seki and Ishio published a paper devoted to implementation techniques [17]. They considered implementation by superimposing two 4-level PSK modulation techniques at different amplitudes to achieve a square QAM constellation and using a similar process in reverse at the demodulator, giving them the advantage of being able to use existing PSK modulator and demodulator circuits. This method of implementing QAM is discussed in Chapter 4. They implemented a prototype system without clock or carrier recovery and concluded that its performance was sufficiently good to merit further investigation. Further groundwork was covered in 1978 by W. Weber, again working on satellite applications, who considered differential (i.e. non-coherent) encoding systems for the various constellations still in favour [18].

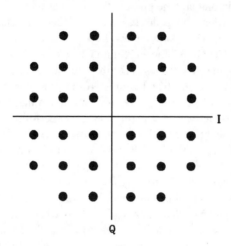

Figure 1.8: Optimum 32-level constellation according to Smith [16] ©IEEE, 1975

His paper essentially added a theoretical basis to the differential techniques that had been in use at that point, although suggesting that non-coherent demodulation techniques deserved more attention.

In late 1979 evidence of the construction of QAM prototype systems worldwide started to emerge. A paper from the CNET laboratories in France by Dupuis *et al.* [19] considered a 140 Mbit/s modem for use in point-to-point links in the 10-11 GHz band. This prototype employed the square 16-level constellation and included carrier recovery, although no details were given. Theoretical calculations of impairments were presented followed by measurements made over a 58 km hop near Paris. Their conclusions were that QAM had a number of restrictions in its use, relating to its sensitivity to non-linearities, and that in the form they had implemented it, PSK offered improved performance. However, they suggested that with further work these problems might be overcome.

The Japanese simultaneously announced results from a prototype 200 Mbit/s 16-QAM system in a paper by Horikawa, Murase and Saito [20]. They used differential coding coupled with a new form of carrier recovery based on a decision feedback method (detailed in Chapter 6). Their modem was primarily designed for satellite applications and their experiment included the use of TWT amplifiers, but was only carried out back-to-back in the laboratory. Their conclusions were that their prototype had satisfactory performance and was an efficient way to increase bandwidth efficiency.

One of the last of the purely theoretical, as opposed to practical papers on QAM appeared in April 1980, marking the progression of QAM from a technical curiosity into a practical system, some twenty years after its introduction. This came from V. Prabhu of Bell Labs [21], further developing the theory to allow calculation of error probabilities in the presence of co-channel interference. Prabhu concluded that 16-QAM had a co-channel interference immunity superior to 16-PSK but inferior to 8-PSK.

1.2.3.1 Non-Linear Amplification

In 1982 there came a turning point in the use of QAM for satellite applications when Feher turned his attention to this problem. His first major publication in this field [22] introduced a new method of generation of QAM signals using highly non-linear amplifiers which he termed non-linear amplified QAM (NLA-QAM). Two separate amplifiers for the 16-QAM case were used, one operating with half the output power of the other. The higher power amplifier coded the two most significant bits of the 4-bit symbol only, and the lower power amplifier coded only the two least significant bits. The amplified coded signals were then summed at full output power to produce the QAM signal. Because both amplifiers were therefore able to use constant envelope modulation they could be run at full power with resulting high efficiency, although with increased complexity due to the need for two amplifiers and a hybrid combiner, compared to previous systems which used only a single amplifier. However, in satellite applications, complexity was relatively unimportant compared to power efficiency, and this NLA technique offered a very substantial 5 dB power gain, considerably increasing the potential of QAM in severely power limited applications.

1.2.3.2 Frequency Selective Fading and Channel Equalisers

This work was soon followed by a performance study of a NLA 64-state system [23] which extended the NLA scheme to 64 levels by using three amplifiers all operating at different power levels. Performance estimates were achieved using computer simulation techniques which included the effects of frequency selective fading.

Frequency selective fading is essentially caused when there are a number of propagation paths between the transmitter and receiver, imposed for example by the reflection of the radio waves from nearby buildings or mountains. When the time delay of the longest path compared to that of the shortest path becomes comparable to a symbol period, intersymbol interference (ISI) arises. Since every time domain effect has an equivalent frequency domain effect, which are related to each other through the Fourier transform, a dispersive channel impulse response results in an undulating frequency-domain channel transfer function, where the corresponding frequency-domain fluctuations are also referred to as 'frequency-selective' fading. This frequency-selective fading phenomenon may be mitigated with the aid of adaptive channel equaliser techniques, which attempt to remove the channel-induced ISI. They do so by calculating the ISI introduced and then subtracting it from the received signals. They are often extremely complex devices and are considered in detail in Chapter 7.

On a historical note, in the context of linear equalisers pioneering work was carried out amongst other researchers by Tufts [24], where the design of the transmitter and receiver was jointly optimised. The optimisation was based on the minimisation of the MSE between the transmitted signal and the equalised signal. This was achieved under the Zero Forcing (ZF) condition, where the ISI was completely mitigated at the sampling instances. Subsequently, Smith [25] introduced a similar optimisation criterion with and without applying the ZF condition. Similar works as a result of these pioneering contributions were achieved by, amongst others, Hänsler [26], Ericson [27] and Forney [28].

The development of the DFE was initiated by the idea of using previously detected symbols to compensate for the ISI in a dispersive channel, which was first proposed by Austin [29]. This idea was adopted by Monsen [30], who managed to optimise the DFE based on minimising the MSE between the equalised symbol and the transmitted symbol.

The optimisation of the DFE based on joint minimisation of both the noise and ISI was undertaken by Salz [31], which was subsequently extended to QAM systems by Falconer and Foschini [32]. At about the same time, Price [33] optimised the DFE by utilizing the so-called ZF criterion, where all the ISI was compensated by the DFE. The pioneering work achieved so far assumed perfect decision feedback and that the number of taps of the DFE was infinite. A more comprehensive history of the linear equaliser and the DFE can be found in the classic papers by Lucky [34] or by Belfiore [35] and a more recent survey was produced by Qureshi [36].

In recent years, there has not been much development on the structure of the linear and decision feedback equalisers. However, considerable effort has been given to the investigation of adaptive algorithms that are used to adapt the equalisers according to the prevalent CIR. These contributions will be elaborated in the next chapter. Nevertheless, some interesting work on merging the MLSE detectors with the DFE has been achieved by Cheung et al. [37,38], Wu et al. [39,40] and Gu et al. [41]. In these contributions, the structure of the MLSE and DFE was merged in order to yield an improved BER performance, when compared to the DFE, albeit at the cost of increased complexity. However, the complexity incurred was less, when compared to that of the MLSE.

In the context of error propagation in the DFE, which will be explained in Chapter 7, this phenomenon has been reported and researched in the past by Duttweiler et al. [42] and more recently by Smee et al. [43] and Altekar et al. [44]. In this respect some solutions have been proposed by, amongst others, Tomlinson [45], Harashima [46], Russell et al. [47] and Chiani [48], in reducing the impact of error propagation.

1.2.3.3   History of Blind Equalisation

The philosophy of channel equalisation is that the transmitter sends known so-called channel-sounding symbols to the receiver. Upon receiving these known symbols the receiver typically evaluates the difference between the pre-agreed transmitted signal as well as the received signal and uses this error signal for adaptively adjusting the response of the equaliser, so that it eliminates the channel-induced ISI. When the wireless channel's impulse response changes rapidly, the channel's response has to be estimated at regular intervals, which requires the frequent transmission of redundant channel-sounding symbols.

By contrast, the principle of blind equalisation is that no known channel-sounding symbols are transmitted over the channel for the sake of estimating its response, which allows the system to maximise its effective throughput. This blind equalisation concept was originally proposed by Sato [49]. Five years after Sato's publication, the blind equalisation problem was further studied for example by Benveniste, Goursat and Ruget in [50], where several blind equalisation issues were clarified and a new algorithm was proposed [51]. At the same time Godard [52] introduced a criterion, namely the so-called "constant modulus" (CM) criterion, leading to a new class of blind equalisers. Following Godard's contribution a range of studies were conducted employing the constant modulus criterion. Foschini [53] was the first researcher studying the convergence properties of Godard's equaliser upon assuming an infinite equaliser length. Later, Ding et al. continued this study [54] and provided an indepth analysis of the convergence issue in the context of a realistic equaliser. Although numerous researchers have studied this issue, nevertheless, a general solution is yet to be found. A plethora of authors have studied Godard's equaliser, rendering it the most widely studied blind equaliser. A well-known algorithm of the so-called Bussgang type [55, 56] was also proposed by Picchi and Prati [57]. Their "Stop-and-Go" algorithm constitutes a combination of the Decision-Directed algorithm [3] with Sato's algorithm [49]. After 1991, a range of different solutions to the blind equalisation problem were proposed. Seshadri [58] suggested the employment of the so-called M-algorithm, as a "substitute" for the Viterbi algorithm [59] for the blind scenario, combined with the so-called "least mean squares (LMS)" based CIR estimation. This CIR estimation was replaced by "recursive least squares (RLS)" estimation by Raheli, Polydoros and Tzou [60], combining the associated convolutional decoding with the CIR estimation, leading to what was termed as *"Per-Survivor Processing"*. Since then a number of papers have focused on this technique [60–69]. At the same time as Seshadri, Tong *et al.* [70] proposed a different approach to blind equalisation, which used oversampling in order to create a so-called "cyclostationary" received signal, and performed CIR estimation by measuring the autocorrelation function of this signal and by exploiting this signal's cyclostationarity. This technique was also applied to the case of 'sampling' the received signals of different antennas (instead of oversampling the signal of a single antenna) and further extended by Moulines *et al.* using a different method of CIR estimation, namely the so-called subspace method in [71]. Furthermore, Tsatsanis and Giannakis suggested that the cyclostationarity can be induced by the transmitter upon transmitting the signal more than once [72]. A number of further contributions have also been published in the context of these techniques [73–88]. Finally, nearly coincidentally with Seshadri [58] and Tong *et al.* [70], Hatzinakos and Nikias [89] proposed a more sophisticated approach to blind equalisation by exploiting the so-called "tricepstrum" of the received signal. Until today, the blind equalisation problem is an open research topic, attracting significant amount of research. A general answer to the fundamental question *"Under what circumstances is it preferable to use a blind equaliser to a trained-equaliser ?"* is yet to be provided. Despite the scarcity of reviews on the topic, in the context of the Global System of Mobile Communications known as GSM an impressive effort was made by Boss, Kammeyer and Petermann [90], who also proposed two novel blind algorithms. We recommend furthermore the fractionally-spaced equalisation review of Endres *et al.* [91] and the Constant Modulus overview of Johnson *et al.* [92] based on a specific type of equalisers, namely on the so-called "fractionally-spaced" equalisers. A review of subspace-ML multichannel blind equalisers was provided by Tong and Perreau [93]. Further important references are the monograph by Haykin [3], the relevant section by Proakis [94] and the blind deconvolution book due to Nandi [95]. Comparative performance studies between various blind equalisers have also been performed. We recommend the second-order statistics-based comparative performance studies of Becchetti *et al.* [96], Kristensson *et al.* [97] and Altuna *et al.* [98], which are based on the mobile environment as well as the second-order statistics and PSP-based comparative study of Skowratanont and Chambers [99]. Furthermore, we recommend the fractionally-spaced Bussgang algorithm based comparative performance study by Shynk *et al.* [100], the CMA comparative performance study of Schirtzinger *et al.* [101] and the comparative convergence study by Endres *et al.* [102].

1.2.3.4   Filtering

In a book published around this time [103], Feher also suggested the use of non-linear filtering (NLF) for QAM satellite communications. Since the I and Q components of the time domain QAM signal change their amplitude abruptly at the signalling intervals their transmission would require an infinite bandwidth. These abrupt changes are typically smoothed by a bandlimiting filter. The design and implementation of such filters were critical, particularly when the NLA-QAM signal is filtered at high power level. In order to alleviate these problems Feher developed the NLF technique along with Huang in 1979 which simplified filter design by simply fitting a quarter raised cosine segment between two initially abruptly changing symbols for both of the quadrature carriers. We have already hinted that filtering is required to prevent abrupt changes in the transmitted signal. This issue is considered in more rigour in Chapter 4. This allowed the generation of jitter-free bandlimited signals, which had previously been a problem, improving clock recovery techniques. Feher's work continued to increase the number of levels used, with a paper on 256-QAM in May 1985 [104] noting the problems that linear group delay distortion caused, and a paper in April 1986 on 512-QAM [105] which came to similar conclusions.

1.2.4   Advanced Prototypes

Work was still continuing in France, Japan and also in New Zealand. CNET were continuing their attempt to overcome the problems they had found in their initial trials reported in 1979. A paper published in 1985 by M. Borgne [106] compared the performance of 16, 32, 64 and 128 level QAM schemes using computer simulation with particular interest in the impairments likely to occur over point-to-point radio links. Borgne concluded that non-linearity cancellers and adaptive equalisers would be necessary for this application. Soon after this, the first major paper on adaptive equalisers for QAM was published by Shafi and Moore [107]. Much of this paper was concerned with clock and carrier recovery without going into detail as to how these operations were performed. Details were provided of a fractional decision feedback equaliser (DFE) which they considered suitable for point-to-point radio links. They concluded that carrier recovery and clock timing was critical and likely to cause major problems, which were somewhat ameliorated by their fractionally spaced system.

Although lagging somewhat behind Feher, the Japanese made up for this delay by publishing a very detailed paper describing the development of a 256-level QAM modem in August 1986. In this paper from Saito and Nakamura [108], the authors further developed the work disseminated in 1979 by Horikawa, Murase and Saito [20] which was discussed earlier. In this new paper they detailed automatic gain control (which he termed automatic threshold control) and carrier recovery methods. The carrier recovery was a slight enhancement to the system announced in 1979 and the AGC system was based on decision-directed methods. Details were given as to how false lock problems were avoided (see Chapter 6) and the back-to-back prototype experiments gave results which the authors considered showed the feasibility of the 256-QAM modem. Evidence of the ever increasing interest in QAM was that in the IEEE special issue on advances in digital communications by radio there was a substantial section devoted to high-level modulation techniques. In a paper by Rustako *et al.* [109] which considered point-to-point applications, the standard times-two carrier recovery method for binary modulation was expanded for QAM. The authors claimed the advantage of not requiring accurate data decisions or interacting with any equaliser. They acknowledged that their system had the disadvantage of slow reacquisition after fades and suggested that it would only be superior in certain situations. This form of carrier recovery is considered in some detail in Chapter 6.

Clearly, until the late 1980s developments were mainly targeted at telephone line and point-to-point radio applications, which led to the definition of the CCITT telephone circuit modem standards V.29 to V.33 based on various QAM constellations ranging from uncoded 16-QAM to trellis coded (TC) 128-QAM. The basic concept of coded modulation is introduced in Chapter 8 along with the members of CCITT standard V-series modem scheme family, including the V.29 - V.33 modems designed for telephone lines.

1.2.5 QAM for Wireless Communications

Another major development occurred in 1987 when Sundberg, Wong and Steele published a pair of papers [110, 111] considering QAM for voice transmission over Rayleigh fading channels, the first major paper considering QAM for mobile radio applications. In these papers, it was recognized that when a Gray code mapping scheme was used, some of the bits constituting a symbol had different error rates from other bits. Gray coding is a method of assigning bits to be transmitted to constellation points in an optimum manner and is discussed in Chapter 5. For the 16-level constellation two classes of bits occurred, for the 64-level three classes and so on. Efficient mapping schemes for pulse code modulated (PCM) speech coding were discussed where the most significant bits (MSBs) were mapped onto the class with the highest integrity. A number of other schemes including variable threshold systems and weighted systems were also discussed. Simulation and theoretical results were compared and found to be in reasonable agreement. They used no carrier recovery, clock recovery or AGC, assuming these to be ideal, and came to the conclusion that channel coding and post-enhancement techniques would be required to achieve acceptable performance.

This work was continued, resulting in a publication in 1990 by Hanzo, Steele and Fortune [112], again considering QAM for mobile radio transmission, where again a theoretical argument was used to show that with a Gray encoded square constellation, the bits encoded onto a single symbol could be split into a number of subclasses, each subclass having a different average BER. The authors then showed that the difference in BER of these different subclasses could be reduced by constellation distortion at the cost of slightly increased total BER, but was best dealt with by using different error correction powers on the different 16-QAM subclasses. A 16 kbit/s sub-band speech coder was subjected to bit sensitivity analysis and the most sensitive bits identified were mapped onto the higher integrity 16-QAM subclasses, relegating the less sensitive speech bits to the lower integrity classes. Furthermore, different error correction coding powers were considered for each class of bits to optimise performance. Again ideal clock and carrier recovery were used, although this time the problem of automatic gain control (AGC) was addressed. It was suggested that as bandwidth became increasingly congested in mobile radio, microcells would be introduced supplying the required high SNRs with the lack of bandwidth as an incentive to use QAM.

In the meantime, CNET were still continuing their study of QAM for point-to-point applications, and Sari and Moridi published a paper [113] detailing an improved carrier recovery system using a novel combination of phase and frequency detectors which seemed promising. However, interest was now increasing in QAM for mobile radio usage and a paper was published in 1989 by J. Chuang of Bell Labs [114] considering NLF-QAM for mobile radio and concluding that NLF offered slight improvements over raised cosine filtering when there was mild intersymbol interference (ISI).

A technique, known as the transparent tone in band method (TTIB) was proposed by

McGeehan and Bateman [115] of Bristol University, UK, which facilitated coherent detection of the square QAM scheme over fading channels and was shown to give good performance but at the cost of an increase in spectral occupancy. This important technique is discussed in depth in Chapter 10. At an IEE colloquium on multilevel modulation techniques in March 1990 a number of papers were presented considering QAM for mobile radio and point-to-point applications. Matthews [116] proposed the use of a pilot tone located in the centre of the frequency band for QAM transmissions over mobile channels.

Huish discussed the use of QAM over fixed links, which was becoming increasingly widespread [117]. Webb *et al.* presented two papers describing the problems of square QAM constellations when used for mobile radio transmissions and introduced the star QAM constellation with its inherent robustness in fading channels [118, 119].

During the 1990s a number of publications emerged, describing various techniques designed for enhancing the achievable performance of QAM transmissions schemes, when communicating over mobile radio channels. All of these techniques are described in detail in Part II of the book, namely in Chapters 9 - 13. In December 1991 a paper appeared in the *IEE Proceedings* [120] which considered the effects of channel coding, trellis coding and block coding when applied to the star QAM constellation. This was followed by another paper in the *IEE Proceedings* [121] considering equaliser techniques for QAM transmissions over dispersive mobile radio channels. A review paper appearing in July 1992 [122] considered areas where QAM could be put to most beneficial use within the mobile radio environment, and concluded that its advantages would be greatest in microcells. Further work on spectral efficiency, particularly of multilevel modulation schemes [123] concluded that variable level QAM modulation was substantially more efficient than all the other modulation schemes simulated. Variable level QAM was first discussed in a paper by Steele and Webb in 1991 [124].

Further QAM schemes for hostile fading channels characteristic of mobile telephony can be found in the following recent references [1, 1, 125–158]. If Feher's previously mentioned NLA concept cannot be applied, then power-inefficient class A or AB linear amplification has to be used, which might become an impediment in lightweight, low-consumption handsets. However, the power consumption of the low-efficiency class A amplifier [134, 135] is less critical than that of the digital speech and channel codecs. In many applications 16-QAM, transmitting 4 bits per symbol reduces the signalling rate by a factor of 4 and hence mitigates channel dispersion, thereby removing the need for an equaliser, while the higher SNR demand can be compensated by diversity reception.

Significant contributions were made by Cavers, Stapleton *et al.* at Simon Fraser University, Burnaby, Canada in the field of pre- and post-distorter design. Out-of-band emissions due to class AB amplifier non-linearities and hence adjacent channel interferences can be reduced by some 15-20 dB using Stapleton's adaptive pre-distorter [1, 1, 136] and a class AB amplifier with 6 dB back-off, by adjusting the pre-distorter's adaptive coefficients using the complex convolution of the pre-distorter's input signal and the amplifier's output signal. Further aspects of linearised power amplifier design are considered in references [137] and [138].

A further important research trend is hallmarked by Cavers' work targeted at pilot symbol assisted modulation (PSAM) [139], where known pilot symbols are inserted in the information stream in order to allow the derivation of channel measurement information. The recovered received symbols are then used to linearly predict the channel's attenuation and phase. This arrangement will be considered in Chapter 10. A range of advanced QAM modems have also been proposed by Japanese researchers doing cutting-edge research in the field, including Sampei *et al.* [129, 130, 140, 141], Adachi [142] *et al.* and Sasaoka [133].

Since QAM research has reached a mature stage, a number of mobile speech, audio and video transmission schemes have been proposed [157–169]. These system design examples demonstrated that substantial system performance benefits accrue, when the entire system is jointly optimised, rather than just a conglomerate of independent system components. A range of digital video broadcasting (DVB) schemes will be the topic of Chapter 30.

1.3 History of Near-Instantaneously Adaptive QAM

A comprehensive overview of adaptive transceivers was provided in [170] and this section is also based on [170]. As we noted in the previous chapters, mobile communications channels typically exhibit a near-instantaneously fluctuating time-variant channel quality [170–173] and hence conventional fixed-mode modems suffer from bursts of transmission errors, even if the system was designed for providing a high link margin. *An efficient approach to mitigating these detrimental effects is to adaptively adjust the modulation and/or the channel coding format as well as a range of other system parameters based on the near-instantaneous channel quality information perceived by the receiver, which is fed back to the transmitter with the aid of a feedback channel [174].* This plausible principle was recognised by Hayes [174] as early as 1968.

It was also shown in the previous sections that these near-instantaneously adaptive schemes require a reliable feedback link from the receiver to the transmitter. However, the channel quality variations have to be sufficiently slow for the transmitter to be able to adapt its modulation and/or channel coding format appropriately. The performance of these schemes can potentially be enhanced with the aid of *channel quality prediction techniques [175].* As an efficient fading counter-measure, Hayes [174] proposed the employment of transmission power adaptation, while *Cavers [176] suggested invoking a variable symbol duration scheme* in response to the perceived channel quality at the expense of a variable bandwidth requirement. A disadvantage of the variable-power scheme is that it increases both the average transmitted power requirements and the level of co-channel interference imposed on other users, while requiring a high-linearity class-A or AB power amplifier, which exhibit a low power-efficiency. As a more attractive alternative, *the employment of AQAM was proposed by Steele and Webb,* which circumvented some of the above-mentioned disadvantages by employing various star-QAM constellations [124, 177].

With the advent of *Pilot Symbol Assisted Modulation (PSAM)* [139, 140, 178], Otsuki *et al.* [179] employed square-shaped AQAM constellations instead of star constellations [180], as a practical fading counter measure. By analysing the channel capacity of Rayleigh fading channels [181], *Goldsmith et al.* [182] and Alouini *et al.* [183] showed that combined variable-power, variable-rate adaptive schemes are attractive in terms of approaching the capacity of the channel and characterised the achievable throughput performance of variable-power AQAM [182]. However, they also found that *the extra throughput achieved by the additional variable-power assisted adaptation over the constant-power, variable-rate scheme is marginal* for most types of fading channels [182, 184].

In 1996 Torrance and Hanzo [185] proposed a set of *mode switching levels* s designed for achieving a high average BPS throughput, while maintaining the target average BER. Their method was based on defining a specific combined BPS/BER cost-function for transmission over narrowband Rayleigh channels, which incorporated both the BPS throughput as well as the target average BER of the system. *Powell's optimisation was invoked for finding a set of mode switching thresholds, which were constant,* regardless of the actual channel Signal to Noise Ratio (SNR) encountered, i.e. irrespective of the prevalent instantaneous channel conditions. However, in 2001 Choi and Hanzo [186] noted that a *higher BPS throughput can be achieved, if under high channel SNR conditions the activation of high-throughput AQAM modes is further encouraged by lowering the AQAM mode switching thresholds. More explicitly, a set of SNR-dependent AQAM mode switching levels was proposed [186],* which keeps the average BER constant, while maximising the achievable throughput. We note furthermore that the set of switching levels derived in [185, 187] is based on Powell's multidimensional optimisation technique [188] and hence the optimisation process may become trapped in a local minimum. This problem was overcome by Choi and Hanzo upon deriving an *optimum set of switching levels [186], when employing the Lagrangian multiplier technique.* It was shown that this set of switching levels results in the global optimum in a sense that the corresponding AQAM scheme obtains the maximum possible average BPS throughput, while maintaining the target average BER. An important further development was Tang's contribution [189] in the area of contriving an *intelligent learning scheme for the appropriate adjustment of the AQAM switching thresholds.*

These contributions demonstrated that *AQAM exhibited promising advantages,* when compared to fixed modulation schemes in terms of spectral efficiency, BER performance and robustness against channel delay spread, etc. Various systems employing AQAM were also characterised in [180]. The *numerical upper bound performance of narrow-band BbB-AQAM* over slow Rayleigh flat-fading channels was evaluated by Torrance and Hanzo [190], while over *wideband channels* by Wong and Hanzo [191, 192]. Following these developments, adaptive modulation was also studied *in conjunction with channel coding and power control techniques by Matsuoka et al. [193] as well as Goldsmith and Chua [194, 195].*

In the early phase of research more emphasis was dedicated to the system aspects of adaptive modulation in a narrow-band environment. A reliable method of transmitting the modulation control parameters was proposed by Otsuki *et al.* [179], where the parameters were embedded in the transmission frame's mid-amble using Walsh codes. Subsequently, at the receiver the Walsh sequences were decoded using maximum likelihood detection. Another technique of *signalling the required modulation mode used was proposed by Torrance and Hanzo [196], where the modulation control symbols were represented by unequal error protection 5-PSK symbols.* Symbol-by-Symbol (SbS) adaptive, *rather than BbB-adaptive systems were proposed by Lau and Maric in [197], where the transmitter is capable of transmitting each symbol in a different modem mode, depending on the channel conditions. Naturally, the receiver has to synchronise with the transmitter in terms of the SbS-adapted mode sequence, in order to correctly demodulate the received symbols and hence the employment of BbB-adaptivity is less challenging, while attaining a similar performance to that of BbB-adaptive arrangements under typical channel conditions.*

The adaptive modulation philosophy was then extended to wideband multi-path environments *amongst others for example by Kamio et al. [198] by utilizing a bi-directional Decision Feedback Equaliser (DFE)* in a micro- and macro-cellular environment. This equalisation technique employed both forward and backward oriented channel estimation based on the pre-amble and post-amble symbols in the transmitted frame. Equaliser tap gain interpolation across the transmitted frame was also utilised for reducing the complexity in conjunction with space diversity [198]. The authors concluded that the cell radius could be enlarged in a macro-cellular system and a higher area-spectral efficiency could be attained for micro-cellular environments by utilizing adaptive modulation. The data transmission latency effect, which occurred when the input data rate was higher than the instantaneous transmission throughput was studied and solutions were formulated using *frequency hopping [199] and statistical multiplexing, where the number of Time Division Multiple Access (TDMA) time-slots allocated to a user was adaptively controlled [200].*

In reference [201] *symbol rate adaptive modulation* was applied, where the symbol rate or the number of modulation levels was adapted by using $\frac{1}{8}$-rate 16QAM, $\frac{1}{4}$-rate 16QAM, $\frac{1}{2}$-rate 16QAM as well as full-rate 16QAM and the criterion used for adapting the modem modes was based on the instantaneous received signal to noise ratio and channel delay spread. The slowly varying channel quality of the uplink (UL) and downlink (DL) was rendered similar by utilizing short frame duration Time Division Duplex (TDD) and the maximum normalised delay spread simulated was 0.1. A *variable channel coding rate* was then introduced by Matsuoka *et al.* in conjunction with adaptive modulation in [193], where the transmitted burst incorporated an outer Reed Solomon code and an inner convolutional code in order to achieve high-quality data transmission. The coding rate was varied according to the prevalent channel quality using the same method, as in adaptive modulation in order to achieve a certain target BER performance. A so-called *channel margin* was introduced in this contribution, which effectively increased the switching thresholds for the sake of preempting *the effects of channel quality estimation errors,* although this inevitably reduced the achievable BPS throughput.

In an effort to improve the achievable performance versus complexity trade-off in the context of AQAM, Yee and Hanzo [202] studied the design of various *Radial Basis Function (RBF) assisted neural network-based schemes, while communicating over dispersive channels.* The advantage of these RBF-aided DFEs is that they are capable of delivering error-free decisions even in scenarios, when the received phasors cannot be error-freely detected by the conventional DFE, since they cannot be separated into decision classes with the aid of a linear decision boundary. In these so-called linearly non-separable decision scenarios the RBF-assisted DFE still may remain capable of classifying the received phasors into decision classes without decision errors. A further improved turbo BCH-coded version of this RBF-aided system was characterised by Yee *et al.* in [203], while a turbo-equalised RBF arrangement was the subject of the investigation conducted by Yee, Liew and Hanzo in [204, 205]. The RBF-aided AQAM research has also been extended to the turbo equalisation of a convolutional as well as space-time trellis coded arrangement proposed by Yee, Yeap and Hanzo [170, 206, 207]. The same authors then endeavoured to reduce the associated implementation complexity of an RBF-aided QAM modem with the advent of employing a separate in-phase / quadrature-phase turbo equalisation scheme in the quadrature arms of the modem.

As already mentioned above, the performance of *channel coding in conjunction with adaptive modulation in a narrow-band environment* was also characterised by Chua and Goldsmith [194]. In their contribution trellis and lattice codes were used without channel interleaving, invoking a feedback path between the transmitter and receiver for modem mode control purposes. Specifically, the simulation and theoretical results by Goldsmith and Chua showed that a 3dB coding gain was achievable at a BER of 10^{-6} for a 4-sate trellis code and 4dB by an 8-state trellis code in the context of the adaptive scheme over Rayleigh-fading channels, while a 128-state code performed within 5dB of the Shannonian capacity limit.

The *effects of the delay in the AQAM mode signalling feedback path* on the adaptive modem's performance were studied and this scheme exhibited a higher spectral efficiency, when compared to the non-adaptive trellis coded performance. Goeckel [208] also contributed in the area of adaptive coding and employed realistic outdated, rather than perfect fading estimates. Further research on adaptive multidimensional coded modulation was also conducted by Hole *et al.* [209] for transmissions over flat fading channels. *Pearce, Burr and Tozer [210] as well as Lau and Mcleod [211] have also analysed the performance trade-offs associated with employing channel coding and adaptive modulation or adaptive trellis coding,* respectively, as efficient fading counter-measures. In an effort to provide a fair comparison of the various coded modulation schemes known at the time of writing, Ng, Wong and Hanzo have also studied Trellis Coded Modulation (TCM), Turbo TCM (TTCM), Bit-Interleaved Coded Modulation (BICM) and Iterative-Decoding assisted BICM (BICM-ID), where TTCM was found to be the best scheme at a given decoding complexity [212].

Subsequent contributions by Suzuki *et al.* [213] incorporated *space-diversity and power-adaptation* in conjunction with adaptive modulation, for example in order to combat the effects of the multi-path channel environment at a 10Mbits/s transmission rate. *The maximum tolerable delay-spread was deemed to be one symbol duration for a target mean BER performance of* 0.1%. This was achieved in a TDMA scenario, where the channel estimates were predicted based on the extrapolation of previous channel quality estimates. As mentioned above, variable transmitted power was applied in combination with adaptive modulation in reference [195], where the transmission rate and power adaptation were optimised for the sake of achieving an increased spectral efficiency. In their treatise a slowly varying channel was assumed and the instantaneous received power required for achieving a certain upper bound performance was assumed to be known prior to transmission. *Power control in conjunction with a pre-distortion type non-linear power amplifier compensator* was studied in the context of adaptive modulation in reference [214]. This method was used to mitigate the non-linearity effects associated with the power amplifier, when QAM modulators were used.

Results were also recorded concerning the performance of adaptive modulation in conjunction with *different multiple access schemes in a narrow-band channel environment.* In a TDMA system, *dynamic channel assignment* was employed by Ikeda *et al.*, where in addition to assigning a different modulation mode to a different channel quality, priority was always given to those users in their request for reserving time-slots, which benefitted from the best channel quality [215]. The performance was compared to fixed channel assignment systems, where substantial gains were achieved in terms of system capacity. Furthermore, a *lower call termination probability was recorded.* However, the probability of intra-cell hand-off increased as a result of the associated dynamic channel assignment (DCA) scheme, which constantly searched for a high-quality, high-throughput time-slot for supporting the actively communicating users. The application of adaptive modulation in packet transmission was introduced by Ue, Sampei and Morinaga [216], where the results showed an improved BPS throughput. The performance of adaptive modulation was also characterised in conjunction with an *automatic repeat request* (ARQ) system in reference [217], where the transmitted bits were encoded using a cyclic redundant code (CRC) and a convolutional punctured code in order to increase the data throughput.

A further treatise was published by Sampei, Morinaga and Hamaguchi [218] on *laboratory test results* concerning the utilization of adaptive modulation in a TDD scenario, where the modem mode switching criterion was based on the signal to noise ratio and on the normalised delay-spread. *In these experimental results, the channel quality estimation errors degraded the performance* and consequently - as already alluded to earlier - a channel estimation error margin was introduced for mitigating this degradation. Explicitly, the channel estimation error margin was defined as the measure of how much extra protection margin must be added to the switching threshold levels for the sake of minimising the effects of the channel estimation errors. The delay-spread also degraded the achievable performance due to the associated irreducible BER, which was not compensated for by the receiver. However, the performance of the adaptive scheme in a delay-spread impaired channel environment was better, than that of a fixed modulation scheme. *These experiments also concluded that the AQAM scheme can be operated for a Doppler frequency of $f_d = 10Hz$ at a normalised delay spread of 0.1 or for $f_d = 14Hz$ at a normalised delay spread of 0.02, which produced a mean BER of 0.1% at a transmission rate of 1 Mbits/s.*

Finally, the *data buffering-induced latency and co-channel interference aspects* of AQAM modems were investigated in [219, 220]. Specifically, the latency associated with storing the information to be transmitted during severely degraded channel conditions was mitigated by frequency hopping or statistical multiplexing. As expected, the latency is increased, when either the mobile speed or the channel SNR are reduced, since both of these result in prolonged low instantaneous SNR intervals. It was demonstrated that as a result of the proposed measures, typically more than 4dB SNR reduction was achieved by the proposed adaptive modems in comparison to the conventional fixed-mode benchmark modems employed. However, the achievable gains depend strongly on the prevalent co-channel interference levels and hence interference cancellation was invoked in [220] on the basis of adjusting the demodulation decision boundaries after estimating the interfering channel's magnitude and phase.

The associated AQAM principles may also be invoked in the context of *multicarrier Orthogonal Frequency Division Multiplex (OFDM) modems [180]*. This principle was first proposed by Kalet [155] for employment in OFDM systems and was then further developed for example by Czylwik *et al.* [221] as well as by Chow, Cioffi and Bingham [222]. The associated concepts were detailed for example in [180] and they will also be augmented in Part III of this monograph. Let us now briefly review the recent history of OFDM-based QAM systems in the next section.

1.4 History of OFDM-based QAM

1.4.1 History of OFDM

The first QAM-related so-called orthogonal frequency division multiplexing (OFDM) scheme was proposed by Chang in 1966 [143] for dispersive fading channels, which has also undergone a dramatic evolution due to the efforts of Weinstein, Peled, Ruiz, Hirosaki, Kolb, Cimini, Schüssler, Preuss, Rückriem, Kalet *et al.* [143–156]. OFDM was standardised as the European digital audio broadcast (DAB) as well as digital video broadcast (DVB) scheme. It constituted also a credible proposal for the recent third-generation mobile radio standard competition in Europe. It was recently selected as the high performance local area network (HIPERLAN) transmission technique.

The system's operational principle is that the original bandwidth is divided to a high number of narrow sub-bands, in which the mobile channel can be considered non-dispersive. Hence no channel equaliser is required and instead of implementing a bank of sub-channel modems they can conveniently be implemented with the help of a single Fast Fourier Transformer (FFT). This scheme will be the topic of Chapters 15 - 20.

These OFDM systems - often also termed as frequency division multiplexing (FDM) or multi-tone systems - have been employed in military applications since the 1960s, for example by Bello [223], Zimmermann [144], Powers and Zimmermann [224], Chang and Gibby [225] and others. Saltzberg [226] studied a multi-carrier system employing orthogonal time-staggered quadrature amplitude modulation (O-QAM) of the carriers.

The employment of the discrete Fourier transform (DFT) to replace the banks of sinusoidal generators and the demodulators was suggested by Weinstein and Ebert [145] in 1971, which significantly reduces the implementation complexity of OFDM modems. In 1980, Hirosaki [156] suggested an equalisation algorithm in order to suppress both intersymbol and intersubcarrier interference caused by the channel impulse response or timing and frequency errors. Simplified OFDM modem implementations were studied by Peled [149] in 1980, while Hirosaki [150] introduced the DFT-based implementation of Saltzberg's O-QAM OFDM system. At Erlangen University, Kolb [151], Schüßler [152], Preuss [153] and Rückriem [154] conducted further research into the application of OFDM. Cimini [146] and Kalet [155] published analytical and early seminal experimental results on the performance of OFDM modems in mobile communications channels.

More recent advances in OFDM transmission were presented in the impressive state-of-the-art collection of works edited by Fazel and Fettweis [227], including the research by Fettweis *et al.* at Dresden University, Rohling *et al.* at Braunschweig University, Vandendorp at Loeven University, Huber *et al.* at Erlangen University, Lindner *et al.* at Ulm University, Kammeyer *et al.* at Brehmen University and Meyr *et al.* [228,229] at Aachen University, but the individual contributions are too numerous to mention.

While OFDM transmission over mobile communications channels can alleviate the problem of multipath propagation, recent research efforts have focused on solving a set of inherent difficulties regarding OFDM, namely the peak-to-mean power ratio, time and frequency synchronisation, and on mitigating the effects of the frequency selective fading channel. These issues are addressed below in slightly more depth, while a treatment is given in Chapters 15 - 20.

1.4.2   Peak-to-Mean Power Ratio

The peak-to-mean power ratio problem of OFDM systems has been detailed along with a range of mitigating techniques in [173]. It is plausible that the OFDM signal - which is the superposition of a high number of modulated sub-channel signals - may exhibit a high instantaneous signal peak with respect to the average signal level. Furthermore, large signal amplitude swings are encountered, when the time domain signal traverses from a low instantaneous power waveform to a high power waveform, which may results in a high out-of-band (OOB) harmonic distortion power, unless the transmitter's power amplifier exhibits an extremely high linearity across the entire signal level range (Section 4.5.1). This then potentially contaminates the adjacent channels with adjacent channel interference. Practical amplifiers exhibit a finite amplitude range, in which they can be considered almost linear. In order to prevent severe clipping of the high OFDM signal peaks - which is the main source of OOB emissions - the power amplifier must not be driven to saturation and hence they are typically operated with a certain so-called back-off, creating a certain "head room" for the signal peaks, which reduces the risk of amplifier saturation and OOB emmission. Two different families of solutions have been suggested in the literature, in order to mitigate these problems, either reducing the peak-to-mean power ratio, or improving the amplification stage of the transmitter.

More explicitly, Shepherd [230], Jones [231], and Wulich [232] suggested different coding techniques which aim to minimise the peak power of the OFDM signal by employing different data encoding schemes before modulation, with the philosophy of choosing block codes whose legitimate code words exhibit low so-called Crest factors or peak-to-mean power envelope fluctuation. Müller [233], Pauli [234], May [235] and Wulich [236] suggested different algorithms for post-processing the time domain OFDM signal prior to amplification, while Schmidt and Kammeyer [237] employed adaptive subcarrier allocation in order to reduce the crest factor. Dinis and Gusmão [238–240] researched the use of two-branch amplifiers, while the clustered OFDM technique introduced by Daneshrad, Cimini and Carloni [241] operates with a set of parallel partial FFT processors with associated transmitting chains. OFDM systems with increased robustness to non-linear distortion have been proposed by Okada, Nishijima and Komaki [242] as well as by Dinis and Gusmão [243].

1.4.3   Synchronisation

Time and frequency synchronisation between the transmitter and receiver is of crucial importance as regards the performance of an OFDM link [244, 245]. A wide variety of techniques have been proposed for estimating and correcting both timing and carrier frequency offsets at the OFDM receiver. Rough timing and frequency acquisition algorithms relying on known pilot symbols or pilot tones embedded into the OFDM symbols have been suggested by Claßen [228], Warner [246], Sari [247], Moose [248], as well as Brüninghaus and Rohling [249]. Fine frequency and timing tracking algorithms exploiting the OFDM signal's cyclic extension were published by Moose [248], Daffara [250] and Sandell [251].

1.4.4   OFDM/CDMA

Combining OFDM transmissions with code division multiple access (CDMA) [172] allows us to exploit the wideband channel's inherent frequency diversity by spreading each symbol across multiple subcarriers. This technique has been pioneered by Yee, Linnartz and Fettweis [252], by Chouly, Brajal and Jourdan [253], as well as by Fettweis, Bahai and Anvari [254]. Fazel and Papke [255] investigated convolutional coding in conjunction with OFDM/CDMA. Prasad and Hara [256] compared various methods of combining the two techniques, identifying three different structures, namely multi-carrier CDMA (MC-CDMA), multi-carrier direct sequence CDMA (MC-DS-CDMA) and multi-tone CDMA (MT-CDMA). Like non-spread OFDM transmission, OFDM/CDMA methods suffer from high peak-to-mean power ratios, which are dependent on the frequency domain spreading scheme, as investigated by Choi, Kuan and Hanzo [257].

1.4.5   Adaptive Antennas in OFDM Systems

The employment of adaptive antenna techniques in conjunction with OFDM transmissions was shown to be advantageous in suppressing co-channel interference in cellular communications systems. Li, Cimini and Sollenberger [258–260], Kim, Choi and Cho [261], Lin, Cimini and Chuang [262] as well as Münster *et al.* [263] have investigated algorithms for multi-user channel estimation and interference suppression. To elaborate a little further, multiple antenna assisted wireless communications systems are discussed in [173] in detail, when incorporated in OFDM systems for the sake of increasing the number of users supported. Furthermore, multiple antenna aided space-time coding arrangements constitute the topic of [171], where the main objective is the mitigation of the channel-induced fading, since space-time codecs are capable of achieving substantial transmit diversity gains. A range of space-time spreading aided CDMA schemes, which were designed with a similar objective to space-time codecs, were characterised in [172]. Finally, multiple antenna-based beamformers are discussed in [264], where the basic design objective is to achieve angular selectivity and hence mitigate the effects of co-channel interference.

1.4.6   Decision-Directed Channel Estimation for OFDM

1.4.6.1   Decision-Directed Channel Estimation for Single-User OFDM

In recent years numerous research contributions have appeared on the topic of channel transfer function estimation techniques designed for employment in single-user, single transmit antenna-assisted OFDM scenarios, since the availability of an accurate channel transfer function estimate is one of the prerequisites for coherent symbol detection with an OFDM receiver. The techniques proposed in the literature can be classified as *pilot-assisted*, *decision-directed* (DD) and *blind* channel estimation (CE) methods.

In the context of pilot-assisted channel transfer function estimation, a subset of the available subcarriers is dedicated to the transmission of specific pilot symbols known to the receiver, which are used for "sampling" the desired channel transfer function. Based on these samples of the frequency domain transfer function, the well-known process of interpolation is used for generating a transfer function estimate for each subcarrier residing between the pilots. This is achieved at the cost of a reduction in the number of useful subcarriers available for data transmission. The family of *pilot-assisted* channel estimation techniques was investigated for example by Chang and Su [288], Höher [265,273,274], Itami *et al.* [277], Li [280], Tufvesson and Maseng [272], Wang and Liu [283], as well as Yang *et al.* [279,284,290].

By contrast, in the context of Decision-Directed Channel Estimation (DDCE) all the sliced and remodulated subcarrier data symbols are considered as pilots. In the absence of symbol errors and also depending on the rate of channel fluctuation, it was found that accurate channel transfer function estimates can be obtained, which often are of better quality, in terms of the channel transfer function estimator's mean-square error (MSE), than the estimates offered by pilot-assisted schemes. This is because the latter arrangements usually invoke relatively sparse pilot patterns.

The family of *decision-directed* channel estimation techniques was investigated for example by van den Beek *et al.* [268], Edfors *et al.* [269, 275], Li *et al.* [260], Li [286], Mignone and Morello [271], Al-Susa and Ormondroyd [278], Frenger and Svensson [270], as well as Wilson *et al.* [267]. Furthermore, the family of *blind* channnel estimation techniques was studied by Lu and Wang [285], Necker and Stüber [289], as well as by Zhou and Giannakis [282]. The various contributions have been summarized in Tables 1.1 and 1.2.

In order to render the various DDCE techniques more amenable to use in scenarios as-

Year	Author	Contribution
'91	Höher [265]	Cascaded 1D-FIR channel transfer factor interpolation was carried out in the frequency- and time-direction for frequency-domain PSAM.
'93	Chow, Cioffi and Bingham [266]	Subcarrier-by-subcarrier-based LMS-related channel transfer factor equalisation techniques were employed.
'94	Wilson, Khayata and Cioffi [267]	Linear channel transfer factor filtering was invoked in the time-direction for DDCE.
'95	van den Beek, Edfors, Sandell, Wilson and Börjesson [268]	DFT-aided CIR-related domain Wiener filter-based noise reduction was advocated for DDCE. The effects of leakage in the context of non-sample-spaced CIRs were analysed.
'96	Edfors, Sandell, van den Beek, Wilson and Börjesson [269]	SVD-aided CIR-related domain Wiener filter-based noise reduction was introduced for DDCE.
	Frenger and Svensson [270]	MMSE-based frequency-domain channel transfer factor prediction was proposed for DDCE.
	Mignone and Morello [271]	FEC was invoked for improving the DDCE's remodulated reference.
'97	Tufvesson and Maseng [272]	An analysis of various pilot patterns employed in frequency-domain PSAM was provided in terms of the system's BER for different Doppler frequencies. Kalman filter-aided channel transfer factor estimation was used.
	Höher, Kaiser and Robertson [273, 274]	Cascaded 1D-FIR Wiener filter channel interpolation was utilised in the context of 2D-pilot pattern-aided PSAM
'98	Li, Cimini and Sollenberger [260]	An SVD-aided CIR-related domain Wiener filter-based noise reduction was achieved by employing CIR-related tap estimation filtering in the time-direction.
	Edfors, Sandell, van den Beek, Wilson and Börjesson [275]	A detailed analysis of SVD-aided CIR-related domain Wiener filter-based noise reduction was provided for DDCE, which expanded the results of [269].
	Tufvesson, Faulkner and Maseng [276]	Wiener filter-aided frequency domain channel transfer factor prediction-assisted pre-equalisation was studied.
	Itami, Kuwabara, Yamashita, Ohta and Itoh [277]	Parametric finite-tap CIR model-based channel estimation was employed for frequency domain PSAM.

Table 1.1: Contributions to channel transfer factor estimation for single-transmit antenna-assisted OFDM; ©John Wiley and IEEE Press, 2003 [173].

Year	Author	Contribution
'99	Al-Susa and Ormondroyd [278]	DFT-aided Burg algorithm-assisted adaptive CIR-related tap prediction filtering was employed for DDCE.
	Yang, Letaief, Cheng and Cao [279]	Parametric, ESPRIT-assisted channel estimation was employed for frequency domain PSAM.
'00	Li [280]	Robust 2D frequency domain Wiener filtering was suggested for employment in frequency domain PSAM using 2D pilot patterns.
'01	Yang, Letaief, Cheng and Cao [281]	Detailed discussions of parametric, ESPRIT-assisted channel estimation were provided in the context of frequency domain PSAM [279].
	Zhou and Giannakis [282]	Finite alphabet-based channel transfer factor estimation was proposed.
	Wang and Liu [283]	Polynomial frequency domain channel transfer factor interpolation was contrived.
	Yang, Cao and Letaief [284]	DFT-aided CIR-related domain one-tap Wiener filter-based noise reduction was investigated, which is supported by variable frequency domain Hanning windowing.
	Lu and Wang [285]	A Bayesian blind turbo receiver was contrived for coded OFDM systems.
	Li and Sollenberger [286]	Various transforms were suggested for CIR-related tap estimation filtering-assisted DDCE.
	Morelli and Mengali [287]	LS- and MMSE-based channel transfer factor estimators were compared in the context of frequency domain PSAM.
'02	Chang and Su [288]	Parametric quadrature surface-based frequency domain channel transfer factor interpolation was studied for PSAM.
	Necker and Stüber [289]	Totally blind channel transfer factor estimation based on the finite alphabet property of PSK signals was investigated.

Table 1.2: Contributions to channel transfer factor estimation for single-transmit antenna-assisted OFDM; ©John Wiley and IEEE Press, 2003 [173].

sociated with a relatively high rate of channel variation expressed in terms of the OFDM symbol normalised Doppler frequency, linear prediction techniques well known from the speech coding literature [169, 291] can be invoked. To elaborate a little further, we will substitute the CIR-related tap estimation filter - which is part of the two-dimensional channel transfer function estimator proposed in [260] - by a CIR-related tap prediction filter. The employment of this CIR-related tap prediction filter enables a more accurate estimation of the channel transfer function encountered during the forthcoming transmission time slot and thus potentially enhances the performance of the channel estimator. We will be following the general concepts described by Duel-Hallen *et al.* [292] and the ideas presented by Frenger and Svensson [270], where frequency domain prediction filter-assisted DDCE was proposed. Furthermore, we should mention the contributions of Tufvesson *et al.* [276, 293], where a prediction filter-assisted frequency domain pre-equalisation scheme was discussed in the context of OFDM. In a further contribution by Al-Susa and Ormondroyd [278], adaptive prediction filter-assisted DDCE designed for OFDM has been proposed upon invoking techniques known from speech coding, such as the Levinson-Durbin algorithm or the Burg algorithm [291, 294, 295] in order to determine the predictor coefficients.

1.4.6.2 Decision-Directed Channel Estimation for Multi-User OFDM

In contrast to the above-mentioned single-user OFDM scenarios, in a multi-user OFDM scenario the signal received by each antenna is constituted by the superposition of the signal contributions associated with the different users or transmit antennas. Note that in terms of the multiple-input multiple-output (MIMO) structure of the channel the multi-user single-transmit antenna scenario is equivalent, for example, to a single-user space-time coded (STC) scenario using multiple transmit antennas. For the latter a Least-Squares (LS) error channel estimator was proposed by Li *et al.* [296], which aims at recovering the different transmit antennas' channel transfer functions on the basis of the output signal of a specific reception antenna element and by also capitalising on the remodulated received symbols associated with the different users. The performance of this estimator was found to be limited in terms of the mean-square estimation error in scenarios, where the product of the number of transmit antennas and the number of CIR taps to be estimated per transmit antenna approaches the total number of subcarriers hosted by an OFDM symbol. As a design alternative, in [297] a DDCE was proposed by Jeon *et al.* for a space-time coded OFDM scenario of two transmit antennas and two receive antennas.

Specifically, the channel transfer function[1] associated with each transmit-receive antenna pair was estimated on the basis of the output signal of the specific receive antenna upon *subtracting* the interfering signal contributions associated with the remaining transmit antennas. These interference contributions were estimated by capitalising on the knowledge of the channel transfer functions of all interfering transmit antennas predicted during the $(n-1)$-th OFDM symbol period for the n-th OFDM symbol, also invoking the corresponding remodulated symbols associated with the n-th OFDM symbol. To elaborate further, the difference between the subtraction-based channel transfer function estimator of [297] and the LS estimator proposed by Li *et al.* in [296] is that in the former the channel transfer functions predicted during the previous, i.e. the $(n-1)$-th OFDM symbol period for the current, i.e.

---

[1]In the context of the OFDM system the set of K different subcarriers' channel transfer factors is referred to as the channel transfer function, or simply as the channel.

Year	Author	Contribution
'99	Li, Seshadri and Ariyavisitakul [296]	The LS-assisted DDCE proposed exploits the cross-correlation properties of the transmitted subcarrier symbol sequences.
'00	Jeon, Paik and Cho [297]	Frequency-domain PIC-assisted DDCE is studied, which exploits the channel's slow variation versus time.
	Li [298]	Time-domain PIC-assisted DDCE is investigated as a simplification of the LS-assisted DDCE of [296]. Optimum training sequences are proposed for the LS-assisted DDCE of [296].
'01	Mody and Stüber [299]	Channel transfer factor estimation designed for frequency-domain PSAM based on CIR-related domain filtering is studied.
	Gong and Letaief [300]	MMSE-assisted DDCE is advocated which represents an extension of the LS-assisted DDCE of [300]. The MMSE-assisted DDCE is shown to be practical in the context of transmitting consecutive training blocks. Additionally, a low-rank approximation of the MMSE-assisted DDCE is considered.
	Jeon, Paik and Cho [301]	2D MMSE-based channel estimation is proposed for frequency-domain PSAM.
	Vook and Thomas [302]	2D MMSE-based channel estimation is invoked for frequency domain PSAM. A complexity reduction is achieved by CIR-related domain-based processing.
	Xie and Georghiades [303]	Expectation maximisation (EM)-based channel transfer factor estimation approach for DDCE.
'02	Li [304]	A more detailed discussion on time-domain PIC-assisted DDCE is provided and optimum training sequences are proposed [298].
	Bölcskei, Heath and Paulraj [305]	Blind channel identification and equalisation using second-order cyclostationary statistics as well as antenna precoding were studied.
	Minn, Kim and Bhargava [306]	A reduced complexity version of the LS-assisted DDCE of [296] is introduced, based on exploiting the channel's correlation in the frequency-direction, as opposed to invoking the simplified scheme of [304], which exploits the channel's correlation in the time-direction. A similar approach was suggested by Slimane [307] for the specific case of two transmit antennas.
	Komninakis, Fragouli, Sayed and Wesel [308]	Fading channel tracking and equalisation were proposed for employment in MIMO systems assisted by Kalman estimation and channel prediction.

Table 1.3: Contributions on channel transfer factor estimation for multiple-transmit antenna assisted OFDM; ⓒJohn Wiley and IEEE Press, 2003 [173].

the n-th OFDM symbol, are employed for both symbol detection *as well as* for obtaining an updated channel estimate for employment during the $(n + 1)$-th OFDM symbol period. In the approach advocated in [297] the subtraction of the different transmit antennas' interfering signals is performed in the frequency domain.

By contrast, in [298] a similar technique was proposed by Li with the aim of simplifying the DDCE approach of [296], which operates in the time domain. A prerequisite for the operation of this parallel interference cancellation (PIC)-assisted DDCE is the availability of a reliable estimate of the various channel transfer functions for the current OFDM symbol, which are employed in the cancellation process in order to obtain updated channel transfer function estimates for the demodulation of the next OFDM symbol. In order to compensate for the channel's variation as a function of the OFDM symbol index, linear prediction techniques can be employed, as was also proposed for example in [298]. However, due to the estimator's recursive structure, determining the optimum predictor coefficients is not as straightforward as for the transversal FIR filter-assisted predictor.

An overview of further publications on channel transfer factor estimation for OFDM systems supported by multiple antennas is provided in Table 1.3, although these topics are beyond the scope of this book. Various multiple antenna aided wireless communications systems are discussed in [173] in detail, when incorporated in OFDM systems. Furthermore, multiple antenna aided space-time coding arrangements constitute the topic of [171], while space-time spreading is addressed in [172]. Finally, multiple antenna-based beamformers are discussed in [264].

1.4.7   Uplink Detection Techniques for Multi-User SDMA-OFDM

The related family of Space-Division-Multiple-Access (SDMA) communication systems has recently attracted wide research interests. In these systems the L different users' transmitted signals are separated at the base-station (BS) with the aid of their unique, user-specific spatial signature, which is constituted by the P-element vector of channel transfer factors between the users' single transmit antenna and the P different receiver antenna elements at the BS, upon assuming flat-fading channel conditions such as those often experienced in the context of each of the OFDM subcarriers.

A whole host of multi-user detection (MUD) techniques known from Code-Division-Multiple-Access (CDMA) communications lend themselves also to an application in the context of SDMA-OFDM on a per-subcarrier basis. Some of these techniques are the Least-Squares (LS) [318, 324, 332, 334], Minimum Mean-Square Error (MMSE) [310–313, 315, 318, 322, 326, 334–336], Successive Interference Cancellation (SIC) [309, 314, 318, 322, 324, 329, 331, 333, 334, 336], Parallel Interference Cancellation (PIC) [330, 334] and Maximum Likelihood (ML) detection [317, 319–323, 325, 328, 334, 336]. A comprehensive overview of recent publications on MUD techniques for MIMO systems is given in Tables 1.4 and 1.5.

1.4.8   OFDM Applications

Due to their implementational complexity, OFDM applications have been scarce until quite recently. Recently, however, OFDM has been adopted as the new European digital audio broadcasting (DAB) standard [147, 148, 337–339] as well as for the terrestrial digital video broadcasting (DVB) system [247, 340]. During this process the design of OFDM systems

Year	Author	Contribution
'96	Foschini [309]	The concept of the BLAST architecture was introduced.
'98	Vook and Baum [310]	SMI-assisted MMSE combining was invoked on an OFDM subcarrier basis.
	Wang and Poor [311]	Robust sub-space-based weight vector calculation and tracking were employed for co-channel interference suppression, as an improvement of the SMI-algorithm.
	Wong, Cheng, Letaief and Murch [312]	Optimisation of an OFDM system was reported in the context of multiple transmit and receive antennas upon invoking the maximum SINR criterion. The computational was reduced by exploiting the channel's correlation in the frequency direction.
	Li and Sollenberger [313]	Tracking of the channel correlation matrix' entries was suggested in the context of SMI-assisted MMSE combining for multiple receiver antenna assisted OFDM, by capitalizing on the principles of [260].
'99	Golden, Foschini, Valenzuela and Wolniansky [314]	The SIC detection-assisted V-BLAST algorithm was introduced.
	Li and Sollenberger [315]	The system introduced in [313] was further detailed.
	Vandenameele, Van der Perre, Engels and de Man [316]	A comparative study of different SDMA detection techniques, namely that of MMSE, SIC and ML detection was provided. Further improvements of SIC detection were suggested by adaptively tracking multiple symbol decisions at each detection node.
	Speth, Senst and Meyr [317]	Soft-bit generation techniques were proposed for MLSE in the context of a coded SDMA-OFDM system.
'00	Sweatman, Thompson, Mulgrew and Grant [318]	Comparisons of various detection algorithms including LS, MMSE, D-BLAST and V-BLAST (SIC detection) were carried out.
	van Nee, van Zelst and Awater [319–321]	The evaluation of ML detection in the context of a Space-Division Multiplexing (SDM) system was provided, considering various simplified ML detection techniques.
	Vandenameele, Van der Perre, Engels, Gyselinckx and de Man [322]	More detailed discussions were provided on the topics of [316].

Table 1.4: Contributions on multi-user detection techniques designed for multiple transmit antenna-assisted OFDM systems; ©John Wiley and IEEE Press, 2003 [173].

Year	Author	Contribution
'00	Li, Huang, Lozano and Foschini [323]	Reduced complexity ML detection was proposed for multiple transmit antenna systems employing adaptive antenna grouping and multi-step reduced-complexity detection.
'01	Degen, Walke, Lecomte and Rembold [324]	An overview of various adaptive MIMO techniques was provided. Specifically, pre-distortion was employed at the transmitter, as well as LS- or BLAST detection were used at the receiver or balanced equalisation was invoked at both the transmitter and receiver.
	Zhu and Murch [325]	A tight upper bound on the SER performance of ML detection was derived.
	Li, Letaief, Cheng and Cao [326]	Joint adaptive power control and detection were investigated in the context of an OFDM/SDMA system, based on the approach of Farrokhi *et al.* [327].
	van Zelst, van Nee and Awater [328]	Iterative decoding was proposed for the BLAST system following the turbo principle.
	Benjebbour, Murata and Yoshida [329]	The performance of V-BLAST or SIC detection was studied in the context of backward iterative cancellation scheme employed after the conventional forward cancellation stage.
	Sellathurai and Haykin [330]	A simplified D-BLAST was proposed, which used iterative PIC capitalizing on the extrinsic soft-bit information provided by the FEC scheme used.
	Bhargave, Figueiredo and Eltoft [331]	A detection algorithm was suggested, which followed the concepts of V-BLAST or SIC. However, multiple symbols states are tracked from each detection stage, where - in contrast to [322] - an intermediate decision is made at intermediate detection stages.
	Thoen, Deneire, Van der Perre and Engels [332]	A constrained LS detector was proposed for OFDM/SDMA, which was based on exploiting the constant modulus property of PSK signals.
'02	Li and Luo [333]	The block error probability of optimally ordered V-BLAST was studied. Furthermore, the block error probability is also investigated for the case of tracking multiple parallel symbol decisions from the first detection stage, following an approach similar to that of [322].

Table 1.5: Contributions on detection techniques for MIMO systems and for multiple transmit antenna-assisted OFDM systems;©John Wiley and IEEE Press, 2003 [173].

has matured and their wide-range employment has become a cost-efficient commercial real-ity. In recent years OFDM schemes have found their way into wireless local area networks (WLANs) as well, such as the 802.11 family.

For fixed-wire applications, OFDM is employed in the asynchronous digital subscriber line (ADSL) and high-bit-rate digital subscriber line (HDSL) systems [341–344] and it has also been suggested for power line communications systems [345, 346] due to its resilience to time-dispersive channels and narrow band interferers.

OFDM applications were studied also within the various European Research projects [347]. The MEDIAN project investigated a 155 Mbps wireless asynchronous transfer mode (WATM) network [348–351], while the Magic WAND group [352,353] developed an OFDM-based WLAN. Hallmann and Rohling [354] presented a range of different OFDM systems that were applicable to the European Telecommunications Standardisation Institute's (ETSI) recent personal communications oriented air interface concept [355].

1.5   History of QAM-Based Coded Modulation

The history of channel coding or Forward Error Correction (FEC) coding dates back to Shannon's pioneering work [356] in 1948, in which he showed that it is possible to design a communication system with any desired small probability of error, whenever the rate of transmission is smaller than the capacity of the channel. While Shannon outlined the theory that explained the fundamental limits imposed on the efficiency of communications systems, he provided no insights into how to actually approach these limits. This motivated the search for codes that would produce arbitrarily small probability of error. Specifically, Hamming [357] and Golay [358] were the first to develop practical error control schemes. Convolutional codes [359] were later introduced by Elias in 1955, while Viterbi [360] invented a maximum likelihood sequence estimation algorithm in 1967 for efficiently decoding convolutional codes. In 1974, Bahl proposed the more complex Maximum A-Posteriori (MAP) algorithm, which is capable of achieving the minimum achievable BER.

The first successful application of channel coding was the employment of convolutional codes [359] in deep-space probes in the 1970s. However, for years to come, error control coding was considered to have limited applicability, apart from deep-space communications. Specifically, this is a power-limited scenario, which has no strict bandwidth limitation. By contrast mobile communications systems constitute a power- and bandwidth-limited scenario. In 1987, a bandwidth efficient Trellis Coded Modulation (TCM) [361] scheme employing symbol-based channel interleaving in conjunction with Set-Partitioning (SP) [362] assisted signal labelling was proposed by Ungerböck. Specifically, the TCM scheme, which is based on combining convolutional codes with multidimensional signal sets, constitutes a bandwidth efficient scheme that has been widely recognised as an efficient error control technique suitable for applications in mobile communications [363]. Another powerful coded modulation scheme utilising bit-based channel interleaving in conjunction with Gray signal labelling, which is referred to as Bit-Interleaved Coded Modulation (BICM), was proposed by Zehavi [364] as well as by Caire, Taricco and Biglieri [365]. Another breakthrough in the history of error control coding is the invention of turbo codes by Berrou, Glavieux and Thitimajshima [366] in 1993. Convolutional codes were used as the component codes and decoders based on the MAP algorithm were employed. The results proved that a performance close to the Shannon limit can be achieved in practice with the aid of binary codes. The attractive properties of turbo codes have stimulated intensive research in this area [367–369]. As a result, turbo coding has reached a state of maturity within just a few years and was standardised in the recently ratified third-generation (3G) mobile radio systems [370].

However, turbo codes often have a low coding rate and hence require considerable bandwidth expansion. Therefore, one of the objectives of turbo coding research is the design of bandwidth-efficient turbo codes. In order to equip the family of binary turbo codes with a higher spectral efficiency, BICM-based Turbo Coded Modulation (TuCM) [371] was proposed in 1994. Specifically, TuCM uses a binary turbo encoder, which is linked to a signal mapper, after its output bits were suitably punctured and multiplexed for the sake of transmitting the desired number of information bits per transmitted symbol. In the TuCM scheme of [371] Gray-coding based signal labelling was utilised. For example, two 1/2-rate Recursive Systematic Convolutional (RSC) codes are used for generating a total of four turbo coded bits and this bit stream may be punctured for generating three bits, which are mapped to an 8PSK modulation scheme. By contrast, in separate coding and modulation scheme, any modulation schemes for example BPSK, may be used for transmitting the channel coded bits. Finally, without puncturing, 16QAM transmission would have to be used for maintaining the original transmission bandwidth. Turbo Trellis Coded Modulation (TTCM) [372] is a more recently proposed channel coding scheme that has a structure similar to that of the family of turbo codes, but employs TCM schemes as its component codes. The TTCM symbols are transmitted alternatively from the first and the second constituent TCM encoders and symbol-based interleavers are utilised for turbo interleaving and channel interleaving. It was shown in [372] that TTCM performs better than the TCM and TuCM schemes at a comparable complexity. In 1998, iterative joint decoding and demodulation assisted BICM referred to as BICM-ID was proposed in [373, 374], which uses SP-based signal labelling. The aim of BICM-ID is to increase the Euclidean distance of BICM and hence to exploit the full advantage of bit interleaving with the aid of soft-decision feedback based iterative decoding [374]. Many other bandwidth efficient schemes using turbo codes have been proposed in the literature [368], but we will focus our study on TCM, BICM, TTCM and BICM-ID schemes in the context of wireless channels in this part of the book.

1.6 QAM in Multiple Antenna-Based Systems

In recent years various smart antenna designs have emerged, which have found application in diverse scenarios, as seen in Table 1.6. The main objective of employing smart antennas is that of combating the effects of multipath fading on the desired signal and suppressing interfering signals, thereby increasing both the performance and capacity of wireless systems [375]. Specifically, in smart antenna-assisted systems multiple antennas may be invoked at the transmitter and/or the receiver, where the antennas may be arranged for achieving spatial diversity, directional beamforming or for attaining both diversity and beamforming. In smart antenna systems the achievable performance improvements are usually a function of the antenna spacing and that of the algorithms invoked for processing the signals received by the antenna elements.

In beamforming arrangements [264] typically $\lambda/2$-spaced antenna elements are used for the sake of creating a spatially selective transmitter/receiver beam. Smart antennas using

Beamforming [264]	Typically $\lambda/2$-spaced antenna elements are used for the sake of creating a spatially selective transmitter/receiver beam. Smart antennas using beamforming have been employed for mitigating the effects of co-channel interfering signals and for providing beamforming gain.
Spatial Diversity [171] and Space-Time Spreading	In contrast to the $\lambda/2$-spaced phased array elements, in spatial diversity schemes, such as space-time block or trellis codes [171] the multiple antennas are positioned as far apart as possible, so that the transmitted signals of the different antennas experience independent fading, resulting in the maximum achievable diversity gain.
Space Division Multiple Access	SDMA exploits the unique, user-specific "spatial signature" of the individual users for differentiating amongst them. This allows the system to support multiple users within the same frequency band and/or time slot.
Multiple Input Multiple Output Systems [309]	MIMO systems also employ multiple antennas, but in contrast to SDMA arrangements, not for the sake of supporting multiple users. Instead, they aim for increasing the throughput of a wireless system in terms of the number of bits per symbol that can be transmitted by a given user in a given bandwidth at a given integrity.

Table 1.6: Applications of multiple antennas in wireless communications

beamforming have widely been employed for mitigating the effects of various interfering signals and for providing beamforming gain. Furthermore, the beamforming arrangement is capable of suppressing co-channel interference, which allows the system to support multiple users within the same bandwidth and/or same time-slot by separating them spatially. This spatial separation becomes, however, only feasible, if the corresponding users are separable in terms of the angle of arrival of their beams. These beamforming schemes, which employ appropriately phased antenna array elements that are spaced at distances of $\lambda/2$ typically result in an improved SINR distribution and enhanced network capacity [264].

In contrast to the $\lambda/2$-spaced phased array elements, **in spatial diversity schemes**, such as space-time coding [171] aided transmit diversity arangements, the multiple antennas are positioned as far apart as possible. A typical antenna element spacing of 10λ [375] may be used, so that the transmitted signals of the different antennas experience independent fading, when they reach the receiver. This is because the maximum diversity gain can be achieved, when the received signal replicas experience independent fading. Although spatial diversity can be achieved by employing multiple antennas at either the base station, mobile station, or both, it is more cost effective and practical to employ multiple transmit antennas at the base station. A system having multiple receiver antennas has the potential of achieving receiver diversity, while that employing multiple transmit antennas exhibits transmit diversity. Recently, the family of transmit diversity schemes based on space-time coding, either space-time block codes or space-time trellis codes, has received wide attention and has been invoked in the 3rd-generation systems [264,376]. The aim of using spatial diversity is to provide both transmit as well as receive diversity and hence enhance the system's integrity/robustness. This typically results in a better physical-layer performance and hence a better network-layer performance, hence space-time codes indirectly increase not only the transmission integrity, but also the achievable spectral efficiency.

A third application of smart antennas is often referred to as **Space Division Multiple Access** (SDMA), which exploits the unique, user-specific "spatial signature" of the individual users for differentiating amongst them. In simple conceptual terms one could argue that both a conventional CDMA spreading code and the Channel Impulse Response (CIR) affect the transmitted signal similarly - they are namely convolved with it. Hence, provided that the CIR is accurately estimated, it becomes known and certainly unique, although - as opposed to orthogonal Walsh-Hadamad spreading codes, for example - not orthogonal to the other CIRs. Nonetheless, it may be used for uniquely identifying users after channel estimation and hence for supporting several users within the same bandwidth. Provided that a powerful multiuser detector is available, one can support even more users than the number of antennas. Hence this method enhances the achievable spectral efficiency directly.

Finally, Multiple Input Multiple Output (MIMO) systems [309, 377–380] also employ multiple antennas, but in contrast to SDMA arrangements, not for the sake of supporting multiple users. Instead, they aim for increasing the throughput of a wireless system in terms of the number of bits per symbol that can be transmitted by a single user in a given bandwidth at a given integrity.

1.7 Outline of the Book

1.7.1 Part I: QAM Basics

- In **Chapter 2** we consider the communications channels over which we wish to send our data. These channels are divided into Gaussian and mobile radio channels, and the characteristics of each are explained.

- **Chapter 3** provides an introduction to modems, considering the manner in which speech or other source waveforms are converted into a form suitable for transmission over a channel, and introducing some of the fundamentals of modems.

- **Chapter 4** provides a more detailed description of modems, specifically that of the modulator, considering QAM constellations, pulse shaping techniques, methods of generating and detecting QAM, as well as amplifier techniques to reduce the problems associated with non-linearities.

- **Chapter 5** elaborates on the details of decision theory and highlights the theoretical aspects of QAM transmission, showing how the BER can be mathematically computed for transmission over Gaussian channels.

- **Chapter 6** considers a range of classic clock and carrier recovery schemes, which are applicable mainly to systems operating over benign Gaussian channels, such as the times-two and the early-late recovery schemes as well as their derivatives.

- **Chapter 7** continues our discourse by considering channel equalisers. First, the family of classic zero-forcing and least mean square equalisers, as well as Kalman filtering based schemes are discussed. Then a large part of this chapter is dedicated to the portrayal of blind channel equalisers.

- **Chapter 8** introduces the concept of classic trellis coded modulation schemes and deals with the historically important family of modems designed for Gaussian channels such as telephone lines.

1.7.2 Part II: Adaptive QAM Techniques for Fading Channels

- **Chapter 9** constitutes the first chapter of Part II of the book, focusing on QAM-based wireless communication by providing a theoretical analysis of QAM transmission over Rayleigh fading mobile radio channels using the so-called maximum minimum distance square-shaped constellation.

- **Chapter 10** introduces the concept of differentially encoded QAM, which was designed for maintaining a low detection complexity, when communicating over hostile wireless channels. These schemes are capable of operating without the employment coherent carrier recovery arrangements, which are prone to false locking in the presence of channel fading. This chapter also considers some of the practicalities of QAM transmissions over these wireless links, including the effects of intentional constellation distortions on the probabilities of the four individual bits of a 4-bit symbol and of hardware imperfections.

- **Chapter 11** details a range of various clock and carrier recovery schemes designed for mobile radio communications using QAM.

- **Chapter 12** proposes a range of various channel equalisers designed for wideband QAM-aided transmissions.

- In **Chapter 13** we consider various orthogonal transmission and pulse shaping techniques in the context of quadrature-quadrature amplitude modulation also referred to as Q^2AQM.

- **Chapter 14** considers the spectral efficiency gains that can be achieved, when using QAM instead of conventional binary modulation techniques, when communicating in interference-limited cellular environments.

1.7.3 Part III: Advanced QAM
Adaptive OFDM Systems

- In **Chapter 15** we focus our attention on the employment of the Fourier transform in order to show mathematically, how orthogonal frequency division multiplexing (OFDM) schemes may be implemented at the cost of a low complexity.

- In **Chapter 16** the BER performance of OFDM modems in AWGN channels is studied for a set of different modulation schemes in the subcarriers. The effects of amplitude limiting of the transmitter's output signal, caused by a simple clipping amplifier model, and of finite resolution D/A and A/D conversion on the system performance are investigated. Oscillator phase noise is considered as a source of inter-subcarrier interference and its effects on the system performance are demonstrated.

- In **Chapter 17** the effects of time-dispersive frequency-selective Rayleigh fading channels on OFDM transmissions are demonstrated. Channel estimation techniques are presented which support the employment of coherent detection in frequency selective channels. Additionally, differential detection is investigated, and the resulting system performance over the different channels is compared.

- **Chapter 18** focuses our attention on the time and frequency synchronisation requirements of OFDM transmissions and the effects of synchronisation errors are demonstrated. Two novel synchronisation algorithms designed for both transmission frame and OFDM symbol synchronisation are suggested and compared. The resultant system performance recorded, when communicating over fading wideband channels is examined.

- In **Chapter 19**, based on the results of Chapter 22 and Chapter 17, the employment of adaptive modulation schemes is suggested for duplex point-to-point links over frequency-selective time-varying channels. Different bit allocation schemes are investigated and a simplified sub-band adaptivity OFDM scheme is suggested for alleviating the associated signalling constraints. A range of blind modulation scheme detection algorithms are also investigated and compared. The employment of long-block-length convolutional turbo codes is suggested for improving the system's throughput and the turbo coded adaptive OFDM modem's performance is compared using different sets of parameters. Then the effects of using pre-equalisation at the transmitter are examined, and a set of different pre-equalisation algorithms is introduced. A joint pre-equalisation and adaptive modulation algorithm is proposed and its BER as well as throughput performance are studied.

- In **Chapter 20** the adaptive OFDM transmission ideas of Chapter 22 and Chapter 19 are extended further, in order to include adaptive error correction coding, based on redundant residual number system (RRNS) and turbo BCH codes. A joint modulation and code rate adaptation scheme is presented.

- **Chapter 21** is dedicated to an OFDM-based system design study, which identifies the benefits and disadvantages of both space-time trellis as well as space-time block codes versus adaptive modulation under various propagation conditions.

- **Chapter 22** provides a detailed mathematical characterisation of adaptive QAM systems, which are capable of appropriately adjusting the number of bits per QAM symbol on the basis of the instantaneous channel conditions. When the instantaneous channel quality is high, a high number of bits is transmitted. By contrast, under hostile channel conditions a low number of bits is transmitted for the sake of maintaining the target integrity. These concepts are also extended to sophisticated space-time coded multi-carrier OFDM and MC-CDMA systems employing multiple transmitters and receivers.

1.7.4 Part IV: Advanced QAM
Turbo-Equalised Adaptive TCM, TTCM, BICM, BICM-ID and Space-Time Coding Assisted OFDM, CDMA and MC-CDMA Systems

- **Chapter 23** deals with an important accomplishment of information theory,namely with the determination of the channel capacity, which quantifies the maximum achievable transmission rate of a system communicating over a bandlimited channel, while maintaining an arbitrarily low probability of error. Given the fact that the available bandwidth of all transmission media is limited, it is desirable to transmit information as bandwidth-efficiently, as possible, which the ultimate goal of this monograph.

- In **Chapter 24** four different coded modulation schemes, namely TCM, TTCM, BICM and BICM-ID are introduced. The conceptual differences amongst these four coded modulation schemes are studied in terms of their coding structure, signal labelling philosophy, interleaver type and decoding philosophy. The symbol-based MAP algorithm operating in the logarithmic domain is also highlighted.

- **Chapter 25** studies the achievable performance of the above-mentioned coded modulation schemes, when communicating over AWGN and narrowband fading channels. Multi-carrier Orthogonal Frequency Division Multiplexing (OFDM) is also combined with the coded modulation schemes designed for communicating over wideband fading channels. With the aid of multi-carrier OFDM the wideband channel is divided into numerous narrowband sub-channels, each associated with an individual OFDM subcarrier. The performance trends of the coded modulation schemes are studied in the context of these OFDM sub-channels and compared in terms of the associated decoding complexity, coding delay and effective throughput under the assumption of encountering non-dispersive channel conditions in each sub-channel.

- In **Chapter 26** the channel equalisation concepts of Chapter 7 are developed further in the context of AQAM and both a conventional Decision Feedback Equaliser (DFE) and a Radial Basis Function (RBF)-based DFE are introduced. These schemes are then combined with various coded modulation schemes communicating over wideband fading channels. The concepts of conventional DFE-based adaptive modulation as well as RBF-based turbo equalisation and a reduced complexity RBF-based In-phase(I)/Quadrature-phase(Q) turbo equalisation scheme are also presented. We will incorporate the various coded modulation schemes considered into these systems and evaluate their performance in terms of the achievable BER, FER and effective throughput, when assuming a similar bandwidth, coding rate and decoding complexity for the various arrangements.

- In **Chapter 27** the performance of the various coded modulation schemes is also evaluated in conjunction with a Direct Sequence (DS) Code-Division Multiple Access (CDMA) system. Specifically, a DFE based Multi-User Detection (MUD) scheme is introduced for assisting the fixed-mode coded modulation schemes as well as the adaptive coded modulation schemes operating in conjunction with DS-CDMA, when communicating over wideband fading channels. The concept of Genetic Algorithm (GA) based MUD is also highlighted, which is invoked in conjunction with the coded modulation schemes for employment in the CDMA system. The performance of this MUD is compared to that of the optimum MUD.

- In **Chapter 28** IQ-interleaved Coded Modulation (IQ-CM) schemes are introduced for achieving IQ diversity. Space Time Block Coding (STBC) is also introduced for attaining additional space/transmit and time diversity. The concept of Double-Spreading aided Rake Receivers (DoS-RR) is proposed for achieving multipath diversity in a CDMA downlink, when transmitting over wideband fading channels. Finally, a STBC-based IQ-CM assisted DoS-RR scheme is proposed for attaining transmit-, time-, IQ- and multipath-diversity, in a CDMA downlink, when communicating over wideband fading channels.

- **Chapter 29** provides a comparative study of the various coded modulation schemes studied in Part IV of the book, including suggestions for future research on coded modulation aided transceivers.

- **Chapter 30** presents a variety of advanced QAM and OFDM assisted turbo-coded DVB schemes. Specifically, it compares the performance of schemes that use blind-equalised QAM modems. It is demonstrated that in comparison to the standard-based DVB systems the employment of turbo coding can provide an extra 5-6 dB channel SNR gain and this can be exploited, for example, to double the number of transmitted bits in a given bandwidth. This then ultimately allows us to improve for example the associated video quality that can be guaranteed in a given bandwidth at the cost of the associated additional implementational complexity.

1.8 Summary

Here we conclude our introduction to QAM and the review of publications concerning QAM as well as its various applications, spanning a period of four decades between the first theoretical study conducted in the context of Gaussian channels, leading to a whole host of applications in various sophisticated wireless systems operating in diverse propagation environments. We now embark on a detailed investigation of the topics highlighted in this introductory chapter.

Chapter **2**

Communications Channels

In this chapter we consider the communication channel which exists between the transmitter and the receiver. Accurate characterisation of this channel is essential if we are to remove the impairments imposed by the channel using signal processing at the receiver. Initially we consider fixed links whereby both terminals are stationary, and then mobile radio communications channels which change significantly with time are investigated.

2.1 Fixed Communication Channels

2.1.1 Introduction

We define fixed communications channels to be those between a fixed transmitter and a fixed receiver. These channels are exemplified by twisted pairs, cables, waveguides, optical fibre and point-to-point microwave radio channels. Whatever the nature of the channel, its output signal differs from the input signal. The difference might be deterministic or random, but it is typically unknown to the receiver. Examples of channel impairments are dispersion, non-linear distortions, delay, and random noise.

Fixed communications channels can often be modelled by a linear transfer function, which describes the channel dispersion. The ubiquitous additive Gaussian noise (AWGN) is a fundamental limiting factor in communications via linear time invariant (LTI) channels.

Although the channel characteristics might change due to ageing, temperature changes, channel switching, etc., these variations will not be apparent over the course of a typical communication. It is this inherent time invariance which characterises fixed channels.

An ideal, distortion-free communications channel would have a flat frequency response and linear phase response over the bandwidth (B) of the signals to be transmitted, as seen in Figure 2.1. In this figure $A(\omega)$ represents the magnitude of the channel response at frequency ω, and ωT represents the phase shift at frequency ω. Practical channels always have some linear distortions, i.e. a non-flat frequency response and a non-linear phase response. In addition, the group delay response of the channel, which is the derivative of the phase response, is often given.

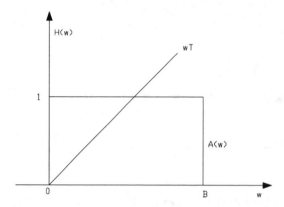

Figure 2.1: Ideal, distortion-free channel model

2.1.2   Fixed Channel Types

Conventional telephony uses twisted pairs to connect subscribers to the local exchange. The
bandwidth is approximately 3.4 kHz, and the waveform distortions are relatively benign.

For applications requiring a higher bandwidth coaxial cables can be used. Their attenuation increases approximately with the square root of the frequency. Hence for wideband,
long-distance operation they require channel equalisation. Typically, coaxial cables can provide a bandwidth of about 50 MHz, and the transmission rate they can support is limited by
the so-called skin effect.

Recent advances in optical communications, such as the invention of the optical amplifiers, render optical fibres more attractive for higher bit rates. Guided optical wave transmission fibres use fine strands of silicon-based, high purity glass. For short-haul systems
the so-called step index fibres are used where the higher refractive index internal core is
surrounded by a lower refractive index external cladding. Their effective bandwidth is limited by their length and the difference between the refractive index of the core and that of
the cladding. For long-distance, high rate applications the so-called graded index fibres are
more suitable, where the refractive index gradually reduces from the central core towards the
cladding. Their relative loss expressed in dB/km versus wavelength characteristic shows a
minimum around 1550 nm. This wavelength also allows the design of low dispersion fibres.

Point-to-point microwave radio channels typically utilise high-gain directional transmit
and receive antennae in a line-of-sight scenario, where free space propagation conditions
may be applicable.

2.1.3   Characterisation of Noise

Irrespective of the communications channel used, random noise is always present. Noise can
be broadly classified as natural or man-made. Examples of man-made noise are those due to
electrical appliances, lighting, etc., and the effects of those sources can usually be mitigated at
the source. Natural noise sources affecting radio transmissions include galactic star radiations
and atmospheric noise. There exists a low noise window in the range of 1–10 GHz, where

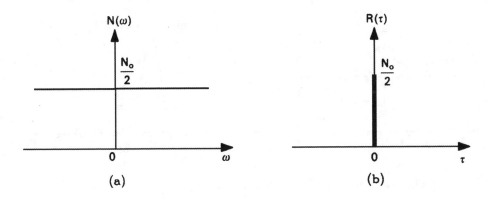

Figure 2.2: Power spectral density and autocorrelation of WN

the effects of these sources are minimised.

Natural thermal noise is ubiquitous. This is due to the random motion of electrons and can be reduced by reducing the temperature. Since thermal noise contains practically all frequency components up to some 10^{13} Hz with equal power, it is often referred to as white noise (WN) in an analogy to white light containing all colours with equal intensity. This WN process can be characterised by its uniform power spectral density (PSD) $N(\omega) = N_0/2$ shown together with its autocorrelation function (ACF) in Figure 2.2. The ACF $R(\tau)$ can be computed using the Wiener-Khintchine theorem, which states that $N(\omega)$ and $R(\tau)$ are Fourier transform pairs such that:

$$
\begin{aligned}
R(\tau) &= \frac{1}{2\pi} \int_{-\infty}^{\infty} N(\omega) e^{j\omega\tau} d\omega = \frac{1}{2\pi} \int_{-\infty}^{\infty} \frac{N_0 e^{j\omega\tau}}{2} d\omega \\
&= \frac{N_0}{2} \frac{1}{2\pi} \int_{-\infty}^{\infty} e^{j\omega\tau} d\omega = \frac{N_0}{2} \delta(\tau),
\end{aligned}
\tag{2.1}
$$

where $\delta(\tau)$ is the Dirac delta function. Clearly, for any time domain shift $\tau > 0$ the noise is uncorrelated.

Bandlimited communications systems bandlimit not only the signal, but the noise as well, and this filtering limits the rate of change of the time domain noise signal, introducing some correlation over the interval of $\pm 1/2B$. The stylised PSD and ACF of bandlimited white noise are displayed in Figure 2.3. With bandlimiting the autocorrelation function becomes

$$
\begin{aligned}
R(\tau) &= \frac{1}{2\pi} \int_{-B}^{B} \frac{N_0}{2} e^{j\omega\tau} d\omega = \frac{N_0}{2} \int_{-B}^{B} e^{j2\pi f\tau} df \\
&= \frac{N_0}{2} \left[\frac{e^{j2\pi f\tau}}{j2\pi\tau} \right]_{-B}^{B} = N_0 B \frac{\sin(2\pi B\tau)}{2\pi B\tau}.
\end{aligned}
\tag{2.2}
$$

In the time domain the amplitude distribution of the white thermal noise has a so-called normal or Gaussian distribution and since it is inevitably added to the received signal, it is usually referred to as additive white Gaussian noise (AWGN). Note that AWGN is therefore

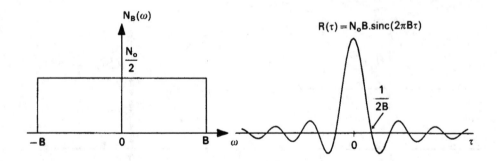

Figure 2.3: Power spectral density and autocorrelation of bandlimited WN

the noise generated in the receiver. The probability density function (PDF) is the well-known bell-shaped curve of the Gaussian distribution, given by:

$$p(x) = \frac{1}{\sigma\sqrt{2\pi}} e^{-(x-m)^2/2\sigma^2}, \tag{2.3}$$

where m is the mean and σ^2 is the variance. The effects of AWGN can be mitigated by increasing the transmitted signal power and thereby reducing the relative effects of noise. The signal-to-noise ratio (SNR) at the receiver's input provides a good measure of the received signal quality. This SNR is often referred to as the channel SNR.

If an analogue channel's bandwidth is B and the prevailing SNR is known, then according to the Shannon-Hartley theorem, the maximum transmission rate at which information can be sent without errors via this channel is given by:

$$C = B \log_2(1 + SNR) \text{ [bit/sec]}, \tag{2.4}$$

where C is referred to as the channel capacity. The Shannon-Hartley law has a number of fundamental consequences, of which we emphasise here only one. Namely, expressing the ratio C/B in terms of SNR we have:

$$\frac{C}{B} = \log_2(1 + SNR) \left[\frac{\text{bit/sec}}{\text{Hz}} \right], \tag{2.5}$$

and selecting the SNR values seen in Table 2.1 we arrive at the near-linear curve seen in Figure 2.4.

It can be seen that the channel capacity for a given bandwidth B is nearly linearly proportional to the SNR. Consequently, using a binary signalling scheme with low C/B ratio when high SNR values are available results in an inefficient use of the channel. This fact provides the main motivation for using the multilevel modulation schemes discussed in this book.

SNR		C/B
Ratio	dB	bit/sec/Hz
1	0	1
3	4.8	2
7	8.5	3
15	11.8	4
31	14.9	5
63	18.0	6
127	21.0	7

Table 2.1: Relative channel capacity versus SNR

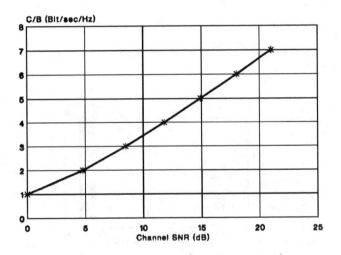

Figure 2.4: Relative channel capacity versus SNR

2.2   Telephone Channels

The ubiquitous telephone lines play an important role as data transmission circuits. Although their attenuation and group delay characteristics vary widely, the international leased telephone lines commonly used for multilevel data transmission must comply with the CCITT recommendations, particularly M1020 and M1025. In general, the overall loss is unknown, but should be less than -13 dB relative (dBr) for four-wire circuits, with a variation of less than ±4 dB. The attenuation and group delay characteristics as specified by the M1020 and M1025 recommendations are given in Figures 2.5, 2.6, 2.7 and 2.8.

When multilevel modems are used for data transmission over these lines, as will be described in Chapter 8, the number of sudden amplitude hits greater than ±2 dB should not exceed 10 in any 15 minutes. The psophometrically weighted noise power for lines longer than 10,000 km must be below -38 dBm, where 1 dBm is related to 1 mW measured across a load resistance of 600 ohms. However, for shorter lines the psophometrically weighted

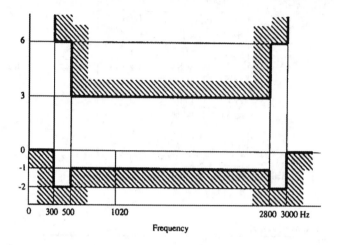

Figure 2.5: Circuit loss as recommended by M1020

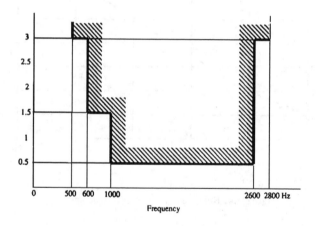

Figure 2.6: Group delay as recommended by M1020

noise power must be proportionally lower, e.g. about -50 dBm at 1,000 km. The number of impulse noise peaks exceeding -21 dBm should not be more than 18 in 15 minutes. The maximum allowed phase jitter is limited to 15° peak-to-peak. The signal-to-total-distortion power ratio should be better than 28 dB when measured with a 1020 Hz -10 dBm sine wave. The level of single tone interference must be at least 3 dB below the random noise level while the maximum frequency error is specified as ± 5 Hz.

Further standards, such as M1030, impose less stringent requirements and are applicable to lower transmission rates or to private switched networks.

Having briefly considered fixed communications channels, we now consider the more complex mobile radio channels.

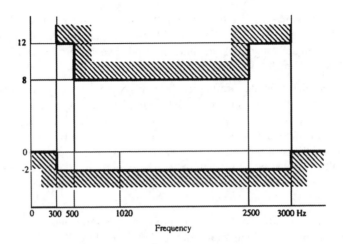

Figure 2.7: Circuit loss as recommended by M1025

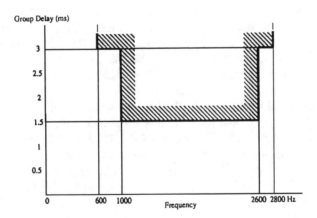

Figure 2.8: Group delay as recommended by M1025

2.3 Mobile Radio Channels

2.3.1 Introduction

The mobile radio channel has been characterised in a number of excellent treatises by Jakes [381], Lee [382], Steele [383], Parsons [384] and others. In this section we follow a similar approach to Steele [383], Greenwood, Hanzo [385] and Cheung [37], while keeping the treatment of mobile channels as simple and practical as possible.

Mobile radio links are established between a fixed base station (BS) and a number of roaming mobile stations (MSs) [381–383]. In order to cover a large area it is necessary to have a number of transmitters. The coverage area of each transmitter is defined as the area in which satisfactory communications between the mobile and the transmitter can be achieved, and this is known as a cell. First- and second-generation mobile systems use cells with radii up to some 35 km, which implies that the existence of a line-of-sight (LOS) path between transmitter and receiver will be rare. Nearby cells are assigned different frequencies, but as the distance between cells increases, the interference between them reduces, and it becomes possible to reuse frequencies, thus increasing system capacity.

In areas where the generated traffic density is high or when only low power handhelds are available, such as the British Cordless Telecommunications system CT-2 or the Digital European Cordless Telecommunications (DECT) system, the cell size is necessarily reduced. If the BS antenna is mounted below rooftop level of surrounding buildings, i.e. fixed to the side of a building, the resulting radio cell is termed a microcell, and typically has a radius of some 200-400 m [386, 387]. In such cells, due to the proximity to the BS, there is an increased likelihood of a LOS path with the resulting effect that multipath fading is often less severe.

Propagation mechanisms can be divided into three types, distance effects, slow and fast fading. In this section we will ignore the slow fading caused by large obstructions. The distance effect is simply that as the separation between a BS and a MS increases, the received mean signal level tends to decrease. Over distances of a few metres the mean signal level is essentially constant, but the instantaneous signal level can vary rapidly by amounts typically up to 40 dB. These rapid variations are known as fast fading. This is because the MS receives a number of versions of the signal transmitted by the BS, which have been reflected and diffracted by buildings, mountains and other obstructions and have a range of delays, attenuations and frequency shifts. These various signals are all summed at the MS antenna. When this summation is constructive, because all paths are received with the same phase change, the received signal level is enhanced. However, when the multipath signals sum destructively, because of phase changes which cause cancellation, the received signal level is much reduced and the receiver is said to be in a fade. Because differences of 180° in the carrier phase could change the interference from constructive to destructive, and at a carrier frequency of 2 GHz this corresponds to some 7 cm, the change in instantaneous signal level can be rapid and so is termed fast fading. As previously suggested, these fades typically occur every half-wavelength.

In mobile communications, digital information is transmitted at a certain bit rate. The propagation phenomena are highly dependent on the ratio of the symbol duration to the delay spread of the time variant radio channel. The delay spread may be defined as the length of the channel's impulse response. We can see that if we transmit data at a slow rate the individual symbols can be resolved at the receiver. This is because the numerous reflected paths all arrive at the receiver before the next symbol is transmitted. However, if we increase the transmitted data rate, a point will be reached where this no longer occurs and each data symbol significantly spreads into adjacent symbols, a phenomenon known as intersymbol interference (ISI). Without the use of channel equalisers to remove the ISI, the bit error rate (BER) may become unacceptably high.

Measurements show that cellular radio networks using large cells, where the excess delay spread may exceed 10 μ s, need equalisers when the bit rate is relatively low, say 64 kb/s, while cordless communications in buildings where the excess delay spread is often significantly below a microsecond may exhibit flat fading when the bit rate exceeds a 1 Mb/s. Very small cells, sometimes referred to as picocells, may support many megabits per second without equalisation because the delay spread is only tens of nanoseconds.

2.3.2   Equivalent Baseband and Passband Systems

When modelling mobile radio systems it is advantageous to model them at baseband rather than radio frequencies as this dramatically simplifies the analysis. In this section we show how baseband equivalent models can be developed, following the approach of Proakis [388].

In general the modulated signal $s(t)$ occupies a relatively narrow bandwidth B, when compared to the carrier frequency f_c, where $f_c \gg B$, and can be represented by:

$$s(t) = a(t) \cos[2\pi f_c t + \Theta(t)] = \Re\{a(t)e^{j[\omega_c t + \Theta(t)]}\}, \tag{2.6}$$

where $\Re$ represents the real part of a complex expression. The modulated signal's amplitude is represented by $a(t)$ and its phase by $\Theta(t)$. The carrier f_c is amplitude modulated by the signal $a(t)$ and phase modulated by the signal $\Theta(t)$. It is common practice to refer to the modulated radio signal $s(t)$ as a narrowband bandpass signal. In computer simulations, however, it is impractical to directly model the above equation. This can be appreciated by the following example.

If we have a signal bandwidth of 16 kHz transmitted at 2 GHz, then for every cycle of our baseband signal, the RF signal goes through $(2 \times 10^9)/(16 \times 10^3) = 1.25 \times 10^5$ cycles. To simulate each cycle of the RF carrier, we must oversample by a minimum of a factor of 2. This means that an order of a quarter of a million operations are required for every baseband symbol transmitted, in addition to the simulation of RF mixers, filters, etc.

Fortunately, we can translate the bandpass signal to an equivalent baseband signal with no loss of accuracy. Expanding Equation 2.6 leads to

$$
\begin{aligned}
s(t) &= a(t) \cos \Theta(t) \cos(2\pi f_c t) - a(t) \sin \Theta(t) \sin(2\pi f_c t) \\
&= u_I \cos(2\pi f_c t) - u_Q \sin(2\pi f_c t) \tag{2.7}
\end{aligned}
$$

where u_I and u_Q are termed the quadrature components of $s(t)$ and modulate quadrature carriers, and therefore are defined as

$$
\begin{aligned}
u_I &= a(t) \cos \Theta(t) \\
u_Q &= a(t) \sin \Theta(t). \tag{2.8}
\end{aligned}
$$

This is a Cartesian or rectangular representation of the baseband signal $u(t)$ with u_I and u_Q corresponding to in-phase (I) and quadrature (Q) signals. There is an equivalent polar representation which is often convenient, given by

$$u(t) = a(t)e^{j\Theta(t)}, \tag{2.9}$$

where $u(t) = u_I(t) + ju_Q(t)$ is the complex baseband envelope or complex lowpass equivalent of $s(t)$. When combining Equations 2.6 and 2.9, this leads to

$$s(t) = \Re\left[u(t)e^{j2\pi f_c t}\right] \tag{2.10}$$

where $\Re[\,]$ represents the real part of $[\,]$. This equation implies that the knowledge of $u(t)$ and f_c uniquely describes the modulated signal $s(t)$, where all the useful information is conveyed by $u(t)$.

In order to establish the validity of the complex baseband model for studying bandpass

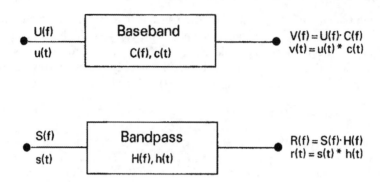

Figure 2.9: Equivalent baseband and passband channels

systems, we will show that the response of a bandpass system to a bandpass signal is identical to the response of the equivalent baseband system to the equivalent baseband signal. This confirms that any operations such as filtering, which are carried out at RF in hardware, can be represented as baseband operations in software, and thus simulations can be carried out at baseband. In order to show this, we take a bandpass modulated signal $s(t)$ with its equivalent lowpass signal $u(t)$ as defined in Equation 2.9. We then allow $s(t)$ to excite a narrowband bandpass system having impulse response $h(t)$ and equivalent baseband impulse response $c(t)$. The baseband equivalent response $c(t)$ is derived from $h(t)$ in the same way as $u(t)$ was derived from $s(t)$. The notation is summarised in Figure 2.9.

The time domain response $r(t)$ of the bandpass system is given by:

$$r(t) = \int_{-\infty}^{\infty} s(\tau)h(t - \tau)\, d\tau. \tag{2.11}$$

This is the well-known convolution integral and is often simply written as

$$r(t) = s(t) * h(t), \tag{2.12}$$

where * indicates convolution. When translating Equation 2.12 to the frequency domain we have:

$$R(f) = S(f)H(f), \tag{2.13}$$

where $S(f)$ is the Fourier transform of $s(t)$, given by:

$$S(f) = \int_{-\infty}^{\infty} s(t)e^{-j2\pi ft}dt. \tag{2.14}$$

Upon substituting $s(t)$ from Equation 2.10 into Equation 2.14 we can write:

$$S(f) = \int_{-\infty}^{\infty} \left\{ \Re\left[u(t)e^{j2\pi f_c t} \right] \right\} e^{-j2\pi ft}dt. \tag{2.15}$$

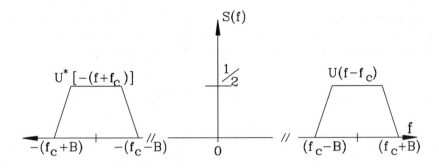

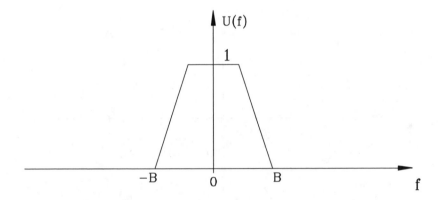

Figure 2.10: Relationship between the baseband spectrum $U(f)$ and the bandpass spectrum $S(f)$

Noting the identity

$$\Re(z) = \frac{1}{2}(z + z^*), \tag{2.16}$$

where the asterisk represents the complex conjugate and taking into account that $\mathcal{F}\{u^*(t)\} = U^*(-f)$, where $\mathcal{F}$ represents the Fourier transform, we can rewrite Equation 2.15 as:

$$
\begin{aligned}
S(f) &= \frac{1}{2} \int_{-\infty}^{\infty} \left[u(t)e^{j2\pi f_c t} + u^*(t)e^{-j2\pi f_c t} \right] e^{-j2\pi f t} dt \\
&= \frac{1}{2} \left[U(f - f_c) + U^*(-f - f_c) \right],
\end{aligned} \tag{2.17}
$$

where $U(f)$ is the Fourier transform of $u(t)$. The spectra $U(f)$ of $u(t)$ and $S(f)$ of $s(t)$ are shown in Figure 2.10. Observe that the multiplication of $u(t)$ by $e^{j2\pi f_c t}$ in Equation 2.15 shifts the baseband spectrum $U(f)$ to $\pm f_c$ and its conjugate complex symmetry implies that the modulated signal $s(t)$ is real. It is worth noting that since $f_c \gg B$, the energy is concentrated in a tight band around the carrier frequency.

In a similar fashion, the conjugate complex symmetric frequency response $H(f)$ of the bandpass system can be described with the help of shifted versions of the equivalent baseband system's frequency response. Since the bandpass system's impulse response $h(t)$ is real, $H(f)$ is conjugate complex symmetric, i.e. $H(f) = H^*(-f)$. Defining

$$C(f - f_c) = \begin{cases} H(f) & \text{if } f > 0 \\ 0 & \text{if } f \leq 0 \end{cases} \tag{2.18}$$

and

$$C^*[-(f + f_c)] = \begin{cases} 0 & \text{if } f \geq 0 \\ H^*(-f) & \text{if } f < 0 \end{cases}, \tag{2.19}$$

as seen in Figure 2.11 allows us to express $H(f)$ as

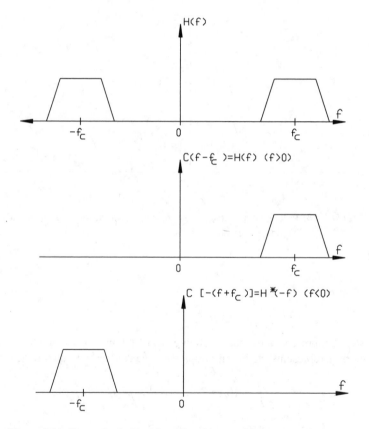

Figure 2.11: Spectral relationship of bandpass and lowpass equivalent systems

$$H(f) = C(f - f_c) + C^*(-f - f_c). \tag{2.20}$$

When returning to the time domain the relationship of the baseband and passband impulse responses is given by:

$$
\begin{aligned}
h(t) &= c(t)e^{j\omega_c t} + c^*(t)e^{-j\omega_c t} \\
&= 2\Re\left\{c(t)e^{j\omega_c t}\right\},
\end{aligned}
\tag{2.21}
$$

indicating that the complex baseband equivalent impulse response $c(t)$ forms the time domain envelope of $h(t)$, where $c(t)$ can be described by its in-phase and quadrature phase components as follows:

$$
c(t) = c_I(t) + jc_Q(t).
\tag{2.22}
$$

The received passband signal $r(t)$ or $R(f)$ can now be expressed also by the help of the baseband equivalents $U(f)$ and $C(f)$ or $u(t)$ and $c(t)$. Substituting Equations 2.17 and 2.20 into Equation 2.13 yields:

$$
R(f) = \frac{1}{2}\left[U(f - f_c) + U^*(-f - f_c)\right]\left[C(f - f_c) + C^*(-f - f_c)\right].
\tag{2.23}
$$

Figures 2.10 and 2.11 reveal that $U(f - f_c) \cdot C^*(f - f_c) = 0$ if $f \leq 0$ and $U^*(-f - f_c) \cdot C(-f - f_c) = 0$ if $f \geq 0$, hence Equation 2.23 simplifies to:

$$
\begin{aligned}
R(f) &= \frac{1}{2}\left[U(f - f_c)C(f - f_c) + U^*(-f - f_c)C^*(-f - f_c)\right] \\
&= \frac{1}{2}\left[V(f - f_c) + V^*(-f - f_c)\right],
\end{aligned}
\tag{2.24}
$$

where

$$
V(f) = U(f)C(f)
\tag{2.25}
$$

is the response of the baseband system $C(f)$ to excitation $U(f)$. But for a real impulse response $v(t)$ we have

$$
V(f) = V^*(-f)
\tag{2.26}
$$

giving that the passband system's response is constituted by the shifted baseband responses, i.e.:

$$
R(f) = \frac{1}{2}\left[V(f - f_c) + V(f + f_c)\right].
\tag{2.27}
$$

The baseband systems' response $v(t)$ to the baseband information signal $u(t)$ can be expressed by the following convolution:

$$
v(t) = u(t) * c(t).
\tag{2.28}
$$

Substituting the quadrature decomposition of $u(t)$ and $c(t)$ from Equations 2.22 and 2.8 into Equation 2.28 gives

$$
\begin{aligned}
v(t) &= [u_I(t) + ju_Q(t)] * [c_I(t) + jc_Q(t)] \\
&= [u_I(t) * c_I(t) - u_Q(t) * c_Q(t)] + \\
&\quad j[u_I(t) * c_Q(t) + u_Q(t) * c_I(t)] \\
&= v_I(t) + jv_Q(t),
\end{aligned}
\tag{2.29}
$$

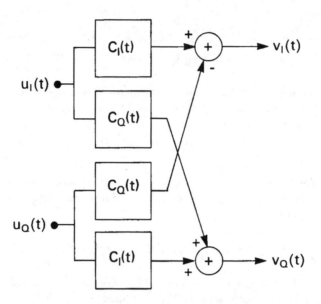

Figure 2.12: Complex convolution in the baseband

where the quadrature components of the equivalent baseband system's response are:

$$v_I(t) = u_I(t) * c_I(t) - u_Q(t) * c_Q(t)$$
$$v_Q(t) = u_I(t) * c_Q(t) + u_Q(t) * c_I(t). \qquad (2.30)$$

These operations are portrayed in Figure 2.12.

In summary, in this section we revealed the equivalence of bandpass and lowpass systems. The relationships portrayed allow us to study bandpass systems in the baseband both by computer simulation and analytical methods. The relevant equivalences are summarised in Table 2.2.

2.3.3   Gaussian Mobile Radio Channel

The simplest type of channel is the Gaussian channel as described in Section 2.1.3. Basically it is a linear time invariant transmission system impaired by the thermal noise generated in the receiver. The noise is assumed to have a constant power spectral density (PSD) over the channel bandwidth, and a Gaussian amplitude probability density function (PDF). This type of channel is occasionally realised in digital mobile radio communications mainly in microcells where it is possible to have a line of sight (LOS) with essentially no multipath propagation. Even when there is multipath propagation, but the mobile is stationary and there are no other moving objects such in its vicinity, as vehicles, the mobile channel may be thought of as Gaussian. This is because, due to the zero mobile speed, the fading is represented by a local path loss.

The Gaussian channel is also important for providing an upper bound on system perfor-

Type	Equations
Inputs	$s(t) = \Re\left\{u(t)e^{j2\pi f_c t}\right\}$ $S(f) = \frac{1}{2}[U(f - f_c) + U^*(-f - f_c)]$
System Description	$r(t) = s(t) * h(t)$ $R(f) = S(f)H(f)$ $v(t) = u(t) * c(t)$ $V(f) = U(f)C(f)$ $h(t) = 2\Re\{c(t)e^{j2\pi f_c t}\}$ $H(f) = C(f - f_c) + C^*(-f - f_c)$ $c(t) = c_I(t) + jc_Q(t)$
Outputs	$r(t) = \Re\left\{v(t)e^{j2\pi f_c t}\right\}$ $R(f) = \frac{1}{2}[V(f - f_c) + V^*(-f - f_c)]$ $u(t) = u_I(t) + ju_Q(t)$ $v(t) = v_I(t) + jv_Q(t)$

Table 2.2: Summary of baseband and bandpass system equations

mance. For a given modulation scheme we may calculate or measure the BER performance in the presence of a Gaussian channel. When multipath fading occurs the BER will increase for a given channel SNR. By using techniques to combat multipath fading, such as diversity, equalisation, channel coding, data interleaving, and so forth, that are described throughout the book, we can observe how close the BER approaches that for the Gaussian channel.

2.3.4 Narrow-Band Fading Channels

Propagation in narrowband mobile radio channels can be divided into three phenomena. These are:

(1) Propagation path loss law;

(2) Slow fading statistics;

(3) Fast fading statistics.

These all vary with the propagation frequency, surrounding natural and man-made objects, vehicular speed, etc. Consequently a deterministic treatment is not possible and so statistical methods are used instead. The calculation of a power budget for a mobile radio link considering the propagation path-loss law, the slow fading and fast fading margins is shown in Figure 2.13.

Since most existing public mobile radio systems use the 0.45-1.8 GHz band, we concentrate on channels in this range. These frequencies fall in the so-called ultra high frequency (UHF) band. At these frequencies the signal level drops quickly once over the radio horizon, limiting co-channel interference when the frequencies are reused in neighbouring cell clusters. At these frequencies, even if there is no line-of-sight path between transmitter and receiver, sufficient signal power may be received by means of wave scattering, reflection and diffraction to allow communications.

The prediction of the expected mean or median received signal power plays a crucial role in planning the coverage area of a specific base station and in determining the closest

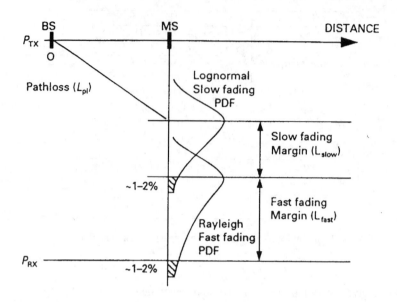

Figure 2.13: Path-loss, slow-fading and fast-fading [383]©1992, Steele

acceptable reuse of the propagation frequency deployed. For high antenna elevations and large rural cells a more slowly decaying power exponent is expected than for low elevations and densely built-up urban areas. As suggested by Figure 2.13, the received signal is also subjected to slow or shadow fading which is mainly governed by terrain and topographical features in the vicinity of the mobile receiver such as small hills and tall buildings. When designing the system's power budget and coverage area pattern, the slow fading phenomenon is taken into account by including a shadow fading margin as demonstrated by Figure 2.13.

Slow fading is compensated for by increasing the transmitted power P_{tx} by the slow fading margin L_{slow}, which is usually chosen to be the $1-2\%$ quantile of the slow fading probability density function (PDF) to minimise the probability of unsatisfactorily low received signal power P_{rx}. Additionally, the short-term fast signal fading due to multipath propagation is taken into account by deploying the so-called fast fading margin L_{fast}, which is typically chosen to be also a few percentage quantile of the fast fading distribution. There remains a certain probability that both of these fading margins are simultaneously exceeded by the superimposed slow and fast envelope fading. This situation is often referred to as "fading margin overload", and results in a very low received signal level which may cause call dropping in mobile telephony. The worst-case probability of these events can be taken to be the sum of the individual fading margin overload probabilities.

Mathematical models can be derived, or measurements made, to determine appropriate margins, which can then be used to analyse likely system performance. For practicality, any models used must consider signals to have been transmitted at baseband through a channel which is equivalent to the mobile radio channel.

2.3.4.1 Propagation Path Loss Law

When considering the propagation path loss, the parameter of prime concern is normally the distance from the BS. Path loss increases with distance due to the increasing area of the circular wavefront expanding from the BS. More details about path loss calculations can be found in [381, 382, 384]. Path loss calculations become significantly more complex when there is any form of obstruction between the transmitter and receiver. In large cells these obstructions will often be hills or undulating terrain. In modelling tools these obstacles are often considered to be knife-edges at the point of the summit of the obstruction, and knife-edge diffraction techniques can then be used [384, 389] to predict the path loss. The analysis becomes increasingly complex if there are a number of hills between the transmitter and receiver. An alternative approach is to determine which Fresnel zone the hill obstructs and perform the prediction accordingly [390].

In urban areas modelling becomes more complex, and in the case of large cells it is rarely possible to model each individual building, although this can be achieved in microcells [391]. Typically this situation is resolved by adding a clutter loss dependent on the density of the buildings in the urban area, and also allowing for a shadowing loss behind buildings when computing the link budget. This approximation leads to sub-optimal network design, with overlapping cells and inefficient frequency reuse. Nevertheless, it seems unlikely that more accurate tools will become available in the near future. In our probabilistic approach it is difficult to give a worst-case path loss exponent for any mobile channel. However, it is possible to specify the most optimistic scenario. This is given by propagation in free space. The free space path loss, L_{pl} is given by [384]:

$$L_{pl} = -10\log_{10} G_T - 10\log_{10} G_R + 20\log_{10} f^{Hz} + 20\log_{10} d^m - 147.6\,\mathrm{dB}, \quad (2.31)$$

where G_T and G_R are the transmitter and receiver antenna gains, f^{Hz} is the propagation frequency in hertz and d^m is the distance from the BS antenna in metres. Observe that the free space path loss is increased by 6 dB every time, the propagation frequency is doubled or the distance from the mobile is doubled. This corresponds to a 20 dB/decade decay and at $d = 1$ km, $f = 1$ GHz and $G_T = G_R = 1$ a path loss of $L_F = 92.4$ dB is encountered. Clearly, not only technological difficulties, but also propagation losses discourage the deployment of higher frequencies. Nevertheless, spectrum is usually only available in these higher frequency bands.

In practice, for UHF mobile radio propagation channels of interest, the free space conditions do not apply. There are, however, a number of useful path loss prediction models that can be adopted to derive other prediction bounds. One such case is the "plane earth" model. This is a two-path model constituted by a direct line-of-sight path and a ground-reflected path which ignores the curvature of the earth's surface. Assuming transmitter base station (BS) and receiver mobile station (MS) antenna heights of $h_{BS}, h_{MS} \ll d$, the plane earth path loss formula [384] can be written as follows:

$$L_{pl} = -10\log_{10} G_T - 10\log_{10} G_R - 20\log_{10} h_{BS}^m - \quad (2.32)$$
$$20\log_{10} h_{MS}^m - 40\log_{10} d^m,$$

where the dependence on propagation frequency is removed. Observe that a 6 dB path loss reduction is the result, when doubling the transmitter or receiver antenna elevations, and there is an inverse fourth power law decay when increasing the BS-MS distance d. In the close vicinity of the transmitter antenna, where the condition h_{BS} or $h_{MS} \ll d$ does not hold, Equation 2.33 is no longer valid.

Hata [392] developed three path loss models for large cells which are widely used, forming the basis for many modelling tools. These were developed from an extensive database derived by Okumura [393] from measurements in and around Tokyo. The typical *urban* Hata model is defined as:

$$
\begin{aligned}
L_{Hu} &= 69.55 + 26.16 \log_{10} f - 13.82 \log_{10} h_{BS} - a(h_{MS}) + \\
&\quad (44.9 - 6.55 \log_{10} h_{BS}) \log_{10} d \; [dB],
\end{aligned}
\tag{2.33}
$$

where f is the propagation frequency in MHz, h_{BS} and h_{MS} are the BS and MS antenna elevations (m), $a(h_{MS})$ is a terrain dependent correction factor, while d is the BS-MS distance in (km). The correction factor $a(h_{MS})$ for small and medium-sized cities was found to be

$$
a(h_{MS}) = (1.1 \log_{10} f - 0.7) h_{MS} - (1.56 \log_{10} f - 0.8),
\tag{2.34}
$$

while for large cities it is frequency-parametrised:

$$
a(h_{MS}) = \begin{cases} 8.29 [\log_{10}(1.54 h_{MS})]^2 - 1.1 & \text{if } f \le 200 MHz \\ 3.2 [\log_{10}(11.75 h_{MS})]^2 - 4.97 & \text{if } f \ge 400 MHz \end{cases}.
\tag{2.35}
$$

The typical *suburban* Hata model applies a correction factor to the urban model, and it yields:

$$
L_{Hsuburban} = L_{Hu} - 2[\log_{10}(f/28)]^2 - 5.4 \; [dB].
\tag{2.36}
$$

The *rural* Hata model modifies the urban formula like this

$$
L_{Hrural} = L_{Hu} - 4.78(\log_{10} f)^2 + 18.33 \log_{10} f - 40.94 \; [dB].
\tag{2.37}
$$

The limitations of its parameters as listed by Hata are as follows:

$$
\begin{aligned}
f &: \quad 150 - 1500 MHz \\
h_{BS} &: \quad 30 - 200m \\
h_{MS} &: \quad 1 - 10m \\
d &: \quad 1 - 20km.
\end{aligned}
$$

For a 900 MHz public land mobile radio (PLMR) system these conditions can usually be satisfied, but for a typical 1.8 GHz personal communications network (PCN) urban microcell all these limits have to be slightly stretched.

Microcellular channels can be described by a four-path model including two more reflected waves from building walls along the streets [394, 395]. In this scenario it is assumed that the transmitter antenna is below the characteristic urban skyline. The four-path model used by Green [394] assumed smooth reflecting surfaces yielding specular reflections with no scattering, finite permittivity and conductivity, vertically polarised waves and half-wave dipole antennas. The resultant path loss profile vs. distance becomes rather erratic with received signal level variations in excess of 20 dB, which renders path loss modelling by a simple power exponent rather inaccurate, however attractive it would appear due to its sim-

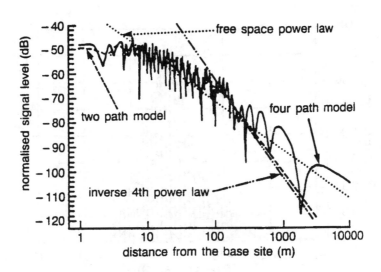

Figure 2.14: Signal level profiles for the two- and four-path models. Also shown are the free space and inverse fourth power laws [394] ©1990, Green

plicity.

Figure 2.14 also allows a comparison of the two-path model, the inverse second law and the inverse fourth law characteristics.

2.3.4.2 Slow Fading Statistics

Having studied a few propagation path loss prediction models we briefly focus our attention on the characterisation of the slow fading phenomena, which constitutes the second component of the overall power budget design of mobile radio links, as portrayed in Figure 2.13. In slow fading analysis the effects of fast fading and path loss have to be removed. The fast fading fluctuations are removed by averaging the signal level over a distance of typically some 20 wavelengths. The slow fading fluctuations are separated by subtracting the best-fit path loss regression estimate from each individual 20 wavelength spaced averaged received signal value. A slow fading histogram derived this way is depicted in Figure 2.15.

Figure 2.15 suggests a lognormal distribution in terms of dBs caused by normally distributed random shadowing effects. Indeed, when subjected to rigorous distribution fitting using the lognormal hypothesis, the hypothesis is confirmed at a high confidence level. The associated standard deviation in this particular case is 6.5 dB.

2.3.4.3 Fast Fading Statistics

Irrespective of the distribution of the numerous individual constituent propagation paths of both quadrature components (a_i, a_q) of the received signal, their distribution will be normal due to the central limit theorem. Then the complex baseband equivalent signal's amplitude

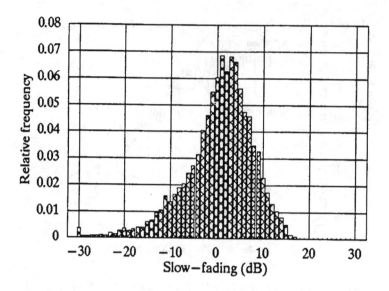

Figure 2.15: Typical microcellular slow fading histogram [383] ©1992, Steele

and phase characteristics are given by:

$$a(k) = \sqrt{a_i^2(k) + a_q^2(k)} \tag{2.38}$$

$$\phi(k) = \arctan[a_q(k)/a_i(k)]. \tag{2.39}$$

Our aim is now to determine the distribution of the amplitude $a(k)$, if $a_i(k)$ and $a_q(k)$ are known to have a normal distribution.

In general, for n normally distributed random constituent processes with means $\overline{a_i}$ and identical variances σ^2, the resultant process $y = \sum_{i=1}^{n} a_i^2$ has a so-called χ^2 distribution with the PDF given below [388]:

$$p(y) = \frac{1}{2\sigma^2} \left(\frac{y}{s^2} \right)^{(n-2)/4} \cdot e^{-(s^2+y)/2\sigma^2} \cdot I_{(n/2)-1} \left(\sqrt{y} \frac{s}{\sigma^2} \right) \tag{2.40}$$

where

$$y \geq 0 \tag{2.41}$$

and

$$s^2 = \sum_{i=1}^{n} (\overline{a_i})^2 \tag{2.42}$$

is the so-called non-centrality parameter computed from the first moments of the $a_1, a_2, \cdots, a_n$ component processes. If the constituent processes have zero means, the χ^2 distribution is central, otherwise non-central. Each process has a variance of σ^2 and $I_k(x)$ is the modified k th order Bessel function of the first kind, given by

$$I_k(x) = \sum_{j=0}^{\infty} \frac{(x/2)^{k+2j}}{j!\Gamma(k+j+1)}, \; x \geq 0. \tag{2.43}$$

The Γ function is defined as

$$\begin{aligned} \Gamma(p) &= \int_0^{\infty} t^{p-1}e^{-t}dt \quad \text{if } p > 0 \\ \Gamma(p) &= (p-1)! \quad \text{if } p > 0 \text{ integer} \\ \Gamma(\tfrac{1}{2}) &= \sqrt{\pi}, \; \Gamma(\tfrac{3}{2}) = \frac{\sqrt{\pi}}{2}. \end{aligned} \tag{2.44}$$

In our case we have two quadrature components, i.e. $n = 2$, $s^2 = (\overline{a_i})^2 + (\overline{a_q})^2$, the envelope is computed as $a = \sqrt{y} = \sqrt{a_i^2 + a_q^2}, a^2 = y, p(a) \, da = p(y) \, dy$, and hence $p(a) = p(y) \, dy/da = 2ap(y)$ yielding the Rician PDF.

$$p_{\text{Rice}}(a) = \frac{a}{\sigma^2} e^{-(a^2+s^2)/2\sigma^2} I_o\left(\frac{as}{\sigma^2}\right) \quad a \geq 0. \tag{2.45}$$

Formally introducing the Rician K-factor as

$$K = s^2/2\sigma^2 \tag{2.46}$$

renders the Rician distribution's PDF to depend on one parameter only:

$$p_{\text{Rice}}(a) = \frac{a}{\sigma^2} \cdot e^{-\frac{a^2}{2\sigma^2}} \cdot e^{-K} \cdot I_o\left(\frac{a}{\sigma} \cdot \sqrt{2K}\right), \tag{2.47}$$

where K physically represents the ratio of the power received in the direct line-of-sight path, to the total power received via indirect scattered paths. Therefore, if there is no dominant propagation path, $K = 0$, $e^{-K} = 1$ and $I_0(0) = 1$ yielding the worst-case Rayleigh PDF:

$$p_{\text{Rayleigh}}(a) = \frac{a}{\sigma^2} e^{-\frac{a^2}{2\sigma^2}}. \tag{2.48}$$

Conversely, in the clear direct line-of-sight situation with no scattered power, $K = \infty$, yielding a "Dirac delta shaped" PDF, representing a step-function-like cumulative distribution function (CDF). The signal at the receiver then has a constant amplitude with a probability of one. Such a channel is referred to as a Gaussian channel. This is because although there is no fading present, the receiver will still see the additive white Gaussian noise (AWGN) referenced to its input. If the K-factor is known, the fast fading envelope's distribution is completely described. The set of Rician PDFs for $K = 0, 1, 2, 4, 10$ and 15 is plotted in Figure 2.16.

Note that the relationship between the variances and means of the Rayleigh and the component Gaussian distributions are given as follows [388]:

$$\sigma_R^2 = (2 - \pi/2) \cdot \sigma^2$$

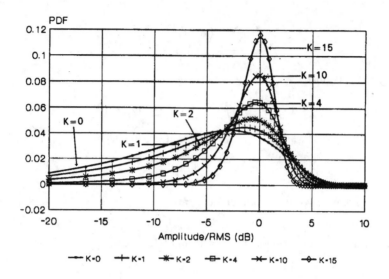

Figure 2.16: Rician PDFs

and
$$m_R = \sqrt{2} \cdot \sigma \cdot \sqrt{\pi}/2 = \sigma\sqrt{\pi/2},$$

Clearly, the Rayleigh distribution's variance is not twice that of the composite Gaussian processes, since we are adding the squares of the Gaussian processes, not the processes themselves.

The Rician CDF takes the shape of [388]:

$$
\begin{aligned}
C_{\text{Rice}}(a) &= 1 - e^{-\left(K + \frac{a^2}{2\sigma^2}\right)} \sum_{m=0}^{\infty} \left(\frac{s}{a}\right)^m \cdot I_m\left(\frac{a\,s}{\sigma^2}\right) \\
&= 1 - e^{-\left(K + \frac{a^2}{2\sigma^2}\right)} \sum_{m=0}^{\infty} \left(\frac{\sigma\sqrt{2K}}{a}\right)^m \cdot I_m\left(\frac{a\sqrt{2K}}{\sigma}\right).
\end{aligned}
$$

(2.49)

Clearly, this formula is more difficult to evaluate than the PDF of Equation 2.47 due to the summation of an infinite number of terms, requiring double or quadruple precision and it is avoided in numerical evaluations, if possible. However, in practical terms it is sufficient to increase m to a value where the last term's contribution becomes less than 0.1%.

A range of Rician CDFs evaluated from Equation 2.49 are plotted on a linear probability scale in Figure 2.17 for $K = 0, 1, 2, 4, 10$ and 15. Figure 2.18 shows the same Rician CDFs plotted on a more convenient logarithmic probability scale, which reveals the enormous difference in terms of deep fades for the K values considered. When choosing the fading margin overload probability, Figure 2.18 is more useful as it expands the tails of the CDFs, where for a Rician CDF with $K = 1$ the 15 dB fading margin overload probability is approximately 10^{-2}.

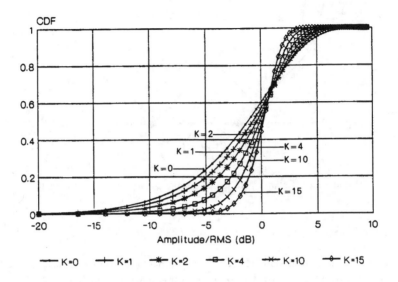

Figure 2.17: Rician CDFs

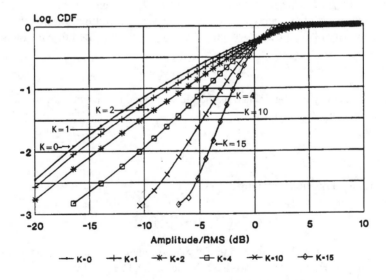

Figure 2.18: Rician logarithmic CDFs

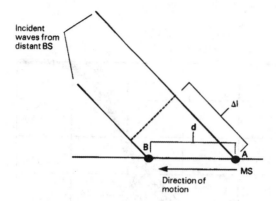

Figure 2.19: Relative path-length change due to the MS's movement

2.3.4.4 Doppler Spectrum

Having described the fading statistics let us now concentrate on the effects of the Doppler shift and consider again the transmission of an unmodulated carrier frequency f_c from a BS. A MS travelling in a direction making an angle α_i with respect to the signal received on the ith path as seen in Figure 2.19 advances a distance of $d = v \cdot \triangle t$ during $\triangle t$, when travelling at a velocity of v. This introduces a relative carrier phase change of $\triangle \Phi = 2\pi$, if the flight of the wave is shortened by an amount given by $\triangle l = \lambda$, where the wavelength can be computed from $\lambda = c/f_c$ with c being the velocity of light.

For an arbitrary $\triangle l$ value we have $\triangle \Phi = -2\pi \cdot \triangle l/\lambda$, where the negative sign implies that the carrier wave's phase delay is reduced if the MS is travelling towards the BS. From the simple geometry of Figure 2.19 we have $\triangle l = d \cdot \cos \alpha_i$ and therefore the phase change becomes:

$$\triangle \Phi = -\frac{2\pi v \triangle t \cos \alpha_i}{\lambda}. \tag{2.50}$$

The Doppler frequency can be defined as the phase change due to the movement of the MS during the infinitesimal interval $\triangle t$:

$$f_D = -\frac{1}{2\pi} \frac{\triangle \Phi}{\triangle t}. \tag{2.51}$$

When substituting Equation 2.50 into 2.51 we get:

$$f_D = \frac{v}{\lambda} \cos \alpha_i = f_m \cos \alpha_i, \tag{2.52}$$

where $f_m = v/\lambda = vf_c/c$ is the maximum Doppler frequency deviation from the transmitted carrier frequency due to the MS's movement. Notice that a Doppler frequency can be positive or negative depending on α_i, and that the maximum and minimum Doppler frequencies are $\pm f_m$. These extreme frequencies correspond to the $\alpha_i = 0°$ and $180°$, when the ray is aligned with the street that the MS is travelling along, and corresponds to the ray coming towards or from behind the MS, respectively. It is analogous to the change in the frequency of a whistle from a train perceived by a person standing on a railway line when the train is bearing down or receding from the person, respectively.

According to Equation 2.52 and assuming that α_i is uniformly distributed, the Doppler frequency has a so-called random cosine distribution. The received power in an angle of $d\alpha$ around α_i is given by $p(\alpha_i)\,d\alpha$, where $p(\alpha_i)$ is the probability density function (PDF) of the received power, which is assumed to be uniformly distributed over the range of $0 \leq \alpha_i \leq 2\pi$, giving $p(\alpha_i) = 1/2\pi$. The Doppler power spectral density $S(f_D)$ can be computed using Parseval's theorem by equating the incident received power $p(\alpha_i)\,d\alpha$ in an angle $d\alpha$ with the Doppler power $S(f_D)\,df_D$, yielding $S(f_D) = d\alpha_i/(2\pi \cdot df_D)$. Upon expressing α_i from $f_D = f_m \cos\alpha_i$ and exploiting that

$$\frac{d\cos^{-1} x}{dx} = -\frac{1}{\sqrt{1 - x^2}}$$

we then have:

$$S(f_D) = \frac{d\alpha_i}{2\pi\,df_D} = \frac{d(\cos^{-1} f_D/f_m)}{2\pi\,df_D} = -\frac{1}{2\pi f_m \sqrt{1 - (f_D/f_m)^2}}. \qquad (2.53)$$

The incident received power at the MS depends on the power gain of the antenna and the polarisation used. Thus the transmission of an unmodulated carrier is received as a "smeared" signal whose spectrum is not a single carrier frequency f_c, but contains frequencies up to $f_c \pm f_m$. In general we can express the received RF spectrum $S(f_D)$ for a particular MS speed, propagation frequency, antenna design and polarisation as,

$$S(f_D) = \frac{C}{\sqrt{1 - (f_D/f_m)^2}} \qquad (2.54)$$

where C is a constant that absorbs the $1/2\pi f_m$ multiplier in Equation 2.53. Notice that the Doppler spectrum of Equation 2.54 becomes $S(f_D = 0) = C$ at $f_D = 0$, while $S(f_D = f_m) = \infty$, when $f_D = f_m$. Between these extreme values $S(f)$ has a U-shaped characteristic, as it is portrayed in a stylised form in the simulation model of Figure 2.20, which will be developed in the next section.

2.3.4.5   Simulation of Narrowband Channels

The Rayleigh fading channel model used in the previous subsection can be represented by the quadrature arrangement shown in Figure 2.20, where the distribution of both received quadrature components (a_i, a_q) is normal due to the central limit theorem. These can be modelled as uncorrelated normally distributed AWGN sources. The outputs from the AWGN sources are applied to lowpass filters having U-shaped frequency domain transfer functions that represent the effects of Doppler frequency shifts, as will be demonstrated.

Observe that the maximum Doppler frequency $f_m = v/\lambda = v \cdot f_c/c$ depends on the product of the speed v of the MS and the propagation frequency f_c. Therefore, the higher the speed or the propagation frequency, the wider the frequency band over which the received carrier is smeared. The effect of deploying the Doppler filter in Figure 2.20 is that the originally uncorrelated quadrature components (a_i, a_q) are now effectively lowpass filtered,

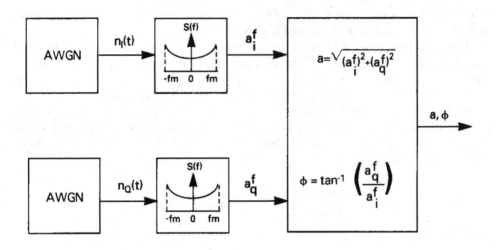

Figure 2.20: Baseband Rayleigh fading simulation model

which introduces some intersample correlation by restricting the maximum rate of change of (a_i, a_q), without altering the Rayleigh distributed envelope fading. This is true, because passing an AWGN process through a linear system, such as a lowpass filter, will reduce the variance of the signal but retain the Gaussian nature of the distribution for both quadrature components. Having no lowpass Doppler filter at all would be equivalent to $f_m = \infty$, implying an infinitely high vehicular speed hence allowing the reception of arbitrary uncorrelated frequencies when transmitting the unmodulated carrier frequency of f_c.

The simulation of the Rayleigh distributed fast fading envelope is based on the concept portrayed in Figure 2.20, but it can be implemented in both the time and the frequency domain.

2.3.4.5.1 Frequency Domain Fading Simulation   In the frequeny domain approach one could exploit the fact that the Fourier transform of an AWGN process is another AWGN process. Hence the frequency domain AWGN sequence generated by the Box-Müller algorithm to be described below can be multiplied by the frequency domain Doppler transfer function of Equation 2.54 and then transformed back by IFFT to the time domain in order to generate the I and Q components (a_i^f, a_q^f) of the filtered complex Gaussian process. Then the Rayleigh distributed magnitude a is given by

$$a = \sqrt{(a_i^f)^2 + (a_q^f)^2},$$

while the uniformly distributed phase ϕ is computed as

$$\phi = \arctan\left[\frac{a_q^f}{a_i^f}\right].$$

The main difficulty associated with this frequency domain approach is that the Doppler filter's bandwidth f_m is typically much lower than the sampling frequency at which the faded samples are produced. The faded samples have to be generated exactly at the signalling rate at which modulation symbols are transmitted in order to supply a fading envelope and phase rotation value for their corruption.

A typical Doppler bandwidth f_m at a carrier frequency of $f_c = 1.8$ GHz and vehicular speed of $v = 30$ mph $= 13.3$ m/s is given by

$$f_m = \frac{v \cdot f_c}{c} = \frac{13.3 \frac{m}{s} \cdot 1.8 \cdot 10^9 \frac{1}{s}}{3 \cdot 10^8 \frac{m}{s}} \approx 80 Hz,$$

where $c = 3 \times 10^8$ m/s is the speed of light. This f_m value becomes about 40 Hz at $f_c = 900$ MHz and 8 Hz at a pedestrian speed of 3 mph. Typical multi-user signalling rates in state-of-art time division multiple access (TDMA) systems are in excess of $f_s = 40$ KHz. Therefore the pedestrian relative Doppler bandwidth D_r becomes $D_r = 8$ Hz/40 KHz $= 2 \times 10^{-4}$ suggesting that only a negligible proportion of the input samples will fall in the Doppler filter's passband. This fact may severely affect the statistical soundness of this approach, unless a very long IFFT is invoked, which will retain a sufficiently high number of frequency domain AWGN samples after removing those outside the Doppler filter's bandwidth.

2.3.4.5.2 Time Domain Fading Simulation When using the time domain approach the AWGN samples generated by the Box-Müller algorithm must be convolved with the impulse response $h_D(t)$ of the Doppler filter $S(f)$ of Equation 2.54 in both quadrature arms of Figure 2.20. Similarly to the frequency domain method, we are faced with a difficult notch filtering problem. Since the filter bandwidth is typically very narrow in comparision to the sampling frequency, the impulse response $h_D(t)$ will be very slowly decaying, requiring tens of thousands of tap values to be taken into account in the convolution in case of a low Doppler bandwidth. This problem is aggravated by the non-decaying U-shaped nature of the Doppler's transfer function $S(f)$. The consequence of these difficulties is that the confidence measures delivered by rigorous goodness-of-fit techniques such as the χ^2 test or the Kolmogorov-Smirnov test described for example in Chapter 2 of [383] might become low, although the resulting fading envelope and phase trajectories are appropriate for simulation studies of mobile radio systems.

2.3.4.5.3 Box-Müller Algorithm of AWGN Generation The previously mentioned Box-Müller algorithm is formulated as follows:

(1) Generate two random variables u_1, u_2 and let:

$$s = u_1^2 + u_2^2.$$

(2) While $s \geq 1$, discard s and re-compute u_1, u_2 and s.

(3) If $s < 1$ is satisfied, compute the I and Q components of the noise as follows:

$$u_I \ = \ u_1 \sqrt{-\frac{2\sigma^2}{s} \cdot \log_e s}$$

$$u_Q \ = \ u_2 \sqrt{-\frac{2\sigma^2}{s} \cdot \log_e s},$$

where σ is the standard deviation of the AWGN. This algorithm can be used both for the generation of the Rayleigh-faded signal envelope as well as for generating the thermal AWGN.

2.3.5   Wideband Channels

2.3.5.1   Modelling of Wideband Channels

The impulse response of the flat Rayleigh fading mobile radio channel consists of a single delta function whose weight has a Rayleigh PDF. This occurs because all the multipath components arrive almost simultaneously and combine to have a Rayleigh PDF. If the signal's transmission bandwidth is narrower than the channel's so-called coherence bandwidth $B_c = 1/(2 \cdot \pi \cdot \Delta)$, where Δ represents the time dispersion interval over which significant multipath components are received, then all transmitted frequency components encounter nearly identical propagation delays. Therefore the so-called narrowband condition is met and the signal is subjected to non-selective or flat envelope fading. The channel's coherence bandwidth (B_c) is defined as the frequency separation, where the correlation of two received signal components' attenuation becomes less than 0.5.

The effect of multipath propagation is to spread the received symbols. If the path delay differences are longer than the symbol duration, several echoes of the same transmitted modulated symbol are received over a number of symbol periods. This is equivalent to saying that in wideband channels the symbol rate is sufficiently high that each symbol is spread into adjacent symbols, causing intersymbol interference (ISI). In order for the receiver to remove the ISI and regenerate the symbols correctly it must determine the impulse response of the mobile radio channel by channel sounding. This response must be frequently measured, since the mobile channel may change rapidly both in time and space.

The magnitude of a typical impulse response is a continuous waveform when plotting received amplitude against time delay. If we partition the time delay axis into equal delay segments, usually called delay bins, then there will generally be, a number of received signals in each bin corresponding to the different paths whose times of arrival are within the bin duration. These signals when vectorially combined can be represented by a delta function occurring in the centre of the bin having a magnitude that is Rayleigh distributed. Impulses which are sufficiently small that they are of no significance to the receiver can then be discarded.

As an example here we describe a set of frequently used typical wideband channel impulses specified by the Group Speciale Mobile committee during the definition of the pan-European mobile radio system known as GSM. These impulse responses describe typical urban, rural and hilly terrain environments, as well as an artificially contrived equaliser test response.

The wideband propagation channel is the superposition of a number of dispersive fading paths, suffering from various attenuations and delays, aggravated by the phenomenon of

Doppler shift caused by the MS's movement. The maximum Doppler shift (f_m) is given by $f_m = v/\lambda = v \cdot f_c/c$, where v is the vehicular speed, λ is the wavelength of the carrier frequency f_c and c is the velocity of light. The momentary Doppler shift f_D depends on the angle of incidence α_i, which is uniformly distributed, i.e. $f_D = f_m \cdot \cos \alpha_i$, hence it has a so-called random cosine distribution with a Doppler spectrum limited to $-f_m < f_D < f_m$. Due to time-frequency duality, this "frequency dispersive" phenomenon results in "time selective" behaviour and the wider the Doppler spread, i.e. the higher the vehicular speed, the faster the time domain impulse response fluctuation.

The set of 6-tap GSM impulse responses is depicted in Figure 2.21. In simple terms the wideband channel's impulse response is measured by transmitting an impulse and detecting the received echoes at the channel's output in every D-spaced so-called delay bin. In some bins no delayed and attenuated multipath component is received, while in others significant energy is detected, depending on the typical reflecting objects and their distance from the receiver. The path delay can easily be related to the distance of the reflecting objects, since radio waves are travelling at the speed of light. For example, at a speed of 300 000 km/s a reflecting object at a distance of 0.15 km yields a multipath component at a round-trip delay of 1 μs.

The typical urban (TU) impulse response spreads over a delay interval of 5 μs, therefore it definitely spills energy into adjacent signalling intervals for signalling rates in excess of 200 ksymbols/s, resulting in serious ISI. In practical terms the transmissions can be considered non-dispersive, if the so-called excess path delay does not exceed 10% of the signalling interval, which in our example would correspond to 20 ksymbols/s or 20 kBaud. The hilly terrain (HT) model has a sharply decaying short-delay section due to local reflections and a long-delay part around 15 μs due to distant reflections. Therefore in practical terms it can be considered a two- or three-path model having reflections from a distance of $3 \cdot 10^8$ m/s $\cdot$ 15 μs = 2.25 km. The rural area (RA) response seems the least hostile amongst all standardised responses, decaying fast within 1 μs and hence up to signalling rates of 100 kBaud can be treated as a flat-fading narrowband channel.

The last standardised GSM impulse response is artificially contrived to test the channel equaliser's performance and is constituted by six equidistant unit-amplitude impulses representing six equal-powered independent Rayleigh fading paths with a delay spread of over 16 μs. With these impulse responses in mind the required channel is simulated by summing the appropriately delayed and weighted received signal components. In all but one case the individual components are assumed to have Rayleigh amplitude distribution. In the RA model the main tap at zero delay is supposed to have Rician distribution with the presence of a dominant path.

We can model a wideband impulse response using the algorithm of Figure 2.22.

The in-phase modulated signal $s_I(t)$ is applied to a series of delays equal to the width of a delay bin. At each delay it is multiplied by the magnitude of the wideband channel at that delay. However, the wideband channel will be changing as the mobile moves, so it is necessary to superimpose fading on each of the wideband responses. This is done using a noise generator and Doppler filter as described earlier. An identical arrangement to that in Figure 2.22 is used for the quadrature component $s_Q(t)$, and the appropriate convolutional terms are combined to obtain the received signal's quadrature components $r_I(t)$ and $r_Q(t)$.

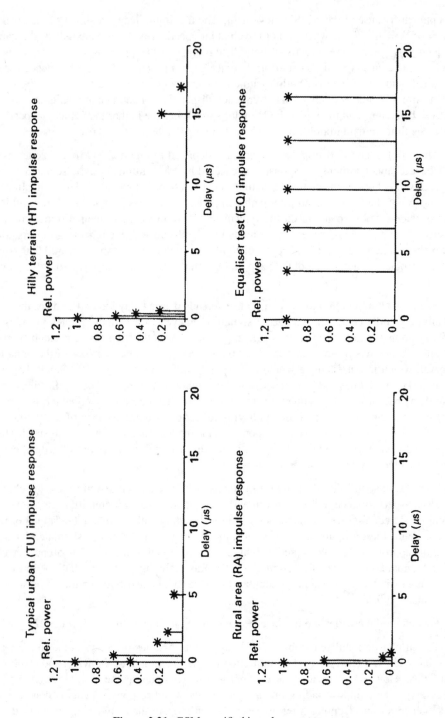

Figure 2.21: GSM specified impulse responses

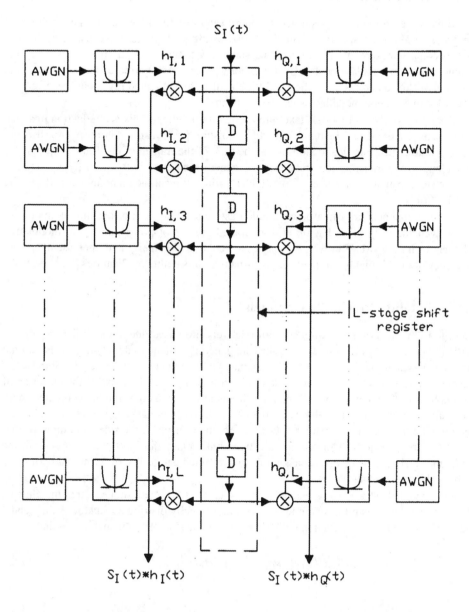

Figure 2.22: Generation of the baseband in-phase received signal via wideband channels [383] ©1992, Steele

2.4 Mobile Satellite Propagation

2.4.1 Fixed-Link Satellite Channels

These links are normally between geostationary satellites and fixed, dish-type receivers on the ground. With the spread of satellite broadcasting systems to home users, such links have become widespread. Because of the stationary nature of both ends of the link, the channel itself is normally modelled by a stationary process. Furthermore, because of the high directivity of the antennas used, there is little likelihood of reflected rays being received. To first order, they can be modelled as Gaussian channels.

More precise channel models take into account rain fading. This is a reduction in signal level caused by the signal being absorbed by moisture present in the atmosphere. It is dependent on the meterological conditions at the time of the transmission. Rain fading is also dependent on the transmission frequency. A general guideline is that rain fading starts to become significant at frequencies above 10 GHz, with a typical fade margin at 20 GHz being 10 dB [396].

Satellite channels are often power limited as the satellite only has a limited amount of power available to it. This has often discouraged the use of power inefficient modulation schemes such as QAM. However, techniques are continually emerging to increase the power efficiency of QAM, making it increasingly attractive for satellite applications.

2.4.2 Satellite-to-Mobile Channels

Satellite-to-mobile communications is still relatively little used due to the difficulties of this particular environment [397–400]. Mobiles are generally unable to deploy highly directional dish antennas, and satellites are normally unable to transmit at high power levels. Some satellite communication systems are restricted to LOS transmission. Because of the elevation of the satellite, LOS transmission will be more frequent than in a conventional cellular system, especially in suburban and rural areas. However, LOS is often masked by trees in rural areas, and by tall buildings in urban areas. Furthermore, the user may be inside a building, further increasing the path loss of the communications link. Once the LOS path is obscured, the signal may arrive at the receiver via a number of reflected paths. The differing phase of these paths will ensure Rician fading.

Few models have been suggested for this channel. Loo [401] suggests that the channel should be modelled as a LOS component with lognormally distributed shadow fading and a multipath component with Rayleigh distribution. This leads to the channel model that

$$p(r) = r/(b_0\sqrt{2\pi d_0}) \int_o^\infty 1/z I_0(rz/b_0)$$

$$\cdot \ \exp[-(ln\ z - mu)^2/2d_0 - (r^2 + z^2)/2b_0)]dz \qquad (2.55)$$

where b_0 is the average scattered power due to multipath, d_0 and μ are the variance and mean due to shadowing, respectively, I_0 is the modified Bessel function, and r is the received signal envelope. The shadowing parameters depend not only on the local environment, but also the satellite elevation, with higher elevations tending to produce lower shadowing losses due to the reduced likelihood of having an obscured LOS path between mobile and satellite.

Because of the hostility of these channels, it is unlikely that the use of multilevel modulation will be widespread.

2.5 Summary

In this chapter we have briefly reviewed the properties of various fixed and mobile communications channels and have shown the equivalence of baseband and passband channels. This allows us to carry out simulations and theoretical analysis in the baseband. In the case of fixed channels the dominant impairment is typically AWGN and linear distortions, as we have seen for example in case of the CCITT standard channels M1020 and M1025 in Figures 2.5 to 2.8. Narrowband mobile radio channels were characterised in terms of path loss, lognormal slow fading and Rician fast fading, as seen in Figure 2.13. The best- and worst-case Rician channels, respectively associated with $K = \infty$ and $K = 0$, are the Gaussian and Rayleigh fading channels, characteristic of the strong LOS and no LOS scenarios. The phenomenon of Doppler shift was introduced, which yielded a smeared spectrum when a single tonal frequency corresponding to an unmodulated carrier was transmitted. A simple simulation model was derived for narrowband channels, which was later extended to dispersive channels. The concept of dispersive wideband channels was portrayed in the context of the GSM wideband channels.

Having considered the non-ideal transmission medium we now embark upon a rudimentary introduction to modems, and review how modems can counteract the channel impairments and facilitate reliable information transmission in a finite bandwidth.

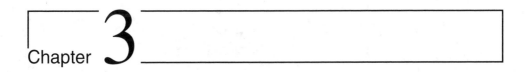

Chapter 3

Introduction to Modems

In this chapter we expand upon our introduction in Chapter 1 by looking in more detail at the basic constituent parts of a modem and considering filtering in more detail.

A typical communications system will input data, perform some form of processing and frequency translation, transmit the data, and perform the converse operations at the receiver. The basic block diagram of such a system employing QAM is shown in Figure 3.1. In a digital communications system the modem's input signal will be a digital signal stream from a digital source or channel encoder. If, however, the modem's input signal is generated by an analogue information source, the signal must be first bandlimited to a bandwidth of B before sampling takes place. According to Nyquist's fundamental theory [402] the sampling frequency must be equal to or higher than twice the bandwidth B, i.e. $f_s \geq 2B$. If this condition is met, the original bandlimited signal can be recovered from its $1/(2B)$ spaced sampled representation with the aid of a lowpass filter having a cut-off frequency of B.

For example, most of the energy of a voice signal is concentrated at frequencies below 4 kHz, hence speech signals are typically lowpass filtered to 4 kHz. As a result, a sampling rate of 8 kHz or higher would be required in order to accurately reconstruct such a signal. In practice, most voice communication systems use a sampling rate of 8 kHz.

We now describe the operation of each of the blocks in the block diagram of Figure 3.1.

3.1 Analogue-to-Digital Conversion

The analogue-to-digital convertor (ADC) takes the bandlimited signal to be transmitted and digitises it by converting the analogue level of each sample to a discrete level. For example, in an 8-bit ADC each discrete level is represented by eight binary output bits. This provides a resolution of 256 distinct digital levels. The action of the ADC is shown in Figure 3.2, where a typical analogue signal is depicted along with its digitised counterpart using only a small number of quantised bits.

The key difference between digital communication systems and the more traditional analogue systems is the signal transmission techniques employed. In an analogue radio, the signal to be transmitted is modulated directly, often by simple multiplication, onto the car-

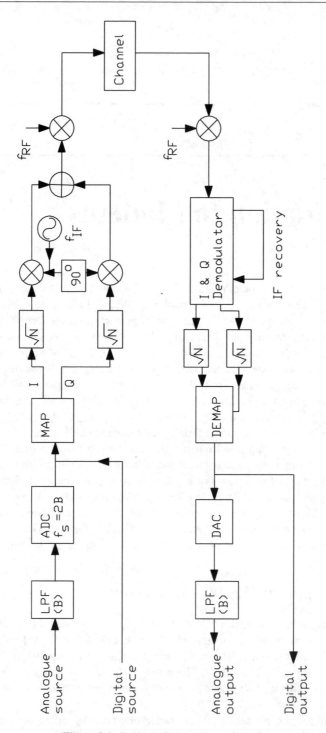

Figure 3.1: Basic QAM modem schematic

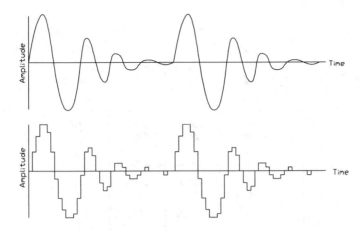

Figure 3.2: Typical ADC input and output waves

rier. On the other hand, digital systems must incorporate modulation schemes which can map the input data onto the carrier in an efficient and readily decoded manner. The benefit gained from using the more sophisticated digital techniques is that the digital link can be rendered error-free by means of signal processing, while in analogue systems the inherent thermal device noise always causes some impairment and hence loss of information.

3.2 Mapping

The process of mapping the information bits onto the streams modulating the I and Q carriers plays a fundamental role in determining the properties of the modem and this is discussed in Chapter 5. The mapping can be represented by the so-called constellation diagram. A constellation is the resulting two-dimensional plot when the amplitudes of the I and Q levels of each of the points which could be transmitted (the constellation points) are drawn in a rectangular coordinate system. For a simple binary amplitude modulation scheme, the constellation diagram would be two points both on the positive x-axis. For a binary PSK (BPSK) scheme the constellation diagram would consist again of two points on the x-axis, but both equidistant from the origin, one to the left and one to the right, as we have seen in Figure 1.2. The "negative amplitude" of the point to the left of the origin represents a phase shift in the transmitted signal of 180°. If we allow phase shifts of angles other than 0 and 180°, then the constellation points move off the x-axis. They can be considered to possess an amplitude and phase, the amplitude representing the magnitude of the transmitted carrier, and the phase representing the phase shift of the carrier relative to the local oscillator in the transmitter. The constellation points may also be considered to have Cartesian, or complex coordinates, which are normally referred to as in-phase (I) and quadrature (Q) components corresponding to the x and y axes, respectively.

In the 16-QAM square constellation of Figure 3.3 each phasor is represented by a 4-bit symbol, constituted by the in-phase bits i_1, i_2 and quadrature bits q_1, q_2, which are interleaved to yield the sequence i_1, q_1, i_2, q_2. The quaternary quadrature components I and Q are Gray

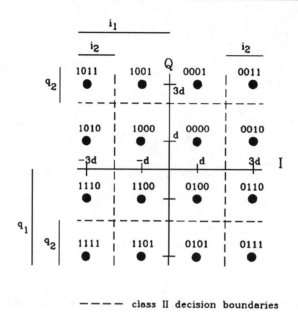

---- class II decision boundaries

Figure 3.3: 16-QAM square constellation

encoded by assigning the bits 01, 00, 10 and 11 to the levels $3d, d, -d$ and $-3d$, respectively. This constellation is widely used because it has equidistant constellation points arranged in a way that the average energy of the phasors is maximised. Using the geometry of Figure 3.3 the average energy is computed as

$$E_0 = (2d^2 + 2 \times 10d^2 + 18d^2)/4 = 10 \times d^2. \tag{3.1}$$

For any other phasor arrangement the average energy will be less and therefore, assuming a constant noise energy, the signal-to-noise ratio required to achieve the same bit error rate (BER) will be higher.

Notice from the mapping in Figure 3.3 that the Hamming distance amongst the constellation points, which are "closest neighbours" with a Euclidean distance of $2d$ is always 1. The Hamming distance between any two points is the difference in the mapping bits for those points, so points labelled 0101 and 0111 would have a Hamming distance of 1, and points labelled 0101 and 0011 would have a Hamming distance of 2. This is a fundamental feature of the Gray coding process and ensures that whenever a transmitted phasor is corrupted by noise sufficiently that it is incorrectly identified as a neighbouring constellation point, the demodulator will choose a phasor with a single bit error. This minimises the error probability.

In Figure 3.4 we plot a typical quaternary I component sequence generated by the mapper. Because of the instantaneous transitions in the time domain this I sequence has an infinite bandwidth and hence would require an infinite channel bandwidth for transmission. The Q component has similar time and frequency domain representations. These signals must be bandlimited before transmission in order to contain the spectrum within a limited band and so minimise interference with other users or systems sharing the spectrum.

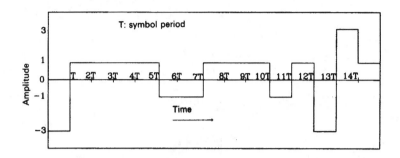

Figure 3.4: Typical I or Q component time-domain representation

3.3 Filtering

An ideal linear phase lowpass filter (LPF) with a cut-off frequency of $f_N = f_s/2$, where $f_s = 1/T$ is the signalling frequency, T is the signalling interval duration and f_N is the so-called Nyquist frequency, would retain all the information conveyed by the quadrature components I and Q within a compact frequency band. Due to the linear phase response of the filter all frequency components would exhibit the same group delay. Because such a filter has a sinc function shaped impulse response with equidistant zero-crossings at the sampling instants, it does not result in intersymbol interference (ISI). This ideal transfer function is referred to as a Nyquist characteristic and is shown in Figure 3.5 along with its impulse response. However, such a filter is unrealisable, as all practical lowpass (LP) filters exhibit amplitude and phase distortions, particularly towards the transition between passband and stopband. Conventional Butterworth, Chebyshev or inverse Chebyshev LP filters have impulse responses with non-zero values at the equispaced sampling instants and hence introduce ISI. They therefore degrade the bit error rate (BER) performance.

Nyquist's fundamental theoretical work [402] suggested that special pulse shaping filters must be deployed, ensuring that the total transmission path, including the channel, has an impulse response with a unity value at the correct signalling instant and zero-crossings at all other sampling instants. He showed that any odd-symmetric frequency domain extension characteristic fitted to the ideal LPF amplitude spectrum yields such an impulse response, and is therefore free from ISI.

A practical odd-symmetric extension characteristic is the so-called raised cosine (RC) characteristic fitted to the ideal LPF of Figure 3.5. The parameter controlling the bandwidth of the Nyquist filter is the so-called roll-off factor α; which is one if the ideal LPF bandwidth is doubled by the extension characteristic. If $\alpha = 0.5$ a total bandwidth of $1.5 \times B$ results, and so on. The lower the value of the roll-off factor, the more compact the spectrum becomes but the higher the complexity of the required filter. The frequency response of these filters is shown in Figure 3.6 for $\alpha = 0.9$ and $\alpha = 0.1$.

Observe in Figures 3.7 and 3.8 how the roll-off factor's value affects the time domain signal in case of 16-QAM for $\alpha = 0.9$ and $\alpha = 0.1$, respectively. With the signal is less sharply filtered and in the time domain it strongly resembles the unfiltered I component of Figure 3.4. The high frequency signal components are removed and hence the sharp time domain transitions are smoothed. The spectrum extends to $1.9 \times B$, i.e. it is virtually doubled

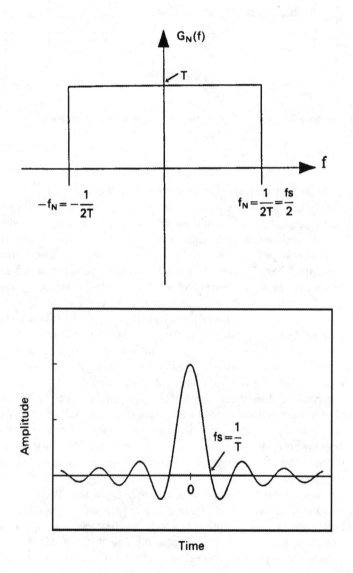

Figure 3.5: Ideal Nyquist filter characteristic and its ISI-free impulse response

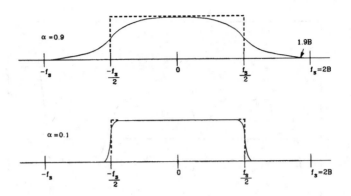

Figure 3.6: Stylised frequency response of two filters with $\alpha = 0.9$ and 0.1

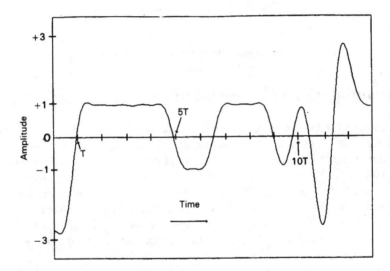

Figure 3.7: Typical filtered I component for $\alpha = 0.9$

when compared to the minimum requirement of B. Waveforms for $\alpha = 0.1$ are shown in Figure 3.8 where it can be seen that the spectrum is confined to $1.1 \times B$, and due to the sharper filtering the time domain waveform is highly smoothed, exhibiting a lesser resemblance to the unfiltered I component of Figure 3.4.

In case of additive white Gaussian noise (AWGN) with a uniform power spectral density (PSD), the noise power admitted to the receiver is proportional to its bandwidth. Therefore it is also necessary to limit the received signal bandwidth at the receiver to a value close to the transmitter's bandwidth. Optimum detection theory [403] shows that the SNR is maximised, if so-called matched filtering is used, where the Nyquist characteristic is divided between two identical filters, each characterised by the square root of the Nyquist shape, as suggested by the filters $\sqrt{N}$ in Figure 3.1.

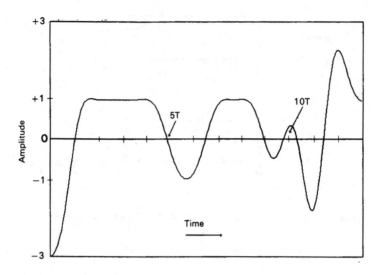

Figure 3.8: Typical filtered I component for $\alpha = 0.1$

3.4   Modulation and Demodulation

Once the analogue I and Q signals have been generated and filtered, they are modulated by an I-Q modulator as shown in Figure 3.1. This modulator essentially consists of two mixers, one for the I channel and another for the Q channel. The I channel is mixed with an intermediate frequency (IF) signal that is in phase with respect to the carrier, and the Q channel is mixed with an IF that is 90° out of phase. This process allows both signals to be transmitted over a single channel within the same bandwidth using quadrature carriers. In a similar fashion, the signal is demodulated at the receiver. Provided signal degradation is kept to a minimum, the orthogonality of the I and the Q channels will be retained and their information sequences can be demodulated.

Following I-Q modulation, the signal is modulated by a radio frequency (RF) mixer, increasing its frequency to that used for transmission. Since the IF signal occurred at both positive and negative frequencies, it will occur at both the sum and difference frequencies when mixed up to the RF. Since there is no reason to transmit two identical sidebands, one is usually filtered out, as seen plotted in dashed lines in Figure 3.9.

The transmission channel is often the most critical factor influencing the performance of any communications system. Here we consider only the addition of noise based on the signal to noise ratio (SNR). The noise is often the major contributing factor to signal degradation and its effect exhibits itself as a noise floor, as portrayed in the received RF spectrum of Figure 3.9.

The RF demodulator mixes the received signal down to the IF for the I-Q demodulator. In order to accurately mix the signal back to the appropriate intermediate frequency, the RF mixer operates at the difference between the IF and RF frequencies. Since the I-Q demodulator includes IF recovery circuits, the accuracy of the RF oscillator frequency is not critical. However, it should be stable, exhibiting a low phase noise, since any noise present in the

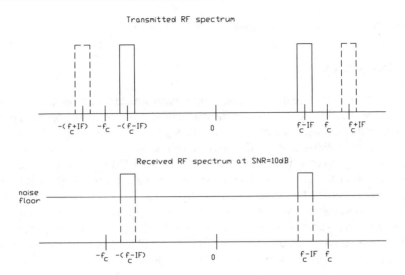

Figure 3.9: Typical transmitted and received RF spectra

downconversion process will be passed on to the detected I and Q baseband signals, thereby adding to the possibility of bit errors. The recovered IF spectrum is similar to the transmitted one but with the additive noise floor seen in the RF spectrum of Figure 3.9.

Returning to Figure 3.1, I-Q demodulation takes place in the reverse order to the modulation process. The signal is split into two paths, with each path being mixed down with IFs that are 90° apart. Since the exact frequency of the original reference must be known to determine absolute phase, IF carrier recovery circuits are used to reconstruct the precise reference frequency at the receiver. The recovered I component should be near identical to that transmitted, with the only differences being caused by noise.

3.5 Data Recovery

Once the analogue I and Q components have been recovered, they must be digitised. This is done by the bit detector. The bit detector determines the most likely bit transmitted by sampling the I and Q signals at the correct sampling instants and comparing them to the legitimate I and Q values of $-3d$, $-d$, d, $3d$ in the case of a square QAM constellation. From each I and Q decision two bits are derived, leading to a 4-bit 16-QAM symbol. The four recovered bits are then passed on to the DAC. Although the process might sound simple, it is complicated by the fact that the "right time" to sample is a function of the clock frequency at the transmitter. The data clock must be regenerated upon recovery of the carrier. Any error in clock recovery will increase the BER.

If there is no channel noise or the SNR is high, the reconstructed digital signal is identical to the original input signal. Provided the DAC operates at the same frequency and with the same number of bits as the input ADC, then the analogue output signal after lowpass filtering with a cut-off frequency of B, is also identical to the output signal of the LPF at the input to the transmitter. Hence it is a close replica of the input signal.

3.6 Summary

In this short chapter we have described the fundamental structure of modems using the basic schematic shown in Figure 3.1. A rudimentary introduction to Gray coding and Nyquist filtering was offered. Nyquist filtering is an essential prerequisite for intersymbol interference (ISI) free communications over bandlimited channels, as we will show in the next chapter.

Accordingly, Chapter 4 will focus on a range of further QAM techniques, including the design of various phasor constellations for channels having a range of impairments. Then an in-depth treatment of ISI-free signalling using Nyquist filtering will be given, which is followed by a detailed discussion on various QAM detection methods, such as threshold detection, matched filtering and correlation receivers. Finally, Chapter 4 will incorporate a section on the linearisation problems of power amplifiers and will be concluded by a discussion on non-differential versus differential coding of QAM signals.

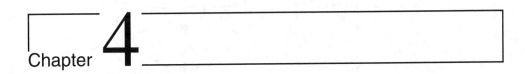
Basic QAM Techniques

In this chapter we consider some of the details of QAM modems in more depth than previously in Chapter 3. We are specifically concerned with constellation types for various channels having different dominant impairments, encoding techniques and suitable forms of pulse shaping for QAM transmission. We will also present a short theoretical discourse on the necessary conditions for interference-free communications over bandlimited channels and derive the optimum transmitter and receiver filters. Methods of QAM detection, such as threshold detection, matched filtering and correlation receivers are also the subject of this chapter, which is concluded with a brief discussion on differential versus non-differential coding of QAM signals.

4.1 Constellations for Gaussian Channels

As discussed in Chapter 1, a large number of different constellations have been proposed for QAM transmissions over Gaussian channels. However, the three constellations shown in Figure 4.1 are often preferred. The essential problem is to maintain a high minimum distance, d_{min}, between points while keeping the average power required for the constellation to a minimum. Calculation of d_{min} and the average power is a straightforward geometric procedure, and has been performed for a range of constellations by Proakis [388]. The results show that the square constellation is optimal for Gaussian channels. The Type I and Type II constellations require a higher energy to achieve the same d_{min} as the square constellation and so are generally not preferred for Gaussian channels. However, there may be implementational reasons for favouring circular constellations over the square ones.

When designing a constellation, consideration must be given to:

(1) The minimum Euclidean distance amongst phasors, which is characteristic of the noise immunity of the scheme.

(2) The minimum phase rotation amongst constellation points, determining the phase jitter immunity and hence the scheme's resilience against clock recovery imperfections and channel phase rotations.

87

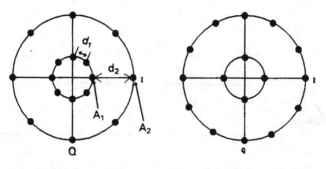

Type I QAM Constellation   Type II QAM Constellation

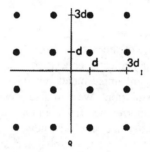

Type III QAM Constellation

Figure 4.1: Variety of QAM constellations

(3) The ratio of the peak-to-average phasor power, which is a measure of robustness against non-linear distortions introduced by the power amplifier.

It is quite instructive to estimate the optimum ring ratio for the Type I constellation of Figure 4.1 in AWGN under the constraint of a constant average phasor energy E_0. Accordingly, a high ring ratio value implies that the Euclidean distance amongst phasors on the inner ring is reduced, while the distance amongst phasors on different rings is increased. In contrast, upon reducing the ring ratio the cross-ring distance is reduced and the distances on the inner ring become larger.

Intuitively, one expects that there will be an optimum ring ratio, where the overall bit error rate (BER) constituted by detection errors on the same ring plus errors between rings is minimised. This problem will be revisited in Figure 10.9 in the context of transmissions over Rayleigh fading channels using simulation experiments. Suffice to say here that the minimum Euclidean distance amongst phasors is maximised if $d_1 = d_2 = A_2 - A_1$ in the

Type I constellation of Figure 4.1. Using the geometry of Figure 4.1 we can write that:

$$\cos 67.5^o = \frac{d_1}{2} \cdot \frac{1}{A_1}$$
$$d_1 = 2 \cdot A_1 \cdot \cos 67.5^o$$

and hence
$$A_2 - A_1 = d_1 = d_2 = 2 \cdot A_1 \cdot \cos 67.5^o.$$

Upon dividing both sides by A_1 and introducing the ring ratio RR we arrive at:

$$RR - 1 = 2 \cdot \cos 67.5^o$$
$$RR \approx 1.77.$$

Simulation results using $1.5 < RR < 3.5$ both over Rayleigh and AWGN channels showed [404] that the BER depends on the channel SNR and has a very flat minimum in the above range, hence in the chapters commencing with Chapter 10 in the systems designed for communicating over Rayleigh fading channels we often opted for RR $= 3$.

Under the constraint of having identical distances amongst constellation points, when $d_1 = d_2 = d$, the average energy E_0 of the Type I constellation can be computed as follows:

$$E_0 = \frac{8 \cdot A_1^2 + 8 \cdot A_2^2}{16} = \frac{1}{2}(A_1^2 + A_2^2)$$

where

$$A_1 = \frac{d}{2 \cdot \cos 67.5^o} \approx \frac{d}{0.765} \approx 1.31d$$

and

$$A_2 \approx 1.77 \cdot A_1 \approx 2.3d$$

yielding

$$E_0 \approx 0.5 \cdot (5.3 + 1.72)d^2 \approx 3.5d^2.$$

The minimum distance of the constellation for an average energy of E_0 becomes:

$$d_{min} \approx \sqrt{E_0/3.5} \approx 0.53 \cdot \sqrt{E_0},$$

while the peak-to-average phasor energy ratio is:

$$r \approx \frac{(2.3d)^2}{3.5d^2} \approx 1.5.$$

The minimum phase rotation θ_{min}, the minimum Euclidean distance d_{min} and the peak-to-average energy ratio r are summarised in Table 4.1 for both the Type I and Type III constellations.

Let us now derive the above characteristic parameters for the Type III constellation. Observe from Figure 4.1 that $\theta_{min} < 45 \deg$, while the distance between phasors is $2 \cdot d$. Hence

Type	θ_{min}	d_{min}	r
I	45°	$0.53\sqrt{E_0}$	1.5
III	$< 45^o$	$0.63 \cdot \sqrt{E_0}$	1.8

Table 4.1: Comparison of Type I and Type III constellations

the average phasor energy becomes:

$$
\begin{aligned}
E_0 &= \frac{1}{16}\left[4 \cdot (d^2 + d^2) + 8(9d^2 + d^2) + 4 \cdot (9d^2 + 9d^2)\right] \\
&= \frac{1}{16}(8d^2 + 80 \cdot d^2 + 72d^2) \\
&= 10d^2.
\end{aligned}
$$

Hence, assuming the same average phasor energy E_0 as for the Type I constellation we now have a minimum distance of

$$
d_{min} = 2d = 2 \cdot \sqrt{E_0/10} = \sqrt{E_0/2.5} \approx 0.63 \cdot \sqrt{E_0}.
$$

Finally, the peak-to-average energy ratio r is given by:

$$
r = \frac{18d^2}{10d^2} = 1.8.
$$

The Type III constellation's characteristics are also summarised in Table 4.1. Observe that the Type I constellation has a higher jitter immunity and a slightly lower peak-to-average energy ratio than the Type III scheme. However, the Type III phasor constellation has an almost 50% higher minimum distance at the same average phasor energy and hence it is very attractive for AWGN channels, where noise is the dominant channel impairment.

The Type III square constellation can only be implemented, when there are N bits per symbol, where $N = 2M$ and M is an integer. If an odd number of bits is to be encoded then the optimum constellation shape is less obvious. This issue was investigated by Smith [16] who determined that the best constellations were those which have come to be known as cross-QAM. An example of such a constellation is shown in Figure 4.2. It is not possible to Gray encode such a cross-shaped constellation so that all nearest neighbour points separated by d_{min} have a Hamming distance of one, and so a small Gray coding penalty must be accepted along with an increased encoding and decoding complexity. For these reasons, even-bit square constellations are significantly more common than odd-bit constellations.

4.2 General Pulse Shaping Techniques

4.2.1 Baseband Equivalent System

In this section we follow the approach proposed by Lucky, Salz and Weldon [9] using the simplified system model of Figure 4.3, which is essentially a simplified version of Figure 3.1, ignoring the frequency translation or mixing sections of the modem schematic, which do not

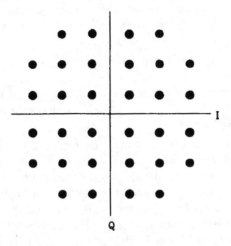

Figure 4.2: 5-bit QAM cross-shaped constellation

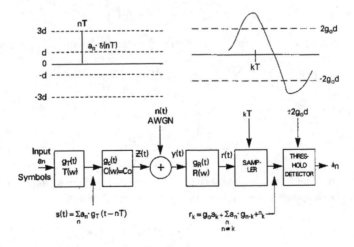

Figure 4.3: Baseband equivalent system model for the I or Q component of square 16-QAM

influence the shape of the signal's spectrum or the signal detection process. In the figure only one of the quadrature arms is portrayed, but we consider a 2^m-ary QAM system, where the I and Q components are $2^{m/2}$-ary. The information to be conveyed is encoded as amplitude values of the signalling pulse shape $g_T(t)$, and these values change at the signalling intervals nT, $n = 1, 2, \ldots, \infty$. Thus the baseband transmitted signal on one of the quadrature channels is

$$s(t) = \sum_n a_n g_T(t - nT) \tag{4.1}$$

where $g_T(t)$ is the transmitter filter's impulse response, implying that the information symbols a_n are idealised zero-width pulses. The demodulator has to infer the transmitted information a_n from the received signal.

We continue to use the baseband equivalent system model introduced in Chapter 2, and try to find that transmitter and receiver filter pair $T(\omega)$ and $R(\omega)$, respectively, which minimises the intersymbol interference (ISI). The legitimate QAM levels are $\pm d$, $\pm 3d$, $\ldots$, $\pm(2^{m/2} - 1)$, as seen, for example, in the Type III phasor constellation of Figure 4.1. The transmitted signal is contaminated by AWGN $n(t)$ and may suffer multipath dispersion (see Chapter 2) via the channel whose transfer function is $C(\omega)$. The receiver filters the signal and after sampling at instants $kT + \tau$, where τ is the channel delay, infers the transmitted information by threshold detection. The signal $r(t)$ at the output of the receiver filter for one quadrature arm is given by:

$$r(t) = \sum_n a_n g(t - nT) + n(t), \tag{4.2}$$

where $g(t)$ represents the impulse response of the total transmission path constituted by the transmitter, receiver and channel transfer functions $T(\omega)$, $R(\omega)$ and $C(\omega)$, respectively. Using their impulse responses of $g_T(t)$, $g_R(t)$ and $g_C(t)$, respectively, we have

$$g(t) = g_T(t) * g_C(t) * g_R(t), \tag{4.3}$$

where * means convolution. The received signal at the receiver's sampling instants is written as:

$$r(kT + \tau) = \sum_n a_n g(kT - nT + \tau) + n(kT + \tau), \tag{4.4}$$

or using a convenient shorthand we have:

$$r_k = \sum_n a_n g_{k-n} + n_k, \tag{4.5}$$

where the system delay τ has been absorbed in the signalling interval index. Equation 4.5 can be rewritten to emphasise the effect of the current symbol of index $k = n$ as:

$$r_k = g_0 a_k + \sum_{\substack{n \\ n \neq k}} a_n g_{n-k} + n_k, \tag{4.6}$$

where g_0 is the main tap of the system's impulse response, while the second and third terms represent the ISI and AWGN, respectively. The receiver compares the signal r_k to the appropriately scaled decision levels of $0, \pm 2g_0d, \ldots, \pm(2^{m/2} - 2)g_0d$, and decides which of the legitimate a_k values is most likely to have been transmitted. An erroneous decision occurs when the (ISI + AWGN) term becomes larger than g_0d, i.e.:

$$\left| \sum_{\substack{n \\ n \neq k}} a_n g_{n-k} + n_k \right| > g_0 d. \tag{4.7}$$

Minimising the left-hand side (LHS) term of Equation 4.7 will minimise the BER. Since the channel's transfer function may be varying with time, we assume a linear phase, distortion-free, equalised channel with $C(\omega) = C_0$ and $\phi_C = \omega T$ within the transmission band. Then the problem is that of finding the filter characteristics $T(\omega)$ and $R(\omega)$ minimising the LHS of Equation 4.7.

4.2.2 Nyquist Filtering

From Equation 4.7 the BER performance of one channel of the QAM system depends on the samples g_{n-k}, $n \neq k$ of the system's impulse response and the noise n_k. It is Nyquist's sampling theorem [402] that enables us to relate these time domain samples to the spectral domain filter characteristic $G(\omega) = T(\omega)R(\omega)$, where $G(\omega)$ is the Fourier transform pair of g_{n-k}. If the system is bandlimited to $[-B, B]$, its behaviour is determined by its samples spaced apart by $T = (1/2B)$. Hence

$$g(nT) = g\left(\frac{n}{2B}\right). \tag{4.8}$$

The system's transfer function $G(\omega)$ is given in terms of these samples by the Fourier series expansion of

$$G(\omega) = \begin{cases} \frac{1}{2B} \sum_n g\left(\frac{n}{2B}\right) e^{-j\omega n/2B} & \text{if } |\omega| \leq 2\pi B \\ 0 & \text{otherwise.} \end{cases} \tag{4.9}$$

Nyquist's sampling theorem also states that the original signal $g(t)$ can be perfectly recovered from its samples by lowpass filtering (LPF) the signal with a filter having a cut-off frequency of B. In time domain this filtering corresponds to convolution with the impulse response

$$h_{LPF}(t) = 2B \frac{\sin(2\pi Bt)}{2\pi Bt} = 2B sinc(2\pi Bt) \tag{4.10}$$

of the LPF. The sampling interval given by $T = 1/2B$ is often referred to as Nyquist interval and the associated frequency of $f_N = 1/2T$ is known as the Nyquist frequency.

The ISI term of Equation 4.7 can only be eliminated if $g_{n-k} = 0$ is met for $n \neq k$, which requires g_n to have zero-crossings at $nT \neq 0$. Since we are only using the T-spaced samples of $g(n)$, this requirement can be formulated by the help of the Fourier transform as:

$$g_n = \int_{-\infty}^{\infty} G(f)e^{j2\pi fnT} df = \delta_n. \tag{4.11}$$

where δ_n is the Kronecker delta. The above integration can be carried out in segments of

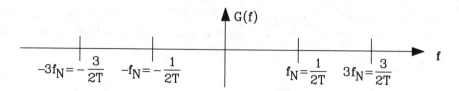

Figure 4.4: Nyquist frequency segments

Nyquist bandwidths given by $f_N = 1/2T$, as seen in Figure 4.4. Then we have:

$$g_n = \sum_k \int_{(2k-1)/2T}^{(2k+1)/2T} G(f)e^{j2\pi fnT}df = \delta_n , \tag{4.12}$$

Instead of integrating $G(f)$ piecewise over the Nyquist frequency segments, as k is increased we can integrate over $[-1/2T, 1/2T]$ and shift $G(f)$ along the frequency axis by $2f_N = 1/T$, as k is increased, giving:

$$g_n = \sum_k \int_{-1/2T}^{1/2T} G\left(f + \frac{k}{T}\right) e^{j2\pi fnT}df = \delta_n . \tag{4.13}$$

Upon interchanging the summation with the integration, g_n can be expressed by the equivalent Nyquist characteristic $G_N(f)$ as follows:

$$g_n = \int_{-1/2T}^{1/2T} G_N(f)e^{j\omega nT}df = \delta_n , \tag{4.14}$$

where

$$G_N(f) = \begin{cases} \sum_k G\left(f + \frac{k}{T}\right) & \text{if } |f| \le \frac{1}{2T} \\ 0 & \text{otherwise.} \end{cases} \tag{4.15}$$

Equation 4.15 reveals that the equivalent Nyquist characteristic $G_N(f)$ is the superposition of all the $1/T$ wide frequency domain slices of $G(f)$ within the band $[-1/2T, 1/2T]$, while it is zero outside this band. Since Equation 4.14 fully specifies $g_n = \delta_n$ in terms of its $T = 1/2B$ spaced samples represented by the Kronecker delta, it also describes $G_N(f)$. The *sinc* function in Equation 4.10 fulfils the criterion of ISI-free signalling and constrains $G_N(f)$ to be an ideal LPF characteristic with zero phase, as seen in Figure 4.5 and given by

$$G_N(f) = \begin{cases} T & \text{if } |f| \le \frac{1}{2T} \\ 0 & \text{if } |f| > \frac{1}{2T}. \end{cases} \tag{4.16}$$

The simplest implementation of the ISI-free Nyquist equivalent characteristic would then be the ideal LPF of Equation 4.16 plotted in Figure 4.5. Unfortunately, this filter characteristic is unrealisable and any approximation to it results in serious ISI penalty. Any practical Nyquist filter must have a wider bandwidth than $f_N = 1/2T$, but a bandwidth in excess of $2f_N$ is usually unacceptable in terms of bandwidth efficiency. This situation is depicted in Figure 4.6, where $G(f)$ is limited to $[-2f_N, 2f_N]$ and hence has only three Nyquist seg-

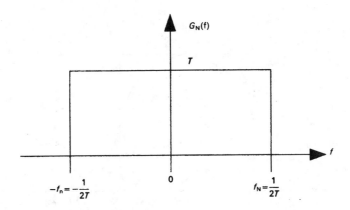

Figure 4.5: Baseband equivalent Nyquist characteristic for ISI-free transmission

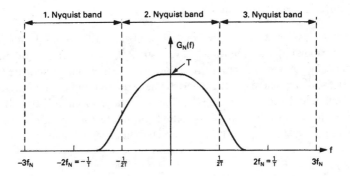

Figure 4.6: Sketch of a practical Nyquist filter characteristic

ments. Then Equation 4.15 has only three terms as seen below:

$$G_N(f) = G\left(f - \frac{1}{T}\right) + G(f) + G\left(f + \frac{1}{T}\right) \qquad (4.17)$$

Observe from Figure 4.6 that the Nyquist segment overlaying specified by Equation 4.17 is equivalent to folding 1 and 3 Nyquist segments back in segment 2 around the frequencies $-f_N$ and f_N, respectively. This is due to the symmetry of the positive and negative wings of the spectrum. However, if the phase characteristic is non-zero, the spectrum's conjugate complex symmetry must be taken into account.

Equivalently, the folding process can be stated as follows. The Nyquist equivalent characteristic is constituted by an ideal LPF with a cut-off frequency of $f_N = 1/2T$ plus an overlaid characteristic having odd symmetry around f_N, but not extending beyond $2f_N$.

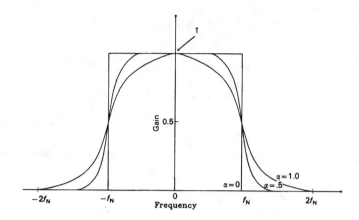

Figure 4.7: Sketch of raised-cosine Nyquist filter characteristics for various roll-off factors

4.2.3   Raised-Cosine Nyquist Filtering

From the above discussion we know that there is a plethora of filters fulfilling the Nyquist criterion, having bandwidths between $f_N = 1/2T$ and $2f_N = 1/T$. In general, the higher the bandwidth efficiency the better. A practical set of convenient Nyquist filters are the so-called raised cosine (RC) filters, where the characteristic fits a quarter period of a frequency domain cosine-shaped curve to the ideal LPF transfer characteristic, as seen in Figure 4.7.

The bandwidth is controlled by the so-called roll-off factor α, defined as the ratio of excess bandwidth above f_N to f_N. The RC characteristic is then defined as [9]:

$$G(f) = \begin{cases} T & \text{for } 0 \leq \mid f \mid \leq \frac{1-\alpha}{2T} \\ \frac{T}{2}\left(1 - \sin\left[\frac{\pi T}{\alpha}\left(f - \frac{1}{2T}\right)\right]\right) & \text{for } \frac{1-\alpha}{2T} < \mid f \mid \leq \frac{1+\alpha}{2T}. \end{cases} \tag{4.18}$$

4.2.4   The Choice of Roll-Off Factor

The impulse response of the RC characteristic plays an important role in deciding upon the choice of α in a particular system. Namely, when the channel becomes non-ideal and hence the ISI is non-zero, then the decay of the impulse response must be as rapid as possible to minimise the duration over which a previous transmitted symbol can influence the received signal.

The set of impulse responses corresponding to the RC curves in Equation 4.18 is given [9] by:

$$g(t) = \frac{\sin(\pi t/T)}{\pi t/T} \frac{\cos(\alpha \pi t/T)}{1 - 4\alpha^2 t^2/T^2}. \tag{4.19}$$

These impulse responses are plotted in Figure 4.8 for roll-off factors of $\alpha = 0, 0.5$ and 1. Observe in Equation 4.19 that for $\alpha = 0$ we have:

$$g(t) = \frac{\sin(\pi t/T)}{\pi t/T}, \tag{4.20}$$

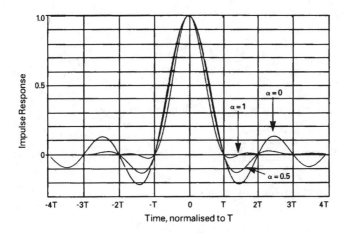

Figure 4.8: Raised-cosine Nyquist filter impulse responses for various roll-off factors

associated with the lowest bandwidth occupancy but also with the most slowly decaying impulse response. The fastest decay is ensured, when $\alpha = 1$, at the price of doubled bandwidth occupancy. In this case the impulse response exhibits an additional set of zero-crossings between those stipulated by the Nyquist condition. A good compromise is represented by $\alpha = 0.5$ in terms of both bandwidth occupancy and impulse response decay. Notice that the impulse response decay plays a crucial role in determining the system's robustness against ISI, because for slowly decaying responses the channel's ISI has a prolonged influence on future and past sampling instants. Therefore higher roll-off factors must be used in high-ISI environments, despite the lower bandwidth efficiency.

In summary, we determined the required transfer characteristic of the total transmission system, which must have an odd-symmetric roll-off around $f_N = 1/2T$. Lower roll-off factors yield spectral compactness, but they are also more vulnerable to ISI via non-ideally equalised channels. This makes clock recovery much more difficult to achieve.

4.2.5 Optimum Transmit and Receive Filtering

In this subsection we highlight briefly how the transfer characteristic $G_N(f)$ must be split between the transmitter filter $G_T(f)$ and the receiver filter $G_R(f)$ in order to maximise noise immunity [9]. Recall from Equation 4.7 that we have minimised the effects of ISI by selecting an overall transfer characteristic $G_N(f)$. The BER due to AWGN is minimised by maximising the SNR at sampling instants for a given transmitted power.

Recall from Equation 4.1 that the transmitted signal was given by

$$s(t) = \sum_{n=-\infty}^{\infty} a_n g_T(t - nT),$$

where a_n represents the Dirac-shaped information symbols, while $g_T(t)$ is the transmitter filter's impulse response. Then the average transmitted power over an interval of $2 \cdot N \cdot T$ is given by the expectation value:

$$P_T = \left\langle \lim_{N \to \infty} \frac{1}{2NT} \int_{-NT}^{NT} \left[\sum_{n=-N}^{N} a_n g_T(t - nT) \right]^2 dt \right\rangle. \tag{4.21}$$

Upon interchanging the expectation and limit computation we have:

$$P_T = \lim_{N \to \infty} \frac{1}{2NT} \sum_{n=-N}^{N} \sum_{m=-N}^{N} \langle a_n a_m \rangle \int_{-NT}^{NT} g_T(t - nT) g_T(t - mT) dt. \tag{4.22}$$

For uncorrelated input symbols we get:

$$\langle a_n a_m \rangle = \begin{cases} 0 & n \neq m \\ \bar{a}^2 & n = m \end{cases} \tag{4.23}$$

and hence

$$
\begin{aligned}
P_T &= \lim_{N \to \infty} \frac{\bar{a}^2}{2NT} \sum_{n=-N}^{N} \int_{-NT}^{NT} g_T^2(t - nT) dt \\
&= \frac{\bar{a}^2}{T} \int_{-\infty}^{\infty} g_T^2(t) dt. \tag{4.24}
\end{aligned}
$$

By exploiting Parseval's theorem:

$$\int_{-\infty}^{\infty} g_T^2(t) dt = \int_{-\infty}^{\infty} |G_T(f)|^2 df \tag{4.25}$$

the channel's input power can also be expressed in terms of the spectral domain transmitter transfer function $G_T(f)$:

$$P_T = \frac{\bar{a}^2}{T} \int_{-\infty}^{\infty} |G_T(f)|^2 df = \frac{\bar{a}^2}{T} \int_{-\infty}^{\infty} \left| \frac{G_N(f)}{G_R(f)} \right|^2 df, \tag{4.26}$$

where we have used the fact that $G_N(f) = G_T(f) \cdot G_R(f)$. The received noise has a spectral density of $N(f)$ and after passing through the receiver filter it has a power of

$$P_N = \int_{-\infty}^{\infty} N(f) |G_R(f)|^2 df. \tag{4.27}$$

The noise power P_N has to be minimised under the constraint of constant channel input power P_T so as to maximise the SNR. Using the technique of variational calculus [9] the functional F defined below must be minimised:

$$F = \lambda P_T + P_N, \tag{4.28}$$

where λ is the so-called Lagrange multiplier.

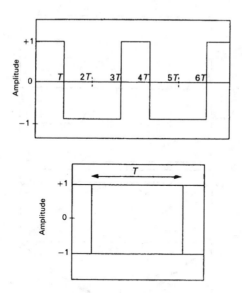

Figure 4.9: A typical binary data sequence and its eye diagram

The result of this minimisation problem is that for a flat noise spectral density N_0 and for a real Nyquist equivalent characteristic, the transmitter and receiver filters must be identical in order to maximise the SNR, i.e.:

$$G_T(f) = G_R(f) = \sqrt{G_N(f)}. \qquad (4.29)$$

This important result was already implicitly exploited in Figure 3.1, where we represented the square root of the Nyquist filter by $\sqrt{N}$.

4.2.6 Characterisation of ISI by Eye Diagrams

When a dispersive channel introduces ISI, previously transmitted symbols will cause non-zero contributions at subsequent sampling instants. The so-called eye diagram is a simple and convenient tool to study the effects of ISI and other channel impairments both by simulation and in hardware experiments.

Let us imagine an ideal noiseless and ISI-free, infinite bandwidth communications system transmitting a binary data sequence having values of ± 1, at a bit rate of $R = 1/T$, as shown in Figure 4.9.

If this system's output signal is displayed on an analogue oscilloscope screen with persistence, when using external synchronisation having a frequency of $R = 1/T$, the beam starts its travel across the screen every $1/T$ second and draws the adjacent signal sections on top of each other, as also shown in Figure 4.9.

If this PRBS is now transmitted via a Nyquist-filtered, bandlimited, but ISI-free channel, the corresponding signals are portrayed in Figure 4.10.

Observe that since no ISI and no noise are present, the eye is "clean" or "fully open" at

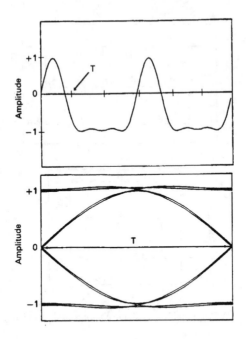

Figure 4.10: Received signal and its eye diagram via bandlimited channel with no noise and ISI

the sampling instants and the zero-crossings occur always at multiples of T, appearing at the same instant in the eye diagram.

Let us now introduce ISI. The signalling waveforms will have typically non-zero values at all sampling instants and the received PRBS and its eye diagram will typically resemble those in Figure 4.11.

When AWGN is also superimposed, we expect waveforms similar to those in Figure 4.12.

As demonstrated by Figure 4.13, nearly all typical channel impairments become recognisable in the eye diagrams, and hence eye diagrams are very useful in quick overall system evaluation. Observe that imperfectly timed sampling has a similar effect to increasing the AWGN. The zero-crossing jitter becomes very crucial in case of clock recovery circuits that operate by detecting zero-crossings (see Chapter 6) and using the time at which these occur to derive the optimal sampling point. When multilevel transmissions are used, the eye diagram becomes much more complicated due to the increased number of possible transitions. The eye opening is narrower, as demonstrated by Figure 4.14, where the transmitter's eye pattern is seen for a 16-QAM Nyquist-filtered modem having a roll-off factor of $\alpha = 1$.

This Nyquist filter with $\alpha = 1$ has the sharpest possible impulse response decay and hence minimises the effects of ISI. Lower α values result in smoother time domain transitions, yielding a more closed eye pattern. This increased sensitivity is the price paid for the higher spectral compactness of narrower Nyquist filters.

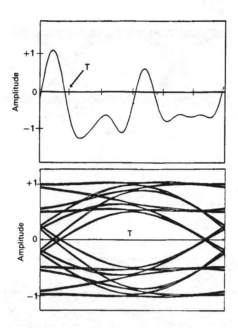

Figure 4.11: Received signal and eye diagram impaired by ISI

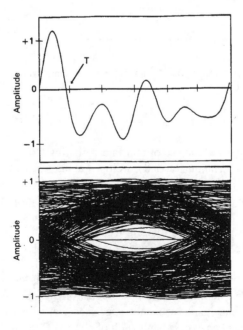

Figure 4.12: Received signal and eye diagram with ISI and AWGN

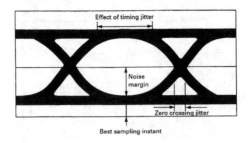

Figure 4.13: Effect of channel impairments on the eye diagram

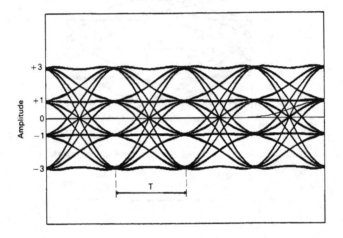

Figure 4.14: 16-QAM transmitter eye diagram using $\alpha = 1$

4.2.7   Non-Linear Filtering

In 1982 Feher [103] suggested the use of non-linear filtering (NLF) for QAM satellite communications. This was a technique he had developed along with Huang in 1979 which simplified filter design by simply fitting a time domain quarter period of a sine wave between two symbols for both of the quadrature carriers. This allowed the generation of jitter-free bandlimited signals, which had previously been a problem, improving clock recovery techniques. Furthermore, unlike partial response filtering techniques which intentionally introduce ISI, NLF produces an ISI-free waveform since there is no contribution from previous symbols at any sample point, which is advantageous when complex high-level QAM constellations are transmitted. The disadvantage of this form of filtering is that it is less spectrally efficient than optimal partial response filtering schemes. We performed some simulation work to compare their spectral efficiency, and our results suggested that while optimal partial response schemes [405] necessitated an excess bandwidth of about 20% above the Nyquist frequency $f_N = 1/(2 \cdot T)$, the NLF scheme required an excess bandwidth of about 40%. Nevertheless, its implementational advantages often render this loss of efficiency acceptable. The power

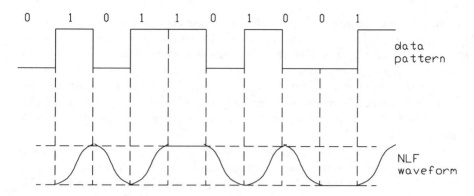

Figure 4.15: Non-linear filtering waveforms

spectrum of a NLF signal is given by [103]

$$S(f) = T \left(\frac{\sin 2\pi fT}{2\pi fT} \frac{1}{1 - 4(fT)^2} \right)^2 \tag{4.30}$$

and some sample waveforms are given in Figure 4.15 for one quadrature arm. The other quadrature arm is identical.

4.3   Methods of Generating QAM

4.3.1   Generating Conventional QAM

As discussed in Chapter 3 and as shown in Figure 4.16, in a typical QAM modem the incom-

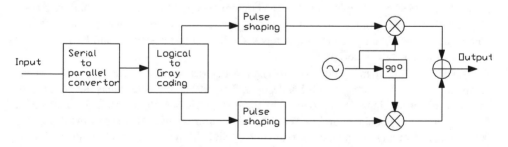

Figure 4.16: Standard QAM modulator

ing binary data stream is passed through a serial-to-parallel convertor of an appropriate width. The data is then passed through a logic block which performs the translation between the binary input data and the constellation points, to allow Gray coding. Typically such a block would be implemented as a look-up table using a programmable memory device. In the case of square QAM the output from the serial-to-parallel convertor can be partitioned into two subsets, and each encoded onto a separate axis using identical mapping blocks. Digital-to-analogue (D/A) conversion is then required, followed by pulse shaping, before upconverting the signal to the carrier frequency. It is often advantageous to perform the pulse shaping prior to the D/A conversion by oversampling the transmitted data and using further look-up tables containing the pulse shaping law in order to produce a digital representation of the smoothed signal. This will still need to be filtered after D/A conversion in order to remove aliasing errors caused by the oversampling ratio, but this filtering is substantially simpler than would otherwise be required.

In order to demodulate the signal, the reverse process is followed. The incoming waveform is sampled at the correct instants provided by the clock recovery circuit, A/D converted and a decision is made as to the most likely constellation point to have been transmitted. The position of this constellation point is then passed through a logic block which performs the reverse operation to the one in the transmitter, leading to the encoded information. This is passed through a parallel-to-serial convertor leading to the output data stream.

While this is by far the most common method for generating QAM signals, an alternative suggestion referred to as superposed QAM was made by Miyauchi *et al.* [17], which will be considered in the next section.

4.3.2 Superposed QAM

In superposed QAM modems the approach is to use a number of PSK modems in tandem as explained below. This had the advantage of being able to use circuitry already developed for PSK modems, allowing ease of implementation. This technique has not gained much popularity, since the conventional QAM generation method described in the previous section has generally proved more efficient. Nevertheless, a brief description is included for completeness.

A diagram showing how a 16-level square QAM constellation could be constructed by this method is given in Figure 4.17. Essentially, the first two bits of the 4-bit symbol are passed to a QPSK modulator operating at a certain power and the second two bits are passed to a second QPSK modulator operating at one-quarter of the power of the first. The two QPSK modulated signals are then superimposed before transmission. A similar philosophy can be employed at the receiver.

Block diagrams of transmitter and receiver structures are given in Figure 4.18. At the receiver the signal is passed to a QPSK decoder. This essentially determines the quadrant of transmission, or the signal level produced by the first QPSK modulator within the transmitter. The decision is remodulated and passed to a subtraction circuit. This forms the difference between the incoming signal and the remodulated signal, giving the signal produced by the second QPSK modulator. This can also be demodulated to produce the remaining output bits.

4.3.3 Offset QAM

Offset, or staggered modulation, is a technique which evolved with PSK transmissions in an effort to reduce the envelope variations of the transmitted signal, and in particular to prevent the envelope of the transmitted signal passing through zero during transitions between symbols as this required the power amplifier to maintain linearity across a wide amplitude range.

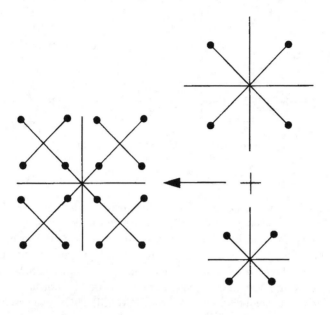

Figure 4.17: Constructing QAM with two PSK modulators

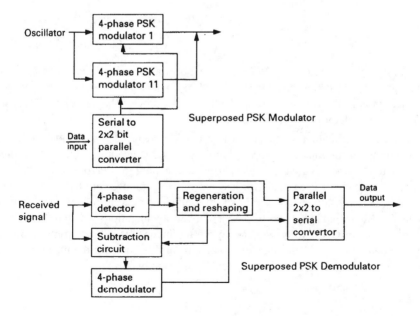

Figure 4.18: Superposed PSK modulator and demodulator

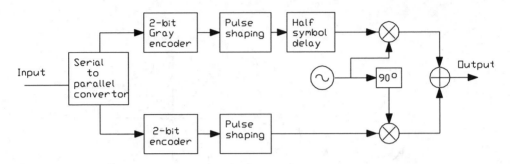

Figure 4.19: O-QAM modulator structure

This was problematic as non-linearities in amplifiers, particularly those used for satellite applications, could cause considerable distortion of the transmitted waveform. Furthermore, large variations in the envelope could result in a requirement for tight filtering if out-of-band spillage was to be avoided. Offset modulation involves delaying one of the quadrature arms by half a symbol period with respect to the other arm. Therefore, when the maximum rate of change of level is taking place on one arm (i.e. in the centre of a transition between levels) then the other arm is near stationary. Offset QAM (O-QAM) can be expressed as:

$$
s(t) = \left[\sum_k a_{2k} h(t - 2kT) \right] \cos(2\pi f_c t) \tag{4.31}
$$

$$
- \left[\sum_k a_{2k-1} h(t - (2k - 1)T) \right] \sin(2\pi f_c t)
$$

where f_c is the carrier frequency, $1/T$ is the symbol rate, a_k is the kth data symbol, for 16-level square QAM taking on values ± 1 and ± 3, and $h(t)$ is the signalling waveform corresponding to the impulse response of the transmitter's pulse shaping filter. The structure of an O-QAM modulator is shown in Figure 4.19.

Offsetting the modulation can make clock recovery more problematic, and decoding more complex. Further uncertainty can exist within the receiver concerning which axis is delayed with respect to the other, and we consider differential coding to overcome this ambiguity in Section 4.7. Claims have been made for less out-of-band spillage for O-QAM compared to QAM [131], with a resulting increase in spectral efficiency due to tighter packing of adjacent channels. Malkamaki [131] performed a number of simulations where he compared the spectral efficiency of QAM with O-QAM. Spectral efficiency is defined in many ways, and we consider some of these definitions in Chapter 14 of this book. In the meantime, we use Malkamaki's definition that the spectral efficiency is given by

$$
S_r = \frac{R_m}{B_m} \left(\frac{C}{I_c} \right)^{-2/\alpha} \tag{4.32}
$$

where R_m is the modulation rate, B_m is the channel spacing, C is the received carrier power,

C/I_c dB	$S_r\ (\alpha = 2)$	$S_r\ (\alpha = 4)$	$S_r\ (\alpha = 6)$
-9	1.07	1.07	1.06
-15	1.05	1.04	1.03
-20	1.04	1.03	1.03

Table 4.2: Relative efficiency of O-QAM compared to QAM

I_c is the interference power, and α is the assumed propagation exponent. Malkamaki performed a number of simulations and reported the relative efficiency results displayed in Table 4.2.

The results suggest that in terms of spectral efficiency O-QAM is only marginally better than QAM, and implementation difficulties may reduce this slight advantage. Furthermore, O-QAM requires a less efficient differential coding system than QAM, as discussed in Section 4.7, and it is not clear whether Malkamaki has taken this into account. It would appear that O-QAM is only advantageous where large envelope variations must be avoided due to amplifier design considerations.

4.3.4 Non-Linear Amplification

QAM had been proposed for satellite applications in the late 1960s but a major implementational stumbling block was that of amplification. Because of limited power within satellites, power efficient non-linear class C amplifiers were preferred. QAM, being a non-constant envelope modulation, requires linear amplifiers. The increase in power required when using low-efficiency linear class A amplifiers precluded QAM from most satellite applications. Feher [22] suggested a new method for generating QAM signals using highly non-linear amplifiers, which he termed non-linearly amplified QAM (NLA-QAM), in order to overcome this problem. Two separate amplifiers for the 16-QAM case were used, one operating with a quarter the output power of the other. This is shown in Figure 4.20.

The higher power amplifier processed the two most significant bits (MSBs) of the 4-bit symbol only, and the lower power amplifier processed only the two least significant bits (LSBs). The two MSBs and LSBs represented the high power and low power phasors, respectively, which always maintained a constant magnitude and hence were class C amplified. The amplified processed signals were then summed at full output power to produce the QAM signal.

Again, in this arrangement each amplifier is producing a QPSK signal, therefore the scheme has certain similarities with that proposed by Miyauchi, except the latter performs summation of QPSK signals prior to amplification, whereas Feher sums the signals after amplification. Because both amplifiers were able to use constant envelope modulation they could be operated at a constant level with resulting high power-efficiency. However, Feher's NLA scheme resulted in an increased complexity due to the need for two amplifiers and a hybrid combiner, compared to previous systems which used only a single amplifier. Nevertheless, in satellite applications, complexity was relatively unimportant compared to power efficiency, and this NLA technique offered a substantial 5 dB power gain, considerably increasing the potential of QAM in severely power-limited applications. This work was soon followed by a performance study of a NLA 64-state system [23] which extended the NLA scheme to 64

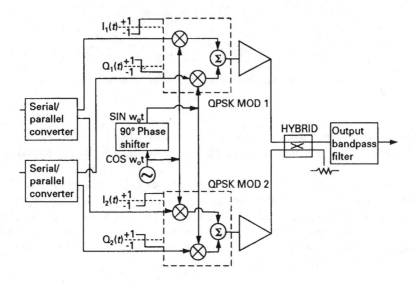

Figure 4.20: NLA structure

levels by using three amplifiers all operating at different power levels.

Non-linear amplification is rarely used in other applications due to the increase in hardware complexity.

4.4   Methods of Detecting QAM Signals

4.4.1   Threshold-Detection of QAM

In most of this book we will be assuming that the filtered received waveform is sampled at the correct point by use of a clock recovery circuit (see Chapter 6) and subjected to threshold detection in order to derive the received information symbol for that symbol instant. A generic receiver designed around these principles is shown in Figure 4.21. The optimum receiver filter maximising the signal-to-noise ratio (SNR) and hence minimising the bit error rate (BER) was derived in Section 4.2, while the BER using a perfectly coherent threshold detector will be analytically derived for 16-QAM and 64-QAM over AWGN channels in Chapter 5.

Unless stated to the contrary, throughout this book we will be considering the receiver of Figure 4.21, which is based on threshold detection.

4.4.2   Matched-Filtered Detection

In the previous section we have briefly summarised the salient features of the low complexity threshold detector receiver, which is constituted by a receiver filter followed by a sampler and threshold detector. In Section 4.2 we optimised the receiver filter under the constraint of constant channel input power or transmitted power P_T and minimised the amount of noise power

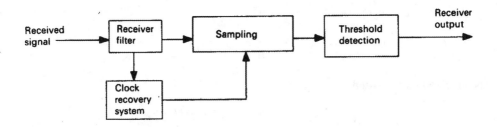

Figure 4.21: Receiver structure using threshold detection

P_N admitted to the threshold detector by appropriately designing the receiver filter. The result was that the Nyquist characteristic had to be split symmetrically between the transmitter and receiver filter, while the channel was assumed to be equalised.

When designing a so-called matched filter, the approach is to maximise the signal-to-noise ratio (SNR) at the output of the matched filter for a given received signal waveform [406]. The system's schematic is identical to that in Figure 4.3. Considering for example one of the quadrature components of a 2^m-ary QAM signal at the input of the receiver we have:

$$y(t) = \sum_n a_n \cdot h(t - nT) + n(t) = z(t) + n(t)$$

where a_n represents the $M = 2^{m/2}$-ary I or Q components, $n(t)$ is the AWGN and $h(t)$ is given by the convolution:

$$h(t) = g_T(t) * g_c(t),$$

with $g_T(t)$ and $g_c(t)$ being the transmitter filter's and the channel's impulse response, respectively. Clearly, signalling is carried out using the signalling waveforms $h(t)$ weighted by the $M = 2^{m/2}$-ary I and Q components. Assuming that there is no intersymbol interference (ISI) the sampling circuit's output signal at time $t = T$ is the sum of a signal contribution r_k and a noise contribution n_k.

Our goal is then to maximise the following SNR term [388, 406, 407]:

$$SNR_T = \frac{r_k^2}{\sigma_0^2},$$

at $t = T$, where σ_0^2 is the noise variance, as a function of the filter transfer function $R(f)$. The information signal $r(t)$ at the filter's output can be formulated by the help of the Fourier transform as:

$$r(t) = \int_{-\infty}^{\infty} R(f)Z(f) \cdot e^{j2\pi ft} df,$$

where $Z(f)$ is the frequency domain representative of the received signal $z(t)$ and $R(f)$ is the transfer function of the filter $g_R(t)$ at the receiver's input.

Following Sklar's approach [406] and assuming a double-sided noise spectral density of

$N_0/2$, the filter's output noise power is given by:

$$\sigma_0^2 = \frac{N_0}{2} \int_{-\infty}^{\infty} |R(f)|^2 df.$$

Then the SNR is expressed as:

$$SNR_T = \frac{|\int_{-\infty}^{\infty} R(f) \cdot Z(f) e^{j2\pi fT} df|^2}{N_0/2 \int_{-\infty}^{\infty} |R(f)|^2 df}. \tag{4.33}$$

In order to find the optimum filter $R(f) = R_{opt}(f)$, Schwartz's inequality can be invoked, which is formulated in general as:

$$\left| \int_{-\infty}^{\infty} g_1(x) \cdot g_2(x) dx \right|^2 \leq \int_{-\infty}^{\infty} |g_1(x)|^2 dx \cdot \int_{-\infty}^{\infty} |g_2(x)|^2 dx.$$

The equality is satisfied, if $g_1(x) = c \cdot g_2^*(x)$, where c is a constant and $*$ denotes the complex conjugate. If we let $g_1(x) = R(f)$ and $g_2(x) = Z(f) \cdot e^{j2\pi fT}$, we arrive at:

$$\left| \int_{-\infty}^{\infty} R(f) \cdot Z(f) e^{j2\pi fT} df \right|^2 \leq \int_{-\infty}^{\infty} |R(f)|^2 df \cdot \int_{-\infty}^{\infty} |Z(f)|^2 df, \tag{4.34}$$

where equality is satisfied if

$$R(f) = c \cdot Z^*(f) e^{-j2\pi fT}. \tag{4.35}$$

Upon substituting Equation 4.34 into Equation 4.33 we have

$$SNR_T \leq \frac{\int_{-\infty}^{\infty} |R(f)|^2 df \cdot \int_{-\infty}^{\infty} |Z(f)|^2 df}{N_0/2 \int_{-\infty}^{\infty} |R(f)|^2 df}$$

giving

$$SNR_T \leq \frac{2}{N_0} \int_{-\infty}^{\infty} |Z(f)|^2 df.$$

If we denote the received signal's energy by E_z and exploit the Parseval theorem, we arrive at:

$$E_z = \int_{-\infty}^{\infty} |z(t)|^2 dt = \int_{-\infty}^{\infty} |Z(f)|^2 df$$

and hence the maximum SNR is given by:

$$SNR_T^{max} = \frac{2E_z}{N_0}. \tag{4.36}$$

Observe that the received signal's waveform shape does not appear in the above equation, indicating that the SNR depends on the energy of $z(t)$, rather than its shape. Recall from Equations 4.35 and 4.36 that the equality condition ensured the maximum SNR and from

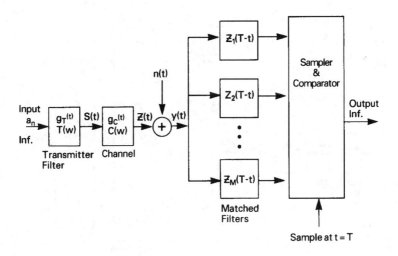

Figure 4.22: Matched filtered receiver

Equation 4.35 this requires in the time domain that

$$\begin{aligned} g_R(t) &= \mathcal{F}^{-1}\{R(f)\} \\ &= \mathcal{F}^{-1}\{c \cdot Z^*(f)e^{-j2\pi fT}\}, \end{aligned}$$

where $\mathcal{F}$ represents the Fourier transform. When taking into account that:

$$\mathcal{F}^{-1}\{Z^*(f)\} = z(-t)$$

and that the factor $e^{-j2\pi fT}$ corresponds to a time domain delay of T, we arrive at

$$g_R(t) = \begin{cases} c \cdot z(T-t) & 0 \le t \le T \\ 0 & \text{otherwise.} \end{cases} \tag{4.37}$$

Explicitly, the impulse response of the filter maximising the SNR at $t = T$ must be the reflected and shifted version of the received signal $z(t)$. Observe from Equation 4.37 that the delay of T renders the impulse response causal or realisable, since $g_R(t) = z(-t)$ would not be realisable.

The matched filtered receiver will then have a bank of filters, each of which is matched to one of the possible $M = 2^{m/2}$ received I or Q waveforms $z_1(t), z_2(t), \ldots, z_M(t)$, as portrayed in Figure 4.22. The received signal is filtered through all matched filters and the output signal of that specific filter will be the largest, which is adequately matched to the received signal. The output signal of the remaining filters will be significantly lower, and hence the comparator finally infers, which $a_n \ n = 1, \ldots, M$ has been transmitted.

Having described the matched filtered receiver we now briefly focus our attention on an alternative implementation of it, the so-called *correlation receiver*.

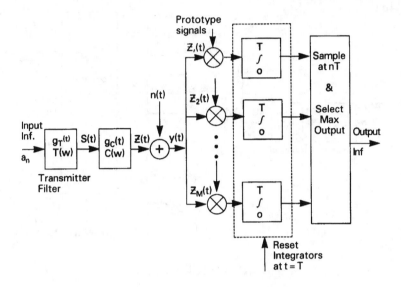

Figure 4.23: Correlation receiver

4.4.3   Correlation Receiver

Observe from Figure 4.22 that the causal filter's output signal can also be described by the help of the convolution:

$$r(t) = y(t) * g_R(t) = \int_0^t y(\tau) \cdot g_R(t - \tau)d\tau \qquad (4.38)$$

and setting arbitrarily $c = 1$ in Equation 4.37 as well as exploiting that:

$$g_R(t - \tau) = c \cdot z[T - (t - \tau)]$$

yields:

$$r(t) = \int_0^t y(\tau) \cdot z(T - t + \tau)d\tau.$$

For $t = T$ this can be written as:

$$r(t = T) = \int_0^{t=T} y(\tau) \cdot z(\tau)d\tau,$$

which allows us to interpret the matched filtered receiver as a *correlation receiver*.

Explicitly, in the correlation receiver of Figure 4.23 the received signal $y(t)$ is correlated with all legitimate prototype signals $z_n(t), n = 1, \ldots, M$ using a bank of M correlators and one of them is assumed to have been transmitted which exhibits the highest correlation at $t = T$. It is important to emphasise, however, that the output of the matched filter is only equal to that of the correlator at $t = T$, which can easily be demonstrated using simple specific received signal waveforms.

Finally, it is important to note that the channel's impulse response $g_c(t)$ can be determined by using channel sounding methods, briefly explained later in Section 7.6. If $g_c(t)$ is known, then the prototype signals $z_n(t), n = 1, \ldots, M$ can be determined. Alternatively, the channel can be equalised initially in order to render $C(f) = C_0$, where C_0 is a constant. This constant transfer function is associated with an ideal Dirac delta-shaped impulse response $g_c(t)$, in which case we have for the prototype signals $z_n(t) = s_n(t), n = 1, \ldots, M = 2^{m/2}$ given by $z_n(t) = s_n(t) = a_n g_T(t - nT)$.

4.5 Linearisation of Power Amplifiers

4.5.1 The Linearisation Problem

A major problem with bandwidth efficient QAM schemes is that their performance is strongly dependent on the linearity of the transmission system. The system designer is faced with the following choices:

(1) Using a low-efficiency linear class A amplifier.

(2) Opting for Feher's non-linear amplification principle [408] relying on separate amplifiers for all amplitude levels.

(3) Employing a power-efficient, weakly non-linear (NL) amplifier and using linearisation techniques to minimise the power amplifier's out-of-band emission and produce a waveform which can be accurately demodulated.

Techniques designed for achieving (3) include active biasing [409], the feedforward method [410], the LINC technique [411], negative feedback [412, 413], pre-distortion [1, 136, 414] or postdistortion [415].

4.5.2 Linearisation by Pre-distortion [1]

4.5.2.1 The Pre-distortion Concept

This subsection concentrates on linearisation by pre-distortion following the approach proposed by Stapleton *et al.* [1, 136, 414, 416–420] whereby a slowly adapting pre-distorter is used [1] to minimise the out-of-band spectral spillage due to power amplifier non-linear (NL) distortions. The simplest method would be to use a non-linear device in front of the power amplifier to produce intermodulation products in antiphase to those of the amplifier in an open-loop configuration. However, this will not adapt to varying transmitter characteristics and needs to be set individually for each amplifier. Instead we use a slowly adapting feedback system, which is shown in Figure 4.24. This method monitors the out-of-band power produced by the non-linear amplifier (NLA) and adjusts the pre-distorter's parameters to minimise it.

As seen in this figure, the pre-distorter creates the distorted signal $v_d(t)$ from the undistorted modulated signal $v_m(t)$. Upon inputting this signal to the amplifier A, its output signal is given by $v_a(t)$.The adaptive feedback path downconverts the bandpass amplifier's output

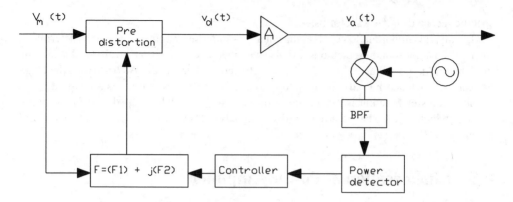

Figure 4.24: Basic pre-distorter block diagram [1] ©IEEE, 1992,
Stapleton *et al.*

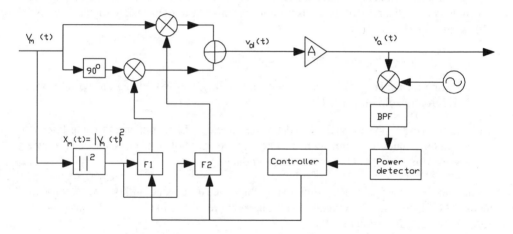

Figure 4.25: Detailed pre-distorter block diagram [1] ©IEEE, 1992, Stapleton *et al.*

signal, which is then bandpass filtered (BPF) to separate the out-of-band (OOB) signal power
from the wanted signal. The OOB signal power is then averaged by the power detector and
used by the controller to adjust the pre-distorter's complex transfer characteristic to minimise
the OOB power.

4.5.2.2 Pre-distorter Description

Based on the above concept Stapleton *et al.* have shown [1] that using the in-phase and
quadrature-phase second-order non-linear functions, F_1 and F_2, which are derived from the
envelope of the pre-distorter's input signal $v_m(t)$, the inverse functions of the power amplifier's amplitude and phase responses can be interpolated. The pre-distorted signal $v_d(t)$
is derived by multiplying the I and Q components of the modulated signal $v_m(t)$ with the
second-order non-linear functions F_1 and F_2, as portrayed in Figure 4.25. The magnitude and
phase characteristics of a typical class AB amplifier are depicted in Figure 4.26 as a function

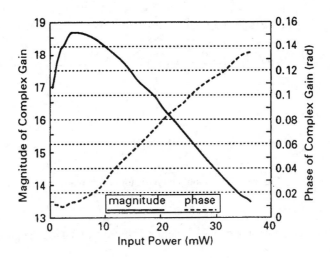

Figure 4.26: Magnitude and phase versus input power for a typical class AB amplifier [1] ©IEEE, 1992,
Stapleton *et al.*

of the input power. Observe the diagrammatic gain and phase variations due to the input power fluctuations. These amplitude characteristics, which depend exclusively on the input power level, must be linearised. It follows that the second-order functions F_1 and F_2 will also only have to depend on the magnitude of the modulated signal, i.e. on $x_m(t) = |v_m(t)|^2$, as shown below:

$$
\begin{aligned}
F_1\{x_m(t)\} &= a_{11} + a_{13}x_m(t) + a_{15}x_m{}^2(t) \\
F_2\{x_m(t)\} &= a_{21} + a_{23}x_m(t) + a_{25}x_m{}^2(t).
\end{aligned}
\tag{4.39}
$$

Note that the coefficient indices have been chosen for later notational convenience. Observe also that the scalar gains a_{11} and a_{21} linearly weight the I and Q components, without introducing any dependency on $v_m(t)$.

The in-phase and quadrature phase weighting functions F_1 and F_2 allow us to express the pre-distorter's complex gain as:

$$
\begin{aligned}
F\{x_m(t)\} &= F_1\{x_m(t)\} + jF_2\{x_m(t)\} \\
&= a_{11} + ja_{21} + (a_{13} + ja_{23})x_m(t) + (a_{15} + ja_{25})x_m{}^2(t) \\
&= a_1 + a_3 x_m(t) + a_5 x_m{}^2(t),
\end{aligned}
\tag{4.40}
$$

where the complex coefficients are given below:

$$
\begin{aligned}
a_1 &= a_{11} + ja_{21} \\
a_3 &= a_{13} + ja_{23} \\
a_5 &= a_{15} + ja_{25}.
\end{aligned}
\tag{4.41}
$$

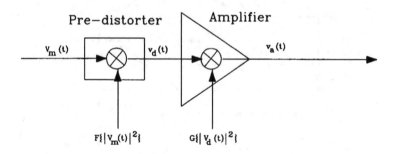

Figure 4.27: Simplified pre-distorter schematic

Using the pre-distorter's complex transfer function F in Equation 4.40, the pre-distorted signal $v_d(t)$ is given by:

$$
\begin{aligned}
v_d(t) &= v_m(t)F\{x_m(t)\} \\
&= a_1 v_m(t) + a_3 v_m(t)x_m(t) + a_5 v_m(t)x_m{}^2(t) \\
&= a_1 v_m(t) + a_3 v_m(t)|v_m(t)|^2 + a_5 v_m(t)|v_m(t)|^4.
\end{aligned}
\tag{4.42}
$$

Observe in the above equation that $v_d(t)$ has a first-, a third- and a fifth-order $v_m(t)$-dependent term, which explains the choice of coefficient indices. Therefore, by appropriately choosing the coefficients a_3 and a_5 the amplifier's third- and fifth-order intermodulation distortion can be reduced.

In a similar approach to our previous deliberations we introduce $x_d(t) = |v_d(t)|^2$ for the envelope of the pre-distorted signal and model the power amplifier also by a simple complex gain expression of the form:

$$
G\{x_d(t)\} = G_1\{x_d(t)\} + jG_2\{x_d(t)\} = G\{|v_d(t)|^2\},
\tag{4.43}
$$

giving the amplifier's output signal $v_a(t)$ in the following form:

$$
v_a(t) = v_d(t)G\{|v_d(t)|^2\}.
\tag{4.44}
$$

The complex amplifier's gain can be modelled by the following complex power series:

$$
G\{x_d(t)\} = g_1 + g_3 x_d(t) + g_5 x_d{}^2(t).
\tag{4.45}
$$

This simple model allows us to portray the whole system as shown in Figure 4.27. Observe that both $F\{|v_m(t)|^2\}$ and $G\{|v_d(t)|^2\}$ depend only on the power of their input signals, but not on their phases. The complex coefficients of G in Equation 4.45 describe both the amplitude modulation to amplitude modulation (AM-AM) and the amplitude modulation to phase modulation (AM-PM) conversion characteristics of the power amplifier.

The system model displayed in Figure 4.27 can be further simplified, as depicted in Figure 4.28, where the overall complex system gain $K\{|v_m(t)|^2\}$ can be derived as follows. First, the amplifier's output signal is expressed as:

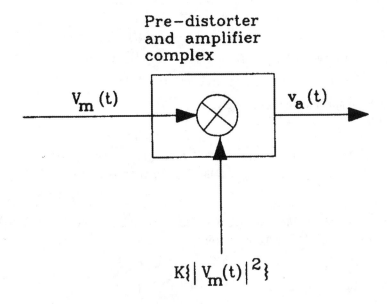

Figure 4.28: Block diagram of pre-distorter and amplifier complex

$$
\begin{aligned}
v_a(t) &= v_d(t).G\{x_d(t)\} \\
&= v_m(t).F\{x_m(t)\}.G\{x_m(t).|F[x_m(t)]|^2\},
\end{aligned}
\tag{4.46}
$$

where we exploited the relationships that $x_d(t) = x_m(t).|F\{x_m(t)\}|^2$ and $v_d = v_m.F\{x_m(t)\}$. From Equation 4.46 the complex system gain is given by:

$$
K\{|v_m(t)|^2\} = \frac{v_a(t)}{v_m(t)} = F\{x_m(t)\}.G\{x_m(t).|F[x_m(t)]|^2\}.
\tag{4.47}
$$

Similarly to the amplifier, the whole system can be modelled by a complex power series, truncated after the fifth-order term, given by:

$$
\begin{aligned}
K\{x_m(t)\} &= K_1\{x_m(t)\} + jK_2\{x_m(t)\} \\
&= k_1 + k_3 x_m(t) + k_5 x_m{}^2(t).
\end{aligned}
\tag{4.48}
$$

Then, in accordance with Equation 4.42 we can write:

$$
\begin{aligned}
v_a(t) &= v_m(t)K\{x_m(t)\} \\
&= k_1 + k_3 v_m(t)x_m(t) + k_5 v_m(t)x_m{}^2(t) \\
&= k_1 + k_3 v_m(t)|v_m(t)|^2 + k_5 v_m(t)|v_m(t)|^4 ,
\end{aligned}
\tag{4.49}
$$

which shows the third- and fifth-order dependence of $v_a(t)$ on $v_m(t)$.

The computation of the coefficients k_1, k_3 and k_5 proceeds by equating the right-hand sides of Equations 4.47 and 4.48 and substituting the appropriate expressions for $F\{\ \}$ and

$G\{\ \}$ to yield:

$$k_1 + k_3 x_m(t) + k_5 x_m{}^2(t)$$
$$= [a_1 + a_3 x_m(t) + a_5 x_m{}^2(t)][g_1 + g_3 x_m(t)|F\{x_m(t)\}|^2$$
$$+ g_5 x_m{}^2(t)|F\{x_m(t)\}|^4]. \tag{4.50}$$

Performing the required multiplications gives:

$$k_1 + k_3 x_m(t) + k_5 x_m{}^2(t)$$
$$= [a_1 + a_3 x_m(t) + a_5 x_m{}^2(t)][g_1 + g_3 x_m(t)|a_1 + a_3 x_m(t) +$$
$$a_5 x_m{}^2(t)|^2 + g_5 x_m{}^2(t)|a_1 + a_3 x_m(t) + a_5 x_m{}^2(t)|^4].$$

$$\tag{4.51}$$

After collecting the zeroth-, first- and second-order terms of $x_m(t)$ on the right-hand side of Equation 4.51 we get the following relationships:

$$
\begin{aligned}
k_1 &= a_1 g_1 \\
k_3 &= a_3 g_1 + a_1 g_3 |a_1|^2 \\
k_5 &= a_5 g_1 + a_3 g_3 |a_1|^2 + a_1 g_5 |a_1|^4 + 2 a_1 g_3 Re\{a_1 a_3{}^*\}.
\end{aligned}
$$

$$\tag{4.52}$$

Observe that the third-order distortion depends on the coefficients a_1 and a_3 of the pre-distorter (PD), while the fifth-order distortion can be controlled by a_1, a_3 and a_5. This suggests that both the third- and fifth-order intermodulation products can be efficiently reduced by the appropriate choice of the coefficients a_1, a_3 and a_5.

4.5.2.3　Pre-distorter Coefficient Adjustment

The multiplicative second-order non-linear functions F_1 and F_2 to be used by the pre-distorter are derived from the amplifier's AM-AM and AM-PM characteristics. A specific 5 W class AB amplifier's characteristics operating in the 800 MHz frequency band have been measured by Stapleton et al. [1] and are shown in Figure 4.29. The inverses of these characteristics are the required F_1 and F_2 non-linear functions, which can be inspected in Figure 4.30 along with the minimum mean squared error (MMSE) fitted second-order curves. The exact coefficients for the initial F_1 and F_2 functions for this particular amplifier were provided by Stapleton et al. [1]:

$$
\begin{aligned}
F_1\{x_m(t)\} &= 0.58315 - 0.08498 x_m(t) + 0.05169 x_m{}^2(t) \\
F_2\{x_m(t)\} &= -0.00395 + 0.0214 x_m(t) - 0.01887 x_m{}^2(t).
\end{aligned}
$$

$$\tag{4.53}$$

These coefficients represent the initial values for the adaptive system. The functions F_1 and F_2 can be implemented for example by the help of operational amplifiers and linear four-quadrant multipliers, using the block diagram of Figure 4.31, where $F\{x_m{}^2(t)\} = a_1 + x_m{}^2(t)[a_3 + a_5 x_m{}^2(t)]$. The function $x_m(t) = |v_m(t)|^2$ can be realised by means

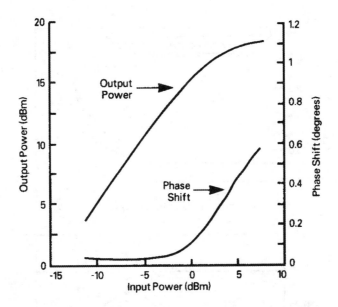

Figure 4.29: AM-AM and AM-PM characteristics of class AB power amplifier [1] ©IEEE 1992, Stapleton *et al.*

of an envelope detector, which must exhibit good linearity across the range of input signal magnitudes.

The adaptive PD coefficient adjustment is based on measuring the OOB signal power at the frequency of the third and fifth upper harmonics. To render this power measurement practical, the output signal of the power amplifier (PA) must be down-converted to a frequency where the third and fifth harmonics can be filtered out. The power of these components must be measured by a peak power detector and then averaged. As seen in Figure 4.24, the power detector's output drives the pre-distorter's controller, which adjusts the coefficients a_1, a_3 and a_5 adaptively to minimise the OOB signal power and hence minimise the detrimental effects of the non-linearity of the power amplifier (PA).

4.5.2.4 Pre-distorter Performance

In general the choice of a specific minimisation algorithm will depend on the initial convergence, tracking accuracy and computational complexity requirements. In our application the initial convergence requirement is not particularly crucial, but low complexity and good tracking accuracy are desirable. Direct search algorithms, such as the Hooke-Jeeve method [421] are attractive in terms of both low complexity and high noise immunity, but converge generally more slowly than gradient techniques.

Stapleton *et al.* [422] studied the performances of three different coefficient PD adaptation methods, namely, the Hooke-Jeeve direct search technique, the steepest descent (SD) gradient algorithm [421] and the Davidson–Fletcher–Powell (DFP) surface fit method [421]. Their convergence results are displayed in Figure 4.32.

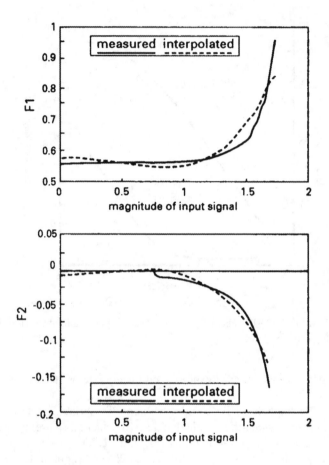

Figure 4.30: Interpolated F_1 and F_2 functions [1] ©IEEE 1992, Stapleton *et al.*

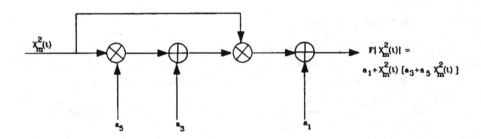

Figure 4.31: F-function implementation [1] ©IEEE, 1992, Stapleton *et al.*

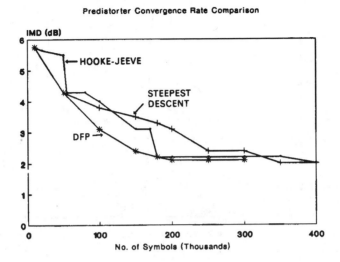

Figure 4.32: Pre-distorter convergence rate comparison [422] ©IEEE 1992, Stapleton *et al.*

As demonstrated by the figure, the DFP method has the fastest convergence at somewhat higher computational complexity. The initial convergence of the SD algorithm appears to be identical to that of the DFP method. However, after the rapid initial intermodulation decay its performance gradually tails off, finally exhibiting similar tracking performance to its other two counterparts.

The overall performance of the pre-distorter can be characterised by the spectral plot of Figure 4.33 [422], where the PA's input spectrum was plotted along with its output spectrum both with and without the PD. In the underlying experiment the previously portrayed 5 W class AB amplifier was used with a back-off power of 6 dB. The modulated signal was generated by a 64 kbps 16–QAM modem having a tight roll-off factor of $\alpha = 0.33$. The spectrally compact sharp roll-off yields slow impulse response decay and increased ISI sensitivity. The PD coefficients were adjusted using the DFP method and the spectrum depicted was generated after a convergence time of 9.6 sec.

In conclusion, pre-distortion is an efficient technique to reduce the non-linear power amplifier's intermodulation distortion, and hence out-of-band spectral spillage, which appears in adjacent channels in frequency division multiplex systems. Stapleton *et al.* demonstrated that the out-of-band emissions of a specific 5 W class AB amplifier deployed in a 16–QAM system can be reduced by some 10 dB, while maintaining high power-efficiency in the amplifier. The complexity of the pre-distorter is moderate.

4.5.3 Postdistortion of NLA-QAM [2]

4.5.3.1 The Postdistortion Concept

In the previous section we considered cancelling the third- and fifth-order OOB intermodulation distortions (IMD) of non-linearly amplified (NLA) QAM signals by pre-distortion.

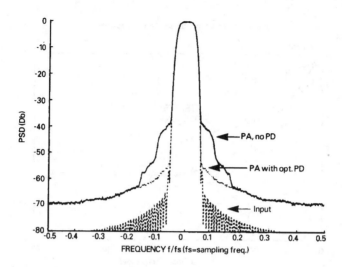

Figure 4.33: Simulated power spectrum after DFP optimisation of PD coefficients [422] ©IEEE 1992

The OOB spectral spillage was reduced by about 15 dB, when using a power-efficient class AB power amplifier. However, the pre-distorter's complexity and power consumption are not negligible.

It may prove more attractive to relax the OOB emission requirement, for instance, from -60 dBc to -45 dBc and remove the resulting unwanted adjacent channel interference at the base station (BS). This eliminates the pre-distorter in front of the QAM modulator and hence reduces the complexity of the hand-held mobile station (MS). The undesirable IMD is reduced at the BS by a postdistorter. The BS transmits the signal to the receiving MS using power inefficient class A linear amplification, since the BS's complexity and power consumption constitute a less severe limitation than that of the MS, and so no postdistortion is required in the mobile.

Best IMD cancellation is achieved using an adaptive postdistorter, which accounts for time variant IMD caused by amplifier ageing, supply voltage variations, temperature fluctuations or other slowly drifting time variant IMD sources. The basic principle of the postdistorter is fairly similar to that of the pre-distorter, in that the AM-AM and AM-PM nonlinearities of the power amplifier are adaptively measured and compensated. The postdistorter measures the envelope of the received signal and generates a non-linear signal, which models the MS power amplifier's third- and fifth-order IMD. Then the postdistorter's output signal is subtracted from the distorted QAM signal in order to cancel the IMD and hence the adjacent channel interference is significantly reduced.

The block diagram of a three-channel, QAM-based, frequency division multiple access (FDMA) scheme is depicted in Figure 4.34. The non-linearly amplified (NLA) QAM signals of three adjacent channels having carrier frequencies of f_1, f_2 and f_3 are received by the BS. Due to the power-efficient class AB or class B amplification at the MSs, channels 1 and 3 spill their third- and fifth-order intermodulation distortions into the adjacent channel 2. The postdistorter's coefficients are initially adjusted to cancel the typical average third- and

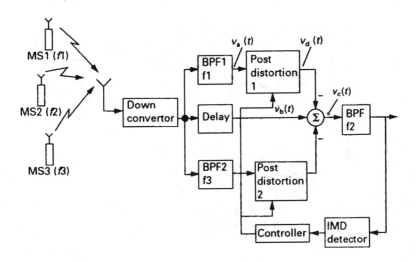

Figure 4.34: QAM-based three-channel FDM mobile radio system using postdistortion [2] ©IEEE, 1993, Quach *et al.*

fifth-order IMD. Then their spectral spillage into channel 2 is measured by the IMD detector during channel 2's idle periods and the postdistorter coefficients are adaptively adjusted to minimise the IMD.

In this simple three-channel scheme only two adjacent channels can spill IMD into channel 2's band. If, however, there are more channels and their third- or fifth-order IMD falls into channel 2, the schematic of Figure 4.34 should be extended to reduce their spillage as well. The downconverted received signal is filtered by the band-pass filters BPF1 and BPF2 to pass only channel 1's signal to postdistorter 1 and that of channel 3 to postdistorter 2. The delay block is provided to account for the processing delays of the BPFs and postdistorters and to pass the received composite signal to the "distortion canceller" block, which is denoted by Σ. The postdistorted channel 1 and channel 3 signals are subtracted from the received composite signal and the "clean" channel 2 signal is bandpass filtered to remove any vestige of its OOB components.

Adaptive coefficient update can be introduced to counteract the slow IMD variations due to ageing, temperature fluctuations, etc. The IMD detector senses the OOB spillage of channel 1 and channel 3 into channel 2 during idle intervals of channel 2 and the controller adaptively adjusts the postdistorter's coefficients to minimise the IMD.

4.5.3.2 Postdistorter Description

Similarly to the previously described pre-distorter, the postdistorter uses a memoryless, non-linear function to model the power amplifier's non-linearity. Using the postdistorter's input and output signal, $v_a(t)$ and $v_d(t)$, respectively, we have:

$$v_d(t) = v_a(t)F(|v_a(t)|^2),$$ (4.54)

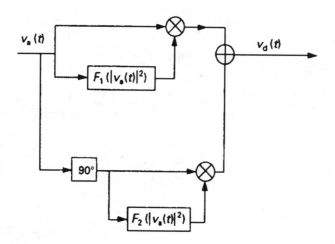

Figure 4.35: Postdistorter block diagram [2] ©IEEE, 1993, Quach *et al.*

where $F(|v_a(t)|^2)$ is the postdistorter's transfer function. Since we want to cancel the third- and fifth-order IMDs, we assume that $v_d(t)$ is a fifth-order function of the input signal $v_a(t)$. The postdistorter's transfer function $F(\ )$ is a second-order function of the envelope of $v_a(t)$, i.e. that of $x_a(t) = |v_a(t)|^2$:

$$
\begin{aligned}
F(|v_a(t)|^2) &= a_1 + a_3|v_a(t)|^2 + a_5(|v_a(t)|^2)^2 \\
&= a_1 + a_3 x_a(t) + a_5 x_a{}^2(t),
\end{aligned}
\tag{4.55}
$$

where the coefficients a_1, a_3 and a_5 are complex. Then the postdistorter's output signal $v_d(t)$ will be a fifth-order function of $v_a(t)$:

$$
\begin{aligned}
v_d(t) &= v_a(t)F(|v_a(t)|^2) \\
&= a_1 v_a(t) + a_3 v_a(t)|v_a(t)|^2 + a_5 v_a(t)|v_a(t)|^4.
\end{aligned}
\tag{4.56}
$$

Again, similarly to the pre-distorter, the postdistorter's complex gain $F(\ )$ can be expressed as a function of the in-phase and quadrature-phase distortion functions as follows:

$$
F[x_a(t)] = F_1[x_a(t)] + jF_2[x_a(t)].
\tag{4.57}
$$

The in-phase and quadrature phase components of the amplified signal $v_a(t)$ are multiplied by the corresponding $v_a(t)$-dependent non-linear functions $F_1(|v_a(t)|^2)$ and $F_2(|v_a(t)|^2)$, respectively, in order to form the complex signal $v_d(t)$, which is also shown in Figure 4.35. The non-linearly class AB or class B amplified signal's expression is given by

$$
v_a(t) = v_m(t)G[|v_m(t)|^2],
\tag{4.58}
$$

where $v_m(t)$ is the modulated signal and $G[|v_m(t)|^2]$ is the non-linear class AB power amplifier's complex gain, which is assumed to be a quadratic non-linear function of the envelope

$x_m(t) = |v_m(t)|^2$. Then we have:

$$
\begin{aligned}
G[x_m(t)] &= g_1 + g_3 x_m(t) + g_5 x_m{}^2(t) \\
&= g_1 + g_3 |v_m(t)|^2 + g_5 |v_m(t)|^4 .
\end{aligned}
\tag{4.59}
$$

Similarly to the pre-distorter we can define the complex system transfer function

$$
\begin{aligned}
K[|v_m(t)|^2] &= \frac{v_d(t)}{v_m(t)} \\
&= K_1[|v_m(t)|^2] + jK_2[|v_m(t)|^2] \\
&= k_1 + k_3 x_m(t) + k_5 x_m{}^2(t) \\
&= k_1 + k_3 |v_m(t)|^2 + k_5 |v_m(t)|^4 .
\end{aligned}
\tag{4.60}
$$

Using Equations 4.56 and 4.58 we then get:

$$
\begin{aligned}
K[|v_m(t)|^2] &= \frac{v_a(t)F[|v_a(t)|^2]}{v_m(t)} \\
&= G[|v_m(t)|^2].F\left\{|v_m(t)G[|v_m(t)|^2]|^2\right\} \\
&= G[x_m(t)]F\{x_m(t).|G[x_m(t)]|^2\},
\end{aligned}
\tag{4.61}
$$

which has the reciprocal structure of the pre-distorter's complex transfer function given in Equation 4.47. Therefore the coefficients k_1, k_3 and k_5 of $K[|v_m(t)|^2]$ are obtained by interchanging a_i with g_i, $i = 1, 3, 5$ in Equation 4.52 yielding:

$$
\begin{aligned}
k_1 &= g_1 a_1 \\
k_3 &= g_3 a_1 + g_1 a_3 |g_1|^2 \\
k_5 &= g_5 a_1 + g_3 a_3 |g_1|^2 + g_1 a_5 |g_1|^4 + 2g_1 a_3 Re\{g_1 g_3{}^*\},
\end{aligned}
\tag{4.62}
$$

where $*$ means the complex conjugate.

From Figure 4.34 we see that the IMD cancellation is carried out by subtracting the OOB IMD components of channel 1 and channel 3 from the delayed composite signal $v_b(t)$. Considering channel 1's IMD only, we obtain the "cleaned" signal $v_c(t)$ as follows:

$$
v_c(t) = v_b(t) - v_d(t) = v_a(t) - v_d(t),
\tag{4.63}
$$

where we have exploited that, apart from a delay component, $v_b(t) = v_a(t)$. Cancelling the IMD produced by channel 3 can be carried out in the same way, as specified by Equation 4.63. When substituting $v_a(t)$ from Equations 4.58 and $v_d(t) = K[|v_m(t)|^2].v_m(t)$ from

Equation 4.60 into Equation 4.63 we have:

$$
\begin{aligned}
v_c(t) &= G[|v_m(t)|^2]v_m(t) - K[|v_m(t)|^2]v_m(t) \\
&= \left\{ G[|v_m(t)|^2] - K[|v_m(t)|^2] \right\} v_m(t) \\
&= [(g_1 - k_1) + (g_3 - k_3)|v_m(t)|^2 + (g_5 - k_5)|v_m(t)|^4]v_m(t).
\end{aligned}
$$

$$(4.64)$$

Substituting the coefficients k_1, k_3 and k_5 from Equation 4.62 into Equation 4.64, the OOB spectral spillage into channel 2's band due to channel 1's IMD can be reduced by minimising the third- and fifth-order coefficients of Equation 4.64, which are given below:

$$
\begin{aligned}
(g_3 - k_3) &= g_3 - (g_3 a_1 + g_1 a_3 |g_1|^2) \\
(g_5 - k_5) &= g_5 - (g_5 a_1 + g_3 a_3 |g_1|^2 + g_1 a_5 |g_1|^4 + \\
& \quad 2g_1 a_3 Re\{g_1 g_3^*\})
\end{aligned}
$$

$$(4.65)$$

Since the linear term $(g_1 - k_1)v_m(t)$ is due to channel 1's desired signal, it is outside of the bandwidth of BPF2, and therefore $(g_1 - k_1)$ can be set to any arbitrary value; it will be removed by BPF2. The remaining two terms, $(g_3 - k_3)$ and $(g_5 - k_5)$, must be minimised by appropriately choosing the postdistorter coefficients a_1, a_3 and a_5.

4.5.3.3 Postdistorter Coefficient Adaptation

Stapleton and Le Quach have shown [415] that the average OOB IMD power is a parabolic function of the postdistorter coefficients a_3 and a_5 and hence it has a global minimum, which can be located by optimisation algorithms. The initial setting of the postdistorter coefficients can be obtained by least squares fitting (LSF) of the second-order non-linear functions F_1 and F_2 to model the power amplifier's AM-AM and AM-PM characteristics, as described earlier in Section 4.5.2.3, in the context of the pre-distorter. This initial set-up can be significantly improved by optimisation methods, such as the Hooke-Jeeve algorithm [421], as also proposed for the pre-distorter's optimisation in Section 4.5.2.3.

4.5.3.4 Postdistorter Performance

Stapleton and Le Quach [415] evaluated the adjacent channel's IMD rejection of a three-channel 16-QAM system with and without postdistortion by means of simulations. Their results are presented in Figure 4.36 for the following scenarios:

(1) Without postdistortion.

(2) With postdistortion using LSF coefficients.

(3) With Hooke-Jeeve optimisation.

Observe the significant OOB IMD reduction in the idle channel 2 band from an average of about -40 dB to about -55 dB.

In summary, similarly to the pre-distorter, the postdistorter also offers significant out-of-band intermodulation distortion reduction, when using non-linear power amplifiers for QAM

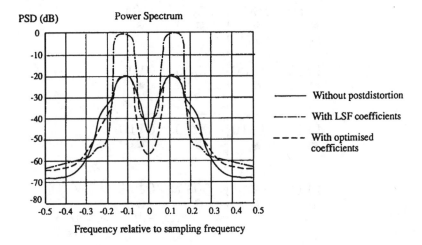

Figure 4.36: Power spectra of a postdistortion system with optimal coefficients, channel 2 is Idle [415]©IEEE 1992, Stapleton *et al.*

transmissions. Its performance and complexity are comparable to that of the pre-distorter, but it is incorporated into the BS, where its additional cost, complexity and power consumption are more readily justified than in the portable station.

Having considered a variety of QAM constellations, along with some pulse shaping and amplification techniques, we now focus our attention on differential versus non-differential QAM transmissions.

4.6 Non-differential Coding for Square QAM

When using a multilevel modulation scheme, it is necessary to map the binary bit stream to be transmitted onto the constellation in an optimal manner. Typically, each of the M constellation points is assigned an N-bit code, where $M = 2^N$. The binary input stream is passed through an N-bit serial-to-parallel convertor and the appropriate constellation point transmitted. The mapping of the data onto the constellation is generally not made in a haphazard fashion, but so as to minimise the number of bit errors given a symbol error. That is, if an incorrect symbol is received, composed of N bits, it is required that the number of bits in error be as small as possible. Since errors are most likely to be made to neighbouring symbol points, this suggests that Gray coding should be employed, whereby the code for each constellation point only differs by one bit from any of its closest neighbours.

An example of an optimally coded 16-level QAM constellation is given in Figure 4.37, where two bits are Gray encoded onto the I-axis and two onto the Q-axis. Assuming that only one bit error occurs per symbol, which is the most likely form of error, we have that

$$P_{E^B} = \frac{1}{N} P_{E^S} \tag{4.66}$$

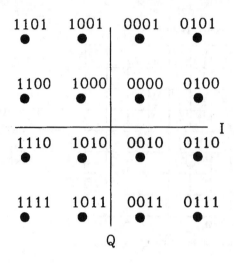

Figure 4.37: Perfect Gray coded 16-level QAM constellation

where P_{EB} and P_{ES} are the probability of bit error and of symbol error, respectively. With some constellations, it is not possible to have perfect Gray coding, (e.g. circular QAM with four points on the inner ring and eight on the outer [18]), and in this case the right-hand side of the above equation must be multiplied by the Gray coding penalty, $g > 1$, to reflect this coding problem.

4.7 Differential Coding for Square QAM

A problem which arises in conjunction with most practical QAM constellations is that the receiver is unable to resolve a number of possible carrier lock positions without additional information from the transmitter. This problem was considered by Weber [18] and we follow his approach here. We define an L-fold rotationally symmetric constellation as a signal set for which the phasor constellation maps onto itself for a rotation of $\pm K \cdot (2\pi/L)$ radians where K and L are integers. Then, given $M = 2^N$ constellation points, the receiver cannot uniquely assign the N bits to each constellation point without resolving the L-fold ambiguity.

Weber suggested that a general algorithm for differentially encoding an arbitrary constellation, where $L = 2^I$, I integer, was to:

(1) Divide the signal space into L equal pie-shaped sectors, and differentially Gray encode the identifiers of these sectors in a rotational manner. For example, in case of square 16-QAM the $L = 2^2 = 4$ pie-shaped sectors have a 90° angle, and are encoded by the identifiers $00, 01, 11, 10$ sequentially when rotating around the constellation.

(2) Gray encode the remaining $(N - I)$ bits within each sector. For our 16-QAM example $(N - I) = 2$ bits belonging to the same quarter of the constellation will be Gray coded by assigning codes of 00 and 11 as well as 01 and 10 to opposite corners of a square. This will allow closest neighbours to differ only in one bit position.

This process has the advantage of differentially encoding as few of the bits as possible, which results in the best performance, by keeping error propagation to a minimum.

If it is assumed that the SNR is sufficiently high that errors are only made to neighbouring symbols then the penalty through using Gray coding can be readily computed. There are now two forms of errors, those within sectors and those between symbols in different sectors. For the former there will be no differential coding penalty, as differential coding has not been used within a sector. There may be a Gray coding penalty, depending on whether perfect Gray coding was possible. When a sector boundary is crossed, a minimum of two bit errors, one in each of two consecutive symbols due to the comparisons used in the differential decoding process, will result. Further errors may result if the sector identifier's Gray coding was such that it was not perfect across boundaries. This can be represented by a Gray coding penalty for boundary symbols. In order to derive a general formula representing this we form the following equation:

$$P_{E^B}(diff\ encoded) = F.P_{E^B}(non\ diff) \tag{4.67}$$

where F is the relative performance of the differentially coded system to the non-differentially coded system. For the non-differentially coded case the general relationship between bit and symbol errors is given by

$$P_{E^B}(non\ diff) = \frac{g}{N}P_{E^S} \tag{4.68}$$

where g is the Gray coding penalty representing the average number of bits in error per symbol error, where only errors between adjacent symbols are considered. A similar expression for the differentially encoded non-ideal Gray coding case is

$$P_{E^B}(diff\ encoded) = \frac{f}{N}P_{E^S} \tag{4.69}$$

where f is the differentially encoded Gray penalty. The penalty for differentially encoding a particular signal set is then given by

$$F = \frac{f}{g}. \tag{4.70}$$

Weber suggests the following algorithm for finding f, which is demonstrated with reference to the square QAM constellation of Figure 4.38:

(a) Within a single sector draw lines between symbol point pairs separated by δ_{min} and let the total number of these lines be N_1.

(b) Write next to each line the Hamming distance, i.e. the number of bits which differ between the two signal points, and let the sum of all the Hamming distances within one sector be H_1.

(c) Draw lines between adjacent constellation points separated by δ_{min} and lying on opposite sides of the sector border, and let the number of these lines be N_2 and the sum of their Hamming distances be H_2.

(d) The differential coding penalty is then given by

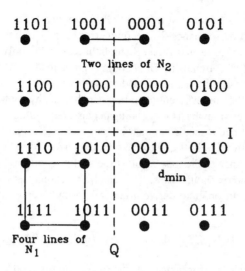

Figure 4.38: Calculating differential coding penalty with a square constellation

$$f = \frac{H_1 + H_2 + 2N_2}{N_1 + N_2} \tag{4.71}$$

where the weighting factor of 2 in the third term of the numerator is due to error propagation in the differentially encoded bits. Note that $g = H_1/N_1$ is the Gray coding penalty within a sector, whereas H_2/N_2 is the Gray coding penalty across a sector.

In order to augment our exposition, an example showing the application of these steps to rectangular 16-level QAM is given in Figure 4.38. From the figure it can be seen that $N_1 = 4$ and $H_1 = 4$ while $N_2 = 2$ and $H_2 = 2$, implying that $g = 1$ and therefore there is no Gray coding penalty with this constellation. Because of this $F = f$ and we have that $F = (4 + 2 + 4)/(4 + 2) = 1.67$, or 2.22 dB. For the general M-level square constellation it can be shown [18] that

$$
\begin{aligned}
N_1 &= 2^{N-1} - 2^{N/2} \\
H_1 &= 2^{N-1} - 2^{N/2} \\
N_2 &= 2^{N/2-1} \\
H_2 &= \left(\frac{N}{2} - 1\right) 2^{(N/2-1)}
\end{aligned}
\tag{4.72}
$$

and since $g = 1$ we have that

$$F = 1 + \frac{N/2}{2^{n/2} - 1}. \tag{4.73}$$

Because only two bits are ever required for differential coding of the four sectors, the coding penalty drops as the number of bits increases, from 2 (3 dB) for $N=2$ (QPSK) to nearly 1 (0 dB) for higher-order systems.

A further problem arises when offset QAM (O-QAM) is employed as not only must the

L-fold rotational ambiguity be resolved, but it is also necessary to determine which axis is delayed relative to the other. Weber proposed overcoming this ambiguity by encoding the I and Q channel transitions independently using separate differential coding systems so that now all bits are being differentially encoded. This will lead to a pattern of I bits and then Q bits, alternating. It will therefore be of no importance which axis is staggered with relation to the other as they are now treated independently. This will increase the differential coding penalty to the full 3 dB, and is an argument against using O-QAM as compared to QAM.

4.8 Summary

In this chapter we have deepened our knowledge as regards to general QAM signal process- ing techniques. After a brief discourse on various QAM constellations for different channels we have analytically derived the transfer functions of the optimum transmitter and receiver filters. Both of these filters must retain a square root Nyquist shape in order to minimise the ISI and maximise the noise immunity. Various methods of generating QAM signals have been proposed and techniques of QAM detection, such as threshold detection, matched filter- ing and the correlation receiver have been reviewed. The chapter was concluded by a short discussion on differential versus non-differential coding.

Equipped with these QAM techniques in the next chapter we now consider the bit error rate versus channel SNR performance of 16-QAM and 64-QAM over AWGN channels.

5

Square QAM

In this chapter we consider the transmission and demodulation of square QAM signals over Gaussian channels. The theory of QAM transmissions over Rayleigh fading channels is given in Chapter 9. In order to arrive at explicit BER versus SNR formulae, let us first review a few fundamental results of decision theory.

5.1 Decision Theory

The roots of decision theory stem from Bayes' theorem formulated as follows:

$$P(X/Y) \cdot P(Y) = P(Y/X) \cdot P(X) = P(X, Y), \tag{5.1}$$

where the random variables X and Y have probabilities of $P(X)$ and $P(Y)$, their joint probability is $P(X, Y)$ and their conditional probabilities are given by $P(X/Y)$ and $P(Y/X)$.

In decision theory the above theorem is invoked in order to infer from the noisy analogue received sample y, what the most likely transmitted symbol was, assuming that the so-called *a priori* probability $P(x)$ of the transmitted symbols $x_n, n = 1, \ldots, M$ is known. Given that the received sample y is encountered at the receiver, the conditional probability $P(x_n/y)$ quantifies the chance that x_n has been transmitted:

$$P(x_n/y) = \frac{P(y/x_n) \cdot P(x_n)}{P(y)}, \quad n = 1, \ldots, N \tag{5.2}$$

where $P(y/x_n)$ is the conditional probability of the continuous noise sample y, given that $x_n, n = 1, \ldots, N$ was transmitted. The probability of encountering a specific y value will be the sum of all possible combinations of receiving y and $x_n, n = 1, \ldots, N$ simultaneously, which is given by:

$$P(y) = \sum_{n=1}^{N} P(y/x_n) \cdot P(x_n) = \sum_{n=1}^{N} P(y, x_n). \tag{5.3}$$

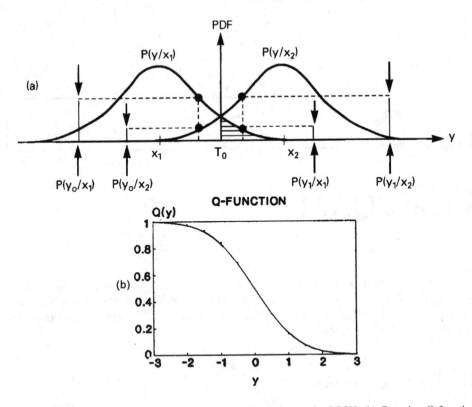

Figure 5.1: (a) Transmitted phasors and noisy received phasors for BPSK, (b) Gaussian Q-function

Let us now consider the case of binary phase shift keying (BPSK), where there are two legitimate transmitted waveforms, x_1 and x_2, which are contaminated by noise, as portrayed in Figure 5.1(a). The conditional probability of receiving any particular analogue sample y given that x_1 or x_2 was transmitted is quantified by the Gaussian probability density functions (PDFs) seen in Figure 5.1(a), which are described by:

$$P(y/x) = \frac{1}{\sigma\sqrt{2\pi}} e^{\frac{-(y-x)^2}{2\sigma^2}}, \tag{5.4}$$

where $x = x_1$ or x_2 is the mean and σ^2 is the variance.

Observe from the figure that the shaded area represents the probability of receiving values larger than the threshold T_0, when x_1 was transmitted and this is equal to the probability of receiving a value below T_0, when x_2 was transmitted. As displayed in the figure, when receiving a specific $y = y_0$ sample, there is an ambiguity, as to which symbol was transmitted. The corresponding conditional probabilities are given by $P(y_0/x_1)$, and $P(y_0/x_2)$ and their values are also marked on Figure 5.1(a). Given the knowledge that x_1 was transmitted, we are more likely to receive y_0 than on the condition that x_2 was transmitted. Hence, upon observing $y = y_0$, statistically speaking it is advisable to decide that x_1 was transmitted. Following similar logic, when receiving y_1 as seen in Figure 5.1(a), it is logical to conclude that x_2 was transmitted.

Indeed, according to optimum decision theory [406], the optimum decision threshold above which x_2 is inferred is given by:

$$T_0 = \frac{x_1 + x_2}{2} \tag{5.5}$$

and below this threshold x_1 is assumed to have been transmitted. If $x_1 = -x_2$ then $T_0 = 0$ is the optimum decision threshold minimising the bit error probability.

In order to compute the error probability in case of transmitting x_1, the PFD $P(y/x_1)$ of Equation 5.4 has to be integrated from x_1 to ∞, which gives the shaded area under the curve in Figure 5.1(a). In other words, the probability of a zero-mean noise sample exceeding the magnitude of x_1 is sought, which is often called the *noise protection distance*, given by the so-called Gaussian Q-function:

$$Q(x_1) = \frac{1}{\sigma\sqrt{2\pi}} \int_{x_1}^{\infty} e^{\frac{-y^2}{2\sigma^2}} \, dy, \tag{5.6}$$

where σ^2 is the noise variance. Notice that since $Q(x_1)$ is the probability of exceeding the value x_1, it is actually the complementary cumulative density function (CDF) of the Gaussian distribution.

Assuming that $x_1 = -x_2$, the probability that the noise can carry x_1 across $T_0 = 0$ is equal to that of x_2 being corrupted in the negative direction. Hence, assuming that $P(x_1) = P(x_2) = 0.5$, the overall error probability is given by:

$$
\begin{aligned}
P_e &= P(x_1) \cdot Q(x_1) + P(x_2) \cdot Q(x_2) \\
&= \frac{1}{2}Q(x_1) + \frac{1}{2}Q(x_1) = Q(x_1)
\end{aligned} \tag{5.7}
$$

The values of the Gaussian Q-function plotted in Figure 5.1(b) are tabulated in many textbooks [406], along with values of the Gaussian PDF in case of zero-mean, unit-variance processes. For abscissa values of $y > 4$ the following approximation can be used:

$$Q(y) \approx \frac{1}{y\sqrt{2\pi}} e^{\frac{-y^2}{2}} \quad \text{for } y > 4. \tag{5.8}$$

Having provided a rudimentary introduction to decision theory, let us now focus our attention on the demodulation of 16-QAM signals in AWGN.

5.2 QAM Modulation and Transmission

In general the modulated signal can be represented by

$$s(t) = a(t)\cos[2\pi f_c t + \Theta(t)] = Re(a(t)e^{j[w_c t + \Theta(t)]}) \tag{5.9}$$

where the carrier $\cos(w_c t)$ is said to be amplitude modulated if its amplitude $a(t)$ is adjusted in accordance with the modulating signal, and is said to be phase modulated if $\Theta(t)$ is varied in accordance with the modulating signal. In QAM the amplitude of the baseband modulating signal is determined by $a(t)$ and the phase by $\Theta(t)$. The in-phase component I is then given by

$$I = a(t) \cos \Theta(t) \tag{5.10}$$

and the quadrature component Q by

$$Q = a(t) \sin \Theta(t). \tag{5.11}$$

This signal is then corrupted by the channel. Here we will only consider AWGN. The received signal is then given by

$$r(t) = a(t) \cos[2\pi f_c t + \Theta(t)] + n(t) \tag{5.12}$$

where $n(t)$ represents the AWGN, which has both an in-phase and quadrature component. It is this received signal which we will attempt to demodulate.

5.3   16-QAM Demodulation in AWGN

The demodulation of the received QAM signal is achieved by performing quadrature amplitude demodulations using the decision boundaries seen in Figure 3.3 for the I and Q components, as shown below for the bits i_1 and q_1:

$$\begin{aligned} \text{if} \quad & I, Q \geq 0 \quad \text{then} \quad i_1, q_1 = 0 \\ \text{if} \quad & I, Q < 0 \quad \text{then} \quad i_1, q_1 = 1 \end{aligned}$$

The decision boundaries for the third and fourth bits i_2 and q_2, respectively, are again shown in Figure 3.3, and thus:

$$\begin{aligned} \text{if} \quad & & I, Q & \geq & 2d \quad & \text{then} \quad & i_2, q_2 = 1 \\ \text{if} \quad -2d \leq & & I, Q & < & 2d \quad & \text{then} \quad & i_2, q_2 = 0 \\ \text{if} \quad -2d > & & I, Q & & & \text{then} \quad & i_2, q_2 = 1. \end{aligned}$$

We will show that in the process of demodulation the positions of the bits in the QAM symbols associated with each point in the QAM constellation have an effect on the probability of them being in error. In the case of the two most significant bits (MSBs) of the 4-bit symbol i_1, q_1, i_2, q_2, i.e. i_1 and q_1, the distance from a demodulation decision boundary of each received phasor in the absence of noise is $3d$ for 50% of the time, and d for 50% of the time; if each phasor occurs with equal probability. The average protection distance for these bits is therefore $2d$ although the bit error probability for a protection distance of $2d$ would be dramatically different from that calculated. Indeed, the average protection distance is never encountered, we only use this term to aid our investigations. The two least significant bits (LSBs), i.e. i_2 and q_2, are always at a distance of d from the decision boundary and consequently the average protection distance is d. We may consider our QAM system as a class 1 ($C1$) and as a class 2 ($C2$) sub-channel, where bits transmitted via the $C1$ sub-channel are received with a lower probability of error than those transmitted via the $C2$ sub-channel.

Observe in the phasor diagram of Figure 3.3 that upon demodulation in the $C2$ sub-channel, a bit error will occur if the noise exceeds d in one direction or $3d$ in the opposite

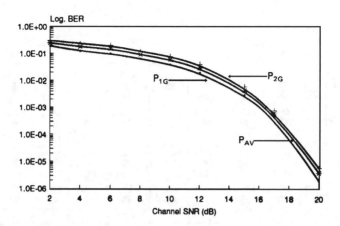

Figure 5.2: 16-QAM BER versus channel SNR curves over AWGN channel

direction, where the latter probability is insignificant. Hence the $C2$ bit error probability becomes

$$P_{2G} = Q \left\{ \frac{d}{\sqrt{N_0/2}} \right\} = \frac{1}{\sqrt{2\pi}} \int_{\frac{d}{\sqrt{N_0/2}}}^{\infty} \exp\left(-x^2/2\right) dx \qquad (5.13)$$

where $N_0/2$ is the double-sided spectral density of the AWGN, $\sqrt{N_0/2}$ is the corresponding noise voltage, and the $Q\{\}$ function was given in Equation 5.6 and Figure 5.1. As the average symbol energy of the 16-level QAM constellation computed for the phasors in Figure 3.3 is

$$E_0 = 10d^2, \qquad (5.14)$$

we have that

$$P_{2G} = Q \left\{ \sqrt{\frac{E_0}{5N_0}} \right\}. \qquad (5.15)$$

For the $C1$ sub-channel data the bits i_1, q_1 are at a protection distance of d from the decision boundaries for half the time, and their protection distance is $3d$ for the remaining half of the time. Therefore the probability of a bit error is

$$
\begin{aligned}
P_{1G} &= \frac{1}{2} Q \left\{ \frac{d}{\sqrt{N_0/2}} \right\} + \frac{1}{2} Q \left\{ \frac{3d}{\sqrt{N_0/2}} \right\} \\
&= \frac{1}{2} \left[Q \left\{ \sqrt{\frac{E_0}{5N_0}} \right\} + Q \left\{ 3\sqrt{\frac{E_0}{5N_0}} \right\} \right].
\end{aligned} \qquad (5.16)
$$

The probabilities P_{1G} and P_{2G} as a function of E_0/N_0 are given by Equations 5.15 and 5.16 and displayed in Figure 5.2.

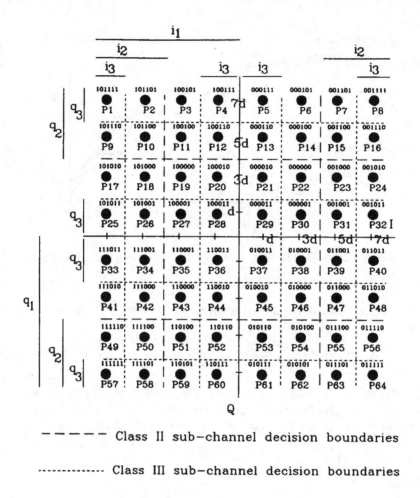

Figure 5.3: Square 64-QAM phasor constellation

Also shown is the average probability P_{AV} of bit error for the 16-level QAM system as

$$P_{AV} = (P_{1G} + P_{2G})/2. \tag{5.17}$$

Our simulation results gave practically identical curves to those in Figure 5.2, where the AWGN performance is seen to have only a small advantage in using the $C1$ sub-channel over using the $C2$ sub-channel.

5.4   64-QAM Demodulation in AWGN

Each constellation point in 64-level QAM (64-QAM) systems is represented by a unique 6-bit symbol, which is Gray coded to minimise the decoded error probability, as seen in Figure 5.3.

The complex phasors of the constellation are decomposed into the 8-level I and Q components. The amplitudes $7d$, $5d$, $3d$, d, $-d$, $-3d$, $-5d$ and $-7d$ of the I and Q AM signals are assigned the 3-bit Gray codes 011, 010, 000, 001, 101, 100, 110 and 111, respectively. The three I and Q bits are denoted by i_1, i_2, i_3 and q_1, q_2, q_3, respectively. These bits are interleaved to give a 6-bit QAM symbol represented by i_1, q_1, i_2, q_2, i_3, q_3. The 64-level QAM phasors are transmitted over the channel, where they become corrupted. They are demodulated using the decision boundaries shown in Figure 5.3. Observe that the construction of the signal constellation is similar to that of a Karnaugh table, where in the left half-plane of the coordinate axis i_1 is always logical 1 and in the bottom half-plane q_1 is logical 1, etc. The bits i_1, q_1, i_2, q_2, i_3, q_3, i_4, q_4 are recovered as follows:

$$
\begin{aligned}
\text{if} \qquad I, Q &\geq 0 \quad \text{then} \quad i_1, q_1 = 0 \\
\text{if} \qquad I, Q &< 0 \quad \text{then} \quad i_1, q_1 = 1,
\end{aligned}
\tag{5.18}
$$

for the most significant bits, and

$$
\begin{aligned}
\text{if} \qquad\qquad I, Q &\geq 4d \quad \text{then} \quad i_2, q_2 = 1 \\
\text{if} \quad -4d \leq I, Q &< 4d \quad \text{then} \quad i_2, q_2 = 0 \\
\text{if} \quad -4d > I, Q &\qquad\quad\; \text{then} \quad i_2, q_2 = 1.
\end{aligned}
\tag{5.19}
$$

for the next most significant bits, while finally for the least significant bits

$$
\begin{aligned}
\text{if} \qquad\qquad I, Q &\geq 6d \quad \text{then} \quad i_3, q_3 = 1 \\
\text{if} \quad 2d \leq I, Q &< 6d \quad \text{then} \quad i_3, q_3 = 0 \\
\text{if} \quad -2d \leq I, Q &< 2d \quad \text{then} \quad i_3, q_3 = 1 \\
\text{if} \quad -6d \leq I, Q &< -2d \quad \text{then} \quad i_3, q_3 = 0 \\
\text{if} \quad -6d > I, Q &\qquad\quad\; \text{then} \quad i_3, q_3 = 1.
\end{aligned}
\tag{5.20}
$$

Similarly to 16-QAM the position of the bits in the 6-bit QAM symbol has an effect on their error probabilities. In case of the i_1, q_1 bits, for example, the phasor can be at a distance d, $3d$, $5d$ or $7d$ from the decision boundary represented by the coordinate axis. Therefore their "average protection distance" is $16d/4 = 4d$. Note again that this average protection distance is never encountered and due to the non-linear nature of the $Q\{\}$ function the actual BER will be quite different from that of a $4d$ average distance PSK constellation. Figure 5.3 reveals that this average protection distance for the i_2, q_2 bits is $2d$, while for the i_3, q_3 bits it is d. We may view the bits (i_1, q_1), (i_2, q_2) and (i_3, q_3) as three sub-channels each having different integrities. These channels will be referred to as $C1$, $C2$ and $C3$, respectively.

We assume that an independent random data sequence is conveyed to each of the $C1$, $C2$ and $C3$ sub-channels, and that perfect coherent detection is carried out at the receiver using Equations 5.18 to 5.20. Let us determine the BER performance of the I component, as the Q component being in quadrature will support data that is subjected to the same BER.

Let us consider first the $C3$ sub-channel (i_3, q_3) with its decision boundaries of $-6d$, $-2d$, $2d$ and $6d$. Each phasor has a different error probability, depending on its position in the constellation of Figure 5.3. The bit i_3 of phasor P_5 (a logical 1) will, for example, be corrupted, if a noise sample with larger amplitude than d is added to it when the decoded phasor is P_6. This probability is represented by the first term of Equation 5.21. However, when P_5 is carried by a noise vector having larger than $5d$ amplitude over the $C3$ decision boundary at

$6d$, the decision becomes error-free again, although the decoded phasor is erroneously considered to be P_8. This is quantified by the third term in Equation 5.21, where the negative sign takes account of the favourable influence of this event on the BER. In the negative direction there is no corruption until the noise amplitude exceeds $-3d$ (phasor P_3) in the negative direction, which has the same probability as exceeding $3d$ in the positive direction, and hence contributes the second term to Equation 5.21. For noise vectors having amplitudes between $-3d$ and $-7d$ there are erroneous decisions. However, once the $-6d$ decision boundary is exceeded, i.e. the negative noise value is larger than $-7d$, the received sample falls into the error-free domain again, although the decoded phasor is P_1 which is reflected by the fourth term of Equation 5.21. The i_3 bit error probability of the P_5 phasor is then given by the following summation of Q-functions:

$$P_{e5} = Q\left[\frac{d}{\sqrt{N_0/2}}\right] + Q\left[\frac{3d}{\sqrt{N_0/2}}\right] - Q\left[\frac{5d}{\sqrt{N_0/2}}\right] - Q\left[\frac{7d}{\sqrt{N_0/2}}\right], \qquad (5.21)$$

where $N_0/2$ denotes the double-sided Gaussian noise spectral density and $\sqrt{N_0/2}$ is the corresponding noise voltage. After averaging the powers of the individual phasors $P_1, \ldots, P_{64}$ the average symbol energy for $64-QAM$ is found to be $E = 42d^2$. Substituting $d = \sqrt{E/42}$ into Equation 5.21, and introducing the average SNR $\gamma = E/N_0$, we get:

$$P_{e5} = Q\left[\sqrt{\frac{\gamma}{21}}\right] + Q\left[3\sqrt{\frac{\gamma}{21}}\right] - Q\left[5\sqrt{\frac{\gamma}{21}}\right] - Q\left[7\sqrt{\frac{\gamma}{21}}\right]. \qquad (5.22)$$

For the phasor P_8 the situation is different, as its $C3$ bit i_3 is not corrupted by positive noise samples of arbitrarily large amplitudes. For negative noise levels it cycles in and out of error as the noise increases past -d, -$5d$, -$9d$ and -$13d$, as seen in Figure 5.3. Therefore the associated error probability is given by

$$P_{e8} = Q\left[\sqrt{\frac{\gamma}{21}}\right] + Q\left[9\sqrt{\frac{\gamma}{21}}\right] - Q\left[5\sqrt{\frac{\gamma}{21}}\right] - Q\left[13\sqrt{\frac{\gamma}{21}}\right]. \qquad (5.23)$$

Following a similar argument the error probability P_{e6} of the i_3 bit of phasor P_6 is

$$P_{e6} = Q\left[\sqrt{\frac{\gamma}{21}}\right] + Q\left[3\sqrt{\frac{\gamma}{21}}\right] - Q\left[5\sqrt{\frac{\gamma}{21}}\right] + Q\left[9\sqrt{\frac{\gamma}{21}}\right], \qquad (5.24)$$

while

$$P_{e7} = Q\left[\sqrt{\frac{\gamma}{21}}\right] + Q\left[3\sqrt{\frac{\gamma}{21}}\right] - Q\left[7\sqrt{\frac{\gamma}{21}}\right] + Q\left[11\sqrt{\frac{\gamma}{21}}\right]. \qquad (5.25)$$

The error probabilities P_{e1}, P_{e2}, P_{e3}, P_{e4} are equivalent to those given by P_{e8}, P_{e7}, P_{e6}, P_{e5}, respectively, and the same holds for all corresponding phasors in the columns of the phasor diagram of Figure 5.3. Furthermore, the q_3 bit error probability is identical to that of i_3, if independent random sequences are transmitted. Averaging the $C3$ bit error probabilities

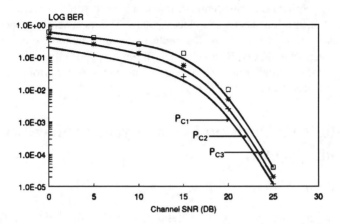

Figure 5.4: Square 64-QAM BER versus channel SNR curves over AWGN channel

yields:

$$P_{C3}(\gamma) = Q\left[\sqrt{\frac{\gamma}{21}}\right] + \frac{3}{4}Q\left[3\sqrt{\frac{\gamma}{21}}\right] - \frac{3}{4}Q\left[5\sqrt{\frac{\gamma}{21}}\right] - \frac{1}{2}Q\left[7\sqrt{\frac{\gamma}{21}}\right] \quad (5.26)$$
$$+ \frac{1}{2}Q\left[9\sqrt{\frac{\gamma}{21}}\right] + \frac{1}{4}Q\left[11\sqrt{\frac{\gamma}{21}}\right] - \frac{1}{4}Q[13\sqrt{\frac{\gamma}{21}}].$$

The last three terms in Equation 5.26 represent extremely unlikely events as the Gaussian noise sample must exceed the $9d$ protection distance of the best-protected $C1$ bit. Consequently we neglect these terms.

Applying similar arguments to those used in the formulation of $P_{C3}(\gamma)$, we have the bit error probabilities for $C2$ and $C1$ as

$$P_{C2}(\gamma) = \frac{1}{2}Q\left[\sqrt{\frac{\gamma}{21}}\right] + \frac{1}{2}Q\left[3\sqrt{\frac{\gamma}{21}}\right] + \frac{1}{4}Q\left[5\sqrt{\frac{\gamma}{21}}\right] + \frac{1}{4}Q\left[7\sqrt{\frac{\gamma}{21}}\right], \quad (5.27)$$

and

$$P_{C1}(\gamma) = \frac{1}{4}Q\left[\sqrt{\frac{\gamma}{21}}\right] + \frac{1}{4}Q\left[3\sqrt{\frac{\gamma}{21}}\right] + \frac{1}{4}Q\left[5\sqrt{\frac{\gamma}{21}}\right] + \frac{1}{4}Q\left[7\sqrt{\frac{\gamma}{21}}\right]. \quad (5.28)$$

The P_{C1}, P_{C2} and P_{C3} error probability curves evaluated using Equations 5.26-5.28 are displayed in Figure 5.4.

Our simulation results differed by less than a decibel from these curves. Observe from the figure the consistent BER advantage obtained using the $C1$ sub-channel. All sub-channels have BERs$< 10^{-3}$ for SNR values in excess of 23 dB. They are therefore suitable for digitised speech transmission. Notice that for SNRs above 23 dB there is only a modest BER advantage in using the $C1$ sub-channel.

As we have seen, both 64-QAM and 16-QAM possess sub-channels with different BERs.

The sub-channel integrities are consistently different and these differences can be used advantageously when transmitting speech and video signals. The most sensitive speech or image bits can be transmitted via the lower BER sub-channel, while the more robust source coded bits can be sent over the higher BER sub-channel. In certain applications this property is very advantageous, because it might remove the need for complex forward error correction codecs with unequal error protection [112, 157–159].

5.5 Recursive Algorithm for the Error Probability Evaluation of M-QAM

L.L. Yang, L. Hanzo

In the previous sections the achievable BER has been directly estimated for 16-QAM and 64-QAM constellations using Gray coded bit mapping. Below we will streamline these equations into a joint formula for directly estimating the BER of arbitrary M-QAM constellations, when communicating over AWGN hannels. Again, in contrast to conventional approaches, which treat M-QAM having different values of M separately, in our current approach, we take advantage of the relationship between different square M-QAM constellations.

This section is organized as follows. Subsection 5.5.1 describes the M-QAM system model. As a special case, the BER of a 16-QAM constellation is briefly revisited in Subsection 5.5.2. The approach used for 16-QAM is then generalized for arbitrary square M-QAM constellations in Subsection 5.5.3. In Subsection 5.5.4 we present numerical examples.

5.5.1 System Model

The square M-QAM signal constellation is exemplified in Figure 5.5 for $M = 16$. The M-QAM signal can be mathematically represented by:

$$s(t) = A_c \cos \omega_c t - A_s \sin \omega_c t, \ 0 \leq t \leq T_s, \tag{5.29}$$

where T_s is the symbol interval, which is related to the bit duration by $T_s = mT_b$ where $m = \log_2 M$ represents the number of bits per symbol, the quadrature amplitudes A_c and A_s range over the set of $\left\{ \pm d, \pm 3d, \ldots, \pm(\sqrt{M} - 1)d \right\}$, in which $2d$ represents the minimum Euclidean distance of constellation points, while d can be computed according to [423]:

$$d = \sqrt{\frac{3m \cdot E_b}{2(M - 1)}}, \tag{5.30}$$

where E_b represents the average energy per bit.

Throughout, we will restrict our considerations to Gray coded bit mapping. With reference to Figure 5.5 as an example, it can be seen that according to the Gray coded bit mapping, two adjacent m-bit symbols differ in a single bit. As a result, an erroneous decision resulting in an adjacent symbol is accompanied by one and only one bit error. Furthermore, we will assume optimum coherent detection with perfect carrier tracking, perfect frequency tracking, and symbol synchronization.

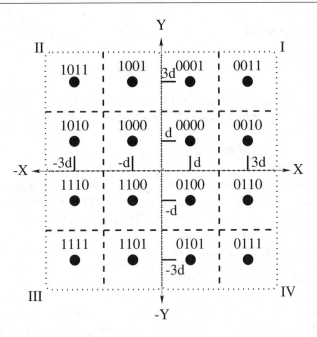

Figure 5.5: 16-QAM square constellation

5.5.2   BER of 16-QAM Constellation

We commence by considering the simple case of a square 16-QAM signal constellation, as shown in Figure 5.5, where each phasor can be represented by a 4-bit symbol, $i_1 q_1 i_2 q_2$, where i_1 and i_2 indicate the inphase (I) bits, while q_1 and q_2 the quadrature-phase (Q) bits. The M-QAM signal can be decomposed into two independent $\sqrt{M}$-AM signals [423]. These two $\sqrt{M}$-AM signals have the same error probability and can be treated independently. Consequently, considering the in-phase (I) 4-AM signal the average bit error probability can be computed by considering the i_1th bit and the i_2th bit, respectively.

Referring to Figure 5.5 and the previous two sections, the error probability of bit i_1 can be expressed as:

$$P_{C1}^{16}(\gamma) = \frac{1}{2}\left[Q\left(\sqrt{\gamma}\right) + Q\left(3\sqrt{\gamma}\right)\right], \tag{5.31}$$

where $Q(\cdot)$ is the Q-function, which is defined as $Q(x) = \frac{1}{\sqrt{2\pi}}\int_x^\infty e^{-t^2/2}dt$, and

$$\gamma = \frac{d^2}{N_0/2}. \tag{5.32}$$

Substituting Equation 5.30 into the above equation, we arrive at $\gamma = \frac{4}{5} \cdot \frac{E_b}{N_0}$ for 16-QAM, where E_b/N_0 represents the average signal-to-noise ratio (SNR) per bit. Similarly, according to Figure 5.5 and following the approach of the previous two section, the error probability of the i_2 bit can be expressed as:

$$P_{C2}^{16}(\gamma) = \frac{1}{2}\left[2Q(\sqrt{\gamma}) + Q(3\sqrt{\gamma}) - Q(5\sqrt{\gamma})\right]. \tag{5.33}$$

Finally, the exact average BER for the square 16-QAM signal is computed by averaging the error probabilities given by Equation 5.31 and Equation 5.33, yielding:

$$P^{16}(\gamma) = \frac{1}{2}\left[P_{C1}^{16}(\gamma) + P_{C2}^{16}(\gamma)\right]. \tag{5.34}$$

In conjunction with Gray coded bit mapping, in addition to two adjacent m-bit symbols differing in a single bit, another important property is that the $M/4$ constellation points of the four quadrants in Figure 5.5 in the mirror-symmetric positions have identical bits assigned to them with respect to the X and Y axes, if we ignore the i_1 and q_1 bits. Hence, without considering bits i_1 and q_1, the points in each quadrant actually constitute an $(M/4)$-QAM constellation. Consequently, the average bit error probability of an M-QAM constellation can be expressed with the aid of the BER expression of an $(M/4)$-QAM constellation, if we ignore the bits i_1 and q_1. Let us now invoke this property in order to investigate the approximate BER expression of 16-QAM. Throughout, let $P^4(\gamma)$ represent the BER expression of a 4-QAM constellation but with γ computed according to the 16-QAM constellation, i.e. using $\gamma = \frac{4}{5} \cdot \frac{E_b}{N_0}$, as shown previously.

5.5.2.1 Approximation 1

In practical terms it is reasonable to assume that a bit error is most frequently caused by a noise sample exceeding d, while the probability of exceeding $3d$ is insignificant, when the signal-to-noise ratio is sufficiently high. Consequently, the BER expression of Equation 5.31 for the bit i_1 can be approximated as:

$$P_{C1}^{16}(\gamma) \approx \frac{1}{2}Q\left(\sqrt{\gamma}\right). \tag{5.35}$$

Upon invoking the BER expression of 4-QAM - which is expressed as $P^4(\gamma) = Q(\sqrt{\gamma})$ - the BER of Equation 5.33 for the i_2 bit can be expressed as:

$$P_{C2}^{16}(\gamma) \approx P^4(\gamma). \tag{5.36}$$

Finally, the average BER of square 16-QAM can be approximated with the aid of Equation 5.34, where $P_{C1}^{16}(\gamma)$ and $P_{C2}^{16}(\gamma)$ were given by Equation 5.35 and Equation 5.36.

5.5.2.2 Approximation 2

The approximate formulae of Equation 5.35 and Equation 5.36 are suitable for 16-QAM upon assuming sufficiently high SNRs. However, if the SNR is low, a more close approximation is required. This can be achieved by considering also the case, when the noise exceeds $3d$. In this case the BER of the i_1 bit is given by Equation 5.31, while that of the i_2 bit can be expressed as:

$$P_{C2}^{16}(\gamma) = Q(\sqrt{\gamma}) + \frac{1}{2}Q(3\sqrt{\gamma}) = P^4(\gamma) + \frac{1}{2}Q(3\sqrt{\gamma}) \tag{5.37}$$

upon neglecting the probability of the noise exceeding $5d$ in Equation 5.33. And finally, the average BER for the square 16-QAM signal can approximated by Equation 5.34 with $P_{C1}^{16}(\gamma)$ and $P_{C2}^{16}(\gamma)$ given by Equation 5.31 and Equation 5.37, respectively.

Above, we have used 16-QAM as an example and investigated its average exact and approximate BER. The above approach of combining the BER expressions of M-QAM having different values of M can be extended to arbitrary square M-QAM constellations and consequently can be used to simplify the associated BER computations. Let us now consider the general algorithm suitable for computing the BER of an arbitrary square M-QAM constellation.

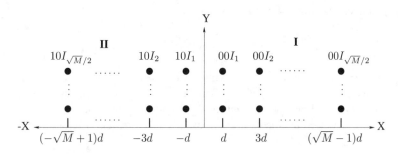

Figure 5.6: Simplified representation of an arbitrary square M-QAM constellation using Gray coded bit mapping

5.5.3   BER of Arbitrary Square M-QAM Constellations

The points of an M-QAM constellation in quadrants I and II can be simply portrayed, as seen in Figure 5.6, where the first two bits indicate the bits i_1 and q_1 encountered in the quadrants I and II, while the remaining $(m-2)$ number of bits of the M-QAM symbol are represented by I_1, I_2, ..., or $I_{\sqrt{M}/2}$. Referring to Figure 5.5, it becomes explicit that I_1, I_2, ..., $I_{\sqrt{M}/2}$ actually represent the corresponding points of a row in the $(M/4)$-QAM constellation. Let $i_1q_1i_2q_2\cdots i_{m/2}q_{m/2}$ represent the m-bit M-QAM symbol. Then, the average BER can be derived as follows.

5.5.3.1   Approximation 1

Based on Approximation 1, where the probability of noise exceeding $3d$ was neglected, the BER of the i_1 bit of an M-QAM constellation can be expressed as:

$$P_{C1}^{M}(\gamma) = \frac{2}{\sqrt{M}}Q(\sqrt{\gamma}), \tag{5.38}$$

where γ is given by Equation 5.32, while d is given by Equation 5.30 for M-QAM. The average BER of bits $i_2, i_3, \ldots, i_{m/2}$ of an M-QAM constellation can be expressed as:

$$P_{C2}^M(\gamma) = P_{M/4}(\gamma). \tag{5.39}$$

Consequently, the average BER of M-QAM can be derived by averaging the BER given by Equation 5.38 and Equation 5.39, yielding:

$$P^M(\gamma) = \frac{1}{m/2} \cdot \frac{2}{\sqrt{M}} Q(\sqrt{\gamma}) + \frac{m/2 - 1}{m/2} \cdot P^{M/4}(\gamma). \tag{5.40}$$

According to Equation 5.40 $P^{M/4}(\gamma)$ has to be determined first, in order to derive $P^M(\gamma)$. Hence the general BER expression of Equation 5.40 can be implemented with the aid of a recursive algorithm, which is described as follows. Let $P^4(\gamma) = \lambda = Q(\sqrt{\gamma})$ and evaluate:

$$P^K(\gamma) = \frac{1}{k/2} \cdot \frac{2}{\sqrt{K}} \lambda + \frac{k/2 - 1}{k/2} \cdot P^{K/4}(\gamma) \tag{5.41}$$

for $K = 16, 4 \times 16, \ldots$, until $K = M$, which is the average BER of the considered square M-QAM scheme, where $k = \log_2 K$ represents the number of bits per symbol of a K-QAM constellation. It can be shown on the basis of Equation 5.41 that we have to evaluate only one integral, in order to determine the resulting BER of an M-QAM constellation.

5.5.3.2   Approximation 2

When using Approximation 2, where the probability of the noise exceeding $5d$ was neglected $P_{C1}^M(\gamma)$ and $P_{C2}^M(\gamma)$ can be expressed as:

$$P_{C1}^M(\gamma) = \frac{2}{\sqrt{M}} [Q(\sqrt{\gamma}) + Q(3\sqrt{\gamma})], \tag{5.42}$$

$$P_{C2}^M(\gamma) = P^{M/4}(\gamma) + \frac{2}{\sqrt{M}} \cdot \frac{1}{m/2 - 1} Q(3\sqrt{\gamma}), \tag{5.43}$$

where the former is the probability of an i_1 bit error, while the latter is the average BER of bits $i_2, i_3, \ldots, i_{m/2}$. Note that in the computation of $P_{C2}^M(\gamma)$ there exists only one term including $Q(3\sqrt{\gamma})$, except those included in $P^{M/4}(\gamma)$. This term is the second part at the right-hand side of Equation 5.43, which is derived from $00I_1$ and $10I_2$ by considering that I_1 and I_2 have a one-bit difference according to Figure 5.6, due to the associated Gray coding.

When invoking Approximation 2, let $P^4(\gamma) = \lambda = Q(\sqrt{\gamma})$, $\beta = Q(3\sqrt{\gamma})$. Then the average BER of the square M-QAM signal can be recursively determined according to:

$$P^K(\gamma) = \frac{1}{k/2} \cdot \frac{2}{\sqrt{K}} [\lambda + \beta] + \frac{k/2 - 1}{k/2} \left[P^{K/4}(\gamma) + \frac{2}{\sqrt{K}} \cdot \frac{1}{k/2 - 1} \beta \right], \tag{5.44}$$

for $K = 16, 4 \times 16, \ldots$, until $K = M$, which is the average BER of the considered square-shaped M-QAM constellation. Equation 5.44 implies that only two integrals have to be evaluated at the first step, in order to obtain the resulting BER of the square M-QAM constellation.

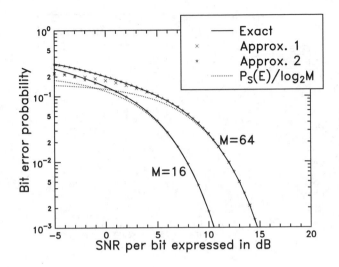

Figure 5.7: BER performance of square M-QAM using Gray coded bit mapping over AWGN channels, where $P_s(E)$ represents the average error probability per M-QAM symbol

5.5.4 Numerical Examples

In this section we use the above algorithms, in order to evaluate the BER performance of various square M-QAM schemes using different values of M. Figure 5.7 is used to show the accuracy of the BER estimated by the proposed approximation approaches, while Figure 5.8 implies that we can readily estimate the BER of an arbitrarily high-order square M-QAM constellation. In Figure 5.7, the BER of 16-QAM and 64-QAM computed according to the recursive expression of Equation 5.41 and Equation 5.44) were compared with the exact BER of 16-QAM - which was obtained from Equation 5.34 - and the exact BER of 64-QAM - which was computed according to the related equations. These results were also compared with the 'standard' approximation derived from the division of the symbol error probability by the number of bits per symbol, $\log_2(M)$ [423] [pp.630]. It can be shown that both Approximation 1 and 2 are closer to the exact BER than the 'standard' approximation within the range of SNR per bit considered. An accurate BER can be achieved by Approximation 1, when the SNR per bit is sufficiently high. However, when the SNR per bit is too low (for example, lower than 4dB for 64-QAM), the more accurate Approximation 2 has to be invoked, in order to achieve a satisfactory BER approximation. As shown in Figure 5.7, we cannot distinguish the exact BER from that computed by Approximation 2 for 16-QAM and 64-QAM over the SNR range of interest.

Figure 5.8 shows the BER performance of various square M-QAM constellations for $M = 4, 16, 64, 256, 1024$ and 4096. Note that for $M = 4$ the BER formula is reduced to that of quadrature phase shift keying (QPSK), yielding $P^4 = Q\left(\sqrt{\frac{2E_b}{N_0}}\right)$, while the BER of the remaining M-QAM constellations was estimated according to Equation 5.41 and Equa-

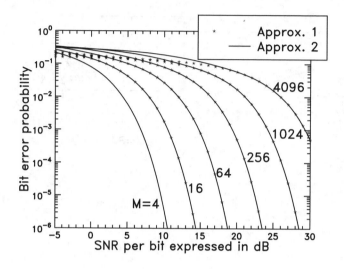

Figure 5.8: BER performance of square M-QAM using Gray coded bit mapping over AWGN channels

tion 5.44, upon invoking Approximations 1 and 2, respectively. It can be shown that at the average BER of 10^{-3}, 4-5dB of SNR per bit has to be invested, in order to transmit an extra 2 bits/symbol by employing a higher-order square M-QAM constellation.

5.6   Summary

In this chapter we have examined the theoretical performance of square QAM transmissions over Gaussian channels. Theoretical calculations showed the presence of different classes of bits with different performance within the constellation, and we will see in Chapter 9 how the relative performance of these bits diverges even further when the constellation is transmitted over mobile radio channels.

Finally, in Section 5.5 a simple recursive algorithm was proposed, which is based on the symmetry of the different square M-QAM constellations using Gray coded bit mapping for evaluating the BER of arbitrarily high-order square M-QAM constellations. The numerical examples show that the BER of any square M-QAM constellation can be accurately estimated by the proposed algorithm.

In the next chapter we consider some of the practical difficulties of receiving modulated signals by examining clock and carrier recovery techniques. These are needed to recover the symbol timing and the carrier phase, respectively.

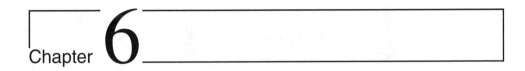

Chapter **6**

Clock and Carrier Recovery Systems for QAM

6.1   Introduction

Both clock and carrier recovery systems attempt to derive information about timing from the received signal, often in a similar manner. While carrier recovery is only necessary in a coherent demodulation system, clock recovery is required in all schemes, and accurate clock recovery is essential for reliable data transmission. Confusion often exists between clock and carrier recovery. Clock recovery attempts to synchronise the receiver clock with the baseband symbol rate transmitter clock, whereas carrier recovery endeavours to align the receiver local oscillator with the transmitted carrier frequency.

It is not our intention here to give a detailed account of all clock and carrier recovery schemes, such background information can be found in the literature [406,424–426]. Instead we describe the particular clock and carrier recovery schemes which have proved suitable for QAM over Gaussian channels. In many cases these are either the same, or modified versions of timing recovery systems which have been used for binary modulation, and in these instances we describe how these schemes have been adapted for QAM. We consider the issue of timing recovery systems for mobile radio transmissions in Chapter 11.

6.2   Clock Recovery

There exist a plethora of different clock recovery techniques. However, many apparently different clock recovery techniques are equivalent, and this is not surprising as they must all use the properties of the received waveform. Basic clock recovery systems which can be used successfully in conjunction with QAM include times-two, early-late, zero-crossing and synchroniser clock recovery systems [424].

We now consider each of these clock recovery systems. We have evaluated the performance of all these systems using computer simulation and found that they all offer similar

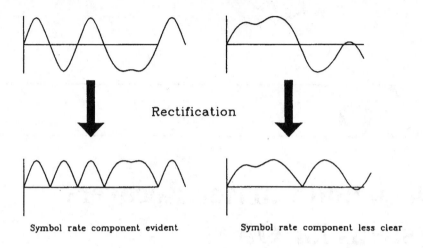

Figure 6.1: Effect of squaring waveforms

performance in a Gaussian environment.

6.2.1   Times-Two Clock Recovery

The most fundamental of all clock recovery schemes is the times-two, or squaring system [424].   Typical BPSK data signals often contain sections of repetitive strings of -1, $+1$, -1, $+1$, ... which after bandlimiting and pulse shaping have a waveform similar to that shown on the left-hand side of Figure 6.1. Observe that this signal has periodic sections having a frequency of half the signalling clock rate. Therefore, if the received demodulated signal is squared, or passed through a non-linear rectifier, then it will possess a periodic frequency domain component at the symbol rate. A bandpass filter tuned close to the symbol rate will extract this periodic signal, allowing derivation of the required timing information. This is shown diagrammatically at baseband for both a binary and a multilevel modulation scheme in Figure 6.1. The times-two clock recovery works best for binary modulation schemes where the -1, $+1$, -1, $+1$, ... sections are frequent and hence the energy of the clock frequency component is high, but not so well for multilevel sections. The reason for this is that the multilevel scheme has a reduced component at the symbol frequency due to the increased possibility of non-zero-crossing transitions such as $+3$, $+1$, -1, ... A schematic of a simple baseband times-two clock recovery system is shown in Figure 6.2 where the clock pulse regenerator could be implemented with a saturating amplifier, leading to a rectangular pulse shape.

6.2.2   Early-Late Clock Recovery

Another well-known form of clock recovery is early-late clock recovery [427], which is portrayed in Figure 6.3. Whilst times-two clock recovery exploits the whole of the incoming waveform, early-late clock recovery works on the peaks in the received waveform. The basic

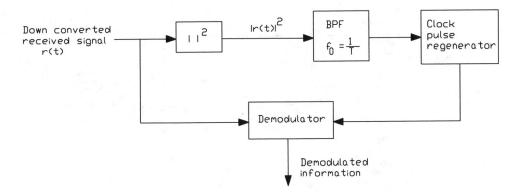

Figure 6.2: Times-two clock recovery schematic

assumptions made by the early-late method are that the peaks in the incoming waveform are at the correct sampling points and that these peaks are symmetrical. This is often true, but for some modulation schemes, such as partial response raised cosine arrangements [405], neither of these assumptions is valid. It may still be possible to use early-late clock recovery with these modulation schemes but the timing jitter will be increased.

The early-late scheme of Figure 6.3 firstly squares the incoming signal in order to render all peaks positive, and then takes two samples of the received waveform, both equispaced around the predicted sampling instant. If the predicted sampling instant is aligned with the correct sample point, and the assumptions made above are correct, then the sample taken just prior to the sample point - the early sample - will be identical to the sample taken just after the sampling instant - the late sample. If the early sample is larger than the late sample, this indicates that the recovered clock is sampling too late, and if the early sample is smaller than the late sample this indicates that the recovered clock is sampling too early. It is common practice to lowpass filter (LPF) the difference of each pair of samples to reduce the effect of random noise on the system. This filtered difference signal adjusts the frequency of a voltage controlled oscillator (VCO) in order to delay or advance the arrival of the next clock impulse, as required by the early and late samples.

The early-late clock recovery works well in conjunction with binary modulation schemes which have peaks in most of the symbol periods, but as with times-two clock recovery, less satisfactorily with multilevel schemes because there are fewer distinctive peaks.

6.2.3 Zero-Crossing Clock Recovery

The zero-crossing clock recovery principle is similar to early-late clock recovery, in that it looks for a specific feature in the received waveform. This works on the premise that with symmetrical signalling pulses, the received waveform will pass through zero exactly midway between the sampling points. The receiver detects a change in the polarity of the received signal, and if this does not occur midway between predicted sampling points, it speeds up or slows down its reconstructed clock appropriately. Again, the zero-crossing clock recovery is a scheme which works well for binary modulation, but the assumption that the waveform will pass through zero midway between sampling instances is not always valid with multilevel

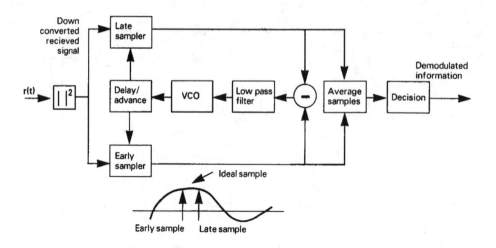

Figure 6.3: Early-late clock recovery schematic

schemes. For example, if the received waveform moves between amplitudes of $+1$ and -3 in a smooth manner, it will pass through zero before the middle of the sample period.

There are two solutions to allow the use of zero-crossing with multilevel modulation. The first is to ignore this problem, as on average the crossing will be in the middle of the symbol period, and to lowpass filter or integrate the observed zero-crossing point in order to produce this average. This increases the timing jitter compared to binary modulation schemes. The second is only to observe the zero-crossings when we expect them to fall midway between sampling instants [428]. If we detect a transition between two symbols of equal magnitude but opposite polarity then we can expect the zero-crossing associated with this transition to be in the middle of a symbol period and can use this to update the timing. If only these transitions are used for timing updates, then, because of the accuracy of the zero-crossing detection, less averaging needs to be performed to remove any noise and so the filtering requirements are reduced, leading to a lower jitter in the recovered clock.

In the case of binary modulation, a similar schematic to that of Figure 6.3 is applicable to zero-crossing clock recovery. However, in the case of multilevel modulation a control logic block must be included in order to enable or disable sampling adjustments, depending on whether zero-crossings fall midway between sampling instances.

6.2.4 Synchroniser

A completely different clock recovery system is the so-called synchroniser. This is the scheme used by the pan-European mobile radio system known as GSM. In this system the transmitter periodically sends a sounding sequence. The receiver searches for this sequence by performing an autocorrelation at a rate significantly faster than the symbol rate (typically four times). The oversampling point at which the maximum correlation occurs is deemed to be the correct sampling point, and is assumed to remain correct until the next sounding sequence is received. This scheme is simple, but requires a sounding sequence to be periodically inserted into the data stream, increasing the bandwidth requirements. It also requires that the transmitter and receiver clocks do not drift significantly over a synchronisation block, which may place constraints on the circuitry that can be employed. This synchronisation scheme has the advantage that it performs equally well regardless of the modulation scheme. The simplified block diagram of such a scheme is shown in Figure 6.4.

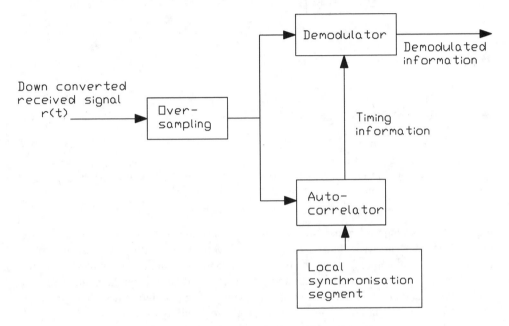

Figure 6.4: Synchroniser clock recovery schematic

6.3 Carrier Recovery

Carrier recovery has received more attention than clock recovery [429, 430] but is not always required as non-coherent detection can often be applied, whereas all modulation schemes require accurate clock recovery. Carrier recovery is always required for square QAM constellations, even when differential coding is employed. For an explanation of differential coding see Chapter 4. This is because the square constellation consists of discrete information encoded on the I and Q axes separately. In order to decode this information it is essential to be able to separate the effects of the I and Q modulation and this can only be performed if a coherent source exists within the receiver.

The effect of coherent detection suffering from inaccurate carrier recovery in the case of a non-differential square 16-QAM system is shown in Figure 6.5, and that on a differential system is shown in Figure 6.6. With the non-differential system a constellation point with a magnitude of (3, 1) on the I and Q axes, respectively, is shown to be transformed to another one with magnitude of approximately (2.2, 2.2) due to the channel-induced carrier phase rotation. When demodulated with reference to the original coordinate system drawn in dashed lines this could easily be decoded as an incorrect symbol. For the differential case shown in

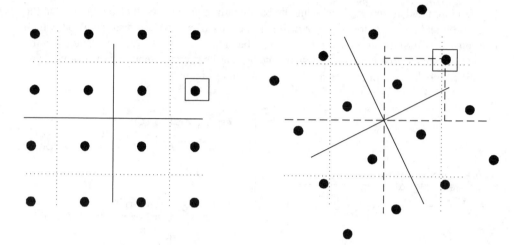

Figure 6.5: Coherent detection suffering from inaccurate carrier recovery in
the case of non-differential QAM

Figure 6.6 let us suppose that the information to be transmitted requires that the constellation
point be moved positively by one quadrant and positively by $90°$ within that quadrant. At the
receiver when demodulating using the original unrotated coordinates shown in dashed lines
we will perceive the correct quadrant move but the received point is on the boundary of the
decision bars, and only a small amount of noise would be required to move it into any of the
four decision areas within the quadrant, giving a high probability of error.

Circular QAM constellations do not need carrier recovery if differential encoding is used.
This is because, with these constellations, the information is encoded onto the phase and amplitude of the signal, corresponding to a polar rather than rectangular encoding, and so only
phase and amplitude differences are required which can be obtained even when the receiver's
carrier reference is at a different frequency from the transmitter reference. In Gaussian channels, the performance of a non-coherent system using full differential encoding will be 3 dB
worse than a coherent system with *perfect* carrier recovery. This is because, with the differential coding, if one symbol is outside the appropriate decision boundary it will cause two
errors, one when it is compared to the previous symbol, and another when the next symbol is
compared with it. In a non-differential system only one error will be experienced.

A common problem for most carrier recovery schemes is an inability to resolve phase
ambiguities encountered in the case of rotationally symmetric, or near-symmetric constellations. So for constellations such as QPSK, phase lock can be established at all multiples of
$90°$, but without additional transmitted information it is not possible to resolve the angle at
which phase lock has been established. This problem can be overcome by sending a sounding
sequence known to the receiver, or through the use of differential coding. The decision as to
whether to employ coherent non-differential, coherent differential, or non-coherent differential modulation depends on the particular environment and the QAM constellation chosen. If
the channel changes only very slowly, which would be the case for a Gaussian channel with
a high SNR, then coherent non-differential transmission is to be preferred, with a training sequence in order to resolve the false locking problem. If the square constellation is employed

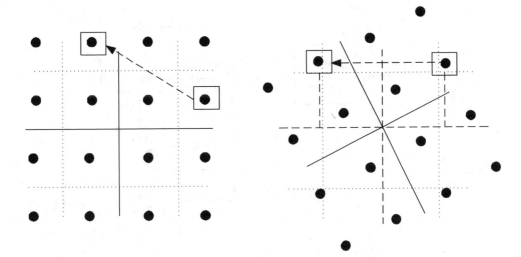

Figure 6.6: Coherent detection suffering from inaccurate carrier recovery in the case of differential QAM

in channels whose phase change can be relatively rapid, then coherent differential modulation will prevent error bursts caused by the possible false locking. If acceptable carrier recovery performance cannot be achieved, then non-coherent differential modulation coupled with a circular QAM constellation is preferable.

After this introduction we now describe some of the more common carrier recovery systems and examine techniques which have been proposed for QAM systems.

6.3.1 Times-n Carrier Recovery

As with clock recovery, a common form of carrier recovery for binary schemes is the times-two or the squared carrier recovery. In this scheme the incoming carrier wave is squared, leading to a periodic component at the carrier frequency which can be isolated using a filter and phase locked loop. Times-two carrier recovery can be implemented using the schematic of Figure 6.7 [388].

Assuming BPSK modulation and noiseless communications for the sake of simplicity, the received signal is given by

$$r(t) = a(t) \cdot \cos(w_c t + \psi), \tag{6.1}$$

where $a(t) = \pm 1$ represents the modulating signal, and ψ the phase of the carrier. After squaring the received signal we get

$$r^2(t) = a^2(t)\cos^2(w_c t + \psi) \tag{6.2}$$
$$= \frac{1}{2}[1 + \cos[2(w_c t + \psi)]].$$

Upon removing the DC component and bandpass filtering, the remaining signal $\cos[2(w_c t + \psi)]$ can be used to drive a phase locked loop (PLL) constituted by a lowpass filter, a multiplier

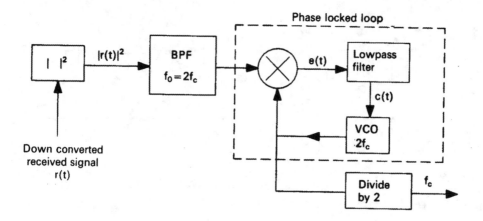

Figure 6.7: Times-two carrier recovery schematic

and a voltage controlled oscillator (VCO).

Assuming that the VCO oscillates at $2w_c$ and has an initial phase of $\hat{\psi}$, its output signal of $\sin[2(w_ct + \hat{\psi})]$ is multiplied by $\cos[2(w_ct + \psi)]$ in order to generate the signal

$$
\begin{aligned}
e(t) &= \cos[2(w_ct + \psi)] . \sin[2(w_ct + \hat{\psi})] \\
&= \frac{1}{2}\sin[4w_ct + 2(\psi + \hat{\psi})] + \frac{1}{2}\sin[2(\hat{\psi} - \psi)].
\end{aligned}
\tag{6.3}
$$

The first term of the above equation is removed by the lowpass loop filter, while the second term of

$$
c(t) = \frac{1}{2}\sin[2(\hat{\psi} - \psi)] \approx \Delta\psi \; if \; \Delta\psi \ll 1
\tag{6.4}
$$

is used to drive the VCO. Finally, the output from the VCO is divided by 2.

Unfortunately, according to Franks [424] the squaring approach does not work with quadrature modulation. This is because when a quadrature signal with equal power on average in each quadrature branch is squared, the signal power in each of the branches tends to cancel, leaving a low signal level on which to perform carrier recovery. However, this can be overcome through the use of a fourth-power device which can generate a significant component at four times the carrier frequency. This can then be divided by 4 to provide carrier recovery. Passing the complex received signal through a fourth-power device has the effect of raising the magnitude of the received phasor to the fourth power and multiplying the phase angle by 4. Times-four carrier recovery was investigated by Rustako *et al.* [109].

Rustako notes that decision directed carrier recovery (see Section 6.3.2) suffers when the occasional receiver outage destroys the accuracy of the data detections. This causes a loss of carrier recovery and considerable time may be required to reacquire the carrier. Furthermore, when adaptive equalisers are used, which update their tap coefficients using a decision directed algorithm, then the equaliser and carrier recovery system can interact in unwanted manner. He notes that times-four carrier recovery does not depend on data decisions or interact with the equaliser. Higher powers than 4 increase carrier jitter due to the increased effects

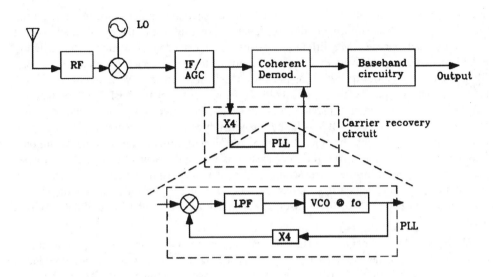

Figure 6.8: Times-four carrier recovery block diagram

of the data pattern.

A schematic diagram of their proposed carrier recovery circuit is shown in Figure 6.8. The received RF signal is downconverted by mixing with the local oscillator (LO). Selectivity and automatic gain control are achieved at the IF stage. The input is raised to the fourth power and input to the PLL. The function of the PLL is to extract the line component at four times the carrier frequency, and to filter out the random component due to the data. Observe that in Figure 6.8 the PLL has an identical structure to that of Figure 6.7, except that it is tuned to f_c rather than $4f_c$, and hence it delivers the required output frequency. The carrier frequency f_c must then be multiplied by four as shown in the $\times 4$ block. This could be achieved by raising the signal to the fourth power. Rustako noted that the times-four function could be performed by a frequency doubler which by virtue of its design also produces frequencies at a fourth power. With suitable filtering such a device could be readily used. In his paper, Rustako gave a thorough mathematical derivation of the theory of times-four carrier recovery. Rustako then went on to construct a times-four carrier recovery and test it using a fading channel simulator. From his experiment Rustako noted that using this form of carrier recovery synchronisation can be maintained even when severe fading causes high error rates. The drawback is that in order to keep pattern-related phase jitter under control, the PLL noise bandwidth must be made small. This will slow down acquisition when loss of carrier does occur. Therefore, times-four carrier recovery will be most suitable in situations where fades are benign and hence loss of lock is not a frequent event.

6.3.2 Decision Directed Carrier Recovery

A carrier recovery scheme which is completely different from all the carrier and clock recovery systems discussed so far is decision directed carrier recovery. In fixed link QAM systems this has proved to be one of the most popular carrier recovery schemes. In this scheme, a decision is made as to which was the most likely constellation point transmitted given the received symbol. This is normally the constellation point closest to the received symbol. It is then assumed that the phase difference between the received symbol and the constellation point is due to carrier recovery drift, and the receiver carrier recovery is updated accordingly. Decision directed carrier recovery has the advantage that it can work with all constellations; however, there is typically a BER threshold. If the prevailing channel BER is less than this threshold, then such methods operate extremely well as they effectively remove additive noise from the input phasor. If the BER is higher than this threshold, then cumulative catastrophic failure can result whereby the constellation point selected in the receiver is the wrong one. The update signal to the carrier recovery system is therefore erroneous, driving the carrier phase further away from the correct value and further increasing the chance of error, etc. Whether decision directed carrier recovery will be suitable for a specific application will depend on this BER threshold, which itself is dependent on the number of modulation levels used.

Whilst simple in principle, decision directed carrier recovery can often require a moderately complex implementation such as that proposed by Horikawa et al. [20] who produced an elegant arrangement for decision directed carrier recovery combined with QAM. The basic premise of the circuit is that the quality of carrier recovery will be acceptable if only half the input signals are used for carrier recovery update.

The 16-QAM phasors can be divided into two subgroups called class I and class II phasors. Note these are not related to the class 1 ($C1$) and class 2 ($C2$) sub-channels derived in Chapter 5. This is shown in Figure 6.9.

Class I consists of the eight phasors lying on the angles:

$$\theta_I = \pi/2 \times (n + 1/2) \quad n = 0, 1, 2, 3 \tag{6.5}$$

and class II consists of all the remaining phasors:

$$\theta_{II} \approx \pi/2 \times (n \pm 1/5) \quad n = 0, 1, 2, 3 \tag{6.6}$$

where the subscripts refer to class I and class II respectively, and $(\pi/2) \cdot (1/5) = (\pi/10) = 18 \deg \approx \tan^{-1}(1/3)$ from the geometry of the figure. In the square 16-QAM constellation shown in Figure 6.9 the class I phasors have $|I| = |Q|$, where $|x|$ means magnitude of x, and I and Q refer to the received voltage on the real and imaginary axes, respectively. The class II signals have $|I| = 3|Q|$ or $|Q| = 3|I|$. The normalised amplitudes for class I phasors are $\sqrt{18}$ and $\sqrt{2}$, and for the class II phasors $\sqrt{10}$.

Horikawa's circuit operates only on class I phasors. It does so by comparing the values of the incoming I and Q signals and if they satisfy the relationship

$$|I|/m < |Q| < n \times |I|, \tag{6.7}$$

where m and n are typically 2, then the circuit concludes that a class I phasor has been sent. For a class I signal received with perfect carrier recovery over a perfect channel we would have $|I| = |Q|$. By subtracting $|I|$ from $|Q|$ we have a signal whose polarity indicates the direction of carrier drift. Determining the quadrant of transmission allows deduction of the required signal to be applied to the VCO. This computation can be carried out by a simple

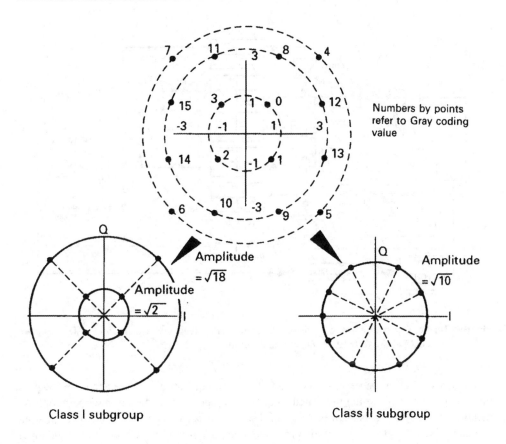

Figure 6.9: 16-level QAM subgroups

algorithm using the following quantities:

$$a = pol(I) \, XOR \, pol(Q) \qquad (6.8)$$
$$b = pol(I + Q)$$
$$c = pol(I - Q)$$
$$d = (a \, XOR \, b) \, XOR \, c$$

where pol (x) is a function which returns 0 if x is negative and 1 if x is positive. The value of d is then used to drive the VCO so that if $d = 1$ the VCO runs slower and if $d = 0$ the VCO runs faster.

For example, say a phasor at $135°$ had been sent but because of an increase in the receiver carrier frequency, this had been received with a phase advance at $140°$. This would give $I = -1.08$ and $Q = 0.91$. So $a = (0 \, XOR \, 1) = 1$, $b =$pol$(-0.17) = 0$, $c =$pol$(-1.99) = 0$ and $d = [(1 \, XOR \, 0) \, XOR \, 0] = 1$. Thus the VCO would be instructed to run slower as required. Examples taken from any quadrant can be seen to function correctly.

The circuit is shown in Figure 6.10. The input phasor is split into real and imaginary

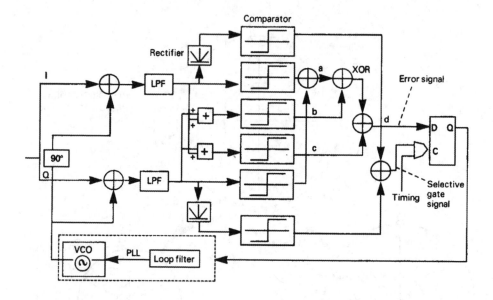

Figure 6.10: Implementation of a decision directed carrier recovery system for 16-level square QAM [113] ©IEEE, 1988, Sari and Moridi

components using the receiver's recovered clock. The quadrature components are lowpass filtered (LPF) to remove the unwanted frequency components of the mixing operation. As mentioned, the decision as to whether a class I signal has been received is based on evaluating whether $|I|/2 < |Q| < 2 \cdot |I|$. Both quadrature components are full-wave rectified and input to a comparator which determines whether the signal level is within the limits expected for a class I signal. The XORed outputs from the two comparators are fed into the control input (c) of an enabling gating arrangement which allows the updating of the VCO if the I and Q amplitudes are within a ratio of 2. The central part of the circuit performs the algorithm described above by computing the parameters a, b, c, d in Equation 6.8, with the loop filter and VCO acting as a phase locked loop (PLL).

This circuit suffers from possible false locking as 90° rotations of the constellation will not change the parameters $a \ldots d$. This must be resolved through the use of a training sequence or some other ambiguity resolving technique.

6.3.2.1 Frequency and Phase Detection Systems

Within amplitude modulation schemes where the phase shift imposed by the modulating signal on the carrier remains constant it is possible to use phase locked loop (PLL) systems. In PLLs phase detectors (PDs) are deployed in order to detect the phase difference between the recovered carrier and the internal reference frequency of the receiver which is used to update the internal reference frequency appropriately. Such a system is referred to as a phase locked loop, which is described in detail in [431].

Such simple systems cannot be used in conjunction with phase or quadrature modulation schemes due to the variation in the carrier phase imposed by the encoded information. These modulation-induced carrier phase variations can either be ignored with the assumption that the average of this phase deviation will be zero, or attempts can be made to remove the phase variation using decision directed methods such as those described in the previous subsection.

Further problems associated with PLL methods for point-to-point microwave links are discussed by Sari and Moridi [113]. Sari notes that the high number of modulation levels used in point-to-point links requires a low phase jitter, which demands narrow PLL bandwidths. However, the frequency uncertainty of the RF oscillators used in the up and down conversion process was often greater than the affordable PLL bandwidth. Possible solutions to this problem include:

(1) Switching of the loop filter. Here a large filter bandwidth is used during the initial phase acquisition process and then a narrow bandwidth is invoked once lock has been achieved. In its simplest form such a system would include two loop filters, and a switch operated by the carrier recovery circuit once the error signal was sufficiently small. The problem with this method is that some means of detecting lock, i.e. a lock detector, is required in order to activate and deactivate the switch optimally.

(2) The use of non-linear elements in the loop filter. Due to this the error signal is no longer proportional to the phase error. The non-linear function is chosen to increase sensitivity (i.e. the phase error seen by the PLL) near lock, and reduce sensitivity further away from lock. Taking the square root or logarithm of the phase error would achieve the desired effect. The chosen non-linear element is placed between the PD and loop filter. Appropriate non-linear devices can often be difficult to construct.

(3) Frequency sweeping. This is a common technique within digital microwave systems whereby the receiver's frequency is slowly changed until lock is achieved using a narrow bandwidth detector. Such a detector should then be able to maintain accurate lock if the channel is relatively benign. In practice the IF oscillator frequency is swept over the uncertainty interval by adding a periodic signal to the filtered PD output. Frequency sweeping is halted once phase lock has been achieved . This process has the disadvantage of taking some time due to theoretical constraints as regards to the speed at which the sweep can take place.

(4) Frequency detectors (FDs). These are devices which produce a signal proportional to the frequency difference between the received signal and local reference carriers. Hence they can readily determine the frequency offset between the received signal and the local reference but as the frequency error does not depend on the phase they cannot be used for phase tracking. They are therefore often used in parallel with PDs, the FD being used during initial acquisition. When the FDs give a near-zero output then a switch is operated such that the FDs are deactivated and the PDs are activated. Sari notes [113] that such a system requires a substantial amount of circuitry.

In his paper, Sari [113] describes an architecture for a new implementation of a combined phase and frequency detector which he refers to as a PFD. A diagram showing the architecture of such a system is given in Figure 6.11. The top PD will be converted into an FD by appropriately manipulating its output. The downconverted received signal is input to a pair of sinusoidal phase detectors (PD) operating in quadrature with each other so that their outputs are $sin\,\phi(t)$ and $cos\,\phi(t)$, where $\phi(t)$ is the phase error, as shown in Figure 6.12. Note that the

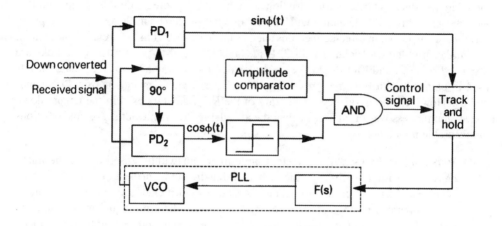

Figure 6.11: PFD block diagram

required stable lock points for a rectangular QAM constellation are at $n \cdot (\pi/2), n = 1, 2, 3, 4$, and these points can be identified as those at the zero crossings of $sin\ \phi$ in Figure 6.12, where $cos\ \phi$ is simultaneously positive. During steady state, the phase error $\phi(t)$ fluctuates around a stable lock point where $sin\ \phi \ll 1$, and only the upper PD and the VCO are being used in a normal PLL configuration. However, during the initial acquisition phase the phase error $\phi(t)$ significantly deviates from zero, and changes at a rate proportional to the frequency offset between the received signal and VCO reference according to $\phi(t) = \Delta wt + \phi(0)$. Clearly, the PD output is a sinusoid with a frequency equal to the difference Δw between the received signal and VCO reference, and therefore has a zero DC content delivering no information about the polarity of the frequency error. In contrast, an FD would provide a DC signal proportional to the frequency offset Δw.

The circuit of Figure 6.11 overcomes this problem by the use of a track and hold device. The device tracks the upper PD when it is in phase lock and so its output is in the range $k\pi/2 \pm \theta$ where $k = 1, 2, 3, 4$ is an integer representing phase lock in four different quadrants, and θ is the maximum allowed phase deviation. In order to track the phase error, the circuitry also requires that the output from the lower quadrant PD is positive, which pinpoints the stable locking domains. Should, however, the phase error $|\phi(t) - n\pi/2|$ be outside the range $\pm\theta$, the circuitry operates in a hold mode and the track and hold device retains the last valid level. The control signal $sin\ \phi$ originating from the upper PD determines whether the phase error $\phi(t)$ is within a predetermined interval $\pm\theta$, and $cos\ \phi$ originating from the lower PD allows recognition of the stable lock points as seen in Figure 6.12. The operation of the device is now explained with reference to Figure 6.13.

The operation of the PFD of the circuit of Figure 6.11 for positive and negative frequency offsets is shown in Figures 6.13(a) and (b), respectively. In these figures which show the situation during initial phase acquisition (i.e. less lock has not been acquired) the dashed line represents the error signal from the upper PD and the solid line represents the output of the track and hold device (and thus the input to the VCO). With a positive frequency offset as shown in Figure 6.13(a) the upper PD tracks the offset until it falls outside the range θ.

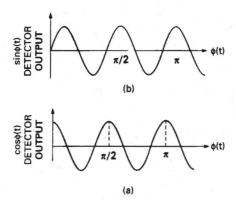

Figure 6.12: In-phase and quadrature phase PD output signal [113] ©IEEE, 1988, Sari *et al.*

At this point the amplitude comparator detects that the PD output has moved outside the allowed range and switches the track and hold device to hold mode. It is maintained in this configuration until the output from the upper PD falls back within the allowed range. Note that the continuous line does not fall back to the dashed line as the latter passes through the axis moving downwards. This is because at this point the track and hold device is maintained in the hold position by the output of the lower PD, which precedes that of the upper by 90° as demonstrated in Figure 6.12. When $cos\,\phi$ is negative, so the upper branch is moving from $+90^o$ to -90°, the track and hold device is maintained at hold. As can be seen, in the positive case portrayed in Figure 6.13(a), the control signal to the VCO is predominantly positive, ensuring that there is sufficient information to allow acquisition. This condition is detected by the comparator at the output of the lower PD. Once lock is acquired, the system will automatically revert to a tracking role as the upper PD output falls within its allowed range. A similar effect can be seen for a negative frequency offset as shown in Figure 6.13(b). The parameter θ should be chosen to maximise the DC at the detector output. For a sinusoidal PD this means that θ should be approximately 16°.

In the case of QAM, according to Sari [113], the phase error associated with each incoming symbol is assessed, and the error angle becomes the PD's output. This is then passed to the track and hold block, which performs its function as before. However, with 16-level square QAM, when the receiver's local oscillator is unsynchronised with that of the transmitter, three circles are apparent, as shown in Figure 6.14. On the inner and outer circles there are four equispaced points forming a QPSK signal set. On the middle ring there are 8 non-regularly spaced points, which can lead to phase ambiguity when used in a circuit of the form proposed by Sari as there will now be 8 possible lock positions, but only in 4 of these can data be correctly demodulated. To overcome this problem, Sari, like Horikawa, proposed to invoke only the class I QAM signals, using a windowing system in order to determine whether a symbol is suitable for carrier recovery. A block diagram of such a system is shown in Figure 6.15.

In this figure the input signal is quadrature downconverted, lowpass filtered (LPF) and analogue-to-digital converted (ADC). It is then input to a practical version of the PFD shown in Figure 6.11. In this scheme the track and hold function is performed by a D-type flip-

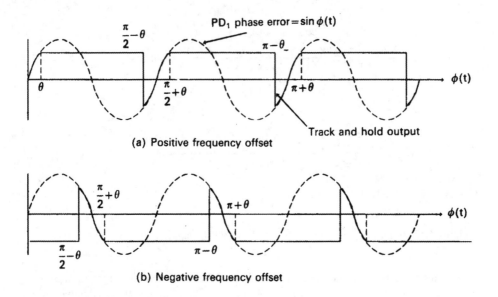

Figure 6.13: PFD waveforms for positive and negative frequency offsets

flop which will only pass the input to the output when clocked. The flip-flop is normally clocked regularly by the clock recovery system (not shown) but can be inhibited by the control logic device. This latter device determines whether the incoming signal is within one of the windows shown as square boxes on the QAM constellation in Figure 6.14. If the signal is within these boxes, then the PD signal is used for carrier recovery purposes, otherwise the input to the loop filter is maintained at the previous level.

Sari claimed a substantially reduced acquisition time for his system compared to a discrete FD and PD system, and a reduced complexity architecture.

6.4   Summary

In this chapter we have considered clock and carrier recovery techniques designed for fixed link QAM systems. Clock recovery techniques used in this environment have tended to be similar or identical to those used for other modulation schemes, with zero-crossing techniques often finding favour. That clock recovery has not been a particular problem for QAM is evidenced by a paucity of publications on the topic.

A considerably more problematic issue has been that of carrier recovery schemes for QAM. Many of the existing schemes proved unsuitable due to the phase-related content of the encoded information and also because a more accurate lock was required due to the higher number of levels used compared to binary modulation. The carrier recovery techniques employed essentially were either times-N or decision directed. Decision directed methods generally were more suitable, where the BER was low and where there were no problems inflicted by a decision feedback equaliser's interactions with a decision directed carrier recovery system.

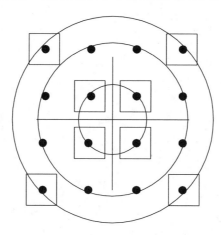

Figure 6.14: Circles and windows within a 16-level square QAM
constellation

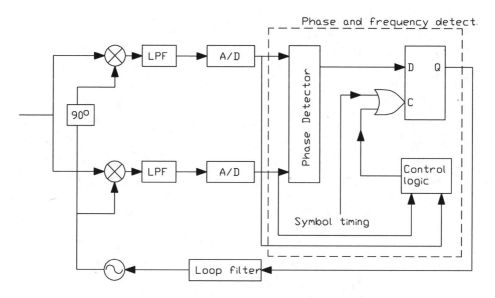

Figure 6.15: General block diagram of a PFD system

Of the decision directed carrier recovery schemes, the one proposed by Sari and Moridi
addressed the issue of both frequency and phase acquisition, whereas that of Horikawa looked
at practical implementation of the discrete time phase detector required. Both decided to use
only the class I QAM symbols for carrier recovery, although for different reasons – Sari
to prevent possible false locking errors and Horikawa to simplify the decoding circuitry required. In the next chapter we consider the important topic of channel equalisation.

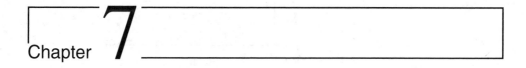

Chapter 7

Trained and Blind Equaliser Techniques

7.1 Introduction

In Chapter 4 we considered the transmission of information via bandlimited channels. It was shown that all bandlimited channels have impulse responses of infinite duration which may result in intersymbol interference (ISI). However, if the Nyquist criterion is met, then the impulse response has zeros at adjacent signalling instants and a unity main tap. Therefore, no ISI is introduced.

In contrast to the approach pursued in Chapter 4, where it was assumed that the channel is distortion-free and hence has a constant frequency domain transfer function and linear phase within the passband, here we consider the more realistic scenario of transmission media having linear distortions. If we characterise the channel by its baseband impulse response $g(t)$, then the baseband received signal $y(t)$ is given by the following convolution:

$$y(t) = x(t) * g(t).$$

A typical continuous channel impulse response is depicted in Figure 7.1. Note that it is no longer symmetric, its zero-crossings are no longer uniformly spaced and this channel introduces linear distortions, which can be seen in the eye pattern.

The aim of channel equalisation is to remove the distortions introduced by the channel. This can be carried out either in the frequency domain by filtering the received signal through the inverse filter of the channel, or in the time domain by convolving it with the appropriate equaliser impulse response.

Linear distortions are introduced, for example, by wideband radio channels described in Chapter 2, when there are a multitude of radio paths between the transmitter and the receiver, and the delay on some of these paths is significantly greater than on others. In extreme cases the longer paths can result from reflections off mountains, more often they occur due to reflections off nearby buildings and sometimes from vehicles such as buses. Wideband

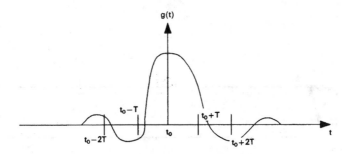

Figure 7.1: Typical channel impulse response

channels can be characterised by their impulse response. This is the signal that would be received were an impulse transmitted (i.e. a pulse of extremely short duration). The impulse response is composed of the superposition of all the paths arriving at the receiver, each path suffering a certain attenuation and delay. Textbooks often show example impulse responses as exponentially decaying curves falling off steeply at first and then becoming more gently undulating. This is because the responses that are significantly delayed are often highly attenuated due to the extra distance which they have travelled.

Historically, wideband channels have been rare in QAM implementations. The only areas where significant multipath fading has been encountered are on high transmission rate (typically > 140 Mbit/s) point-to-point radio links. These are characterised by slowly varying impulse responses extending over a few symbol periods only. Furthermore, the high transmission rate has implied relatively simple equalisers if they are to be implemented with existing circuitry. Two equalisers which are frequently used are the linear and the decision feedback equaliser, while the Viterbi equaliser has rarely been used for QAM mainly due to its high complexity. For that reason we do not describe the Viterbi equaliser here, but refer the reader to Proakis [388]. In the following two sections we consider linear and decision feedback equalisers. In the final section we consider aspects of equalisers specific to QAM, and consider some actual implementations. For more in-depth treatment of equalisation the interested reader is referred to the literature [3, 9, 37, 383, 408, 432].

7.2   Linear Equalisers

7.2.1   Zero-Forcing Equalisers

The basic structure of a linear equaliser (LE) system is given in Figure 7.2.

It is a linear transversal filter with an impulse response of

$$c(t)_{t=jT} = c(jT) = \sum_{j=-N}^{N} c(j)z^{-jT}, \tag{7.1}$$

where $c(t)$ is represented by its T-spaced samples. Equalisation is performed by convolving the T-spaced samples of the received signal, $y(t) = y(kT) = y_k$, with $c(t)$. In practice it is

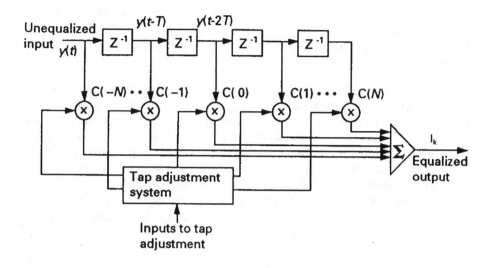

Figure 7.2: General linear equaliser system

performed by passing y_k through a delay chain and the equalised output I_k is derived by a multiplication using taps with variable coefficients $c(j)$. Thus

$$I_k = \sum_{j=-N}^{N} c(j)y(k-j), \qquad (7.2)$$

where there are $(2N+1)$ taps.

In order to satisfy the Nyquist criterion, the equaliser taps can be chosen to drive the T-spaced samples of the combined impulse response of the concatenated channel and equaliser towards zero, allowing only for the main tap to be non-zero. Due to its operating principle this scheme is referred to as a zero-forcing equaliser. Naturally, zero-forcing is only possible over the length of the transversal filter's memory. Nevertheless, as the filter length tends to infinity, the ISI at the equaliser's output tends to zero for all values of jT, $j = 1, 2, \ldots, \infty$. The fundamental limitation of the zero-forcing equaliser is due to the fact that the channel transfer function's inverse filter $C(f)$ constituted by the equaliser when cascaded with the channel transfer function $G(f)$ enhances the channel noise in those frequency intervals where $G(f)$ has a high attenuation. This is because the flat noise spectrum $N(f)$ is boosted along with the signal, where the channel inflicted a high attenuation, which becomes particularly adverse for transmissions over frequency selective fading radio channels.

There are numerous ways of deriving the coefficients $c(j)$. If the channel impulse response can be considered time invariant as a leased telephone line, then the coefficients are constant. In most cases, however, the channel impulse response is time variant and so the tap coefficients need to be variable. A large class of linear equalisers adjust these coefficients by comparing the equalised output with the nearest constellation point and adjusting the coefficients so as to reduce the error. A form of recursive least squares algorithm is normally used in making these adjustments. This is adequate for slowly varying channels with low BERs.

If the BER exceeds a certain threshold, then catastrophic failure will occur due to the wrong symbol being decoded and hence the wrong adjustment being applied to the tap coefficients. These maladjusted coefficients further distort the signal, increasing the probability of error, and so on, until complete failure ensues.

Robustness can be increased by the periodic use of a training sequence to ensure that the coefficient values are close to optimum. The data to be transmitted is divided into blocks, and at the start of each block a known sequence is transmitted. This method can be further subdivided into convergence training and channel sounding, depending on the known sequence transmitted and the processing at the receiver. In the former, any known sequence can be sent. Because the receiver knows this sequence there can be no decoding errors during this period and so the coefficients can be driven towards the optimal value with little risk of catastrophic failure. During normal data transmission the coefficients can either be left unchanged, or updated according to a more cautious update rule compared to that used during the training period.

Another way that the transmission of a known sequence can be used is to periodically sound the channel and then set the coefficients accordingly, eliminating the need for complicated convergence algorithms. The channel may be sounded by sending a maximal length pseudorandom binary sequence (PRBS) having a flat, almost "white" spectrum as the known sequence, and correlating the received sequence against a replica of itself held in the receiver. This correlation process is invoked for each of the symbol intervals over which the expected delay spread of the incoming signal occurs. Maximal length PRBSs have the property that the peak amplitude of the autocorrelation function with zero delay is equal to the length of the code, but for any offset delay the autocorrelation function is equal to -1. The autocorrelation is performed with the XOR function giving a result of either 1 or -1 (rather than 1 or 0). This method is used for example in the CCITT V.27, V.29, V.32 and V.33 standard high speed data modems proposed for telephone lines having linear distortions. These schemes will be described in Chapter 8. Once we have sounded the channel and its impulse response is known, we can calculate the tap coefficients for the LE.

The optimal coefficients for an infinite length LE are the T-spaced samples of the inverse filter. Again, if the baseband equivalent transmitted signal is $x(t)$, and the baseband impulse response of the channel is $g(t)$, then the baseband equivalent input to our receiver is

$$y(t) = g(t) * x(t) \tag{7.3}$$

where * symbolises convolution. An equaliser at the receiver having an impulse response $c(t)$ results in an equalised output

$$h(t) = c(t) * y(t) = c(t) * g(t) * x(t) \tag{7.4}$$

which, to be equal to the input $x(t)$, requires

$$c(t) * g(t) = \delta(t) \tag{7.5}$$

where $\delta(t)$ is the unit pulse sequence [1, 0, 0, ..]. This system is shown in Figure 7.3. Applying the z-domain transfer function to Equation 7.5 gives the z-transform of the filter

Figure 7.3: System model with equaliser

required to equalise the channel having transfer function $G(z)$ as:

$$C(z) = \frac{1}{G(z)}. \tag{7.6}$$

In practice the channel response is of finite length and we represent it by

$$G(z) = g(0) + g(1)z^{-1} + \ldots + g(n)z^{-n} \tag{7.7}$$

where $g(0)$, $g(1)$, ..., $g(n)$ are the coefficients of the finite impulse response (FIR) filter modelling $G(z)$. Its inverse filter has a z-transform computed as follows:

$$
\begin{aligned}
C(z) &= \frac{1}{g(0) + g(1)z^{-1} + \ldots + g(n)z^{-n}} \\
C(z) &= c(0) + c(1)z^{-1} + c(2)z^{-2} + \ldots
\end{aligned} \tag{7.8}
$$

where $c(0)$, $c(1)$,... form an infinite length sequence of equaliser coefficients. If the ISI is mild, then this sequence will decay sharply and the equaliser coefficients, i.e. $c(n)$, are insignificant for large n. Then we can set the coefficients, $c(n)$, of the LE by the long division of $1/G(z)$, which requires little computational effort and can easily be implemented in real time. Thus, if our channel filter has a response of

$$G(z) = g(0) + g(1)z^{-1} + \ldots + g(n)z^{-n} \tag{7.9}$$

where $g(n)$ is a complex number, then the complex equaliser coefficients are given by

$$
\begin{aligned}
c(0) &= \frac{1}{g(0)} \\
c(1) &= \frac{[-c(0) \times g(1)]}{g(0)} \\
c(2) &= \frac{[(-c(0) \times g(2)) + (-c(1) \times g(1))]}{g(0)}
\end{aligned}
$$

and generally

$$c(n) = \frac{\sum_{i=0}^{n-1}(-c(i) \times g(n-i))}{g(0)}. \tag{7.10}$$

If there is no single dominant tap in the impulse response, then the coefficients of the LE may fail to decay for a reasonable value of n. In this case the coefficients calculated by this long division method are no longer the optimal coefficients for the finite length equaliser. Although optimal values for the coefficients of the finite length linear equaliser can be found by more complex processes [388], the linear equaliser is no longer suitable.

7.2.2   Least Mean Squared Equalisers

Least mean squared (LMS) equalisers are considered superior to zero-forcing (ZF) equalisers, because they can minimise the joint effects of ISI and noise at the equaliser's output over their memory length. In our analysis, however, we assume a high signal-to-noise ratio and neglect the effect of noise for the sake of simplicity. We also note that LMS equalisers in general can tolerate higher ISI and have faster convergence than ZF equalisers.

Lucky *et al.* [9] define the mean squared distortion (MSD) as

$$MSD = \frac{1}{h^2(0)} \sum_{\substack{n=-\infty \\ n \neq 0}}^{\infty} h^2(n), \tag{7.11}$$

which is simply the normalised sum of the square of symbol period spaced system response samples at the equaliser's output. Recall from Equation 7.4 that $h(n)$ is the convolution of the received signal $y(n)$ and the equaliser's impulse response, giving:

$$h(n) = y(n) * c(n) = \sum_{j=-N}^{N} c(j)y(n-j), \tag{7.12}$$

and hence

$$h(0) = \sum_{j=-N}^{N} c(j)y(-j). \tag{7.13}$$

Minimising the effects of ISI requires minimising the mean squared error (MSE) term

$$\epsilon = \sum_{\substack{n=-\infty \\ n \neq 0}}^{\infty} h^2(n) = \left(\sum_{n=-\infty}^{\infty} h^2(n) \right) - h^2(0), \tag{7.14}$$

which due to Equation 7.12 and Equation 7.13 is a quadratic function of the equaliser coefficients $c(j)$ and hence it has a minimum. Then from Equation 7.14 we get:

$$\frac{\partial \epsilon}{\partial c(j)} = \left(\sum_{k=-\infty}^{\infty} 2h(n) \frac{\partial h(n)}{\partial c(j)} \right) - 2h(0) \frac{\partial h(0)}{\partial c(j)} = 0, \tag{7.15}$$

and exploiting Equations 7.12 and 7.13 yields:

$$\sum_{n=-\infty}^{\infty} h(n)y(n-j) = h(0)y(-j), \quad j = -N \dots N. \tag{7.16}$$

The main tap value $h(0)$ does not play any role in minimising the (MSE) of the system response's ISI contributions. Hence for the sake of simplicity we set the main tap to $h(0) = 1$ and fulfil this condition by appropriately scaling all equaliser taps by a constant at a later stage.

Substituting Equation 7.12 into Equation 7.16 gives:

$$\sum_{n=-\infty}^{\infty} \sum_{k=-N}^{N} c(k)y(n-k)y(n-j) = y(-j). \tag{7.17}$$

After exchanging the order of summations we arrive at:

$$\sum_{k=-N}^{N} c(k) \sum_{n=-\infty}^{\infty} y(n-k)y(n-j) = y(-j), \tag{7.18}$$

where

$$b(j,k) = \sum_{n=-\infty}^{\infty} y(n-k)y(n-j) \tag{7.19}$$

are recognised as the correlation coefficients of the received signal $y(n)$. Therefore Equation 7.18 simplifies to:

$$\sum_{k=-N}^{N} c(k)b(j,k) = y(-j) \quad j = -N \ldots N. \tag{7.20}$$

This is a set of $(2N+1)$ simultaneous linear equations, from which the $(2N+1)$ equaliser coefficients $c(k)$, $k = -N \ldots N$, minimising the (MSE) term of Equation 7.11 can be determined. Subsequently all coefficients must be scaled by a constant in order to arrive at $h(0) = 1$, which does not affect the shape of the equaliser transfer function. In a more convenient matrix form Equation 7.20 is expressed as:

$$\mathbf{B} \cdot \mathbf{c} = \mathbf{y} \tag{7.21}$$

where $\mathbf{B}$ is a $[2N+1] \times [2N+1]$ dimensional matrix of correlations, while $\mathbf{c}$ and $\mathbf{y}$ are $[2N+1]$ dimensional column vectors. Assuming a stationary received signal $y(n)$, its correlation coefficients will not be dependent on the actual values of the indices (j,k), only their differences, hence Equation 7.21 can be written as follows:

$$\begin{bmatrix} b(0) & b(1) & \ldots & b(2N) \\ b(1) & b(0) & \ldots & b(2N-1) \\ b(2) & b(1) & \ldots & b(2N-2) \\ \vdots & \vdots & & \vdots \\ b(2N) & b(2N-1) & \ldots & b(0) \end{bmatrix} \begin{bmatrix} c(-N) \\ c(-N+1) \\ c(-N+2) \\ \vdots \\ c(N) \end{bmatrix} = \begin{bmatrix} y(N) \\ y(N-1) \\ y(N-2) \\ \vdots \\ y(-N) \end{bmatrix}. \tag{7.22}$$

Observe that the correlation matrix in Equation 7.22 is not only symmetric but all the elements along the main diagonals are identical, hence it has a so-called Töplitz structure. This set of equations are often referred to as the Wiener–Hopf equations in the literature of LMS spectral estimation.

An important consequence of Equation 7.20 can be inferred as follows [9]. The ideal system response has a unity value at $n = 0$, i.e. $h(0) = 1$ and $h(n) = 0$ otherwise. Any deviation from these desired values is actually the error sample $e(n)$ associated with sampling instant n, therefore

$$
\begin{aligned}
e(0) &= h(0) - 1 \text{ for } n = 0 \\
e(n) &= h(n) \ \text{ for } n \neq 0.
\end{aligned}
\tag{7.23}
$$

From Equation 7.16 we infer that

$$
\sum_{\substack{n=-\infty \\ n \neq 0}}^{\infty} h(n)y(n-j) + h(0)y(-j) = h(0)y(-j),
\tag{7.24}
$$

and hence

$$
\sum_{\substack{n=-\infty \\ n \neq 0}}^{\infty} h(n)y(n-j) = \sum_{\substack{n=-\infty \\ n \neq 0}}^{\infty} e(n)y(n-j) = 0 \quad j = -N \ldots N.
\tag{7.25}
$$

Equation 7.25 implies that the optimum equaliser setting will result in an error sequence $e(n)$ which is uncorrelated with the equaliser's input signal $y(n)$ within its range of $[-N, \ldots, N]$.

Similar sets of simultaneous equations are frequently encountered in various spectral estimation problems and a plethora of computationally efficient recursive algorithms are available for their solution. For example, when computing the optimum predictor coefficients in linear predictive speech coding, a similar set of equations must be solved, which is often carried out using the Levinson–Durbin algorithm [383]. Another example is that of designing an optimum autoregressive filter in the decoding of Bose–Chaudhuri–Hocquenghem block codes, which generates the required sequence of syndromes and facilitates the computation of the error locations. In this application the Berlekamp–Massey algorithm is normally employed [383]. The specific solution to be used will depend on the final formulation of Equation 7.20 as regards to the computation of $b(j,k)$ in Equation 7.19. Namely, the error minimisation in Equation 7.14 extends over $[-\infty, \ldots, \infty]$, but in practical terms it must be set to zero outside a finite interval length. Depending on how this windowing is carried out, different solutions arise, because the matrix $\mathbf{B}$ containing the elements $b(j,k)$ will have different properties. Efficient solutions avoid straight matrix inversion, because their complexity is usually proportional to N^3. Instead, recursive solutions exploiting the symmetry of $\mathbf{B}$ or the Töplitz structure are generally favoured due to their lower complexity, which is typically proportional to N^2.

The solution of Equation 7.21 in terms of matrix inversion is given in the form

$$
\mathbf{c}^{\mathrm{opt}} = \mathbf{B}^{-1}\mathbf{y}
\tag{7.26}
$$

but as mentioned, iterative solutions are preferable.

A frequently used straightforward iterative procedure is the *steepest descent or gradient algorithm*, where an arbitrary initial coefficient vector $\mathbf{c}_0$ is assumed. As seen from Equation 7.12 and Equation 7.14, the (MSE) is a quadratic function of the equaliser coefficients comprised by $\mathbf{c}$, and hence exhibits a minimum. The initial $\mathbf{c}_0$ equaliser setting corresponds to an arbitrary momentary MSE ϵ_0 on a $(2N+1)$ dimensional paraboloid surface. The

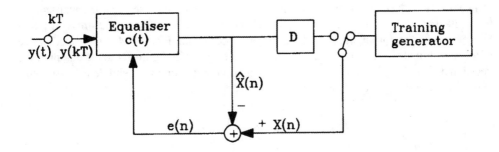

Figure 7.4: Adaptive equaliser schematic

$(2N + 1)$ dimensional gradient vector $\mathbf{g}_0$ of the MSE at this instant is given by:

$$\mathbf{g}_0 = \frac{\partial \epsilon}{\partial \mathbf{c}_0}. \tag{7.27}$$

With the $(2N + 1)$ gradient components $g_{-N}, \ldots, g_N$ known, every equaliser coefficient is modified by a step value $d_0(n)$, $n = -N, \ldots, N$, proportional to its magnitude in the direction opposite to its gradient. For example, if the point $\mathbf{c}_0$ was at a location of negative gradient on the $(2N + 1)$ dimensional paraboloid in terms of all $(2N + 1)$ components, all components must be increased by the common step size d and vice versa, giving the relationship

$$c_{k+1}(n) = c_k(n) - d \cdot g_k(n), \quad n = -N \ldots N. \tag{7.28}$$

Observe that the convergence speed and accuracy of the algorithm depends on the choice of the step size $d_k(n)$ that was fixed to d in our approach. In Section 7.4 we will show that faster convergence can be achieved using a technique referred to as Kalman filtering [388]. In general, the larger the step size, the faster the initial convergence and the larger the residual convergence error. The minimum MSE cannot be attained with this algorithm, but due to its simplicity the method is quite popular.

7.2.3 Decision Directed Adaptive Equalisers

The simplified schematic of a decision directed automatic adaptive equaliser is portrayed in Figure 7.4, where the term "decision" directed implies that after a training period the equaliser is required to update its coefficients on the basis of the signal at the output of the receiver's decision device (D), which is not always an error-free signal. Initially a training sequence known to both the transmitter and the receiver is sent, allowing the equaliser to derive the error signal $e(n) = x(n) - \hat{x}(n)$ from the transmitted signal $x(n)$ and the equalised received signal $\hat{x}(n)$. Observe that in this formulation $e(n)$ can now contain both ISI and noise. Then the MSE term of

$$\epsilon = \sum_{n=-\infty}^{\infty} e^2(n) \tag{7.29}$$

can be used to adjust the equaliser's coefficients. After the training segment has elapsed, the transmitted signal $x(n)$ becomes unknown. However, assuming that no transmission errors have occurred, the output signal of the decision device (D) in Figure 7.4 is the exact replica of the transmitted signal $x(n)$. Therefore flipping the switch in Figure 7.4 to the output of the decision device (D) allows decision directed equalisation to continue unhindered. Using this approach the error signal is given by:

$$e(n) = x(n) - \sum_{k=-N}^{N} c(k)y(n-k). \tag{7.30}$$

Then the MSE is yielded as:

$$\epsilon = \sum_{n=-\infty}^{\infty} \left[x(n) - \sum_{k=-N}^{N} c(k)y(n-k) \right]^2 \tag{7.31}$$

which we want to minimise as a function of the equaliser coefficients $c(k)$. Upon setting the partial derivatives to zero we arrive at:

$$\frac{\partial \epsilon}{\partial c(i)} = -2 \sum_{n=-\infty}^{\infty} \left\{ \left[x(n) - \sum_{k=-N}^{N} c(k)y(n-k) \right] y(n-i) \right\} = 0. \tag{7.32}$$

After rearranging Equation 7.32 we get:

$$\sum_{n=-\infty}^{\infty} x(n)y(n-i) = \sum_{n=-\infty}^{\infty} \sum_{k=-N}^{N} c(k)y(n-k)y(n-i) \tag{7.33}$$

which on exchanging the order of summations yields:

$$\sum_{n=-\infty}^{\infty} x(n)y(n-i) = \sum_{k=-N}^{N} c(k) \sum_{n=-\infty}^{\infty} y(n-k)y(n-i). \tag{7.34}$$

In Equation 7.31 it was implicitly assumed that the error is minimised over the interval $-\infty < n < \infty$, which cannot be implemented in practice. Depending on how the input data undergoing spectral estimation is windowed to a finite interval, two frequently used methods arise, the so-called autocorrelation method and the covariance method [383]. Pursuing the former we implicitly suppose that the data outside the equaliser's range $[-N, \ldots, N]$ is set to zero. Then recognising that in Equation 7.34 for a stationary received signal $y(n)$

$$r(i,k) = r(|i-k|) = \sum_{n=-N}^{N} y(n-k)y(n-i) \tag{7.35}$$

is the autocorrelation of the received samples, while

$$\psi(-i) = \sum_{n=-N}^{N} x(n)y(n-i) \qquad (7.36)$$

is the cross-correlation of the received samples with the transmitted samples, which in the case of error-free decisions are known to the receiver, allows us to express Equation 7.34 as follows:

$$\sum_{k=-N}^{N} c(k)r(|i-k|) = \psi(-i), \quad i = -N \ldots N. \qquad (7.37)$$

Similarly to Equation 7.20, Equation 7.37 also represents a set of $(2N+1)$ simultaneous equations, from which the equaliser coefficients $c(k)$ can be determined. The sole difference arises on the right-hand side of Equation 7.37, where now we have the cross-correlation between the received samples $y(n)$ and the transmitted samples $x(n)$ that are assumed to be locally known. This implies either that the received signal $y(n)$ is an error-free replica of $x(n)$ or that an equaliser training pattern is transmitted, which is known to both the transmitter and the receiver. The solution of Equation 7.37 ensues similarly to that of Equation 7.20, i.e. either by Gauss-Jordan elimination or by some recursive method.

For the success of the equaliser training process it is vital to establish initial synchronisation before the training segment is received in order to ensure that the error signal $e(n)$ is derived from the appropriately synchronised received and locally stored training patterns. If the bit error rate (BER) is low, the decision directed equaliser is capable of tracking slowly varying channel characteristics, such as slow drift in telephone lines, etc. High BERs, however, result in catastrophic error precipitation. In order to attain fast initial convergence, usually a larger step size is used during the training process, which is reduced to a smaller one during decision directed operation to maintain a low tracking error.

7.3 Decision Feedback Equalisers

The structure of a general decision feedback equaliser (DFE) is shown in Figure 7.5.

The equaliser is composed of a feedforward section and a feedback section. The feedforward section is identical to the linear equaliser but the feedback section differs in its input. Instead of using the received input signal, a decision is made as to the transmitted symbol, and this value is passed into the feedback section. This leads to a number of advantages. The data passed into the feedback section is the output signal of the detector and hence no longer contains any noise, thus increasing the accuracy of the interference cancellation. The structure of the feedback system implies that the length of the feedback register need only be as long as the delay spread of the received signal. This is in contrast to the linear equaliser where the equaliser length is governed by the effective length of the infinite impulse response inverse filter, and can easily be an order of magnitude greater than the delay spread.

From Figure 7.5, following Cheung's approach [37], we infer the equalised signal at the

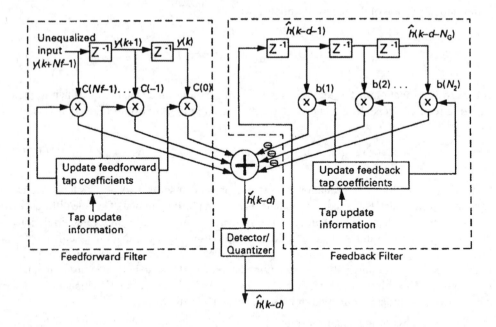

Figure 7.5: General decision feedback equaliser structure

detector's input as:

$$\check{h}(k-d) = \sum_{m=0}^{N_f-1} c(m)y(k+m) - \sum_{p=1}^{N_b} b(p)\hat{h}(k-d-p), \qquad (7.38)$$

which has been delayed by d sampling intervals. The error signal $e(k-d)$ between the transmitted signal $x(k-d)$ and the equalised signal $\check{h}(k-d)$ can be expressed with the help of Equation 7.38 as follows:

$$
\begin{aligned}
e(k-d) &= x(k-d) - \check{h}(k-d) \qquad\qquad (7.39)\\
&= x(k-d) - \sum_{m=0}^{N_f-1} c(m)y(k+m) + \sum_{p=1}^{N_b} b(p).\hat{h}(k-d-p)
\end{aligned}
$$

Using the minimum MSE criterion requires minimisation of the expected value

$$\epsilon = E\{|e(k-d)|^2\}. \qquad (7.40)$$

As we have seen in the previous section, this is achieved by that equaliser setting which renders the error signal $e(k-d)$ uncorrelated with the equaliser's input signal. As regards to the forward filter this requires

$$E\{e(k-d) \cdot y^*(k+m)\} = 0 \quad m = 0\dots(N_f-1) \qquad (7.41)$$

and for the backward filter it requires

$$E\{e(k-d) \cdot \hat{h}^*(k-d-q)\} = 0 \quad q = 1 \dots N_b \tag{7.42}$$

where the superscript * represents conjugate complex computation. Supposing that no decision errors have been encountered, we have

$$\hat{h}(k-d) = x(k-d) \tag{7.43}$$

i.e. the detector's output signal is an identical replica of the transmitted signal. Upon substituting Equation 7.40 into Equation 7.42 we arrive at:

$$E\{ \left[x(k-d) - \sum_{m=0}^{N_f-1} c(m)y(k+m) + \sum_{p=1}^{N_b} b(p)\hat{h}(k-d-p) \right] \tag{7.44}$$
$$\cdot \hat{h}^*(k-d-q)\} = 0 \quad q = 1 \dots N_b$$

and taking into account Equation 7.43 yields:

$$E\{ \left[x(k-d) - \sum_{m=0}^{N_f-1} c(m)y(k+m) + \sum_{p=1}^{N_b} b(p)x(k-d-p) \right] \tag{7.45}$$
$$\cdot x^*(k-d-q)\} = 0 \quad q = 1 \dots N_b.$$

For the random uncorrelated transmitted signal samples we have:

$$E\{x(k-d-p) \cdot x^*(k-d-q)\} = \begin{cases} 1 & \text{for } p = q \\ 0 & \text{otherwise} \end{cases}. \tag{7.46}$$

The received signal $y(k)$ is the convolution of the L_c sample's long channel impulse response $g(k)$ and the transmitted signal $x(k)$ contaminated by the additive noise $n(k)$, as shown below:

$$y(k) = \sum_{i=0}^{L_c-1} g(i) \cdot x(k-i) + n(k). \tag{7.47}$$

Taking Equations 7.46 and 7.47 into account, Equation 7.46 can be written as:

$$\sum_{m=0}^{N_f-1} c(m) \sum_{i=0}^{L_c-1} E\{[g(i)x(k+m-i) + n(k+m)] \cdot x^*(k-d-q)\}$$
$$= \sum_{p=1}^{N_b} b(p) E\{x(k-d-p) \cdot x^*(k-d-q)\} \quad q = 1 \dots N_b \tag{7.48}$$

Exploiting the orthogonality of the additive noise $n(k+m)$ and transmitted signal $x(k-d-q)$

we have:

$$E\{n(k+m) \cdot x^*(k-d-q)\} = E\{n^*(k+m)x(k-d-p)\} = 0 \qquad (7.49)$$

and therefore Equation 7.50 simplifies to:

$$\sum_{m=0}^{N_f-1} c(m)g(m+d+q) = b(q) \quad q = 1 \dots N_b. \qquad (7.50)$$

Explicitly, Equation 7.50 is a set of N_b simultaneous equations delivering the feedback filter's coefficients $b(q)$, $q = 1, \dots, N_b$ as a function of the feedforward filter coefficients $c(m)$, $m = 0, \dots, (N_f - 1)$, and that of the channel impulse response $g(m + d + q)$. The feedforward coefficients can be determined using Equation 7.20, while the channel impulse response is derived by means of channel sounding methods.

7.4 Fast Converging Equalisers

As we have seen in Section 7.2.2, the stochastic gradient method is a low complexity technique for adjusting the equaliser coefficients and it exhibits slow convergence properties. Considerably faster convergence can be attained using a computationally more demanding recursive coefficient adjustment technique, which is referred to as Kalman filtering [433].

Kalman filtering is a fast-converging spectral estimation algorithm for the solution of Wiener–Hopf type normal equations and it has been well documented in references on adaptive filtering and channel equalisation [3, 388, 408]. In our deliberations we follow the approach of Proakis [388] and Haykin [3].

7.4.1 Least Squares Method

The minimum mean squared estimation method used in Sections 7.2.2 and 7.2.3 was based on a statistical ensemble average minimisation of the square of the error signal, which led to the standard Wiener–Hopf equations given in Equation 7.20 or Equation 7.37. The set of simultaneous equations was then solved by matrix inversion, recursive solutions or by the computationally simple steepest descent algorithm, which yielded slow convergence due to applying a common step size d, as seen in Equation 7.28.

In contrast, the least squares (LS) method's optimisation criterion is expressed in terms of a time average, rather than ensemble average. The LS estimation problem is then formulated as estimating the required equaliser output sequence $h(i)$ by $\hat{h}(i)$ from its input sequence $y(i)$ by the help of the equaliser coefficients $c(t)$, using a linear relationship similar to that in Equation 7.30,

$$h(i) = \sum_{t=0}^{T-1} c(t)y(i-t) + e(i) \qquad (7.51)$$

where the index t is used to emphasise the notion of time dependence and $e(i)$ represents the random estimation error. This scheme is depicted in Figure 7.6. In other words, Equation 7.51 states that the required equaliser output sample $h(i)$ can be estimated from the previous T

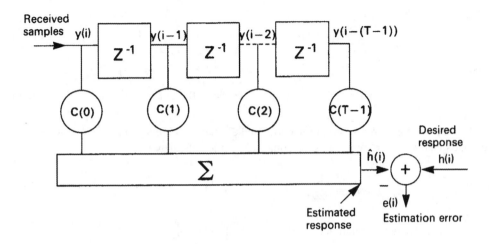

Figure 7.6: LS equaliser structure

received samples by a linear combination, where the random estimation error $e(i)$ has zero mean and a variance of σ^2. This formulation of the estimation problem is identical to the linear regression model often used in source encoding theory.

Our approach is now to find the equaliser coefficients $c(t)$, which minimise the cost function defined as the sum of error squares over a finite interval. The estimation error is expressed from Equation 7.51 as follows:

$$
\begin{aligned}
e(i) &= h(i) - \hat{h}(i) \\
&= h(i) - \sum_{t=0}^{T-1} c(t)y(i-t).
\end{aligned}
\tag{7.52}
$$

Then the optimisation cost function ϵ depends on the equaliser coefficients $c(t)$, $t = 0, \ldots, T-1$ and can be expressed as:

$$
\begin{aligned}
\epsilon &= \sum_i |e(i)|^2 \\
&= \sum_i |h(i) - \sum_{t=0}^{T-1} c(t)y(i-t)|^2,
\end{aligned}
\tag{7.53}
$$

where the interval over which minimisation is carried out depends on the data windowing method invoked. Observing that our cost function ϵ to be minimised in Equation 7.53 is similar to that of Equation 7.31 in Section 7.2.3 and setting the range over which optimisation is carried out in accordance with the covariance method to $i = 1, \ldots, I$, the equaliser coefficients can be computed.

Setting the partial derivatives of Equation 7.53 with respect to the equaliser coefficients $c(t)$, $t = 0, \ldots, T-1$ to zero gives a similar set of equations to Equation 7.37, as shown below:

$$\sum_{t=0}^{T-1} c(t)\Phi(t,k) = \psi(-k), \;\; k = 0\ldots(T-1) \tag{7.54}$$

Here

$$\psi(-k) = \sum_{i=1}^{I} h(i)y(i-k), \;\; k = 0\ldots(T-1) \tag{7.55}$$

is the time-averaged cross-correlation between the equaliser's input signal $y(i)$ and its desired output signal $h(i)$, $c(t)$ represents the equaliser coefficients, while

$$\phi(t,k) = \sum_{i=1}^{I} y(i-k)y(i-t), \;\; 0 \leq t, \; k \leq T - 1 \tag{7.56}$$

is the time-averaged correlation of $y(i)$. In matrix form we have:

$$\mathbf{\Phi} \cdot \mathbf{c} = \mathbf{\Psi} \tag{7.57}$$

or explicitly:

$$\begin{pmatrix} \phi(0,0) & \phi(1,0) & \phi(2,0) & \cdots & \phi(T-1,0) \\ \phi(0,1) & \phi(1,1) & \phi(2,1) & \cdots & \phi(T-1,1) \\ \phi(0,2) & \phi(1,2) & \phi(2,2) & \cdots & \phi(T-1,2) \\ \vdots & \vdots & \vdots & & \vdots \\ \phi(0,T-1) & \phi(1,T-1) & \phi(2,T-1) & \cdots & \phi(T-1,T-1) \end{pmatrix}$$
$$\begin{bmatrix} c(0) \\ c(1) \\ c(2) \\ \vdots \\ c(T-1) \end{bmatrix} = \begin{bmatrix} \psi(0) \\ \psi(-1) \\ \psi(-2) \\ \vdots \\ \psi(-T+1) \end{bmatrix}.$$

The elements of ϕ can be computed from Equation 7.56 as follows:

$$
\begin{aligned}
\phi(0,0) &= y(1)y(1) + \ldots + y(I)y(I) && (7.58) \\
&= \sum_{i=1}^{I} y(i)y(i) \\
\phi(0,1) &= y(0)y(1) + \ldots + y(I-1)y(I) \\
&= \sum_{i=1}^{I} y(i-1)y(i)
\end{aligned}
$$

$$\vdots$$

$$
\begin{aligned}
\phi(0, T-1) &= y(2-T)y(1) + \ldots + y(I-T+1)y(I)y(I) \\
&= \sum_{i=1}^{I} y(i-T+1)y(i) \\
\phi(1,0) &= y(1)y(0) + \ldots + y(I)y(I-1) \\
&= \sum_{i=1}^{I} y(i)y(i-1) \\
\phi(1,1) &= y(0)y(0) + \ldots + y(I-1)y(I-1) \\
&= \sum_{i=1}^{I} y(i-1)y(i-1)
\end{aligned}
$$

$$\vdots$$

$$
\begin{aligned}
\phi(1, T-1) &= y(2-T)y(0) + \ldots + y(I-T+1)y(I-1) \\
&= \sum_{i=1}^{I} y(i-T+1)y(i-1)
\end{aligned}
$$

$$\vdots$$

$$
\begin{aligned}
\phi(T-1,0) &= y(1)y(2-T) + \ldots + y(I)y(I-T+1) \\
&= \sum_{i=1}^{I} y(i)y(i-T+1)
\end{aligned}
$$

$$\vdots$$

$$
\begin{aligned}
\phi(T-1, T-1) &= y(2-T)y(2-T) + \ldots + y(I-T+1)y(I-T+1) \\
&= \sum_{i=1}^{I} y(i-T+1)y(i-T+1)
\end{aligned}
$$

The diagonal elements of the matrix $\boldsymbol{\Phi}$ are now not identical, so it does not have a Töplitz structure. The equaliser coefficients optimised for the interval $i = 1, \ldots, I$ can be found from Equation 7.57 by simply inverting the matrix $\boldsymbol{\Phi}$ or using more efficient recursive algorithms. Having presented the fundamental principles of LS estimation, in the next section we present a recursive LS (RLS) algorithm for equaliser coefficient adjustment.

7.4.2   Recursive Least Squares Method [3]

7.4.2.1   Cost Function Weighting

The principle of the RLS [3] algorithm is that given the current LS equaliser coefficient set, it is possible to update this set, when new equaliser input samples arrive. The RLS algorithm converges quickly, but has a high complexity, as we will see. The cost function ϵ of RLS estimation usually contains an exponentially decaying multiplicative weighting factor β^{n-i} in order to reduce the effect of "old" error terms and emphasise more recent ones:

$$\epsilon = \sum_{i=1}^{n} \beta^{n-i} |e(i)|^2, \tag{7.59}$$

where the error term $e(i)$ was given in Equation 7.52. The factor β is typically close to, but less than unity, and the number of error terms n in the cost function above is variable. This is because in the RLS method the computations commence from a known initial state determined for example by the LS method described in the previous subsection, and new received samples are then used to update previous estimates. For the observation interval n the equaliser coefficients $c(t), t = 0, \ldots, (T-1)$ remain unchanged.

For our further deliberations it is useful to introduce the vector of equaliser input samples, which can be complex valued:

$$\mathbf{y}(i) = [y(i), y(i-1), y(i-2) \ldots y(i-(T-1))]^T, \tag{7.60}$$

where the superscript T means transposition. Then we can express the correlation matrix $\mathbf{\Phi}(\mathbf{n})$ for the observation interval n as follows:

$$\mathbf{\Phi}(\mathbf{n}) = E\left\{\mathbf{y}(i) \cdot \mathbf{y}^{*T}(i)\right\} = \sum_{i=1}^{n} \mathbf{y}(i) \cdot \mathbf{y}^{*T}(i) =$$

$$\begin{bmatrix} \sum_{i=1}^{n} y(i)y^*(i) & \cdots & \sum_{i=1}^{n} y(i)y^*(i-T+1) \\ \sum_{i=1}^{n} y(i-1)y^*(i) & \cdots & \sum_{i=1}^{n} y(i-1)y^*(i-T+1) \\ \vdots & & \vdots \\ \sum_{i=1}^{n} y(i-T+1)y^*(i) & \cdots & \sum_{i=1}^{n} y(i-T+1)y^*(i-T+1) \end{bmatrix}. \tag{7.61}$$

Observe that apart from the upper summation limit n, this matrix has the same elements as in the previous section.

The key equation in Equation 7.54 or 7.57 is now defined in terms of the modified correlation matrix incorporating the exponential scaling factor β^{n-i}:

$$\mathbf{\Phi}(n) = \sum_{i=1}^{n} \beta^{n-i} \mathbf{y}(i) \cdot \mathbf{y}^{*T}(i), \tag{7.62}$$

while the cross-correlation $\boldsymbol{\Psi}(n)$ takes the form

$$\boldsymbol{\Psi}(n) = \sum_{i=1}^{n} \beta^{n-i} \mathbf{y}(i) \cdot h^*(i). \tag{7.63}$$

7.4.2.2 Recursive Correlation Update

Since we intend to derive a recursive update formula for $\boldsymbol{\Phi}(n)$ based on its previous value, we separate the last term from the summation in Equation 7.62 yielding:

$$\boldsymbol{\Phi}(n) = \beta \left[\sum_{i=1}^{n-1} \beta^{n-i-1} \mathbf{y}(i) \cdot \mathbf{y}^{*T}(i) \right] + \mathbf{y}(n)\mathbf{y}^{*T}(n), \tag{7.64}$$

where the square bracketed term can be recognised as $\boldsymbol{\Phi}(n-1)$. Hence Equation 7.64 can be simplified to the required recursive relationship

$$\boldsymbol{\Phi}(n) = \beta\boldsymbol{\Phi}(n-1) + \mathbf{y}(n)\mathbf{y}^{*T}(n), \tag{7.65}$$

where the matrix given by the outer product $\mathbf{y}(n)\mathbf{y}^{*T}(n)$ constitutes the correction term in the update process. Analogously, the update formula for $\boldsymbol{\Psi}(n)$ accrues as follows:

$$\boldsymbol{\Psi}(n) = \beta\boldsymbol{\Psi}(n-1) + \mathbf{y}(n)h^*(n). \tag{7.66}$$

Again, similarly to the LS method, the normal equations given by Equation 7.57 have to be solved either by matrix inversion, or preferably by computationally more efficient recursive methods for $n = 1, 2, \ldots, \infty$. According to Haykin [3] this can be achieved using Woodbury's identity [432], which is also known as the matrix inversion lemma, if the matrix $\boldsymbol{\Phi}(n)$ is non-singular, i.e. it has a non-zero determinant and hence it is invertible.

7.4.2.3 The Ricatti Equation of RLS Estimation

Upon exploiting Woodbury's identity the following recursive formula accrues for the inverted matrix $\boldsymbol{\Phi}^{-1}$ [3]:

$$\boldsymbol{\Phi}^{-1}(n) = \beta^{-1}\boldsymbol{\Phi}^{-1}(n-1) - \frac{\beta^{-2}\boldsymbol{\Phi}^{-1}(n-1)\mathbf{y}(n)\mathbf{y}^{*T}(n)\boldsymbol{\Phi}^{-1}(n-1)}{1 + \beta^{-1}\mathbf{y}^{*T}(n)\boldsymbol{\Phi}^{-1}(n-1)\mathbf{y}(n)} \tag{7.67}$$

This is already an explicit recursive expression for updating $\boldsymbol{\Phi}^{-1}(n-1)$, when new received signal samples $y(n)$ become available, which can be further streamlined using the following shorthand:

$$\mathbf{k}(n) = \frac{\beta^{-1}\boldsymbol{\Phi}^{-1}(n-1)\mathbf{y}(n)}{1 + \beta^{-1}\mathbf{y}^{*T}(n)\boldsymbol{\Phi}^{-1}(n-1)\mathbf{y}(n)}, \tag{7.68}$$

where $k(n)$ is a $T \times 1$ vector, since the numerator is a $T \times 1$ vector, while the denominator is a scalar. Then we have:

$$\boldsymbol{\Phi}^{-1}(n) = \beta^{-1}\boldsymbol{\Phi}^{-1}(n-1) - \beta^{-1}\mathbf{k}(n)\mathbf{y}^{*T}(n)\boldsymbol{\Phi}^{-1}(n-1), \tag{7.69}$$

which is the so-called Ricatti equation of RLS estimation [3]. The vector expression $\mathbf{k}(n)$ in Equation 7.68 can further be simplified by multiplying both sides of it with the denominator, giving:

$$\mathbf{k}(n) + \beta^{-1}\mathbf{k}(n)\mathbf{y}^{*T}(n)\mathbf{\Phi}^{-1}(n-1)\mathbf{y}(n) = \beta^{-1}\mathbf{\Phi}^{-1}(n-1)\mathbf{y}(n)$$

which leads to:

$$
\begin{aligned}
\mathbf{k}(n) &= \beta^{-1}\mathbf{\Phi}^{-1}(n)\mathbf{y}(n) - \beta^{-1}\mathbf{k}(n)\mathbf{y}^{*T}(n)\mathbf{\Phi}^{-1}(n-1)\mathbf{y}(n) \\
&= \left[\beta^{-1}\mathbf{\Phi}^{-1}(n) - \beta^{-1}\mathbf{k}(n)\mathbf{y}^{*T}(n)\mathbf{\Phi}^{-1}(n-1)\right]\mathbf{y}(n).
\end{aligned}
$$

(7.70)

Closer scrutiny of Equation 7.69 reveals that the square bracketed term of Equation 7.70 is equal to $\mathbf{\Phi}^{-1}(n)$, hence Equation 7.70 simplifies to:

$$\mathbf{k}(n) = \mathbf{\Phi}^{-1}(n)\mathbf{y}(n),\qquad(7.71)$$

which defines the vector $\mathbf{k}(n)$ as a transformed version of the received signal vector $\mathbf{y}(n)$, where the transformation is carried out by rotating $\mathbf{y}(n)$ with the aid of the inverse of the correlation matrix.

7.4.2.4 Recursive Equaliser Coefficient Update

Based on the normal equations in Equation 7.57 and the recursive update formula in Equation 7.66 the estimated equaliser coefficients can be expressed by means of $\mathbf{\Phi}^{-1}(n)$ as follows:

$$
\begin{aligned}
\hat{\mathbf{c}}(n) &= \mathbf{\Phi}^{-1}(n)\mathbf{\Psi}(n) \\
&= \beta\mathbf{\Phi}^{-1}(n)\mathbf{\Psi}(n-1) + \mathbf{\Phi}^{-1}(n)\mathbf{y}(n)h^*(n).
\end{aligned}
$$

(7.72)

Substituting $\mathbf{\Phi}^{-1}(n)$ in the first term of Equation 7.72 by Equation 7.69 leads to:

$$
\begin{aligned}
\hat{\mathbf{c}}(n) &= \mathbf{\Phi}^{-1}(n-1)\mathbf{\Psi}(n-1) - \mathbf{k}(n)\mathbf{y}^{*T}(n)\mathbf{\Phi}^{-1}(n-1)\mathbf{\Psi}(n-1) \\
&\quad + \mathbf{\Phi}^{-1}(n)\mathbf{y}(n)h^*(n).
\end{aligned}
$$

(7.73)

Upon substituting the vector

$$\mathbf{\Phi}^{-1}(n-1)\mathbf{\Psi}(n-1) = \hat{\mathbf{c}}(n-1)\qquad(7.74)$$

into Equation 7.73 we arrive at:

$$\hat{\mathbf{c}}(n) = \hat{\mathbf{c}}(n-1) - \mathbf{k}(n)\mathbf{y}^{*T}(n)\hat{\mathbf{c}}(n-1) + \mathbf{\Phi}^{-1}(n)\mathbf{y}(n)h^*(n).\qquad(7.75)$$

Exploiting Equation 7.71 simplifies Equation 7.75 even further to:

$$
\begin{aligned}
\hat{\mathbf{c}}(n) &= \hat{\mathbf{c}}(n-1) - \mathbf{k}(n)\mathbf{y}^{*T}(n)\hat{\mathbf{c}}(n-1) + \mathbf{k}(n)h^*(n) \\
&= \hat{\mathbf{c}}(n-1) + \mathbf{k}(n)\left[h^*(n) - \mathbf{y}^{*T}(n)\hat{\mathbf{c}}(n-1)\right] \\
&= \hat{\mathbf{c}}(n-1) + \mathbf{k}(n)i(n)
\end{aligned}
$$

(7.76)

where

$$i(n) = h^*(n) - \mathbf{y}^{*T}(n)\hat{\mathbf{c}}(n-1) \tag{7.77}$$

represents an estimation error term given by the difference of the desired equaliser output $h^*(n)$ and its estimate $\mathbf{y}^{*T}(n)\hat{\mathbf{c}}(n-1)$. In this light the estimation error term of Equation 7.77 is weighted in Equation 7.76 by the gain vector $\mathbf{k}(n)$ in order to update the equaliser coefficient set $\hat{\mathbf{c}}(n-1)$ to arrive at the RLS estimate $\hat{\mathbf{c}}(n)$. Since the estimate $\mathbf{y}^{*T}(n)\hat{\mathbf{c}}(n-1)$ of $h^*(n)$ is based on the previous RLS estimate $\hat{\mathbf{c}}(n-1)$, the estimation error term $i(n)$ in Equation 7.77 is an a priori estimation error, as opposed to the a posteriori estimation error of

$$e(n) = h(n) - \hat{\mathbf{c}}^{*T}(n)\mathbf{y}(n). \tag{7.78}$$

In other words, $i(n)$ is the estimation error, which would arise without updating the equaliser coefficients from $\hat{\mathbf{c}}(n-1)$ to $\hat{\mathbf{c}}(n)$. Due to updating the error reduces to $e(n)$, which was our cost function given in Equation 7.52 in the RLS estimation procedure.

In summary, the RLS equaliser coefficient estimation algorithm is specified by Equations 7.68, 7.77, 7.76 and 7.69, repeated here for convenience:

$$\mathbf{k}(n) = \frac{\beta^{-1}\mathbf{\Phi}^{-1}(n-1)\mathbf{y}(n)}{1 + \beta^{-1}\mathbf{y}^{*T}(n)\mathbf{\Phi}^{-1}(n-1)\mathbf{y}(n)}$$

$$i(n) = h^*(n) - \mathbf{y}^{*T}(n)\hat{\mathbf{c}}(n-1)$$

$$\hat{\mathbf{c}}(n) = \hat{\mathbf{c}}(n-1) + \mathbf{k}(n)i(n)$$

$$\mathbf{\Phi}^{-1}(n) = \beta^{-1}\mathbf{\Phi}^{-1}(n-1) - \beta^{-1}\mathbf{k}(n)\mathbf{y}^{*T}(n)\mathbf{\Phi}^{-1}(n-1).$$

Observe that Equation 7.77 describes how the received signal $\mathbf{y}(n)$ is equalised by the coefficient set $\hat{\mathbf{c}}(n-1)$, giving an equalised output of $\hat{\mathbf{c}}(n-1) \cdot \mathbf{y}^{*T}(n)$, as seen in Figure 7.7. The estimation error $i(n)$ is computed according to Equation 7.77 and it is weighted by the so-called Kalman gain factor $\mathbf{k}(n)$ in the computation of the new set of equaliser coefficients, as portrayed by Equation 7.76. Before the recursion enters a new iteration cycle, the inverse of the correlation matrix $\mathbf{\Phi}(n)$ is updated in harmony with Equation 7.69 and used to compute the next value of $\mathbf{k}(n)$ according to Equation 7.68. The sequence of operations is best illustrated by Haykin's signal flow diagram [3] seen in Figure 7.8.

The fast convergence of the RLS Kalman algorithm is particularly important in data transmission modems for telephone links, where the length of the equaliser training pattern is limited. If adequate equalisation is not established within the training window, a new training procedure is initiated, which might be prolonged indefinitely. The Kalman algorithm can also be used in time division multiple access (TDMA) mobile radio systems, where the channel is sounded in every TDMA burst in order to acquire an accurate estimate of the non-stationary channel impulse response [37]. In this scenario the Kalman algorithm is fed with the locally stored original and the received sounding sequence in order to derive the estimate of the channel impulse response, which can be used for example in a DFE. Cheung [37] followed this approach in a TDMA mobile scenario, evaluating the channel estimation error for various SNR values and tabulated the computational complexity associated with various steps of the algorithm. The convergence rates of the steepest descent or stochastic gradient

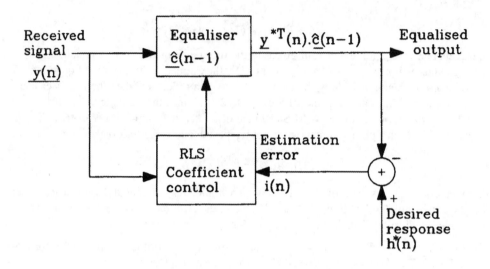

Figure 7.7: Kalman filtering block diagram [3] ©Prentice Hall, 1984, Haykin

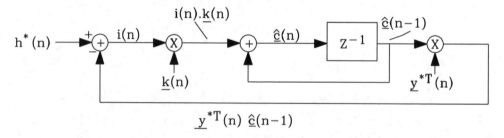

Figure 7.8: Kalman filtering signal flow diagram [3] ©Prentice Hall, 1984, Haykin

and the Kalman algorithm can be characterised by the typical MSE versus iteration index curves [388], as seen in Figure 7.9 where an 11-tap equaliser was used.

7.5 Adaptive Equalisers for QAM

The equalisers discussed so far are suitable in their original form for baseband pulse amplitude modulated (PAM) schemes. If the channel SNR allows the deployment of QAM, higher bit rates can be achieved within the same bandwidth using QAM modems [107,408].

  Figure 7.10 shows the schematic diagram of a QAM system with the square root Nyquist filters $\sqrt{N(\omega)}$ and an adaptive, complex baseband equaliser $C(\omega)$. In Chapter 2 on communications channels we have shown the equivalence of passband and baseband channels. The

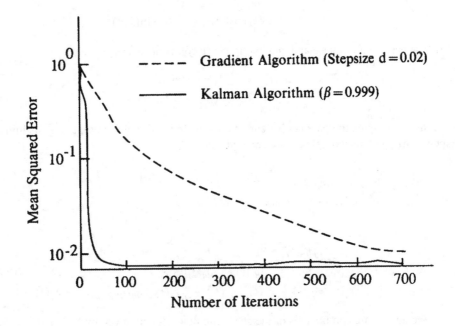

Figure 7.9: Comparison of convergence rate for the Kalman and gradient algorithm [388] ©McGraw-Hill, 1989, Proakis

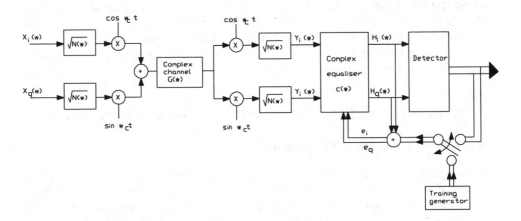

Figure 7.10: QAM system with complex adaptive baseband equaliser

passband modulated signal can be portrayed as follows:

$$
\begin{aligned}
x(t) &= A(t)\cos(\omega_c t + \phi(t)) \\
&= x_i(t)\cos(\omega_c t) - x_q(t)\sin(\omega_c t)
\end{aligned}
\tag{7.79}
$$

where the baseband in-phase and quadrature phase components are given by:

$$
\begin{aligned}
x_i(t) &= A(t)\cos[\phi(t)] \\
x_q(t) &= A(t)\sin[\phi(t)]
\end{aligned}
\tag{7.80}
$$

Exciting the complex lowpass equivalent communication channel represented by its impulse response $g(t)$ with the complex baseband equivalent signal

$$
x(t) = x_i(t) + jx_q(t)
\tag{7.81}
$$

the complex baseband system response becomes:

$$
\begin{aligned}
y(t) &= x(t) * g(t) \\
&= y_i(t) + jy_q(t) \\
&= [x_i(t) + jx_q(t)] * [g_i(t) + jg_q(t)] \\
&= [x_i(t) * g_i(t) - x_q(t) * g_q(t)] + j[x_i(t) * g_q(t) + g_i(t) * x_q(t)]
\end{aligned}
\tag{7.82}
$$

where the quadrature components of the channel output $y(t)$ are given by:

$$
\begin{aligned}
y_i(t) &= x_i(t) * g_i(t) - x_q(t) * g_q(t) \\
y_q(t) &= x_i(t) * g_q(t) + g_i(t) * x_q(t).
\end{aligned}
\tag{7.83}
$$

Using the above complex baseband representation the system model seen in Figure 7.3 is redrawn in terms of complex QAM signals in Figure 7.11. As seen in the figure, the real in-phase equaliser paths $c_i(t)$ cancel the ISI in both of the received components $y_i(t)$ and $y_q(t)$, while the cross-coupling equaliser paths with impulse responses $c_q(t)$ combat the cross-interference between $y_i(t)$ and $y_q(t)$ caused by channel asymmetries, or imperfections in clock and carrier recovery. The price paid for the higher spectral efficiency of the QAM modem is its quadruple equalisation complexity.

7.6   Viterbi Equalisers for Partial Response Modulation

7.6.1   Partial Response Modulation

Although the full treatment of partial response (PR) modulation techniques and that of their equalisation is beyond the scope of this book, for the sake of completeness we briefly highlight the principles of maximum likelihood sequence estimation (MLSE) used for the detection and equalisation of PR systems. The most frequently used MLSE scheme is the Viterbi algorithm (VA), which can also be invoked for the maximum likelihood detection of trellis coded modulation (TCM) that will be the subject of the next chapter. For a full treatment of partial response modulation, Viterbi decoding and Viterbi equalisation, the reader is referred

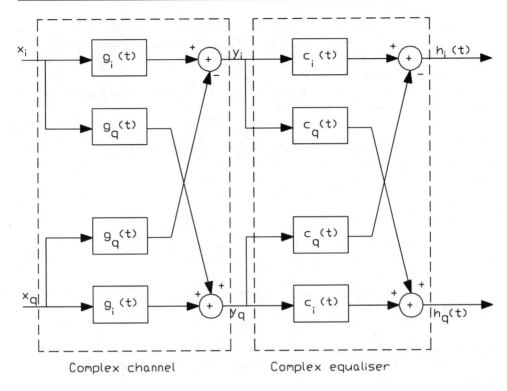

Figure 7.11: Complex baseband channel and equaliser schematic

to Chapters 4 and 6 of [383] or to [369].

The class of constant envelope, continous phase modulation (CPM) schemes is widely used over fading mobile radio channels due to their robustness against signal fading and interference, while maintaining good spectral efficiency. Many of their attractive properties are attributable to the fact that the transmitted signal has a constant envelope, which facilitates the employment of highly power efficient class C amplifiers. In contrast, QAM systems have to use linearised class AB amplifiers, as we highlighted in Chapter 4. The immunity of CPM to fading is also a consequence of the constant signal envelope, since only the phase changes carry information.

The simplest such constant envelope CPM arrangement is minimum shift keying (MSK), where the modulated signal's phase is constrained to change linearly between adjacent signalling instants over the duration of one signalling interval. Hence MSK is a so-called full-response system. However, the derivative of the phase trajectory has discontinuities at the sampling instants due to its step-wise linear nature, indicating that further spectral efficiency gains are possible, when introducing smoother phase changes.

It is plausible that the slower and smoother the phase changes engendered by the modulating signal, the narrower the modulated signal's spectrum and the better the spectral efficiency. Spreading phase changes with maximally flat zero initial and final slopes over several modulation intervals yields a so-called partial response system. Although this PR scheme introduces intersymbol interference (ISI), this can be removed using appropriately designed

channel equalisers, which can cope with the additional deliberately introduced controlled ISI
(CISI). An attractive representative of this PR CPM family is called Gaussian minimum shift
keying (GMSK), where the phase changes are spread typically over two to four signalling
intervals using a Gaussian filter [383].

The schematic of the GMSK modulator is portrayed in Figure 7.13. The impulse response
$g(t)$ of the Gaussian filter is given [434] by:

$$g(t) = \frac{1}{2T} \left[Q \left(2\pi B_b \frac{t - T/2}{\sqrt{\ell n2}} \right) - Q \left(2\pi B_b \frac{t + T/2}{\sqrt{\ell n2}} \right) \right] \tag{7.84}$$

for

$$0 \le B_b T \le \infty$$

where $Q(t)$ is the Q-function

$$Q(t) = \int_t^\infty \frac{1}{\sqrt{2\pi}} \exp(-\tau^2/2) d\tau \tag{7.85}$$

B_b is the 3 dB down bandwidth of the Gaussian phase shaping filter, T is the signalling
interval duration and

$$B_N = B_b T \tag{7.86}$$

is the normalised bandwidth.

The Gaussian impulse response $g(t)$ is shown in Figure 7.12 for different normalised
bandwidths B_N.

Observe in Figure 7.13 that the Gaussian filter is preceeded by an integrator, which converts the data-induced input frequency changes to phase changes $\phi(t, \alpha_n)$. Taking the sin
and cos of this phase yields the base band equivalent in-phase and quadrature-phase modulating signals, which modulate two quadrature carriers. Finally, the orthogonal quadrature
components are added and transmitted in the same frequency band.

7.6.2 Viterbi Equalisation

A set of standardised dispersive wideband mobile radio channels used in the pan-European
mobile radio system, known as GSM was introduced in Figure 2.21. Since dispersive wideband channels can be described by impulse responses extending over a number of signaling
intervals, their uncontrolled ISI must be counteracted by the channel equaliser. Clearly, the
equaliser now has to have a longer memory in order to be able to remove the superimposed
effects of the channel impulse response plus the CISI introduced by the PR GMSK scheme.

If, however, the transmitted signal's bandwidth is narrow compared to the channel's coherence bandwidth (B_c), all transmitted frequency components encounter nearly identical
propagation delays, i.e. the so-called narrowband condition is met and the signal is subjected to non-frequency-selective or flat envelope fading. In this case the channel's impulse
response is an ideal Dirac delta, implying no channel-induced ISI and the equaliser has to
remove solely the CISI.

Both during and after the definition of the GSM standards a number of Viterbi equaliser
(VE) implementations have been proposed in the literature, which have different complexities
and performances [435, 436]. A simple general VE block diagram is shown in Figure 7.14.

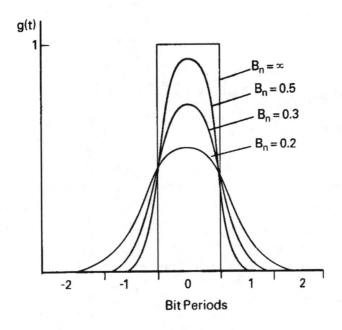

Figure 7.12: Gaussian impulse response $g(t)$ of GMSK for various normalised bandwidths B_N [383] ©Pentech Press, 1992, Steele

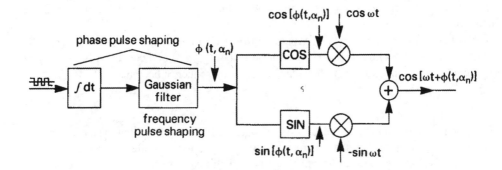

Figure 7.13: GMSK modulator schematic diagram

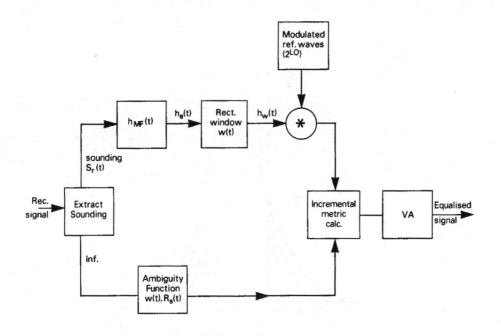

Figure 7.14: MLSE block diagram

In conventional time division multiple access (TDMA) mobile systems, such as the GSM system, communication is maintained using transmission bursts, incorporating a so-called midamble sequence in the centre of the burst. This periodically transmitted Dirac delta-like sequence has a sharply decaying autocorrelation sequence, which allows the system to measure the channel's impulse response and hence assists in the equalisation process. Furthermore, since the midamble is typically a 16-bit unique word, its detection using autocorrelation facilitates frame synchronisation. This synchroniser concept was briefly introduced in Section 6.2.4.

Explicitly, the modulated TDMA burst with the channel sounding sequence $s(t)$ in its centre is convolved with the channel's impulse response $h_c(t)$ and corrupted by noise. Neglecting the noise for simplicity, the received sounding sequence becomes:

$$s_r(t) = s(t) * h_c(t). \tag{7.87}$$

Then the received sounding sequence $s_r(t)$ is matched filtered (MF) using the causal impulse response h_{MF}. (For further notes on matched filtering the reader is referred to Section 4.4.2.) In order to derive the estimate $h_e(t)$ of the channel's impulse response we can then use the following convolution:

$$h_e(t) = s_r(t) * h_{MF}(t) = s(t) * h_c(t) * h_{MF}(t) = R_s(t) * h_c(t), \tag{7.88}$$

where $R_s(t)$ is the sounding sequence's autocorrelation function. If $R_s(t)$ is a highly peaked

Dirac impulse-like function, then its convolution with $h_c(t)$ becomes $h_e(t) \approx h_c(t)$. With $s(t)$ in the middle of the TDMA burst, the estimated channel impulse response $h_c(t)$ can be considered quasi-stationary for the burst duration, and can be used to equalise the useful information bits at both sides of it. The time variant channel typically precipitates higher bit error rates towards the burst edges, where the impulse response estimate becomes less accurate due to the channel's quasi-stationarity.

Since the complexity of the VE increases exponentially with the number of signalling intervals over which the received signal can be spread due to ISI and CISI, the estimated channel response $h_c(t)$ has to be windowed to a computationally affordable length using the rectangular function $w(t)$. However, it has to retain a sufficiently long memory to compensate for the typical channel impulse responses encountered. Specifically, in addition to the duration L_{CISI} of the controlled ISI, the channel's delay spread L_c also has to be considered in calculating the required observation interval $L_o = L_{CISI} + L_c$ of the 2^{L_o-1}-state VE.

In order to augment our exposition, we invoke the GSM system's example. When using a bit interval of 3.69 μs corresponding to the GSM TDMA burst rate of about 270 kbit/s and a maximum channel impulse response duration of around 15–20 μs, as seen in Figure 2.21, a VE with a memory of 4-6 bit intervals is a good compromise. Note that $h_c(t)$ might have an excessively long decay time and hence for practical reasons it is retained over that particular 4-6 bit interval of its total time domain support length, where it is exhibiting the highest energy. In general, L_o consecutive transmitted bits give rise to 2^{L_o} possible transmitted sequences, which are first input to a local modulator to generate the modulated waveforms, and then convolved with the windowed estimated channel response $h_w(t)$ in order to derive the legitimate reference waveforms for metric calculation, as portrayed in Figure 7.14.

Recall that the condition $h_e(t) = h_c(t)$ is met, i.e. the estimated impulse response is identical to the true channel impulse response only, if $R_s(t)$ is the Dirac delta function. This condition cannot be met when finite-length sounding sequences are used. The true channel response $h_c(t)$ could only be computed by deconvolution from Equation 7.88 upon neglecting the rectangular window $w(t)$, which does not give a unique decomposition. Alternatively, the received signal can be convolved for the sake of metric calculation with the known windowed autocorrelation function $w(t) \cdot R_s(t)$ often referred to as ambiguity function, as seen in the lower branch of Figure 7.14 after extracting the sounding sequence from the received TDMA burst. Clearly, this way the received signal is "pre-distorted" using the ambiguity function, identically to the estimated impulse response in Equation 7.88. This filtered signal is then compared to all possible locally generated reference signals and the incremental metrics m_i, $i = 0, \ldots, (2^{L_o-1})$ are computed. Finally, the incremental metrics are utilised by the Viterbi algorithm (VA) in order to determine the maximum likelihood transmitted sequence, as explained in references [383, 436]. Suffice to note here that the maximum likelihood transmitted sequence is identified on the basis of the metrics by finding the specific reference sequence that has the lowest deviation from the currently processed received sequence. The same procedure can be used in order to infer the maximum likelihood transmitted trellis coded sequence in the next chapter.

7.7 Overview of Blind Equalisers

S. Vlahoyiannatos and L. Hanzo

7.7.1   Introduction

Blind channel equalisation has been attracting scientific interest since 1975, when Sato published his original study of a scheme now known as the "Sato algorithm" [49]. Following Sato's study, the topic of blind equalisation attracted more academic than practical attention. In recent years researchers have intensified their efforts in this field due to the definition of a range of high-importance video broadcast standards. The Pan-European Satellite-based Digital Video Broadcast (DVB-S) [437], the terrestrial DVB-T [438] and the cable-based DVB-C [439] systems constitute a family of harmonised systems, where both the DVB-S and the DVB-C schemes recommend blind equalisation. In this review, we offer a comprehensive overview of these techniques, focusing on two particular families of methods, namely on the so-called Bussgang algorithms and on the joint channel impulse response (CIR) and data sequence estimation. This section is structured as follows. In Subsection 7.7.2 we provide a brief historical perspective. The basic principles of blind equalisation are discussed in Subsection 7.7.3, followed by an overview of the so-called Bussgang equalisers in Subsection 7.7.4. Subsection 7.7.6 is dedicated to convergence results. In Subsection 7.7.7 joint data and channel estimation techniques are discussed. The so-called second-order statistics-based algorithms' principles are highlighted in Section 7.7.8. The basic principles of the fourth-order statistics-based channel estimation and equalisation algorithms using the so-called "polycepstra" are summarised in Subsection 7.7.9. Complexity comparisons are provided in Subsection 7.7.10. Finally, in Subsection 7.7.11 the performance of these equalisers is studied comparatively, while decision-directed improvements are reviewed in Subsection 7.7.12. Let us now commence our discourse with a historical perspective.

7.7.2   Historical Background

As mentioned above, blind equalisation was originally proposed by Sato [49]. Five years after Sato's publication, the blind equalisation problem was further studied for example by Benveniste, Goursat and Ruget in [50], where several blind equalisation issues were clarified and a new algorithm was proposed [51]. At the same time Godard [52] introduced a criterion, namely the so-called "constant modulus" (CM) criterion, leading to a new class of blind equalisers. Following Godard's contribution a range of studies were conducted employing the constant modulus criterion. Foschini [53] was the first researcher studying the convergence properties of Godard's equaliser upon assuming an infinite equaliser length. Later, Ding *et al.* continued this study [54] and provided an indepth analysis of the convergence issue in the context of a realistic equaliser. Although numerous researchers studied this issue, nevertheless, a general solution is yet to be found. A plethora of authors have studied Godard's equaliser, rendering it the most widely studied blind equaliser. A well-known algorithm of the so-called Bussgang type [55,56] was also proposed by Picchi and Prati [57]. Their "Stop-and-Go" algorithm constitutes a combination of the Decision-Directed algorithm [3] with Sato's algorithm [49]. After 1991, a range of different solutions to the blind equalisation problem were proposed. Seshadri [58] suggested the employment of the so-called M-algorithm, as a "substitute" for the Viterbi algorithm [59] for the blind scenario, combined with the so-called "least mean squares (LMS)" based CIR estimation. This CIR estimation was replaced by "recursive least squares (RLS)" estimation by Raheli, Polydoros and Tzou [60], combining the associated convolutional decoding with the CIR estimation, leading to what was termed as *"Per-Survivor Processing"*. Since then a number of papers have focused on this technique [60–69]. At the same time as Seshadri, Tong *et al.* [70] proposed a different approach to blind equalisation, which used oversampling in order to create a so-called "cyclostationary" received signal, and performed CIR estimation by measuring the autocorrelation function of this signal and by exploiting this signal's cyclostationarity. This technique was also applied to the case of 'sampling' the received signals of different antennas (instead of oversampling the signal of a single antenna) and further extended by Moulines *et al.* using a different method of CIR estimation, namely the so-called subspace method in [71]. Furthermore, Tsatsanis and Giannakis suggested that the cyclostationarity can be induced by the transmitter upon transmitting the signal more than once [72]. A number of further contributions have also been published in the context of these techniques [73–88]. Finally, nearly coincidentally with Seshadri [58] and Tong *et al.* [70], Hatzinakos and Nikias [89] proposed a more sophisticated approach to blind equalisation by exploiting the so-called "tricepstrum" of the received signal. Despite its long history, at the time of writing blind equalisation still constitutes a burgeoning research topic, especially in the context of novel applications, such as blind space-time equalisation, for example. A general answer to the fundamental question *"Under what circumstances is it preferable to use a blind equaliser to a trained-equaliser ?"* is yet to be provided. Despite the scarcity of reviews on the topic, in the context of the Global System of Mobile Communications known as GSM an impressive effort was made by Boss, Kammeyer and Petermann [90], who also proposed two novel blind algorithms. We recommend furthermore the fractionally-spaced equalisation review of Endres *et al.* [91] and the Constant Modulus overview of Johnson *et al.* [92] based on a specific type of equalisers, namely on the so-called "fractionally-spaced" equalisers. A review of subspace-ML multichannel blind equalisers was provided by Tong and Perreau [93]. Further important references are the monograph by Haykin [3], the relevant section by Proakis [94] and the blind deconvolution book due to Nandi [95]. Comparative performance studies between various blind equalisers have also been performed. We recommend the second-order statistics-based comparative performance studies of Becchetti *et al.* [96], Kristensson *et al.* [97] and Altuna *et al.* [98], which is based on the mobile environment as well as the second-order statistics and PSP-based comparative study of Skowratanont and Chambers [99]. Furthermore, we recommend the fractionally-spaced Bussgang algorithm based comparative performance study by Shynk *et al.* [100], the CMA comparative performance study of Schirtzinger *et al.* [101] and the comparative convergence study by Endres *et al.* [102]. Let us now review the basic principles of blind equalisation in the next section.

7.7.3 Blind Equalisation Principles

In this section we will introduce and discuss a range of blind equalisation principles, mainly focusing on the Bussgang techniques [55, 56]. Since Sato's original study in 1975 [49], blind equalisation has attracted significant scientific interest due to its potentials in terms of:

- Overhead reduction. Training sequences sacrifice bandwidth in order to assist in determining the CIR, hence assisting equalisation. Blind equalisers do not need training sequences and therefore conserve bandwidth.

- Simplification of point to multipoint communications systems or broadcast. When a communications link is reset, equaliser adjustment "from scratch" is necessary. In this

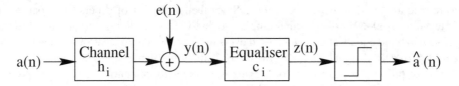

Figure 7.15: Equalised communications system

case, using a training sequence is inefficient, since the transmitter has to retransmit the training sequence specifically for each receiver, which is reset.

We will now proceed to describe the problem of blind equalisation. In our discussion the communication system parameters will be assumed to be time-invariant. This constraint is not necessary in general, although it is beneficial in terms of complexity reduction. Furthermore, the additive noise will be assumed to be Gaussian and white. Let us assume that the input bits, resulting from any encoder prior to modulation, are mapped into complex Quadrature Amplitude Modulation (QAM) [440] input symbols $a(n)$ transmitted at time instant n. These transmitted symbols are filtered by the CIR h_i and then the noise $e(n)$ is added to them, resulting in the received symbols $y(n)$ at time instant n in the form of:

$$y(n) = \sum_{i=-L_1}^{L_2} h_i \cdot a(n-i) + e(n),$$
(7.89)

where L_1 and L_2 are the length of the CIR's pre- and post-cursor sections surrounding the main tap (measured in terms of the number of transmitted symbols), respectively. An equaliser is typically placed after the channel in order to remove the channel-induced dispersion, as shown in Figure 7.15. Blind equalisation involves finding the 'best' equaliser filter, which regenerates the input symbols $a(n)$ at the receiver, without any knowledge of the CIR upon exploiting the knowledge of the distribution of the input QAM symbols. If the equaliser has N_1 feedback and N_2 feedforward taps $\{c_i\}$, then the equalised symbols will be of the form:

$$z(n) = \sum_{i=-N_1}^{N_2} c_i \cdot y(n-i)$$
(7.90)

and upon using Equation 7.89 we have:

$$z(n) = \sum_{i=-L_1-N_1}^{L_2+N_2} t_i \cdot a(n-i) + \sum_{i=-N_1}^{N_2} c_i \cdot e(n-i),$$
(7.91)

where $\{t_i\} = \{h_i\} * \{c_i\}$ is the convolution of the CIR with the equaliser filter, representing the total transfer function of the cascaded system constituted by the channel plus the equaliser. Assuming that the noise power is low compared to the power of the received signal, we can observe that the blind equalisation problem corresponds to estimating the suitable equaliser impulse response, which reduces the first term at the right-hand side of Equation 7.91 to only one of the summation terms. In this case, the cascaded system's impulse response tap-vector $\mathbf{t}$ takes the form of:

$$\mathbf{t} = (0, \cdots, 0, A, 0, \cdots, 0), \tag{7.92}$$

where A is a constant. This is the only case that corresponds to zero inter-symbol interference (ISI). Assuming that the strongest signal path is located at time instant 0, Equation 7.91 can also be expressed as:

$$z(n) = t_0 \cdot a(n) + \sum_{i=-L_1-N_1,\ i\neq 0}^{L_2+N_2} t_i \cdot a(n-i) + \sum_{i=-N_1}^{N_2} c_i \cdot e(n-i). \tag{7.93}$$

The first term of Equation 7.93 is the useful one, including the one and only path at time 0. The second term is the ISI term, which is also referred to as *convolutional noise* [441]. This is because this term is a noise term, as far as the receiver is concerned and since it is the result of the convolution of the CIR h_i and the equaliser's impulse response c_i with the input signal $a(n)$. This is usually the main noise contributor at the initialisation of the equalisation process and it is reduced further during the stages of the equalisation process, leaving only the real noise term as the sole signal impairment, when the equalisation is perfect. Finally, the third term of Equation 7.93 is the noise term $e(n)$, convolved with the equaliser's impulse response c_i, since the noise has been filtered by the equaliser.

A problem similar to blind equalisation is the problem of blind deconvolution [442]. In blind deconvolution the aim is to perform joint channel and data estimation in a non-real time fashion, which is a fundamental difference with respect to the blind equalisation problem. The received signal is stored and then deconvolved in order to produce the CIR together with the input signal. This is the case, for example, when seismic signals are considered [443]. We have no knowledge of the signal emerging from the crust of the earth or of the channel that intermediates between the source of this signal and our receiver, hence we attempt to record the signal in order to deconvolve it later. When a channel estimation is available to us, we can estimate the input signal in two ways. A feasible approach is to use a sequence estimation technique, such as the Viterbi algorithm [59]. Another approach is to produce an inverse filter of the channel and to perform filtering with the aid of this inverse filter. We will discuss this technique in more detail in Section 7.7.7.

Let us now point out an important aspect of channel equalisation, which was observed by Benveniste *et al.* in [50], noting that if the distribution of the equaliser's input signal is Gaussian, then the equalised signal of Equation 7.91 is also Gaussian distributed. This is true even if there are more than one non-zero terms in the sum, i.e. even if ISI exists. Conversely, provided that the distribution of the equaliser's input signal is Gaussian, the equalised signal constituted by a sum of "independent identically distributed (i.i.d.)" Gaussian variables is also a Gaussian variable irrespective of whether there is residual ISI at its output. In this case, equalisation is impossible, since the equaliser cannot distinguish the zero-ISI equalised signal sequence from the other possible candidate sequences, which contain ISI. In other words, it is impossible to equalise the received signal, since the zero-ISI candidate received signal sequence is indistinguishable from the other ISI-contaminated candidate sequences. Hence blind equalisation cannot be performed, if the distribution of the equaliser's input signal is Gaussian. Fortunately, the uniform distribution of typical QAM data sources is far from the Gaussian, which renders the received signal amenable to blind equalisation.

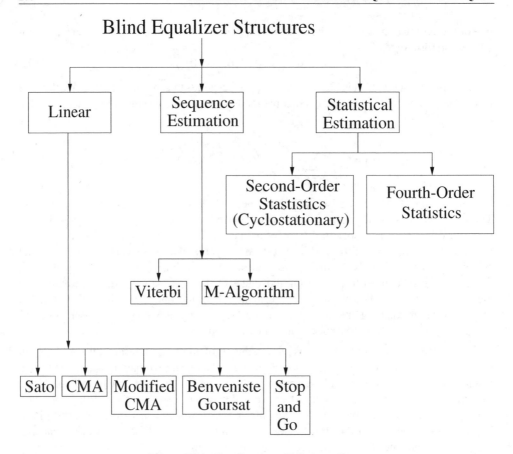

Figure 7.16: Classification of blind equalisers

The equalisers alluded to in this section are classified in Figure 7.16. Blind channel identifiability issues have been discussed in [50] and [444]. Following the above general discussion on blind equalisibility, let us now proceed with the overview of the family of Bussgang techniques in the next section.

7.7.4   Bussgang Blind Equalisers

In this section we set out to characterise a range of basic equaliser schemes involving the class of "*Bussgang*" techniques dating back to 1952 [56]. In our discussion we will consider the following signal vectors, which obey the notations of Figure 7.15:

$$\mathbf{a}(n) = [a(n + N_1 + L_1), \ldots, a(n - N_2 - L_2)]^T \tag{7.94}$$
$$\mathbf{y}(n) = [y(n + N_1), \ldots, y(n - N_2)]^T \tag{7.95}$$
$$\mathbf{c} = [c_{-N_1}, \ldots, c_{N_2}]^T \tag{7.96}$$
$$\mathbf{t} = [t_{-N_1-L_1}, \ldots, t_{N_2+L_2}]^T \tag{7.97}$$
$$\mathbf{e}(n) = [e(n + N_1 + L_1), \ldots, e(n - N_2 - L_2)]^T \tag{7.98}$$

$$\mathbf{H} = \begin{pmatrix} h_{-L_1} & \cdots & h_{L_2} & 0 & \cdots & 0 & 0 \\ 0 & h_{-L_1} & \cdots & h_{L_2} & 0 & \cdots & 0 \\ \vdots & \ddots & \ddots & \ddots & \ddots & \ddots & \vdots \\ 0 & \cdots & 0 & h_{-L_1} & \cdots & h_{L_2} & 0 \\ 0 & 0 & \cdots & 0 & h_{-L_1} & \cdots & h_{L_2} \end{pmatrix} \tag{7.99}$$

where T denotes transpose, $*$ denotes conjugate and H denotes the Hermitian matrix. From these definitions the following matrix relationships hold:

$$\mathbf{y}(n) = \mathbf{H} \cdot \mathbf{a}(n) \tag{7.100}$$
$$\mathbf{t} = \mathbf{H}^T \cdot \mathbf{c} \tag{7.101}$$
$$z(n) = \mathbf{c}^T \cdot \mathbf{y}(n) \tag{7.102}$$
$$z(n) = \mathbf{t}^T \cdot \mathbf{a}(n) + \mathbf{c}^T \cdot \mathbf{e}(n). \tag{7.103}$$

Explicitly, Equation 7.100 describes the received signal $y(n)$ as the convolution of the transmitted signal $a(n)$ and the CIR h_i. Equation 7.101 reflects the convolution of the CIR with the equaliser's impulse response and Equation 7.102 characterises the equaliser's output signal $z(n)$ as the convolution of the received signal $y(n)$ with the equaliser's impulse response c_i. Following these definitions, a general form of the Bussgang equaliser update formulae can be expressed as [56]:

$$\mathbf{c}^{(n+1)} = \mathbf{c}^{(n)} - \lambda \cdot \mathbf{y}^*(n) \cdot [z(n) - g\{z(n)\}], \tag{7.104}$$

where $g\{z(n)\}$ is a non-linear zero-memory function of the equalised output $z(n)$ and λ is the so-called "step-size" parameter, controlling the speed and the accuracy of the equaliser's convergence. The condition for attaining convergence in the mean value for these algorithms is [3]:

$$E[z(n) \cdot \mathbf{y}^*(n)] = E[g\{z(n)\} \cdot \mathbf{y}^*(n)] \tag{7.105}$$

or

$$E[z(n) \cdot y^*(n - i)] = E[g\{z(n)\} \cdot y^*(n - i)], \quad i = -N_1, \cdots, N_2. \tag{7.106}$$

Upon multiplying each side of this equation by the relevant equaliser tap coefficient c_i^* and summing the results for $i = -N_1, \cdots, N_2$ as in [3], we obtain:

$$E[z(n) \cdot z^*(n)] = E[g\{z(n)\} \cdot z^*(n)]. \tag{7.107}$$

If instead of multiplying by c_i^* we had multiplied by c_{i-k}^*, assuming that the equaliser has an infinite number of taps, then Equation 7.107 would take the form of:

$$E[z(n) \cdot z^*(n-k)] \simeq E[g\{z(n)\} \cdot z^*(n-k)], \tag{7.108}$$

which reflects the so-called *Bussgang property* that is satisfied by the Bussgang algorithms, when the equaliser length is doubly infinite. When both the feedforward and feedback equaliser lengths are sufficiently high, then the Bussgang property is approximately satisfied. In the strict sense, however, only Equation 7.107 is satisfied. When the equaliser has converged, then the equalised symbols $z(n)$ approximate the transmitted symbols $a(n)$ and Equation 7.107 becomes:

$$E[z(n) \cdot g\{z(n)\}] = E[|a(n)|^2] = 1, \tag{7.109}$$

provided that the input power is normalised.

The problem of finding the optimum Bussgang equaliser corresponds to finding the function $g\{z(n)\}$, which provides the best estimate of the corresponding transmitted symbol $a(n)$ for each equalised symbol $z(n)$, while satisfying Equation 7.109. According to the Maximum Likelihood (ML) criterion, this can be achieved by setting $g\{z(n)\} = E[a(n)|z(n)]$, i.e. setting $g\{z(n)\}$ equal to the expected (or most probable) value of the transmitted symbol, given the equalised symbol $z(n)$. In order to find this expected value, we have to estimate the distribution $f_z(z)$ of the equalised symbols $z(n)$. During the equaliser's initialisation, in general the equalised signal contains ISI. If during this initialisation phase we ignore the channel noise by assuming that the ISI is the main signal impairment at this stage, then this means that the equalised signal consists of a number of transmitted signal replicas, each having a different delay and weight. When the number of these replicas is sufficiently high, according to the central limit theorem we can approximate the distribution of the equalised symbols $z(n)$ with a Gaussian distribution. In our analysis in this section we shall assume that the central limit theorem condition can be invoked since there is a sufficiently high number of ISI terms. Assuming also M-level QAM transmissions, the expected value giving the estimate of $g\{z(n)\}$ obeys the following form, which is similar to the one given in [442] for pulse-amplitude modulation (PAM):

$$g_{ML}\{z(n)\} = E[a(n)|z(n)] = \frac{\sum_{i=1}^{M} A_i e^{-|z(n)-A_i|^2/2\sigma^2}}{\sum_{i=1}^{M} e^{-|z(n)-A_i|^2/2\sigma^2}} \tag{7.110}$$

where the coefficients A_i constitute the signal amplitudes associated with the QAM constellation and σ^2 is the variance of the noise, consisting of two components, namely the convolutional noise and the Additive White Gaussian Noise (AWGN) induced by the channel. An estimate of the noise variance σ^2 must be available for the evaluation of Equation 7.110. However, the function $g\{z(n)\}$ is only the optimum one under the plain assumption of a Gaussian distribution for the above-mentioned composite noise, which is produced by the ISI plus the channel's additive noise. Depending on this distribution, different Bussgang algorithms exist. The well-known Godard (or CMA) [52], Sato [49], Benveniste-Goursat [50] or Stop-and-Go [57] algorithms constitute a few such algorithms. We will describe each of them in the forthcoming paragraphs and discuss their characteristics. As an illustration, in Figure 7.17 we have plotted the real part of the function $g\{(z)\}$, evaluated from Equation 7.110, for

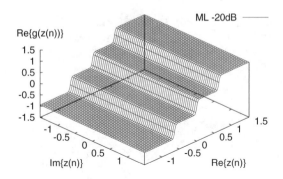

Figure 7.17: The real part of the function $g\{z(n)\}$ plotted against the complex $z(n)$−plane for ML detection assuming Gaussian noise with -20dB variance for 16-QAM

the ML algorithm of $g_{ML}\{z(n)\} = E[a(n)|z(n)]$ under the assumption of $-20dB$ additive Gaussian noise power. The real part of a 16-QAM signal can take four discrete values, symmetrically distributed around the origin. The estimated signal, which approximates the most likely transmitted signal, should be close to these legitimate constellation points. We observe that this is the case, when the above ML algorithm is used, under the assumption of low noise ($-20dB$). In this low-noise scenario, the ML algorithm estimates the transmitted signal as the constellation symbol, which is closest to the equalised symbol at time instant n $z(n)$. This is illustrated in Figure 7.17 by the four levels, corresponding to the four discrete values that the real (or the imaginary) part of a 16-QAM signal can assume. If the noise variance is not sufficiently low, however, then the surface of Figure 7.17 loses its resemblance to the 16-QAM constellation. This is shown in Figure 7.18, where the noise variance was assumed to be $-10dB$, i.e. $10dB$ higher than in Figure 7.17. When no CIR information is available, it might be extremely optimistic to assume that, even without channel noise, the power of the ISI-induced noise would be as low as $-20dB$. In fact it might be well over $0dB$, thus rendering the 16-QAM pattern unrecognisable in the received signal. The approximation of the estimation function $g\{z(n)\}$ for each Bussgang algorithm is given in a 2D plot in [442] for 8-level PAM, where the signals are real-valued. Similar figures can be generated by extending the approach of [442] to the case of 16-QAM, where the signals assume complex values, as was seen in the 3D plots of Figures 7.17 and 7.18.

An alternative interpretation of the function $g\{z(n)\}$ can be observed by considering a "cost-function" $J\{z(n)\}$. The minimisation of this cost-function leads to the desired equaliser tap values, according to the wide-spread steepest descent algorithm [3]:

$$\mathbf{c}^{(n+1)} = \mathbf{c}^{(n)} - \lambda \cdot \frac{\partial J\{z(n)\}}{\partial \mathbf{c}}, \tag{7.111}$$

which physically implies that the taps $\mathbf{c}^{(n)}$ at instant n are modified by the derivative of the cost-function – after weighting by the step-size λ – in the direction of minimizing the cost-

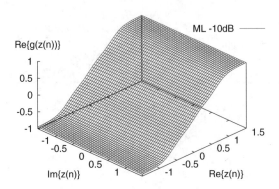

Figure 7.18: The real part of the function $g\{z(n)\}$ plotted against the complex $z(n)-$plane for ML detection assuming Gaussian noise with -10dB variance for 16-QAM

function. According to Equations 7.111 and 7.104 the following relationship holds between $g\{z(n)\}$ and $J\{(z(n)\}$:

$$\frac{\partial J\{z(n)\}}{\partial \mathbf{c}} = \mathbf{y}^*(n) \cdot (z(n) - g\{z(n)\}). \tag{7.112}$$

Again, in simple, but conceptually feasible terms Equations 7.111 and 7.112 can be interpreted as updating each tap of the equaliser on the basis of the gradient of the error term

$$\epsilon(n) = z(n) - g\{z(n)\} \tag{7.113}$$

with respect to (wrt) a specific tap. Depending on the polarity of the cost-function's derivative wrt a specific tap, this tap is updated according to the step-size λ, such that in the next step it reduces $\epsilon(n)$ – hence the negative sign in Equation 7.111. Upon using the differentiation rules with respect to a vector given in Appendix 7.9, we can readily arrive at:

$$g\{z(n)\} = z(n) - J'\{z^*(n)\}, \tag{7.114}$$

where $'$ denotes the derivative and the associated difference quantifies the discrepancy of the equalised output $z(n)$ and the derivative of the cost-function given by Equation 7.112. Equation 7.114 will be useful, when we consider the Bussgang cost-functions individually and derive the corresponding equaliser tap update algorithms.

Again, in all of our discussions, we employ QAM [440], [441] and the general structure of the equaliser is shown in Figure 7.19. As we can see from this figure, the blind equaliser coefficients are updated using the knowledge of the received signal vector $\mathbf{y}(n)$, the equalised signal $z(n)$, the phase-corrected equalised signal $\tilde{z}(n) = z(n)e^{-j\phi}$ and the estimated signal $\hat{a}(n)$. We note, however, that a specific Bussgang equaliser may not make use of all of these signals in order to update its equaliser tap coefficients. It depends on the algorithmic implementation, which of these signals are invoked in the tap update process. What is common,

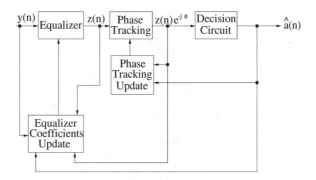

Figure 7.19: Equaliser structure used in Bussgang techniques

however, to all Bussgang algorithms is that the error estimate in Equation 7.113 will be a function of the equalised symbol $z(n)$ at time n only, i.e. they are based on a zero-memory estimation.

In the equalisers presented in this section the sampling rate is identical to the signalling-rate – or Baud-rate, that is we use only one sample per symbol period. These equalisers are referred to as *symbol-spaced* schemes, as opposed to *fractionally-spaced equalisers*, which use more than one sample per symbol in order to equalise the channel. A typical example of fractionally-spaced equalisers is constituted by the family of second-order cyclostationary statistics based blind channel estimation algorithms [445]. In the context of Bussgang schemes, the extension of these equalisers to fractionally spaced arrangements is relatively straightforward. Such algorithms have been reported in the literature for example by Pei and Shih in [446] or by Dogancay and Kennedy in [447]. They have been further studied for example by Endres, Johnson and Green in [448], by LeBlanc, Fijalkow and Johnson in [449], by Endres, Halford, Johnson and Giannakis in [91], by Magarini *et al.* in [450] and by Papadias and Slock in [451].

Before proceeding to the discussion of Sato's algorithm, we note that the Bussgang zero-memory function $g\{z(n)\}$ of Equation 7.104 can be extended to the non-zero-memory case, if we take into consideration more than one equalised symbols in generating the error function of Equation 7.113. This was proposed by Yang for the CMA in [452]. Let us now consider a range of Bussgang algorithms in a little more depth in the forthcoming subsections.

7.7.4.1 Sato's Algorithm [49]

Sato's pioneering contribution in 1975 [49] described the first blind equalisation algorithm proposed, which was designed for real-valued signals and PAM. However, its extension to complex-valued signals and QAM is straightforward, especially in the spirit of Godard's publication of the well-known CMA [52], which was derived for complex-valued QAM signals. Sato's algorithm dedicated to real valued signals $z(n)$ uses the following cost-function [49]:

$$J^S(n) = E\left[(|z(n)| - \gamma)^2\right], \tag{7.115}$$

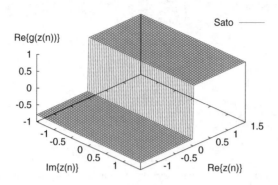

Figure 7.20: The real part of function $g\{z(n)\}$ plotted against the real part of the equalised symbol $z(n)$ for Sato's algorithm. The modulation used is 16-QAM

where γ is Sato's scaling coefficient and $E[]$, again, represents the expectation over all possible transmitted data sequences. It is clear that this cost-function is forcing the absolute value of the equalised signal to a fixed value γ. This is a plausible policy to pursue, when Binary Phase Shift Keying (BPSK) is used, but not for any other multilevel PAM scheme. For these multilevel constellations the minimisation of the Sato cost-function of Equation 7.115 may not seem to lead to correct update of the equaliser taps at each iteration. Nonetheless, experimental experience shows that the minimization of Equation 7.115 may still lead to convergence to the desired zero-ISI equilibrium, although not in all cases, as we shall see in Section 7.7.6. The associated complex Sato cost-function can be defined as in [51]:

$$J^S(n) = E\left[(|Re\{z(n)\}| - \gamma)^2\right] + E\left[(|Im\{z(n)\}| - \gamma)^2\right]. \qquad (7.116)$$

The steepest descent algorithm – which results from the cost-function of Equation 7.116 – can be found by determining the gradient of the cost-function with respect to the equaliser tap vector **c**. Alternatively, using Equation 7.114 we obtain:

$$g_{Sato}\{z(n)\} = csgn(z(n)), \qquad (7.117)$$

where the complex "signum" function is given by:

$$csgn(z(n)) = sgn(Re\{z(n)\}) + j \cdot sgn(Im\{z(n)\}). \qquad (7.118)$$

In Figure 7.20 a plot similar to these in Figures 7.17 and 7.18 is given for Sato's algorithm. Clearly, this surface does not follow the four legitimate values of the 16-QAM constellation pattern of Figure 7.17, it is constituted by two planes. All the other Bussgang algorithms follow a pattern similar to that of Sato's algorithm and none of them follows the specific 16-QAM constellation pattern of the ML algorithm seen in Figure 7.17. Sato's algorithm [49] is thus given, according to Equation 7.104 by:

$$\mathbf{c}^{(n+1)} = \mathbf{c}^{(n)} - \lambda \cdot \mathbf{y}^*(n) \cdot \epsilon^{Sato}(n), \tag{7.119}$$

where $\epsilon^{Sato}(n)$ is the Sato-error defined as:

$$\epsilon^{Sato}(n) = z(n) - \gamma \cdot csgn(z(n)). \tag{7.120}$$

Explicitly, the tap vector $\mathbf{c}$ in Equation 7.119 is adjusted according to the correction term $\lambda \cdot \mathbf{y}^*(n) \cdot \epsilon^{Sato}(n)$, where Sato's error term $\epsilon^{Sato}(n)$ depends on the cost-function of Equation 7.116. As it can be seen from Equation 7.120, this algorithm uses only the sign of the equalised output values $z(n)$ in order to update the equaliser coefficients. This implies that the exact value of $z(n)$ is ignored. Clearly, an error in the polarity of $z(n)$ is less probable than an error in its exact value, when compared to the actual transmitted value $a(n)$. Therefore, Sato's algorithm has the advantage that it avoids using the generally error-prone exact value of $z(n)$, in favour of invoking the less spurious polarity of it. In the case of symmetric multilevel PAM transmissions, for which the algorithm was originally proposed, $Re\{z(n)\}$ assumes equi-probable positive and negative values. Using suitable coding and taking into account only the sign of $z(n)$ implies ignoring the fine-resolution channel effects. The same idea can be adopted for QAM transmissions, where $Re\{z(n)\}$ and $Im\{z(n)\}$ can be treated as two independent PAM constellations.

Setting the value of the scaling coefficient γ in Equation 7.119 is very important, since it actually directs the signal $z(n)$ to the point of its convergence, i.e. to the original constellation points. A way of achieving this is by constraining the mean value of the error update term of Equation 7.119 to zero. Therefore, the optimum value is set for γ in the minimum mean squared error sense. This optimum value was set by Sato [49] to:

$$\gamma = \frac{E[a(n)^2]}{E[|a(n)|]} \tag{7.121}$$

for real valued signals. This is the only value of γ, which sets the mean value of the error term in Equation 7.120 to zero. For complex-valued transmitted signals γ is given by a similar relationship:

$$\gamma = \frac{E[Re\{a(n)^2\}]}{E[Re\{|a(n)|\}]} = \frac{E[Im\{a(n)^2\}]}{E[Im\{|a(n)|\}]}. \tag{7.122}$$

Having described Sato's algorithm, let us now consider the so-called "*Constant Modulus Algorithm*", which was proposed by Godard in [52].

7.7.4.2 Constant Modulus Algorithm [52]

A more general algorithm for blind equalisation was proposed by Godard in [52], which was further generalized later by Shalvi and Weinstein in [453]. Its related cost-function is defined as:

$$J^{(p,q)}(n) = \frac{1}{pq} E[||z(n)|^p - R_p|^q], \tag{7.123}$$

which will be elaborated on below. Godard's original algorithm and all other related algorithms use only $q = 2$ and this is the case that we will focus on from now on. One can observe that Sato's algorithm in Equation 7.116 is similar to the CMA of Equation 7.123, when we have $p = 1$ and $q = 2$. The reason for considering the difference betweeen the amplitude of $z(n)$ and a constant R_p instead of the actually transmitted symbol $a(n)$, is that in blind equalisation we attempt to match the equalised signal not to the actually transmitted sequence, which is never available, but to its statistics, which of course is known at the receiver. Given this cost-function, the coefficient adjustment algorithm using the steepest descent technique of Equation 7.111 can then be invoked. The value of the parameter R_p has to be matched to the constellation in a way similar to the setting of γ for Sato's algorithm in Equation 7.116. Clearly, this algorithm forces only the amplitude of the received signal to match a desired mean value, but ignores the phase. Consequently, the phase of the received signal might have arbitrary variations. Godard suggested that equalising only the magnitude of the signal should be adequate. The associated phase ambiguity can be removed by using differential phase encoding, which is congenial to the nature of blind equalisation. Indeed, a sign ambiguity is related to any blind equaliser, stemming from the fact that the QAM constellations are symmetric with respect to the x and y-axes. As it is clear from the definition of the CMA's cost-function in Equation 7.123, in the case of pure phase modulation, the equaliser's output $z(n)$ will be constrained to a constant value and the algorithm will readily converge [52]. A variation of this algorithm, which solves the problem of the arbitrary phase rotation is the so-called Modified-CMA, presented in the next section. Using Equations 7.114 and 7.123 we obtain:

$$g_{CMA}(z(n)) = z(n) \cdot \left[1 - |z(n)|^2 + R_2^2\right]. \tag{7.124}$$

Substituting this formula into Equation 7.104 we readily arrive at the equaliser tap update equation of the CMA [52]:

$$\mathbf{c}^{(n+1)} = \mathbf{c}^{(n)} - \lambda \cdot \mathbf{y}^*(n) \cdot z(n) \cdot \left[|z(n)|^2 - R_2\right]. \tag{7.125}$$

The value of R_2 can be found by constraining the mean value of the update term of Equation 7.125 equal to zero, assuming that the equalised signal $z(n)$ is equal to the transmitted signal $a(n)$ with its phase rotated by a random value, i.e. assuming that the state of perfect equalisation has been reached. The procedure of determining R_2 is exactly the same as the one, which was used to compute Sato's scaling coefficient γ, yielding [52] :

$$R_2 = \frac{E[|a(n)|^4]}{E[|a(n)|^2]}. \tag{7.126}$$

A generalization of Godard's algorithm was proposed by Shalvi and Weinstein in [453]. For this algorithm the cost-function to be minimised is the so-called Kurtosis of the equalised signal $z(n)$, defined as [89]:

$$K(z) = E[|z(n)|^4] - 2 \cdot E^2[|z(n)|^2] - |E[z^2(n)]|^2 \tag{7.127}$$

for complex-valued signals $z(n)$. The algorithm resulting from this cost-function is given in [453], where it is shown that this generalized algorithm results in Godard's algorithm as a special case. Following this algorithm, Shalvi and Weinstein proposed their so-called super-exponential algorithm [454], which uses 4-th order cumulants and converges at a nearly super-exponential speed. Cumulant-based blind equalisation algorithms have been proposed for example in [455]- [456].

As an important extension of Godard's algorithm, Wesolowsky's modified CMA is presented in the next section.

7.7.5 Modified Constant Modulus Algorithm [457]

Another modified version of Godard's CMA [52] was proposed by Wesolowsky in [457], employing a cost-function, which relies on both the real and imaginary parts of the equalised signal $z(n)$ [52]:

$$
J(n) = E[\left(|Re\{z(n)\}|^2 - R_{2,R}\right)^2 + \left(|Im\{z(n)\}|^2 - R_{2,I}\right)^2].
$$
(7.128)

The idea behind this cost-function, as compared to the CMA cost-function of Equation 7.123 is that both the real and imaginary parts of the signal are forced to a constant value and, therefore, the random phase ambiguity of the CMA now becomes only 90°. This is meaningful in pure phase modulation, in which case the CMA may converge to an arbitrarily phase-shifted solution. For QAM, though, the 90° symmetry of the constellation makes it possible for both algorithms to converge to a 90° phase-shifted solution. Following the same procedure as in the context of the other algorithms, based on Equations 7.111 and 7.112 we obtain:

$$
\mathbf{c}^{(n+1)} = \mathbf{c}^{(n)} -
$$
$$
-\lambda \cdot \mathbf{y}^*(n) \cdot [Re[z(n)] \cdot \left((Re[z(n)])^2 - R_{2,R}\right)
$$
$$
+j \cdot Im\{z(n)\} \cdot \left((Im\{z(n)\})^2 - R_{2,I}\right)]
$$
(7.129)
$$
g_{MCMA}(z) = z - (Re\{z\} \cdot \left((Re\{z\})^2 - R_{2,R}\right) +
$$
$$
j\left(Im\{z\} \cdot \left((Im\{z\})^2 - R_{2,I}\right)\right)).
$$
(7.130)

The values of $R_{2,R}$ and $R_{2,I}$ can be found using the same method as for the other algorithms, namely by constraining the mean value of the update terms in Equation 7.129, yielding [52]:

$$
R_{2,R} = \frac{E\left[(Re\{a(n)\})^4\right]}{E\left[(Re\{a(n)\})^2\right]}
$$
(7.131)

$$
R_{2,I} = \frac{E\left[(Im\{a(n)\})^4\right]}{E\left[(Im\{a(n)\})^2\right]}.
$$
(7.132)

It was shown by Wesolowsky [457] that the MCMA exhibits slightly faster convergence than the classical CMA, in particular for medium-distortion channels. It also offers slightly better steady-state performance, as we will show in Section 7.7.11. Let us now, however, turn our attention to considering the *Benveniste-Goursat* algorithm [51].

7.7.5.1   Benveniste-Goursat Algorithm [51]

In Sato's algorithm [49], the error signal was expressed in Equation 7.120, which is repeated here for convenience:

$$\epsilon^{Sato}(n) = z(n) - \gamma \cdot csgn(z(n)).$$

The error term is used to update the equaliser coefficients, according to Equation 7.119. This error signal is, however, non-zero, when the signal is perfectly equalised, except in the case of QPSK, since γ is a constant value, reflecting the statistics of $a(n)$, while $z(n)$ generally takes its values from a multilevel constellation, when the SNR is sufficiently high and the equaliser has converged. This results in inaccurate steady-state behaviour along with small error fluctuations around the point of equilibrium associated with the minimum error. In other words, even near the optimum equaliser setting, not every tap update drives the equaliser towards the desired equilibrium. In order to remedy these deficiencies, Benveniste and Goursat [51] considered the decision-directed error signal expressed as:

$$\epsilon^{DD}(n) = z(n) - \hat{a}(n), \tag{7.133}$$

which becomes zero, when equalisation has been accomplished, giving a good-steady state performance. On the other hand, this error signal cannot be employed during the equaliser's initial convergence phase, since at the beginning of the equalisation process the decisions concerning $z(n)$ are often erroneous and this would drive the coefficient update equation to an ill-conditioned state. Combining the error signals in Equation 7.120 and 7.133, each one scaled by a certain weight, Benveniste and Goursat [51] formulated a new error signal as:

$$\epsilon^{BG}(n) = k_1 \cdot \epsilon^{DD}(n) + k_2 \cdot |\epsilon^{DD}(n)| \cdot \epsilon^{Sato}(n), \tag{7.134}$$

where k_1 and k_2 are the corresponding weighting factors. This error signal is zero, when equalisation is perfect and, at the same time, it is not as error-prone as a purely blind decision-directed (DD) approach would be at start up, since then the influence of Sato's error term $\epsilon^{Sato}(n)$ in Equation 7.120 offers better error estimation. Using this combined error signal we readily arrive at the Benveniste-Goursat (B-G) algorithm [51], adjusting the equaliser taps according to:

$$\mathbf{c}^{(n+1)} = \mathbf{c}^{(n)} - \lambda \cdot \mathbf{y}^*(n) \cdot \epsilon^{BG}(n). \tag{7.135}$$

A good choice for k_1 and k_2 in Equation 7.134 would be to initialise the algorithm with a large k_2/k_1 ratio and decrease the ratio, when the equaliser is close to convergence in order to render the steady-state equalisation more accurate. This philosophy is similar to the idea of switching to decision-directed mode, when the equaliser has converged.

The related $g\{z(n)\}$ function in this case can be found by comparing Equations 7.104 and 7.135, yielding [442]:

$$g(z) = z(n) - k_1 \cdot \epsilon^{DD}(n) + k_2 \cdot |\epsilon^{DD}(n)| \cdot \epsilon^{Sato}(n). \tag{7.136}$$

Having discussed the Benveniste-Goursat algorithm, we will now consider another DD-like algorithm in the next section, namely the *stop-and-go* algorithm by Picchi and Prati [57].

7.7.5.2 Stop-and-Go Algorithm [57]

In the previous algorithms the equaliser coefficient update is inevitably occasionally wrong due to the statistical nature of the algorithms. This leads to a reduced convergence speed and also to a degradation of the steady-state performance of the equaliser. In order to avoid this impediment to some degree, Picchi and Prati [57] suggested an algorithm, which decides whether a specific received symbol should contribute to the update process and updates the equaliser coefficients, only when it has decided that this would bring their values closer to their steady-state ones. The algorithm used for updating the coefficients is the classic error feedback algorithm with the decision-directed error expressed as in Equation 7.133, which is repeated here for convenience:

$$\epsilon^{DD}(n) = z(n) - \hat{a}(n).$$

Two variables, namely $f_{n,R}$ and $f_{n,I}$ are introduced in [57], each of which defines a measure of the probability that the update of the real or imaginary part of the equaliser coefficients is correct. Naturally, the actual probability is unknown at the receiver, but it can be estimated using the philosophy of the Sato-type error of Equation 7.120, setting $f_{n,R}$ and $f_{n,I}$ to 1 and 0, depending on our confidence in the success of the update, as [57]:

$$f_{n,R} = \begin{cases} 1 & \text{if } sgn(Re[\epsilon^{DD}(n)]) = sgn(Re[\epsilon^{Sato}(n)]) \\ 0 & \text{if } sgn(Re[\epsilon^{DD}(n)] \neq sgn(Re[\epsilon^{Sato}(n)]) \end{cases} \qquad (7.137)$$

and

$$f_{n,I} = \begin{cases} 1 & \text{if } sgn(Im\{\epsilon^{DD}(n)\}) = sgn(Im\{\epsilon^{Sato}(n)\}) \\ 0 & \text{if } sgn(Im\{\epsilon^{DD}(n)\}) \neq sgn(Im\{\epsilon^{Sato}(n)\}). \end{cases} \qquad (7.138)$$

Practically this implies that the update of the real part of the equaliser coefficients only takes place when:

$$\left. \begin{array}{rclclcl} & Re[z(n)] > \gamma & \text{and} & Re[\hat{a}(n)] < Re[z(n)] \\ \text{or} & 0 < Re[z(n)] < \gamma & \text{and} & Re[\hat{a}(n)] > Re[z(n)] \\ \text{or} & -\gamma < Re[z(n)] < 0 & \text{and} & Re[\hat{a}(n)] < Re[z(n)] \\ \text{or} & Re[z(n)] < -\gamma & \text{and} & Re[\hat{a}(n)] > Re[z(n)] \end{array} \right\}. \qquad (7.139)$$

From this we can suggest that the "correct" values for $z(n)$ are those ones that are expected to bring $\hat{a}(n)$ closer to γ, which was richly illustrated in [57] in geometrical terms. This is a plausible, but certainly imperfect criterion. The probabilities of making a false update decision also are calculated in [57].

With these definitions we can form the algorithm using the classical error feedback algorithm of Equation 7.104, but involving both the real and imaginary parts of the error, enabled or disabled by $f_{n,R}$ and $f_{n,I}$, as follows:

$$\mathbf{c}^{(n+1)} = \mathbf{c}^{(n)} - \lambda \cdot \mathbf{y}^*(n) \cdot \left[f_{n,R} Re\{\epsilon^{DD}(n)\} + j f_{n,I} Im\{\epsilon^{DD}(n)\} \right]. \qquad (7.140)$$

This algorithm is expected to have an advantage over the previous ones, since it uses Equations 7.138 - 7.139 to reject unreliable coefficient updates and to render convergence more steady and accurate. Nevertheless, the algorithm's convergence is hampered to a certain degree, since it does not use all the incoming symbols to update the equaliser coefficients.

A modification of this algorithm, which was suggested by Choi *et al.* [458], invokes a CMA-type error term, defined as [458]:

$$e^{CMA}(n) = z(n) \cdot (|z(n)|^2 - R_2), \tag{7.141}$$

instead of the Sato error of Equation 7.120, in order to form the decisions concerning the validity of the equaliser update at each symbol, where the symbols $z(n)$ and the constant R_2 were defined in Section 7.7.4.2. This error term corresponds to circular geometric regions in the equalised symbol domain, since it only considers the square of the magnitude of the equaliser's output $z(n)$, ignoring the phase.

Finally, again, we present the symbol estimation function $g\{z(n)\}$ corresponding to this algorithm, which is readily found by observing Equations 7.104 and 7.140:

$$g\{z(n)\} = z(n) - \left[f_{n,R} \cdot Re\{\epsilon^{DD}(n)\} + j f_{n,I} \cdot Im\{\epsilon^{DD}(n)\} \right]. \tag{7.142}$$

Having reviewed a range of Bussgang algorithms, let us now consider some of the associated convergence issues in the next section.

7.7.6   Convergence Issues

In this section, the convergence properties of the Bussgang equalisers are discussed and the problems associated with them are explored. These issues attracted the attention of researchers as early as 1980, when Godard studied the convergence of the algorithm he proposed [52]. Later, in 1985, Foschini [53] gave a proof of the convergence of the CMA, when the length of the equaliser is doubly infinite, i.e. when both the number of the feedforward when it has and the feedback taps of the equaliser was infinite. It was not until ten years later that Ding *et al.* [54] proved that when the equaliser is not of infinite length, then there can be undesirable stable local minima, depending on the CIR. Ding *et al.* arrived at this conclusion by considering a special class of channels, namely the so-called "autoregressive" channels, and by finding the local minima of a CMA equaliser for these channels. Their theory presented in [459] also revealed that the local minima of the CMA based Baud-rate spaced Bussgang equaliser correspond to local minima of all other Baud-rate spaced Bussgang equalisers for the same channel, arising from the fact that the equalisers do not have an infinite length. These minima are thus referred to as length-dependent minima [460]. Algorithm-dependent minima do not exist in the family of CMAs, but do exist in Sato's algorithm [461], in the context of the Benveniste-Goursat algorithm [51] and in conjunction with the Stop-and-Go algorithm [460]. The above two CMAs and also the Shalvi-Weinstein algorithms [454] exhibit only channel-dependent local minima. It has to be mentioned that a general solution for the convergence of the Bussgang equalisers is still an open research issue. The regions around undesirable local minima have also been studied by Ding *et al.* [462, 463], while initialisation strategies have been proposed by Li and Ding in [464]. Moreover, it has been indicated that the convergence performances of the CMA and the Shalvi-Weinstein algorithms are similar to each other and also similar to the performance of the LMS (or Wiener) receiver [465–467]. Finally, dynamic convergence issues have been treated, for example, in [468–470]. Similar studies have recently been conducted also for fractionally-spaced equalisers [83, 448, 471, 472]. In this overview we give a basic analysis model for the convergence of Bussgang equalisers and interpret some well-established results.

We commence this analysis with the convergence analysis of the CMA. We consider a noiseless environment, which simplifies our discussion. We recall the error term of the CMA's equaliser tap update formula from Equation 7.125:

$$\epsilon^{\mathbf{CMA}}(n) = \mathbf{y}^*(n) \cdot z(n) \cdot (|z(n)|^2 - R_2). \tag{7.143}$$

This error term is basically the derivative of the CMA's cost-function in Equation 7.123 with respect to the equaliser tap vector. The points at which the mean value of this error term becomes zero define the local minima, maxima and saddle points of this algorithm. Therefore, in order to find the possible local minima, we have to evaluate the local minima of the following equation:

$$\mathbf{y}^*(n) \cdot z(n) \cdot (|z(n)|^2 - R_2) = \mathbf{0}. \tag{7.144}$$

By substituting the vectors from Equation 7.94 to 7.99 into Equation 7.144 and taking the expectation we have:

$$
\begin{aligned}
E &\left[\mathbf{y}^*(n) \cdot z(n) \cdot (|z(n)|^2 - R_2) \right] \\
&= E\left[\mathbf{H}^* \cdot \mathbf{a}^*(n) \cdot (\mathbf{a}^T(n) \cdot \mathbf{t}) \cdot (z^*(n)z(n) - R_2) \right] \\
&= E\left[\mathbf{H}^* \cdot (\mathbf{a}^*(n) \cdot \mathbf{a}^T(n)) \cdot \mathbf{t} \cdot ((\mathbf{t}^H \mathbf{a}^*(n)) \cdot (\mathbf{a}^T(n)\mathbf{t}) - R_2) \right] \\
&= E\left[\mathbf{H}^* \cdot ((\mathbf{a}^* \cdot \mathbf{a}^T)(\mathbf{t} \cdot \mathbf{t}^H)(\mathbf{a}^* \cdot \mathbf{a}^T) - R_2 \cdot (\mathbf{a}^* \cdot \mathbf{a}^T)) \cdot \mathbf{t} \right] \\
&= \mathbf{H}^* \cdot \mathbf{T} \cdot \mathbf{t}
\end{aligned}
\tag{7.145}
$$

where

$$
\begin{aligned}
\mathbf{T} = \ & \mu_4 diag(|t_i|^2 - 1) + \\
& \mu_2^2 \cdot (\mathbf{t}\mathbf{t}^H + \mathbf{t}^T \mathbf{t}^* \mathbf{I} - \\
& -2 \cdot diag(|t_{-K_1}|^2, \cdots, |t_{K_2}|^2))
\end{aligned}
\tag{7.146}
$$

or

$$
[\mathbf{T}]_{i,j} = \begin{cases} \mu_4 \cdot (|t_i|^2 - 1) + \mu_2^2 \sum_{k=-K_1, k \neq i}^{K_2} |t_k|^2 & i = j \\ \mu_2^2 t_i \cdot t_j^* & i \neq j \end{cases}
\tag{7.147}
$$

and $\mu_j = E[|a(n)|^j]$, while $K_1 = N_1 + L_1$, $K_2 = N_2 + L_2$. The candidate stationary points will satisfy the set of equations:

$$\mathbf{H}^* \cdot \mathbf{T} \cdot \mathbf{t} = \mathbf{0}. \tag{7.148}$$

The resulting equations may have two types of solutions. We can assume that the null-space of matrix $\mathbf{H}^*$ is trivial, i.e. we assume that:

$$\mathbf{H}^* \cdot \mathbf{x} = \mathbf{0} \Rightarrow \mathbf{x} = \mathbf{0}. \tag{7.149}$$

As long as the number of the taps is finite, the channel matrix $\mathbf{H}^*$ is an $(N_1 + N_2 + 1) \times (N_1 + N_2 + L_1 + L_2)$ dimensional matrix, which has less rows than columns. A system characterised by $\mathbf{H}^* \cdot \mathbf{x} = \mathbf{0}$ in conjunction with such a matrix $\mathbf{H}^*$ always has an infinite number of non-trivial solutions. If the equaliser has an infinite number of feedforward and feedback taps, though, then the situation changes. We can see that the channel matrix $\mathbf{H}$ of Equation 7.99 now has an infinite number of rows and columns, thus being a square matrix having linearly independent rows, which implies that the system characterised by Equation 7.149 has only the trivial solution. This was shown differently by Ding *et al.* in [459]. For the moment, we will assume that the matrix has a trivial nullspace. In this case we can find the stationary equilibrium points of the algorithm by finding the solution of the following set of equations:

$$\mathbf{T} \cdot \mathbf{t} = \mathbf{0}. \tag{7.150}$$

The solution of these equations can be found to be any vector $\mathbf{t}$, which has some zero entries and some non-zero values, all exhibiting the same magnitude. Since the vector $\mathbf{t}$ is constituted by the convolution of the CIR with the equaliser's impulse response, ideally $\mathbf{t}$ would be a Dirac delta function. The vectors $\mathbf{t}$ having more than one non-zero components represent saddle points, yielding unstable equilibria, which do not affect the equaliser's performance, since they do not constitute convergence points. This can be shown by examining the second derivative of the tap update formulae in Equation 7.125 with respect to the equaliser tap vector $\mathbf{c}$. The second derivatives can be found in the same way as the first derivatives, with the exception that now the derivative is with respect to the conjugate of the equaliser tap vector $\mathbf{c}$, yielding:

$$\frac{\partial^2 J_{CMA}(n)}{\partial \mathbf{c} \partial \mathbf{c}^H} = \mathbf{H}^* \cdot \mathbf{T}' \cdot \mathbf{H}^T, \tag{7.151}$$

where

$$\begin{aligned}
\mathbf{T}' &= \mu_4 diag(2|t_i|^2 - 1) + \\
&\quad 2\mu_2^2 \cdot (\mathbf{t}\mathbf{t}^H + \mathbf{t}^T \mathbf{t}^* \mathbf{I} - \\
&\quad -2 \cdot diag(|t_{-K_1}|^2, \cdots, |t_{K_2}|^2))
\end{aligned} \tag{7.152}$$

or

$$[\mathbf{T}']_{i,j} = \begin{cases} \mu_4 \cdot (2|t_i|^2 - 1) + 2\mu_2^2 \sum_{k=-K_1, k \neq i}^{K_2} |t_k|^2 & i = j \\ 2\mu_2^2 \cdot t_i \cdot t_j^* & i \neq j. \end{cases} \tag{7.153}$$

The positive definiteness of a matrix $\mathbf{A}$ can be verified by considering the term $\mathbf{x}^T \cdot \mathbf{A} \cdot \mathbf{x}^*$. In our case, this term becomes:

$$\begin{aligned}
\mathbf{x}^T \cdot \mathbf{H}^* \cdot \mathbf{T}' \cdot \mathbf{H}^T \cdot \mathbf{x}^* \quad &= \\
(\mathbf{H}^H \cdot \mathbf{x})^T \cdot \mathbf{T}' \cdot (\mathbf{H}^H \cdot \mathbf{x})^*. &
\end{aligned} \tag{7.154}$$

From this relationship we observe that if the matrix $\mathbf{T}'$ is positive (negative) definite, then the matrix $\mathbf{H}^* \cdot \mathbf{T}' \cdot \mathbf{H}^T$ is also positive (negative) definite. Therefore, it is sufficient to estimate the positive definiteness of $\mathbf{T}'$. By examining the sub-determinants of $\mathbf{T}'$, we can easily see that if only one entry of the vector $\mathbf{t}$ is non-zero, then all of the sub-determinants have the form μ_4^i and are positive, which implies that the matrix is positive definite and the associated error surface point is a local minimum. Finally, all the other solutions, for which more than one component of the vector $\mathbf{t}$ is non-zero, have positive and negative subdeterminants for the matrix $\mathbf{T}'$, thus constituting saddle points.

The stationary points of the Modified CMA [457] of Section 7.7.5 have a similar form to the stationary points of the CMA [52] of Section 7.7.4.2, as explored by Wesolowsky in [457], since the cost-functions of these two algorithms are similar.

Above we have found the stationary and saddle points of the CMA under the assumption that the rows of the channel matrix are linearly dependent. In general, this assumtion is not true, unless the number of feedforward and feedback taps of the equaliser is infinite. In practice, though, for a high number of taps this is approximately true. Ding *et el.* showed in [54,459] that if the number of equaliser taps is not infinite, depending on the form of the CIR, undesirable equilibria will be present. The authors considered a specific channel, namely the autoregressive channel and found the undesirable equilibria associated with this channel. These are channel-dependent equilibria, which are likely to exist in a similar form for every other Bussgang equaliser. This can be seen, if we recall Equation 7.145. The multiplicative matrix $\mathbf{H}^*$ in Equation 7.145, which is responsible for the undesirable equilibria explored by Ding *et al.* [54], stems from the $\mathbf{y}^*$ factor multiplying the error term in Equation 7.104. This equation is common to all Bussgang equalisers, which implies that the term $\mathbf{H}^*$ will exist for any other algorithm of this type and therefore all these algorithms will exhibit these unstable equilibria. The stationary points of the error surface in Sato's algorithm were explored in [461], while the stationary points of the modified CMA are similar to those of the CMA, which were explored in [457]. A different type of analysis, suggested by Shynk and Chan in [473], assumed that the convolutional noise is normally distributed and provides similar results. Fractionally-spaced equalisers of this type have also been studied. It has been shown by Ding *et al.* [83,472] that a sufficient condition for the existence of only one desirable global minimum is that the equaliser's length is at least equal to the channel's delay spread and, at the same time, there are no common zeros between the subchannels created by considering the z-transform of the oversampled channel. In other words this means that the associated subchannels are reasonably diverse. Endres *et al.* [448] have explored the scenario, where the first of these conditions was not met. In closing, we note that the extension of Bussgang equalisers to fast RLS estimation based schemes has been proposed by Douglas *et al.* in [474] as well as by Papadias and Slock in [475]. Having highlighted the most salient Bussgang algorithms, let us now consider joint CIR and data estimation techniques.

7.7.7 Joint Channel and Data Estimation Techniques

Joint channel and data detection techniques constitute the blind equivalent of sequence estimation algorithms. These techniques were originally used for data estimation, when the CIR was known. Here, instead of equalising the received signal, we estimate the CIR together with the ML data sequence, assuming that the CIR estimate is sufficiently accurate. Originally, Seshadri [58] observed that it may be risky to invoke the Viterbi algorithm [59] for the blind scenario, since by assuming perfect CIR knowledge, we would discard all the surviving paths but one. This would be a bad tactic at the initial stages of the algorithm, when the CIR estimation is poor. In fact, no path should be eliminated in favour of another in the blind detection scenario. This would correspond to a computationally intensive algorithm. Instead, Seshadri suggested that the so-called M-algorithm [476] should be used, retaining M number of surviving paths at each trellis state, rather than just one. He also suggested that the CIR estimation should be initialised and updated at any symbol interval using an LMS estimator. Each survivor path of the trellis should keep its own data plus CIR estimate, as

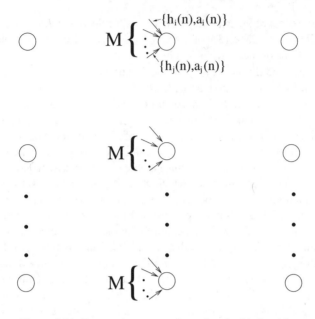

Figure 7.21: Per-survivor processing using the M-algorithm

illustrated in Figure 7.21. Here only the stylised trellis states and the associated M number of trellis transitions are indicated. For a deeper exposure to the M-algorithm the interested reader is referred to [476]. The LMS estimation was extended to RLS-based estimation by Xie *et al.* [477] for the case of multiuser communications, where the interference was due to K users, each having p distinct propagation paths of the dispersive channel. Raheli *et al.* [60] introduced the term *Per-Survivor Processing (PSP)* and showed the applicability of this technique to fast fading channels, owing to its fast convergence.

In this section, we give a brief overview of joint CIR and data estimation using the PSP technique. The definitions of the variables are the same as in Figure 7.15 and Section 7.7.4. Additionally, we define $\hat{\mathbf{h}} = [\hat{h}_{-L_1}, \cdots, \hat{h}_{L_2}]^T$ as the estimated CIR. Any transition from state μ_n to state μ_{n+1} in the trellis of Figure 7.21, describing the evolution of the state-machine modelling the channel, is due to the estimated transmitted symbol vector $\hat{\mathbf{a}}(\mu_n \rightarrow \mu_{n+1}) = \hat{\mathbf{a}}(n)$. The estimated received symbol $\hat{y}(n)$ for this transition at time n is then given by the convolution of the tentatively assumed transmitted symbol vector $\hat{\mathbf{a}}(n)$ and the CIR $\hat{\mathbf{h}}^{(n)}$ assumed, which is formulated as $\hat{\mathbf{a}}^T(n) \cdot \hat{\mathbf{h}}^{(n)}$. Each surviving path in the trellis of Figure 7.21 will have its own CIR and data estimate. The LMS CIR estimation algorithm for each surviving path will then be [58]:

$$\hat{\mathbf{h}}^{(n+1)} = \hat{\mathbf{h}}^{(n)} - \lambda \cdot \hat{\mathbf{a}}^*(n) \cdot \left(y(n) - \underbrace{\sum_{i=0}^{L} \hat{\mathbf{a}}^T(n) \cdot \hat{\mathbf{h}}^{(n)}(n)}_{\hat{y}(n)} \right), \qquad (7.155)$$

which is similar in its philosophy to the tap-update formula of Equation 7.104, updating $\mathbf{c}^{(n)}$ using the step-size λ depending on the error term in the round brackets (). The LMS CIR and data estimator can be replaced by an RLS estimator according to the following equations [3, 60]:

$$\mathbf{\Phi}(\mu_n) = \sum_{i=0}^{n} w^{n-i}\hat{\mathbf{a}}^*(i)\hat{\mathbf{a}}^T(i) = \mathbf{P}^{-1}(\mu_n)$$

$$\mathbf{k}(\mu_{n+1}) = \frac{\mathbf{P}(\mu_n)\hat{\mathbf{a}}^*(n)}{w + \hat{\mathbf{a}}^T(n)\mathbf{P}(\mu_n)\hat{\mathbf{a}}^*(n)}$$

$$\mathbf{P}(\mu_{n+1}) = \frac{1}{w}\left(\mathbf{P}(\mu_n) - \mathbf{k}(\mu_{n+1})\hat{\mathbf{a}}^T(n)\mathbf{P}(\mu_n)\right)$$

$$\epsilon(\mu_n \to \mu_{n+1}) = \sum_{i=0}^{n} w^{n-i}|y(i) - \hat{\mathbf{a}}^T(n)\hat{\mathbf{h}}^{(n)}(i)|^2$$

$$\hat{\mathbf{h}}(\mu_{n+1}) = \hat{\mathbf{h}}(\mu_n) + \mathbf{k}(\mu_{n+1})\epsilon(\mu_n \to \mu_{n+1}). \tag{7.156}$$

This CIR and data estimator provides fast adaptation and renders the estimator applicable to fast fading channels.

The blind data and CIR estimator presented in this section has some significant differences in comparison to the Bussgang blind equalisers of the previous section. Namely:

- In contrast to stand-alone equalisation, there is no filter simulating the inverse of the CIR.

- The convergence is significantly faster.

- This technique can be combined with ML sequence estimation techniques and it is also amenable to amalgamation with ML trellis-based channnel decoding.

As for the other blind equalisation techniques discussed, numerous studies have provided insight into this method. The associated practical considerations and a range of modifications were studied by Chugg and Polydoros in [61], while in [62] the associated performance was studied both theoretically and by simulations. In [63] and [64], Hidden Markov Model (HMM) theory was applied to the problem of joint data detection as well as CIR estimation and performance results were given for the so-called Gaussian Minimum Shift Keying (GMSK) modulation invoked in a Time Division Multiple Access (TDMA) scenario. In [65] the employment of fuzzy logic was proposed for the calculation of the associated decision metrics. In [66] and [67] the acquisition performance of a PSP detector was evaluated and "smart" initialisation strategies were explored. In [68] the algorithm was adapted to make use of soft statistics and to include error prediction. Finally, in [69] a genetic algorithm was applied for estimating the CIR, and the Viterbi algorithm was then invoked.

Let us now consider the family of second-order statistics based techniques.

7.7.8 Blind Equalisation Using Second-Order Cyclostationary Statistics

7.7.8.1 Second-order statistics-based blind channel identifiability

Using second-order statistics [478], in general, is significantly more efficient than invoking higher-order statistics. This is because less samples are needed, in order to generate the required statistics, implying higher convergence speed and better stability. Nevertheless, second-order statistics are not applicable in all cases of interest. Specifically, in the case of blind equalisation, the CIR is not identifiable from second-order statistics, when we sample at the Baud-rate, i.e. at one sample per symbol period. This is because the phase information of the CIR is not contained in the second-order statistics of the received signal sampled at the Baud-rate. This can be readily seen, if we consider the autocorrelation function of the received signal in Figure 7.15, namely:

$$R_{yy}(\tau) = E[y(t + \tau) \cdot y^*(t)] \tag{7.157}$$

and the autocorrelation of the transmitted signal sequence, namely:

$$R_{aa}(\tau) = E[a(t + \tau) \cdot a^*(t)], \tag{7.158}$$

as well as the channel transfer function $H(j\omega)$ in the frequency domain. These considerations assume that the input and the received signals are wide-sense stationary (WSS), implying that the signals' statistical characteristics are insensitive to the instant of capturing them [445]. This is typically true for the input QAM signal. For the received signal, though, it is true only when the channel is static; when it is not, then often we can consider it as slowly varying, in which case the above WSS hypothesis can be regarded as met. For example, when TDMA is used, it is adequate to assume that the CIR remains unchanged within one TDMA frame period. It was verified in [479] or [478] that when both the sequences $\{a(n)\}$ and $\{y(n)\}$ are wide-sense stationary, then the Fourier transforms of these autocorrelation functions, namely $\Phi_{yy}(\omega)$ and $\Phi_{aa}(\omega)$, respectively, are related to each other with the aid of:

$$\Phi_{yy}(\omega) = |H(j\omega)|^2 \cdot \Phi_{aa}(\omega), \tag{7.159}$$

indicating that only the CIR's magnitude information can be extracted, when Baud-rate based sampling is employed. In 1991 Tong *et al.* [70] exploited the fact that by employing over-sampling, the received signal becomes so-called *cyclostationary*, implying that the associated statistical properties become periodic in time [478]. Exploiting this cyclostationarity, Tong *et al.* proposed a method of CIR estimation, using second-order statistics. More recently, in 1995 Mulines *et al.* [71] observed that the same happens, when we are taking the input of multiple sensors – such as multiple antennas for example – at Baud-rate, i.e. without resorting to oversampling. Finally, in 1997, Tsatsanis and Giannakis [78], [72] proposed another substitute for oversampling at the receiver, namely inducing cyclostationarity at the transmitter, instead of oversampling at the receiver. All these ideas lead to a range of similar methods applicable to CIR estimation, which are based on second-order statistics.

7.7.8.2   Second-order statistics-based blind CIR estimation algorithms

Let us invoke oversampling and denote the symbol interval by T, while the sampling interval by T_S, so that $T = T_S \cdot \Delta$, where Δ is the oversampling factor. We also assume in the rest of this chapter that the channel's memory extends from $-L_1T$ to L_2T. The channel memory limits do not have to be multiples of the symbol interval T, it is simply notationally convenient. We observe the system over the time interval $[nT, (n+1) \cdot T]$, using the linear model of:

$$\mathbf{y}(n) = \mathbf{H}(n) \cdot \mathbf{a}(n) + \mathbf{e}(n), \qquad (7.160)$$

where

$$\mathbf{y}(n) = [y(nT), y(nT + T_S), \ldots, y(nT + (\Delta - 1)T_S)]^T \qquad (7.161)$$

represents the received signal vector at time instant n,

$$\mathbf{H}(n) = \begin{bmatrix} h_{-L_1T}(n) & \cdots & h_{L_2T}(n) \\ h_{-L_1T+T_S}(n) & \cdots & h_{L_2T+T_S}(n) \\ \vdots & \vdots & \vdots \\ h_{-L_1T+(\Delta-1)T_S}(n) & \cdots & h_{L_2T+(\Delta-1)T_S}(n) \end{bmatrix} \qquad (7.162)$$

is constituted by the channel's impulse response taps $\{h_i(n)\}$ for $i = -L_1T, \cdots, L_2T$ $+(\Delta - 1)T_S$ at time nT,

$$\mathbf{a}(n) = [a((n + L_1)T), \ldots, a((n - L_2)T)]^T \qquad (7.163)$$

is the transmitted data vector and

$$\mathbf{e}(n) = [e(nT), e(nT + T_S), \ldots, e(nT + (\Delta - 1)T_S)]^T \qquad (7.164)$$

is the channel noise vector.

We now point out the equivalence of oversampling the received signal with rendering the transmitted signal cyclostationary and also its equivalence to involving the outputs of Δ number of 'sensors' or antennae. The former [72] involves transmitting the same output Δ times sequentially, instead of just once. This way the transmitted signal becomes cyclostationary and the same happens to the corresponding received signal sampled at the Baud-rate, when the channel is stationary for the duration of processing. If we consider the transmitter's Baud-rate to be Δ-times lower than the receiver's Baud-rate, then the received signal would be the same as in the previous scenario.

Upon using the outputs of multiple sensors or antennae [71], we can observe that the output of each sensor – say, for example that of the i-th sensor – is equivalent to the output of oversampling at time instants of $iT_S, i = 0, \cdots, \Delta - 1$ in the $[0, (\Delta - 1)T_S]$ space.

Having shown the above equivalences, we will now proceed to discuss the formation of second-order statistics based algorithms. Let us consider the autocorrelation matrix of the received signal $y(nT + iT_S)$ over the time interval iT at time instant nT:

$$
\begin{aligned}
\mathbf{R_{yy}}(i;n) \quad &= \quad E[\mathbf{y}(n) \cdot \mathbf{y}^H(n-i)] \\
&\overset{7.160}{=} \quad E[(\mathbf{H}(n) \cdot \mathbf{a}(n) + \mathbf{e}(n)) \cdot \\
& \qquad (\mathbf{a}^H(n-i) \cdot \mathbf{H}^H(n-i) + \mathbf{e}^H(n-i))] \\
&= \quad E[\mathbf{H}(n)\left(\mathbf{a}(n)\mathbf{a}^H(n-i)\right)\mathbf{H}^H(n-i) + \\
& \qquad +\mathbf{H}(n)\mathbf{a}(n)\mathbf{e}^H(n-i) + \\
& \qquad +\mathbf{e}(n)\mathbf{a}^H(n-i)\mathbf{H}^H(n-i) + \\
& \qquad +\mathbf{e}(n)\mathbf{n}^H(n-i)],
\end{aligned}
\tag{7.165}
$$

where $\mathbf{y}(n)$ was defined in Equation 7.161. The two terms in the middle of this expression have a zero mean value, since they are the product of independent variables. Thus, Equation 7.165 becomes:

$$
\mathbf{R_{yy}}(i;n) = \mathbf{H}(n) \cdot \mathbf{R_{aa}}(i;n) \cdot \mathbf{H}^H(n-i) + \mathbf{R_{ee}}(i;n).
\tag{7.166}
$$

The CIR estimation process consists of estimating the CIR matrix $\mathbf{H}$ using measurements of the autocorrelation matrix $\mathbf{R_{yy}}(n)$ according to Equations 7.165 and 7.166, assuming that the autocorrelation of the transmitted signal $\mathbf{R_{aa}}(i;k)$ is known. Tong *et al.* [445] invoked the assumption that the input data source produces completely uncorrelated symbols $\mathbf{a}(n)$, while Hua *et al.* [82] adapted the method to work with correlated input signals. Furthermore, while in Tong's algorithm [445] the noise was assumed to have some known properties, in [73] a variation was proposed, which assumed no knowledge concerning the channel noise. In order to estimate the CIR matrix $\mathbf{H}$ in Equation 7.162, we must also have knowledge of the noise's correlation. The noise samples cannot be assumed to be uncorrelated (white), since they are generated at a higher rate than the Baud-rate and since the receiver's lowpass filtering imposed autocorrelation upon it. This autocorrelation function has the same shape as the impulse response of the receiver filter, since the noise was white before this filtering and hence its autocorrelation function was a $\delta()$ function.

There are various ways of performing CIR estimation at this stage. Some algorithms, such as that proposed by Tong *et al.*, [70, 445], estimate the CIR by assuming that the noise is white – not a valid assumption in general – and the transmitted sequence is also white. In our forthcoming deliberations the channel is assumed to be static. The CIR matrix $\mathbf{H}$ of Equation 7.162 is then estimated by performing the so-called "singular value decomposition (SVD)" [480] of $\mathbf{R_{yy}}(0)$ and $\mathbf{R_{yy}}(1)$ and then by performing a range of further algebraic calculations. In other algorithms [72] iterative procedures have been proposed, exhibiting lower complexity, but having an asymptotically similar performance. After estimating the CIR matrix $\mathbf{H}$ of Equation 7.162 and ignoring the effect of noise in Equation 7.160, we can invert $\mathbf{H}$ in order to extract the original information symbols $\mathbf{a}(n)$ from the received sequence $\mathbf{y}(n)$, using the following LMS estimator:

$$
\mathbf{c}_{LMS} = \mathbf{H}^t \cdot \mathbf{y}(n),
\tag{7.167}
$$

where $\mathbf{H}^t$ satisfies:

$$
\mathbf{H}^t \cdot \mathbf{H} = \mathbf{I}
\tag{7.168}
$$

and it is the "pseudo-inverse" [465] of the CIR matrix. The input signals have been assumed to be i.i.d. variables with a normalised power of unity, while the noise has been neglected.

An important advantage of the second-order statistics based algorithms – compared to the family of the linear equalisers – is that they are asymptotically accurate. Simulations presented in [445] demonstrate this accuracy in performance terms and also show that the second-order statistics-based algorithms exhibit fast convergence, typically within 100 symbols. This concludes our discussions on the second-order statistics-based algorithms. For more detailed discussions on this issue the reader is referred to [70,445] and also to [73–75]. In [76] a different method of estimating the CIR matrix $\mathbf{H}$ of Equation 7.162 is proposed for the "multiple channel" scenario, using so-called "outer-product matrix decomposition". In [77,78], the transmitter-induced cyclostationarity algorithms are explored, while in [79] the cyclostationarity-based method is applied to an Orthogonal Frequency Modulation (OFDM) receiver. In [80], a modification of Tong's original method [445] is proposed, while in [81] a general study of the cyclostationary method is given. The effects of non-i.i.d. signal distributions are studied in [82] and in [83]. In [84] the second-order statistics-based methods are adapted to the so-called source separation problem, where for example the wanted and interfering signals are separated. In [85] the subspace method is applied to the suppression of both inter-symbol interference and multiple-access interference. In [86] the case of unknown noise distributions is considered. Finally, in [87], the so-called "*Column-Anchored Zeroforcing Equalisation (CAZE)*" is proposed and studied. Having reviewed the family of second-order statistics-based techniques, let us now consider a range of so-called "polycepstra-oriented" algorithms in the next section.

7.7.9 Blind Channel Estimation and Equalisation Using Polycepstra

The last class of blind equalisation algorithms discussed here involves fourth-order – rather than second-order – statistics of the received signal, in order to estimate the inverse of the CIR and to equalise its effects. The algorithms belonging to this class were initially proposed by Hatzinakos and Nikias in [89]. This algorithm, referred to as the *Tricepstrum Equalisation Algorithm (TEA)*, employs the complex cepstrum of the so-called fourth-order cumulants (the tricepstrum is defined in Appendix 7.10) of the received signal sequence sampled at the Baud-rate. These algorithms are capable of identifying both so-called "minimum and maximum-phase" channels, which we will characterise more explicitly during our further discourse. Below we briefly introduce the TEA using the fourth-order cumulants and tricepstra as defined in Appendix 7.10.

In this context we aim to equalise a channel having a z-domain channel transfer function of $H(z)$. We consider $H(z)$ as the product of a minimum-phase [3] section – where the z-domain transfer function of the channel has all its zeros inside the unit circle – and a maximum-phase section – where the z-domain transfer function of the channel has all its zeros outside the unit circle. This is formulated as [89]:

$$H(z) = A \cdot I(z^{-1}) \cdot O(z), \tag{7.169}$$

where A is a constant and $I(z^{-1})$ and $O(z)$ are polynomials of the form $\Pi_{i=1}^{M}(1 - a_i z^{-1})$ and $\Pi_{j=1}^{K}(1 - a_j z)$ respectively, with $|a_i| < 1$, M is the number of channel zeros inside the unit circle and K is the number of channel zeros outside the unit circle. A simple manifestation of

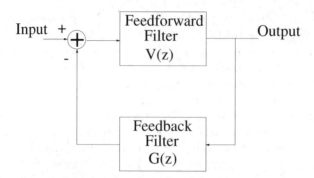

Figure 7.22: The DFE filter structure

a minimum-phase channel exhibits a CIR, where the main tap associated with the time instant 0 is the largest one. The polynomial $I(z^{-1})$ is a minimum-phase polynomial, while $O(z)$ is a maximum-phase polynomial. In order to equalise this channel under the so-called Zero-Forcing (ZF) constraint – implying that the combined CIR and equaliser impulse response is forced to zero at sampling instants $n \neq 0$ – we use an equaliser, having a transfer function of $C(z)$, which is the inverse of that of the channel [89]:

$$C(z) = \frac{1}{H(z)} = \frac{1}{A \cdot I(z^{-1}) \cdot O(z)}. \tag{7.170}$$

Then the cascaded channel and equaliser transfer function constitutes an ideal channel. If instead of the ZF equaliser we use a decision-feedback equaliser (DFE) having a feedforward transfer function of $V(z)$ and a feedback transfer function of $G(z)$, then the equaliser's transfer function will have the form [89]:

$$C(z) = \frac{V(z)}{1 + G(z)}, \tag{7.171}$$

where $V(z)$ and $G(z)$ are the feedforward and feedback section's transfer function, respectively, as shown in Figure 7.22. The feedback filter must be realizable. A possible choice for $V(z)$ and $G(z)$ is [89]:

$$V(z) = \frac{O^*(z^{-1})}{AO(z)} \tag{7.172}$$

$$1 + G(z) = I(z^{-1}) \cdot O^*(z^{-1}). \tag{7.173}$$

Then the system $C(z) \cdot H(z)$ is perfectly equalised and the feedforward section $V(z)$ of the DFE is an all-pass filter having zeros inside and poles outside the unit circle, while the feedback filter $G(z)$ is a minimum-phase filter. In order to construct the DFE we have to find an estimate of the coefficients of the feedforward and feedback filters. Equally, we have to find an estimate of the poles and zeros of the two equaliser filters. This can be achieved using the fourth-order cumulants of the received signal.

We form the received signal $y(n)$ as the convolution of the transmitted signal $a(n)$ with the CIR $\{h_n\}$, plus a zero-mean additive Gaussian stochastic process $e(n)$, as follows:

$$y(n) = a(n) * h_n + e(n). \tag{7.174}$$

We recall from Appendix 7.10 that the tricepstrum $c_y(m, n, l)$ of the received signal $y(n)$ is related to the tricepstrum $c_h(m, n, l)$ of the CIR by [89]:

$$c_y(m, n, l) = c_h(m, n, l), \qquad (m, n, l) \neq (0, 0, 0). \tag{7.175}$$

This implies that in the tricepstrum domain the received signal is equal to the CIR. Therefore, estimating the received signal's tricepstrum directly gives us the CIR's tricepstrum. The problem then becomes that of estimating the CIR in terms of its tricepstrum representation. It was shown in [89] that it is sufficient to consider tricepstra in the form of $c_y(K, 0, 0)$ (K integer). We recall from Appendix 7.10 the following relationship:

$$\sum_{I=1}^{p} A^{(I)}[L_y(m - I, n, l) - L_y(m + I, n + I, l + I)] +$$

$$+ \sum_{J=1}^{q} B^{(J)}[L_y(m - J, n - J, l - J) - L_y(m + J, n, l)] =$$

$$= -mL_y(m, n, l)$$

where

$$K \cdot c_y(K, 0, 0) = \begin{cases} -A^{(K)}, & K = 1, \ldots, p \\ B^{(-K)}, & K = -1, \ldots, -q. \end{cases}$$

This relationship is in fact a system of linear equations, which can be solved by iterative methods, as shown in Appendix 7.10. However, the estimation of the CIR and the equaliser filters is quite an elaborate task and hence the associated derivation was relegated to Appendix 7.10, following the approach of [89].

An estimate of these algorithms' complexity compared to the Bussgang algorithms' complexity was also given by Hatzinakos and Nikias in [89], where the complexity of the former appears to be significantly higher, a fact which was mentioned before. As a trade-off, the simulations presented in [89] indicate the superiority of the TEAs in terms of convergence speed, as well as in terms of their ability to equalise non-minimum phase channels, which are often encountered in fading mobile channels. For more detailed investigations of these techniques the reader is referred to [89, 455, 481–484].

Having presented an overview of a range of basic blind equaliser structures, we will now provide a summary of their complexity.

7.7.10 Complexity Evaluation

In this section the complexity of the blind equalisers presented is evaluated. We commence our discussions by considering the complexity of the Bussgang algorithms. The complexity of the Bussgang techniques is similar for all of them and it is relatively low. For simplicity, in this section we will assume that the number of equaliser taps is $2N + 1$, i.e. $N_1 = N_2 = N$,

Algorithm	Additions	Multiplications	Memory
CMA [52]	$16L + 10$	$16L + 13$	$4L + 4$
Sato [49]	$16L + 8$	$16L + 10$	$4L + 4$
B-G [51]	$16L + 17$	$16L + 19$	$4L + 4$
Modified-CMA [457]	$16L + 8$	$16L + 14$	$4L + 4$
Stop-and-Go [57]	$16L + 8$	$20L + 10$	$4L + 4$
Super-Exponential $(p = 2, q = 1)$ [454]	$O(L^3)$	$O(L^3)$	$O(L)$

Table 7.1: Complexity estimate of the Bussgang techniques of Section 7.7.4, assuming $2N + 1 \approx 2L + 1$, where L is the channel's memory

according to the notation we have used so far. The complexity depends only on the number of equaliser taps, $2N + 1$, and it is on the order of N, which is indicated as $O(N)$. An estimate of the number of real additions and multiplications required for each algorithm per equalised symbol interval is presented in Table 7.1. Here, we have assumed that the complex variables are represented in the memory of the associated arithmetic unit in terms of their real and imaginary parts. We have also assumed that the number of equaliser taps $2N + 1$ is approximately equal to the channel memory $2L + 1$, i.e. $2N + 1 \approx 2L + 1$ and that the square root evaluation required for the computation of $|\epsilon^{DD}(n)|$ in the Benveniste-Goursat algorithm of Section 7.7.5.1 is performed with the aid of 4 real additions and 2 real multiplications. Finally, in Table 7.1 we have also included an estimate of the memory requirements of each algorithm. In addition to the Bussgang algorithms the table incorporated the super-exponential algorithm, which does not fall into the Bussgang category, but exhibits similarities with the CMA. A complexity estimate of the PSP-based algorithms of Section 7.7.7 is obtained similarly to the previously introduced Bussgang algorithms of Section 7.7.4 by calculating the total number of additions and multiplications as well as the associated memory requirements. Considering that the convolution of the CIR with the estimated sequence in Equation 7.155 is only calculated once for each survivor transition and stored in memory (thus saving unnecessary calculations), the associated complexity results are summarised in Table 7.2. These complexity figures refer to one symbol interval. The complexity due to convolutional decoding - provided that channel coding is used - is ignored in these calculations.

In Table 7.2 M is the number of survivors seen in Figure 7.21 that we retain at each step and Q is the number of possible signal constellation points, i.e. $Q = 2^K$ for K-bit per symbol modulation. By comparing Tables 7.1 and 7.2 we observe that the complexity of the sequence estimation techniques (aided by LMS-based CIR estimation) is exponentially increasing with the channel order L, i.e. it is of $O(Q^L)$. By contrast, the complexity of the Bussgang techniques is only $O(L)$ and it is only linearly increasing, implying that channels having long CIRs cannot be equalised by this sequence estimation technique. However, the

	Additions	Mult/ions	Memory
$M \neq 1$	$MQ^{2L}(M/2 + 6$ $Q^{-1}(10L - 1))$	$MQ^{2L}(6+$ $Q^{-1}(20L + 3))$	$2MQ^{2L-1}(L + 2)$
$M = 1$	$Q^{2L}(6.5+$ $Q^{-1}(10L - 1))$	$Q^{2L}(6+$ $Q^{-1}(20L + 3))$	$2Q^{2L-1}(L + 2)$

Table 7.2: Complexity estimate of the Sequence Estimation algorithms of Section 7.7.7

associated complexity also depends on the number of phasors in the modulation constellation as well as on the number M of survivors per state.

From these tables we can also observe that the Bussgang algorithms are attractive for long CIRs, since in this case the sequence estimation techniques exhibit excessively high computational complexity. By contrast, for short CIRs the sequence estimation techniques offer significant advantages at an affordable complexity. However, as will be seen in Section 7.7.11, channels exhibiting long CIRs cannot be equalised using Bussgang equalisers due to the enhancement of convolutional noise associated with the large number of equaliser taps required for equalisation.

Having presented a rudimentary overview of various blind equalisation methods, we note the emergence of a recent approach, based on Neural Networks (NN). For more details concerning NN-based blind methods the reader is referred to [485–490].

7.7.11 Performance Results

In this section we present the associated comparative performance results for the algorithms described in Sections 7.7.4 and 7.7.7. Two different types of results are presented, commencing with Bit Error Rate (BER) learning curves, which offer a measure of the algorithms' convergence speed. The second set of results is concerned with the average BER curves of the algorithms over a dispersive Gaussian channel. The modulation schemes involved are 16-QAM and 64-QAM [440, 441]. The associated signal constellations are shown in Figures 7.23 and 7.24.

7.7.11.1 Channel Models

Three different channels were used in our comparative study. The first one is a typical worst-case Wireless Asynchronous Transfer Mode (WATM) channel, while the second one is a Shortened WATM (SWATM) CIR, both of which are presented in Figure 7.25. These indoor CIRs were generated with the aid of finding the line-of-sight path and the four longest-delay paths in a 100x100x3m^3 hall at a WATM transmission rate of 155 Mbit/s. The third channel used in our comparative study is a simple one-symbol-delay channel, which is shown in Figure 7.26. This channel may characterise a satellite link, where due to the directional parabolic antenna used, only one or two multipath components may arrive at the receiver. The difference between the WATM, SWATM and the one-symbol-delay channel is that while the former two channels carry multipath components at several symbols' delay, the latter exhibites only one additional multipath component at a delay of one symbol. These different

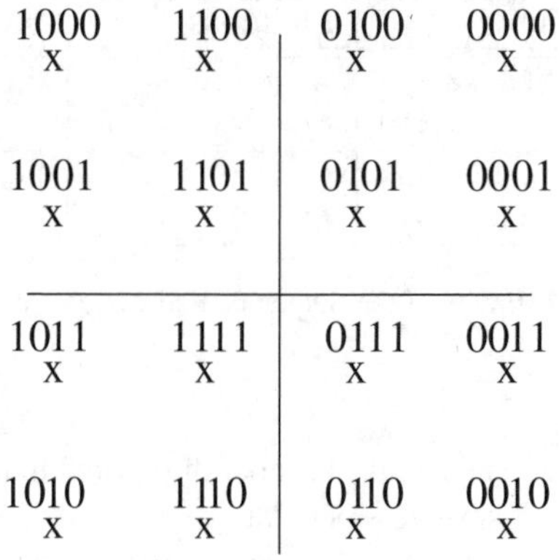

1000	1100	0100	0000
X	X	X	X
1001	1101	0101	0001
X	X	X	X
1011	1111	0111	0011
X	X	X	X
1010	1110	0110	0010
X	X	X	X

Figure 7.23: 16-QAM constellation

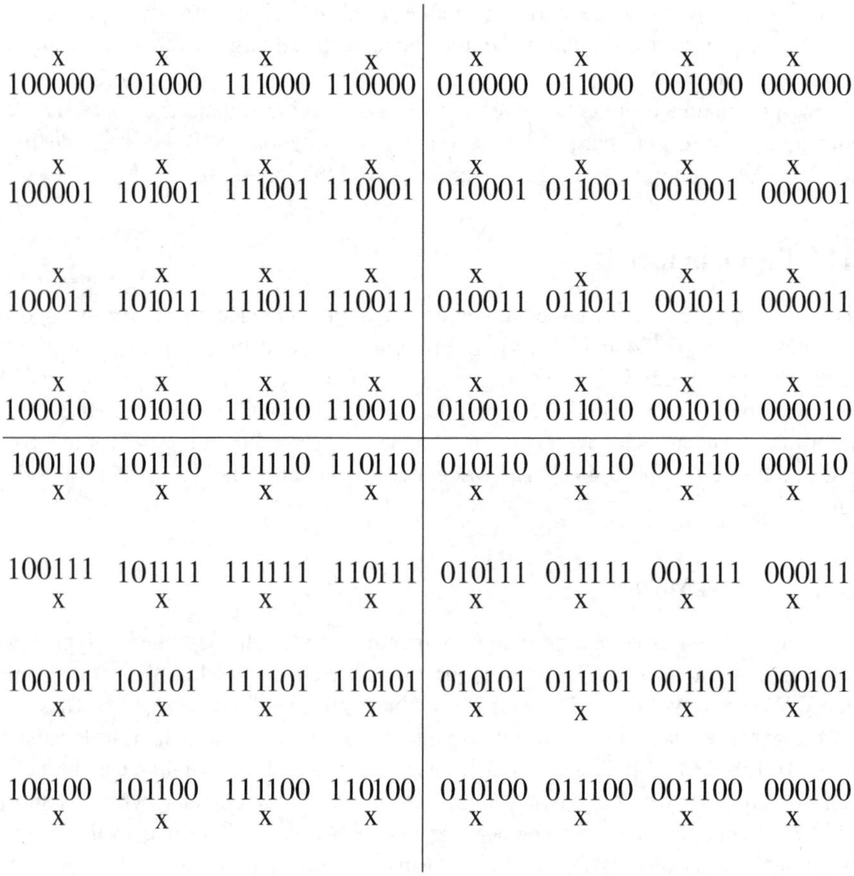

X	X	X	X	X	X	X	X
100000	101000	111000	110000	010000	011000	001000	000000
X	X	X	X	X	X	X	X
100001	101001	111001	110001	010001	011001	001001	000001
X	X	X	X	X	X	X	X
100011	101011	111011	110011	010011	011011	001011	000011
X	X	X	X	X	X	X	X
100010	101010	111010	110010	010010	011010	001010	000010
100110	101110	111110	110110	010110	011110	001110	000110
X	X	X	X	X	X	X	X
100111	101111	111111	110111	010111	011111	001111	000111
X	X	X	X	X	X	X	X
100101	101101	111101	110101	010101	011101	001101	000101
X	X	X	X	X	X	X	X
100100	101100	111100	110100	010100	011100	001100	000100
X	X	X	X	X	X	X	X

Figure 7.24: 64-QAM constellation

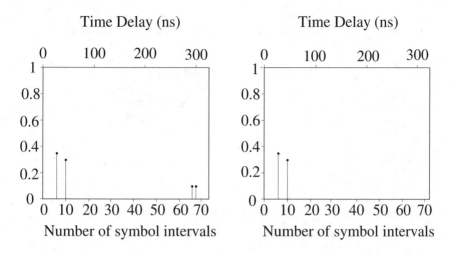

Figure 7.25: (a) The WATM and (b) Shortened WATM channels

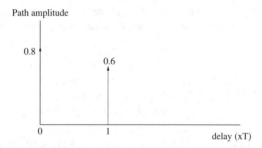

Figure 7.26: The one-symbol-delay channel used in the simulations

CIRs will be used to demonstrate the fact that different equalisers may be appropriate for different channels.

7.7.11.2 Learning Curves

In this section, the associated BER learning curves of the blind equalisers of Sections 7.7.4 and 7.7.7 are presented for the CIRs of Figures 7.25 and 7.26. The step-size parameter λ is common for all the Bussgang algorithms and equal to $5 \cdot 10^{-4}$, while the Benveniste-Goursat parameters of Equation 7.135 are $k_1 = 1$, $k_2 = 5$. For the PSP-based sequence estimation algorithm, the step-size in Equation 7.155 is 10^{-2} and only $M = 1$ survivor was retained, owing to the fact that the CIR was time-invariant. The WATM channel simulations associated with the CIR of Figure 7.25(a) are presented in Figure 7.27 for 16-QAM and in Figure 7.30 for 64-QAM. An equaliser length of 68 taps was employed and the Signal-to-Noise Ratio

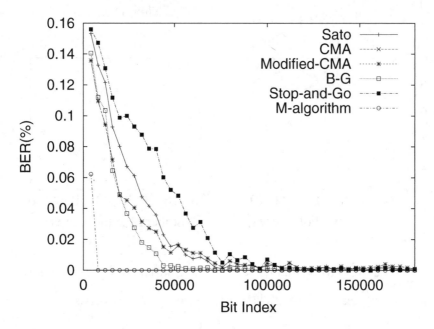

Figure 7.27: BER learning curves over the WATM channel of Figure 7.25(a) for 16-QAM at SNR = 30dB using a 68-tap equaliser obeying the schematic of Figure 7.19

(SNR) was $30dB$. The corresponding SWATM and one-symbol-delay CIR based results are presented in Figures 7.28 and 7.29 as well as 7.31 and 7.32 for 16-QAM as well as 64-QAM, respectively. As we see, the CIR spread was gradually shortened from 68 to 10 and then to 2. Observe that for the M-algorithm of Figure 7.21 we only presented BER results in Figures 7.29 and 7.32 over the shortest one-symbol-delay CIR. In order to quantify the BER associated with the learning curves, we have used the values of BER in variable-length intervals and we have performed smoothing of the curves, which was necessary due to the statistical nature of the blind equalisers of this type; not all of the incoming data symbols drive the equaliser to the point of convergence. From these curves we can infer some observations concerning the convergence speed of each of the tested algorithms.

- The M-Algorithm converges at a higher speed than any of the Bussgang algorithms considered. A reason for this is the employment of a larger step-size value, which could not have been used for the Bussgang algorithms, since this would result in poor tracking performance.

- Sato's algorithm converges at a medium to slow speed.

- Godard's algorithm converges with about the same speed as Sato's and its convergence is faster for higher-order QAM.

- The MCMA algorithm converges at a medium speed, but its convergence is faster for higher-order QAM, as we can see from Figures 7.30 and 7.31 for 64-QAM.

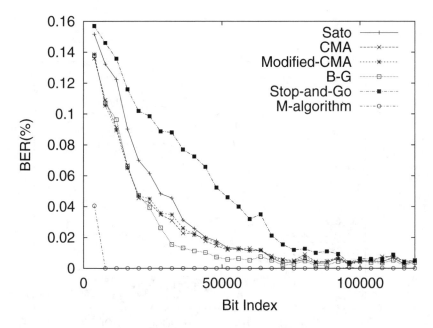

Figure 7.28: BER learning curves over the Shortened WATM channel of Figure 7.25(b) for 16-QAM at SNR = 20dB using a 30-tap equaliser obeying the schematic of Figure 7.19

- The Benveniste-Goursat algorithm converges rapidly only for low-order QAM, and we can see from Figures 7.27 and 7.28 that for 16-QAM it converges faster than any of the other Bussgang algorithms. For higher-order constellations, such as 64-QAM, it converges significantly slower.

- The Stop-and-Go algorithm is definitely the slowest algorithm in all cases, since it is not always enabled to iterate.

In Tables 7.3 and 7.4 an estimate of the number of symbols needed for the convergence of each algorithm is given. The converged state was defined as the state, where the BER has 'just' reached its steady state value and does not change significantly thereafter. The channel SNR is kept at $30dB$. Having presented a comparative simulation study of the convergence speed of the blind equalisers of Sections 7.7.4 and 7.7.7, we will now give an illustration of the blind equalisers' convergence by means of the associated phasor diagrams [440, 441].

7.7.11.3   Phasor Diagrams

In this section some phasor diagrams are presented for the various algorithms considered. We are observing the phasor diagram at the equaliser's output at different instants, each giving an idea of how the equaliser is converging. In Figures 7.33 and 7.34, the phasor constellation is plotted in the complex plane for a CMA-based equaliser having 10 taps, equalising the one-symbol-delay channel of Figure 7.26, when 16-QAM is used. Two snapshots are shown.

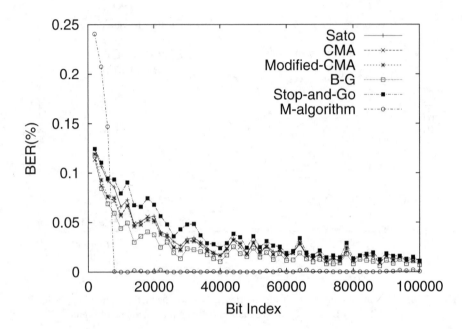

Figure 7.29: BER learning curves over the one-symbol-delay channel of Figure 7.26 for 16-QAM at SNR = 20dB using a 10-tap equaliser obeying the schematic of Figure 7.19

	WATM	S-WATM	TC
M-Algorithm	550	550	3700
Sato	30000	25000	80000
CMA	35000	40000	55000
MCMA	35000	20000	60000
Benveniste-Goursat	20000	15000	-
Stop-and-Go	75000	25000	-

Table 7.3: The number of symbols required for each algorithm to converge for 16-QAM over the WATM and Shortened WATM channels of Figures 7.25(a) and (b) as well as over the simple one-symbol-delay channel of Figure 7.26, which is labelled as TC (Test Channel). The SNR is $30dB$.

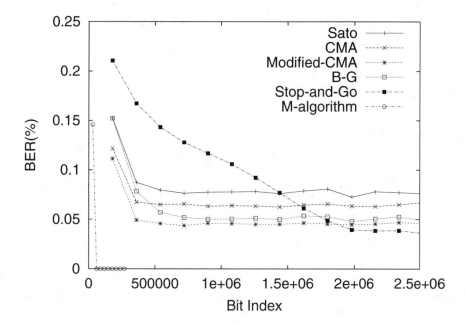

Figure 7.30: BER learning curves over the WATM channel of Figure 7.25(a) for 64-QAM at SNR = 30dB using a 68-tap equaliser obeying the schematic of Figure 7.19

	WATM	S-WATM	TC
M-Algorithm	4150	3000	6000
Sato	50000	50000	120000
CMA	55000	35000	110000
MCMA	45000	45000	115000
Benveniste-Goursat	75000	55000	150000
Stop-and-Go	320000	200000	600000

Table 7.4: The number of symbols required for each algorithm to converge for 64-QAM over the WATM and Shortened WATM channels of Figures 7.25(a) and (b) as well as over the simple one-symbol-delay channel of Figure 7.26, which is labelled as TC (Test Channel). The SNR is $30dB$.

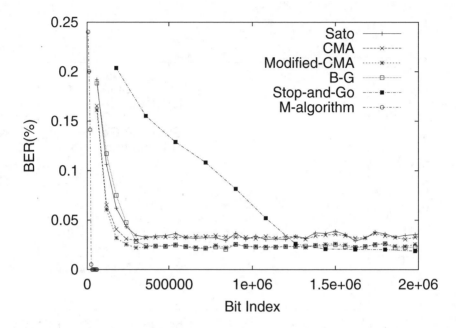

Figure 7.31: BER learning curves over the Shortened WATM channel of Figure 7.25(b) for 64-QAM at SNR = 20dB using a 30-tap equaliser obeying the schematic of Figure 7.19

In the first the equaliser is reaching convergence, but still exhibits residual ISI. In the second snapshot, the equaliser has almost converged. These snapshots demonstrate how the equaliser is adapting and slowly approaching the state of convergence. When it has converged, the residual impairments are the convolutional noise, which is rather small, and the additive channel noise, which has also been chosen to be low, so that the convergence can be better observed. At this state, the equalised signal is confined to small regions around the legitimate QAM constellation points and the diameter of these regions depends basically on the SNR and on the residual ISI, which produces convolutional noise. Finally, in Figures 7.35 and 7.36 the phasor diagram is shown for the case of 64-QAM.

Having presented our comparative results for the convergence speed of the blind equalisers, we will now investigate the accuracy of convergence.

7.7.11.4 Gaussian Channel

In this section the steady-state average BER curves are presented as a function of the bit-SNR over the WATM and Shortened WATM channels of Figure 7.25, as well as over the one-symbol delay channel of Figure 7.26 using 16-QAM and 64-QAM. The bit-SNR is defined as the SNR per bit, i.e. the signal power per bit over the noise power per bit. In mathematical terms we have:

$$\text{Bit-SNR} = \frac{SNR}{\text{Number of bits per symbol}}. \tag{7.176}$$

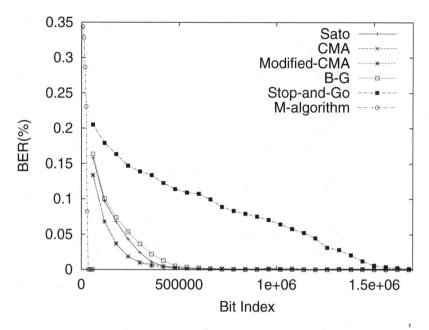

Figure 7.32: BER learning curves over the one-symbol-delay channel of Figure 7.26 for 64-QAM at SNR = 30dB using a 10-tap equaliser obeying the schematic of Figure 7.19

The equaliser characteristics are the same as in Section 7.7.11.2. For the M-algorithm, the channel estimator takes into consideration only the CIR taps that indeed exist, thus reducing the number of calculations. The associated curves are presented in Figures 7.37 to 7.42.

Note here that following the classical initialisation strategy – that is initialising the equaliser tap vector to $\mathbf{c}^o = (1.2, 0, \cdots, 0)^T$ as suggested by Godard [52] – is inadequate for the Stop-and-Go equaliser to converge to the desired equilibrium. Instead, the equaliser converges to an undesirable equilibrium associated with the combined CIR plus equaliser tap vector of $\mathbf{t}^1 = (0.74, 0.48, 0, \cdots, 0)^T$. For the equaliser to converge to the correct equilibrium, the initialisation has to be closer to the desired equilibrium. For example, initialising the equaliser with the tap vector $(1.2, -0.9, 0.6, 0, \cdots, 0)^T$ is adequate. This phenomenon is directly related to the nature of this algorithm and the undesirable equilibrium is clearly an algorithm-dependent equilibrium. This is why this phenomenon is not observed in the other equalisers. From these curves we can infer some observations concerning the accuracy of convergence for each of the tested algorithms, which exhibits itself in terms of the residual BER at a given SNR after reaching the steady-state. We note furthermore that at the SNR concerned, namely at $30dB$, the converged-state-accuracy of the various techniques becomes explicit also from Figures 7.27 to 7.31.

- The M-algorithm provides the best performance.

- Sato's algorithm converges with a medium accuracy.

- Godard's algorithm converges with a medium convergence accuracy, which is im-

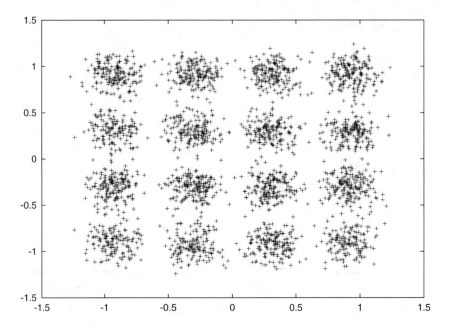

Figure 7.33: The 16-QAM phasor constellation over the one-symbol-delay channel of Figure 7.26 after
40000 symbols at an SNR = 30dB using a CMA 10-tap equaliser obeying the schematic
of Figure 7.19

proved for low-order QAM.

- The Modified-CMA algorithm offers a good accuracy, which is even better for higher-order QAM.

- The Benveniste-Goursat algorithm exhibits excellent accuracy, especially in conjunction with higher-order QAM.

- The Stop-and-Go algorithm has a good convergence accuracy, especially in the context of BPSK.

- While for 16-QAM the BER curves do not tend to exhibit residual errors for high SNRs, the same does not hold for 64-QAM over the WATM channel. This is, because these channels contain multipath components spread to several symbols' delays. This, in turn, means that the equaliser should also have a high number of taps. However, a blind equaliser having a high number of taps is usually not feasible due to their limited accuracy. We can extend the order of these equalisers and improve the associated BER, but only up to the point where the equaliser starts to enhance the ISI and the associated convolutional noise.

- The WATM channel is a channel containing multipath components scattered over a delay of 67 symbols. The equalisation of this channel would involve a very long equaliser

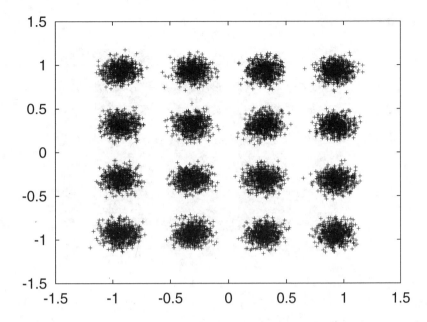

Figure 7.34: The 16-QAM phasor constellation over the one-symbol-delay channel of Figure 7.26 after 60000 symbols at an SNR = 30dB using a CMA 10-tap equaliser obeying the schematic of Figure 7.19

of about 80 taps. In the blind scenario, this is impractical, since the equaliser's resolution is not sufficiently high. Hence beyond a certain length, the taps of these equalisers which are located far from the centre tap do not effectively contribute to the equalisation process. Furthermore, they enhance the convolutional noise by increasing the ISI. Besides that, the speed of convergence is linearly dependent on the equaliser order and it becomes low for such high-order equalisers. This can be viewed in the Figure 7.43, where the MSE is plotted against the equaliser order for a specific example. In this case, the algorithms used were the CMA and the LMS or Wiener filter, which a trained benchmarker used for comparison. The modulation was 16-QAM and the environment was noiseless.

7.7.12   Simulations with Decision-Directed Switching

In this section, we briefly explore the possibility of switching to decision-directed equalisation after the convergence of the blind equaliser. It is expected that for low SNR values this would drive the equaliser away from convergence – since the blind decision-directed equaliser is generally unstable – and therefore, at low SNRs switching to DD mode would be disastrous. Nevertheless, when the SNR is sufficiently high, switching to DD mode is expected to assist the equaliser in converging with a better accuracy. This is indeed what we observed in our investigations. Explicitly, this technique improved the performance of blind

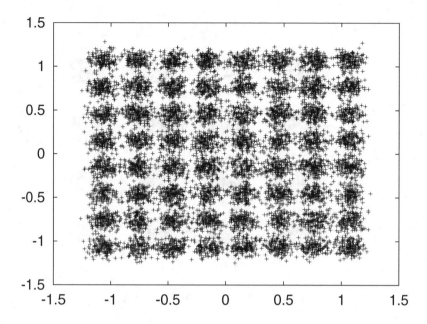

Figure 7.35: The 64-QAM phasor constellation over the one-symbol-delay channel of Figure 7.26 after 70000 symbols at SNR = 35dB using a CMA 10-tap equaliser obeying the schematic of Figure 7.19

equalisation and the improvement was higher, when the order of the QAM constellation was higher. This can be explained by the fact that when the SNR is high, the DD technique is powerful. This property can be exploited more readily in the context of higher-order QAM constellations than in lower-order schemes, since the SNR is typically higher for the higher-order QAM schemes, where the distances between constellation points are smaller. In Figures 7.44 and 7.45 the associated improvement is shown for the case of the CMA and for 16-QAM as well as for 64-QAM, respectively. It can be observed that the improvement is modest for 16-QAM, but it is around $1dB$ for 64-QAM at a BER of 10^{-3}.

Having studied the range of blind equalisation solutions in the previous subsections, a prominent application of the algorithms in the context of the Pan-European Satellite-based Digital Video Broadcast (DVB-S) [437] system [491] will be presented in Chapter 30.

7.8 Summary

This chapter has dealt with the issues of channel equalisation for QAM transmissions, commencing with a summary of conventional equalisation techniques. The principles of Kalman filtering were introduced, followed by an overview of the family of blind channel equalisers, which have found wide-ranging applications in data transmission modems and broadcast systems.

The blind equalisers investigated were classified in Figure 7.16. The Bussgang equalis-

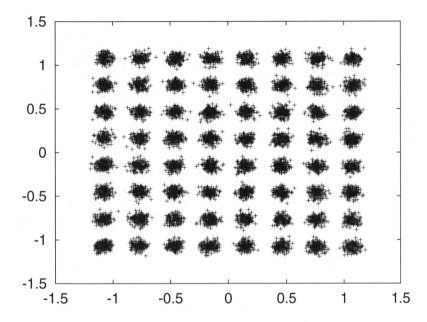

Figure 7.36: The 64-QAM phasor constellation over the one-symbol-delay channel of Figure 7.26 after 180000 symbols at an SNR = 35dB using a CMA 10-tap equaliser obeying the schematic of Figure 7.19

ers exhibit a low complexity, but require a considerable number of symbols to converge. Moreover, their convergence accuracy is mediocre. By contrast, the M-Algorithm appears to be the best choice, giving the best convergence speed and tracking performance at the cost of a higher complexity, depending on the channel's delay spread and the number of survivors M.

For advanced QAM equalisation strategies the reader is referred to Chapters 7, 17 and 26, where equalisers designed for QAM transmissions over dispersive mobile radio channels are considered. Furthermore, a range of blind equalised video broadcast systems will be the topic of Chapter 30.

In the next chapter we consider a transmission technique whereby the design of the constellation used in the modulation scheme and the error correction coding used to improve the channel's bit error rate are combined into one system. Such a system can be shown to have distinct advantages and is known as trellis coded modulation or also often referred to as 'codulation'.

7.9 Appendix: Differentiation with Respect to a Vector

In this appendix we give the definitions of differentiation with respect to complex vectors along with their associated properties. Based on these, we then highlight how we can minimise the cost-functions encountered in our theoretical considerations. First of all we define

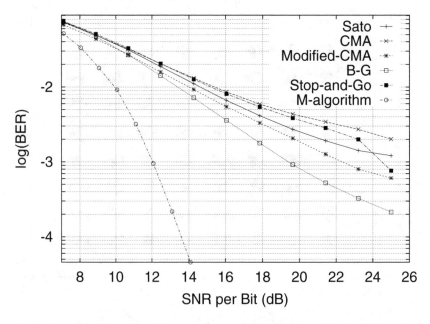

Figure 7.37: Gaussian BER versus SNR curves for the WATM channel of Figure 7.25(a) for 16-QAM
using a 68-tap equaliser obeying the schematic of Figure 7.19 and, for the M-Algorithm,
a 5-path channel estimation obeying the schematic of Figure 7.21

the gradient vector [3]:

$$\nabla_{\mathbf{z}} = \begin{bmatrix} \frac{\partial}{\partial x_1} + j\frac{\partial}{\partial y_1} \\ \frac{\partial}{\partial x_2} + j\frac{\partial}{\partial y_2} \\ \vdots \\ \frac{\partial}{\partial x_N} + j\frac{\partial}{\partial y_N} \end{bmatrix} \tag{7.177}$$

and

$$\nabla_{\mathbf{z}^*} = \begin{bmatrix} \frac{\partial}{\partial x_1} - j\frac{\partial}{\partial y_1} \\ \frac{\partial}{\partial x_2} - j\frac{\partial}{\partial y_2} \\ \vdots \\ \frac{\partial}{\partial x_N} - j\frac{\partial}{\partial y_N} \end{bmatrix}, \tag{7.178}$$

where $\mathbf{z} = [x_1 + jy_1 \cdots , x_N + jy_N]^T$ is a complex vector. The properties of the gradient
vectors, which we will exploit are :

$$\nabla_{\mathbf{z}}\mathbf{z} = \mathbf{O} \tag{7.179}$$

$$\nabla_{\mathbf{z}}\mathbf{z}^* = \mathbf{I} \tag{7.180}$$

$$\nabla_{\mathbf{z}^*}\mathbf{z} = \mathbf{I} \tag{7.181}$$

$$\nabla_{\mathbf{z}^*}\mathbf{z}^* = \mathbf{O} \tag{7.182}$$

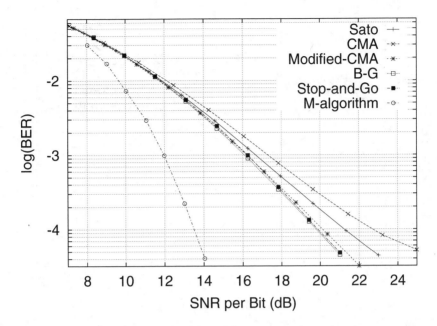

Figure 7.38: Gaussian BER versus SNR curves for the Shortened WATM channel of Figure 7.25(b) for 16-QAM using a 30-tap equaliser obeying the schematic of Figure 7.19 and, for the M-Algorithm, a 3-path channel estimation obeying the schematic of Figure 7.21

$$\nabla_{\mathbf{z}}(\mathbf{a}^T \cdot \mathbf{b}) = \nabla_{\mathbf{z}}\mathbf{b} \cdot \mathbf{a} + \nabla_{\mathbf{z}}\mathbf{a} \cdot \mathbf{b}, \qquad (7.183)$$

where $\mathbf{O}$ is the zero matrix and $\mathbf{I}$ is the unit matrix. These properties can be readily verified by applying the definition of the gradient vector of Equations 7.177 and 7.178 and the properties of differentiation of real functions. We will also highlight a few more properties, which we will invoke in order to derive the analytical expressions of the cost-function derivatives with

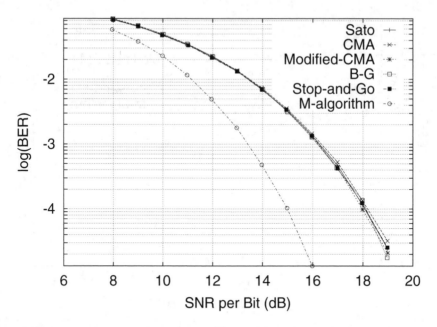

Figure 7.39: Gaussian BER versus SNR curves for the one-symbol-delay channel of Figure 7.26 for 16-QAM using a 10-tap equaliser, obeying the schematic of Figure 7.19 and, for the M-Algorithm, a 2-path channel estimation obeying the schematic of Figure 7.21

respect to the equaliser tap vector of $\mathbf{c}$:

$$
\nabla_{\mathbf{z}}(f(\mathbf{a}))^k = \begin{bmatrix} \frac{\partial}{\partial x_1} + j\frac{\partial}{\partial y_1} \\ \frac{\partial}{\partial x_2} + j\frac{\partial}{\partial y_2} \\ \vdots \\ \frac{\partial}{\partial x_N} + j\frac{\partial}{\partial y_N} \end{bmatrix} (f(\mathbf{a}))^k
$$

$$
= \begin{bmatrix} \frac{\partial (f(\mathbf{a}))^k}{\partial x_1} + j\frac{\partial (f(\mathbf{a}))^k}{\partial y_1} \\ \frac{\partial (f(\mathbf{a}))^k}{\partial x_2} + j\frac{\partial (f(\mathbf{a}))^k}{\partial y_2} \\ \vdots \\ \frac{\partial (f(\mathbf{a}))^k}{\partial x_N} + j\frac{\partial}{\partial y_N} \end{bmatrix}
$$

$$
= k \cdot (f(\mathbf{a}))^{k-1} \begin{bmatrix} \frac{\partial f(\mathbf{a})}{\partial x_1} + j\frac{\partial f(\mathbf{a})}{\partial y_1} \\ \frac{\partial f(\mathbf{a})}{\partial x_2} + j\frac{\partial f(\mathbf{a})}{\partial y_2} \\ \vdots \\ \frac{\partial f(\mathbf{a})}{\partial x_N} + j\frac{\partial f(\mathbf{a})}{\partial y_N} \end{bmatrix}
$$

$$
= k \cdot (f(\mathbf{a}))^{k-1}\nabla_{\mathbf{z}}f(\mathbf{a}). \tag{7.184}
$$

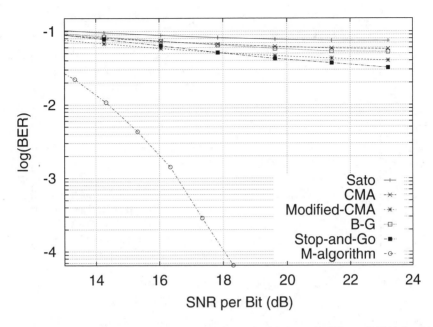

Figure 7.40: Gaussian BER versus SNR curves for the WATM channel of Figure 7.25(a) for 64-QAM using a 68-tap equaliser obeying the schematic of Figure 7.19 and, for the M-Algorithm, a 5-path channel estimation obeying the schematic of Figure 7.21

These properties are useful for the minimisation of the blind equaliser cost-functions given in Equations 7.123, 7.116, 7.128 for the CMA, for Sato's algorithm and for the Modified CMA, respectively. In the next section we will deal with the derivation of the tap-update equations of a specific blind equalisation algorithm, namely the CMA and we will give the minimisation procedure for this algorithm. Three very useful relationships which we will invoke in the next

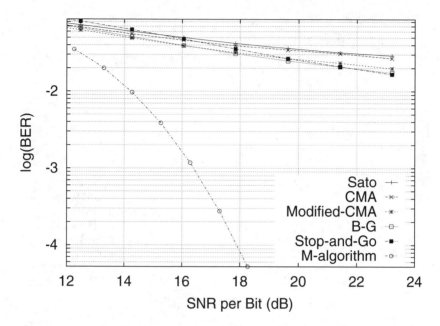

Figure 7.41: Gaussian BER versus SNR curves for the Shortened WATM channel of Figure 7.25(b) for 64-QAM using a 30-tap equaliser obeying the schematic of Figure 7.19

section come from the direct application of Equation (7.184), yielding:

$$
\begin{aligned}
\nabla_{\mathbf{c}} |z(n)| &= \nabla_{\mathbf{c}} \sqrt{|z(n)|^2} \\
&= \frac{1}{2\sqrt{|z(n)|^2}} \nabla_{\mathbf{c}} |z(n)|^2 \\
&\overset{(7.184)}{=} \frac{1}{2|z(n)|} \nabla_{\mathbf{c}} (\mathbf{c}^T \mathbf{y}(n))(\mathbf{c}^H \mathbf{y}^*(n)) \\
&= \frac{1}{2|z(n)|} [(\mathbf{c}^T \mathbf{y}(n)) \nabla_{\mathbf{c}} (\mathbf{c}^H \mathbf{y}^*(n)) + \\
&\quad + (\mathbf{c}^H \mathbf{y}^*(n)) \nabla_{\mathbf{c}} (\mathbf{c}^T \mathbf{y}(n))] \\
&\overset{(7.183)}{=} \frac{1}{2|z(n)|} [(\mathbf{c}^T \mathbf{y}(n)) \{(\nabla_{\mathbf{c}} \mathbf{y}^*(n)) \cdot \mathbf{c}^* + \\
&\quad + (\nabla_{\mathbf{c}} \mathbf{c}^*) \cdot \mathbf{y}^*(n)\} \\
&\quad + (\mathbf{c}^H \mathbf{y}^*(n)) \{(\nabla_{\mathbf{c}} \mathbf{c}) \cdot \mathbf{y}(n) + \\
&\quad + (\nabla_{\mathbf{c}} \mathbf{y}^T(n)) \cdot \mathbf{c}\}] \\
&\overset{(7.179)-(7.182)}{=} \frac{(\mathbf{c}^T \cdot \mathbf{y}(n)) \mathbf{y}^*(n)}{2|z(n)|} \\
&= \frac{z(n) \cdot \mathbf{y}^*(n)}{2|z(n)|} \tag{7.185}
\end{aligned}
$$

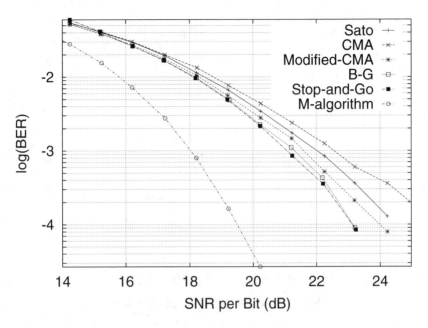

Figure 7.42: Gaussian BER versus SNR curves for the one-symbol-delay channel of Figure 7.26 for 64-QAM using a 10-tap equaliser, obeying the schematic of Figure 7.19 and, for the M-Algorithm, a 2-path channel estimation obeying the schematic of Figure 7.21

$$
\begin{aligned}
\nabla_{\mathbf{c}} Re\{z(n)\} &= \nabla_{\mathbf{c}} \frac{z(n) + z^*(n)}{2} \\[2mm]
&= \nabla_{\mathbf{c}} \frac{\mathbf{c}^T \mathbf{y}(n) + \mathbf{c}^H \mathbf{y}^*(n)}{2} \\[2mm]
&\stackrel{(7.183)}{=} \frac{(\nabla_{\mathbf{c}} \mathbf{c}) \cdot \mathbf{y}(n) + (\nabla_{\mathbf{c}} \mathbf{c}^*) \cdot \mathbf{y}^*(n)}{2} \\[2mm]
&\stackrel{(7.179)-(7.183)}{=} \frac{1}{2} \cdot \mathbf{y}^*(n)
\end{aligned}
\tag{7.186}
$$

$$
\begin{aligned}
\nabla_{\mathbf{c}} Im\{z(n)\} &= \nabla_{\mathbf{c}} \frac{z(n) - z^*(n)}{2j} \\[2mm]
&= \nabla_{\mathbf{c}} \frac{\mathbf{c}^T \mathbf{y}(n) - \mathbf{c}^H \mathbf{y}^*(n)}{2j} \\[2mm]
&\stackrel{(7.183)}{=} \frac{(\nabla_{\mathbf{c}} \mathbf{c}) \cdot \mathbf{y}(n) - (\nabla_{\mathbf{c}} \mathbf{c}^*) \cdot \mathbf{y}^*(n)}{2j} \\[2mm]
&\stackrel{7.179-7.183}{=} \frac{j}{2} \cdot \mathbf{y}^*(n),
\end{aligned}
\tag{7.187}
$$

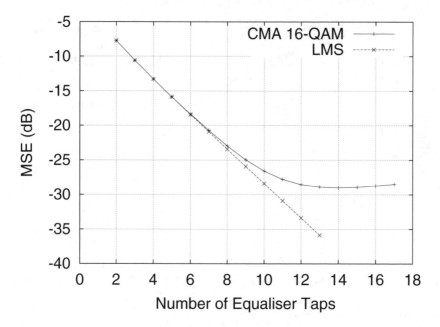

Figure 7.43: MSE as a function of the equaliser order for 16-QAM using the CMA-based equaliser and a noiseless scenario. The channel model used is the one-symbol delay channel as given in Figure 7.26. The LMS-based MSE is given as a lower limit, as this is a non-blind equaliser which gives a benchmark performance. Clearly, for small number of taps the blind equaliser curve coincides with the LMS benchmark curve, while for larger number of taps the CMA produces large MSE, attaining a local minimum at about 14 taps, which is the ideal setting for this scenario.

where $z(n) = \mathbf{c}^T \mathbf{y}(n)$ is the equalised signal in Figure 7.15, expressed as the product of the equaliser tap vector $\mathbf{c}$ and the received signal vector $\mathbf{y}(n)$, which forms the convolution of the received signal with the equaliser's impulse response.

7.9.1 An Illustrative Example: CMA Cost-Function Minimisation

The CMA's cost-function was defined in Equation (7.123), which is repeated here for convenience:

$$J(n) = E[(|z(n)|^2 - R_2)^2]$$

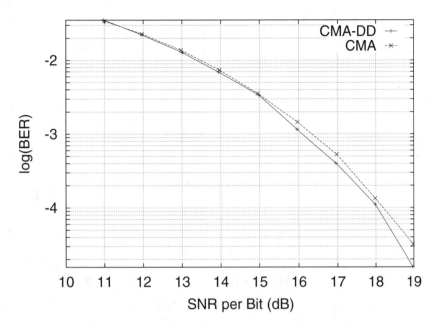

Figure 7.44: Gaussian BER versus SNR curves for the one-symbol-delay channel of Figure 7.26 for 16-QAM using a 10-tap equaliser obeying the schematic of Figure 7.19 for the CMA using switching to decision-directed equalisation after convergence

If we omit the expectation, then its gradient vector, with respect to **c**, is :

$$
\begin{aligned}
\nabla_{\mathbf{c}} J(n) &= \nabla_{\mathbf{c}} \left(|z(n)|^2 - R_2 \right)^2 \\
&\overset{(7.184)}{=} 2 \cdot \left(|z(n)|^2 - R_2 \right) \cdot \nabla_{\mathbf{c}} \left(|z(n)|^2 - R_2 \right) \\
&= \left(|z(n)|^2 - R_2 \right) \cdot \nabla_{\mathbf{c}} z(n) \cdot z^*(n) \\
&= 2 \cdot \left(|z(n)|^2 - R_2 \right) \cdot z(n)' \cdot \nabla_{\mathbf{c}} z^*(n) \\
&= 2 \cdot \left(|z(n)|^2 - R_2 \right) \cdot z(n) \cdot \nabla_{\mathbf{c}} \mathbf{c}^H \mathbf{y}^*(n) \\
&\overset{(7.183)}{=} 2 \cdot \left(|z(n)|^2 - R_2 \right) \cdot z(n) \cdot \mathbf{y}^*(n).
\end{aligned}
\tag{7.188}
$$

This is exactly the equaliser coefficients' update term of Equation (7.125).

7.10 Appendix: Polycepstra Definitions

Based on [3, 89, 442, 454], in this appendix we give some fourth-order statistics definitions and properties, which are useful for the blind equalisation algorithms of Section 7.7.9. The definitions are based on [89], in which the polycepstra-based algorihms first appeared and on [454], where the super-exponential algorithm was proposed, using fourth-order cumulants.

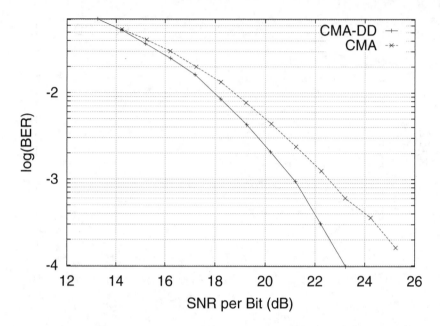

Figure 7.45: Gaussian BER versus SNR curves for the one-symbol-delay channel of Figure 7.26 for 64-QAM using a 10-tap equaliser obeying the schematic of Figure 7.19 for the CMA using switching to decision-directed equalisation after convergence

In general, the joint cumulant of the random variables $x_{n_1}, x_{n_2}, \ldots x_{n_m}$ is given by [454]:

$$L(x_{n_1}; x_{n_2}; \cdots ; x_{n_m}) = (-j)^m \frac{\partial^m ln\phi(\omega)}{\partial \omega_{n_1} \ldots \partial \omega_{n_m}|_{\omega=0}}, \tag{7.189}$$

where $\phi(\omega)$ is the so-called joint characteristic function of the random variables x_i, defined as [478]:

$$\phi(\omega) = E\left[e^{j\omega \mathbf{x}}\right] = E\left[e^{j\sum_{i=1}^{m} \omega_i x_i}\right]. \tag{7.190}$$

For zero-mean random variables and cumulants up to fourth-order, the definition is equivalent to [89]:

$$
\begin{aligned}
L(x_1) &= 0 & (7.191)\\
L(x_1; x_2) &= E[x_1 \cdot x_2] & (7.192)\\
L(x_1; x_2; x_3) &= E[x_1 \cdot x_2 \cdot x_3] & (7.193)\\
L(x_1; x_2; x_3; x_4) &= E[x_1 \cdot x_2 \cdot x_3 \cdot x_4] - \\
& \quad E[x_1 \cdot x_2] \cdot E[x_3 \cdot x_4] - \\
& \quad E[x_1 \cdot x_3] \cdot E[x_2 \cdot x_4] - \\
& \quad E[x_1 \cdot x_4] \cdot E[x_2 \cdot x_3] & (7.194)
\end{aligned}
$$

When the random variables are actually samples from a zero-mean stochastic process, such as the received signal $y(k)$, then the fourth-order cumulant is given by:

$$\begin{aligned}
L_y(m,n,l) &= E[y(k)y(k+m)y(k+n)y(k+l)] - \\
&\quad E[y(k)y(k+m)] \cdot E[y(k)y(k+l-n)] - \\
&\quad E[y(k)y(k+n)] \cdot E[y(k)y(k+l-m)] - \\
&\quad E[y(k)y(k+l)] \cdot E[y(k)y(k+m-n)].
\end{aligned}$$

(7.195)

Following the above definition, we will now summarise some of the associated cumulant properties. First of all, the cumulant of the sum of two independent and zero-mean stationary signals $x_1(k)$ and $x_2(k)$ is equal to the sum of the cumulants of $x_1(k)$ and $x_2(k)$, i.e.:

$$L_{x_1+x_2}(m,n,l) = L_{x_1}(m,n,l) + L_{x_2}(m,n,l).$$

(7.196)

Furthermore, the cumulant of the convolution of two zero-mean stationary signals corresponds to the convolution of their cumulants in the cumulant domain:

$$L_{x_1 * x_2}(m,n,l) = L_{x_1}(m,n,l) * L_{x_2}(m,n,l),$$

(7.197)

where $*$ stands for convolution. Two further properties related to the signal distributions:

- If $x(k)$ is Gaussian, then the cumulant of $x(k)$ is 0

- If $x(k)$ is i.i.d., then the cumulant of $x(k)$ is given by:

$$L_x(m,n,l) = K(x(k)) \cdot \delta(m,n,l),$$

(7.198)

where $K(x(k))$ is the Kurtosis of $x(k)$ defined as:

$$K(x(k)) = E[|x(k)|^4] - 3\left(E[x^2(k)]\right)^2$$

(7.199)

for real-valued signals Equation (7.127) defined the Kurtosis for complex-valued signals, which is repeated here:

$$K(z) = E[|z(n)|^4] - 2 \cdot E^2[|z(n)|^2] - |E[z^2(n)]|^2.$$

Finally, below we will now give a cumulant property of stochastic signals passing through linear systems. We assume that the i.i.d. signal $x(k)$ is input to a system having an impulse response h_k. The output of the system is given by:

$$y(k) = \sum_i h_i \cdot a(k-i).$$

(7.200)

The cumulant of the output is then given by:

$$L_y(n,k,l) = K(a(k)) \cdot L_h(m,n,l),$$

(7.201)

where $L_h(m, n, l)$ is the fourth-order moment of the CIR. Here, the moment has replaced the cumulant because $h(k)$ is a deterministic signal. We can now form the cumulant of the received signal $y(k)$, which is given by the convolution of the i.i.d. input signal $a(k)$ with the CIR h_k plus the additive Gaussian noise:

$$y(k) = h_k * x(k) + e(k) \Rightarrow \qquad (7.202)$$

$$L_y(m, n, l) = K(x(k)) \cdot L_h(m, n, l), \qquad (7.203)$$

where the Gaussian noise term vanishes, since its fourth-order cumulant is zero. We will now define the tricepstrum of a signal $x(k)$ as:

$$c_x(m, n, l) = Z_{(3)}^{-1}\{ln(Z_{(3)}\{L_x(m, n, l)\})\}, \qquad (7.204)$$

where $Z_{(3)}$ stands for the 3D z-transform. Complicated as this definition may seem, it simplifies the cumulant-based relationship of Equation (7.203) to the tricepstrum domain as:

$$c_y(m, n, l) = c_h(m, n, l) \qquad (m, n, l) \neq (0, 0, 0), \qquad (7.205)$$

which means that the tricepstrum of the received signal is the same as the tricepstrum of the CIR, provided that the input is i.i.d. and the noise is Gaussian distributed. By measuring the tricepstrum of the received signal, we can now have an estimate of the tricepstrum of the CIR. It can be shown that the following identity holds [89]:

$$\sum_{I=1}^{p} A^{(I)}[L_y(m - I, n, l) - L_y(m + I, n + I, l + I)] +$$

$$+ \sum_{J=1}^{q} B^{(J)}[L_y(m - J, n - J, l - J) - L_y(m + J, n, l)] =$$

$$= -mL_y(m, n, l),$$

$$(7.206)$$

where

$$K \cdot c_y(K, 0, 0) = \begin{cases} -A^{(K)}, & K = 1, \dots, p \\ B^{(-K)}, & K = -1, \dots, -q. \end{cases} \qquad (7.207)$$

This relationship holds for any (m, n, l). Nonetheless, a rule of thumb, suggested in [89] is to define $w = max(p, q)$, $z \leq w/2$, $s \leq z$ and then select $m = -w, \dots, -1, 1, \dots, w$, $n = -z, \dots, 0, \dots, z$ and $l = -s, \dots, 0, \dots, s$. It is shown in [89] that $A^{(I)}, B^{(J)}$ are the minimum- and maximum-phase differential cepstrum coefficients related to the zeros of the CIR. Thus, estimating $A^{(I)}, B^{(J)}$ from the system of Equations (7.206) means that we can invert the CIR apart from an uncertainty of the magnitude A. Theoretically, it is required that $p, q \to \infty$, but in practice we can use a sufficiently large discrete value for each of them. Equations (7.206) can then be rewritten in matrix form as:

$$\mathbf{P} \cdot \mathbf{a} = \mathbf{p}, \qquad (7.208)$$

where $\mathbf{P}$ is the matrix containing the L_y-elements of the left-hand side of Equation (7.206), $\mathbf{a}$ is the vector containing $A^{(I)}$ and $B^{(J)}$ and $\mathbf{p}$ is the vector containing the L_y-elements of the right-hand side of Equation (7.206). We have to estimate $\mathbf{a}$, in order to invert the CIR. This can be carried out upon invoking the LMS solution of Equation (7.208), as in [89], yielding:

$$\hat{\mathbf{a}}^{(i+1)} = \hat{\mathbf{a}}^{(i)} + \mu^{(i)} \cdot (\hat{\mathbf{P}}^{(i)})^H \cdot (\hat{\mathbf{p}}^{(i)} - \hat{\mathbf{P}}^{(i)} \cdot \hat{\mathbf{a}}^{(i)}) \tag{7.209}$$

with

$$0 < \mu^{(i)} < 2/tr\{(\hat{\mathbf{P}}^{(i)})^H \cdot \hat{\mathbf{P}}^{(i)}\},$$

where $tr()$ stands for the trace of a matrix. When $\mu^{(i)}$ is in this region, stability is ensured, as shown in [3].

The exact relationship of the coefficients $A^{(I)}$ and $B^{(I)}$ with the poles and zeros of the equaliser filters' polynomials is as follows. If the z-domain channel transfer function polynomial $H(z)$ is given by $H(z) = I(z^{-1}) \cdot O(z)$, as in Equation (7.169), where

$$I(z^{-1}) = \frac{\Pi_{i=1}^{L_1}(1 - a_i z^{-1})}{\Pi_{i=1}^{L_3}(1 - c_i z^{-1})} \tag{7.210}$$

with $|a_i| < 1$ and $|c_i| < 1$ and

$$O(z) = \Pi_{i=1}^{L_2}(1 - b_i z) \tag{7.211}$$

with $b_i < 1$, then the coefficients $A^{(I)}$ and $B^{(I)}$ are given by:

$$A^{(I)} = \sum_{i=1}^{L_1} a_i^I - \sum_{i=1}^{L_3} c_i^I \tag{7.212}$$

$$B^{(I)} = \sum_{i=1}^{L_2} b_i^I \tag{7.213}$$

Having estimated $A^{(I)}$ and $B^{(I)}$ we can now use Equation (7.213) to estimate the time-domain coefficients $\hat{i}_{inv}^{(i)}(k)$ and $\hat{o}_{inv}^{(i)}(k)$, which are the coefficients of $1/I(z^{-1})$ and $1/O(z)$, respectively, as in [89] :

$$\hat{i}_{inv}^{(i)}(k) = -\frac{1}{k} \sum_{n=2}^{k+1} [-A^{(n-1),(i)}] \cdot \hat{i}_{inv}^{(i)}(k - n + 1), \quad k = 1, \ldots, N_1 \tag{7.214}$$

$$\hat{o}_{inv}^{(i)}(k) = \frac{1}{k} \sum_{n=k+1}^{0} [-B^{(1-n),(i)}] \cdot \hat{o}_{inv}^{(i)}(k - n + 1), \quad k = -1, \ldots, -N_2, \tag{7.215}$$

where the superscript i denotes the i-th iteration of the algorithm. The equaliser's impulse response is finally given by the convolution $\hat{i}_{inv}^{(i)}(k) * \hat{o}_{inv}^{(i)}(k)$. In the case of a DFE we also

$$\sum_{I=1}^{p} A^{(I)}[L_y(m-I,n,l) - L_y(m+I,n+I,l+I)] +$$
$$+\sum_{J=1}^{q} B^{(J)}[L_y(m-J,n-J,l-J) - L_y(m+J,n,l)] =$$
$$= -mL_y(m,n,l)$$

$$\hat{\mathbf{a}}^{(i)} + \mu^{(i)} \cdot (\hat{\mathbf{P}}^{(i)})^H \cdot (\hat{\mathbf{p}}^{(i)} - \hat{\mathbf{P}}^{(i)} \cdot \hat{\mathbf{a}}^{(i)}))$$
$$\mathbf{a} = (A^{(1)}, \dots, A^{(p)}, B^{(1)}, \dots, B^{(q)})^T$$
$$[\mathbf{p}]_i = -m \cdot L_y(m,n,l)$$
$$[\mathbf{P}]_{ij} = L_y(m-I,n,l) - L_y(m+I,n+I,l+I)$$
$$i = 1, \dots, p$$
$$[\mathbf{P}]_{ij} = L_y(m-J,n-J,l-J) - L_y(m+J,n,l)$$
$$i = p+1, \dots, p+q$$
$$w = max(p,q),\ z \le w/2,\ s \le z$$
$$m = -w, \dots, -1, 1, \dots, w$$
$$n = -z, \dots, 0, \dots, z$$
$$l = -s, \dots, 0, \dots, s$$

$$\hat{i}_{inv}^{(i)}(k) = -\frac{1}{k}\sum_{n=2}^{k+1}[-A^{(n-1),(i)}] \cdot \hat{i}_{inv}^{(i)}(k-n+1)$$
$$k = 1, \dots, N_1$$

$$\hat{o}_{inv}^{(i)}(k) = \frac{1}{k}\sum_{n=k+1}^{0}[-B^{(1-n),(i)}] \cdot \hat{o}_{inv}^{(i)}(k-n+1)$$
$$k = -1, \dots, -N_2$$

$$\hat{i}_o^{(i)}(k) = \frac{1}{k}\sum_{n=2}^{k+1}[A^{(n-1),(i)} + (B^{(n-1),(i)})^*]\hat{i}_o^{(i)}(k-n+1)$$
$$k = 1, \dots, N_3$$

$$\hat{o}_o^{(i)}(k) = -\frac{1}{k}\sum_{n=2}^{k+1}[B^{(n-1),(i)}]^* \cdot \hat{o}_o^{(i)}(k-n+1)$$
$$k = 1, \dots, N_4$$

Table 7.5: A synopsis of the equations related to the TEA

have to estimate the inverse z-transforms of $I(z^{-1})$ and $O^*(z^{-1})$, which is given by [89]:

$$\hat{i}_o^{(i)}(k) = \frac{1}{k}\sum_{n=2}^{k+1}[A^{(n-1),(i)} + (B^{(n-1),(i)})^*]\hat{i}_o^{(i)}(k-n+1)$$
$$k = 1, \dots, N_3 \tag{7.216}$$

$$\hat{o}_o^{(i)}(k) = -\frac{1}{k}\sum_{n=2}^{k+1}[B^{(n-1),(i)}]^* \cdot \hat{o}_o^{(i)}(k-n+1), \quad k = 1, \dots, N_4. \tag{7.217}$$

Finally, the feedforward and feedback equaliser impulse responses are now extracted as $\hat{o}_o^{(i)}(k) * \hat{o}_{inv}^{(i)}(k)$ and $\hat{i}_{inv}^{(i)}(k) * \hat{o}_o^{(i)}(k) - \delta(k)$, respectively. The channel attenutation parameter A can be estimated using Automatic Gain Control (AGC). A synopsis of the equations involved in the TEA is given in Table 7.10.

Chapter **8**

Classic QAM Modems for Bandlimited AWGN Channels

8.1 Introduction

The most widely used AWGN communication channel is the conventional telephone line. Since its amplitude and group delay characteristics vary slowly, mainly due to ageing, humidity, etc., telephone lines can be considered as linear and time invariant. They are typically bandlimited to 300-3000 Hz and have SNR values around 30 dB. The international lines must comply with the CCITT M1020 Recommendation, the amplitude and group delay characteristics of which were shown in Figure 2.5.

The early history of modems designed for AWGN channels is comprehensively documented for example in [492]. Early data transmission modems designed in the 1950s used frequency shift keying (FSK) at 300 bps for switched telephone lines and at 1200 bps for private or leased lines. From the 1960s onward, in conjunction with the upgrading of the telephone network, the equalised bandwidth was gradually extended to 1200 Hz, 1600 Hz and 2400 Hz. This allowed signalling rates to increase from 1200 Hz to 2400 Hz, and the higher SNR values facilitated the transmission of several bits per sample.

An important milestone was the launch of the first four-phase 2400 bps modem in 1962, which was later standardised in the V.26 CCITT Recommendation. This was followed in 1967 by the introduction of the eight-phase differential phase shift keying (8-DPSK), 4800 bps modem using a signalling rate of 1600 Bd and 3 bits/sample. To achieve this signalling rate over leased lines a manual channel equaliser (EQ) was needed, as specified by the V.27 CCITT Recommendation. However, with rapid advances in digital equalisation, the V.27bis and V.27ter Recommendations quickly followed, proposing adaptive channel equalisation, which allowed operation via switched telephone lines.

In 1971 the first 4 bits/symbol 16-QAM modem was developed, which deployed an advanced channel equaliser to achieve a signalling rate of 2400 Bd. This modem transmitted at 9600 bps over point-to-point four-wire leased lines and it was standardised in 1976 in the V.29 CCITT Recommendation. This development was further improved to facilitate 9600 bps

Rate (bps)	Baud rate (Bd)	Constellation	Standard	Features
2400	1200	4-DPSK	V.26, 1968	fixed EQ
4800	1600	8-DPSK	V.27, 1972	manual EQ
9600	2400	16-QAM	V.29, 1976	adapt. EQ
9600	2400	32-QAM, TC	V.32, 1984	adapt. EQ
14 400	2400	128-QAM, TC	V.33, 1988	adapt. EQ

Table 8.1: Summary of modem features

transmissions via two-wire switched telephone lines and it was standardised in the V.32 Recommendation in 1984. Although the basic configuration uses a square-shaped 16-QAM constellation, it also has a 32-point trellis coded option to improve its robustness against channel errors. The most advanced QAM scheme at the time of writing, standardised in 1988 in the V.33 Recommendation, is the 14 400 bps point-to-point four-wire modem, which is recommended for leased lines. It uses trellis coded (TC) 128-QAM for 14 400 bps transmissions, but it also allows trellis coded 64-QAM transmissions at a fallback rate of 12 000 bps over lower quality channels. The history of point-to-point multilevel modems is summarised in Table 8.1.

8.2 Trellis Coding Principles

Historically, modulation and forward error correction (FEC) coding [171] have been treated as distinct subjects, where FEC is typically invoked in order to reduce the BER of the modem. However, in his visionary paper of 1974 Massey [493] surmised that treating modulation and FEC as an integrated entity would allow significant coding gains to be achieved over power and bandlimited transmission media. The first practical coded modulation scheme was proposed by Imai and Hirakawa [494], shortly followed by Ungerböck's more frequently referenced papers [495, 496], where trellis coded modulation (TCM) is introduced. Here we refrain from providing a detailed discourse on the family of coded modulation schemes, since they are discussed in great detail in Chapter 24 and adopt a simple practical approach for characterising their employment in classic telephone-line modems.

During the 1980s TCM reached a state of maturity and has been incorporated into the CCITT V. series multilevel modems designed for Gaussian telephone channels. With the emergence of widespread digital communications via hostile fading mobile radio channels, researchers turned their attention towards adapting TCM principles for these applications, which led to the invention of block coded modulation (BCM) [497]. The importance of coded modulation is evidenced in the large number of conferences [497], special issues [498], and books [499] devoted to this subject.

The motivation for TCM is based on the disadvantages of the conventional approach of using separate modulation and coding blocks. In power-limited applications, such as satellite channels, FEC codes can be invoked in order to improve power efficiency at the cost of expanding the bandwidth required. In a bandwidth-limited scenario, more bits/symbol can be transmitted but at the cost of increased power requirements. In trellis coded modulation the advantages of higher-level modulation are combined with those of FEC, where modulation/encoding and demodulation/FEC decoding form an integral operation.

TCM and BCM were initially deployed over telephone channels. The telephone channel has historically been the scene of the earliest applications of many efficient modulation systems. This is because this ubiquitous medium constitutes a benign channel, being relatively time invariant with a high SNR, normally modelled as 28 dB. However, the channel is severely bandlimited over the range 300-3000 Hz. Calculations of channel capacity [492] showed that uncoded modulation techniques were not realising the full potential of the channel, as also suggested by Figure 2.4. For example, if one-dimensional pulse amplitude modulation (PAM) is transmitted with m bits per dimension using one of the 2^m equally likely levels, normally $\pm 1, \pm 3, \pm 5, ..., \pm(2^m - 1)$, then the average energy of each coordinate is given by

$$S_m = (4^m - 1)/3 \tag{8.1}$$

and it then follows that

$$S_{m+1} = 4S_m + 1. \tag{8.2}$$

Therefore, it takes approximately four times as much energy (6 dB) to send a further 1 bit/dimension, or 2 bits/symbol. The maximum possible information transmission rate over a noisy channel was determined by Shannon [500] by defining the so-called channel capacity C (in bits per dimension) as

$$C = \frac{1}{2} \log_2(1 + S/N) \ \ bits/dimension, \tag{8.3}$$

where S is the signal energy, N is the noise energy, and so S/N is the SNR. When $S/N \gg 1$ it takes approximately four times as much energy to increase the capacity by 1 bit/dimension. Therefore, the ratio of bits/dimension to channel capacity, which is a good measure of how efficiently the channel is being exploited, approaches 1 as S becomes large. How this high efficiency is achieved is not stated by Shannon, but adding more modulation levels until the maximum channel capacity is reached would initially seem the best approach. Unfortunately, high-level modulation schemes are prone to error and hence joint modulation and coding might contribute better towards approaching the predicted maximum channel efficiency.

However, many authors [493, 501] regard a parameter R_0, defined below, as a better estimate of the maximum rate that is practically achievable [502] where

$$R_0 = \frac{1}{2} \log_2(1 + S/2N) \ \ bits/dimension. \tag{8.4}$$

According to Equation 8.4 it therefore takes 3 dB more power to signal at a certain number of bits/dimension. The practical potential saving using coding is therefore probably about 3 dB/dimension, or 6 dB for two-dimensional QAM.

This gain was not achievable using standard coding techniques. It was this which led workers to investigate alternative coding techniques that would actually realise this coding gain.

While considerable work has been performed on channel coding [503–507], here we consider a different approach where coding implies the introduction of interdependencies between sequences of signal points such that not all sequences are possible. The surprising consequence of this is that the minimum distance d_0 between two uncoded sequences in two-dimensional space is increased to d_{min} in an expanded N-dimensional space when coding

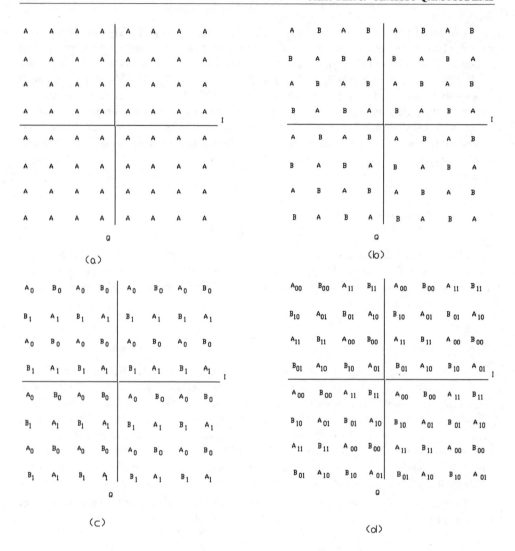

Figure 8.1: Mapping by set partition

is used. Use of maximum likelihood sequence detection at the receiver yields a coding gain factor of d_{min}^2/d_0^2 in energy efficiency, less the amount of extra energy needed for the transmission of the additional redundant coding bits. Conventional channel coding cannot realise this gain. Instead it is necessary to use a principle described by Ungerböck [496] as "mapping by set partition".

The basis of set partitioning is shown in Figure 8.1 and described below. The original constellation shown in Figure 8.1(a), which is 64-level square QAM in this example, can be initially divided into two subsets (Figure 8.1(b)) by assigning alternate points A and B to each subset. This has the result that the minimum squared distance between points within a subset is twice that between points in the original constellation given by d_0^2. Observe that the

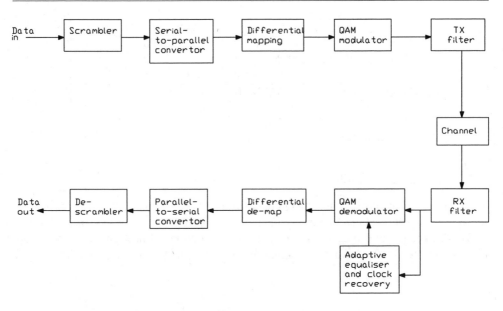

Figure 8.2: General modem schematic for bandlimited AWGN channels

points in each subset now lie on a rectangular grid rotated by 45° with respect to the original grid. This procedure can be repeated to yield 4, 8, 16, ... subsets with similar properties and intrasubset squared distances of 4, 8, 16, ... times d_0^2. This is illustrated in Figure 8.1 for 2 subsets (A, B) of 32 points, 4 subsets (A_0, A_1, B_0, B_1) of 16 points and 8 subsets (A_{00}, ..., A_{11}, B_{00}, ..., B_{11}) of 8 points. A relatively simple coding technique can then be employed. Some of the incoming six data bits of a modulation symbol are encoded in a channel coder resulting in a larger number of bits due to the coding process. The additional protection bits are used to select which subset the phasor representing the original input bits will be assigned to. The remaining uncoded bits select the specific constellation point to be transmitted within the subset selected. The coding can be either block or convolutional, although convolutional coding has consistently proved both more effective and simpler to implement for Gaussian channels, and so we concentrate our efforts on convolutionally coded TCM.

In the following sections we consider how modems have evolved towards high capacity systems by considering the family of classic CCITT, or synonymously referred to as ITU recommendations for modems [508], commencing with the V.29 modem.

8.3   V.29 Modem

The 9600 bps V.29 modem was recommended for use over point-to-point four-wire leased telephone lines, but its users are not precluded from using it over lower quality links. There are provisions for optional fallback rates of 7200 and 4800 bps when the channel conditions are hostile. A generic block diagram for all the modems we consider in this chapter is shown in Figure 8.2.

Absolute phase	Q_1	Relative signal element amplitude
0 deg, 90 deg, 180 deg, 270°	0	3
	1	5
45 deg, 135 deg, 225 deg, 315°	0	$\sqrt{2}$
	1	$3\sqrt{2}$

Table 8.2: Phasor amplitude mapping for the V.29 modem

8.3.1 Signal Constellation

The 9600 bps input data stream is "scrambled" using a linear feedback shift register in order to assist the carrier and clock recovery operation by preventing a long sequence of logical 1's or 0's. The scrambled data sequence is then converted into 4 bits/symbol called "quad-bits", which are Gray coded onto the signal constellation depicted in Figure 8.3. The first bit

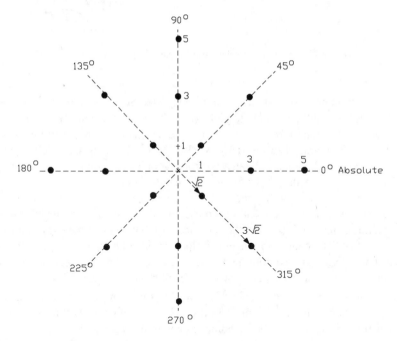

Figure 8.3: Phasor constellation for 9600 bps operation

(Q_1) of every symbol determines the amplitude of the signal element to be transmitted, but the amplitude is also dependent on the instantaneous absolute phase, and the two different sets of amplitudes used are given in Table 8.2. The remaining three bits Q_2, Q_3 and Q_4 are differentially Gray coded onto the phase positions, as summarised in Table 8.3. After differential mapping of the bits, 16-QAM modulation is carried out using the signal constellation of Figure 8.3 at a signalling rate of 2400 Bd.

At the fallback rate of 7200 bps the bit Q_1 is permanently set to zero, limiting the ampli-

Q2	Q3	Q4	Phase change
0	0	1	0°
0	0	0	45°
0	1	0	90°
0	1	1	135°
1	1	1	180°
1	1	0	225°
1	0	0	270°
1	0	1	315°

Table 8.3: Phasor phase mapping for the V.29 modem

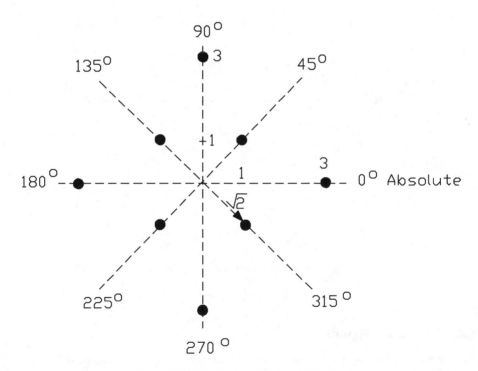

Figure 8.4: Signal space diagram at 7200 bps

Data bits		Quadbits				Phase change
		Q1	Q2	Q3	Q4	
0	0	0	0	0	1	0°
0	1	0	0	1	0	90°
1	1	0	1	1	1	180°
1	0	0	1	0	0	270°

Table 8.4: Mapping table for 4800 bps operation

tude to either $\sqrt{2}$ or 3, depending on the phase information, as seen in Figure 8.4. Otherwise the mapping is identical to that of the 9600 bps constellation. The bit rate is 2400 $Bd \times 3$ bits/symbol = 7200 bps. In case of 4800 bps transmissions $Q_1 = 0$ and $Q_4 = \overline{(Q_2 \oplus Q_3)}$ are set, where $\oplus$ means modulo 2 addition and ($\overline{\bullet}$) is the inverse of ($\bullet$), yielding the mapping summarised in Table 8.4. The resulting constellation is shown in Figure 8.5 and the 2 bits/symbol transmissions at 2400 Bd yield a bit rate of 4800 bps.

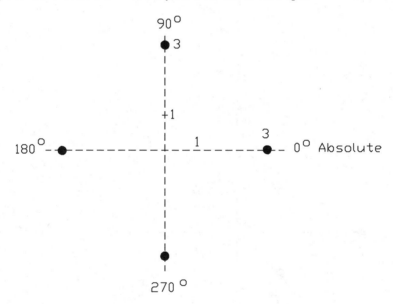

Figure 8.5: Signal space diagram at 4800 bps

8.3.2 Training Signals

The operation of the adaptive channel equaliser, carrier recovery and timing recovery circuits is aided by the transmission of specific training signals which allow characterisation of the telephone line in terms of its impulse response, and the establishment of timing references. These are summarised in Table 8.5 for the V.29 modem. The training sequence is split into four segments, each containing different signals. Segment 1 contains no energy and forms a so-called precursor. Segment 2 consists of the alternate sequence of phasors A and B for

Training segments	Type of line signal	No. sym. intervals	Approx. time (ms)
Segment 1	No transmitted energy	48	20
Segment 2	Alternations	128	53
Segment 3	Equaliser conditioning pattern	384	160
Segment 4	Scrambled all binary 1's	48	20
Total signal	Total synchronising signal	608	253

Table 8.5: Training signals for the V.29 modem

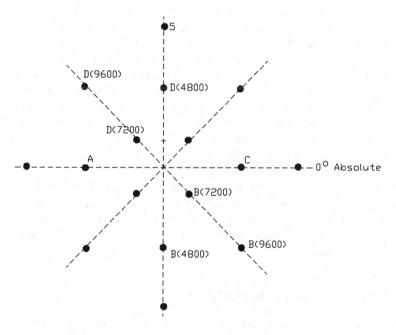

Figure 8.6: Signal space diagram showing synchronising signal points

the duration of 128 symbol intervals, where phasors A and B are depicted in Figure 8.6 for the transmission rates of 9600, 7200 and 4800 bps, respectively. Observe in the figure that symbol A determines the absolute phase reference of 180°. The segment 2 sequence toggles between two distinct phasor states at a signalling rate of 2400 Bd, hence it is periodic at 1200 Hz, therefore having a strong 1200 Hz frequency domain component which assists timing recovery, as explained in Chapter 6. By the time that segment 2 is finished the timing

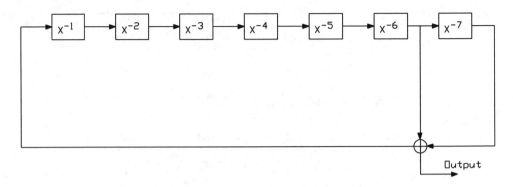

Figure 8.7: Pseudorandom sequence generator

recovery must be established.

Segment 3 of the training pattern is designed to initialise the adaptive channel equaliser by transmitting a pseudorandom (PR) sequence of phasors C and D. This pseudorandom sequence can be used to sound the channel. Observe in Figure 8.6 that the phasor D depends on the transmission rate. The PR sequence generator uses the polynomial $p(x) = 1 + x^{-6} + x^{-7}$, and is shown in Figure 8.7, where $\oplus$ represents modulo 2 addition. Every time the PR sequence contains a logical 0, phasor C is transmitted; for a logical 1, phasor D is sent. The seven-stage PR sequence generator is initialised to 0101010, yielding an initial sequence of CDCDCDC and continues for 384 symbol durations. Since these phasors are widely spaced in the signal constellation, the transmitted signal becomes similar to that of the 2400 Bd BPSK transmissions. These phasors have large Euclidean distances between them and hence are unlikely to be corrupted by AWGN. The same phasors can be used to assist initial equaliser convergence as explained in Chapter 7.

In segment 4 of the training sequence, 48 scrambled binary 1's are transmitted for further updating and fine tuning of the clock and carrier recovery circuits. This is because the initial clock and carrier recovery during segment 2 was performed without equalisation. Now that equalisation is present it is possible to fine-tune these timing recovery circuits. The total training lasts 608 symbol durations, amounting to about 253 ms.

8.3.3 Scrambling and Descrambling

Data scrambling and descrambling are invoked in order to assist the clock and carrier recovery circuits by preventing the transmission of long sequences of all 1's or all 0's. Its operation is based on the generator polynomial

$$g(x) = 1 + x^{-18} + x^{-23} . \tag{8.5}$$

The scrambler performs a logical operation on the data bits to be transmitted using the generator polynomial $g(x)$. Figure 8.8 shows a suitable circuit structure where D_i are the input data bits and D_s the output data stream. At the beginning of transmissions the switch on the scrambler is set to 0, causing the scrambler to be filled with logical 0's. During this period the input data bits D_i are part of the training sequence and can be sent unscrambled since the

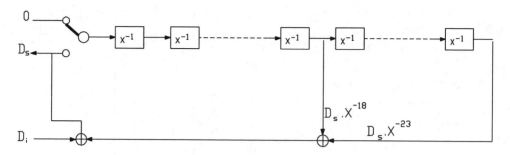

Figure 8.8: V.29 scrambler

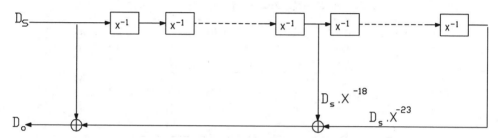

Figure 8.9: V.29 descrambler

input 0's do not affect the sequence D_s. During normal data transmission the scrambled data is fed back to the shift register by setting the switch to D_s. Descrambling is carried out by the circuit of Figure 8.9 by performing the inverse operation using the generator polynomial of Equation 8.5.

8.3.4   Channel Equalisation and Synchronisation

The use of channel equalisation is compulsory in the modem and the receiver has to be provided with appropriate means of detecting whether the channel has changed significantly since the equaliser training sequence was sent. In this instance the equalisation will no longer be correct, and the receiver sends a training sequence to the original transmitter which indicates to this transmitter that in response it should send another training sequence to the receiver. Upon resynchronisation the modem also has to be able to detect the four-segment initial synchronisation signal sequence described earlier in Table 8.5. When detecting the reception of the training sequence, say, from modem A to modem B, then modem B has to initiate the transmission of its own training pattern.

If modem A does not receive a synchronising signal from modem B within the interval of the maximum expected turnaround delay, which is typically 1.2 sec, it sends another synchronising signal. Also, if modem A fails to synchronise on its received sequence, it transmits another four-segment sequence. The carrier frequency is 1700 Hz and the transmitted spectrum measured when transmitting continuous scrambled logical 1 must have an essentially linear phase over the range of 200–2700 Hz. The energy density has to be attenuated at the frequencies of 500 Hz and 2900 Hz by 4.5 ± 2.5 dB with respect to its level at the carrier frequency.

Inputs		Previous outputs		16-QAM outputs		32-QAM outputs		Signal state
$Q1_n$	$Q2_n$	$Y1_{n-1}$	$Y2_{n-1}$	$Y1_n$	$Y2_n$	$Y1_n$	$Y2_n$	
0	0	0	0	0	1	0	0	B
0	0	0	1	1	1	0	1	C
0	0	1	0	0	0	1	0	A
0	0	1	1	1	0	1	1	D
0	1	0	0	0	0	0	1	A
0	1	0	1	0	1	0	0	B
0	1	1	0	1	0	1	1	D
0	1	1	1	1	1	1	0	C
1	0	0	0	1	1	1	0	C
1	0	0	1	1	0	1	1	D
1	0	1	0	0	1	0	1	B
1	0	1	1	0	0	0	0	A
1	1	0	0	1	0	1	1	D
1	1	0	1	0	0	1	0	A
1	1	1	0	1	1	0	0	C
1	1	1	1	0	1	0	1	B

Table 8.6: Differential coding for the V.32 modem

8.4 V.32 Modem

8.4.1 General Features

The V.32 Recommendation [508] proposes a family of two-wire duplex modems for transmission rates of up to 9600 bps for the general switched telephone network as well as for point-to-point leased telephone circuits. The signalling rate is 2400 Bd and transmission rates of 9600 bps as well as 4800 bps are supported using a variety of QAM constellations. At 9600 bps either square QAM or trellis coded 32-QAM are used. There is provision for manual as well as for automatic transmission rate selection. The carrier frequency is 1800 Hz and the spectrum, when transmitting logical 1's, has to be attenuated by 4.5 ± 2.5 dB at frequencies of 600 and 3000 Hz, when compared to the carrier frequency.

8.4.2 Signal Constellation and Bitmapping

8.4.2.1 Non-Redundant 16-QAM

The 9600 bps non-redundant square 16-QAM uses a partially differential bit mapping onto the phasor constellation. The scrambled data stream is divided into groups of four bits and the first two bits $Q1_n$ and $Q2_n$ of the nth symbol are differentially encoded to generate the bits $Y1_n$ and $Y2_n$ according to Table 8.6 where the signal state applies to 4800 bps. Following this the bits $Y1_n$, $Y2_n$, $Q3_n$ and $Q4_n$ are then mapped onto the square 16-QAM constellation, as shown in Figure 8.10. In case of 4800 bps transmissions the 2-bit symbols are generated by differentially encoding $Q1_n$ and $Q2_n$ into $Y1_n$ and $Y2_n$, as specified in

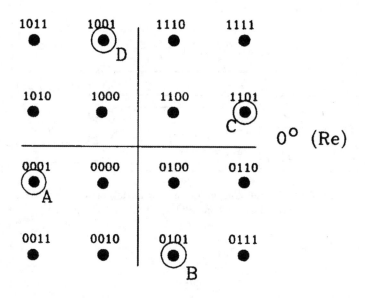

The binary numbers denote Y1$_n$ Y2$_n$ Q3$_n$ Q4$_n$

Figure 8.10: V.32 16-point phasor constellation with non-redundant coding for 9600 bps and subset A B C D of states used at 4800 bps and for training

Table 8.6. The corresponding phasor subset designated by states A, B, C and D coincides with four of the 16-QAM phasors and is also shown in Figure 8.10.

8.4.2.2   Trellis Coded 32-QAM

The fundamentals of trellis coded modulation (TCM) were discussed briefly earlier in this chapter - here we focus our attention on its employment in the V.32 modem. Trellis coded modulation will be revisited in significantly more detail in Chapter 24.1. The general structure of a trellis coded 32-QAM modulator is shown in Figure 8.11 [496]. When using $(m + 1) = 5$ bits per QAM symbol, the signal constellation is constituted by $2^{m+1} = 32$ phasors as displayed in Figure 8.12.

In our specific case the 32 phasors are subdivided into eight subsets $D0, \ldots, D7$, each constituted by four phasors, which are as far apart from each other, as possible. This is shown in Figure 8.13, where the three most significant bits $Y0_n$, $Y1_n$ and $Y2_n$ are the subset identifiers. Phasors of a subset are encoded by the two least significant bits $Q3_n$ and $Q4_n$.

Phasors of different subsets might be close to each other, hence they are easily corrupted and therefore they need error correction coding. The two most significant bits $Q1_n$ and $Q2_n$ of the 4-bit input symbol are therefore $R = 2/3$ rate, constraint length $\nu = 3$ convolutionally coded to yield three bits $Y0_n$, $Y1_n$, and $Y2_n$, which select a specific subset, i.e. D-group, of the 32-QAM constellation. Accordingly, the constraint length $\nu = 3$ code has $\nu = 3$ storage elements and $2^\nu = 8$ trellis states.

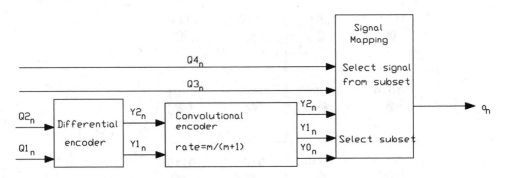

Figure 8.11: General structure of encoder modulator for trellis coded 32-QAM modulation [496] ©IEEE, 1987, Ungerböck

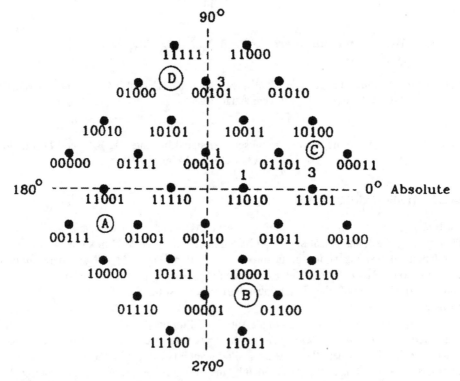

The binary numbers denote $Y0_n$ $Y1_n$ $Y2_n$ $Q3_n$ $Q4_n$

Figure 8.12: V.32 32-point signal structure with trellis coding for 9600 bps and states A, B, C, D used at 4800 bps and for training

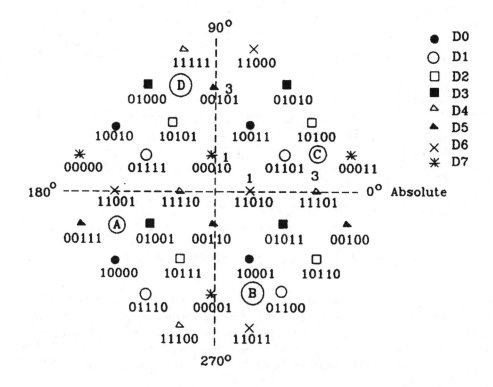

Figure 8.13: V.32 Trellis coded signal subsets $D0, \ldots, D7$

TCM schemes have been shown to have significant coding gains over uncoded constellations [496], but Ungerböck also demonstrated (see Section 8.5) that they are more sensitive to carrier phase offsets than the less dense uncoded constellations. Therefore, in general it is advantageous to design TCM codes having as many rotationally invariant, phase symmetric phasor positions as possible. This would ensure shorter bursts of errors and faster recovery after temporary loss of synchronisation. However, the overall TCM scheme must be rendered transparent to the data, i.e. insensitive to phase rotations, by means of differential encoding and decoding which typically incurs a BER penalty.

A TCM scheme was found by Wei [509] for the 32-QAM constellation of Figure 8.12 that could achieve a coding gain of 4 dB using an eight-state, constraint length $\nu = 3$ code, while retaining 90° phase symmetry. This convolutional encoder contained non-linear elements, as seen in Figure 8.14.

The 32-QAM trellis coded scheme generates the bits $Y1_n$ and $Y2_n$ from the first two input bits $Q1_n$ and $Q2_n$ by differential coding as seen in Table 8.6. The differentially encoded bits $Y1_n$ and $Y2_n$ are then input to the $R = 2/3$ rate systematic convolutional encoder depicted in Figure 8.14. The figure also shows the differential encoding and the mapping onto the 32-QAM signal constellation.

Since the convolutional encoder is systematic, its input bits $Y1_n$ and $Y2_n$ are copied to its output and the simple Boolean logic seen in Figure 8.14 specifies the necessary operations to generate the parity bit $Y0_n$. Finally, the bits $Y0_n$, $Y1_n$, $Y2_n$, $Q3_n$ and $Q4_n$ are mapped onto

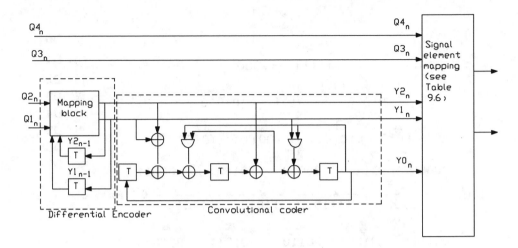

Figure 8.14: Trellis coding at 9600 bps [509] ©IEEE, 1984, Wei

the trellis coded 32-QAM constellation displayed in Figure 8.12. Observe that at the fallback rate of 4800 bps the same states A, B, C and D are used, as in the non-redundant constellation of Figure 8.10 but these now do not coincide with any of the legitimate 32-QAM phasors. The I and Q quadrature component coordinates of these states are identical to those of the corresponding states in Figure 8.10 and hence they are distinctly different from the 32-QAM constellation points.

8.4.3   Scrambler and Descrambler

In the V.32 modem each transmission direction uses a different scrambler. The calling modem uses the same scrambler as the V.29 Recommendation shown in Figure 8.8, specified by the generator polynomial of Equation 8.5, namely $g_1(x) = 1 + x^{-18} + x^{-23}$. At the calling modem, scrambling is performed by dividing the message data polynomial by the generator polynomial, and the coefficients of the quotient constitute the scrambled data. The answering modem computes the descrambled data signal by multiplying the received data with the generator polynomial $g_1(x)$, while scrambling is carried out using:

$$g_2(x) = 1 + x^{-5} + x^{-23} . \tag{8.6}$$

This long division can be carried out by means of a similar circuit to that of Figure 8.8; however, the feedback taps of the linear shift register are at the positions given by the generator polynomial $g_2(x)$.

The initial set-up and training procedures of the V.32 modem are much more complicated than for the V.29 modem, which were briefly described in the previous section. These training procedures tend to use the signalling states A, B, C and D or some subset of them to assist the initial set-up of the carrier and clock recovery circuitries, as well as that of the adaptive channel equaliser.

Since this book is concerned with QAM modulation, we will not consider further standardised features of the V.32 modem. The interested reader is referred to the newest edition of the CCITT V-Series Recommendations [508].

8.5   V.33 Modem

8.5.1   General Features

The 14 400 bps V.33 modem is designed primarily for special quality four-wire leased telephone lines, which comply with the CCITT M.1020 Recommendation shown in Chapter 2. It has a 12 000 bps fallback rate and includes an eight-state trellis-coded modulator and an optional multiplexer to combine data rates of 2400 bps, 4800 bps, 7200 bps, 9600 bps and 12 000 bps. The carrier frequency is 1800 Hz and the signalling rate is 2400 Baud. The generator polynomial for the scrambler and descrambler is identical to that of the V.29 modem, namely $g(x) = 1 + x^{-18} + x^{-23}$.

8.5.2   Signal Constellations and Bitmapping

The 14 400 bps scrambled data stream is divided into groups of six consecutive bits and the first two bits $Q1_n$ and $Q2_n$ are differentially encoded into $Y1_n$ and $Y2_n$, as we have seen in Table 8.6 for the V.32 modem. The differentially encoded bits $Y1_n$ and $Y2_n$ are convolutionally encoded using the same systematic, non-linear convolutional encoder as in case of the V.32 modem, which was portrayed in Figure 8.14. The $R = 2/3$ rate systematic convolutional encoder generates the parity bit $Y0_n$, which is concatenated to the differential encoder's output bits $Y1_n$ and $Y2_n$ and to the uncoded input bits $Q3_n$, $Q4_n$, $Q5_n$ and $Q6_n$. This 7-bit sequence $Q6_n$, $Q5_n$, $Q4_n$, $Q3_n$, $Y2_n$, $Y1_n$, $Y0_n$ is then mapped onto the 128-QAM signal constellation seen in Figure 8.15.

Similarly to the V.32 trellis-coded constellation, the three least significant protected bits $Y2_n$, $Y1_n$ and $Y0_n$ are used as subset identifiers to select one of the eight trellis subsets. For example, in Figure 8.15 the circled phasors belong to the subset identified by $[Y2_n, Y1_n, Y0_n] = [110]$. The remaining four unprotected bits are used to encode the 16 phasor positions within the trellis-coded subset. Since these phasors are sufficiently far apart in the signal constellation, they are less prone to noise corruption and hence they can be transmitted unprotected.

At the 12 000 bps fallback rate the modem operates using 6 bits/symbol and it has the phasor constellation depicted in Figure 8.16. The input data rate of 12 000 bps would only require 5 bits/symbol at 2400 Bd, but the previously described trellis-coding scheme generates and attaches the parity bit $Y0_n$ to protect the differentially coded subset partitioning bits $Y2_n$ and $Y1_n$. Apart from removing the bit $Q6_n$ and using the constellation of Figure 8.16, the 12 000 bps scheme is identical to the 14 400 bps scheme, but with its increased noise protection distances amongst constellation points allows the modem to operate at lower signal-to-noise ratios. The 12 000 bps set partitioning scheme is based again on the three convolutionally coded least significant bits $Y2_n$, $Y1_n$ and $Y0_n$ of the circled phasors in Figure 8.16.

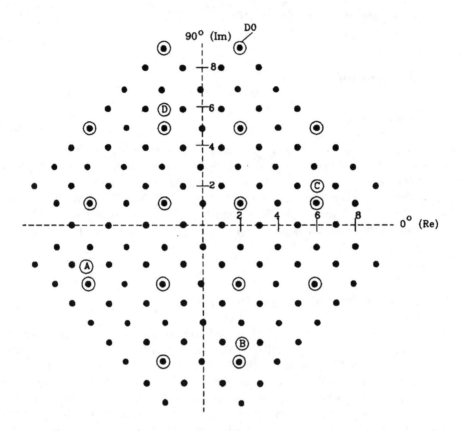

Figure 8.15: Signal space diagram and mapping for trellis-coded modulation at 14 400 bps

8.5.3 Synchronising Signals

Before the transmission of useful data, the modem requires a variety of training signals in order to condition its carrier and clock recovery circuits, as well as the adaptive channel equaliser. The standardised initial synchronising signals are summarised in Table 8.7. Segment 1 of the training pattern is constituted by 256 alternations between the signal constellation states A and B, depicted in Figures 8.15 and 8.16, which occupy the same positions as in the V.32 Recommendations and hence coincide with four of the legitimate non-redundant 16-QAM phasors. This segment assists the initial carrier and clock recovery, as detailed in Chapter 6. This is followed by segment 2, consisting of a scrambled pseudorandom sequence of the states A, B, C and D, shown in Figures 8.15 and 8.16, which is transmitted at 4800 bps and designed to support the initial set-up of the adaptive equaliser as highlighted in Chapter 7. Segment 3 is designed to convey a 16-bit binary sequence, which is scrambled and transmitted eight times at 4800 bps using differentially encoded bits. This sequence carries transmission rate, multiplex set-up or other configuration information. Finally, segment 4 contains scrambled binary 1's transmitted at the rate specified in segment 3. This segment is used to fine-tune the clock and carrier recovery circuits because the linear distortions of

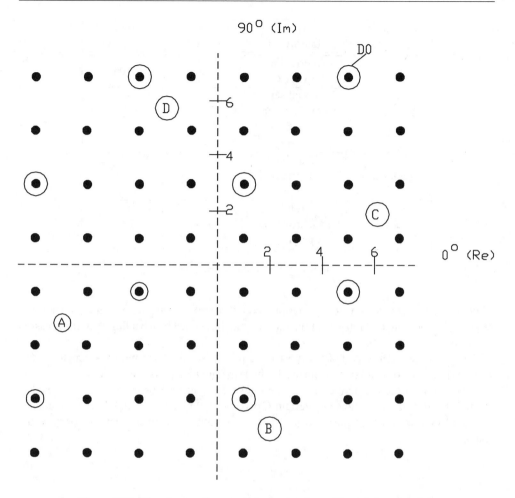

Figure 8.16: Signal space diagram and mapping for trellis-coded modulation at 12 000 bps

the channel should now be equalised and so the clock and carrier references established over the unequalised channel can now be more reliably set. Scrambling is invoked in order to introduce signal transitions into long strings of logical 1's or 0's.

V.33 modems are provided with adaptive channel equalisers. The receiver must be able to detect loss of equalisation and initiate a resynchronisation procedure. Upon initiating a resynchronisation sequence the initiating modem expects to receive a synchronising signal from the remote modem within the maximum turnaround delay.

8.6 Summary

In this chapter we have examined a range of classic QAM schemes designed for AWGN telephone lines, which were standardised by the CCITT – now known as the ITU. The more

Training segments	Type of line signal	No. sym. intervals	Approx. time (ms)
Segment 1	Alternations ABAB	256	106
Segment 2	Equaliser conditioning A,B,C,D	2976	1240
Segment 3	Rate sequence	64	27
Segment 4	Scrambled all binary 1's	48	20
Total	Total synchronising signal time	3344	1393

Table 8.7: V.33 training sequence

recent modems use high-order constellations, complex trellis-coded systems and long training sequences in order to maximise the transmission rate facilitated by a bandlimited transmission medium.

This consideration of one of the most widespread uses of QAM modems concludes our section on QAM for Gaussian channels. In the remaining chapters we concentrate on transmissions over hostile fading mobile channels. We commence the next section by considering the theoretical performance of certain QAM constellations over mobile radio channels and then continue to investigate means means of mitigating the impairments imposed by the channel.

Part II

Adaptive QAM Techniques for Fading Channels

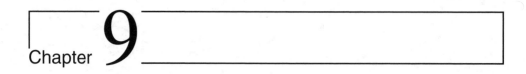

Chapter **9**

Square QAM for Rayleigh Fading Channels

9.1 Square 16-QAM Performance over Rayleigh Fading Channels

Having considered the BER versus channel SNR performance of QAM transmissions via AWGN channels in Chapter 5, we now turn to the analysis of the square 16-QAM constellation of Figure 3.3 over Rayleigh fading channels [112, 157, 158]. Our discussion in this chapter assumes that the reader has already consulted Chapter 5. Assuming a so-called non-frequency-selective, narrowband fading channel, where the signal bandwidth is much lower than the coherence bandwidth of the channel, all frequency components of the transmitted signal undergo the same attenuation and phase shift. The multipath components in the received signal are not resolvable and the channel transfer function is given by

$$c(t) = \alpha(t) \cdot e^{j\Phi(t)}, \tag{9.1}$$

where $\alpha(t)$ represents the Rayleigh fading envelope and $\Phi(t)$ the phase of the channel, which is uniformly distributed over $[-\pi, \pi]$. If additionally the fading is slow, so that $\alpha(t) = \alpha$ and $\Phi(t) = \Phi$ for the duration of one signalling interval, the received signal $r(t)$ at the channel's output is:

$$r(t) = \alpha \cdot e^{j \cdot \Phi} \cdot m(t) + n(t), \tag{9.2}$$

where $m(t)$ is the transmitted modulated signal and $n(t)$ is an AWGN process.

In Chapter 5 we derived the following $C1$ and $C2$ BER results for the AWGN channel using the Gaussian Q-function [510]:

$$
\begin{aligned}
P_{1G} &= \frac{1}{2}Q\left\{\frac{d}{\sqrt{N_0/2}}\right\} + \frac{1}{2}Q\left\{\frac{3d}{\sqrt{N_0/2}}\right\} \\
&= \frac{1}{2}\left[Q\left\{\sqrt{\frac{E_0}{5N_0}}\right\} + Q\left\{3\sqrt{\frac{E_0}{5N_0}}\right\}\right].
\end{aligned}
\tag{9.3}
$$

and

$$
P_{2G} = Q\left\{\sqrt{\frac{E_0}{5N_0}}\right\}.
\tag{9.4}
$$

where $E_0 = 10d^2$ is the average constellation energy.

Assuming perfectly phase coherent reception, no fade compensation, and rewriting Equations 9.3 and 9.4 in terms of the instantaneous SNR γ via Rayleigh fading channels, we obtain:

$$
P_{1R}(\gamma) = \frac{1}{2}\left[Q\left\{\sqrt{\gamma/5}\right\} + Q\left\{3\sqrt{\gamma/5}\right\}\right],
\tag{9.5}
$$

and

$$
P_{2R}(\gamma) = Q\left\{\sqrt{\gamma/5}\right\}.
\tag{9.6}
$$

As the fading envelope fluctuates, the instantaneous SNR γ is also changing. The average BER is found by calculating the bit error probability at a given instantaneous SNR γ and averaging it over all possible SNRs. This is carried out by multiplying the bit error probability $P(\gamma)$ at the instantaneous SNR γ with the occurrence probability of the specific SNR γ expressed in terms of its probability density function (PDF) $C(\gamma)$ and then averaging, i.e. integrating, this product over all possible values of γ:

$$
P_{be} = \int_0^\infty P(\gamma) \cdot C(\gamma)d\gamma.
\tag{9.7}
$$

The random variable $\gamma = \alpha^2 \cdot E_0/N_0$ is a transformed form of the fading envelope α and the PDF of α, $C(\alpha)$ is known to be Rayleigh [383,511]:

$$
C(\alpha) = \left(\frac{\alpha}{\alpha_0^2}\right)e^{-\alpha^2/2\alpha_0^2},
\tag{9.8}
$$

where α_0^2 is the identical variance of the independent quadrature components constituting the fading envelope α, while the second moment of α is given by $E(\alpha^2) = \overline{\alpha^2} = 2\alpha_0^2$ giving the variance of the envelope itself.

Let us now compute the PDF of γ, $C(\gamma)$, given that $C(\alpha)$ is Rayleigh distributed, and $\gamma = \alpha^2 . E_o/N_o$ is a second-order function of α. The following theorem can be invoked [510].

Theorem
Let α be a continuous random variable with the PDF $C(\alpha)$, and let $\gamma = \alpha^2 . E_o/N_o$ be a transformation of α. In order to determine the PDF of γ, $C(\gamma)$, we solve the equation

$\gamma = \alpha^2.E_o/N_o$ for α in terms of γ. If α_1 and α_2 are the only real solutions then

$$C(\gamma) = C(\alpha_1)|\frac{d\alpha_1}{d\gamma}| + C(\alpha_2)|\frac{d\alpha_2}{d\gamma}|. \tag{9.9}$$

As $\gamma = \alpha^2 \cdot E_0/N_0$ has two real roots, α_1 and α_2, but the Rayleigh distribution does not exist in the negative amplitude domain, α_2 can be eliminated and hence $C(\gamma)$ is given by

$$C(\gamma) = C(\alpha_1) \cdot |\frac{d\alpha_1}{d\gamma}|. \tag{9.10}$$

When taking into account that

$$\alpha = \sqrt{\frac{N_0}{E_0}} \cdot \gamma^{\frac{1}{2}}$$

and that

$$\frac{d\alpha_1}{d\gamma} = \frac{1}{2}\sqrt{\frac{N_0}{E_0\gamma}}$$

we arrive at

$$C(\gamma) = \frac{1}{2\alpha_0{}^2}\frac{N_0}{E_0\gamma} \cdot e^{\frac{-N_0\gamma}{2E_0\alpha_0^2}}. \tag{9.11}$$

If we define the average SNR Γ as:

$$\Gamma = \overline{\gamma} = E\{\alpha^2\frac{E_0}{N_0}\} = \overline{\alpha^2}\frac{E_0}{N_0} = 2\alpha_0^2\frac{E_0}{N_0}, \tag{9.12}$$

where we exploited the fact that $E(\alpha^2) = \overline{\alpha^2} = 2\alpha_0{}^2$ for a Rayleigh PDF, then from Equations 9.11 and 9.12 the transformed PDF $C(\gamma)$ is given by

$$C(\gamma) = \frac{1}{\Gamma}e^{-\gamma/\Gamma}, \tag{9.13}$$

which is known as the chi-square distribution [383], with the instantaneous SNR $\gamma \geq 0$.

The $C1$ BER over Rayleigh fading channels is found by substituting into Equation 9.7 the PDF $C(\gamma)$ from Equation 9.13 and the instantaneous bit error probability $P_{1R}(\gamma)$ over Rayleigh fading channels from Equation 9.5, yielding:

$$P_{1R}(\Gamma) = \frac{1}{2\Gamma}\int_0^\infty \left[Q\left\{\sqrt{\gamma/5}\right\} + Q\left\{3\sqrt{\gamma/5}\right\}\right] \cdot e^{-\gamma/\Gamma}d\gamma. \tag{9.14}$$

The $C2$ BER cannot be formulated without considering the effect of a particular level of channel attenuation on the demodulation process. The demodulation is now carried out on an attenuated signal constellation. Therefore we multiply the decision boundaries specified in Equation 4.6 for the $C2$ bits in the I and Q components by the average attenuation $\overline{\alpha}$:

$$
\begin{array}{llllll}
\text{if} & & I,Q & \geq & 2\overline{\alpha}d & \text{then} \quad i_2, q_2 = 1 \\
\text{if} & -2\overline{\alpha}d \leq & I,Q & < & 2\overline{\alpha}d & \text{then} \quad i_2, q_2 = 0 \\
\text{if} & -2\overline{\alpha}d > & I,Q & & & \text{then} \quad i_2, q_2 = 1.
\end{array}
$$

Considering the case when the transmitted bit i_2 or q_2 is a logical 0 and assuming that $\alpha < 2\overline{\alpha}$, the protection distance between the received phasor αd and the decision boundary $2\overline{\alpha}d$ is $d_1 = (2\overline{\alpha}d - \alpha d)$. Upon substituting d_1 into Equation 4.7 instead of d, we get the $C2$ BER for the transmission of a logical 0 as

$$P_{2,0,<2} = Q\left\{\frac{d_1}{\sqrt{N_0/2}}\right\}$$

$$= Q\left\{\frac{d(2\overline{\alpha} - \alpha)}{\sqrt{N_0/2}}\right\}. \tag{9.15}$$

From $E_0 = 10d^2$ we have $d = \sqrt{E_0/10}$, and whence:

$$P_{2,0,<2} = Q\left\{(2\overline{\alpha} - \alpha)\sqrt{\frac{E_0}{5N_0}}\right\}. \tag{9.16}$$

From $\gamma = \alpha^2 E_0/N_0$ we get $E_0/N_0 = \gamma/\alpha^2$, and therefore:

$$P_{2,0,<2} = Q\left\{\left(\frac{2\overline{\alpha} - \alpha}{\alpha}\right)\sqrt{\frac{\gamma}{5}}\right\}$$

$$= Q\left\{\left(\frac{2\overline{\alpha}}{\alpha} - 1\right)\sqrt{\frac{\gamma}{5}}\right\}. \tag{9.17}$$

Upon substituting Equations 9.17 and 9.13 into Equation 9.7 we get one component of the total $C2$ error probability. First the upper integration boundary of Equation 9.7 is derived for $\alpha \leq 2\overline{\alpha}$ as

$$\gamma = \alpha^2 E_0/N_0 \leq 4(\overline{\alpha})^2 E_0/N_0. \tag{9.18}$$

For a Rayleigh distribution we have [511] that $\overline{\alpha} = \alpha_0\sqrt{\pi/2}$, and this means we arrive at:

$$\gamma = 4\left(\alpha_0\sqrt{\frac{\pi}{2}}\right)^2 \frac{E_0}{N_0} = 2\alpha_0{}^2\frac{E_0}{N_0}\pi = \Gamma\pi, \tag{9.19}$$

and

$$P_{2,0,<2}(\Gamma) = \int_0^\infty P(\gamma)C(\gamma)d\gamma = \int_0^{\Gamma\pi} Q\left\{\left(\frac{2\overline{\alpha}}{\alpha} - 1\right)\sqrt{\frac{\gamma}{5}}\right\} \cdot \frac{1}{\Gamma}e^{-\gamma/\Gamma}d\gamma. \tag{9.20}$$

By contrast, for $d_1 = (2\overline{\alpha}d - \alpha d) < 0$, i.e. for $\alpha > 2\overline{\alpha}$ there is no protection distance at all, because the instantaneous attenuation α is higher than twice the average attenuation $\overline{\alpha}$ computed by the AGC over k consecutive samples. Hence, if there is no noise, or if the instantaneous noise sample is not sufficiently negative to carry the received phasor back to the error-free zone, the decisions are always erroneous. Accordingly, over noisy channels the error probability is less than one, namely the probability of one is decreased by the probability of those events, when the noise carries the phasor back to its original domain. Hence we have:

$$P_{2,0,>2} = Q \left\{ \frac{d(2\bar{\alpha} - \alpha)}{\sqrt{N_0/2}} \right\}. \tag{9.21}$$

Exploiting $d = \sqrt{E_0/10}$ we have

$$P_{2,0,>2} = Q \left\{ \sqrt{\frac{E_0}{5N_0}} (2\bar{\alpha} - \alpha) \right\}, \tag{9.22}$$

and by taking into account that $E_0/N_0 = \gamma/\alpha^2$ we arrive at:

$$P_{2,0,>2} = Q \left\{ (\frac{2\bar{\alpha}}{\alpha} - 1) \sqrt{\frac{\gamma}{5}} \right\}. \tag{9.23}$$

The lower integration limit is now identical to the previously computed upper limit, i.e. $\gamma = \pi\Gamma$, and the total probability of receiving a logical 1 in the $C2$ sub-channel instead of the transmitted logical 0 is given by:

$$P_{2,0}(\Gamma) = P_{2,0,<2}(\Gamma) + P_{2,0,>2}(\Gamma). \tag{9.24}$$

Upon substituting Equations 9.20 and 9.23 into Equation 9.24 we get:

$$
\begin{aligned}
P_{2,0}(\Gamma) &= \frac{1}{\Gamma} \int_0^{\pi\Gamma} Q \left\{ (\frac{2\bar{\alpha}}{\alpha} - 1) \sqrt{\frac{\gamma}{5}} \right\} e^{-\gamma/\Gamma} d\gamma \\
&+ \frac{1}{\Gamma} \int_{\pi\Gamma}^{\infty} \left[Q \left\{ (\frac{2\bar{\alpha}}{\alpha} - 1) \sqrt{\frac{\gamma}{5}} \right\} \right] e^{-\gamma/\Gamma} d\gamma \\
&= \frac{1}{\Gamma} \int_0^{\infty} \left[Q \left\{ (\frac{2\bar{\alpha}}{\alpha} - 1) \sqrt{\frac{\gamma}{5}} \right\} \right] e^{-\gamma/\Gamma} d\gamma.
\end{aligned}
\tag{9.25}
$$

In the reverse situation, when a transmitted $C2$ logical 1 is corrupted into a logical 0, the protection distance is $d_1 = 3\alpha d - 2\bar{\alpha}d$, which is positive if $3\alpha d > 2\bar{\alpha}d$, i.e. $\alpha/\bar{\alpha} > 2/3$. The AWGN has to overcome this protection distance to cause an error, and hence the error probability for a positive protection distance d_1 is

$$P_{2,1,>2/3} = Q \left\{ \left(\frac{3\bar{\alpha}}{\alpha} - 2 \right) \sqrt{\frac{\gamma}{5}} \right\}. \tag{9.26}$$

If the protection distance d_1 is negative, then $3\alpha d < 2\bar{\alpha}d$, i.e. $\alpha/\bar{\alpha} < 2/3$. Again, in this situation the decisions are always erroneous, when the additive noise sample is insufficiently high for it to carry the faded sample back to the error-free domain. Hence the corresponding error probability is expressed with the aid of the Gaussian Q-function as:

$$P_{2,1,<2/3} = Q \left\{ (\frac{3\bar{\alpha}}{\alpha} - 2) \sqrt{\frac{\gamma}{5}} \right\}. \tag{9.27}$$

Before computing the total $C2$ bit error probability for the transmission of a logical 1, we determine the integration limit for Equation 9.7. Since $\alpha > 2\overline{\alpha}/3$,

$$\gamma = \alpha^2 \frac{E_0}{N_0} = \left(\frac{2}{3}\overline{\alpha}\right)^2 \frac{E_0}{N_0} = \frac{4}{9}(\overline{\alpha})^2 \frac{E_0}{N_0}. \tag{9.28}$$

For a Rayleigh distribution we have [511] that $\overline{\alpha} = \alpha_0 \sqrt{\pi/\gamma}$, so

$$\gamma = \frac{4}{9}\alpha_0^2 \frac{\pi}{\gamma} \frac{E_0}{N_0} = \Gamma \frac{\pi}{9}. \tag{9.29}$$

Therefore:

$$P_{2,1}(\Gamma) = P_{2,1,<2/3}(\Gamma) + P_{2,1,>2/3}(\Gamma), \tag{9.30}$$

or

$$
\begin{aligned}
P_{2,1}(\Gamma) &= \frac{1}{\Gamma} \int_0^{\pi\Gamma/9} Q\left\{(\frac{3\overline{\alpha}}{\alpha} - 2)\sqrt{\frac{\gamma}{5}}\right\} e^{-\gamma/\Gamma} d\gamma \\
&+ \frac{1}{\Gamma} \int_{\pi\Gamma/9}^{\infty} \left[Q\left\{(\frac{3\overline{\alpha}}{\alpha} - 2)\sqrt{\frac{\gamma}{5}}\right\}\right] e^{-\gamma/\Gamma} d\gamma \\
&= \frac{1}{\Gamma} \int_0^{\infty} \left[Q\left\{(\frac{3\overline{\alpha}}{\alpha} - 2)\sqrt{\frac{\gamma}{5}}\right\}\right] e^{-\gamma/\Gamma} d\gamma.
\end{aligned} \tag{9.31}
$$

For random transmitted data the overall $C2$ sub-channel error probability is given by:

$$P_{2R}(\Gamma) = \frac{1}{2}[P_{2,1}(\Gamma) + P_{2,0}(\Gamma)]. \tag{9.32}$$

When we evaluated the $C1$ and $C2$ bit error probabilities from Equations 9.14 and 9.32, respectively, we received nearly identical curves to those in Figure 9.1 denoted by C1,Ray and C2,Ray, for our simulations. The average $C1$ and $C2$ BER curve is represented by Av,Ray in the figure.

For comparison we also included the corresponding AWGN curves denoted by C1,AWGN and C2,AWGN as well as the average BER Av,AWGN, of the two subchannels. We noted in Chapter 5 that the $C1$ and $C2$ BER differences are consistent but not drastic when transmitting square 16-QAM signals via the AWGN channel. The discrepancy becomes more dramatic over the Rayleigh fading channel. While the $C1$ BER is monotonically decreasing, the $C2$ curve shows a high residual BER. This is because due to the received signal envelope fluctuations the multilevel QAM constellation collapses towards the origin of the coordinate system, causing excessive BERs for the low protection $C2$ sub-channel. The high protection $C1$ sub-channel withstands the channel fading better and hence its behaviour is mostly governed by the AWGN. Automatic gain control (AGC) techniques with and without channel sounding side information for fading compensation are reported in references [112, 157, 159] allowing the reduced residual BERs to be efficiently combated by appropriate FEC methods.

In the final part of this section, we briefly note that there exists a closed-form solution for the integral of the product of the Gaussian Q-function and the PDF $C(\alpha)$. The closed-form

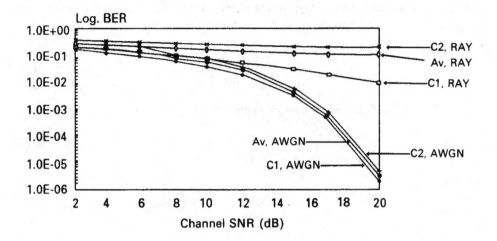

Figure 9.1: Square 16-QAM BER versus channel SNR curves for AWGN and Rayleigh channels

BPSK bit error probability can be derived as follows:

$$
\begin{aligned}
P_{BPSK-R}(\alpha_0) &= \int_0^\infty Q\left\{\frac{\alpha d}{\sqrt{N_o/2}}\right\} C(\alpha)d\alpha \\
&= \int_0^\infty Q\left\{\frac{\alpha d}{\sqrt{N_o/2}}\right\} \frac{\alpha}{\alpha_0^2}\cdot e^{\frac{-\alpha^2}{2\alpha_0^2}}\,d\alpha.
\end{aligned}
\tag{9.33}
$$

We exploit the following fact [512]:

$$
\int_0^\infty [1-erf(\beta x)]e^{-\mu x^2}\cdot x\,dx = \frac{1}{2\mu}\left(1-\frac{\beta}{\sqrt{\mu+\beta^2}}\right)
\tag{9.34}
$$

where the relationship between the Gaussian error function $erf(x)$ and the Q-function is given as follows [510]:

$$
1 - erf(x) = 2\cdot Q\left\{\sqrt{2}x\right\}.
\tag{9.35}
$$

Upon rewriting Equation 9.33 in a more convenient form we have

$$
P_{BPSK-R}(\alpha_0) = \frac{1}{2\alpha_0^2}\int_0^\infty 2\cdot Q\left\{\sqrt{2}\frac{d}{\sqrt{N_o}}\alpha\right\}\cdot\alpha\cdot e^{\frac{-\alpha^2}{2\alpha_0^2}}\,d\alpha
\tag{9.36}
$$

and exploiting Equation 9.35 gives

$$
P_{BPSK-R}(\alpha_0) = \frac{1}{2\alpha_0^2}\int_0^\infty 2\cdot\left(1-erf\left\{\frac{d}{\sqrt{N_o}}\alpha\right\}\right)\cdot\alpha\cdot e^{\frac{-\alpha^2}{2\alpha_0^2}}\,d\alpha
\tag{9.37}
$$

Assigning $\beta = d/\sqrt{N_o}$ and $\mu = 1/2\alpha_0^2$ in order to enable us to use the closed-form formula of Equation 9.34 we get

$$P_{BPSK-R}(\alpha_0) = \frac{1}{2\alpha_0^2}\frac{1}{2\mu}\left(1 - \frac{\beta}{\sqrt{\mu + \beta^2}}\right) \tag{9.38}$$

and after substituting back $\beta = d/\sqrt{N_0}$ and $\mu = 1/2\alpha_0^2$ we arrive at the analytical bit error probability of BPSK in a Rayleigh flat fading environment :

$$
\begin{aligned}
P_{2R}(\alpha_0) &= \frac{\alpha_0^2}{2\alpha_0^2}\left(1 - \frac{\frac{d}{\sqrt{N_o}}}{\sqrt{\frac{1}{2\alpha_0^2} + \frac{d^2}{N_o}}}\right) \\
&= \frac{1}{2}\left[1 - \sqrt{\frac{2\alpha_0^2\frac{d^2}{N_o}}{1 + 2\alpha^2\frac{d^2}{N_o}}}\right]
\end{aligned}
\tag{9.39}
$$

which may be written in a more compact form as:

$$P^b_{\text{BPSK}_{\text{FR}}}(\bar{\gamma}) = 0.5\left(1 - \sqrt{\frac{\bar{\gamma}}{1+\bar{\gamma}}}\right). \tag{9.40}$$

The same derivation can be extended to the higher-order modulation schemes. Therefore the theoretical BER performance of 4QAM, 16QAM and 64QAM over flat Rayleigh-fading channels can be written as:

$$P^b_{\text{4QAM}_{\text{FR}}}(\bar{\gamma}) = 0.5(1 - \sqrt{\frac{\bar{\gamma}}{2+\bar{\gamma}}}) \tag{9.41}$$

$$P^b_{\text{16QAM}_{\text{FR}}}(\bar{\gamma}) = 0.5\left[0.5(1 - \sqrt{\frac{\bar{\gamma}}{10+\bar{\gamma}}}) + 0.5(1 - \sqrt{\frac{9\bar{\gamma}}{10+9\bar{\gamma}}})\right] \tag{9.42}$$

The results of the theoretical Rayleigh flat fading solutions and the experimental simulations are shown in Figure 9.2, where a close correspondence was observed between the analytical and experimental solutions. In the next section, we will categorize the effects of the dispersive channel, termed as (ISI).

9.2 Square 64-QAM Performance over Rayleigh Fading Channels

Adopting an approach similar to that used for the 16-QAM in Section 10.1, the formula for the $C1$ BER can be obtained by substituting the Gaussian channel result of Equation 4.22, repeated here for convenience:

$$P_{C1}(\gamma) = \frac{1}{4}Q\left[\sqrt{\frac{\gamma}{21}}\right] + \frac{1}{4}Q\left[\sqrt{\frac{3\gamma}{21}}\right] + \frac{1}{4}Q\left[\sqrt{\frac{5\gamma}{21}}\right] + \frac{1}{4}Q\left[\sqrt{\frac{7\gamma}{21}}\right] \tag{9.43}$$

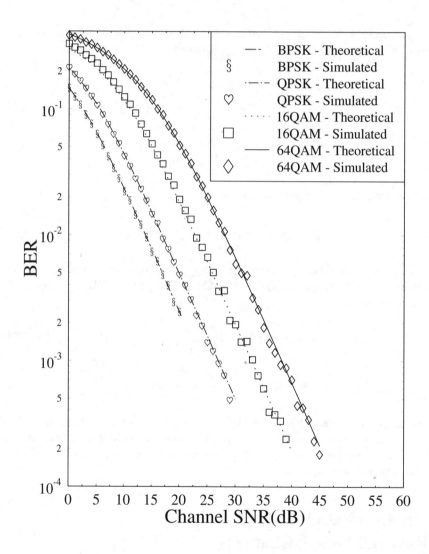

Figure 9.2: Theoretical and simulated BER performance of a single tap optimum DFE for BPSK, 4QAM, 16QAM and 64QAM over a flat fading Rayleigh channel using perfect channel estimation.

into Equation 9.7 along with the transformed PDF $C(\gamma)$ in Equation 9.13, in order to arrive at:

$$P_{C1}(\Gamma) = \frac{1}{4\Gamma} \int_0^{\infty} \left\{ Q\left[\sqrt{\frac{\gamma}{21}}\right] + Q\left[3\sqrt{\frac{\gamma}{21}}\right] + Q\left[5\sqrt{\frac{\gamma}{21}}\right] + Q\left[7\sqrt{\frac{\gamma}{21}}\right] \right\} e^{-\gamma/\Gamma} d\gamma. \tag{9.44}$$

This simple derivation is achieved because the $C1$ decision boundaries constituted by the coordinate axis and shown in Figure 5.3 are unaltered by the fading. However, the probability of bit error for the $C2$ and $C3$ sub-channels cannot be derived without considering the effects of fading on each phasor and decision boundary individually.

When calculating the probability of bit error for $C2$, we acknowledge that although the received phasors are attenuated by the instantaneous α value, the receiver is arranged to demodulate with reference to the expected value $2\alpha_0$. The effect of fading must be computed individually for each phasor. For P_6 the modified protection distance d_1 measured from the decision boundary $4\overline{\alpha}d$ is seen from the phasor constellation depicted in Figure 4.2 to be $d_1 = 4\overline{\alpha}d - 3\alpha d$. If the modified protection distance is overcome by noise, erroneous decisions occur. If the modified protection distance $d_1 < 0$ then in the absence of noise the decisions are consistently erroneous. In the presence of noise the phasors can be carried back into the error-free decision domain by sufficiently large negative noise samples. In both cases the error probability is equivalent to the noise samples exceeding the protection distance, irrespective of its actual value. In terms of the $Q[\,]$ function we have

$$P_{e6}(\gamma) = Q\left[\left(\frac{4\overline{\alpha}}{\alpha} - 3\right)\sqrt{\frac{\gamma}{21}}\right]. \tag{9.45}$$

For the phasor P_5 the modified protection distance from the $C2$ decision boundary $4\overline{\alpha}d$ is $d_1 = 4\overline{\alpha}d - \alpha d$. Observe that the effect of the instantaneous attenuation α is now less profound than for P_6, where it is multiplied by a factor of 3. Using this d_1 value in the argument of the $Q[\,]$ function gives

$$P_{e5}(\gamma) = Q\left[\left(\frac{4\overline{\alpha}}{\alpha} - 1\right)\sqrt{\frac{\gamma}{21}}\right]. \tag{9.46}$$

The modified protection distance from the $C2$ decision boundary at $4\overline{\alpha}d$ for P_7 is seen from the phasor constellation of Figure 5.3 to be $d_1 = 4\overline{\alpha}d - 5\alpha d$, which yields an error probability of

$$P_{e7}(\gamma) = Q\left[\left(\frac{4\overline{\alpha}}{\alpha} - 5\right)\sqrt{\frac{\gamma}{21}}\right]. \tag{9.47}$$

The phasor P_8 has the highest distance from the coordinate axis, therefore it is profoundly affected by the instantaneous fading. It has an associated protection distance of $d_1 = 4\overline{\alpha}d - 7\alpha d$ which results in an error probability of

$$P_{e8}(\gamma) = Q\left[\left(\frac{4\overline{\alpha}}{\alpha} - 7\right)\sqrt{\frac{\gamma}{21}}\right]. \tag{9.48}$$

The partial $C2$ error probabilities from Equations 9.45 to 9.48 have to be substituted in Equation 9.7 along with Equation 9.13 in order to deliver the average square 64-QAM $C2$

BER for Rayleigh fading channels:

$$
\begin{aligned}
P_{C2}(\Gamma) &= \frac{1}{4\Gamma}\left\{\int_0^\infty Q\left[(\frac{4\overline{\alpha}}{\alpha} - 1)\sqrt{\frac{\gamma}{21}}\right] e^{-\gamma/\Gamma} d\gamma\right. \\
&+ \int_0^\infty Q\left[(\frac{4\overline{\alpha}}{\alpha} - 3)\sqrt{\frac{\gamma}{21}}\right] e^{-\gamma/\Gamma} d\gamma \\
&+ \int_0^\infty Q\left[(\frac{4\overline{\alpha}}{\alpha} - 5)\sqrt{\frac{\gamma}{21}}\right] e^{-\gamma/\Gamma} d\gamma \\
&+ \left.\int_0^\infty Q\left[(\frac{4\overline{\alpha}}{\alpha} - 7)\sqrt{\frac{\gamma}{21}}\right] e^{-\gamma/\Gamma} d\gamma\right\}.
\end{aligned}
\tag{9.49}
$$

The $C3$ sub-channel performance is computed similarly, but for each phasor there are two modified protection distances, d_1 and d_2. For the phasor P_6 in Figure 5.3 we find that $d_1 = 6\overline{\alpha}d - 3\alpha d$ and $d_2 = 3\alpha d - 2\overline{\alpha}d$, representing the distances from the $C3$ decision boundaries at $6\overline{\alpha}d$ and $2\overline{\alpha}d$, respectively. When exposed to AWGN, the decisions are erroneous if the modified protection distance d_1 is exceeded by the noise samples. This holds for negative d_1 values as well. Erroneous decisions are also encountered if the phasor P_6 is carried across the $C3$ decision boundary at $2\overline{\alpha}d$ by sufficiently large negative noise samples. This happens whenever the noise value resides below the level $-d_2$. Accordingly, the $C3$ bit error probability for P_6 is

$$
P_{e6}(\gamma) = Q\left[(\frac{6\overline{\alpha}}{\alpha} - 3)\sqrt{\frac{\gamma}{21}}\right] + Q\left[(3 - \frac{2\overline{\alpha}}{\alpha})\sqrt{\frac{\gamma}{21}}\right].
\tag{9.50}
$$

Considering the case of the phasor P_5 the modified protection distances are $d_1 = 2\overline{\alpha}d - \alpha d$ and $d_2 = 6\overline{\alpha}d - \alpha d$, respectively. Erroneous decisions are engendered if the protection distance d_1 is exceeded by noise, but noise samples larger than d_2 carry the phasors into another error-free decision domain. Whence the error probability is given by

$$
P_{e5}(\gamma) = Q\left[(\frac{2\overline{\alpha}}{\alpha} - 1)\sqrt{\frac{\gamma}{21}}\right] - Q\left[(\frac{6\overline{\alpha}}{\alpha} - 1)\sqrt{\frac{\gamma}{21}}\right]
\tag{9.51}
$$

When the phasor P_7 is transmitted the situation is characterised by the distances $d_1 = 6\overline{\alpha}d - 5\alpha d$ and $d_2 = 5\alpha d - 2\overline{\alpha}d$, and hence

$$
P_{e7}(\gamma) = Q\left[(\frac{6\overline{\alpha}}{\alpha} - 5)\sqrt{\frac{\gamma}{21}}\right] + Q\left[(5 - \frac{2\overline{\alpha}}{\alpha})\sqrt{\frac{\gamma}{21}}\right].
\tag{9.52}
$$

Finally, for the phasor P_8 we find $d_1 = 7\alpha d - 6\overline{\alpha}d$ and $d_2 = 7\alpha d - 2\overline{\alpha}d$, where errors are caused by noise samples below the level $-d_1$. Should, however, the noise vector fall below $-d_2$, the phasors are switched to another error-free decision domain across the decision boundary at $2\overline{\alpha}d$, therefore

$$
P_{e8}(\gamma) = Q\left[(7 - \frac{6\overline{\alpha}}{\alpha})\sqrt{\frac{\gamma}{21}}\right] - Q\left[(7 - \frac{2\overline{\alpha}}{\alpha})\sqrt{\frac{\gamma}{21}}\right].
\tag{9.53}
$$

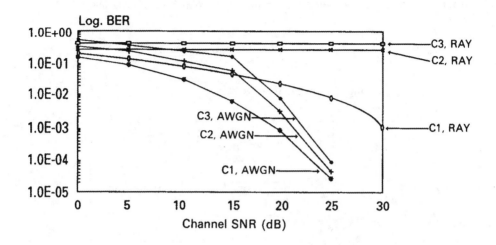

Figure 9.3: Square 64-QAM BER versus channel SNR curves for AWGN and Rayleigh channels

Let us assume again that random sequences are sent, i.e. the occurrence probability of each phasor P_1, ..., P_{64} in Figure 5.3 is identical, and further, let us exploit the symmetry of the signal constellation. On substituting Equations 9.50 to 9.53, along with Equation 9.13, into Equation 9.7 we obtain the average $C3$ bit error probability as a function of the average SNR Γ:

$$
\begin{aligned}
P_{C3}(\Gamma) \;=\; & \frac{1}{4\Gamma}\Big\{\int_0^\infty [Q[(\frac{6\overline{\alpha}}{\alpha}-3)\sqrt{\frac{\gamma}{21}}] + Q[(3-\frac{2\overline{\alpha}}{\alpha})\sqrt{\frac{\gamma}{21}}]]e^{-\gamma/\Gamma}d\gamma \\
& + \int_0^\infty [Q[(\frac{2\overline{\alpha}}{\alpha}-1)\sqrt{\frac{\gamma}{21}}] - Q[(\frac{6\overline{\alpha}}{\alpha}-1)\sqrt{\frac{\gamma}{21}}]]e^{-\gamma/\Gamma}d\gamma \\
& + \int_0^\infty [Q[(\frac{6\overline{\alpha}}{\alpha}-5)\sqrt{\frac{\gamma}{21}}] + Q[(5-\frac{2\overline{\alpha}}{\alpha})\sqrt{\frac{\gamma}{21}}]]e^{-\gamma/\Gamma}d\gamma \\
& + \int_0^\infty [Q[(\frac{7-6\overline{\alpha}}{\alpha})\sqrt{\frac{\gamma}{21}}] - Q[(7-\frac{2\overline{\alpha}}{\alpha})\sqrt{\frac{\gamma}{21}}]]e^{-\gamma/\Gamma}d\gamma\Big\}. \quad (9.54)
\end{aligned}
$$

In summary, the square 64-QAM $C1$, $C2$ and $C3$ sub-channel performances over Rayleigh fading channels are given by Equations 9.44, 9.49 and 9.54, respectively. The equivalent BER vs. channel SNR curves are plotted in Figure 9.3, which again coincide within a fraction of a dB with our simulation results. The BER differences between the three classes become more profound in Rayleigh channels than in the AWGN channel. This is in harmony with our experience for the $16-QAM$ BER curves displayed in Figure 9.1. The BER of the $C1$ sub-channel is well suited for FEC-coded speech or image transmission, while the $C2$ and $C3$ performances have to be improved by fading-compensating automatic gain control and optional diversity reception for FEC techniques to become efficient [112, 157, 159].

Finally, following our arguments outlined in the context of Equations 9.40 - 9.42, a closed-form solution may be found for the bit error probability of 64QAM, which is expressed as:

$$P^b_{64QAM_{FR}}(\bar{\gamma}) = \frac{1}{3}(P^b_{64QAM_1} + P^b_{64QAM_2} + P^b_{64QAM_3}), \tag{9.55}$$

where $P^b_{64QAM_1}$, $P^b_{64QAM_2}$ and $P^b_{64QAM_3}$ are defined as:

$$P^b_{64QAM_1} = \frac{1}{4}\left[0.5(1 - \sqrt{\frac{\bar{\gamma}}{42+\bar{\gamma}}}) + 0.5(1 - \sqrt{\frac{9\bar{\gamma}}{42+9\bar{\gamma}}})\right.$$
$$\left. +0.5(1 - \sqrt{\frac{25\bar{\gamma}}{42+25\bar{\gamma}}}) + 0.5(1 - \sqrt{\frac{49\bar{\gamma}}{42+49\bar{\gamma}}})\right] \tag{9.56}$$

$$P^b_{64QAM_2} = \frac{1}{2}\left[0.5(1 - \sqrt{\frac{\bar{\gamma}}{42+\bar{\gamma}}}) + 0.5(1 - \sqrt{\frac{9\bar{\gamma}}{42+9\bar{\gamma}}})\right.$$
$$\left. +0.25(1 - \sqrt{\frac{25\bar{\gamma}}{42+25\bar{\gamma}}}) + 0.25(1 - \sqrt{\frac{49\bar{\gamma}}{42+49\bar{\gamma}}})\right] \tag{9.57}$$

$$P^b_{64QAM_3} = \frac{1}{2}\left[(1 - \sqrt{\frac{\bar{\gamma}}{42+\bar{\gamma}}}) + 0.75(1 - \sqrt{\frac{9\bar{\gamma}}{42+9\bar{\gamma}}})\right.$$
$$-0.75(1 - \sqrt{\frac{25\bar{\gamma}}{42+25\bar{\gamma}}}) - 0.5(1 - \sqrt{\frac{49\bar{\gamma}}{42+49\bar{\gamma}}})$$
$$+0.5(1 - \sqrt{\frac{81\bar{\gamma}}{42+81\bar{\gamma}}}) + 0.25(1 - \sqrt{\frac{121\bar{\gamma}}{42+121\bar{\gamma}}})$$
$$\left. -0.25(1 - \sqrt{\frac{169\bar{\gamma}}{42+169\bar{\gamma}}})\right]. \tag{9.58}$$

The corresponding performance results were compared to those of BPSK, 4QAM and 16QAM in Figure 9.2. The above results were derived for systems having either perfectly coherent carrier phase recovery but no fading magnitude estimate or perfect channel estimate. These systems were communicating over Rayleigh fading channels, although in recent years the employment of the Nakagami channel model has also become popular in the context of characterising the achievable performance of wireless systems [513], because a range of Rayleigh, Rician and Gaussian channels can be characterised with their aid using a single generalised formula. Furthermore, a recursive algorithm for the exact BER computation of generalized hierarchical QAM constellations was disseminated by Vitthaladevuni and Alouini in [514], while general BER expressions for one- and two-dimensional amplitude modulation schemes were provided by Cho and Yoon in [515]. Finally, the BER computation of hierarchical QAM constellations was studied in [516].

9.3 Reference Assisted Coherent QAM

Two powerful methods have been proposed in order to ensure coherent QAM operation in fading environments. Both these techniques deliver channel measurement information in terms of attenuation and phase shift due to fading. The first is transparent tone in band (TTIB) assisted modulation, where a pilot carrier is inserted typically in the centre of the modulated spectrum. At the receiver the signal is extracted and used to estimate the channel-induced attenuation and phase rotation.

The other technique is called pilot symbol assisted modulation (PSAM) where known channel sounding phasors are periodically inserted into the transmitted time domain signal sequence. Similarly to the frequency domain pilot tone, these known symbols deliver channel measurement information.

9.3.1 Transparent-Tone-in-Band Modulation [115]

9.3.1.1 Introduction

Transparent tone in band (TTIB) assisted modulation has been proposed for various mobile radio applications by McGeehan, Bateman *et al.* [115, 517–523]. TTIB schemes place a pilot tone within the spectrum of the transmitted signal which can be used for a range of applications from in-band signalling in telephone systems, through fading compensation in mobile radio systems [523], to coherent data transmissions [521, 524] to aid the operation of direct conversion receivers. Further related contributions can be found in the literature [525–531].

In the following discussion of TTIB schemes we follow McGeehan and Bateman's approach [115] and concentrate on the use of TTIB in a mobile radio environment. TTIB schemes constitute a frequency domain alternative to pilot symbol assisted modulation (PSAM) [529] (see Section 9.3.2.1). In TTIB systems for mobile radio applications a spectral gap of a certain bandwidth is created in the centre of the signal spectrum to allow insertion of the pilot tone. The pilot tone is often chosen to be the carrier frequency as this can then assist the receiver in the process of coherent demodulation. The use of coherent demodulation overcomes the carrier recovery problems highlighted in Chapter 6 and allows the use of the maximum least square distance constellations discussed in the previous two sections over mobile radio channels.

Further advantages can be achieved if the separation of the two sub-bands created by dividing the spectrum is made to be a subharmonic of the symbol timing. This allows relatively straightforward clock recovery procedures to be employed. Finally, since the amplitude and phase of the transmitted pilot symbol are known to the receiver, extrication of this pilot symbol allows an estimate of the complex fading channel envelope, which removes the need for AGC. However, this will only be possible if the channel is exhibiting flat fading, i.e. all the other frequency components of the signal suffer the same amplitude and phase shift.

The disadvantages of this system are that the bandwidth requirements are slightly increased, and higher transmitter and receiver complexity are required to perform the necessary spectral operations. Furthermore, increased transmitted power is required for the transmission of the pilot tone, and although this is typically at a lower level than the power level of the transmitted signal, at very low levels the pilot signal can become corrupted by noise, reducing the reliability of the fading estimates. Furthermore, the inclusion of the pilot tone increases the signal envelope fluctuations, which results in increased out-of-band emissions due to the amplifier non-linearities, as discussed in Chapter 5.

9.3.1.2   Principles of TTIB

The basic principles of TTIB transmission are shown in Figure 9.4, while the corresponding frequency domain manipulations are portrayed in Figure 9.5, with the 10 stages in processing shown in each of the figures.

The input waveform is firstly filtered by the LPF in order to remove high frequencies which might lead to aliasing. In Figure 9.5 both the lower sub-band (L) and upper sub-band (U) occupy half the total bandwidth B, but there is no reason why the signal could not be split into a number of unequal sidebands apart from the implementational practicalities. The ideal LPF, LPF2 removes U and passes L to the transmitter. Then U is frequency translated using a carrier frequency f_T leading to the spectrum shown at stage 3. This is then bandpass filtered by BPF1 before being recombined with L for transmission. At stage 5, the frequency gap between L and U is f_G, extending the total transmission bandwidth to $B + f_G$. Next the pilot carrier frequency f_P is added to the composite signal. If the total bandwidth $(B + f_G)$ is less than the coherence bandwidth of the channel, then both the received spectrum and the received pilot tone suffer the same attenuation and phase shift, so the same correction factor can be used to remove the fading from both the pilot and the signal spectrum.

At the receiver, the pilot signal is extricated using BPF3, while L is retained by LPF3 and U by BPF2. If the frequency f_T, is known to the receiver, and there is perfect alignment between nominal and actual frequencies giving $f_R = f_T$ then the original signal spectrum can be recovered at stage 10. Before a decision is made as to the transmitted signal, the extricated pilot tone is used to remove the estimated effect of the complex channel fading by deriving an amplitude and phase correction factor. The bandwidth of BPF3 is an important system parameter; if it is made too small then Doppler shift and oscillator drift may prevent recovery of the pilot tone; if it is too large then it will process an excessive amount of channel noise, degrading the reliability of the channel estimate and hence the system performance.

The TTIB receiver typically relies on signals 7 and 8 in order to recover the TTIB subcarrier f_T used by the transmitter in the frequency translation process. The manner in which this is performed is discussed in the next section.

In practice there is usually a frequency offset Δw between the TTIB subcarrier f_T at the transmitter, and that at the receiver, denoted f_R. The effect of this frequency offset of $\Delta w = 2\pi \Delta f = 2\pi (f_T - f_R)$ will also be addressed in the next section, while a practical sub-band splitting technique known as quadrature mirror filtering (QMF) which does not require infinitely steep cut-off filters will be discussed in Section 9.3.1.4.

9.3.1.3   TTIB Subcarrier Recovery

In order to eliminate interferences between the L and U sub-bands and achieve perfect signal reconstruction, it is crucial to combine their frequency and phase coherently. Because the received pilot tone has been subject to up and down conversions in the transmitter and receiver, respectively, and the multipath channel might have introduced some random frequency offset, the pilot will be prone to some frequency error and so cannot be relied upon in producing the receiver's TTIB subcarrier f_R. The channel effects and RF up and down conversions have an

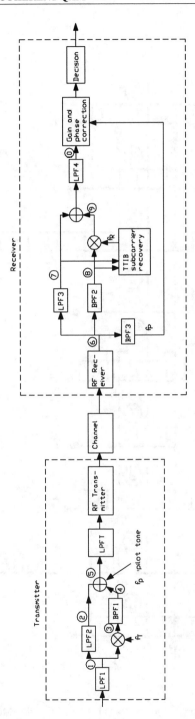

Figure 9.4: TTIB system schematic [115] ©IEEE, 1984, McGeehan,
Bateman

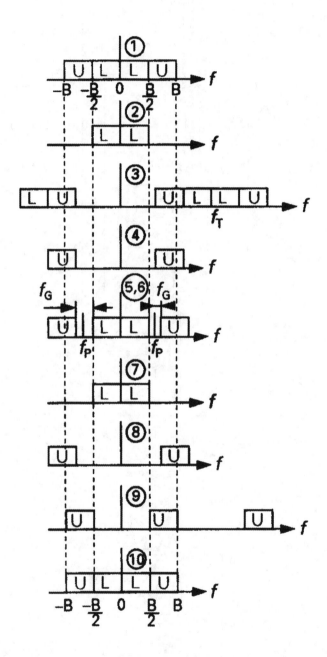

Figure 9.5: TTIB processed spectra [115] ©IEEE, 1984, McGeehan, Bateman

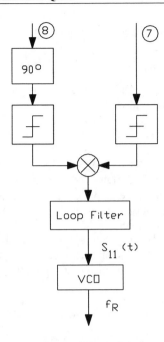

Figure 9.6: TTIB subcarrier recovery [115] ©IEEE, 1984, McGeehan, Bateman

identical influence over L and U, whereas the TTIB subcarrier offset affects only U, as shown in Figure 9.5.

To overcome this problem the TTIB subcarrier recovery method shown in Figure 9.6 is used [115]. Its operation relies on the characteristics of the signals labelled 7 and 8, which are identical to those shown in Figures 9.4 and 9.5. Observe in Figure 9.6 that U is phase shifted by 90° and then both signals 7 and 8 are input to a limiter or null comparator in order to remove any modulation or channel imposed phase variations. The two signals are then multiplied together to create the difference and sum of their frequency components. The loop filter removes the sum of the frequencies but it retains the difference, which is then input to a voltage controlled oscillator (VCO) that generates the frequency and phase coherent local TTIB carrier $f_R = f_T$.

The operation of the TTIB subcarrier recovery system is best understood if a single tone of

$$i(t) = a_i \cos(w_i t + \alpha_i) \qquad (9.59)$$

is used to excite the transmitter, where a_i, w_i and α_i are the amplitude, frequency and initial phase of the tone, respectively. We also assume that overlapping regions of the L and U sub-bands do contain some signal energy. We choose w_i specifically to satisfy this condition and assume that the L and U sub-band filters have attenuations of A_1 and A_2, respectively, at this transition band frequency.

Initially, the U sub-band tone is frequency translated using the TTIB subcarrier of

$$C_T(t) = \cos(w_T t + \alpha_T) \tag{9.60}$$

having a frequency of w_T and initial phase of α_T. It is then added to the L tone giving a signal at point 5 of Figure 9.4 as follows:

$$S_5(t) = A_1 a_i \cos(w_i t + \alpha_i) + A_2 a_i \cos[(w_i + w_T)t + \alpha_i + \alpha_T] \tag{9.61}$$

where due to the filter delay τ_1 introduced by both LPF2 and BPF1, and the delay τ_2 introduced by the transmitter filter LPFT in Figure 9.4, the time axis must be translated such that t becomes $t - \tau_1 - \tau_2$. While it might seem unlikely that LPF2 and BPF1 would have identical delays, we will show how this might be arranged using QMF in the next section.

The received signal is corrupted by the complex fading introducing an envelope factor of $\alpha(t)$ and a phase factor of $\psi(t)$. It also has a residual oscillator error of Δw due to up and down conversion. The signal received at stage 6 of Figure 9.4 is therefore given by

$$\begin{aligned} S_6(t) = {} & A_1 a_i \alpha(t) \cos[w_i t + \alpha_i + \psi(t) + \Delta wt] \\ & + A_2 a_i \alpha(t) \cos[(w_i + w_T)t + \alpha_i + \alpha_T + \psi(t) + \Delta wt]. \end{aligned}$$

The TTIB subcarrier at the receiver is given by

$$C_R(t) = \cos(w_R t + \alpha_R) \tag{9.62}$$

where α_R is its initial phase and w_R is its angular frequency. Then the signal at stage 7 in Figure 9.4, the output of LPF3, becomes

$$S_7(t) = A_3 A_1 a_i \alpha(t) \cos[w_i t + \alpha_i + \psi(t) + \Delta wt] \tag{9.63}$$

where A_3 is the attenuation of LPF3 at $(w_i + \Delta w)$ and the time t is now redefined as $t - \tau_1 - \tau_2 - \tau_3$, with τ_3 being the identical delay of LPF3 and BPF2. After passing $S_6(t)$ through BPF2 we get

$$S_8(t) = A_4 A_2 a_i \alpha(t) \cos[(w_i + w_T)t + \alpha_i + \alpha_T + \psi(t) + \Delta wt] \tag{9.64}$$

where A_4 is the attenuation of BPF2 at $(w_i + w_T + \Delta w)$. By mixing $S_8(t)$ with the output signal from the VCO having a free running frequency of w_R and an initial phase of α_R, components are generated at the difference frequency of $(w_i + w_T - w_R)$ and the sum frequency $(w_i + w_T + w_R)$, with the latter term being removed by LPF4. Ignoring, the latter term, therefore, the difference frequency term at position 9 is then given by

$$\begin{aligned} S_9^{diff}(t) = {} & A_4 A_2 a_i \alpha(t) \cos[(w_i + w_T - w_R)t + \alpha_i + \alpha_T - \alpha_R \\ & + \psi(t) + \Delta wt - w_R(\tau_1 + \tau_2 + \tau_3)]. \end{aligned}$$

The last term of $w_R(\tau_1 + \tau_2 + \tau_3)$ in Equation 9.65 is introduced now because the time variable t defined for the received components as $[t - (\tau_1 + \tau_2 + \tau_3)]$ due to the filtering delays incurred, but the time associated with the receiver's frequency is not delayed. To obtain coherent phase and frequency combination for L and U given by $S_7(t)$ and $S_9^{diff}(t)$

then $w_T = w_R$ and $\alpha_T = \alpha_R - w_R(\tau_1 + \tau_2 + \tau_3) = 0$ must be satisfied. As portrayed in Figure 9.6, initially $S_8(t)$ is phase shifted by 90° in order to rotate the cos() function and generate a sin() function. Then both $S_7(t)$ and the phase-shifted $S_8(t)$ are input to a limiter which removes any envelope fluctuations, and the amplitude-limited signals are multiplied together. Both the sum and differences of their arguments are generated, but the loop filter of Figure 9.6 admits only their difference $S_11(t)$ to the VCO, hence we arrive at:

$$S_{11}(t) = \sin[(w_T - w_R)t + (\alpha_T - \alpha_R) - w_R(\tau_1 + \tau_2 + \tau_3)] \tag{9.65}$$

Observe that $S_{11}(t)$ is independent of any channel fading since $\psi(t)$ was cancelled, but retains the frequency and phase error terms between the transmitter and receiver.

After the addition of $S_7(t)$ and $S_9(t)$ and the removal of the unwanted mixer output component by LPF4, the recovered signal can be derived from Equations 9.63 and 9.65 as

$$S_{10}(t) = (A_1 A_3 + A_2 A_4)\alpha(t)\, a_i \cos(w_i t + \alpha_i + \Delta w t). \tag{9.66}$$

If appropriate filter design is able to maintain $A_1 A_3 + A_2 A_4 = 1$, the recovered signal is corrupted by the channel amplitude fading $\alpha(t)$ and the frequency offset Δw only. However, the TTIB processing itself appears perfectly transparent. Since the pilot signal and the information signal suffer the same attenuation in a flat fading narrowband channel, the channel sounding information derived by the pilot can be used to remove the effects of the fading from the information sequence.

9.3.1.4 TTIB Schemes Using Quadrature Mirror Filters

Quadrature mirror filters (QMFs) [532] allow relatively straightforward TTIB implementations. A variety of suitable QMF designs have been proposed by Johnston [533]. The fundamental property of QMFs is that the lower (L) and upper (U) sidebands are symmetrical around the joint cut-off frequency of the LPF and HPF stages. Hence, despite their overlapping transition region, they do not introduce aliasing distortions and automatically ensure identical signal delays for both sub-band signals. Furthermore, since both stages retain signal energy from the transition band, their output signals in this frequency band will be highly correlated with each other. Through the use of correlation techniques we can combine these two correlated sub-bands perfectly after the removal of the pilot tone at the receiver.

The number of sub-bands is not limited to 2, m cascaded stages of QMFs can be used to provide 2^m sub-bands. With an increased number of sub-bands the system becomes more complex, but as each sub-band becomes narrower in comparison to the coherence bandwidth of the mobile radio channel, and the number of pilots which can be inserted increases, the channel is effectively sounded across the signal's bandwidth, making it possible to dispense with a channel equaliser. However, the increased number of pilots reduces the spectral efficiency of the transmission. In Part III of the book we will introduce a technique based on an orthogonal transformation, and this band splitting might employ 128 sub-channels and a large number of pilot tones.

In this section we consider the twin-band, QMF-based TTIB processing shown in Figure 9.7. Here a complex baseband spectrum is split by the QMF HP and LP stages. In part (c) the subbands are split apart to make room for the pilot tone.

In order for distortion-free signal reconstruction to occur using QMFs, the QMF structure

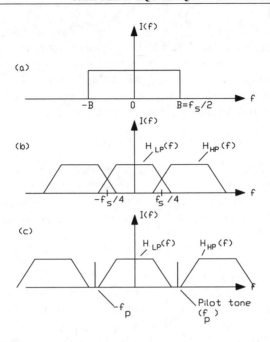

a) Complex baseband spectrum
b) QMF high pass and low pass transfer functions
c) Addition of pilot tone to spectral gap created

Figure 9.7: QMF splitting of a TTIB signal

must have an all-pass transfer function given by

$$|H_{LP}(f)|^2 + |H_{HP}(f)|^2 = 1. \tag{9.67}$$

In order to maintain linear phase, both the HP and LP impulse responses must be symmetric. Finally, the alias, free condition will be met if the impulse responses satisfy the condition

$$h_{LP}(n) = (-1)^n h_{HP}(n). \tag{9.68}$$

Assuming a complex baseband input spectrum of $I(f)$ the QMF outputs can be written as

$$\begin{aligned} I_{LP}(f) &= I(f) \cdot H_{LP}(f) \\ I_{HP}(f) &= I(f) \cdot H_{HP}(f). \end{aligned}$$

In order to shift $I_{LP}(f)$ and $I_{HP}(f)$ sufficiently far apart to create a spectral gap for the pilot symbol, we multiply both transfer functions by $e^{\pm j w_T t}$, where $2w_T$ is the required width of the spectral notch. This leads to a transmitted signal of

$$T(f) = I(f)[H_{LP}(f)e^{-j w_T t} + H_{HP}(f)e^{j w_T t}]. \tag{9.69}$$

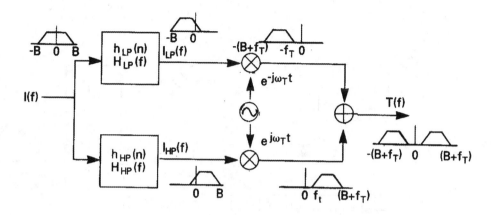

Figure 9.8: TTIB transmitter scheme [524] ©IEEE, 1990, Bateman

These operations are summarised in Figure 9.8 where the complex baseband spectra are also shown at the various stages of the processing.

Note that, due to the half-band splitting, the sub-band signals must be decimated by 2 and in general the connections between blocks represent complex signals. Bateman [524] suggested a QMF-based TTIB receiver structure, but Cavers [529] proposed an improved alternative. Here we explain Caver's structure, but use Bateman's frequency domain approach in our descriptive analysis.

As previously detailed, the transmitted signal $T(f)$ is subjected to amplitude fading and phase rotation by the mobile radio channel. However, in this analysis we ignore these effects since they are considered in the previous section, and concentrate purely on the signal reconstruction. For perfect, alias-free signal regeneration, the sidebands must be shifted back to their original position, and this operation is typically performed in two steps. This is shown in Figure 9.9 where the QMF-based TTIB receiver's schematic is shown along with the typical baseband spectral representation of the various stages of processing.

Initially, after removing the pilot tone, the sub-band components of the received signal $R(f)$ are translated back close to their original positions by being shifted using a frequency of $\pm w_R = \pm(w_T + \Delta w)$, where Δw represents the frequency difference $(w_R - w_T)$ due to misalignment between the transmitter and receiver oscillators. These translated signals can be expressed in terms of the transmitted signal $T(f)$ and noise $N(f)$ using

$$R(f) = T(f) + N(f) \tag{9.70}$$

so that

$$
\begin{aligned}
R_{LP}(f) &= R(f).e^{j(w_T+\Delta w)t} &= [T(f)+N(f)]e^{j(w_T+\Delta w)t} \\
R_{HP}(f) &= R(f).e^{-j(w_T+\Delta w)t} &= [T(f)+N(f)]e^{-j(w_T+\Delta w)t}.
\end{aligned}
$$

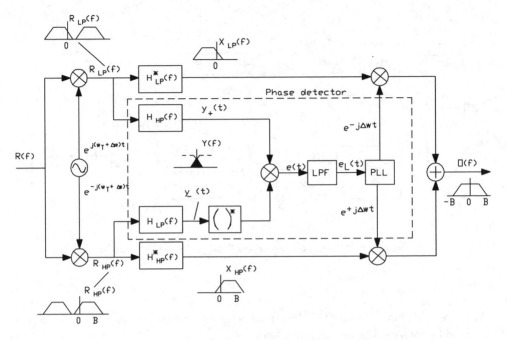

Figure 9.9: QMF-based TTIB receiver structure [524] ©IEEE, 1990, Bateman

Upon substituting $T(f)$ from Equation 9.69 into Equation 9.71 we obtain

$$
\begin{aligned}
R_{LP}(f) &= I(f).H_{LP}(f)e^{j\Delta wt} + I(f).H_{HP}(f).e^{j(2w_T+\Delta w)t} + \\
&\quad N(f)e^{j(w_T+\Delta w)t} \\
R_{HP}(f) &= I(f).H_{LP}(f).e^{-j(2w_T+\Delta w)t} + I(f).H_{HP}(f)e^{-j\Delta wt} + \\
&\quad N(f)e^{-j(w_T+\Delta w)t}.
\end{aligned}
$$

It can be seen in Figure 9.9 that one of the side lobes of each of the spectra $R_{LP}(f)$ and $R_{HP}(f)$ is now shifted back to their original positions, apart from a small frequency error Δw. Filtering $R_{LP}(f)$ and $R_{HP}(f)$ with the filter characteristics $H_{LP}(f)$ and $H_{HP}(f)$, respectively, removes the incorrectly positioned side lobes and retains the required ones, yielding

$$
\begin{aligned}
X_{LP} &= R_{LP}(f).H_{LP}^*(f) \\
&= I(f).|H_{LP}(f)|^2e^{j\Delta wt} + N(f).H_{LP}^*(f)e^{j(w_T+\Delta w)t} \\
X_{HP} &= R_{HP}(f).H_{HP}^*(f) \\
&= I(f).|H_{HP}(f)|^2e^{-j\Delta wt} + N(f).H_{HP}^*(f)e^{j(w_T+\Delta w)t}.
\end{aligned}
$$

The remaining operations in Figure 9.9 are concerned with the removal of the residual frequency error Δw and the reconstruction of the original full-band signal. Assuming that the frequency error can be extricated from $R_{LP}(f)$ and $R_{HP}(f)$ by exploiting the correlations inherent in their overlap transition bands (which we will address later), then the output signal

$O(f)$ can be written as

$$O(f) = X_{LP}(f).e^{-j\Delta wt} + X_{HP}(f).e^{j\Delta wt}. \tag{9.71}$$

Substituting Equation 9.71 into Equation 9.71 yields

$$\begin{aligned} O(f) &= I(f)\left[|H_{LP}(f)|^2 + |H_{HP}(f)|^2\right] + \\ &\quad N(f).H_{LP}^*(f).e^{jw_Tt} + N(f).H_{HP}^*(f).e^{-jw_Tt} \end{aligned}$$

where the filtered and phase-shifted noise term of

$$N_F(f) = N(f).H_{LP}^*(f).e^{jw_Tt} + N(f).H_{HP}^*(f).e^{-jw_Tt} \tag{9.72}$$

is identical to $N(f)$ since it has the same total bandwidth, the filters $H_{LP}(f)$ and $H_{HP}(f)$ have complementary transfer functions and the noise in the two sidebands is uncorrelated due to their narrow overlapping regions. Taking the all-pass nature of the QMFs given in Equation 9.67 into account, Equation 9.72 can be written as

$$O(f) = I(f) + N(f) \tag{9.73}$$

which is the original input signal plus the additive noise. Therefore, the QMF-based TTIB transceiver ensures perfectly transparent baseband processing.

9.3.1.5   Residual Frequency Error Compensation [529]

As mentioned earlier, the compensation of the residual frequency error Δw is based on the correlation of the spectral components of the HP and LP transition regions [529]. If this region does not contain any significant energy, some deterministic signal can be added, although this reduces the overall power efficiency [518].

The filters $H_{HP}(f)$ and $H_{LP}(f)$ in the phase detector section of Figure 9.9 extract the transition band components of the L and U band signals $R_{LP}(f)$ and $R_{HP}(f)$, respectively, which were given in Equation 9.71. Due to the antisymmetric filter arrangement in the phase detector compared to the filters in the corresponding branches of the transmitter, the combined frequency response $H(f)$ of the TTIB transmitter and the receiver's concatenated phase detector becomes identical in both branches, and is described by

$$H(f) = H_{LP}(f) \cdot H_{HP}(f). \tag{9.74}$$

This antisymmetric filter arrangement blocks all frequency components outside the transition band, and the input signal with spectrum $I(f)$ generates an output having a spectrum of

$$\begin{aligned} Y(f) &= I(f) \cdot H(f). \\ &= I(f).H_{LP}(f) \cdot H_{HP}(f). \end{aligned}$$

In the time domain this corresponds to

$$y(t) = i(t) * h(t) = \int_{-\infty}^{\infty} i(t-\tau)h(\tau)d\tau \tag{9.75}$$

where $y(t)$, $i(t)$ and $h(t)$ are the time domain representations of $Y(f)$, $I(f)$ and $H(f)$, respectively.

If the transmission channel is non-ideal, the received signal $r(t)$ differs from the transmitted signal $t(t)$ which is described by the following relationship

$$r(t) = \alpha(t) \cdot e^{j\psi(t)} \cdot t(t) + n(t) \tag{9.76}$$

where $\alpha(t)$, $\psi(t)$ and $n(t)$ represent the fading envelope, the fading phase, and the additive noise, respectively. These quantities are considered to be slowly varying with time, which allows us to drop their dependency on t within a symbol period. The signal $y(t)$ in Equation 9.75 is subjected to the channel effects portrayed in Equation 9.76, but due to the oscillator frequency error $\Delta w = (w_T - w_R)$ between the transmitter and receiver, a further phase component Δwt must be taken into account, yielding the phase-shifted signals $y_+(t)$ and $y_-(t)$ as follows:

$$
\begin{aligned}
y_+(t) &= \alpha \cdot e^{j(\psi + \Delta wt)} \cdot y(t) + n_+(t) \\
&= \alpha \cdot e^{j\psi} \cdot e^{j\Delta wt} \cdot y(t) + n_+(t) \\
y_-(t) &= \alpha \cdot e^{j(\psi - \Delta wt)} \cdot y(t) + n_-(t) \\
&= \alpha \cdot e^{j\psi} \cdot e^{-j\Delta wt} \cdot y(t) + n_-(t)
\end{aligned}
$$

where $n_\pm(t)$ now represents the phase shifted additive noise components. After taking the conjugate of $y_-(t)$ in Equation 9.77 and multiplying $y_+(t)$ and $y_-^*(t)$ as shown in Figure 9.9 we derive the error signal

$$
\begin{aligned}
e(t) &= y_+(t) \cdot y_-^*(t) \\
&= \alpha^2 \cdot e^{j2\Delta wt} \cdot |y(t)|^2 + \alpha \cdot e^{j\psi} \cdot e^{j\Delta wt} \cdot y(t) \cdot n_-^*(t) + \\
&\quad \alpha \cdot e^{-j\psi} \cdot e^{j\Delta wt} \cdot y^*(t) \cdot n_+(t) + n_+(t) \cdot n_-^*(t).
\end{aligned}
$$

In the case of low channel noise we can approximate this equation by

$$e(t) \approx \alpha^2 \cdot e^{j2\Delta wt} \cdot |y(t)|^2 \tag{9.77}$$

which upon evaluating the phase $\Phi = 2\Delta wt$ and halving the frequency allows the receiver to coherently recombine the L and U sub-bands by shifting the signals $X_{LP}(f)$ and $X_{HP}(f)$ in Figure 9.9 back to their original positions. Note, however, that there is a phase ambiguity of π in the phase detection process since shifts of both Φ and $\pi + \Phi$ produce the same signal $e(t)$. In this situation the TTIB receiver could erroneously invert the polarity of the output signal $O(t)$. This problem can be overcome through the use of differential coding, although this does incur a performance penalty as discussed in Chapter 4.

9.3.1.6 TTIB System Parameters [531]

The performance of a TTIB-based 16-level QAM scheme has been reported in Reference [531] where initially an AWGN channel was assumed and the hardware was constructed using a TMS320C25 digital signal processing device. With a perfectly coherent, synchronous implementation the SNR degradation due to device imperfections was about 0.5 dB when compared to the theoretical performance. When TTIB notch-width-based timing recovery was used along with perfect carrier recovery, a further 0.5 dB degradation was incurred.

When evaluating the TTIB performance using pilot-based carrier recovery, further degradations were introduced which were dependent on the level of the pilot signal transmitted. At higher pilot levels the system is less prone to channel noise, but also less power efficient. If a constant total power level is assumed, there is an optimum trade-off between pilot and information signal power. Furthermore, by narrowing the pilot filter bandwidth, less noise is introduced and the pilot signal's SNR is improved, but the maximum fading rate which can be followed is reduced. In their implementation, Martin and Bateman [531] found a bandwidth of 300 Hz to be most appropriate.

They also found that at a pilot power level of -3 dB relative to the signal, the 16-QAM BER against SNR performance was very similar to the perfectly coherent scheme, if the pilot power was not included when calculating the E_b/N_0 ratio. When the pilot power was included, a performance penalty of about 2.5 dB was incurred. At a pilot level of -6 dB relative to the mean information power a further 1 dB loss in SNR was added, which was deemed acceptable by the authors.

When the performance of the TTIB modem was evaluated over a fading channel at a frequency of 70 MHz, the pilot level of -3 dB was still adequate and yielded a 2.5 dB E_b/N_0 penalty for fading rates between 1 and 100 Hz. However, for higher fading rates the pilot filter bandwidth limited the possible range of fading fluctuations and impaired the performance unless a higher notch width was introduced. The BER versus channel SNR performance of the TTIB assisted 16-QAM modem proposed by Martin and Bateman [531] is shown in Figure 9.10 for Rayleigh fading channels exhibiting Doppler frequencies of 1, 10 and 100 Hz.

Clearly, TTIB assisted coherent QAM ensures high-integrity information transmission via fading channels by providing both amplitude and phase references for the receiver. These references assist the receiver in attaining coherent demodulation against a pilot scaled set of demodulation thresholds at a moderate complexity. Having discussed the insertion of channel sounding information in the spectral domain we now focus our attention on a time domain technique called pilot symbol assisted modulation.

9.3.2 Pilot Symbol Assisted Modulation [139]

9.3.2.1 Introduction

As we have seen in Chapter 5, the BER of the square constellation 16-QAM and 64-QAM is sufficiently low for speech transmission over AWGN channels, but it was demonstrated in this chapter that it becomes prohibitively high for transmission over Rayleigh fading channels. This is particularly so for the lower integrity $C2$ sub-channel in case of 16-QAM and for the $C2$ and $C3$ sub-channels of 64-QAM, resulting in the average BER becoming unacceptably high.

As shown in the previous section, one way to improve the BER of the square constellation over fading channels is to use transparent tone in band (TTIB) transmission. This results in higher out-of-band emissions and hence higher adjacent channel interferences, while also increasing the associated complexity. An alternative technique which is analogous to TTIB but, instead of transmitting constantly in part of the available spectrum, transmits periodically in all the available spectrum is referred to as pilot symbol assisted modulation (PSAM) [157, 178, 534, 535]. This method relies upon the insertion of known phasors into the

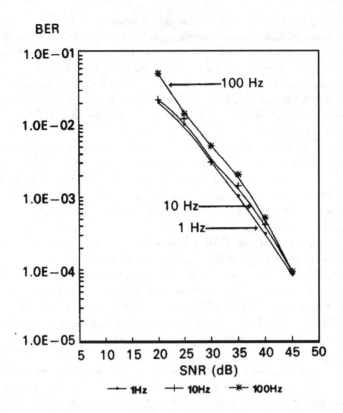

Figure 9.10: BER versus channel SNR performance of TTIB-assisted 16-QAM with f_P=70 MHz, $f_{Doppler}$=1, 10 and 100 Hz. [531] ©IEEE, 1991, Martin, Bateman

stream of useful information symbols for the purpose of channel sounding. These pilot symbols allow the receiver to extract channel attenuation and phase rotation estimates for each received symbol, facilitating the compensation of the fading envelope and phase. Closed-form formulae for the BER of PSAM were provided by Cavers [139] for binary phase shift keying (BPSK), and quadrature phase shift keying (QPSK), while for 16-QAM he derived a tight upper bound of the symbol error rate (SER).

9.3.2.2 PSAM System Description

Following Caver's approach [139], the block diagram of a general PSAM scheme is depicted in Figure 9.11, where the pilot symbols p are cyclically inserted into the data sequence prior to pulse shaping, as demonstrated by Figure 9.12. A frame of data is constituted by M symbols, and the first one in every frame is assumed to be the pilot symbol $b(0)$, followed by $(M-1)$ useful data symbols $b(1), b(2), \ldots, b(M-1)$.

Detection can be carried out by matched filtering, and the output of the matched filter is split into data and pilot paths, as seen in Figure 9.11. The set of pilot symbols can be extracted by decimating the matched filter's sampled output sequence using a decimation

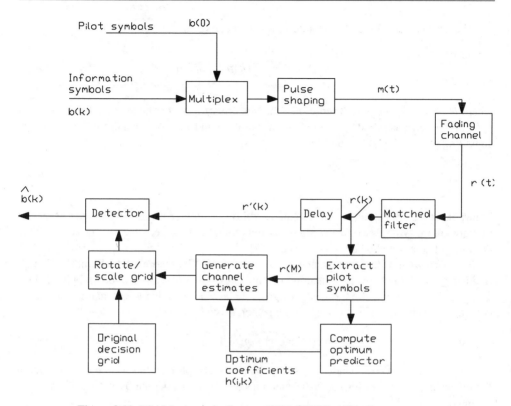

Figure 9.11: PSAM schematic diagram [139] ©IEEE, 1991, Cavers

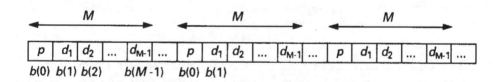

Figure 9.12: Insertion of pilot symbols

factor of M. The extracted sequence of pilot symbols must then be interpolated in order to derive a channel estimate $v(k)$ for every useful received information symbol $r(k)$. Decision is carried out against a decision level reference grid, scaled and rotated according to the instantaneous channel estimate $v(k)$.

Observe in Figure 9.11 that the received data symbols must be delayed according to the interpolation and prediction delay incurred. This delay becomes longer if interpolation is carried out using a longer history of the received signal to yield better channel estimates. Consequently, there is a trade-off between processing delay and accuracy. The interpolation coefficients can be kept constant over a whole pilot period of length M, but better channel estimates can be obtained if the interpolator's coefficients are optimally updated for every received symbol.

The complex envelope of the modulated signal can be formulated as:

$$m(t) = \sum_{k=-\infty}^{\infty} b(k)p(t - kT), \tag{9.78}$$

where $b(k) = -3, -1, 1$ or 3 represents the quaternary I or Q components of the 16-QAM symbols to be transmitted, T is the symbol duration and $p(t)$ is a bandlimited unit-energy signalling pulse, for which we have:

$$\int_{-\infty}^{\infty} |p(t)|^2 dt = 1. \tag{9.79}$$

The value of the pilot symbols $b(kM)$ can be arbitrary, although sending a sequence of known pseudorandom symbols avoids the transmission of a periodic tone, which would increase the detrimental adjacent channel interference [178].

The narrowband Rayleigh channel is assumed to be "flat" fading, which implies that all frequency components of the transmitted signal suffer the same attenuation and phase shift. This condition is met if the transmitted signal's bandwidth is much lower than the channel's coherence bandwidth. The received signal is then given by:

$$r(t) = c(t) \cdot m(t) + n(t), \tag{9.80}$$

where $n(t)$ is the AWGN and $c(t)$ is the channel's complex gain. Assuming a Rayleigh fading envelope $\alpha(t)$, uniformly distributed phase $\phi(t)$ and a residual frequency offset of f_0, we have:

$$c(t) = \alpha(t)e^{j\phi(t)} \cdot e^{j\omega_0 t} \tag{9.81}$$

The matched filter's output symbols at the sampling instant kT are then given as:

$$r(k) = b(k) \cdot c(k) + n(k). \tag{9.82}$$

Without imposing limitations on the analysis, Cavers [139] assumed that in every channel sounding block $b(0)$ was the pilot symbol and considered the detection of the useful information symbols in the range $\lfloor -M/2 \rfloor \leq k \leq \lfloor (M - 1)/2 \rfloor$, where $\lfloor \bullet \rfloor$ is the integer of $\bullet$. Optimum detection is achieved if the corresponding channel gain $c(k)$ is estimated for every received symbol $r(k)$ in the above range. The channel gain estimate $v(k)$ can be derived as a weighted sum of the surrounding K received pilot symbols $r(iM)$, $\lfloor -K/2 \rfloor \leq i \leq \lfloor K/2 \rfloor$, as shown below:

$$v(k) = \sum_{i=\lfloor -K/2 \rfloor}^{\lfloor K/2 \rfloor} h(i, k) \cdot r(iM), \tag{9.83}$$

and the weighting coefficients $h(i, k)$ explicitly depend on the symbol position k within the frame of M symbols.

The estimation error $e(k)$ associated with the gain estimate $v(k)$ is computed as:

$$e(k) = c(k) - v(k). \tag{9.84}$$

9.3.2.3   Channel Gain Estimation

While previously proposed PSAM schemes used either a lowpass interpolation filter [178] or an approximately Gaussian filter [535], Cavers deployed an optimum Wiener filter to minimise the channel estimation error variance $\sigma^2{}_e(k) = E\{e^2(k)\}$, where $E\{\ \}$ represents the expectation. This well-known estimation error variance minimisation problem can be formulated as follows:

$$
\begin{aligned}
\sigma_e^2(k) &= E\{e^2(k)\} = E\{[c(k) - v(k)]^2\} \\
&= E\left\{ \left[c(k) - \sum_{i=\lfloor -K/2 \rfloor}^{\lfloor K/2 \rfloor} h(i,k) \cdot r(iM) \right]^2 \right\}.
\end{aligned}
$$

In order to find the optimum interpolator coefficients $h(i,k)$, minimising the estimation error variance $\sigma^2{}_e(k)$ we consider estimating the kth sample and set:

$$
\frac{\partial \sigma_e^2(k)}{\partial h(i,k)} = 0 \quad \text{for} \quad \lfloor -K/2 \rfloor \leq i \leq \lfloor K/2 \rfloor. \tag{9.85}
$$

Then using Equation 9.85 we have:

$$
\frac{\partial \sigma_e^2(k)}{\partial h(i,k)} = E\left\{ 2\left[c(k) - \sum_{i=\lfloor -K/2 \rfloor}^{\lfloor K/2 \rfloor} h(i,k) \cdot r(iM) \right] \cdot r(jM) \right\} = 0. \tag{9.86}
$$

After multiplying both square bracketed terms with $r(jM)$, and computing the expected value of both terms separately, we arrive at

$$
E\{c(k) \cdot r(jM)\} = E\left\{ \sum_{i=\lfloor -K/2 \rfloor}^{\lfloor K/2 \rfloor} h(i,k) \cdot r(iM) \cdot r(jM) \right\}. \tag{9.87}
$$

Observe that

$$
\Phi(j) = E\{c(k) \cdot r(jM)\} \tag{9.88}
$$

is the cross-correlation of the received pilot symbols and complex channel gain values, while

$$
R(i,j) = E\{r(iM) \cdot r(jM)\} \tag{9.89}
$$

represents the pilot symbol autocorrelations, hence Equation 9.87 yields:

$$
\sum_{i=\lfloor -K/2 \rfloor}^{\lfloor K/2 \rfloor} h(i,k) \cdot R(i,j) = \Phi(j), \quad j = \lfloor -\frac{k}{2} \rfloor \ldots \lfloor \frac{k}{2} \rfloor. \tag{9.90}
$$

If the fading statistics can be considered stationary, the autocorrelations $R(i,j)$ will only depend on the difference $|i - j|$, giving $R(i,j) = R(|i - j|)$. Therefore Equation 9.90 can be written as:

$$\sum_{i=\lfloor -K/2 \rfloor}^{\lfloor K/2 \rfloor} h(i,k) \cdot R(|i-j|) = \Phi(j), \quad j = \lfloor -K/2 \rfloor \dots \lfloor K/2 \rfloor, \tag{9.91}$$

which is a form of the well-known Wiener-Hopf equations [3], often used in estimation and prediction theory [383].

This set of K equations contains K unknown prediction coefficients $h(i,k)$, $i = \lfloor -K/2 \rfloor, \dots, \lfloor K/2 \rfloor$, which must be determined in order to arrive at a minimum error variance estimate of $c(k)$ by $v(k)$. First, the correlation terms $\Phi(j)$ and $R(|i-j|)$ must be computed and to do this the expectation value computations in Equations 9.88 and 9.89 need to be restricted to a finite duration window. This approach is referred to as the *autocorrelation method* [383].

The set of Equations 9.91 can also be expressed in a convenient matrix form as:

$$\begin{bmatrix} R(0) & R(1) & R(2) & \dots & R(K) \\ R(1) & R(0) & R(1) & \dots & R(K-1) \\ R(2) & R(1) & R(0) & \dots & R(K-2) \\ \vdots & \vdots & \vdots & \dots & \vdots \\ R(K) & R(K-1) & R(K-2) & \dots & R(0) \end{bmatrix} \tag{9.92}$$

$$\cdot \begin{bmatrix} h\left(\lfloor -\frac{K}{2} \rfloor, k\right) \\ h\left(\lfloor -\frac{K}{2}+1 \rfloor, k\right) \\ h\left(\lfloor -\frac{K}{2}+2 \rfloor, k\right) \\ \vdots \\ h\left(\lfloor \frac{K}{2} \rfloor, k\right) \end{bmatrix} = \begin{bmatrix} \Phi\left(\lfloor -\frac{K}{2} \rfloor\right) \\ \Phi\left(\lfloor -\frac{K}{2}+1 \rfloor\right) \\ \Phi\left(\lfloor -\frac{K}{2}+2 \rfloor\right) \\ \vdots \\ \Phi\left(\lfloor \frac{K}{2} \rfloor\right) \end{bmatrix},$$

which can be solved for the optimum predictor coefficients $h(i,k)$ by matrix inversion using Gauss-Jordan elimination or any appropriate recursive algorithm.

Once the optimum predictor coefficients $h(i,k)$ are known, the minimum error variance channel estimate $v(k)$ can be derived from the received pilot symbols using Equation 9.83, as also demonstrated by Figure 9.11.

9.3.2.4 PSAM Parameters

Cavers [139] provided a range of results for a variety of PSAM parameters. As regards to the pilot spacing M, the optimum value is determined by the Nyquist rate of the fading envelope. Inserting pilot symbols less frequently results in a significant increase in BER due to the inability to track the fading, while an unnecessarily frequent insertion wastes energy without significantly improving the BER performance.

The typical energy loss and Doppler frequency interdependence can be highlighted by the following simple example. Let us assume a fast-moving vehicle with a Doppler frequency of $f_D = 100$ Hz, and a TDMA multi-user signalling rate of 100 ksymbols/s, yielding a normalised Doppler shift of $f_d \cdot T = 10^{-3}$. This would require a Nyquist frequency of $f_N = 2 \cdot 10^{-3}$, implying a pilot spacing of $M = 500$, which results in an energy loss of only 0.2%. If the signalling rate reduces to 10 ksymbols/s, the relative frequency of the pilot symbol insertion must be increased tenfold in order for the pilot symbols to appear in the same positions in terms of absolute time, facilitating equally frequent channel sounding, as in the case of 100 ksymbols/s signalling. The corresponding pilot insertion frequency of $M = 50$ implies an energy loss of about 2%. However, when using the PSAM scheme in a single-user scenario with a low signalling rate of, say, 2 ksymbols/s, the energy loss due to five times more frequent channel sounding associated with $M = 10$ becomes 10%.

As seen in Figure 9.11, the operation of the PSAM scheme is prone to a delay of $KM/2$ symbols, which might become an impediment in case of speech transmissions. Due to this start-up delay the first $KM/2$ modulation symbols are corrupted by fading and hence must be discarded. This might affect, for example, some $7 \cdot 10/2 = 35$ start-up symbols for the specific case of $K = 7$ and $M = 10$.

Clearly, the design of the predictor coefficients $h(i, k)$ depends on a variety of PSAM system parameters, such as K and M, as well as on channel conditions, such as the prevailing SNR and normalised Doppler shift. The symbol error rate performance upper bounds derived by Cavers [139] for 16-QAM cannot be approached if the above conditions are drastically different from those for which the predictor coefficients were optimised. Best results can be attained using adaptive predictors updating their parameters online, as signalling and channel conditions vary.

A further advantage of PSAM is that it generates an absolute phase reference and hence differential encoding of the data is not required. In contrast, TTIB transmissions require a perfect phase lock to be maintained [529], which assumes differential coding, associated with some SNR penalty. In a more recent paper [536] Cavers investigated the performance of PSAM schemes under dispersive channel conditions introduced by frequency selective fading channels in case of BPSK and QPSK. He found that the performance gains are sensitive to delay spread, but PSAM always ensured a better performance than differential coding. Recently Seymour and Fitz [537] recognised that ignoring the transmitted information symbols in the predictor is wasteful. This is because the modulated received information symbols also carry channel-specific information. Exploiting this property allows for more sparse pilot insertion, thereby saving modulation energy, or achieving slightly better BER performance.

9.3.2.5 PSAM Performance

In this section we embarked on the performance evaluation of two low-complexity square-constellation pilot-assisted modems, namely a 16-QAM/PSAM scheme and a 64-QAM/PSAM scheme. The pilot separation of the 16-QAM scheme was $M = 10$, while that of the 64-QAM modem was $M = 5$ at a pedestrian speed of 4 mph, propagation frequency of 1.9 GHz and signalling rate of 100 kBd. The channel magnitude and phase fluctuation was linearly interpolated between two adjacent pilot symbols. The performance of these scheme is displayed in Figures 9.13. and 9.14. This low complexity approach typically failed to track the amplitude and phase trajectory upon emerging from a deep fade, which manifests itself in a residual bit error rate, but overall provided a significant performance improvement when compared to the scenarios portrayed earlier in Figures 9.1 and 5.3 without PSAM.

The performance of a combined 16-QAM/PSAM/TCM scheme over Rayleigh fading channels is reported in Figure 11.8, where three PSAM modulation schemes having a net throughput of 2 bits/symbol are studied comparatively. Specifically, uncoded QPSK, 2/3 rate coded 8-PSK/TCM and 1/2 rate square 16-QAM/TCM are compared and the 16-QAM/TCM

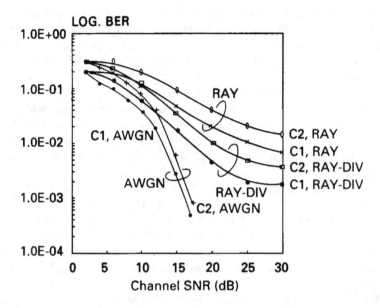

Figure 9.13: Square 16-QAM/PSAM BER versus channel SNR curves for AWGN and Rayleigh channels with and without diversity at 4 mph, 1.9 GHz, and 100 kBd

scheme has the best performance, since it has no unprotected bits.

9.4 Summary

In this chapter we started by considering the theoretical performance of 16-level and 64-
level square QAM over mobile radio channels. The results showed a significant drop in
performance compared to transmission over the Gaussian channels considered earlier in this
book.

We then continued to consider two particular schemes of QAM transmission over mobile
radio channels whereby an attempt was made to transmit some information to the receiver
concerning the instantaneous state of the mobile radio channel. Transparent tone in band
attempted to split the transmitted spectrum in two, to increase the spacing between these
two parts, and to insert a continuous tone which the receiver could use to estimate the channel response. Such a scheme was shown to be complex and a number of problems with it
were detailed. Pilot symbol assisted modulation inserted a known symbol periodically into
the data stream. Since the receiver had been pre-programmed as to what this symbol was
and when it would appear, it could use this information to estimate the channel response at
that particular instant. Advanced extrapolation techniques could then be used to predict the
channel response between any two of these periodic symbols. Both schemes were shown to
have some success in overcoming the worst problems of the mobile radio channel although
both had the disadvantage of increasing the required bandwidth due to the transmission of
redundant information.

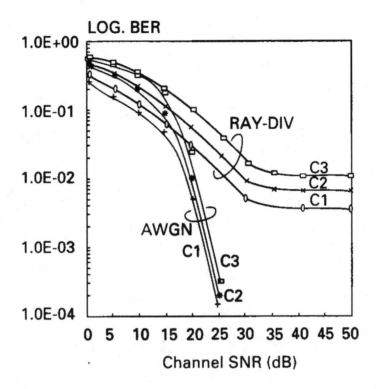

Figure 9.14: Square 64-QAM/PSAM BER versus channel SNR curves for AWGN and Rayleigh channels with diversity at 4 mph, 1.9 GHz and 100 kBd

In the next chapter we consider a completely different means of overcoming the problem
of a rapidly changing channel. This is through the use of a different QAM constellation
known as star QAM which, through its symmetry and use of differential coding, removes the
need for the receiver to estimate the channel response at all.

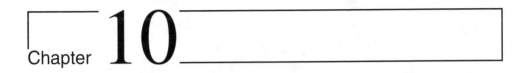

Chapter **10**

Star QAM Schemes for Rayleigh Fading Channels

10.1 Introduction

In Chapter 9 we considered the transmission of coherently detected maximum least distance square QAM signals over Rayleigh fading channels. We have shown that adequate BER performance can only be achieved through deploying either frequency domain pilots (TTIB) or using time domain pilot signals (PSAM). The channel attenuation and phase rotation information derived using these techniques facilitates fading compensation and coherent detection at moderate implementational complexity. In many applications, such as light-weight hand-held portable telephones, power consumption, weight and low cost construction are crucial issues, and hence lower complexity modem schemes are desirable. In this case it may be preferable to reduce the system performance slightly in order to be able to employ a low-complexity non-coherent differentially encoded QAM constellation, such as the circular or star QAM scheme proposed in this chapter. We can then dispense with TTIB, PSAM, or AGC reference schemes, and no longer require carrier recovery. Consequently, the result is an attractive low complexity system. In this chapter we investigate the properties and performance of a variety of star QAM systems and then consider constellation distortion and practical issues.

10.2 Star 16-QAM Transmissions over Mobile Channels

In Section 4.1 we presented a brief discussion on the most appropriate choice of constellation for a range of channel conditions, and showed the superiority of the Type III rectangular constellation of Figure 4.1 over the Type I circular or star constellation for AWGN channels. The QAM constellation which we advocate for Rayleigh fading channels is shown in Figure 10.1. This constellation does not have a minimum least free distance between points in the manner of the square constellation of Chapter 9, but does allow efficient differential encoding and decoding methods to be used which remove the need for AGC and carrier recovery and

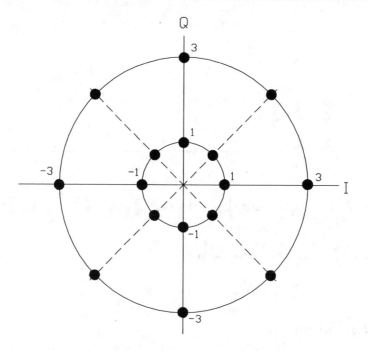

Figure 10.1: Transmitted differential star QAM constellation

thus go some way towards mitigating the effects of Rayleigh fading. The structure of this constellation will be further investigated throughout this section.

10.2.1 Differential Coding

In the case of PLL type carrier tracking, some form of differential encoding is essential as the Rayleigh fading channel can introduce phase shifts in excess of 50° between consecutive symbols in the case of slow signalling rates, making it extremely difficult to establish an absolute phase reference.The encoding method used in the star QAM system is as follows. Of the four bits in each symbol, b_1, b_2, b_3 and b_4, the first is differentially encoded onto the QAM phasor amplitude so that a 1 causes a change to the amplitude ring which was not used in the previous symbol, and a 0 causes the current symbol to be transmitted at the same amplitude as the previous symbol. The remaining three bits are differentially Gray encoded onto the phase so that, for example,000 would cause the current symbol to be transmitted with the same phase as the previous one, 001 would cause a 45° phase shift relative to the previous symbol, "011" a 90° shift relative to the previous symbol, and so on.

10.2.2 Differential Decoding

Using the above differential encoding technique, decoding data is now reduced to a comparison test between the previous and current received symbols. Suppose we fix the transmitted rings at amplitude levels A_1 and A_2. Let the received phasor amplitudes be Z_t and Z_{t+1}

at times t and $t+1$, respectively. The demodulator must identify whether there has been a significant change in amplitude in order to regenerate a logical 1. This is equivalent to asking whether the amplitude of the phasor at the transmitter changed rings at time $t+1$ compared to time t. The algorithm employed at the demodulator uses two adaptive thresholds to carry out its decisions. If

$$Z_{t+1} \geq \left(\frac{A_1 + A_2}{2} \right) Z_t \qquad (10.1)$$

or if

$$Z_{t+1} < \left(\frac{2}{A_1 + A_2} \right) Z_t \qquad (10.2)$$

then a significant change in amplitude is deemed to have occurred and bit b_1 is set to logical 1 at time $t+1$. Notice that the thresholds are dependent on Z_t, and as the amplitudes of the phasors change in fading conditions so do the thresholds. Should both Equations 10.1 and 10.2 fail to be satisfied, b_1 is assigned logical 0.

If the received symbol phases are θ_t and θ_{t+1} at time t and t+1, respectively, the demodulated angle is

$$\theta_{dem} = (\theta_{t+1} - \theta_t) \bmod 2\pi. \qquad (10.3)$$

This angle is then quantised to the nearest multiple of 45° and a lookup table is consulted in order to derive the remaining three output bits, b_2, b_3 and b_4. This differential system can considerably improve the BERs compared to the square constellation because it eliminates the long error bursts that occur when a false lock was inflicted. It is considerably important that with differential amplitude encoding there is no longer any need for AGC. This not only simplifies the system, but also removes errors caused by an inability of the AGC to follow the fading envelope.

10.2.3 Effect of Oversampling

In mobile radio channels the fading continues to cause problems because there are changes in channel amplitude and phase during symbol periods, which in general move the differentially decoded phasors nearer to the decision boundaries. The most likely cause of error in the star QAM system is when both the noise, and the change in the phasor's amplitude or phase due to fading, combine to drive the incoming signal level over a decision boundary. As the Rayleigh fading envelope is bandlimited by the normalised Doppler frequency, it is highly correlated and predictable, particularly at low vehicular speeds. Hence a correction factor may be applied to the incoming signal in order to compensate for the changes in the fading envelope during the last symbol period. Such a system must be fast acting so that the sudden change from an amplitude decrease to an amplitude increase experienced at the bottom of a fade can immediately be detected and compensated. Methods such as using the differential of the PLL error signal in order to change the step size of the phase correction signal tend to overshoot at sudden changes and exhibit damped second-order system behaviour.

In order to overcome this problem a simple oversampling receiver can be used. In this system n observations equally spaced in time are made per symbol period. At SNRs above 30 dB the current phasor can be modified to compensate for the fading. This is done by finding the change in the incoming symbol phase and amplitude over the current symbol period. We do this by subtracting the phase at the end of the symbol from the phase at the

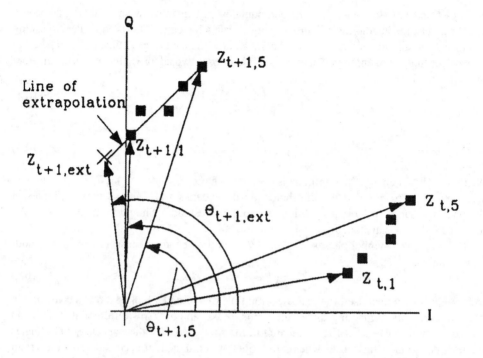

Figure 10.2: Correction for changes in fading envelope

beginning of the symbol, and subtracting the amplitude at the end of the symbol from the amplitude at the beginning of the symbol. The situation is shown in Figure 10.2.

We use the same notation as before, but with the addition of subscript n to the symbols, to signify the nth observation of the symbol at time t. Thus

$$Z_{t+1,diff} = Z_{t+1,n} - Z_{t+1,1} \tag{10.4}$$

where *diff* means differential, and

$$\theta_{t+1,diff} = \theta_{t+1,n} - \theta_{t+1,1}. \tag{10.5}$$

These changes can be used to extrapolate from the first observation in the current symbol back to the time when the last observation from the previous symbol was made.

With the aim of improving the accuracy of the references used in Equations 10.1 and 10.2 we replace Z_t by its last oversampled value, namely $Z_{t,n}$ and we use an estimate of Z_{t+1} which is formed by extrapolation as

$$Z_{t+1,ext} = Z_{t+1,1} - \frac{Z_{t+1,diff}}{n} \tag{10.6}$$

rather than Z_t. Notice that $Z_{t+1,1}$ is the first sample of Z_{t+1} and is closest in time to $Z_{t,n}$. By subtracting the average change in Z_{t+1} over a symbol period, i.e. $\frac{Z_{t+1,diff}}{n}$ from $Z_{t+1,1}$, we obtain an estimate of Z_{t+1} had it been transmitted at time t. This is beneficial as there may have been significant amplitude and phase changes between times t and $t+1$ as can be seen in Figure 10.2.

After determining bit b_1 with the aid of Equations 10.1 and 10.2 using the modified Z_t and Z_{t+1} values, we determine bits b_2, b_3 and b_4 by formulating

$$\theta_{dem} = (\theta_{t+1,ext} - \theta_{t,n}) \bmod 2\pi. \tag{10.7}$$

where $\theta_{t+1,ext}$ and $\theta_{t,n}$ are the phase angles associated with $Z_{t+1,ext}$ and $Z_{t,n}$. Again θ_{dem} is quantised and used to address a look-up table which provides values of b_2, b_3 and b_4.

In the above description it has been assumed that no pulse shaping or filtering was employed so that square pulse shapes result. This is impractical because of the spectral spillage caused by such modulation. The oversampling system can work with practical pulse shaping systems. With pulse shaping the oversampling system needs to perform an inverse operation before predicting the change undergone in the channel. At the receiver an attempt is made to reconstruct the transmitted waveform. This is done by passing the estimated received and regenerated data in square waveform through an identical filter to the transmitter filter. The difference between this reconstructed waveform and the received waveform can then be used to estimate the average change that has taken place over the symbol period, and thus form some estimate of the Rayleigh fading. This estimate can then be used in the decoding of the next symbol, with extrapolation techniques instigated as required. Simulations showed the degradation in BER over unfiltered transmissions to be negligible.

10.2.4 Star 16-QAM Performance

Previous QAM systems tended to exhibit a residual BER at high SNRs due to the rapidly changing Rayleigh channel, rather than the additive noise. With this system the residual BER is reduced by approximately two orders of magnitude. The simulation results of this system when the 16-QAM carrier was 1.9 GHz, the symbol rate was 16kSym/s, the mobile's speed was 30 mph and the channel exhibited Rayleigh fading with additive white Gaussian noise (AWGN) are given in Figure 10.3. The receiver operated as described above, with the oversampling ratio set at $n = 8$. Non-linear filtering as introduced in Chapter 4 was employed at the transmitter. Both transmitter and receiver used a fourth-order Butterworth lowpass filter with a 3dB point at 1.5 times the Baud rate, i.e. 24 kHz.

The variation of BER versus channel SNR is shown in curve (c) in Figure 10.3.

Also shown as benchmarks are the results for the similar complexity coherent non-differential square QAM constellation using no fading compensation AGC, TTIB or PSAM (curve (a)) and the star 16-QAM with differential encoding (curve (b)) both over a Rayleigh fading channel, and the star 16-QAM with differential encoding over an AWGN channel (curve (e)). The performance of the system having differential encoding and oversampling (curve(c)) can be considerably enhanced by the use of spatial diversity, where two antennas and receiver circuits are used. For these simulations, switched diversity was used, whereby for each phasor received, the receiver with the incoming phasor of largest magnitude was selected. Both receivers must have their own differential decoders. Curve (d) shows the performance of this second, order diversity assisted system.

As can be seen, a very substantial improvement in BER has been obtained over the coher-

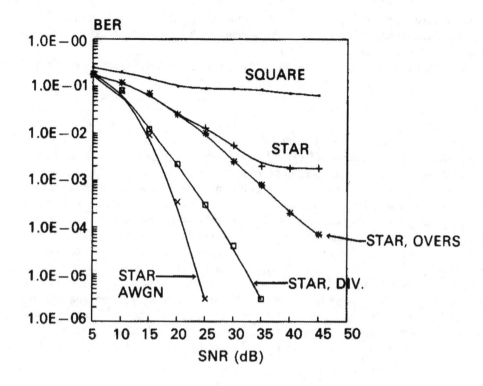

Figure 10.3: BER versus channel SNR results for various 16-QAM systems

ent non-compensated square 16-QAM scheme having similar complexity by using star QAM, and at high SNRs oversampling gave a further significant gain. By introducing second-order switched diversity the system operated with a channel SNR of only 5 dB above that for an AWGN channel for a BER of 10^{-3}.

10.3 Trellis Coded Modulation for QAM

In an attempt to further reduce the BER achieved by the star 16-QAM schemes proposed in the previous section, we turned our attention towards various FEC coding schemes. Papers published by Ungerböck and others [496] suggest that, for an AWGN channel, significant coding gains can be achieved by expanding the symbol set size and using the extra bit(s) gained for channel coding. This technique is referred to as trellis coded modulation (TCM). TCM schemes employ non-binary modulation in combination with a finite state convolutional encoder, as described briefly in Chapter 8 and in more detail in Chapter 24.

Historically TCM was conceived to operate with modems over AWGN channels. Here it could be assumed that an accurate phase reference was established and maintained and so there was no problem with false locking. One of the properties of TCM systems is that they exhibit invariance to $180°$ rotations, i.e. if lock is established $180°$ from the correct lock

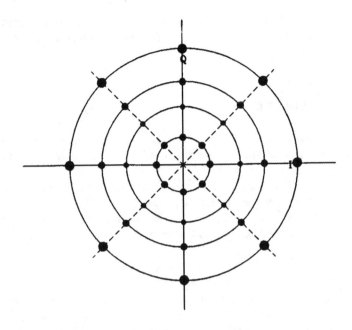

Figure 10.4: Star 32-TCM constellation

position, their operation will be unaffected. Systems have been suggested [538] having non-linear convolutional encoders which exhibit invariance to all 90° rotations, but no systems exhibiting invariance to 45° rotations have been published. This means that TCM cannot be used as it stands with the star 16-QAM system, as half of the possible lock points will lead to incorrect system operation.

In order to maintain phase invariance to 45° rotations it is necessary to introduce differential coding whereby we associate each TCM word with a phase and amplitude change, rather than with an absolute phase and amplitude as is normally done. The constellation we opt for is a four-ring star scheme with 8 points on each level, i.e. the 32-level star QAM shown in Figure 10.4.

This arrangement does not have a constant free distance amongst constellation points, but it does enable the TCM codewords to be optimally and differentially mapped onto the constellation. Based on our experiments we discovered that sub-optimal mappings involving constellations, for example, that have 4, 6, 10, and 12 points on the first, second, third and fourth rings, respectively, where the rings are numbered starting with the inner ring, have poorer results than a constellation having 8 points on each of its four rings. Diagonal interleaving over approximately three fades was used in order to randomise errors. Simulations of TCM revealed marginal improvements in BER at very high SNRs compared to 16-level star QAM, although significant degradation in BER was experienced at the lower SNRs encountered in practice. This degradation at lower SNRs was because the BERs on which the convolutional decoder was operating were so high that the decoder often chose the incorrect path through the trellis and thus increased the number of errors. It was not until the BER fell below a certain level that the convolutional coding system was able to reduce the number of errors. TCM systems only proved beneficial when switched diversity was used, achieving a residual BER of 5×10^{-5} at SNRs above 35 dB. This was because current TCM systems have been optimised for AWGN.

10.4 Block Coding

In our next endeavour in attempting to reduce the star 16-QAM BER, block coding [507] rather than convolutional coding [507] was employed by expanding the signal set to cope with the extra bits required for the FEC code. A 2/3 coding rate was considered to be appropriate, causing the number of QAM levels to increase from 16 to 64. An extension of the star constellation was selected having four amplitude rings with 16 points equispaced on each ring. Again, this was not optimal as regards to the minimum distance criterion, but simulations showed it to perform better in fading environments than constellations which were optimal for AWGN.

For block codes to perform well in the Rayleigh fading environment it is necessary to add interleaving to the system in order to randomly distribute the errors in time. The block code and the interleaving process introduce a delay, and the maximum permissible delay depends on the type of information to be transmitted. The integrity constraints of computer data transmissions are at least three orders of magnitude higher than those of digital speech transmissions. Fortunately, data channels can accept longer interleaving delays which allow effective randomisation of the bursty error statistics of the Rayleigh fading channel. This considerably increases the ability of the forward error correction (FEC) decoder to decrease the BER. For each transmission rate, propagation frequency, vehicular speed and interleaver algorithm there is a minimum interleaving depth or delay to transform the BER statistics of a fading channel into a good approximation of those encountered in a Gaussian channel. For a specific combination of system parameters, an interleaved FEC code word has to span a number of channel fades.

For the propagation frequencies 1.7 to 1.9 GHz used for personal communications networks (PCN) the wavelength is about 17 cm and there is approximately one fade every 8.5 cm distance. If the data transmission rate is 64 kbit/s, yielding a signalling rate of 16 ksamples/s for the uncoded 16-level QAM, then when the mobile station is travelling at 30 mph or 13.3 m/s, there are approximately 100 QAM samples transmitted between two deep fades which are about half a wavelength apart. Diagonal interleaving over six fades randomises the bursty error statistics and has a delay of approximately 600 QAM samples corresponding to about 40 ms. This delay is typically acceptable for speech communications [539]. By increasing the interleaving depth to much higher values associated with higher delay, which is acceptable for data transmission, the BER is further improved.

We consider Bose-Chaudhuri-Hocquenghem (BCH) block codes [507] to have favourable properties for PCN transmissions. They can correct both random and bursty errors and also their error detection capability allows the BCH decoder to know when the received code word contains more errors than the correcting power of the code can cope with. Provided a systematic BCH code is used, the information part of the coded word can be separated from the parity bits without attempting error correction which would precipitate more errors due to the fact that the codec's error correction power was exceeded and hence the decoding operation would be erroneous.

A special subclass of BCH codes are the maximum minimum distance Reed-Solomon(RS) codes [507]. These codes operate on non-binary symbols and have identical error locator and symbol fields. The non-binary RS codes are optimum due to their maximum distance property, and may also be sufficiently long to span channel fades without additional interleaving. However, they are more complex to implement than binary BCH codes. To explore both ends of the complexity/performance trade-off we selected an extremely long and complex RS code, the $RS(252, 168, 42)$ code over Galois field $GF(256)$ using 8-bit symbols, the moderately long and less complex $RS(44, 30, 7)$ code over $GF(64)$, as well as the low complexity short binary $BCH(63, 45, 3)$ code. All these codes have an approximately equal coding rate of $R = 0.7$. The nomenclature used here is $RS(m, n, k)$ where m is the number of encoded symbols, n is the number of information symbols and k is the number of corrected symbols in a code word. Every field element of the finite field $GF(2^l)$ is represented by an l-bit non-binary symbol in the case of RS codes.

10.5 64-level TCM

As a benchmark scheme for the block-coded arrangements a 64-level 2/3 rate convolutional coding system was devised. However, TCM as outlined in Chapters 8 and 24 cannot be used directly, since it doubles the symbol set size, and here we have a quadrupling of the set size from 16 to 64 levels. Nevertheless, the principles of TCM can be retained. The 64 possible code words were divided into 16 groups of four so-called D-groups [496]. TCM only adds FEC coding to those bits that have the highest probability of error. With four input bits and six bits per symbol we can add half-rate convolutional coding to the more vulnerable two bits and leave uncoded the other two that are more widely spaced. The two coded bits plus the two FEC parity bits can specify one of 16 groups, and the uncoded two bits can specify one of four points within this group. The four points in each group are spaced as far apart across the constellation as possible so that there is little chance of one of these four points being mistakenly received as another in the presence of noise.

In Figure 10.5 we display the constellation of 64 points representing the 64 possible transmitted phasors. Figure 10.5 may also be viewed as a look-up table for the differential encoding. The six bits span the decimal range 0-63 as they vary from 000000-111111. If this decimal number is in the range 0-3, 4-7, 8-11, ..., 60-63 we select subgroup D0, D1, D2, ..., D15, respectively. In each subgroup we map each of the four possible numbers onto one of the four points in that subgroup. This is done by starting on the positive x-axis and the innermost circle, and rotating anticlockwise. The first point in the subgroup that we come across is given the lowest number for that subgroup. Thus if we consider subgroup D0, we would assign the point on the positive x-axis the value 0, i.e. it is the point we map to when the incoming six bits are 000000. The point on the positive y-axis and the second circle becomes 1 corresponding to incoming bits 000001, the point on the negative x-axis becomes 2 corresponding to 000010, and the point on the negative y-axis becomes 3 corresponding to 000011. This is repeated for all subgroups. Thus the D4 point on the negative y-axis corresponds to 19 or 010011. Once we have identified the phasor for the incoming code word, we can use this to derive the differential signal which we must add onto the previous constellation point transmitted as a reference in order to derive the constellation point that we transmit. The representation is as follows.

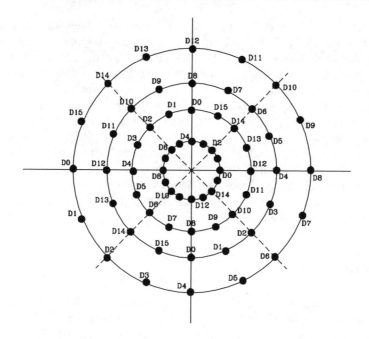

Figure 10.5: Star 64-TCM constellation

A point on the inner ring causes us to transmit the current symbol with an identical amplitude to the previous symbol transmitted.

A point on the second ring causes us to transmit the current symbol with an amplitude on the next amplitude ring up from the previous symbol transmitted. If the previous symbol amplitude was that of the outermost ring, then a change is made to the innermost ring, i.e. wrap-around occurs.

A point on the third ring causes an increase of two rings in amplitude from the previous amplitude transmitted again with wrap-around, and so on.

A point at 0° on the positive x-axis represents no phase change from the previous point transmitted. A point at 22.5^o represents a phase change of $+22.5^o$ relative to the previous point transmitted. A point at 45° represents a phase change of $+45^o$ relative to the previous point transmitted, and so on.

As an example, if the last phasor transmitted was that corresponding to the point labelled D13 at 22.5^o on the second ring and the input bits were 000010, we decode these input bits as the D0 point on the negative x-axis. Thus we increase the ring size by three, wrapping round from the D13 point to the innermost ring, and rotate this point by 180° to end up at the unlabelled D9 point between the corresponding D8 and D10 points on the innermost ring at -157.5^o. This is the phasor transmitted.

This heuristic representation has been used as it allows set partitioning into the D-groups to be carried out in the same way as for conventional non-differential TCM.

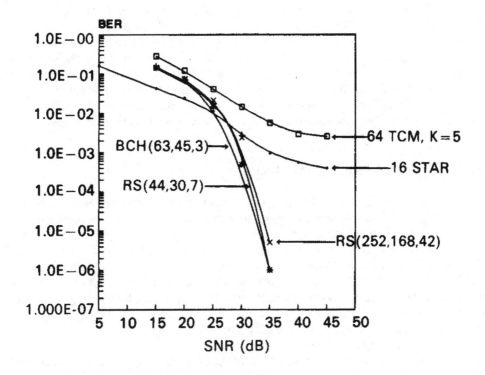

Figure 10.6: TCM and block coded performance

The half-rate coder chosen was of constraint length $K = 5$ with generator polynomials

$$g1 = 1 + X^3 + X^4$$

and

$$g2 = 1 + X + X^2 + X^4$$

as recommended for the GSM system [540]. These generator polynomials are optimal for random error statistics. The encoded bits are diagonally interleaved prior to the mapping shown in Figure 10.5. This interleaving was the same as used with the block codes.

10.6   Bandwidth Efficient Coding Results

The above 64-level $R = 2/3$, $K = 5$ TCM system contrived in the previous section performed worse than the star 16-QAM at all SNRs, as shown in Figure 10.6.

When using separate block coding and absorbing the parity bits by constellation expansion at the cost of higher channel BERs and implementational complexity, but without bandwidth expansion as described in Section 10.4, the BER decreased dramatically compared to both our 64-TCM $K = 5$ system and to the uncoded star 16-QAM scheme. The $BCH(63, 45, 3)$ code and the $RS(44, 30, 7)$ code had nearly identical performances, in spite of the considerably higher block length and complexity of the RS code. The $RS(252, 168, 42)$ code offers an extra 2 dB SNR gain for a large increase in complexity. We therefore favour the BCH code which provides virtually error-free communications for channel SNR values in excess of 30-35 dB, a value that may be realisable in the small microcells to be ultimately found in a fully developed PCN. By using error correction coding and compensating for the increased bit rate by deploying higher-level modulation, we are able to provide transmissions of higher integrity, but with identical bandwidth compared to an uncoded system. The transmitted power and system complexity are also increased.

10.7 Overall Coding Strategy

We now consider 16-level and 64-level QAM schemes which are error protected, where the FEC coding now decreases the useful information transmission rate while increasing the integrity of the transmitted data. The previous 64-level QAM scheme having a 2/3 rate code had its code rate reduced to 1/2 while maintaining constant transmission rate, but decreasing the primary information rate. We selected this overall 1/2 coding rate as it is widely used in mobile radio [507, 541]. A further coding rate reduction does not bring substantial coding gains and squanders channel capacity which cannot be compensated for by using more modulation levels. We also employed a 3/4 rate code in conjunction with the star 16-QAM in order to provide the same overall bit rate as the coded 64-level QAM.

Our results are depicted in Figure 10.7.

The uncoded star 16-QAM and the 64-level QAM/$RS(44, 30, 7)$ curves are repeated for comparison along with the 64-level QAM/$RS(60, 30, 15)$ and 16-level QAM/$RS(60, 44, 8)$ arrangements. Observe that the error correction power of the 64-QAM/$RS(60, 30, 15)$ scheme is about twice as strong as that of the 16-QAM/$RS(60, 44, 8)$ system. However, 64-QAM has typically more than twice the symbol error rate of 16-QAM and this is why the 16-QAM arrangement is likely to outperform the coded 64-QAM scheme. There is a consistent and remarkable improvement in both the 16-level QAM and 64-level QAM performance. For SNRs in excess of 25 dB there is an almost 5 dB extra SNR gain improvement due to the stronger RS code in case of the 64-level QAM scheme. By employing error correction coding, the performance of the 16-level QAM arrangement is dramatically improved and becomes superior to that of the 1/2 rate coded 64-level QAM system. Depending on the integrity required, an SNR value of 25 dB is sufficient for reliable signalling via the 16-QAM/$RS(60, 44, 8)$ system.

10.7.1 Square 16-QAM/PSAM/TCM Scheme

The previously considered differential 32-TCM and 64-TCM star QAM schemes failed to improve the star 16-QAM performance partly due to the lack of fading compensation, and partly because of the ambitious goal of providing a 4 bit/symbol net transmission rate. Clearly, the expanded 64-TCM constellation had reduced the ability of the scheme to withstand fading compared to the 16-QAM schemes.

Ho, Cavers and Varaldi [542] embarked on a comparison of three modulation schemes with a net throughput of 2 bits/symbol. These were uncoded QPSK, rate 2/3 8-PSK/TCM as proposed by Ungerböck [496], and rate 1/2 square 16-QAM/TCM combined with PSAM.

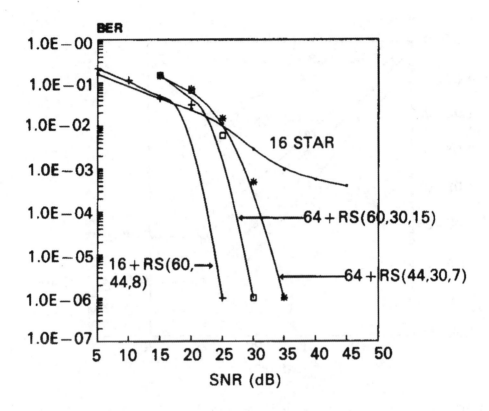

Figure 10.7: Overlaid block coding performance

The pilot symbol assisted fading estimates allowed the authors to improve the power efficiency over uncoded QPSK via fading channels.

The proposed square 16-QAM/TCM scheme was based on separate four-state TCM of the I and Q components. Ho *et al.* compared the BER versus channel SNR performance of their three schemes for a variety of fading rates, using a pilot periodicity of $M = 7$ and a $k = 11$ channel predictor. The authors also provided best-case performance estimates when using the actual complex fading sample for channel estimation. Their results are shown in Figure 10.8, where P represents a perfect channel estimate.

Observe that the best performance was achieved by 16-QAM/TCM/PSAM, since there are no unprotected bits transmitted over the channel. For BER values below 10^{-6} 16-QAM was about 5 dB more power efficient than 8-PSK, and about 25 dB more efficient than uncoded QPSK. As regards power amplifier design, it is important to compare these schemes on the basis of peak power. If all three schemes are allocated the same peak power, their corresponding peak-to-average power ratios can be used in order to derive their resulting average power. Ho *et al.* report that for a 50% excess bandwidth raised cosine pulse shaping, the corresponding peak-to-average ratios are 2.45 dB, 2.75 dB and 5.31 dB for QPSK, 8-PSK and 16-QAM, respectively. For a fair comparison the BER curves of Figure 10.8 must

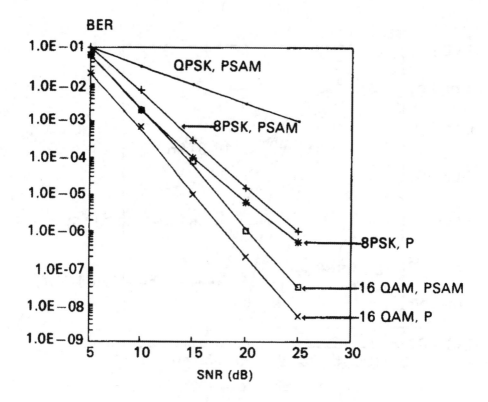

Figure 10.8: BER performance of coded 8-PSK and coded 16-QAM [542] ©IEEE, 1992, Ho *et al.*

be shifted to the right by the above SNR values. After such a shift the 16-QAM still has a 2.44 dB peak power advantage at a BER of 10^{-6}. These figures were confirmed by other researchers [178, 543].

10.8   Distorted Constellation Star QAM

10.8.1   Introduction

In Chapters 4 and 9 we have shown that the maximum distance square 16-QAM scheme has two sub-channels with different integrities, which we have referred to as the $C1$ and $C2$ sub-channels. The BER advantage of the $C1$ channel over the $C2$ channel is not dramatic for AWGN channels, but is more significant over Rayleigh fading channels. The different sub-channel integrities can be equalised using matched FEC codes. However, in the case of coded speech it is often advantageous to exploit the unequal BERs by transmitting the more vulnerable speech data via the $C1$ channel, while the more robust speech data is sent via the $C2$ subchannel [112, 157–159]

In the 16-level star QAM systems each bit in the 4-bit symbol has an approximately equal probability of being in error. In general this is a desirable feature as data can be transmitted without applying any special mapping strategy and this normally yields the best overall bit error rate (BER). There are situations, however, where equal BERs for all bits is disadvantageous. For example, when the data applied to the star QAM modem is from an n-bit PCM source, where the most significant bit corresponds to a 2^{n-1} higher voltage than the least significant bit, then the MSB needs to be received with a concomitant lower BER than the least significant bit (LSB). For information sources having a hierarchical structure in their data it is necessary to employ source-matched channel coding strategies in order to ensure that the recovered data has the requisite performance after decoding. This procedure, employing a range of channel codecs or an adaptive channel codec, is complex and expensive. We seek a simpler solution here, investigating whether distorting the star constellation may yield an acceptable performance without the need for complex source-matched channel coding.

10.8.2 Distortion of the Star-Constellation

As it is desirable for the average of the four individual BERs to remain as low as possible, only minor distortions to the star QAM constellation are considered, as it was found to be well suited to the Rayleigh fading environment. Accordingly the twin-level circular constellation is retained, along with the differential Gray coding of the constellation points.

Within these confines two forms of distortion are considered. One is to change the spacing between the inner and outer rings, and the other is to no longer equi-space the eight points around each ring. Simulations were performed using random data, encoded onto the star QAM constellation and transmitted on a 1.9 GHz carrier at a Baud rate of 16 ksymbols/sec corresponding to 64 kbit/s. The mobile's speed was a constant 30 mph. Appropriate Rayleigh fading and Gaussian noise samplews were generated leading to an average channel SNR of 30 dB. The data was differentially decoded and the BER was calculated.

10.8.2.1 Amplitude Distortion

Bearing in mind the structure of the star QAM constellation of Figure 10.1, increasing the spacing between the two rings of the star QAM constellation potentially improves the performance of bit b_1 at the expense of the other three bits. Bit b_1 is expected to have improved integrity because more noise is required to move it across the decision boundary. The remaining three bits will suffer because they are moved closer together on the inner ring, increasing the probability of phase error for inner-inner, inner-outer and outer-inner ring transitions, and decreasing the probability of error only for the outer-outer ring transition. In the case of random data all transitions are equally likely.

Figure 10.9 shows the simulation results for the variation of the BER of each of the four bits and the average BER as the ratio of the distance between the rings was varied. While bits b_2, b_3 and b_4 generally exhibited increasing BER as would be expected, bit b_1 had a minimum BER at a ring ratio of 3, where the inner ring has a magnitude of 1 and the outer ring a magnitude of 3. The reason that the BER did not continue to fall with increasing ring ratio is that as the inner ring shrinks at the expense of the growth of the outer ring, constellation points on the inner ring become increasingly susceptible to noise, considerably distorting their amplitude and phase value. So although inner-outer and outer-inner transitions have a low probability of being decoded incorrectly, there is a relatively high probability that an inner-inner ring transition will be incorrectly decoded as either an inner-outer or outer-

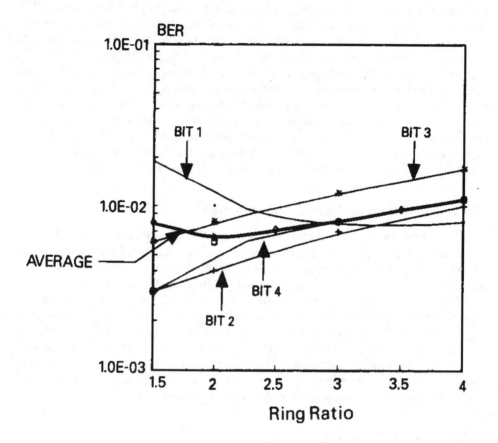

Figure 10.9: BER performance of star 16-QAM for ring ratio variations between 1.5 and 4 at 30 mph, 1.9 GHz, 16 ksymbols/s and SNR = 30 dB

inner transition. Decoding this change as an outer-inner transition is possible because the receiver does not decide upon which ring the previous symbol sent was on, but looks for a significant change in amplitude between previous and current symbols. This avoids error propagation due to incorrect decisions. Simulations with the receiver attempting to make a decision as to the ring on which the previous symbol resided in order to eliminate the incorrect decoding of an inner-inner ring transition as an outer-inner ring transition showed that this scheme exhibited worse performance than the original system because of the error propagation associated with incorrect decisions.

It can be seen that there is little scope for changing the ring ratio. Variations between 1.5 and 3 would be acceptable. A ring ratio of 1.5 allows the largest variation in terms of BER between bits, with bit b_1 having a BER about 6 times greater than bits b_2 and b_4.

The average BER exhibited a minimum at a ring ratio of 2, rising slowly as the ring ratio was increased, and rising more quickly as the ring ratio was decreased. A theoretical analysis performed by Chow *et al.* [404] for AWGN also came to similar conclusions. However, since

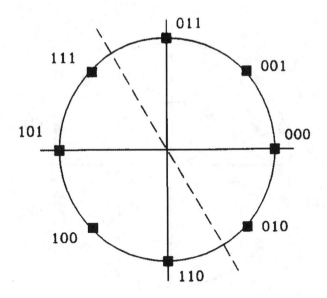

Figure 10.10: Mapping for standard star system

there was a modest 0.9 dB difference between the ring ratios of 2 and 3, the performance gains achieved by optimising this ring ratio will be small.

The second form of constellation distortion considered is to change the angles between each of the points on a ring. The current system has a spacing of 45° between each adjacent point. By increasing this angular spacing for some points and decreasing it for others, the BER of certain bits can be altered. The bitmapping onto the phasor points for one of the rings on the current system is shown in Figure 10.10.

In all cases the bit mapping on the other ring has constellation points at identical angular spacing. Only three bits are shown against each point, corresponding to bits b_2, b_3 and b_4 as b_1 is used to differentiate between the rings. It is important to remember that this diagram represents a differential mapping system. That is, a phase point at an angle of 45°, say, implies that the current symbol should be transmitted with a phase shift of $+45^o$ relative to the previous symbol transmitted, interpreted as modulo 360°. In all subsequent simulations the ring ratio was held at a value of 3.

10.8.2.2 Phase Variations

As can be seen in Figure 10.10 the 8 points on a circle can be split into two groups of 4 points each. In each of these subgroups, separated by the dashed line in the figure, the first bit, namely bit b_2 in the 4 bits mapping scheme, is the same, i.e. a logical 1 to the left of the dashed line and a logical 0 to the right of the line. The integrity of b_2 can be increased by further separating these subgroups as shown in Figure 10.11.

Here the angle θ is our variable parameter. We note from the geometry of Figure 10.11

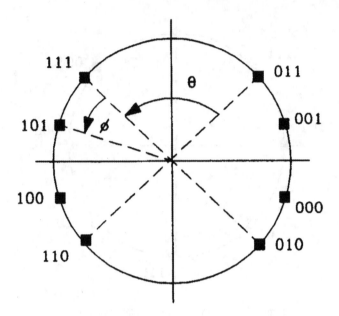

Figure 10.11: Bit b_2 distortion system

that the angle ϕ is given by

$$\phi = 60 \deg - \frac{\theta}{3} \tag{10.8}$$

where ϕ, in common with all the angles quoted throughout, is in degrees. Hence, for example, for the undistorted case of $\theta = 45°$ we have that $\phi = 45°$, as expected. The variation of BER against θ is given in Figure 10.12 for our initial conditions of SNR = 30 dB, 30 mph, 1.9 GHz and 16 ksymbols/s.

Bit b_1 is not shown because it remained constant with variations in θ as would be expected. It is apparent that as θ increased above about $100°$, the average BER rose steeply and this factor probably precludes operation in this region. Indeed, the average BER was at a minimum for $\theta = 45°$, and the price paid for a different BER on each bit was a higher average BER. As desired, the BER performance of bit b_2 was improved at the expense of bits b_3 and b_4. For values of θ in excess of $80°$ there was no longer any significant improvement in the BER of bit b_2. At this point there is a ratio of about 4 between the BER of bit b_2 and bits b_3 as well as b_4. With the ring ratio held at 3, bit b_1 had a performance similar to bits b_3 and b_4.

There is a further distortion that can be applied to the constellation. This is shown in Figure 10.13 where each of the two groups of 4 points are divided into two further subgroups.

In each of these subgroups bit b_4 has a constant value. Thus we would expect to increase the performance of bit b_4 at the expense of bit b_3 by increasing the angle α. This is because the protection distance between opposite values of b_4 is now increased, while that between

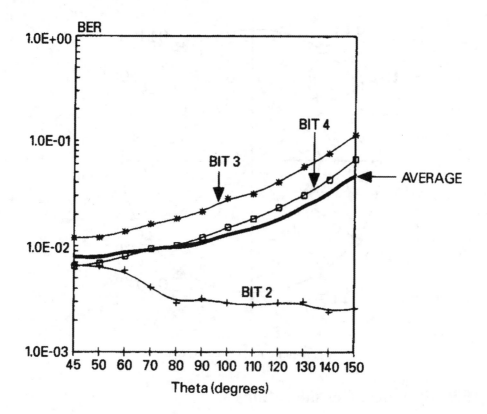

Figure 10.12: BER versus distortion angle θ for bits 2, 3 and 4 of the star 16-QAM constellation at 30 mph, 1.9 GHz, 16 ksymbols/s and SNR = 30 dB

opposite values of b_3 is decreased. The angles in Figure 10.13 are related by

$$\phi = 90 \deg - \frac{\theta + \alpha}{2}. \tag{10.9}$$

Again, for the undistorted case of $\theta = \alpha = 45^o$, Equation 10.9 gives $\phi = 45^o$. The BER performance of this system for different values of θ is shown in Figures 10.14 to 10.16.

The BER performance is displayed for only bits b_3 and b_4, as the BER for bits b_1 and b_2 remained constant. The trend in all graphs is the same, a sharply rising average BER, and only a slowly falling BER for b_4. This suggests that only small increases in terms of α are acceptable. For example a value of $\alpha = 50^o$ when $\theta = 90^o$ gives a ratio of 6.5 between the BER of bit b_4 and that of bits b_2 and b_3. Similarily, for the same distortion a BER ratio of 1.5 accrues between bit b_4 and bit b_1. In this instance the average BER is degraded by a factor of 1.75 which corresponds to about 2.5 dB.

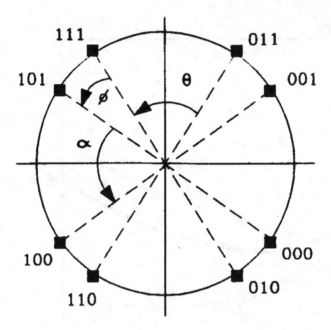

Figure 10.13: Bit b_4 distortion system

10.9   Practical Considerations

10.9.1   Introduction

In previous chapters, it was assumed that no hardware imperfections were introduced. This gave an optimistic estimate of the performance. Possible imperfections include a finite number of quantisation levels, I-Q crosstalk, AM-AM and AM-PM conversion non-linearities and a finite oversampling ratio. Even if it appeared possible to implement a modem which, as far as the QAM technique employed is concerned, appears perfect, it may be sensible to reduce the performance in order to achieve a simpler and cheaper design. In this section the effects of a number of imperfections which are likely to occur in a hardware modem are investigated, in order to establish hardware parameters.

10.9.2   Hardware Imperfections

10.9.2.1   Quantisation Levels

Typical QAM modems employ digital baseband circuitry, and so analogue-to-digital (A-D) conversion is required at the input of the receiver. It is important that an adequate number of bits is chosen for this converter. If too low a number of bits are chosen then the performance will suffer badly, and if too many are chosen then the hardware complexity and cost will rise substantially while the maximum operating speed will fall. Initial quantisation level

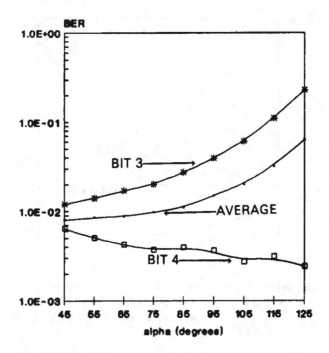

Figure 10.14: BER versus distortion angle α for bits 3 and 4 of the star 16-QAM constellation with $\theta = 45°$, at 30 mph, 1.9 GHz, 16 ksymbols/s and SNR = 30 dB

simulations assumed that the received signal level after pre-amplification would fall within a certain range. The quantisation levels are spread evenly throughout this dynamic range, and the incoming signal is rounded to the nearest quantisation level. Peak clipping occurs if the signal exceeds this range. Results of simulations carried out using a variety of quantisation levels are shown in Figure 10.17.

For these simulations data was transmitted at 128 ksymbols/s, at a mobile speed of 30 mph, and a propagation frequency of 1.9 GHz. Non-linear filtering (NLF) was used to shape the transmitted data and a fourth-order Butterworth lowpass filter with the 3 dB cut-off point at 1.5 times the data rate (i.e. 192 kHz) was used in the receiver. Clock recovery was implemented using the modified early-late recovery system discussed in Chapter 11, and 16 times oversampling was implemented throughout the system. Unless stated otherwise, these simulation parameters can be assumed to hold true for all subsequent simulations in this chapter.

The results show that at least 14 bits will be required if the receiver is not to suffer a drop in terms of BER performance. In practice this means that 16-bit A-D converters will be necessary, since 14-bit converters are rare. Operation with only 12 bits would be feasible but does entail a BER penalty at high SNRs.

Consideration of the waveforms into and out of the A-D converter showed that the full dynamic range of the converter was not used over a short time span, but it was necessary to

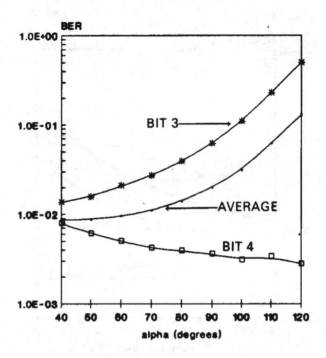

Figure 10.15: BER versus distortion angle α for bits 3 and 4 of the star 16-QAM constellation with $\theta = 60^\circ$, at 30 mph, 1.9 GHz, 16 ksymbols/s and SNR = 30 dB

accommodate the peaks and troughs in the Rayleigh fading pattern. This suggested that performance could be improved by the use of a simple AGC prior to the A-D converter. Because of the differential detection employed, the absolute signal level is unimportant. However, it is critical that the AGC must be sufficiently slow acting in order that its gain does not change significantly from one symbol to the next, as this would make the differential detection more error prone. In practice this is not a problem because there are many symbols in each fade and so the AGC can be slow acting with respect to the symbol period and still track the envelope adequately. It is stressed that this AGC does not attempt to maintain a constant or near constant input envelope, but merely to compress the fading range in order to reduce the A-D converter's resolution requirements. Simulations suggested that a suitable system was a slow-acting AGC, whose dynamic range could suddenly be increased if the input signal exceeded 95% of its allowed range. This tended to keep the fades within the input clipping range without unduly distorting the waveform when it was around the average value.

Results of simulations using this system are shown in Figure 10.18.

Here it can be seen that adequate performance is obtained with only 12 bit conversion, and that with only 8 bits the degradation is not unduly severe.

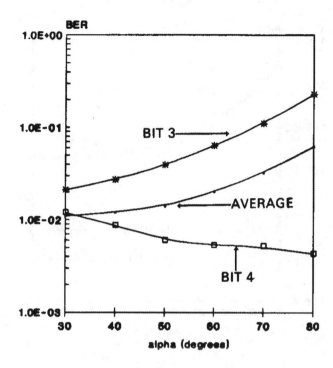

Figure 10.16: BER versus distortion angle α for bits 3 and 4 of the star 16-QAM constellation with $\theta = 90^{\circ}$, at 30 mph, 1.9 GHz, 16 ksymbols/s and SNR = 30 dB

10.9.2.2 I-Q Crosstalk

Another problem that is often encountered in the RF side of the modem is that of crosstalk between the quadrature branches. This problem can be reduced by balanced RF system design, but here an attempt is made to discover the maximum level of crosstalk which can be tolerated before unacceptable BER performance degradation occurs. No baseband counter measures are proposed, merely adequate RF system design.

Results of simulations employing 8 A-D converter quantisation levels and the same parameters as the previous section are shown in Figure 10.19.

Here it can be seen that levels of I-Q crosstalk up to -40 dB would seem acceptable, that is the leakage power from one quadrature branch to another should be 40 dB less than the average power in that branch.

10.9.2.3 Oversampling Ratio

In order for the clock recovery system to operate satisfactorily, it is necessary to sample the received waveform a number of times during each symbol period. This allows accurate location of waveform peaks. Ideally, the modem should work with as low an oversampling ratio as possible in order to keep internal clock speeds low. In this section the complete star

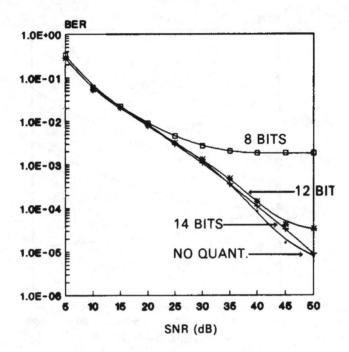

Figure 10.17: Star 16-QAM BER versus channel SNR performance using different input A-D resolutions at 128 ksymbols/s, 30 mph, 1.9 GHz

QAM system was simulated as the oversampling ratio was varied. For accuracy, a 16 times oversampling ratio was maintained in the simulations, but we only made a certain number of these samples available to the receiver.

The results for three different oversampling ratios are shown in Figure 10.20.

Here it can be seen that the ratio can be satisfactorily reduced to 8 samples per symbol, but a reduction to only 4 causes a significant performance degradation.

10.9.2.4 AM-AM and AM-PM Distortion

We have already touched upon the problems of amplitude modulation to both amplitude modulation (AM-AM) and to phase modulation (AM-PM) conversion, when discussing the characteristics of power amplifiers in Section 5.6. A typical pair of AM-AM and AM-PM conversion curves was given in Figure 5.27. In this subsection we briefly attempt to quantify the effects of amplifier imperfections on decoded BER. This impairment is difficult to model as it is not known exactly what form the imperfections will take. In this section a number of assumptions are made about possible forms of distortion in order to assess the effect they may have on a QAM system. The AM-AM distortion is modelled by non-linear, compressed operation due to the limited dynamic range of the amplifier. This means that while the maximum output is maintained, the pulse shape is distorted. It was found that this had very little effect

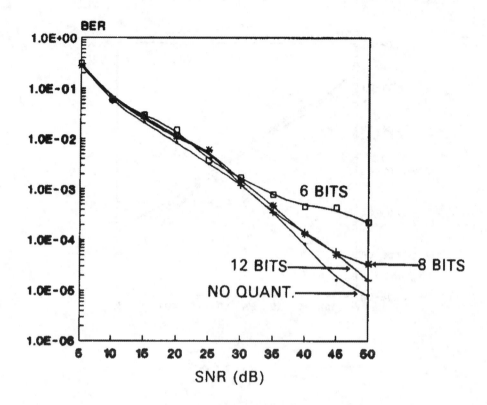

Figure 10.18: Star 16-QAM BER versus channel SNR performance using AGC and a variety of A-D converter resolutions at 128 ksymbols/s, 30 mph, 1.9 GHz

on the system, causing a slight but acceptable degradation in the clock recovery system. The degradation could be completely removed by regenerating the original pulse shape. Since the pulse shape is typically derived from a look-up table, this does not increase the complexity of the design.

The AM-PM distortion is modelled by introducing a phase shift of P degrees when the received phasor is on the outer ring. So when moving from the inner to the outer ring an undesirable differential phase shift of $\theta_{enc} + P$ is experienced where θ_{enc} is the differentially encoded transmitted angle. When moving from the outer to the inner ring a detrimental phase shift of $\theta_{enc} - P$ is experienced.

In the first set of simulations, no corrective action was attempted at the receiver, and the maximum permissible distortion which did not unduly degrade the BER performance was found. BER versus channel SNR performance curves are shown in Figure 10.21 for various AM-PM distortion values, which suggest that a maximum uncompensated phase shift of up to 10° could be tolerated.

However, since this phase shift is time invariant, it should be possible to take some sort of corrective action. In the differential decoder at the receiver, we can monitor the amplitude change. If there appears to have been a change from the inner to the outer ring, then we

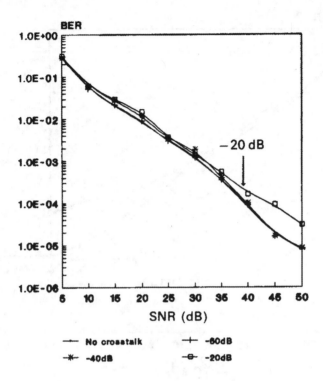

Figure 10.19: Star 16-QAM BER versus channel SNR performance using AGC and 8-bit A-D conversion when using different levels of I-Q crosstalk, at 128 ksymbols/s, 30 mph, 1.9 GHz

subtract the known AM-PM distortion, and if there appears to have been a change from the outer to the inner ring, then the known AM-PM distortion is added. This method is expected to remove most of the AM-PM distortion, but can precipitate errors if an error is made in decoding the amplitude change. BER versus channel SNR performance curves for this system are shown in Figure 10.22 for 0° and 40° conversion errors, where we can see that satisfactory performance is achieved for large values of phase distortion.

10.10 Summary

In this chapter we have discussed various signal processing techniques to facilitate reliable QAM transmissions in Rayleigh fading environments. The well-known 16-level square constellation was found to be unsuitable for the mobile radio environment. The sub-optimal star scheme with differential encoding and oversampling signal estimation dramatically improved the BER performance, rendering the channel appropriate for speech transmissions. The lower BERs essential for data transmissions were achieved by expanding the 16-level QAM signal set to 64 levels and using the extra channel capacity acquired for error correction coding. For SNRs in excess of 25 dB the performance of the coded 64-level QAM was superior to that of

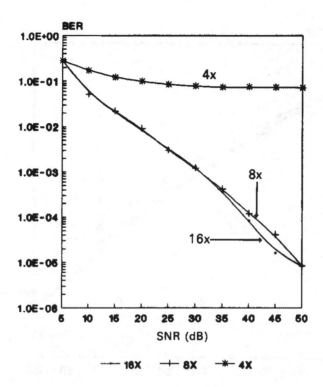

Figure 10.20: Star 16-QAM BER versus channel SNR performance using AGC, 8-bit A-D conversion, when oversampling at rates of 4, 8 and 16, 128 ksymbols/s, 30 mph, 1.9 GHz

the 16-level QAM and at values of SNR in excess of 30 dB it was virtually error-free. When the overall coding rate was lowered to allow the 16-level QAM scheme to incorporate a 3/4 rate RS code and the 64-level QAM system had a 1/2 rate RS code, the 16-QAM arrangement outperformed the more complex 64-level scheme.

Unlike square QAM modems, star QAM has an approximately equal BER for each of the bits in its symbols. This is normally advantageous, but there are some instances where a differing BER for each bit can be effectively used for unequal source-matched FEC coding of speech and video sources. Ways of rendering the BER of each bit different were considered. Unfortunately, in all cases the average BER was increased, and large ratios between the BER performances of certain bits were not achieved for the scenarios considered. However, the BER ratios of approximately 6 were achieved when the average BER was degraded by a factor of 1.75.

The effects of hardware imperfections on star 16-QAM modem performance was then considered. It was established that 8-bit A-D converters could be employed if a slow-acting AGC was deployed prior to the converter. Mild levels of IQ crosstalk were found acceptable and the minimum oversampling ratio was found to be 8. Finally, it was determined that small AM-AM and AM-PM distortions could be tolerated, especially if correction could be applied.

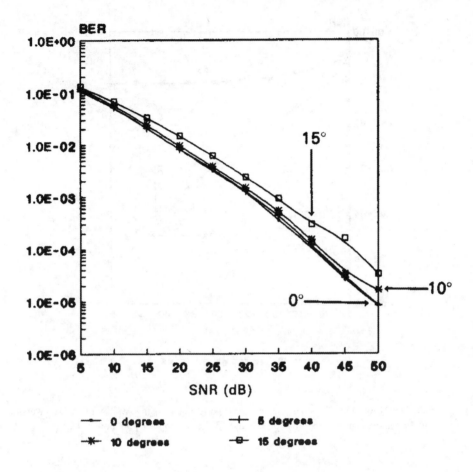

Figure 10.21: Star 16-QAM BER versus channel SNR performance at 128 ksymbols/s, 30 mph, 1.9 GHz for different values of AM-PM distortion without correction

In the next chapter we investigate further the use of star QAM transmission over mobile radio channels by considering the application of the clock and carrier recovery techniques introduced in Chapter 6. We show that a new clock recovery technique can be introduced which is advantageous for the particular combination of star QAM and Rayleigh fading channels.

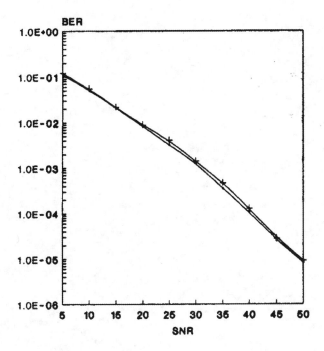

Figure 10.22: Star 16-QAM BER versus channel SNR performance at 128 ksymbols/s, 30 mph, 1.9 GHz for different values of AM-PM distortion with correction

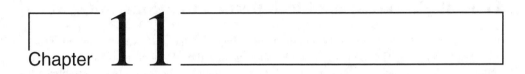

Chapter 11

Timing Recovery for Fading Channels

11.1 Introduction

Although the differential star QAM scheme proposed in Chapter 10 does not need carrier recovery, the characteristic 3 dB penalty associated with non-coherent detection need not be incurred if low complexity timing recovery can be implemented. Accurate clock recovery is always essential, irrespective of whether or not non-coherent detection is possible.

In this chapter clock and carrier recovery schemes for 16-level star QAM transmitted over mobile radio channels are considered, and the effect of ISI is simulated. We commence our investigations by considering the two most pertinent classic clock recovery methods, namely, the times-two and the early-late timing recovery techniques. These two techniques were discussed in Chapter 6 in the context of AWGN channels; here their performance over mobile radio channels with star QAM is considered.

11.2 Times-two Clock Recovery for QAM

In order to implement times-two clock recovery in conjunction with QAM the received signal is split into in-phase (I) and quadrature (Q) waveforms and each is processed separately, before being combined in the clock recovery system. In the case of QAM, however, the presence of high energy frequency components at half the data rate is less likely than with binary data schemes. This is because instead of changes in the baseband modulated signal about zero whenever we have a change of transmitted symbol, as with binary modulation, there can be swings between symbols on the positive side and to and from those on the negative side of the constellation resulting in fewer zero-crossings. For example, in the case of 16-level QAM and transmitted levels of ±1 and ±3 on the I and Q channels, a data sequence of -1, -3, -1, 1, 3, 1, -1, -3, -1 ..., -1, -3 yields a component at 1/6 of the data rate. Simulations were performed using times-two clock recovery which showed the QAM performance to be unacceptably poor. For more details on times-two clock recovery the reader is referred to Chapter 6.

11.3   Basic Early-Late Clock Recovery for Star 16-QAM

The early-late (EL) clock recovery technique described in Chapter 6 was introduced into a star QAM receiver. The simulation results showed that the EL clock recovery performed better than the times-two system. Correct locking of the clock was eventually established, and maintained moderately well during fades. The locking time, however, was unacceptably long.

The major problem with EL clock recovery for QAM transmissions is that not all QAM sequences result in time domain waveform peaks occurring every sampling period. An even more detrimental effect is that half the peaks are of the wrong polarity for the clock recovery technique. This problem is illustrated in Figure 11.1 where the polarity of the early signal minus the late signal is considered for three different received 16-QAM sequences, assuming in all cases that the sampling point is too early. Due to the squaring operation identical waveform segments having opposite polarity yield identical PLL control sequences. The early and late sampling points are shown by dashed vertical lines on the squared signal. The chosen sampling point is assumed to lie in the centre of these lines, whereas the correct sampling instant is seen to be nearer to the late sample, or the rightmost dashed line. When the sampling point is early, the early signal minus the late signal must be negative for conventional early-late schemes to function correctly, and this is represented in Figure 11.1 as "E-L Negative". In this example only the I channel is considered for simplicity. The practical implementation of this scheme could have independent EL recovery systems for both the I and Q channels, which would drive the same PLL using control logic.

In Figure 11.1(a) the QAM sequence 1, 3, 1 is shown where the first number in the sequences refers to the modulation level on the I channel transmitted in the symbol period prior to the one in which peak detection is to be performed. The second number in the sequences refers to the modulation level in the symbol period in which peak detection is attempted, while the third number refers to the modulation level in the symbol period after the one in which peak detection is attempted. Clearly not an oversampled waveform, but three independent symbols are considered. Also shown is the sequence -1, -3, -1 which leads to an identical waveform after squaring, where the squaring is an integral part of the EL process. Both of these sequences result in correct EL polarity. In Figure 11.1(b) the received sequences 3, 1, 3 and -3, -1, -3 are shown, both of which lead to an identical waveform after squaring. It can be seen that despite sampling occurring early as in (a), the EL value is of the opposite polarity to that in (a). Also problematic are sequences such as -1, 1, 3 and 1, -1, -3, both of which again lead to an identical waveform after squaring as shown in Figure 11.1(c). Observe that the monotonic waveform leads to negative early-late difference signals regardless of whether the sampling is early or late, although we only show it early.

The consequence is that the early-late difference signal, which should be negative for all the waveforms in Figure 11.1, is correct for only a fraction of the transmitted symbols. The dips of the type in Figure 11.1(b) will cause an incorrect symbol regeneration. Since the monotonic waveforms of Figure 11.1(c) always indicate that a timing retardation is required, they will cause an equal distribution of correct and incorrect PLL control signals. These

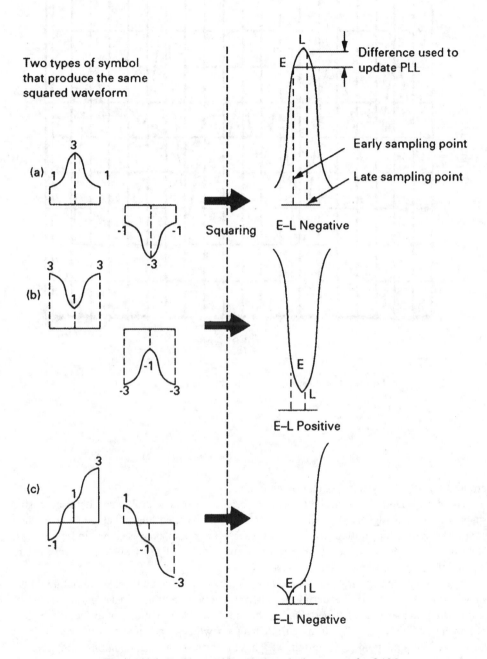

Figure 11.1: Problems with early-late clock recovery for QAM

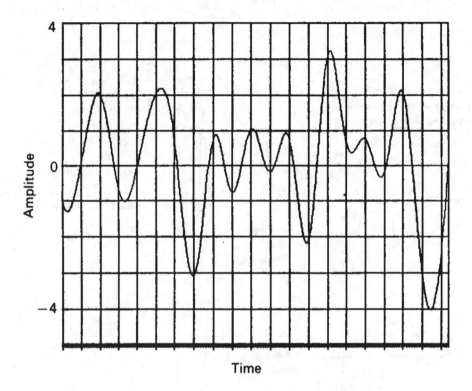

Figure 11.2: Raised cosine pulse shaping on the I channel

problems render the system's operation unreliable.

The pulse shaping used at the transmitter can cause further problems. If raised cosine shaping is used, as suggested by Kingsbury [405], with matched filtering at the receiver, then a typical resulting transmitted star 16-QAM waveform I component is shown in Figure 11.2. Note that for 16-level star QAM the possible magnitudes of the I channel waveform are $0, \pm 0.707, \pm 1, \pm 2.12$ and ± 3 for the valid symbols.

As it can be seen, the peaks do not always coincide with the equispaced sampling points shown by the vertical lines, rendering EL clock recovery inaccurate. This suggests the use of alternative pulse shaping schemes. One of the most suitable schemes would seem to be that of non-linear filtering (NLF) discussed in Chapter 4 which guarantees that the peaks are at the correct sampling point. In addition, NLF has the virtue of implementational simplicity. A typical NLF waveform is shown in Figure 11.3, where it can be seen that the peaks coincide with the optimal sampling points even though the zero-crossings are not equispaced. Again, the correct sampling times are given by the vertical grid lines. Based on the above arguments, it was decided to use NLF for our further investigations.

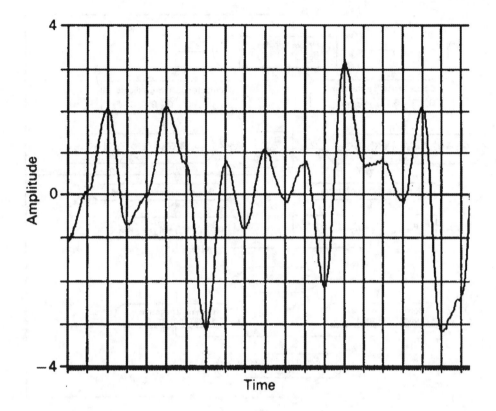

Figure 11.3: Non-linear filtered I channel waveform for star 16-QAM

11.4 Modified Early-Late Clock Recovery for QAM

The basis of modified EL clock recovery is to search the current symbol period for a peak, and if one is found to assess its suitability for clock recovery. Only if a suitable peak is found is update information sent to the PLL. This improved system, which is based on the use of oversampling, is now described in more detail with reference to Figure 11.4.

The clock recovery system waits until all n of the oversampled observations for a symbol period have been made, where the n observations are equispaced around the current sampling point. The maximum and the minimum samples during the observation period are then identified. If the maximum is at either end of the sampling period, the search for a positive pulse is discontinued since this implies that there cannot be a valid positive peak in the current sampling period. The same applies if the minimum is at either end of the sampling period.

If there is a valid maximum, sample gradients on both sides of the peak are calculated and the peak is rated depending on these gradients as described below. A similar rating is calculated for the negative pulse if there is still a valid minimum. The algorithm used for

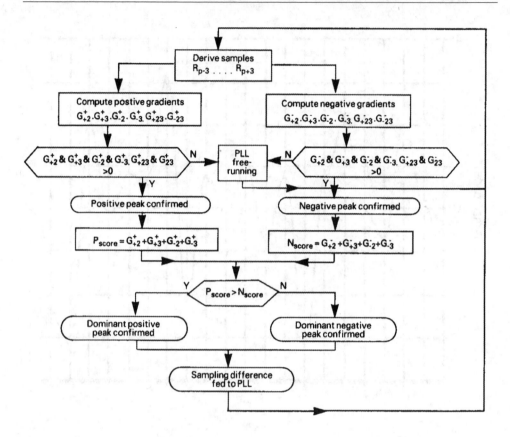

Figure 11.4: Modified early-late clock recovery flow chart

rating the positive peaks is based on evaluating the following gradients:

$$
\begin{array}{rcl}
G^+_{+2} &=& R_p - R_{p+2} \\
G^+_{+3} &=& R_p - R_{p+3} \\
G^+_{-2} &=& R_p - R_{p-2} \\
G^+_{-3} &=& R_p - R_{p-3} \\
G^+_{+23} &=& R_{p+2} - R_{p+3} \\
G^+_{-23} &=& R_{p-2} - R_{p-3}
\end{array}
$$

where R_p is the amplitude of the incoming waveform at the peak p and R_{p+k} is the amplitude at the point k oversamples offset from p, where k takes the values $\pm 2, \pm 3$. The presence of a positive peak is only confirmed if all of these gradients are positive. This implies that, moving away from the peak in both directions, the waveform falls away and that it does so with increasing steepness. The oversampled values of R_{p+1} and R_{p-1} are not used as, due to the nature of the pulse shaping, these are very similar to R_p and so the gradient being small can be easily falsified by noise. If all these quantities are positive, then the peak is rated according to

$$P_{score} = G_{+2}^+ + G_{+3}^+ + G_{-2}^+ + G_{-3}^+ \tag{11.1}$$

where P_{score} is the positive peak score.

Similarly for the negative peaks

$$
\begin{aligned}
G_{+2}^- &= R_{p+2} - R_p \\
G_{+3}^- &= R_{p+3} - R_p \\
G_{-2}^- &= R_{p-2} - R_p \\
G_{-3}^- &= R_{p-3} - R_p \\
G_{+23}^- &= R_{p+3} - R_{p+2} \\
G_{-23}^- &= R_{p-3} - R_{p-2}
\end{aligned}
$$

and if all these quantities are positive then the negative peak is rated as:

$$N_{score} = G_{+2}^- + G_{+3}^- + G_{-2}^- + G_{-3}^- \tag{11.2}$$

where N_{score} is the negative peak score. If $P_{score} > N_{score}$ then the positive peak is deemed more dominant and the negative peak is discarded, otherwise the negative peak is retained and the positive peak is discarded. If a valid peak is identified, the difference between the current sampling time and the peak just identified is used to update the variable internal clock, normally a phase locked loop (PLL) system. Otherwise the PLL is allowed to continue running at its current level.

The BER performance of the star 16-QAM scheme using this clock recovery system operating without any ISI in Rayleigh fading is shown in Figure 11.5, compared to that of a system operating with perfect clock recovery.

As can be seen in the figure the performance of the two systems is almost identical over a wide range of channel SNR values, indicating correct operation of the clock recovery system.

11.5 Clock Recovery in the Presence of ISI

11.5.1 Wideband Channel Models

In our earlier studies we used low signalling rates, which were significantly below the channel's coherence bandwidth, and hence no ISI was introduced. When the signalling rate is increased to a few MSymbols/s, in most mobile radio environments distortion due to distant echos is introduced. In this section, unless stated, we use a vehicular speed of 30 mph, a propagation frequency of 1.9 GHz, and a signalling rate of 2 Msymbols/s. Intersymbol interference (ISI) distorts the received QAM pulse shape, complicating correct clock recovery. ISI can take many forms depending on the number of local radio scatterers and the terrain in the vicinity of the mobile station (MS). Let us consider two extreme cases.

The first extreme scenario is when the MS is in the presence of a single large reflector. Here there is a main path incident upon local scatterers leading to a flat Rayleigh fading path, and one single reflected path with a delay dependent on the position of the reflector, which is again incident upon the local scatterers, giving another independent Rayleigh distribution. This is the classic two-path dispersive channel with two delay bins. In our simulations we studied the effect of the delay τ between the direct path and the reflected path. In most

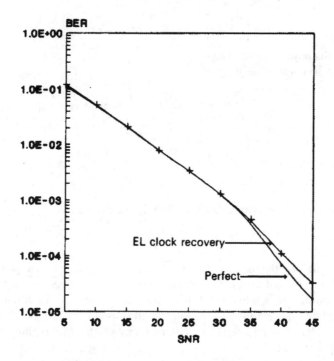

Figure 11.5: BER versus SNR performance of star 16-QAM using perfect and modified early-late clock recovery with no ISI, a mobile speed of 30 mph, and a transmission frequency of 1.9 GHz

practical cases τ will not be exactly equal to any multiple of the symbol period T, i.e. $\tau \neq nT$ where n represents an integer, hence the received signal peak due to the reflected path component arrives somewhere between two peaks spaced by T. There is the alternative, less likely two-path case where $\tau = nT$, that is the peak due to the reflected path component coincides with the peak of a different symbol which arrives at the receiver via the main path.

The second extreme scenario does not use the two-ray model. Instead a large number of indistinguishable incoming paths occur leading to a "peak smearing", or smoothing of the peak. We term this condition *smearing*. This situation was simulated using 16 delay bins, each one comprising 1/16 of a symbol period, and using 16-fold oversampling.

In our simulations Bultitude's wideband impulse response measurements [544] made in a street microcell in Canada were used. If the results presented graphically in his paper are divided into delay bins for a transmission rate of 2 Msymbols/s with an oversample ratio of 16 then significant responses occur in bins 0, 12 and 32, each bin being 31.25 ns wide. Each path was then subjected to independent Rayleigh fading as seen in Figures 2.21 and 2.22. The approximate relative powers in these delay bins are 1, 0.28 and 0.01 respectively, i.e. if the first path is assumed to have a power of 1 then the path in delay bin 12 has a power of 0.28 times that of the first path. These results can be used directly for the $\tau \neq nT$ scenario. Furthermore, by altering the transmission rate to $16/12 \times 2$ MHz, i.e. 2.66 MHz, and ignoring the relatively insignificant path in delay bin 32, then the results can be used for the $\tau = nT$ scenario. For the smearing scenario an exponentially decaying impulse response

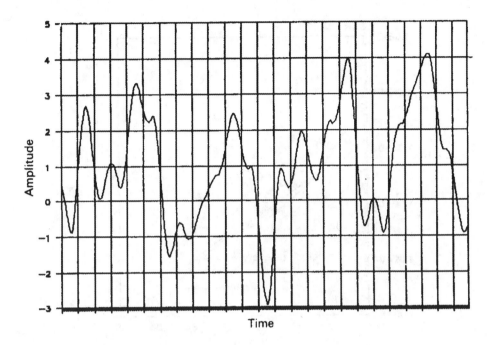

Figure 11.6: Typical star 16-QAM NLF I channel waveform with $\tau \neq nT$

over one symbol period was assumed. Based on measurements the smearing scenario seems less likely than the other scenarios considered here, nevertheless it is also included for the sake of completeness.

11.5.2   Clock Recovery in Two-Path Channels

11.5.2.1   Case of $\tau \neq nT$

A typical star 16-QAM NLF waveform for the ISI situation is shown in Figure 11.6.

Let us compare Figure 11.6 with the ISI-free case shown in Figure 11.3. There is considerable distortion of the received waveform due to ISI, and although the peaks of the main pulses are still distinguishable, there are other peaks at different sampling points. Operation of the clock recovery system was closely monitored for this $\tau \neq nT$ scenario, and it was found that provided the interfering signal remained weaker than the wanted signal, the ISI did not hamper clock recovery as any additional peaks were either ignored, or given a lower peak score by the pulse rating system described previously. Once the delayed multipath signal became stronger than the main signal, which would be the case when the main signal was in a fade, the clock recovery scheme switched to the delayed signal. The delayed signal had effectively become the main signal in that situation. The BER versus channel SNR performance of the star 16-QAM modified EL clock recovery system exposed to this form of ISI is shown in Figure 11.7.

In this simulation no channel equalisation has been performed in order to keep the simu-

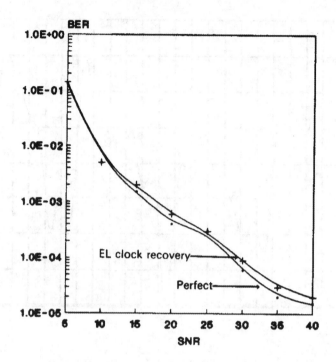

Figure 11.7: Clock recovery performance with $\tau \neq nT$, 30 mph mobile speed, 1.9 GHz carrier frequency, 2 Msymbols/s data rate

lations tractable. A suitable equaliser for the star 16-QAM system is discussed in Chapter 12.

It can be seen from the figure that the performance of the system with the modified EL clock recovery scheme was almost identical to that using perfect clock recovery, showing that the clock recovery system worked adequately in this ISI-exposed scenario.

11.5.2.2 Case of $\tau = nT$

A typical star 16-QAM waveform for this ISI-impaired case is shown in Figure 11.8. Again additional erroneous waveform peaks occur.

Although the waveform is considerably distorted compared to the situation seen in Figure 11.3, the clock recovery system was not unduly affected. Figure 11.9 shows the BER versus channel SNR performance curves obtained with the modified EL clock recovery system and when ideal clock recovery was used. The EL clock recovery is seen to operate almost perfectly in the presence of this form of ISI.

11.5.3 Clock Recovery Performance in Smeared ISI

A typical waveform for the star 16-QAM NLF smeared ISI condition is shown in Figure 11.10.

Again, considerable distortion of the original waveform was experienced, but interest-

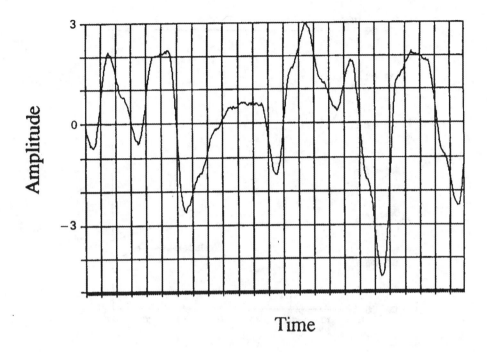

Figure 11.8: Typical star 16-QAM NLF I channel waveform with $\tau = nT$, 30 mph mobile speed, 1.9 GHz carrier frequency, 2 Msymbols/s signalling rate

ingly the peaks became more distinguishable, although they were often shifted from the sampling point. This shifting of peaks caused some problems to the clock recovery system, the performance of which is shown in Figure 11.11.

The performance of the system with clock recovery was about 3 dB worse in terms of channel SNR compared to ideal clock recovery. This decrease in performance may be acceptable, especially as measurements suggest that such channels are unlikely to be encountered in practice. The reason for the performance of the clock recovery system diverging from that of perfect clock recovery as the SNR increases is likely to be due to random phase jitter caused by the ISI. This will tend to cause a certain number of errors regardless of the SNR, and as the BER falls, the effect of jitter-induced errors will become more significant.

11.6 Early-Late Clock Recovery Implementation Details

A possible implementation arrangement of the modified EL clock recovery system is shown in Figure 11.12.

The system is seen to consist of simple building blocks of comparators, adder/subtractors, selector switches and low capacity memory. In Figure 11.12 only the positive peak calculation system has been shown, an almost identical negative peak calculation system is also required. The received input signal passes to a comparator which compares it with the previous maximum value. If a new maximum has occurred, then the value of this, along with

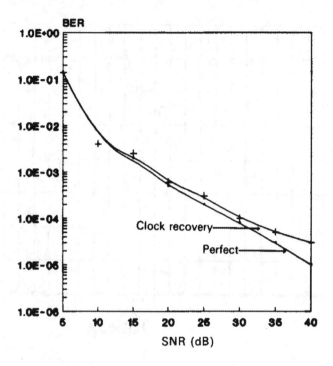

Figure 11.9: Star 16-QAM modified EL clock recovery performance with $\tau = nT$, 30 mph, 1.9 GHz, 2.66 Msymbols/s data rate

its oversample address, is stored. The received input signal is also passed to a shift register which stores all the oversampled inputs occurring during the symbol period, typically eight. A bank of n-way switches is then used to latch the appropriate values into the subtractors which perform the calculations of Equation 11.1. The results are then summed according to Equation 11.1, leading to the positive peak score P_{score}. This is then compared with the negative peak score N_{score} which has been derived in an almost identical system, and the larger of the two is passed to a switch. According to whether the score exceeds the threshold and whether it was a maximum or minimum, the switch is set appropriately to select the sample point and update the PLL.

Simplifications to the implementation can be made by reducing the number of gradient calculations in Equation 11.1; this reduces performance slightly but considerably simplifies the circuit's construction.

11.7   Carrier Recovery

The problem of clock recovery is closely related to that of carrier recovery. In the case of simple modulation schemes such as BPSK, the carrier recovery operation is straightforward as there is a transmitted component at the carrier frequency for the PLL to lock onto. QAM, however, is a suppressed carrier form of modulation. This means that we either have to

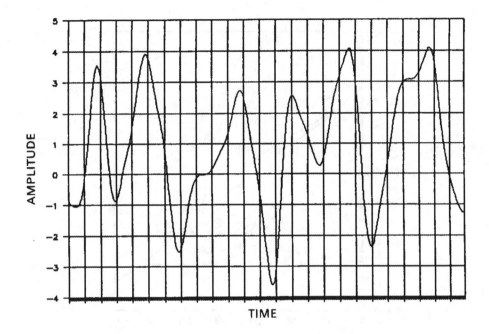

Figure 11.10: Star 16-QAM NLF I channel waveforms with smeared ISI

deliberately insert a carrier as described in the TTIB section in Chapter 9, or we employ more sophisticated recovery schemes [109]. The problem with carrier insertion methods, such as transparent tone in band (TTIB), is that they expand the required spectrum, and these defeat the underlying premise that spectral efficiency is the reason that QAM is employed. They also significantly increase the complexity. At low symbol rates, e.g. below 64 Ksymbols/s, experience has shown that it is very difficult to resolve the phase ambiguities inherent in QAM resulting in false locking, especially during fades. This is because at low signalling rates the fading envelope and phase change more dramatically between signalling instances than in the case of high symbol rates. In this respect, doubling the symbol rate is equivalent to halving the vehicular speed since they both mitigate the rapid envelope fluctuations. At these low symbol rates, differential detection is preferred and it removes the need for carrier recovery. At high symbol rates, the inherent ability of an adaptive equaliser to remove phase ambiguity renders non-differential detection viable.

Simulations with decision directed methods showed little promise. The use of the circular star constellation suggests an interesting and effective form of carrier recovery. One of the systems which has been used for QAM is *times-four carrier recovery* [109]. In this system the complex received incoming signal is raised to the fourth power. This has the effect of raising the amplitude to the fourth power and multiplying the phase angle by 4. In the case of the square QAM constellation this rotates the phase of half the constellation points (those where the magnitude of I is equal to the magnitude of Q) to 180°, and distributes the other half of the constellation points randomly in phase. This operation generates a carrier component at four times the carrier frequency which can be used to facilitate carrier recovery. This can

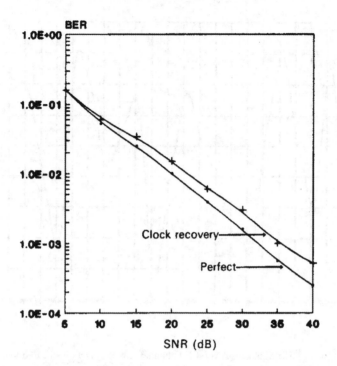

Figure 11.11: Star 16-QAM modified EL clock recovery performance with smeared ISI, 30 mph, 1.9 GHz, 2 Msymbols/s

easily be shown using simple trigonometrical equivalences and a QAM carrier. The accuracy of this recovery is limited by the constellation points which do not map onto 180°. A long run of these can cause significant jitter and even loss of lock.

The circular QAM constellation offers the intriguing possibility of times-eight carrier recovery. This would have the effect of rotating *all* the constellation point to 0°, eliminating the jitter inherent in the previous system. This system, however, cannot resolve ambiguity between lock at any of the eight 45° positions and so it is necessary to periodically send a sounding sequence. The performance of this system compared to the differential system is shown in Figure 11.13. It is apparent that the system operating with carrier recovery performs slightly better than that without, but only by a small margin. The gains are lower than the expected 3 dB because of operation in Rayleigh fading rather than Gaussian noise. Here the probability of the next symbol being in error is highly correlated with the probability of the current symbol being in error. There is also the problem of catastrophic failure during deep fades. It can be concluded that the extra complexity involved in adding carrier recovery is not compensated for by the gains in performance and so differential, non-coherent reception is generally preferred.

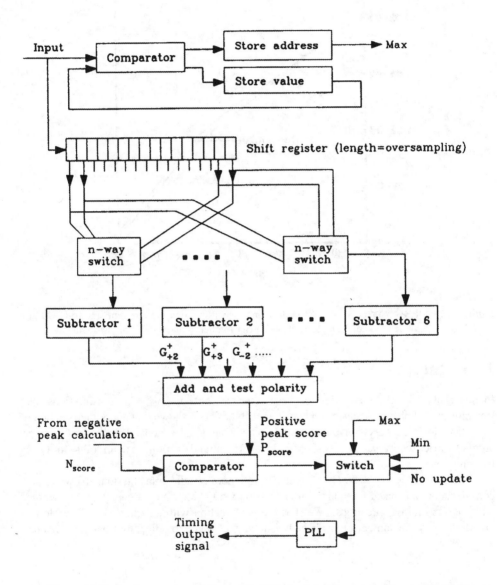

Figure 11.12: Implementation of the modified EL clock recovery system for star 16-QAM

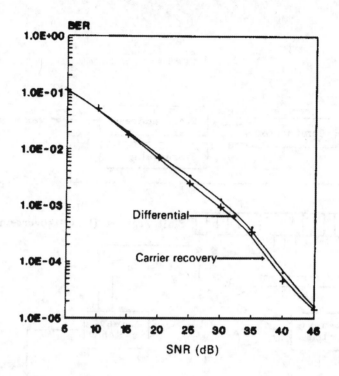

Figure 11.13: BER versus channel SNR performance of differential and coherent star 16-QAM at a mobile speed of 30 mph, 1.9 GHz carrier frequency and 64 Ksymbols/s data rate

11.8   Summary

In this chapter clock and carrier recovery schemes for star 16-QAM transmissions over Rayleigh fading channels have been considered. It has been shown that a modified version of early-late clock recovery works well both in narrowband and wideband channels. Consideration of carrier recovery schemes showed that differential detection was preferable for QAM transmission.

This concludes the basic QAM schemes for mobile radio transmission. Now we will consider more advanced methods of transmission via QAM, which are generally concerned with mobile radio applications. We start the next section with an analysis of variable rate QAM whereby the number of levels is dynamically changed to reflect the changing channel.

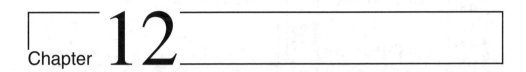

Chapter **12**

Wideband QAM Transmissions over Fading Channels

12.1 Introduction

In current communications systems the majority of teletraffic is typically generated in office buildings, hence the capacity of mobile systems in these buildings must be very high. The high bit rates required to accommodate video, audio, computer data and other traffic may necessitate the symbol rate of the multilevel modulation systems being so high that the mobile radio channels exhibit dispersion. Performing equalisation of multilevel signals that are subjected to dispersion and fading is not an easy task, as we argued in detail in the context of Chapter 7. In this chapter methods whereby high bit rates can be conveyed via star 16-level QAM over the wideband mobile radio channels of Chapter 2 are proposed and investigated, which we consider to be composed of three independent Rayleigh fading paths. A range of other sophisticated channel-equalised coded modulation aided QAM schemes will be studied in Chapter 26, while various adaptive modulation assisted channel-equalised QAM-aided systems using both Kalman filters and classification based Radial Basis Function (RBF) equalisers were investigated in full detail in [170].

The wideband channel response used in our studies is shown in Figure 12.1, which is based on Bultitude's indoor and outdoor measurements [544, 545] of impulse responses in microcells.

Again, in Chapter 7 we provided a rudimentary introduction to various channel equalisers. Many of them were designed to operate in conjunction with binary modulation, but most of them can be modified to operate in conjunction with QAM. In this chapter we propose a novel channel equaliser scheme for employment in star 16-QAM systems, which is based on a combination of various signal processing blocks, such as FEC codecs, the so-called RAKE diversity combiner to be described in the next section, and a linear equaliser (LE). We firstly detail the RAKE combiner.

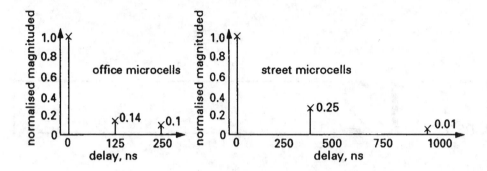

Figure 12.1: Microcellular impulse responses used in the simulations

12.2   The RAKE Combiner

The RAKE receiver is not an equaliser, but a combiner for diversity systems. The RAKE system inherits its name from the way it "rakes" in all the incoming pulses to form an equalised signal. A block diagram of the basic system is shown in Figure 12.2. The RAKE system is identical in structure to a maximum ratio combiner. It is designed to operate in a situation where the signal to be transmitted has a narrow bandwidth, but the bandwidth available is wide, and so the transmitted signal's bandwidth can be expanded in order to use the multipath diversity inherent in the wideband system. Multipath diversity exhibits itself in the fact that in wideband channels several echoes of the transmitted signal can be observed due to far-field reflections. The channel equaliser can remove the dispersion introduced by the channel, and by locking onto the echoes appearing in the impulse response, taps within the equaliser can exploit this so-called time domain diversity effect. In order to achieve the required bandwidth expansion the baseband signal is transmitted as a series of narrow pulses with a gap between each pulse as shown in Figure 12.3. Suppose, for example, that the signalling interval duration is T and the bandwidth required when these pulses are appropriately shaped is B, but the available system bandwidth is $N \times B$. In this case an N times narrower signalling pulse of duration T/N can be used to exploit the whole bandwidth. Accordingly the RAKE receiver of Figure 12.2 must be clocked at an N times higher rate.

Because of the delay spread, the receiver perceives each transmitted pulse as a series of received pulses. In order for the RAKE combiner to operate correctly it needs to know the expected attenuation and phase change suffered by each path. Each incoming pulse is then amplified and de-rotated by these expected factors, and then the resulting partially equalised pulses are superimposed to form the input to the symbol decision section. This is shown diagrammatically in Figure 12.3. Adaptive coefficient update systems can be used instead of channel sounding methods, but these tend to be less robust due to error propagation in the coefficient update section.

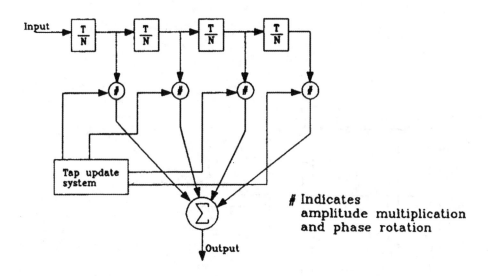

Figure 12.2: RAKE diversity combining system

12.3 The Proposed Equaliser

12.3.1 Linear Equaliser

When transmitting in the Rayleigh fading environment, there are a number of counter-measures that are often employed against fading. Mobile subscribers may become stationary in a deep fade with consequent system failure. In order to combat this condition frequency hopping is often used. Accordingly, the data is divided into blocks, and at the end of each block a hop is made to a new frequency outside the coherence bandwidth of the channel. The channel is sounded at the start of the block in order to establish the impulse response of the channel at this new frequency. Sounding in the centre of the block will give better results because the impulse response of the channel cannot change so dramatically during half a block as during a whole block, and is often employed. The next block of data is then sent. The size of each block must be chosen so that the channel does not change significantly during the time the block is transmitted. However, the channel should be sounded as infrequently as possible in order to keep the data throughput high. The optimum block size is related to the mobile speed.

Simulations of adaptive linear and DFEs showed very poor performance due to catastrophic failure in fading environments. A similar problem was experienced with DFEs employing channel sounding. The linear equaliser with channel sounding offered the most promising performance.

A few slight changes are necessary as regards the channel sounding method used for binary modulation schemes in the case of QAM as the pseudo random binary sequence (PRBS) used for sounding must be mapped onto the non-binary symbol set for transmission, and both the amplitude and phase shift of each of the propagation paths must be recovered. A logical 1 from the PRBS is mapped onto the outermost phasor of the star 16-QAM constellation at 0°

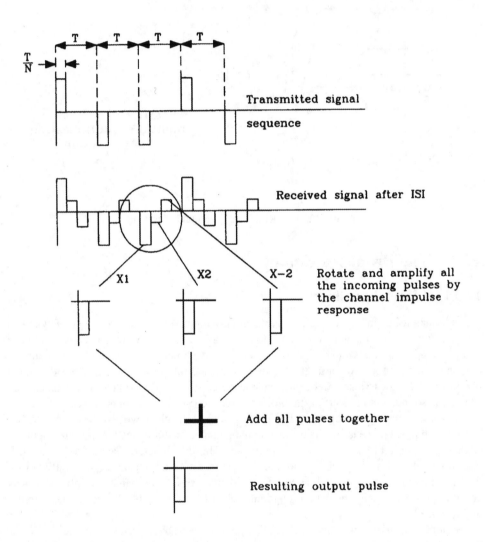

Figure 12.3: Time domain operation of the RAKE combiner

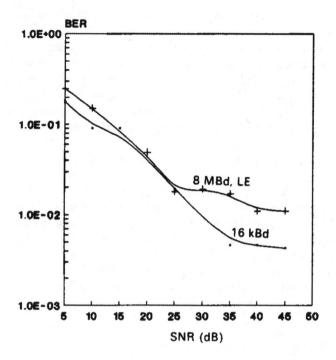

Figure 12.4: BER versus channel SNR performance of narrowband and wideband star 16-QAM at 30 mph and 1.9 GHz for 16 kBd and 8 MBd, respectively

and a logical 0 from the PRBS onto the outermost symbol at 180°. The use of the outermost symbol maximises the received SNR and by only transmitting on the I-axis, measurement of the introduced phase shift is simplified. This principle is similar to those used in the generation of equaliser training sequences in the CCITT V. series modems discussed in Chapter 8. Separate correlation is performed on both the I and Q axes, and the phase shift is found by taking the arctan of the results. A comparison of the BER performance of the differential 16-level star QAM modem operating at 16 kBd over a flat fading, i.e. narrowband channel, and the 8 MBd (32 Mbps) system with a LE over a frequency selective wideband channel is presented in Figure 12.4 where it can be seen that the performance of the wideband system is inferior to the narrowband system, although the difference is not substantial.

12.3.2 Iterative Equaliser System

Upon studying error events using the linear equaliser it was observed in the simulations that the performance of the system would be considerably improved if it could be made more tolerant to fades which prevented the LE coefficients from decaying, as was mentioned in Chapter 7. Furthermore, the DFE is a potentially powerful equaliser if it can be prevented from failing catastrophically due to error propagation effects in the case of erroneous decisions, when incorrect information is fed back into the coefficient adjustment mechanism. A solution to this problem is to use an equaliser which exploits the advantages of both the LE and DFE but avoids their shortcomings.

The linear equaliser works satisfactorily most of the time, as previously described, except when the coefficients fail to decay. This failure to decay can be detected by comparing the values of the last few equaliser coefficients against an appropriate threshold value. If this threshold value is exceeded by the coefficient values, the decoded data from the linear equaliser will have a high error rate. A robust initial equaliser system is required which will not fail catastrophically during fades and whose output can then be further processed to reduce the BER.

In those situations where the LE failed to converge, a DFE was therefore invoked which was constrained in order to prevent catastrophic failure, and a multi-stage equalisation algorithm was contrived. The first pass at equalising the channel was carried by a non-optimal equaliser based on the RAKE equaliser [388] that was described in the previous section. Although the RAKE combiner's operation is prone to errors, experience showed that these will be typically confined to neighbouring constellation points. The data regenerated by the RAKE nows has a lower BER than the received data and it can be used to keep the DFE from failing catastrophically. We refer to this non-optimal equaliser as a one symbol window equaliser (OSWE). A block diagram of the complete system proposed is shown in Figure 12.5 and its operation is discussed in depth in the following sections.

12.3.2.1 The One-Symbol Window Equaliser

This equaliser utilises the principle that the ISI will generally exhibit itself as random noise, and calculating an appropriately weighted average of the magnitude and phase of the incoming phasors over the duration of the channel impulse response will enable it to remove this ISI. Its principle of operation is discussed below and shown diagrammatically in Figure 12.6. First the channel impulse response is estimated from the received channel sounding sequence and in Figure 12.6 three significant paths are resolved, each having an associated magnitude and phase. Next the delay bin is determined, m symbol periods after the first significant received multipath component, in which the last significant path component was received. This informs the receiver about the duration of the channel's delay spread. For each of the m delay bins a copy of the locally stored transmitted constellation is rotated by the measured rotation of the impulse response for that particular bin and held in memory as portrayed in the figure. By this means any rotation of the constellation caused by the channel can be compensated. In order to make a decision as to the rth received symbol all m received QAM phasors which include a contribution from the rth symbol are taken into account. These phasors are mapped onto the appropriately rotated constellation. Thus the QAM phasor due to the first path signal is held with the constellation rotated by the phase shift of the first path. Similarly, all other paths up to path m generate a received phasor indicated by the arrows at the bottom of Figure 12.6 for $m = 3$ consecutive signalling periods. The QAM phasor comprising the first path from the $(r+1)$th transmitted symbol plus ISI contributions from the rth symbol would be held with the constellation rotated by the phase shift which the channel imposes on the second bin delay, and so on.

Then for each of the M possible transmitted phasors a distance summation is performed over all m rotated constellations. Specifically the distance between each rotated point of the m rotated constellations to which the mth phasor has been mapped and the received symbol

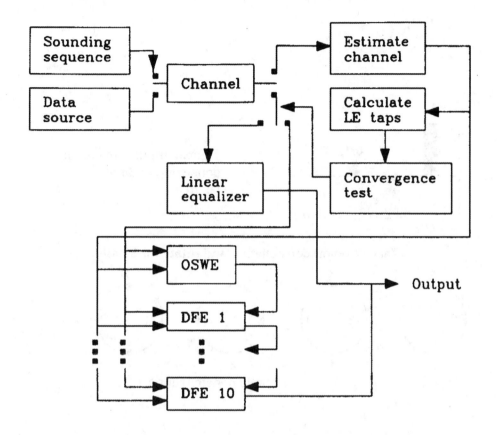

Figure 12.5: System block diagram

are found. Each distance is scaled by the ratio of the path amplitude for that delay to the path amplitude of the main tap. This scaling has the effect of reducing the importance of significantly attenuated delayed signals when regenerating the symbol and is necessary as these delayed signals will be highly corrupted by noise and ISI from other symbols.

This distance calculation is demonstrated at the foot of Figure 12.6 for $M = 4$ and $m = 3$. The Euclidean distances between the received phasors in the $m = 3$ consecutive signalling intervals, and the appropriately rotated phasor 1 are computed, giving distances of a, b and c. Weighting them with their associated path magnitude introduces the required scaling based on their significance, giving a distance of $d_1 = a + 0.5b + 0.1c$. Similarly, for phasor 2 the distance becomes $d_2 = d + 0.5e + 0.1f$, etc., up to $d_M = d_4$. Then the rotated constellation point yielding the lowest distance d is deemed to be the most likely to have been transmitted. In general terms the distance d_n, $n = 1$, ..., M associated with any particular received symbol $S_n = x_t + jy_t$ at time t is given by

$$d_n = \sum_{i=1}^{m} \frac{a_i}{a_0} \sqrt{(x_{t+i} - u_i)^2 + (y_{t+i} - v_i)^2} \qquad (12.1)$$

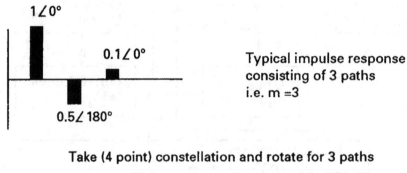

$1\angle 0°$

$0.1\angle 0°$

$0.5\angle 180°$

Typical impulse response
consisting of 3 paths
i.e. m =3

Take (4 point) constellation and rotate for 3 paths

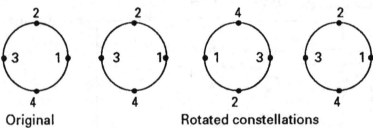

Original   Rotated constellations

The full lines
represent the
received
phasors for
3 consecutive
symbol periods.

We calculate $d_1 = a+0.5b+0.1c$ $d_2 = d+0.5e+0.1f$ and so on for each of
the (4) constellation points. We choose the constellation with the
smallest distance.

Figure 12.6: OWSE operation for $M = 4$ and $m = 3$

where m is the last significant delay bin, x_t and y_t are the Cartesian values of the received signal at time t, a_i and a_0 are the amplitudes of the measured impulse response at tap i and the main tap respectively, and u and v are the Cartesian values of the rotated constellation points and are given by

$$u_i = r_n \cos(\theta_n + \phi_i)$$
$$v_i = r_n \sin(\theta_n + \phi_i) \tag{12.2}$$

r_n and θ_n are the polar amplitude and phase associated with symbol n at the transmitter, and ϕ_i is the phase shift of the ith component in the estimated channel impulse response.

Having found the total distance from between the m received phasors which include components of the rth sample and each of the M constellation points, the constellation point associated with the lowest total distance according to Equation 12.1 is selected as the most likely one to have been transmitted. This symbol along with the actual received signal levels x_t and y_t is passed to the DFE. Observe the underlying similarity between the operation of the RAKE receiver described in Section 12.2 and that of the OSWE proposed.

12.3.2.2   The Limited Correction DFE

The power of the DFE can be used in order to significantly reduce the error rate from the OSWE. When the LE coefficients have failed to decay, and this is detected by the convergence test block of Figure 12.5, the DFE acts on the received signal level as per normal and decides on the transmitted symbol. If this decision concurs with that of the OSWE the symbol is input to the feedback taps of the DFE and is stored in a decision register maintained for the duration of the decoding of one block. If the DFE's decision represents a neighbouring symbol in the QAM constellation to the one decoded by the OSWE, then again the symbol decoded by the DFE is passed back to the feedback taps and stored as the new decoded symbol. However, should the DFE's regenerated symbol *not* be a neighbour of the OWSE-estimated symbol, the OSWE's estimated symbol is changed to one of its neighbours in such a way that it is moved nearer to the symbol position regenerated by the DFE, is input to the DFE feedback taps and stored as the new decoded symbol. This movement towards the DFE both tends to bring the DFE back towards lock and accounts for the likely inaccuracy of the OSWE during a fade.

This restriction on the symbols fed back by the DFE prevents it from failing catastrophically in all but the most severe fades. In order to improve the performance the DFE can be run iteratively a number of times using the same input signals but different feedback signals as the constraining output register changes slightly on each pass. Simulations revealed that when the LE failed completely, this iterative system typically produced a very low or zero symbol error rate. Using a knowledge of the transmitted symbol in the simulation showed that each iteration produced a lower or identical BER than the previous iteration. After approximately 10 iterations a minimum BER was reached.

The OSWE will automatically lock onto and operate on the strongest multipath component in the channel impulse response. DFEs normally have a feedforward section to allow for the strongest component not being the first path. This feedforward section is essentially a LE and the tap coefficients are set by the inverse filter process as previously explained in Chapter 7. With the combination of equalisers proposed here, it is possible to remove most of the feedforward LE section of the DFE as the OSWE has already decoded the data that would

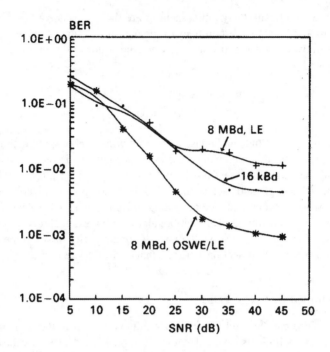

Figure 12.7: BER versus channel SNR comparison of star 16-QAM in the case of 16 kBd and 8 MBd transmission using the LE and OSWE at 30 mph and 1.9 GHz

normally be decoded in the feedforward part of the DFE. The DFE operates on the untampered OSWE output, the received signal itself and the feedback data derived as previously explained. This equaliser does not have as good iterative convergence properties as when the first path is the strongest, as it has to use data from the OSWE before it can modify it, but again it shows a significant improvement in performance over the LE.

The results for this system are shown in Figure 12.7 where the performance is significantly better than that of the LE alone, and also shows an improvement over the narrowband system. This improvement over the narrowband system is because this equaliser is successfully using the inherent multipath time diversity present in the wideband channel.

12.3.3   Employing Error Correction Coding

The results shown in Figure 12.7 show the presence of a residual BER caused by impulse responses which cannot be equalised satisfactorily by this system. For the indoor environment, the system fails when the first component in the channel impulse response is in a fade of about 10 dB, when its magnitude becomes approximately equal to that of the other multipath components.

Error correction coding can be used to overcome the residual BER. Conventional FECs encode the bits, interleave and transmit them, perform equalisation, de-interleave the bits and then decode and correct them. The results of this strategy overlaid onto the system described

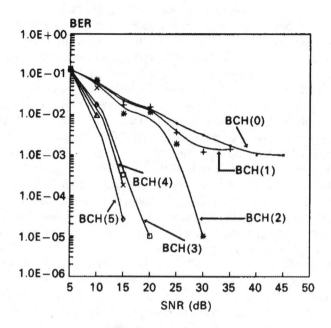

Figure 12.8: BER versus channel SNR performance of star 16-QAM at 8 MBd using the LE, the OSWE and a range of BCH FEC codes at 30 mph and 1.9 GHz

in Section 12.3.2.2 are shown in Figure 12.8.

The bracketed numbers in the caption indicate the bit correction ability for a block length of 63 bits when low complexity binary BCH codecs are used. Explicitly, the codes employed are $BCH(0) =$ NoFEC,

$$BCH$$

$(1) = BCH(63,57,1)$, $BCH(2) = BCH(63,51,2)$, $BCH(3) = BCH(63,45,3)$, $BCH(4) = BCH(63,39,4)$, $BCH(5) = BCH(63,36,5)$ and $BCH(6) = BCH(63,30,6)$.

As it can be seen in the figure, even mild error correcting powers proved to be effective, particularly with SNRs in excess of about 20 dB. The residual BER is removed and the channel is rendered suitable for computer data transmission. Coding powers in excess of the $BCH(63, 45, 3)$ offer little extra gain for the trade-off in spectral efficiency. Note that coding reduces spectral efficiency but it has proved necessary to take this measure in most transmissions over Rayleigh fading channels.

With the system described in Section 12.3.2.2, error correction codes can be used in a more efficient manner than by merely decoding after the QAM symbol regeneration. In the same way that we use a number of passes through the DFE, we can pass the partially equalised data through the BCH decoder a number of times. If, as occurs most of the time, the LE has been used, the decoded data is stored in an array ready for de-interleaving. If the OSWE has been used then in addition to storing the decoded data in this array, both the received signal and a flag to identify that this block has been decoded by the OSWE are stored. The blocks of data, whether from the LE or the OSWE, are de-interleaved and BCH decoded.

Most de-interleaved blocks will have too many errors at this stage to allow their correction by the BCH decoder, but if any correction does occur then a marker is placed beside the corrected bits. The non-interleaved blocks which were decoded by the OSWE are then passed through the DFE as previously described but with the small modification that any bits that were marked as correct are not permitted to be altered by the DFE. Notice that differential coding cannot be used because if we have corrected a differentially decoded symbol we would know the correct difference between the previous and current symbol, but not the absolute values of these symbols. This is not a severe limitation, since differential coding is often avoided in high integrity systems because of the 3 dB SNR penalty associated with its use.

After being passed through the DFE the data is again de-interleaved and another BCH decoding pass is attempted. This process continues for as many passes as are necessary. Since a good estimate of the number of errors can be gained from the overload detector of the BCH decoder, this iterative process can be continued until either there are no errors in the array, or no more errors have been removed since the previous pass. Convergence of the DFE is improved by marking of the correct bits, helping to prevent it from making incorrect decisions. The performance of this system for the same BCH codes as before is shown in Figure 12.9. As can be seen, the overall effect is very similar to the more conventional decoding method of Figure 12.8. Comparison of the two systems shows that there is a small gain of the order of 2-3 dB for most coding powers and SNRs. This is not a particularly large reward for the considerable increase in receiver complexity. This system would only be worth implementing if SNR or BER were at an absolute premium.

The reason for the gain being small is probably that although the equaliser is now forced into a better convergence, the previous system was able to remove most of the errors from an unconverged equaliser, reducing the benefit of the more complex system.

12.4   Diversity in the Wideband System

In Chapter 10 it was shown that second-order spatial diversity is beneficial in narrowband QAM. The selection or combination of the received signals from diversity receivers depends on their relative amplitudes. In the wideband system considered here, the scenario is different. Less benefit would be expected from space diversity because the time diversity inherent in a multipath environment is used to some extent in the equalisation process. If there are two channels with different impulse responses, then to implement switched diversity reception a decision must be made as to which channel has the most favourable response. This decision is governed by the equaliser's properties.

It is necessary to produce some figure of merit that relates to the likely number of errors for a given impulse response. With switched diversity the channel with the least likelihood of errors is chosen, and with maximal ratio combining the incoming channels are weighted by the likelihood of errors. In order to identify the types of impulse responses that are more likely to yield relatively low symbol error rates, a series of simulations were performed using a channel whose impulse response has three coefficients, but in contrast to Figure 12.1 the taps weights were now varied.

The magnitude of the strongest path is designated by a_0, that of the next strongest path by a_1 and that of the weakest path by a_2. The average power of the impulse response is arranged to be a constant, and the lowest error rate was achieved when there was a dominant path a_0,

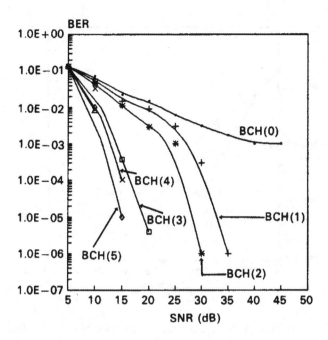

Figure 12.9: BER versus channel SNR performance of star 16-QAM at 8 MBd using the LE, the OSWE and a range of iteratively decoded BCH FEC codes at 30 mph and 1.9 GHz

and a_1 was significantly larger than a_2. For the situation when two paths were approximately equal, fewer errors were incurred when the two smaller paths were equal compared to the two larger ones being equal. The highest error rate inflicted was when all the paths had the same magnitude.

The diversity system was therefore arranged to rate an impulse response according to the system parameter

$$r = ((a_0/a_1) + (a_0/a_2)) \cdot F + (a_1/a_2) \tag{12.3}$$

where F is another system parameter which must exceed unity. The first term in r weighted by the factor F provides a measure of the dominance of the strongest path a_0, while the second unweighted term (a_1/a_2) represents the dominance of a_1 over a_2. For different channels, the one having the largest r parameter will have the most favourably shaped impulse response, and if all impulse responses have the same average power, it will give the lowest BER. In general, the impulse responses will have differing average powers and so the channel rating is modified to

$$R = a_0 r \tag{12.4}$$

as r deals in ratios and is only concerned with the shape of the impulse response. Due to multiplying r by the dominant coefficient a_0, r is scaled appropriately. The BER was examined as the parameter F varied from 1 to 10, and was found to be relatively insensitive to the value selected.

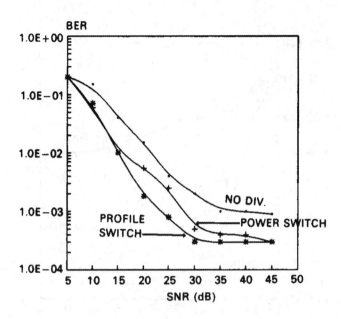

Figure 12.10: BER versus channel SNR performance of star 16-QAM at 8 MBd using the LE, the OSWE and a range of diversity combining techniques at 30 mph and 1.9 GHz

Armed with an algorithm R to rate the quality of a channel's impulse response, second-order diversity simulations can be performed. The two diversity channels had impulse responses with three coefficients each. All three coefficients had magnitudes and phases which conformed to independent Rayleigh fading statistics. The PRBS sounding sequence was transmitted over both channels for each block of data, and the channel impulse responses were estimated. The diversity branch whose impulse response had the larger R parameter was used in our switch diversity arrangements.

The variation of BER as a function of channel SNR is displayed in Figure 12.10 for this diversity system with a data rate of 8 MBd, i.e. 32 Mbps. A benchmark for the non-diversity condition is also shown. For a BER of 10^{-2} (no FEC is used here) the diversity yielded a reduction in SNR of approximately 6 dB. For comparison it shows the performance of the system which selects the incoming channel not by using R of Equation 12.4, but merely by selecting the impulse response with the strongest average power, i.e. rating the impulse responses on the basis of

$$\hat{R} = a_0^2 + a_1^2 + a_2^2 \tag{12.5}$$

The curves in Figure 12.10 show that despite the inherent diversity in the multipath signal which was exploited by the equaliser system of Figure 12.5, there were further significant gains to be made by introducing second-order spatial diversity. The method of selecting the impulse response using Equation 12.4 yielded the profile switch curve. This performed better than the conventional method of selecting the signal with the highest average power using $\hat{R}$ of Equation 12.5, the performance of which is labelled as the power switch curve in Figure 12.10.

12.5 Summary

At a transmission frequency of 1.9 GHz, data at a rate of 8 Msymbols/s, corresponding to 32 Mbits/s, can be sent via star 16-level QAM over wideband dispersive channels at an integrity acceptable in many applications. In most instances a linear equaliser worked satisfactorily, but when the main component of the impulse response was in a deep fade it was advantageous to carry out a first-pass equalisation with a non-optimal equaliser and then use the partially equalised signal to prevent a DFE from failing catastrophically. When this system was used in an iterative fashion, considerable improvements in BER were achieved. The system discussed incorporates three types of equalisers in a concatenated arrangement with switching between them as conditions in the fading channel change. Improvements in BER of an order of magnitude over a linear equaliser system were obtained for SNRs in excess of 15 dB.

Second-order spatial diversity gave about 5-7 dB gain in terms of channel SNR over an equivalent uncoded system using no diversity providing that it was implemented in an intelligent way, leading to a BER of 3×10^{-4} at a SNR of 30 dB with no channel coding.

The use of channel coding provided very significant reductions in BER. With an approximately 2/3 rate BCH code, BERs as low as 1×10^{-5} were achieved for SNRs as low as 20 dB, and when an iterative channel decoding approach was adopted, BERs of 1×10^{-6} were realised for the same coding power and SNR, but at the cost of a considerable increase in complexity.

In the next chapter we consider an alternative means of transmitting at high bit rates over mobile radio channels. This involves dividing the message to be transmitted into a number of streams and sending each of these streams simultaneously. It is known as orthogonal multiplexing.

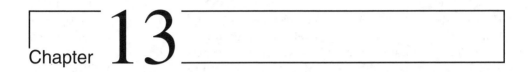

Chapter **13**

Quadrature-Quadrature AM

13.1 Introduction

A variety of other orthogonal signalling functions were proposed in the literature [546–548], where a number of messages can be multiplexed for transmission in the same bandwidth without interfering with each other. Orthogonal multiplexing can potentially improve the bandwidth efficiency of a communications system. One of the simplest manifestations of orthogonal multiplexing is conventional QAM, where information is transmitted using two orthogonal carriers – the in-phase and quadrature channels – within the same bandwidth. The in-phase and quadrature channels constitute two dimensions of an orthogonal space. There are a plethora of methods that can be invoked for generating *orthogonal dimensions*, which can be used for uniquely recoverable telecommunications signalling. Using a set of orthogonal modulation elements or signalling pulses $m_T(t)$ can create a further dimension for bandwidth efficient orthogonal signalling. In conventional modulation schemes, bandlimited pulses are used, but there is considerable freedom in the choice of the pulse shape. These issues will form the topic of this chapter in the context of quadrature-quadrature amplitude modulation (Q^2AM). As an introduction to these systems, Q^2PSK [549] is examined first.

13.2 Quadrature-Quadrature Phase Shift Keying

We have seen in previous chapters that by moving from BPSK to QPSK, the number of bits per Hertz is increased. This is achieved by transmitting two bits per modulation symbol, which allows us to double the BPSK signalling interval of T to $2T$, thereby halving the bandwidth requirement, while maintaining the same bit rate. Naturally, this is achieved at an increased SNR as well as hardware complexity. Alternatively, one could double the bit rate in the original BPSK bandwidth. This is because the number of signalling dimensions used in the modulation scheme is increased. BPSK can be considered to constitute a scheme having only a single dimension – the in-phase or I channel. Following this line of argument, QPSK then has two dimensions, the in-phase and quadrature or I and Q channels. In general, given a time-limited signalling pulse of duration τ which may be viewed as having a bandwidth of W, then there are $N - 2\tau W$ possible signalling dimensions [549, 550]. If a channel is strictly bandlimited to $W_0 = 1/\tau$, such as for example a system having a Nyquist roll-off factor of $\alpha = 0$, then the number of signalling dimensions is given by $N = 2\tau W_0 = 2$. Hence, when the channel is strictly bandlimited to $W_0 = 1/\tau$, the bit duration of BPSK and QPSK is given by $T_b^{BPSK} = \tau = 1/W_0$ and $T_b^{QPSK} = \tau/2 = 1/2W_0$, respectively. Therefore the bandwidth efficiency of QPSK and BPSK is $1/W_0 T_b^{QPSK} = 2$ bit/s/Hz and $1/W_0 T_b^{BPSK} = 1$ bit/s/Hz, respectively. However, if a system having a Nyquist roll-off factor of $\alpha = 1$ was used, where an excess bandwidth of 100% is imposed, then the channel is no longer strictly bandlimited to W_0, but bandlimited to $W = 2/\tau$. In this scenario, the number of signalling dimensions is $N = 2\tau W = 4$ and the bit duration of BPSK and QPSK becomes $T_b^{BPSK} = \tau = 2/W$ and $T_b^{QPSK} = \tau/2 = 1/W$, respectively. Hence, the bandwidth efficiency of QPSK and BPSK is given by 1 bit/s/Hz and 0.5 bit/s/Hz, respectively. Clearly, an $N = 4$-dimensional signalling scheme is needed in order to provide a bandwidth efficiency of 2 bit/s/Hz when the channel's bandwidth is $W = 2/\tau$.

In theory, the bound of $N = 2\tau W$ on the number of dimensions can be reached by using the so-called prolate spheroidal wave functions, but these have been shown to be unrealisable [551]. Fortunately, there may be other ways of approaching this bound, ways which can be readily realised. As an example, by moving from QPSK to Q^{2}PSK, it becomes possible to exploit all four available dimensions. More specifically, Q^{2}PSK utilises two orthogonal cosinusoidal pulse shaping functions instead of a rectangular pulse shaping function that may be employed by the QPSK scheme, together with two different carrier frequencies in order to create 4-dimensional Q^{2}PSK symbols. More specifically, it was argued in [549] and we will also show this equivalence mathematically in Equations 13.3 and 13.8, demonstrating that Q^{2}PSK signalling may be viewed as the superposition of two Minimum Shift Keying (MSK) [370, 388] signals. The associated bandwidth is given by $W = 2/\tau = 1/T$, where τ is the QPSK signalling period and T is the MSK signalling interval. The bit duration of Q^{2}PSK is given by $T_b^{Q^2 PSK} = T/2 = 1/2W$ and hence a bandwidth efficiency of $1/W T_b^{Q^2 PSK} = 2$ bit/s/Hz is attained. Therefore, the Q^{2}PSK scheme may be viewed as being potentially twice as bandwidth efficient as QPSK, when the channel is not strictly bandlimited to $W_0 = 1/\tau$, although this doubled bandwidth-potential may not be fully attained, when taking into account the practical realisability of the associated filters, as we will see in the context of Q^{2}AM in Section 13.3.

For unshaped infinite-bandwidth QPSK the input data consists of pulses $a(t)$ having a period of T and amplitude of ± 1, which is equivalent to using the rectangular modulation elements $m_T(t) = rect(t/T)$ of the previous chapter. The incoming data stream of pulse duration T is demultiplexed into two parallel streams, $a_1(t)$ and $a_2(t)$, consisting of alternative pulses of duration $2T$. These streams are multiplied by the orthogonal sine and cosine carriers and summed to form the QPSK signal

$$S_{QPSK}(t) = \frac{1}{\sqrt{2T}} a_1(t) \cos\left(2\pi f_c t\right) + \frac{1}{\sqrt{2T}} a_2(t) \sin\left(2\pi f_c t\right), \qquad (13.1)$$

$$= \frac{1}{\sqrt{T}} \cos\left(2\pi f_c t + \phi(t)\right), \qquad (13.2)$$

where $a_1(t)$ and $a_2(t)$ may assume the value of either $+1$ or -1, while $\phi(t)$ may assume the value of either 0°, $\pm 90^o$ or 180° depending on the values of $a_1(t)$ and $a_2(t)$. Note that there

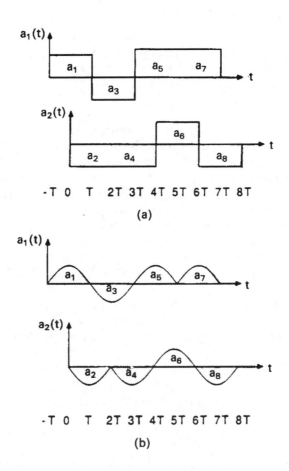

Figure 13.1: Pulse shaping: (a). rectangular, (b). sinusoidal and cosinusoidal [549] ©IEEE, 1989, Saha

is only a single carrier frequency f_c involved in Equation 13.2. In offset QPSK (OQPSK) the bit stream $a_2(t)$ is delayed by a fraction of the symbol period $2T$, normally by half, i.e. by T, in order to mitigate the instantaneous changes in the carrier phase, as portrayed by Figure 13.1(a).

Although the power spectral densities of QPSK and OQPSK are identical, they react to bandlimiting and non-linear amplification differently. As opposed to QPSK, OQPSK cannot have a $180°$ phase shift, which could be caused by the coincident changes in $a_1(t)$ and $a_2(t)$. Thus OQPSK has a lower envelope fluctuation than QPSK and hence non-linear amplification will not regenerate the high frequency modulated signal side lobes removed by filtering. The above-mentioned instantaneous rectangular transition can be completely avoided in MSK [370, 388], where the data streams $a_1(t)$ and $a_2(t)$ are shaped by a sinusoidal pulse shaping function instead of the rectangular pulse shaping function of OQPSK. The MSK signal can be written as [549]:

$$
\begin{aligned}
S_{MSK}(t) &= \frac{1}{\sqrt{T}}a_1(t)cos\left(\frac{\pi t}{2T}\right)cos(2\pi f_c t) + \frac{1}{\sqrt{T}}a_2(t)sin\left(\frac{\pi t}{2T}\right)sin(2\pi f_c t), \\
&= \frac{1}{\sqrt{T}}cos\left[2\pi\left(f_c + \frac{b(t)}{4T}\right)t + \phi(t)\right],
\end{aligned}
\tag{13.3}
$$

where $b(t) = -a_1(t)a_2(t)$ and $\phi(t) = 0$ or π according to whether we have $a_1(t) = +1$ or -1. Although the same sinusoidal pulse shaping function is used for both $a_1(t)$ and $a_2(t)$, as illustrated in Figure 13.1(b), the pulse shaping function assigned for $a_2(t)$ is treated as a cosinusoidal function because of the time offset of T. Note, however, that the signalling pulses of $cos(\frac{\pi t}{2T})$ and $sin(\frac{\pi t}{2T})$ shown in Figure 13.1(b) are different from the more conventional time-domain raised cosine pulse introduced as non-linear filtering (NLF) in Chapter 4. Furthermore, the MSK signals require two carrier frequencies which are given in Equation 13.3 as $f_c \pm 1/4T$. More explicitly, MSK belongs to the Frequency Shift Keying (FSK) [388] family. The minimum frequency separation that is necessary for ensuring the orthogonality of two FSK signals over a signalling interval of T is given by [388, p. 197] $\Delta F = (f_c + 1/4T) - (f_c - 1/4T) = 1/2T$. Therefore, the bandwidth required for MSK signalling having $M = 2$ different carrier frequencies is given by [388, p. 283] $W = M\Delta F = 1/T$. However, MSK utilises only two out of the $N = 2\tau W = 4$ available signal dimensions. The issue of signalling space dimensionality will be revisited in greater detail in Chapter 23.

Observe in Equation 13.3 that we have introduced two orthogonal pulse shaping functions, namely:

$$
p_1(t) = sin(\pi t/2T)
\tag{13.4}
$$

and

$$
p_2(t) = cos(\pi t/2T)
\tag{13.5}
$$

respectively. Armed with these functions the following set of orthogonal basis functions can be formulated:

$$
\begin{aligned}
s_1(t) &= p_1(t)sin2\pi f_c t \\
s_2(t) &= p_2(t)sin2\pi f_c t \\
s_3(t) &= p_1(t)cos2\pi f_c t \\
s_4(t) &= p_2(t)cos2\pi f_c t.
\end{aligned}
\tag{13.6}
$$

It can be seen that between any two signals in the above set there is a common factor, which is either a common data shaping pulse or a common carrier component. The remaining factor is orthogonal to the corresponding factor in the other. In other words, these second factors are in quadrature with each other. With the advent of the four orthogonal basis functions seen

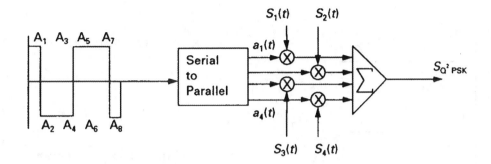

Figure 13.2: Structure of a quadrature-quadrature phase shift keying modulator

in Equation 13.6, the Q^2PSK modulated signal is formulated as [549]:

$$S_{Q^2PSK}(t) = \sum_{i=1}^{4} a_i(t)s_i(t),$$

(13.7)

$$= cos\left[2\pi\left(f_c + \frac{b_{14}(t)}{4T}\right)t + \phi_{14}(t)\right] + sin\left[2\pi\left(f_c + \frac{b_{23}(t)}{4T}\right)t + \phi_{23}(t)\right],$$

(13.8)

where $b_{14}(t) = -a_1(t)a_4(t)$ and $\phi_{14}(t) = 0$ or π, according to whether we have $a_1(t) = +1$ or -1. Similarly, $b_{23}(t) = +a_2(t)a_3(t)$ and $\phi_{23}(t) = 0$ or π, according to whether we have $a_3(t) = +1$ or -1. Unlike in MSK, there is no time offset between $a_1(t)$ and $a_4(t)$ (or $a_2(t)$ and $a_3(t)$), but two pulse shaping functions are employed, namely the sinusoidal function of Equation 13.4 and the cosinusoidal function of Equation 13.5. We infer from Equations 13.3 and 13.8 that Q^2PSK signalling may be viewed as the superposition of two MSK signals. Specifically, Q^2PSK consists of two signals, one of them is sinusoidal and assumes one of two legitimate frequencies given by $f_c \pm 1/4T$ as well as another one, which is cosinusoidal and also assumes one of two legitimate frequencies, namely $f_c \pm 1/4T$. The minimum frequency separation between the two frequencies of $f_c + 1/4T$ and $f_c - 1/4T$ is $\triangle f = 1/2T$ and hence the bandwidth required is $W = 2\triangle f = 1/T$. The number of signalling dimensions becomes $N = 2\tau W = 4$, which is fully exploited by the 4-dimensional Q^2PSK signalling scheme of Equation 13.7. The block diagram of the Q^2PSK modulator is shown in Figure 13.2.

The waveforms of the components $a_i(t)s_i(t), i = 1, 2, 3, 4$, are displayed in Figure 13.3 for the input pulses $A_1, A_2, ..., A_8$, where A_n is the nth binary input pulse with value -1 or 1, and whose polarities are shown in Figure 13.2.

The waveforms are displayed for two symbol periods, remembering that this modulation scheme encodes four bits per symbol. For the first symbol period the output from the serial-to-parallel converter $a_1(t)$ is A_1. In the second symbol interval $a_1(t)$ is A_5, and in general $a_1(t) = A_{4i+1}, i = 0, 1, ..., .$ The waveform at the top of Figure 13.3, $a_1(t)p_1(t)$, is composed of two positive sine half-cycles, corresponding to positive pulses A_1 and A_5. The waveform for $a_2(t)p_2(t)$, is composed of a negative cosine half-cycle as A_2 is -1, followed by a positive cosine half-cycle as A_6 is $+1$. Since the serial-to-parallel converter processes

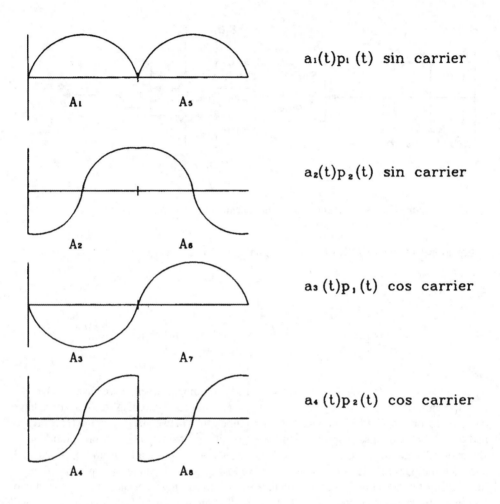

Figure 13.3: Pulse shapes for the Q^2PSK modulator

four bits at a time, compared to two for QPSK, twice the amount of data per modulation interval is accommodated. Indeed, without any bandlimiting, Q²PSK can obtain twice the throughput of QPSK at the same SNR and BER [549]. However, due to the sudden changes in the modulating signal as for example $a_4(t)p_2(t)$ in Figure 13.3, when the signal is bandlimited this statement is invalid. Bandlimiting Q²PSK transmissions over AWGN channels requires an increase of 1.6 dB transmitted power in order to provide an identical BER performance to that of QPSK for the same channel bandwidth [549].

a_n	a_{n+1}	Output
0	0	-1
0	1	-3
1	0	+1
1	1	+3

Table 13.1: Encoding scheme for Figure 13.4

13.3 Quadrature-Quadrature Amplitude Modulation

13.3.1 Square 16-QAM

In this section the modifications necessary to include orthogonal data shaping pulses in square and star QAM are discussed.

In our approach, which is a direct expansion of Q^2PSK, we take four data bits for each of the I and Q channels. The structure of the modulator for this scheme is shown in Figure 13.4 and the corresponding waveforms in Figure 13.5. The four bits for the I channel, which are A_1 to A_4 for the first symbol and A_9 to A_{12} for the second symbol in Figure 13.4 are split into two groups. The first group consists of $a_1(t) + a_2(t)$ where $a_1(t) = A_1$ and $a_2(t) = A_2$, representing the first two bits of an eight bit symbol. Each group of two adjacent input bits passes in parallel through a block labelled ENCODE in Figure 13.4, which takes the two binary bits and encodes them onto one of four levels, ± 1 or ± 3, corresponding to the valid amplitude levels in a square 16-QAM constellation. The mapping is Gray encoded and shown in Table 13.1, where $n = 1, 3, 5, 7$.

Extending the assignments for $s_i(t)$ in Equation 13.6 to 16-QAM by allowing not only the sign but also the magnitude of the modulation pulses to change, the amplitude derived from $a_1(t)$ and $a_2(t)$ can be encoded onto the sinusoidal pulse shape on the sinusoid, i.e. the I carrier, while that derived from $a_3(t)$ and $a_4(t)$ can be encoded onto the cosinusoidal pulse shape on the I carrier as suggested by the waveforms of Figure 13.5. The same procedure applies to the Q channel. Namely, the amplitude derived from bits $a_5(t)$ and $a_6(t)$ is encoded onto the sinusoidal pulse shape on the cosinusoid i.e., the Q carrier, and that derived from bits $a_7(t)$ and $a_8(t)$ is encoded onto the cosinusoidal pulse shape on the Q carrier.

In Figure 13.5 the same principle has been used as for Figure 13.3. Here $A_1 + A_2$ has been used to signify the amplitude derived by the mapping described previously from the bits A_1 and A_2, etc. Therefore, considering the uppermost waveform, from Figure 13.4 it can be seen that $A_1 = 1$ and $A_2 = 0$, where negative pulses are considered to have a logical value of 0. Consulting Table 13.1 shows that the output of the encoder has an amplitude of +1, consequently a positive half-cycle of a sine wave with amplitude 1 results. The next symbol period uses A_9 and A_{10}, which are both logical 0; Table 13.1 shows the output is -1, and accordingly a sine half-wave of amplitude -1 is observed. This allows 8 bits per symbol to be encoded onto a 16-level QAM constellation.

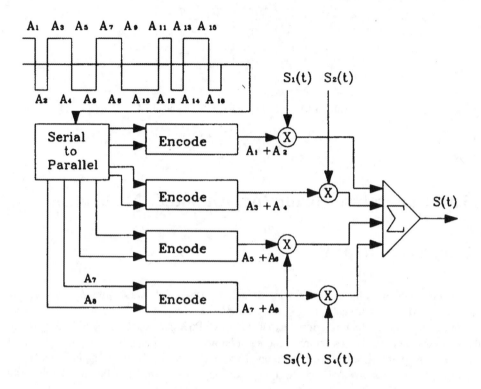

Figure 13.4: Structure of a quadrature-quadrature amplitude modulator

13.3.2   Star 16-QAM

The square constellation Q^2AM with 8 bits/symbol maps almost directly onto the star 16-QAM system. Instead of considering the I and Q channels separately, the system is considered to have two separate differential encoders. This is shown diagrammatically in Figure 13.6. The incoming pulse stream is split into two streams, each of which has 4 bits per symbol. Both streams are fed to identical differential encoders which both internally derive an amplitude and phase angle representing their differentially encoded data. The amplitude is designated $Z_1(t)$ and $Z_2(t)$ for differential coders 1 and 2, respectively, and the corresponding phases are $\theta_1(t)$ and $\theta_2(t)$.

The encoding scheme differentially encodes the first bit onto the amplitude and differentially Gray encodes the remaining three bits onto the phase. This information is then converted into two pairs of values representing I and Q magnitude for each differential coder. Thus $I_1(t) = Z_1(t) \sin \theta_1(t)$, $Q_1(t) = Z_1(t) \cos \theta_1(t)$ and so on. Note that the star 16-QAM system allows nine possible values of $I(t)$ and $Q(t)$. $I_1(t)$ is then encoded onto a sinusoidal pulse while $I_2(t)$ from differential encoder 2 is encoded onto a cosinusoidal pulse. These are then added and multiplied by $\cos w_c t$ to give the I carrier. A similar procedure is carried out for the Q channel.

At the receiver there are four matched filters, two with sinusoidal impulse responses and two with cosinusoidal impulse responses. These form two pairs used for the I and Q channels,

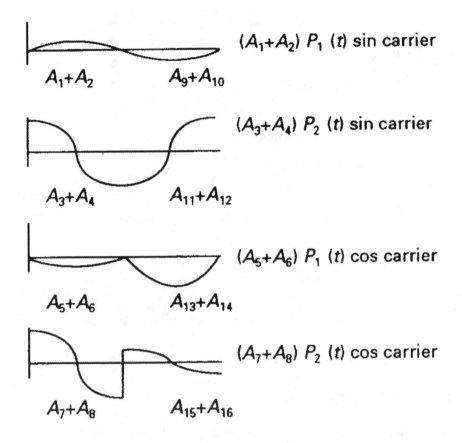

(A_1+A_2) P_1 (t) sin carrier

A_1+A_2 A_9+A_{10}

(A_3+A_4) P_2 (t) sin carrier

A_3+A_4 $A_{11}+A_{12}$

(A_5+A_6) P_1 (t) cos carrier

A_5+A_6 $A_{13}+A_{14}$

(A_7+A_8) P_2 (t) cos carrier

A_7+A_8 $A_{15}+A_{16}$

Figure 13.5: Pulse shapes for the Q^2AM modulator

each pair consisting of a sine and cosine filter. After the I and Q channels are separated by the quadrature mixers, each is simultaneously applied to the matched sine and cosine filter pairs for the appropriate quadrature arms. Because the encoding consisted of a sine and cosine pulse on each channel, by this matched filtering the pulses with their appropriate magnitude and polarity can be separated. The outputs of these matched filters are fed to the appropriate differential decoders which are configured so as to reconstruct the original data stream.

Simulations using this 16-Q^2AM system based on different orthogonal modulation pulses in an infinite bandwidth channel gave an identical BER to a 16-QAM system using a specific sinusoidal pulse shaping over one symbol alone. It should be noted that such sinusoidal pulse shaping would not normally be used; we would employ some form of raised cosine pulse shaping [405]. Conventional pulse shaping techniques have been considered in Chapter 4.

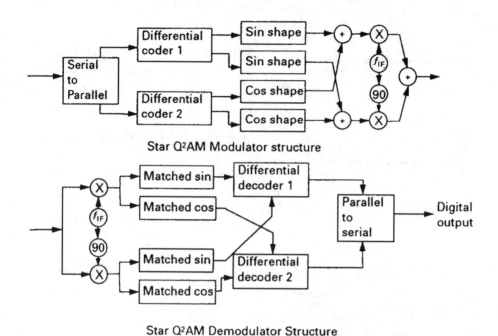

Star Q²AM Modulator structure

Star Q²AM Demodulator Structure

Figure 13.6: Star Q^2AM modulator and demodulator structure

13.4   Spectral Efficiency

As expected, there is a penalty to pay for the doubling of the data throughput. The sharp changes caused mainly by the step change in the cosinusoidal pulse, as can be seen in a typical time domain waveform shown in Figure 13.7, cause considerable spectral spreading.

The spectral efficiency of Q^2AM can be improved by using different orthogonal pulse shapes, and then independently filtering these pulse shapes in order to reduce the sharp changes at modulation instants necessary in order to achieve pulse shape orthogonality. In [549] a number of transmit and receiver filter pairs are suggested for this purpose. It is easy to smooth the sinusoidal pulse, but there are severe difficulties with the cosinusoidal pulse due to the sharp changes that occur at the end of some symbol periods. When bandlimiting filtering is carried out, it "smears" these changes significantly, considerably distorting the modulated signal waveform and increasing the BER. Essentially this filtering reduces the orthogonality between the data pulses. QAM is more sensitive than QPSK as more information has been encoded onto each pulse, and severe bandlimiting renders this information inaccurate.

13.5   Bandlimiting 16-Q^2AM

In order to make an accurate comparison between previous 16-QAM systems and 16-Q^2AM systems we need to compare their bandlimited performance.

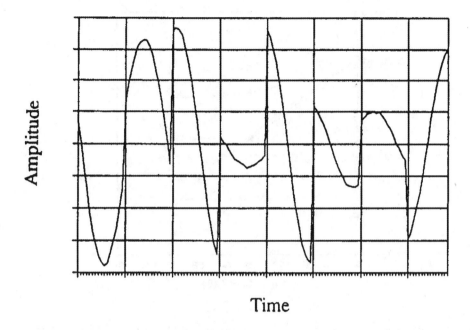

Figure 13.7: Q^{2}AM I channel time domain waveforms

Three bandlimited filtering systems are examined. The first system was the raised cosine pulse shaping of the data bits over five symbols prior to modulation as suggested in [405]. This makes use of a generalised equation for the filter impulse response given as

$$g(t) = 1 - A\cos(wt) + B\cos(2wt) - C\cos(3wt)$$
$$0 < t < LT \qquad (13.9)$$

where $w = 2\pi/LT$ and LT is the duration of the impulse response. The recommended coefficients for the transmit filter are given in Reference [405] as $A = 2.0$, $B = 1.823$, $C = 0.823$ and $L = 5$. The receiver filter is the corresponding matched filter for this response, but as the response is symmetrical about $t = 0$ it has an identical impulse response. Also considered was the non-linear filtering (NLF) system as suggested by Feher [103] and discussed in Chapter 4. This system fits a raised cosine segment between adjacent symbols if they have differing amplitudes, and a fixed level if they are of the same magnitude. Finally Q^{2}AM with the previously described simple unsmoothed sinusoidal and cosinusoidal pulse shapes was also included as a benchmark.

Since Q^{2}AM has twice the throughput of QAM, it can be considered a superior system if it can operate with the same BER as QAM but in less than *twice* the bandwidth. Indeed, it could be argued, especially at high bit rates, that Q^{2}AM is superior if it can operate with the same BER in just over twice the bandwidth, as the lower symbol rate will tend to reduce the effects of intersymbol interference (ISI) and thus dispense with equalisation or render it simpler and more accurate.

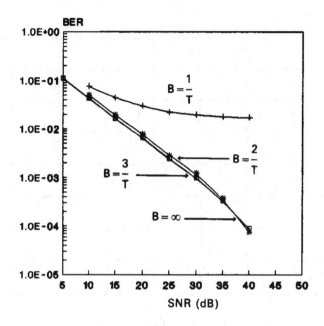

Figure 13.8: BER versus channel SNR performance of star 16-QAM at 30 mph, 1.9 GHz, 2 Msymbols/s using raised cosine filtering and various lowpass bandwidths

13.6   Performance of Bandlimited 16-QAM and 16-Q^{2}AM

Initial simulations using our star QAM modem were performed at a transmission rate of 2 MSymbols/s over a flat fading channel where intersymbol interference (ISI) did not occur. This situation would correspond to a microcellular or indoor environment. Lower transmission rates than 2 Msymbol/s merely had the effect of increasing the residual BER as shown in previous chapters due to the inability of the differential detector to track the rapidly changing fading envelope. The vehicular speed of the mobile station (MS) was assumed to be 30 mph. Higher MS speeds increased the residual BER, while lower speeds reduced it. The MS speed could also be directly traded off against transmission rate, so if we were to reduce our MS speed by half to 15 mph, we would be able to reduce our transmission rate by half to 1 Msymbol/s, and get identical results (assuming no ISI). The carrier frequency was 1.9 GHz.

The performance of the raised cosine filtering is shown in Figure 13.8. In Reference [405] the bandwidth of QPSK with this pulse shaping is quoted as being $1.2/T$, where T is the symbol rate. That this bandwidth is approximately true for star 16-QAM is substantiated by the results shown. When the bandwidth of the equipment filters was 3 dB down at $B = 1/T$, the performance decreased dramatically from the infinite bandwidth case. When the 3 dB point was moved to $B = 2/T$, the performance was only about 1 dB worse than the infinite bandwidth case, and by increasing the 3 dB point to $B = 3/T$, the performance of the infinite bandwidth system was essentially achieved, as shown in Figure 13.8.

Figure 13.9 shows the performance of the NLF 16-QAM system under the previous propagation conditions with various amounts of bandlimiting. Note that the BER of the non-

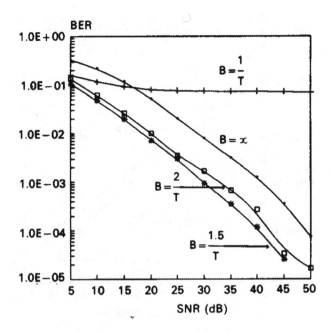

Figure 13.9: BER versus channel SNR performance of star 16-QAM at 30 mph, 1.9 GHz, 2 Msymbols/s using raised NLF filtering and various lowpass bandwidths

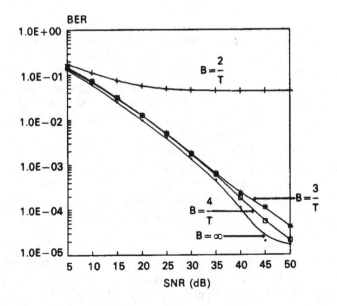

Figure 13.10: BER versus channel SNR performance of star 16-Q^{2}AM at 30 mph, 1.9 GHz, 2 Msymbols/s using various lowpass bandwidths

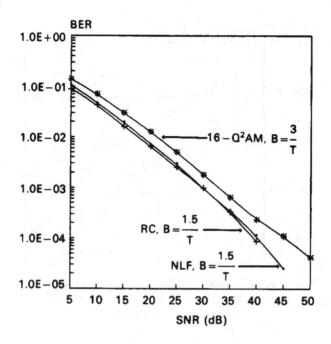

Figure 13.11: BER versus channel SNR comparison of star 16-QAM and 16-Q^{2}AM at 30 mph, 1.9 GHz, 2 Msymbols/s using filtering at $B = 1.5/T$ and $B = 3/T$

linearly filtered but otherwise unbandlimited 16-QAM scheme was significantly higher than the case of raised cosine (RC) pulse shaping. This was because of the use of a matched filter in the raised cosine filtered receiver compared to simple sampling in the NLF 16-QAM scheme. NLF schemes cannot employ matched filtering, instead they must rely on lowpass filtering of the noise at the receiver. In our simulations we used the same receive filter as transmit filter. There was an optimum filter bandwidth for the case where the lowpass filter removes most of the noise without smearing the pulses significantly. It can be seen from the figure that this occurred at a 3 dB down cut-off frequency of about $B = 1.5/T$. When the 3 dB point was decreased to $B = 1/T$, significant performance degradation occurred.

Figure 13.10 shows the performance of the 16-Q^{2}AM system with various amounts of bandlimiting. The performance was approximately that of the infinite bandwidth channel for the 3 dB point above $B = 2/T$, but when this point was reduced below $B = 2/T$ the BER became unacceptable.

In Figures 13.11 and 13.12 we compare the performance of the three different filtering systems all operating at *identical bit rates*. This means that the symbol rate of Q^{2}AM is half that of the other two systems. Hence in Figure 13.11 we compare the performance of raised cosine and NLF with the 3 dB point at $B = 1.5/T$ to Q^{2}AM with the 3 dB point at $B = 3/T$. This comparison is carried out in order to show how Q^{2}AM performs relative to a near-optimally filtered 16-QAM modem using raised cosine pulse shaping and a practically filtered QAM modem using the NLF system. The same format is adhered to in Figure 13.12

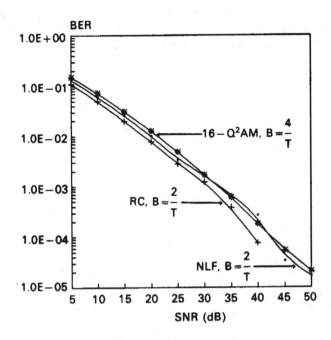

Figure 13.12: BER versus channel SNR comparison of star 16-QAM and 16-Q^{2}AM at 30 mph, 1.9 GHz, 2 Msymbols/s using filtering at $B = 2/T$ and $B = 4/T$

where we consider 3 dB bandwidths of $B = 2/T$ and $B = 4/T$ for the various systems. When bandlimited to $B = 2/T$ and $B = 4/T$ the performance of all three systems is virtually identical, although the Q^{2}AM scheme is slightly down on the other two. When the bandlimiting is tightened to $B = 1.5/T$ and $B = 3/T$, a level at which our preferred NLF scheme performs best, the performance of the Q^{2}AM scheme deteriorates to some 3 dB below that of the other two schemes. This is more significant and it suggests that Q^{2}AM is only worth implementing if ISI is causing serious problems.

13.7   Summary

This chapter considered how quadrature-quadrature amplitude modulation can be successfully applied for transmissions over Rayleigh fading channels. This scheme is termed Q^{2}AM, and for a given symbol rate Q^{2}AM requires approximately twice the channel bandwidth of QAM but operates at twice the bit rate. Alternatively, for the same channel bandwidth, Q^{2}AM operates at half the transmission rate compared to QAM, causing any ISI to be spread over fewer symbols. Decreasing the symbol rate by a factor of 2 may result in Q^{2}AM not needing equalisation when QAM does. In this case Q^{2}AM is advantageous. Otherwise it can be concluded that the necessary increase in complexity and decrease in performance are not warranted.

   In the next chapter we go on to further examine the issue of spectral efficiency which was touched upon in this chapter when the need to compare different schemes within the same bandwidth was described. The next chapter considers the spectral efficiency of QAM when employed in cellular communications and compares it to other modulation schemes.

Chapter **14**

Area Spectral Efficiency of Adaptive Cellular QAM Systems

14.1 Introduction

Within fixed communications links where co-channel interference is typically insignificant, the achievable bandwidth efficiency can be directly increased upon increasing the number of bits per symbol, provided that the channel's SNR is sufficiently high for maintaining the required target integrity quantified in terms of the BER. Hence in propagation environments, where no co-channel interference is encountered, the number of bits per symbol transmitted determines the achievable bandwidth efficiency. For example, when considering a BPSK system using a 1 bit/symbol modulation scheme and a 4 bits/symbol QAM arrangement, the attainable bandwidth efficiency may be expected to increase by a factor of four by the QAM scheme. It is this increased bandwidth efficiency, which has promoted the widespread use of QAM in fixed link based systems.

However, in mobile radio applications the situation is less clear. Here we define a different efficiency which we term as area spectral efficiency. We define this as the number of channels that can be provided in a cellular system relaying on frequency reuse. For cellular systems, the system's user capacity is directly related to the area spectral efficiency, but it is only loosely related to the above-mentioned bandwidth efficiency. This is because changing from 1 bit/symbol to 4 bits/symbol modulation will typically require an increased signal to co-channel interference ratio. To obtain this increased signal to co-channel interference ratio, the same carrier frequency must be assigned to BSs more sparsely in the cellular re-use structure. This implies that the frequency reuse cluster size (i.e. the number of orthogonal frequency sets available within the complete system) will have to be increased, assuming a constant overall system bandwidth. Increasing the cluster size will reduce the number of carrier frequencies available within each cell. This will reduce the number of channels available in each cell. Therefore, in order to have a clear picture of the true effect of a change of modulation scheme in mobile radio applications we must consider the area spectrum efficiency, rather than only the bandwidth efficiency.

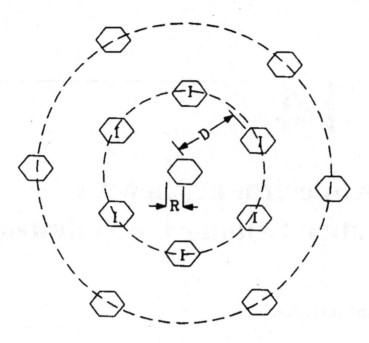

Figure 14.1: Idealised cell structure with interferers

According to Lee [552] the calculation of spectrum efficiency for cellular systems proceeds as follows. Firstly, an idealised cell structure with six significant interferers is postulated as shown in Figure 14.1. Regardless of the cluster size there will always be six significant interferers. These may not all be equidistant but for simplicity it is assumed they are all at their average distance from the transmitter. This leads to a small inaccuracy which will not significantly affect the result. With this assumption the signal-to-interference ratio SIR is given by

$$SIR = \frac{S}{\sum_{k=1}^{6} I_k + n} \tag{14.1}$$

where S is the signal power, I_k is the interference power from the kth interfering cell, and n is the local noise. Based on a logarithmic path loss model, neglecting noise, and assuming that the signal level from all the interferers is approximately equal, Equation 14.1 may be rewritten as

$$SIR = \frac{R^{-\gamma}}{6.D^{-\gamma}} \tag{14.2}$$

where R is the radius of a cell, D the distance between interfering cells and γ the path loss law. For more details on propagation path loss models the interested reader is referred to Chapter 2. A diagrammatic representation of the interferers is shown in Figure 14.1 where the centre hexagon is the wanted cell and the significant interferers are labelled with the letter I. It then follows that

$$\frac{D}{R} = (6.SIR)^{1/\gamma}. \tag{14.3}$$

For a hexagonal cell configuration [552] we have

$$\frac{D}{R} = \sqrt{3K} \tag{14.4}$$

where K is the number of cells in a frequency reuse pattern. Equating Equations 14.3 and 14.4 yields

$$K = \frac{1}{3}(6.SIR)^{2/\gamma}. \tag{14.5}$$

The radio capacity in terms of the number of channels available is defined by

$$m = \frac{B_t}{B_c.K} \tag{14.6}$$

where B_t is the total bandwidth available and B_c is the bandwidth required per channel. Substituting for K from Equation 14.5 in terms of SIR and assuming that $\gamma = 4$, gives

$$m = \frac{B_t}{B_c\sqrt{\frac{2}{3}SIR}} \tag{14.7}$$

which is known as the radio capacity equation for a hexagonal cellular structure with a cluster size of K.

14.2 Spectrum Efficiency in Conventional Cells

We have evaluated Equation 14.7 for a wide variety of modulation schemes using a computer simulation. This simulation considered frequency shift keying (FSK), phase shift keying (PSK), and fixed-level and variable-level QAM. The QAM simulations are for differential star QAM introduced in Chapter 10, and the variable-level QAM schemes are for variable rate star QAM introduced in Chapter 22. The 8-level and 16-level star QAM constellations have two amplitude rings, whereas the 32-level and 64-level constellations have four rings.

The simulation program used includes clock and carrier recovery, co-channel and adjacent channel interferers using a range of modulation schemes, Rayleigh fading at a variety of symbol rates, and a wide range of convolutional and block coding systems. Shadow fading is not included as it does not affect the results in any way. For the simulations, near-optimal root raised cosine filtering at transmitter, receiver and co-channel transmitter was used. The interference was generated using a single interferer with the same modulation scheme and filtering as the transmitter. This was subject to a Rayleigh fading path between interferer and receiver, and arranged such that the average interference power was as selected in the simulation parameters. A carrier frequency of 1.8 GHz, a mobile speed of $15\ m/s \approx 30$ mph and a data rate of 25 kBd were assumed while the SNR was fixed at 40 dB.

Using this tool the BER against SIR performance for a range of modulation schemes, each using a range of coding techniques, was established. This led to over 60 performance curves, the most important of which are shown in Figures 14.2 and 14.3. In these figures BCH coders have been used, where the numbers in brackets refer to the number of bits corrected per 63-bit block, and the coding rate is 0.9, 0.81, 0.71, and 0.62 for correction powers of 1 to 4 respectively. Curves for square QAM have been omitted due to its poor performance, although we note that proposed schemes using pilot symbol assisted modulation (PSAM) as discussed in Chapter 9 are expected to provide similar efficiency levels to that of star QAM, albeit at a higher complexity. The curves for 16 FSK have also been omitted, again due to poor performance.

Using these curves, spectrum efficiency for a variety of BERs was established. In order to achieve this, nine BER values spanning the range 5×10^{-2} to 1×10^{-4} were selected. For each modulation scheme and each FEC coding scheme, the SIR at which these error rates were achieved, if it was possible to achieve them, was read off the performance curves drawn previously. The SIR and bandwidth required after FEC coding were then fed into Equation 14.7. The resulting number, m, is the number of channels available within an idealised cellular system, neglecting frequency-domain guard band space between adjacent channels. For any BER and any given modulation scheme there were five curves relating to the four error control codes used, and to no error control coding. Each of these led to a different channel bandwidth requirement B_c and SIR, leading to a different number of channels supported, or radio capacity m. The highest value of m was selected. This meant that at different BERs, different error control codes are optimal in spectral efficiency terms for any one particular modulation scheme. Explicit specification of the error control code favoured for every condition would be unduly complicated and unnecessary as can be readily inferred from the graphs supplied. In general, low power error control was best for high BERs and high power error control for low BERs.

The graphs showing the radio capacity in terms of the number of channels m supported by different modems at different BER values are given in Figure 14.4. In the figures "nPSK", where n is integer, refers to n-level PSK [406], similarly for FSK, and for QAM "n fixed" refers to n-level star QAM where the number of levels are fixed over the duration of the transmission, whereas the variable-level scheme employed an average of 4 bits per symbol. From these graphs we can see that the performance of all the schemes is broadly similar. The 4PSK scheme is superior for all values of BER, closely followed by the variable rate scheme. There is a tendency for the schemes which use fewest modulation levels to give the best performance, so that 8-PSK is superior to 16-PSK and 16-QAM. Schemes which use the same number of levels, such as 16-PSK and 16-QAM, had broadly similar performances.

14.3 Spectrum Efficiency in Microcells

While the above equations hold for idealised hexagonal cellular systems, they are not applicable to microcellular systems. This is because in microcells the clusters are different from those in conventional cells, and the interference is not so straightforward to calculate. This necessitates a change in Equation 14.7. It was with this in mind that an investigation was undertaken into spectrum efficiency in microcells. With this incentive a number of microcell clusters were constructed on an idealised grid pattern based on maps of New York City and the interference resulting from these clusters was established. The results were then applied to Lee's formula which allowed conclusions to be drawn as to the most appropriate modulation schemes for microcellular systems. This approach is now described in detail.

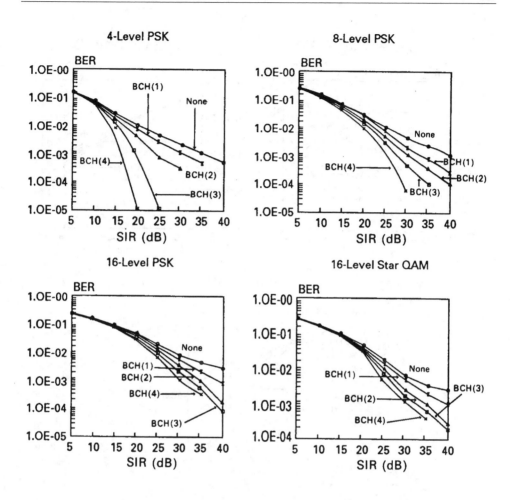

Figure 14.2: BER versus SIR performance of 4, 8, 16 PSK and 16-QAM at 30 mph, 1.8 GHz, 25 kBd, using SNR = 40 dB and various BCH codes

14.3.1 Microcellular Clusters

In a microcell, the interference from any given co-channel interferer varies throughout the microcell, due to the small size and rapid decrease in signal level as the mobile moves away from the BS. As a result of this, it is difficult to calculate SIRs for given clusters. Instead we resorted to a microcellular prediction tool developed at Multiple Access Communications (MAC) and known as Microcellular Design System (MIDAS). This tool can predict both signal and interference levels for any microcellular cluster in both practical and idealised situations and has been proven by comparison with a large database of microcellular measurement data established at MAC.

An idealised grid map based on the layout of New York City, with 100 m blocks and 20 m wide streets was loaded into MIDAS and a range of predictions were performed at both 900 MHz and 1.8 GHz. In order to establish the microcell size, the cell boundary was set

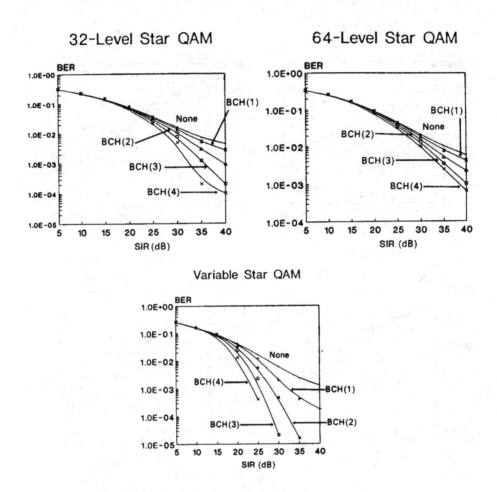

Figure 14.3: BER versus SIR performance of 32, 64 and variable rate star QAM at 30 mph, 1.8 GHz, 25 kBd, using SNR = 40 dB and various BCH codes

based approximately on the minimum received power of the CT-2 cordless telephone system. With a 10 mW transmitter and 0 dB gain antennas, the effective radiated power was 10 dBm. Given a minimum signal level of -95 dBm at which the data can be received and a fast fading margin of 20 dB, the minimum received signal level was -75 dBm, relating to a path loss of 85 dB. This process was visualised in Figure 2.13. With microcell BSs placed midway along blocks, which is a suitable set of locations to produce a cluster with regular frequency assignment, this path loss of 85 dB gives a range of 180 m at 1.8 GHz, giving a microcell length of 360 m. The path loss versus distance relationship was established using the MIDAS prediction system with its six-breakpoint model rather than the less accurate $\gamma = 4$ fourth power law. This requires a microcell every third block as shown in Figure 14.5, which also illustrates the BS pattern for a four-cell cluster. In this figure, which is not drawn to scale, the square areas represent city blocks and the letters represent a BS location and its specific

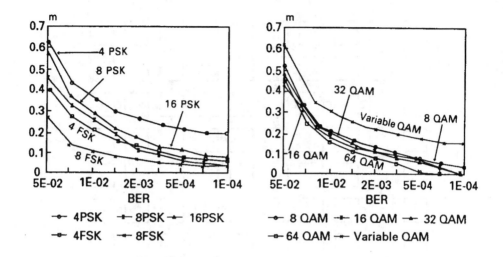

Figure 14.4: Spectrum efficiency versus BER performance of a range of modulation schemes at 30 mph, 1.8 GHz, 25 kBd, using SNR = 40 dB

frequency assignment.

At 900 MHz the same path loss leads to a range of 300 m giving a microcell length of 600 m and a BS every fifth block as shown in Figure 14.6, which also shows the frequency assignment for a six-cell cluster, where the same representation as in Figure 14.5 applies.

Using these BS patterns we considered the SIRs. This was simply achieved by placing the BSs appropriately within the MIDAS modelling package, and then selecting the SIR option. The package produces a visual display of the SIR on a $1\ m^2$ basis which is used to determine the worst-case SIR. At 1.8 GHz using four cells per cluster only the BSs along the same street cause significant interference at a level of 25-30 dB close to the microcell edge. In contrast the use of six-cell clusters leads to there being no significant interference, since with three frequency sets available for the horizontal streets, BSs can be spaced further apart on the same street and also staggered on neighbouring streets to cause less interference. With $K = 6$ the microcell becomes SNR limited as opposed to SIR limited.

At 900 MHz the interference from BSs using the same frequency on parallel adjacent streets becomes more significant, due to the increase in reflection and diffraction around corners at this frequency. A similar method to that used to construct the cluster at 1.8 GHz was used at 900 MHz, but with the wider microcell spacing caused by increased propagation distances at the lower frequency. With $K = 4$, areas where the SIR dropped to 20 dB were experienced, this SIR was lower than that for 1.8 GHz, due to the higher diffraction and reflection of energy around corners. With $K = 6$, the system again became SNR limited.

14.3.2 System Design for Microcells

Communications systems are generally designed for worst-case propagation. Consequently, despite the fact that microcells only experience significant interference in certain places, a communications system should nevertheless be designed to cope with this level of inter-

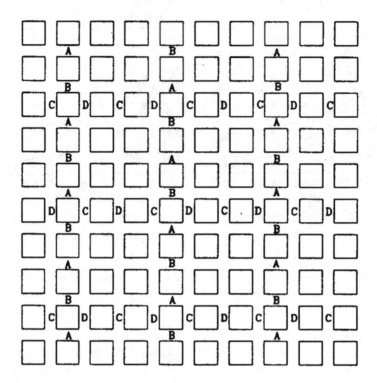

Figure 14.5: Microcellular BS pattern at 1.8 GHz for a four-cell cluster using MIDAS propagation
prediction

ference. It is possible that future designers will find ways of overcoming or reducing the
interference even in four-cell clusters, by a combination of dynamic channel allocation, and
switching the frequencies of mobiles experiencing significant interference to different frequencies within the same set. However, here we only consider a simple system with fixed
allocation, and investigate its results. More intelligent systems will tend to reduce the cluster
size.

14.3.3   Microcellular Radio Capacity

When deriving his equation for spectrum capacity, Lee used a formula for conventional cells
linking SIR with the number of cells per cluster, K. We performed the above work in an
effort to produce a similar equation for microcells, but because of the limited range of K
that it is practical to use, we feel it inappropriate to produce a formula of this sort. Instead
we substitute our results directly into Equation 14.6 for the values of K for which we have
results.

Hence, for $K = 4$ we use the simulation results pertaining to 20 dB SIR, whereas for
$K = 6$, where transmission is SNR limited, we use the simulation results for a range of SNRs,
particularly 20, 30 and 40 dB. The BSs are maintained in the same position for each of the

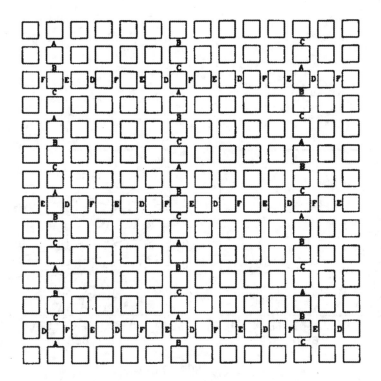

Figure 14.6: Microcellular BS pattern at 900 MHz for a six-cell cluster using MIDAS propagation prediction

SNRs, and the operation of the mobiles with each SNR is examined.

Equation 14.6 is evaluated for both $K = 4$ and $K = 6$, and the cluster giving the highest number of channels is selected for each BER and each modulation scheme. In most cases a cluster size of 4 was selected. The microcellular SNR is of great importance in deciding upon the optimum modulation scheme and depends on transmitter power and receiver sensitivity for any given system. Here we give results for a range of SNRs, always using the most efficient cluster size for each SNR value to derive the number of available channels.

14.3.4   Modulation Schemes for Microcells

Given SNRs of 20, 30 and 40 dB and assuming an SIR of 20 dB for the four-cell cluster and infinite for the six-cell cluster, we can consider the most efficient modulation schemes to use within the microcellular environment. Graphs showing the BER against number of channels are given in Figure 14.7.

These graphs have been derived in the same way as the graphs for the conventional cells, but with appropriate SIRs assumed. The same procedure was followed as with macrocells in that for each BER along the x-axis, the SIR at which this was achieved was read off Figures 14.2 and 14.3. However, if this SIR was greater than that available within the microcell

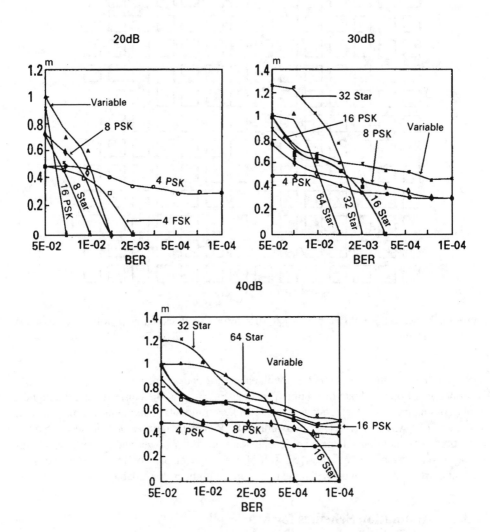

Figure 14.7: Spectrum efficiency versus BER performance in microcells of a range of modulation schemes at 30 mph, 1.8 GHz, 25 kBd, using SNR = 20 dB, 30 dB and 40 dB

for a given cluster size the scheme was disallowed. If it was less, the scheme was allowed at the cluster size under consideration. Therefore, achieving the BER with an SIR significantly lower than that available was disadvantageous compared to achieving it with a SIR closer to that available.

This procedure was followed for $K = 4$ and $K = 6$, and the cluster size giving the best result used at each point. The QAM schemes were star QAM, using two concentric circles with four and eight amplitude points per ring for the 8-level and 16-level schemes, respectively. Curves with very poor performance have been omitted for clarity. In all cases a Rayleigh fading channel is assumed with transmission at 1.8 GHz, mobile speed of 15 m/s and data rate of 25 kBd. The results at 900 MHz are similar because almost the same SIR and SNR values apply to both frequencies. Differential transmission and reception have been used along with near-optimal root raised cosine filtering, modified early-late clock recovery and with the interferer using the same modulation scheme as the transmitter. The same nomenclature has been used as in Figure 14.4.

For the 20 dB SNR case, variable QAM is most effective for high BERs, with 4-PSK giving the best performance for BERs below 1×10^{-2}. When the SNR is 30 dB the variable-level QAM is superior for all but the highest BERs. If only fixed-level schemes are considered then 8-PSK proves slightly better than 4-PSK and 4-FSK for low BERs. For the 40 dB SNR case for high BERs the higher-level modulation schemes, namely 32-QAM and 64-QAM prove best. At BERs below 2×10^{-4} the variable rate QAM schemes are in line with the 32-level QAM, with the 4-level and 8-level schemes giving substantially worse performance due to their inability to make the best use of the SNR and SIR available. We note that if variable rate QAM giving an average of 5 bits/symbol had been employed, we would have expected it to outperform 32-level QAM in all scenarios.

14.4 Summary

In this chapter the interference levels that might be expected in a range of conventional cellular and microcellular clusters have been considered. Simulation work has suggested that for conventional cell sizes the lower-level modulation schemes such as 4-PSK are best, with variable rate QAM schemes only incurring a slight penalty. For microcells with communication at both 900 MHz and 1.8 GHz, four-cell or six-cell clusters are advocated depending on the SNR expected. Based on the expected SNRs and interference levels we then surmised that variable rate QAM schemes are often superior to the other modulation schemes considered here. For low BERs and SNR = 20 dB, 4-PSK provides the best performance; whereas for high BERs, particularly when the SNR is as high as 30 or 40 dB, then 32-level and 64-level star QAM are most suitable.

Having considered the performance of the modulation scheme in isolation, in the next chapter we proceed to investigate QAM and other modulation schemes when used for the transmission of speech, and investigate means whereby the different bit error rates of different bits on the same QAM constellation can be linked advantageously with the properties of modern speech codecs.

Part III

Advanced QAM:
Adaptive versus Space-Time
Block- and Trellis-Coded OFDM

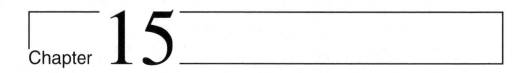

Chapter **15**

Introduction to Orthogonal
Frequency Division Multiplexing

15.1 Introduction

In this chapter we examine orthogonal frequency division multiplexing (OFDM) as a means of dealing with the channel-induced linear distortion problems encountered when transmitting over a dispersive radio channel. The fundamental principle of orthogonal multiplexing originates from Chang [143], and over the years a number of researchers have investigated this technique [144–146, 149–154, 224–226]. Despite its conceptual elegance, until recently its deployment has been mostly limited to military applications due to implementational difficulties. However, it has recently been adopted as the new European digital audio broadcasting (DAB) standard, and this consumer electronics application underlines its significance as a broadcasting technique [147, 148, 155, 337, 338].

In the OFDM scheme of Figure 15.1 the serial data stream of a traffic channel is passed through a serial-to-parallel convertor which splits the data into a number of parallel channels. The data in each channel is applied to a modulator, such that for N channels there are N modulators whose carrier frequencies are f_0, f_1, ..., f_{N-1}. The difference between adjacent channels is Δf and the overall bandwidth W of the N modulated carriers is $N\Delta f$, as shown in Figure 15.2.

These N modulated carriers are then combined to give an OFDM signal. We may view the serial-to-parallel convertor as applying every Nth symbol to a modulator. This has the effect of interleaving the symbols into each modulator, e.g. symbols S_0, S_N, S_{2N}, ... are applied to the modulator whose carrier frequency is f_1. At the receiver the received OFDM signal is demultiplexed into N frequency bands, and the N modulated signals are demodulated. The baseband signals are then recombined using a parallel-to-serial convertor.

In the more conventional approach the traffic data is applied directly to the modulator operating with a carrier frequency at the centre of the transmission band f_0, ..., f_{N-1}, i.e. at $(f_{N-1} + f_0)/2$, and the modulated signal occupies the entire bandwidth W. When the data is applied sequentially the effect of a deep fade in a mobile channel is to cause burst

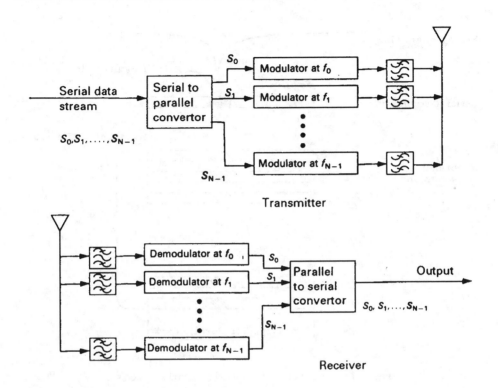

Figure 15.1: Basic OFDM system

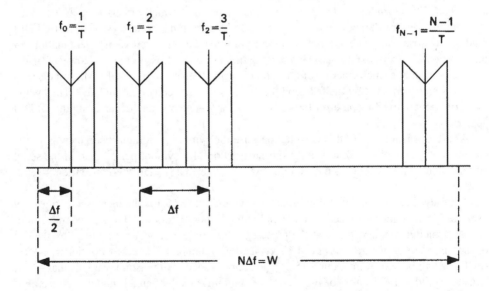

Figure 15.2: FDM frequency assignment

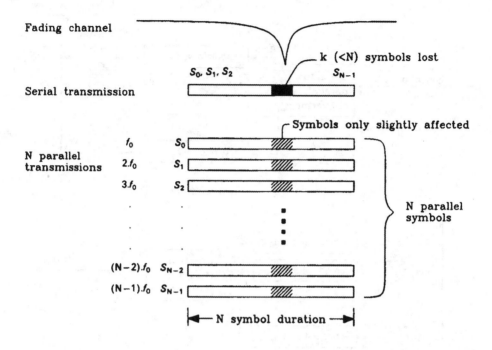

Figure 15.3: Effect of a fade on serial and parallel systems

errors. Figure 15.3 shows the serial transmission of symbols S_0, S_1, ..., S_{N-1}, while the solid shaded block indicates the position of the error burst which affects $k < N$ symbols. By contrast, during the N-symbol period of the conventional serial system, each OFDM modulator carries only one symbol, and the error burst causes severe signal degradation for the duration of k serial symbols. This degradation is shown cross-hatched. However, if the error burst is only a small fraction of the symbol period then each of the OFDM symbols may only be slightly affected by the fade and they can still be correctly demodulated. Thus while the serial system exhibits an error burst, no errors or few errors may occur using the OFDM approach.

A further advantage of OFDM is that because the symbol period has been increased, the channel delay spread is a significantly shorter fraction of a symbol period than in the serial system, potentially rendering the system less sensitive to ISI than the conventional serial system.

A disadvantage of the OFDM approach shown in Figure 15.1 is the increased complexity over the conventional system caused by employing N modulators and filters at the transmitter and N demodulators and filters at the receiver. This complexity can be reduced by the use of the discrete Fourier transform (DFT), typically implemented as a fast Fourier transform (FFT), as will be shown in Section 15.3. The simple conceptual reason for this is that – as it will be argued in the context of Figure 15.4 – the OFDM sub-channel modulators use a set of harmonically related sinusoidal and cosinusoidal sub-channel carriers, just like the basis functions of the DFT.

15.2 Principles of QAM-OFDM

The simplest version of the basic OFDM system has N sub-bands, each separated from its neighbour by a sufficiently large guard band in order to prevent interference between signals in adjacent bands. However, the available spectrum can be used much more efficiently if the spectra of the individual sub-bands are allowed to overlap. By using coherent detection and orthogonal sub-band tones the original data can be correctly recovered.

In the system shown in Figure 15.4, the input serial data stream is rearranged into a sequence $\{d_n\}$ of N QAM symbols at baseband. Each serial QAM symbol is spaced by $\Delta t = 1/f_s$ where f_s is the serial signalling or symbol rate. At the nth symbol instant, the QAM symbol $d(n) = a(n) + jb(n)$ is represented by an in-phase component $a(n)$ and a quadrature component $b(n)$. A block of N QAM symbols are applied to a serial-to-parallel convertor and the resulting in-phase symbols $a(0), a(1), ..., a(N-1)$, and quadrature symbols $b(0), b(1), ..., b(N-1)$ are applied to N pairs of balanced modulators. The quadrature components $a(n)$ and $b(n)$, $n = 0, 1, ..., N-1$, modulate the quadrature carriers $\cos w_n t$ and $\sin w_n t$, respectively. Notice that the signalling interval of the sub-bands in the parallel system is N times longer than that of the serial system giving $T = N\Delta t$, which corresponds to an N-times lower signalling rate. The sub-band carrier frequencies $w_n = 2\pi n f_0$ are spaced apart by $w_0 = 2\pi/T$.

The modulated carriers $a(n)\cos w_n t$ and $b(n)\sin w_n t$ when added together constitute a QAM signal, and as $n = 0, 1, ..., N-1$ we have N QAM symbols at RF, where the nth QAM signal is given by:

$$X_n(t) = a(n)\cos w_n t + b(n)\sin w_n t \qquad (15.1)$$
$$= \gamma(n)\cos(w_n t + \psi_n)$$

where

$$\gamma(n) = \sqrt{a^2(n) + b^2(n)} \qquad (15.2)$$

and

$$\psi_n = \tan^{-1}\left(\frac{b(n)}{a(n)}\right) \qquad (15.3)$$
$$n = 0, 1, ..., N-1.$$

By adding the outputs of each sub-channel signal $X_n(t)$ whose carriers are offset by $w_0 = 2\pi/T$ we obtain the FDM/QAM signal

$$D(t) = \sum_{n=0}^{N-1} X_n(t). \qquad (15.4)$$

This set of N FDM/QAM signals is transmitted over the mobile radio channel. At the receiver this OFDM signal is demultiplexed using a bank of N filters to regenerate the N QAM signals. The QAM baseband signals $a(n)$ and $b(n)$ are recovered and turned into serial form $\{d_n\}$. Recovery of the data ensues using the QAM baseband demodulator and differential decoding.

In theory, such a system is capable of achieving the maximum transmission rate of

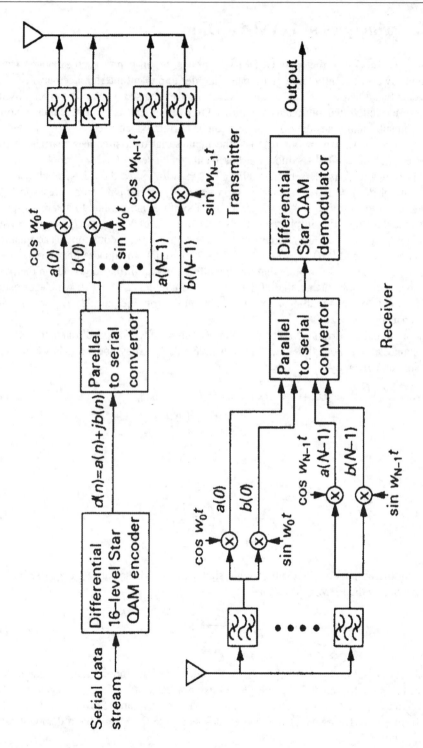

Figure 15.4: Detailed OFDM system

$log_2 \, Q \, bits/s/Hz$, where Q is the number of QAM levels. In practice, there is some spectral spillage due to adjacent frequency sub-bands which reduces this efficiency. Spectral spillage due to the sub-bands at the top and bottom of the overall frequency band requires a certain amount of guard space between adjacent users. Furthermore, spectral spillage between OFDM sub-bands due to the imperfections of each of the sub-band filters requires that the sub-bands be spaced further apart than the theoretically required minimum amount, decreasing spectral efficiency. In order to obtain the highest efficiency, the block size should be kept high and the sub-band filters made to meet stringent specifications.

One of the most attractive features of this scheme is that the bandwidth of the sub-channels is very narrow when compared to the communications channel's coherence bandwidth. Therefore, flat-fading narrowband propagation conditions apply. The sub-channel modems can use almost any modulation scheme, and QAM is an attractive choice in some situations.

15.3 Modulation by Discrete Fourier Transform [553, 554]

A fundamental problem associated with the OFDM scheme described is that in order to achieve high resilience against fades in the channels we consider, the block size, N, has to be of the order of 100, requiring a large number of sub-channel modems. Fortunately, it can be shown mathematically that taking the discrete Fourier transform (DFT) of the original block of N QAM symbols and then transmitting the DFT coefficients serially is exactly equivalent to the operations required by the OFDM transmitter of Figure 15.4. Substantial hardware simplifications can be made with OFDM transmissions if the bank of sub-channel modulators/demodulators is implemented using the computationally efficient pair of inverse fast Fourier transform and fast Fourier transform (IFFT/FFT).

The modulated signal $m(t)$ is given by

$$m(t) = \Re \left\{ b(t)e^{j2\pi f_0 t} \right\}, \tag{15.5}$$

where $b(t)$ is the equivalent baseband information signal and f_0 is the carrier frequency, as introduced in Equations 2.6, 2.9 and 2.10. Using the rectangular full-response modulation elements $m_T(t - kT) = rect \frac{(t-kT)}{T}$ "weighted" by the complex QAM information symbols $X(k) = I(k) + jQ(k)$ to be transmitted, where $I(k)$ and $Q(k)$ are the quadrature components, the equivalent baseband information signal is given by:

$$b(t) = \sum_{k=-\infty}^{\infty} X(k)m_T(t - kT), \tag{15.6}$$

where k is the signalling interval index and T is its duration. On substituting Equation 15.6 into Equation 15.5 we have:

$$m(t) = \Re \left\{ \sum_{k=-\infty}^{\infty} X(k)m_T(t - kT)e^{j2\pi f_0 t} \right\}. \tag{15.7}$$

Without loss of generality let us consider the signalling interval $k = 0$:

$$m_0(t) = m(t) rect \frac{t}{T},\tag{15.8}$$

where adding the modulated signals of the sub-channel modulators yields

$$m_0(t) = \sum_{n=0}^{N-1} m_{0n}(t).\tag{15.9}$$

Observe that the stream of complex baseband symbols $X(k)$ to be transmitted can be described both in terms of in-phase $I(k)$ and quadrature phase $Q(k)$ components, as well as by magnitude and phase. In case of a square-shaped constellation which was used in Chapter 9, $X(k) = I(k) + jQ(k)$ might be a more convenient formalism; for the star QAM constellation introduced in Chapter 10, $X(k) = |X(k)|e^{j\Phi(k)}$ appears to be more attractive.

If $X_n = X_{n,k=0}$ is the complex baseband QAM symbol to be transmitted via sub-channel n in signalling interval $k = 0$, then

$$m_0(t) = \begin{cases} \sum_{n=0}^{N-1} \Re\left\{X_{n,0}e^{j2\pi f_{0,n}t}\right\} & \text{for } |t| < \frac{T}{2}, \\ 0 & \text{otherwise.} \end{cases}\tag{15.10}$$

Bearing in mind that $m_0(t)$ is confined to the interval $|t| < \frac{T}{2}$ we drop the sampling interval index $k = 0$ and simplify our formalism to

$$m_0(t) = \sum_{n=0}^{N-1} \Re\left\{X_n e^{j2\pi f_{0n}t}\right\}.\tag{15.11}$$

When computing the $\Re$ part using the complex conjugate we have:

$$\begin{aligned} m_0(t) &= \sum_{n=0}^{N-1} \frac{1}{2}\left\{X_n e^{j2\pi f_{0n}t} + X_n{}^* e^{-j2\pi f_{0n}t}\right\} \\ &= \sum_{n=-(N-1)}^{N-1} \frac{1}{2} X_n e^{j2\pi f_{0n}t}, \end{aligned}\tag{15.12}$$

where for $n = 0, \ldots, N - 1$ we have:

$$X_{-n} = X_n{}^*, \quad X_0 = 0, \quad f_{0(-n)} = -f_{0n}, \quad f_{00} = 0.$$

This conjugate complex symmetric sequence is shown in Figure 15.5 in case of the square 16-QAM constellation of Figure 3.3, where both the I and Q components can assume values of ± 1 and ± 3.

We streamline our formalism in Equation 15.12 by introducing the Fourier coefficients F_n given by

$$F_n = \begin{cases} \frac{1}{2}X_n & if & 1 \le n \le N - 1 \\ \frac{1}{2}X_n{}^* & if & -(N-1) \le n \le -1 \\ 0 & if & n = 0 \end{cases}\tag{15.13}$$

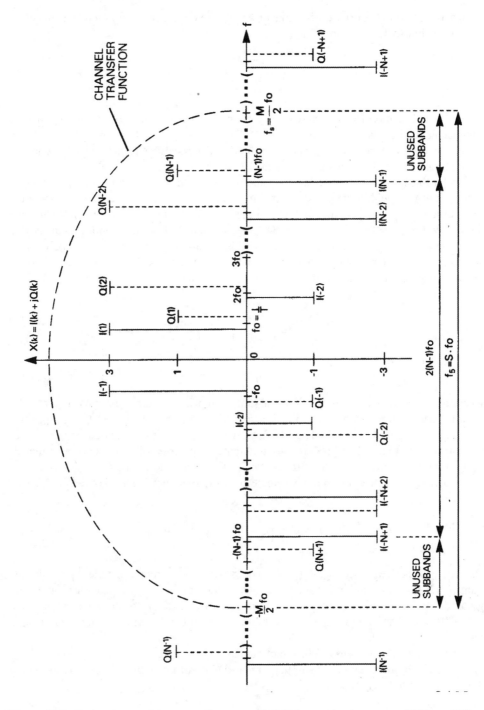

Figure 15.5: Conjugate complex symmetric square 16-QAM transmitted sequence $X(k) = I(k) + jQ(k)$

emphasising that the Fourier coefficients of a real signal are conjugate complex symmetric. Then Equation 15.12 can be rewritten as:

$$m_0(t) = \sum_{n=-(N-1)}^{N-1} F_n \cdot e^{j2\pi f_{0n}t}. \tag{15.14}$$

Observe that Equation 15.14 already bears close resemblance to the DFT. Assuming that the sub-channel carriers take values of $f_{0n} = nf_0$, $n = 0, \ldots, N-1$, where again $f_0 = 1/T$ represents the subcarrier spacing chosen to be the reciprocal of the sub-channel signalling interval, then the total one-sided bandwidth is $B = (N-1)f_0$.

So far the sub-channel modulated signals $m_{0n}(t)$ have been assumed to be continuous functions of time within the parallel signalling interval $k = 0$. In order to assist sampled, time-discrete processing by the DFT within the signalling interval $k = 0$ we introduce the discretised time $t = i\Delta t$, where $\Delta t = 1/f_s$ is the reciprocal of the sampling frequency f_s which must be chosen in accordance with Nyquist's sampling theorem to adequately represent $m_0(t)$. Recall furthermore from the previous section that $\Delta t = 1/f_s$ is the original serial QAM symbol spacing and f_s is the serial QAM symbol rate. Then we have:

$$m_0(i\Delta t) = \sum_{n=-(N-1)}^{N-1} F_n e^{j2\pi n f_0 i\Delta t}. \tag{15.15}$$

The Nyquist criterion is met if

$$f_s > 2(N-1)f_0, \tag{15.16}$$

where for practical reasons $f_s = Mf_0$ is assumed, implying that the sampling frequency f_s is an integer multiple of the subcarrier spacing f_0, with $M > 2(N-1)$ being a positive integer, implying also that $f_s = 1/\Delta t = Mf_0$, i.e. $f_0\Delta t = 1/M$. Again, these frequencies are portrayed in Figure 15.5. Bearing in mind that the spectrum of a sampled signal repeats itself at multiples of the sampling frequency f_s with a periodicity of $M = f_s/f_0$ samples, and exploiting the conjugate complex symmetry of the spectrum, for the Fourier coefficients F_n we have:

$$F_n = \begin{cases} F_{n-M} = F_{M-n}^* & if\ (\frac{M}{2}+1) \le n \le M-1 \\ 0 & if\ N-1 < n \le \frac{M}{2}. \end{cases} \tag{15.17}$$

Observe that the frequency region $(N-1)f_0 < f_n \le \frac{M}{2}f_0$ represents the typical unused transition band of the communications channel, where it exhibits significant amplitude and group delay distortion as suggested by Figure 15.5. However, in the narrow sub-bands the originally wide band frequency selective fading channel can be rendered flat upon using a high number of sub-channels.

The set of QAM symbols can be interpreted as a spectral domain sequence, which due to its conjugate complex symmetry will have a real IDFT pair in the time domain, representing the real modulated signal, which can be written as:

$$m_0(i\Delta t) = \sum_{n=0}^{M-1} F_n e^{j\frac{2\pi}{M}ni}, \quad i = 0\ldots M-1. \tag{15.18}$$

This is the standard IDFT that can be computed by the IFFT if the transform length M is an integer power of 2. The rectangular modulation elements $m_T(t) = rect\frac{t}{T}$ introduced in Equation 15.6 have an infinite bandwidth requirement since in effect they rectangularly window the set of orthogonal basis functions constituted by the carriers $rect\frac{t}{T} \cdot \cos w_{0l}t$ and $rect\frac{t}{T} \cdot \sin w_{0l}t$. It is possible to use a time-domain raised cosine pulse instead of the $rect\frac{t}{T}$ function [145], but this will impose further slight impairments on the time-domain modulated signal which is also exposed to the hostile communications channel.

Observe that the representation of $m_0(t)$ by its $\Delta t = 1/f_s$-spaced samples is only correct if $m_0(t)$ is assumed to be periodic and bandlimited to $2(N - 1)f_0$. This is equivalent to saying that $m_0(t)$ can only have a bandlimited frequency domain representation if in the time domain it expands from $-\infty$ to ∞. Due to sampling at a rate of $f_s = 1/\Delta t$ the spectral lobes become periodic at the multiples of f_s, but if the Nyquist sampling theorem is observed, no aliasing occurs. In order to fulfil these requirements, the modulating signal $m_0(t)$ derived by IFFT from the conjugate complex symmetric baseband information signal X_l has to be quasi-periodically extended before transmission via the channel, at least for the duration of the channel's memory. The effects of bandlimited transmission media will be discussed in the following section.

15.4 Transmission via Bandlimited Channels

The DFT/IDFT operations assume that the input signal is periodic in both time and frequency domain with a periodicity of M samples. If the modulated sample sequence of Equation 15.18 is periodically repeated and transmitted via the lowpass filter (LPF) preceding the channel, the channel is excited with a continuous, periodic signal. However, in order not to waste precious transmission time and hence channel capacity we would like to transmit only one period of $m_0(t)$ constituted by M samples. Assuming a LPF with a cut-off frequency of $f_c = 1/2\Delta t = f_s/2$ and transmitting only one period of $m_0(i\Delta t)$ the channel's input signal becomes:

$$\begin{aligned} m_{0,LPF}(t) &= m_0(i\Delta t) * \frac{1}{\Delta t}\frac{\sin(\pi t/\Delta t)}{\pi t/\Delta t} \\ &= m_0(i\Delta t) * \frac{1}{\Delta t} sinc\frac{\pi t}{\Delta t}, \end{aligned} \tag{15.19}$$

where the LPF's impulse response is given by the *sinc* function and hence $m_{0,LPF}(t)$ has an infinite time domain duration. One period of the periodic modulated signal $m_{0,p}(i\Delta t)$, is given by

$$m_0(i\Delta t) = m_{0,p}(i\Delta t)rect\frac{t}{T}. \tag{15.20}$$

The convolution in Equation 15.19 can be written as:

$$m_{0,LPF}(t) = \sum_{i=0}^{M-1} m_0(i\Delta t)\frac{1}{\Delta t} sinc\frac{\pi(t - i\Delta t)}{\Delta t}. \tag{15.21}$$

In the spectral domain this is equivalent to writing

$$M_{0,LPF}(f) = M_0(f)rect\frac{f}{f_c}, \tag{15.22}$$

where $M_0(f) = FFT\{m_0(i\Delta t)\}$ and $H_{LPF}(f) = rect\frac{f}{f_c}$ is the LPF's frequency domain transfer function. Transforming Equation 15.20 into the frequency domain yields:

$$\begin{aligned} M_0(f) &= FFT\{m_{0,p}(i\Delta t)rect\frac{t}{T}\} \\ &= M_{0,p}(f) * \frac{1}{f_0}sinc\frac{\pi f}{f_0}, \end{aligned} \tag{15.23}$$

where $M_{0,p}(f)$ is the frequency domain representation of $m_{0,p}(i\Delta t)$, which is convolved with the Fourier transform of the $rect\frac{t}{T}$ function. Now $M_0(f)$ of Equation 15.23 is lowpass filtered according to Equation 15.22, giving:

$$\begin{aligned} M_{0,LPF}(f) &= M_0(f)rect\frac{f}{f_c} \\ &= \left[M_{0,p}(f) * \frac{1}{f_0}sinc\frac{\pi f}{f_0}\right]rect\frac{f}{f_c}. \end{aligned} \tag{15.24}$$

So the effect of the time domain truncation of the periodic modulated signal $m_{0,p}(i\Delta t)$ to a single period as in Equation 15.20 manifests itself in the frequency domain as the convolution of Equation 15.23, generating the infinite bandwidth signal $M_0(f)$. When $M_0(f)$ is lowpass filtered according to Equation 15.24, it becomes bandlimited to f_c, and its Fourier transform pair $m_{0,LPF}(t)$ in Equation 15.19 has an infinite time domain support due to the convolution with $sinc(\pi t/\Delta t)$. This phenomenon results in interference due to time domain overlapping between consecutive transmission blocks, which can be mitigated by quasi-periodically extending $m_0(i\Delta t)$ for the duration of the memory of the channel before transmission. At the receiver only the unimpaired central section is used for signal detection.

In order to portray a practical OFDM scheme and to aid the exposition here and in the previous section, in Figure 15.6 we plotted a few characteristic signals. More specifically, Figure 15.6(a) shows the transmitted spectrum using $M = 128$ and rectangular 16-QAM having $I, Q = \pm1, \pm3$.

In contrast to Figure 15.5, where the I and Q components were portrayed next to each other in order to emphasise that they belong to the same frequency, here the I and Q components associated with the same frequency are plotted with the same spacing as adjacent sub-bands. In Figure 15.6(a) there are 64 legitimate frequencies between 0 and 4 kHz, corresponding to 128 lines. Observe, however, that for frequencies of 0-300 Hz and 3.4-4.0 kHz we have allocated no QAM symbols. This is because this signal was transmitted after conjugate complex extension and IFFT over the M1020 CCITT telephone channel simulator, which has a high attenuation in these frequency slots. The real modulated signal after IFFT is plotted in Figure 15.6(b), which is constituted by 128 real samples. At the receiver this signal is demodulated by FFT in order to derive the received signal $\tilde{X}(k)$, which is then subjected to hard decision delivering the sequence $\hat{X}(k)$ seen in Figure 15.6(c). Finally, Figure 15.6(d) portrays the error signal $\Delta(k) = X(k) - \tilde{X}(k)$, where large errors can be observed towards

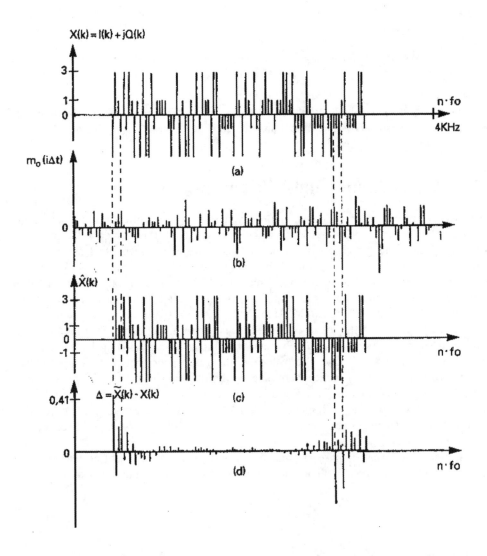

Figure 15.6: Characteristic OFDM signals for $M = 128$ and 16-QAM

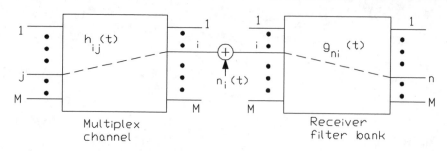

Figure 15.7: Multidimensional interference model [556] ©IEEE, 1975, Van
Etten

the transmission band edges due to the M1020 channel's attenuation. The interpretation of
these characteristic signals will be further augmented in Section 15.6, as further details of the
OFDM system are unravelled.

15.5   Generalised Nyquist Criterion

Using frequency division or orthogonal multiplexing (OFDM) two different sources of interference can be identified [555–559]. Intersymbol interference (ISI) is defined as the crosstalk
between signals within the same sub-channel of consecutive FFT frames, which are separated
in time by the signalling interval T. Inter-Channel Interference (ICI) is the cross-talk between
adjacent sub-channels or frequency slots of the same FFT frame. Since the effects of these
interference sources and their mitigation methods are similar, it is convenient to introduce the
term multidimensional interference (MDI) [557].

In order to describe MDI we assume the linear system model depicted in Figure 15.7,
which is the concatenation of the M-sub-channel multiplex transmission medium to the M-
sub-channel receiver filter bank. The channel's impulse response between its input j and
output i is denoted by $h_{ij}(t)$, while that of the receiver filter bank is $g_{ni}(t)$. With the assumption of this linear system model the output signal of the receiver filter is the superposition
of the system responses due to all input signals $1, \ldots, M$. Our objective is to optimise the
linear receiver in order to minimise the effects of AWGN and MDI, which will reduce the
BER of the receiver.

Assuming a sub-channel noise spectral density of N_i and receiver filter transfer functions
of $G_{ni}(f)$ between the input i and output n, the noise variance at output n becomes:

$$\sigma_n{}^2 = \sum_{i=1}^{M} N_i \int_0^\infty G_{ni}^2(f)df, \tag{15.25}$$

which can also be expressed in terms of $g_{ni}(t)$ using Parseval's theorem:

$$\sigma_n{}^2 = \sum_{i=1}^{M} N_i \int_0^\infty g_{ni}^2(\tau)d\tau. \tag{15.26}$$

Following the approaches in [557–559] the noise variance can be minimised as a function of the receiver filter impulse response $g_{ni}(t)$. Supposing that the sub-channel input signals a_j $j = 1, \ldots, M$ can take any value from the mutually independent, equiprobable set of L-ary alphabet, then the channel's input vector in the kth signalling interval is given by:

$$\mathbf{u}^{\mathbf{T}}(\mathbf{k}) = \left[a_1^k, a_2^k, \ldots, a_M^k \right]. \tag{15.27}$$

Observe that the number of possible input vectors is L^M.

From Figure 15.7 we can see that the system's response at the receiver output can be computed through convolution of the channel and receiver filter impulse responses. The input signal of sub-channel j generates an output at every receiver filter input $i = 1, \ldots, M$, each of which gives rise to a response at the nth receiver filter output. In order to derive the nth output, the contributions must be summed first fixing the jth sub-channel input over the receiver filter input index i. Then, since the input signal vector $\mathbf{u}^{\mathbf{T}}(\mathbf{k})$ excites all sub-channel inputs $j = 1, \ldots, M$, summing over the nth index j gives the output signal of the receiver filter:

$$s_n{}^k(t) = \sum_{j=1}^{M} a_j{}^k \sum_{i=1}^{M} \int_0^\infty g_{ni}(\tau) h_{ij}(t - \tau) d\tau. \tag{15.28}$$

At the received sampling instants $(t_s + lT)$ we have

$$s_n{}^k(t_s + lT) = \sum_{j=1}^{M} a_j{}^k \sum_{i=1}^{M} \int_0^\infty g_{ni}(\tau) h_{ij}(t_s + lT - \tau) d\tau. \tag{15.29}$$

The optimisation problem now is to find the system transfer functions, which minimise the noise variance at the output of the nth receiver filter under the constraint of constant $s_n{}^k(t_s + lT)$, for all l and k. Using the technique of variational calculus, the following functional must be minimised [556]:

$$F_n = \sigma_n^2 - 2 \sum_{k=1}^{L^M} \sum_l \lambda_{nkl} s_n{}^k(t_s + lT), \tag{15.30}$$

where λ_{nkl} is the so-called Lagrange multiplier. On substituting Equations 15.26 and 15.29 into Equation 15.30 we have:

$$
\begin{aligned}
F_n =\ & \sum_{i=1}^{M} N_i \int_0^\infty g_{ni}{}^2(\tau) d\tau \\
& - 2 \sum_{k=1}^{L^M} \sum_l \lambda_{nkl} \sum_{j=1}^{M} a_j{}^k \sum_{i=1}^{M} \int_0^\infty g_{ni}(\tau) h_{ij}(t_s + lT - \tau) d\tau.
\end{aligned}
$$

The minimisation [556] of F_n in Equation 15.31 yields:

$$g_{ni}(t) = \frac{1}{N_i} \sum_{k=1}^{L^M} \sum_l \lambda_{nkl} \sum_{j=1}^{M} a_j{}^k h_{ij}(t_s + lT - t). \tag{15.31}$$

Assuming identical noise spectral density in all sub-channels, i.e. that $N_i = N$ for all i and introducing the shorthand

$$c_{njl} = \frac{1}{N} \sum_{k=1}^{L^M} a_j{}^k \lambda_{nkl}, \tag{15.32}$$

then Equation 15.31 can be simplified to:

$$g_{ni}(t) = \sum_{j=1}^{M}\sum_{l} c_{njl}h_{ij}(t_s + lT - t).$$

(15.33)

As seen from Equation 15.33, the set of optimum receiver filters $g_{ni}(t)$ depends on the sub-channel impulse responses $h_{ij}(t)$. The receiver filter impulse responses $g_{ni}(t)$ from the ith filter input to all filter outputs have to be matched to the responses $h_{ij}(t)$ from all sub-channel inputs $j = 1, \ldots, M$ to the ith subchannel output or received filter input, as detailed in section 4.5.2.

In order to derive the matched receiver filter response $g_{ni}(t)$ according to Equation 15.33, we have to superimpose the responses of all receiver filters that are matched to the OFDM sub-channel outputs due to the jth OFDM sub-channel input. The summation over the index l weighted by the coefficients c_{njl} in Equation 15.33 represents the ISI due to the T-spaced h_{ij} contributions from previous sampling instants.

Using our general MDI model in Figure 15.7, the overall impulse response at the sampling instant (lT) between the jth OFDM sub-channel input and the nth receiver filter output can be denoted by $f_{nj}(lT), n, j = 1, \ldots, M$ or in matrix form:

$$\mathbf{F}_1 = \begin{bmatrix} f_{11}(lT) & f_{12}(lT) & \cdots & f_{1M}(lT) \\ f_{21}(lT) & f_{22}(lT) & \cdots & f_{2M}(lT) \\ \vdots & & & \\ f_{M1}(lT) & f_{M2}(lT) & \cdots & f_{MM}(lT) \end{bmatrix}.$$

(15.34)

The total accumulated MDI at the nth receiver filter output due to both ISI and ICI can be defined [556] as:

$$MDI_n = \frac{|f_{nn}(0)| - \sum_l \sum_{i=1}^{M} |f_{ni}(l)|}{f_{nn}(0)}.$$

(15.35)

It is plausible that the condition of MDI-free transmission is met if there is no cross-talk amongst the sub-channels, i.e. the matrix of Equation 15.34 is the diagonal identity matrix for any arbitrary sampling instant (lT). In formal terms this means that for $n, j = 1, 2, \ldots, M$, $l = 0, \pm 1, \pm 2, \ldots, \pm\infty$ the generalised Nyquist criterion proposed by Schnidman [555] must be satisfied:

$$\sum_l f_{nj}(lT) = \begin{cases} 1 & \text{if } n = j \\ 0 & \text{otherwise.} \end{cases}$$

(15.36)

If Equation 15.36 is satisfied, the MDI term in Equation 15.35 becomes zero and the generalised Nyquist criterion can be fulfilled by the appropriate choice of the coefficients c_{njl}.

Quite clearly, the generalised Nyquist criterion in Equation 15.36 requires not only that the conventional Nyquist criterion shall be met, i.e. $f_{nn}(lT) = \delta_l, l = 0, 1, \ldots, \infty$, where δ_l is the Kronecker delta, in order to render the ISI from other signalling intervals zero. It also requires that all the other $M(M-1)$ sub-channels with their matched receiver filters shall have a zero-crossing in their impulse responses, i.e. for $n, j = 1, 2, \ldots, M$ and $l = 0, \pm 1, \ldots, \pm\infty$ we have that

$$F_{nj}(lT) = \delta_{nj}\delta_l$$

(15.37)

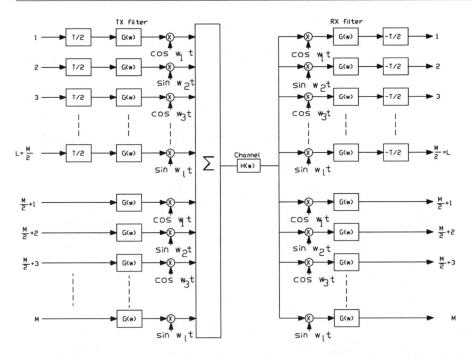

Figure 15.8: Filter-bank implementation of the FDM modem [150] ©IEEE, 1981, Hirosaki

shall be satisfied, where

$$\delta_{nj} = \begin{cases} 1 & \text{if } n = j \\ 0 & \text{otherwise.} \end{cases} \qquad (15.38)$$

Therefore, the generalised Nyquist criterion requires the equivalent folded-in baseband transfer characteristic introduced in Chapter 4 to be an ideal lowpass characteristic for $n = j$ and zero otherwise.

15.6 Basic OFDM Modem Implementations

A historic OFDM modem implementation was proposed by Hirosaki in Reference [150] and is shown in Figure 15.8. This implementation is based on staggered QAM (SQAM) or offset QAM (OQAM), where the quadrature components are delayed by half a signalling interval with respect to each other in order to reduce the signal envelope fluctuation (see Chapter 4). The system is constituted by M synchronised baseband channels operating with a Baud rate of $f_0 = 1/T$. The baseband modulating signals of sub-channels i and $(i + M/2)$ are amplitude modulated onto the carriers $f_i = [f_0 + (i - 1)f_0]$, $i = 1, \ldots, M/2$, with the carrier suppressed. This implies that the sub-channels are spaced according to the Baud rate of $f_0 = 1/T$. Then the sum of the sub-channel signals i and $(i + M/2)$, $i = 1, \ldots, M/2$ form the ith SQAM sub-channel, where the in-phase and quadrature phase modulating signals are shifted by $T/2$ with respect to each other. The transmit and receive filters $G(\omega)$ are identical square-root raised-cosine Nyquist filters and an equaliser implementation is also proposed

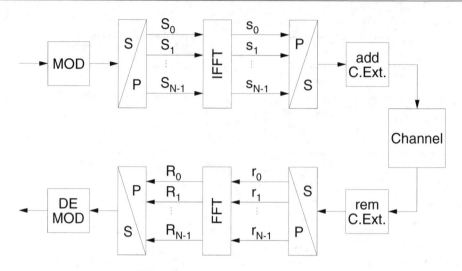

Figure 15.9: Schematic of N-subcarrier OFDM transmission system

in [150].

However, Hirosaki's conceptually simple OFDM implementation may become implementationally prohibitive in terms of complexity and cost [150], especially for a high number of sub-channels. Weinstein [145] suggested the digital implementation of OFDM subcarrier modulators/demodulators based on the discrete Fourier transform (DFT).

The DFT and its more efficient implementation, the fast Fourier transform (FFT) are typically employed in practice in the baseband OFDM modulation/demodulation process, as can be seen in the schematic of Figure 15.9. The serial data stream is mapped to data symbols with a symbol rate of $1/T_s$, employing a general phase and amplitude modulation scheme, and the resulting symbol stream is demultiplexed into a vector of N data symbols S_0 to S_{N-1}. The parallel data symbol rate is $1/N \cdot T_s$, ie. the parallel symbol duration is N times longer than the serial symbol duration T_s. The inverse FFT (IFFT) of the data symbol vector is computed and the coefficients s_0 to s_{N-1} constitute an OFDM symbol. The s_n are the time domain samples of the OFDM symbol and are transmitted sequentially over the channel at a symbol rate of $1/T_s$. At the receiver, a spectral decomposition of the received time domain samples r_n is computed employing an N-tap FFT, and the recovered data symbols R_n are restored in serial order and demultiplexed.

If we assume that the bandwidth of the OFDM spectrum is finite, then simple Fourier theory dictates that the corresponding time domain signal has an infinite duration. The underlying assumption upon invoking the IFFT for modulation is that although N frequency domain samples produce N time domain samples, the time domain signal is assumed to be periodically repeated, theoretically, for an infinite duration. In practice, however, it is sufficient to repeat the time domain signal periodically for the duration of the channel's memory, i.e. for a duration that is comparable to the length of the CIR. Hence, for transmission over time-dispersive channels, each time domain OFDM symbol is extended by the so-called cyclic extension (C. Ext. in Figure 15.9) or a guard interval of N_g samples duration, in order to overcome the inter-OFDM symbol interference due to the channel's memory.

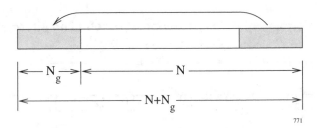

Figure 15.10: Stylised plot of N-subcarrier OFDM time domain signal with a cyclic extension of N_g samples

The samples of the cyclic extension are copied from the end of the time domain OFDM symbol, generating the transmitted time domain signal $(s_{N-N_g-1}, \ldots, s_{N-1}, s_0, \ldots, s_{N-1})$ depicted in Figure 15.10. At the receiver, the samples of the cyclic extension are discarded. Clearly, the need for a cyclic extension in time-dispersive environments reduces the efficiency of OFDM transmissions by a factor of $N/(N + N_g)$. Since the duration N_g of the necessary cyclic extension depends only on the channel's memory, OFDM transmissions employing a high number of carriers N are desirable for efficient operation. Typically a guard interval length of not more than 10% of the OFDM symbol's duration is employed.

15.7 Cyclic OFDM Symbol Extension

As we have seen in Section 15.4, the response of the low pass filtered OFDM channels to a block of modulated signal is theoretically infinite. In Section 15.5 we introduced the generalised Nyquist criterion, which allowed infinite-length system responses, as long as their periodic zero-crossings yielded no MDI. If, however, the zero-crossings or the sampling instants fluctuate, the MDI soon becomes prohibitively high.

An alternative method of combating the previously characterised multi-dimensional interference (MDI) is to transmit quasi-periodically extended time domain blocks, i.e. OFDM symbols, after modulation by IFFT [151]. The length of the added quasi-periodic extension depends on the memory length of the channel, in other words, on the length of the transient response of the channel to the quasi-periodic excitation constituted by the modulated signal. The method is best explained with reference to Figure 15.11. Every block of length-T modulated signal segment is quasi-periodically extended by a length of T_T transient duration simply repeating N_T samples of the useful information block. Then the total sequence length becomes $(M + N_T)$ samples, corresponding to a duration of $(T + T_T)$. Trailing and leading samples of this extended block are corrupted by the channel's transient response, hence the receiver is instructed to ignore the first j number of samples of the received block and also disregard $(M + N_T - j)$ trailing samples. Only the central M number of samples are demodulated by FFT at the receiver, which are essentially unaffected by the channel's transient response, as seen in Figure 15.11.

The number of extension samples N_T required depends on the length of the channel's transient response and the number of modulation levels. If the number of modulation levels is high, the maximum acceptable MDI due to channel transients must be kept low in order to maintain sufficient noise margins before the data is corrupted. This then requires a

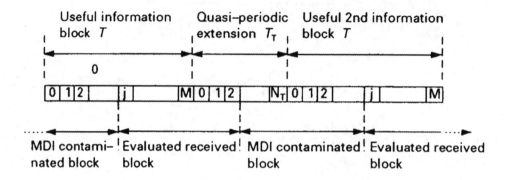

Figure 15.11: Reducing MDI by quasi-periodic block extension

longer quasi-periodic extension, i.e. N_T must be also higher. It must also be appreciated that this MDI-corrupted extension actually wastes channel capacity as well as transmitted power. However, if the useful information blocks are long, i.e. $M \geq 128$, the extension length can be kept as low as 10% of the useful information block length.

15.8 Reducing MDI by Compensation [151]

15.8.1 Transient System Analysis

Following Kolb's [151] and Schüssler's [152] approach, we can describe the OFDM transmission system by using input and output vectors $\mathbf{u}(k)$ and $\mathbf{y}(k)$, respectively, that are related by

$$\mathbf{y}(k) = \mathbf{S}\mathbf{u}(k), \tag{15.39}$$

where the convolution matrix $\mathbf{S}$ [560] is given by:

$$\mathbf{S} = \begin{bmatrix} h_0 & 0 & \cdots & \cdots & 0 \\ h_1 & h_0 & \cdots & \cdots & 0 \\ h_2 & h_1 & h_0 & \cdots & 0 \\ \vdots & \vdots & \vdots & & \vdots \end{bmatrix}. \tag{15.40}$$

Due to causality the vectors $\mathbf{u}(k)$ and $\mathbf{y}(k)$ are of infinite length for positive sample indices only. Let us assume furthermore that the input vector $\mathbf{u}(k)$ represents $L \geq 2$ periods of an M-sample periodic sequence and that the channel's impulse response is shorter than M samples. Then, for example, for $L = 3$ periods and a channel impulse response length of $n = M - 1$ samples the convolution in Equation 15.39 can be written as:

$$\begin{bmatrix} \mathbf{y}_1 \\ \mathbf{y}_2 \\ \mathbf{y}_3 \\ \mathbf{y}_4 \end{bmatrix} = \begin{bmatrix} \mathbf{S}_0 & 0 & 0 \\ \mathbf{S}_1 & \mathbf{S}_0 & 0 \\ 0 & \mathbf{S}_1 & \mathbf{S}_0 \\ 0 & 0 & \mathbf{S}_1 \end{bmatrix} \begin{bmatrix} \mathbf{u} \\ \mathbf{u} \\ \mathbf{u} \end{bmatrix}, \tag{15.41}$$

where

$$\mathbf{S}_0 = \begin{bmatrix} h_0 & 0 & \cdots & \cdots & 0 \\ h_1 & h_0 & \cdots & \cdots & 0 \\ h_2 & h_1 & h_0 & \cdots & 0 \\ \vdots & \vdots & \vdots & & \vdots \\ h_n & h_{n-1} & h_{n-2} & \cdots & h_0 \end{bmatrix} \quad (15.42)$$

and

$$\mathbf{S}_1 = \begin{bmatrix} 0 & h_n & h_{n-1} & \cdots & h_1 \\ 0 & 0 & h_n & \cdots & h_2 \\ \vdots & \vdots & \vdots & & \vdots \\ 0 & 0 & 0 & \cdots & h_n \\ 0 & 0 & 0 & \cdots & 0 \end{bmatrix} . \quad (15.43)$$

Upon exciting our transmission channel with a periodic signal, its response is constituted by a transient response $\mathbf{y}_{in}(k)$ plus a stationary periodic response $\mathbf{y}_{st}(k)$. The stationary response has the periodicity M of the excitation, while the channel's transient response dies down after $n = M - 1$ samples.

The system's response during the first period of M samples is then given by:

$$\begin{aligned} \mathbf{y}_1 &= \mathbf{y}_{in} + \mathbf{y}_{st} = \mathbf{S}_0\mathbf{u} \quad (15.44) \\ &= [h_0u_0; (h_1u_0 + h_0u_1); (h_2u_0 + h_1u_1 + h_0u_2); \\ &\quad \ldots; (h_nu_0 + h_{n-1}u_1 + h_{n-2}u_2 + \ldots + h_0u_n)]^T, \end{aligned}$$

where we also exploited Equation 15.41 and $[\ ]^T$ means the transpose of $[\ ]$. During the second cycle of the excitation the transient does not affect the system's response since $u < M$, and hence by the help of Equation 15.41 we get:

$$\begin{aligned} \mathbf{y}_2 &= \mathbf{y}_{st} = [\mathbf{S}_1\mathbf{S}_0][\mathbf{uu}]^T \quad (15.45) \\ &= [\mathbf{S}_1 + \mathbf{S}_0]\mathbf{u} = \mathbf{Hu}, \end{aligned}$$

where the matrix $\mathbf{H} = [\mathbf{S}_1 + \mathbf{S}_0]$ is computed from Equations 15.42 and 15.43 as follows:

$$\mathbf{H} = \mathbf{S}_1 + \mathbf{S}_0 = \begin{bmatrix} h_0 & h_n & h_{n-1} & \cdots & h_1 \\ h_1 & h_0 & h_n & \cdots & h_2 \\ \vdots & & & & \\ h_n & h_{n-1} & h_{n-2} & \cdots & h_0 \end{bmatrix} . \quad (15.46)$$

Then the system's response for the second cycle from Equation 15.46 can be written as:

$$\mathbf{y}_2 = \begin{bmatrix} h_0u_0 + h_nu_1 + \ldots + h_1u_n \\ h_1u_0 + h_0u_1 + \ldots + h_2u_n \\ \vdots \\ h_nu_0 + h_{n-1}u_1 + \ldots + h_0u_n \end{bmatrix} . \quad (15.47)$$

Similarly, the system's response during the third period of the excitation is unaffected by transients, and hence given by:

$$\mathbf{y}_3 = \mathbf{H}\mathbf{u}. \tag{15.48}$$

The switch-off transient $\mathbf{y}_4 = \mathbf{y}_{out}$ of the system describes the output signal after the excitation died down. From Equation 15.41 we get:

$$\begin{aligned}
\mathbf{y}_{out} = \mathbf{y}_4 &= \mathbf{S}_1\mathbf{u} = \mathbf{S}_1\mathbf{u} + \mathbf{S}_0\mathbf{u} - \mathbf{S}_0\mathbf{u} \\
&= (\mathbf{S}_1 + \mathbf{S}_0)\mathbf{u} - \mathbf{S}_0\mathbf{u}.
\end{aligned} \tag{15.49}$$

Combining Equations 15.45 and 15.46 gives:

$$\mathbf{y}_{out} = \mathbf{y}_4 = \mathbf{y}_{st} - (\mathbf{y}_{in} + \mathbf{y}_{st}) = -\mathbf{y}_{in}, \tag{15.50}$$

which implies that the trailing transient is the "inverse" of the leading transient.

Based on the previous analysis our aim is to transmit each information block only once $(L = 1)$ and remove the effects of MDI by cancellation, rather than reduce the system's effective transmission rate by the factor $M/(M + N_T)$, where N_T is the number of samples used for the quasi-periodic transmission block expansion. The kth received signal vector $\mathbf{y}^k$ is constituted by an initial transient due to the trailing transient $\mathbf{y}_{out}^{k-1}$ of the $(k - 1)$th transmitted vector, a stationary response $\mathbf{y}_{st}^k$ due to the kth transmitted vector plus the initial transient $\mathbf{y}_{in}^{k-1}$ due to the kth transmitted vector:

$$\mathbf{y}^k = \mathbf{y}_{out}^{k-1} + \mathbf{y}_{st}^k + \mathbf{y}_{in}^k. \tag{15.51}$$

Equation 15.50, for the trailing transient from block $(k - 1)$, gives us:

$$\mathbf{y}_{out}^{k-1} = -\mathbf{y}_{in}^{k-1}, \tag{15.52}$$

which can be substituted in Equation 15.51 to yield:

$$\mathbf{y}^k = \mathbf{y}_{in}^k + \mathbf{y}_{st}^k - \mathbf{y}_{in}^{k-1}. \tag{15.53}$$

15.8.2 Recursive MDI Compensation

Our objective is now to determine the transmitted vector $\mathbf{u}^k$, which can be inferred if $\mathbf{y}^k$ is known. Hence the compensation of MDI ensues as follows:

(1) Initially a known preamble sequence $\mathbf{u}_0$ is transmitted, sending $L \geq 2$ repetitions to determine the channel's impulse response, i.e. the matrix $\mathbf{H}$. This is possible using Equations 15.45 and 15.46, where the system's response is in its stationary phase during the second cycle of the periodic preamble. The system's impulse response can be conveniently measured by transmitting a specific preamble $\mathbf{u}_0$, which is constituted by a "white" spectrum with the real part set to one and the imaginary part set to zero. This input signal modulates the subchannel modulators, where the modulated signal after IFFT becomes the Kronecker delta. When this preamble signal excites the transmission system, its response is the impulse response itself. Once the impulse response is known, the matrices $\mathbf{H}$, $\mathbf{S}_0$ and $\mathbf{S}_1$ are also known.

(2) Since the system responses $\mathbf{y}_1^0$ and $\mathbf{y}_2^0$ due to the preamble $\mathbf{u}_0$ during the first two periods are now known, the system's initial transient can be computed from Equations 15.45

and 15.46 using:

$$\mathbf{y}_1^0 = \mathbf{y}_{in}^0 + \mathbf{y}_{st}^0 = \mathbf{S}_0 \mathbf{u} \tag{15.54}$$

and

$$\mathbf{y}_2^0 = \mathbf{y}_{st}^0 = [\mathbf{S}_1 + \mathbf{S}_0] \mathbf{u} \tag{15.55}$$

as follows:

$$\mathbf{y}_{in}^0 = \mathbf{y}_1^0 - \mathbf{y}_{st}^0 = \mathbf{y}_1^0 - \mathbf{y}_2^0 = -\mathbf{S}_1 \mathbf{u}. \tag{15.56}$$

With the knowledge of this initial transient response the trailing effects corrupting the consecutive blocks can be recursively compensated.

(3) The preamble $\mathbf{u}_0$ is followed by the useful information blocks $\mathbf{y}^k$, $k = 1, 2, \ldots, \infty$, transmitted only once ($L = 1$). Upon receiving the system's response $\mathbf{y}^k$ due to $\mathbf{u}^k$, $k = 1, 2, \ldots, \infty$, the trailing transient of the previous block can be subtracted as follows:

$$\mathbf{y}^k - \mathbf{y}_{out}^{k-1} = \mathbf{y}^k + \mathbf{y}_{in}^{k-1}. \tag{15.57}$$

When considering Equations 15.51 and 15.45 we get:

$$\mathbf{y}^k - \mathbf{y}_{out}^{k-1} = \mathbf{y}_{st}^k + \mathbf{y}_{in}^k = \mathbf{S}_0 \tilde{\mathbf{u}}^k, \tag{15.58}$$

where we used $\tilde{\mathbf{u}}^k$ which is different from $\mathbf{u}^k$ due to channel effects. The estimated transmitted signal $\tilde{\mathbf{u}}^k$ can be recovered using Equations 15.57 and 15.58:

$$\tilde{\mathbf{u}}^k = \mathbf{S}_0^{-1}(\mathbf{y}^k - \mathbf{y}_{out}^{k-1}) = \mathbf{S}_0^{-1}(\mathbf{y}^k + \mathbf{y}_{in}^{k-1}). \tag{15.59}$$

From Equation 15.49 we have $\mathbf{y}_{out}^{k-1} = \mathbf{S}_1 \mathbf{u}^{k-1}$ that can be substituted in Equation 15.59, giving

$$\tilde{\mathbf{u}}^k = \mathbf{S}_0^{-1}(\mathbf{y}^k - \mathbf{S}_1 \mathbf{u}^{k-1}), \tag{15.60}$$

which is an explicit formula for the estimated transmitted vector $\tilde{\mathbf{u}}^k$ in terms of the received vector $\mathbf{y}^k$, the previously recovered transmitted vector $\mathbf{u}^{k-1}$ and the matrices $\mathbf{S}_0$ and $\mathbf{S}_1$, which depend only on the system's impulse response.

(4) Now the compensated vector $\tilde{\mathbf{u}}^k$ in Equation 15.60 can be demodulated by FFT and the demodulated signal $\tilde{\mathbf{U}}^k$ is subjected to hard decisions, which are represented as $D\{\ \}$, giving the recovered information signal in the following form:

$$\mathbf{U}^k = D\{\tilde{\mathbf{u}}^k\} = D\{FFT[\tilde{\mathbf{u}}^k]\}. \tag{15.61}$$

(5) Since $\mathbf{u}^k$ is needed in Equation 15.60 for the next recursive compensation step,

$$\mathbf{u}^k = IFFT[\mathbf{U}^k] \tag{15.62}$$

is computed to conclude the compensation process.

This method bears a strong resemblance to the conventional partial response technique, allowing signals belonging to adjacent signalling intervals to overlap. The controlled ISI can then be recursively compensated, if the channel can be considered slowly changing.

15.9   Adaptive Channel Equalisation

In this section it will be assumed that the MDI was removed by compensation, quasi-periodic block extension or by obeying the generalised Nyquist criterion. The linear distortions introduced by the unequalised channel transfer function $H(f)$ can be removed by estimating $H(if_0)$ and then dividing the received signal spectrum by $H(if_0)$, before hard decision decoding is performed using the function $D\{\ \}$.

In the previous section we mentioned how $H(if_0)$ can be measured using a preamble having a real-valued "white" spectrum. By setting all real spectral lines to unity and all imaginary lines to zero in the preamble data frame, after modulation by the IFFT the transmitted signal is the Kronecker delta. Therefore the received signal is the channel's impulse response, which after demodulation by FFT gives the channel's frequency domain transfer function $H(if_0)$. The received signal's spectrum after demodulation by FFT becomes:

$$\tilde{X}_i = H(if_0)X_i, \; i = 1, \ldots, M, \tag{15.63}$$

where X_i is the ith transmitted spectral line at frequency if_0. After equalisation by dividing the received spectral line $\tilde{X}_i$ with the estimated channel transfer function $H(if_0)$ and taking hard decisions, we obtain the recovered sequence:

$$\hat{X}_i = D\left\{\frac{\tilde{X}_i}{H(if_0)}\right\}, \; i = 1, \ldots, M. \tag{15.64}$$

If the recovered sequence $\hat{X}_i$ is error-free, it lends itself to the recursive recomputation of the channel's frequency response in order to cope with slowly time-varying transmission media [151]. The up-dated transfer function is given by:

$$H_a(if_0) = \frac{\tilde{X}_i}{\hat{X}_i} = \frac{\tilde{X}_i}{D\left\{\frac{\tilde{X}_i}{H(if_0)}\right\}}, \; i = 1, \ldots, M. \tag{15.65}$$

In order to retain robustness for transmissions over channels having high bit error rate a leaky algorithm can be introduced to generate a weighted average of the previous and current transfer function using the leakage factor β in the following fashion:

$$H_{\text{adaptive}}(if_0) = \beta H_a^k(if_0) + (1 - \beta)H_a^{k-1}(if_0). \tag{15.66}$$

The leakage factor β is a parameter determined by the prevailing channel bit error rate.

An interesting aspect of our OFDM scheme is that if the channel varies slowly, then using differential coding between corresponding sub-channels of consecutive OFDM transmission frames removes the requirement for a channel equaliser as long as the difference is computed before hard decision takes place [145]. This is due to the fact that spectral lines of the same sub-channel or same frequency will suffer the same attenuation due to the channel's linear distortion. Hence the effect of channel attenuation and phase shift drops out before hard decision takes place. Similar arguments can be exploited between adjacent lines of the same OFDM transmission frame as well, if the channel's transfer characteristic is sufficiently smooth.

If the frequency domain transfer function $H(if_0)$ is more erratic as a function of frequency or time, a number of known pilot tones can be included in the transmitted spectrum. These pilots facilitate the more accurate estimation and equalisation of the channel transfer function. This technique was proposed by Cimini for wide-band, frequency-selective multipath mobile channels in reference [146] and it will be widely used throughout the forthcoming chapters. In case of narrowband Rayleigh-fading mobile radio channels the fading is "frequency flat fading", hence each transmitted frequency component suffers the same attenuation and phase shift. In this case time domain pilot symbols employed as in the PSAM schemes of Chapter 9 are useful in supporting the tracking of the Rayleigh fading envelope [139].

15.10   OFDM Bandwidth Efficiency

In the OFDM system each symbol to be transmitted modulates an assigned carrier of a set of wide-sense orthogonal basis functions and these modulated sub-channel signals are superimposed for transmission via the communications channel. The received signal can be demodulated for example by the correlation receiver described in the previous section or by FFT.

A set of suitable wide-sense orthogonal basis functions of gradually increasing length T, $3T$ and $5T$, similar to those used in our OFDM schemes, is depicted in Figure 15.12 for a one-carrier, three-carrier and five-carrier system, respectively [546, 547].

Both their time domain waveforms and stylised spectra are shown in Figure 15.12. In a simplistic approach here we assume that the signal spectra can be band-limited to the bandwidth of its main spectral lobe, as suggested by the figure.

Using an essentially serial system with one carrier, as seen in Figure 15.12(a), the minimum bandwidth required is $f_B = 1/T$ and the bandwidth efficiency is $\eta = 1$ Bd/Hz because the spectrum of this pulse is represented by the sinc function whose first zero is at $f_B = 1/T$. The three-carrier system of Figure 15.12(b) expands the length of the basis functions to $3T$, thereby reducing the bandwidth requirement to $B = \frac{2}{3}f_B$, giving $\eta = 1.5$ Bd/Hz. This is because the rectangularly windowed sin and cos spectra are represented by the convolution of a tonal spectral line and a frequency domain sinc function describing the spectrum of the rectangular time domain window. The five-carrier scheme using basis functions of $5T$ length further reduces the bandwidth to $B = \frac{3}{5}f_B$ and increases the spectral efficiency to 1.67 Bd/Hz.

Similarly, the approximate bandwidth of a $(2M + 1)$-carrier system using an impulse as well as M sine and M cosine carriers of length $(2M + 1)T$ becomes

$$B = \frac{M+1}{2M+1}\frac{1}{T},$$
(15.67)

yielding a bandwidth efficiency of

$$\eta = \frac{2M+1}{M+1} \text{ Bd/Hz.}$$
(15.68)

When $M \to \infty$, we have $\lim_{M\to\infty} \eta = 2$ Bd/Hz, which for a typical value of $M = 64$ gives $\eta = 129/65 = 1.98$ Bd/Hz. The interferences caused by the above band-limitation are

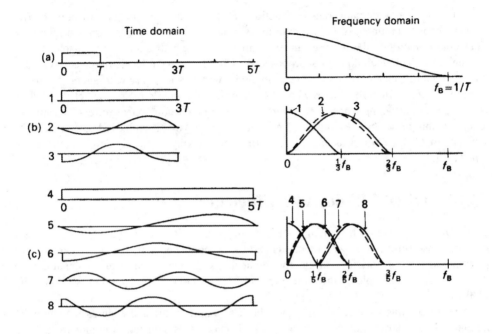

Figure 15.12: OFDM carriers, their stylised spectra and bandwidth requirement: (a) one-carrier system, (b) three-carrier system, (c) five-carrier system [546] ©Springer, 1969, Harmuth

given in closed form in Reference [547] for a variety of carriers, but for the more attractive scenarios using a higher number of carriers it can only be estimated by simulation studies.

15.11 Summary

In this chapter we have investigated orthogonal multiplexing as an alternative means of transmitting high-rate information over highly dispersive mobile radio channels. The essential premise of orthogonal multiplexing is to divide the serial input data stream into a number of parallel streams and to transmit these low-rate parallel streams simultaneously. This offered two main advantages. Firstly, by increasing the time – due to the low sub-channel signalling rate – over which a symbol was transmitted, the probability of it being completely destroyed in a fade was reduced. Secondly, by decreasing the bandwidth of each channel, the need for a channel equaliser was diminished. Modulation on each of these orthogonal channels could be either binary or multilevel, and both were considered.

However, orthogonal multiplexing also has a number of disadvantages. It requires an increase in the complexity of the transmitting and receiving equipment, and may also engender an increased transmission delay. It will typically also require pilot symbols or some other similar redundancy such that the receiver can be informed of how each individual channel has been modified by the mobile radio channel.

A number of sections were concerned with the issue of multidimensional interference, whereby interference might occur not only between consecutive symbols of the same sub-channel, but also from one sub-channel to the adjacent sub-channels, and methods to ameliorate this were suggested.

Following this rudimentary introduction to OFDM, in the forthcoming chapters we will delve into OFDM more deeply, considering a range of associated research and implementational aspects.

16

OFDM Transmission over Gaussian Channels

High data rate communications over additive white Gaussian noise (AWGN) channels are limited not only by noise, but especially with increasing symbol rates, often more significantly by the intersymbol interference (ISI) due to the memory of the dispersive wireless communications channel. Explicitly, this channel memory is caused by the dispersive channel impulse response (CIR) due to the different length propagation paths between the transmitting and the receiving antennae. This dispersion effect could theoretically be measured by transmitting an infinitely short impulse and "receiving" the channel impulse response itself. On this basis, several measures of the effective duration of the impulse response can be calculated, one being the delay spread. The multipath propagation of the channel manifests itself by different echoes of possibly different transmitted symbols overlapping at the receiver, which leads to error rate degradation.

This effect occurs not only in wireless communications, but also over all types of electrical and optical waveguides, although for these media the relative time differences are comparatively small, mostly due to multimode transmission or incorrect electrical or optical adaption at interfaces.

In wireless communications systems the duration and the shape of the channel impulse response depend heavily on the propagation environment of the communications system in question. While indoor wireless networks typically exhibit only short relative delays, outdoor networks, like the Global System of Mobile communications known as GSM may have maximum delays spreads on the order of 15 μs.

As a general rule, the effects of ISI on the transmission error statistics are negligible, as long as the delay spread is significantly shorter than the duration of one transmitted symbol. This implies that the symbol rate of communications systems is practically limited by the channel's memory. For higher symbol rates, there is typically significant deterioration of the system's error rate performance.

If symbol rates exceeding this limit are to be transmitted over the channel, mechanisms must be implemented in order to combat the effects of ISI. Channel equalisation techniques can be used to suppress the echoes caused by the channel. To perform this operation, the channel impulse response must be estimated. Significant research efforts were invested into the development of such channel equalisers, and most wireless systems in operation use equalisers to combat ISI.

There is, however, an alternative approach towards transmitting data over a multipath channel. Instead of attempting to cancel the effects of the channel's echoes, orthogonal frequency division multiplexing (OFDM) [173] modems employ a set of harmonically related carriers in order to transmit information symbols in parallel over the channel. Since the system's data throughput is the sum of all the parallel channels' throughputs, the data rate per sub-channel is only a fraction of the data rate of a conventional single-carrier system having the same throughput. This allows us to design a system supporting high data rates, while maintaining symbol durations much longer than the channel's memory, thus circumventing the need for channel equalisation.

16.1 Orthogonal Frequency Division Multiplexing

16.1.1 History

Frequency division multiplexing (FDM) or multi-tone systems have been employed in military applications since the 1960s, for example by Bello [223], Zimmerman [144], Powers and Zimmerman [224], and others. Orthogonal frequency division multiplexing (OFDM), which employs multiple carriers overlapping in the frequency domain, was pioneered by Chang [143,225]. Saltzberg [226] studied a multicarrier system employing orthogonal time-staggered quadrature amplitude modulation (O-QAM) on the carriers.

The use of the discrete Fourier transform (DFT) to replace the banks of sinusoidal generators and the demodulators was suggested by Weinstein and Ebert [145] in 1971, which significantly reduces the implementation complexity of OFDM modems. In 1980, Hirosaki [156] suggested an equalisation algorithm in order to suppress both intersymbol and intersubcarrier interference caused by the channel impulse response or timing and frequency errors. Simplified OFDM modem implementations were studied by Peled [149] in 1980, while Hirosaki [150] introduced the DFT based implementation of Saltzberg's O-QAM OFDM system. From Erlangen University, Kolb [151], Schüßler [152], Preuss [153] and Rückriem [154] conducted further research into the application of OFDM. Cimini [146] and Kalet [155] published analytical and early seminal experimental results on the performance of OFDM modems in mobile communications channels.

More recent advances in OFDM transmission are presented in the impressive state-of-the-art collection of works edited by Fazel and Fettweis [227], including the research by Fettweis *et al.* at Dresden University, Rohling *et al.* at Braunschweig University, Vandendorp at Loeven University, Huber *et al.* at Erlangen University, Lindner *et al.* at Ulm University, Kammeyer *et al.* at Bremen University and Meyr *et al.* [228, 229] at Aachen University, but the individual contributions are too numerous to mention.

While OFDM transmission over mobile communications channels can alleviate the problem of multipath propagation, recent research efforts have focused on solving a set of inherent difficulties regarding OFDM, namely the peak-to-meanpower ratio, time and frequency synchronisation, and on mitigating the effects of the frequency selective fading channel. These issues are addressed below in more depth.

16.1.1.1 Peak-to-Mean Power Ratio

It is plausible that the OFDM signal – which is the superposition of a high number of modulated sub-channel signals – may exhibit a high instantaneous signal peak with respect to the average signal level. Furthermore, large signal amplitude swings are encountered, when the time domain signal traverses from a low instantaneous power waveform to a high power waveform, which may result in a high out-of-band (OOB) harmonic distortion power, unless the transmitter's power amplifier exhibits an extremely high linearity across the entire signal level range, as discussed in Section 4.5.1. This then potentially contaminates the adjacent channels with adjacent channel interference. Practical amplifiers exhibit a finite amplitude range, in which they can be considered near-linear. In order to prevent severe clipping of the high OFDM signal peaks, which is the main source of OOB emissions, the power amplifier must not be driven into saturation and hence they are typically operated with a certain so-called back-off, creating a certain "headroom" for the signal peaks, which reduces the risk of amplifier saturation and OOB emission. Two different families of solutions have been suggested in the literature, in order to mitigate these problems, either reducing the peak-to-mean power ratio, or improving the amplification stage of the transmitter.

More explicitly, Shepherd [230], Jones [231], and Wulich [232] suggested different coding techniques which aim to minimise the peak power of the OFDM signal by employing different data encoding schemes before modulation, with the philosophy of choosing block codes whose legitimate code words exhibit low so-called crest factors or peak-to-mean power envelope fluctuation. Müller [233], Pauli [234], May [235] and Wulich [236] suggested different algorithms for post-processing the time domain OFDM signal prior to amplification, while Schmidt and Kammeyer [237] employed adaptive subcarrier allocation in order to reduce the crest factor. Dinis and Gusmão [238–240] researched the use of two-branch amplifiers, while the clustered OFDM technique introduced by Daneshrad, Cimini and Carloni [241] operates with a set of parallel partial FFT processors with associated transmitting chains. OFDM systems with increased robustness to non-linear distortion have been proposed by Okada, Nishijima and Komaki [242] as well as by Dinis and Gusmão [243].

16.1.1.2 Synchronisation

Time and frequency synchronisation between the transmitter and receiver are of crucial importance as regards to the performance of an OFDM link [244, 245]. A wide variety of techniques have been proposed for estimating and correcting both timing and carrier frequency offsets at the OFDM receiver. Rough timing and frequency acquisition algorithms relying on known pilot symbols or pilot tones embedded into the OFDM symbols have been suggested by Claßen [228], Warner [246], Sari [247], Moose [248], as well as Brüninghaus and Rohling [249]. Fine frequency and timing tracking algorithms exploiting the OFDM signal's cyclic extension were published by Moose [248], Daffara [250] and Sandell [251].

16.1.1.3 OFDM/CDMA

Combining OFDM transmissions with code division multiple access (CDMA) allows us to exploit the wideband channel's inherent frequency diversity by spreading each symbol across multiple subcarriers. This technique has been pioneered by Yee, Linnartz and Fettweis [252], by Chouly, Brajal and Jourdan [253], as well as by Fettweis, Bahai and Anvari [254].

Fazel and Papke [255] investigated convolutional coding in conjunction with OFDM/CDMA. Prasad and Hara [256] compared various methods of combining the two techniques, identifying three different structures, namely multicarrier CDMA (MC-CDMA), multicarrier direct sequence CDMA (MC-DS-CDMA) and multi-tone CDMA (MT-CDMA). Like non-spread OFDM transmission, OFDM/CDMA methods suffer from high peak-to-mean power ratios, which are dependent on the frequency domain spreading scheme, as investigated by Choi, Kuan and Hanzo [257].

16.1.1.4   Adaptive Antennas

Combining adaptive antenna techniques with OFDM transmissions was shown to be advantageous in suppressing co-channel interference in cellular communications systems. Li, Cimini and Sollenberger [258–260], Kim, Choi and Cho [261], Lin, Cimini and Chuang [262] as well as Münster *et al.* [263] have investigated algorithms for multi-user channel estimation and interference suppression.

16.1.1.5   OFDM Applications

Due to their implementational complexity, OFDM applications have been scarce until quite recently. Recently, however, OFDM has been adopted as the new European digital audio broadcasting (DAB) standard [147, 148, 337–339] as well as for the terrestrial digital video broadcasting (DVB) system [247, 340].

For fixed-wire applications, OFDM is employed in the asynchronous digital subscriber line (ADSL) and high-bit-rate digital subscriber line (HDSL) systems [341–344] and it has also been suggested for power line communication systems [345, 346] due to its resilience to time dispersive channels and narrowband interferers.

More recently, OFDM applications were studied within the European *4th* Framework Advanced Communications Technologies and Services (ACTS) programme [347]. The ME-DIAN project investigated a 155 Mbps wireless asynchronous transfer mode (WATM) network [348–351], while the Magic WAND group [352, 353] developed a wireless local area network (LAN). Hallmann and Rohling [354] presented a range of different OFDM-based system proposals that were applicable to the European Telecommunications Standardisation Institute's (ETSI) recent personal communications oriented air interface concept [355].

16.2   The Frequency Domain Modulation

Modulation of the OFDM subcarriers is analogous to the modulation in conventional serial systems. The modulation schemes of the subcarriers are generally quadrature amplitude modulation (QAM) or phase shift keying (PSK) in conjunction with both coherent and non-coherent detection. The differentially encoded star-QAM (DSQAM) scheme of Chapter 10 may also be employed. If coherently detected modulation schemes are employed, then the reference phase of the OFDM symbol must be known, which can be acquired with the aid of pilot tones [139] embedded in the spectrum of the OFDM symbol, as will be discussed in Chapter 17. For differential detection the knowledge of the absolute subcarrier phase is not necessary, and differentially coded signalling can be invoked either between neighbouring subcarriers or between the same subcarriers of consecutive OFDM symbols.

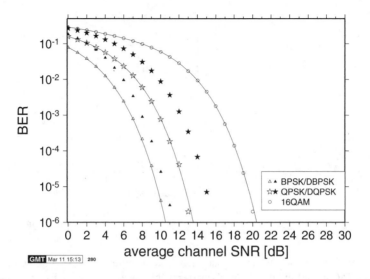

Figure 16.1: BER versus SNR curves for the OFDM modem in AWGN channel using BPSK, DBPSK, QPSK, DQPSK, and 16-QAM. The lines indicate the theoretical performance of the coherently detected modulation schemes in a serial modem over AWGN channels

16.3 OFDM System Performance over AWGN Channels

As the additive white Gaussian noise (AWGN) in the time domain channel corresponds to AWGN of the same average power in the frequency domain, an OFDM modem's performance in an AWGN channel is identical to that of a serial modem. Analogously to a serial system, the bit error rate (BER) versus signal-to-noise rate (SNR) characteristics are determined by the frequency domain modulation scheme used. In Figure 16.1 the simulated BER versus SNR curves for binary phase shift keying (BPSK), differential BPSK (DBPSK), quaternary phase shift keying (QPSK), differential QPSK (DQPSK) and coherent 16-quadrature amplitude modulation (16-QAM) are shown, together with the theoretical BER curves of serial modems, as derived in [94]:

$$p_{e,BPSK}(\gamma) = Q(\sqrt{2\gamma}), \tag{16.1}$$

$$p_{e,QPSK}(\gamma) = Q(\sqrt{\gamma}), \tag{16.2}$$

$$p_{e,16QAM}(\gamma) = 0.75 \cdot Q\left(\sqrt{\frac{\gamma}{5}}\right) 0.25 \cdot Q\left(\sqrt{\frac{\gamma}{5}}\right), \tag{16.3}$$

where γ is the SNR and the Gaussian $Q()$-function is defined as

$$Q(y) = \frac{1}{\sqrt{2\pi}} \int_y^\infty e^{-x^2/2} dx = \text{erfc}\left(\frac{y}{\sqrt{2}}\right).$$

It can be seen from the figure that the experimental BER performance of the OFDM modem is in accordance with the theoretical BER curves of conventional serial modems in AWGN

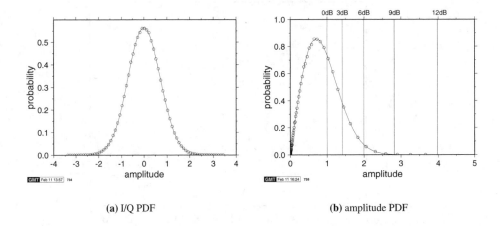

(a) I/Q PDF   (b) amplitude PDF

Figure 16.2: Statistics of OFDM time domain signal: (a) amplitude histogram of the I component of a 256-subcarrier OFDM signal (markers) obeying a Gaussian PDF (continuous line) with $\sigma = 1/\sqrt{2}$; (b) two-dimensional amplitude histogram (markers) following a Rayleigh PDF (continuous line) with $\sigma = 1/\sqrt{2}$. The vertical lines in (b) indicate the corresponding powers above mean in decibels

channels.

16.4   Clipping Amplification

16.4.1   OFDM Signal Amplitude Statistics

The time domain OFDM signal is constituted by the sum of complex exponential functions, whose amplitudes and phases are determined by the data symbols transmitted over the different carriers. Assuming random data symbols, the resulting time domain signal exhibits an amplitude probability density function (PDF) approaching the two-dimensional or complex Gaussian distribution for a high number of subcarriers. Figure 16.2(a) explicitly shows that the measured amplitude histogram of the in-phase (I) component of a 256-subcarrier OFDM signal obeys a Gaussian distribution with a standard deviation of $\sigma = 1/\sqrt{2}$. The mean power of the signal is $2\sigma^2 = 1$. As the amplitude distribution in both the in-phase and the quadrature phase (Q) component is Gaussian, the two-dimensional amplitude histogram follows a complex Gaussian, or in other words, a Rayleigh distribution [94] with the same standard deviation. The observed amplitude histogram of the 256-subcarrier OFDM signal and the corresponding Rayleigh probability density function are depicted in Figure 16.2(b). The vertical grid lines in the figure indicate the relative amplitude above the mean value in decibels. It can be seen that unlike full response serial modulation schemes – which have a more limited range of possible output amplitudes – the parallel modem's output signal exhibits strong amplitude fluctuations. Note that the standard deviation σ of the probability density functions depicted in Figure 16.2(a) is independent of the number of subcarriers employed, since the mean power of the signal is normalised to 1.

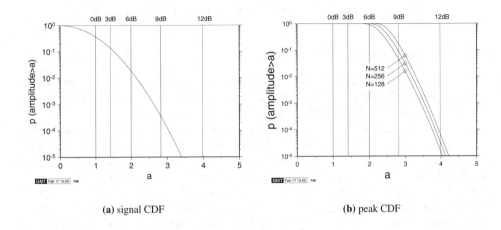

(a) signal CDF   (b) peak CDF

Figure 16.3: OFDM time domain signal: probability of (a) the instantaneous signal amplitude and (b) the peak amplitude per OFDM symbol being above a given threshold value. The lines indicate the corresponding power levels above the mean. The peak value CDF depends on the number of subcarriers, and curves are shown for 128, 256 and 512 carriers

The probability of the instantaneous signal amplitude being above a given threshold, the cumulative density function (CDF), is depicted in Figure 16.3(a). It can be noted that the probability of the signal amplitude exceeding the 6 dB mark is about 1.1%, while the 9dB mark is exceeded with a probability of about $3.5 \cdot 10^{-4}$. The signal amplitude is higher than the average by 10.6 dB with a probability of 10^{-5}.

16.4.2 Clipping Amplifier Simulations

In order to evaluate the effects of a non-linear amplifier on the performance of an OFDM system, simulations have been conducted employing a simple clipping amplifier model. This clipping amplifier limits the amplitude of the transmitted signal to a given level, without perturbing the phase information. This amplitude limitation of the time domain signal affects both the received symbols on all subcarriers in the OFDM symbol, as well as the frequency domain out-of-band emissions, and therefore increases the interference inflicted on adjacent carriers.

In Figure 16.3(a) the probability of the instantaneous signal amplitude exceeding a given level was shown. As amplitude limitation in the time domain affects all the subcarriers in the OFDM symbol, the probability of the time domain peak amplitude per OFDM symbol period being clipped by the amplifier is the probability of at least one of the N time domain samples exceeding a given amplitude limit. This clipping probability for a given maximal amplifier amplitude a is displayed in Figure 16.3(b). It can be observed that the clipping probability per OFDM symbol is dependent on the number of subcarriers employed.

Given the information in Figure 16.3(b), the necessary amplifier back-off for an OFDM transmitter can be determined. If the acceptable clipping probability per OFDM symbol is 10^{-5}, then the necessary amplifier back-off values would be 12.1 dB, 12.3 dB, and 12.5 dB for a 128, 256, and 512 subcarrier OFDM modem, respectively. If a clipping probability of

10^{-4} is acceptable, then these back-off values can be reduced by about 0.6 dB.

16.4.2.1 Peak-Power Reduction Techniques

Two main types of peak-to-mean power ratio reduction techniques have been investigated in the literature, which rely on either introducing redundancy in the data stream or on post-processing the time domain OFDM signal before amplification, respectively.

Shepherd [230], Jones [231], and Wulich [232] suggest different coding techniques which aim to ensure that only low peak power OFDM symbols are chosen for transmission, while excluding the transmission of the specific combinations of modulating bits that are known to result in the highest peak factors. These schemes can be viewed as simple k out of n block codes. Depending on the tolerable peak to mean power ratio, the set of acceptable OFDM symbols is computed and the data sequences are mapped onto OFDM symbols from this set only. These techniques reduce the data throughput of the system, but mitigate the severity of clipping amplification.

Müller [233] proposes techniques based on generating a set of OFDM signals by multiplying the modulating data vector in the frequency domain with a set of different phase vectors known to both the transmitter and the receiver, before applying the IDFT. The transmitter will then choose the resulting OFDM symbol exhibiting the lowest peak factor and transmits this together with the chosen phase vector's identification. This technique reduces the average peak factors and requires only a low signalling overhead, but cannot guarantee a given maximum peak factor.

Another group of peak amplitude limiting techniques is based on modifying the time domain signal, where its amplitude exceeds the given peak amplitude limit. Straightforward clipping falls into this category, but induces strong frequency domain out-of-band emissions. In order to avoid the spectral domain spillage of hard clipping, both multiplicative and additive time domain modifications of the OFDM signal have been investigated. Pauli [234] proposes a multiplicative correction of the peak values and their adjacent samples using smooth Gaussian functions for limiting out-of-band emissions. However, multiplying the time domain signal with the time-varying amplitude limiting function introduces intersubcarrier interference.

A similar technique, using additive instead of multiplicative amplitude limiting, is described by May [235]. A smoothly varying time domain signal is added to the time domain signal, which has been optimised for low out-of-band emissions. As the DFT is a linear operation, adding a correction signal in the time domain will overlay the frequency domain data symbols with the DFT of the correction signal, introducing additional noise.

As an alternative solution, Wulich [236] suggests measuring the peak amplitude in each OFDM time domain symbol and scaling the amplitude for the whole OFDM symbol, so that the peak value becomes constant from symbol to symbol. At the receiver, an amplitude correction similar to pilot-assisted scaling over fading channels has to be performed. The advantage of this technique is that the system's throughput is not reduced by introducing redundant information, nor is the orthogonality of the subcarriers impaired. As the OFDM symbol energy is fluctuating with the peak power exhibited by the different OFDM symbols, however, the BER performance of the system is impaired.

A further approach to reducing the amplifier linearity requirements, and which relies on the findings of Figure 16.3(b), is to split an OFDM symbol into groups of subcarriers, which

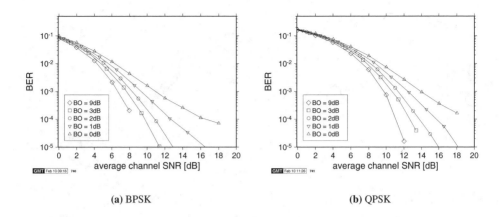

(a) BPSK   (b) QPSK

Figure 16.4: Influence of amplitude clipping on BER performance of an OFDM modem

are processed by IFFT, amplified and transmitted over separate transmit chains. This system, referred to as clustered OFDM, has been proposed by Daneshrad, Cimini and Carloni [241], and combines the advantages of reduced peak power in each transmitter chain with additional spatial diversity from the use of multiple transmit antennas, which can be exploited by coding or OFDM-CDMA techniques.

16.4.2.2   BER Performance Using Clipping Amplifiers

If lower back-off values are chosen and no peak power reduction techniques are employed in the OFDM transmission system, then the clipping amplifier will influence both the performance of the OFDM link as well as the out-of-band emissions. The effects of the amplifier back-off value on the BER performance in an AWGN channel for coherently detected 512-subcarrier BPSK and QPSK OFDM transmission are depicted in Figure 16.4.

The investigated amplifier back-off values range from 9 dB down to 0 dB. According to Figure 16.3(b), an amplifier back-off of 9 dB results in a clipping probability per OFDM symbol of about 17%, while an amplifier back-off of 0 dB results in nearly certain clipping of every transmitted OFDM symbol.

It can be observed in Figure 16.4 for both BPSK and QPSK that for an amplifier back-off value of 9 dB the BER performance is indistinguishable from the non-limited case, which was portrayed in Figure 16.1. For smaller back-off values, however, the effects of clipping on the BER performance are more severe. A 3 dB amplifier back-off results in a SNR penalty of about 1.3 dB at a BER of 10^{-4}, and a back-off of 0 dB results in more than 8 dB SNR penalty at the same BER.

The BER performance of the OFDM modem can be improved slightly if the receiver's decision boundaries are adjusted according to the amplitude loss in each subcarrier due to the overall signal power loss, which was observed by O'Neill and Lopes [561]. In fading channels this amplitude adjustment would be performed in conjunction with the channel estimation.

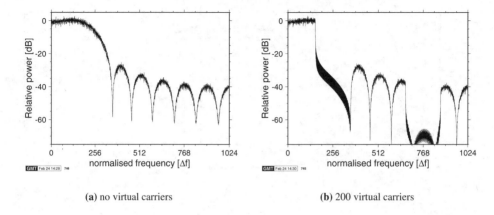

(a) no virtual carriers   (b) 200 virtual carriers

Figure 16.5: Spectrum of a 512-subcarrier OFDM signal using raised-cosine Nyquist pulse shaping filter with an excess bandwidth of $\alpha = 0.35$: (a) no virtual subcarriers, (b) 2×100 subcarriers at the edges of the spectrum. The characteristic sinc-signal shaped curve of this figure is a consequence of rectangular windowing used in the spectral estimation. This effect can be mitigated by more smooth windowing.

16.4.2.3   Signal Spectrum with Clipping Amplifier

The simulated spectrum of a 512-subcarrier OFDM signal is shown in Figure 16.5. Figure 16.5(a) depicts the spectrum of the transmitted signal, when all 512 subcarriers are used for data transmission. A raised cosine Nyquist filter with an excess bandwidth of $\alpha = 0.35$ was used for pulse shaping, resulting in a relatively slow fall-off of the power spectral density outside the FFT bandwidth of $\pm 256 \Delta f$. In order to avoid adjacent channel interference due to spectral overlapping, an adjacent carrier must be placed at a frequency distance considerably higher than the data symbol rate. If a tighter spectral packing of adjacent carriers is important, then pulse shaping filters with a steeper frequency transfer function have to be employed, which are more complex to implement. Since in this scenario all the subcarriers are used for data transmission, the time domain samples are uncorrelated and therefore the spectrum in Figure 16.5(a) is equivalent to that of a serial modem transmitting at the same sample rate and employing the same pulse shaping filter.

Figure 16.5(b) depicts the signal spectrum of a 512-subcarrier modem employing 200 virtual subcarriers at the edges of the bandwidth, which are effectively disabled and carry no energy. Although the same pulse shaping filter was employed as in Figure 16.5(a), the resulting spectral shape is nearly rectangular, with a drop of approximately 20 dB within one subcarrier distance Δf outside the signal bandwidth. Comparing this spectrum with Figure 16.5(a) it can be observed that the overall shape of the spectrum is determined by the pulse shaping filter, while the employment of disabled virtual subcarriers cuts out parts of the spectrum. It can be seen from the spectrum that by employing virtual subcarriers a very tight power spectrum can be obtained with a simple pulse shaping filter. This rectangular signal spectrum allows tight packing of adjacent carriers, therefore enhancing the spectral efficiency of an OFDM system.

The effects of a clipping amplifier on the OFDM signal spectrum are demonstrated in

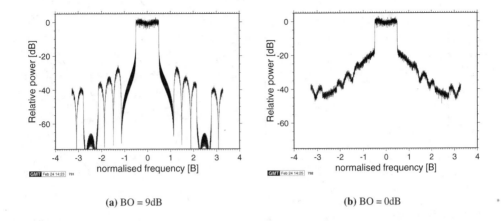

(a) BO = 9dB   **(b)** BO = 0dB

Figure 16.6: Power spectra of a 512-subcarrier OFDM modem employing 200 virtual carriers and Nyquist filtering with an excess bandwidth of 0.35. The frequencies are normalised to the signal bandwidth of $B = 312 \cdot \Delta f$: (a) 9 dB amplifier back-off (BO), (b) 0 dB amplifier back-off

Figure 16.6. Specifically, Figure 16.6(a) shows the spectrum corresponding to the system employed in the context of Figure 16.5(b), but with the amplifier back-off set to 9 dB. The frequency axis is normalised to the signal bandwidth of $B = 312 \cdot \Delta f$ due to the disabled carriers and the power spectral density is normalised to the average signal power in the signal band. Comparison of Figure 16.6(a) and 16.5(b) shows no discernible change in the spectral domain.

Reducing the amplifier back-off results in increased out-of-band emissions. Figure 16.6(b) shows the averaged power spectral density of the same system with an amplifier back-off of 0 dB. In this case, the spectral spillage is apparent directly outside the signal band.

Since the amplitude limiting of the transmitted time domain signal is due to the power amplifier, no bandlimiting filtering is performed after the amplitude distortion. Li and Cimini [562] proposed to employ a hard limiter and a lowpass filter before the non-linear amplification stage. In this case, the out-of-band emissions can be reduced, while the BER penalty due to hard limiting the time domain signal still applies.

In order to evaluate the amount of adjacent channel interference caused by non-linear amplification of the OFDM signals, the total interference power in a frequency domain window of width B having a variable frequency offset was integrated and normalised by the in-band power for varying values of amplifier back-off. Figure 16.7(b) shows the results of this integration for amplifier back-off values up to 9 dB and adjacent carrier separations from B to $2B$. A carrier separation of B corresponds to directly adjacent signal bands, and such a set-up would approach the maximal possible system throughput of 2 symbols/s/Hz or 2 Baud/Hz.

It can be observed from the figure that the integrated interference levels are indistinguishable for amplifier back-off values of 6 dB and 9 dB. The integrated interference level in this case is below -30 dB with directly adjacent carriers and below -33 dB for all carrier separation values above $1.03B$. The drop in interference for carrier separations above $1.7B$ is

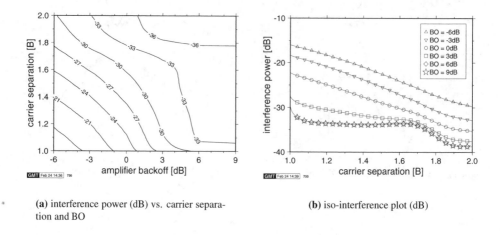

(a) interference power (dB) vs. carrier separation and BO

(b) iso-interference plot (dB)

Figure 16.7: Integrated adjacent channel interference power (dB) in a signal bandwidth of B for different values of normalised carrier separation and amplifier back-off

due to the frequency domain image of the virtual carriers that can be seen in Figure 16.6(a) at frequencies between $2.2B$ and $3.8B$. For a back-off value of 3 dB the interference is approximately 1 dB worse than for the higher back-off values for carrier separations above $1.5B$, and this difference is up to 5 dB for smaller carrier separations.

For a back-off value of 0 dB, which corresponds to the spectrum shown in Figure 16.6(b), the integrated interference is below -22.5 dB for all carrier separations. The figure also shows two graphs for negative amplifier back-off values for comparison, which exhibit considerably higher interference values. Since negative back-off values imply amplifier clipping below the nominal mean power of the signal, this would not be a likely case at the transmitter's power amplifier.

16.4.3 Clipping Amplification – Summary

We have seen that for OFDM transmission a considerable amplifier back-off is necessary, in order to maintain the BER and spectral characteristics of the ideal OFDM system. This complicates the transmitter design and leads to inefficient power amplifier applications.

Although a 512-subcarrier OFDM modem is capable of producing a peak-to-average power ratio of up to 27 dB, we have seen that much lower amplifier back-off values result in acceptable system performance. A back-off of 12.5 dB results in only one of 10^5 OFDM symbols suffering clipping at all, and we have seen for a 6 dB back-off that both the BER performance as well as the averaged spectrum are indistinguishable from the ideally amplified scenario.

16.5 Analogue-to-Digital Conversion

We have seen in Figure 16.2 that the time domain OFDM signal follows a two-dimensional or complex Gaussian distribution. The distribution of signal amplitudes determines the nec-

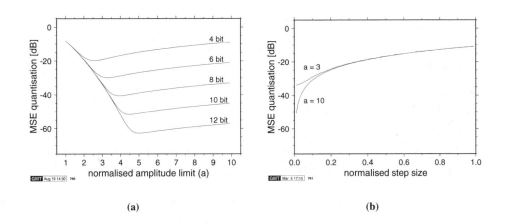

Figure 16.8: Mean square quantisation error for linearly spaced ADC at OFDM receiver: (a) quantisation noise for different amplitude limits normalised to the average input signal amplitude σ versus ADC resolution; (b) quantisation noise for amplitude limits of $a = 3$ and $a = 10$ versus the linear step size

essary parameters of the analogue-to-digital convertor (ADC) at the receiver. We have investigated the effects of the receiver input ADC on the BER performance of an OFDM system in an AWGN channel environment.

Since an ADC transforms the continuous analogue input signal to discrete-valued digital information, the original input signal is not preserved perfectly. As the dynamic range of an ADC is limited, there are amplitude clipping effects if the input signal exceeds the ADC's maximal input amplitude.

A linear analogue-to-digital convertor quantises the input signal into a set of n equidistant output values, which span an interval with clipping values of $-a$ and a. In this case, the ADC error stems from quantisation error due to the finite step size of $2a/n$ for input values within the dynamic range $[-a, a]$, as well as from amplitude clipping for input values outside this range.

Taking into account the Gaussian nature of the I and Q components of the OFDM signal, the mean square ADC error can be determined for different ADC set-ups. Figure 16.8(a) shows the mean square ADC error normalised to the average input signal power versus clipping amplitude for different quantisation resolution values. For this and the following figures, the input signal is modelled by a white Gaussian noise function with a standard deviation σ. The mean square quantisation error is relative to the mean input power σ^2, and the amplitude limit a is in units of σ.

The mean square ADC error value can be interpreted as the mean power of an additional random noise superimposed on the quantised signal. This ADC noise is not Gaussian, hence estimating the BER performance degradation with the Gaussian noise approximation is inadequate.

It can be seen from the figure that a small dynamic range of $2a$ results in high ADC noise irrespective of the number of ADC bits, which is due to predominant clipping. For $a = 1$, the normalised ADC noise power is about -8 dB, again independently of the resolution of the

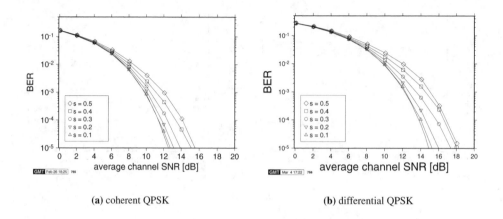

(a) coherent QPSK   (b) differential QPSK

Figure 16.9: BER versus channel SNR for a QPSK modem with coherent and differential detection using different normalised receiver ADC resolution values in an AWGN channel. The maximal normalised amplitude is $a = 3$

ADC. For higher values of a, the total ADC noise is dominated by the effects of increasing step size. Depending on the AGC's given resolution, there is an optimal value of a, for which the ADC noise, constituted by clipping and granular noise, is minimal.

In order to separate the effects of the amplitude limitation and the quantisation step size, BER simulations have been conducted with a fixed amplitude limiter and a range of ADC step sizes. The linear ADC step size is given for a range of ADC resolutions b and two amplitude limits a in Table 16.1. The effects of clipping on the OFDM performance have been discussed in the context of non-linear amplification, and Figure 16.4 showed that an amplifier back-off of 9 dB results in no BER performance degradation due to amplifier clipping in an AWGN channel for a 512-subcarrier OFDM modem. Setting the amplitude limits of the simulated ADC to $a = 3$ corresponds to a 9.5 dB dynamic range, hence the amplitude clipping contribution to the ADC noise does not significantly impair the modem's BER performance.

$a \backslash b$	4	5	6	7	8	9	10	11	12
3	0.375	0.188	0.094	0.0469	0.0234	0.0117	0.00586	0.00293	0.00146
10	0.625	0.313	0.156	0.0781	0.039	0.0195	0.00977	0.00488	0.00244

Table 16.1: Linear ADC step size for amplitude limits of $a = 3$ and $a = 10$ and ADC resolutions from $b = 4$ bits/sample to $b = 12$ bits/sample

The average ADC noise power for the normalised amplitude limits of $a = 3$ and $a = 10$ are depicted in Figure 16.8(b). The amplitude limit of $a = 3$ results in a residual ADC noise of about -34 dB due to amplitude clipping, while the amplitude clipping noise contribution for a limit of $a = 10$ is below -50 dB. For step sizes above 0.2, which corresponds to ADC resolutions of 5 bits or less, the gap between the two curves is closed, and the noise power is dominated by the quantisation noise.

The BER performance of a 512-subcarrier OFDM modem employing coherently as well as differentially detected QPSK is depicted in Figure 16.9. These experiments have been conducted for a maximal amplitude of $a = 3$ with step sizes from 0.1 to 0.5, which roughly corresponds to ADC resolutions between 6 and 3 bits. In both graphs the curve without markers represents the modem's BER performance in the absence of ADC quantisation and clipping noise.

It can be observed in both figures that the SNR penalty at a BER of 10^{-4} with an ADC step size of 0.1 is 0.15 dB and 0.2 dB for QPSK and DQPSK, respectively. This resolution corresponds to 60 steps in the range from -3 to 3, and therefore would require an ADC of 6-bit resolution. For higher step sizes the SNR penalty is more severe; a step size of 0.5 results in a SNR penalty of about 2.8 dB and 3.2 dB for coherent and differential detection, respectively. Let us now consider the effects of oscillator noise.

16.6   Phase Noise

The presence of phase noise is an important limiting factor for an OFDM system's performance [244, 245, 563], and depends on the quality and the operating conditions of the system's RF hardware. In conventional mobile radio systems around a carrier frequency of 2GHz the phase noise constitutes typically no severe limitation, however, in the 60GHz carrier frequency, 225MHz bandwidth system considered in Section 17.1.1 its effects were less negligible and hence had to be investigated in more depth. Oscillator noise stems from oscillator inaccuracies in both the transmitter and receiver and manifests itself in the baseband as additional phase and amplitude modulation of the received samples [564]. The oscillator noise influence on the signal depends on the noise characteristics of the oscillators in the system and on the signal bandwidth. It is generally split in amplitude noise $A(t)$ and phase noise $\Phi(t)$, and the influence of the amplitude noise $A(t)$ on the data samples is often neglected. The time domain functions $A(t)$ and $\Phi(t)$ have Gaussian histograms, and their time domain correlation is determined by their respective long-term power spectra through the Wiener-Khintchine theorem.

If the amplitude noise is neglected, imperfect oscillators are characterised by the long-term power spectral density (PSD) $N_p(f')$ of the oscillator output signal's phase noise, which is also referred to as the phase noise mask. The variable f' represents the frequency distance from the oscillator's nominal carrier frequency in a bandpass model, or equivalently, the absolute frequency in the baseband. An example of this phase noise mask for a practical oscillator is given in Figure 16.10(a). If the phase noise PSD $N_p(f')$ of a specific oscillator is known, then the variance of the phase error $\Phi(t)$ for noise components in a frequency band $[f_1, f_2]$ is the integral of the phase noise spectral density over this frequency band as in [564]:

$$\bar{\Phi}^2 = \int_{f_1}^{f_2} \left(\frac{2N_p(f')}{C} \right) df', \tag{16.4}$$

where C is the carrier power and the factor 2 represents the double-sided spectrum of the phase noise. The phase noise variance $\bar{\Phi}^2$ is also referred to as the integrated phase jitter, which is depicted in Figure 16.10(b).

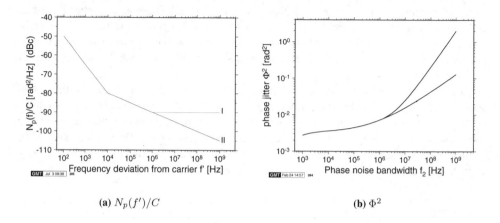

(a) $N_p(f')/C$ (b) Φ^2

Figure 16.10: Phase noise characterisation: (a) spectral phase noise density (phase noise mask), (b) integrated phase jitter for two different phase noise masks

16.6.1   Effects of Phase Noise

The phase noise contribution of both the transmitter and receiver can be viewed as an additional multiplicative effect of the radio channel, like fast and slow fading. In serial modulation schemes phase noise manifests itself as random phase errors of the received samples. The effects of this additional random phase component on the received samples depend on the modulation scheme employed, and as expected, they are more pronounced in differential detection schemes than in coherently detected arrangements invoking carrier recovery. The carrier recovery, however, is affected by the phase noise, which in turn degrades the performance of a coherently detected scheme.

For OFDM schemes, multiplication of the received time domain signal with a time-varying channel transfer function is equivalent to convolving the frequency domain spectrum of the OFDM signal with the frequency domain channel transfer function. Since the phase noise spectrum's bandwidth is wider than the subcarrier spacing, this results in energy spillage into other sub-channels and therefore in intersubcarrier interference, an effect which will be quantified below.

16.6.2   Phase Noise Simulations

In our studies two different models of the phase noise in an OFDM communications system have been investigated: the simple bandlimited white phase noise model, which is solely based on the value of the integrated phase jitter of Equation 16.4, and a coloured phase noise model, which also takes into account the phase noise mask of Figure 16.10(a). Both of these models will be described below.

16.6.2.1   White Phase Noise Model

The simplest way of modelling the effect of phase noise in simulations is to assume uncorrelated Gaussian distributed phase errors with a standard deviation of $\bar{\Phi}$ at the signal sampling instants. This corresponds to the phase noise exhibiting a uniform PSD $N_p(f') = (\bar{\Phi})^2/B$ throughout the signal bandwidth B.

As the phase noise is assumed to be white, the integrated phase jitter, or the variance of the phase noise given in Equation 16.4 constitute sufficient information for generating the phase error signal $\Phi(t)$. Clearly, since no correlation between the noise samples is assumed, this is the worst-case scenario with the highest possible differences of $\Phi(t)$ between two consecutive samples.

Simulations have been performed for both a 512-subcarrier OFDM modem, as well as for a serial modem for comparison. The serial modem's BER performance curves are depicted in Figure 16.11 for BPSK and QPSK employing both coherent and differential detection, for integrated phase jitter values of $\bar{\Phi}^2 = 0.05 \, rad^2$ to $\bar{\Phi}^2 = 0.25 \, rad^2$. For coherently detected BPSK, as shown in Figure 16.11(a), white phase noise of $\bar{\Phi}^2 = 0.05 \, rad^2$ already causes a measurable SNR penalty of about 0.6 dB at a BER of 10^{-4}, while a phase jitter of $0.1 \, rad^2$ results in a SNR penalty of 1.8 dB at this BER. For phase jitter values above $0.1 \, rad^2$, residual bit error rates of $6 \cdot 10^{-5}$, $5 \cdot 10^{-4}$ and $2 \cdot 10^{-3}$ were observed at phase jitters of 0.15, 0.2 and 0.25 rad^2, respectively.

16.6.2.1.1 Serial Modem The serial QPSK modem's performance is more vulnerable to the effects of phase noise than that of BPSK, as is shown in Figure 16.11(a) and 16.11(b). For all simulated values of $\bar{\Phi}^2$, the bit error rate exceeds 10^{-4}. For a phase jitter of 0.25 rad^2, the observed residual bit error rate is 6%. The curves for differential detection of BPSK and QPSK are shown in Figure 16.11(c) and 16.11(d), respectively. Since for differentially detected PSK schemes both the received modulated symbol as well as the reference phase are corrupted by phase noise, the phase noise effects are severe. The DBPSK modem exhibits residual bit errors of more than 10^{-4} for all simulated phase jitter values above 0.05 rad^2, while the DQPSK modem's BER performance is worse than 0.9% for all simulated values of $\bar{\Phi}^2$.

16.6.2.1.2 OFDM Modem While for serial modems the phase noise results in phase errors of the received samples, the FFT modem's performance is affected by the inter-subcarrier interference caused by the convolution of the phase noise spectrum with the OFDM spectrum. The average amount of power spillage from one subcarrier into the rest of the OFDM spectrum observed for different values of integrated phase jitter is depicted in Figure 16.12(a). It can be seen that the white phase noise results in uniform spreading of inter-subcarrier interference over all subcarriers, and a loss of signal power in the desired subcarrier number 256. It can be noted that doubling the variance of the phase noise results approximately in doubling the interference power in the other subcarriers, as demonstrated by the 3 dB differences seen in Figure 16.12(a). The received signal power in the central subcarrier drops accordingly.

The total interference power received in each subcarrier is the sum of all active subcarriers' contributions, which is shown in Figure 16.12(b) for a 512-subcarrier OFDM modem. It can be seen that the signal energy drops while the interference energy rises for rising values of the integrated phase jitter, with the signal-to-interference ratio (SIR) dropping to about 1.5, which corresponds to 2 dB for $\bar{\Phi}^2 = 0.5$. For small values of phase jitter up to approximately 0.1, the interference power rises linearly with the phase jitter, and more slowly for higher values of phase jitter. The interchannel-interference (ICI) experienced in each subcarrier for the white phase noise channel depending on the phase jitter is shown in Figure 16.13(a). The

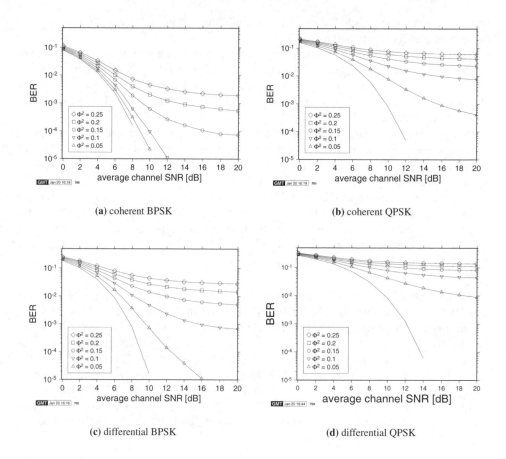

(a) coherent BPSK

(b) coherent QPSK

(c) differential BPSK

(d) differential QPSK

Figure 16.11: Bit error rate versus channel SNR for a serial modem in an AWGN channel and exposed to white phase noise for integrated phase jitter values of $0.05\ rad^2$ to $0.25\ rad^2$: (a) BPSK, (b) QPSK, (c) DBPSK, (d) DQPSK

SIR is above 10 dB for phase jitter values below approximately 0.1, and 20 dB for values phase jitter values of 0.01. This is in accordance to the linear relationship of interference power and phase jitter for small values observed above.

Figure 16.14 shows the BER performance results observed for a 512-subcarrier OFDM modem in the white phase noise channels. For coherently detected OFDM BPSK the BER performance is generally worse than for the serial modem, characterised in Figure 16.11(a). For differential detection, however, the OFDM modem's BER performance is better than the serial modem's depicted in Figure 16.11(c). This different behaviour can be explained by the different effects of the phase noise on the different systems. It has been observed in Figure 16.12(b) that for the investigated values of the integrated phase jitter the mean interference power in each subcarrier of an OFDM modem increases approximately linearly with the mean square of the phase error in the serial modem. For the same values of noise variance, coherent serial BPSK is more robust to the phase errors in the received serial symbols than

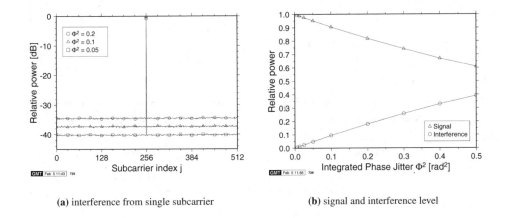

(a) interference from single subcarrier   (b) signal and interference level

Figure 16.12: The effect of white phase noise on OFDM: (a) power spillage due to white phase noise, (b) average signal and interference levels for one subcarrier

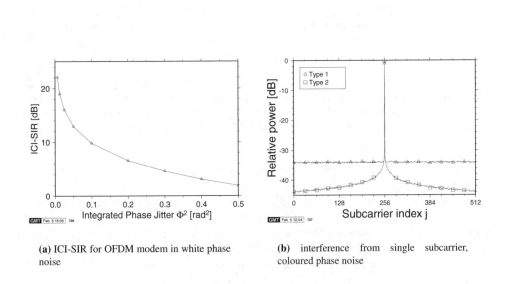

(a) ICI-SIR for OFDM modem in white phase noise   (b) interference from single subcarrier, coloured phase noise

Figure 16.13: OFDM in phase noise channel: (a) ICI-SIR for OFDM modem in white phase noise channel; (b) average signal and interference levels for one subcarrier in the coloured phase noise channels corresponding to the phase noise masks shown in Figure 16.10(a)

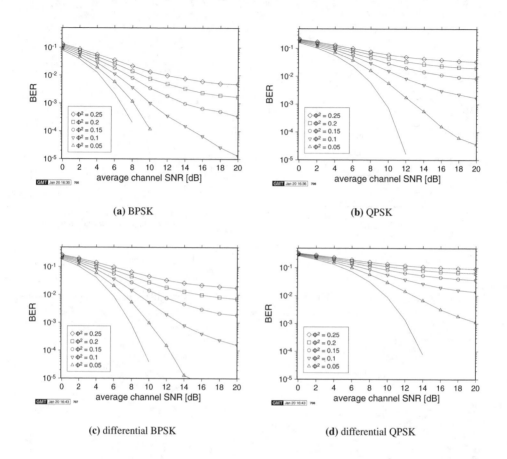

(a) BPSK   (b) QPSK

(c) differential BPSK   (d) differential QPSK

Figure 16.14: Bit error rate versus channel SNR for a 512 subcarrier OFDM modem in an AWGN channel and white phase noise for different values of integrated phase jitter of $0.05\ rad^2$ to
$0.25\ rad^2$: (a) BPSK, (b) QPSK, (c) DBPSK, (d) DQPSK (compared with Figure 16.11)

DBPSK, while the noise-like interference observed in the OFDM system is less detrimental
to the differentially detected BPSK than to coherent BPSK.

For both differential and coherent QPSK the OFDM modem performs better than the serial modem, as seen by comparing Figure 16.14(b) with 16.11(b) and 16.14(d) with 16.11(d).
There is, however, considerable degradation of the BER performance in both cases. Having
considered the worst-case scenario of uncorrelated white phase noise, let us now focus our
attention on the more practical case of coloured phase noise sources.

16.6.2.2   Coloured Phase Noise Model

The integral $\bar{\Phi}^2$ of Equation 16.4 characterises the long-term statistical properties of the oscillator's phase and frequency errors due to phase noise. In order to create a time domain
function satisfying the standard deviation $\bar{\Phi}^2$, a white Gaussian noise spectrum was filtered
with the phase noise mask $N_p(f')$ depicted in Figure 16.10(a), which was transformed into

f'[Hz]	100	1k	10k	100k	1M
N_p/C[dB]	-50	-65	-80	-85	-90

Table 16.2: Two-sided phase noise mask used for simulations. f' = frequency distance from carrier, N_p/C = normalised phase noise density

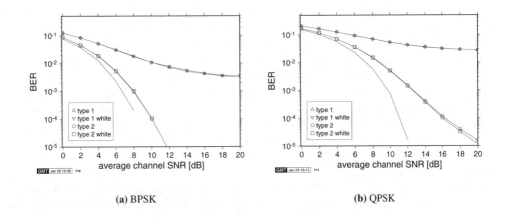

(a) BPSK   (b) QPSK

Figure 16.15: Bit error rate versus channel SNR for a 512-subcarrier OFDM modem in the presence of phase noise. Type 1 represents the coloured phase noise channel with the phase noise mask depicted in Figure 16.10(a) assuming a noise floor of 90 rad^2/Hz, while Type 2 is the channel without phase noise floor. The curves designated "white" are the corresponding white phase noise results. The lines without markers give the corresponding results in the absence of phase noise

the time domain. A frequency resolution of about 50 Hz was assumed in order to model the shape of the phase noise mask at low frequencies, which led to a FFT transform length of $2^{22} = 4194304$ samples for the frequency range of Figure 16.10(a).

The resulting time domain phase noise channel data is a stream of phase error samples, which were used to distort the incoming signal at the receiver. The double-sided phase noise mask used for the simulations is given in Table 16.2. Between the points given in Table 16.2, a log-linear interpolation is assumed, as shown in Figure 16.10(a). As the commercial oscillator's phase noise mask used in our investigations was not specified for frequencies beyond 1 MHz, two different cases were considered for frequencies beyond 1 MHz: (I) a phase noise floor at -90 dB, and (II) an $f^{-1/2}$ law. Both of these extended phase noise masks are shown in Figure 16.10(a). The integrated phase jitter has been calculated using Equation 16.4 for both scenarios, and the value of the integral for different noise bandwidths is depicted in Figure 16.10(b).

For the investigated 155 Mbits/s wireless ATM (WATM) system's [348–351] double-sided bandwidth of 225 MHz, the integration of the phase noise masks results in phase jitter values of $\bar{\Phi}^2 = 0.2303\ rad^2$ and $\bar{\Phi}^2 = 0.04533\ rad^2$ for the phase noise mask with and without noise floor, respectively.

The simulated BER performance of a 512-subcarrier OFDM system with a subcarrier distance $\Delta f = 440\, kHz$ over the two different phase noise channels is depicted in Figure 16.15 for coherently detected BPSK and QPSK. In addition to the BER graphs corresponding to the coloured phase noise channels described above, graphs of the modems' BER performance over white phase noise channels with the equivalent integrated phase jitter values was also plotted in the figures. It can be observed that the BER performance for both modulation schemes and for both phase noise masks is very similar for the coloured and the white phase noise models.

The interference caused by one OFDM subcarrier in the two coloured phase noise channels is depicted in Figure 16.13(b). It can be seen that the Type 1 channel exhibits a virtually white interference spectrum, very similar to the white phase noise channels depicted in Figure 16.12(a). Only the subcarriers directly adjacent to the signal bearer show higher interference influence than in the white phase noise channel. This is due to the Type 1 phase noise density given in the phase noise mask; it is non-flat only for frequencies below 1 MHz, which corresponds to approximately 2.3 subcarrier distances at a separation of 440 kHz. It can be observed in Figure 16.13(b) that higher interference was measured in the two subcarriers adjacent to the signal carrier. If the interference caused by all subcarriers is combined, then all but the very closest subcarriers have equal contributions to the interference signal, resulting in interference like Gaussian noise.

The Type 2 channel, which exhibits no noise floor in the phase noise mask, results in interference that is dependent on the distance from the interfering subcarrier. The summation of the interference is dominated by the interference contribution of the carriers in close vicinity, hence the resulting interference is less Gaussian than for the Type 1 channel.

The simulated BER results shown in Figure 16.15 show virtually indistinguishable performance for the modems in both the coloured and the white phase noise channels. Only a slight difference can be observed for QPSK between the Type 1 and Type 2 phase noise masks, where the corresponding white phase noise results in a better performance than the coloured noise. This difference can be explained with the interference being caused by fewer interferers compared to the white phase noise scenario, resulting in a non-Gaussian error histogram.

16.6.3 Phase Noise – Summary

Phase noise, like all time-varying channel conditions experienced by the time-domain signal, results in intersubcarrier interference in OFDM transmissions. If the bandwidth of the phase noise is high compared to the OFDM subcarrier spacing, then this interference is caused by a high number of contributions from different subcarriers, resulting in a Gaussian noise interference. Besides this noise inflicted upon the received symbols, the signal level in the subcarriers drops by the amount of energy spread over the adjacent subcarriers.

The integral over the phase noise mask, termed phase jitter, is a measure of the signal-to-interference ratio that can be expected in the received subcarriers, if the phase noise has a wide bandwidth and is predominantly white. The relationship between the phase jitter and the SIR is shown in Figure 16.13(a). For narrowband phase noise this estimation is pessimistic.

16.7 Summary

Following a brief historical perspective on the development of OFDM schemes in this chapter we commenced our discourse by characterising the system's achievable performance, when communicating over AWGN channels. It was shown that the various combinations of the subcarrier symbols result in a widely fluctuating transmitted power, which may result in high out-of-band spurious emissions, when the transmitter's amplifier exhibits an amplitude-dependent gain, which is referred to as amplifier non-linearity. Then finite dynamic range amplification was introduced and the effects of so-called clipping amplifiers on the system's BER were quantified. The effects of various analogue-to-digital converter resolutions were also evaluated. Finally, the influence of phase-noise was investigated.

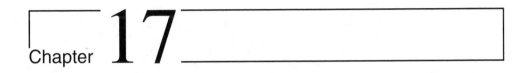

17

OFDM Transmission over Wideband Channels

Orthogonal frequency division multiplexing modems were originally conceived in order to transmit data reliably in time-dispersive or frequency-selective channels without the need for a complex time domain channel equaliser. In this chapter, the techniques employed for the transmission of quadrature amplitude modulated (QAM) OFDM signals over a time dispersive channel are discussed and channel estimation methods are investigated.

17.1 The Channel Model

The channel model assumed in this chapter is that of a finite impulse response (FIR) filter with time-varying tap values. Every propagation path i is characterised by a fixed delay τ_i and a time-varying amplitude $A_i(t) = a_i \cdot g_i(t)$, which is the product of a complex amplitude a_i and a Rayleigh fading process $g_i(t)$. The Rayleigh processes g_i are independent from each other, but they all exhibit the same normalised Doppler frequency f'_d, depending on the parameters of the simulated channel.

The ensemble of the p propagation paths constitutes the impulse response

$$h(t,\tau) = \sum_{i=1}^{p} A_i(t) \cdot \delta(\tau - \tau_i) = \sum_{i=1}^{p} a_i \cdot g_i(t) \cdot \delta(\tau - \tau_i), \qquad (17.1)$$

which is convolved with the transmitted signal.

All the investigations carried out in this chapter were based on one of the system and channel models characterised below. Each of the models represents a framework for a class of similar systems, grouped in three categories: wireless asynchronous transfer mode (WATM), wireless local area networks (WLAN) and a time division multiple access (TDMA) OFDM form of a personal communications type scheme, in a similar framework to the universal mobile telecommunication system (UMTS).

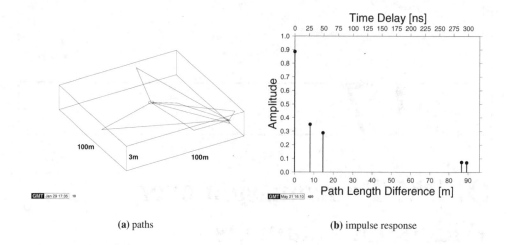

(a) paths (b) impulse response

Figure 17.1: The 60GHz, 225MHz bandwidth WATM channel: five-path model and resulting impulse response

17.1.1 The Wireless Asynchronous Transfer Mode System

The wireless asynchronous transfer mode (WATM) system parameters used for our investigations follow closely the specifications of the Advanced Communications Technologies and Services (ACTS) Median system, which is a proposed wireless extension to fixed-wire ATM-type networks. In the Median system, the OFDM FFT length is 512, and each symbol is padded with a cyclic prefix of length 64. The sampling rate of the Median system is 225 Msamples/s, and the carrier frequency is 60 GHz. The uncoded target data rate of the Median system is 155 MBps.

17.1.1.1 The WATM Channel

The WATM channel employed here is a pessimistic model of the operating environment of an indoor wireless ATM network similar to that of the ACTS Median system. For our simulations, we assumed a vehicular velocity of about 50 km/h or 13.9 m/s, resulting in a normalised Doppler frequency of $f'_d = 1.235 \cdot 10^{-5}$. The impulse response was determined by simple ray tracing in a warehouse-type environment of 100 m × 100 m × 3 m, which is shown schematically in Figure 17.1(a). The resulting impulse response, shown in Figure 17.1(b) exhibits a maximum path delay of 300 ns, which corresponds to 67 sampling intervals. The five taps of the impulse response were derived assuming free space propagation, using the inverse second power law.

The Fourier transform of this impulse response leads to the static frequency domain channel transfer function depicted in Figure 17.2(a). Note that the central frequency of the bandwidth is in FFT bin 0, hence the spectrum appears wrapped around in this graph. Throughout this work the subcarrier index 0 contains the central frequency, with the subcarriers 1 to $N/2$ spanning the positive relative frequencies, and the subcarriers $N/2 - 1$ to $N - 1$ containing the negative relative frequency range. Since the assumed impulse response is real, the result-

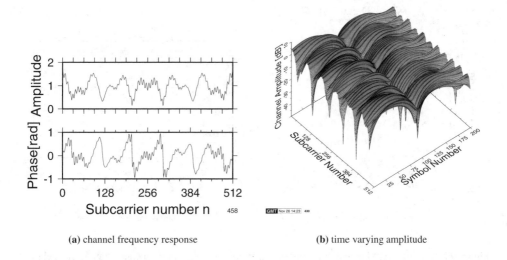

(a) channel frequency response   **(b)** time varying amplitude

Figure 17.2: WATM channel: (a) unfaded frequency domain channel transfer function $H(n)$; (b) time-varying channel amplitude for 200 consecutive OFDM symbols

ing channel transfer function is conjugate complex symmetric around the central frequency bin.

If each of the impulses in Figure 17.1(b) is faded according to a Rayleigh fading process with the normalised Doppler frequency f'_d, then the resulting time-varying impulse response will lead to a time-varying frequency domain channel transfer function, which is depicted in Figure 17.2(b). In this figure, the amplitude of the channel transfer function has been plotted over the bandwidth of the OFDM system for 200 consecutive OFDM symbol time slots.

Figure 17.2(b) reveals a relatively high correlation of the channel transfer function amplitude both along the time and the frequency axes. The correlation in the frequency domain is dependent on the impulse response; the longer the tap delays, the lower the correlation in the transfer function. The similarity of the channel transfer functions along the time axis stems from the slowly changing nature, i.e. the low normalised Doppler frequency, of the assumed narrowband channels. The impulse response changes only little between consecutive OFDM symbol intervals; the correlation between neighbouring channel transfer functions is therefore affected by the Doppler frequency of the constituting narrowband fading channels relative to the duration of one OFDM symbol. For this graph, the impulse response was kept constant for the duration of each OFDM symbol. We will show in Section 17.2.2 that the notion of defining a frequency domain channel transfer function is problematic in environments exhibiting rapid changes of the impulse response.

As the channel impulse response depicted in Figure 17.1(b) is of the line-of-sight (LOS) type with one dominant path, fading of this LOS path results in an amplitude variation across the whole signal bandwidth. This is seen in Figure 17.2(b) at around symbol number 100, where a substantial amplitude fade across all subcarriers occurs.

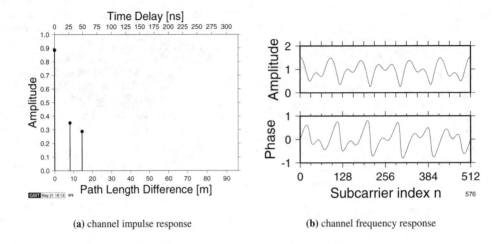

(a) channel impulse response   **(b)** channel frequency response

Figure 17.3: Short WATM channel: (a) impulse response, (b) unfaded frequency domain channel transfer function $H(n)$

17.1.1.2   The Shortened WATM Channel

Since the WATM channel discussed above is the worst-case scenario for an indoor wireless
high data rate network, we have introduced a truncated version of the original WATM channel impulse response depicted in Figure 17.1(b), by only retaining the first three impulses.
This reduces the total length of the impulse response, with the last path arriving at a delay
of 48.9 ns due to the reflection with an excess path length of about 15 m with respect to
the line-of-sight path, which corresponds to 11 sample periods. Omitting the two long delay
pulses in the impulse response does not affect the total impulse response energy significantly,
lowering the power by only 0.045 dB, hence no renormalisation of the impulse response was
necessary. The resulting impulse response exhibits a root mean squared (RMS) delay spread
of $1.5276 \cdot 10^{-8}$s, and it is shown in Figure 17.3(a). The resulting frequency domain transfer
function for the unfaded short WATM impulse response is given in Figure 17.3(b), and comparison with Figure 17.2(a) reveals a close correspondence with the original WATM channel
transfer function. As the first three low delay paths are the same for both channel models, the
general shape of the channel transfer function is very similar; the last two paths in the WATM
impulse response result in fast ripple on top of the dominant lower frequency components of
the channel transfer function of Figure 17.2(a), which is absent in the shortened channel of
Figure 17.3(b).

17.1.2   The Wireless Local Area Network System

The main wireless local area network (WLAN) system parameters we opted for were loosely
based on the high performance local area network (HIPERLAN) system [565], leading to a
sampling rate of 20 MHz and a carrier frequency of 17 GHz. The assumed data symbols
utilise a 1024-point FFT and a cyclic extension of 168 samples.

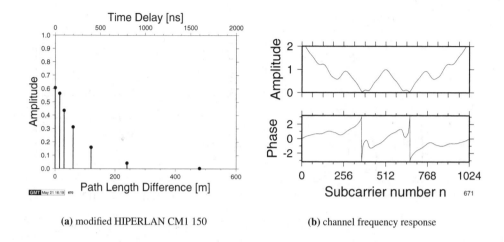

(a) modified HIPERLAN CM1 150   **(b)** channel frequency response

Figure 17.4: WLAN channel: (a) modified HIPERLAN CM1 150 ns impulse response, (b) unfaded frequency domain channel transfer function $H(n)$

17.1.2.1 The WLAN Channel

The WLAN channel's impulse response is based on the HIPERLAN CM1 impulse response with a RMS delay spread of 150 ns, as described by Tellambura *et al.* [565]. The second and third path of the original impulse response were combined to achieve a symbol-spaced impulse response pattern. The resulting impulse response is shown in Figure 17.4(a). The corresponding frequency domain channel transfer function is displayed in Figure 17.4(b). Assuming a worst-case vehicular velocity of 50km/h, fading channel simulations were conducted employing Rayleigh fading narrowband channels with a normalised Doppler frequency of $f'_d = 3.94 \cdot 10^{-5}$.

17.1.3 The UMTS System

The set of parameters employed for the UMTS-type investigations was based on a version of the ACTS FRAMES Mode 1 proposal [566], resulting in a sampling rate of 2.17 MHz, a carrier frequency of 2 GHz, and a bandwidth of 1.6 MHz. A 1024-point FFT OFDM symbol with 410 virtual subcarriers is employed, in order to comply with the spectral constraints. The data segment of the OFDM symbol is padded with a 168-sample cyclic extension.

17.1.3.1 The UMTS Type Channel

A discretised COST 207 bad urban (BU) impulse response, as described in the COST 207 final report [567], was chosen for the UMTS-type channel. The symbol-spaced discretised impulse response employed for our investigations is depicted in Figure 17.5(a). This impulse response results in a strongly frequency selective channel, as shown in Figure 17.5(b). The area shaded in grey shows the location of the virtual subcarriers in the OFDM spectrum. Since no signal is transmitted in this range of frequencies, the channel transfer function for the

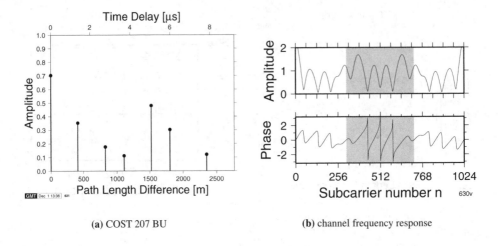

(a) COST 207 BU   (b) channel frequency response

Figure 17.5: UMTS channel: (a) COST 207 BU impulse response, (b) unfaded frequency domain channel transfer function $H(n)$

virtual subcarriers does not affect the modem's performance. For simulations under Rayleigh
fading conditions, a carrier frequency of 1.9 GHz and a vehicular velocity of 50 km/h was
assumed, leading to a normalised Doppler frequency of $f'_d = 4.0534 \cdot 10^{-5}$.

	f_c	$1/T_s$	$f_{d,max}$	$f'_{d,max}$	τ_{max}	RMS(τ)
UMTS	2 GHz	2.17 MHz	87.9 Hz	$4.05 \cdot 10^{-5}$	7.83 μs	3.28 μs
WLAN	17 GHz	20 MHz	787 Hz	$3.94 \cdot 10^{-5}$	1.6 μs	0.109 μs
WATM	60 GHz	225 MHz	2278 Hz	$2.345 \cdot 10^{-5}$	300 ns	34.3 ns
WATM - short	-"-	-"-	-"-	-"-	48.9 ns	16.9 ns

Table 17.1: Carrier frequency f_c, sample rate $1/T_s$, maximal Doppler frequency $f_{d,max}$, normalised
maximal Doppler frequency $f'_{d,max}$, maximal path delay τ_{max}, and channel RMS delay
spread RMS(τ) of the various system frameworks

17.2 Effects of Time Dispersive Channels on OFDM

The effects of the time variant and time dispersive channels on the data symbols transmitted
in an OFDM symbol's subcarriers are diverse. Firstly, if the impulse response of the channel
is longer than the duration of the OFDM guard interval, then energy will spill over between
consecutive OFDM symbols, leading to inter-OFDM symbol interference. We will not investigate this effect here, as the length of the guard interval is generally chosen to be longer than
the longest anticipated channel impulse response.

   If the channel is changing only slowly compared to the duration of an OFDM symbol,
then a time invariant impulse response can be associated with each transmitted OFDM symbol, which may change between different OFDM symbols. In this case, the frequency se-

lective effects of the channel result in a frequency-dependent multiplicative distortion of the received symbols, very similar to the effects of a time domain fading channel envelope in a serial modem. If the channel impulse response duration is shorter than the OFDM guard interval, then no intersubcarrier interference (ISI) is experienced, a case corresponding to a narrowband fading channel in the case of a serial modem. This will be investigated in Section 17.2.1.

A non-stationary channel, however, will introduce inter subcarrier interference due to the time variant impulse response. The effects of this ISI will be studied in Section 17.2.2.

17.2.1   Effects of the Stationary Time-Dispersive Channel

Here a channel is referred to as stationary if the impulse response does not vary significantly over the duration of one OFDM symbol, but it is time variant over longer periods of time. In this case, the time domain convolution of the transmitted time domain signal with the channel impulse response corresponds simply to the multiplication of the spectrum of the signal with the channel frequency transfer function $H(f)$:

$$s(t) * h(t) \longleftrightarrow S(f) \cdot H(f), \tag{17.2}$$

where the channel's frequency domain transfer function $H(f)$ is the Fourier transform of the impulse response $h(t)$:

$$h(t) \longleftrightarrow H(f). \tag{17.3}$$

Since the information symbols $S(n)$ are encoded into the amplitude of the transmitted spectrum at the subcarrier frequencies f_n, the received symbols $r(n)$ are the product of the transmitted symbol with the channel's frequency domain transfer function $H(n)$ plus the additive complex Gaussian noise samples $n(n)$:

$$r(n) = S(n) \cdot H(n) + n(n). \tag{17.4}$$

17.2.2   Non-Stationary Channel

A channel is classified here as non-stationary if the impulse response changes significantly over the duration of an OFDM symbol. In this case, the frequency domain transfer function is time variant during the transmission of an OFDM symbol and this time-varying frequency domain transfer function leads to the loss of orthogonality between the OFDM symbol's subcarriers. The amount of this intersubcarrier interference depends on the rate of change in the impulse response.

The simplest environment to study the effects of non-stationary channels is the narrowband channel, whose impulse response consists of only one fading path. If the amplitude of this path is varying in time, then the received OFDM symbol's spectrum will be the original OFDM spectrum convolved with the spectrum of the channel variation during the transmission of the OFDM symbol. As this short-term channel spectrum is varying between different transmission bursts, we will investigate the effects of the time-varying narrowband channel averaged over a high number of transmission bursts.

Since the interference is caused by the variation of the channel impulse response during the transmission of each OFDM symbol, we introduce the "OFDM symbol normalised"

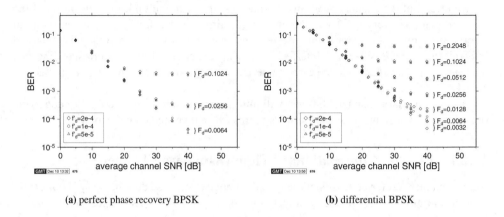

(a) perfect phase recovery BPSK   (b) differential BPSK

Figure 17.6: BPSK OFDM modem performance in a fading narrowband channel for normalised Doppler frequencies of $f'_d = 5 \cdot 10^{-5}$, $1 \cdot 10^{-4}$ and $2 \cdot 10^{-4}$ and FFT lengths between 16 and 4096. The FFT length for a given normalised Doppler frequency f'_d and an OFDM symbol normalised Doppler frequency F_d can be obtained from Table 17.2

Doppler frequency F_d:

$$F_d = f_d \cdot NT_s = f'_d \cdot N, \qquad (17.5)$$

where N is the FFT length, $1/T_s$ is the sampling rate, f_d is the Doppler frequency characterising the fading channel and $f'_d = f_d \cdot T_s$ is the conventional normalised Doppler frequency.

f'_d / N	16	32	64	128	256	512	1024	2048	4096
5×10^{-5}	-	-	0.0032	0.0064	0.0128	0.0256	0.0512	0.1024	0.2048
1×10^{-4}	-	0.0032	0.0064	0.0128	0.0256	0.0512	0.1024	0.2048	-
2×10^{-4}	0.0032	0.0064	0.0128	0.0256	0.0512	0.1024	0.2048	-	-

Table 17.2: OFDM symbol normalised Doppler frequency $F_d = f_d NT_s = f'_d N$ for FFT lengths between 16 and 4096 and the set of system parameters employed for non-stationary channel experiments

The BER performance for the OFDM modem configurations shown in Table 17.2 was determined by simulation and the simulation results for BPSK are given in Figure 17.6. Figure 17.6(a) depicts the BER performance of an OFDM modem employing BPSK with perfect narrowband fading channel estimation, where it can be observed that for any given value of F_d the different FFT lengths and channels behave similarly. For an F_d value of 0.0256, a residual bit error rate of about $2.8 \cdot 10^{-4}$ is observed, while for $F_d = 0.1024$ the residual BER is about 0.37%.

Figure 17.6(b) shows the performance of an OFDM system without the assumption of perfect phase recovery. Instead, differential detection is employed at the receiver, and, as in Figure 17.6(a), there is good correspondence between the BER curves for the same value of F_d. It can be seen that for F_d values from 0.0128 on, there is already a visible performance penalty at a SNR value of 40 dB. The highest observed residual BER is 4.2% for $f_d = 0.2048$.

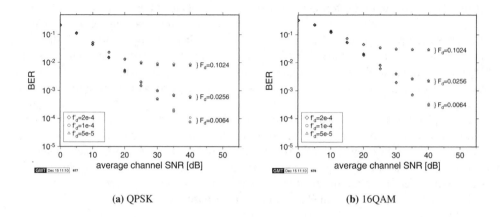

(a) QPSK　　　　　　　　　　　　　　　　**(b) 16QAM**

Figure 17.7: OFDM modem performance assuming perfect phase recovery in a fading narrowband channel with normalised Doppler frequencies of $f'_d = 5 \cdot 10^{-5}$, $1 \cdot 10^{-4}$ and $2 \cdot 10^{-4}$ and FFT lengths between 16 and 2048 symbols. The FFT length for a given normalised Doppler frequency f'_d and an OFDM symbol normalised Doppler frequency F_d can be obtained from Table 17.2

Higher-order modulation schemes, such as QPSK and 16-QAM, are more affected by the intersubcarrier interference caused by the time-varying narrowband fading channel characteristics, as shown in Figure 17.7. Again, the correspondence between different set-ups resulting in the same value of F_d is very good. Since the BER performance is limited by the intersubcarrier interference instead of noise, the BER curves of Figures 17.6(a), 17.7(a) and 17.7(b) do not exhibit the expected fixed SNR shift of 3 dB between BPSK and QPSK and 9.5 dB between BPSK and 16-QAM, as experienced for noise-only limited systems. From the BER curves and residual BER values of Figures 17.6 and 17.7, it can be deduced that the effects of the intersubcarrier interference for $F_d = 0.1024$ and $F_d = 0.0256$ are equivalent to the effects of noise at SNR values of $\gamma = 17$ dB and $\gamma = 28$ dB, respectively. These values are long-term averages and appear to be valid for all the modulation schemes used.

17.2.2.1 Summary of Time-Variant Channels

Time variant channels are of major influence on the BER performance of OFDM systems. The intersubcarrier interference caused by the time-varying nature of the transmission channel limits the attainable bit error rates for the UMTS and the WLAN scenario, where normalised Doppler frequencies of around $4 \cdot 10^{-5}$ and FFT lengths of 1024 are assumed, leading to $F_d \approx 0.04$. Therefore, an error floor of about 10^{-3} has to be faced by these systems, even if perfect channel estimation is assumed.

17.2.3 Signalling Over Time-Dispersive OFDM Channels

Analogously to the case of serial modems in narrowband fading channels, the amplitude and phase variations inflicted by the channel's frequency domain transfer function $H(n)$ upon the received symbols will severely affect the bit error probabilities, where different modulation schemes suffer to different extents from the effects of the channel transfer function. Coherent modulation schemes rely on the knowledge of the symbols' reference phase, which will be distorted by the phase of $H(n)$. If such a modulation scheme is to be employed, then this phase distortion has to be estimated and corrected. For multilevel modulation schemes, where also the magnitude of the received symbol bears information, the magnitude of $H(n)$ will affect the demodulation. Clearly, the performance of such a system depends on the quality of the channel estimation.

A simpler approach to signalling over fading channels is to employ differential modulation, where the information is encoded in the difference between consecutive symbols. Differential phase shift keying (DPSK) employs the phase of the previously received symbol as phase reference, encoding information in the phase difference between symbols. DPSK is thus only affected by the differential channel phase distortion between two consecutive symbols, rather than by the channel phase distortion's absolute value.

17.3   Channel Estimation

The issue of channel estimation was discussed in great detail in [173] both analytically as well as by simulation, hence here only a light-hearted introduction of this topic is offered. Frequency domain channel estimation algorithms generate the channel transfer function estimates $\hat{H}(n)$ for subsequent correction of the received symbols prior to demodulation. The accuracy of the algorithm influences the total system performance to a great extent, especially for systems employing multilevel modulation and coherent detection. We have investigated several different wideband channel estimation techniques, which can be split into two groups: those operating on the spectrum of the received OFDM symbol and those employing time domain correlation algorithms. This topic was discussed in great detail in analytical terms in [173], but here we restrict our discussions to a rudimentary conceptual treatment.

17.3.1   Frequency Domain Channel Estimation

Frequency domain channel estimation algorithms operate on the basis of the received samples in the frequency domain, that is after the receiver's FFT processing stage. At this stage, the frequency domain channel transfer function $H(n)$ can be estimated by using known frequency domain pilot symbols embedded in the OFDM symbol's spectrum and exploiting the strong correlation between consecutive samples of the frequency domain channel transfer function $H(n)$.

17.3.1.1   Pilot Symbol Assisted Schemes

Pilot symbol assisted modulation (PSAM) schemes obtain a channel transfer function estimate on the basis of known frequency domain pilot symbols that are interspersed with the transmitted data symbols [139]. Conventionally, PSAM schemes are utilised in narrowband fading environments for serial modems. For each received pilot symbol the instantaneous channel fading value is estimated as the quotient of the received and the expected symbol, and the channel fading estimation for the data symbols is derived from these pilot fading estimates by means of interpolation. A range of different interpolation techniques were comparatively studied by Torrance et al. [568] for serial modems in Rayleigh fading narrowband channels, ultimately favouring linear interpolation due to its low complexity and good performance. We will address the issues of interpolation schemes in the context of OFDM transmission in fading wideband channels below.

In parallel modems, PSAM can be utilised for estimating the frequency domain channel transfer function $H(n)$ in a time dispersive environment, if no inter-subcarrier interference is assumed. Accordingly, n_p pilot symbols P_i are transmitted in the subcarriers with indices $p_i, i = 1, \ldots, n_p$ within the total OFDM symbol bandwidth of N subcarriers. At the receiver, the channel transfer function $\hat{H}(p_i)$ at the pilot subcarriers is estimated from the received samples $r(p_i)$:

$$\hat{H}(p_i) = r(p_i)/P_i. \qquad (17.6)$$

In a second step, the values of the channel transfer function are estimated for the unknown data symbols by interpolation using the $\hat{H}(p_i)$ values of Equation 17.6. Clearly, the placement of the pilots and the interpolation technique will influence the quality of the channel estimation. In this chapter we will investigate two different interpolation techniques, linear and lowpass interpolation, both with a varying number of pilot subcarriers.

17.3.1.1.1 Linear Interpolation for PSAM   The simplest interpolation technique that can be used to estimate the channel transfer function for the subcarriers between two neighbouring channel estimation samples $\hat{H}(p_i)$ and $\hat{H}(p_{i+1})$ is a linear function:

$$\hat{H}(n) = \hat{H}(p_i) + \frac{\hat{H}(p_{i+1}) - \hat{H}(p_i)}{p_{i+1} - p(i)} \cdot (n - p_i) \quad \text{for} \quad p_i \leq n \leq p_{i+1}. \qquad (17.7)$$

Figure 17.8 shows the amplitude and the phase of the stationary WATM channel's frequency domain transfer function $H(n)$ for different numbers of equidistant pilot symbols in the frequency domain in an ideal noiseless environment. It can be seen that the channel estimation accuracy improves with increasing number of pilot tones in the spectrum. The number of pilot tones in the OFDM spectrum necessary to sample the channel transfer function can be determined on the basis of the sampling theorem as follows.

The frequency domain channel's transfer function $H(f)$ is the Fourier transform of the channel impulse response $h(t)$. Each of the impulses in the impulse response will result in a complex exponential function $e^{-(j2\pi\tau/T_s)f}$ in the frequency domain, depending on its time delay τ. In order to sample this contribution to $H(f)$ according to the sampling theorem, the maximum pilot spacing Δp in the OFDM symbol is:

$$\Delta p \leq \frac{N}{2\tau/T_s} \Delta f. \qquad (17.8)$$

In the case of the WATM channel impulse response with a maximum delay of $\tau/T_s = 67$, the resulting pilot spacing in the OFDM symbol is three subcarriers, requiring 171 pilot carriers per 512-subcarrier OFDM symbol. If the channel estimation is to resolve only the effects of the first three dominant paths with a maximum delay of $\tau/T_s = 11$ in Figure 17.1(b), then the minimal pilot spacing can be increased from 3 to 23 subcarriers.

Inspection of Figure 17.8 underlines these points: in order to resolve the full detail of the original transfer function, as depicted at the top of the figure, 171 pilots must be used. Furthermore, even using 171 pilots in the OFDM symbol, the estimated curve is not a very

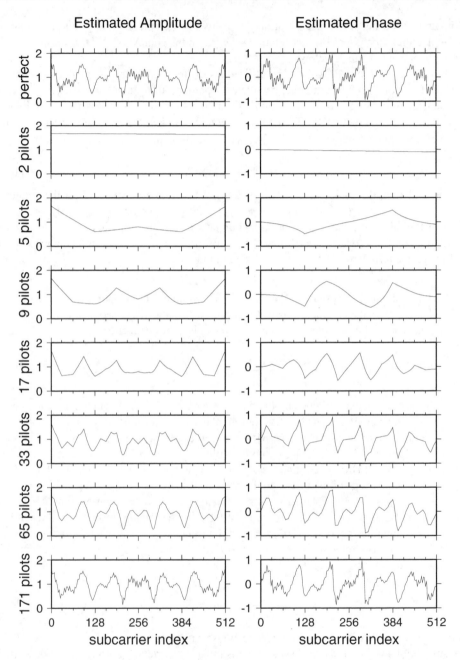

Figure 17.8: Estimation of the stationary WATM channel of Figure 17.2, a noiseless environment using linear interpolation between pilots employing 2 . . . 171 pilots per 512-subcarrier OFDM symbol

close match of the original. This is due to the linear interpolation algorithm used.

Nonetheless, the estimations performed with 33 and 65 pilots, respectively, show very similar levels of detail. Both schemes can appropriately sample the effects of the first three paths, but they are undersampled for capturing information concerning the two long-delay paths. If the pilot spacing is higher than the calculated number of 23, then the accuracy of the estimation declines further.

The shortened WATM channel, as characterised in Figure 17.3(a), can be sampled adequately utilising any pilot spacing of less than 23 subcarriers, which for our 512-subcarrier system corresponds to a minimum number of 23 pilots per OFDM symbol. Figure 17.9 shows the linearly interpolated channel estimation values for PSAM channel estimation using 2, ..., 171 pilots in the 512-subcarrier OFDM symbol for the noiseless shortened WATM channel. Again, it can be seen that the accuracy of the channel transfer function estimation improves with the number of pilots, but that a total of 171 pilots is necessary to yield a channel estimation that is indistinguishable from the perfect channel estimation curve in this scale. As 33 and 65 pilot subcarriers already exceed the 23 pilots required to sample the frequency domain channel transfer function according to the Nyquist sampling theorem, it becomes clear that the estimation errors are due to the linear interpolation algorithm and that – for optimal channel estimation accuracy – substantial oversampling of the channel has to be employed, thus deteriorating the system's bandwidth efficiency.

17.3.1.1.2 Ideal Lowpass Interpolation for PSAM The perfect interpolator for the measured channel transfer function samples $\hat{H}(p_i)$ is the ideal lowpass filter with a cut-off frequency of $1/\Delta p$, where Δp is the spectral distance between consecutive pilot subcarriers. Unlike the linear interpolation, this requires equidistant pilot placement in the OFDM symbol.

While an ideal rectangular filter transfer function is impossible to implement perfectly in the time domain due to its infinitely long impulse response, we employed an ideal rectangular frequency domain filter based on a FFT/IFFT operation. A perfect FFT-based lowpass filter can be implemented if its bandwidth coincides with a FFT bin of the chosen Fourier transform, which limits the set of possible pilot numbers. In our case, a 512-point FFT/IFFT operation was employed, therefore allowing for sets of 2^n with $1 \leq n \leq 9$ pilots.

For comparison with the linear interpolation, the same set of pilot distances Δp has been employed for the experiments. Since the lowpass interpolator does not require a pilot in the last subcarrier, the corresponding number of pilots is reduced by one for the lowpass interpolator, when compared to the linear interpolator.

Figure 17.10 gives an overview of the estimation accuracy of the PSAM scheme with ideal lowpass interpolation in a noiseless environment. Again, the resolution of the estimation depends on the number of pilots per OFDM symbol. The estimated transfer functions for 32 and for 64 pilots are identical and exactly correspond to the effect of the first three impulses in $h(t)$. In order to resolve $H(f)$ more finely, at least 170 pilot subcarriers must be employed. Closer inspection of the estimated $\hat{H}(n)$ for 170 pilots reveals relatively high errors in the estimation. This is due to non-optimal lowpass filtering; the optimum bandwidth of the lowpass filter corresponding to $\Delta p = 3\Delta f$ does not line up with the filter's FFT bins, hence the passband is too wide. This leads to considerable estimation errors, especially at the highest subcarrier indices, since there is no pilot symbol at the last subcarrier. If a perfect lowpass interpolation scheme was to be used over the WATM channel, then a pilot distance of $\Delta p = 2\Delta f$ would have to be used, resulting in 256 pilot subcarriers per 512-subcarrier OFDM symbol.

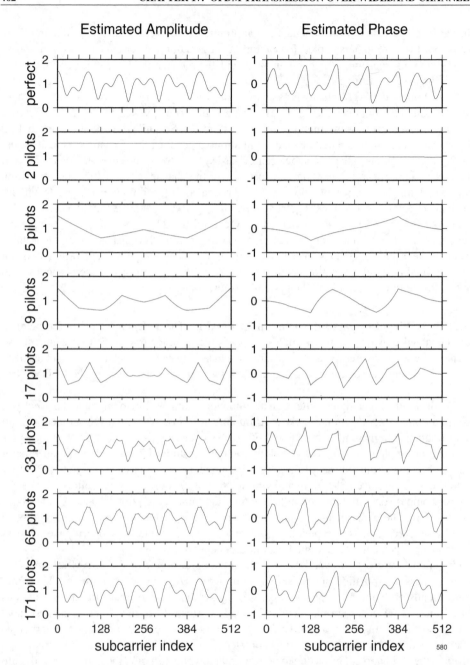

Figure 17.9: Estimation of the stationary shortened WATM channel of Figure 17.3, a noiseless environment using linear interpolation between pilots employing 2 ..., 171 pilots per 512-subcarrier OFDM symbol

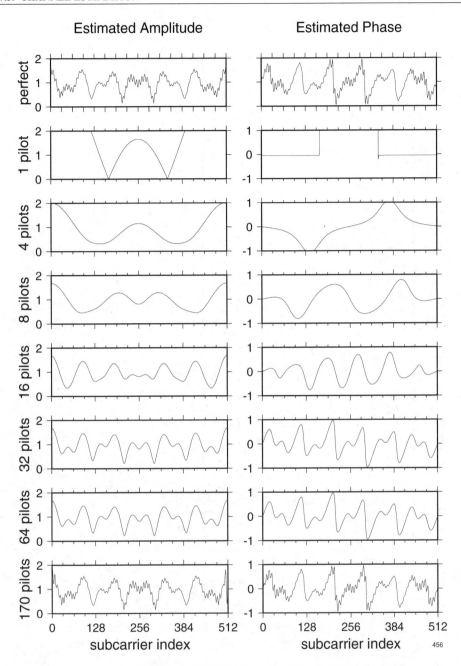

Figure 17.10: Estimation of the stationary WATM channel of Figure 17.2, a noiseless environment using ideal lowpass interpolation between pilots employing 2, ..., 171 pilots per 512-subcarrier OFDM symbol

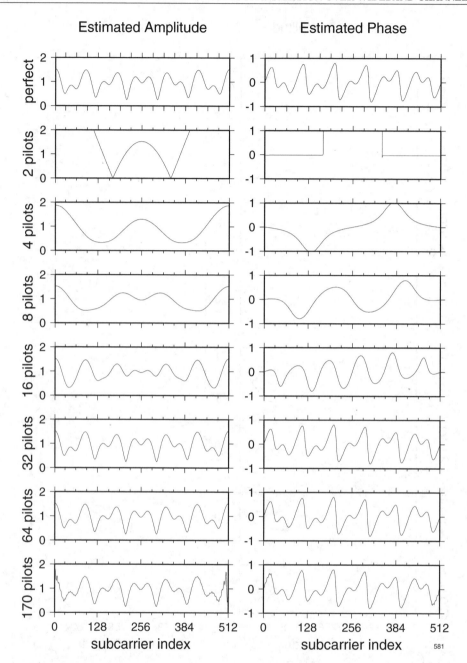

Figure 17.11: Estimation of the stationary shortened WATM channel of Figure 17.3, a noiseless environment using ideal lowpass interpolation between pilots employing 2, ..., 171 pilots per 512-subcarrier OFDM symbol

The equivalent estimated channel transfer functions for the case of the shortened WATM channel are depicted in Figure 17.11. It can be seen from the figure that the channel estimation is indistinguishable from the perfect case for the 32-pilot scenario and the 64-pilot scenario. Since the number of pilots necessary for the perfect sampling of this channel is 23, which corresponds to a pilot spacing of 23, the channel estimation utilising less than 23 pilots yields inaccurate results. If 170 pilot subcarriers are employed in each OFDM symbol, then the same effect as observed for the full-length WATM channel applies; because the frequency resolution of the FFT employed for the lowpass filter implementation is insufficient, the filter passband does not correspond exactly to the pilot frequency.

17.3.1.1.3 Summary   The Nyquist sampling theorem applies to the channel estimation in the frequency domain utilising pilot subcarriers in the OFDM symbol, but delivers only a lower limit for the number of pilots necessary, which is valid for perfect lowpass interpolation. If linear interpolation is employed, then oversampling beyond the Nyquist rate is necessary, in order to achieve high estimation accuracy. The number of pilot subcarriers that has to be used is dependent on the modulation scheme employed for signalling over the subcarriers.

The ideal lowpass interpolator yields better estimation accuracy than the linear interpolator for a given number of pilots, but the choice of pilot sets is limited by the implementation of the lowpass filter. If a FFT/IFFT based filter implementation is chosen, then only a limited set of possible pilot distances is available. This, in the WATM channel, leads to the need for a substantially higher number of pilots than would be necessary for the more flexible linear interpolator.

17.3.2   Time Domain Channel Estimation

An alternative approach to channel estimation in OFDM transmission systems is to estimate the channel directly from the time domain received signal. Analogous to wideband estimation in serial modems, a training sequence in the transmitted data stream can be employed in order to perform a correlation-based impulse response estimation. The corresponding frequency domain channel transfer function can be computed by FFT from this impulse response. A further attractive design alternative is to employ frequency-domain pilots for estimating the channel's frequency-domain tranfer function and invoke the previous estimates of the channel for improving the achievable channel estimation quality with the aid of predictive techniques, as suggested in [173]. The prediction itself may be carried out both in the time and the frequency domain [173].

17.4   System Performance

The various channel estimation techniques discussed above have been studied in a perfectly noiseless environment and without considering the effects of the channel estimation upon the system's bit error rate performance. In this section, we will consider the attainable performance of data transmission systems over the time dispersive channel, for both non-fading and stationary channels.

17.4.1   Static Time-Dispersive Channel

The static time dispersive channel exhibits a time invariant channel impulse response, which inflicts no intersubcarrier interference as described in Section 17.2.2. Therefore, the OFDM subcarriers are independent and each subcarrier n corresponds to an AWGN channel with a signal-to-noise ratio of γ_n. If the average channel SNR is γ, then the sub-channel SNR values γ_n depend on the magnitude of the frequency domain channel transfer function $H(n)$:

$$\gamma_n = \gamma \cdot |H(n)|^2. \tag{17.9}$$

If we assume perfect channel estimation, corresponding to $\hat{H}(n) = H(n)$, then the bit error rate $p_{e,n}$ for each subcarrier n depends exclusively on the sub-channel SNR γ_n, which allows the calculation of the overall system bit error rate p_e by simply averaging over the sub-channel SNR values:

$$p_e = \frac{1}{N_u} \sum_n p_{e,n} = \frac{1}{N_u} \sum_n p_e(\gamma_n), \tag{17.10}$$

where N_u is the number of subcarriers used for data signalling and the use of the same modulation scheme for all subcarriers is assumed.

17.4.1.1   Perfect Channel Estimation

Perfect channel estimation is the best-case scenario for OFDM transmission over time dispersive channels, as the performance is only limited by the SNR of the subcarriers, not by the reference phase and amplitude estimation errors in the receiver. Simulations have been performed for both coherent and non-coherent modulation schemes, where the frequency domain channel transfer function $H(n)$ was calculated at the receiver by applying the Fourier transform to the perfectly known noiseless impulse response of the channel, as depicted in Figure 17.1(b). The resulting channel transfer function is shown in Figure 17.2(a).

Figure 17.13(a) shows the measured BER per subcarrier for different levels of average OFDM SNR for the simulated 512-subcarrier modem over the WATM channel employing coherent binary phase shift keying (BPSK) in the subcarriers. It is apparent that the bit error probability varies significantly between different subcarriers, from about 2% in good subchannel conditions, up to 40% in the deep fades of the channel transfer function $H(f)$ at an average OFDM SNR of 0 dB. At an average SNR of 5 dB, groups of virtually errorfree subcarriers can be observed, interspersed with bundles of carriers exhibiting high bit error probabilities. At 15 dB, only a very small number of carriers experience measurable transmission error rates at all.

The recorded bit error rates in the simulation correspond very closely to the theoretical results obtained by calculating the sub-channel SNR γ_n for each subcarrier index n following Equation 17.9 and evaluating the appropriate bit error probability function for the chosen modulation scheme. Specifically, the BER over the AWGN channel was calculated employing Equations 16.1 to 16.3 For coherent modulation, the bit error probabilities $p_e(\gamma)$ for BPSK, QPSK and 16-QAM are depicted in Figure 16.1, and Figure 17.12 shows good correspondence between the obtained theoretical bit error rates and the simulation results.

The corresponding graph for QPSK, as shown in Figure 17.13(b), is similar to the BPSK case, only exhibiting higher bit error probabilities. Again, the bit error rate per subcarrier

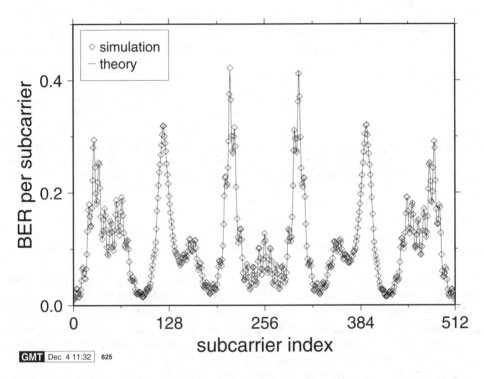

Figure 17.12: Simulated and theoretical BER per subcarrier for coherently detected BPSK transmission with perfect channel estimation over the static WATM channel of Figure 17.1(b) at an average OFDM SNR of 0 dB

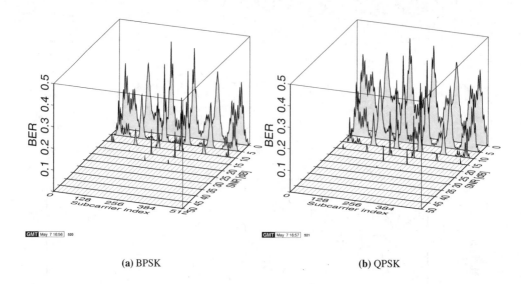

(a) BPSK   (b) QPSK

Figure 17.13: Simulated BER per subcarrier over the static WATM channel of Figure 17.1(b) with perfect channel estimation for (a) coherent BPSK transmission and (b) coherent QPSK transmission. The symmetry of the BER curves indicates that the unfaded CIR was real.

varies strongly with the channel transfer function, which is due to the variation in sub-channel SNR.

The OFDM bit error probability averaged over all subcarriers versus the channel SNR for BPSK, QPSK and 16-QAM transmission over the non-fading WATM channel is shown in Figure 17.14(a). The QPSK curve is shifted to the right by 3 dB compared to the BPSK performance curve, which is in accordance with the performance in a narrowband channel, as demonstrated in Section 16.3. The same observation can be made for the performance difference between 16-QAM and the other signalling schemes.

Although the relative BER performance relationship between the different modulation schemes corresponds to the BER results found in the narrowband AWGN channel, the absolute bit error rate values depend on the channel transfer function. Figure 17.14(b) depicts the corresponding BER versus channel SNR graph for the shortened WATM channel of Figure 17.3. The relative BER performance relationship between the different modulation schemes is the same as over the other channels, but the shape of the BER curve is different from that of the WATM channel. Comparing the frequency domain channel transfer function of the shortened WATM channel of Figure 17.3(b) and that of the original WATM channel of Figure 17.2(a), it can be seen that the fading is less severe in the shortened WATM channel, leading to better BER performance at high SNR values.

The UMTS channel depicted in 17.5(b) exhibits even stronger fading in the frequency domain than the two WATM channels, and consequently the BER performance is worse than that of both the WATM systems. At a BER of 10^{-3} the required SNR for the corresponding modulation schemes is about 9 dB higher for the UMTS system than for the WATM system.

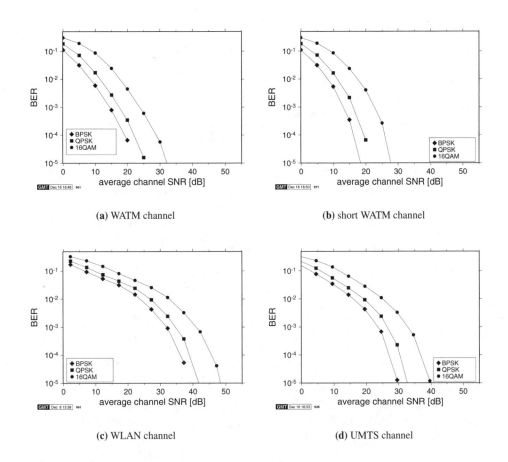

(a) WATM channel

(b) short WATM channel

(c) WLAN channel

(d) UMTS channel

Figure 17.14: Bit error rate versus channel SNR for static (a) WATM channel (Figure 17.1(b)), (b) shortened WATM channel (Figure 17.3), (c) WLAN channel (Figure 17.4), (d) UMTS channel (Figure 17.5(a)); perfect channel estimation and coherent detection

17.4.1.2 Differentially Coded Modulation

Differentially coded modulation enables the receiver to detect the data symbols without knowledge of the reference phase, therefore no channel estimation is required, if differential phase shift keying (DPSK) is used for modulating the subcarriers. Since the transmitted information is encoded in the phase difference between two consecutive subcarriers, there needs to be one phase reference subcarrier per OFDM symbol, typically in the first subcarrier.

While the absolute phase of the frequency domain channel transfer function $H(n)$ at subcarrier n does not affect the reception, the phase change of $H(n)$ between two consecutive subcarriers does influence the reception of the symbols. Note that the BER performance of both the WATM system in Figure 17.15(a) as well as that of the UMTS system in Figure 17.15(d) exhibit a residual bit error rate for high SNR levels for certain modulation schemes. These noise-independent errors are caused by channel effects and are investigated for the

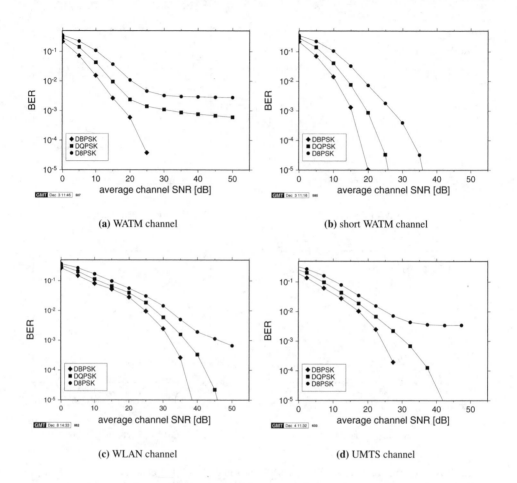

(a) WATM channel

(b) short WATM channel

(c) WLAN channel

(d) UMTS channel

Figure 17.15: Bit error rate versus channel SNR for static (a) WATM channel (Figure 17.1(b)), (b) shortened WATM channel (Figure 17.3), (c) WLAN channel (Figure 17.4), (d) UMTS channel (Figure 17.5(a)), using differential detection

UMTS system below.

In differential phase shift keying, the information to be transmitted is mapped to the phase difference between two consecutively received symbols. As all the symbols transmitted over the wideband OFDM channel suffer different phase rotations from the channel's frequency domain transfer function $H(n)$, the channel phase rotation difference $(\angle H(n) - \angle H(n-1))$ between subcarrier $(n-1)$ and subcarrier n offsets the receiver's decision boundaries for detected signals. Figure 17.16(a) shows the phase difference $(\angle H(n) - \angle H(n-1))$ between adjacent subcarriers for the UMTS channel of Figure 17.5. The grey area in the graph marks the positions of the virtual subcarriers, which are not used for transmission, and therefore do not have an impact on the system's performance. The horizontal lines at $\pm\pi/8$ and $\pm\pi/4$ mark the receiver's decision boundaries for DQPSK and D8PSK, respectively. If the channel transfer function's differential phase shift crosses the appropriate decision boundary for a

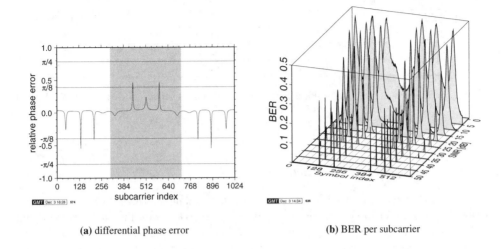

(a) differential phase error   (b) BER per subcarrier

Figure 17.16: Differential modulation over the static UMTS channel of Figure 17.5. (a) Phase difference between consecutive frequency domain samples of the frequency domain channel
transfer function $H(n)$; the grey area designates the virtual subcarriers, which are not
used for data transmission. (b) Simulated BER per subcarrier in static UMTS channel
for D8PSK without channel estimation

given subcarrier index, then the data symbol transmitted over this subcarrier suffers from
residual bit errors. In our example of Figure 17.16(a), there are four visible peaks comprising
actually 6 out of 612 data-bearing subcarriers crossing the $\pi/8$ decision boundary for D8PSK.
As the data is grey mapped on the PSK symbols, exactly one residual bit error will occur on
each of the six subcarriers, resulting in a residual bit error rate of $6/1836 = 3.268 \cdot 10^{-3}$.
Figure 17.16(b) gives an overview of the resulting BER per subcarrier for the data-bearing
subcarriers for SNR values from 0 dB up to 50 dB. It can be seen that the residual bit errors
are concentrated in four bursts, comprising the six symbols with subcarrier indices 136, 137,
312, 401, 477, and 478. At 50 dB SNR the symbols in the outer two bursts have a bit error
rate of $1/3$, corresponding to one bit error per D8PSK symbol. The bit error rate of the inner
two error bursts is lower, at about 23%, but is rising with higher SNR values to $1/3$. This
rising BER with higher SNR values for the two inner error bursts is explained by the channel
transfer function shown in Figure 17.16(a). The channel induces exactly one bit error for the
symbols in each of the corresponding subcarriers in the absence of noise. Since the phase
error due to the channel's phase rotation is very close to the decision 8-DPSK boundary, the
presence of noise can influence the reception towards the correct decision, therefore reducing
the long-term bit error rate.

In addition to residual bit errors, the decision boundary offset caused by the differential
phase shift of the channel transfer function as depicted in Figure 17.16(a) will also compromise the noise sensitivity of the subcarriers that are not directly affected by residual errors.

System	FFT length	Max. delay $[T_s]$	max. $\Delta p\,[\Delta f]$	min. N_p
WATM	512	67	3	171
short WATM	512	11	23	22
WLAN	1024	32	16	64
UMTS	1024	17	30	34

Table 17.3: Theoretical maximum pilot distance and minimal number of pilots for the WATM, short WATM, WLAN and UMTS systems

17.4.1.3 Pilot Symbol Assisted Modulation

As discussed in Section 17.3.1.1, pilot symbol assisted modulation (PSAM) schemes acquire a channel transfer function estimate $\hat{H}(n)$ by sampling $H(n)$ with the aid of known pilot symbols embedded in the transmission burst and interpolating between the received pilots. We have seen that the minimal number of pilots needed to sample the channel transfer function follows the sampling theorem and that the maximal pilot distance in the OFDM symbol is given by Equation 17.8. Furthermore, we have seen that this minimal number of pilots only applies to the ideal lowpass filter interpolation algorithm, and more pilot symbols have to be employed, if linear rather than lowpass interpolation is to be used, therefore trading receiver complexity for transmission overhead.

Here we will characterise the performance of the four different OFDM systems in their respective non-fading channel environments. Each system was investigated using a range of pilot numbers. Using Equation 17.8, we can predict the number of pilot symbols necessary for each scenario, if ideal interpolation is assumed, and these numbers are given in Table 17.3. The performance evaluation for all these scenarios was carried out for 8, 16, 32 and 64 pilots with ideal lowpass interpolation and with the equivalent 9, 17, 33 and 65 pilot sets for linear interpolation. We will concentrate on our most robust and least robust modulation schemes, namely BPSK and 16-QAM.

Figure 17.17 shows the BER performance for all the discussed systems in their respective non-fading channels employing PSAM-16-QAM with ideal lowpass interpolation as modulation scheme. It can be seen in Figure 17.17(a) that the WATM system faces high residual bit error rates for all sets of pilot symbols investigated. This is expected, because the number of pilot subcarriers employed in these simulations was lower than the minimum number of 171 stated in Table 17.3. Therefore, the channel transfer function is not adequately sampled in the frequency domain and the resulting channel estimation is of insufficient accuracy for coherently detecting 16-QAM. Note that the residual bit error rate drops with increasing number of pilots employed for the channel estimation, but that the curves for 32-pilot and 64-pilot PSAM exhibit virtually equal performance. Figure 17.17(a) reveals that the estimated channel transfer function for these two cases is the same, owing to the twin-burst structure of the WATM impulse response of Figure 17.1(b). Therefore, increasing the number of pilots from 32 to 64 will not increase the system's performance, since this does not allow the modem to resolve the channel effects due to the two paths around 90 m.

According to Table 17.3, the short WATM channel can be adequately sampled with 22 pilot symbols equidistantly interspersed with the data symbols. This is in accordance with the system's performance curves in Figure 17.17(b). The BER curves for 32 and 64 subcarriers

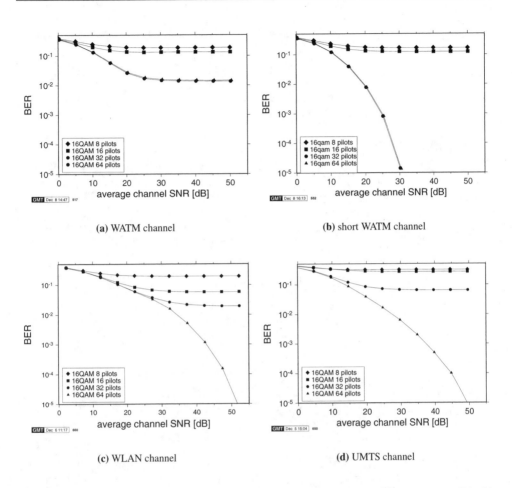

(a) WATM channel

(b) short WATM channel

(c) WLAN channel

(d) UMTS channel

Figure 17.17: Bit error rate versus channel SNR for static (a) WATM channel (Figure 17.1(b)), (b) shortened WATM channel (Figure 17.3), (c) WLAN channel (Figure 17.4), (d) UMTS channel (Figure 17.5(a)), PSAM-16QAM with ideal lowpass interpolation

exhibit essentially identical performance, while for 8 and 16 pilot symbols the residual bit error rate is above 10%. For the WLAN channel, at least 64 pilots have to be employed in order to sample all the frequency components of $H(f)$. We can see in Figure 17.17(c) that all but the 60 pilot symbol systems exhibit residual bit error rates and that the BER performance increases with an increased number of pilots.

If the more robust BPSK modulation scheme is used in the subcarriers, then the effects of inaccurate channel estimation are less pronounced. Figure 17.18 gives an overview of the BER performance for BPSK modems in the four operational environments, employing a perfect lowpass filter interpolator. The BER performance curves for the WATM system, shown in Figure 17.18(a), exhibit no residual errors. Like the 16-QAM curves, the performance is identical, irrespective of whether 32 or 64 pilot symbols are employed. Unlike in the 16-QAM case, however, the performance is best for 16-pilot subcarriers employed in the

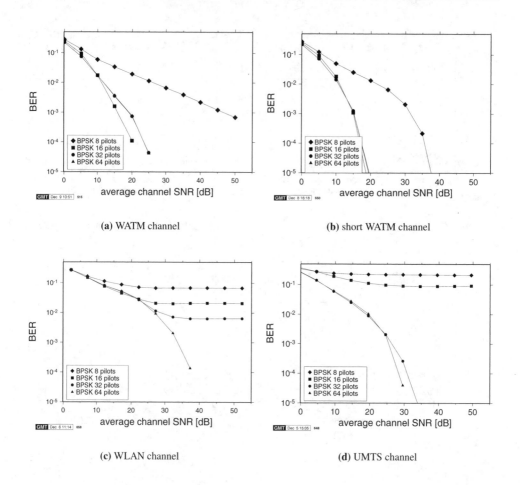

(a) WATM channel

(b) short WATM channel

(c) WLAN channel

(d) UMTS channel

Figure 17.18: Bit error rate versus channel SNR for static (a) WATM channel (Figure 17.1(b)), (b) shortened WATM channel (Figure 17.3), (c) WLAN channel (Figure 17.4), (d) UMTS channel (Figure 17.5(a)), PSAM-BPSK with ideal lowpass interpolation

OFDM symbol. This oddity can be explained by comparing the channel phase estimation errors for 16 and for 32 subcarriers, which are depicted in Figure 17.19. The phase estimation error for the WATM channel transfer function estimated with 16 pilot symbols is shown in Figure 17.19(a), which exhibits high fluctuations of the phase error of up to 0.7 radians. The phase estimation error for 32 pilot symbols, as shown in Figure 17.19(b), while substantially smaller on average, exhibits a peak error value of about 1.0 rad, which is closer to the BPSK decision boundary of $\pm\pi/2$ than the maximum error of the 16-pilot estimation. For higher SNR values the overall BER will be dominated by the bit errors occurring in the subcarriers with the highest phase estimation errors, therefore leading to the 16-pilot system performing better than the overall more accurate 32 and 64 pilot symbol estimation systems.

The WATM system's performance over the short WATM channel of Figure 17.3 is depicted in Figure 17.18(b). No residual bit error rates are observed and the BER performance

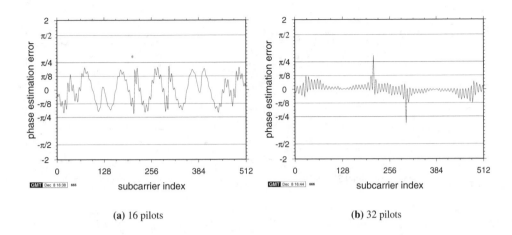

(a) 16 pilots

(b) 32 pilots

Figure 17.19: Channel transfer function phase estimation error for the WATM channel of Figure 17.1(b) employing ideal lowpass filter interpolator: (a) 16 pilot symbols, (b) 32 pilot symbols

with 32 and 64 pilots is identical. The 16-pilot BER curve is close to the 32- and 64-pilot results, and the modem performs about 18 dB worse with 8 pilot symbols. Both the WLAN as well as the UMTS channels exhibit residual bit errors for 8 and 16 pilot symbols and show an improving BER performance with increasing numbers of pilots. The WLAN system, whose performance is shown in Figure 17.18(c), performs without residual errors only when using 64 pilot symbols. For the UMTS channel, however, 32-pilot channel estimation results in a BER performance comparable to that of the 64-pilot estimation.

If the linear interpolation algorithm is employed instead of the ideal lowpass filter, then the BER performance of all the systems investigated is degraded. This is in accordance with the channel estimation experiments, as evidenced for example by the WATM channel estimation plots in Figures 17.8 and 17.10. The BER performance curves for 16-QAM transmission employing linearly interpolated PSAM are given in Figure 17.20, and it can be seen that for all modems, the performance is worse than that associated with the lowpass interpolation curves shown in Figure 17.17. The least differences between the two interpolation algorithms are visible for the WATM modem. In both cases, 64 pilot symbols are insufficient for combating residual bit errors and the residual bit error rates for both schemes are similar. In the short UMTS channel, the differences are more obvious. While the performance of the 32- and 64-pilot estimation systems was the same for lowpass interpolation, the linear interpolation algorithm shows residual errors for 32 pilot symbols. Compared to the lowpass interpolation, the BER results for the linear interpolator using 64 pilots are about 3 dB worse.

The BER performance for the WLAN system, as depicted in Figures 17.20(c) and 17.17(c), shows relatively minor differences between lowpass and linear interpolation. The residual bit error rates for 8, 16 and 32 pilots are similar, while for 64 pilots there is a SNR loss of about 5 dB for the linearly interpolated system at a BER of 10^{-3}. The UMTS, whose performance curves are plotted in Figure 17.20(d), exhibits a residual BER of about 1% for 64 pilots.

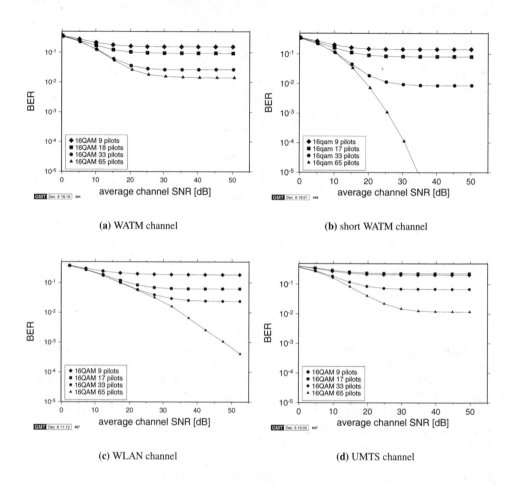

Figure 17.20: Bit error rate versus channel SNR for static (a) WATM channel (Figure 17.1(b)), (b) shortened WATM channel (Figure 17.3), (c) WLAN channel (Figure 17.4), (d) UMTS channel (Figure 17.5(a)), using PSAM 16-QAM with linear interpolation

For BPSK-PSAM transmission over the OFDM systems, the differences in performance between lowpass and linear interpolation algorithms are much less apparent than for 16-QAM. Figure 17.21 gives an overview of the BER performance of the simulated systems in their respective environments using BPSK with linear interpolation-based channel estimation. The BER performance of the WATM system, as depicted in Figure 17.21(a), does not vary significantly with the number of pilots employed. This differs from the curves observed for the lowpass interpolator in Figure 17.18(a), where much higher bit error rates were observed, if only 8 pilots were employed for the channel estimation. Both the linear as well as the lowpass interpolator results show best performance for 17 and 16 pilot symbols, respectively. Over the short WATM channel, systems employing either interpolator exhibit an essentially identical performance for all but 8 and 9 pilots, respectively. In the latter case, there is a performance difference of 13-dB SNR for a BER of 10^{-3} in favour of the linear interpolation.

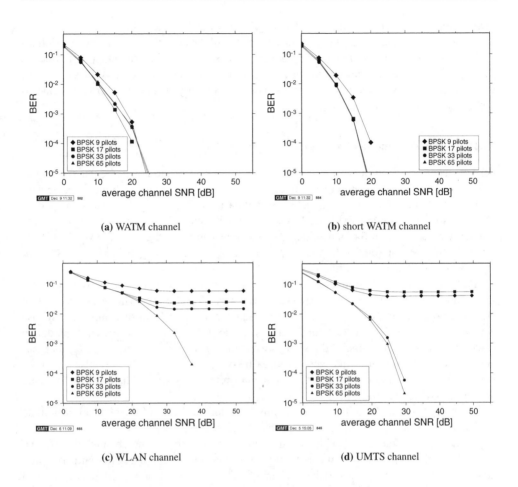

Figure 17.21: Bit error rate versus channel SNR for static (a) WATM channel (Figure 17.1(b)), (b) shortened WATM channel (Figure 17.3), (c) WLAN channel (Figure 17.4), (d) UMTS channel (Figure 17.5(a)), using PSAM-BPSK with linear interpolation

For the WLAN and the UMTS, the BER performance with lowpass and linear interpolation is fairly similar, the difference being slightly lower residual bit error rates for the linearly interpolated channel estimation.

17.4.2 Slowly Varying Time-Dispersive Channel

The slowly varying time-dispersive channel is characterised by an impulse response that is time-varying, but which is assumed constant for the duration of one OFDM symbol. Hence we also refer to this channel model as an OFDM symbol time-invariant channel. This stationary property prevents intersubcarrier interference and renders the wideband channel estimation possible. In this case, the channel can be viewed as a succession of static impulse responses generated upon fading the constituting paths. The rate of change for each of the paths' fading is described by the Doppler frequency of the fading channel.

As seen in Section 17.2.2, the time-varying nature of the channel transfer function will introduce intersubcarrier interference depending on the symbol normalised Doppler frequency $F_d = f_d \cdot T_s \cdot N$. Performance degradations can be observed even for small values of f_d. We will investigate the validity of the assumption of the stationary channel for our four scenarios in the case of perfect channel estimation.

17.4.2.1 Perfect Channel Estimation

Perfect channel estimation is the best-case scenario for transmission over wideband channels and the system performance in these circumstances constitutes an upper limit to the achievable performance with realistic channel estimation algorithms. As the results for perfect channel estimation are not influenced by sub-optimal channel estimation, the effects of the time-varying channel on the transmitted data symbols can be studied and the validity of the static channel assumption for the four operating frameworks, namely WATM, short WATM, WLAN and UMTS, can be verified.

Perfect channel estimation of the time-varying wideband channel was achieved by taking a snapshot of the channel impulse response in the centre of the OFDM symbol and calculating the corresponding frequency domain channel transfer function from this impulse response with the aid of the FFT. This transfer function does not fully characterise the channel, unlike in the static channel case, since it does not give any information concerning the intersubcarrier interference and only takes into account the time-varying nature of the impulse response. Nonetheless, the achieved channel estimation constitutes a best-case benchmark for realistic channel estimation algorithms.

In order to isolate the influence of the time-domain variations of the channel during the transmission of an OFDM symbol, experiments have been conducted for all four operating frameworks of Table 17.1 with two different models of the fading channel. Firstly, the channel impulse response was updated after every transmitted sample, resulting in a realistic model for a time-varying wideband channel. The second set of experiments employed a stationary channel model, where the impulse response was kept constant for the duration of each transmitted OFDM symbol. The resulting BER results from these experiments are shown in Figure 17.22. A BER performance degradation due to the time-varying channel can be observed for all the systems. The BER degradation is low for the WATM system over both the WATM and the shortened WATM channel with $F_d = 0.0063$, exhibiting a residual BER of under $2 \cdot 10^{-4}$ for 16-QAM. For the WLAN system and the UMTS, whose F_d values are about 0.04, the BER degradation is much more evident. In both cases, the residual BER for 16-QAM transmission is about $6 \cdot 10^{-3}$ and for BPSK the BER residual is $7.4 \cdot 10^{-4}$. Clearly, for the UMTS and WLAN at the given vehicular speed, the channel cannot be assumed to be stationary.

17.4.2.2 Pilot Symbol Assisted Modulation

In order to investigate the effects of imperfect channel estimation on the system's BER performance in wideband fading channels separately from the effects of time-domain variations of the impulse response, the PSAM experiments have been conducted in the stationary channel. Therefore, the observed performance degradation compared to the perfect channel estimation

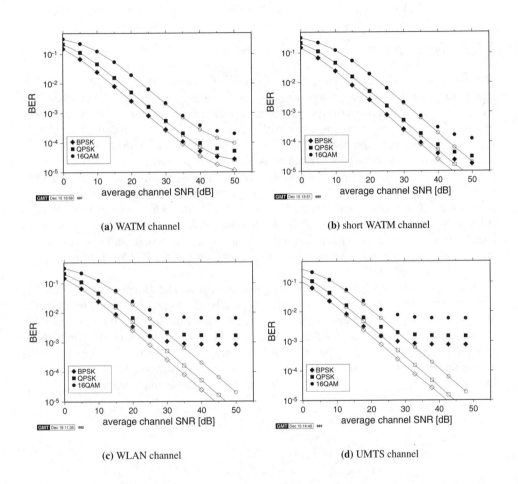

(a) WATM channel

(b) short WATM channel

(c) WLAN channel

(d) UMTS channel

Figure 17.22: Bit error rate versus channel SNR for fading (a) WATM channel (Figure 17.1(b)), (b) shortened WATM channel (Figure 17.3), (c) WLAN channel (Figure 17.4), (d) UMTS channel (Figure 17.5(a)), using coherent detection and perfect channel estimation. The open symbols correspond to the stationary channel model and the filled symbols correspond to the continuously fading channel model

results depicted in Figure 17.22 are caused by incorrect amplitude and phase estimations. The experimental results for 16-QAM transmission employing PSAM in conjunction with ideal lowpass and linear interpolation techniques are given in Figures 17.23 and 17.24, respectively. The estimation accuracy of the PSAM interpolation algorithms depends on each faded impulse response actually encountered by the modem, with some impulse amplitude combinations resulting in very poor estimation quality for the linear interpolation algorithm.

17.5 Summary

The performance of OFDM transmission over time-varying and time-dispersive channels is mainly limited by two factors, long-term shape and the time variability of the channel's impulse response. The impulse response determines the frequency domain channel transfer function, which manifests itself as frequency domain fading in the bandwidth of the OFDM symbol. The effects of this fading can be combated by the same methods as employed in serial modems, namely by pilot symbol assisted modulation (PSAM) or differential detection of the received data symbols. Analogous to time domain transmission systems, PSAM requires the transmission of additional pilot symbols and hence increases the system's overhead, while differential detection typically shows a 3 dB SNR performance loss compared to coherent schemes. For both differential detection and PSAM the structure of the channel's impulse response determines the performance. Long channel delays result in rapid fading in the frequency domain, requiring more pilots per data symbol for PSAM or decreasing the BER performance for differential detection.

Variations of the channel's impulse response, caused by fast fading of the constituting paths, result in interference between the OFDM symbol's subcarriers. This intersubcarrier interference is low, if the channel varies only insignificantly during one OFDM symbol, but causes severe performance degradation, if the rate of the channel variation is not much slower than the OFDM symbol rate. The effect of intersubcarrier interference therefore limits the maximum number of subcarriers employed in an OFDM system, depending on the channel's Doppler frequency and the robustness of the modulation scheme employed.

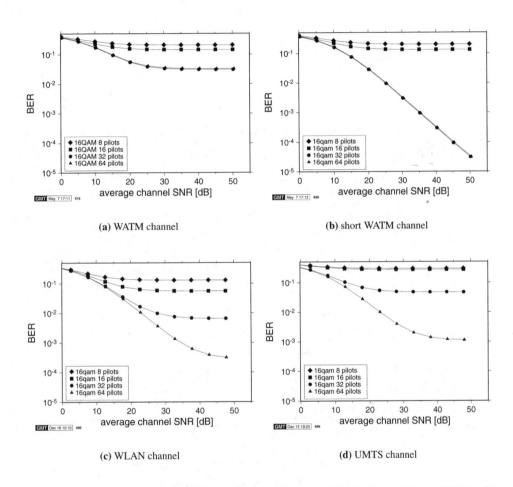

(a) WATM channel

(b) short WATM channel

(c) WLAN channel

(d) UMTS channel

Figure 17.23: Bit error rate versus channel SNR for fading (a) WATM channel (Figure 17.1(b)), (b) shortened WATM channel (Figure 17.3), (c) WLAN channel (Figure 17.4), (d) UMTS channel (Figure 17.5(a)), using coherent detection and pilot symbol assisted channel estimation with ideal lowpass filter interpolation for 16-QAM

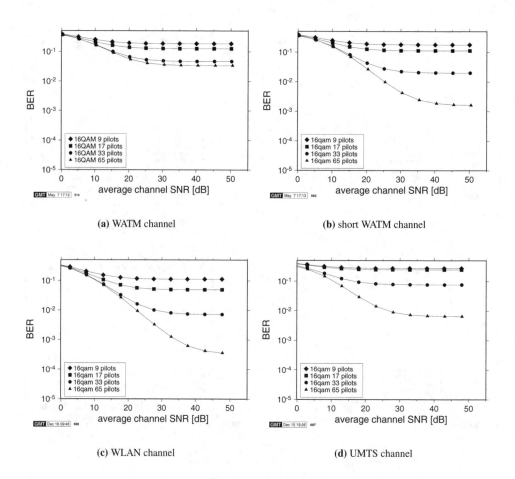

Figure 17.24: Bit error rate versus channel SNR for fading (a) WATM channel (Figure 17.1(b)), (b) shortened WATM channel (Figure 17.3), (c) WLAN channel (Figure 17.4), (d) UMTS channel (Figure 17.5(a)), using coherent detection and pilot symbol assisted channel estimation with linear interpolation for 16-QAM

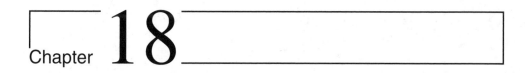

Chapter **18**

Time and Frequency Domain Synchronisation for OFDM

In this chapter we will investigate the effects of time and frequency domain synchronisation errors on the performance of an OFDM system, and two different synchronisation algorithms will be presented for time-domain burst-based OFDM communications systems.

18.1 System Performance with Frequency and Timing Errors

The performance of the synchronisation subsystem, in particular the accuracy of the frequency and timing error estimations, is of major influence on the overall OFDM system performance. In order to investigate the effects of carrier frequency and time domain FFT window alignment errors, a series of investigations has been performed over different channels.

18.1.1 Frequency Shift

Carrier frequency errors result in a shift of the received signal's spectrum in the frequency domain. If the frequency error is an integer multiple n of the subcarrier spacing Δf, then the received frequency domain subcarriers are shifted by $n \cdot \Delta f$. The subcarriers are still mutually orthogonal, but the received data symbols, which were mapped to the OFDM spectrum, are in the wrong position in the demodulated spectrum, resulting in a bit error rate of 0.5.

If the carrier frequency error is not an integer multiple of the subcarrier spacing, then energy is spilling over between the subcarriers, resulting in loss of their mutual orthogonality. In other words, interference is observed between the subcarriers, which deteriorates the bit error rate of the system. The amount of this intersubcarrier interference can be evaluated by investigating the spectrum of the OFDM symbol.

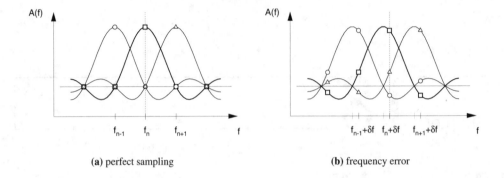

(a) perfect sampling   (b) frequency error

Figure 18.1: Stylised plot of OFDM symbol spectrum with sampling points for three subcarriers. The symbols on the curves signify the contributions of the three subcarriers to the sum at the sampling point: (a) no frequency offset between transmitter and receiver, (b) frequency error δf present

18.1.1.1   Spectrum of the OFDM Signal

The spectrum of the OFDM signal is derived from its time domain representation transmitted over the channel. A single OFDM symbol in the time domain can be described as:

$$u(t) = \left[\sum_{n=0}^{N-1} a_n e^{j\omega_n \cdot t} \right] \times \text{rect}\left(\frac{t}{N \cdot T_s} \right), \tag{18.1}$$

which is the sum of N subcarriers $e^{\omega_n \cdot t}$, each modulated by a QAM symbol a_n and windowed by a rectangular window of the OFDM symbol duration T_s. The Fourier transform of this rectangular window is a frequency domain sinc function, which is convolved with the Dirac delta subcarriers, determining the spectrum of each of the windowed complex exponential functions, leading to the spectrum of the nth single subcarrier in the form of

$$A_n(\omega) = \frac{\sin(N \cdot T_s \cdot \omega/2)}{N \cdot T_s \cdot \omega/2} * \delta(\omega - \omega_n).$$

Replacing the angular velocities ω by frequencies and using the relationship $N \cdot T_s = 1/\Delta f$, the spectrum of a subcarrier can be expressed as:

$$A_n(f) = \frac{\sin(\pi \frac{f - f_n}{\Delta f})}{\pi \frac{f - f_n}{\Delta f}} = \text{sinc}\left(\frac{f - f_n}{\Delta f} \right).$$

The OFDM receiver samples the received time domain signal, demodulates it by invoking the FFT and in case of a carrier frequency shift it generates the sub-channel signals in the frequency domain at the sampling points $f_n + \delta f$, which are spaced from each other by the subcarrier spacing Δf and misaligned by the frequency error δf. This scenario is shown in Figure 18.1. Figure 18.1(a) shows the sampling of the subcarrier at frequency f_n at the optimum frequency raster, resulting in a maximum signal amplitude and no intersubcarrier interference. If the frequency reference of the receiver is offset with respect to that of the transmitter by a frequency error of δf, then the received symbols suffer from intersubcarrier interference, as depicted in Figure 18.1(b).

The total amount of intersubcarrier interference experienced by subcarrier n is the sum of the interference amplitude contributions of all the other subcarriers in the OFDM symbol:

$$I_n = \sum_{j,j\neq n} a_j \cdot A_j(f_n + \delta f).$$

Since the QAM symbols a_j are random variables, the interference amplitude in subcarrier n, I_n, is also a random variable, which cannot be calculated directly. If the number of interferers is high, however, then the power spectral density of I_n can be approximated with that of a Gaussian process, according to the central limit theorem. Therefore, the effects of the intersubcarrier interference can be modelled by additional white Gaussian noise superimposed on the frequency domain data symbols.

The variance of this Gaussian process σ_{ISI_n} is the sum of the variances of the interference contributions,

$$\sigma_{ISI_n}^2 = \sum_{j,j\neq n} \sigma_{a_j}^2 \cdot |A_j(f_n + \delta f)|^2 .$$

The quantities $\sigma_{a_j}^2$ are the variances of the data symbols, which are the same for all j in a system that is not varying the average symbol power across different subcarriers. Additionally, because of the constant subcarrier spacing of Δf, the interference amplitude contributions can be expressed more conveniently as:

$$A_j(f_n + \delta f) = A_j n(\delta f) = sinc((n - j) + \frac{\delta f}{\Delta f}).$$

The sum of the interferer powers leads to the intersubcarrier interference variance expression:

$$\sigma_{ISI}^2 = \sigma_a^2 \cdot \sum_{i=-N/2-1}^{N/2} \left| sinc(i + \frac{\delta f}{\Delta f}) \right| . \tag{18.2}$$

The value of the intersubcarrier interference (ISI) variance for FFT lengths of $N = 64$, 512 and 4096 and for a range of frequency errors δf is shown in Figure 18.2. It can be seen that the number of subcarriers does not influence the ISI noise variance for OFDM symbol lengths of more than 64 subcarriers. This is due to the rapid decrease of the interference amplitude with increasing frequency separation, so that only the interference from close subcarriers contributes significantly to the interference load on the subcarriers.

In order to investigate the accuracy of the Gaussian approximation, simulations were conducted and histograms of the measured interference amplitude were produced for QPSK and 16-QAM modulation of the subcarriers. The triangles in Figure 18.3 depict the histograms of ISI noise magnitudes recorded for a 512-subcarrier OFDM modem employing QPSK and 16-QAM in a system having a frequency error of $\delta f = 0.3\Delta f$. The continuous line drawn in the same graph is the corresponding approximation of the histogram by a Gaussian probability density function (PDF) of the variance calculated using Equation 18.2. It can be observed that the Gaussian curve is a reasonable approximation for both histograms in the central region,

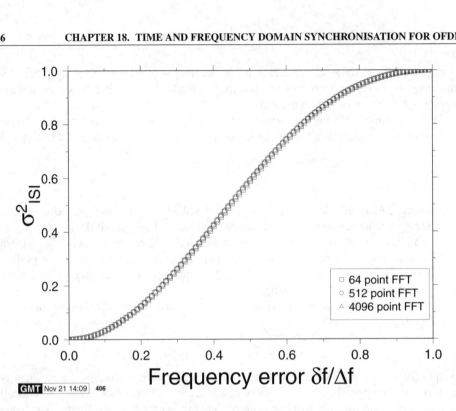

Figure 18.2: Intersubcarrier interference variance due to a frequency shift δf FFT lengths of $N = 64$, 512 and 4096 for normalised frequency errors $\delta f/\Delta f$ between 0 and 1

but that for the tails of the distributions the Gaussian function exhibits high relative errors. The histogram of the interference caused by the 16-QAM signal is, however, closer to the Gaussian curve than the QPSK interference histogram.

However, the frequency mismatch between the transmitter and receiver of an OFDM system not only results in intersubcarrier interference, but it also reduces the useful signal amplitude at the frequency domain sampling point by a factor of $f(\delta f) = \text{sinc}(\delta f/\Delta f)$. Using this and σ_{ISI}^2, the theoretical influence of the inter-subcarrier interference, approximated by a Gaussian process, can be calculated for a given modulation scheme in an AWGN channel. In the case of coherently detected QPSK, the closed-form expression for the BER $P_e(\gamma)$ at a channel SNR γ is given [94] by:

$$P_e(\gamma) = Q(\sqrt{\gamma}),$$

where the Gaussian $Q()$-function is defined as

$$Q(y) = \frac{1}{\sqrt{2\pi}} \int_y^\infty e^{-x^2/2} dx = \text{erfc}\left(\frac{y}{\sqrt{2}}\right).$$

Assuming that the effects of the frequency error can be approximated by white Gaussian noise of variance σ_{ISI}^2 and taking into account the attenuated signal magnitude $f(\delta f) =$

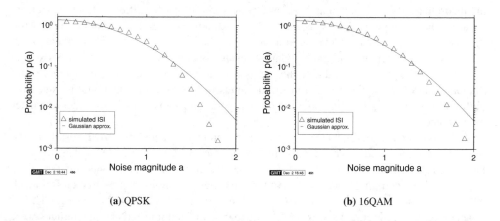

(a) QPSK   **(b) 16QAM**

Figure 18.3: Histogram of the ISI magnitude for a simulated 512-subcarrier OFDM modem using QPSK or 16-QAM for $\delta f = 0.3\Delta f$; the line represents the Gaussian approximation having the same variance

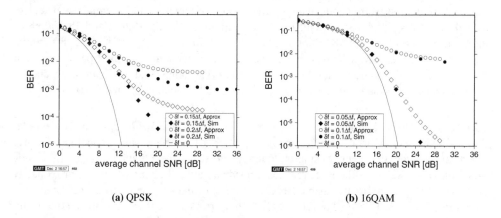

(a) QPSK   **(b) 16QAM**

Figure 18.4: The effect of intersubcarrier interference due to frequency synchronisation error on the BER over AWGN channels. (a) Bit error probability versus channel SNR for frequency errors of $0.15\Delta f$ and $0.2\Delta f$ for a QPSK modem. (b) BER versus channel SNR for frequency errors of $0.05\Delta f$ and $0.1\Delta f$ for a 16-QAM modem. In both graphs, the filled symbols are simulated BER results, while the open symbols are the predicted BER curves using the Gaussian intersubcarrier interference model

$sinc(\pi\delta f/\Delta f)$, we can adjust the equivalent SNR to:

$$\gamma' = \frac{f(\delta f) \cdot \sigma_a^2}{\sigma_{ISI}^2 + \sigma_a^2/\gamma},$$

where σ_a^2 is the average symbol power and γ is the real channel SNR. Comparison between the theoretical BER calculated using γ' and simulation results for different frequency errors δf are shown in Figure 18.4(a). While for both frequency errors the theoretical BER using the Gaussian approximation fits the simulation results well for channel SNR values of up to 12 dB, the predictions and the simulation results diverge for higher values of SNR. The pessimistic BER prediction is due to the pronounced discrepancy between the histogram and the Gaussian curve in Figure 18.3 at the tail ends of the amplitude histograms, since for high noise amplitudes the Gaussian model is a poor approximation for the intersubcarrier interference.

The equivalent experiment, conducted for coherently detected 16-QAM, results in the simulated and predicted bit error rates depicted in Figure 18.4(b). For 16-QAM transmission, the noise resilience is much lower than for QPSK, hence for our experiments smaller values of δf have been chosen. It can be observed that the Gaussian noise approximation is a much better fit for the simulated BER in a 16-QAM system than for a 4-QAM modem. This is in accordance with Figure 18.3, where the histograms of the interference magnitudes were depicted.

18.1.1.2   Effects of Frequency Mismatch on Different Modulation Schemes

In order to investigate the effects of frequency mismatch on different modulation schemes, a series of simulations were conducted employing both coherently and differentially detected, as well as pilot symbol assisted QPSK systems. Figures 18.5(a), 18.6(a) and 18.7(a) show the performance of the QPSK, pilot symbol assisted QPSK (PSA-QPSK) and differential QPSK (DQPSK) OFDM schemes, respectively. As a benchmark, the BER performance of the equivalent serial modulation scheme is plotted [94] as a line on all the graphs, which also represents the performance that is achieved by the parallel modem, when $\delta f = 0$.

18.1.1.2.1   Coherent Modulation   Figure 18.5(a) reveals the BER performance degradation due to increasing the carrier frequency offset δf. The adjacent sub-channel interference effects are considerable, even for small frequency errors. It is clear that for a carrier frequency offset of $\delta f = 0.2 \cdot \Delta f$, the BER reaches a residual value of about 10^{-3} at approximately 26 dB of SNR. The minimal possible BER for different values of carrier frequency mismatch are plotted in Figure 18.5(b). For relative frequency errors above $0.18 \cdot \Delta f$, the attainable bit error rate is worse than 10^{-4}.

18.1.1.2.2   PSAM   Figure 18.6(a) shows the bit error rate performance of the pilot-assisted QPSK system over an AWGN channel, which is consistently worse than that of an equivalent coherently demodulated system without pilot assistance. This is in accordance with the performance of PSAM schemes over AWGN channels, where PSAM systems are generally handicapped by errors in the channel estimation, which is unnecessary for non-fading channels. Since the frequency error mismatch results in additional noise in the received frequency domain pilots, the quality of the channel estimation deteriorates. This reduces the BER performance of a PSAM system compared to a coherently demodulated system refraining from using pilots.

The bit error rate curves given in the figure were computed for three pilot subcarriers per 512-subcarrier OFDM symbol, which were invoked to mitigate the effects of the channel's fading envelope in a time or frequency fading environment. The corresponding relationship

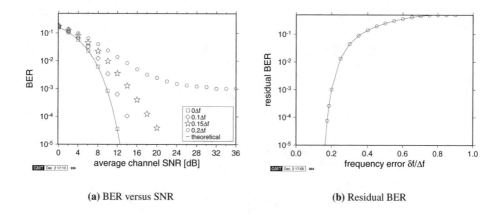

(a) BER versus SNR   **(b)** Residual BER

Figure 18.5: Bit error rate versus channel SNR performance for a non-pilot-assisted QPSK OFDM modem in an AWGN channel: (a) bit error rate versus signal-to-noise ratio plot for different constant frequency errors, (b) plot of residual bit error rate versus the frequency error

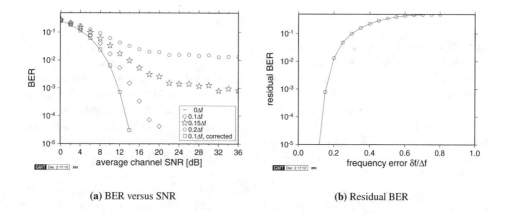

(a) BER versus SNR   **(b)** Residual BER

Figure 18.6: Bit error rate versus channel SNR performance for a 3-pilot-assisted QPSK OFDM modem in an AWGN channel: (a) bit error rate versus signal-to-noise ratio plot for different fixed frequency errors, (b) plot of residual bit error rate versus the frequency error

between the residual bit error rate and the frequency error is given in Figure 18.6(b). In the simulations presented here linear interpolation was employed between the pilots. Clearly, the BER performance of the system depends on the number of pilots employed as well as on the interpolation method, and these effects are discussed in Section 17.3.1.1 in more depth.

18.1.1.2.3 Differential Modulation The corresponding simulation results for differentially encoded QPSK are shown in Figure 18.7. Again, the impact of the intersubcarrier interference is severe even for small relative frequency errors. Figure 18.7(b) shows that a frequency error of only $0.12\Delta f$ results in a BER residual of 10^{-4}.

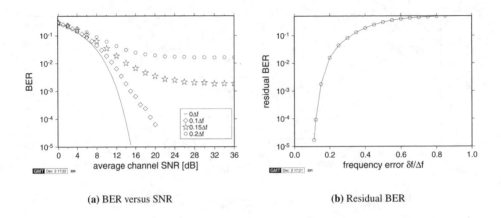

(a) BER versus SNR   (b) Residual BER

Figure 18.7: Bit error rate versus channel SNR performance for differential QPSK modulation scheme in AWGN channel. The OFDM FFT length is 512. (a) Bit error rate versus signal-to-noise ratio plot for different fixed frequency errors, (b) plot of residual bit error rate versus the frequency error

18.1.1.2.4   Frequency Error - Summary   A frequency error in an OFDM system results in a shift of the received frequency domain symbols relative to the receiver's raster. This leads to intersubcarrier interference, whose nature is noise-like, owing to the great number of contributing interfering subcarriers. Because of the frequency domain aliasing, the variance of this interference is constant for all subcarriers in the OFDM symbol, if all subcarriers carry the same average power. The variance of the interference can be computed and used to model the interference effects by additional white noise superimposed on the data symbols. Depending on the modulation scheme employed, this model typically yields a good estimation of the actual BER.

Different modulation schemes are affected differently by the presence of frequency errors in the system, analogously to their performance in purely AWGN environments. Coherent detection suffers the least penalty, followed by differential detection and PSAM schemes. Let us now consider the effects of time domain synchronisation errors.

18.1.2   Time-Domain Synchronisation Errors

Unlike frequency mismatch, as discussed above, time synchronisation errors do not result in intersubcarrier interference. Instead, if the receiver's FFT window spans samples from two consecutive OFDM symbols, inter-OFDM symbol interference occurs.

Additionally, even small misalignments of the FFT window result in an evolving phase shift in the frequency domain symbols, leading to BER degradation. Initially, we will concentrate on these phase errors.

If the receiver's FFT window is shifted with respect to that of the transmitter, then the time shift property of the Fourier transform, formulated as:

$$f(t) \quad \longleftrightarrow \quad F(\omega)$$
$$f(t - \tau) \quad \longleftrightarrow \quad e^{-j\omega\tau} F(\omega)$$

describes its effects on the received symbols. Any misalignment τ of the receiver's FFT window will introduce a phase error of $2\pi\Delta f\tau/T_s$ between two adjacent subcarriers. If the time shift is an integer multiple m of the sampling time T_s, then the phase shift introduced between two consecutive subcarriers is $\delta\phi = 2\pi m/N$, where N is the FFT length employed. This evolving phase error has a considerable influence on the BER performance of the OFDM system, clearly depending on the modulation scheme used.

18.1.2.1 Coherent Demodulation

Coherent modulation schemes suffer the most from FFT window misalignments, since the reference phase evolves by 2π throughout the frequency range for every sampling time misalignment. Clearly, this results in a total loss of the reference phase, and hence coherent modulation cannot be employed without phase correction mechanisms, if imperfect time synchronisation has to be expected.

18.1.2.2 Pilot Symbol Assisted Modulation

Pilot symbol assisted modulation (PSAM) schemes can be employed in order to mitigate the effects of spectral attenuation and the phase rotation throughout the FFT bandwidth. Pilots are interspersed with the data symbols in the frequency domain and the receiver can estimate the evolving phase error from the received pilots' phases.

This operation is performed with the aid of the wideband channel estimation discussed in Section 17.3.1.1, and the number of pilot subcarriers necessary for correctly estimating the channel transfer function depends on the maximum anticipated time shift τ. Following the notion of the frequency domain channel transfer function $H(n)$ introduced in Chapter 17, the effects of phase errors can be written as:

$$H(f) = e^{-j2\pi f\tau}. \tag{18.3}$$

Replacing the frequency variable f by the subcarrier index n, where $f = n\Delta f = n/(NT_s)$ and normalising the time misalignment τ to the sampling time T_s, so that $\tau = m \cdot T_s$, the frequency domain channel transfer function can be expressed as:

$$H(n) = e^{-j2\pi \frac{nm}{N}}. \tag{18.4}$$

The number of pilots necessary for correctly estimating this frequency domain channel transfer function $H(n)$ is dependent on the normalised time delay m. Following the Nyquist sampling theorem, the distance Δp between two pilot tones in the OFDM spectrum must be less than or equal to half the period of $H(n)$, so that

$$\Delta p \leq \frac{N}{2m}. \tag{18.5}$$

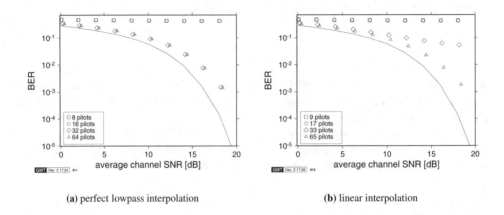

(a) perfect lowpass interpolation   (b) linear interpolation

Figure 18.8: Bit error rate versus channel SNR performance for 16-level PSA-QAM in an AWGN channel for different pilot subcarrier spacings in the presence of a fixed FFT window misalignment of $\tau = 10T_s$. The OFDM FFT length is 512. (a) PSAM interpolation using ideal lowpass interpolator, (b) PSAM using linear interpolator. In both graphs, the line marks the coherently detected 16-QAM performance in the absence of both FFT window misalignment and PSAM

The simulated performance of a 512-subcarrier 16-QAM PSAM modem in the presence of a constant timing error of $\tau = 10T_s$ in an AWGN channel is depicted in Figure 18.8 for both of the PSAM interpolation algorithms investigated in Section 17.3.1.1. Following Equation 18.5, the maximum acceptable pilot subcarrier distance required for resolving a normalised FFT window misalignment of $m = \tau/T_s = 10$ is $\Delta p = N/20 = 512/20 = 25.6$, requiring at least 20 pilot subcarriers equidistantly spaced in the OFDM symbol. We can see in both graphs of Figure 18.8 that the bit error rate is 0.5 for both schemes if less than 20 pilot subcarriers are employed in the OFDM symbol. For pilot numbers above the required minimum of 20, however, the performance of the ideal lowpass interpolated PSAM scheme does not vary with the number of pilots employed, while the linearly interpolated PSAM scheme needs higher numbers of pilot subcarriers for achieving a similar performance to the lowpass interpolator scheme. The continuous lines in the graphs show the BER curve for a coherently detected 16-QAM OFDM modem in the absence of timing errors, while utilising no PSAM. The BER penalty of PSAM in a narrowband AWGN channel as well as the performance differences between the two PSAM schemes are in accordance with the results found in Section 17.3.1.1.

18.1.2.3 Differential Modulation

Differential encoding of OFDM symbols can be implemented both between corresponding subcarriers of consecutive OFDM symbols or between adjacent subcarriers of the same OFDM symbol. We found the latter more advantageous in TDMA environments and hence this principle is employed here. The BER performance of differentially encoded modulation schemes is affected by the phase shift $\delta\phi$ between adjacent subcarriers introduced by timing errors and this influence can be evaluated for example for DPSK systems, as will be shown below.

The two-dimensional probability density function of a noisy phasor in polar coordinates is calculated in Appendix 18.6 and is given by Equation 18.49, if we assume the transmitted phase to be zero. Integration of this function over the magnitude r gives the phase probability function $p_\phi(\phi)$:

$$p_\phi(\phi) = \int_0^\infty \frac{r}{2\pi\sigma^2} \cdot e^{-(r^2+\mathcal{A}^2-2r\mathcal{A}\cos\phi)/2\sigma^2} dr, \qquad (18.6)$$

where $\mathcal{A}$ is the amplitude of the noiseless phasor. For differential phase modulation, the error of the difference between the phases of two consecutive symbols, $\Phi = \phi_k - \phi_{k-1}$, determines the symbol error rate (SER). The PDF $p_\Phi(\Phi)$ of this difference can be expressed by a variable transform, which results in the following integral:

$$p_\Phi(\Phi) = \int_{-\pi}^{\pi} p_\phi(\psi - \Phi) \cdot p_\phi(\psi) d\psi, \qquad (18.7)$$

where ψ is an auxiliary variable. The received symbol will be demodulated correctly, if the difference error $\Delta\phi$ is within the decision boundaries of the PSK constellation. For M-DPSK, the symbol error rate (SER) is given by:

$$SER = 1 - \int_{-\pi/M}^{\pi/M} p_\Phi(\Phi) d\Phi. \qquad (18.8)$$

If there is an FFT window misalignment induced phase shift $\delta\phi$ between consecutive symbols, then the integration limits in Equation 18.8 are biased by this shift:

$$SER = 1 - \int_{-\pi/M+\delta\phi}^{\pi/M+\delta\phi} p_\Phi(\Phi) d\Phi. \qquad (18.9)$$

Simulations have been performed for a 512-subcarrier OFDM system, employing DBPSK and DQPSK for different FFT window misalignment values. The BER performance curves for timing errors up to six sampling intervals are displayed in Figure 18.9. Note that one sample interval misalignment represents a phase error of $2\pi/512$ between two consecutive samples, which explains why the BER effects of the simulated positive timing misalignments marked by the hollow symbols are negligible for DBPSK. Specifically, a maximum SNR degradation of 0.5 dB was observed for DQPSK.

Positive FFT window time shifts correspond to a delayed received data stream and hence all samples in the receiver's FFT window belong to the same quasi-periodically extended OFDM symbol. In the case of negative time shifts, however, the effects on the bit error rate are much more severe due to inter-OFDM symbol interference. Since the data is received prematurely, the receiver's FFT window contains samples of the forthcoming OFDM symbol, not from the cyclic extension of the wanted symbol.

This non-symmetrical behaviour of the OFDM receiver with respect to positive and negative relative timing errors can be mitigated by adding a short postamble, consisting of copies of the OFDM symbol's first samples. Figure 18.10 shows the BER versus SNR curves for the same offsets, while using a 10-sample postamble. Now, the behaviour for positive and negative timing errors becomes symmetrical. Clearly, the required length of this postamble

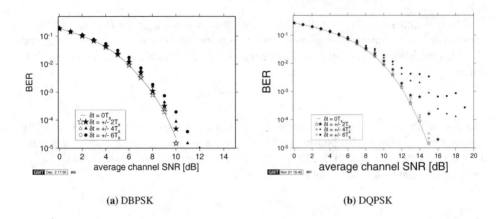

(a) DBPSK   (b) DQPSK

Figure 18.9: Bit error rate versus SNR over AWGN channels for a 512-subcarrier OFDM modem employing DBPSK and DQPSK, respectively. Positive time shifts imply time-advanced FFT window or delayed received data

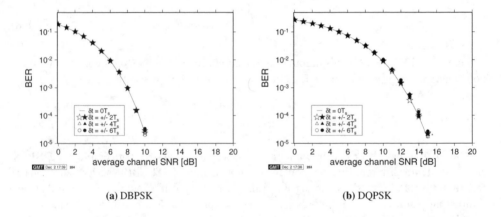

(a) DBPSK   (b) DQPSK

Figure 18.10: Bit error rate versus SNR over AWGN channels for a 512-subcarrier OFDM modem employing a postamble of 10 symbols for DBPSK and DQPSK, respectively. Positive time shifts correspond to time-advanced FFT window or delayed received data

depends on the largest anticipated timing error, which adds further redundancy to the system. This postamble can be usefully employed, however, to make an OFDM system more robust to time misalignments and thus to simplify the task of the timedomain FFT window synchronisation system.

18.1.2.3.1   Time-Domain Synchronisation Errors - Summary   Misalignment of the receiver's FFT window relative to the received sample stream leads to possible inter-OFDM symbol interference as well as to an evolving shift of the reference phase throughout the received frequency domain OFDM symbol. While the effects of inter OFDM symbol inter-

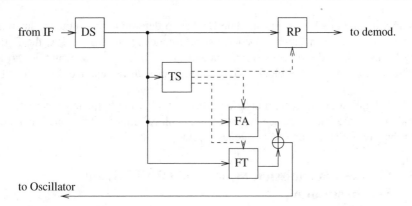

Figure 18.11: Block diagram of the synchronisation system: **DS**-downsampling and clock recovery, **TS**-time synchronisation, **FA**-frequency acquisition, **FT**-frequency tracking, **RP**-remove cyclic extension prefix

ference can be mitigated for moderate misalignments by appending a cyclic postamble to the OFDM symbol, the phase errors in the frequency domain make it impossible to use coherently detected modulation schemes without phase recovery methods. Instead, differentially detected schemes can be employed, which nonetheless suffer from performance degradation due to the phase errors. Alternatively, pilot symbol assisted channel estimation schemes can be employed in conjunction with coherent detection.

18.2 Synchronisation Algorithms

The results of Section 18.1 indicate that the accuracy of a modem's time and frequency domain synchronisation system dramatically influences the overall BER performance. We have seen that carrier frequency differences between the transmitter and the receiver of an OFDM system will introduce additional impairments in the frequency domain caused by intersubcarrier interference, while FFT window misalignments in the time domain will lead to phase errors between the subcarriers. Both of these effects will degrade the system's performance and have to be kept to a minimum by the synchronisation system.

In a TDMA-based OFDM system, the frame synchronisation between a master station – in cellular systems generally the base station – and the portable stations has also to be maintained. For these systems, a reference symbol marking the beginning of a new time frame is commonly used. This added redundancy can be exploited for both frequency synchronisation and FFT window alignment, if the reference symbol is correctly chosen.

In order to achieve synchronisation with a minimal amount of computational effort at the receiver, while also minimising the amount of redundant information added to the data signal, the synchronisation process is normally split into an acquisition phase and a tracking phase, if the characteristics of the random frequency and timing errors are known. In the acquisition phase, an initial estimate of the errors is acquired, using more complex algorithms and possibly a higher amount of synchronisation information in the data signal, whereas later the tracking algorithms only have to correct for small short-term deviations.

The block diagram of a possible synchronisation system is shown in Figure 18.11. The down sampling and clock recovery module DS has to determine the optimum sampling instant. The time synchronisation TS controls the frequency acquisition FA, the frequency tracking FT as well as the timedomain alignment of the FFT window, which is carried out by the "remove prefix" or RP block. These operations will be detailed during our further discourse.

At the beginning of the synchronisation process neither the frequency error nor the timing misalignment are known, hence synchronisation algorithms must be found that are sufficiently robust to initial timing and frequency errors.

18.2.1   Coarse Transmission Frame and OFDM Symbol Synchronisation

Coarse frame and symbol synchronisation algorithms presented in the literature all rely on additional redundancy inserted in the transmitted data stream. The pan-European digital video broadcasting (DVB) system uses a so-called null symbol as the first OFDM symbol in the time frame. No energy is transmitted [569], during the null symbol and it is detected by monitoring the received baseband power in the time domain, without invoking FFT processing. Claßen [228] proposed an OFDM synchronisation burst of at least three OFDM symbols per time frame. Two of the OFDM symbols in the burst would contain synchronisation subcarriers bearing known symbols along with normal data transmission carriers, but one of the OFDM symbols would be the exact copy of one of the other two, thus resulting in more than one OFDM symbol synchronisation overhead per synchronisation burst. For the ALOHA environment, Warner [246] proposed the employment of a power detector and subsequent correlation-based detection of a set of received synchronisation subcarriers embedded in the data symbols. The received synchronisation tones are extracted from the received time domain signal using an iterative algorithm for updating the synchronisation tone values once per sampling interval. For a more detailed discussion on these techniques the interested reader is referred to the literature [228, 246], while for a variety of further treatises to the contributions by Mammela and his team [570–572].

18.2.2   Fine Symbol Tracking

Fine symbol tracking algorithms are generally based on correlation operations either in the time or in the frequency domain. Warner [246] and Bingham [573] employed frequency domain correlation of the received synchronisation pilot tones with known synchronisation sequences, while de Couasnon [574] utilised the redundancy of the cyclic prefix by integrating over the magnitude of the difference between the data and the cyclic extension samples. Sandell [251] proposed exploiting the autocorrelation properties of the received time domain samples imposed by the cyclic extension for fine time domain tracking.

18.2.3   Frequency Acquisition

The frequency acquisition algorithm has to provide an initial frequency error estimate, which is sufficiently accurate for the subsequent frequency tracking algorithm to operate reliably. Generally the initial estimate must be accurate to half a subcarrier spacing. Sari [247] proposed the use of a pilot tone embedded into the data symbol, surrounded by zero-valued virtual subcarriers, so that the frequency-shifted pilot can be located easily by the receiver. Moose [248] suggested a shortened repeated OFDM symbol pair, analogous to his frequency tracking algorithm to be highlighted in the next section. By using a shorter DFT for this reference symbol pair, the subcarrier distance is increased and thus the frequency error estimation range is extended. Claßen [228, 229] proposed using binary pseudo-noise (PN) or so-called CAZAC training sequences carried by synchronisation subcarriers, which are also employed for the frequency tracking. The frequency acquisition, however, is performed by a search for the training sequence in the frequency domain. This is achieved by means of frequency domain correlation of the received symbol with the training sequence.

18.2.4 Frequency Tracking

Frequency tracking generally relies on an already established coarse frequency estimation having a frequency error of less than half a subcarrier spacing. Moose [248] suggested the use of the phase difference between subcarriers of repeated OFDM symbols in order to estimate frequency deviations of up to one-half of the subcarrier spacing, while Claßen [228] employed frequency domain synchronisation subcarriers embedded into the data symbols, for which the phase shift between consecutive OFDM symbols can be measured. Daffara [250] and Sandell [251] used the phase of the received signal's autocorrelation function, which represents a phase shift between the received data samples and their repeated copies in the cyclic extension of the OFDM symbols.

Following the above brief literature survey, we will investigate two different synchronisation algorithms, both making use of a reference symbol marking the beginning of a new time frame. This limits the use of both algorithms to systems whose channel access scheme is based on time division multiple access (TDMA) frames.

18.2.5 Time- and Frequency-Domain Synchronisation Based on Autocorrelation

Both the frequency and the time domain synchronisation control signals can be derived from the received signal samples' cyclic nature upon exploiting the OFDM symbols' cyclic time domain extension by means of correlation techniques. A range of symbol timing and fine frequency tracking algorithms were proposed by Mandarini and Falaschi [575]. Originally, Moose [248] proposed a synchronisation algorithm using repeated data symbols, and methods for the frequency error estimation using the cyclic extension of OFDM symbols were presented by Daffara et al. [250] and Sandell et al. [251]. The frequency acquisition and TDMA frame synchronisation proposed here are based on similar principles, employing a dedicated reference symbol exploited in the time domain [576].

No added redundancy in the data symbols and no a priori knowledge of the synchronisation sequences constituting the reference symbol are required, since only the repetitive properties of the OFDM symbols and those of the reference symbol (REF) in the proposed time division duplex (TDD) frame structure seen in Figure 18.13 are exploited. All the processing is carried out in the time domain, hence no FFT-based demodulation of the reference symbol is necessary.

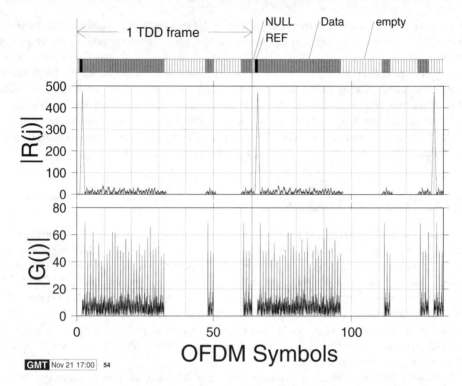

Figure 18.12: Time synchronisation: plots of the correlation terms $R(j)$ and $G(j)$ from Equation 18.10 and 18.11 for two consecutive 64-slot TDD frames under perfect channel conditions. The peaks indicate the correct TDD frame and OFDM symbol synchronisation instants, respectively

18.2.6   Multiple Access Frame Structure

The proposed 64-slot TDMA/TDD frame structure is depicted at the top of Figure 18.12, which is constituted by a null symbol, a reference symbol and 62 data symbols. Let us initially consider the role of the reference symbol.

18.2.6.1   The Reference Symbol

The reference symbol shown in Figure 18.13 was designed to assist in the operation of the synchronisation scheme and consists of repetitive copies of a synchronisation pattern SP of N_s pseudorandom complex samples. The synchronisation algorithm at the receiver needs no knowledge of the employed synchronisation pattern, hence this sequence could be used for channel-sounding training sequences or for base station identification signals. Note therefore that there are three hierarchical periodic time domain structures in the proposed framing scheme: the short-term intrinsic periodicity in the reference symbol of Figure 18.13, the medium-term periodicity associated with the quasi-periodic extension of the OFDM symbols and the long-term periodicity of the OFDM TDMA/TDD frame structure, repeating the reference symbol every 64 OFDM symbols, as portrayed in Figure 18.12. The long-term

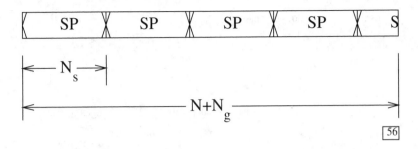

Figure 18.13: Reference symbol consisting of consecutive copies of a synchronisation pattern (SP) in the time domain

reference symbol periodicity is exploited in order to maintain OFDM frame synchronisation, while the medium-term synchronism of the cyclic extension assists in the process of OFDM symbol synchronisation. A detailed discussion of this figure will be provided during our further discourse. Let us begin with the macroscopic structure.

18.2.6.2   The Correlation Functions

The synchronisation algorithms rely on the evaluation of the following correlation functions $G(j)$ and $R(j)$, where j is the index of the most recent input sample:

$$G(j) = \sum_{m=0}^{N_g-1} z(j-m) \cdot z(j-m-N)^* \qquad (18.10)$$

$$R(j) = \sum_{m=0}^{N+N_g-N_s-1} z(j-m) \cdot z(j-m-N_s)^*, \qquad (18.11)$$

where $z(j)$ are the received complex signal samples, N is the number of subcarriers per OFDM symbol, N_g is the length of the cyclic extension and N_s is the periodicity within the reference symbol, as seen in Figure 18.13. The asterisk $*$ denotes the conjugate of a complex value.

$G(j)$ is used for both frequency tracking and OFDM symbol synchronisation, expressing the correlation between two sequences of N_g samples length spaced by N in the received sample stream, as shown in Figure 18.14. The second function, $R(j)$, is the corresponding expression for the reference symbol, where the period of the repetitive synchronisation pattern is N_s, as seen in Equation 18.11 and Figure 18.13. In this case, $(N_g + N - N_s)$ samples are taken into account for the correlation computation, and they are spaced by a distance of N_s samples. Having defined the necessary correlation functions for quantifying the time and frequency synchronisation error, let us now concentrate on how the synchronisation algorithms rely on their evaluation.

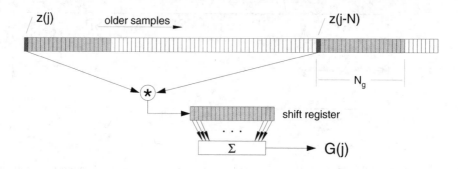

Figure 18.14: Schematic plot of the computation of the correlation function $G(j)$. The grey area represents the memory of the shift register

18.2.7   Frequency Tracking and OFDM Symbol Synchronisation

In this section we consider details of the frequency tracking and OFDM symbol synchronisation algorithms, which make use of $G(j)$, as defined by Equation 18.10.

18.2.7.1   OFDM Symbol Synchronisation

The magnitude of $G(j_{max})$ is maximum, if $z(j_{max})$ is the last sample of the current OFDM symbol, since then the guard samples constituting the cyclic extension and their copies in the current OFDM symbol are perfectly aligned in the summation windows. Figure 18.12 shows the simulated magnitude plots of $G(j)$ and $R(j)$ for two consecutive WATM frames, with $N = 512$ and $N_g = N_s = 50$, under perfect channel conditions. The observed correlation peaks of $|G(j)|$ can be easily identified at the last sample of an OFDM symbol. The amplitudes of the correlation peaks fluctuate, since the transmitted OFDM data symbols differ. The correlation peak magnitude is equal to the energy contained in the N_g samples of the cyclic extension and averages 50 for our system with $N_g = 50$ and an average sample power of unity.

The simulated accuracy of the OFDM symbol synchronisation in an AWGN channel is characterised in Figure 18.15. Observe in the figure that for SNRs in excess of about 7 dB the histogram is tightly concentrated around the perfect estimate, typically resulting in OFDM symbol timing estimation errors below $\pm 20 T_s$. However, as even slightly misaligned time domain FFT windows cause phase errors in the frequency domain, this estimation accuracy is insufficient. In order to improve the OFDM symbol timing synchronisation, the estimates must be lowpass filtered. Let us now concentrate on the issues of fine frequency tracking.

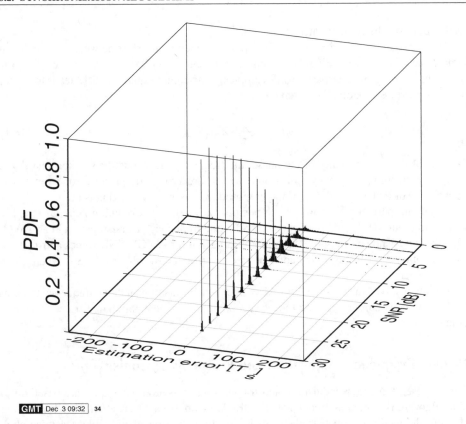

Figure 18.15: Histogram of the symbol timing estimation errors normalised to the sample interval T_s in an AWGN channel for $N = 512$ and $N_g = 50$ with no lowpass filtering of the estimates

18.2.7.2 Frequency Tracking

A carrier frequency error of δf results in an evolving phase error $\Psi(j)$ of the received samples $z(j)$:

$$\Psi(\delta f, j) \;=\; 2\pi\delta f \cdot j \cdot T_s \tag{18.12}$$

$$\;=\; 2\pi\frac{j\delta f}{N\Delta f}. \tag{18.13}$$

Clearly, the phase error difference between two received time domain samples $z(j_1)$ and $z(j_2)$ is a function of the frequency error and their time delay, and is given by $\Psi(\delta f, j_2) - \Psi(\delta f, j_1) = \Psi(\delta f, |j_2 - j_1|)$. If the original phase difference between the two received symbols $z(j_1)$ and $z(j_2)$ is known, and all other phase distortion is absent, then the phase difference error can be used to determine the frequency error δf.

Since the time domain samples of the cyclic extension or the guard interval are known to be a copy of the last N_g data samples of the OFDM symbol, the frequency error can be estimated using each of these N_g pairs of identical samples. In order to improve the estimation accuracy when exposed to noise and other channel impairments, averaging can be carried out over the N_g estimates.

The phase of $G(j)$ at $j = j_{max}$ equals the averaged phase shift between the guard time samples and the corresponding data samples of the current OFDM symbol. Since the corresponding sample pairs are spaced by N samples, rearranging Equation 18.13 leads to the fine frequency error estimation δf_t given by:

$$\delta f_t = \frac{\Delta f}{2\pi} \cdot \angle G(j_{max}). \tag{18.14}$$

Because of the 2π ambiguity of the phase, the frequency error must be smaller than $\Delta f/2$. Therefore, the initial frequency acquisition must ensure a rough frequency error estimate with an accuracy of better than $\Delta f/2$, if the proposed fine frequency tracking is used.

Assuming perfect estimation of the position j_{max} of the correlation peak $G(j_{max})$, as we saw in Figure 18.12, the performance of the fine frequency error estimation in an AWGN environment is shown in Figure 18.16(a). Observe that for AWGN SNR values above 10 dB, the estimation error histogram is concentrated to errors below about $0.02\Delta f$, where Δf is the subcarrier spacing.

Having resolved the issues of OFDM symbol synchronisation and fine frequency tracking, let us now focus our attention on the aspects of frequency acquisition and frame synchronisation.

18.2.8 Frequency Acquisition and Frame Synchronisation

Our proposed frequency acquisition and frame synchronisation techniques are based on the same algorithms as the fine frequency and OFDM symbol synchronisation. However, instead of using the medium-term periodicity of the cyclic extension of the OFDM data symbols, the dedicated reference symbol with shorter cyclic period $N_s < N$ is exploited in order to improve the frequency capture range, as it will be highlighted below.

18.2.8.1 Frame Synchronisation

Similarly to the OFDM symbol synchronisation, the magnitude of $R(j)$ in Equation 18.11 and Figure 18.12 is maximum when the periodic synchronisation segments SP of length N_s of the reference symbol shown in Figure 18.13 perfectly overlap. Again, the magnitude of $R(j)$ for two simulated TDD frames is shown at the top of Figure 18.12. The OFDM frame timing is synchronised with the peak of $R(j)$, which can additionally be taken into account for the OFDM symbol synchronisation. The peak height is constant under perfect channel conditions, owing to the fixed reference symbol.

18.2.8.2 Frequency Acquisition

The frequency acquisition algorithm uses the same principle as the frequency tracking scheme of Section 18.2.7.2. Specifically, the phase of $R(j)$ at the last sample of the reference symbol j_{max} contains information on the frequency error:

$$\angle R(j_{max}) = 2\pi \cdot \delta f_a \cdot N_s \cdot T_s = 2\pi \cdot \delta f_a \cdot \frac{N_s}{N \cdot \Delta f} \tag{18.15}$$

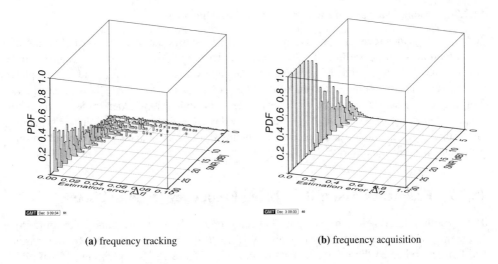

(a) frequency tracking   (b) frequency acquisition

Figure 18.16: Histogram of the simulated frequency estimation error for the frequency acquisition and frequency tracking algorithms using $N = 512$ and $N_s = N_g = 50$ over AWGN channels

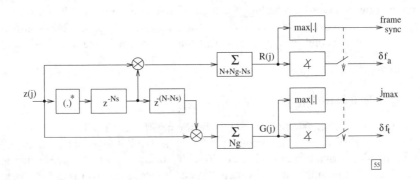

Figure 18.17: Block diagram of the synchronisation algorithms

leading to

$$\delta f_a = \frac{N}{N_s} \frac{\Delta f}{2\pi} \cdot \angle R(j_{max}).$$ (18.16)

Since the spacing between the sample pairs used in the computation of $R(j)$ is smaller than in the case of $G(j)$, which was used for the frequency tracking ($N_s < N$), the maximum detectable frequency error is now increased from $\Delta f/2$ to $N/N_s \cdot \Delta f/2$, where Δf is the subcarrier spacing of the OFDM symbols. The simulated performance of the frequency acquisition algorithm for $N = 512$, $N_s = 50$ and perfect time synchronisation is shown in Figure 18.16(b) for transmissions over an AWGN channel. As seen in the figure, the scheme maintains an acquisition error below $\Delta f/2$ even for SNR values down to 0 dB, which exceeds the system's operational specifications.

18.2.8.3   Block Diagram of the Synchronisation Algorithms

To summarise our previous elaborations, Figure 18.17 shows the detailed block diagram of the synchronisation algorithms. The received samples are multiplied with the complex conjugate of the delayed input sequences and summed up over $(N + N_g - N_s)$ and N_g samples, respectively. The magnitude maxima of the two sequences $G(j)$ and $R(j)$ are detected and they trigger the sampling of the phase estimates $\angle G$ and $\angle R$ in order to derive the two frequency error estimates δf_a and δf_t. Let us now describe an alternative frequency synchronisation technique, which can be used as a benchmark in order to assess the potential of our previously described technique.

18.2.9   Frequency Acquisition Using Frequency-Domain Pilots

The algorithm described in this section has been proposed by Mandarini and Falaschi [575]. In contrast to the reference symbol of Figure 18.13, the technique advocated by Mandarini and Falaschi relies on the reference symbol shown in Figure 18.18 containing a set of pilot tones, which is transmitted once per TDD frame. The reference symbol is processed in the frequency domain, after demodulation by FFT.

The idea of the pilot tone algorithm is that a frequency mismatch between the transmitter and the receiver will result in a shift of the pilot tones in the received spectrum. This shift is measured at the receiver by searching for the FFT frequency bin with the maximum amplitude, resulting in an estimation accuracy of half a subcarrier spacing. This estimation is improved further in a second stage by considering the relative amplitudes of the adjacent FFT bins.

18.2.9.1   The Reference Symbol

The reference symbol is transmitted once at the beginning of every TDD frame, as depicted in Figure 18.12. This reference symbol consists of a set of M pilot tones spanning the OFDM signal's bandwidth, spaced from each other by ΔN subcarriers or by a frequency gap of $\Delta F = \Delta N \cdot \Delta f$. Accordingly, the maximum detectable frequency error is $\Delta F/2 = \Delta N/2 \cdot \Delta f$ and hence the higher the number of pilots, the lower the frequency capture range. A stylised plot of the reference symbol in the frequency domain is shown in Figure 18.18.

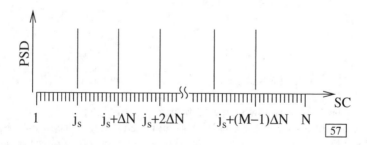

Figure 18.18: Stylised plot of the power spectral density (PSD) of the pilot-tone-based reference symbol using M pilot tones in the N subcarrier (SC) spectrum, spaced by $(\Delta N - 1)$ blank subcarriers

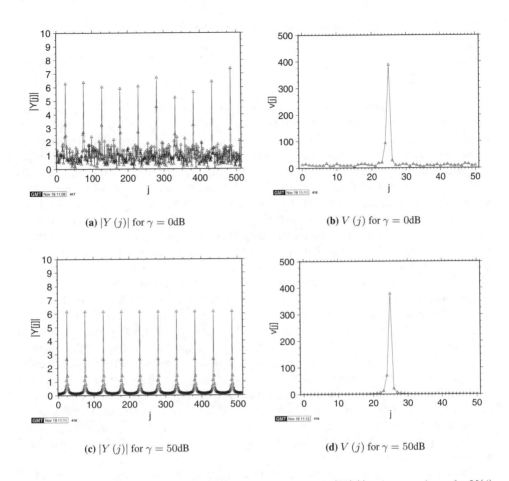

Figure 18.19: Magnitude of the received frequency domain samples $|Y(j)|$ and averaged samples $V(j)$ for SNR values of $\gamma = 0$ dB and $\gamma = 50$ dB for a 512-subcarrier OFDM modem employing 10 pilot tone subcarriers in the reference symbol. The simulated frequency mismatch between the transmitter and the receiver was $\delta f = 0.3\Delta f$

In order to keep the overall OFDM symbol energy constant, the total energy of the reference symbol is set equal to the average OFDM symbol energy and split equally between the M pilot tones.

18.2.9.2 Frequency Acquisition

The frequency acquisition algorithm estimates the frequency error by searching for the position of the pilot tones in the spectrum of the received reference symbol. In order to accomplish this, the received time domain signal of the reference symbol is demodulated by FFT. In order to minimise the influence of noise, the sum of the received power spectral amplitudes over all the M frequency ranges depicted in Figure 18.18 is then calculated as follows:

$$V(j) = \sum_{m=0}^{M-1} |Y(j + j_s + m\Delta N)|^2 \quad \text{for} \quad -\frac{\Delta N - 1}{2} \le j \le \frac{\Delta N - 1}{2}, \quad (18.17)$$

where $Y(j)$ represents the frequency domain samples of the received demodulated reference symbol. Observe that for the sample position identified by j $V(j)$ corresponds to the super-position of the powers of the ΔN-spaced $Y(j)$ samples, one from each of the M frequency ranges. In Figure 18.19 simulated values for both $|Y(j)|$ and $V(j)$ are given for SNR values of 0 dB and 50 dB, respectively, for a 10-pilot reference symbol in a 512-subcarrier OFDM system with $\delta f = 0.3\Delta f$ frequency error. It can be seen that the positions of the pilot tones in the received spectrum are easily determined even for SNR values of 0 dB, thanks to the high power of the pilot tones relative to the average data symbol power. The total energy of the reference is 512, leading to a pilot tone amplitude of $\sqrt{512/10} = 7.15$. Because of the subcarrier energy spillover caused by the frequency error, the amplitude of the peaks of $Y(j)$ in the figure is lowered to about 6.2. Accordingly, the peak value of $V(j)$ is about 390 instead of 512, with the rest of the received energy located in the adjacent FFT bins.

The frequency error is determined by estimating the position of the highest peak in $V(j)$ over the range of j given in Equation 18.17. First, the index j_{max} of the maximum value of $V(j)$ is found. This gives a rough estimate of the frequency error with a frequency resolution of the subcarrier spacing Δf. As can be observed in Figure 18.19, this rough frequency error estimation is very noise resilient, resulting in a reliable detection of the frequency error. The accuracy of this estimation, however, is insufficient for low BER modem operation. If a subsequent frequency error tracking algorithm like the proposed time domain correlation-based approach is employed in the system, then the $0.5\Delta f$ estimation accuracy can be an adequate starting point for the tracking algorithm. If a better estimate is needed, however, then the amplitudes in the neighbouring bins around the peak value in $V(j)$ can be exploited to refine the first estimate, as follows.

Since the frequency error is not generally an integer multiple of the resolution Δf of the FFT, a better frequency error estimate can be derived by determining the position of the peak value of $V(j)$ around j_{max}. The proposed algorithm exploits the amplitude of $V(j)$ at $j = j_{max}$ and the two adjacent values $V(j_{max} \pm 1)$ in order to derive the following quantities:

$$\rho_1 = \frac{\sqrt{V(j_{max} + 1)}}{\sqrt{V(j_{max})}} \quad (18.18)$$

$$\rho_{-1} = \frac{\sqrt{V(j_{max} - 1)}}{\sqrt{V(j_{max})}}, \quad (18.19)$$

which represent the normalised frequency domain pilot values at the positions adjacent to the $V(j)$ peak at $j = j_{max}$. Their difference $(\rho_1 - \rho_{-1})$ is computed and the rough frequency error estimate of $j_{max} \cdot \Delta f$ is corrected by the help of the estimated deviation d as follows:

$$d = \frac{\rho_1 - \rho_{-1}}{2} \quad (18.20)$$

$$\delta f = \Delta f \cdot (j_{max} + \text{sgn}\{d\} \cdot \sqrt{|d|}), \quad (18.21)$$

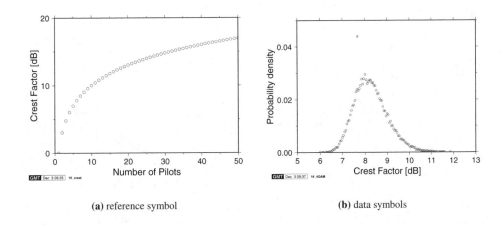

(a) reference symbol   (b) data symbols

Figure 18.20: Characterisation of the coarse frequency synchronisation using the reference symbol of Figure 18.18: (a) crest factor versus number of pilot tones for non-optimised reference symbols, (b) histogram of simulated data symbol crest factors for a 512-subcarrier, 4-QAM OFDM system

where the value of $\sqrt{|d|}$ is the estimation of the true peak position between the peak $V(j_{max})$ and the higher of the two adjacent values of $V(j_{max} \pm 1)$.

Clearly, the number of pilot tones included in the reference symbol, M, determines the maximal frequency deviation that can be detected using this algorithm, which is given by $\delta f_{max} = \Delta f \cdot \lfloor N/2M \rfloor$, where $\lfloor x \rfloor$ means the highest integer number smaller or equal to x. Not only the frequency capture range, but also the peak-to-average power ratio or crest factor (CF) of the reference symbol in the time domain depends on the number of pilot tones used. If the phase of all the M pilot tones is equal, then the upper bound for the resulting crest factor equals M, which would be the result of all M sinusoids adding constructively to a peak value. Experimental results in Figure 18.20(b) show that the actually measured CF for pilot symbols with a varying number of pilot tones is very close to the stated upper bound. For comparison, the histogram of the simulated CF values for pseudorandom OFDM data symbols is shown in Figure 18.20(b). This PDF shows that the average CF value for a 512-subcarrier 4-QAM OFDM system is about 8.5 dB, although peak values up to 12 dB have been occasionally observed. For small numbers of pilot tones, the crest factor of the reference symbol is therefore not worse than that of the average data symbols.

18.2.9.3 Performance of the Pilot-Based Frequency Acquisition in AWGN Channels

A series of simulations was performed in order to characterise the performance of the pilot tone based frequency acquisition algorithm in an AWGN environment. For different fixed frequency errors and noise levels, the frequency error estimation was invoked and the difference between the estimated and the actual error was recorded. Figure 18.21 shows the histograms of the frequency mismatch estimation errors for 10, 20, 30 and 40 pilots and for a range of SNR values in the case of an actual frequency error of $\delta f = 0.0$.

Inspection of the figure reveals that the estimation accuracy is not sensitive to the number of pilot tones used in the reference symbol; in fact, the differences in the simulated histograms

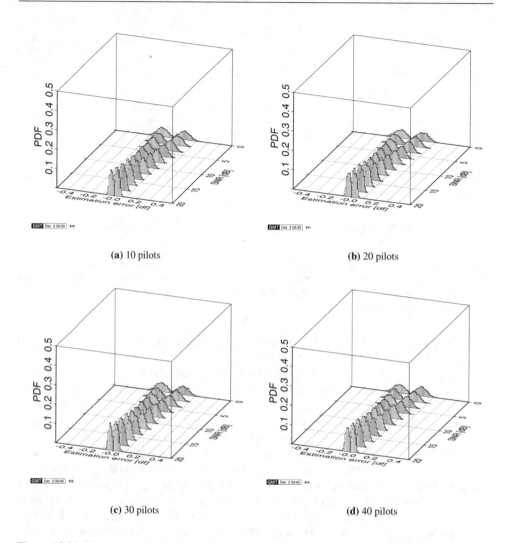

(a) 10 pilots

(b) 20 pilots

(c) 30 pilots

(d) 40 pilots

Figure 18.21: Histograms of simulated frequency mismatch estimation error for a 512-subcarrier OFDM modem in a narrowband AWGN channel for 10, 20, 30 and 40 pilot tones. The simulated frequency error was $\delta f = 0.0$

are hardly visible. The estimation accuracy is significantly better than $0.5\Delta f$ throughout the simulated SNR range and hence meets the requirements of the subsequent fine frequency synchronisation for all simulated signal-to-noise ratios. It can be observed, however, that there is a bifurcation of the estimation error histogram which leads to two distinct probability maxima on both sides of the true value, with a low probability for a perfect estimation. This behaviour can be explained by the fine peak location estimation, which exhibits a high noise sensitivity for low δf values and hence small differences in the noise samples on either side of the correlation peak lead to considerable etimation errors.

The gap between the maxima closes with increasing SNR, but even at 20 dB there are frequency estimation errors of up to 8% of the subcarrier spacing, which underlines the need for subsequent fine frequency synchronisation. Alternatively, averaging the estimated frequency errors for several estimations would improve the estimation accuracy, but it would slow down the system's response time to frequency deviations. The histogram is symmetric around the perfect estimation value, hence a longer-term averaging would yield the correct frequency error estimate.

The estimation accuracy of the pilot tone frequency acquisition algorithm depends on the actual frequency error. In order to illustrate this, frequency estimation error histograms for a given frequency mismatch of $\delta f = 0.3\Delta f$ are shown in Figure 18.22, and they exhibit rather different characteristics from the previous case with $\delta f = 0$. The histograms are offset with respect to the perfect estimation value, hence averaging over multiple estimations will result in a residual estimation error. Subsequent fine frequency sychronisation is therefore necessary for low SNR values. For low noise levels, however, the estimation accuracy is greatly improved with respect to the earlier case. For SNRs in excess of 10 dB, the estimation errors do not exceed 5% of the subcarrier spacing, and will therefore not affect the system performance to a great extent. It can be observed that the asymmetry of the estimation errors is dependent on the number of pilots employed.

The frequency deviation estimation error histogram for $\delta f = 0.5\Delta f$ is shown in Figure 18.23, where it can be observed that in this case the histogram is split into two distinct parts, and that each of the groups is centred at 8% of the subcarrier distance above and below the correct value. Again, fine frequency synchronisation would be necessary to ensure optimal system performance.

In order to evaluate the residual estimation errors, Figure 18.24 depicts the estimated versus the actual frequency error for a perfect, noiseless channel. For frequency errors close to an integer multiple of Δf the estimation accuracy is very good, but values close to $(n + 1/2)\Delta f$ are estimated with errors of up to 8% of the subcarrier spacing. This observation is in accordance with the results of the above experiments, which showed a residual estimation error of about 1% of the carrier separation at $\delta f = 0.3\Delta f$ and of $\pm 8\%$ at $\delta f = 0.5\Delta f$.

18.2.9.4 Alternative Frequency Error Estimation for Frequency-Domain Pilot Tones

In order to improve the frequency error dependent performance of the frequency domain pilot-based frequency synchronisation algorithm outlined above, an alternative peak position estimation has been investigated in order to enhance the algorithm based on Equation 18.21. As we have seen earlier, the continuous spectrum of each OFDM subcarrier follows a sinc function centred around the subcarrier frequency. In the presence of a frequency error between the transmitter and receiver, the receiver's sampling raster in the frequency domain is not aligned with the received spectrum's maximum and nulls, but shifted by the frequency error δf.

The spectrum of the received reference symbol – which contains M pilot tones at the frequencies $(j_s + m\Delta N) \cdot \Delta f$ with $0 \le m \le (M-1)$ – in the presence of a frequency error δf in a noiseless environment can be expressed as:

$$Y(j) = \sqrt{\frac{N}{M}} \cdot \sum_{m=0}^{M-1} \text{sinc}\left(j - j_s - m\Delta N + \frac{\delta f}{\Delta f}\right). \qquad (18.22)$$

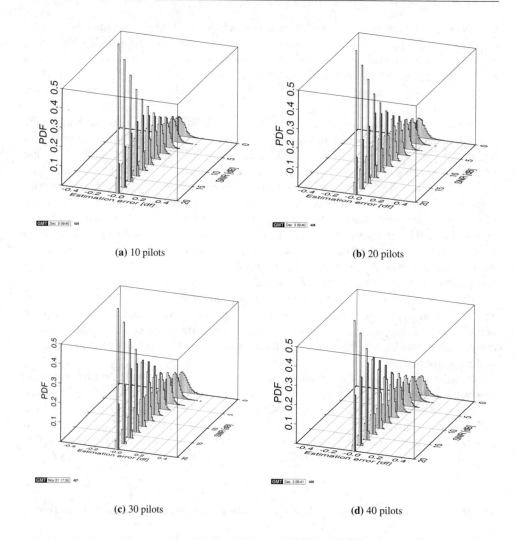

(a) 10 pilots

(b) 20 pilots

(c) 30 pilots

(d) 40 pilots

Figure 18.22: Histograms of simulated frequency mismatch estimation error for a 512-subcarrier
OFDM modem in a narrowband AWGN channel for 10, 20, 30 and 40 pilot tones. The
simulated frequency error was $\delta f = 0.3$

The factor $\sqrt{N/M}$ is the amplitude of each pilot tone, ensuring that the overall energy of
the reference symbol is equal to the average OFDM symbol energy. If the frequency distance
ΔN between two consecutive pilot tones is sufficiently large so that the received spectra
of the different pilot tones do not significantly overlap, then the vector $V(j)$, as defined by

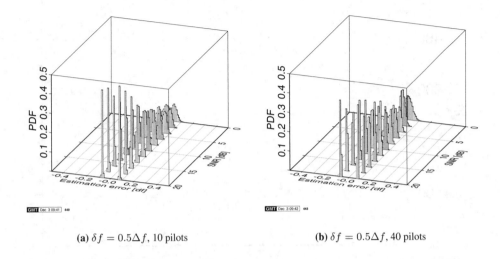

(a) $\delta f = 0.5\Delta f$, 10 pilots   **(b)** $\delta f = 0.5\Delta f$, 40 pilots

Figure 18.23: Histograms of simulated frequency mismatch estimation error for a 512-subcarrier OFDM modem in a narrowband AWGN channel for 10 and 40 pilot tones. The simulated frequency error was $\delta f = 0.5\Delta f$

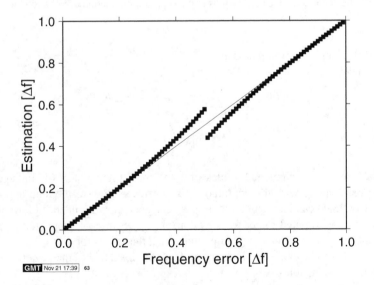

Figure 18.24: Estimated versus actual frequency error for the pilot-tone-based coarse frequency synchronisation algorithm in a perfect noiseless channel

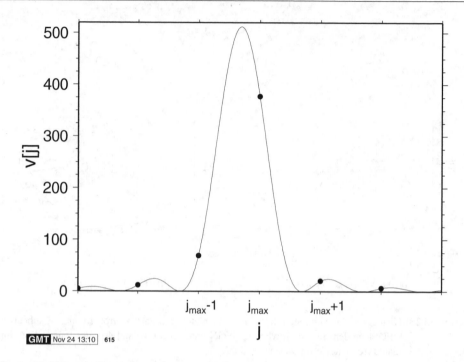

Figure 18.25: Detail from Figure 18.19(d): $V(j)$ for an SNR value of $\gamma = 50$ dB for a 512-subcarrier OFDM modem employing 10 pilot tones in the reference symbol. The simulated frequency mismatch is $0.3\Delta f$. The symbols represent the simulated $V(j)$ values, while the continuous line corresponds to the sinc^2 approximation of Equation 18.23

Equation 18.17, can be approximated as:

$$V(j) = \sum_{m-0}^{M-1} |Y(j + j_s + m\Delta N)|^2 \tag{18.23}$$

$$\approx N \cdot \text{sinc}^2 \left(j + \frac{\delta f}{\Delta f} \right). \tag{18.24}$$

Figure 18.25 shows the simulated values of $V(j)$ from Figure 18.19(d) in greater detail around the peak, along with the continuous line corresponding to the sinc approximation of Equation 18.23. It can be seen that there is a very good correspondence between the simulated and the approximated values for the pilot spacing of $M = 50$ employed in this case.

Using the sinc^2 approximation of Equation 18.23, the values for $V(j_{max})$, $V(j_{max} + 1)$ and $V(j_{max} - 1)$ can be expressed in terms of the normalised fine frequency error estimation $\nu = \delta f / \Delta f - j_{max}$ as follows:

$$V(j_{max}) \approx N\text{sinc}^2(\nu), \tag{18.25}$$
$$V(j_{max} + 1) \approx N\text{sinc}^2(1 + \nu), \quad \text{and} \tag{18.26}$$
$$V(j_{max} - 1) \approx N\text{sinc}^2(-1 + \nu). \tag{18.27}$$

Then the following normalised terms can be defined:

$$\rho_1 = \frac{\sqrt{V(j_{max}+1)}}{\sqrt{V(j_{max})}} \tag{18.28}$$

$$\approx \frac{\sqrt{N}\,|\sin(\pi(1+\nu))|}{|\pi(1+\nu)|} \cdot \frac{|\pi\nu|}{\sqrt{N}\,|\sin(\pi\nu)|} \tag{18.29}$$

$$= \frac{|\nu|}{|1+\nu|} \quad \text{and} \tag{18.30}$$

$$\rho_{-1} = \frac{\sqrt{V(j_{max}-1)}}{\sqrt{V(j_{max})}} \tag{18.31}$$

$$\approx \frac{\sqrt{N}\,|\sin(\pi(-1+\nu))|}{|\pi(-1+\nu)|} \cdot \frac{|\pi\nu|}{\sqrt{N}\,|\sin(\pi\nu)|} \tag{18.32}$$

$$= \frac{|\nu|}{|-1+\nu|}. \tag{18.33}$$

The value d, defined as half the difference between ρ_1 and ρ_{-1} in Equation 18.20, can therefore be approximated as:

$$d = \frac{\rho_1 - \rho_{-1}}{2} \approx \frac{1}{2}\left(\frac{|\nu|}{|1+\nu|} - \frac{|\nu|}{|-1+\nu|}\right). \tag{18.34}$$

Solving Equation 18.34 for ν values smaller than one subcarrier distance yields:

$$\nu = \begin{cases} -\sqrt{\frac{d}{d+1}} & \text{for} \quad -1 < \nu < 0 \quad (\text{or} \quad (d>0) \\ \sqrt{\frac{d}{d-1}} & \text{for} \quad 0 \le \nu < 1 \quad (\text{or} \quad d \le 0) \end{cases}. \tag{18.35}$$

An alternative pilot-based frequency synchronisation algorithm can therefore be contrived by calculating the peak position estimate ν from d, as defined in Equation 18.20, using Equation 18.35 and replacing the δf estimate in Equation 18.21 by:

$$\delta f = \Delta f \cdot (j_{max} + \nu). \tag{18.36}$$

A series of simulations were conducted in order to investigate the performance of this modified peak position estimation algorithm in noisy conditions. All the investigations have been performed for a 512-subcarrier system employing a 10-pilot reference symbol in a narrowband Gaussian white noise channel.

Histograms of the estimation errors for fixed frequency errors of $\delta f = 0$, $\delta f = 0.3\Delta f$ as well as for $\delta f = 0.5\Delta f$ are given in Figure 18.26. In all cases, the estimation accuracy was better than $0.5\Delta f$, therefore allowing the subsequent use of the OFDM data symbol based tracking algorithm of Section 18.2.7. Comparison of Figure 18.26(a) with the corresponding results for the original peak position estimation algorithm in Figure 18.21(a) reveals a similar performance for both algorithms. This was expected, since the modified estimation terms $\sqrt{d/(d-1)}$ and $-\sqrt{d/(d+1)}$ are close to the term $\sqrt{d}$ in Equation 18.21 of the original algorithm for small values of d.

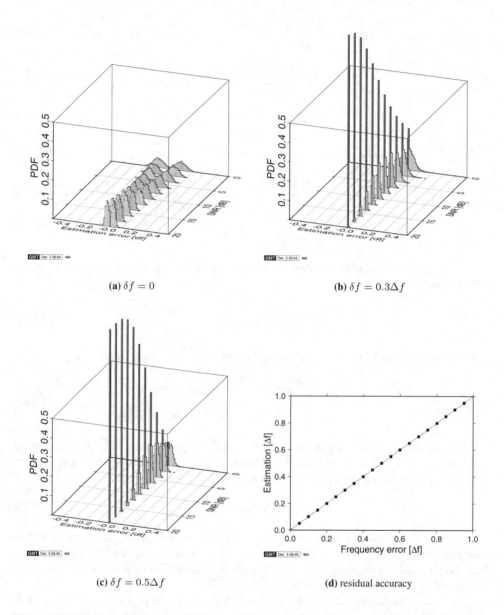

(a) $\delta f = 0$

(b) $\delta f = 0.3\Delta f$

(c) $\delta f = 0.5\Delta f$

(d) residual accuracy

Figure 18.26: (a, b, c): Histograms of the simulated frequency mismatch estimation error for a 512-subcarrier OFDM modem in a narrowband AWGN channel for 10 pilot tones employing the alternative peak position estimation algorithm. The simulated frequency error was $\delta f = 0.3$. (d) Estimated versus actual frequency error for pilot-tone-based frequency synchronisation employing the alternative peak position estimation algorithm

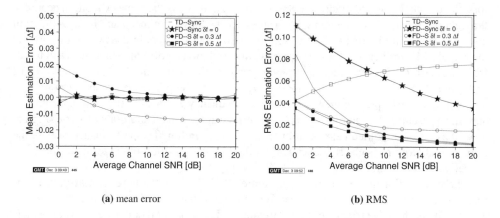

(a) mean error

(b) RMS

Figure 18.27: Comparison of the frequency acquisition algorithms over AWGN channels: (a) mean estimation error, (b) RMS estimation error. TD = time domain estimation algorithm, FD = frequency domain estimation algorithm employing a 10-pilot reference symbol. The open symbols show the performance associated with the original peak position estimation of Equation 18.21, while the closed symbols signify the alternative peak position estimation algorithm of Equation 18.36 for given frequency errors δf of 0, 0.3 and 0.5 times the subcarrier distance Δf

For a frequency error of $\delta f = 0.3\Delta f$, as depicted in Figure 18.26(b), the estimation accuracy was dramatically improved, when compared to the case of $\delta f = 0$. This was in accordance with the results achieved using the original peak position estimation, as shown in Figure 18.22(a). Although both schemes achieved accurate estimation results, the modified algorithm of Equation 18.36 exhibited a lower estimation bias and it was slightly more accurate than the original algorithm of Equation 18.21.

Imposing the third simulated frequency error of $\delta f = 0.5\Delta f$ revealed the different behaviour of the two investigated algorithms. While the original version of the algorithm exhibited a split histogram with systematic estimation errors of about 8% of the subcarrier distance, as can be seen in Figure 18.23(a), the modified algorithm delivered a significantly more accurate frequency error estimate. The estimation accuracy of the modified algorithm in a noiseless environment is demonstrated in Figure 18.26(d). There are no systematic estimation errors for the whole frequency error range that the peak position estimation is expected to handle.

18.3 Comparison of the Frequency Acquisition Algorithms

Both the frequency acquisition algorithms examined achieve the minimum requirements of an estimation accuracy better than half a subcarrier spacing for AWGN channel SNR levels down to 0 dB and hence ensure reliable operation of the subsequent fine frequency synchronisation algorithm.

There is, however, a difference in their absolute accuracy. While the pilot tone algorithm of Section 18.2.9 suffers from residual estimation errors for certain ranges of frequency errors, the correlation-based algorithm of Section 18.2.5 delivers the same estimation accuracy throughout its frequency capture range. Values for the long-term mean and RMS estimation error are shown in Figure 18.27. The solid lines without markers in the two graphs are the mean and RMS error curves for the time domain correlation algorithm, which are valid for all frequency errors within the frequency capture range of the algorithm ($\pm 5.12\Delta f$ for $N = 512$ and $N_s = 50$). The curves with markers correspond to the pilot tone synchronisation system employing a 10-pilot reference symbol, for frequency errors of 0, 0.3, and 0.5 times the subcarrier distance Δf. For these frequency error values, curves are drawn for the original as well as for the modified peak position estimation algorithm. The performance of both the pilot tone algorithms is clearly dependent on the frequency error, both in terms of the mean as well as the RMS errors.

The mean estimation error curves in Figure 18.27(b) show the long-term averaged estimation errors for the different schemes. While none of the algorithms exhibits mean errors above 0.5% of the subcarrier distance for frequency deviations of 0 or 0.5% Δf over the SNR range investigated, frequency errors of $0.3\Delta f$ cause both the pilot tone estimation algorithms to deliver biased estimates. The amount of this mean estimation error depends on the SNR, and the noise level dependency of the estimation is the same for both the original as well as for the alternative peak search algorithm. The residual estimation error in the absence of noise is different for the two peak position estimators, however. While the alternative estimation algorithm delivers a non-biased estimation for high SNR values, the original algorithm exhibits an estimation error of about 1.5% of the subcarrier distance. This is in accordance with Figure 18.24, which shows the residual estimation errors for the original peak position estimation. It is interesting to note that the original peak search algorithm does not result in a biased non-zero mean error estimate for frequency errors of $\delta f = 0.5\Delta f$, although the instantaneous estimation errors at this point are at a maximum, as can be seen from Figures 18.24 and 18.23(a). This is because of the symmetry of the histogram, which will result in a non-biased mean estimation value. For all other frequency errors between 0 and $0.5\Delta f$, however, the histogram of estimation errors will be asymmetrical and the estimation will therefore be biased, as seen for example in Figure 18.22.

The modified peak search algorithm of Section 18.2.9.4, too, delivers biased estimations for frequency errors between 0 and $0.5\Delta f$, as seen in Figure 18.27. Since these are due to the noise affecting the estimation, there is no residual estimation error in noiseless environments. For SNR values in excess of 8 dB, the mean estimation error is smaller than 0.5% of the subcarrier distance for $\delta f = 0.3\Delta f$. The time domain frequency synchronisation algorithm of Section 18.2.5 does not exhibit any significant bias in the estimation error for the range of SNR values investigated. Therefore, long-term averaging of the time domain algorithm's estimates will result in correct frequency error estimation at the cost of reduced agility.

Figure 18.27(b) depicts the RMS estimation error curves for the investigated synchronisation algorithms for the same set of frequency errors. It can be seen that for both frequency domain pilot-based error estimation algorithms the estimation quality varies greatly with the frequency error to be estimated. The original peak position estimation algorithm exhibits residual RMS values of $0.015\Delta f$ and about $0.8\Delta f$ for frequency errors δf of $0.3\Delta f$ and $0.5\Delta f$, respectively. For $\delta f = 0.3\Delta f$ this corresponds to the bias of 1.5% of the subcarrier distance, which has been observed in Figure 18.27(a). For $\delta f = 0.5\Delta f$, this residual estimation error can be observed in the histogram, as shown in Figure 18.23(a). Although the long-term mean of the estimation is the correct value, each single estimation exhibits an error of about 8% of the subcarrier distance. This value corresponds to the residual level of RMS estimation error for $\delta f = 0.5\Delta f$ in Figure 18.27(b).

The frequency domain pilot-based algorithm employing the alternative peak position estimation of Section 18.2.9.4 does not exhibit residual RMS estimation errors. The estimation accuracy does, however, vary with the frequency error to be estimated. For $\delta f = 0$, its performance is virtually the same as that of the original peak position estimation algorithm, with high estimation errors for the whole AWGN SNR range. At $\delta f = 0.3\Delta f$ and $0.5\Delta f$, the estimation accuracy is much better, at RMS estimation errors of around 1% of the subcarrier distance for SNR values in excess of 12 dB and 10 dB, respectively.

The time domain correlation-based algorithm's accuracy does not depend on the frequency error to be estimated. Its performance is comparable to that of the frequency domain algorithm employing the alternative peak search algorithm of Section 18.2.9.4 for frequency errors δf of $0.3\Delta f$ or $0.5\Delta f$ for SNR values above 10 dB, but it performs significantly better at $\delta f = 0$.

Comparing frequency error the estimation accuracy of the different algorithms it becomes clear that all the algorithms would satisfy the estimation accuracy of $0.5\Delta f$, necessary if subsequent fine frequency tracking mechanisms are used data symbol by data symbol. If no fine frequency tracking is to be employed, however, then the constraints on the estimation accuracy are significantly tighter. We have seen in Section 18.1.1.2 that, depending on the modulation scheme used, frequency errors of 5–10% of the subcarrier spacing lead to degradation of the system's SNR performance. In this case – and if no lowpass filtering of the estimated value is to be employed – the time domain correlation-based synchronisation algorithm of Section 18.2.5 would be the only applicable algorithm, since its accuracy is consistently high and its estimation is unbiased throughout the frequency error range.

If the pilot tone reference symbol based frequency synchronisation algorithms are to be employed without subsequent frequency tracking, then the estimation accuracy has to be improved using lowpass filtering techniques of the estimates. Let us now consider the system's BER performance when subjected to frequency synchronisation errors.

18.4 BER Performance with Frequency Synchronisation

In order to investigate the effects of the proposed frequency synchronisation algorithm on an OFDM modem, a series of experiments were conducted. The synchronisation algorithm chosen was the time domain correlation algorithm of Section 18.2.5, because of its frequency error independent estimation accuracy. We modelled a system employing one reference symbol and one data symbol per 64-slot TDMA frame. The frequency error estimates were not lowpass filtered, instead every estimate was used directly to correct the frequency error of the subsequent data symbol. Coherently detected BPSK, QPSK and 16-QAM were employed as modulation schemes in the subcarriers and the BER performance of the modem was investigated in both narrowband AWGN as well as in fading and non-fading wideband channels. In all the simulations, a 512-subcarrier OFDM scheme employing no virtual carriers and perfect wideband channel estimation was used.

The BER performance of the simulated modems in all the channels is given in Figure 18.28. Figure 18.28(a) depicts the BER versus channel SNR for BPSK, QPSK and 16-QAM in an AWGN channel with a frequency error of $\delta f = 0.3\Delta f$. The white symbols in the graph

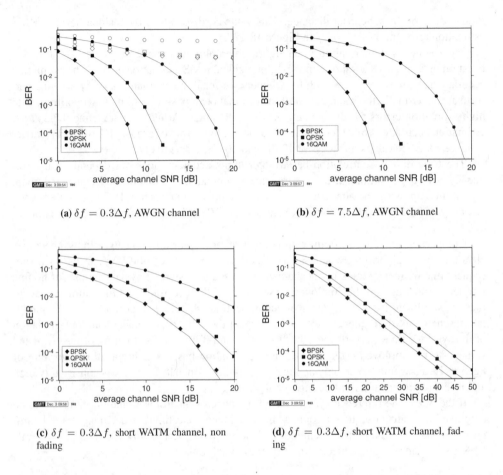

(a) $\delta f = 0.3\Delta f$, AWGN channel

(b) $\delta f = 7.5\Delta f$, AWGN channel

(c) $\delta f = 0.3\Delta f$, short WATM channel, non fading

(d) $\delta f = 0.3\Delta f$, short WATM channel, fading

Figure 18.28: BER versus channel SNR performance curves for the 512-subcarrier OFDM system in the presence of fixed frequency errors. The lines indicate the performance for perfectly corrected frequency error, and the open symbols show the performance for uncorrected frequency errors. The filled symbols indicate simulations with our proposed frequency error estimation using the time domain correlation technique. The short WATM channel impulse response is shown in Figure 17.3(a).

portray the BER performance of an OFDM modem employing no frequency synchronisation. It can be seen that the corresponding frequency error results in heavy intersubcarrier interference, which manifests itself by a high residual bit error rate of about 5% for BPSK and QPSK and about 20% for 16-QAM. The lines in the graph characterise the performance of the modem in the absence of frequency errors. The black markers correspond to the BER recorded with the frequency synchronisation algorithm in operation. It can be seen that the performance of the modem employing the proposed frequency synchronisation algorithm of Section 18.2.5 is nearly indistinguishable from the perfectly synchronised case. In Figure 18.28(b), the modem's BER curves for an AWGN channel at a frequency error of $7.5\Delta f$ are depicted. Since the synchronisation algorithm's accuracy does not vary with varying frequency errors, the modem's BER performance employing the proposed synchronisation algorithm at $\delta f = 7.5\Delta f$ is the same as at $\delta f = 0.3\Delta f$. The BER for the non-synchronised modem is, however, 50% and the corresponding markers are off the graph.

The synchronised modem's BER performance in wideband channels is given in Figure 18.28(c) and 18.28(d). The impulse response used in these investigations was the short WATM impulse response, which is depicted in Figure 17.3(a). Perfect knowledge of the channel impulse response was assumed for perfect phase and amplitude correction of the data symbols with coherent detection. Again, the BER curves for both the non-fading and the fading channels show a remarkable correspondence between the ideal performance lines and the performance of the synchronised modems. In all the investigated environments, the modem's performance was unaffected by the estimation accuracy of the time domain reference symbol synchronisation algorithm.

18.5 Summary

In this chapter, the effects of frequency and timing errors in OFDM transmissions have been characterised. While frequency errors result in frequency domain inter-subcarrier interference, timing errors lead to time domain inter-OFDM symbol interference and to frequency domain phase rotations.

In order to overcome the effects of moderate timing errors, a cyclic postamble and the use of pilot symbol assisted modulation or differential detection were proposed. Different frequency and timing error estimation algorithms were portrayed, and their performance was investigated. A new combined frequency and timing synchronisation algorithm based on a dedicated reference symbol exploited in the time domain was proposed and the system's performance employing this algorithm was investigated in AWGN and fading channels. The solution of Section 18.2.5 was finally advocated for implementation in a 34Mbit/s real-time demonstration testbed.

18.6 Appendix: Theoretical Performance of OFDM Synchronisation Algorithms

18.6.1 Frequency Synchronisation in an AWGN Channel

The correlation operation used in the frequency synchronisation algorithm is based on the conjugate complex multiplication of the noisy input sample with its noisy and phase–shifted copy. To derive the theoretical performance of this algorithm, the influence of the noise on the phase of this product must be known. Throughout the calculations, we will assume no frequency– and time invariant phase errors, in order to simplify the notation.

18.6.1.1 One Phasor in AWGN Environment

18.6.1.1.1 Cartesian Coordinates Every received sample is a superposition of the transmitted phasor and two statistically independent quadrature noise samples; if the received phasor is real and of magnitude $\mathcal{A}$, the noisy signal is described by the complex stochastic variable Z:

$$Z \quad = \quad \mathcal{A} + N \tag{18.37}$$

$$= \quad X + jY \tag{18.38}$$

As the two quadrature noise processes are statistically independent, the joint probability density function (PDF) of the two quadrature components x and y is given by the product of two one–dimensional Gaussian PDFs:

$$p_{x,y}(x,y) \quad = \quad p(x) \cdot p(y) \tag{18.39}$$

$$= \quad \left(\frac{1}{\sqrt{2\pi}\sigma}\right)^2 \cdot e^{-(x-\mathcal{A})^2/2\sigma^2} \cdot e^{-y^2/2\sigma^2} \tag{18.40}$$

$$= \quad \frac{1}{2\pi\sigma^2} \cdot e^{-((x-\mathcal{A})^2+y^2)/2\sigma^2} \tag{18.41}$$

18.6.1.1.2 Polar Coordinates The stochastic variable Z can be expressed in polar coordinates:

$$Z \quad = R \cdot e^{j\Phi} \tag{18.42}$$

$$with R \quad = \sqrt{X^2 + Y^2} \tag{18.43}$$

$$\Phi \quad = \tan^{-1}\frac{Y}{X} \tag{18.44}$$

To derive the PDF in polar coordinates $p_{r,\phi}(r,\phi)$ from the expression in Cartesian coordinates $p_{x,y}(x,y)$, given in Equation 18.41, the variable transform $(x,y) \Rightarrow (r,\phi)$ is performed:

$$x(r,\phi) \quad = r \cdot \cos\phi \tag{18.45}$$

$$y(r,\phi) \quad = r \cdot \sin\phi \tag{18.46}$$

The determinant J of the corresponding Jacobean matrix is:

$$J = \begin{vmatrix} \frac{\partial x(r,\phi)}{\partial r} & \frac{\partial y(r,\phi)}{\partial r} \\ \frac{\partial x(r,\phi)}{\partial \phi} & \frac{\partial y(r,\phi)}{\partial \phi} \end{vmatrix} = \begin{vmatrix} \cos\phi & \sin\phi \\ -r \cdot \sin\phi & r \cdot \cos\phi \end{vmatrix} = r \tag{18.47}$$

This leads to the probability density function of one noisy phasor in polar coordinates:

$$p_{r,\phi}(r,\phi) \quad = |J| \cdot p_{x,y}(x(r,\phi), y(r,\phi)) \tag{18.48}$$

$$= \frac{r}{2\pi\sigma^2} \cdot e^{-(r^2+\mathcal{A}^2-2r\mathcal{A}\cos\phi)/2\sigma^2} \tag{18.49}$$

18.6.1.2 Product of Two Noisy Phasors

18.6.1.2.1 Joint Probability Density To derive the probability density function of the product of two noisy phasors, we will assume that both of the phasors are stochastic variables with a PDF given in Equation 18.49, with the same signal amplitude $\mathcal{A}$ and zero signal phase. The new stochastic variable Π represents the product of the two noisy phasors Z_1 and Z_2:

$$\Pi = \Xi * e^{j\Psi} = Z_1 \cdot Z_2 \tag{18.50}$$

The joint probability density function $p_\Pi(\Xi = \xi, \Psi = \psi)$ is derived from Equation 18.49 using complex multiplication:

$$\Xi = R_1 \cdot R_2 \tag{18.51}$$

$$\Psi = \Phi_1 + \Phi_2 \tag{18.52}$$

Two auxiliary variables must be introduced in order to solve the system. We choose $\Lambda = R_2$ and $\Omega = \Phi_2$. The resulting transformation functions and their inverse functions are:

$$\left.\begin{array}{rcl} \xi &=& r_1 \cdot r_2 \\ \psi &=& \phi_1 + \phi_2 \\ \lambda &=& r_2 \\ \omega &=& \phi_2 \end{array}\right\} \Leftrightarrow \left\{\begin{array}{rcl} r_1 &=& \xi/\lambda \\ \phi_1 &=& \psi - \omega \\ r_2 &=& \lambda \\ \phi_2 &=& \omega \end{array}\right. \tag{18.53}$$

The determinant J of the corresponding Jacobean matrix is:

$$J = \begin{vmatrix} \frac{\partial r_1}{\partial \xi} & \frac{\partial r_2}{\partial \xi} & \frac{\partial \phi_1}{\partial \xi} & \frac{\partial \phi_2}{\partial \xi} \\ \frac{\partial r_1}{\partial \psi} & \frac{\partial r_2}{\partial \psi} & \frac{\partial \phi_1}{\partial \psi} & \frac{\partial \phi_2}{\partial \psi} \\ \frac{\partial r_1}{\partial \lambda} & \frac{\partial r_2}{\partial \lambda} & \frac{\partial \phi_1}{\partial \lambda} & \frac{\partial \phi_2}{\partial \lambda} \\ \frac{\partial r_1}{\partial \omega} & \frac{\partial r_2}{\partial \omega} & \frac{\partial \phi_1}{\partial \omega} & \frac{\partial \phi_2}{\partial \omega} \end{vmatrix} = \begin{vmatrix} 1/\lambda & 0 & 0 & 0 \\ 0 & 1 & 0 & 0 \\ \xi \ln \lambda & 0 & 1 & 0 \\ 0 & 0 & 0 & 1 \end{vmatrix} = 1/\lambda \tag{18.54}$$

The resulting probability density function depending on the four new variables is:

$$p_\Pi(\xi, \psi, \lambda, \omega) = |J| \cdot p_Z(r_1 = \xi/\lambda, \phi_1 = \psi - \omega, r_2 = \lambda, \phi_2 = \omega) \tag{18.55}$$

The two probability density functions $p_Z(r1, \phi_1)$ and $p_Z(r2, \phi_2)$ are statistically independent, therefore

$$p_Z(r_1, \phi_1, r_2, \phi_2) = p_Z(r1, \phi_1) \cdot p_Z(r2, \phi_2) \tag{18.56}$$

Using the Equations 18.49 and 18.56 in 18.55, we find the expression for the PDF of the product of two noisy phasors as a function of the four transformed variables ξ, ψ, λ and ω:

$$p_\Pi(\xi, \psi, \lambda, \omega) = \frac{\xi}{\psi} \cdot \frac{1}{4\pi^2 \sigma^4} \cdot e^{-\frac{\xi^2/\lambda^2 + \lambda^2 + 2A^2 - 2A(\lambda \cos \omega + \xi/\lambda \cos(\psi - \omega)))}{2\sigma^2}} \tag{18.57}$$

Eliminating the auxiliary variables λ and ω by integration of 18.57 yields:

$$p_\Pi(\Xi = \xi, \Psi = \psi) = \int_0^\infty \int_{-\pi}^\pi p_\Pi(\xi, \psi, \lambda, \omega) d\omega d\lambda \tag{18.58}$$

18.6.1.2.2   Phase Distribution   The distribution of the phase of Π is obtained by integrating Equation 18.58 over the amplitude ξ:

$$p_\Pi(\Psi = \psi) = \int_0^\infty p_\Pi(\xi, \psi) d\xi \tag{18.59}$$

18.6.1.2.3   Numerical Integration   The integrals in Equations 18.58 and 18.59 cannot be solved analytically, therefore numerical integration has to be employed to determine the prob-

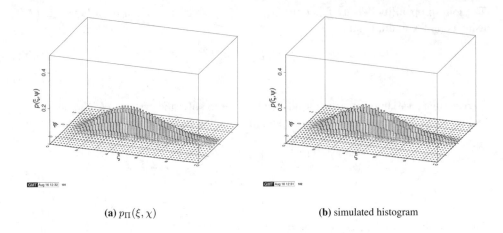

(a) $p_\Pi(\xi, \chi)$ (b) simulated histogram

Figure 18.29: Product of two noisy phasors with $\mathcal{A} = 2$ and $\gamma = 6\text{dB}$: (a): numerical solution of Equation 18.58, (b): histogram of 100000 simulated products

ability density functions $p_\Pi(\xi, \psi)$ and $p_\Pi(\psi)$. This numerical integral has been evaluated for a range of different signal-to-noise (SNR) values γ:

$$\gamma = \frac{\mathcal{A}}{2\sigma^2} \qquad (18.60)$$

Simulations have been performed for the same SNR values to verify the expressions 18.58 and 18.59. The simulated histograms have been employed to perform a χ^2–test with the numerical results. The χ^2 value gives a distance measure between an expected probability distribution and a N–binned set of experimental data:

$$\chi^2 = \sum_{i=0}^{N-1} \frac{(p(i) - h(i))^2}{p(i)} \qquad (18.61)$$

where $h(i)$ is the value of the ith data bin, and $p(i)$ its expected value. As an example, Figure 18.6.1.2.3 illustrates both the numerical evaluation of Equation 18.58 and the histogram of the simulation for a SNR value of 6dB and a phasor amplitude of $\mathcal{A} = 2$. The figures reflect a high degree of correspondence between the numerical and the simulation results. This is confirmed by a quantitative measure, the χ^2 test, which yields a confidence–level in excess of 99% when testing the hypothesis that the simulated results are of the same distribution as given in equation 18.58. Table 18.1 displays the values of χ^2 and the resulting confidence level $\Gamma(2100, \chi^2)$ for SNR values between 0 and 10dB. The numerically integrated phase probability density function Equation 18.59 and the histogram of the simulated values are shown in Figure 18.30(a). Figure 18.30(b) depicts the phase PDF for different SNR values.

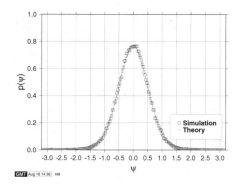
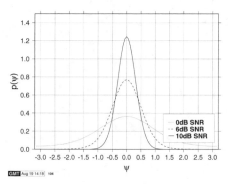

(a) phase PDF, theory and simulation   (b) phase PDFs for different SNR values

Figure 18.30: Product of two noisy phasors: phase probability density functions. (a): theoretical PDF and simulated histogram of the phase distribution for SNR=6dB (b): theoretical PDFs for different SNR values

SNR	χ^2	Confidence level
0	3.959337e-1	$> 99\%$
2	3.344832e-1	–"–
4	3.012187e-1	–"–
6	2.110520e-1	–"–
8	1.835500e-1	–"–
10	1.718824e-1	–"–

Table 18.1: χ^2 values and confidence levels for $p_\Pi(\xi, \psi)$ for different SNR values

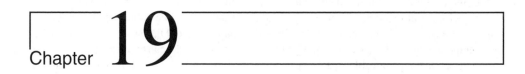

Chapter 19

Adaptive Single- and Multi-user OFDM Techniques

19.1 Introduction

Steele and Webb [124] proposed adaptive modulation for exploiting the time-variant Shannonian channel capacity of fading narrowband channels, and further research was conducted at Osaka University by Sampei *et al.* [193], at the University of Stanford by Goldsmith *et al.* [577], by Pearce, Burr and Tozer at the University of York [210], Lau and McLeod at the University of Cambridge [211], and at Southampton University [219, 220]. The associated principles can also be invoked in the context of parallel modems, as has been demonstrated by Kalet [155], Czylwik [221] as well as by Chow, Cioffi and Bingham [222].

19.1.1 Motivation

We saw in Figure 17.13 that the bit error probability of different OFDM subcarriers transmitted in time dispersive channels depends on the frequency domain channel transfer function. The occurrence of bit errors is normally concentrated in a set of severely faded subcarriers, while in the rest of the OFDM spectrum often no bit errors are observed. If the subcarriers that will exhibit high bit error probabilities in the OFDM symbol to be transmitted can be identified and excluded from data transmission, the overall BER can be improved in exchange for a slight loss of system throughput. As the frequency domain fading deteriorates the SNR of certain subcarriers, but improves that of others' above the average SNR value, the potential loss of throughput due to the exclusion of faded subcarriers can be mitigated by employing higher-order modulation modes on the subcarriers exhibiting high SNR values.

In addition to excluding sets of faded subcarriers and varying the modulation modes employed, other parameters such as the coding rate of error correction coding schemes can be adapted at the transmitter according to the perceived channel transfer function. This issue will be addressed in Chapter 20.

Adaptation of the transmission parameters is based on the transmitter's perception of the channel conditions in the forthcoming time slot. Clearly, this estimation of future channel parameters can only be obtained by extrapolation of previous channel estimations, which are acquired upon detecting each received OFDM symbol. The channel characteristics therefore have to be varying sufficiently slowly compared to the estimation interval.

Adapting the transmission technique to the channel conditions on a time slot by time slot basis for serial modems in narrowband fading channels has been shown to considerably improve the BER performance [578] for time division duplex (TDD) systems assuming duplex reciprocal channels. However, the Doppler fading rate of the narrowband channel has a strong effect on the achievable system performance: if the fading is rapid, then the prediction of the channel conditions for the next transmit time slot is inaccurate, and therefore the wrong set of transmission parameters may be chosen. If, however, the channel varies slowly, then the data throughput of the system is varying dramatically over time, and large data buffers are required at the transmitters in order to smoothen the bit rate fluctuation. For time critical applications, such as interactive speech transmission, the potential delays can become problematic. A given single-carrier adaptive system in narrowband channels will therefore operate efficiently only in a limited range of channel conditions.

Adaptive OFDM modem channels can ease the problem of slowly time-varying channels, since the variation of the signal quality can be exploited in the time domain and the frequency domain. The channel conditions still have to be monitored based on the received OFDM symbols, and relatively slowly varying channels have to be assumed, since we saw in Section 17.2.2 that OFDM transmissions are not well suited to rapidly varying channel conditions.

19.1.2   Adaptive Modulation Techniques

Adaptive modulation is only suitable for duplex communication between two stations, since the transmission parameters have to be adapted using some form of two-way transmission in order to allow channel measurements and signalling to take place.

Transmission parameter adaptation is a response of the transmitter to time-varying channel conditions. In order to efficiently react to the changes in channel quality, the following steps have to be taken:

(1) *Channel quality estimation:* In order to appropriately select the transmission parameters to be employed for the next transmission, a reliable estimation of the channel transfer function during the next active transmit time slot is necessary.

(2) *Choice of the appropriate parameters for the next transmission:* Based on the prediction of the channel conditions for the next time slot, the transmitter has to select the appropriate modulation modes for the subcarriers.

(3) *Signalling or blind detection of the employed parameters:* The receiver has to be informed, as to which demodulator parameters to employ for the received packet. This information can either be conveyed within the OFDM symbol itself, with the loss of effective data throughput, or the receiver can attempt to estimate the parameters employed by the remote transmitter by means of blind detection mechanisms.

19.1.2.1 Channel Quality Estimation

The transmitter requires an estimate of the expected channel conditions for the time when the next OFDM symbol is to be transmitted. Since this knowledge can only be gained by prediction from past channel quality estimations, the adaptive system can only operate efficiently in an environment exhibiting relatively slowly varying channel conditions.

The channel quality estimation can be acquired from a range of different sources. If the communication between the two stations is bidirectional and the channel can be considered reciprocal, then each station can estimate the channel quality on the basis of the received OFDM symbols, and adapt the parameters of the local transmitter to this estimation. We will refer to such a regime as *open-loop adaptation*, since there is no feedback between the receiver of a given OFDM symbol and the choice of the modulation parameters. A time division duplex (TDD) system in absence of interference is an example of such a system, and hence a TDD regime is assumed for generating the performance results below. Channel reciprocity issues were addressed for example in [579, 580].

If the channel is not reciprocal, as in a frequency division duplex (FDD) system, then the stations cannot determine the parameters for the next OFDM symbol's transmission from the received symbols. In this case, the receiver has to estimate the channel quality and explicitly signal this perceived channel quality information to the transmitter in the reverse link. Since in this case the receiver explicitly instructs the remote transmitter as to which modem modes to invoke, this regime is referred to as *closed-loop adaptation*. With the aid of this technique the adaptation algorithms cas take into account effects such as interference as well as non-reciprocal channels. If the communication between the stations is essentially unidirectional, then a low-rate signalling channel must be implemented from the receiver to the transmitter. If such a channel exists, then the same technique as for non-reciprocal channels can be employed.

Different techniques can be employed to estimate the channel quality. For OFDM modems, the bit error probability in each subcarrier is determined by the fluctuations of the channel's instantaneous frequency domain channel transfer function H_n, if no interference is present. The estimate of the channel transfer function $\hat{H}_n$ can be acquired by means of pilot tone-based channel estimation, as demonstrated in Section 17.3.1.1. More accurate measures of the channel transfer function can be gained by means of decision directed or time domain training sequence based techniques. The estimate of the channel transfer function $\hat{H}_n$ does not take into account effects, such as co-channel or intersubcarrier interference. Alternative channel quality measures that include interference effects can be devised on the basis of the error correction decoder's soft output information or by means of decision feedback local SNR estimations.

The delay between the channel quality estimate and the actual transmission of the OFDM symbol in relation to the maximal Doppler frequency of the channel is crucial to the adaptive system's performance. If the channel estimate is obsolete at the time of transmission, then poor system performance will result. For a closed-loop adaptive system the delays between channel estimation and transmission of the packet are generally longer than for an open-loop adaptive system, and therefore the Doppler frequency of the channel is a more critical parameter for the system's performance than in the context of open-loop adaptive systems.

19.1.2.2   Parameter Adaptation

Different transmission parameters can be adapted to the anticipated channel conditions, such as the modulation and coding modes. Adapting the number of modulation levels in response to the anticipated local SNR encountered in each subcarrier can be employed, in order to achieve a wide range of different trade-offs between the received data integrity and through-put. Corrupted subcarriers can be excluded from data transmission and left blank or perhaps used for crest factor reduction.

The adaptive channel coding parameters entail code rate, adaptive interleaving and punc-turing for convolutional and turbo codes, or varying block lengths for block codes [170,171]. These techniques can be combined with adaptive modulation mode selection and will be discussed in Chapter 20.

Based on the estimated frequency domain channel transfer function, spectral pre-distortion at the transmitter of one or both communicating stations can be invoked, in order to partially or fully counteract the frequency selective fading of the time dispersive channel. Unlike frequency domain equalisation at the receiver – which corrects for the amplitude and phase errors inflicted upon the subcarriers by the channel but cannot improve the signal-to-noise ratio in poor quality channels – spectral predistortion at the OFDM transmitter can deliver near-constant signal-to-noise levels for all subcarriers and can be thought of as power control on a subcarrier-by-subcarrier basis.

In addition to improving the system's BER performance in time dispersive channels, spec-tral predistortion can be employed in order to perform all channel estimation and equalisation functions at only one of the two communicating duplex stations. Low cost, low power con-sumption mobile stations can communicate with a base station that performs the channel estimation and frequency domain equalisation of the uplink, and uses the estimated channel transfer function for predistorting the down-link OFDM symbol. This set-up would lead to different overall channel quality on the uplink and the downlink, and the superior downlink channel quality could be exploited by using a computationally less complex channel decoder having weaker error correction capabilities in the mobile station than in the base station.

If the channel's frequency domain transfer function is to be fully counteracted by the spectral predistortion upon adapting the subcarrier power to the inverse of the channel transfer function, then the output power of the transmitter can become excessive, if heavily faded subcarriers are present in the system's frequency range. In order to limit the transmitter's maximal output power, hybrid channel predistortion and adaptive modulation schemes can be devised, which would deactivate transmission in deeply faded sub-channels, while retaining the benefits of predistortion in the remaining subcarriers.

19.1.2.3   Signalling the AQAM Parameters

Signalling plays an important role in adaptive systems and the range of signalling options is summarised in Figure 19.1 for both open-loop and closed-loop signalling, as well as for blind detection. If the channel quality estimation and parameter adaptation have been performed at the transmitter of a particular link, based on open-loop adaptation, then the resulting set of parameters has to be communicated to the receiver in order to successfully demodulate and decode the OFDM symbol. If the receiver itself determines the requested parameter set to be used by the remote transmitter, the closed-loop scenario, then the same amount of information has to be transported to the remote transmitter in the reverse link. If this

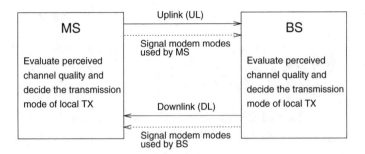

(a) Reciprocal channel, open-loop control

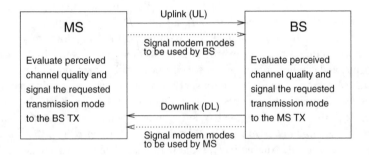

(b) Non-reciprocal channel, closed-loop signalling

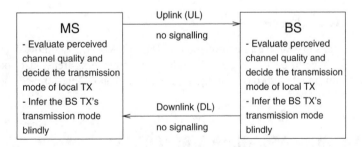

(c) Reciprocal channel, blind modem-mode detection

Figure 19.1: Signalling scenarios in adaptive modems

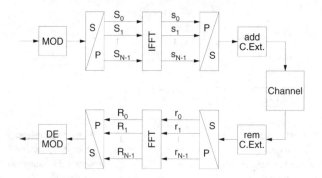

Figure 19.2: Schematic model of the OFDM system

signalling information is corrupted, then the receiver is generally unable to correctly decode the OFDM symbol corresponding to the incorrect signalling information.

Unlike adaptive serial systems, which employ the same set of parameters for all data symbols in a transmission packet [219, 220], adaptive OFDM systems have to react to the frequency selective nature of the channel, by adapting the modem parameters across the sub-carriers. The resulting signalling overhead may become significantly higher than that for serial modems, and can be prohibitive for subcarrier-by-subcarrier modulation mode adapta-tion. In order to overcome these limitations, efficient and reliable signalling techniques have to be employed for practical implementation of adaptive OFDM modems.

If some flexibility in choosing the transmission parameters is sacrificed in an adaptation scheme, as in the sub-band adaptive OFDM schemes described below, then the amount of signalling can be reduced. Alternatively, blind parameter detection schemes can be devised, which require little or no signalling information, respectively. Two simple blind modulation scheme detection algorithms are investigated in Section 19.2.6.2 [581].

19.1.3   System Aspects

The effects of transmission parameter adaptation for OFDM systems on the overall commu-nication system have to be investigated in at least the following areas: data buffering and latency due to varying data throughput, the effects of co-channel interference and bandwidth efficiency.

19.2   Adaptive Modulation for OFDM

19.2.1   System Model

The system model of the N-subcarrier orthogonal frequency division multiplexing (OFDM) modem is shown in Figure 19.2 [143]. At the transmitter, the modulator generates N data symbols S_n, $0 \leq n \leq N - 1$, which are multiplexed to the N subcarriers. The time do-main samples s_n transmitted during one OFDM symbol are generated by the inverse fast Fourier transform (IFFT) and transmitted over the channel after the cyclic extension (C. Ext.)

has been inserted. The channel is modelled by its time-variant impulse response $h(\tau, t)$ and AWGN. At the receiver, the cyclic extension is removed from the received time domain samples, and the data samples r_n are fast Fourier transformed, in order to yield the received frequency-domain data symbols R_n.

The channel's impulse response is assumed to be time invariant for the duration of one OFDM symbol, therefore it can be characterised for each OFDM symbol period by the N-point Fourier transform of the impulse response, which is referred to as the frequency domain channel transfer function H_n. The received data symbols R_n can be expressed as:

$$R_n = S_n \cdot H_n + n_n,$$

where n_n is an AWGN sample. Coherent detection is assumed for the system, therefore the received data symbols R_n have to be defaded in the frequency domain with the aid of an estimate of the channel transfer function H_n. This estimate $\hat{H}_n$ can be obtained by the use of pilot subcarriers in the OFDM symbol, or by employing time domain channel sounding training sequences embedded in the transmitted signal. Since the noise energy in each subcarrier is independent of the channel's frequency domain transfer function H_n, the "local" signal-to-noise ratio SNR in subcarrier n can be expressed as

$$\gamma_n = |H_n|^2 \cdot \gamma,$$

where γ is the overall SNR. If no signal degradation due to intersubcarrier interference (ISI) or interference from other sources appears, then the value of γ_n determines the bit error probability for the transmission of data symbols over the subcarrier n.

The goal of adaptive modulation is to choose the appropriate modulation mode for transmission in each subcarrier, given the local SNR γ_n, in order to achieve a good trade-off between throughput and overall BER. The acceptable overall BER varies depending on other systems parameters, such as the correction capability of the error correction coding and the nature of the service supported by this particular link.

The adaptive system has to fulfil these requirements:

(1) Channel quality estimation;

(2) Choice of the appropriate modulation modes;

(3) Signalling or blind detection of the modulation modes.

We will examine these three points with reference to Figure 19.1 in the following sections for the example of a 512-subcarrier OFDM modem in the shortened WATM channel of Section 17.1.1.2.

19.2.2 Channel Model

The impulse response $h(\tau, t)$ used in our experiments was generated on the basis of the symbol-spaced impulse response shown in Figure 19.3(a) by fading each of the impulses obeying a Rayleigh distribution of a normalised maximal Doppler frequency of $f'_d = 1.235 \cdot 10^{-5}$, which corresponds to the WLAN channel experienced by a modem transmitting at a carrier frequency of 60 GHz with a sample rate of 225 MHz and a vehicular velocity of

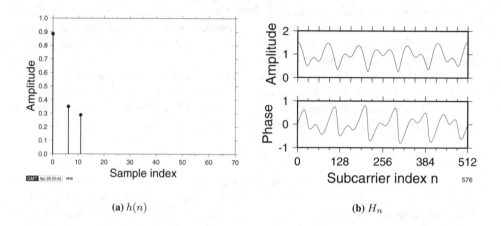

(a) $h(n)$ **(b)** H_n

Figure 19.3: WATM wideband channel: (a) unfaded symbol spaced impulse response, (b) the corresponding frequency domain channel transfer function

50 km/h. The complex frequency domain channel transfer function H_n corresponding to the unfaded impulse response is shown in Figure 19.3(b).

19.2.3   Channel Estimation

The most convenient setting for an adaptive OFDM (AOFDM) system is a time division duplex (TDD) system in a slowly varying reciprocal channel, allowing open-loop adaptation. Both stations transmit an OFDM symbol in turn, and at each station the most recent received symbol is used for the channel estimation employed for the modulation mode adaptation for the next transmitted OFDM symbol. The channel estimation on the basis of the received symbol can be performed by PSAM (see Section 17.3.1.1), or upon invoking more sophisticated methods, such as decision directed channel estimation. Initially, we will assume perfect knowledge of the channel transfer function during the received time slot.

19.2.4   Choice of the AQAM Modes

The two communicating stations use the open-loop predicted channel transfer function acquired from the most recent received OFDM symbol, in order to allocate the appropriate modulation modes to the subcarriers. The modulation modes were chosen from the set of binary phase shift keying (BPSK), quadrature phase shift keying (QPSK), 16-quadrature amplitude modulation (16-QAM), as well as "no transmission", for which no signal was transmitted. These modulation modes are denoted by M_m, where $m \in (0, 1, 2, 4)$ is the number of data bits associated with one data symbol of each mode.

In order to keep the system complexity low, the modulation mode is not varied on a subcarrier-by-subcarrier basis, but instead the total OFDM bandwidth of 512 subcarriers is split into blocks of adjacent subcarriers, referred to as sub-bands, and the same modulation scheme is employed for all subcarriers of the same sub-band. This substantially simplifies the task of signalling the modem mode and renders the employment of alternative blind detection

	l_0	l_1	l_2	l_4
speech system	$-\infty$	3.31	6.48	11.61
data system	$-\infty$	7.98	10.42	16.76

Table 19.1: Optimised switching levels for adaptive modulation over Rayleigh fading channels for the "speech" and "data" systems, shown in instantaneous channel SNR [dB] (from [185])

mechanisms feasible, which will be discussed in Section 19.2.6.

Three modulation mode allocation algorithms were investigated in the sub-bands: a fixed threshold controlled algorithm, an upper bound BER estimator and a fixed-throughput adaptation algorithm.

19.2.4.1 Fixed Threshold Adaptation Algorithm

The fixed threshold algorithm was derived from the adaptation algorithm proposed by Torrance for serial modems [578]. In the case of a serial modem, the channel quality is assumed to be constant for all symbols in the time slot, and hence the channel has to be slowly varying in order to allow accurate channel quality prediction. Under these circumstances, all data symbols in the transmit time slot employ the same modulation mode, chosen according to the predicted SNR. The SNR thresholds for a given long-term target BER were determined by Powell optimisation [185]. Torrance assumed two uncoded target bit error rates: 1% for a high data rate "speech" system, and 10^{-4} for a higher integrity, lower data rate "data" system. The resulting SNR thresholds l_n for activating a given modulation mode M_n in a slowly Rayleigh fading narrowband channel for both systems are given in Table 19.1. Specifically, the modulation mode M_n is selected if the instantaneous channel SNR exceeds the switching level l_n.

This adaptation algorithm originally assumed a constant instantaneous SNR over all of the block's symbols, but in the case of an OFDM system in a frequency selective channel the channel quality varies across the different subcarriers. For sub-band adaptive OFDM transmission, this implies that if the sub-band width is wider than the channel's coherence bandwidth, then the original switching algorithm cannot be employed. For our investigations, we have therefore employed the lowest quality subcarrier in the sub-band for the adaptation algorithm based on the thresholds given in Table 19.1. The performance of the 16 sub-band adaptive system over the shortened WATM Rayleigh fading channel of Figure 17.3 is shown in Figure 19.4.

Adjacent or consecutive time slots have been used for the uplink and downlink slots in these simulations, so that the delay between channel estimation and transmission was rendered as short as possible. Figure 19.4 shows the long-term average BER and throughput of the studied modem for the "speech" and "data" switching levels of Table 19.1 as well as for a subcarrier-by-subcarrier adaptive modem employing the "data" switching levels. The results show the typical behaviour of a variable throughput AOFDM system, which constitutes a trade-off between the best BER and best throughput performance. For low SNR values, the system achieves a low BER by transmitting very few bits and only when the channel conditions allow. With increasing long-term SNR, the throughput increases without significant change in the BER. For high SNR values the BER drops as the throughput approaches its

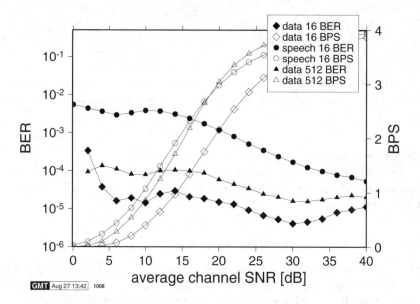

Figure 19.4: BER and BPS throughput performance of the 16 sub-band, 512 subcarrier switching level
adaptive OFDM modem employing BPSK, QPSK, 16-QAM and "no transmission" over
the Rayleigh fading time dispersive channel of Figure 17.3 using the switching thresholds
of Table 19.1

maximum of 4 bits per symbol, since the highest-order constellation was 16-QAM.

It can be seen from the figure that the adaptive system performs better than its target bit
error rates of 10^{-2} and 10^{-4} for the "speech" and "data" systems, respectively, resulting
in measured bit error rates lower than the targets. This can be explained by the adaptation
regime, which was based on the conservative principle of using the lowest quality subcarrier
in each sub-band for channel quality estimation, leading to a pessimistic channel quality
estimate for the entire sub-band. For low values of SNR, the throughput in bits per data
symbol is low and exceeds the fixed BPSK throughput of 1 bit/symbol only for SNR values
in excess of 9.5 dB and 14 dB for the "speech" and "data" systems, respectively.

The upper bound performance of the system with subcarrier-by-subcarrier adaptation is
also portrayed in the figure, shown as 512 independent sub-bands, for the "data" optimised
set of threshold values. It can be seen that in this case the target BER of 10^{-4} is closely met
over a wide range of SNR values from about 2 dB to 20 dB, and that the throughput is considerably higher than in the case of the 16 sub-band modem. This is the result of more accurate
subcarrier-by-subcarrier channel quality estimation and fine-grained adaptation, leading to
better exploitation of the available channel capacity.

Figure 19.5 shows the long-term modulation mode histograms for a range of channel
SNR values for the "data" switching levels in both the 16 sub-band and the subcarrier-by-
subcarrier adaptive modems using the switching thresholds of Table 19.1. Comparison of the
graphs shows that higher-order modulation modes are used more frequently by the subcarrier-
by-subcarrier adaptation algorithm, which is in accordance with the overall throughput performance of the two modems in Figure 19.4.

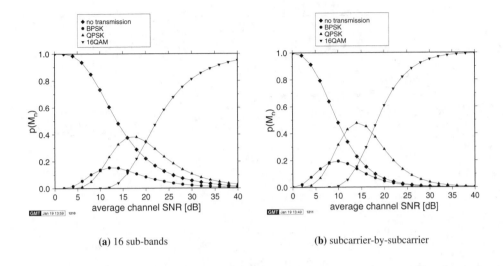

(a) 16 sub-bands   (b) subcarrier-by-subcarrier

Figure 19.5: Histograms of modulation modes versus channel SNR for the "data" switching level adaptive 512 subcarrier, 16 sub-band OFDM modem over the Rayleigh fading time dispersive channel of Figure 17.3 using the switching thresholds of Table 19.1

The throughput penalty of employing sub-band adaptation depends on the frequency domain variation of the channel transfer function. If the sub-band bandwidth is lower than the channel's coherence bandwidth, then the assumption of constant channel quality per sub-band is closely met, and the system performance is equivalent to that of a subcarrier-by-subcarrier adaptive scheme.

19.2.4.2 Sub-band BER Estimator Adaptation Algorithm

We have seen above that the fixed switching level based algorithm leads to a throughput performance penalty, if used in a sub-band adaptive OFDM modem, when the channel quality is not constant throughout each sub-band. This is due to the conservative adaptation based on the subcarrier experiencing the most hostile channel in each sub-band.

An alternative scheme taking into account the non-constant SNR values γ_j across the N_s subcarriers in the jth sub-band can be devised by calculating the expected overall bit error probability for all available modulation modes M_n in each sub-band, which is denoted by $\bar{p}_e(n) = 1/N_s \sum_j p_e(\gamma_j, M_n)$. For each sub-band, the mode having the highest throughput, whose estimated BER is lower than a given threshold, is then chosen. While the adaptation granularity is still limited to the sub-band width, the channel quality estimation includes not only the lowest quality subcarrier, which leads to an improved throughput.

Figure 19.6 shows the BER and throughput performance for the 16 sub-band adaptive OFDM modem employing the BER estimator adaptation algorithm in the Rayleigh fading time dispersive channel of Figure 17.3. The two sets of curves in the figure correspond to target bit error rates of 10^{-2} and 10^{-1}, respectively. Comparing the modem's performance for a target BER of 10^{-2} with that of the "speech" modem in Figure 19.4, it can be seen

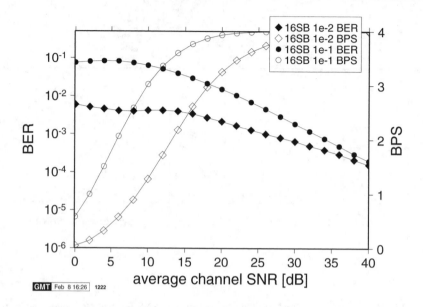

Figure 19.6: BER and BPS throughput performance of the 16 sub-band, 512 subcarrier BER estimator adaptive OFDM modem employing BPSK, QPSK, 16-QAM and "no transmission" over the Rayleigh fading time dispersive channel of Figure 17.3

that the BER estimator algorithm results in significantly higher throughput while meeting the BER requirements. The BER estimator algorithm is readily adjustable to different target bit error rates, which is demonstrated in the figure for a target BER of 10^{-1}. Such adjustability is beneficial when combining adaptive modulation with channel coding, as will be discussed in Section 19.2.7.

19.2.5   Constant-Throughput Adaptive OFDM

The time-varying data throughput of an adaptive OFDM modem operating with either of the two adaptation algorithms discussed above makes it difficult to employ such a scheme in a variety of constant rate applications. Torrance [578] studied the system implications of variable-throughput adaptive modems in the context of narrowband channels, stressing the importance of data buffering at the transmitter, in order to accommodate the variable data rate. The required length of the buffer is related to the Doppler frequency of the channel, and a slowly varying channel – as required for adaptive modulation – results in slowly varying data throughput and therefore the need for a high buffer capacity. Real-time interactive audio or video transmission is sensitive to delays, and therefore different modem mode adaptation algorithms are needed for such applications.

The constant throughput AOFDM scheme proposed here exploits the frequency selectivity of the channel, while offering a constant bit rate. Again, sub-band adaptivity is assumed, in order to simplify the signalling or the associated blind detection of the modem schemes.

The modulation mode allocation of the sub-bands is performed on the basis of a cost

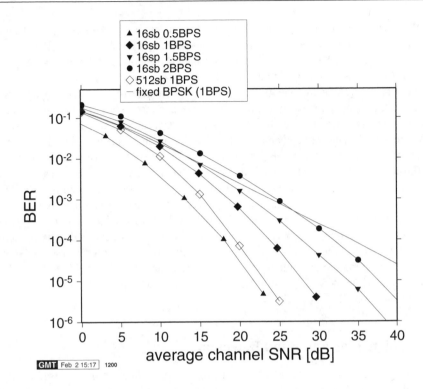

Figure 19.7: BER performance versus SNR for the 512 subcarrier, 16 sub-band constant throughput
adaptive OFDM modem employing BPSK, QPSK, 16-QAM and "no transmission" in
the Rayleigh fading time dispersive channel of Figure 17.3 for 0.5, 1, 1.5 and 2 bits per
symbol (BPS) target throughput

function to be introduced below, based on the expected number of bit errors in each sub-band.
The expected number of bit errors $e_{n,s}$, for each sub-band n and for each possible modulation
mode index s, is calculated on the basis of the estimated channel transfer function $\hat{H}$, taking
into account the number of bits transmitted per sub-band and per modulation mode, $b_{n,s}$.

Each sub-band is assigned a state variable s_n holding the index of a modulation mode.
Each state variable is initialised to the lowest-order modulation mode, which in our case is 0
for "no transmission". A set of cost values $c_{n,s}$ is calculated for each sub-band n and state s:

$$c_{n,s} = \frac{e_{n,s+1} - e_{n,s}}{b_{n,s+1} - b_{n,s}} \tag{19.1}$$

for all but the highest modulation mode index s. This cost value is related to the expected
increase in the number of bit errors, divided by the increase of throughput, if the modulation
mode having the next higher index is used instead of index s in sub-band n. In other words,
Equation 19.1 quantifies the expected incremental bit error rate of the state transition $s \rightarrow
s + 1$ in sub-band n.

The modulation mode adaptation is performed by repeatedly searching for the block n

having the lowest value of c_{n,s_n}, and incrementing its state s_n. This is repeated until the total number of bits in the OFDM symbol reaches the target number of bits. Because of the granularity in bit numbers introduced by the sub-bands, the total number of bits may exceed the target. In this case, the data is padded with dummy bits for transmission.

Figure 19.7 gives an overview of the BER performance of the fixed-throughput 512-subcarrier OFDM modem over the time dispersive channel of Figure 17.3 for a range of target bit numbers. The curve without shapes represents the performance of a fixed BPSK OFDM modem over the same channel and where the modem transmits, i.e. 1 bit over each data subcarrier per OFDM symbol. The diamond shapes give the performance of the equivalent throughput adaptive scheme for the 16 sub-band arrangement filled shapes and for the subcarrier-by-subcarrier adaptive scheme open shapes. It can be seen that the 16 sub-band adaptive scheme yields a significant improvement in BER terms for SNR values above 10 dB. The SNR gain for a bit error rate of 10^{-4} is 8 dB compared to the non-adaptive case. Subcarrier-by-subcarrier adaptivity increases this gain by a further 4 dB. The modem can readily be adapted to the system requirements by adjusting the target bit rate, as shown in Figure 19.7. Halving the throughput to 0.5 BPS, the required SNR is reduced by 6 dB for a BER of 10^{-4}, while increasing the throughput to 2 BPS deteriorates the noise resilience by 8 dB at the same BER.

19.2.6   Signalling and Blind Detection

The adaptive OFDM receiver has to be informed of the modulation modes used for the different sub-bands. This information can either be conveyed using signalling subcarriers in the OFDM symbol itself, or the receiver can employ blind detection techniques in order to estimate the transmitted symbols' modulation modes, as seen in Figure 19.1.

19.2.6.1   Signalling

The simplest way of signalling the modulation mode employed in a sub-band is to replace one data symbol by an M-PSK symbol, where M is the number of possible modulation modes. In this case, reception of each of the constellation points directly signals a particular modulation mode in the current sub-band. In our case, for four modulation modes and assuming perfect phase recovery, the probability of a signalling error $p_s(\gamma)$, when employing one signalling symbol, is the symbol error probability of QPSK. Then the correct sub-band mode signalling probability is:

$$(1 - p_s(\gamma)) = (1 - p_{b,QPSK}(\gamma))^2,$$

where $p_{b,QPSK}$ is the bit error probability for QPSK:

$$p_{b,QPSK}(\gamma) = Q(\sqrt{\gamma}) = \frac{1}{2} \cdot \text{erfc}\left(\sqrt{\frac{\gamma}{2}}\right),$$

which leads to the expression for the modulation mode signalling error probability of

$$p_s(\gamma) = 1 - \left(1 - \frac{1}{2} \cdot \text{erfc}\left(\sqrt{\frac{\gamma}{2}}\right)\right)^2.$$

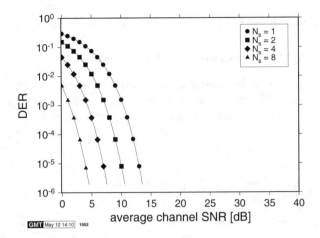

Figure 19.8: Modulation mode detection error ratio (DER) if signalling with maximum ratio combining is employed for QPSK symbols in an AWGN channel for 1, 2, 4 and 8 signalling symbols per sub-band, evaluated from Equation 19.2

The modem mode signalling error probability can be reduced by employing multiple signalling symbols and maximum ratio combining of the received signalling symbols $R_{s,n}$, in order to generate the decision variable R'_s prior to decision:

$$R'_s = \sum_{n=1}^{N_s} R_{s,n} \cdot \hat{H}^*_{s,n},$$

where N_s is the number of signalling symbols per sub-band, the quantities $R_{s,n}$ are the received symbols in the signalling subcarriers, and $\hat{H}_{s,n}$ represents the estimated values of the frequency domain channel transfer function at the signalling subcarriers. Assuming perfect channel estimation and constant values of the channel transfer function across the group of signalling subcarriers, the signalling error probability for N_s signalling symbols can be expressed as:

$$p'_s(\gamma, N_s) = 1 - \left(1 - \frac{1}{2} \cdot \mathrm{erfc}\left(\sqrt{\frac{N_s \gamma}{2}}\right)\right)^2. \tag{19.2}$$

Figure 19.8 shows the signalling error rate in an AWGN channel for 1, 2, 4 and 8 signalling symbols per sub-band, respectively. It can be seen that doubling the number of signalling subcarriers improves the performance by 3 dB. Modem mode detection error ratios (DER) below 10^{-5} can be achieved at 10 dB SNR over AWGN channels if two signalling symbols are used. The signalling symbols for a given sub-band can be interleaved across the entire OFDM symbol bandwidth, in order to benefit from frequency diversity in fading wideband channels.

As seen in Figure 19.1, blind detection algorithms aim to estimate the employed modulation mode directly from the received data symbols, therefore avoiding the loss of data capacity due to signalling subcarriers. Two algorithms have been investigated, one based on geometrical SNR estimation and another incorporating error correction coding.

19.2.6.2 Blind AQAM Mode Detection by SNR Estimation

The receiver has no a-priori knowledge of the modulation mode employed in a particular received sub-band and estimates this parameter by quantising the defaded received data symbols $R_n/\hat{H}_n$ in the sub-band to the closest symbol $\hat{R}_{n,m}$ for all possible modulation modes M_m for each subcarrier index n in the current sub-band. The decision directed error energy e_m for each modulation mode is calculated:

$$e_m = \sum_n \left(R_n/\hat{H}_n - \hat{R}_{n,m} \right)^2$$

and the modulation mode M_m which minimises e_m is chosen for the demodulation of the sub-band.

The DER of the blind modulation mode detection algorithm described in this section for a 512-subcarrier OFDM modem in an AWGN channel is depicted in Figure 19.9. It can be seen that the detection performance depends on the number of symbols per sub-band, with fewer sub-bands and therefore longer symbol sequences per sub-band leading to a better detection performance. It is apparent, however, that the number of available modulation modes has a more significant effect on the detection reliability than the block length. If all four legitimate modem modes are employed, then reliable detection of the modulation mode is only guaranteed for AWGN SNR values of more than 15-18 dB, depending on the number of sub-bands per OFDM symbol. If only M_0 and M_1 are employed, however, the estimation accuracy is dramatically improved. In this case, AWGN SNR values above 5-7 dB are sufficient to ensure reliable detection. The estimation accuracy could be improved by using the estimate of the channel quality, in order to predict the modulation mode, which is likely to have been employed at the transmitter. For example, at an estimated channel SNR of 5 dB it is unlikely that 16-QAM was employed as the modem mode and hence this a priori knowledge can be exploited, in order increase our confidence in the corresponding decision.

Figure 19.10 shows the BER performance of the fixed threshold "data" 16 sub-band adaptive system in the fading wideband channel of Figure 17.3 for both sets of modulation modes, namely for (M_0, M_1) and (M_0, M_1, M_2, M_4) with blind modulation mode detection. Erroneous modulation mode decisions were assumed to yield a BER of 50% in the received block. This is optimistic, since in a realistic scenario the receiver would have no knowledge of the number of bits actually transmitted, leading to loss of synchronisation in the data stream. This problem is faced by all systems having a variable throughput and not employing an ideal reliable signalling channel. This impediment must be mitigated by data synchronisation measures.

It can be seen from Figure 19.10 that while blind modulation mode detection yields poor performance for the quadruple-mode adaptive scheme, the twin-mode scheme exhibits BER results consistently better than 10^{-4}.

19.2.6.3 Blind AQAM Mode Detection by Multi-Mode Trellis Decoder

If error correction coding is invoked in the system, then the channel decoder can be employed to estimate the most likely modulation mode per sub-band. Since the number of bits per

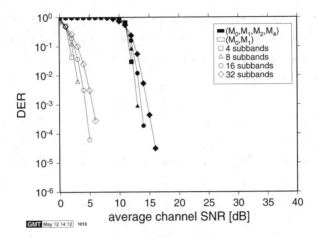

Figure 19.9: Blind modulation mode detection error ratio (DER) for 512-subcarrier OFDM systems employing (M_0, M_1) as well as for (M_0, M_1, M_2, M_4) for different numbers of subbands in an AWGN channel

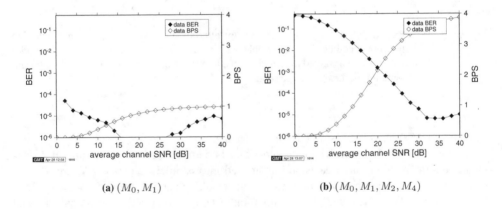

(a) (M_0, M_1) **(b)** (M_0, M_1, M_2, M_4)

Figure 19.10: BER and BPS throughput performance of a 16 sub-band, 512 subcarrier adaptive OFDM modem employing (a) no transmission (M_0) and BPSK (M_1), or (b) (M_0, M_1, M_2, M_4), both using the data-type switching levels of Table 19.1 and the SNR-based blind modulation mode detection of Section 19.2.6.2 over the Rayleigh fading time dispersive channel of Figure 17.3

OFDM symbol is varying in this adaptive scheme, and the channel encoder's block length is therefore not constant, for the sake of implementational convenience we have chosen a convolutional encoder at the transmitter. Once the modulation modes to be used are decided upon at the transmitter, the convolutional encoder is employed to generate a zero-terminated code word having the length of the OFDM symbol's capacity. This code word is modulated on the subcarriers according to the different modulation modes for the different sub-bands,

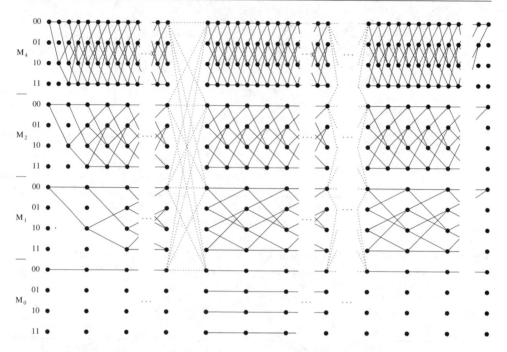

Figure 19.11: Schematic plot of the parallel trellises for blind modulation mode detection employing
convolutional coding. In this example, a four-state 00-terminated convolutional encoder
was assumed. The dotted lines indicate the intersub-band transitions for the 00 state, and
are omitted for the other three states

and the OFDM symbol is transmitted over the channel.

At the receiver, each received data subcarrier is demodulated by all possible demodulators, and the resulting hard decision bits are fed into parallel trellises for Viterbi decoding. Figure 19.11 shows a schematic sketch of the resulting parallel trellis if 16-QAM (M_4),
QPSK (M_2), BPSK (M_1), and "no transmission" (M_0) are employed, for a convolutional
code having four states. Each sub-band in the adaptive scheme corresponds to a set of four
parallel trellises, whose inputs are generated independently by the four demodulators of the
legitimate modulation modes. The number of transitions in each of the trellises depends on
the number of output bits received from the different demodulators, so that the 16-QAM (M_4)
trellis contains four times as many transitions as the BPSK and "no transmission" trellises.
Since in the case of "no transmission" no coded bits are transmitted, the state of the encoder
does not change. Therefore, legitimate transitions for this case are only horizontal ones.

At sub-band boundaries, transitions are allowed between the same state of all the parallel
trellises associated with the different modulation modes. This is not a transition due to a
received bit, and therefore preserves the metric of the originating state. Note that in the figure
only the possible allowed transitions for the state 00 are drawn; all other states originate the
equivalent set of transitions. The initial state of the first sub-band is 00 for all modulation
modes, and since the code is terminated in 00, the last sub-band's final states are 00.

The receiver's Viterbi decoder calculates the metrics for the transitions in the parallel

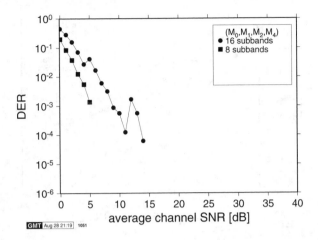

Figure 19.12: Blind modulation mode detection error ratio (DER) using the parallel trellis algorithm of Section 19.2.6.3 with a $K = 7$ convolutional code in an AWGN channel for a 512-subcarrier OFDM modem

trellises, and once all the data symbols have been processed, it traces back through the parallel trellis on the surviving path. This backtracing commences at the most likely 00 state at the end of the last sub-band. If no termination was used at the decoder, then the backtracing would start at the most likely of all the final states of the last block.

Figure 19.12 shows the modulation mode detection error ratio (DER) for the parallel trellis decoder in an AWGN channel for 16 and 8 sub-bands, if a convolutional code of constraint length 7 is used. Comparison with Figure 19.9 shows considerable improvements relative to the BER estimation-based blind detection scheme of Section 19.2.6.2, both for 16 and 8 sub-bands. Higher sub-band lengths improve the estimation accuracy by a greater degree than has been observed for the BER estimation algorithm of Figure 19.9. A DER of less than 10^{-5} was observed for AWGN SNR values of 6 dB and 15 dB in the 8 and 16 sub-band scenarios, respectively. The use of stronger codes could further improve the estimation accuracy, at the cost of higher complexity.

19.2.7   Sub-band Adaptive OFDM and Turbo Coding

Adaptive modulation can reduce the BER to a level where channel decoders can perform well. Figure 19.13 shows both the uncoded and coded BER performance of a 512-subcarrier OFDM modem in the fading wideband channel of Figure 17.3, assuming perfect channel estimation. The channel coding employed in this set of experiments was a turbo coder [366] with a data block length of 1000 bits, employing a random interleaver and 8 decoder iterations. The log-MAP decoding algorithm was used [582]. The constituent half-rate convolutional encoders were of constraint length 3, with octally represented generator polynomials of $(7, 5)$ [171]. It can be seen that the turbo decoder provides a considerable coding gain for the different fixed modulation schemes, with a BER of 10^{-4} for SNR values of 13.8 dB, 17.3 dB and 23.2 dB for BPSK, QPSK and 16-QAM transmission, respectively.

Figure 19.14 depicts the BER and throughput performance of the same decoder employed

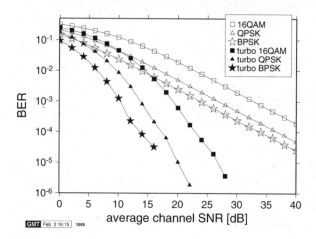

Figure 19.13: BER performance of the 512-subcarrier OFDM modem in the fading time dispersive
channel of Figure 17.3 for both uncoded and half-rate turbo-coded transmission, using
8-iteration log-MAP turbo decoding, 1000-bit random interleaver, and a constraint length
of 3

in conjunction with the adaptive OFDM modem for different adaptation algorithms. Figure 19.14(a) shows the performance for the "speech" system employing the switching levels
listed in Table 19.1. As expected, the half-rate channel coding results in a halved throughput
compared to the uncoded case, but offers low BER transmission over the channel of Figure
17.3 for SNR values of down to 0 dB, maintaining a BER below 10^{-6}.

Further tuning of the adaptation parameters can ensure a better average throughput, while
retaining error-free data transmission. The switching level based adaptation algorithm of
Table 19.1 is difficult to control for arbitrary bit error rates, since the set of switching levels
was determined by an optimisation process for uncoded transmission. Since the turbo codec
has a non-linear BER versus SNR characteristic, direct switching level optimisation is an
arduous task. The sub-band BER predictor of Section 19.2.4.2 is easier to adapt to a channel
codec, and Figure 19.14(b) shows the performance for the same decoder, with the adaptation
algorithm employing the BER prediction method having an upper BER bound of 1%. It can
be seen that the less stringent uncoded BER constraints when compared to Figure 19.14(a)
lead to a significantly higher throughput for low SNR values. The turbo-decoded data bits are
error-free, hence a further increase in throughput is possible while maintaining a high degree
of coded data integrity.

The second set of curves in Figure 19.14(b) shows the system's performance, if an uncoded target BER of 10% is assumed. In this case, the turbo decoder's output BER is below
10^{-5} for all the SNR values plotted, and shows a slow decrease for increasing values of SNR.
The throughput of the system, however, exceeds 0.5 data bits per symbol for SNR values of
more than 2 dB.

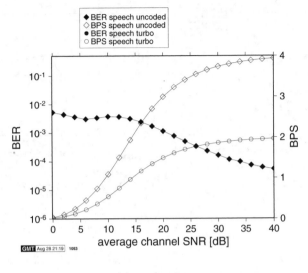

(a) speech system

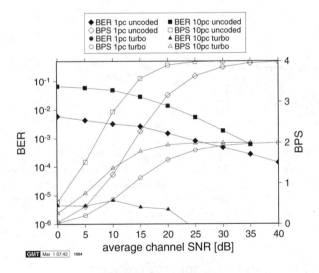

(b) maximal BER 1% and 10%

Figure 19.14: BER and BPS throughput performance of 16 sub-band, 512 subcarrier adaptive turbo coded and uncoded OFDM modem employing (M_0, M_1, M_2, M_4) for (a) speech-type switching levels of Table 19.1 and (b) a maximal estimated sub-band BER of 1% and 10% over the channel of Figure 17.3. The turbo-coded transmission over the speech system and the 1% maximal BER system are error-free for all examined SNR values and therefore the corresponding BER curves are omitted from the graphs, hence the lack of black circles on (a) and (b)

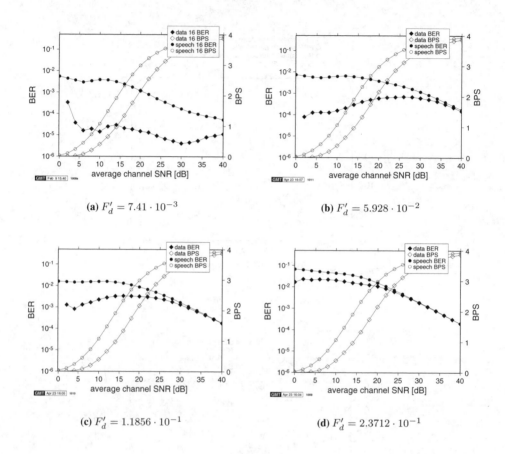

(a) $F_d' = 7.41 \cdot 10^{-3}$

(b) $F_d' = 5.928 \cdot 10^{-2}$

(c) $F_d' = 1.1856 \cdot 10^{-1}$

(d) $F_d' = 2.3712 \cdot 10^{-1}$

Figure 19.15: BER and BPS throughput performance of 16 sub-band, 512 subcarrier adaptive OFDM modem employing (M_0, M_1, M_2, M_4) for both data-type and speech-type switching levels with perfect modulation mode detection and different frame normalised Doppler frequencies F_d' over the channel of Figure 17.3. The triangular markers in (a) show the performance of a subcarrier-by-subcarrier adaptive modem using the data-type switching levels of Table 19.1 for comparison

19.2.8   Effect of Channel's Doppler Frequency

Since the adaptive OFDM modem employs the most recently received OFDM symbol in order to predict the frequency domain transfer function of the reverse channel for the next trans-mission, the quality of this prediction suffers from the time variance of the channel transfer function between the uplink and downlink time slots. We assume that the time delay between the uplink and downlink slots is the same as the delay between the downlink and uplink slots, and we refer to this time as the frame duration T_f. We normalise the maximal Doppler fre-quency f_d of the channel to the frame duration T_f, and define the *frame normalised Doppler frequency* F_d' as $F_d' = f_d \cdot T_f$. Figure 19.15 depicts the fixed switching level (see Table 19.1) modem's BER and throughput performance in bits per symbol (BPS) for values of F_d'

between $7.41 \cdot 10^{-3}$ and $2.3712 \cdot 10^{-1}$. These values stem from the studied WATM system with a time slot duration of 2.67 μs and up/downlink delays of 1, 8, 16 and 32 time slots at a channel Doppler frequency of 2.78 kHz. As mentioned in Section 17.1.1.1, this corresponds to a system employing a carrier frequency of 60 GHz, a sampling rate of 225 Msamples/s and a vehicular velocity of 50 km/h or $13.\overline{8}$ m/s.

Figure 19.15(a) shows the BER and BPS throughput of the studied modems in a framework with consecutive uplink and downlink time slots. This corresponds to $F'_d = 7.41 \cdot 10^{-3}$, while the target bit error rates for the speech and data system are met for all SNR values above 4 dB, and the BER performance is generally better than the target error rates. This was explained above with the conservative choice of modulation modes based on the most corrupted subcarrier in each sub-band, resulting in lower throughput and lower bit error rates for the switching level based sub-band adaptive modem.

Comparing Figure 19.15(a) with the other performance curves, it can be seen that the bit error rate performance for both the speech and the data system suffer from increasing decorrelation of the predicted and actual channel transfer function for increasing values of F'_d. In Figure 19.15(b) an 8 time slot delay was assumed between uplink and downlink time slots, which corresponds to $F'_d = 5.928 \cdot 10^{-2}$, and therefore the BER performance of the modem was significantly deteriorated. The "speech" system still maintains its target BER, but the "data" system delivers a BER of up to 10^{-3} for SNR values between 25 and 30 dB. It is interesting to observe that the delayed channel prediction mainly affects the higher-order modulation modes, which are employed more frequently at high SNR values. This explains the shape of the BER curve for the "data" system, which is rising from below 10^{-4} at 2 dB SNR up to 10^{-3} at 26 dB SNR. The average throughput of the modem is mainly determined by the statistics of the estimated channel transfer function at the receiver, and this is therefore not affected by the delay between the channel estimation and the packet transmission.

19.2.9   Channel Estimation

All the adaptive modems above rely on the estimate of the frequency domain channel transfer function, both for equalisation of the received symbols at the receiver, as well as for the modem mode adaptation of the next transmitted OFDM symbol. Figure 19.16 shows the BER versus SNR curves for the 1% target BER modem, as it was presented above, if pilot symbol assisted channel estimation [139] is employed instead of the previously used delayed, but otherwise perfect, channel estimation.

Comparing the curves for perfect channel estimation and for the 64-pilot lowpass interpolation algorithm, it can be seen that the modem falls short of the target bit error rate of 1% for channel SNR values of up to 20 dB. More noise resilient channel estimation algorithms can improve the modem's performance. If the passband width of the interpolation lowpass filter (see Section 17.3.1.1) is halved, which is indicated in Figure 19.16 as the reduced bandwidth (red. bw.) scenario, then the BER gap between the perfect and the pilot symbol assisted channel estimation narrows, and a BER of 1% is achieved at a SNR of 15 dB. Additionally, employing pairs of pilots with the above bandwidth-limited interpolation scheme further improves the modem's performance, which results in BER values below 1% for SNR values above 5 dB. The averaging of the pilot pairs improves the noise resilience of the channel estimation, but introduces estimation errors for high SNR values. This can be observed in the residual BER in the figure.

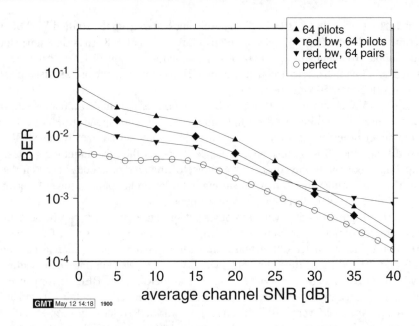

Figure 19.16: BER versus channel SNR performance for the 1% target BER adaptive 16 sub-band, 512 subcarrier OFDM modem employing pilot symbol assisted channel transfer function estimation over the channel of Figure 17.3

Having studied a range of different AOFDM modems, let us now embark on a system design study in the context of an adaptive interactive speech system.

19.3   Adaptive OFDM Speech System[1]

19.3.1   Introduction

In this section we introduce a bidirectional high-quality audio communications system, which will be used to highlight the systems aspects of adaptive OFDM transmissions over time dispersive channels. Specifically, the channel-coded adaptive transmission characteristics and a potential application for joint adaptation of modulation, channel coding and source coding are studied.

The basic principle of adaptive modulation is to react to the anticipated channel capacity for the next OFDM symbol transmission burst, by employing modulation modes of different robustness to channel impairments and of different data throughput. The trade-off between data throughput and integrity can be adapted to different system environments. For data transmission systems, which are not dependent on a fixed data rate and do not require low transmission delays, variable-throughput adaptive schemes can be devised that operate effi-

---

[1]T. Keller, M. Münster, L. Hanzo: A Turbo-coded Burst-by-burst Adaptive Wideband Speech Transceiver, ©IEEE JSAC, November 2000, Vol. 18, No. 11, pp 2363-2372 [583]. Details concerning the source codec and its performance are discussed in the above publication.

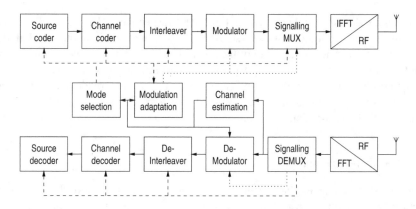

Figure 19.17: Schematic model of the multimode adaptive OFDM system

ciently with powerful error correction coders, such as long block length turbo codes [584]. Real-time audio or video communications employing source codecs, which allow variable bit rates, can also be used in conjunction with variable-rate adaptive schemes, but in this case block-based error correction coders cannot be readily employed.

Fixed-rate adaptive OFDM systems, which sacrifice a guaranteed BER performance for the sake of a fixed data throughput, are more readily integrated into interactive communications systems, and can coexist with long block based channel coders in real-time applications.

For these investigations, we propose a hybrid adaptive OFDM scheme, based on a multimode constant throughput algorithm, consisting of two adaptation loops: an inner constant throughput algorithm, having a bit rate consistent with the source and channel coders, and an outer mode switching control loop, which selects the target bit rate of the whole system from a set of distinct operating modes. These issues will become more explicit during our further discourse.

19.3.2 System Overview

The structure of the studied adaptive OFDM modem is depicted schematically in Figure 19.17. The top half of the diagram is the transmitter chain, which consists of the source and channel coders, a channel interleaver decorrelating the channel's frequency domain fading, an adaptive OFDM modulator, a multiplexer adding signalling information to the transmit data, and an IFFT/RF OFDM block. The receiver, at the lower half of the schematic, consists of a RF/FFT OFDM receiver, a demultiplexer extracting the signalling information, an adaptive demodulator, a de-interleaver/channel decoder and the source decoder. The parameter adaptation linking the receiver and transmitter chain consists of a channel estimator, and the throughput mode selection as well as the modulation adaptation blocks.

The open-loop control structure of the adaptation algorithms can be seen in the figure: the receiver's operation is controlled by the signalling information that is contained in the received OFDM symbol, while the channel quality information generated by the receiver is employed to determine the remote transmitter's matching parameter set by the modulation adaptation algorithms. The two distinct adaptation loops distinguished by the dotted and dashed lines are the inner and outer adaptation loops, respectively. The outer adaptation loop controls the overall throughput of the system, which is chosen from a finite set of predefined modes, so that a fixed delay decoding of the received OFDM data packets becomes possible. This outer loop controls the block length of the channel encoder and interleaver, and the target throughput of the inner adaptation loop. The operation of the adaptive modulator, controlled by the inner loop, is transparent to the rest of the system. The operation of the adaptation loops is described in more detail below.

19.3.2.1   System Parameters

The transmission parameters have been adopted from the TDD mode of the UMTS system of Section 17.1.3, with a carrier frequency of 1.9 GHz, a time frame and time slot duration of 4.615 ms and 122 μs, respectively. The sampling rate is assumed to be 3.78 MHz, leading to a 1024-subcarrier OFDM symbol with a cyclic extension of 64 samples in each time slot. For spectral shaping of the OFDM signal, there are a total of 206 virtual subcarriers at the bandwidth boundaries.

The 7 kHz bandwidth PictureTel audio codec[2] has been chosen for this system because of its good audio quality, robustness to packet dropping and adjustable bit rate. The channel encoder/interleaver combination is constituted by a convolutional turbo codec [366] employing block turbo interleavers in conjunction with a subsequent pseudorandom channel interleaver. The constituent half-rate recursive systematic convolutional (RSC) encoders are of constraint length 3, with octal generator polynomials of $(7, 5)$ [171]. At the decoder, 8 iterations are performed, utilising the so-called maximum a posteriori (MAP) [582] algorithm and log-likelihood ratio soft inputs from the demodulator.

The channel model consists of a four-path COST 207 typical urban impulse response [567], where each impulse is subjected to independent Rayleigh fading with a normalised Doppler frequency of $2.25 \cdot 10^{-6}$, corresponding to a pedestrian scenario with a walking speed of 3 mph.

The unfaded impulse response and the time- and frequency-varying amplitude of the channel transfer function are depicted in Figure 19.18.

19.3.3   Constant-Throughput Adaptive Modulation

The constant throughput adaptive algorithm attempts to allocate a given number of bits for transmission in subcarriers exhibiting a low BER, while the use of high BER subcarriers is minimised. We employ the open-loop adaptive regime of Figure 19.1, basing the decision concerning the next transmitted OFDM symbol's modulation scheme allocation on the channel estimation gained at the reception of the most recently received OFDM symbol by the local station. Sub-band adaptive modulation [585] – where the modulation scheme is adapted not on a subcarrier-by-subcarrier basis, but for sub-bands of adjacent subcarriers – is employed in order to simplify the signalling requirements. The adaptation algorithm was highlighted in Section 19.2.5. For these investigations we employed 32 sub-bands of 32 subcarriers in each OFDM symbol. Perfect channel estimation and signalling were used.

---

[2]see http://www.picturetel.com

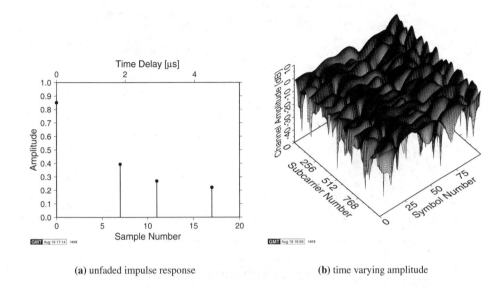

(a) unfaded impulse response   (b) time varying amplitude

Figure 19.18: Channel for PictureTel experiments: (a) unfaded channel impulse response (b) time-varying channel amplitude for 100 OFDM symbols

19.3.3.1 Constant-Rate BER Performance

Figure 19.19 characterises the fixed-throughput adaptive modulation scheme's performance under the channel conditions characterised above, for a block length of 578 coded bits. As a comparison, the BER curve of a fixed BPSK modem transmitting the same number of bits in the same channel, employing 578 out of 1024 subcarriers, is also depicted. The number of useful audio bits per OFDM symbol was based on a 200-bit target data throughput, which corresponds to a 10 kbps data rate, padded with 89 bits, which can contain a checksum for error detection and high-level signalling information. Furthermore, half-rate channel coding was used.

The BER plotted in the figure is the hard-decision based bit error rate at the receiver before channel decoding. It can be seen that the adaptive modulation scheme yields a significantly improved performance, which is also reflected in the frame error rate (FER). This FER approximates the probability of a decoded block containing errors, in which case it is unusable for the audio source decoder and hence it is dropped. This error event can be detected by using the checksum in the OFDM data symbol.

As an example, the modulation mode allocation for the 578 data bit adaptive modem at an average channel SNR of 5 dB is given in Figure 19.20(a) for 100 consecutive OFDM symbols. The unused sub-bands with indexes 15 and 16 contain the virtual carriers, and therefore do not transmit any data. It can be seen that the constant throughput adaptation algorithm of Section 19.2.5 allocates data to the higher quality subcarriers on a symbol-by-symbol basis, while keeping the total number of bits per OFDM symbol constant. As a comparison, Figure 19.20(b) shows the equivalent overview of the modulation modes employed for a fixed bit rate of 1458 bits per OFDM symbol. It can be seen that in order to meet this increased

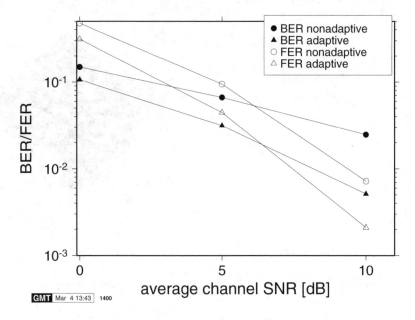

Figure 19.19: BER and FER performance for the fixed-throughput adaptive and non-adaptive OFDM
modems in the fading time dispersive channel of Section 17.1.3 for a block length of 578
coded bits

throughput target, hardly any sub-bands are in "no transmission" mode, and overall higher-
order modulation schemes have to be employed.

19.3.4   Multimode Adaptation

While the fixed-throughput adaptive algorithm described above copes with the frequency
domain fading of the channel, there is a medium-term variation of the overall channel capacity
due to time domain channel quality fluctuations as indicated in Figure 19.18(b). While it is
not straightforward to employ powerful block-based channel coding schemes, such as turbo
coding, in variable-throughput adaptive OFDM schemes for real-time applications like voice
or video telephony, a multimode adaptive system can be designed that allows us to switch
between a set of different source and channel codecs as well as transmission parameters,
depending on the overall channel quality. We have investigated the use of the estimated
overall BER at the output of the receiver, which is the sum of all the $e(j, s_j)$ quantities of
Equation 19.1 after adaptation. On the basis of this expected bit error rate at the input of the
channel decoder, the probability of a frame error (FER) must be estimated and compared with
the estimated FER of the other modem modes. Then the mode having the highest throughput
exhibiting an estimated FER of less than 10^{-6} – or alternatively the mode exhibiting the
lowest FER – is selected and the source encoder, the channel encoder and the adaptive modem
are set up accordingly.

   We have defined four different operating modes, which correspond to unprotected audio
data rates of 10, 16, 24, and 32 kbps at the source codec's interface. With half-rate channel

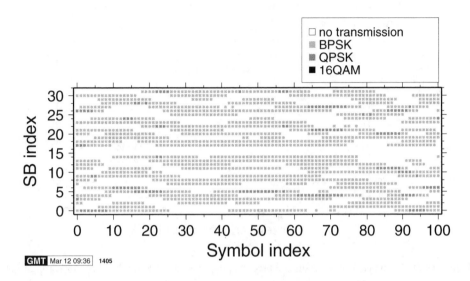

(a) 578 data bits per OFDM symbol

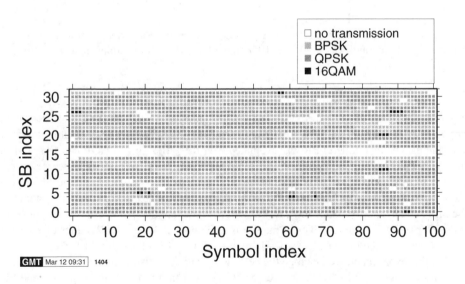

(b) 1458 data bits per OFDM symbol

Figure 19.20: Overview of modulation mode allocation for fixed-throughput adaptive modems over the fading time dispersive channel of Figure 19.18(b) at 5 dB average channel SNR

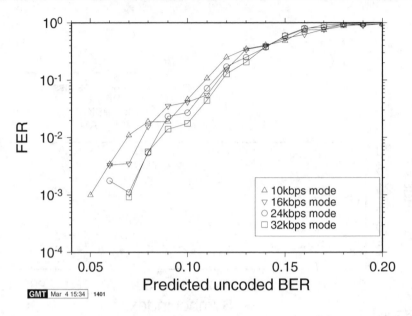

Figure 19.21: Frame error rate versus the predicted unprotected BER for 10 kbps, 16 kbps, 24 kbps and 32 kbps modes

coding and allowing for checksum and signalling overhead, the number of transmitted coded bits per OFDM symbol was 578, 722, 1058 and 1458 for the four modes, respectively.

19.3.4.1   Mode Switching

Figure 19.21 shows the *observed* FER for all four modes versus the unprotected BER that was *predicted* at the transmitter. The predicted unprotected BER was discretised into intervals of 1%, and the channel-coded FER was evaluated over these BER intervals. It can be seen from the figure that for estimated protected BER values below 5% no frame errors were observed for any of the modes. For higher estimated unprotected BER values, the higher throughput modes exhibited a lower FER than the lower throughput modes, which was consistent with the turbo coder's performance increase for longer block lengths. A FER of 1% was observed for a 7% predicted unprotected error rate for the 10 kbps mode, and BER values of 8% to 9% were allowed for the longer blocks, while maintaining a FER of less than 1%

For this experiment, we assumed the best-case scenario of using the actual measured FER statistics of Figure 19.21 for the mode switching algorithm rather than estimating the FER on the basis of the estimated uncoded BER. In this case, the previously observed FER corresponding to the predicted overall BER values for the different modes were compared, and the mode having the lowest FER was chosen for transmission. The mode switching sequence for the first 500 OFDM symbols at 5 dB channel SNR over the channel of Figure 19.18(b) is depicted in Figure 19.22. It can be seen that in this segment of the sequence 32 kbps transmission is the most frequently employed mode, followed by the 10 kbps mode. The intermediate modes are mostly transitory, as the improving or deteriorating channel conditions makes it

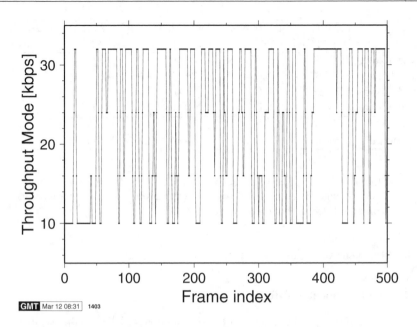

Figure 19.22: Mode switching pattern at 5 dB channel SNR over the channel of Figure 19.18(b)

necessary to switch between the 10 kbps and 32 kbps modes. This behaviour is consistent with Table 19.2, for the Switch-I scheme.

19.3.5 Simulation Results

The comparison between the different adaptive schemes will be based on a channel SNR of 5 dB over the channel of Figure 19.18(b), since the audio codec's performance is unacceptable for SNR values around 0 dB, and as the adaptive modulation is most effective for channel SNR values below 10 dB.

19.3.5.1 Frame Error Rate Results

The audio experiments [583] have shown that the audio quality is acceptable for frame dropping rates of about 5%, and that the perceived audio quality increases with increasing throughput. Table 19.2 gives an overview of the frame error rates and mode switching statistics of the system for a channel SNR of 5 dB over the channel of Figure 19.18(b). It can be seen that for the fixed modes the FER increases with the throughput, from 4.45% in the 10 kbps mode up to 18.65% for the 32 kbps mode. This is because the turbo codec performance improves for longer interleavers, the OFDM symbol had to be loaded with more bits, hence there was a higher unprotected BER. The time-variant bit rate mode-switching schemes, referred to as Switch-I and Switch-II for the four- and three-mode switching regimes used, deliver frame dropping rates of 4.44% and 5.58%, respectively. Both these FER values are acceptable for the audio transmission. It can be seen that upon incorporating the 10 kbps mode in the switching regime Switch-I of Table 19.2 the overall FER is lowered only by an insignificant amount,

Scheme	FER [%]	Rate-10kbps [%]	Rate-16kbps [%]	Rate-24kbps [%]	Rate-32kbps [%]
Fixed-10kbps	4.45	95.55	0.0	0.0	0.0
Fixed-16kbps	5.58	0.0	94.42	0.0	0.0
Fixed-24kbps	10.28	0.0	0.0	89.72	0.0
Fixed-32kbps	18.65	0.0	0.0	0.0	81.35
Switch-I	4.44	21.87	13.90	11.59	48.20
Switch-II	5.58	0.0	34.63	11.59	48.20

Table 19.2: FER and relative usage of different bit rates in the fixed bit rate schemes and the variable schemes Switch-I and Switch-II (successfully transmitted frames) for a channel SNR of 5 dB over the channel of Figure 19.18(b)

while the associated average throughput is reduced considerably.

19.3.5.2 Audio Segmental SNR

Figure 19.23 displays the cumulative density function (CDF) of the segmental SNR (SEGSNR) [169] obtained from the reconstructed signal of an audio test sound for all the modes of Table 19.2 discussed above at a channel SNR of 5 dB over the channel of Figure 19.18(b).

Focusing our attention on the figure, we can draw a whole range of interesting conclusions. As expected, for any given SEGSNR it is desirable to maintain as low a proportion as possible of the audio frames' SEGSNRs below a given abscissa value. Hence we conclude that the best SEGSNR CDF was attributable to the Switch-II scheme, while the worst performance was observed for the fixed 10 kbps scheme. Above a SEGSNR of 15 dB the CDFs of the fixed 16, 24 and 32 kbps modes follow our expectations. Viewing matters from a different perspective, the Switch-II scheme exhibits a SEGSNR of less than 20 dB with a probability of 0.8, compared to 0.95 for the fixed 10 kbps scheme.

Before concluding we also note that the CDFs do not have a smoothly tapered tail, since for the erroneous audio frames a SEGSNR of 0 dB was registered. This results in the step-function-like behaviour associated with the discontinuities corresponding to the FER values in the FER column of Table 19.2.

19.4 Pre-Equalisation

We have seen above how the receiver's estimate of the channel transfer function can be employed by the transmitter in order to dramatically improve the performance of an OFDM system by adapting the subcarrier modulation modes to the channel conditions. For sub-channels exhibiting a low signal-to-noise ratio, robust modulation modes were used, while for subcarriers having a high SNR, high throughput multi-level modulation modes can be employed. An alternative approach to combating the frequency selective channel behaviour is to apply pre-equalisation to the OFDM symbol prior to transmission on the basis of the anticipated channel transfer function. The optimum scheme for the power allocation to the

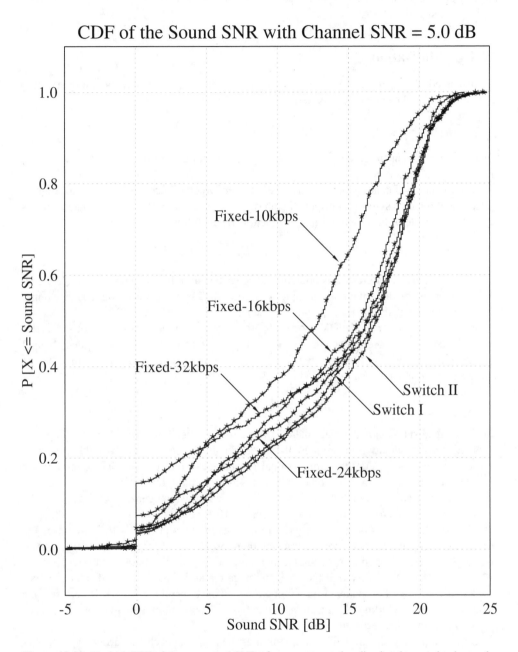

Figure 19.23: Typical CDF of the segmental SNR of a reconstructed audio signal transmitted over the fading time dispersive channel of Section 17.1.3 at a channel SNR of 5 dB (from [583])

subcarriers is Shannon's water-pouring solution [586]. We will investigate a range of related topics in this section.

19.4.1   Motivation

As discussed above, the received data symbol R_n of subcarrier n over a stationary time dispersive channel can be characterised by:

$$R_n = S_n \cdot H_n + n_n,$$

where S_n is the transmitted data symbol, H_n is the channel transfer function of subcarrier n, and n_n is a noise sample.

The frequency domain equalisation at the receiver, which is necessary for non-differential detection of the data symbols, corrects the phase and amplitude of the received data symbols using the estimate of the channel transfer function $\hat{H}_n$ as follows:

$$R'_n = R_n/\hat{H}_n = S_n \cdot H_n/\hat{H}_n + n_n/\hat{H}_n.$$

If the estimate $\hat{H}_n$ is accurate, this operation defades the constellation points before decision. However, upon defading, the noise sample n_n is amplified by the same amount as the signal, therefore preserving the SNR of the received sample.

Pre-equalisation for the OFDM modem operates by scaling the data symbol of subcarrier n, S_n, by a predistortion function E_n, computed from the inverse of the anticipated channel transfer function, prior to transmission. At the receiver, no equalisation is performed, hence the received symbols can be expressed as:

$$R_n = S_n \cdot E_n \cdot H_n + n_n.$$

Since no equalisation is performed, there is no noise amplification at the receiver. Similarly to the adaptive modulation techniques illustrated above, pre-equalisation is only applicable to a duplex link, since the transmitted signal is adapted to the specific channel conditions perceived by the receiver. As in other adaptive schemes, the transmitter needs an estimate of the current frequency domain channel transfer function, which can be obtained from the received signal in the reverse link, as seen in Figure 19.1.

The simplest choice of the pre-equalisation transfer function E_n is the inverse of the estimated frequency domain channel transfer function, $E_n = 1/\hat{H}_n$. If the estimation of the channel transfer function is accurate, then perfect channel inversion would result in an AWGN-like channel perceived at the receiver, since the anticipated time and frequency dependent behaviour of the channel is precompensated at the transmitter. The BER performance of such a system, accordingly, is identical to that of the equivalent modem in an AWGN channel with respect to the received signal power.

Since the pre-equalisation algorithm amplifies the power in each subcarrier by the corresponding estimate of the channel transfer function, the transmitter's output power fluctuates in an inverse fashion with respect to time variant channel. The fades in the frequency domain channel transfer function can be deep, hence the transmit power in the corresponding subcarriers may be high. Figure 19.24 shows the histogram of the total OFDM symbol energy at the transmitter's output for the short WLAN channel of Figure 17.3 in conjunction with

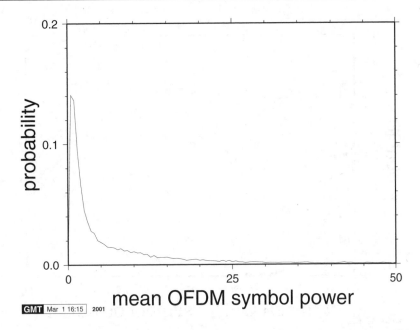

Figure 19.24: OFDM symbol energy histogram for 512-subcarrier 16-QAM with full channel inversion over the short WATM channel of Figure 17.3

"full channel inversion", normalised to the fixed average output energy. It can be seen that the OFDM symbol energy fluctuates widely, with observed peak values in excess of 55. The long-term mean symbol energy was measured to be 22.9, which corresponds to an average output power increase of 13.6 dB. This naturally would impose unacceptable constraints on the required perfectly linear dynamic range of the power amplifier. Hence in practice only limited dynamic scenarios can be considered.

Explicitly, in order to limit the associated transmit power fluctuations, the dynamic range of the pre-equalisation algorithm can be limited to a value l, so that the following relations apply:

$$E_n = a_n \cdot e^{-j\phi_n}, \text{with} \tag{19.3}$$

$$\phi_n = \angle \hat{H}_n \quad \text{and} \tag{19.4}$$

$$a_n = \begin{cases} \left| \hat{H}_n \right| & \text{for } \left| \hat{H}_n \right| \le l \\ l & \text{otherwise.} \end{cases} \tag{19.5}$$

Limiting the values of E_n to the value of l does not affect the phase of the channel pre-equalisation. Depending on the modulation mode employed for transmission, reception of the symbols affected by the amplitude limitation is still possible, perhaps for phase shift keying. Multilevel modulation modes exploiting the received symbol's amplitude will be affected by the imperfect pre-equalisation. The associated mean OFDM symbol power histogram is shown in Figure 19.25. Given - for example - a maximum allowed amplification factor

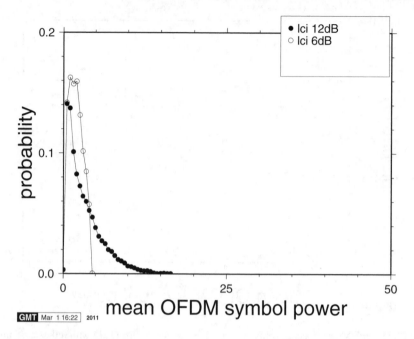

Figure 19.25: OFDM symbol energy histogram for 512-subcarrier 16-QAM transmission with limited dynamic range channel inversion (lci) over the WATM channel of Figure 17.3.

of 6 and 12 dB, the normalised transmitted OFDM symbol power in the figure should be limited to 4 and 16, respectively. However, in practice even higher values may be observed, which is due to the OFDM symbol's energy fluctuation as a function of the specific data sequence, if multilevel modulation modes are used. In order to circumvent the above peak-to-mean envelope fluctuation problems, it is in practical terms more attractive to combine pre-equalisation with sub-band blocking, which is the topic of the next section.

19.4.2   Pre-Equalisation Using Sub-Band Blocking

Limiting the maximal amplification of the subcarriers leads to a reduced BER performance compared to the full channel inversion, but the system performance can be improved by identifying the subcarriers that cannot be fully pre-equalised and by disabling subsequent transmission in these subcarriers. This "blocking" of the transmission in certain subcarriers can be seen as adaptive modulation with two modulation modes, which introduces the problem of modulation mode signalling. As has been discussed in the context of Figure 19.1 for the adaptive modulation modems above, this signalling task can be solved in different ways, namely by blind detection of blocked subcarriers, or by transmitting explicit signalling information contained in the data block. We have seen above that employing sub-band adaptivity - rather than subcarrier-by-subcarrier adaptivity - simplifies both the modem mode detection as well as the mode signalling, at the expense of a lower system throughput. In order to keep the system's complexity low and to allow for simple modem mode signalling or blind detection,

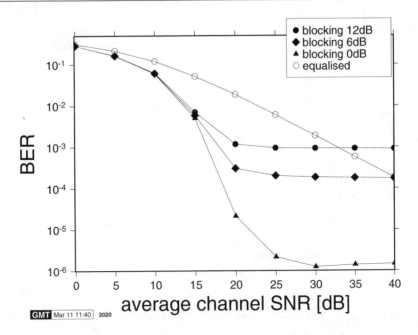

Figure 19.26: BER performance of the 512-subcarrier 16-QAM OFDM modem over the fading short WATM channel of Figure 17.3 employing 16 sub-band pre-equalisation with blocking and a delay of 1 time slot between the instants of perfect channel estimation and reception. Note that the transmit power is not shown in this figure

we will assume a 16 subband adaptive scheme here.

Analogously to the adaptive modulation schemes above, the transmitter decides for all subcarriers in each sub-band, whether to transmit data or not. If pre-equalisation is possible under the power constraints, then the subcarriers are modulated with the pre-equalised data symbols. The information on whether or not a sub-band is used for transmission, this is signalled to the receiver.

Since no attempt is made to transmit in the sub-bands that cannot be power-efficiently pre-equalised, the power conserved in the blank subcarriers can be used for "boosting" the data-bearing sub-bands. This scheme allows for a more flexible pre-equalisation algorithm than the fixed threshold method described above; here is a summary:

(1) Calculate the necessary transmit power p_n for each sub-band n, assuming perfect pre-equalisation.

(2) Sort sub-bands according to their required transmit power p_n.

(3) Select sub-band n with the lowest power p_n, and add p_n to the total transmit power. Repeat this procedure with the next lowest power, until no further sub-bands can be added without the total power $\sum p_j$ exceeding power limit l.

Figure 19.26 depicts the 16-QAM BER performance over the short WATM channel of Figure 17.3. The BER floor stems from the channel's time-variant nature, since there is a

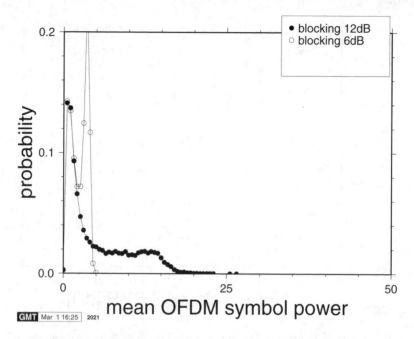

Figure 19.27: OFDM symbol energy histogram for 512 subcarrier, 16 sub-band pre-equalisation with blocking over the short WATM channel of Figure 17.3 using 16-QAM. The corresponding BER curves are given in Figure 19.26

delay between the channel estimation instant and the instant of transmission. The average throughput figures for the 6 dB and 12 dB symbol energy limits are 3.54 and 3.92 bits per data symbol, respectively. It can be noted that the BER floor is lower for $l = 6$ dB than for $l = 12$ dB. This is because the effects of the channel variation due to the delay between the instants of channel estimation and reception in the faded subcarriers on the equalisation function are more dramatic, than in the higher quality subcarriers. The lower the total symbol energy limit l, the smaller the number of low quality subcarriers used for transmission. If the symbol energy is limited to 0 dB, then the BER floor drops to $1.5 \cdot 10^{-6}$ at the expense of the throughput, which attains 2.5 BPS. Figure 19.27 depicts the mean OFDM symbol energy histogram for this scenario. It can be seen that, compared with the limited channel inversion scheme of Figure 19.25, the allowable symbol energy is more efficiently allocated, with a higher probability of high energy OFDM symbols. This is the result of the flexible reallocation of energy from blocked sub-bands, instead of limiting the output power on a subcarrier-by-subcarrier basis.

19.4.3 Adaptive Modulation Using Spectral Pre-Distortion

The pre-equalisation algorithms discussed above invert the channel's anticipated transfer function, in order to transform the resulting channel into a Gaussian-like non-fading channel, whose SNR is dependent only on the path loss. Sub-band blocking has been introduced above, in order to limit the transmitter's output power, while maintaining the near-constant SNR across the used subcarriers. The pre-equalisation algorithms discussed above do not cancel out the channel's path loss, but rely on the receiver's gain control algorithm to automatically account for the channel's average path loss.

We have already seen that maintaining Gaussian channel characteristics is not the most efficient way of exploiting the channel's time variant capacity. If maintaining a constant data throughput is not required by the rest of the communications system, then a fixed BER scheme in conjunction with error correction coding can assist in maximising the system's throughput. The results presented for the target BER adaptive modulation scheme in Figure 19.14(b) showed that, for the particular turbo coding scheme used, an uncoded BER of 1% resulted in error-free channel coded data transmission, and that for an uncoded target BER of 10% the turbo decoded data BER was below 10^{-5}. We have seen that it is impossible to exactly reach the anticipated uncoded target BER with the adaptive modulation algorithm, since the adaptation algorithm operates in discrete steps between modulation modes.

Combining the target BER adaptive modulation scheme and spectral pre-distortion allows the transmitter to react to the channel's time and frequency variant nature, in order to fine-tune the behaviour of the adaptive modem in fading channels. It also allows the transmitter to invest the energy that is not used in "no transmission" sub-bands into the other sub-bands without affecting the equalisation at the receiver.

The combined algorithm for adaptive modulation with spectral predistortion described here does not aim for inverting the channel's transfer function across the OFDM symbol's range of subcarriers, it is therefore not a pure pre-equalisation algorithm. Instead, the aim is to transmit a sub-band's data symbols at a power level which ensures a given target SNR at the receiver, that is constant for all subcarriers in the sub-band, which in turn results in the required BER. Clearly, the receiver has to anticipate the different relative power levels for the different modulation modes, so that error-free demodulation of the multilevel modulation modes employed can be ensured.

The joint adaptation algorithm requires the estimates of the noise floor level at the receiver as well as the channel transfer function, which includes the path loss. On the basis of these values, the necessary amplitude of E_n required to transmit a data symbol over the subcarrier n for a given received SNR of γ_n can be calculated as follows:

$$|E_n| = \frac{\sqrt{N_0 \cdot \gamma_n}}{\left|\hat{H}_n\right|},$$

where N_0 is the noise floor at the receiver. The phase of E_n is used for the pre-equalisation, and hence:

$$\angle E_n = -\angle \hat{H}_n.$$

The target SNR, of subcarrier n, γ_n, is dependent on the modulation mode that is signalled over the subcarrier, and determines the system's target BER. We have identified three sets of target SNR values for the modulation modes, with uncoded target BER values of 1% and 10% for use in conjunction with channel coders, as well as 10^{-4} for transmission without channel coding. Table 19.3 gives an overview of these levels, which have been read from the BER performance curves of the different modulation modes in a Gaussian channel.

Figure 19.28 shows the performance of the joint predistortion and adaptive modulation

target BER	10^{-4}	1%	10%
SNR(BPSK)[dB]	8.4	4.33	−0.85
SNR(QPSK)[dB]	11.42	7.34	2.16
SNR(16QAM)[dB]	18.23	13.91	7.91

Table 19.3: Required target SNR levels for 1% and 10% target BER for the different modulation schemes over an AWGN channel

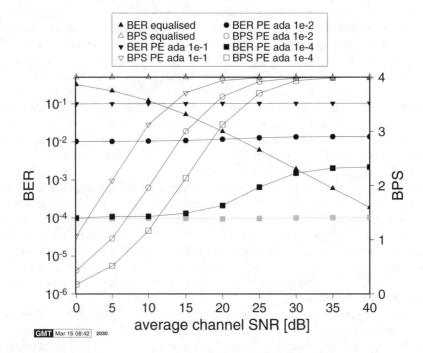

Figure 19.28: BER performance and BPS throughput of the 512 subcarrier, 16 sub-band adaptive OFDM modem with spectral predistortion over the Rayleigh fading time dispersive short WATM channel of Figure 17.3, and that of the perfectly equalised 16-QAM modem. The light grey BER curve gives the performance of the adaptive modem for a target BER of 10^{-4} with no delay between channel estimation and transmission, while the other results assume 1 time slot delay between uplink and downlink.

algorithm over the fading time-dispersive short WATM channel of Figure 17.3 for the set of different target BER values of Table 19.3, as well as the comparison curves of the perfectly equalised 16-QAM modem under the same channel conditions. It can be seen that the BER achieved by the system is close to the BER targets. Specifically, for a target BER of 10%, no perceptible deviation from the target has been recorded, while for the lower BER targets the deviations increase for higher channel SNRs. For a target BER of 1%, the highest measured deviation is at the SNR of 40 dB, where the recorded BER is 1.36%. For the target BER of 10^{-4}, the BER deviation is small at 0 dB SNR, but at an SNR of 40 dB the experimental

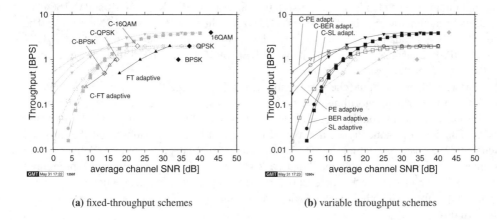

(a) fixed-throughput schemes   (b) variable throughput schemes

Figure 19.29: BPS throughput versus average channel SNR for non-adaptive and adaptive modulation as well as for pre-equalised adaptive techniques, for a data bit error rate of 10^{-4}. Note that for the coded schemes the achieved BER values are lower than 10^{-4}. (a) Fixed-throughput systems: coded (C-) and uncoded BPSK, QPSK, 16-QAM, and fixed-throughput (FT) adaptive modulation. (b) Variable-throughput systems: coded (C-) and uncoded switching level adaptive (SL), target BER adaptive (BER) and pre-equalised adaptive (PE) systems. Note that the separately plotted variable-throughput graph also shows the light grey benchmark curves of the complementary fixed-rate schemes and vice versa

BER is $2.2 \cdot 10^{-3}$. This increase of the BER with increasing SNR is due to the rapid channel variations in the deeply faded subcarriers, which are increasingly used at higher SNR values. The light grey curve in the figure denotes the system's performance if no delay is present between the channel estimation and the transmission. In this case, the simulated BER shows only very little deviation from the target BER value. This is consistent with the behaviour of the full channel inversion pre-equalising modem.

19.5   Comparison of the Adaptive Techniques

Figure 19.29 compares the different adaptive modulation schemes discussed in this chapter. The comparison graph is split into two sets of curves, depicting the achievable data throughput for a data BER of 10^{-4} highlighted for the fixed-throughput systems in Figure 19.29(a), and for the time-variant -throughput systems in Figure 19.29(b).

The fixed-throughput systems, highlighted in black in Figure 19.29(a), comprise the non-adaptive BPSK, QPSK and 16-QAM modems, as well as the fixed-throughput adaptive scheme, both for coded and uncoded applications. The non-adaptive modems' performance is marked on the graph as diamonds, and it can be seen that the uncoded fixed schemes require the highest channel SNR of all examined transmission methods to achieve a data BER of 10^{-4}. Channel coding employing the advocated turbo coding schemes dramatically improves the SNR requirements, at the expense of half the data throughput. The uncoded fixed-throughput (FT) adaptive scheme, marked by filled triangles, yields consistently worse data throughput than the coded (C-) fixed modulation schemes C-BPSK, C-QPSK and C-16QAM, with its throughput being about half the coded fixed scheme's at the same SNR values. The coded FT (C-FT) adaptive system, however, delivers very similar throughput to the C-BPSK and C-QPSK transmission, and can deliver a BER of 10^{-4} for SNR values down to about 9 dB.

The variable throughput schemes, highlighted in Figure 19.29(b), outperform the comparable fixed-throughput algorithms. For high SNR values, all uncoded schemes' performance curves converge to a throughput of 4 bits/symbol, which is equivalent to 16-QAM transmission. The coded schemes reach a maximal throughput of 2 bits/symbol. Of the uncoded schemes, the "data" switching level (SL) and target BER adaptive modems deliver a very similar BPS performance, with the target-BER scheme exhibiting slightly better throughput than the SL adaptive modem. The adaptive modem employing pre-equalisation (PE) significantly outperforms the other uncoded adaptive schemes and offers a throughput of 0.18 BPS at an SNR of 0 dB.

The coded transmission schemes suffer from limited throughput at high SNR values, since the half-rate channel coding limits the data throughput to 2 BPS. For low SNR values, however, the coded schemes offer better performance than the uncoded schemes, with the exception of the "speech" SL adaptive coded scheme, which is outperformed by the uncoded PE adaptive modem. The poor performance of the coded SL scheme can be explained by the lower uncoded target BER of the "speech" scenario, which was 1%, in contrast to the 10% uncoded target BER for the coded BER and PE adaptive schemes. The coded PE adaptive modem outperforms the target-BER adaptive scheme, thanks to its more accurate control of the uncoded BER, leading to a higher throughput for low SNR values.

It is interesting to observe that for the given set of four modulation modes the uncoded PE adaptive scheme is close in performance to the coded adaptive schemes, and that for SNR values of more than 14 dB it outperforms all other studied schemes. It is clear, however, that the coded schemes would benefit from higher-order modulation modes, which would allow these modems to increase the data throughput further when the channel conditions allow. Before concluding this chapter, in the next section let us consider the generic problem of optimum power and bit allocation in the context of uncoded OFDM systems.

19.6   A Fast Algorithm for Near-Optimum Power and Bit Allocation in OFDM Systems [587]

19.6.1   State-of-the-Art

In this section the problem of efficient OFDM symbol-by-symbol based power and bit allocation is analysed in the context of highly dispersive time-variant channels. A range of solutions published in the literature is reviewed briefly and Piazzo's [587] computationally efficient algorithm is discussed in somewhat more detail.

When OFDM is invoked over highly frequency selective channels, each subcarrier can be allocated a different transmit power and a different modulation mode. This OFDM symbol-by-symbol based "resource" allocation can be optimised with the aid of an algorithm which, if the channel is time variant, has to be repeated on an OFDM symbol-by-symbol basis. Some of the existing algorithms [155, 588] are mainly of theoretical interest due to their high complexity. Amongst the practical algorithms [221, 222, 587, 589, 590] the Hughes-Hartog algorithm (HHA) [590–592] is perhaps best known, but its complexity is somewhat high, especially for real-time OFDM symbol-by-symbol based applications at high bit rates. Hence the HHA has stimulated extensive research for computationally more efficient algorithms [221, 222, 587, 589, 590]. The most efficient appears to be that of Lai *et al.* [590], which is a fast version of the HHA and that of Piazzo [587].

19.6.2 Problem Description

Piazzo [587] considered an OFDM system using N subcarriers, each employing a potentially different modulation mode and transmit power. Below we follow the notation and approach proposed by Piazzo [587]. The different modes use different modem constellations and thus carry a different number of bits per subcarrier, ranging from 1 to I bits per subcarrier, corresponding to BPSK and 2^I-ary QAM. We denote the transmit power and the number of bits allocated to subcarrier k ($k = 0, \ldots, N - 1$) by p_k and b_k, respectively. If $b_k = 0$, subcarrier k is allocated no power and no bits, hence it is disabled. The total transmit power is $P = \frac{1}{N} \sum_{k=0}^{N-1} p_k$ and the number of transmitted bits per OFDM symbol is $B = \sum_{k=0}^{N-1} b_k$. The i-bit modulation mode is characterised by the function $R_i(S)$, denoting the SNR required at the input of the detector, in order to achieve a target bit error rate (BER) equal to S. Finally, we denote the channel's power attenuation at subcarrier k by a_k, and the power of the Gaussian noise by P_N, so that the SNR of subcarrier k is $r_k = p_k/(a_k \cdot P_N)$.

We consider the problem of minimising the transmit power for a fixed target BER of S and for a fixed number of transmitted bits B per OFDM symbol. We impose an additional constraint, namely that the BER of every carrier has to be equal to S. This constraint simplifies the problem, while producing a system close to the unconstrained optimum system [222, 590, 592], while [588] considers an unconstrained system. Furthermore, from an important practical point of view, it produces a near-constant BER at the input of the channel decoder, if FEC is used, which maximises the achievable coding gain, since the channel does not become overwhelmed by the plethora of transmission errors, which would be the case for a more bursty error statistics without this constraint. In order to satisfy this constraint, the power transmitted on subcarrier k has to be $p_k = P_N a_k R_{b_k}(S)$, and the total power to be minimised is given by the sum of the N subcarriers' powers across the OFDM symbol:

$$P = \frac{P_N}{N} \sum_{k=0}^{k=N-1} a_k \cdot R_{b_k}(S). \tag{19.6}$$

We now state a property of the optimum system. Namely, in the optimum system if a subcarrier has a lower attenuation than another one - i.e. it exhibits a higher frequency domain transfer function value and hence experiences a higher received SNR - then it must carry at least as many bits as the lower SNR subcarrier. More explicitly:

$$a_k < a_h \ \Rightarrow \ b_k \geq b_h. \tag{19.7}$$

The above property in Equation 19.7 can be readily proven. Let us briefly consider a system, which does not satisfy Equation 19.7, where for subcarriers k and h the above condition is violated and hence we have $a_k < a_h$ and $b_k = i_1 < b_h = i_2$. In other words, although the attenuation a_k is lower than a_h, $i_1 < i_2$. Consider now a second system, where the lower attenuation subcarrier was assigned the higher number of bits, i.e. $b_k = i_2$ and $b_h = i_1$. Since the required SNR for maintaining the target BER of S is lower for a lower number of bits, - i.e. we have $R_{i_1}(S) < R_{i_2}(S)$, upon substituting these SNR values in Equation 19.6 we can infer that the second system requires a lower total power P per OFDM symbol for maintaining the target BER S. Thus the first system is not optimum in this sense.

Equation 19.7 states a necessary condition of optimality, which was also exploited by Lai *et al.* in [590], but it can be exploited further, as we will demonstrate below. From now on, we consider the channel's transfer function or attenuation vector sorted in the order of $a_0 \leq a_1 \leq a_2 \ldots$, which simplifies our forthcoming discussions.

19.6.3 Power- and Bit-Allocation Algorithm

Piazzo's algorithm [587] solves the above resource allocation problem for the general system by repeatedly solving the problem for a simpler system. Explicitly, the simpler system employs only two modulation modes, those carrying J and $J - 1$ bits. This system can be termed the twin-mode system (TMS). On the basis of Equation 19.7 and since the channel's frequency -domain attenuation vector was sorted in the order of $a_0 \leq a_1 \leq a_2 \ldots$, for the optimum TMS (OTMS) the OFDM subcarriers will be partitioned in three groups:

(1) Group J comprises the first or lowest attenuation OFDM subcarriers using a J-bit modulation mode.

(2) Group 0 is constituted by the last or highest attenuation OFDM subcarriers transmitting zero bits.

(3) Group $(J-1)$ hosts the remaining OFDM subcarriers using a $(J-1)$-bit modem mode.

In order to find the OTMS - minimising the required transmit power of the OFDM symbol for a fixed target BER of S and for a fixed number of transmitted bits B per OFDM symbol - we initially assign all the B bits of the OFDM symbol to the highest quality, i.e. lowest attenuation, group J. This of course would be a sub-optimum scheme, leaving the medium quality subcarriers of group $J - 1$ unused, since even the highest quality subcarriers would require an excessive SNR, i.e. transmit power, for maintaining the target BER, when transmitting B bits per OFDM symbol. We note furthermore that the above bit allocation may require padding of the OFDM symbol with dummy bits if J is not an integer divisor of B.

Following the above initial bit allocation, Piazzo suggested performing a series of *bit re-allocations, reducing the transmit power upon each reallocation.* Specifically, in each power and bit reallocation step we move the $J \cdot (J - 1)$ bits allocated to the last, i.e. highest attenuation or lowest quality, $(J-1)$ OFDM subcarriers of group J to group $(J-1)$. For example, if 1 bit/symbol BPSK and 2 bit/symbol 4-QAM are used, then we move $2 \cdot 1 = 2$ bits, which were allocated to the highest attenuation 4-QAM subcarrier to two BPSK modulated subcarriers. The associated trade-off is that while previously the lowest quality subcarrier of class J had to carry 2 bits, it will now be conveying only 1 bit and additionally the highest quality and previously unused subcarrier has to be assigned 1 bit. This reallocation was motivated by the fact that before reallocation the lowest quality subcarrier of class J would have required a higher power for meeting the target BER requirement of S upon carrying 2 bits, than the regime generated by the reallocation step.

In general, for the sake of performing this power-reducing bit reallocation we have to add J subcarriers to group $(J-1)$. Hence we assign the last, i.e. lowest quality, $(J-1)$ OFDM subcarriers of group J and the first, i.e. highest quality, unused subcarrier to group $(J-1)$. Based on Equation 19.6 and upon denoting the index of the last, i.e. lowest quality, subcarrier of group J before reallocation by M_J and the index of the first, i.e. highest quality, unused subcarrier before reallocation by M_0, the condition of successful power reduction after the tentative bit reallocation can be formulated. Specifically, the bit reallocation results in a system using less power, if the sum of the subcarriers' attenuations carrying J bits weighted by their SNR $R_J(S)$ required for the J-bit modem mode for maintaining the target BER of S is higher than that of the corresponding constellation after the above bit reallocation process, when an extra previously unused subcarrier was invoked for transmission. This can be expressed in a more compact form as:

$$R_J(S) \sum_{k=0}^{J-2} a_{M_J-k} > R_{J-1}(S)(a_{M_0} + \sum_{k=0}^{J-2} a_{M_J-k}). \tag{19.8}$$

If Equation 19.8 is satisfied, the reallocation is performed and another tentative reallocation step is attempted. Otherwise the process is terminated, since the optimum twin-mode power and bit allocation scheme has been found.

According to Piazzo's proposition [587] the above procedure can be further accelerated. Since the attenuation vector was sorted, we have $a_{M_J-k} \approx a_{M_J}$ in Equation 19.8. Upon replacing a_{M_J-k} by a_{M_J}, after some manipulations we can reformulate Equation 19.8, i.e. the condition for the modem mode allocation after the bit reallocation to become more efficient as:

$$K_J(S)a_{M_J} - a_{M_0} > 0, \tag{19.9}$$

where $K_J(S) = (J-1)(\frac{R_J(S)}{R_{J-1}(S)} - 1)$ and $K_J(S) > 0$ holds, since $R_J(S) > R_{J-1}(S)$. Piazzo denoted the values of M_J and M_0 after m reallocation steps by $M_J(m)$ and $M_0(m)$. Since initially all the bits were allocated to group J, we have for the index of the last subcarrier of group J at the commencement of the bit reallocation steps $M_J(0) = \lfloor B/J \rfloor - 1$, while for the index of the first unused subcarrier is $M_0(0) = \lfloor B/J \rfloor$, where $\lfloor x \rfloor$ is the smallest integer greater than or equal to x. Furthermore, since in each bit reallocation step the last $(J-1)$ OFDM subcarriers of group J and the first subcarrier of group 0 are moved to group $(J-1)$, after m reallocations we have $M_J(m) = \lfloor B/J \rfloor - 1 - (J-1)m$ and $M_0(m) = \lfloor B/J \rfloor + m$. Upon substituting these values in Equation 19.9 the left-hand side becomes a function of m, namely $f(m) = K_J(S)a_{M_J(m)} - a_{M_0(m)}$. Because the frequency domain channel transfer function's attenuation vector was ordered and since $K_J(S) > 0$, hence it is readily seen that $f(m)$ is a monotonically decreasing function of the reallocation index m. Therefore the method presented above essentially attempts to find the specific value of the reallocation index m, for which we have $f(m) > 0$ and $f(m+1) < 0$. In other words, when we have $f(m+1) < 0$, the last reallocation step resulted in a power increment, not a decrement, hence the reallocation procedure is completed.

The search commences from $m = 0$ and increases m by one at each bit reallocation step. In order to accelerate the search procedure, Piazzo replaced the linearly incremented search by a logarithmic search. This is possible, since the $f(m)$ function is monotonically decreasing upon increasing the reallocation index m. Piazzo [587] stipulated the search range by commencing from the minimum value of the reallocation index m, namely from $m_0 = 0$.

The maximum value, denoted by m_1, is determined by the number of OFDM subcarriers N or by the number of bits B to be transmitted per OFDM symbol, as will be argued below. There are two limitations, which determine the maximum possible number of reallocation steps. Namely, the reallocation steps have to be curtailed when there are no more bits left in the group of subcarriers associated with the J-bit modem mode group or when there are no more unused carriers left after iteratively invoking the best unused carrier from the group of disabled carriers. These limiting factors, which determine the maximum possible number of bit reallocation steps, are discussed further below.

Recall that at the commencement of the algorithm all the bits were assigned to the sub-carrier group associated with the J-bit modem mode and hence there were $\lceil B/J \rceil$ subcarriers in group J. Upon reallocating the $J \cdot (J - 1)$ bits allocated to the last, i.e. highest attenuation or lowest quality, $(J - 1)$ OFDM subcarriers of group J to group $(J - 1)$ until no more bits were left in the subcarrier group associated with the J-bit modem mode naturally constitutes an upper limit for the maximum number of reallocation steps m_1, which is given by $\lceil B/J \rceil / (J - 1)$. Again, the other limiting factor of the maximum number of bit reallocation steps is the number of originally unused carriers, which was $N - \lceil B/J \rceil$). Hence the maximum possible number of reallocations is given by $m_1 = min(\lfloor \lfloor B/J \rfloor / (J - 1) \rfloor; N - \lfloor B/J \rfloor)$, where $\lceil x \rceil$ is the highest integer smaller than or equal to x.

The accelerated logarithmic search proposed by Piazzo [587] halves the maximum possible range at each bit reallocation step, by testing the value of $f(m)$ at the centre of the range and by updating the range accordingly. Piazzo's proposed algorithm can be summarised in a compact form as follows [587]:

Algorithm 1 $OTMS(B, S, J, N, a_k)$
(1) Initialise $m_0 = 0$, $m_1 = min(\lfloor \lfloor B/J \rfloor / (J - 1) \rfloor, N - \lfloor B/J \rfloor)$.
(2) Compute $m_x = m_0 + \lfloor m_1 - m_0 \rfloor / 2$.
(3) If $f(m_x) \geq 0$ let $m_0 = m_x + 1$; else let $m_1 = m_x$.
(4) If $m_1 = m_0 + 1$ goto step 5); else goto step 2).
(5) Stop. The number of carriers in group J is $N_J = \lfloor B/J \rfloor - m_0 \cdot (J - 1)$.

When the algorithm is completed, the value N_J becomes known and this specifies the number of OFDM subcarriers in the group J associated with the J-bit modem mode.

Having generated the optimum twin-mode system, Piazzo also considered the problem of finding the optimum general system (OGS) employing OFDM subcarrier modulation modes carrying $1, ..., I$ bits. The procedure proposed initially invoked Algorithm 1 in order to find the optimum twin-mode system carrying a total of B bits per OFDM symbol using the I-bit and the $(I - 1)$-bit per subcarrier modulation modes. At the completion of Algorithm 1 we know N_I, the number of OFDM subcarriers carrying I bits. These subcarriers are now confirmed. These OFDM subcarriers as well as the associated $I \cdot N_I$ bits can now be eliminated from the resource allocation problem, and the optimum system transmitting the remaining $B - I \cdot N_I$ bits of the remaining $(N - N_I)$ subcarriers can be sought, using subcarrier modulation modes transmitting $(I - 1), (I - 2), ...$ bits.

Again, Algorithm 1 can be applied to this new system, now using the modulation modes with $(I-1)$ and $(I-2)$ bits per subcarrier and repeating the procedure. After each application of Algorithm 1 a new group of subcarrier is confirmed. Piazzo's general algorithm can be summarised in a compact form as follows [587]:

Algorithm 2 $OGS(B, S, I, N, a_k)$
(1) Initialise $\hat{B} = B$, $\hat{N} = N$, $\hat{a}_k = a_k$, $J = I$.
(2) Perform $OTMS(\hat{B}, S, J, \hat{N}, \hat{a}_k)$ to compute N_J.
(3) If $J = 2$, let $N_1 = \hat{B} - 2 \cdot N_2$ and stop.
*(4) Remove the first N_J carriers from $\hat{a}_k$, let $\hat{B} = \hat{B} - J \cdot N_J$, $\hat{N} = \hat{N} - N_J$, $J = J - 1$
and goto step 2).*

When the algorithm is completed, the values N_i specifying the number of OFDM subcarriers conveying i bits, become known for all the legitimate modes carrying $i = I, (I-1), ..., 1$ bits per subcarrier. Hence we know the number of bits allocated to subcarrier k ($k = 0, ..., N - 1$) expressed in terms of the b_k values as well as the associated minimum power requirements. Hence the system is specified in terms of $p_k = P_N a_k R_{b_k}(S)$. In closing it is worthwile noting that the algorithm can be readily modified to handle the case where the two modes of the twin-mode system carry J and $K < J - 1$ bits.

Having considered the above near-optimum power- and bit allocation algorithm, in the next section we will consider a variety of OFDM systems supporting multiple users and invoking adaptive beam-steering.

19.7 Multi-User AOFDM

M. Münster, T. Keller and L. Hanzo [3]

19.7.1 Introduction

Signal fading as well as co-channel interference is known to have a severe impact on the system performance in multi-cellular mobile environments. *Adaptive modulation* as a method of matching the system to fading induced variations of the channel quality was originally proposed for single carrier transmission, but its potential was also soon discovered in the context of multicarrier transmissions, with the aim of concentrating the throughput on subcarriers least affected by frequency selective fading [585]. On the other hand, adaptive antenna array techniques have been shown to be effective in reducing co-channel interference at the receiver side [335, 593]. One of the most prominent schemes for performing the combining operation is the *Sample Matrix Inversion* (SMI) technique, which has recently attracted wide interest [259, 260, 310]. We commence our discussions in the next section with a description of a system amalgamating adaptive modulation and co-channel interference suppression. Initial performance results will be presented in Subsection 19.7.3 assuming perfect knowledge of all channel parameters, whilst in Subsection 19.7.4 the problem of channel parameter estimation will be addressed by means of orthogonal pilot sequences, leading to our conclusions.

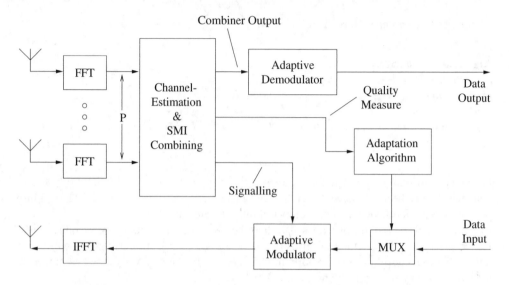

Figure 19.30: Schematic structure of the adaptive transceiver with interference suppression at the receiver

19.7.2   Adaptive Transceiver Architecture

An Overview - The transceiver schematic is shown in Figure 19.30, where the receiver employs a multiple-antenna assisted front end. The signal received by each individual antenna element is fed to an FFT block, and the resulting parallel received OFDM symbols are combined on a subcarrier-by-subcarrier basis. The combining is accomplished on the basis of the weight vector, which has been obtained by solving the Wiener equation, constituting the core of the sample matrix inversion algorithm [335, 593]. After combining the signal is fed into the adaptive demodulator of Figure 19.30, which delivers the output bits in the form of soft-decision information to an optional channel decoder. The demodulator operates in one of a set of four modes, namely 'no transmission', BPSK, QPSK and 16-QAM. Since an interfered channel cannot be considered to constitute a reciprocal system, the modem mode adaptation operates in a closed-loop fashion, where each of the receivers instructs the remote transmitter as to the required set of modulation modes for the next AOFDM symbol, which is necessary for maintaining a given target Bits per Symbol (BPS) performance. On reception of a packet the adaptation algorithm computes the set of modulation modes to be employed by the remote transmitter for the next transmitted AOFDM symbol on the basis of a channel quality measure, namely the subcarrier SNR, which can be estimated by the interference suppression algorithm. The set of requested modulation modes is signalled to the remote receiver along with the next transmitted AOFDM symbol, which is then used by the remote transmitter in its next transmission.

The Signal Model - The $P \times 1$ vector of complex signals, $\mathbf{x}[n, k]$, received by the antenna array in the k-th subcarrier of the n-th OFDM symbol is constituted by a superposition of the

---

[3]This section is based on M. Münster, T. Keller and L. Hanzo; Co-Channel Interference Suppression Assisted Adaptive OFDM In Interference Limited Environments, ©IEEE, VTC'99, Amsterdam, NL, 17-19 Sept. 1999

independently faded signals associated with the *desired* user and the L *undesired* users plus the Gaussian noise at the array elements:

$$\mathbf{x}[n,k] = \mathbf{d}[n,k] + \mathbf{u}[n,k] + \mathbf{n}[n,k], \quad \text{with} \qquad (19.10)$$

$$\mathbf{d}[n,k] = \mathbf{H}^{(0)}[n,k]s_0[k]$$

$$\mathbf{u}[n,k] = \sum_{l=1}^{L} \mathbf{H}^{(l)}[n,k]s_l[k],$$

where $\mathbf{H}^{(l)}[n,k]$ for $l = 0, \ldots, L$ denotes the $P \times 1$ vector of complex channel coefficients between the l-th user and the P antenna array elements. We assume that the vector components $H_m^{(l)}[n,k]$ for different array elements m or users l are independent, stationary, complex Gaussian distributed processes with zero-mean and different variance σ_l^2, $l = 0, \ldots, L$. The variable $s_l[n,k]$ - which is assumed to have zero-mean and unit variance - represents the complex data of the l-th user and $\mathbf{n}[n,k]$ denotes the aforementioned $P \times 1$ vector of additive white Gaussian noise contributions with zero mean and variance σ^2 [259].

The SMI Algorithm - The idea behind minimum mean-square error (MMSE) beamforming [335] is to adjust the antenna weights, so that the power of the differential signal between the combiner output and a reference signal - which is characteristic of the desired user - is minimised. The solution to this problem is given by the well-known Wiener equation, which can be directly solved by means of the SMI algorithm in order to yield the optimum weight vector $\mathbf{w}[n,k]$ of dimension $P \times 1$. Once the instantaneous correlation between the received signals, which is represented by the $P \times P$ matrix $\mathbf{R}[n,k]$ - and the channel vector $\mathbf{H}^{(0)}[n,k]$ of the desired user become known, the weights are given by [259, 310, 335]:

$$\mathbf{w}[n,k] = (\mathbf{R}[n,k] + \gamma\mathbf{I})^{-1}\mathbf{H}^{(0)}[n,k], \qquad (19.11)$$

where γ represents the so-called diagonal augmentation factor [259]. Assuming knowledge of all channel parameters and the noise variance σ^2, the correlation matrix can be determined by:

$$\mathbf{R}[n,k] \triangleq E_c\{\mathbf{x}[n,k]\mathbf{x}^H[n,k]\} \qquad (19.12)$$

$$= \mathbf{R}_d[n,k] + \mathbf{R}_u[n,k] + \mathbf{R}_n[n,k], \quad \text{with}$$

$$\mathbf{R}_d[n,k] = \mathbf{H}^{(0)}[n,k]\mathbf{H}^{(0)^H}[n,k]$$

$$\mathbf{R}_u[n,k] = \sum_{l=1}^{L} \mathbf{H}^{(l)}[n,k]\mathbf{H}^{(l)^H}[n,k]$$

$$\mathbf{R}_n[n,k] = \sigma^2\mathbf{I},$$

which is a superposition of the correlation matrices $\mathbf{R}_d[n,k]$, $\mathbf{R}_u[n,k]$ and $\mathbf{R}_n[n,k]$ of the desired and undesired users as well as of the array element noise, respectively. The combiner output can now be inferred from the array output vector $\mathbf{x}[n,k]$ by means of:

$$y[n,k] = \mathbf{w}^H[n,k]\mathbf{x}[n,k]. \qquad (19.13)$$

The Signal-to-Noise Ratio (SNR) at the combiner output - which is of vital importance for the modulation mode adaptation - is given by [335]:

$$SNR = \frac{E\{|\mathbf{w}^H[n,k]\mathbf{d}[n,k]|^2\}}{E\{|\mathbf{w}^H[n,k]\mathbf{n}[n,k]|^2\}} \quad (19.14)$$

$$= \frac{\mathbf{w}^H[n,k]\mathbf{R}_d[n,k]\mathbf{w}[n,k]}{\mathbf{w}^H[n,k]\mathbf{R}_n[n,k]\mathbf{w}[n,k]}$$

and correspondingly the Signal-to-Interference+Noise Ratio (SINR) is given by [335]:

$$SINR = \frac{\mathbf{w}^H[n,k]\mathbf{R}_d[n,k]\mathbf{w}[n,k]}{\mathbf{w}^H[n,k](\mathbf{R}_u[n,k]+\mathbf{R}_n)\mathbf{w}[n,k]}. \quad (19.15)$$

Equation 19.12 is the basis for our initial simulations, where *perfect channel knowledge* has been assumed. Since in a real environment the receiver does not have perfect knowledge of the channel, its parameters have to be estimated, a problem which we will address in Section 19.7.4 on an OFDM symbol-by-symbol basis by means of orthogonal pilot sequences.

The Adaptive Bit-assignment Algorithm - The adaptation performed by the modem is based on the choice between a set of four modulation modes, namely 4, 2, 1 and 0 bit/subcarrier, where the latter corresponds to 'no transmission'. The modulation mode could be assigned on a subcarrier-by-subcarrier basis, but the signalling overhead of such a system would be prohibitive. Hence, we have grouped adjacent subcarriers into 'sub-bands' and assign the same modulation mode to all subcarriers in a sub-band. Note that the frequency domain channel transfer function is typically not constant across the subcarriers of a sub-band, hence the modem mode adaptation will be sub-optimal for some of the subcarriers. The Signal-to-Noise Ratio (SNR) of the subcarriers will be shown to be in most cases an effective measure for controlling the modulation assignment. The modem mode adaptation is hence achieved by calculating in the first step for each sub-band and for all four modulation modes the expected overall sub-band bit error rate (BER) by means of averaging the estimated individual subcarrier BERs. Throughout the second step of the algorithm - commencing with the lowest modulation mode in all sub-bands - in each iteration the number of bits/subcarrier of that sub-band is increased, which provides the best compromise in terms of increasing the number of expected bit errors compared to the number of additional data bits accommodated, until the target number of bits is reached.

The Channel Models - Simulations have been conducted for the indoor Wireless Asynchronous Transfer Mode (WATM) channel impulse response (CIR) of Section 17.1 [585]. This three-path impulse response exhibits a maximal dispersion of 11 time-domain OFDM samples, with each path faded according to a Rayleigh distribution of a normalised maximal Doppler frequency of $f_d' = 1.235 \cdot 10^{-5}$, where the normalisation interval was the OFDM symbol duration. This model corresponds to the channel experienced by a mobile transmitting at a carrier frequency of 60 GHz with a sampling rate of 225 MHz and travelling at a vehicular velocity of 50 km/h. An alternative channel model, which we considered in our simulations is a Wireless Local Area Network (WLAN) model associated with a seven-path impulse response having a maximal dispersion of 32 samples. However, for this more dispersive and higher Doppler frequency channel adaptive modulation has turned out to be less effective due to its significantly increased normalised Doppler frequency of $f_d' = 3.935 \cdot 10^{-5}$ corresponding to a carrier frequency of 17 GHz, sampling rate of 20 MHz and vehicular velocity of 50 km/h.

19.7.3   Simulation Results - Perfect Channel Knowledge

General Remarks - In our initial simulations we assumed that the receiver had perfect knowledge of all channel parameters, which enabled the estimation of the correlation matrix required by the SMI algorithm upon using Equation 19.12. Furthermore, we initially assumed that the receiver was capable of signalling the modulation modes to the transmitter without any additional delay. Throughout our discussions we will gradually remove the above idealistic assumptions. In all simulations we assumed a partitioning of the 512-subcarrier OFDM symbol's total bandwidth into 16 equal-sized 32-subcarrier sub-bands. This has been shown to provide a reasonable compromise between signalling overhead and performance degradation compared to a subcarrier-by-subcarrier based modulation mode assignment.

Two-Branch Maximum–Ratio Combining - Initial simulations were conducted in the absence of co-channel interference. In this scenario the SMI equations take the form of the MMSE maximum-ratio combiner, resulting in a high diversity gain even with the minimal configuration of only two reception elements. Adaptive modulation was performed on the basis of the estimated SNR of each subcarrier, which is given by Equation 19.14. Since due to diversity reception the dramatic fades of the channel frequency response have been mitigated, the performance advantage of adaptive modulation is more modest, as illustrated by Figure 19.31 for the 'transmission frame-invariant' WATM channel, for which the fading profile is kept constant for the OFDM symbol duration, in order to avoid inter-subcarrier interference. For the equivalent simulations in the 'transmission frame-invariant' WLAN channel environment we observed a more distinct performance gain due to adaptive modulation, which is justified by the higher degree of frequency selectivity introduced by the WLAN channel's seven-path impulse response.

SMI Co-Channel Interference Suppression -   In these simulations we considered first of all the case of a single dominant co-channel interferer of the same signal strength as the desired user. It is well-known that if the total number of users - whose signals arrive at the antenna array - is less or equal to the number of array elements, the unwanted users are suppressed quite effectively. Hence, for our modulation adaptation requirements we can assume that $SNR \approx SINR$, which enables us to use the algorithm described in Section 19.7.2, on the basis of the SNR estimated with the aid of Equation 19.14. Figure 19.32 illustrates the impact of adaptive modulation in the WATM channel environment under the outlined conditions. At a given SNR the performance gain due to adaptive modulation decreases with an increasing bitrate, since the higher bitrate imposes a more stringent constraint on the modulation mode assignment, invoking a higher number of low-SNR subcarriers. Upon comparing Figure 19.32 and 19.33 we observe that AOFDM attains a significantly higher SNR gain in the presence of co-channel interference, than without interference. As alluded to in the previous section this is because under co-channel interference the SMI scheme exploits most of its diversity information extracted from the antenna array for suppressing the unwanted signal components, rather than mitigating the frequency domain channel fades experienced by the wanted user. For decreasing values of the Interference-to-Noise Ratio (INR) at the antenna array output, the system performance will gradually approach the performance observed for the MRC system. In order to render our investigations more realistic in our next experiment we allow a continous, i.e. 'frame-variant' fading across the OFDM symbol du-

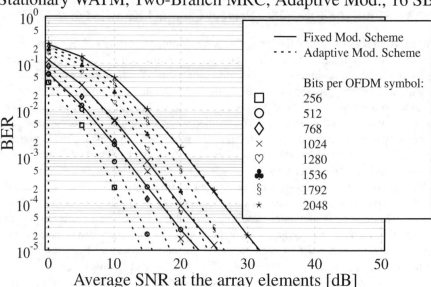

Figure 19.31: BER of 16-sub-band AOFDM modem with *two-branch maximum-ratio combining* in a *'frame-invariant' indoor WATM* environment, assuming *perfect channel knowledge* and *zero-delay signalling* of the modulation modes

ration. The system performance corresponding to this scenario is illustrated in Figure 19.33. At low SNRs the observed performance is identical to that recorded in Figure 19.32 for the 'frame-invariant' channel model, whereas at high SNRs we experience a residual BER due to inter-subcarrier interference. Again, for a low number of bits per OFDM symbol the adaptive scheme is capable of reducing the 'loading' of subcarriers with low SNR values, which are particularly impaired by inter-subcarrier interference. Hence AOFDM exhibits a BER improvement in excess of an order of magnitude. So far we have assumed that the receiver is capable of instantaneously signalling the required modulation modes for the next OFDM symbol to the transmitter. This assumption cannot be maintained in practice. Here we assume a time division duplexing (TDD) system with identical transceivers at both ends of the link, which communicate with each other using adjacent up-link and down-link slots. Hence we have to account for this by incorporating an additional delay of at least one OFDM symbol, while neglecting the finite signal processing delay. Simulation results for this scenario are depicted in Figure 19.34. We observe that the performance gain attained by adaptive modulation is reduced compared to that associated with the zero-delay assumption in Figure 19.33. This could partly be compensated for by channel prediction. When employing a higher number of array elements, the performance gain achievable by adaptive modulation will mainly depend on the number of users and their signal strength. If the number of users is lower, than

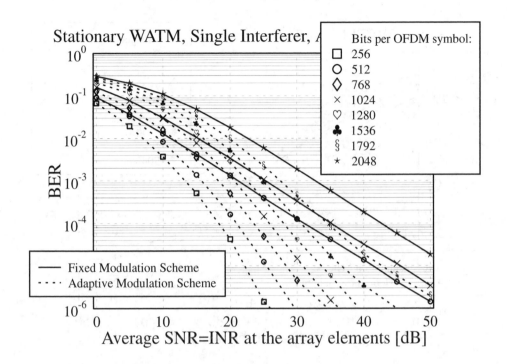

Figure 19.32: BER of 16-sub-band AOFDM modem with *two-branch SMI* and 2 *users* in a *'frame-invariant' indoor WATM* environment, assuming *perfect channel knowledge* and *zero-delay signalling* of the modulation modes

the number of array elements, or if the interferers are predominantly weak, the remaining degrees of freedom for influencing the array response are dedicated by the SMI scheme to providing diversity for the reception of the wanted user and hence adaptive modulation proves less effective. If the number of users exceeds the number of array elements, the system becomes incapable of suppressing the undesired users effectively, resulting in a residual BER at high SNRs due to the residual co-channel interference. Since for a relatively high number of users the residual interference exhibits Gaussian-noise like characteristics, the SINR given by Equation 19.15 could be a suitable measure for performing the modulation mode assignment. By contrast, for a low number of interferers it is difficult to predict the impact on the system performance analytically. A possible approach would be to use the instantaneous number of errors in each sub-band (e.g. at the output of a turbo decoder) as a basis for the modulation assignment, which constitutes our future work. Let us now consider the issues of channel parameter estimation.

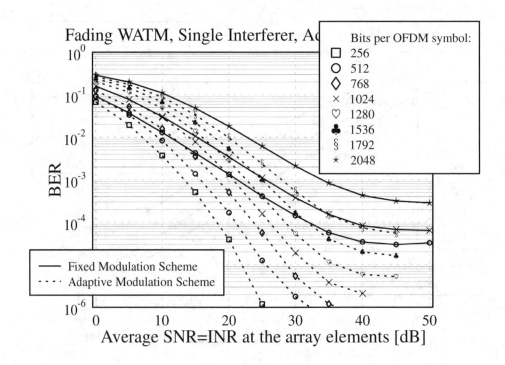

Figure 19.33: Performance results of Figure 19.32 repeated for a *fading indoor WATM* environment

code/bit	0	1	2	3
0	1	1	1	1
1	1	1	-1	-1
2	1	-1	-1	1
3	1	-1	1	-1

Table 19.4: Orthogonal Walsh codes with a length of 4 bit

19.7.4   Pilot-Based Channel Parameter Estimation

System Description - Vook and Baum [310] have proposed SMI parameter estimation for OFDM by means of orthogonal reference sequences carrying pilot slots, which are transmitted over several OFDM symbol durations. This principle can also be applied on an OFDM symbol-by-symbol basis, as required for adaptive modulation. Upon invoking the idea of pilot based channel estimation by means of sampling and low-pass interpolating the channel transfer function, we replace each single pilot subcarrier by a group of pilots, which carries a replica of the user's unique reference sequence. This is illustrated in Figure 19.35 for a reference sequence having a length of 4 bit, and for a pilot group distance of 16 subcarriers, which

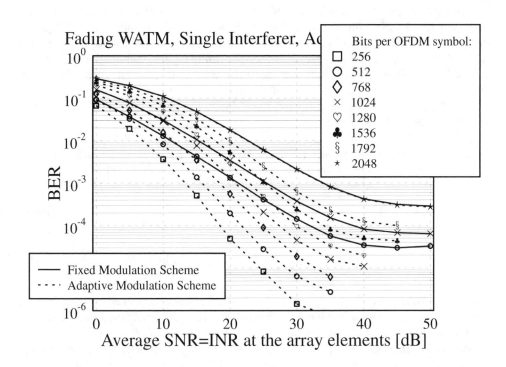

Figure 19.34: Performance results of Figure 19.33 repeated for *one OFDM symbol delayed signalling*

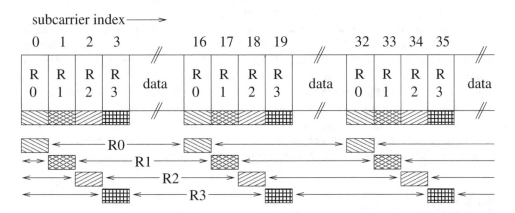

Figure 19.35: Pilot arrangement in each OFDM symbol for a *reference length of 4 bit* and a *group distance of 16 subcarriers*; interpolation is performed between pilots associated with the same bit position within the reference sequence

corresponds to the frequency required for sampling the WATM channel's transfer function. The corresponding 4-bit orthogonal Walsh code-based reference sequences are listed in Table 19.4. Each of these 4 bits is assigned using BPSK to one of the 4 pilots in a pilot-group. The complex signal received by the m-th antenna in a pilot subcarrier at absolute index k and local index i within the reference sequence is constituted by a contribution of all users, each of which consists of the product of the Walsh code value associated with the user at bit position i of the reference sequence of Table 19.4 and the complex channel coefficient between the transmitter and the m-th antenna. MMSE lowpass interpolation is performed between all pilot symbols of the same relative index i within the k-spaced pilot blocks - as seen in Figure 19.35 - in order to generate an interpolated estimate of the reference for each subcarrier. An estimate of the channel vector $\hat{\mathbf{H}}^{(0)}[n, k]$ and the correlation matrix $\hat{\mathbf{R}}[n, k]$ for the k-th subcarrier of the n-th OFDM symbol is then given by [310, 335, 593]:

$$\hat{\mathbf{H}}^{(0)}[n, k] = \frac{1}{N} \sum_{i=0}^{N-1} r^{(0)^*}(i)\mathbf{x}_{LP}[n, k](i) \tag{19.16}$$

$$\hat{\mathbf{R}}[n, k] = \frac{1}{N} \sum_{i=0}^{N-1} \mathbf{x}_{LP}[n, k](i)\mathbf{x}_{LP}^{H}[n, k](i), \tag{19.17}$$

where $r^{(0)}(i)$ denotes the i-th value of the reference sequence associated with the desired user, $\mathbf{x}_{\mathbf{LP}}[n, k](i)$ represents the low-pass interpolated received signal at sequence position i and N denotes the total reference length.

Simulation Results - The performance of this scheme is characterised by the simulation results presented in Figure 19.36. Compared to the results presented in Figure 19.34 for 'perfect channel knowledge', in Figure 19.36 we observe that besides the reduced range of supported bitrates there is an additional performance degradation, which is closely related to the choice of the reference length. Specifically, there is a reduction in the number of useful data subcarriers due to the pilot overhead, which reduces the 'adaptively' exploitable diversity potential. Secondly, a relatively short reference sequence results in a limited accuracy of the estimated channel parameters - an effect which can be partly compensated for by a technique referred to as diagonal loading [259]. However, the effect of short reference sequences becomes obvious for a higher number of antenna elements, since more signal samples are required, in order to yield a reliable estimate of the correlation matrix. Hence our scheme proposed here is attractive for a scenario having 2-3 reception elements, where the interference is due to 1-2 dominant interferers and an additional Gaussian noise like contribution of background interferers, which renders the SINR of Equation 19.15 an effective measure of channel quality. In conclusion, the proposed adaptive array assisted AOFDM scheme resulted typically in an order of magnitude BER reduction due to employing adaptive modulation.

19.8   Summary

A range of adaptive modulation and spectral predistortion techniques have been presented in this chapter, all of which aim to react to the time and frequency dependent channel transfer function experienced by OFDM modems in fading time dispersive channels. It has been demonstrated that by exploiting the knowledge of the channel transfer function at the trans-

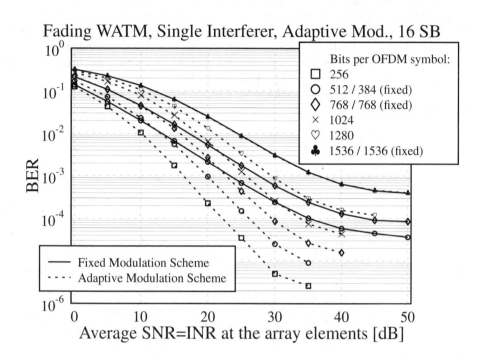

Figure 19.36: BER of 16-sub-band adaptive OFDM modem with *two-branch SMI* and 2 *users* in a *fading indoor WATM* environment, with *pilot based channel parameter estimation* and *one OFDM symbol delayed signalling* of the modulation assignment using a *diagonal loading* of $\gamma = 1.0$

mitter, the overall system performance can be increased substantially over the non-adaptive case. It has been pointed out that the prediction of the channel transfer function for the next transmission time slot and the signalling of the parameters are the main practical problems in the context of employing adaptive techniques in duplex communications.

The channel prediction accuracy is dependent on the quality of the channel estimation at the receiver, as well as on the temporal correlation of the channel transfer function between the uplink and downlink time slots. Two-dimensional channel estimation techniques [260, 594] can be invoked in order to improve the channel prediction at the receivers.

It has been demonstrated that sub-band adaptivity instead of subcarrier-by-subcarrier adaptivity can significantly decrease the necessary signalling overhead, with a loss of system performance that is dependent on the channel's coherence bandwidth. We have seen that sub-band adaptivity allows the employment of blind detection techniques in order to minimise the signalling overhead. Further work into blind detection algorithms as well as new signalling techniques is needed for improving the overall bandwidth efficiency of adaptive OFDM systems.

Pre-equalisation or spectral predistortion techniques have been demonstrated to significantly improve an OFDM system's performance in time dispersive channels, while not increasing the system's output power. It has been shown that spectral predistortion can integrate well with adaptive modulation techniques, improving the system's performance significantly. We have seen in Figure 19.1 that a data throughput of 0.5 bits/symbol has been achieved at 0 dB average channel SNR with a BER of below 10^{-4}.

An adaptive power and bit allocation algorithm was also highlighted and the performance of AOFDM in a beam-forming assisted multi-user scenario was characterised.

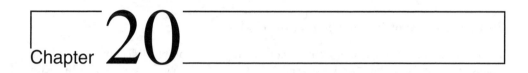

Chapter **20**

Block-Coded Adaptive OFDM

T. Keller, T-H. Liew, L.L. Yang and L. Hanzo

20.1 Introduction

20.1.1 Motivation

We saw in Chapter 19 that adaptive modulation techniques can be used in order to achieve a given target level of bit error rate (BER) in a duplex communications link by choosing the appropriate modulation modes for the instantaneous channel conditions. The various adaptive modem mode signalling regimes were portrayed in Figure 19.1. employing recursive systematic Convolutional (RSC) codes as the constituent codes in turbo coding in conjunction with a turbo decoder allowed us to dramatically improve the effective data throughput for a given target bit error rate of 10^{-4} at low SNR values. The performance comparison in Figure 19.29 showed that the observed throughput for the coded, target BER adaptive modem was higher than in the uncoded case for SNR values up to 18 dB. For higher SNR values, however, the redundant information overhead of the FEC limits the system's effective data throughput if only a limited set of modulation modes is employed by the adaptation algorithm. In the case of the half-rate convolutional coding based turbo code employed in Section 19.2.7, half the data bandwidth was unnecessarily absorbed at high SNR values.

In this chapter, we will investigate the employment of variable rate channel coding schemes within an OFDM transmission system, in order to improve the system's throughput at high SNR values. Analogously to the adaptive modulation modes discussed in Chapter 19, in addition to the modem modes, in this chapter the code rate is also adapted in response to the time and frequency dependent estimated channel conditions. We note, however, that this variable code rate scheme can operate along with time-invariant non-adaptive modulation or in conjunction with additional adaptive OFDM (AOFDM) modulation techniques, as will be demonstrated here.

Similarly to the adaptive modulation schemes discussed in Chapter 19, adaptive coding relies on the same fundamental principles of channel quality estimation, parameter adaptation

and signalling of the modem parameters employed. Like adaptive modulation, it can operate
in open and closed adaptation loops, which were portrayed in Figure 19.29. In this chapter,
we will concentrate on the upper bound performance study of adaptive error correction coding
for OFDM modems, and assume perfect channel estimation and signalling.

20.1.2   Choice of Error Correction Codes

The behaviour of the adaptive modems in conjunction with long-block-length convolutional
coding based turbo coding has been studied in Chapter 19. We have seen that the error
correction performance of the turbo decoder was high, resulting in output bit error rates of
below 10^{-4} for input bit error rates of up to 1%. The disadvantage of this RSC-based turbo
codec was the long block length of 2000 bits, which prevented code rate adaptation on a
short-term basis. Radically shortening the interleaving block length of turbo codes results in
significantly reduced performance, as does additional puncturing of the code word for code
rate adaptation.

Similarly to the adaptive modulation algorithms, the adaptive error correction codec has
to be able to vary its code rate according to the frequency dependent channel transfer function observed for the OFDM system in a time dispersive channel. Ideally, the error correction
capability of the code would be adjustable for each data bit's expected BER independently.
For our experiments, short block length codes of less than 100 bits per code word were employed, in order to allow flexible adaptation of code parameters, while delivering reasonable
error protection for the data bits.

Two different block codes have been investigated for adaptive OFDM transmission,
namely variable-length redundant residue number system (RRNS) based codes [171,595,596]
and turbo BCH codes [597,598].

20.2   Redundant Residue Number System Codes

Residue number system (RNS) [595,596] based algorithms have been studied in the context
of digital filtering, spectral analysis, correlation and matrix operations as well as in image
processing [599–602]. Until quite recently RRNSs have only been proposed for fault tolerant
processing of signals, for example in digital arithmetics. RRNSs represent each operand with
the aid of a set of residues. The residues are generated as the remainders upon dividing the
operand to be represented with the aid of the residues by each of the so-called moduli of
the RRNS, which have to be relative primes, i.e. do not have a common divisor. Amongst
others the following two advantages accrue for RRNS-based processing [603]: they have
the ability to use carry-free arithmetic, since all residue-based operations can be processed
independently of each other. This is amenable to high-speed parallel processing. A related
RRNS property is the lack of ordered significance among residue digits, which implies that as
long as a sufficient number of residues was retained, in order to unambiguously represent the
results of the computations, any erroneous residue digit can be discarded without affecting
the result.

Error detection and correction algorithms based on the RNS have been proposed by Szabo and Tanaka [595], and by Watson and Hastings [596], which exploited the properties
of the redundant residue number system (RRNS). More recently, a computationally efficient

procedure was described in [604] for correcting a single error. In [605], the procedure was extended to correcting double errors as well as simultaneously correcting single errors and detecting multiple errors. Efficient soft-decision multiple error correcting algorithms were suggested in [606]. Furthermore, a RNS-based M-ary modulation scheme has been proposed and analysed in [607], while an RRNS-based CDMA system was the topic of [608]. For a detailed discourse on the application and turbo decoding algorithms of RRNS, please refer to [172].

RRNS(n, k) codes are akin to the well-known family of Reed-Solomon (RS) codes, and both families achieve the so-called maximum minimum distance of $d = n - k + 1$, provided that the moduli of the RRNS code obey certain conditions. It can also be readily shown that the RRNS code's weight distribution can be approximated by the weight distribution of RS codes if all the moduli assume values close to their average value [604, 605, 609]. For further details concerning the construction, coding theory, encoding and decoding of RRNS codes the interested reader is referred to the literature [604–609].[1] Based on the above arguments a similar coding performance is achieved by an identical rate RS code and RRNS code, provided that they both use the same number of bits per symbol or bits per residue. However, their encoding and decoding algorithms are distinctly different [606], requiring further research into their implementation, coding theory and application. Since in [606] the authors have developed a soft-decision decoding algorithm, here we favoured RRNS codes, with the intention of stimulating further research in this interesting novel field. Typically there exists a larger variety of RRNS codes for a given code rate and code word length than in the family of RS codes.

The RRNS codes employed in our investigations are systematic, which means that k of the n code residues contain the original data bits, and the additional $(n - k)$ redundant residues can be employed for error correction at the decoder. The error correction capability of the code is $t = \lfloor \frac{n-k}{2} \rfloor$ residues [606].

The code rate, and accordingly the error correction capability of the code, can be readily varied by transmitting only a fraction of the generated redundant residues. If the channel conditions are favourable, then only the systematic information-bearing residues are transmitted, resulting in a unity rate code with no added redundancy and no error correction capability. Upon transmitting two redundant residues with the data bits, the resulting code can correct one residue error for a code rate of $\frac{n}{n+2}$. More of the redundant residues can be transmitted, lowering the code rate and improving the code's error resilience at the cost of a lower effective information throughput, when the channel quality degrades.

In our investigations RRNS codes employing 8 bits per residue have been chosen. Three or six systematic information-bearing residues – corresponding to 24 or 48 useful data bits per code word – and up to six redundant residues have been employed. The code parameters for these codes are shown in Table 20.1. As can be seen from the table, the code rates vary from 0.33 to 0.75. The uncoded case, corresponding to a $(3, 3)$ code, is not shown in the table.

Code	(5, 3)	(7, 3)	(9, 3)	(8, 6)	(10, 6)	(12, 6)
Sys. residues	3	3	3	6	6	6
Red. residues	2	4	6	2	4	6
Correction capability t	1	2	3	1	2	3
Code rate	0.6	0.43	0.33	0.75	0.6	0.5
Data bits per word	24	24	24	48	48	48
Red. bits per word	16	32	48	16	32	48

Table 20.1: RRNS codes employed in our investigations, each using 8-bit residues

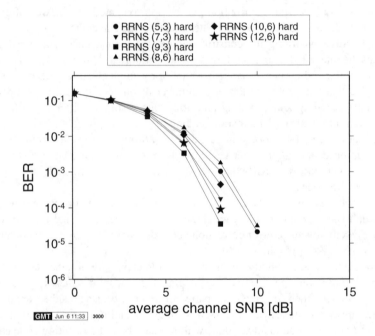

Figure 20.1: BER performance for RRNS-coded QPSK transmission in AWGN channel using hard decision decoding

20.2.1   Performance in an AWGN Channel

The BER performance of an OFDM system employing QPSK and the RRNS codes of Table 20.1 is depicted in Figure 20.1. It can be seen that the relative BER performance of the different codes is largely in line with their respective code rates. The $(9, 3)$ code, with a code rate of 0.33, exhibits the strongest error correction properties; the $(8, 6)$ code, with a code rate of 0.75, is the weakest code of the set. Comparing the performance of the $(5, 3)$ code with that of the $(10, 6)$, code both having a code rate of 0.6, shows that the longer code exhibits a superior performance. The $(12, 6)$ code, having a code rate of 0.5, outperforms the shorter

---

[1]A range of related RRNS-coding oriented papers by the authors can be found under http://www-mobile.ecs.soton.ac.uk/lly/

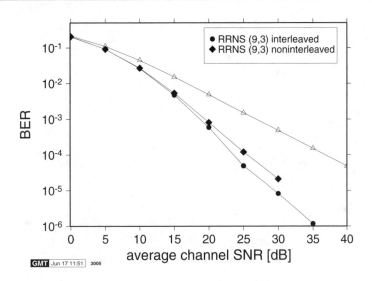

Figure 20.2: BER performance and throughput in bits per symbol for the RRNS-coded OFDM system in the Rayleigh fading shortened WATM channel of Section 17.1.1.2, with 512-FFT QPSK OFDM transmission

$(7, 3)$ code for SNR values in excess of 6 dB.

20.2.1.1 Performance in a Fading Time-Dispersive Channel

Figure 20.2 portrays the BER performance of a RRNS-coded 512-subcarrier modem using a 512-point FFT-based (512-FFT) OFDM modem employing QPSK transmission over the Rayleigh fading short WATM channel described in Section 17.1.1.2. The associated BER curves for uncoded, coded non-interleaved and coded interleaved transmission are shown. The interleaver employed in acquiring the interleaved results was a shortened block interleaver of the structure shown in Figure 20.3. This interleaver structure was chosen since it is sufficiently flexible for varying numbers of residues if adaptive modulation is invoked.

We note that residue-based interleaving has a better performance than bit interleaving, since bit interleaving would increase the probability of residue errors due to spreading bursts of erroneous bits across residues. Since the RRNS decoding algorithm is symbol based, the increased residue error rate would degrade the system's performance.

It can be seen from Figure 20.2 that the coded schemes deliver a gain of about 12 dB in SNR terms. Interleaving of the residues within the OFDM symbol improves the BER performance by about 2 dB.

20.2.1.2 Adaptive RRNS-Coded OFDM

Analogously to the adaptive modulation schemes discussed in Chapter 19, the code rate adaptation reacts to the time- and frequency-varying channel conditions experienced in a duplex link. Each station exploits the channel quality information extracted from the last received OFDM symbol for determining the coding parameters of the next transmitted frame.

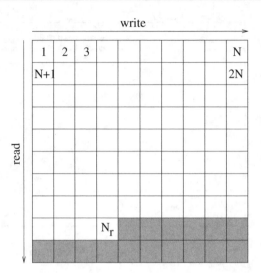

Figure 20.3: Structure of the interleaver employed for the ARRNS system. The interleaver is a square $N \times N$ interleaver for N_r residues, with $N = \lceil \sqrt{N_r} \rceil$; only the first N_r positions are used, where N_r is the number of residues in the OFDM symbol

Although the longer codes exhibit better error correction properties than the shorter codes investigated, we will concentrate on the group of $(9,3)$, $(7,3)$, $(5,3)$ codes and on uncoded $(3,3)$ transmission for the variable code rate application, since a short code word length allows for increased flexibility in the adaptation to the channel conditions. These codes exhibit an error correction capability of $t = 3, 2, 1$, and 0 residues per code word, respectively. Correspondingly, four error correction coding modes c are defined, which are shown in Table 20.2.

Mode c	0	1	2	3
n_c	3	5	7	9
k_c	3	3	3	3
t_c	0	1	2	3
R	1	0.6	0.43	0.33

Table 20.2: RRNS coding modes used for the code rate adaptation employing 8-bit residues

The choice of the coding mode for each code word in the AOFDM symbol is determined on the basis of the estimated channel transfer function. As discussed for the adaptive modulation scheme of Section 19.2.4.2, the predicted bit error probabilities p_e are calculated for all bits to be transmitted in an OFDM symbol, based on the estimated subcarrier SNR and the modulation mode to be employed. More explicitly, the expected overall bit error probability for the given OFDM symbol is computed by averaging the individual subcarrier BERs for all legitimate modulation modes M_n, yielding $\bar{p}_e(M_n) = 1/N_s \sum_j p_e(\gamma_j, M_n)$ where γ_j is the subcarrier SNR for $j = 1, \ldots, 512$. The specific subcarrier modem mode allocation with the highest bits per symbol (BPS) throughput, whose estimated BER is lower than the required value is then confirmed. Currently, however, only QPSK/OFDM is used, which simplifies the associated bit allocation and BER computation procedure. AOFDM will be the topic of Section 20.2.2. This algorithm allows the direct adjustment of the desired maximum BER.

If OFDM is to be employed in conjunction with adaptive RRNS coding, then the number of bits per OFDM symbol and the mapping of bits to subcarriers can change from one OFDM symbol to the next. Exlicitly, this implies using a variable bitrate, near-constant BER regime. Hence the RRNS coding scheme adaptation algorithm operates on the basis of the estimated BER, rather than directly relying on the estimated channel transfer function. Once the vector of estimated bit error probabilities $p_e(n)$ for the specific number of bits N_b - which has to be conveyed by the OFDM symbol - is known, the total number of bits to be transmitted is split into blocks of K bits, where K is the number of bits per residue. Again, the error correction capability of the code in each RRNS code word is a given number of residues, not bits. Hence, as argued before, interleaving of bits would increase the residue error rate at the decoder's input and lower the system's performance.

From the bit error probability $p_e(n)$ of bit n, $n = 1 \ldots N_b$ - where N_b is the number of bits allocated to the OFDM symbol - the estimated residue error rate $p_r(r)$ for residue index r - which is defined as the proportion of residues in error - for the $N_r = \lfloor N_b/K \rfloor$ number of residues in the OFDM symbol can be calculated as:

$$p_r(r) = 1 - \left(\prod_{n=0}^{K-1} (1 - p_e(r \cdot K + n)) \right). \tag{20.1}$$

The remaining $N_b - K \cdot N_r$ data bits of the OFDM symbol that are not allocated to any residue are filled with padding bits and hence contain no useful data. The mapping of the residues with index r to the RRNS code words is based on the estimated residue error probabilities $p_r(r)$. The interleaver $I(r)$ of Figure 20.3 is used to map the stream of residues to the residue positions in the transmitted OFDM symbol. The interleaver function used in our experiments is defined more explicitly below.

A received code word of the codec mode c_w is irrecoverable if more than t_c of the received residues are in error. The RRNS code word error probability p_w for word w can be calculated as:

$$p_w(w) = p(R_r(w) > t_{c_w}) = 1 - P(R_r(w) \leq t_{c_w}), \tag{20.2}$$

where $R_r(w)$ is the number of residue errors in code word w, and $p_w(w)$ can be calculated from the residue error probabilities $p_r(r)$ as:

$$p_w(w) = 1 - p[R_r(w) = 0] - p[R_r(w) = 1] - \ldots - p[R_r(w) = t_{c_w}]. \tag{20.3}$$

Upon elaborating further:

$$p(R_r(w) = 0) \quad = \quad \prod_{r=0}^{n_{cw}-1} (1 - p_r(I(r_{0,w} + r))) \tag{20.4}$$

$$p(R_r(w) = 1) \quad = \quad \sum_{r=0}^{n_{cw}-1} p_r(I(r_{0,w} + r)) \prod_{s=0,s \neq r}^{n_{cw}-1} (1 - p_r(I(r_{0,w} + s))) \tag{20.5}$$

$$= \quad p(R_r(w) = 0) \cdot \sum_{r=0}^{n_{cw}-1} \frac{p_r(I(r_{0,w} + r))}{1 - p_r(I(r_{0,w} + r))} \tag{20.6}$$

$$p(R_r(w) = 2) \quad = \quad \frac{1}{2} \cdot p(R_r(w) = 0)$$

$$\cdot \quad \sum_{r=0}^{n_{cw}-1} \left[\frac{p_r(I(r_{0,w} + r))}{1 - p_r(I(r_{0,w} + r))} \cdot \sum_{s=0,s \neq r}^{n_{cw}-1} \frac{p_r(I(r_{0,w} + s))}{1 - p_r(I(r_{0,w} + s))} \right] \tag{20.7}$$

$$p(R_r(w) = 3) \quad = \quad \frac{1}{3!} \cdot p(R_r(w) = 0) \cdot \sum_{r=0}^{n_{cw}-1} \left[\frac{p_r(I(r_{0,w} + r))}{1 - p_r(I(r_{0,w} + r))} \right. \tag{20.8}$$

$$\left. \cdot \quad \sum_{s=0,s \neq r}^{n_{cw}-1} \left[\frac{p_r(I(r_{0,w} + s))}{1 - p_r(I(r_{0,w} + s))} \cdot \sum_{t=0,t \neq q,r}^{n_{cw}-1} \frac{p_r(I(r_{0,w} + t))}{1 - p_r(I(r_{0,w} + t))} \right] \right],$$

where $r_{0,w}$ is the index of the first residue in code word w.

The code rate adaptation algorithm calculates the word error probability $p_w(w)$ for the RRNS code word index w for the lowest-power codec mode of Table 20.2, $c = 0$. If the word error probability exceeds a certain threshold, i.e. $p_w(w) > \alpha$ for $c = 0$, then the next stronger coding mode $c = 1$ is selected, and the word error probability is evaluated again. If the new RRNS code word error probability exceeds the threshold α, then the next codec mode is evaluated, until the estimated RRNS code word error probability falls below the threshold α, or until the highest power codec mode is selected.

The parameter α is supplied to the algorithm and it can be used to control the adaptation process, similarly to the target BER in the adaptive modulation algorithm of Section 19.2.4.2.

The effect of different settings of the parameter α on the effective data throughput and BER of a QPSK modem in an AWGN channel is shown in Figure 20.4. The word error rate (WER) threshold α was varied from 10^{-1} up to 10^{-4}, and it can be seen that the system performance varies only in a limited range of SNR values, approximately between 7 and 12 dB, with the WER α. The three RRNS-coded BER curves of Figure 20.1 are easily identified in this figure as segments of the adaptive scheme's BER performance curve together with the BER curve of uncoded QPSK. The different codec modes are easily identifiable in this figure, since in the AWGN channel the channel quality is time invariant, and therefore the same codec mode is chosen for all transmitted RRNS code words at a given SNR point. The choice of the RRNS codec mode clearly depends on the value of the WER α. For all four values of α, the system performance is identical for SNR values up to 6 dB. For higher SNR values, it can be observed that the WER α influences the codec mode switching algorithm. For $\alpha = 10^{-1}$, the codec mode is switched to the next lower-power higher-rate mode of

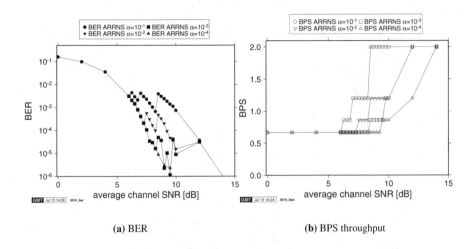

(a) BER   (b) BPS throughput

Figure 20.4: BER performance and throughput in bits per symbol for the adaptive RRNS (ARRNS) coded OFDM system in AWGN channels, with QPSK transmission in the subcarriers

Table 20.2 at SNR values of 6.25, 7 and 8.5 dB, resulting in a BER of about 0.7% after each reconfiguration as seen in Figure 20.4. The maximal throughput of 2 bits/symbol is reached at an SNR of 8.5 dB by the $\alpha = 10^{-1}$ curve. For lower values of α, the codec mode switching is more conservative hence lower bit error rates and throughputs are observed. For $\alpha = 10^{-4}$ the modem employs codec mode 3 of Table 20.2 for SNR values of up to 9.5 dB, reaching mode 0 at 14 dB SNR.

Over fading time dispersive channels the ARRNS codec mode is adapted to the time and frequency variant channel conditions. In order to demonstrate the RRNS code allocation process, in Figure 20.5 we captured the expected subcarrier SNR versus the subcarrier index for one specific OFDM symbol. It can be seen that the channel SNR varies by about 18 dB across the set of subcarriers, which will result in a varying BER across the OFDM symbol. In our illustrative example a specific OFDM symbol was selected that contained at least one RRNS code word corresponding to each of the RRNS codec modes of Table 20.2. In order to augment our understanding, in this example the residues of only one RRNS code word per codec mode are shown in the figure, corresponding to a total of $3 + 5 + 7 + 9 = 24$ residues, signified by vertical bars in the figure. Each vertical bar corresponds to one transmitted residue, and the shading of each bar relates it to one of the four ARRNS code words. The position of a bar gives the residue index r, while its height signifies the expected residue error rate (RER) $p_r(r)$ evaluated from Equation 20.1 for each of the residues of the four chosen ARRNS code words, which depends on the subcarrier SNR experienced at index r.

The channel model used for the fading experiments was described in Section 17.1.1.2. The word error probability threshold α was set to 10^{-1}.

The four code words were chosen so that one of each codec mode of Table 20.2 is demonstrated: the light grey bars mark the position and the residue error probabilities for an uncoded (mode 0) (3, 3) RRNS code word. The positions of the residues in the transmitted OFDM

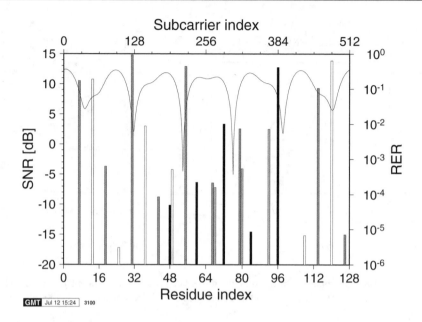

Figure 20.5: Predicted subcarrier SNR (continuous line) for an OFDM symbol and residue error probabilities (vertical bars) for four selected ARRNS code words for a 512-subcarrier OFDM
modem employing QPSK in the short WATM channel of Figure 17.3. The average SNR
is 10 dB, and $\alpha = 10^{-1}$. Key: light grey = (3, 3) code at indices of 68, 80, 92; black = (5,
3) at 48, 60, 72, 84, 96; white = (7, 3); dark grey (9, 3)

symbol are determined by the residue-based interleaver function. It can be seen that, because
of the good channel quality in the subcarriers used for transmission, all of the three residues
of the uncoded code word exhibit low RER values. The word error probability for this code
word is mainly dominated by the highest RER residue transmitted at index 92, which exhibits
an error probability of about $7.2 \cdot 10^{-3}$. The resulting word error probability due to all three
residues is $7.9 \cdot 10^{-3}$, which is below the threshold of $\alpha = 10^{-1}$, and therefore uncoded
transmission was chosen by the algorithm.

The black bars in Figure 20.5 mark the residues for a mode 1 (5, 3) code word, with
five transmitted residues, while the white marked residues form a mode 2 (7, 3) code word,
consisting of seven transmitted residues. The first of this code word's residues at residue
index $r = 0$ is not visible on the figure, as its expected RER is $2.1 \cdot 10^{-7}$.

The dark grey bars of Figure 20.5, represent a mode 3 (9, 3) RRNS code word with 9
transmitted residues. The word error probability for this mode 3 RRNS code word is 11.5%,
which exceeds the set threshold. Since there is no stronger code in the set of coder modes of
Table 20.2, the highest possible code rate was chosen.

Figure 20.6 shows the BER performance and data throughput of a 512-subcarrier
ARRNS-coded OFDM modem employing QPSK in the SWATM channel of Figure 17.3.
Figure 20.6(a) depicts the BER and BPS performance for the modem employing no residue
interleaving, while for Figure 20.6(b) a residue interleaver obeying the structure introduced in
Figure 20.3 was employed. This comparison is relevant and necessary, since it is not intuitive

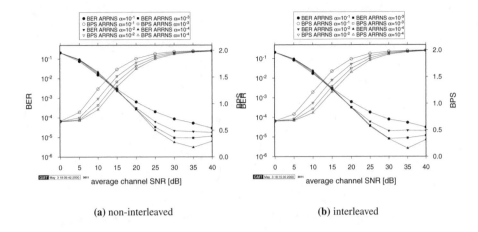

(a) non-interleaved   (b) interleaved

Figure 20.6: BER and BPS throughput versus average channel SNR for ARRNS-coded 512-subcarrier OFDM transmission employing QPSK over the shortened WATM channel of Figure 17.3. The stipulated WER values were $\alpha = 10^{-1}, 10^{-2}, 10^{-3}$ and 10^{-4}

as to whether residue interleaving results in performance improvements. This is because the residue interleaving disperses the bursty channel errors across the OFDM subcarriers, which is expected to improve the coding performance of conventional fixed rate coding. We, however, proposed adaptive RRNS-coding, in order to combat the bursty error distribution across the OFDM subcarriers, which was designed to counteract the frequency selective fading and its bursty errors. Hence interleaving and adaptive RRNS coding may disadvantageously interfere with each other, necessitating a comparison of the interleaved and non-interleaved results.

It can be seen that the BER performance is fairly similar for the interleaved and non-interleaved modems. Specifically, although the non-interleaved modem slightly outperforms the interleaved one in BER terms for WERs of $\alpha = 10^{-3}$ and for $\alpha = 10^{-4}$ at SNR values in excess of 25 dB, the BER difference is not significant. Upon comparing the achieved throughput, however, it is clear that the non-interleaved modem offers an average throughput benefit of about 0.1 bit/symbol for all target WERs α in the SNR region up to 25 dB. Since the range of code rates is limited, the BER performance at low values of SNR cannot be significantly lowered. Similarly to the adaptive modulation systems of Chapter 19, transmission blocking would have to be introduced in order to guarantee a target bit error rate. Let us now consider the combination of ARRNS coding and AOFDM modulation in the next section.

20.2.2 ARRNS/AOFDM Transceivers

In this section we will demonstrate that upon combining adaptive modulation with adaptive coding, the low SNR performance can be dramatically improved by amalgamating transmission blocking with adaptive error correction coding. We have advocated here the target BER adaptive modulation algorithm of Section 19.2.4.2 due to its high performance and easy adjustability to different target bit error rates.

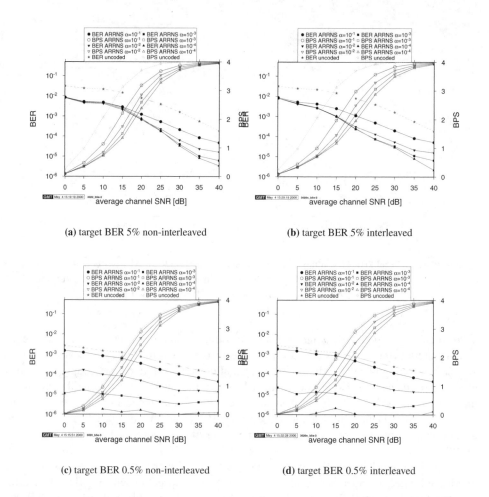

(a) target BER 5% non-interleaved   (b) target BER 5% interleaved

(c) target BER 0.5% non-interleaved   (d) target BER 0.5% interleaved

Figure 20.7: BER and BPS throughput versus average channel SNR for hard decision ARRNS-coded 512-subcarrier OFDM transmission employing adaptive modulation over the shortened WATM channel of Figure 17.3: (a, b) uncoded adaptive modulation target BER 5%, (c, d) uncoded target BER 0.5%, (a, c) non-interleaved, (b, d) interleaved. The stipulated WER values were $\alpha = 10^{-1}, 10^{-2}, 10^{-3}$ and 10^{-4}. The light grey curves show the uncoded BER and BPS throughput

The transmission parameter adaptation is performed in two steps. First, the modulation modes are allocated to the subcarriers according to the algorithm outlined in Section 19.2.4.2. Recall that for AOFDM the expected overall bit error probability for the given OFDM symbol was computed by averaging the individual subcarrier BERs for all legitimate modulation modes M_n, yielding $\bar{p}_e(M_n) = 1/N_s \sum_j p_e(\gamma_j, M_n)$ where γ_j is the subcarrier SNR for $j = 1, \ldots, 512$. The specific subcarrier modem mode allocation with the highest bits per symbol (BPS) throughput, whose estimated BER is lower than the required value, is then confirmed. According to this step, the number of bits N_b transmitted in the next OFDM symbol and their estimated bit error probabilities $p_e(n)$ become known. On the basis of this, the code rate adaptation algorithm calculates the residue error rates $p_r(r)$ from Equation 20.1, constructs the interleaver $I(r)$ for the correct number of residues and invokes the appropriate codec modes for the RRNS code words, as outlined above.

Figure 20.7 gives an overview of the ARRNS/AOFDM system's BER and throughput performance over the short WATM channel of Figure 17.3. Two target BER values have been stipulated, both with and without interleaving of the transmitted residues. Figure 20.7(b) and 20.7(a) portray the system's performance if a target BER of 5% is assumed for the adaptive modulation with and without interleaving, respectively. As has been observed in Chapter 19, the uncoded BER, represented by the grey curves, is lower than the AOFDM target BER, which is due to the operation of the AOFDM algorithm. It can be seen that the coded BER is below 1% for all simulated modem configurations, and that the SNR gain is much higher than for fixed QPSK transmission in Figure 20.2. The BER performance is limited, however, by the limited error correction capability of the RRNS (9, 3) mode when the SNR is very low.

Lowering the adaptive modulation's target BER to 0.5%, the BER of the system can be influenced over the whole SNR range by varying the WER α. Figures 20.7(d) and 20.7(c) depict the corresponding BER and BPS throughput. For a WER of $\alpha = 10^{-1}$ the achieved BER is better than $2 \cdot 10^{-3}$, while for $\alpha = 10^{-2}$ a BER of $2 \cdot 10^{-4}$ is never exceeded and for $\alpha = 10^{-3}$ a BER of $2 \cdot 10^{-5}$.

Comparing the interleaved performance with the non-interleaved results, it can be seen that the BER performance of comparable modems is fairly similar. The throughput is slightly higher, however, for the non-interleaved systems, demonstrating the efficiency of the hard decision ARRNS/AOFDM schemes in terms of combating the bursty errors of frequency selective fading. Let us now consider the effects of invoking soft decisions.

20.2.3 Soft-Decision Aided RRNS Decoding

The performance of the ARRNS codes can be increased by soft decision decoding [606]. Previously, the ARRNS decoder assumed that the outputs of the demodulator were binary hard decision values. However, the transceiver is capable of exploiting soft outputs provided by the demodulator at the receiver. Soft decoding of the ARRNS codes can be implemented by combining the classic Chase algorithm [610] with the hard decision ARRNS decoder.

Figure 20.8 shows the soft decision decoded performance of the AOFDM/ARRNS system. Comparison with Figure 20.7 shows an improved BER performance for the soft decoder. This is especially significant for the target bit error rate of 5%, where the soft decoded AOFDM/ARRNS system achieves BER values of below 0.3% at 0 dB SNR. Since the adaptation algorithm is unchanged, the same throughput is observed for the hard and soft decoded systems. A BER of below 10^{-4} was registered for the 0.5% target BER system for a WER of $\alpha = 10^{-2}$. Under these circumstances the interleaved system exhibits a lower throughput and worse BER performance than the non-interleaved system. Let us now compare the performance results to those of turbo BCH codecs in the next section.

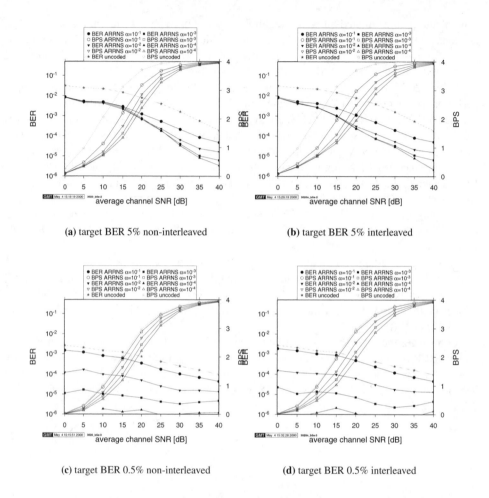

(a) target BER 5% non-interleaved   (b) target BER 5% interleaved

(c) target BER 0.5% non-interleaved   (d) target BER 0.5% interleaved

Figure 20.8: BER and BPS throughput versus average channel SNR for soft decision ARRNS-coded 512-subcarrier OFDM transmission employing adaptive modulation over the shortened WATM channel of Figure 17.3. (a, b) - uncoded adaptive modulation target BER 5%, (c, d) uncoded target BER 0.5%, (a, c) non-interleaved, (b, d) interleaved. The stipulated WER values were $\alpha = 10^{-1}, 10^{-2}, 10^{-3}$ and 10^{-4}. The light grey curves show the uncoded BER and BPS throughput. The corresponding hard decision results are plotted in Figure 20.7

20.3   Turbo BCH Codes

As an alternative to the above ARRNS codes, we have investigated the employment of turbo BCH codes for the adaptive modem. In Section 19.2.7, half-rate convolutional turbo codes were employed in the adaptive modem. However, the convolutional constituent codes can also be replaced by block codes [597,598], which have been shown for example by Hagenauer [611, 612] to perform impressively even at near-unity coding rates. Generally, block codes

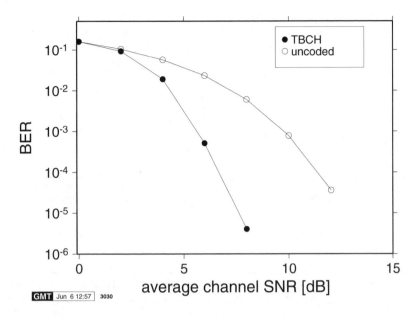

Figure 20.9: BER performance of the 0.72 rate $TBCH(36, 26)$ coded OFDM system over the AWGN channel, employing QPSK transmission in the subcarriers using 6 iterations and a 4×13 turbo interleaver

are appropriate for high code rates and convolutional codes for lower code rates, typically below 2/3, as argued by Hagenauer *et al.* [611]. Turbo block codes have since been studied in a variety of applications [203, 613–615].

For our adaptive coding system, a turbo (T) code with constituent $BCH(31, 26)$ coders was advocated. The parity bits of the two constituent codes were not punctured, resulting in a $TBCH(36, 26)$ code. The resulting code rate was therefore 0.72. The turbo interleaver employed was a 4×13 bit interleaver, leading to a block size of 52 uncoded or 72 coded bits.

Figure 20.9 shows the BER performance of the TBCH code over an AWGN channel with QPSK transmission using 6 iterations and a 4×13 turbo interleaver. Comparison with the AWGN performance of the hard decision decoded RRNS codes in Figure 20.1 reveals that the TBCH code outperforms all the investigated RRNS codes. The SNR gain of the 0.72 rate TBCH code at a BER of 10^{-4} is about 4.5 dB, which is 0.8 dB better than that of the 1/3 rate $(9, 3)$ RRNS code. We note, however, that the RRNS coding performance can be improved by invoking turbo coding. This issue is the subject of further research.

For transmission over the short WATM channel of Figure 17.3, a block channel interleaver obeying the structure depicted in Figure 20.3 was employed. In the case of the 0.72 rate $TBCH(26, 26)$ coder, the interleaving was performed bit-by-bit, instead of the residue-based interleaving of the RRNS system. Figure 20.10 shows the BER performance of the studied TBCH code over the short WATM channel of Figure 17.3. It can be seen that channel interleaving of the TBCH-coded bits significantly improves the performance of the codec. For a BER of 10^{-4} a SNR gain of 10 dB was recorded for the non-interleaved, while an SNR gain of 16.5 dB was observed for the interleaved system.

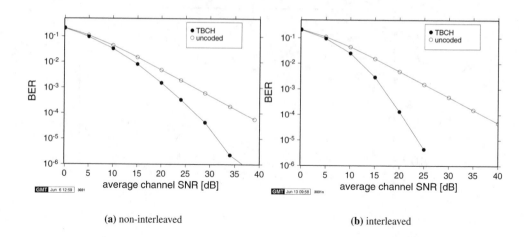

(a) non-interleaved   (b) interleaved

Figure 20.10: BER versus average channel SNR for the 0.72 rate $TBCH(36, 26)$ coded 512-subcarrier OFDM transmission employing QPSK over the short WATM channel of Figure 17.3 with and without channel interleaving, using 6 iterations and a 4×13 turbo interleaver

Figure 20.11 gives a comparison between the 0.72 rate TBCH performance of Figure 20.10 and the long block length half-rate turbo convolutional code of Figure 19.13 in E_b/N_0 terms. It can be seen that the TBCH code with additional channel interleaving performs only about 1 dB worse in E_b/N_0 than the convolutional turbo code at a BER of 10^{-4}.

20.3.1   Adaptive TBCH Coding

Unlike the ARRNS code rate adaptation algorithm, which employed four code rates, the proposed TBCH rate adaptation regime operates as a simple on/off switch. If the channel quality is lower than a threshold, then N_d data bits are encoded into N_c coded bits and transmitted in the bit positions allocated for the code word. If the channel quality is sufficiently high for uncoded transmission, then N_c data bits are directly mapped to the allocated bit positions.

Since binary BCH codes are employed, the estimated bit error rate in each code word can be used directly as the adaptation criterion. Based on the estimated bit error probability $p_e(n)$, the uncoded bit error probability in the code word w of length N_c can be calculated as:

$$\bar{p}_e(w) = \frac{1}{N_c} \cdot \sum_{n=0}^{N_c} p_e\left(I\left(n + n_0(w)\right)\right), \qquad (20.9)$$

where $I()$ is the channel interleaver, and $n_o(w)$ is the index of the first bit of the code word w.

The adaptation algorithm compares the estimated BER in the code word with a bit error rate threshold β, and if $\bar{p}_e(w) \leq \beta$ then N_c data bits are transmitted in the word w. Otherwise, coded transmission is selected.

Figure 20.12 shows the BER and throughput of the ATBCH modem employing QPSK over the short WATM channel of Figure 17.3 for BER thresholds of $\beta =$

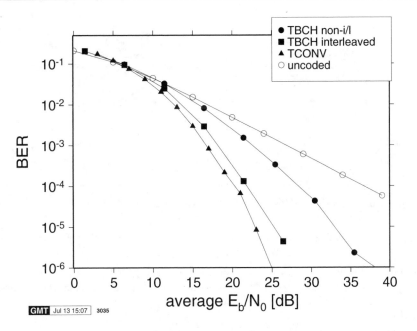

Figure 20.11: Comparison of the BER versus E_b/N_0 performance of the 0.72 rate $TBCH(36, 26)$ coded with and without channel interleaver and the $1/2$ rate turbo convolutional coded OFDM system with 2000-bit turbo interleaver of Section 19.2.7 (Figure 19.13) employing QPSK transmission over the short WATM channel of Figure 17.3

$10^{-2}, 10^{-3}$ and 10^{-4}. It can be seen that for both the interleaved and non-interleaved cases the BER curves follow the trends found in Figure 20.10 for low SNR values before the adaptation algorithm gradually increases the average code rate towards unity, approaching in a throughput of 2 BPS. Comparing the throughput of the interleaved and non-interleaved modems, it can be seen that the non-interleaved scheme yields a slightly higher throughput across the SNR range. Let us now combine ATBCH coding with AOFDM in the next section.

20.3.2   Joint ATBCH/AOFDM Algorithm

As in the ARRNS/AOFDM system of Section 20.2.2, the ATBCH code has been combined with the adaptive modulation scheme selection regime discussed in Section 19.2.4.2. The BER and BPS throughput performance of the ATBCH/AOFDM system is determined by the combination of the target BER for the adaptive modulation scheme and the BER threshold β for the code rate adaptation. Figure 20.13 depicts the resulting system performance for uncoded AOFDM target BER values of 10% and 1%, and for ATBCH activation threshold BER values of $\beta = 10^{-2}, 10^{-3}$ and 10^{-4}. For the uncoded AOFDM target BER of 10%, neither the interleaved nor the non-interleaved modems can reduce the data BER to acceptable levels for low SNR values. The interleaved system shows a better BER performance, but exhibits lower throughput than the non-interleaved system. For an uncoded AOFDM target BER of 1%, the ATBCH decoder can efficiently reduce the data BER below 10^{-4} for $\beta =$

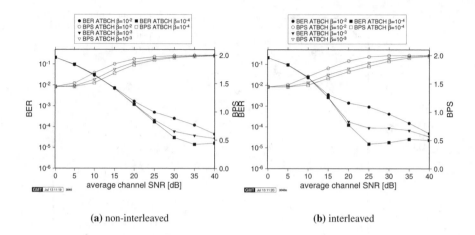

(a) non-interleaved   (b) interleaved

Figure 20.12: BER and BPS throughput versus average channel SNR for adaptive $TBCH(36, 26)$
coded 512-subcarrier OFDM transmission employing QPSK over the short WATM channel of Figure 17.3 with and without channel interleaving

10^{-4}. This meets the requirements of the adaptive modem's BER performance, as set out in
Chapter 19.

20.4   Signalling

Signalling the codec modes employed for the code words of each OFDM symbol requires
a significant amount of bandwidth, analogously to the modulation mode signalling for
AOFDM. The effects of erroneous mode detection at the receiver are the same as for the
modulation mode signalling, resulting in loss of synchronisation in the data stream.

Blind detection of the codec mode at the transmitter in the case of open-loop coding
adaptation could be employed for longer block-length codes, by tentatively decoding the
received data sequence and exploiting soft outputs of the decoder as a reliability measure. If
decoding of the received sequence increases the reliability measure for the received block's
data bits, then the corresponding coder mode is chosen. However, the short block lengths,
necessitated by the adaptation algorithms discussed in this chapter, make it impossible to use
blind detection of the codec mode based on the decoder's soft outputs.

The two different adaptive error correction schemes discussed in this chapter both require
signalling of the codec mode for each of the transmitted code words, but the scenarios are
different. The ARRNS scheme, which employs four codec modes with code words of variable
length between 24 and 72 bits, requires a signalling scheme similar to that of the AOFDM
scheme discussed in Section 19.2.6.1.

The ATBCH scheme, with only two different codec modes and a fixed block length, is
easier to signal. Depending on the required mode detection error rate, and assuming that the
achieved uncoded BER is the uncoded AOFDM target BER, then the number of signalling
bits necessary for majority voting signalling can be calculated. If a target BER of 1% is

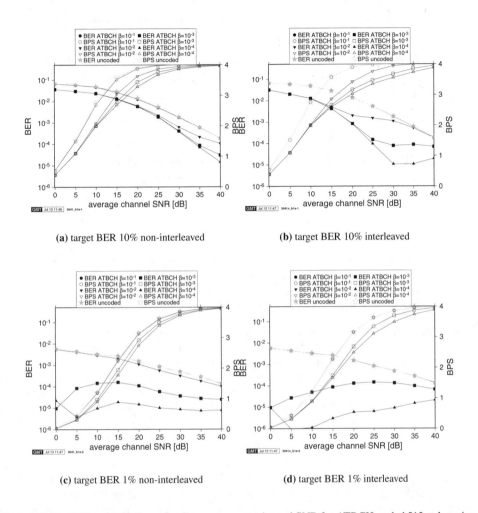

Figure 20.13: BER and BPS throughput versus average channel SNR for ATBCH-coded 512-subcarrier AOFDM over the shortened WATM channel of Figure 17.3: (a, b) - uncoded AOFDM target BER 10%, (c, d) target BER 1%, (a, c) non-interleaved (b, d) - interleaved. The stipulated BER threshold values were $\beta = 10^{-1}, 10^{-2}, 10^{-3}$ and 10^{-4}. The light grey curves show the uncoded BER and BPS throughput

assumed, then a 5-bit signalling sequence results in a detection error rate of 10^{-5}. If 7 signalling bits are used, then a DER of $3.5 \cdot 10^{-7}$ can be achieved.

20.5 Comparison of Coded Adaptive OFDM Schemes

Figure 20.14 shows the throughput of the various adaptive transmission systems studied for a target data BER of 10^{-4}. The lightly shaded curves represent the variable throughput systems' performance graphs from Figure 19.29. The ARRNS/AOFDM and ATBCH/AOFDM

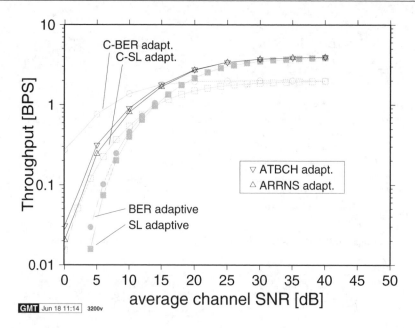

Figure 20.14: BPS throughput versus average channel SNR for AOFDM for a maximum data bit error rate of 10^{-4}. The light grey curves present the performance of the variable-throughput adaptive modulation schemes Figure 19.29 with and without convolutional turbo coding. The variable-throughput systems include convolutionally turbo coded (C-) and uncoded switching level adaptive (SL) and target BER adaptive (BER) systems, as well as the joint adaptive RRNS/AOFDM and ATBCH/AOFDM systems. The short WATM channel of Figure 17.3 was employed

systems employed no interleaving and had an AOFDM target BER of 1%. The ARRNS adaptation algorithm used a WER threshold of $\alpha = 10^{-2}$, while the ATBCH system used a data BER threshold of $\beta = 10^{-4}$. The pre-equalisation results from Figure 19.29 were omitted in this figure. It should be noted that the above performance figures do not take into account the signalling overhead required for adaptive modulation and adaptive code rates and hence constitute the upper bound performance of the system.

Let us now in the next section summarise our discourse on various OFDM schemes.

20.6 Summary

20.6.1 Summary of the OFDM-related Chapters in Part III

In the six OFDM-related chapters constituting Part III of the book we have discussed the implementational, algorithmic and performance aspects of orthogonal frequency division multiplexing in predominantly duplex mobile communications environments.

Following a rudimentary introduction to OFDM in Chapter 15, in Chapter 16 we have further studied the structure of an OFDM modem and we have investigated the problem of the high peak-to-mean power ratio observed for OFDM signals, and that of clipping amplification caused by insufficient amplifier back-off. We have investigated the BER performance and the spectrum of the OFDM signal in the presence of clipping, and we have seen that for an amplifier back-off of 6 dB the BER performance was indistinguishable from the perfectly amplified case. We have investigated the effects of quantisation of the time-domain OFDM signal. The effects of phase noise on the OFDM transmission were studied, and two phase noise models were suggested. One model was based on white phase noise, only relying on the integrated phase jitter, while a second model used coloured noise, which was generated from the phase noise mask.

In Chapter 17 we studied OFDM transmissions over time dispersive channels. The spectrum of the transmitted frequency-domain symbols is multiplied with the channel's frequency domain channel transfer function, hence the amplitude and phase of the received subcarriers are distorted. If the channel is varying significantly during each OFDM symbol's duration, then additional inter-subcarrier interference occurs, affecting the modem's performance. We have seen the importance of channel estimation on the performance of coherently detected OFDM, and we have studied two simple pilot-based channel estimation schemes. Differentially detected modulation can operate without channel estimation, but exhibits lower BER performance than coherent detection. We have seen that the signal-to-noise ratio is not constant across the OFDM symbol's subcarriers, and that this translates into a varying bit error probability across the different subcarriers.

The effects of timing- and frequency errors between transmitter and receiver were studied in Chapter 18. We saw that a timing error results in a phase rotation of the frequency domain symbols, and possibly inter-OFDM-symbol interference, while a carrier frequency error leads to inter-subcarrier-interference. We suggested a cyclic postamble, in order to suppress inter-OFDM-symbol interference for small timing errors, but we saw that frequency errors of more than 5% of the subcarrier distance lead to severe performance losses. In order to combat this, we investigated a set of frequency- and timing-error estimation algorithms. We suggested a time-domain based joint time- and frequency-error acquisition algorithm, and studied the performance of the resulting system over fading time dispersive channels.

Based on the findings of Chapter 17, we investigated adaptive modulation techniques to exploit the frequency diversity of the channel. Specifically, in Chapter 19, three adaptive modulation algorithms were proposed and their performance was investigated. The issue of signalling was discussed, and we saw that adaptive OFDM systems require a significantly higher amount of signalling information than adaptive serial systems. In order to limit the amount if signalling overhead, a sub-band adaptive scheme was suggested, and the performance trade-offs against a subcarrier-by-subcarrier adaptive scheme were discussed. Blind modulation mode detection schemes were investigated, and combined with an error correction decoder. We saw that by combining adaptive modulation techniques with a strong convolutional turbo channel codec significant system throughput improvements were achieved for low SNR values. Finally, frequency-domain pre-distortion techniques were investigated in order to pre-equalise the time-dispersive channel's transfer function. We observed that by incorporating pre-distortion in adaptive modulation significant throughput performance gains were achieved compared to adaptive modems without pre-equalisation.

We highlighted in Chapter 19 that although channel coding significantly improves the achievable throughput for low SNR values in adaptive modems, with increasing average channel quality the coding overhead limits the system's throughput. In order to combat this problem, in Chapter 20 we investigated coding schemes which offer readily adjustable code rates. RRNS and BCH codes were employed to implement an adaptive coding scheme allowing us to adjust the code rate across the subcarriers of an OFDM symbol. Combinations of adaptive coding and the adaptive modulation techniques introduced in Chapter 20 were studied, and we saw that a good compromise between the coded and uncoded transmission characteristics can be found. Let us now in the next subsection draw some conclusions.

20.6.2   Conclusions Concerning the OFDM Chapters in Part III

(1) Based on the implementation-oriented characterisation of OFDM modems, leading to a real-time testbed implementation and demonstration at 34 Mbps we concluded that OFDM is amenable to the implementation of high bit rate wireless ATM networks, which is underlined by the recent ratification of the HIPERLAN II standard.

(2) The range of proposed joint time- and frequency synchronisation algorithms efficiently supported the operation of OFDM modems in a variety of propagation environments, resulting in virtually no BER degradation in comparison to the perfectly synchronised modems. For implementation in the above-mentioned 34 Mbps real-time testbed simplified versions of these algorithms were invoked.

(3) Symbol-by-symbol adaptive OFDM substantially increases the BPS throughput of the system at the cost of moderately increased complexity. It was demonstrated in the context of an adaptive real-time audio system that this increased modem throughput can be translated in improved audio quality at a given channel quality.

(4) The proposed blind symbol-by-symbol adaptive OFDM modem mode detection algorithms were shown to be robust against channel impairments in conjunction with twin-mode AOFDM. However, it was necessary to combine it with higher-complexity channel coding based mode detection techniques, in order to maintain sufficient robustness, when using quadruple-mode AOFDM.

(5) The combination of frequency-domain pre-equalisation with AOFDM resulted in further performance benefits at the cost of a moderate increase in the peak-to-mean envelope flluctuation and system complexity.

(6) The combination of adaptive RRNS and turbo BCH FEC coding with AOFDM provided further flexibility for the system, in order to cope with hostile channel conditions, in particular in the low SNR region.

20.6.3   Suggestions for Further OFDM Research

In our investigations we occasionally assumed perfect knowledge of the channel's frequency-domain transfer function at the receiver and a co-channel interference-free environment. Clearly, these assumptions do not apply to a cellular communications system, and the adaptive schemes in the suggested form will not operate optimally in a co-channel interference limited cellular environment. Interference will generally render the channel non-reciprocal, and will make the channel estimation more arduous. The problem of the non-reciprocal channel could be circumvented by using a closed-loop adaptation system, as it was suggested in Chapter 19. Combining antenna arrays, interference cancellation or joint-detection techniques combined with space-time coding and channel estimation techniques in conjunction with the adaptive OFDM ideas of Chapters 19 and 20 may result in an adaptive OFDM scheme suitable for cellular systems. These issues constitute our current research.

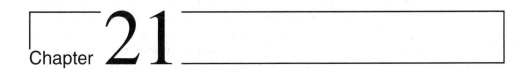

Chapter **21**

Space-Time Coded Versus Adaptive QAM-Aided OFDM

T.H. Liew and L. Hanzo[1]

21.1 Introduction

In [171] the encoding and decoding processes as well as the various design trade-offs of *space-time block codes* [616–618] were reviewed. More explicitly, various previously proposed space-time block codes [617, 618] have been discussed and their performance was investigated over perfectly interleaved, non-dispersive Rayleigh fading channels. A range of systems consisting of space-time block codes and different channel codecs were investigated. The performance versus estimated complexity trade-off of the different systems was investigated and compared.

In an effort to provide as comprehensive a technology road-map as possible and to identify the most promising schemes in the light of their performance versus estimated complexity, in this chapter we shall explore the family of *space-time trellis codes* [619–624], which were proposed by Tarokh *et al.* Space-time trellis codes incorporate jointly designed channel coding, modulation, transmit diversity and optional receiver diversity. The performance criteria for designing space-time trellis codes were outlined in [619], under the assumption that the channel is fading slowly and that the fading is frequency non-selective. It was shown in [619] that the system's performance is determined by matrices constructed from pairs of distinct code sequences. Both the *diversity gain* and *coding gain* of the codes are determined by the minimum rank and the minimum determinant [619, 625] of the matrices, respectively. The results were then also extended to fast fading channels. The space-time trellis codes proposed in [619] provide the best tradeoff between data rate, diversity advantage and trellis

---

[1]Space-Time Trellis Coding and Space-Time Block Coding versus Adaptive Modulation: An Overview and Comparative Study for Transmission over Wideband Channels, submitted to IEEE Tr. on Vehicular Technology, 2001 ©IEEE

complexity.

The performance of both space-time trellis and block codes over narrowband Rayleigh fading channels was investigated by numerous researchers [616, 618, 619, 623, 626]. The investigation of space-time codes was then also extended to the class of practical wideband fading channels. The effect of multiple paths on the performance of space-time trellis codes was studied in [624] for transmission over slowly varying Rayleigh fading channels. It was shown in [624] that the presence of multiple paths does not decrease the diversity order guaranteed by the design criteria used to construct the space-time trellis codes. The evidence provided in [624] was then also extended to rapidly fading dispersive and non-dispersive channels. As a further performance improvement, turbo equalisation was employed in [627] in order to mitigate the effects dispersive channels. However space-time coded turbo equalisation involved an enormous complexity. In addressing the complexity issues, Bauch and Al-Dhahir [628] derived finite-length multi-input multi-output (MIMO) channel filters and used them as prefilters for turbo equalisers. These prefilters significantly reduce the number of turbo equaliser states and hence mitigate the decoding complexity. As an alternative solution, the effect of Inter Symbol Interference (ISI) could be eliminated by employing Orthogonal Frequency Division Multiplexing (OFDM) [180]. A system using space-time trellis coded OFDM is attractive, since the decoding complexity reduced, as demonstrated by the recent surge of research interest [626, 629–631]. In [626, 629, 631], non-binary Reed-Solomon (RS) codes were employed in the space-time trellis coded OFDM systems for improving its performance.

Similarly, the performance of space-time block codes was also investigated over frequency selective Rayleigh fading channels. In [632], a multiple input multiple output equaliser was utilised for equalising the dispersive multipath channels. Furthermore, the advantages of OFDM were also exploited in space-time block coded systems [626, 633, 634].

We commence our discussion with a detailed description of the encoding and decoding processes of the space-time trellis codes in Section 21.2. The state diagrams of a range of other space-time trellis codes are also given in Section 21.2.3. In Section 21.3, a specific system is introduced, which enables the comparison of space-time trellis codes and space-time block codes over wideband channels. Our simulation results are then given in Section 21.4. We continue our investigations by employing space-time coded adaptive modulation based OFDM in Section 21.5. Finally, we conclude in Section 21.6.

21.2   Space-Time Trellis Codes

In this section, we will detail the encoding and decoding processes of space-time trellis codes. Space-time trellis codes are defined by the number of transmitters p, by the associated state diagram and the modulation scheme employed. For ease of explanation, as an example we shall use the simplest 4-state, 4-level Phase Shift Keying (4PSK) space-time trellis code, which has $p = 2$ two transmit antennas.

21.2.1   The 4-State, 4PSK Space-Time Trellis Encoder

At any time instant k, the 4-state 4PSK space-time trellis encoder transmits symbols $x_{k,1}$ and $x_{k,2}$ over the transmit antennas $Tx\,1$ and $Tx\,2$, respectively. The output symbols at time

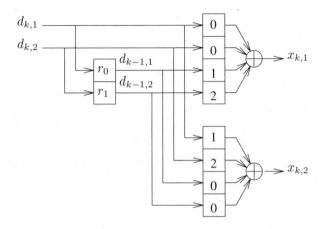

Figure 21.1: The 4-state, 4PSK space-time trellis encoder

Input queue	Instant k	Input bits $(d_{k,1}; d_{k,2})$	Shift register $(d_{k-1,1}; d_{k-1,2})$	State S_k	Transmitted symbols $(x_{k,1}; x_{k,2})$
00011110	0	- -	0 0	0	- -
000111	1	0 1(2)	0 0	0	0 2
0001	2	1 1(3)	0 1	2	2 3
00	3	1 0(1)	1 1	3	3 1
	4	0 0(2)	1 0	1	1 0
	5	- -	0 0	0	- -

Table 21.1: Operation of the space-time encoder of Figure 21.1

instant k are given by [619]:

$$x_{k,1} = 0.d_{k,1} + 0.d_{k,2} + 1.d_{k-1,1} + 2.d_{k-1,2} \qquad (21.1)$$
$$x_{k,2} = 1.d_{k,1} + 2.d_{k,2} + 0.d_{k-1,1} + 0.d_{k-1,2} \qquad (21.2)$$

where $d_{k,i}$ represents the current input bits, whereas $d_{k-1,i}$ the previous input bits and $i = 1, 2$. More explicitly, we can represent Equation 21.2 with the aid of a shift register, as shown in Figure 21.1, where $\oplus$ represents modulo 4 addition. Let us explain the operation of the shift register encoder for the random input data bits 01111000. The shift register stages r_0 and r_1 must be reset to zero before the encoding of a transmission frame starts. They represent the state of the encoder. The operational steps are summarised in Table 21.1. Again, given the register stages $d_{k-1,1}$ and $d_{k-1,2}$ as well as the input bits $d_{k,1}$ and $d_{k,2}$, the output symbols seen in the table are determined according to Equation 21.2 or Figure 21.1. Note that the last two binary data bits in Table 21.1 are intentionally set to zero in order to force the 4-state 4PSK trellis encoder back to the zero state which is common practice at the end of a transmission frame. Therefore, the transmit antenna Tx 1 will transmit symbols 0, 2, 3, 1. By contrast, symbols 2, 3, 1, 0 are then transmitted by the antenna Tx 2.

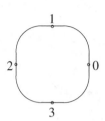

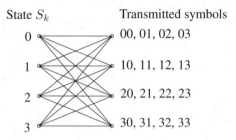

State S_k Transmitted symbols

Figure 21.2: The 4PSK constellation points

Figure 21.3: The 4-state, 4PSK space-time trellis code

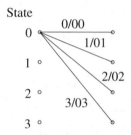

Figure 21.4: The trellis transitions from state $S_k = 0$ to various states

According to the shift register encoder shown in Figure 21.1, we can find all the legitimate subsequent states, which result in transmitting the various symbols $x_{k,1}$ and $x_{k,2}$, depending on a particular state of the shift register. This enables us to construct the state diagram for the encoder. The 4PSK constellation points are seen in Figure 21.2, while the corresponding state diagram of the 4-state 4PSK space-time trellis code [619] is shown in Figure 21.3. In Figure 21.3, we can see that for each current state there are four possible trellis transitions to the states $0, 1, 2$ and 3, which correspond to the legitimate input symbols of $0(d_{k,1} = 0, d_{k,2} = 0), 1(d_{k,1} = 1, d_{k,2} = 0), 2(d_{k,1} = 0, d_{k,2} = 1)$ and $3(d_{k,1} = 1, d_{k,2} = 1)$, respectively. Correspondingly, there are four sets of possible transmitted symbols associated with the four trellis transitions, shown at right of the state diagram. Each trellis transition is associated with two transmitted symbols, namely with x_1 and x_2, which are transmitted by the antennas Tx 1 and Tx 2, respectively. In Figure 21.4, we have highlighted the trellis transitions from state zero $S_k = 0$ to various states. The associated input symbols and the transmitted symbols of each trellis transitions are shown on top of each trellis transition. If the input symbol is 0, then the symbol $x_1 = 0$ will be sent by the transmit antenna Tx 1, and symbol $x_2 = 0$ by the transmit antenna Tx 2 as seen in Figure 21.4 or Figure 21.3. The next state remains $S_{k+1} = 0$. However, if the input symbol is 2 associated with $d_{k,1} = 0$, $d_{k,2} = 1$ in Table 21.1 then, the trellis traverses from state $S_k = 0$ to state $S_{k+1} = 2$ and the symbols $x_1 = 0$ and $x_2 = 2$ are transmitted over the antennas Tx 1 and Tx 2, respectively. Again, the encoder is required to be in the zero state both at the beginning and at the end of the encoding process.

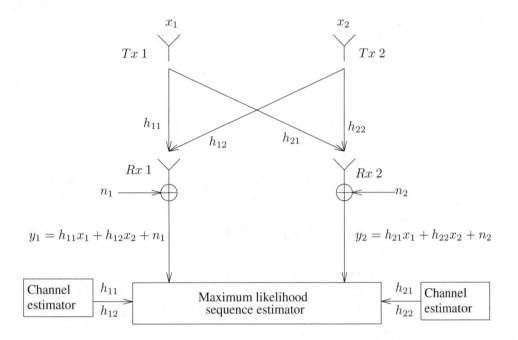

Figure 21.5: Baseband representation of the 4-state, 4PSK space-time trellis code using two receivers

21.2.2   The 4-State, 4PSK Space-Time Trellis Decoder

In Figure 21.5 we show the baseband representation of the 4-state, 4PSK space-time trellis code using two receivers. At any transmission instant, we have symbols x_1 and x_2 transmitted by the antennas $Tx\,1$ and $Tx\,2$, respectively. At the receivers $Rx\,1$ and $Rx\,2$, we would have:

$$y_1 = h_{11}x_1 + h_{12}x_2 + n_1 \qquad\qquad (21.3)$$
$$y_2 = h_{21}x_1 + h_{22}x_2 + n_2 \;, \qquad\qquad (21.4)$$

where h_{11}, h_{12}, h_{21} and h_{22} represent the corresponding complex time-domain channel transfer factors. Aided by the channel estimator, the Viterbi Algorithm based maximum likelihood sequence estimator [619] first finds the branch metric associated with every transition in the decoding trellis diagram, which is identical to the state diagram shown in Figure 21.3. For each trellis transition, we have two estimated transmit symbols, namely $\tilde{x}_1$ and $\tilde{x}_2$, for which the branch metric BM is given by:

$$
\begin{aligned}
BM & = \left| y_1 - h_{11}\tilde{x}_1 - h_{12}\tilde{x}_2 + y_2 - h_{21}\tilde{x}_1 - h_{22}\tilde{x}_2 \right|^2 \\
& = \sum_{i=1}^{2} \left| y_i - h_{i1}\tilde{x}_1 - h_{i2}\tilde{x}_2 \right|^2 \\
& = \sum_{i=1}^{2} \left| y_i - \sum_{j=1}^{2} h_{ij}\tilde{x}_j \right|^2 .
\end{aligned}
\tag{21.5}
$$

We can, however, generalise Equation 21.5 to p transmitters and q receivers, as follows:

$$
BM = \sum_{i=1}^{p} \left| y_i - \sum_{j=1}^{q} h_{ij}\tilde{x}_j \right|^2 .
\tag{21.6}
$$

When all the transmitted symbols were received and the branch metric of each legitimate transition was calculated, the maximum likelihood sequence estimator invokes the Viterbi Algorithm (VA) in order to find the maximum likelihood path associated with the best accumulated metric.

21.2.3 Other Space-Time Trellis Codes

In Section 21.2.1, we have shown the encoding and decoding process of the simple 4-state, 4PSK space-time trellis code. More sophisticated 4PSK space-time trellis codes were designed by increasing the number of trellis states [619], which are reproduced in Figures 21.6 to 21.8. With an increasing number of trellis states the number of tailing symbols required for terminating the trellis at the end of a transmitted frame is also increased. Two zero-symbols are needed to force the trellis back to state zero for the space-time trellis codes shown in Figures 21.6 and 21.7. By contrast, three zero-symbols are required for the space-time trellis code shown in Figure 21.8.

Space-time trellis codes designed for the higher-order modulation scheme of 8PSK were also proposed in [619]. In Figure 21.9, we showed the constellation points employed in [619]. The trellises of the 8-state, 16-state and 32-state 8PSK space-time trellis codes were reproduced from [619] in Figures 21.10, 21.11 and 21.12, respectively. One zero-symbol is required to terminate the 8-state, 8PSK space-time trellis code, whereas two zero-symbols are needed for both the 16-state and 32-state 8PSK space-time trellis codes.

21.3 Space-Time Coded Transmission over Wideband Channels

In Section 21.2, we have detailed the concept of space-time trellis codes. Let us now elaborate further by investigating the performance of space-time codes over dispersive wideband fading channels. As mentioned in Section 21.1, Bauch's approach [627, 628] of using turbo equalisation for mitigating the ISI exhibits a considerable complexity. Hence we argued that

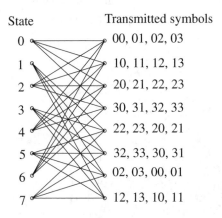

Figure 21.6: The 8-state, 4PSK space-time trellis code ©IEEE [619]

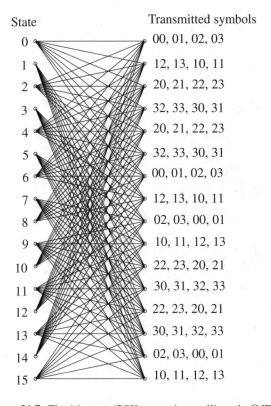

Figure 21.7: The 16-state, 4PSK space-time trellis code ©IEEE [619]

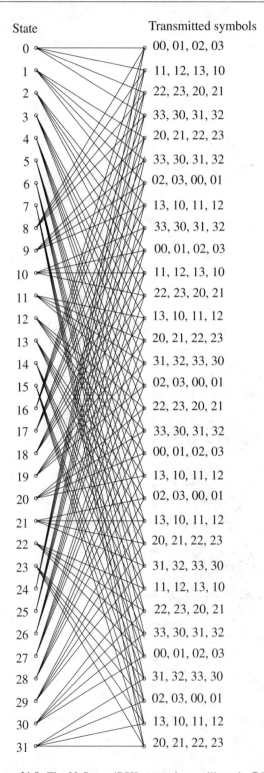

State   Transmitted symbols

0   00, 01, 02, 03
1   11, 12, 13, 10
2   22, 23, 20, 21
3   33, 30, 31, 32
4   20, 21, 22, 23
5   33, 30, 31, 32
6   02, 03, 00, 01
7   13, 10, 11, 12
8   33, 30, 31, 32
9   00, 01, 02, 03
10   11, 12, 13, 10
11   22, 23, 20, 21
12   13, 10, 11, 12
13   20, 21, 22, 23
14   31, 32, 33, 30
15   02, 03, 00, 01
16   22, 23, 20, 21
17   33, 30, 31, 32
18   00, 01, 02, 03
19   13, 10, 11, 12
20   02, 03, 00, 01
21   13, 10, 11, 12
22   20, 21, 22, 23
23   31, 32, 33, 30
24   11, 12, 13, 10
25   22, 23, 20, 21
26   33, 30, 31, 32
27   00, 01, 02, 03
28   31, 32, 33, 30
29   02, 03, 00, 01
30   13, 10, 11, 12
31   20, 21, 22, 23

Figure 21.8: The 32-State, 4PSK space-time trellis code ©IEEE [619]

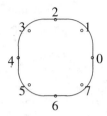

Figure 21.9: The 8PSK constellation points ©IEEE [619]

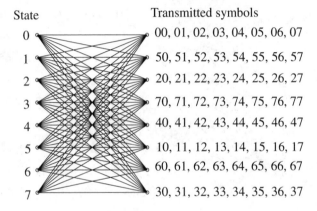

Figure 21.10: The 8-state, 8PSK space-time trellis code ©IEEE [619]

using space-time coded OFDM constitutes a more favourable approach to transmission over dispersive wireless channels, since the associated decoding complexity is significantly lower. Therefore, in this chapter OFDM is employed for mitigating the effects of dispersive channels.

It is widely recognised that space-time trellis codes [619] perform well at the cost of high complexity. However, Alamouti's G_2 space-time block code [616] could be invoked instead of space-time trellis codes. The space-time block code G_2 is appealing in terms of its simplicity, although there is a slight loss in performance. Therefore, we concatenate the space-time block code G_2 with Turbo Convolutional (TC) codes in order to improve the performance of the system. The family of TC codes was favoured, because it was shown in [635, 636] that TC codes achieve an enormous coding gain at a moderate complexity, when compared to convolutional codes, turbo BCH codes, trellis coded modulation and turbo trellis coded modulation. The performance of concatenated space-time block codes and TC codes will then be compared to that of space-time trellis codes. Conventionally, Reed-Solomon (RS) codes have been employed in conjunction with the space-time trellis codes [626,629,631] for improving the performance of the system. In our forthcoming discussion, we will concentrate on comparing the performance of space-time block and trellis codes in conjunction with various channel coders.

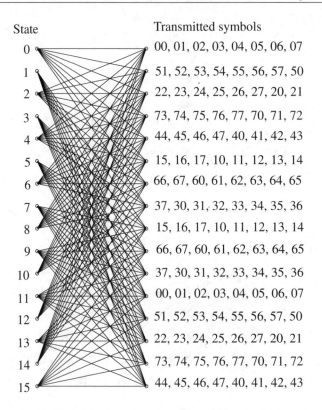

State   Transmitted symbols

0   00, 01, 02, 03, 04, 05, 06, 07
1   51, 52, 53, 54, 55, 56, 57, 50
2   22, 23, 24, 25, 26, 27, 20, 21
3   73, 74, 75, 76, 77, 70, 71, 72
4   44, 45, 46, 47, 40, 41, 42, 43
5   15, 16, 17, 10, 11, 12, 13, 14
6   66, 67, 60, 61, 62, 63, 64, 65
7   37, 30, 31, 32, 33, 34, 35, 36
8   15, 16, 17, 10, 11, 12, 13, 14
9   66, 67, 60, 61, 62, 63, 64, 65
10   37, 30, 31, 32, 33, 34, 35, 36
11   00, 01, 02, 03, 04, 05, 06, 07
12   51, 52, 53, 54, 55, 56, 57, 50
13   22, 23, 24, 25, 26, 27, 20, 21
14   73, 74, 75, 76, 77, 70, 71, 72
15   44, 45, 46, 47, 40, 41, 42, 43

Figure 21.11: The 16-State, 8PSK space-time trellis code ©IEEE [619]

21.3.1   System Overview

Figure 21.13 shows the schematic of the system employed in our performance study. At the transmitter, the information source generates random information data bits. The information bits are then encoded by TC codes, RS codes or left uncoded. The coded or uncoded bits are then channel interleaved, as shown in Figure 21.13. The output bits of the channel interleaver are then passed to the Space-Time Trellis (STT) or Space-Time Block (STB) encoder. We will investigate all the previously mentioned space-time trellis codes proposed by Tarokh, Seshadri and Calderbank in [619], where the associated state diagrams are shown in Figures 21.3, 21.6, 21.7, 21.10, 21.11 and 21.12. The modulation schemes employed are 4PSK as well as 8PSK and the corresponding trellis diagrams were shown in Figures 21.2 and 21.9, respectively. On the other hand, from the family of space-time block codes only Alamouti's $\mathbf{G}_2$ code is employed in the system, since it was shown in [636] that the best performance is achieved by concatenating the space-time block code $\mathbf{G}_2$ with TC codes. For convenience, the transmission matrix of the space-time block code $\mathbf{G}_2$ is reproduced here as follows:

$$\mathbf{G}_2 = \begin{pmatrix} x_1 & x_2 \\ -\bar{x}_2 & \bar{x}_1 \end{pmatrix}. \tag{21.7}$$

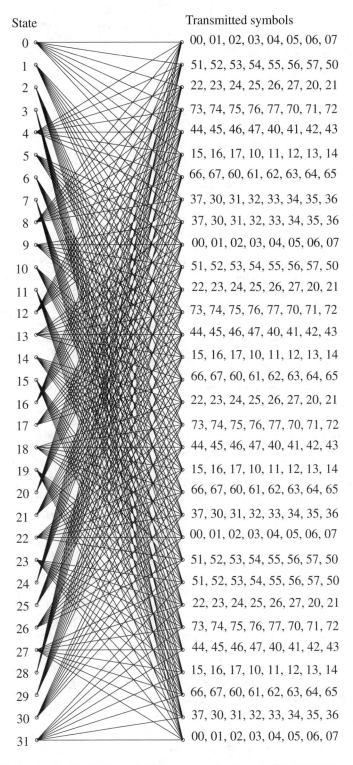

State — Transmitted symbols

State	Transmitted symbols
0	00, 01, 02, 03, 04, 05, 06, 07
1	51, 52, 53, 54, 55, 56, 57, 50
2	22, 23, 24, 25, 26, 27, 20, 21
3	73, 74, 75, 76, 77, 70, 71, 72
4	44, 45, 46, 47, 40, 41, 42, 43
5	15, 16, 17, 10, 11, 12, 13, 14
6	66, 67, 60, 61, 62, 63, 64, 65
7	37, 30, 31, 32, 33, 34, 35, 36
8	37, 30, 31, 32, 33, 34, 35, 36
9	00, 01, 02, 03, 04, 05, 06, 07
10	51, 52, 53, 54, 55, 56, 57, 50
11	22, 23, 24, 25, 26, 27, 20, 21
12	73, 74, 75, 76, 77, 70, 71, 72
13	44, 45, 46, 47, 40, 41, 42, 43
14	15, 16, 17, 10, 11, 12, 13, 14
15	66, 67, 60, 61, 62, 63, 64, 65
16	22, 23, 24, 25, 26, 27, 20, 21
17	73, 74, 75, 76, 77, 70, 71, 72
18	44, 45, 46, 47, 40, 41, 42, 43
19	15, 16, 17, 10, 11, 12, 13, 14
20	66, 67, 60, 61, 62, 63, 64, 65
21	37, 30, 31, 32, 33, 34, 35, 36
22	00, 01, 02, 03, 04, 05, 06, 07
23	51, 52, 53, 54, 55, 56, 57, 50
24	51, 52, 53, 54, 55, 56, 57, 50
25	22, 23, 24, 25, 26, 27, 20, 21
26	73, 74, 75, 76, 77, 70, 71, 72
27	44, 45, 46, 47, 40, 41, 42, 43
28	15, 16, 17, 10, 11, 12, 13, 14
29	66, 67, 60, 61, 62, 63, 64, 65
30	37, 30, 31, 32, 33, 34, 35, 36
31	00, 01, 02, 03, 04, 05, 06, 07

Figure 21.12: The 32-State, 8PSK space-time trellis code ©IEEE [619]

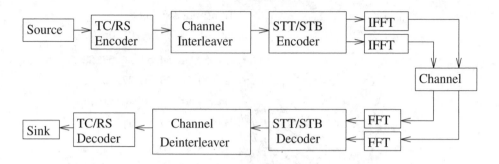

Figure 21.13: System overview

Different modulation schemes could be employed [370], such as Binary Phase Shift Keying (BPSK), Quadrature Phase Shift Keying (QPSK), 16-level Quadrature Amplitude Modulation (16QAM) and 64-level Quadrature Amplitude Modulation (64QAM). Gray-mapping of the bits to symbols was applied and this resulted in different protection classes in higher-order modulation schemes [180]. The mapping of the data bits and parity bits of the TC encoder was chosen such that it yielded the best achievable performance along with the application of the random separation channel interleaver [635]. The output of the space-time encoder was then OFDM [180] modulated and transmitted by the corresponding antenna. The number of transmit antennas was fixed to two, while the number of receive antennas constituted a design parameter. Dispersive wideband channels were used and the associated channels' profiles will be discussed later.

At the receiver the signal of each receive antenna is OFDM demodulated. The demodulated signals of the receiver antennas are then fed to the space-time trellis or space-time block decoder. The space-time decoders apply the MAP [582] or Log-MAP [637, 638] decoding algorithms for providing soft outputs for the channel decoders. If no channel codecs are employed in the system, the space-time decoders apply the VA [619, 639], which gives similar performance to the MAP decoder at a lower complexity. The decoded bits are finally passed to the sink for the calculation of the Bit Error Rate (BER) or Frame Error Rate (FER).

21.3.2   Space-Time and Channel Codec Parameters

In Figure 21.13, we have given an overview of the system studied. In this section, we present the parameters of the space-time codes and the channel codecs employed in the system. We will employ the set of various space-time trellis codes shown in Figures 21.3, 21.6, 21.7, 21.8, 21.10, 21.11 and 21.12. The associated space-time trellis coding parameters are summarised in Table 21.2. On the other hand, from the family of space-time block codes only Alamouti's G_2 code is employed, since we have shown in [636] that the best performance in the set of investigated schemes was yielded by concatenating the space-time block code G_2 with TC codes. The transmission matrix of the code is shown in Equation 21.7, while the number of transmitters used by the space-time block code G_2 is two, which is identical to the number of transmitters in the space-time trellis codes shown in Table 21.2.

Let us now briefly consider the TC channel codes used. In this chapter we will concen-

Modulation scheme	BPS	Decoding algorithm	No. of states	No. of transmitters	No. of termination symbols
4PSK	2	VA	4	2	1
			8	2	2
			16	2	2
			32	2	3
8PSK	3	VA	8	2	1
			16	2	2
			32	2	2

Table 21.2: Parameters of the space-time trellis codes shown in Figures 21.3, 21.6, 21.7, 21.8, 21.10, 21.11 and 21.12

Code	Octal generator polynomial	No. of states	Decoding algorithm	Puncturing pattern	No. of iterations
TC(2, 1, 3)	7,5	4	Log-MAP	10,01	8

Table 21.3: The associated parameters of the TC(2, 1, 3) code

trate on using the simple half-rate TC(2, 1, 3) code. Its associated parameters are shown in Table 21.3. As seen in Table 21.4, different modulation schemes are employed in conjunction with the concatenated space-time block code G_2 and the TC(2,1,3) code. Since the half-rate TC(2,1,3) code is employed, higher-order modulation schemes such as 16QAM and 64QAM were chosen, so that the throughput of the system remained the same as that of the system employing the space-time trellis codes without channel coding. It is widely recognised that the performance of TC codes improves upon increasing the turbo interleaver size and near-optimum performance can be achieved using large interleaver sizes exceeding 10,000 bits. However, this performance gain is achieved at the cost of high latency, which is impractical for a delay-sensitive real-time system. On the other hand, space-time trellis codes offer impressive coding gains [619] at low latency. The decoding of the space-time trellis codes is carried out on a transmission burst-by-burst basis. In order to make a fair comparison between the systems investigated, the turbo interleaver size was chosen so that all the coded bits were hosted by one transmission burst. This enables burst-by-burst turbo decoding at the receiver. Since we employ an OFDM modem, latency may also be imposed by a high number of subcarriers in an OFDM symbol. Therefore, the turbo interleaver size was increased, as the number of sub-carriers increased in our investigations. In Table 21.4, we summarised the modulation schemes and interleaver sizes used for different number of OFDM subcarriers in the system. The random separation based channel interleaver of [635] was used. The mapping of the data bits and parity bits into different protection classes of the higher-order modulation scheme was carried out so that the best possible performance was attained. This issue was addressed in [635].

Reed-Solomon codes were employed in conjunction with the space-time trellis codes.

Code	Code Rate R	Modulation Mode	BPS	Random turbo interleaver depth	Random separation interleaver depth
TC(2, 1, 3)	0.50	**128 carriers**			
		16QAM	2	256	512
		64QAM	3	384	768
		512 carriers			
		QPSK	1	512	1024
		16QAM	2	1024	2048

Table 21.4: The simulation parameters associated with the TC(2, 1, 3) code

Code	Galois Field	Rate	Correctable symbol errors
RS(105,51)	2^{10}	0.49	27
RS(153,102)	2^{10}	0.67	25

Table 21.5: The coding parameters of the Reed-Solomon codes employed

Hard decision decoding was utilised and the coding parameters of the Reed-Solomon codes employed are summarised in Table 21.5.

21.3.3 Complexity Issues

In this section, we will address the implementational complexity issues of the systems studied. We will, however, focus mainly on the relative complexity of the systems, rather than attempting to quantify their exact complexity. In order to simplify our comparative study, several assumptions were stipulated. In our simplified approach, the estimated complexity of the system is deemed to depend only on that of the space-time trellis decoder and turbo decoder. In other words, the complexity associated with the modulator, demodulator, space-time block encoder and decoder as well as that of the space-time trellis encoder and turbo encoder are assumed to be insignificant compared to the complexity of space-time trellis decoder and turbo decoder.

In [636], we have detailed our complexity estimates for the TC decoder and the reader is referred to the paper for further details. The estimated complexity of the TC decoder is assumed to depend purely on the number of trellis transitions per information data bit and this simple estimated complexity measure was also used in [636] as the basis of our comparisons. Here, we adopt the same approach and evaluate the estimated complexity of the space-time trellis decoder on the basis of the number of trellis transitions per information data bit.

In Figures 21.3, 21.6, 21.7, 21.8, 21.10, 21.11 and 21.12, we have shown the state diagrams of the 4PSK and 8PSK space-time trellis codes. From these state diagrams, we can see that the number of trellis transitions leaving each state is equivalent to 2^{BPS}, where BPS denotes the number of transmitted bits per modulation symbol. Since the number of information bits is equal to BPS, we can approximate the complexity of the space-time trellis decoder as:

$$comp\{STT\} = \frac{2^{BPS} \times \text{No. of States}}{BPS}$$

$$= 2^{BPS-1} \times \text{No. of States} . \qquad (21.8)$$

Applying Equation 21.8 and assuming that the Viterbi decoding algorithm was employed, we tabulated the approximated complexities of the space-time trellis decoder in Table 21.6.

Modulation scheme	BPS	No. of states	Complexity
4PSK	2	4	8
		8	16
		16	32
		32	64
8PSK	3	8	21.33
		16	42.67
		32	85.33

Table 21.6: Estimated complexity of the space-time trellis decoders shown in Figures 21.3, 21.6, 21.7, 21.8, 21.10, 21.11 and 21.12

21.4 Simulation Results

In this section, we will present our simulation results characterising the OFDM-based system investigated. As mentioned earlier, we will investigate the system's performance over dispersive wideband Rayleigh fading channels. We will commence our investigations using a simple two-ray channel impulse response (CIR) having equal tap weights, followed by a more realistic Wireless Asynchronous Transfer Mode (WATM) channel [180]. The CIR of the two-ray model is shown in Figure 21.14. From the figure we can see that the reflected path has the same amplitude as the Line Of Sight (LOS) path, although arriving $5\mu s$ later. However, in our simulations we also present results over two-ray channels separated by various delay spreads, up to $40\mu s$. Jakes' model [381] was adapted for modelling the fading channels. In Figure 21.15, we portray the 128-subcarrier OFDM symbol employed, having a guard period of $40\mu s$. The guard period of $40\mu s$ or cyclic extension of 32 samples was employed to overcome the inter-OFDM symbol interference due to the channel's memory.

In order to obtain our simulation results, several assumptions were stipulated:

- The average signal power received from each transmitter antenna was the same;

- All multipath components undergo independent Rayleigh fading;

- The receiver has a perfect knowledge of the CIR.

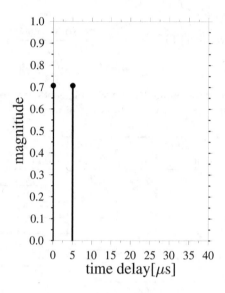

Figure 21.14: Two-ray channel impulse response having equal amplitudes

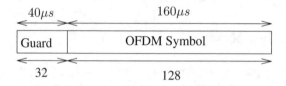

Figure 21.15: Stylised plot of 128-subcarrier OFDM time-domain signal using a cyclic extension of 32 samples

We note that the above assumptions are unrealistic, yielding the best-case performance, nonetheless, facilitating the performance comparison of the various techniques under identical circumstances.

21.4.1   Space-Time Coding Comparison – Throughput of 2 BPS

In Figure 21.16, we show our frame error rate (FER) performance comparison between 4PSK space-time trellis codes and the space-time block code G_2 concatenated with the TC(2,1,3) code using one receiver and the 128-subcarrier OFDM modem. The CIR had two equal-power rays separated by a delay spread of $5\mu s$ and the maximum Doppler frequency was 200 Hz. The TC(2, 1, 3) code is a half-rate code and hence 16QAM was employed, in order to support the same 2 BPS throughput, as the 4PSK space-time trellis codes using no channel

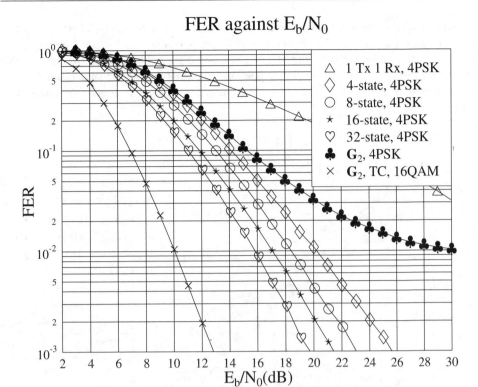

Figure 21.16: **FER** performance comparison between various 4PSK space-time trellis codes and the space-time block code $\mathbf{G}_2$ concatenated with the TC(2,1,3) code using **one receiver** and the 128-subcarrier OFDM modem over a channel having a CIR characterised by two equal-power rays separated by a delay spread of $5\mu s$. The maximum Doppler frequency was 200 Hz. The effective throughput was **2 BPS** and the coding parameters are shown in Tables 21.2, 21.3 and 21.4

codes. We can clearly see that at FER=10^{-3} the performance of the concatenated scheme is at least 7 dB better than that of the space-time trellis codes.

The performance of the space-time block code $\mathbf{G}_2$ without TC(2, 1, 3) channel coding is also shown in Figure 21.16. It can be seen in the figure that the space-time block code $\mathbf{G}_2$ does not perform well, exhibiting a residual BER. Moreover, at high E_b/N_0 values, the performance of the single-transmitter, single-receiver system is better than that of the space-time block code $\mathbf{G}_2$. This is because the assumption that the fading is constant over the two consecutive transmission instants is no longer valid in this situation. Here, the two consecutive transmission instants are associated with two adjacent subcarriers in the OFDM symbol and the fading variation is relatively fast in the frequency domain. Therefore, the orthogonality of the space-time code has been destroyed by the frequency-domain variation of the fading envelope. At the receiver, the combiner can no longer separate the two different transmitted signals, namely x_1 and x_2. More explicitly, the signals interfere with each other. The increase in SNR does not improve the performance of the space-time block code G_2, since this also increases the power of the interfering signal. We will address this issue more explicitly in Section 21.4.4. By contrast, the TC(2, 1, 3) channel codec succeeds in overcoming this problem. However, we will show later in Section 21.4.4 that the concatenated channel coded scheme exhibits the same residual BER problem, if the channel's variation becomes more rapid.

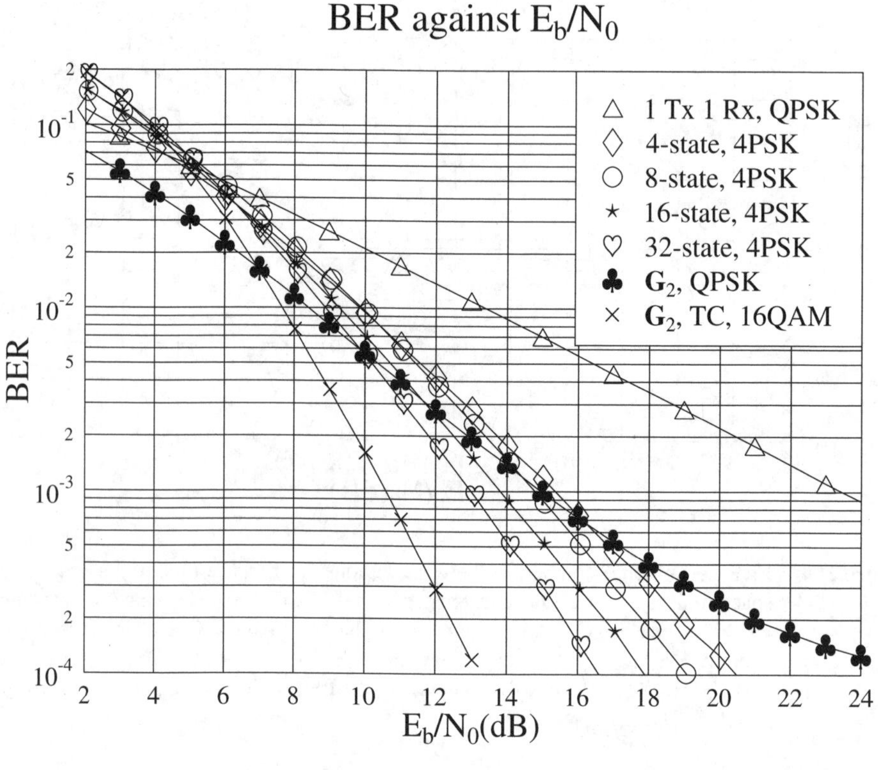

BER against E_b/N_0

Figure 21.17: **BER** performance comparison between various 4PSK space-time trellis codes and the space-time block code G_2 concatenated with the TC(2,1,3) code using **one receiver** and the 128-subcarrier OFDM modem over a channel having a CIR characterised by two equal-power rays separated by a delay spread of $5\mu s$. The maximum Doppler frequency was 200 Hz. The effective throughput was **2 BPS** and the coding parameters are shown in Tables 21.2, 21.3 and 21.4

In Figure 21.17, we provide the corresponding BER performance comparison between the 4PSK space-time trellis codes and the space-time block code G_2 concatenated with the TC(2,1,3) code using one receiver and the 128-subcarrier OFDM modem over a channel characterised by two equal-power rays separated by a delay spread of $5\mu s$ and having a maximum Doppler frequency of 200 Hz. Again, we show in the figure that the 2 BPS throughput concatenated G_2/TC(2,1,3) scheme outperforms the 2 BPS space-time trellis codes using no channel coding. At a BER of 10^{-4}, the concatenated channel coded scheme is at least 2 dB superior in SNR terms to the space-time trellis codes using no channel codes. At high E_b/N_0 values, the space-time block code $\mathbf{G}_2$, again exhibits a residual BER. On the other hand, at low E_b/N_0 values the latter outperforms the concatenated $\mathbf{G}_2$/TC(2,1,3) channel coded scheme as well as the space-time trellis codes using no channel coding.

FER against E_b/N_0

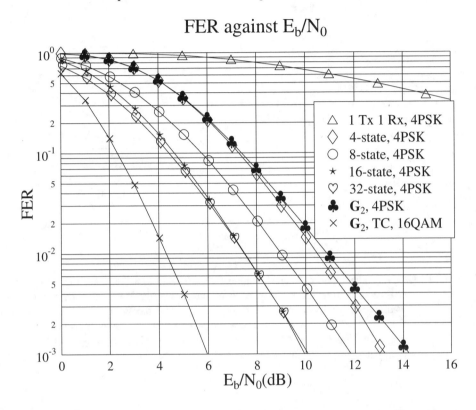

Figure 21.18: FER performance comparison between various 4PSK space-time trellis codes and the space-time block code $\mathbf{G}_2$ concatenated with the TC(2,1,3) code using **two receivers** and the 128-subcarrier OFDM modem over a channel having a CIR characterised by two equal-power rays separated by a delay spread of $5\mu s$. The maximum Doppler frequency was 200 Hz. The effective throughput was **2 BPS** and the coding parameters are shown in Tables 21.2, 21.3 and 21.4

Following the above investigations, the number of receivers was increased to two. In Figure 21.18, we show our FER performance comparison between the various 4PSK space-time trellis codes and the space-time block code $\mathbf{G}_2$ concatenated with the TC(2,1,3) code using two receivers and the 128-subcarrier OFDM modem. As before, the CIR had two equal-power rays separated by a delay spread of $5\mu s$. Again, we can see that the concatenated $\mathbf{G}_2$/TC(2,1,3) channel coded scheme outperforms the space-time trellis codes using no channel coding. However, the associated difference is lower and at a FER of 10^{-3} the concatenated channel coded scheme is about 4 dB better in E_b/N_0 terms than the space-time trellis codes using no channel codes. On the other hand, by employing two receivers the per-

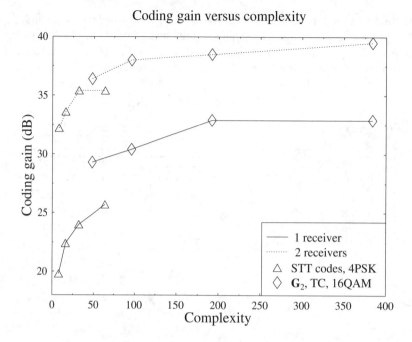

Figure 21.19: Coding gain versus estimated complexity for the various 4PSK space-time trellis codes and the space-time block code G_2 concatenated with the TC(2,1,3) code **using one as well as two receivers** and the 128-subcarrier OFDM modem over a channel having a CIR characterised by two equal-power rays separated by a delay spread of $5\mu s$. The maximum Doppler frequency was 200 Hz. The effective throughput was **2 BPS** and the coding parameters are shown in Tables 21.2, 21.3 and 21.4

formance of the space-time block code G_2 improved and the performance flattening effect happens at a lower FER.

In [636] and 21.3.3, we have derived the complexity estimates of the TC decoders and space-time trellis decoders, respectively. By employing Equation 21.8 and equations in [636], we compare the performance of the schemes studied, while considering their approximate complexity. Our performance comparison of the various schemes was carried out on the basis of the coding gain defined as the E_b/N_0 difference, expressed in decibels (dB), at FER= 10^{-3} between the schemes studied and the uncoded single-transmitter, single-receiver system having the same throughput of 2 BPS. In Figure 21.19, we show our coding gain versus estimated complexity comparison for the various 4PSK space-time trellis codes and the space-time block code G_2 concatenated with the TC(2,1,3) code using one as well as two receivers. The 128-subcarrier OFDM modem transmitted over the channel having a CIR of two equal-power rays separated by a delay spread of $5\mu s$ and a maximum Doppler frequency of 200 Hz. The estimated complexity of the space-time trellis codes was increased by increasing the number of trellis states. By contrast, the estimated complexity of the TC(2, 1, 3) code was increased by increasing the number of turbo iterations. The coding gain of the concatenated G_2/TC(2,1,3) scheme using one, two, four and eight iterations is shown in Figure 21.19. It

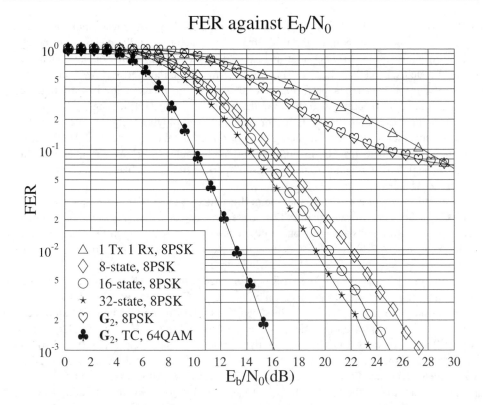

Figure 21.20: **FER** performance comparison between various 8PSK space-time trellis codes and the
space-time block code G_2 concatenated with the TC(2,1,3) code using **one receiver** and
the 128-subcarrier OFDM modem over a channel having a CIR characterised by two
equal-power rays separated by a delay spread of $5\mu s$. The maximum Doppler frequency
was 200 Hz. The effective throughput was **3 BPS** and the coding parameters are shown
in Tables 21.2, 21.3 and 21.4

can be seen that the concatenated scheme outperforms the space-time trellis codes using no
channel coding, even though the number of turbo iterations was only one. Moreover, the
improvement in coding gain was obtained, at an estimated complexity comparable to that of
the 32-state 4PSK space-time trellis code using no channel coding. From the figure we can
also see that the performance gain of the concatenated G_2/TC(2,1,3) channel coded scheme
over the space-time trellis codes becomes lower, when the number of receivers is increased
to two.

21.4.2   Space-Time Coding Comparison – Throughput of 3 BPS

In Figure 21.20, we show our FER performance comparison between the various 8PSK
space-time trellis codes of Table 21.2 and space-time block code G_2 concatenated with the
TC(2,1,3) code using one receiver and the 128-subcarrier OFDM modem. The CIR exhibited
two equal-power rays separated by a delay spread of $5\mu s$ and a maximum Doppler frequency

of 200 Hz. Since the TC$(2, 1, 3)$ scheme is a half-rate code, 64QAM was employed in order to ensure the same 3 BPS throughput, as the 8PSK space-time trellis codes using no channel coding. We can clearly see that at FER $= 10^{-3}$ the performance of the concatenated channel coded scheme is at least 7 dB better in terms of the required E_b/N_0 than that of the space-time trellis codes. The performance of the space-time block code $\mathbf{G}_2$ without the concatenated TC$(2, 1, 3)$ code is also shown in the figure. In Table 21.4, we can see that although there is an increase in the turbo interleaver size, due to employing a higher-order modulation scheme, nonetheless, no performance gain is observed for the concatenated TC$(2, 1, 3)$-$\mathbf{G}_2$ scheme over the space-time trellis codes using no channel coding. We speculate that this is because the potential gain due to the increased interleaver size has been offset by the vulnerable 64QAM scheme.

We also show in Figure 21.20 that the performance of the 3 BPS 8PSK space-time block code $\mathbf{G}_2$ without the concatenated TC$(2, 1, 3)$ scheme is worse than that of the other schemes investigated. It exhibits the previously noted flattening effect, which becomes more pronounced near FER$= 10^{-1}$. The same phenomenon was observed near FER$= 10^{-2}$ for the corresponding $\mathbf{G}_2$-coded 4PSK scheme, which has a throughput of 2 BPS.

In Figure 21.21, we portray our BER performance comparison between the various 8PSK space-time trellis codes and the space-time block code $\mathbf{G}_2$ concatenated with the TC(2,1,3) scheme using one receiver and the 128-subcarrier OFDM modem. The CIR exhibited two equal-power rays separated by a delay spread of $5\mu s$ and the maximum Doppler frequency was 200 Hz. Again, we observe in the figure that the concatenated $\mathbf{G}_2$/TC(2,1,3)-coded scheme outperforms the space-time trellis codes using no channel coding. At a BER of 10^{-4}, the concatenated scheme is at least 2 dB better in terms of its required E_b/N_0 value than the space-time trellis codes. The performance of the space-time block code $\mathbf{G}_2$ without TC(2,1,3) channel coding is also shown in Figure 21.21. As before, at high E_b/N_0 values, the space-time block code $\mathbf{G}_2$ exhibits a flattening effect. On the other hand, at low E_b/N_0 values it outperforms the concatenated $\mathbf{G}_2$/TC(2,1,3) scheme as well as the space-time trellis codes.

In Figure 21.22, we compare the FER performance of the 8PSK space-time trellis codes and the space-time block code $\mathbf{G}_2$ concatenated with the TC(2,1,3) channel codec using two receivers and the 128-subcarrier OFDM modem. As before, the CIR has two equal-power rays separated by a delay spread of $5\mu s$ and exhibits maximum Doppler frequency of 200 Hz. Again, with the increase in the number of receivers the performance gap between the concatenated channel coded scheme and the space-time trellis codes using no channel coding becomes smaller. At a FER of 10^{-3} the concatenated channel coded scheme is only about 2 dB better in terms of its required E_b/N_0 than the space-time trellis codes using no channel coding.

With the increase in the number of receivers, the previously observed flattening effect of the space-time block code $\mathbf{G}_2$ has been substantially mitigated, dipping to values below FER$= 10^{-3}$. However, it can be seen in Figure 21.22 that its performance is about 10 dB worse than that of the 8-state 8PSK space-time trellis code. In the previous system characterised in Figure 21.18, which had an effective throughput of 2 BPS, the performance of the space-time block code $\mathbf{G}_2$ was only about 1 dB worse in E_b/N_0 terms than that of the 4-state 4PSK space-time trellis code, when the number of receivers was increased to two. This observation clearly shows that higher-order modulation schemes have a tendency to saturate the channel's capacity and hence result in a poorer performance than the identical-throughput

BER against E_b/N_0

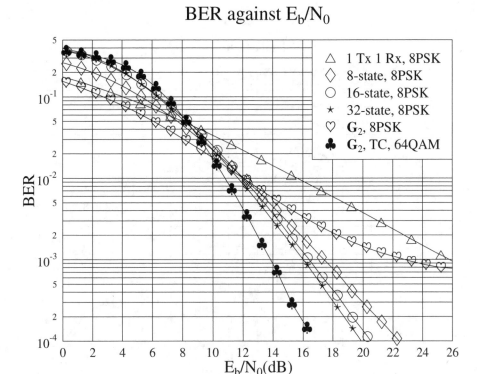

Figure 21.21: BER performance comparison between various 8PSK space-time trellis codes and the space-time block code $\mathbf{G}_2$ concatenated with the TC(2,1,3) code using **one receiver** and the 128-subcarrier OFDM modem over a channel having a CIR characterised by two equal-power rays separated by a delay spread of $5\mu s$. The maximum Doppler frequency was 200 Hz. The effective throughput was **3 BPS** and the coding parameters are shown in Tables 21.2, 21.3 and 21.4

space-time trellis codes using no channel coding.

Similarly to the 2 BPS schemes of Figure 21.19, we compare the performance of the 3 BPS throughput schemes by considering their approximate decoding complexity. The derivation of the estimated complexity has been detailed in [636] and in Section 21.3.3. As mentioned earlier, the performance comparison of the various schemes was made on the basis of the coding gain defined as the E_b/N_0 difference, expressed in decibels, at a FER= 10^{-3} between the schemes investigated and the uncoded single-transmitter, single-receiver system having a throughput of 3 BPS. In Figure 21.23, we show the associated coding gain versus estimated complexity curves for the 8PSK space-time trellis codes using no channel coding and the space-time block code $\mathbf{G}_2$ concatenated with the TC(2,1,3) code using one and two receivers and the 128-subcarrier OFDM modem. For the sake of consistency, the CIR, again exhibited two equal-power rays separated by a delay spread of $5\mu s$ and a maximum Doppler frequency of 200 Hz. Again, the estimated complexity of the space-time trellis codes was

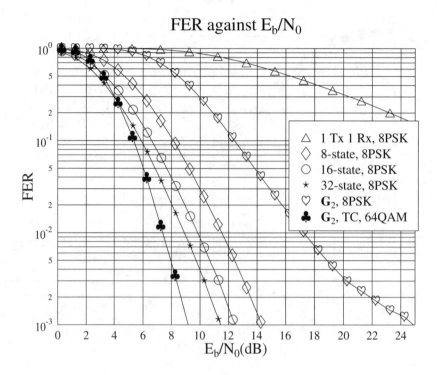

Figure 21.22: **FER** performance comparison between various 8PSK space-time trellis codes and the
space-time block code G_2 concatenated with the TC(2,1,3) code using **two receivers**
and the 128-subcarrier OFDM modem over a channel having a CIR characterised by two
equal-power rays separated by a delay spread of $5\mu s$. The maximum Doppler frequency
was 200 Hz. The effective throughput was **3 BPS** and the coding parameters are shown
in Tables 21.2, 21.3 and 21.4

increased by increasing the number of states. On the other hand, the estimated complexity of
the TC(2, 1, 3) code was increased by increasing the number of iterations. The coding gain of
the concatenated channel coded scheme invoking one, two, four and eight iterations is shown
in Figure 21.23. Previously in Figure 21.19 we have shown that the concatenated TC(2,1,3)-
coded scheme using one iteration outperformed the space-time trellis codes using no channel
coding. However, in Figure 21.23 the concatenated scheme does not exhibit the same performance trend. For the case of one receiver, the concatenated scheme using one iteration has a
negative coding gain and exhibits a saturation effect. This is again due to the employment of
the high-order 64QAM scheme, which has a preponderance to exceed the channel's capacity.
Again, we can also see that the performance gain of the concatenated G_2/TC(2,1,3)-coded
scheme over the space-time trellis codes using no channel coding becomes smaller, when
the number of receivers is increased to two. Having studied the performance of the various

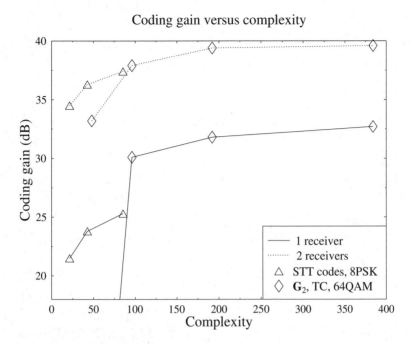

Figure 21.23: Coding gain versus estimated complexity for the various 8PSK space-time trellis codes and the space-time block code $\mathbf{G}_2$ concatenated with the TC(2,1,3) code **using one and two receivers** and the 128-subcarrier OFDM modem over a channel having a CIR characterised by two equal-power rays separated by a delay spread of $5\mu s$. The maximum Doppler frequency was 200 Hz. The effective throughput was **3 BPS** and the coding parameters are shown in Tables 21.2, 21.3 and 21.4

schemes over the channel characterised by the two-path, 5μ-dispersion CIR at a fixed Doppler frequency of 200Hz, let us in the next section study the effects of varying the Doppler frequency.

21.4.3 The Effect of Maximum Doppler Frequency

In our further investigations we have generated the FER versus E_b/N_0 curves similar to those in Figure 21.16, when the Doppler frequency was fixed to 5, 10, 20, 50 and 100 Hz. In order to present these results in a compact form, we then extracted the required E_b/N_0 values for maintaining a FER of 10^{-3}. In Figure 21.24, we show the E_b/N_0 crossing point at FER = 10^{-3} versus the maximum Doppler frequency for the 32-state 4PSK space-time trellis code using no channel coding and for the space-time block code $\mathbf{G}_2$ concatenated with the TC(2,1,3) code using one receiver and the 128-subcarrier OFDM modem. As before, the CIR exhibited two equal-power rays separated by a delay spread of $5\mu s$. We conclude from the near-horizontal curves shown in the figure that the maximum Doppler frequency does not significantly affect the performance of the space-time trellis codes and the concatenated scheme. Furthermore, the performance of the concatenated scheme is always better than that

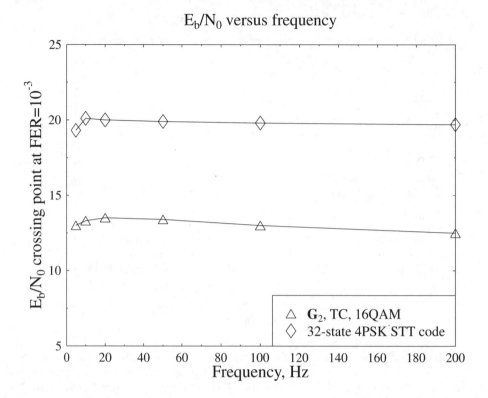

Figure 21.24: The E_b/N_0 value required for maintaining FER $= 10^{-3}$ versus the maximum Doppler
frequency for the 32-state 4PSK space-time trellis code and for the space-time block code
G_2 concatenated with the TC(2,1,3) code using **one receiver** and the 128-subcarrier
OFDM modem. The CIR exhibited two equal-power rays separated by a delay spread
of $5\mu s$. The effective throughput was **3 BPS** and the coding parameters are shown in
Tables 21.2, 21.3 and 21.4

of the space-time trellis codes using no channel coding. Having studied the effects of various
Doppler frequencies, let us now consider the impact of varying the delay spread.

21.4.4   The Effect of Delay Spreads

In this section, we will study how the variation of the delay spread between the two paths
of the channel affects the system performance. By varying the delay spread, the channel's
frequency-domain response varies as well. In Figure 21.25, we show the fading amplitude
variation of the 128 subcarriers in an OFDM symbol for a delay spread of (a) $5\mu s$, (b) $10\mu s$,
(c) $20\mu s$ and (d) $40\mu s$. It can be seen from the figure that the fading amplitudes vary more
rapidly, when the delay spread is increased. For the space-time block code G_2 the fading envelopes of the two consecutive transmission instants of antennas Tx 1 and Tx 2 are
assumed to be constant [616]. In Figure 21.25 (d), we can see that the variation of the
frequency-domain fading amplitudes is so dramatic that we can no longer assume that the

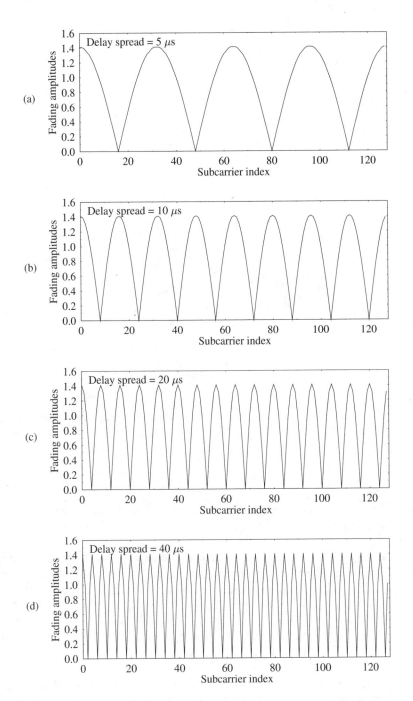

Figure 21.25: Frequency-domain fading amplitudes of the 128 subcarriers in an OFDM symbol for a delay spread of (a) $5\mu s$, (b) $10\mu s$, (d) $20\mu s$ and (c) $40\mu s$

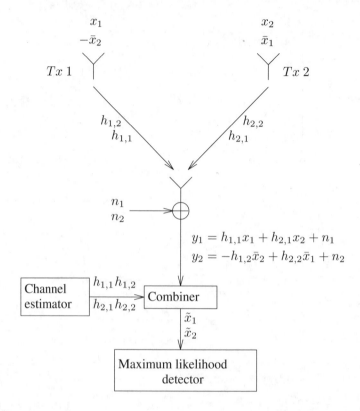

Figure 21.26: Baseband representation of the simple twin-transmitter space-time block code $\mathbf{G}_2$ of Equation 21.7 using one receiver over varying fading conditions

fading envelopes are constant for two consecutive transmission instants. The variation of the frequency-domain fading envelope will eventually destroy the orthogonality of the space-time block code $\mathbf{G}_2$.

We show in Figure 21.26 that the two transmission instants are no longer assumed to be associated with the same complex transfer function values. The figure shows the baseband representation of the simple twin-transmitter space-time block code $\mathbf{G}_2$ of Equation 21.7 using one receiver. At the receiver, we have

$$y_1 = h_{1,1}x_1 + h_{2,1}x_2 + n_1 \tag{21.9}$$

$$y_2 = -h_{1,2}\bar{x}_2 + h_{2,2}\bar{x}_1 + n_2 , \tag{21.10}$$

where y_1 is the first received signal and y_2 is the second. Both signals y_1 and y_2 are passed to the combiner in order to extract the signals x_1 and x_2. Aided by the channel estimator, which in this example provides perfect estimation of the diversity channels' frequency-domain transfer functions, the combiner performs simple signal processing in order to separate the signals x_1 and x_2. Specifically, in order to extract the signal x_1, it combines signals

y_1 and y_2 as follows:

$$
\begin{aligned}
\tilde{x}_1 &= \bar{h}_{1,1}y_1 + h_{2,2}\bar{y}_2 && (21.11)\\
&= \bar{h}_{1,1}h_{1,1}x_1 + \bar{h}_{1,1}h_{2,1}x_2 + \bar{h}_{1,1}n_1 - h_{2,2}\bar{h}_{1,2}x_2 + h_{2,2}\bar{h}_{2,2}x_1 + h_{2,2}\bar{n}_2\\
&= \left(|h_{1,1}|^2 + |h_{2,2}|^2\right)x_1 + \left(\bar{h}_{1,1}h_{2,1} - h_{2,2}\bar{h}_{1,2}\right)x_2 + \bar{h}_{1,1}n_1 + h_{2,2}\bar{n}_2.
\end{aligned}
$$

Similarly, for signal x_2 the combiner generates:

$$
\begin{aligned}
\tilde{x}_2 &= \bar{h}_{2,1}y_1 - h_{1,2}\bar{y}_2 && (21.12)\\
&= \bar{h}_{2,1}h_{1,1}x_1 + \bar{h}_{2,1}h_{2,1}x_2 + \bar{h}_{2,1}n_1 + h_{1,2}\bar{h}_{1,1}x_2 - h_{1,2}\bar{h}_{2,2}x_1 - h_{1,2}\bar{n}_2\\
&= \left(|h_{2,1}|^2 + |h_{1,2}|^2\right)x_2 + \left(\bar{h}_{2,1}h_{1,1} - h_{1,2}\bar{h}_{2,2}\right)x_1 + \bar{h}_{2,1}n_1 - h_{1,2}\bar{n}_2.
\end{aligned}
$$

In contrast to the prefect cancellation scenario of [616], we can see from Equations 21.12 and 21.13 that the signals x_1 and x_2 now interfere with each other. We can no longer cancel the cross-coupling of signals x_2 and x_1 in Equations 21.12 and 21.13, respectively, unless the fading envelopes satisfy the condition of $h_{1,1} = h_{1,2}$ and $h_{2,1} = h_{2,2}$.

At high SNRs the noise power is insignificant compared to the transmitted power of the signals x_1 and x_2. Therefore, we can ignore the noise terms n in Equations 21.12 and 21.13. However, the interference signals' power increases, as we increase the transmission power. Assuming that both the signals x_1 and x_2 have an equivalent signal power, we can then express the signal to interference ratio (SIR) for signal x_1 as:

$$
SIR = \frac{|h_{1,1}|^2 + |h_{2,2}|^2}{\bar{h}_{1,1}h_{2,1} - h_{2,2}\bar{h}_{1,2}}, \tag{21.13}
$$

and similarly for signal x_2 as:

$$
SIR = \frac{|h_{2,1}|^2 + |h_{1,2}|^2}{\bar{h}_{2,1}h_{1,1} - h_{1,2}\bar{h}_{2,2}}. \tag{21.14}
$$

In Figure 21.27, we show the FER performance of the space-time block code $\mathbf{G}_2$ concatenated with the TC(2,1,3) code using one receiver and the 128-subcarrier 16QAM OFDM modem. The CIR has two equal-power rays separated by various delay spreads and a maximum Doppler frequency of 200 Hz. As we can see in Equations 21.13 and 21.14, we have $SIR \to \infty$, if $h_{1,1} = h_{1,2}$ and $h_{2,1} = h_{2,2}$. On the other hand, we encounter $SIR \to 1$, if $h_{1,1} = \delta h_{1,2}$ and $h_{2,1} = \delta h_{2,2}$, where $\delta \to \infty$. Since the SIR decreases, when the delay spread increases due to the rapidly fluctuating frequency-domain fading envelopes, as shown in Figure 21.25, we can see in Figure 21.27 that the performance of the concatenated scheme degrades, when increasing the delay spread. When the delay spread is more than $15\mu s$, we can see from the figure that the concatenated scheme exhibits the previously observed flattening effect. Furthermore, the error floor of the concatenated scheme becomes higher, as the delay spread is increased.

Similarly to Figure 21.24, where the Doppler frequency was varied, we show in Figure 21.28 the E_b/N_0 value required for maintaining FER=10^{-3} versus the delay spread for the 32-state 4PSK space-time trellis code and for the space-time block code $\mathbf{G}_2$ concatenated

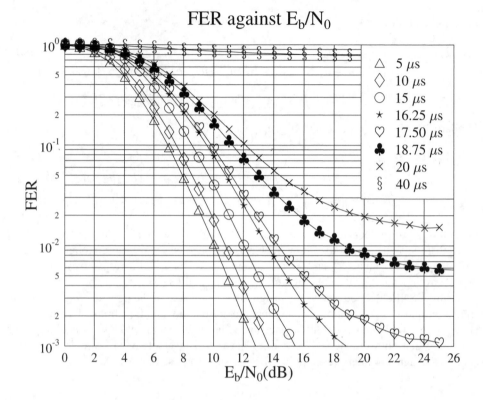

Figure 21.27: FER performance of the space-time block code $\mathbf{G}_2$ concatenated with the TC(2,1,3) code using one receiver, the 128-subcarrier OFDM modem and 16QAM. The CIR exhibits two equal-power rays separated by various delay spreads and a maximum Doppler frequency of 200 Hz. The coding parameters are shown in Tables 21.2, 21.3 and 21.4

with the TC(2,1,3) code using one receiver and the 128-subcarrier OFDM modem. The CIR exhibited two equal-power rays separated by various delay spreads. The maximum Doppler frequency was 200 Hz. We can see in the figure that the performance of the 32-state 4PSK space-time trellis code does not vary significantly with the delay spread. However, the concatenated TC(2,1,3)-coded scheme suffers severe performance degradation upon increasing the delay spread, as evidenced by the associated error floors shown in Figure 21.27. The SIR associated with the various delay spreads was obtained using computer simulations and the associated SIR values are also shown in Figure 21.28, denoted by the hearts. As we have expected, the calculated SIR decreases with the delay spread. We can see in the figure that the performance of the concatenated $\mathbf{G}_2$/TC(2,1,3) scheme suffers severe degradation, when the delay spread is in excess of $15\mu s$, as indicated by the near-vertical curve marked by triangles. If we relate this curve to the SIR curve marked by the hearts, we can see from the figure that the SIR is approximately 10 dB. Hence the SIR of the concatenated $\mathbf{G}_2$/TC(2,1,3) scheme has to be more than 10 dB, in order for it to outperform the space-time trellis codes using no channel coding.

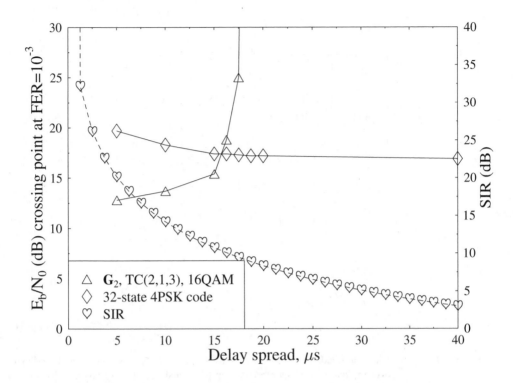

Figure 21.28: The E_b/N_0 values required for maintaining FER=10^{-3} versus delay spreads for the 32-state 4PSK space-time trellis code and for the space-time block code $\mathbf{G}_2$ concatenated with the TC(2,1,3) code using **one receiver** and the 128-subcarrier OFDM modem. The CIR exhibited two equal-power rays separated by various delay spreads and a maximum Doppler frequency of 200 Hz. The effective throughput was **2 BPS** and the coding parameters are shown in Tables 21.2, 21.3 and 21.4. The SIR of various delay spreads are shown as well

21.4.5 Delay Non-sensitive System

Previously, we have provided simulation results for a delay-sensitive, OFDM symbol-by-symbol decoded system. More explicitly, the received OFDM symbol had to be demodulated and decoded on a symbol-by-symbol basis, in order to provide decoded bits for example for a low-delay source decoder. Therefore, the two transmission instants of the space-time block code $\mathbf{G}_2$ had to be in the same OFDM symbol. They were allocated to the adjacent subcarriers in our previous studies. Moreover, we have shown in Figure 21.25 that the variation of the frequency-domain fading amplitudes along the subcarriers becomes more severe, as we increase the delay spread of the two rays. In Figure 21.29 we show both the frequency-domain and time-domain fading amplitudes of the channels' fading amplitudes for a fraction of the subcarriers in the 128-subcarrier OFDM symbols over the previously used two-path channel having two equal-power rays separated by a delay spread of $40\mu s$. The maximum Doppler

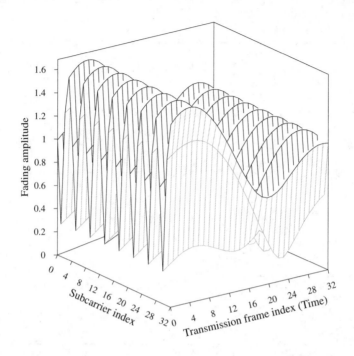

Figure 21.29: The fading amplitude versus time and frequency for various 128-subcarrier OFDM symbols over the two-path channel exhibiting two equal-power rays separated by a delay spread of $40\mu s$ and maximum Doppler frequency of 100 Hz

frequency was set here to 100 Hz. It can be clearly seen from the figure that the fading amplitude variation versus time is slower than that versus the subcarrier index within the OFDM symbols. This implies that the SIR attained would be higher, if we were to allocate the two transmission instants of the space-time block code $\mathbf{G}_2$ to the same subcarrier of consecutive OFDM symbols. This increase in SIR is achieved by doubling the delay of the system, since in this scenario two consecutive OFDM symbols have to be decoded, before all the received data becomes available.

In Figure 21.30, we show our FER performance comparison for the above two scenarios, namely using two adjacent subcarriers and the same subcarrier in two consecutive OFDM symbols for the space-time block code $\mathbf{G}_2$ concatenated with the TC(2,1,3) code using one 128-subcarrier 16QAM OFDM receiver. As before, the CIR exhibited two equal-power rays separated by a delay spread of $40\mu s$ and a maximum Doppler frequency of 100 Hz. It can be seen from the figure that there is a severe performance degradation, if the two transmission instants of the space-time block $\mathbf{G}_2$ are allocated to two adjacent subcarriers. This is evidenced by the near-horizontal curve marked by diamonds across the figure. On the other hand, upon assuming that having a delay of two OFDM-symbol durations does not pose any problems in terms of real-time interactive communications, we can allocate the two transmission instants of the space-time block code $\mathbf{G}_2$ to the same subcarrier of two consecutive OFDM symbols.

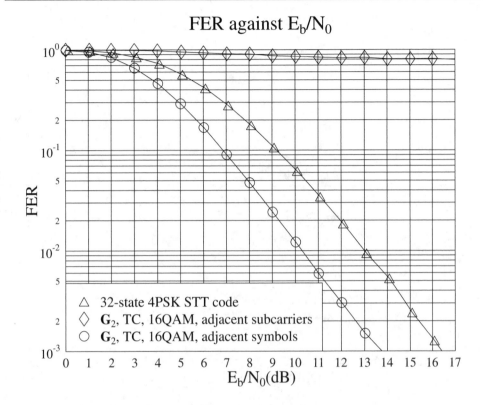

Figure 21.30: **FER** performance comparison between adjacent subcarriers and adjacent OFDM symbols allocation for the space-time block code G_2 concatenated with the TC(2,1,3) code using one receiver, the 128-subcarrier OFDM modem and 16QAM over a channel having a CIR characterised by two equal-power rays separated by a delay spread of $40\mu s$. The maximum Doppler frequency was 100 Hz. The coding parameters are shown in Tables 21.2, 21.3 and 21.4

From Figure 21.30, we can observe a dramatic improvement over the previous allocation method. Furthermore, the figure also indicates that by tolerating a two OFDM-symbol delay, the concatenated G_2/TC(2,1,3) scheme outperforms the 32-state 4PSK space-time trellis code by approximately 2 dB in terms of the required E_b/N_0 value at a FER of 10^{-3}.

Since the two transmission instants of the space-time block code G_2 are allocated to the same subcarrier of two consecutive OFDM symbols, it is the maximum Doppler frequency that would affect the performance of the concatenated scheme more gravely, rather than the delay spread. Hence we extended our studies to consider the effects of the maximum Doppler frequency on the performance of the concatenated G_2/TC(2,1,3) scheme. Specifically, Figure 21.31 shows the E_b/N_0 values required for maintaining FER=10^{-3} versus the Doppler frequency for the 32-state 4PSK space-time trellis code, and for the space-time block code G_2 concatenated with the TC(2,1,3) code using one 128-subcarrier 16QAM OFDM receiver, when mapping the two transmission instants to the same subcarrier of two consecutive

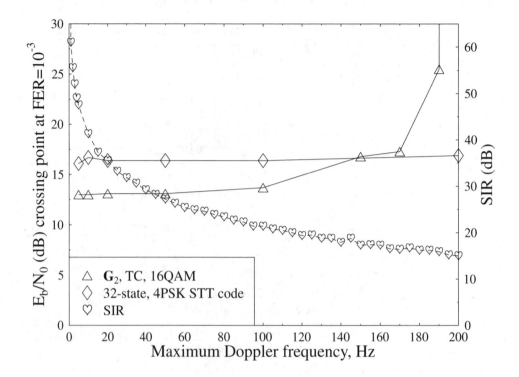

Figure 21.31: The E_b/N_0 value required for maintaining FER=10^{-3} versus the maximum Doppler frequency for the 32-state 4PSK space-time trellis code and for the adjacent OFDM symbols allocation of the space-time block code $\mathbf{G}_2$ concatenated with the TC(2,1,3) code using **one receiver** and the 128-subcarrier OFDM modem. The CIR exhibited two equal-power rays separated by a delay spread of $40\mu s$. The effective throughput was **2 BPS** and the coding parameters are shown in Tables 21.2, 21.3 and 21.4. The SIR of various maximum Doppler frequencies are shown as well

OFDM symbols. The channel exhibited two equal-power rays separated by a delay spread of $40\mu s$ and various maximum Doppler frequencies. The SIR achievable at various maximum Doppler frequencies is also shown in Figure 21.31. Again, we can see that the performance of the concatenated $\mathbf{G}_2$/TC(2,1,3) scheme suffers severely, if the maximum Doppler frequency is above 160 Hz. More precisely, we can surmise that in order for the concatenated scheme to outperform the 32-state 4PSK space-time trellis code, the SIR should be at least 15 dB, which is about the same as the required SIR in Figure 21.28. From Figures 21.28 and 21.31, we can conclude that the concatenated $\mathbf{G}_2$/TC(2,1,3) scheme performs better, if the SIR is in excess of about 10-15 dB.

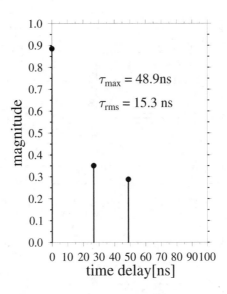

Figure 21.32: Short WATM channel impulse response

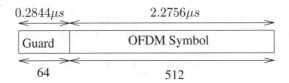

Figure 21.33: Short WATM plot of 512-subcarrier OFDM time domain signal with a cyclic extension of 64 samples

21.4.6 The Wireless Asynchronous Transfer Mode System

We have previously investigated the performance of different schemes over two-path channels having two equal-power rays. In this section, we investigate the performance of the various systems over indoor Wireless Asynchronous Transfer Mode (WATM) channels. The WATM system used 512 subcarriers and each OFDM symbol was extended with a cyclic prefix of length 64. The sampling rate was 225 Msamples/s and the carrier frequency was 60 GHz. In [180] two WATM CIRs were used, namely a five-path and a three-path model, where the latter one was referred to as the shortened WATM CIR. This CIR was used also in our investigations here.

The shortened WATM channel's impulse response is depicted in Figure 21.32, where the longest-delay path arrived at a delay of 48.9 ns, which corresponds to 11 sample periods

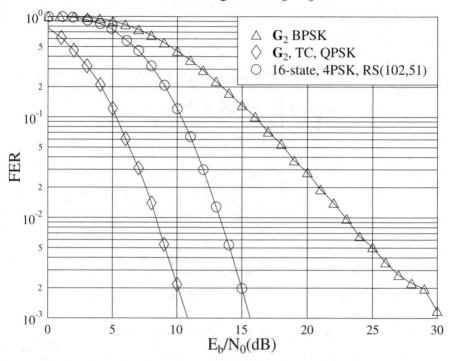

Figure 21.34: FER performance comparison between the TC(2,1,3) coded space-time block code G_2 and the RS(102,51) GF(2^{10}) coded 16-state 4PSK space-time trellis code using one 512-subcarrier OFDM receiver over the shortened WATM channel at an effective throughput of **1 BPS**. The coding parameters are shown in Tables 21.2, 21.3, 21.4 and 21.5

at the 225 Msamples/s sampling rate. The 512-subcarrier OFDM time-domain transmission frame having a cyclic extension of 64 samples is shown in Figure 21.33.

21.4.6.1   Channel Coded Space-Time Codes – Throughput of 1 BPS

Previously, we have compared the performance of the space-time trellis codes to that of the TC(2,1,3)-coded space-time block code G_2. We now extend our comparisons to Reed-Solomon (RS) coded space-time trellis codes, which were used in [626, 629, 631]. In Figure 21.34 we show our FER performance comparison between the TC(2,1,3) coded space-time block code G_2 and the RS(102,51) GF(2^{10}) coded 16-state 4PSK space-time trellis code using one 512-subcarrier OFDM receiver over the shortened WATM CIR of Figure 21.32 at an effective throughput of 1 BPS. We can see from the figure that the TC(2,1,3) coded space-time block code G_2 outperforms the RS(102,51) GF(2^{10}) coded 16-state 4PSK space-time trellis code by approximately 5 dB in E_b/N_0 terms at a FER of 10^{-3}. The performance of the RS(102,51) GF(2^{10}) coded 16-state 4PSK space-time trellis code would be improved by about 2 dB, if the additional complexity of maximum likelihood decoding were affordable.

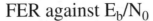

FER against E_b/N_0

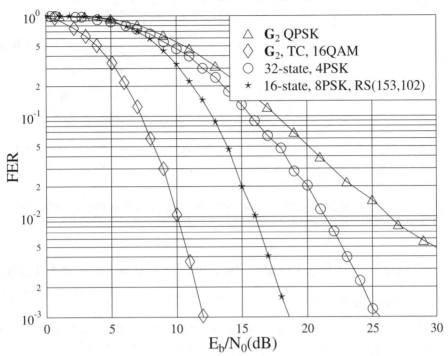

Figure 21.35: FER performance comparison between the TC(2,1,3) coded space-time block code $\mathbf{G}_2$ and the RS(153,102) GF(2^{10}) coded 16-state 8PSK space-time trellis code using one 512-subcarrier OFDM receiver over the shortened WATM channel at an effective throughput of 2 BPS. The coding parameters are shown in Tables 21.2, 21.3, 21.4 and 21.5

However, even assuming this improvement, the TC(2,1,3) coded space-time block code $\mathbf{G}_2$ would outperform the RS(102,51) GF(2^{10}) coded 16-state 4PSK space-time trellis code.

21.4.6.2 Channel Coded Space-Time Codes – Throughput of 2 BPS

In our next experiment, the throughput of the system was increased to 2 BPS by employing a higher-order modulation scheme. In Figure 21.35 we show our FER performance comparison between the TC(2,1,3) coded space-time block code $\mathbf{G}_2$ and the RS(153,102) GF(2^{10}) coded 16-state 8PSK space-time trellis code using one 512-subcarrier OFDM receiver over the shortened WATM channel of Figure 21.32 at an effective throughput of 2 BPS. Again, we can see that the TC(2,1,3) coded space-time block code $\mathbf{G}_2$ outperforms the RS(153,102) GF(2^{10}) coded 16-state 8PSK space-time trellis code by approximately 5 dB in terms of E_b/N_0 at a FER of 10^{-3}. The corresponding performance of the 32-state 4PSK space-time trellis code is also shown in the figure. It can be seen that its performance is about 13 dB worse in E_b/N_0 terms than that of the TC(2,1,3) coded space-time block code $\mathbf{G}_2$. Let us

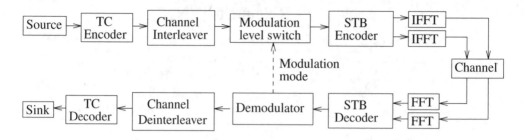

Figure 21.36: System overview of the turbo-coded and space-time-coded adaptive OFDM

now continue our investigations by considering, whether channel-quality controlled adaptive space-time coded OFDM is capable of providing further performance benefits.

21.5 Space-Time Coded Adaptive Modulation for OFDM

21.5.1 Introduction

Adaptive modulation was proposed by Steele and Webb [124, 640], in order to combat the time-variant fading of mobile channels. The main idea of adaptive modulation is that when the channel quality is favourable, higher-order modulation modes are employed, in order to increase the throughput of the system. On the other hand, more robust but lower-throughput modulation modes are employed, if the channel quality is low. This simple but elegant idea has motivated a number of researchers to probe further [180,211,219,220,577,581,585,641, 642].

Recently adaptive modulation was also invoked in the context of OFDM, which was termed adaptive OFDM (AOFDM) [155, 180, 581, 585]. AOFDM exploits the variation of the signal quality both in the time domain as well as in the frequency domain. In what is known as sub-band adaptive OFDM transmission, all subcarriers in an AOFDM symbol are split into blocks of adjacent subcarriers, referred to as sub-bands. The same modulation scheme is employed for all subcarriers of the same sub-band. This substantially simplifies the task of signalling the modulation modes, since there are typically four modes and for example 32 sub-bands, requiring a total of 64 AOFDM mode signalling bits.

21.5.2 Turbo-Coded and Space-Time-Coded Adaptive OFDM

In this section, the adaptive OFDM philosophy portrayed for example in [180,581,585,643] is extended, in order to exploit the advantages of multiple transmit and receive antennas. Additionally, turbo coding is employed for improving the performance of the system. In Figure 21.36, we show the system schematic of the turbo-coded and space-time-coded adaptive OFDM system. Similarly to Figure 21.13, random data bits are generated and encoded by the TC(2,1,3) encoder using an octal generator polynomial of $(7, 5)$. Various TC(2,1,3) coding rates were used for the different modulation schemes. The encoded bits were channel interleaved and passed to the modulator. The choice of the modulation scheme to be used by the transmitter for its next OFDM symbol is determined by the channel quality estimate of the receiver based on the current OFDM symbol. In this study, we assumed perfect channel quality estimation and perfect signalling of the required modem mode of each sub-band based on the channel quality estimate acquired during the current OFDM symbol. Aided by the perfect channel quality estimator, the receiver determines the highest-throughput modulation mode to be employed by the transmitter for its next transmission while maintaining the system's target BER. Five possible transmission modes were employed in our investigations, which are no transmission (NoTx), BPSK, QPSK, 16QAM and 64QAM. In order to simplify the task of signalling the required modulation modes, we employed the sub-band adaptive OFDM transmission scheme detailed for example in [180, 581, 585, 643]. The modulated signals were then passed to the encoder of the space-time block code G_2. The space-time encoded signals were OFDM modulated and transmitted by the corresponding antennas. The shortened WATM channel was used, where the CIR profile and the OFDM transmission frame are shown in Figures 21.32 and 21.33, respectively.

The number of receivers invoked constitutes a design parameter. The received signals were OFDM demodulated and passed to the space-time decoders. Log-MAP [638] decoding of the received space-time signals was performed, in order to provide soft-outputs for the TC(2,1,3) decoder. Assuming that the demodulator of the receiver has perfect knowledge of the instantaneous channel quality, this information is passed to the transmitter in order to determine its next AOFDM modulation mode allocation. The received bits were then channel deinterleaved and passed to the TC decoder, which again, employs the Log-MAP decoding algorithm [637]. The decoded bits were finally passed to the sink for calculation of the BER.

21.5.3 Simulation Results

As mentioned earlier, all the AOFDM-based simulation results were obtained over the shortened WATM channel. The channels' profile and the OFDM transmission frame structure are shown in Figures 21.32 and 21.33, respectively. Again, Jakes' model [381] was adopted for modelling the fading channels.

In order to obtain our simulation results, several assumptions were stipulated:

- The average signal power received from each transmitter antenna was the same;

- All multipath components undergo independent Rayleigh fading;

- The receiver has a perfect knowledge of the CIR;

- Perfect signalling of the AOFDM modulation modes.

Again, we note that the above assumptions are unrealistic, yielding the best-case performance, nonetheless, they facilitate the performance comparison of the various techniques under identical circumstances.

21.5.3.1 Space-Time Coded Adaptive OFDM

In this section, we employ the fixed threshold based modem mode selection algorithm, which was also used in [180], adapting the techniques of [185, 641, 644] for serial modems. It was

System	NoTx	BPSK	QPSK	16QAM	64QAM
Speech	$-\infty$	3.31	6.48	11.61	17.64
Data	$-\infty$	7.98	10.42	16.76	26.33

Table 21.7: Optimised switching levels quoted from [641] for adaptive modulation over Rayleigh fading channels for the speech and data systems, shown in instantaneous channel SNR (dB)

assumed that the channel quality was constant for all the symbols in a transmission burst, i.e. that the channel's fading envelope varied slowly across the transmission burst. Under these conditions, all the transmitted symbols are modulated using the same modulation mode, chosen according to the predicted SNR. Torrance optimised the modem mode switching thresholds [185, 641, 644] for the target BERs of 10^{-2} and 10^{-4}, which will be appropriate for a high-BER speech system and for a low-BER data system, respectively. The resulting SNR switching thresholds for activating a given modulation mode in a slowly Rayleigh fading narrowband channel are given in Table 21.7 for both systems. Assuming perfect channel quality estimation, the instantaneous channel SNR is measured by the receiver and the information is passed to the modulation mode selection switch at the transmitter, as shown in Figure 21.36 using the system's control channel. This side-information signalling does not constitute a problem, since state-of-the-art wireless systems, such as for example IMT-2000 [370] have a high-rate, low-delay signalling channel. This modem mode signalling feedback information is utilised by the transmitter for selecting the next modulation mode. Specifically, a given modulation mode is selected, if the instantaneous channel SNR perceived by the receiver exceeds the corresponding switching levels shown in Figure 21.7, depending on the target BER.

As mentioned earlier, the adaptation algorithm of [185, 641, 644] assumes constant instantaneous channel SNR over the whole transmission burst. However, in the case of an OFDM system transmitting over frequency selective channels, the channels' quality varies across the different subcarriers. In [180, 643] employing the lowest-quality sub-carrier in the sub-band for controlling the adaptation algorithm based on the switching thresholds given in Table 21.7. Again, this approach significantly simplifies the signalling and therefore it was also adopted in our investigations.

In Figure 21.37, we show the BER and BPS performance of the 16 sub-band AOFDM scheme employing the space-time block code $\mathbf{G}_2$ in conjunction with multiple receivers and a target BER of 10^{-4} over the shortened WATM channel shown in Figure 21.32. The switching thresholds are shown in Table 21.7. The performance of the conventional AOFDM scheme using no diversity [180] is also shown in the figure. From Figure 21.37, we can see that the BPS performance of the space-time coded AOFDM scheme using one receiver is better than that of the conventional AOFDM scheme. The associated performance gain improves, as the throughput increases. At a throughput of 6 BPS, the space-time coded scheme outperforms the conventional scheme by at least 10 dB in E_b/N_0 terms. However, we notice in Figure 21.37 that as a secondary effect, the BER performance of the space-time coded AOFDM scheme using one receiver degrades, as we increase the average channel SNR. Again, this problem is due to the interference of signals x_1 and x_2 caused by the rapidly varying frequency-domain fading envelope across the subcarriers. At high SNRs, predominantly 64QAM was employed. Since the constellation points in 64QAM are densely packed,

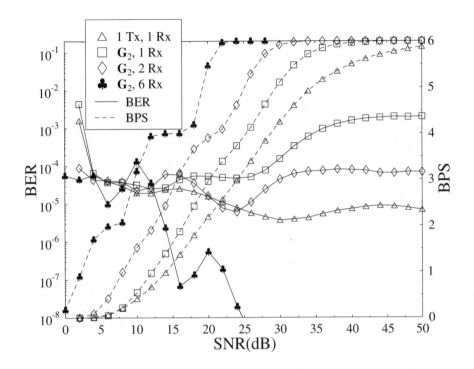

Figure 21.37: BER and BPS performance of 16 sub-band AOFDM employing the space-time block code G_2 using multiple receivers for a target BER of 10^{-4} over the shortened WATM channel shown in Figure 21.32 and the transmission format of Figure 21.33. The switching thresholds are shown in Table 21.7

this modulation mode is more sensitive to the 'cross-talk' of the signals x_1 and x_2. This limited the BER performance to 10^{-3} even at high SNRs. However, at SNRs lower than 30 dB typically more robust modulation modes were employed and hence the target BER of 10^{-4} was readily met. We will show in the next section that this problem can be overcome by employing turbo channel coding in the system.

In Figure 21.37 we also observe that the BER and BPS performance improves, as we increase the number of AOFDM receivers, since the interference between the signals x_1 and x_2 is eliminated. Upon having six AOFDM receivers, the BER of the system drops below 10^{-8}, when the average channel SNR exceeds 25 dB and there is no sign of the BER flattening effect. At a throughput of 6 BPS, the space-time coded AOFDM scheme using six receivers outperforms the conventional system by more than 30 dB.

Figure 21.38 shows the probability of each AOFDM sub-band modulation mode for (a) conventional AOFDM and for space-time coded AOFDM using (b) 1, (c) 2 and (d) 6 receivers over the shortened WATM channel shown in Figure 21.32. The transmission format obeyed

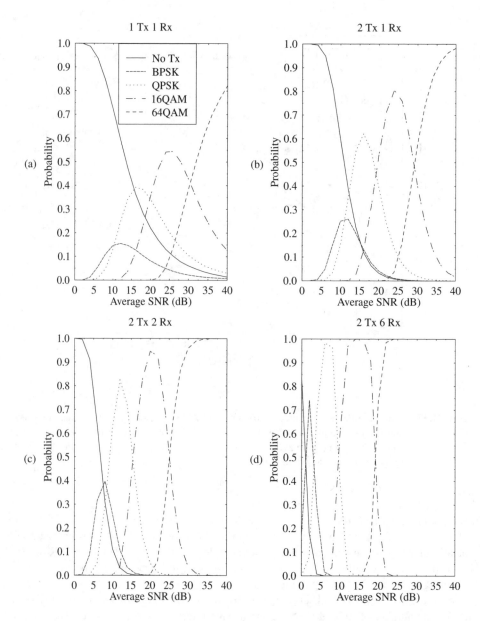

Figure 21.38: Probability of each modulation mode for (a) conventional AOFDM and for space-time coded AOFDM using (b) 1, (c) 2 and (d) 6 receivers over the shortened WATM channel shown in Figure 21.32 and using the transmission frame of Figure 21.33. The thresholds were optimised for the data system having a BER of 10^{-4} and they are shown in Table 21.7. All sub-figures share the legends shown in Figure 21.38 (a)

Figure 21.33. The switching thresholds were optimised for the data system having a target BER of 10^{-4} and they are shown in Table 21.7. By employing multiple transmitters and receivers, we increase the diversity gain and we can see in the figure that this increases the probability of the most appropriate modulation mode at a certain average channel SNR. This is clearly shown by the increased peaks of each modulation mode at different average channel SNRs. As an example, in Figure 21.38 (d) we can see that there is an almost 100% probability of transmitting in the QPSK and 16QAM modes at average channel SNR of approximately 6 dB and 15 dB, respectively. This strongly suggests that it is a better solution, if fixed modulation based transmission is employed in space-time coded OFDM, provided that we can afford the associated complexity of using six receivers. We shall investigate these issues in more depth at a later stage.

On the other hand, the increased probability of a particular modulation mode at a certain average channel SNR also means that there is less frequent switching amongst the various modulation modes. For example, we can see in Figure 21.38 (b) that the probability of employing 16QAM increased to 0.8 at an average channel SNR of 25 dB compared to 0.5 in Figure 21.38 (a). Furthermore, there are almost no BPSK transmissions at SNR = 25 dB in Figure 21.38 (b). This situation might be an advantage in the context of the AOFDM system, since most of the time the system will employ 16QAM and only occasionally switches to the QPSK and 64QAM modulation modes. This can be potentially exploited to reduce AOFDM modem mode the signalling traffic and to simplify the system.

The characteristics of the modem mode probability density functions in Figure 21.38 in conjunction with multiple transmit antennas can be further explained with the aid of Figure 21.39. In Figure 21.39 we show the instantaneous channel SNR experienced by the 512-subcarrier OFDM symbols for a single-transmitter, single-receiver scheme and for the space-time block code G_2 using one, two and six receivers over the shortened WATM channel. The average channel SNR is 10 dB. We can see in Figure 21.39 that the variation of the instantaneous channel SNR for a single transmitter and single receiver is fast and severe. The instantaneous channel SNR may become as low as 4 dB due to deep fades of the channel. On the other hand, we can see that for the space-time block code G_2 using one receiver the variation in the instantaneous channel SNR is slower and less severe. Explicitly, by employing multiple transmit antennas as shown in Figure 21.39, we have reduced the effect of the channels' deep fades significantly. This is advantageous in the context of adaptive modulation schemes, since higher-order modulation modes can be employed, in order to increase the throughput of the system. However, as we increase the number of receivers, i.e. the diversity order, we observe that the variation of the channel becomes slower. Effectively, by employing higher-order diversity, the fading channels have been converted to AWGN-like channels, as evidenced by the space-time block code G_2 using six receivers. Since adaptive modulation only offers advantages over fading channels, we argue that using adaptive modulation might become unnecessary, as the diversity order is increased.

To elaborate a little further, from Figure 21.38 and 21.39 we surmise that fixed modulation schemes might become more attractive, when the diversity order increases, which is achieved in this case by employing more receivers. This is because for a certain average channel SNR, the probability of a particular modulation mode increases. In other words, the fading channel has become an AWGN-like channel, as the diversity order is increased. In Figure 21.40 we show our throughput performance comparison between AOFDM and fixed modulation based OFDM in conjunction with the space-time block code G_2 employing (a) one receiver and

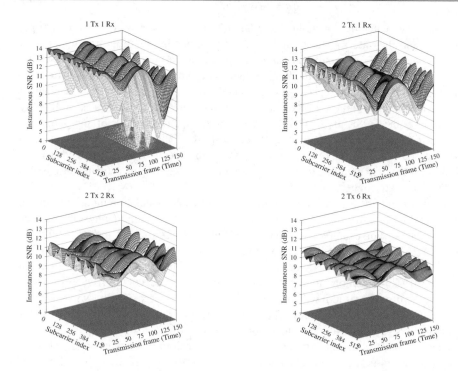

Figure 21.39: Instantaneous channel SNR of 512-subcarrier OFDM symbols for single-transmitter single-receiver and for the space-time block code $\mathbf{G}_2$ using one, two and six receivers over the shortened WATM channel shown in Figure 21.32 and using the transmission format of Figure 21.33. The average channel SNR is 10 dB

(b) two receivers over the shortened WATM channel. The throughput of fixed OFDM was 1, 2, 4 and 6 BPS and the corresponding E_b/N_0 values were extracted from the associated BER versus E_b/N_0 curves of the individual fixed-mode OFDM schemes. It can be seen from Figure 21.40 (a) that the throughput performance of the adaptive and fixed OFDM schemes is similar for a 10^{-2} target BER system. However, for a 10^{-4} target BER system, there is an improvement of 5-10 dB in E_b/N_0 terms at various throughputs for the adaptive OFDM scheme over the fixed OFDM scheme. At high average channel SNRs the throughput performance of both schemes converged, since 64QAM became the dominant modulation mode for AOFDM.

On the other hand, if the number of receivers is increased to two, we can see in Figure 21.40 (b) that the throughput performance of both adaptive and fixed OFDM is similar for both the 10^{-4} and 10^{-2} target BER systems. We would expect similar trends, as the number of receivers is increased, since the fading channels become AWGN-like channels. From Figure 21.40, we conclude that AOFDM is only beneficial for the space-time block code $\mathbf{G}_2$ using one receiver in the context of the 10^{-2} target BER system.

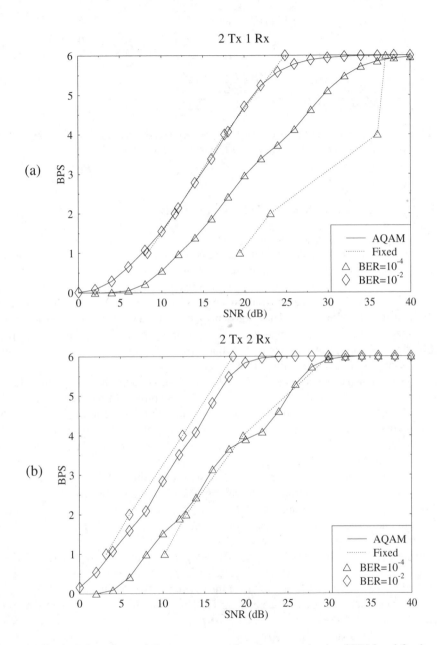

Figure 21.40: BPS throughput performance comparison between adaptive OFDM and fixed modulation based OFDM using the space-time block code $\mathbf{G}_2$ employing **(a)** one receiver and **(b)** two receivers over the shortened WATM channel shown in Figure 21.32 and the transmission format of Figure 21.33

	NoTx	BPSK	QPSK	16QAM	64QAM
Half-rate TC(2,1,3)					
Rate	–	0.50	0.50	0.50	0.50
Thresholds (dB)	$-\infty$	-4.0	-1.3	5.4	9.8
Variable-rate TC(2,1,3)					
Rate	–	0.50	0.67	0.75	0.90
Thresholds (dB)	$-\infty$	-4.0	2.0	9.70	21.50

Table 21.8: Coding rates and switching levels (dB) for TC(2,1,3) and space-time coded adaptive OFDM over the shortened WATM channel of Figure 21.32 for a target BER of 10^{-4}

21.5.3.2   Turbo and Space-Time Coded Adaptive OFDM

In the previous section we have discussed the performance of space-time coded adaptive OFDM. Here we extend our study by concatenating turbo coding with the space-time coded AOFDM scheme in order to improve both the BER and BPS performance of the system. As earlier, the turbo convolutional code TC(2,1,3) having a constraint length of 3 and octal generator polynomial of $(7, 5)$ was employed. Since the system was designed for high-integrity, low-BER data transmission, it was delay non-sensitive. Hence a random turbo interleaver size of approximately 10,000 bits was employed. The random separation channel interleaver [635] was utilised in order to disperse the bursty channel errors. The Log-MAP [637] decoding algorithm was employed, using eight iterations.

We proposed two TC coded schemes for the space-time coded AOFDM system. The first scheme is a fixed half-rate turbo and space-time coded adaptive OFDM system. It achieves a high BER performance, but at the cost of a maximum throughput limited to 3 BPS due to half-rate channel coding. The second one is a variable-rate turbo and space-time coded adaptive OFDM system. This scheme sacrifices the BER performance in exchange for an increased system throughput. Different puncturing patterns are employed for the various code rates R. The puncturing patterns were optimised experimentally by simulations. The design procedure for punctured turbo codes was proposed by Acikel *et al.* [645] in the context of BPSK and QPSK. The optimum AOFDM mode switching thresholds were obtained by computer simulations over the shortened WATM channel of Figure 21.32 and they are shown in Table 21.8.

In Figure 21.41, we show the BER and BPS performance of 16 sub-band AOFDM employing the space-time block code $\mathbf{G}_2$ concatenated with both half-rate and variable-rate TC(2,1,3) coding at a target BER of 10^{-4} over the shortened WATM channel of Figure 21.32. We can see in the figure that by concatenating fixed half-rate turbo coding with the space-time coded adaptive OFDM scheme, the BER performance of the system improves tremendously, indicated by a steep dip of the associated BER curve marked by the solid line and diamonds. There is an improvement in the BPS performance as well, exhibiting an E_b/N_0 gain of approximately 5 dB and 10 dB at an effective throughput of 1 BPS, compared to space-time coded AOFDM and conventional AOFDM, respectively. However, again, the maximum throughput of the system is limited to 3 BPS, since half-rate channel coding was employed. In Figure 21.41, we can see that at an E_b/N_0 value of about 30 dB the maximum throughput of the turbo coded and space-time adaptive OFDM system is increased

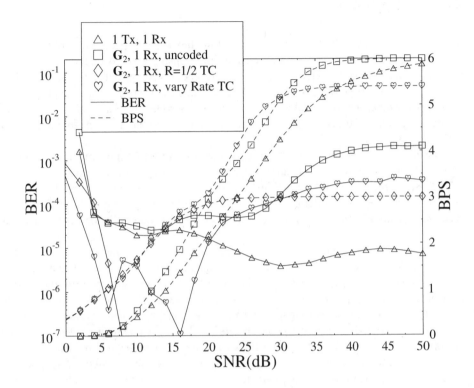

Figure 21.41: BER and BPS performance of 16 sub-band AOFDM employing the space-time block code G_2 concatenated with both half-rate and variable-rate TC(2,1,3) at a target BER of 10^{-4} over the shortened WATM channel shown in Figure 21.32 and using the transmission format of Figure 21.33. The switching thresholds and coding rates are shown in Table 21.7

from 3.0 BPS to 5.4 BPS by employing the variable-rate TC(2,1,3) code. Furthermore, the BPS performance of the variable-rate turbo coded scheme is similar to that of the half-rate turbo coded scheme at an average channel SNR below 15 dB. The BER curve marked by the solid line and clubs drops, as the average channel SNR is increased from 0 dB to 15 dB. Due to the increased probability of the 64QAM transmission mode, the variable-rate turbo coded scheme was overloaded by the plethora of channel errors introduced by the 64QAM mode. Therefore, we can see in Figure 21.41 that the BER increases and stabilises at 10^{-4}. Again, the interference of the signals x_1 and x_2 in the context of the space-time block code G_2 prohibits further improvements in the BER performance, as the average channel SNR is increased. However, employing the variable-rate turbo codec has lowered the BER floor, as demonstrated by the curve marked by the solid line and squares.

21.6   Summary

Space-time trellis codes [619–624] and space-time block codes [616–618] constitute state-of-the-art transmission schemes based on multiple transmitters and receivers. Both codes were introduced in Section 21.1. Space-time trellis codes were introduced in Section 21.2 by utilising the simplest possible 4-state, 4PSK space-time trellis code as an example. The state diagrams for other 4PSK and 8PSK space-time trellis codes were also given. The branch metric of each trellis transition was derived, in order to facilitate their maximum likelihood (ML) decoding.

In Section 21.3, we proposed the employment of an OFDM modem for mitigating the effects of dispersive multipath channels due to its simplicity compared to other approachs [627, 628]. Turbo codes and Reed-Solomon codes were invoked in Section 21.3.1 for concatenation with the space-time block code G_2 and the various space-time trellis codes, respectively. The estimated complexity of the various space-time trellis codes was derived in Section 21.3.3.

We presented our simulation results for the proposed schemes in Section 21.4. The first scheme studied was the TC(2,1,3) coded space-time block code G_2, whereas the second one was based on the family of space-time trellis codes. It was found that the FER and BER performance of the TC(2,1,3) coded space-time block G_2 was better than that of the investigated space-time trellis codes at a throughput of 2 and 3 BPS over the channel exhibiting two equal-power rays separated by a delay spread of $5\mu s$ and having a maximum Doppler frequency of 200 Hz. Our comparison between the two schemes was performed by also considering the estimated complexity of both schemes. It was found that the concatenated G_2/TC(2,1,3) scheme still outperformed the space-time trellis codes using no channel coding, even though both schemes exhibited a similar complexity.

The effect of the maximum Doppler frequency on both schemes was also investigated in Section 21.4.3. It was found that the maximum Doppler frequency had no significant impact on the performance of both schemes. By contrast, in Section 21.4.4 we investigated the effect of the delay spread on the system. Initially, the delay-spread dependent SIR of the space-time block code G_2 was quantified. It was found that the performance of the concatenated TC(2,1,3)-G_2 scheme degrades, as the delay spread increases due to the decrease in the associated SIR. However, varying the delay spread had no significant effect on the space-time trellis codes. We proposed in Section 21.4.5 an alternative mapping of the two transmission instants of the space-time block code G_2 to the same subcarrier of two consecutive OFDM symbols, a solution which was applicable to a delay non-sensitive system. By employing this approach, the performance of the concatenated scheme was no longer limited by the delay spread, but by the maximum Doppler frequency. We concluded that a certain minimum SIR has to be maintained for attaining the best possible performance of the concatenated scheme.

The shortened WATM channel was introduced in Section 21.4.6. In this section, space-time trellis codes were concatenated with Reed-Solomon codes, in order to improve the performance of the system. Once again, both channel coded space-time block and trellis codes were compared at a throughput of 1 and 2 BPS. It was also found that the TC(2,1,3) coded space-time block code G_2 outperforms the RS coded space-time trellis codes.

Space-time block coded AOFDM was studied in the Section 21.5, which is the last section of this chapter. It was shown in Section 21.5.3.1 that only the space-time block code G_2 using one AOFDM receiver was capable of outperforming the conventional single-transmitter, single-receiver AOFDM system designed for a data transmission target BER of 10^{-4} over the shortened WATM channel. We also confirmed that upon increasing the diversity order, the fading channels become AWGN-like channels. This explains, why fixed-mode OFDM transmission constitutes a better trade-off than AOFDM, when the diversity order is increased. In Section 21.5.3.2, we continued our investigations into AOFDM by concatenating turbo coding with the system. Two schemes were proposed: half-rate turbo and space-time coded AOFDM as well as variable-rate turbo and space-time coded AOFDM. Despite the impressive BER performance of the half-rate turbo and space-time coded scheme, the maximum throughput of the system was limited to 3 BPS. However, by employing the variable-rate turbo and space-time coded scheme, the BPS performance improved, achieving a maximum throughput of 5.4 BPS. However, the improvement in BPS performance was achieved at the cost of a poorer BER performance. In conclusion, this is a burgeoning research area, which is expected to achieve further advances in the field of turbo space-time coding.

Chapter 22

Adaptive QAM Optimisation for OFDM and MC-CDMA

B.J. Choi and L. Hanzo

22.1 Motivation

In Chapter 21 we considered the design trade-offs of adaptive versus space-time coded transmissions. Although both methods aim for mitigating the effects of fading-induced channel quality fluctuations, their approach is fundamentally different. Adaptive QAM aims for adjusting the modulation modes for the sake of maintaining the target integrity, despite the channel quality fluctuations. By contrast, space-time coding [171] employs multiple transmitters and receivers for the sake of directly mitigating the channel quality undulations. It transpired that both techniques were capable of attaining a similar performance, although AQAM imposed a lower complexity owing to employing a single transmitter and receiver.

In this chapter our discourse evolves further by providing a comparative study of the various approaches developed for optimising the AQAM switching thresholds and then employing the thresholds in the context of both AQAM and space-time coded OFDM [173] as well as frequency-domain spread MC-CDMA [173].

Let us commence our detailed discourse with a glimpse of history. In recent years the concept of intelligent multi-mode, multimedia transceivers (IMMT) has emerged in the context of wireless systems [168–173,180,640,646,647]. The range of various existing solutions that have found favour in already operational standard systems was summarised in the excellent overview by Nanda *et al.* [647]. *The aim of these adaptive transceivers is to provide mobile users with the best possible compromise amongst a number of contradicting design factors, such as the power consumption of the hand-held portable station (PS), robustness against transmission errors, spectral efficiency, teletraffic capacity, audio/video quality and so forth [646].*

The fundamental limitation of wireless systems is constituted by their time- and frequency-domain channel fading, as illustrated in Figure 22.1 [171] in terms of the Signal-to-Noise Ratio (SNR) fluctuations experienced by a modem over a dispersive channel. The violent SNR fluctuations observed as a function of both time and frequency suggest that over these channels no fixed-mode transceiver can be expected to provide an attractive performance, complexity and delay trade-off. Motivated by the above mentioned performance limitations of fixed-mode transceivers, IMMTs have attracted considerable research interest in the past decade [168, 169, 180, 640, 646, 647]. Some of these research results are collated in this monograph.

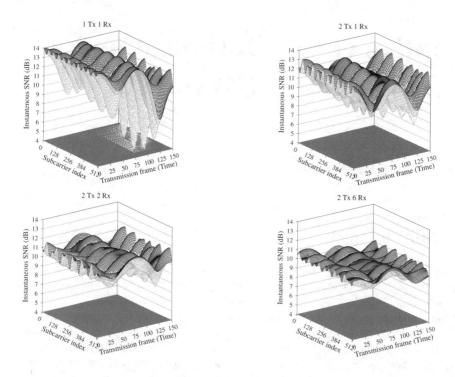

Figure 22.1: Instantaneous channel SNR versus time and frequency for a 512-subcarrier OFDM modem in the context of a single-transmitter single-receiver as well as for the space-time block code G_2 [616] using one, two and six receivers when communicating over an indoor wireless channel. The average channel SNR is 10 dB. ©IEEE, Liew and Hanzo [171, 648], 2001

In Figure 22.1 we show the instantaneous channel SNR experienced by the 512-subcarrier OFDM symbols for a single-transmitter, single-receiver scheme and for the space-time block code G_2 [616] using one, two and six receivers over the shortened WATM channel. The average channel SNR is 10 dB. We can see in Figure 22.1 that the variation of the instantaneous channel SNR for a single transmitter and single receiver is severe. The instantaneous channel SNR may become as low as 4 dB due to deep fades of the channel. On the other hand, we can see that for the space-time block code G_2 using one receiver the variation in the instantaneous channel SNR is slower and less severe. Explicitly, by employing multiple transmit antennas as shown in Figure 22.1, we have reduced the effect of the channels' deep fades significantly. This is advantageous in the context of adaptive modulation schemes, since higher-order modulation modes can be employed, in order to increase the throughput of the system. However, as we increase the number of receivers, i.e. the diversity order, we observe that the variation of the channel becomes slower. Effectively, by employing higher-order diversity, the fading channels have been converted to AWGN-like channels, as evidenced by the scenario employing the space-time block code G_2 using six receivers. Since adaptive modulation only offers advantages over fading channels, we argue that using adaptive modulation might become unnecessary, as the diversity order is increased. Hence, adaptive modulation can be viewed as a lower-complexity alternative to space-time coding, since only a single transmitter and receiver are required.

The above mentioned calamities inflicted by the wireless channel can be mitigated by contriving a suite of near-instantaneously adaptive or Burst-by-Burst Adaptive (BbBA) wideband single-carrier [180], multi-carrier or Orthogonal Frequency Division Multiplex [173] (OFDM) as well as Code Division Multiple Access (CDMA) [172] transceivers. The aim of these IMMTs is to communicate over hostile mobile channels at a higher integrity or higher throughput than conventional fixed-mode transceivers. A number of existing wireless systems already support some grade of adaptivity and future research is likely to promote these principles further by embedding them into the already existing standards. For example, due to their high control channel rate and with the advent of the well-known Orthogonal Variable Spreading Factor (OVSF) codes, the third-generation UTRA/IMT2000 systems are amenable not only to long-term spreading factor reconfiguration, but also to near-instantaneous reconfiguration on a 10ms transmission burst-duration basis. The High-Speed Data Packet Access (HSDPA) mode of the third-generation wireless systems has also opted for using adaptive modulation [170] and adaptive channel coding [171].

With the advent of BbBA QAM, OFDM or CDMA transmissions it becomes possible for mobile stations (MS) to invoke for example in indoor scenarios or in the central propagation cell region - where typically benign channel conditions prevail - a high-throughput modulation mode, such as 4 bit/symbol Quadrature Amplitude Modulation (16QAM). By contrast, a robust, but low-throughput modulation mode, such as 1 bit/symbol Binary Phase Shift Keying (BPSK) can be employed near the edge of the propagation cell, where hostile propagation conditions prevail. The BbBA QAM, OFDM or CDMA mode switching regime is also capable of reconfiguring the transceiver at the rate of the channel's slow- or even fast-fading. This may prevent premature hand-overs and - more importantly - unnecessary powering up, which would inflict an increased interference upon co-channel users, resulting in further potential power increments. This detrimental process could result in all mobiles operating at unnecessarily high power levels.

A specific property of these transceivers is that their bit rate fluctuates, as a function of time. This is not an impediment in the context of data transmission. However, in interactive speech [169] or video [168] communications appropriate source codecs have to be designed, which are capable of promptly reconfiguring themselves according to the near-instantaneous bitrate budget provided by the transceiver.

The expected performance of our BbBA transceivers can be characterised with the aid of a whole plethora of performance indicators. In simple terms, adaptive modems outperform their individual fixed-mode counterparts, since given an average number of transmitted bits per symbol (BPS), their average BER will be lower than that of the fixed-mode modems. From a different perspective, at a given BER their BPS throughput will be always higher. In general, the higher the tolerable BER, the closer the performance to that of the Gaussian channel capacity. Again, this fact underlines the importance of designing programmable-rate, error-resilient source codecs - such as the Advanced Multi-Rate (AMR) speech codec to be employed in UMTS - which do not expect a low BER.

Similarly, when employing the above BbBA or AQAM principles in the frequency domain in the context of OFDM [180] or in conjunction with OVSF spreading codes in CDMA systems, attractive system design trade-offs and a high overall performance can be attained [168]. However, despite the extensive research in the field by the international community, there is a whole host of problems that remain to be solved and this monograph intends to contribute towards these efforts.

22.2 Adaptation Principles

AQAM is suitable for duplex communication between the MS and BS, since the AQAM modes have to be adapted and signalled between them, in order to allow channel quality estimates and signalling to take place. The AQAM mode adaptation is the action of the transmitter in response to time-varying channel conditions. In order to efficiently react to the changes in channel quality, the following steps have to be taken:

- *Channel quality estimation:* In order to appropriately select the transmission parameters to be employed for the next transmission, a reliable estimation of the channel transfer function during the next active transmit time-slot is necessary.

- *Choice of the appropriate parameters for the next transmission:* Based on the prediction of the channel conditions for the next timeslot, the transmitter has to select the appropriate modulation and channel coding modes for the subcarriers.

- *Signalling or blind detection of the employed parameters:* The receiver has to be informed, as to which demodulator parameters to employ for the received packet. This information can either be conveyed within the OFDM symbol itself, at the cost of loss of effective data throughput, or the receiver can attempt to estimate the parameters employed by the remote transmitter by means of blind detection mechanisms [180].

22.3 Channel Quality Metrics

The most reliable channel quality estimate is the bit error rate (BER), since it reflects the channel quality, irrespective of the source or the nature of the quality degradation. The BER can be estimated invoking a number of approaches.

Firstly, the BER can be estimated with a certain granularity or accuracy, provided that the system entails a channel decoder or - synonymously - Forward Error Correction (FEC) decoder employing algebraic decoding [171].

Secondly, if the system contains a soft-in-soft-out (SISO) channel decoder, the BER can be estimated with the aid of the Logarithmic Likelihood Ratio (LLR), evaluated either at the input or the output of the channel decoder. A particularly attractive way of invoking LLRs is employing powerful turbo codecs, which provide a reliable indication of the confidence associated with a particular bit decision in the context of LLRs.

Thirdly, in the event that no channel encoder / decoder (codec) is used in the system, the channel quality expressed in terms of the BER can be estimated with the aid of the mean-squared error (MSE) at the output of the channel equaliser or the closely related metric of Pseudo-Signal-to-Noise-Ratio (Pseudo-SNR) [168]. The MSE or pseudo-SNR at the output of the channel equaliser have the important advantage that they are capable of quantifying the severity of the inter-symbol-interference (ISI) and/or Co-channel Interference (CCI) experienced, in other words quantifying the Signal to Interference plus Noise Ratio (SINR).

As an example, let us consider OFDM. In OFDM modems [180] the bit error probability in each subcarrier can be determined by the fluctuations of the channel's instantaneous frequency domain channel transfer function H_n, if no co-channel interference is present. The estimate $\hat{H}_n$ of the channel transfer function can be acquired by means of pilot-tone based channel estimation [180]. For CDMA transceivers similar techniques are applicable, which constitute the topic of this monograph.

The delay between the channel quality estimation and the actual transmission of a burst in relation to the maximal Doppler frequency of the channel is crucial as regards the adaptive system's performance. If the channel estimate is obsolete at the time of transmission, then poor system performance will result [168].

22.4 Transceiver Parameter Adaptation

Different transmission parameters – such as the modulation and coding modes – of the AQAM single- and multi-carrier as well as CDMA transceivers can be adapted to the anticipated channel conditions. For example, adapting the number of modulation levels in response to the anticipated SNR encountered in each OFDM subcarrier can be employed, in order to achieve a wide range of different trade-offs between the received data integrity and throughput. Corrupted subcarriers can be excluded from data transmission and left blank or used for example for Crest-factor reduction. A range of different algorithms for selecting the appropriate modulation modes have to be investigated by future research. *The adaptive channel coding parameters entail code rate, adaptive interleaving and puncturing for convolutional and turbo codes, or varying block lengths for block codes [170, 171].*

Based on the estimated frequency-domain channel transfer function, *spectral pre-distortion at the transmitter of one or both communicating stations can be invoked, in order to partially of fully counteract the frequency-selective fading of the time-dispersive channel.* Unlike frequency-domain equalisation at the receiver – which corrects for the amplitude- and phase-errors inflicted upon the subcarriers by the channel, but which cannot improve the SNR in poor quality OFDM subchannels – spectral pre-distortion at the OFDM transmitter can deliver near-constant signal-to-noise levels for all subcarriers and can be viewed as power control on a subcarrier-by-subcarrier basis.

In addition to improving the system's BER performance in time-dispersive channels, spectral pre-distortion can be employed in order to perform all channel estimation and equalisation functions at only one of the two communicating duplex stations. Low-cost, low power consumption mobile stations can communicate with a base station that performs the channel estimation and frequency-domain equalisation of the uplink, and uses the estimated channel

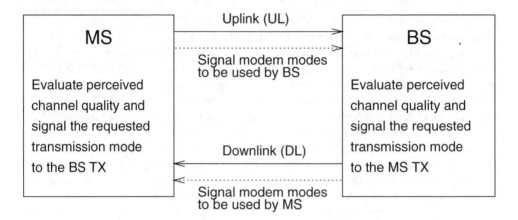

Figure 22.2: Parameter signalling in BbBA OFDM, CDMA and AQAM modems, IEEE Press-John Wiley, 2000, Hanzo, Webb, Keller [180]

transfer function for pre-distorting the down-link OFDM symbol. This setup would lead to different overall channel quality on the up- and down-link, and the superior pre-equalised downlink channel quality could be exploited by using a computationally less complex channel decoder, having weaker error correction capabilities in the mobile station than in the base station.

If the channel's frequency-domain transfer function is to be fully counteracted by the spectral pre-distortion upon adapting the subcarrier power to the inverse of the channel transfer function, then the output power of the transmitter can become excessive, if heavily faded subcarriers are present in the system's frequency range. In order to limit the transmitter's maximal output power, hybrid channel pre-distortion and adaptive modulation schemes can be devised, which would de-activate transmission in deeply faded subchannels, while retaining the benefits of pre-distortion in the remaining subcarriers.

BbBA mode signalling plays an important role in adaptive systems and the range of signalling options is summarised in Figure 22.2 for *closed-loop signalling*. If the channel quality estimation and parameter adaptation have been performed at the transmitter of a particular link, based on open-loop adaptation, then the resulting set of parameters has to be communicated to the receiver in order to successfully demodulate and decode the OFDM symbol. Once the receiver has determined the requested parameter set to be used by the remote transmitter, then this information has to be signalled to the remote transmitter in the reverse link. If this signalling information is corrupted, then the receiver is generally unable to correctly decode the OFDM symbol corresponding to the incorrect signalling information, yielding an OFDM symbol error.

Unlike adaptive serial systems, which employ the same set of parameters for all data symbols in a transmission packet [180], adaptive OFDM systems [180] have to react to the frequency selective nature of the channel, by adapting the modem parameters across the subcarriers. The resulting signalling overhead may become significantly higher than that for serial modems, and can be prohibitive for example for subcarrier-by-subcarrier based modulation mode adaptation. In order to overcome these limitations, efficient and reliable signalling techniques have to be employed for practical implementation of adaptive OFDM modems.

If some flexibility in choosing the transmission parameters is sacrificed in an adaptation scheme, as in sub-band adaptive OFDM schemes [180], then the amount of signalling can be reduced. Alternatively, blind parameter detection schemes can be devised, which require little or no OFDM mode signalling information, respectively [180].

In conclusion, fixed mode transceivers are incapable of achieving a good trade-off in terms of performance and complexity. The proposed BbB adaptive system design paradigm is more promising in this respect. A range of problems and solutions were highlighted in conceptual terms with reference to an OFDM-based example, indicating the areas, where substantial future research is required. A specific research topic, which raised substantial research interest recently is invoking efficient channel quality prediction techniques [175]. Before we commence our in-depth discourse in the forthcoming chapters, in the next section we provide a brief historical perspective on adaptive modulation.

22.5 Milestones in Adaptive Modulation History

22.5.1 Adaptive Single- and Multi-carrier Modulation

A comprehensive overview of adaptive transceivers was provided in [170] and this section is also based on [170]. As we noted in the previous chapters, mobile communications channels typically exhibit a near-instantaneously fluctuating time-variant channel quality [170–173] and hence conventional fixed-mode modems suffer from bursts of transmission errors, even if the system was designed for providing a high link margin. *An efficient approach to mitigating these detrimental effects is to adaptively adjust the modulation and/or the channel coding format as well as a range of other system parameters based on the near-instantaneous channel quality information perceived by the receiver, which is fed back to the transmitter with the aid of a feedback channel [174].* This plausible principle was recognised by Hayes [174] as early as 1968.

It was also shown in the previous sections that these near-instantaneously adaptive schemes require a reliable feedback link from the receiver to the transmitter. However, the channel quality variations have to be sufficiently slow for the transmitter to be able to adapt its modulation and/or channel coding format appropriately. The performance of these schemes can potentially be enhanced with the aid of *channel quality prediction techniques [175]*. As an efficient fading counter-measure, Hayes [174] proposed the employment of transmission power adaptation, while *Cavers [176] suggested invoking a variable symbol duration scheme* in response to the perceived channel quality at the expense of a variable bandwidth requirement. A disadvantage of the variable-power scheme is that it increases both the average transmitted power requirements and the level of co-channel interference imposed on other users, while requiring a high-linearity class-A or AB power amplifier, which exhibit a low power-efficiency. As a more attractive alternative, *the employment of AQAM was proposed by Steele and Webb*, which circumvented some of the above-mentioned disadvantages by employing various star-QAM constellations [124, 177].

With the advent of *Pilot Symbol Assisted Modulation (PSAM)* [139, 140, 178], Otsuki *et al.* [179] employed square-shaped AQAM constellations instead of star constellations [180], as a practical fading counter measure. With the aid of analysing the channel capacity of Rayleigh fading channels [181], *Goldsmith et al. [182] and Alouini et al. [183] showed that combined variable-power, variable-rate adaptive schemes are attractive* in terms of approaching the capacity of the channel and characterised the achievable throughput performance of variable-power AQAM [182]. However, they also found that *the extra throughput achieved by the additional variable-power assisted adaptation over the constant-power, variable-rate scheme is marginal* for most types of fading channels [182, 184].

In 1996 Torrance and Hanzo [185] proposed a set of *mode switching levels* s designed for achieving a high average BPS throughput, while maintaining the target average BER. Their method was based on defining a specific combined BPS/BER cost-function for transmission over narrowband Rayleigh channels, which incorporated both the BPS throughput as well as the target average BER of the system. *Powell's optimisation was invoked for finding a set of mode switching thresholds, which were constant*, regardless of the actual channel Signal to Noise Ratio (SNR) encountered, i.e. irrespective of the prevalent instantaneous channel conditions. However, in 2001 Choi and Hanzo [649] noted that a *higher BPS throughput can be achieved, if under high channel SNR conditions the activation of high-throughput AQAM modes is further encouraged by lowering the AQAM mode switching thresholds. More explicitly, a set of SNR-dependent AQAM mode switching levels was proposed [649]*, which keeps the average BER constant, while maximising the achievable throughput. We note furthermore that the set of switching levels derived in [185, 187] is based on Powell's multidimensional optimisation technique [188] and hence the optimisation process may become trapped in a local minimum. This problem was overcome by Choi and Hanzo upon deriving an *optimum set of switching levels [649]*, when employing the *Lagrangian multiplier technique*. It was shown that this set of switching levels results in the global optimum in a sense that the corresponding AQAM scheme obtains the maximum possible average BPS throughput, while maintaining the target average BER. An important further development was Tang's contribution [189] in the area of contriving an *intelligent learning scheme for the appropriate adjustment of the AQAM switching thresholds*.

These contributions demonstrated that *AQAM exhibited promising advantages*, when compared to fixed modulation schemes in terms of spectral efficiency, BER performance and robustness against channel delay spread, etc. Various systems employing AQAM were also characterised in [180]. The *numerical upper bound performance of narrow-band BbB-AQAM* over slow Rayleigh flat-fading channels was evaluated by Torrance and Hanzo [190], while over *wide-band channels* by Wong and Hanzo [191, 192]. Following these developments, adaptive modulation was also studied *in conjunction with channel coding and power control techniques by Matsuoka et al. [193] as well as Goldsmith and Chua [194, 195]*.

In the early phase of research more emphasis was dedicated to the system aspects of adaptive modulation in a narrow-band environment. A reliable method of transmitting the modulation control parameters was proposed by Otsuki, Sampei and Morinaga [179], where the parameters were embedded in the transmission frame's mid-amble using Walsh codes. Subsequently, at the receiver the Walsh sequences were decoded using maximum likelihood detection. Another technique of *signalling the required modulation mode used was proposed by Torrance and Hanzo [196], where the modulation control symbols were represented by unequal error protection 5-PSK symbols. Symbol-by-Symbol (SbS) adaptive, rather than BbB-adaptive systems were proposed by Lau and Maric in [197], where the transmitter is capable of transmitting each symbol in a different modem mode, depending on the channel conditions. Naturally, the receiver has to synchronise with the transmitter in terms of the SbS-adapted mode sequence, in order to correctly demodulate the received symbols and hence the employment of BbB-adaptivity is less challenging, while attaining a similar performance to that of BbB-adaptive arrangements under typical channel conditions.*

The adaptive modulation philosophy was then extended to wideband multi-path environments *amongst others for example by Kamio et al. [198] by utilising a bi-directional Decision Feedback Equaliser (DFE)* in a micro- and macro-cellular environment. This equalisation technique employed both forward and backward oriented channel estimation based on the pre-amble and post-amble symbols in the transmitted frame. Equaliser tap gain interpolation across the transmitted frame was also utilised for reducing the complexity in conjunction with space diversity [198]. The authors concluded that the cell radius could be enlarged in a macro-cellular system and a higher area-spectral efficiency could be attained for micro-cellular environments by utilising adaptive modulation. The data transmission latency effect, which occurred when the input data rate was higher than the instantaneous transmission throughput was studied and solutions were formulated using *frequency hopping [199] and statistical multiplexing, where the number of Time Division Multiple Access (TDMA) timeslots allocated to a user was adaptively controlled [200].*

In Reference [201] *symbol rate adaptive modulation* was applied, where the symbol rate or the number of modulation levels was adapted by using $\frac{1}{8}$-rate 16QAM, $\frac{1}{4}$-rate 16QAM, $\frac{1}{2}$-rate 16QAM as well as full-rate 16QAM, and the criterion used for adapting the modem modes was based on the instantaneous received signal to noise ratio and channel delay spread. The slowly varying channel quality of the uplink (UL) and downlink (DL) was rendered similar by utilising short frame duration Time Division Duplex (TDD) and the maximum normalised delay spread simulated was 0.1. A *variable channel coding rate* was then introduced by Matsuoka *et al.* in conjunction with adaptive modulation in Reference [193], where the transmitted burst incorporated an outer Reed Solomon code and an inner convolutional code in order to achieve high-quality data transmission. The coding rate was varied according to the prevalent channel quality using the same method, as in adaptive modulation in order to achieve a certain target BER performance. A so-called *channel margin* was introduced in this contribution, which effectively increased the switching thresholds for the sake of pre-empting *the effects of channel quality estimation errors*, although this inevitably reduced the achievable BPS throughput.

In an effort to improve the achievable performance versus complexity trade-off in the context of AQAM, Yee and Hanzo [202] studied the design of various *Radial Basis Function (RBF) assisted neural network-based schemes, while communicating over dispersive channels.* The advantage of these RBF-aided DFEs is that they are capable of delivering error-free decisions even in scenarios, when the received phasors cannot be error-freely detected by the conventional DFE, since they cannot be separated into decision classes with the aid of a linear decision boundary. In these so-called linearly non-separable decision scenarios the RBF-assisted DFE still may remain capable of classifying the received phasors into decision classes without decision errors. A further improved turbo BCH-coded version of this RBF-aided system was characterised by Yee *et al.* in [203], while a turbo-equalised RBF arrangement was the subject of the investigation conducted by Yee, Liew and Hanzo in [204, 205]. The RBF-aided AQAM research has also been extended to the turbo equalisation of a convolutional as well as space-time trellis coded arrangement proposed by Yee, Yeap and Hanzo [170, 206, 207]. The same authors then endeavoured to reduce the associated implementation complexity of an RBF-aided QAM modem with the advent of employing a separate in-phase / quadrature-phase turbo equalisation scheme in the quadrature arms of the modem.

As already mentioned above, the performance of *channel coding in conjunction with adaptive modulation in a narrow-band environment* was also characterised by Chua and Goldsmith [194]. In their contribution trellis and lattice codes were used without channel interleaving, invoking a feedback path between the transmitter and receiver for modem mode control purposes. Specifically, the simulation and theoretical results by Goldsmith and Chua showed that a 3dB coding gain was achievable at a BER of 10^{-6} for a 4-sate trellis code and 4dB by an 8-state trellis code in the context of the adaptive scheme over Rayleigh-fading channels, while a 128-state code performed within 5dB of the Shannonian capacity limit. The *effects of the delay in the AQAM mode signalling feedback path* on the adaptive modem's performance were studied and this scheme exhibited a higher spectral efficiency, when compared to the non-adaptive trellis coded performance. Goeckel [208] also contributed in the area of adaptive coding and employed realistic outdated, rather than perfect fading estimates. Further research on adaptive multidimensional coded modulation was also conducted by Hole *et al.* [209] for transmissions over flat fading channels. *Pearce, Burr and Tozer [210] as well as Lau and Mcleod [211] have also analysed the performance trade-offs associated with employing channel coding and adaptive modulation or adaptive trellis coding,* respectively, as efficient fading counter measures. In an effort to provide a fair comparison of the various coded modulation schemes known at the time of writing, Ng, Wong and Hanzo have also studied Trellis Coded Modulation (TCM), Turbo TCM (TTCM), Bit-Interleaved Coded Modulation (BICM) and Iterative-Decoding assisted BICM (BICM-ID), where TTCM was found to be the best scheme at a given decoding complexity [212].

Subsequent contributions by Suzuki *et al.* [213] incorporated *space-diversity and power-adaptation* in conjunction with adaptive modulation, for example in order to combat the effects of the multi-path channel environment at a 10Mbits/s transmission rate. *The maximum tolerable delay-spread was deemed to be one symbol duration for a target mean BER performance of* 0.1%. This was achieved in a TDMA scenario, where the channel estimates were predicted based on the extrapolation of previous channel quality estimates. As mentioned above, variable transmitted power was applied in combination with adaptive modulation in Reference [195], where the transmission rate and power adaptation was optimised for the sake of achieving an increased spectral efficiency. In their treatise a slowly varying channel was assumed and the instantaneous received power required for achieving a certain upper bound performance was assumed to be known prior to transmission. *Power control in conjunction with a pre-distortion type non-linear power amplifier compensator* was studied in the context of adaptive modulation in Reference [214]. This method was used to mitigate the non-linearity effects associated with the power amplifier, when QAM modulators were used.

Results were also recorded concerning the performance of adaptive modulation in conjunction with *different multiple access schemes in a narrow-band channel environment*. In a TDMA system, *dynamic channel assignment* was employed by Ikeda *et al.*, where in addition to assigning a different modulation mode to a different channel quality, priority was always given to those users in their request for reserving time-slots, which benefitted from the best channel quality [215]. The performance was compared to fixed channel assignment systems, where substantial gains were achieved in terms of system capacity. Furthermore, a *lower call termination probability was recorded.* However, the probability of intra-cell hand-off increased as a result of the associated dynamic channel assignment (DCA) scheme, which constantly searched for a high-quality, high-throughput time-slot for supporting the actively communicating users. The application of adaptive modulation in packet transmission was introduced by Ue, Sampei and Morinaga [216], where the results showed an improved BPS throughput. The performance of adaptive modulation was also characterised in conjunction with an *automatic repeat request* (ARQ) system in Reference [217], where the transmitted bits were encoded using a cyclic redundant code (CRC) and a convolutional punctured code in order to increase the data throughput.

A further treatise was published by Sampei, Morinaga and Hamaguchi [218] on *laboratory test results* concerning the utilisation of adaptive modulation in a TDD scenario, where the modem mode switching criterion was based on the signal to noise ratio and on the normalised delay-spread. *In these experimental results, the channel quality estimation errors degraded the performance* and consequently - as already alluded to earlier - a channel estimation error margin was introduced for mitigating this degradation. Explicitly, the channel estimation error margin was defined as the measure of how much extra protection margin must be added to the switching threshold levels for the sake of minimising the effects of the channel estimation errors. The delay-spread also degraded the achievable performance due to the associated irreducible BER, which was not compensated by the receiver. However, the performance of the adaptive scheme in a delay-spread impaired channel environment was better than that of a fixed modulation scheme. *These experiments also concluded that the AQAM scheme can be operated for a Doppler frequency of $f_d = 10Hz$ at a normalised delay spread of 0.1 or for $f_d = 14Hz$ at a normalised delay spread of 0.02, which produced a mean BER of 0.1% at a transmission rate of 1 Mbits/s.*

Finally, the *data buffering-induced latency and co-channel interference aspects* of AQAM modems were investigated in [219, 220]. Specifically, the latency associated with storing the information to be transmitted during severely degraded channel conditions was mitigated by frequency hopping or statistical multiplexing. As expected, the latency is increased, when either the mobile speed or the channel SNR are reduced, since both of these result in prolonged low instantaneous SNR intervals. It was demonstrated that as a result of the proposed measures, typically more than 4dB SNR reduction was achieved by the proposed adaptive modems in comparison to the conventional fixed-mode benchmark modems employed. However, the achievable gains depend strongly on the prevalant co-channel interference levels and hence interference cancellation was invoked in [220] on the basis of adjusting the demodulation decision boundaries after estimating the interfering channel's magnitude and phase.

The associated principles can also be invoked in the context of *multicarrier Orthogonal Frequency Division Multiplex (OFDM) modems [180]*. This principle was first proposed by Kalet [155] and was then further developed for example by Czylwik *et al.* [221] as well as by Chow, Cioffi and Bingham [222]. The associated concepts were detailed for example in [180] and will be also augmented in this monograph. Let us now briefly review the recent history of the BbB adaptive concept in the context of CDMA in the next section.

22.5.2 Adaptive Code Division Multiple Access

The techniques described in the context of single- and multi-carrier modulation are conceptually similar to multi-rate transmission [650] in CDMA systems. However, in BbB adaptive CDMA the transmission rate is modified according to the near-instantaneous channel quality, instead of the service required by the mobile user. BbB-adaptive CDMA systems are also useful for employment in arbitrary propagation environments or in hand-over scenarios, such as those encountered, when a mobile user moves from an indoor to an outdoor environment or in a so-called 'birth-death' scenario, where the number of transmitting CDMA users changes frequently [651], thereby changing the interference dramatically. Various methods of multi-rate transmission have been proposed in the research literature. Below we will briefly discuss some of the recent research issues in multi-rate and adaptive CDMA schemes.

Ottosson and Svensson compared various multi-rate systems [650], including multiple spreading factor (SF) based, multi-code and multi-level modulation schemes. According to the multi-code philosophy, the SF is kept constant for all users, but multiple spreading codes transmitted simultaneously are assigned to users requiring higher bit rates. In this case – unless the spreading codes's perfect orthogonality is retained after transmission over the channel – the multiple codes of a particular user interfere with each other. This inevitably reduces the system's performance.

Multiple data rates can also be supported by a variable SF scheme, where the chip rate is kept constant, but the data rates are varied, thereby effectively changing the SF of the spreading codes assigned to the users; at a fixed chip rate the lower the SF, the higher the supported data rate. Performance comparisons for both of these schemes have been carried out by Ottosson and Svensson [650], as well as by Ramakrishna and Holtzman [652], demonstrating that both schemes achieved a similar performance. Adachi, Ohno, Higashi, Dohi and Okumura proposed the employment of multi-code CDMA in conjunction with pilot symbol-assisted channel estimation, RAKE reception and antenna diversity for providing multi-rate capabilities [653, 654]. The employment of multi-level modulation schemes was also investigated by Ottosson and Svensson [650], where higher-rate users were assigned higher-order modulation modes, transmitting several bits per symbol. However, it was concluded that the performance experienced by users requiring higher rates was significantly worse than that experienced by the lower-rate users. The use of M-ary orthogonal modulation in providing variable rate transmission was investigated by Schotten, Elders-Boll and Busboom [655]. According to this method, each user was assigned an orthogonal sequence set, where the number of sequences, M, in the set was dependent on the data rate required – the higher the rate required, the larger the sequence set. Each sequence in the set was mapped to a particular combination of $b = (\log_2 M)$ bits to be transmitted. The M-ary sequence was then spread with the aid of a spreading code of a constant SF before transmission. It was found [655] that the performance of the system depended not only on the MAI, but also on the Hamming distance between the sequences in the M-ary sequence set.

Saquib and Yates [656] investigated the employment of the decorrelating detector in conjunction with the multiple-SF scheme and proposed a modified decorrelating detector, which utilised soft decisions and maximal ratio combining, in order to detect the bits of the different-rate users. Multi-rate transmission schemes involving interference cancellation receivers have previously been investigated amongst others by Johansson and Svensson [657,658], as well as by Juntti [659]. Typically, multiple users transmitting at different bit rates are supported in the same CDMA system invoking multiple codes or different spreading factors. SIC schemes and multi-stage cancellation schemes were used at the receiver for mitigating the MAI [657–659], where the bit rate of the users was dictated by the user requirements. The performance comparison of various multiuser detectors in the context of a multiple-SF transmission scheme was presented for example by Juntti [659], where the detectors compared were the decorrelator, the PIC receiver and the so-called group serial interference cancellation (GSIC) receiver. It was concluded that the GSIC and the decorrelator performed better than the PIC receiver, but all the interference cancellation schemes including the GSIC, exhibited an error floor at high SNRs due to error propagation.

The bit rate of each user can also be adapted according to the near-instantaneous channel quality, in order to mitigate the effects of channel quality fluctuations. Kim [660] analysed the performance of two different methods of combating the near-instantaneous quality variations of the mobile channel. Specifically, Kim studied the adaptation of the transmitter power or the switching of the information rate, in order to suit the near-instantaneous channel conditions. It was demonstrated using a RAKE receiver that rate adaptation provided a higher average information rate than power adaptation for a given average transmit power and a given BER [660]. Abeta, Sampei and Morinaga [661] conducted investigations into an adaptive packet transmission based CDMA scheme, where the transmission rate was modified by varying the channel code rate and the processing gain of the CDMA user, employing the carrier to interference plus noise ratio (CINR) as the switching metric. When the channel quality was favourable, the instantaneous bit rate was increased and conversely, the instantaneous bit rate was reduced when the channel quality dropped. In order to maintain a constant overall bit rate, when a high instantaneous bit rate was employed, the duration of the transmission burst was reduced. Conversely, when the instantaneous bit rate was low, the duration of the burst was extended. This resulted in a decrease in interference power, which translated to an increase in system capacity. Hashimoto, Sampei and Morinaga [662] extended this work also to demonstrate that the proposed system was capable of achieving a higher user capacity with a reduced hand-off margin and lower average transmitter power. In these schemes the conventional RAKE receiver was used for the detection of the data symbols. A variable-rate CDMA scheme – where the transmission rate was modified by varying the channel code rate and, correspondingly, the M-ary modulation constellations – was investigated by Lau and Maric [197]. As the channel code rate was increased, the bit-rate was increased by increasing M correspondingly in the M-ary modulation scheme. Another adaptive system was proposed by Tateesh, Atungsiri and Kondoz [663], where the rates of the speech and channel codecs were varied adaptively. In their adaptive system, the gross transmitted bit rate was kept constant, but the speech codec and channel codec rates were varied according to the channel quality. When the channel quality was low, a lower rate speech codec was used, resulting in increased redundancy and thus a more powerful channel code could be employed. This resulted in an overall coding gain, although the speech quality dropped with decreasing speech rate. A variable rate data transmission scheme was proposed by Okumura and Adachi [664], where the fluctuating transmission rate was mapped to discontinuous transmission, in order to reduce the interference inflicted upon the other users, when there was no transmission. The transmission rate was detected blindly at the receiver with the help of cyclic redundancy check decoding and RAKE receivers were employed for coherent reception, where pilot-symbol-assisted channel estimation was performed.

The information rate can also be varied according to the channel quality, as will be demonstrated shortly. However, in comparison to conventional power control techniques - which again, may disadvantage other users in an effort to maintain the quality of the links considered - the proposed technique does not disadvantage other users and increases the network capacity [170, 665]. The instantaneous channel quality can be estimated at the receiver and the chosen information rate can then be communicated to the transmitter via explicit signalling in a so-called closed-loop controlled scheme. Conversely, in an open-loop scheme - provided that the downlink and uplink channels exhibit a similar quality - the information rate for the downlink transmission can be chosen according to the channel quality estimate related

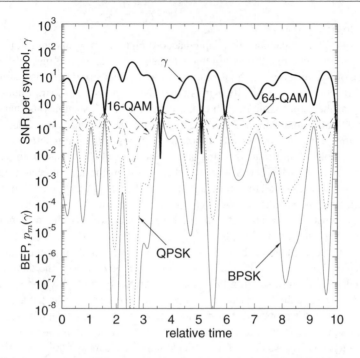

Figure 22.3: Instantaneous SNR per transmitted symbol, γ, in a flat Rayleigh fading scenario and the associated instantaneous bit error probability, $p_m(\gamma)$, of a fixed-mode QAM. The average SNR is $\bar{\gamma} = 10$dB. The fading magnitude plot is based on a normalised Doppler frequency of $f_N = 10^{-4}$ and for the duration of $100ms$, corresponding to a mobile terminal travelling at the speed of $54km/h$ and operating at $f_c = 2GHz$ frequency band at the sampling rate of $1MHz$

to the uplink and vice versa. The validity of the uplink/downlink similarity in TDD-CDMA systems has been studied by Miya *et al.* [579], Kato *et al.* [580] and Jeong *et al.* [666].

22.6   Increasing the Average Transmit Power as a Fading Counter-Measure

The radio frequency (RF) signal radiated from the transmitter's antenna takes different routes, experiencing defraction, scattering and reflections, before it arrives at the receiver. Each multi-path component arriving at the receiver simultaneously adds constructively or destructively, resulting in fading of the combined signal. When there is no line-of-sight component amongst these signals, the combined signal is characterised by Rayleigh fading. The instantaneous SNR (iSNR), γ, per transmitted symbol[1] is depicted in Figure 22.3 for a typical Rayleigh fading using the thick line. The Probability Density Function (PDF) of γ is given

---

[1]When no diversity is employed at the receiver, the SNR per symbol, γ, is the same as the channel SNR, γ_c. In this case, we will use the term "SNR" without any adjective.

as [388]:

$$f_{\bar{\gamma}}(\gamma) = \frac{1}{\bar{\gamma}} e^{\gamma/\bar{\gamma}} , \tag{22.1}$$

where $\bar{\gamma}$ is the average SNR and $\bar{\gamma} = 10$dB was used in Figure 22.3.

The instantaneous Bit Error Probability (iBEP), $p_m(\gamma)$, of BPSK, QPSK, 16-QAM and 64-QAM is also shown in Figure 22.3 with the aid of four different thin lines. These probabilities are obtained from the corresponding bit error probability over AWGN channel conditioned on the iSNR, γ, which are given as [180]:

$$p_m(\gamma) = \sum_i A_i Q(\sqrt{a_i \gamma}) , \tag{22.2}$$

where $Q(x)$ is the Gaussian Q-function defined as $Q(x) \triangleq \frac{1}{\sqrt{2\pi}} \int_x^\infty e^{-t^2/2} dt$ and $\{A_i, a_i\}$ is a set of modulation mode dependent constants. For the Gray-mapped square QAM modulation modes associated with $m = 2, 4, 16, 64$ and 256, the sets $\{A_i, a_i\}$ are given as [180,667]:

$$
\begin{array}{lll}
m = 2, & \text{BPSK} & \{(1,2)\} \\
m = 4, & \text{QPSK} & \{(1,1)\} \\
m = 16, & \text{16-QAM} & \left\{ \left(\frac{3}{4}, \frac{1^2}{5}\right), \left(\frac{2}{4}, \frac{3^2}{5}\right), \left(-\frac{1}{4}, \frac{5^2}{5}\right) \right\} \\
m = 64, & \text{64-QAM} & \left\{ \left(\frac{7}{12}, \frac{1^2}{21}\right), \left(\frac{6}{12}, \frac{3^2}{21}\right), \left(-\frac{1}{12}, \frac{5^2}{21}\right), \left(\frac{1}{12}, \frac{9^2}{21}\right), \left(-\frac{1}{12}, \frac{13^2}{21}\right) \right\} \\
m = 256, & \text{256-QAM} & \left\{ \left(\frac{15}{32}, \frac{1^2}{85}\right), \left(\frac{14}{32}, \frac{3^2}{85}\right), \left(\frac{5}{32}, \frac{5^2}{85}\right), \left(-\frac{6}{32}, \frac{7^2}{85}\right), \left(-\frac{7}{32}, \frac{9^2}{85}\right), \right. \\
& & \left. \left(\frac{6}{32}, \frac{11^2}{85}\right), \left(\frac{9}{32}, \frac{13^2}{85}\right), \left(\frac{8}{32}, \frac{15^2}{85}\right), \left(-\frac{7}{32}, \frac{17^2}{85}\right), \left(-\frac{6}{32}, \frac{19^2}{85}\right), \right. \\
& & \left. \left(-\frac{1}{32}, \frac{21^2}{85}\right), \left(\frac{2}{32}, \frac{23^2}{85}\right), \left(\frac{3}{32}, \frac{25^2}{85}\right), \left(-\frac{2}{32}, \frac{27^2}{85}\right), \left(-\frac{1}{32}, \frac{29^2}{85}\right) \right\} .
\end{array}
\tag{22.3}
$$

As we can observe in Figure 22.3, $p_m(\gamma)$ exhibits high values during the deep channel envelope fades, where even the most robust modulation mode, namely BPSK, exhibits a bit error probability $p_2(\gamma) > 10^{-1}$. By contrast even the error probability of the high-throughput 16-QAM mode, namely $p_{16}(\gamma)$, is below 10^{-2}, when the iSNR γ exhibits a high peak. This wide variation in the communication link's quality is a fundamental problem in wireless radio communication systems. Hence, numerous techniques have been developed for combating this problem, such as increasing the average transmit power, invoking diversity, channel inversion, channel coding and/or adaptive modulation techniques. In this section we will investigate the efficiency of employing an increased average transmit power.

As we observed in Figure 22.3, the instantaneous Bit Error Probability (BEP) becomes excessive, when sustaining an adequate service quality during instances, when the signal experiences a deep channel envelope fade. Let us define the cut-off BEP p_c, below which the Quality Of Service (QOS) becomes unacceptable. Then the outage probability P_{out} can be defined as:

$$P_{out}(\bar{\gamma}, p_c) \triangleq \Pr[p_m(\gamma) > p_c] , \tag{22.4}$$

where $\bar{\gamma}$ is the average channel SNR dependent on the transmit power, p_c is the cut-off BEP and $p_m(\gamma)$ is the instantaneous BEP, conditioned on γ, for an m-ary modulation mode, given for example by (22.2). We can reduce the outage probability of (22.4) by increasing the

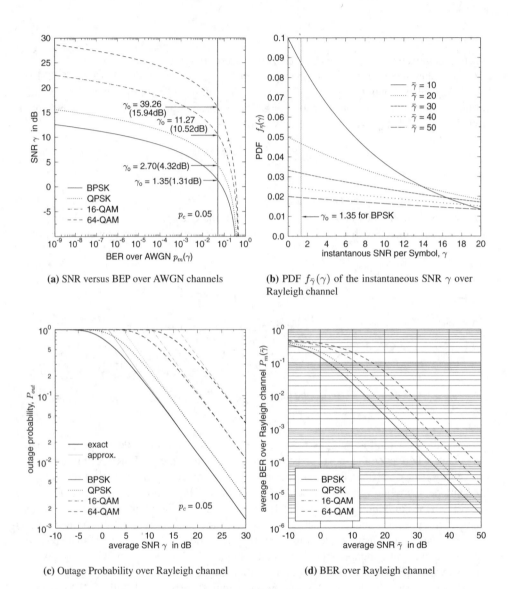

(a) SNR versus BEP over AWGN channels

(b) PDF $f_{\bar{\gamma}}(\gamma)$ of the instantaneous SNR γ over Rayleigh channel

(c) Outage Probability over Rayleigh channel

(d) BER over Rayleigh channel

Figure 22.4: The effects of an increased average transmit power. (a) The cut-off SNR γ_o versus the cut-off BEP p_c for BPSK, QPSK, 16-QAM and 64-QAM. (b) PDF of the iSNR γ over Rayleigh channel, where the outage probability is given by the area under the PDF curve surrounded the two lines given by $\gamma = 0$ and $\gamma = \gamma_o$. An increased transmit power increases the average SNR $\bar{\gamma}$ and hence reduces the area under the PDF proportionately to $\bar{\gamma}$. (c) The exact outage probability versus the average SNR $\bar{\gamma}$ for BPSK, QPSK, 16-QAM and 64-QAM evaluated from (22.7) confirms this observation. (d) The average BEP is also inversely proportional to the transmit power for BPSK, QPSK, 16-QAM and 64-QAM

transmit power, and hence increasing the average channel SNR $\bar{\gamma}$. Let us briefly investigate the efficiency of this scheme.

Figure 22.4(a) depicts the instantaneous BEP as a function of the instantaneous channel SNR. Once the cut-off BEP p_c is determined as a QOS-related design parameter, the corresponding cut-off SNR γ_o can be determined, as shown for example in Figure 22.4(a) for $p_c = 0.05$. Then, the outage probability of (22.4) can be calculated as:

$$P_{out} = \Pr[\gamma < \gamma_o], \tag{22.5}$$

and in physically tangible terms its value is equal to the area under the PDF curve of Figure 22.4(b) surrounded by the left y-axis and $\gamma = \gamma_o$ vertical line. Upon taking into account that for high SNRs the PDFs of Figure 22.4(b) are near-linear, this area can be approximated by $\gamma_o/\bar{\gamma}$, considering that $f_{\bar{\gamma}}(0) = 1/\bar{\gamma}$. Hence, the outage probability is inversely proportional to the transmit power, requiring an approximately 10-fold increased transmit power for reducing the outage probability by an order of magnitude, as seen in Figure 22.4(c). The exact value of the outage probability is given by:

$$P_{out} = \int_0^{\gamma_o} f_{\bar{\gamma}}(\gamma)\, d\gamma \tag{22.6}$$

$$= 1 - e^{-\gamma_o/\bar{\gamma}}, \tag{22.7}$$

where we used the PDF $f_{\bar{\gamma}}(\gamma)$ given in (22.1). Again, Figure 22.4(c) shows the exact outage probabilities together with their linearly approximated values for several QAM modems recorded for the cut-off BEP of $p_c = 0.05$, where we can confirm the validity of the linearly approximated outage probability,[2] when we have $P_{out} < 0.1$.

The average BEP $P_m(\bar{\gamma})$ of an m-ary Gray-mapped QAM modem is given by [180, 388, 668]:

$$P_m(\bar{\gamma}) = \int_0^{\infty} p_m(\gamma)\, f_{\bar{\gamma}}(\gamma)\, d\gamma \tag{22.8}$$

$$= \frac{1}{2} \sum_i A_i \{1 - \mu(\bar{\gamma}, a_i)\}, \tag{22.9}$$

where a set of constants $\{A_i, a_i\}$ is given in (22.3) and $\mu(\bar{\gamma}, a_i)$ is defined as:

$$\mu(\bar{\gamma}, a_i) \triangleq \sqrt{\frac{a_i \bar{\gamma}}{1 + a_i \bar{\gamma}}}. \tag{22.10}$$

In physical terms (22.8) implies weighting the BEP $p_m(\gamma)$ experienced at an iSNR γ by the probability of occurrence of this particular value of γ - which is quantified by its PDF $f_{\bar{\gamma}}(\gamma)$ - and then averaging, i.e. integrating, this weighted BEP over the entire range of γ. Figure 22.4(d) displays the average BER evaluated from (22.9) for the average SNR rage of $-10\text{dB} \geq \bar{\gamma} \geq 50\text{dB}$. We can observe that the average BEP is also inversely proportional to the transmit power.

---

[2]The same approximate outage probability can be derived by taking the first term of the Taylor series of e^x of (22.7).

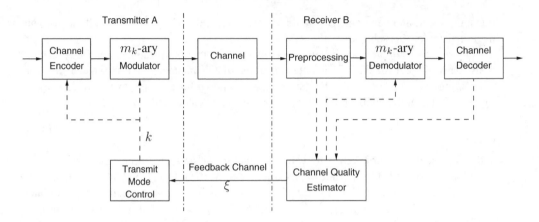

Figure 22.5: Stylised model of near-instantaneous adaptive modulation scheme

In conclusion, we studied the efficiency of increasing the average transmit power as a fading counter-measure and found that the outage probability as well as the average bit error probability are inversely proportional to the average transmit power. Since the maximum radiated powers of modems are regulated in order to reduce the co-channel interference and transmit power, the acceptable transmit power increase may be limited and hence employing this technique may not be sufficiently effective for achieving the desired link performance. We will show that the AQAM philosophy of the next section is a more attractive solution to the problem of channel quality fluctuation experienced in wireless systems.

22.7 System Description

A stylised model of our adaptive modulation scheme is illustrated in Figure 22.5, which can be invoked in conjunction with any power control scheme. In our adaptive modulation scheme, the modulation mode used is adapted on a near-instantaneous basis for the sake of counteracting the effects of fading. Let us describe the detailed operation of the adaptive modem scheme of Figure 22.5. Firstly, the channel quality ξ is estimated by the remote receiver B. This channel quality measure ξ can be the instantaneous channel SNR, the Radio Signal Strength Indicator (RSSI) output of the receiver [177], the decoded BER [177], the Signal to Interference-and-Noise Ratio (SINR) estimated at the output of the channel equaliser [192], or the SINR at the output of a CDMA joint detector [669]. The estimated channel quality perceived by receiver B is fed back to transmitter A with the aid of a feedback channel, as seen in Figure 22.5. Then, the transmit mode control block of transmitter A selects the highest-throughput modulation mode k capable of maintaining the target BEP based on the channel quality measure ξ and the specific set of adaptive mode switching levels **s**. Once k is selected, m_k-ary modulation is performed at transmitter A in order to generate the transmitted signal $s(t)$, and the signal $s(t)$ is transmitted through the channel.

The general model and the set of important parameters specifying our constant-power adaptive modulation scheme are described in the next subsection in order to develop the

k	0	1	2	3	4
m_k	0	2	4	16	64
b_k	0	1	2	4	6
c_k	0	1	1	2	2
modem	No Tx	BPSK	QPSK	16-QAM	64-QAM

Table 22.1: The parameters of five-mode AQAM system.

underlying general theory. Then, in Subsection 22.7.2 several application examples are introduced.

22.7.1 General Model

A K-mode adaptive modulation scheme adjusts its transmit mode k, where $k \in \{0, 1 \cdots K - 1\}$, by employing m_k-ary modulation according to the near-instantaneous channel quality ξ perceived by receiver B of Figure 22.5. The mode selection rule is given by:

$$\text{Choose mode } k \text{ when } s_k \leq \xi < s_{k+1}, \tag{22.11}$$

where a switching level s_k belongs to the set $\mathbf{s} = \{s_k \mid k = 0, 1, \cdots, K\}$. The Bits Per Symbol (BPS) throughput b_k of a specific modulation mode k is given by $b_k = \log_2(m_k)$ if $m_k \neq 0$ otherwise $b_k = 0$. It is convenient to define the incremental BPS c_k as $c_k = b_k - b_{k-1}$, when $k > 0$ and $c_0 = b_0$, which quantifies the achievable BPS increase, when switching from the lower-throughput mode $k-1$ to mode k.

22.7.2 Examples

22.7.2.1 Five-Mode AQAM

A five-mode AQAM system has been studied extensively by many researchers, which was motivated by the high performance of the Gray-mapped constituent modulation modes used. The parameters of this five-mode AQAM system are summarised in Table 22.1. In our investigation, the near-instantaneous channel quality ξ is defined as instantaneous channel SNR γ. The boundary switching levels are given as $s_0 = 0$ and $s_5 = \infty$. Figure 22.6 illustrates operation of the five-mode AQAM scheme over a typical narrow-band Rayleigh fading channel scenario. Transmitter A of Figure 22.5 keeps track of the channel SNR γ perceived by receiver B with the aid of a low-BER, low-delay feedback channel – which can be created for example by superimposing the values of ξ on the reverse direction transmitted messages of transmitter B – and determines the highest-BPS modulation mode maintaining the target BEP depending on which region γ falls into. The channel-quality related SNR regions are divided by the modulation mode switching levels s_k. More explicitly, the set of AQAM switching levels $\{s_k\}$ is determined so that the average BPS throughput is maximised, while satisfying the average target BEP requirement, P_{target}. We assumed a target BEP of $P_{target} = 10^{-2}$ in Figure 22.6. The associated instantaneous BPS throughput b is also depicted using the

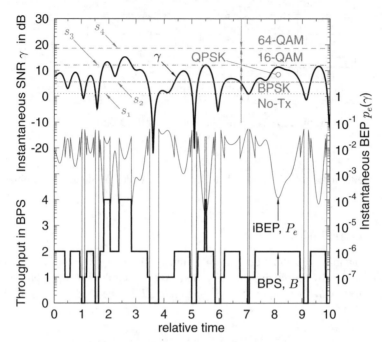

Figure 22.6: The operation of the five-mode AQAM scheme over a Rayleigh fading channel. The instantaneous channel SNR γ is represented as a thick line at the top part of the graph, the associated instantaneous BEP $P_e(\gamma)$ as a thin line at the middle, and the instantaneous BPS throughput $b(\gamma)$ as a thick line at the bottom. The average SNR is $\bar{\gamma} = 10$dB, while the target BEP is $P_{target} = 10^{-2}$.

thick stepped line at the bottom of Figure 22.6. We can observe that the throughput varied from 0 BPS, when the no transmission (No-Tx) QAM mode was chosen, to 4 BPS, when the 16-QAM mode was activated. During the depicted observation window the 64-QAM mode was not activated. The instantaneous BEP, depicted as a thin line using the middle trace of Figure 22.6, is concentrated around the target BER of $P_{target} = 10^{-2}$.

22.7.2.2 Seven-Mode Adaptive Star-QAM

Webb and Steele revived the research community's interest on adaptive modulation, although a similar concept was initially suggested by Hayes [174] in the 1960s. Webb and Steele reported the performance of adaptive star-QAM systems [177]. The parameters of their system are summarised in Table 22.2.

22.7.2.3 Five-Mode APSK

Our five-mode Adaptive Phase-Shift-Keying (APSK) system employs m-ary PSK constituent modulation modes. The magnitude of all the constituent constellations remained constant, where adaptive modem parameters are summarised in Table 22.3.

k	0	1	2	3	4	5	6
m_k	0	2	4	8	16	32	64
b_k	0	1	2	3	4	5	6
c_k	0	1	1	1	1	1	1
modem	No Tx	BPSK	QPSK	8-QAM	16-QAM	32-QAM	64-QAM

Table 22.2: The parameters of a seven-mode adaptive star-QAM system [177], where 8-QAM and 16-QAM employed four and eight constellation points allocated to two concentric rings, respectively, while 32-QAM and 64-QAM employed eight and 16 constellation points over four concentric rings, respectively.

k	0	1	2	3	4
m_k	0	2	4	8	16
b_k	0	1	2	3	4
c_k	0	1	1	1	1
modem	No Tx	BPSK	QPSK	8-PSK	16-PSK

Table 22.3: The parameters of the five-mode APSK system.

22.7.2.4 Ten-Mode AQAM

Hole, Holm and Øien [209] studied a trellis coded adaptive modulation scheme based on eight-mode square- and cross-QAM schemes. Upon adding the No-Tx and BPSK modes, we arrive at a ten-mode AQAM scheme. The associated parameters are summarised in Table 22.4.

22.7.3 Characteristic Parameters

In this section, we introduce several parameters in order to characterise our adaptive modulation scheme. The constituent mode selection probability (MSP) $\mathcal{M}_k$ is defined as the probability of selecting the k-th mode from the set of K possible modulation modes, which can be calculated as a function of the channel quality metric ξ, regardless of the specific

k	0	1	2	3	4	5	6	7	8	9
m_k	0	2	4	8	16	32	64	128	256	512
b_k	0	1	2	3	4	5	6	7	8	9
c_k	0	1	1	1	1	1	1	1	1	1
modem	No Tx	BPSK	QPSK	8-Q	16-Q	32-C	64-Q	128-C	256-Q	512-C

Table 22.4: The parameters of the ten-mode adaptive QAM scheme based on [209], where m-Q stands for m-ary square QAM and m-C for m-ary cross QAM.

metric used, as:

$$\mathcal{M}_k = \Pr[\, s_k \leq \xi < s_{k+1} \,] \tag{22.12}$$

$$= \int_{s_k}^{s_{k+1}} f(\xi)\, d\xi \,, \tag{22.13}$$

where s_k denotes the mode switching levels and $f(\xi)$ is the probability density function (PDF) of ξ. Then, the average throughput B expressed in terms of BPS can be described as:

$$B = \sum_{k=0}^{K-1} b_k \int_{s_k}^{s_{k+1}} f(\xi)\, d\xi \tag{22.14}$$

$$= \sum_{k=0}^{K-1} b_k \, \mathcal{M}_k \,, \tag{22.15}$$

which in simple verbal terms can be formulated as the weighted sum of the throughput b_k of the individual constituent modes, where the weighting takes into account the probability $\mathcal{M}_k$ of activating the various constituent modes. When $s_K = \infty$, the average throughput B can also be formulated as:

$$B = \sum_{k=0}^{K-1} b_k \int_{s_k}^{s_{k+1}} f(\xi)\, d\xi \tag{22.16}$$

$$= \sum_{k=0}^{K-1} c_k \int_{s_k}^{\infty} f(\xi)\, d\xi \tag{22.17}$$

$$= \sum_{k=0}^{K-1} c_k \, F_c(s_k), \tag{22.18}$$

where $F_c(\xi)$ is the complementary Cumulative Distribution Function (CDF) defined as:

$$F_c(\xi) \triangleq \int_{\xi}^{\infty} f(x)\, dx \,. \tag{22.19}$$

Let us now assume that we use the instantaneous SNR γ as the channel quality measure ξ, which implies that no co-channel interference is present. By contrast, when operating in a co-channel interference limited environment, we can use the instantaneous SINR as the channel quality measure ξ, provided that the co-channel interference has a near-Gaussian distribution. In such scenario, the mode-specific average BEP P_k can be written as:

$$P_k = \int_{s_k}^{s_{k+1}} p_{m_k}(\gamma)\, f(\gamma)\, d\gamma \,, \tag{22.20}$$

where $p_{m_k}(\gamma)$ is the BEP of the m_k-ary constituent modulation mode over the AWGN channel and we used γ instead of ξ in order to explicitly indicate the employment of γ as the channel quality measure. Then, the average BEP P_{avg} of our adaptive modulation scheme can be represented as the sum of the BEPs of the specific constituent modes divided by the average adaptive modem throughput B, formulated as [190]:

$$P_{avg} = \frac{1}{B} \sum_{k=0}^{K-1} b_k \, P_k \, , \qquad (22.21)$$

where b_k is the BPS throughput of the k-th modulation mode, P_k is the mode-specific average BEP given in (22.20) and B is the average adaptive modem throughput given in (22.15) or in (22.18).

The aim of our adaptive system is to transmit as high a number of bits per symbol as possible, while providing the required Quality of Service (QOS). More specifically, we are aiming for maximising the average BPS throughput B of (22.14), while satisfying the average BEP requirement of $P_{avg} \leq P_{target}$. Hence, we have to satisfy the constraint of meeting P_{target}, while optimising the design parameter of s, which is the set of modulation-mode switching levels. The determination of optimum switching levels will be investigated in Section 22.8. Since the calculation of the optimum switching levels typically requires the numerical computation of the parameters introduced in this section, it is advantageous to express the parameters in a closed form, which is the objective of the next section.

22.7.3.1   Closed Form Expressions for Transmission over Nakagami Fading Channels

Fading channels often are modelled as Nakagami fading channels [670]. The PDF of the instantaneous channel SNR γ over a Nakagami fading channel is given as [670]:

$$f(\gamma) = \left(\frac{m}{\bar{\gamma}} \right)^m \frac{\gamma^{m-1}}{\Gamma(m)} \, e^{-m\gamma/\bar{\gamma}} \, , \quad \gamma \geq 0 \, , \qquad (22.22)$$

where the parameter m governs the severity of fading and $\Gamma(m)$ is the Gamma function [671]. When $m = 1$, the PDF of (22.22) is reduced to the PDF of γ over Rayleigh fading channel, which is given in (22.1). As m increases, the fading behaves like Rician fading, and it becomes the AWGN channel, when m tends to ∞. Here we restrict the value of m to be a positive integer. In this case, the Nakagami fading model of (22.22), having a mean of $\bar{\gamma}_s = m \, \bar{\gamma}$, will be used to describe the PDF of the SNR per symbol γ_s in an m-antenna based diversity assisted system employing Maximal Ratio Combining (MRC).

When the instantaneous channel SNR γ is used as the channel quality measure ξ in our adaptive modulation scheme transmitting over a Nakagami channel, the parameters defined in Section 22.7.3 can be expressed in a closed form. Specifically, the mode selection probability $\mathcal{M}_k$ can be expressed as:

$$\mathcal{M}_k = \int_{s_k}^{s_{k+1}} f(\gamma) \, d\gamma \qquad (22.23)$$

$$= F_c(s_k) - F_c(s_{k+1}) \, , \qquad (22.24)$$

where the complementary CDF $F_c(\gamma)$ is given by:

$$F_c(\gamma) = \int_{\gamma}^{\infty} f(x)\, dx \tag{22.25}$$

$$= \int_{\gamma}^{\infty} \left(\frac{m}{\bar{\gamma}}\right)^m \frac{x^{m-1}}{\Gamma(m)} e^{-mx/\bar{\gamma}}\, dx \tag{22.26}$$

$$= e^{-m\gamma/\bar{\gamma}} \sum_{i=0}^{m-1} \frac{(m\gamma/\bar{\gamma})^i}{\Gamma(i+1)}\,. \tag{22.27}$$

In deriving (22.27) we used the result of the indefinite integral of [672]:

$$\int x^n e^{-ax}\, dx = -(e^{-ax}/a) \sum_{i=0}^{n} x^{n-i}/a^i\, n!/(n-i)!\,. \tag{22.28}$$

In a Rayleigh fading scenario, i.e. when $m = 1$, the mode selection probability $\mathcal{M}_k$ of (22.24) can be expressed as:

$$\mathcal{M}_k = e^{-s_k/\bar{\gamma}} - e^{-s_{k+1}/\bar{\gamma}}\,. \tag{22.29}$$

The average throughput B of our adaptive modulation scheme transmitting over a Nakagami channel is given by substituting (22.27) into (22.18), yielding:

$$B = \sum_{k=0}^{K-1} c_k\, e^{-ms_k/\bar{\gamma}} \left\{ \sum_{i=0}^{m-1} \frac{(ms_k/\bar{\gamma})^i}{\Gamma(i+1)} \right\}\,. \tag{22.30}$$

Let us now derive the closed form expressions for the mode specific average BEP P_k defined in (22.20) for the various modulation modes when communicating over a Nakagami channel. The BER of a Gray-coded square QAM constellation for transmission over AWGN channels was given in (22.2) and it is repeated here for convenience:

$$p_{m_k, QAM}(\gamma) = \sum_i A_i\, Q(\sqrt{a_i \gamma})\,, \tag{22.31}$$

where the values of the constants A_i and a_i were given in (22.3). Then, the mode specific average BEP $P_{k,QAM}$ of m_k-ary QAM over a Nakagami channel can be expressed as seen in

Appendix 31.3 as follows:

$$P_{k,QAM} = \int_{s_k}^{s_{k+1}} p_{m_k,QAM}(\gamma) f(\gamma) \, d\gamma \tag{22.32}$$

$$= \sum_i A_i \int_{s_k}^{s_{k+1}} Q(\sqrt{a_i\gamma}) \left(\frac{m}{\bar{\gamma}}\right)^m \frac{\gamma^{m-1}}{\Gamma(m)} e^{-m\gamma/\bar{\gamma}} \, d\gamma \tag{22.33}$$

$$= \sum_i A_i \left\{ \left[-e^{-m\gamma/\bar{\gamma}} Q(\sqrt{a_i\gamma}) \sum_{j=0}^{m-1} \frac{(m\gamma/\bar{\gamma})^j}{\Gamma(j+1)} \right]_{s_k}^{s_{k+1}} + \left[\sum_{j=0}^{m-1} X_j(\gamma, a_i) \right]_{s_k}^{s_{k+1}} \right\}, \tag{22.34}$$

where $g(\gamma)]_{s_k}^{s_{k+1}} \triangleq g(s_{k+1}) - g(s_k)$ and $X_j(\gamma, a_i)$ is given by:

$$X_j(\gamma, a_i) = \frac{\mu^2}{\sqrt{2a_i\pi}} \left(\frac{m}{\bar{\gamma}}\right)^j \frac{\Gamma(j+\frac{1}{2})}{\Gamma(j+1)} \sum_{k=1}^{j} \left(\frac{2\mu^2}{a_i}\right)^{j-k} \frac{\gamma^{k-\frac{1}{2}}}{\Gamma(k+\frac{1}{2})} e^{-a_i\gamma/(2\mu^2)}$$

$$+ \left(\frac{2\mu^2 m}{a_i\bar{\gamma}}\right)^j \frac{1}{\sqrt{\pi}} \frac{\Gamma(j+\frac{1}{2})}{\Gamma(j+1)} \mu \, Q\left(\sqrt{a_i\gamma}/\mu\right), \tag{22.35}$$

where, again, $\mu \triangleq \sqrt{\frac{a_i\bar{\gamma}}{2+a_i\bar{\gamma}}}$ and $\Gamma(x)$ is the Gamma function.

On the other hand, the high-accuracy approximated BEP formula of a Gray-coded m_k-ary PSK scheme ($k \geq 3$) transmitting over an AWGN channel is given as [673]:

$$p_{m_k,PSK} \simeq \frac{2}{k} \left\{ Q\left(\sqrt{2\gamma} \sin(\pi/2^k)\right) + Q\left(\sqrt{2\gamma} \sin(3\pi/2^k)\right) \right\} \tag{22.36}$$

$$= \sum_i A_i Q(\sqrt{a_i\gamma}), \tag{22.37}$$

where the set of constants $\{(A_i, a_i)\}$ is given by $\{(2/k, 2\sin^2(\pi/m_k)), (2/k, 2\sin^2(3\pi/m_k))\}$. Hence, the mode-specific average BEP $P_{k,PSK}$ can be represented using the same equation, namely (22.34), as for $P_{k,QAM}$.

22.8 Optimum Switching Levels

In this section we restrict our interest to adaptive modulation schemes employing the SNR per symbol γ as the channel quality measure ξ. We then derive the optimum switching levels as a function of the target BEP and illustrate the operation of the adaptive modulation scheme. The corresponding performance results of the adaptive modulation schemes communicating over a flat-fading Rayleigh channel are presented in order to demonstrate the effectiveness of the schemes.

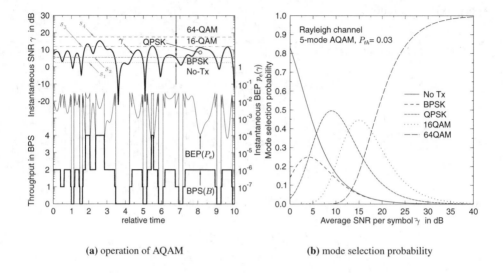

(a) operation of AQAM   (b) mode selection probability

Figure 22.7: Various characteristics of the five-mode AQAM scheme communicating over a Rayleigh fading channel employing the specific set of switching levels designed for limiting the peak instantaneous BEP to $P_{th} = 3 \times 10^{-2}$. (a) The evolution of the instantaneous channel SNR γ is represented by the thick line at the top of the graph, the associated instantaneous BEP $p_e(\gamma)$ by the thin line in the middle and the instantaneous BPS throughput $b(\gamma)$ by the thick line at the bottom. The average SNR is $\bar{\gamma} = 10$dB. (b) As the average SNR increases, the higher-order AQAM modes are selected more often

22.8.1 Limiting the Peak Instantaneous BEP

The first attempt at finding the optimum switching levels that are capable of satisfying various transmission integrity requirements was made by Webb and Steele [177]. They used the BEP curves of each constituent modulation mode, obtained from simulations over an AWGN channel, in order to find the Signal-to-Noise Ratio (SNR) values, where each modulation mode satisfies the target BEP requirement [180]. This intuitive concept of determining the switching levels has been widely used by researchers [179, 184] since then. The regime proposed by Webb and Steele can be used for ensuring that the instantaneous BEP always remains below a certain threshold BEP P_{th}. In order to satisfy this constraint, the first modulation mode should be "no transmission". In this case, the set of switching levels s is given by:

$$\mathbf{s} = \{ s_0 = 0, \ s_k \mid p_{m_k}(s_k) = P_{th} \ k \geq 1 \}. \tag{22.38}$$

Figure 22.7 illustrates how this scheme operates over a Rayleigh channel, using the example of the five-mode AQAM scheme described in Section 22.7.2.1. The average SNR was $\bar{\gamma} = 10$dB and the instantaneous target BEP was $P_{th} = 3 \times 10^{-2}$. Using the expression given in (22.2) for p_{m_k}, the set of switching levels can be calculated for the instantaneous target BEP, which is given by $s_1 = 1.769$, $s_2 = 3.537$, $s_3 = 15.325$ and $s_4 = 55.874$. We can observe that the instantaneous BEP represented as a thin line by the middle of trace of Figure 22.7(a) was limited to values below $P_{th} = 3 \times 10^{-2}$.

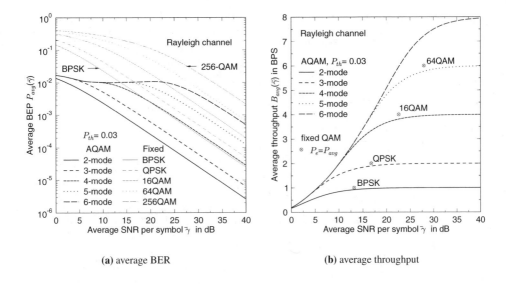

(a) average BER **(b)** average throughput

Figure 22.8: The performance of AQAM employing the specific switching levels defined for limiting the peak instantaneous BEP to $P_{th} = 0.03$. (a) As the number of constituent modulation modes increases, the SNR region where the average BEP remains around $P_{avg} = 10^{-2}$ widens. (b) The SNR gains of AQAM over the fixed-mode QAM scheme required for achieving the same BPS throughput at the same average BEP of P_{avg} are in the range of 5dB to 8dB

At this particular average SNR predominantly the QPSK modulation mode was invoked. However, when the instantaneous channel quality is high, 16-QAM was invoked in order to increase the BPS throughput. The mode selection probability $\mathcal{M}_k$ of (22.24) is shown in Figure 22.7(b). Again, when the average SNR is $\bar{\gamma} = 10$dB, the QPSK mode is selected most often, namely with the probability of about 0.5. The 16-QAM, No-Tx and BPSK modes have had the mode selection probabilities of 0.15 to 0.2, while 64-QAM is not likely to be selected in this situation. When the average SNR increases, the next higher order modulation mode becomes the dominant modulation scheme one by one and eventually the highest order of 64-QAM mode of the five-mode AQAM scheme prevails.

The effects of the number of modulation modes used in our AQAM scheme on the performance are depicted in Figure 22.8. The average BEP performance portrayed in Figure 22.8(a) shows that the AQAM schemes maintain an average BEP lower than the peak instantaneous BEP of $P_{th} = 3 \times 10^{-2}$ even in the low SNR region, at the cost of a reduced average throughput, which can be observed in Figure 22.8(b). As the number of the constituent modulation modes employed of the AQAM increases, the SNR regions, where the average BEP is near constant around $P_{avg} = 10^{-2}$ expands to higher average SNR values. We can observe that the AQAM scheme maintains a constant SNR gain over the highest-order constituent fixed QAM mode, as the average SNR increases, at the cost of a negligible BPS throughput degradation. This is because the AQAM activates the low-order modulation modes or disables transmissions completely, when the channel envelope is in a deep fade, in order to avoid inflicting bursts of bit errors.

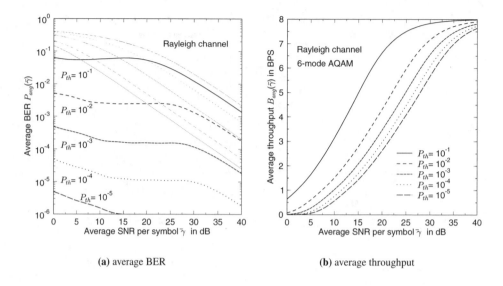

(a) average BER   **(b)** average throughput

Figure 22.9: The performance of the six-mode AQAM employing the switching levels of (22.38) designed for limiting the peak instantaneous BEP

Figure 22.8(b) compares the average BPS throughput of the AQAM scheme employing various numbers of AQAM modes and those of the fixed QAM constituent modes achieving the same average BER. When we want to achieve the target throughput of $B_{avg} = 1$ BPS using the AQAM scheme, Figure 22.8(b) suggest that 3-mode AQAM employing No-Tx, BPSK and QPSK is as good as four-mode AQAM, or in fact any other AQAM schemes employing more than four modes. In this case, the SNR gain achievable by AQAM is 7.7dB at the average BEP of $P_{avg} = 1.154 \times 10^{-2}$. For the average throughputs of $B_{avg} = 2, 4$ and 6, the SNR gains of the 6-mode AQAM schemes over the fixed QAM schemes are 6.65dB, 5.82dB and 5.12dB, respectively.

Figure 22.9 shows the performance of the six-mode AQAM scheme, which is an extended version of the five-mode AQAM of Section 22.7.2.1, for the peak instantaneous BEP values of $P_{th} = 10^{-1}, 10^{-2}, 10^{-3}, 10^{-4}$ and 10^{-5}. We can observe in Figure 22.9(a) that the corresponding average BER P_{avg} decreases as P_{th} decreases. The average throughput curves seen in Figure 22.9(b) indicate that as anticipated the increased average SNR facilitates attaining an increased throughput by the AQAM scheme and there is a clear design trade-off between the achievable average throughput and the peak instantaneous BEP. This is because predominantly lower-throughput, but more error-resilient AQAM modes have to be activated, when the target BER is low. By contrast, higher-throughput but more error-sensitive AQAM modes are favoured, when the tolerable BEP is increased.

In conclusion, we introduced an adaptive modulation scheme, where the objective is to limit the peak instantaneous BEP. A set of switching levels designed for meeting this objective was given in (22.38), which is independent of the underlying fading channel and the average SNR. The corresponding average BEP and throughput formulae were derived in Section 22.7.3.1 and some performance characteristics of a range of AQAM schemes for transmitting over a flat Rayleigh channel were presented in order to demonstrate the effectiveness of the adaptive modulation scheme using the analysis technique developed in Section 22.7.3.1. The main advantage of this adaptive modulation scheme is in its simplicity regarding the design of the AQAM switching levels, while its drawback is that there is no direct relationship between the peak instantaneous BEP and the average BEP, which was used as our performance measure. In the next section a different switching-level optimization philosophy is introduced and contrasted with the approach of designing the switching levels for maintaining a given peak instantaneous BEP.

22.8.2 Torrance's Switching Levels

Torrance and Hanzo [185] proposed the employment of the following cost function and applied Powell's optimization method [188] for generating the optimum switching levels:

$$\Omega_T(\mathbf{s}) = \sum_{\bar{\gamma}=0\text{dB}}^{40\text{dB}} [10 \log_{10}(\max\{P_{avg}(\bar{\gamma};\mathbf{s})/P_{th},1\}) + B_{max} - B_{avg}(\bar{\gamma};\mathbf{s})] , \qquad (22.39)$$

where the average BEP P_{avg} is given in (22.21), $\bar{\gamma}$ is the average SNR per symbol, $\mathbf{s}$ is the set of switching levels, P_{th} is the target average BER, B_{max} is the BPS throughput of the highest order constituent modulation mode and the average throughput B_{avg} is given in (22.14). The idea behind employing the cost function Ω_T is that of maximising the average throughput B_{avg}, while endeavouring to maintain the target average BEP P_{th}. Following the philosophy of Section 22.8.1, the minimisation of the cost function of (22.39) produces a set of constant switching levels across the entire SNR range. However, since the calculation of P_{avg} and B_{avg} requires the knowledge of the PDF of the instantaneous SNR γ per symbol, in reality the set of switching levels $\mathbf{s}$ required for maintaining a constant P_{avg} is dependent on the channel encountered and the receiver structure used.

Figure 22.10 illustrates the operation of a five-mode AQAM scheme employing *Torrance*'s SNR-independent switching levels designed for maintaining the target average BEP of $P_{th} = 10^{-2}$ over a flat Rayleigh channel. The average SNR was $\bar{\gamma} = 10\text{dB}$ and the target average BEP was $P_{th} = 10^{-2}$. *Powell*'s minimisation [188] involved in the context of (22.39) provides the set of optimised switching levels, given by $s_1 = 2.367$, $s_2 = 4.055$, $s_3 = 15.050$ and $s_4 = 56.522$. Upon comparing Figure 22.10(a) to Figure 22.7(a) we find that the two schemes are nearly identical in terms of activating the various AQAM modes according to the channel envelope trace, while the peak instantaneous BEP associated with Torrance's switching scheme is not constant. This is in contrast to the constant peak instantaneous BEP values seen in Figure 22.7(a). The mode selection probabilities depicted in Figure 22.10(b) are similar to those seen in Figure 22.7(b).

The average BEP curves, depicted in Figure 22.11(a) show that *Torrance*'s switching levels support the AQAM scheme in successfully maintaining the target average BEP of $P_{th} = 10^{-2}$ over the average SNR range of 0dB to 20dB, when five or six modem modes are employed by the AQAM scheme. Most of the AQAM studies found in the literature have applied *Torrance*'s switching levels owing to the above mentioned good agreement between the design target P_{th} and the actual BEP performance P_{avg} [674].

Figure 22.11(b) compares the average throughputs of a range of AQAM schemes employing various numbers of AQAM modes to the average BPS throughput of fixed-mode QAM

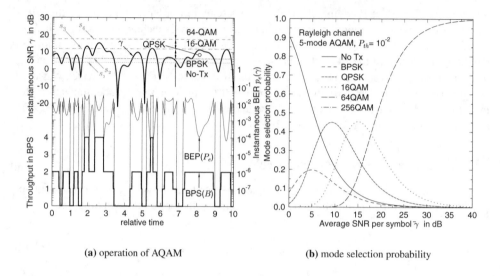

(a) operation of AQAM   (b) mode selection probability

Figure 22.10: Performance of the five-mode AQAM scheme over a flat Rayleigh fading channel employing the set of switching levels derived by Torrance and Hanzo [185] for achieving the target average BEP of $P_{th} = 10^{-2}$. (a) The instantaneous channel SNR γ is represented as a thick line at the top part of the graph, the associated instantaneous BEP $p_e(\gamma)$ as a thin line at the middle, and the instantaneous BPS throughput $b(\gamma)$ as a thick line at the bottom. The average SNR is $\bar{\gamma} = 10$dB. (b) As the SNR increases, the higher-order AQAM modes are selected more often

arrangements achieving the same average BEP, i.e. $P_e = P_{avg}$, which is not necessarily identical to the target BEP of $P_e = P_{th}$. Specifically, the SNR values required by the fixed mode scheme in order to achieve $P_e = P_{avg}$ are represented by the markers '$\otimes$', while the SNRs, where the target average BEP of $P_e = P_{th}$ is achieved, are denoted by the markers '$\odot$'. Compared to the fixed QAM schemes achieving $P_e = P_{avg}$, the SNR gains of the AQAM scheme were 9.06dB, 7.02dB, 5.81dB and 8.74dB for the BPS throughput values of 1, 2, 4 and 6, respectively. By contrast, the corresponding SNR gains compared to the fixed QAM schemes achieving $P_e = P_{th}$ were 7.55dB, 6.26dB, 5.83dB and 1.45dB. We can observe that the SNR gain of the AQAM arrangement over the 64-QAM scheme achieving a BEP of $P_e = P_{th}$ is small compared to the SNR gains attained in comparison to the lower-throughput fixed-mode modems. This is due to the fact that the AQAM scheme employing *Torrance*'s switching levels allows the target BEP to drop at a high average SNR due to its sub-optimum thresholds, which prevents the scheme from increasing the average throughput steadily to the maximum achievable BPS throughput. This phenomenon is more visible for low target average BERs, as can be observed in Figure 22.12.

In conclusion, we reviewed an adaptive modulation scheme employing Torrance's switching levels [185], where the objective was to maximise the average BPS throughput, while maintaining the target average BEP. Torrance's switching levels are constant across the entire SNR range and the average BEP P_{avg} of the AQAM scheme employing these switching levels shows good agreement with the target average BEP P_{th}. However, the range of average

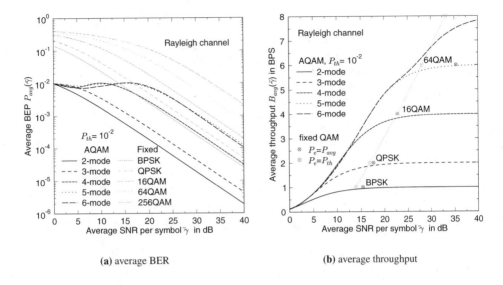

(a) average BER   **(b)** average throughput

Figure 22.11: The performance of various AQAM systems employing *Torrance*'s switching levels [185] designed for the target average BEP of $P_{th} = 10^{-2}$. (a) The actual average BEP P_{avg} is close to the target BEP of $P_{th} = 10^{-2}$ over an average SNR range which becomes wider, as the number of modulation modes increases. However, the five-mode and six-mode AQAM schemes have a similar performance across much of the SNR range. (b) The SNR gains of the AQAM scheme over the fixed-mode QAM arrangements, while achieving the same throughput at the same average BEP, i.e. $P_e = P_{avg}$, range from 6dB to 9dB, which corresponds to a 1dB improvement compared to the SNR gains observed in Figure 22.8(b). However, the SNR gains over the fixed mode QAM arrangement achieving the target BEP of $P_e = P_{avg}$ are reduced, especially at high average SNR values, namely for $\bar{\gamma} > 25$dB.

SNR values, where $P_{avg} \simeq P_{th}$ was limited up to 25dB.

22.8.3 Cost Function Optimization as a Function of the Average SNR

In the previous section, we investigated *Torrance*'s switching levels [185] designed for achieving a certain target average BEP. However, the actual average BEP of the AQAM system was not constant across the SNR range, implying that the average throughput could potentially be further increased. Hence here we propose a modified cost function $\Omega(\mathbf{s}; \bar{\gamma})$, putting more emphasis on achieving a higher throughput and optimise the switching levels for a given SNR, rather than for the whole SNR range [187]:

$$\Omega(\mathbf{s}; \bar{\gamma}) = 10 \log_{10}(\max\{P_{avg}(\bar{\gamma}; \mathbf{s})/P_{th}, 1\}) + \rho \log_{10}(B_{max}/B_{avg}(\bar{\gamma}; \mathbf{s})) , \quad (22.40)$$

where $\mathbf{s}$ is a set of switching levels, $\bar{\gamma}$ is the average SNR per symbol, P_{avg} is the average BEP of the adaptive modulation scheme given in (22.21), P_{th} is the target average BEP of the adaptive modulation scheme, B_{max} is the BPS throughput of the highest order constituent modulation mode. Furthermore, the average throughput B_{avg} is given in (22.14) and ρ is a

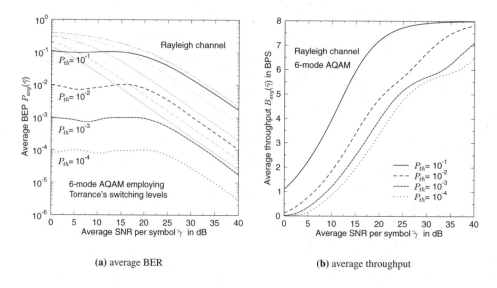

(a) average BER   (b) average throughput

Figure 22.12: The performance of the six-mode AQAM scheme employing Torrance's switching levels [185] for various target average BERs. When the average SNR is over 25dB and the
target average BEP is low, the average BEP of the AQAM scheme begins to decrease,
preventing the scheme from increasing the average BPS throughput steadily

weighting factor, facilitating the above-mentioned BPS throughput enhancement. The first
term at the right-hand side of (22.40) corresponds to a cost function, which accounts for the
difference, in the logarithmic domain, between the average BEP P_{avg} of the AQAM scheme
and the target BEP P_{th}. This term becomes zero, when $P_{avg} \leq P_{th}$, contributing no cost
to the overall cost function Ω. On the other hand, the second term of (22.40) accounts for
the logarithmic distance between the maximum achievable BPS throughput B_{max} and the
average BPS throughput B_{avg} of the AQAM scheme, which decreases, as B_{avg} approaches
B_{max}. Applying Powell's minimisation [188] to this cost function under the constraint of
$s_{k-1} \leq s_k$, the optimum set of switching levels $s_{opt}(\bar{\gamma})$ can be obtained, resulting in the
highest average BPS throughput, while maintaining the target average BEP.

Figure 22.13 depicts the switching levels versus the average SNR per symbol optimised in
this manner for a five-mode AQAM scheme achieving the target average BEP of $P_{th} = 10^{-2}$
and 10^{-3}. Since the switching levels are optimised for each specific average SNR value, they
are not constant across the entire SNR range. As the average SNR $\bar{\gamma}$ increases, the switching
levels decrease in order to activate the higher-order mode modulation modes more often in
an effort to increase the BPS throughput. The low-order modulation modes are abandoned
one by one, as $\bar{\gamma}$ increases, activating always the highest-order modulation mode, namely
64-QAM, when the average BEP of the fixed-mode 64-QAM scheme becomes lower, than
the target average BEP P_{th}. Let us define the *avalanche SNR* $\bar{\gamma}_\alpha$ of a K-mode adaptive
modulation scheme as the lowest SNR, where the target BEP is achieved, which can be

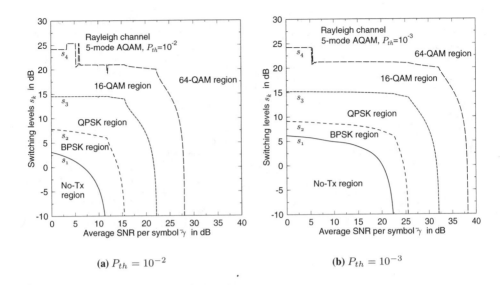

(a) $P_{th} = 10^{-2}$

(b) $P_{th} = 10^{-3}$

Figure 22.13: The switching levels optimised at each average SNR value in order to achieve the target average BEP of (a) $P_{th} = 10^{-2}$ and (b) $P_{th} = 10^{-3}$. As the average SNR $\bar{\gamma}$ increases, the switching levels decrease in order to activate the higher-order mode modulation modes more often in an effort to increase the BPS throughput. The low-order modulation modes are abandoned one by one as $\bar{\gamma}$ increases, activating the highest-order modulation mode, namely 64-QAM, all the time when the average BEP of the fixed-mode 64-QAM scheme becomes lower than the target average BEP P_{th}

formulated as:

$$P_{e,m_K}(\bar{\gamma}_\alpha) = P_{th} \,, \tag{22.41}$$

where m_K is the highest order modulation mode, P_{e,m_K} is the average BEP of the fixed-mode m_K-ary modem activated at the average SNR of $\bar{\gamma}$ and P_{th} is the target average BEP of the adaptive modulation scheme. We can observe in Figure 22.13 that when the average channel SNR is higher than the avalanche SNR, i.e. $\bar{\gamma} \geq \bar{\gamma}_\alpha$, the switching levels are reduced to zero. Some of the optimised switching level versus SNR curves exhibit glitches, indicating that the multi-dimensional optimization might result in local optima in some cases.

The corresponding average BEP P_{avg} and the average throughput B_{avg} of the two to six-mode AQAM schemes designed for the target average BEP of $P_{th} = 10^{-2}$ are depicted in Figure 22.14. We can observe in Figure 22.14(a) that now the actual average BEP P_{avg} of the AQAM scheme is exactly the same as the target BEP of $P_{th} = 10^{-2}$, when the average SNR $\bar{\gamma}$ is less than or equal to the avalanche SNR $\bar{\gamma}_\alpha$. As the number of AQAM modulation modes K increases, the range of average SNRs where the design target of $P_{avg} = P_{th}$ is met extends to a higher SNR, namely to the avalanche SNR. In Figure 22.14(b), the average BPS throughputs of the AQAM modems employing the 'per-SNR optimised' switching levels introduced in this section are represented by thick lines, while the BPS throughput of the six-mode AQAM arrangement employing Torrance's switching levels [185] is represented using a solid thin

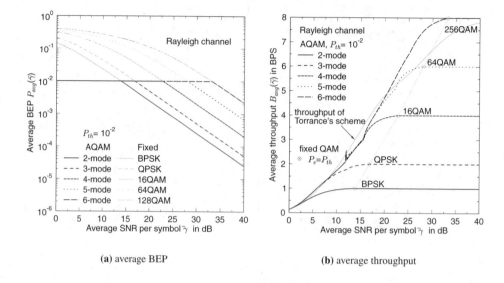

(a) average BEP   (b) average throughput

Figure 22.14: The performance of K-mode AQAM schemes for $K = 2, 3, 4, 5$ and 6, employing the switching levels optimised for each SNR value designed for the target average BEP of $P_{th} = 10^{-2}$. (a) The actual average BEP P_{avg} is exactly the same as the target BER of $P_{th} = 10^{-2}$, when the average SNR $\bar{\gamma}$ is less than or equal to the so-called avalanche SNR $\bar{\gamma}_\alpha$, where the average BEP of the highest-order fixed-modulation mode is equal to the target average BEP. (b) The average throughputs of the AQAM modems employing the 'per-SNR optimised' switching levels are represented in the thick lines, while that of the six-mode AQAM scheme employing Torrance's switching levels [185] is represented by a solid thin line.

line. The average SNR values required by the fixed-mode QAM scheme for achieving the target average BEP of $P_{e,m_K} = P_{th}$ are represented by the markers '⊙'. As we can observe in Figure 22.14(b) the new per-SNR optimised scheme produces a higher BPS throughput than the scheme using Torrance's switching regime, when the average SNR $\bar{\gamma} > 20$dB. However, for the range of 8dB $< \bar{\gamma} < 20$dB, the BPS throughput of the new scheme is lower than that of *Torrance*'s scheme, indicating that the multi-dimensional optimization technique might reach local minima for some SNR values.

Figure 22.15(a) shows that the six-mode AQAM scheme employing 'per-SNR optimised' switching levels satisfies the target average BEP values of $P_{th} = 10^{-1}$ to 10^{-4}. However, the corresponding average throughput performance shown in Figure 22.15(b) also indicates that the thresholds generated by the multi-dimensional optimization were not satisfactory. The BPS throughput achieved was heavily dependent on the value of the weighting factor ρ in (22.40). The glitches seen in the BPS throughput curves in Figure 22.15(b) also suggest that the optimization process might result in some local minima.

We conclude that due to these problems it is hard to achieve a satisfactory BPS throughput for adaptive modulation schemes employing the switching levels optimised for each SNR value based on the heuristic cost function of (22.40), while the corresponding average BEP exhibits a perfect agreement with the target average BEP.

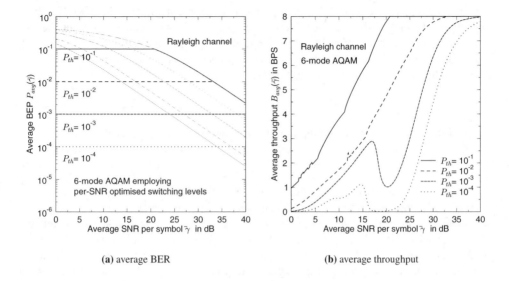

(a) average BER   **(b)** average throughput

Figure 22.15: The performance of six-mode AQAM employing 'per-SNR optimised' switching levels
for various values of the target average BEP. (a) The average BEP P_{avg} remains constant
until the average SNR $\bar{\gamma}$ reaches the avalanche SNR, then follows the average BEP curve
of the highest-order fixed-mode QAM scheme, i.e. that of 256-QAM. (b) For some
SNR values the BPS throughput performance of the six-mode AQAM scheme is not
satisfactory due to the fact that the multi-dimensional optimization algorithm becomes
trapped in local minima and hence fails to reach the global minimum

22.8.4 Lagrangian Method

As argued in the previous section, Powell's minimisation [188] of the cost function often leads
to a local minimum, rather than to the global minimum. Hence, here we adopt an analytical
approach to finding the globally optimised switching levels. Our aim is to optimise the set
of switching levels, $\mathbf{s}$, so that the average BPS throughput $B(\bar{\gamma}; \mathbf{s})$ can be maximised under
the constraint of $P_{avg}(\bar{\gamma}; \mathbf{s}) = P_{th}$. Let us define P_R for a K-mode adaptive modulation
scheme as the sum of the mode-specific average BEP weighted by the BPS throughput of the
individual constituent mode:

$$P_R(\bar{\gamma}; \mathbf{s}) \triangleq \sum_{k=0}^{K-1} b_k \, P_k \,, \qquad (22.42)$$

where $\bar{\gamma}$ is the average SNR per symbol, $\mathbf{s}$ is the set of switching levels, K is the number of
constituent modulation modes, b_k is the BPS throughput of the k-th constituent mode and the
mode-specific average BEP P_k is given in (22.20) as:

$$P_k = \int_{s_k}^{s_{k+1}} p_{m_k}(\gamma) \, f(\gamma) \, d\gamma \,, \qquad (22.43)$$

where again, $p_{m_k}(\gamma)$ is the BEP of the m_k-ary modulation scheme over the AWGN channel and $f(\gamma)$ is the PDF of the SNR per symbol γ. Explicitly, (22.43) implies weighting the BEP $p_{m_k}(\gamma)$ by its probability of occurrence quantified in terms of its PDF and then averaging, i.e. integrating it over the range spanning from s_k to s_{k+1}. Then, with the aid of (22.21), the average BEP constraint can also be written as:

$$P_{avg}(\bar{\gamma}; \mathbf{s}) = P_{th} \iff P_R(\bar{\gamma}; \mathbf{s}) = P_{th}\, B(\bar{\gamma}; \mathbf{s}) \,. \tag{22.44}$$

Another rational constraint regarding the switching levels can be expressed as:

$$s_k \leq s_{k+1} \,. \tag{22.45}$$

As we discussed before, our optimization goal is to maximise the objective function $B(\bar{\gamma}; \mathbf{s})$ under the constraint of (22.44). The set of switching levels $\mathbf{s}$ has $K+1$ levels in it. However, considering that we have $s_0 = 0$ and $s_K = \infty$ in many adaptive modulation schemes, we have $K-1$ independent variables in $\mathbf{s}$. Hence, the optimization task is a $K-1$ dimensional optimization under a constraint [675]. It is a standard practice to introduce a modified object function using a Lagrangian multiplier and convert the problem into a set of one-dimensional optimization problems. The modified object function Λ can be formulated employing a Lagrangian multiplier λ [675] as:

$$\Lambda(\mathbf{s}; \bar{\gamma}) = B(\bar{\gamma}; \mathbf{s}) + \lambda \left\{ P_R(\bar{\gamma}; \mathbf{s}) - P_{th}\, B(\bar{\gamma}; \mathbf{s}) \right\} \tag{22.46}$$

$$= (1 - \lambda P_{th})\, B(\bar{\gamma}; \mathbf{s}) + \lambda P_R(\bar{\gamma}; \mathbf{s}) \,. \tag{22.47}$$

The optimum set of switching levels should satisfy:

$$\frac{\partial \Lambda}{\partial \mathbf{s}} = \frac{\partial}{\partial \mathbf{s}} \left(B(\bar{\gamma}; \mathbf{s}) + \lambda \left\{ P_R(\bar{\gamma}; \mathbf{s}) - P_{th}\, B(\bar{\gamma}; \mathbf{s}) \right\} \right) = 0 \quad \text{and} \tag{22.48}$$

$$P_R(\bar{\gamma}; \mathbf{s}) - P_t\, B(\bar{\gamma}; \mathbf{s}) = 0 \,. \tag{22.49}$$

The following results are helpful in evaluating the partial differentiations in (22.48) :

$$\frac{\partial}{\partial s_k} P_{k-1} = \frac{\partial}{\partial s_k} \int_{s_{k-1}}^{s_k} p_{m_{k-1}}(\gamma)\, f(\gamma)\, d\gamma = p_{m_{k-1}}(s_k)\, f(s_k) \tag{22.50}$$

$$\frac{\partial}{\partial s_k} P_k = \frac{\partial}{\partial s_k} \int_{s_k}^{s_{k+1}} p_{m_k}(\gamma)\, f(\gamma)\, d\gamma = -p_{m_k}(s_k)\, f(s_k) \tag{22.51}$$

$$\frac{\partial}{\partial s_k} F_c(s_k) = \frac{\partial}{\partial s_k} \int_{s_k}^{\infty} f(\gamma)\, d\gamma = -f(s_k) \,. \tag{22.52}$$

Using (22.50) and (22.51), the partial differentiation of P_R defined in (22.42) with respect to s_k can be written as:

$$\frac{\partial P_R}{\partial s_k} = b_{k-1}\, p_{m_{k-1}}(s_k)\, f(s_k) - b_k\, p_{m_k}(s_k)\, f(s_k) \,, \tag{22.53}$$

where b_k is the BPS throughput of an m_k-ary modem. Since the average throughput is given by $B = \sum_{k=0}^{K-1} c_k\, F_c(s_k)$ in (22.18), the partial differentiation of B with respect to s_k can be written as, using (22.52) :

$$\frac{\partial B}{\partial s_k} = -c_k \, f(s_k) \,, \qquad (22.54)$$

where c_k was defined as $c_k \triangleq b_k - b_{k-1}$ in Section 22.7.1. Hence (22.48) can be evaluated as:

$$\left[-c_k(1 - \lambda \, P_{th}) + \lambda \left\{ b_{k-1} \, p_{m_{k-1}}(s_k) - b_k p_{m_k}(s_k) \right\} \right] f(s_k) = 0 \text{ for } k = 1, 2, \cdots, K - 1 \,. \qquad (22.55)$$

A trivial solution of (22.55) is $f(s_k) = 0$. Certainly, $\{s_k = \infty, \; k = 1, 2, \cdots, K-1\}$ satisfies this condition. Again, the lowest throughput modulation mode is 'No-Tx' in our model, which corresponds to no transmission. When the PDF of γ satisfies $f(0) = 0$, $\{s_k = 0, \; k = 1, 2, \cdots, K - 1\}$ can also be a solution, which corresponds to the fixed-mode m_{K-1}-ary modem. The corresponding avalanche SNR $\bar{\gamma}_\alpha$ can obtained by substituting $\{s_k = 0, \; k = 1, 2, \cdots, K - 1\}$ into (22.49), which satisfies:

$$p_{m_{K-1}}(\bar{\gamma}_\alpha) - P_{th} = 0 \,. \qquad (22.56)$$

When $f(s_k) \neq 0$, Equation (22.55) can be simplified upon dividing both sides by $f(s_k)$, yielding:

$$-c_k(1 - \lambda \, P_{th}) + \lambda \left\{ b_{k-1} \, p_{m_{k-1}}(s_k) - b_k p_{m_k}(s_k) \right\} = 0 \text{ for } k = 1, 2, \cdots, K - 1 \,. \qquad (22.57)$$

Rearranging (22.57) for $k = 1$ and assuming $c_1 \neq 0$, we have:

$$1 - \lambda \, P_{th} = \frac{\lambda}{c_1} \left\{ b_0 \, p_{m_0}(s_1) - b_1 p_{m_1}(s_1) \right\} \,. \qquad (22.58)$$

Substituting (22.58) into (22.57) and assuming $c_k \neq 0$ for $k \neq 0$, we have:

$$\frac{\lambda}{c_k} \left\{ b_{k-1} \, p_{m_{k-1}}(s_k) - b_k p_{m_k}(s_k) \right\} = \frac{\lambda}{c_1} \left\{ b_0 \, p_{m_0}(s_1) - b_1 p_{m_1}(s_1) \right\} \,. \qquad (22.59)$$

In this context we note that the Lagrangian multiplier λ is not zero because substitution of $\lambda = 0$ in (22.57) leads to $-c_k = 0$, which is not true. Hence, we can eliminate the Lagrangian multiplier dividing both sides of (22.59) by λ. Then we have:

$$y_k(s_k) = y_1(s_1) \text{ for } k = 2, 3, \cdots K - 1 \,, \qquad (22.60)$$

where the function $y_k(s_k)$ is defined as:

$$y_k(s_k) \triangleq \frac{1}{c_k} \left\{ b_k p_{m_k}(s_k) - b_{k-1} \, p_{m_{k-1}}(s_k) \right\} \,, \; k = 2, 3, \cdots K - 1 \,, \qquad (22.61)$$

which does not contain the Lagrangian multiplier λ and hence it will be referred to as the 'Lagrangian-free function'. This function can be physically interpreted as the normalised

BEP difference between the adjacent AQAM modes. For example, $y_1(s_1) = p_2(s_1)$ quantifies the BEP increase, when switching from the No-Tx mode to the BPSK mode, while $y_2(s_2) = 2p_4(s_2) - p_2(s_2)$ indicates the BEP difference between the QPSK and BPSK modes. These curves will be more explicitly discussed in the context of Figure 22.16. The significance of (22.60) is that the relationship between the optimum switching levels s_k, where $k = 2, 3, \cdots K - 1$, and the lowest optimum switching level s_1 is independent of the underlying propagation scenario. Only the constituent modulation mode related parameters, such as b_k, c_k and $p_{m_k}(\gamma)$, govern this relationship.

Let us now investigate some properties of the Lagrangian-free function $y_k(s_k)$ given in (22.61). Considering that $b_k > b_{k-1}$ and $p_{m_k}(s_k) > p_{m_{k-1}}(s_k)$, it is readily seen that the value of $y_k(s_k)$ is always positive. When $s_k = 0$, $y_k(s_k)$ becomes:

$$y_k(0) \triangleq \frac{1}{c_k} \left\{ b_k p_{m_k}(0) - b_{k-1} p_{m_{k-1}}(0) \right\} = \frac{1}{c_k} \left\{ \frac{b_k}{2} - \frac{b_{k-1}}{2} \right\} = \frac{1}{2}. \tag{22.62}$$

The solution of $y_k(s_k) = 1/2$ can be either $s_k = 0$ or $b_k p_{m_k}(s_k) = b_{k-1} p_{m_{k-1}}(s_k)$. When $s_k = 0$, $y_k(s_k)$ becomes $y_k(\infty) = 0$. We also conjecture that

$$\frac{d\,s_k}{d\,s_1} = \frac{y_1'(s_1)}{y_k'(s_k)} > 0 \text{ when } y_k(s_k) = y_1(s_1), \tag{22.63}$$

which states that the k-th optimum switching level s_k always increases, whenever the lowest optimum switching level s_1 increases. Our numerical evaluations suggest that this conjecture appears to be true.

As an example, let us consider the five-mode AQAM scheme introduced in Section 22.7.2.1. The parameters of the five-mode AQAM scheme are summarised in Table 22.1. Substituting these parameters into (22.60) and (22.61), we have the following set of equations.

$$y_1(s_1) = p_2(s_1) \tag{22.64}$$

$$y_2(s_2) = 2p_4(s_2) - p_2(s_2) \tag{22.65}$$

$$y_3(s_3) = 2p_{16}(s_3) - p_4(s_3) \tag{22.66}$$

$$y_4(s_4) = 3p_{64}(s_4) - 2p_{16}(s_4) \tag{22.67}$$

The Lagrangian-free functions of (22.64) through (22.67) are depicted in Figure 22.16 for Gray-mapped square-shaped QAM. As these functions are basically linear combinations of BEP curves associated with AWGN channels, they exhibit waterfall-like shapes and asymptotically approach 0.5, as the switching levels s_k approach zero (or $-\infty$ expressed in dB). While $y_1(s_1)$ and $y_2(s_2)$ are monotonic functions, $y_3(s_3)$ and $y_4(s_4)$ cross the $y = 0.5$ line at $s_3 = -7.34$ dB and $s_4 = 1.82$ dB respectively, as can be observed in Figure 22.16(b). One should also notice that the trivial solutions of (22.60) are $y_k = 0.5$ at $s_k = 0$, $k = 1, 2, 3, 4$, as we have discussed before.

For a given value of s_1, the other switching levels can be determined as $s_2 = y_2^{-1}(y_1(s_1))$, $s_3 = y_3^{-1}(y_1(s_1))$ and $s_4 = y_4^{-1}(y_1(s_1))$. Since deriving the analytical inverse function of y_k is an arduous task, we can rely on a graphical or a numerical method. Figure 22.16(b) illustrates an example of the graphical method. Specifically, when $s_1 = \alpha_1$, we first find the point on the curve y_1 directly above the abscissa value of α_1 and then draw a horizontal

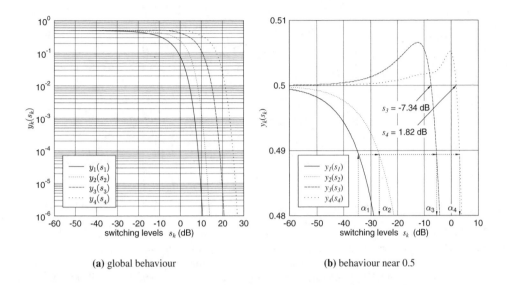

(a) global behaviour   **(b)** behaviour near 0.5

Figure 22.16: The Lagrangian-free functions $y_k(s_k)$ of (22.64) through (22.67) for Gray-mapped square-shaped QAM constellations. As s_k becomes lower $y_k(s_k)$ asymptotically approaches 0.5. Observe that while $y_1(s_1)$ and $y_2(s_2)$ are monotonic functions, $y_3(s_3)$ and $y_4(s_4)$ cross the $y = 0.5$ line

line across the corresponding point. From the crossover points found on the curves of y_2, y_3 and y_4 with the aid of the horizontal line, we can find the corresponding values of the other switching levels, namely those of α_2, α_3 and α_4. In a numerical sense, this solution corresponds to a one-dimensional (1-D) root finding problem [188] (Chapter 9). Furthermore, the $y_k(s_k)$ values are monotonic, provided that we have $y_k(s_k) < 0.5$ and this implies that the roots found are unique. The numerical results shown in Figure 22.17 represent the direct relationship between the optimum switching level s_1 and the other optimum switching levels, namely s_2, s_3 and s_4. While the optimum value of s_2 shows a near-linear relationship with respect to s_1, those of s_3 and s_4 asymptotically approach two different constants, as s_1 becomes smaller. This corroborates the trends observed in Figure 22.16(b), where $y_3(s_3)$ and $y_4(s_4)$ cross the $y = 0.5$ line at $s_3 = -7.34$ dB and $s_4 = 1.82$ dB, respectively. Since the low-order modulation modes are abandoned at high average channel SNRs in order to increase the average throughput, the high values of s_1 on the horizontal axis of Figure 22.17 indicate encountering a low channel SNR, while low values of s_1 suggest that high channel SNRs are experienced, as it transpires for example from Figure 22.13.

Since we can relate the other switching levels to s_1, we have to determine the optimum value of s_1 for the given target BEP, P_{th}, and the PDF of the instantaneous channel SNR, $f(\gamma)$, by solving the constraint equation given in (22.49). This problem also constitutes a 1-D root finding problem, rather than a multi-dimensional optimization problem, which was the case in Sections 22.8.2 and 22.8.3. Let us define the constraint function $Y(\bar{\gamma}; \mathbf{s}(s_1))$ using

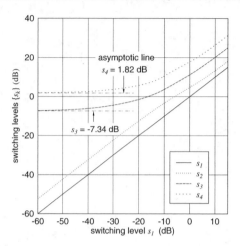

Figure 22.17: Optimum switching levels as functions of s_1, where the linear relationship of s_1 versus s_1 was also plotted for completeness. Observe that while the optimum value of s_2 shows a linear relationship with respect to s_1, those of s_3 and s_4 asymptotically approach constant values as s_1 is reduced

(22.49) as:

$$Y(\bar{\gamma}; \mathbf{s}(s_1)) \triangleq P_R(\bar{\gamma}; \mathbf{s}(s_1)) - P_{th}\, B(\bar{\gamma}; \mathbf{s}(s_1)), \qquad (22.68)$$

where we represented the set of switching levels as a vector, which is the function of s_1, in order to emphasise that s_k satisfies the relationships given by (22.60) and (22.61).

More explicitly, $Y(\bar{\gamma}; \mathbf{s}(s_1))$ of (22.68) can be physically interpreted as the difference between $P_R(\bar{\gamma}; \mathbf{s}(s_1))$, namely the sum of the mode-specific average BEPs weighted by the BPS throughput of the individual AQAM modes, as defined in (22.42) and the average BPS throughput $B(\bar{\gamma}; \mathbf{s}(s_1))$ weighted by the target BEP P_{th}. Considering the equivalence relationship given in (22.44), (22.68) reflects just another way of expressing the difference between the average BEP P_{avg} of the adaptive scheme and the target BEP P_{th}.

Even though the relationships implied in $\mathbf{s}(s_1)$ are independent of the propagation conditions and the signalling power, the constraint function $Y(\bar{\gamma}; \mathbf{s}(s_1))$ of (22.68) and hence the actual values of the optimum switching levels are dependent on propagation conditions through the PDF $f(\gamma)$ of the SNR per symbol and on the average SNR per symbol $\bar{\gamma}$.

Let us find the initial value of $Y(\bar{\gamma}; \mathbf{s}(s_1))$ defined in (22.68), when $s_1 = 0$. An obvious solution for s_k when $s_1 = 0$ is $s_k = 0$ for $k = 1, 2, \cdots, K - 1$. In this case, $Y(\bar{\gamma}; \mathbf{s}(s_1))$ becomes:

$$Y(\bar{\gamma}; 0) = b_{K-1}\left(P_{m_{K-1}}(\bar{\gamma}) - P_{th}\right), \qquad (22.69)$$

where b_{K-1} is the BPS throughput of the highest-order constituent modulation mode, while $P_{m_{K-1}}(\bar{\gamma})$ is the average BEP of the highest-order constituent modulation mode for transmission over the underlying channel scenario and P_{th} is the target average BEP. The value of $Y(\bar{\gamma}; 0)$ could be positive or negative, depending on the average SNR $\bar{\gamma}$ and on the target average BEP P_{th}. Another solution exists for s_k when $s_1 = 0$, if $b_k\, p_{m_k}(s_k) = b_{k-1}\, p_{m_{k-1}}(s_k)$.

The value of $Y(\bar{\gamma}; 0^+)$ using this alternative solution turns out to be close to $Y(\bar{\gamma}; 0)$. However, in the actual numerical evaluation of the initial value of Y, we should use $Y(\bar{\gamma}; 0^+)$ for ensuring the continuity of the function Y at $s_1 = 0$.

In order to find the minima and the maxima of Y, we have to evaluate the derivative of $Y(\bar{\gamma}; \mathbf{s}(s_1))$ with respect to s_1. With the aid of (22.50) to (22.54), we have:

$$
\begin{aligned}
\frac{dY}{ds_1} &= \sum_{k=1}^{K-1} \frac{\partial Y}{\partial s_k} \frac{ds_k}{ds_1} \\
&= \sum_{k=1}^{K-1} \frac{\partial}{\partial s_k} \{P_R - P_{th} B\} \frac{ds_k}{ds_1} \\
&= \sum_{k=1}^{K-1} \{b_{k-1} p_{m_{k-1}}(s_k) - b_k p_{m_k}(s_k) + P_{th} c_k\} f(s_k) \frac{ds_k}{ds_1} \\
&= \sum_{k=1}^{K-1} \left[\frac{c_k}{c_1} \{b_0 p_{m_0}(s_1) - b_1 p_{m_1}(s_1)\} + P_{th} c_k \right] f(s_k) \frac{ds_k}{ds_1} \\
&= \frac{1}{c_1} \{b_0 p_{m_0}(s_1) - b_1 p_{m_1}(s_1) + P_{th}\} \sum_{k=1}^{K-1} c_k f(s_k) \frac{ds_k}{ds_1} .
\end{aligned}
\tag{22.70}
$$

Considering $f(s_k) \geq 0$ and using our conjecture that $\frac{ds_k}{ds_1} > 0$ given in (22.63), we can conclude from (22.70) that $\frac{dY}{ds_1} = 0$ has roots, when $f(s_k) = 0$ for all k or when $b_1 p_{m_1}(s_1) - b_0 p_{m_0}(s_1) = P_{th}$. The former condition corresponds to either $s_i = 0$ for some PDF $f(\gamma)$ or to $s_k = \infty$ for all PDFs. By contrast, when the condition of $b_1 p_{m_1}(s_1) - b_0 p_{m_0}(s_1) = P_{th}$ is met, $dY/ds_1 = 0$ has a unique solution. Investigating the sign of the first derivative between these zeros, we can conclude that $Y(\bar{\gamma}; s_1)$ has a global minimum of Y_{min} at $s_1 = \zeta$ so that $b_1 p_{m_1}(\zeta) - b_0 p_{m_0}(\zeta) = P_{th}$ and a maximum of Y_{max} at $s_1 = 0$ and another maximum value at $s_1 = \infty$.

Since $Y(\bar{\gamma}; s_1)$ has a maximum value at $s_1 = \infty$, let us find the corresponding maximum value. Let us first consider $\lim_{s_1 \to \infty} P_{avg}(\bar{\gamma}; \mathbf{s}(s_1))$, where upon exploiting (22.21) and (22.42) we have:

$$
\lim_{s_1 \to \infty} P_{avg}(\bar{\gamma}; s_k) = \frac{\lim_{s_1 \to \infty} P_R}{\lim_{s_1 \to \infty} B}
\tag{22.71}
$$

$$
= \frac{0}{0} .
\tag{22.72}
$$

When applying l'Hopital's rule and using Equations (22.50) through (22.54), we have:

$$
\frac{\lim_{s_1 \to \infty} P_R}{\lim_{s_1 \to \infty} B} = \frac{\lim_{s_1 \to \infty} \frac{d}{ds_1} P_R}{\lim_{s_1 \to \infty} \frac{d}{ds_1} B}
\tag{22.73}
$$

$$
= \lim_{s_1 \to \infty} \frac{1}{c_1} b_1 p_{m_1}(s_1) - b_0 p_{m_0}(s_1)
\tag{22.74}
$$

$$
= 0^+ ,
\tag{22.75}
$$

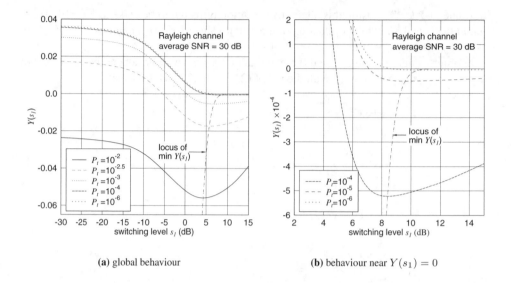

(a) global behaviour   **(b)** behaviour near $Y(s_1) = 0$

Figure 22.18: The constraint function $Y(\bar{\gamma}; \mathbf{s}(s_1))$ defined in (22.68) for our five-mode AQAM scheme employing Gray-mapped square-constellation QAM operating over a flat Rayleigh fading channel. The average SNR was $\bar{\gamma} = 30$ dB and it is seen that Y has a single minimum value, while approaching 0^-, as s_1 increases. The solution of $Y(\bar{\gamma}; \mathbf{s}(s_1)) = 0$ exists, when $Y(\bar{\gamma}; 0) = 6\{p_{64}(\bar{\gamma}) - P_{th}\} > 0$ and is unique.

implying that $P_{avg}(\bar{\gamma}; s_k)$ approaches zero from positive values, when s_1 tends to ∞. Since according to (22.21), (22.42) and (22.68) the function $Y(\bar{\gamma}; \mathbf{s}(s_1))$ can be written as $B(P_{avg} - P_{th})$, we have:

$$\lim_{s_1 \to \infty} Y(\bar{\gamma}; s_1) = \lim_{s_1 \to \infty} B(P_{avg} - P_{th}) \tag{22.76}$$

$$= \lim_{s_1 \to \infty} B(0^+ - P_{th}) \tag{22.77}$$

$$= 0^-, \tag{22.78}$$

Hence $Y(\bar{\gamma}; \mathbf{s}(s_1))$ asymptotically approaches zero from negative values, as s_1 tends to ∞. From the analysis of the minimum and the maxima, we can conclude that the constraint function $Y(\bar{\gamma}; \mathbf{s}(s_1))$ defined in (22.68) has a unique zero only if $Y(\bar{\gamma}; 0^+) > 0$ at a switching value of $0 < s_1 < \zeta$, where ζ satisfies $b_1 \, p_{m_1}(\zeta) - b_0 \, p_{m_0}(\zeta) = P_{th}$. By contrast, when $Y(\bar{\gamma}; 0^+) < 0$, the optimum switching levels are all zero and the adaptive modulation scheme always employs the highest-order constituent modulation mode.

As an example, let us evaluate the constraint function $Y(\bar{\gamma}; s_1)$ for our five-mode AQAM scheme operating over a flat Rayleigh fading channel. Figure 22.18 depicts the values of $Y(s_1)$ for several values of the target average BEP P_{th}, when the average channel SNR is 30dB. We can observe that $Y(s_1) = 0$ may have a root, depending on the target BEP P_{th}.

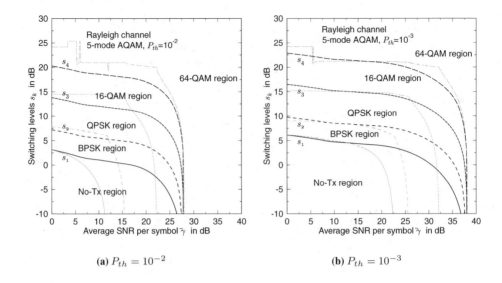

(a) $P_{th} = 10^{-2}$ (b) $P_{th} = 10^{-3}$

Figure 22.19: The switching levels for our five-mode AQAM scheme optimised at each average SNR
value in order to achieve the target average BEP of (a) $P_{th} = 10^{-2}$ and (b) $P_{th} = 10^{-3}$
using the Lagrangian multiplier based method of Section 22.8.4. The switching levels
based on Powell's optimization are represented in thin grey lines for comparison

When $s_k = 0$ for $k < 5$, according to (22.21), (22.42) and (22.68) $Y(s_1)$ is reduced to

$$Y(\bar{\gamma}; 0) = 6(P_{64}(\bar{\gamma}) - P_{th}),\tag{22.79}$$

where $P_{64}(\bar{\gamma})$ is the average BEP of 64-QAM over a flat Rayleigh channel. The value of
$Y(\bar{\gamma}; 0)$ in (22.79) can be negative or positive, depending on the target BEP P_{th}.

We can observe in Figure 22.18 that the solution of $Y(\bar{\gamma}; \mathbf{s}(s_1)) = 0$ is unique, when it exists. The locus of the minimum $Y(s_1)$, i.e. the trace curve of points $(Y_{min}(s_{1,min}), s_{1,min})$,
where Y has the minimum value, is also depicted in Figure 22.18. The locus is always below
the horizontal line of $Y(s_1) = 0$ and asymptotically approaches this line, as the target BEP
P_{th} becomes smaller.

Figure 22.19 depicts the switching levels optimised in this manner for our five-mode
AQAM scheme maintaining the target average BEPs of $P_{th} = 10^{-2}$ and 10^{-3}. The switching
levels obtained using Powell's optimization method in Section 22.8.3 are represented as the
thin grey lines in Figure 22.19 for comparison. In this case all the modulation modes may
be activated with a certain probability, until the average SNR reaches the avalanche SNR
value, while the scheme derived using Powell's optimization technique abandons the lower
throughput modulation modes one by one, as the average SNR increases.

Figure 22.20 depicts the average throughput B expressed in BPS of the AQAM scheme
employing the switching levels optimised using the Lagrangian method. In Figure 22.20(a),
the average throughput of our six-mode AQAM arrangement using Torrance's scheme discussed in Section 22.8.2 is represented as a thin grey line. The Lagrangian multiplier based
scheme showed SNR gains of 0.6dB, 0.5dB, 0.2dB and 3.9dB for a BPS throughput of 1, 2, 4

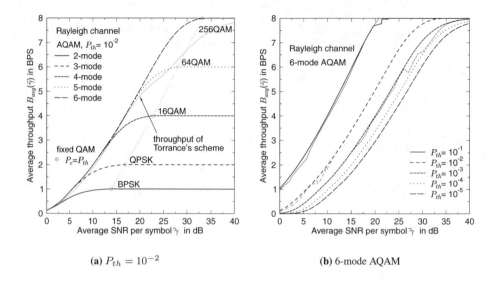

(a) $P_{th} = 10^{-2}$ (b) 6-mode AQAM

Figure 22.20: The average BPS throughput of various AQAM schemes employing the switching levels
optimised using the Lagrangian multiplier method (a) for $P_{th} = 10^{-2}$ employing two
to six-modes and (b) for $P_{th} = 10^{-2}$ to $P_{th} = 10^{-5}$ using six-modes. The average
throughput of the six-mode AQAM scheme using Torrance's switching levels [185] is
represented for comparison as the thin grey line in figure (a). The average throughput of
the six-mode AQAM scheme employing per-SNR optimised thresholds using Powell's
optimization method are represented by the thin lines in figure (b) for the target average
BEP of $P_{th} = 10^{-1}$, 10^{-2} and 10^{-3}.

and 6, respectively, compared to Torrance's scheme. The average throughput of our six-mode
AQAM scheme is depicted in Figure 22.20(b) for the several values of P_{th}, where the corresponding BPS throughput of the AQAM scheme employing per-SNR optimised thresholds
determined using Powell's method are also represented as thin lines for $P_{th} = 10^{-1}$, 10^{-2}
and 10^{-3}. Comparing the BPS throughput curves, we can conclude that the per-SNR optimised Powell method of Section 22.8.3 resulted in imperfect optimization for some values of
the average SNR.

In conclusion, we derived an optimum mode-switching regime for a general AQAM
scheme using the Lagrangian multiplier method and presented our numerical results for various AQAM arrangements. Since the results showed that the Lagrangian optimization based
scheme is superior in comparison to the other methods investigated, we will employ these
switching levels in order to further investigate the performance of various adaptive modulation schemes.

22.9   Results and Discussions

The average throughput performance of adaptive modulation schemes employing the globally
optimised mode-switching levels of Section 22.8.4 is presented in this section. The mobile

channel is modelled as a Nakagami-m fading channel. The performance results and discussions include the effects of the fading parameter m, that of the number of modulation modes, the influence of the various diversity schemes used and the range of Square QAM, Star QAM and MPSK signalling constellations.

22.9.1 Narrow-Band Nakagami-m Fading Channel

The PDF of the instantaneous channel SNR γ of a system transmitting over the Nakagami fading channel is given in (22.22). The parameters characterising the operation of the adaptive modulation scheme were summarised in Section 22.7.3.1.

22.9.1.1 Adaptive PSK Modulation Schemes

Phase Shift Keying (PSK) has the advantage of exhibiting a constant envelope power, since all the constellation points are located on a circle. Let us first consider the BEP of fixed-mode PSK schemes as a reference, so that we can compare the performance of adaptive PSK and fixed-mode PSK schemes. The BEP of Gray-coded coherent M-ary PSK (MPSK), where $M = 2^k$, for transmission over the AWGN channel can be closely approximated by [673]:

$$p_{MPSK}(\gamma) \simeq \sum_{i=1}^{2} A_i \, Q(\sqrt{a_i \gamma}) \,, \tag{22.80}$$

where $M \geq 8$ and the associated constants are given by [673]:

$$A_1 = A_2 = 2/k \tag{22.81}$$

$$a_1 = 2\sin^2(\pi/M) \tag{22.82}$$

$$a_2 = 2\sin^2(3\pi/M) \,. \tag{22.83}$$

Figure 22.21(a) shows the BEP of BPSK, QPSK, 8PSK, 16PSK, 32PSK and 64PSK for transmission over the AWGN channel. The differences of the required SNR per symbol, in order to achieve the BER of $p_{MPSK}(\gamma) = 10^{-6}$ for the modulation modes having a throughput difference of 1 BPS are around 6dB, except between BPSK and QPSK, where a 3dB difference is observed.

The average BEP of MPSK schemes over a flat Nakagami-m fading channel is given as:

$$P_{MPSK}(\bar{\gamma}) = \int_0^\infty p_{MPSK}(\gamma) \, f(\gamma) \, d\gamma \,, \tag{22.84}$$

where the BEP $p_{MPSK}(\gamma)$ for a transmission over the AWGN channel is given by (22.80) and the PDF $f(\gamma)$ is given by (22.22). A closed form solution of (22.84) can be readily obtained for an integer m using the results given in Section (14-4-15) of [388], which can be expressed as:

$$P_{MPSK}(\bar{\gamma}) = \sum_{i=1}^{2} A_i \left[\tfrac{1}{2}(1 - \mu_i)\right]^m \sum_{j=0}^{m-1} \binom{m-1+j}{j} \left[\tfrac{1}{2}(1 + \mu_i)\right]^j \,, \tag{22.85}$$

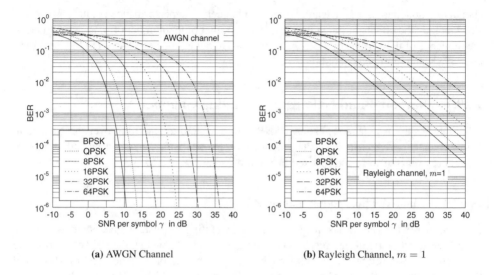

(a) AWGN Channel   (b) Rayleigh Channel, $m = 1$

Figure 22.21: The average BEP of various MPSK modulation schemes

k	0	1	2	3	4	5	6	7	8
m_k	0	2	4	8	16	32	64	128	256
b_k	0	1	2	3	4	5	6	7	8
c_k	0	1	1	1	1	1	1	1	1
mode	No Tx	BPSK	QPSK	8PSK	16PSK	32PSK	64PSK	128PSK	256PSK

Table 22.5: Parameters of a nine-mode adaptive PSK scheme following the definitions of the generic adaptive modulation model developed in Section 22.7.1

where μ_i is defined as:

$$\mu_i \triangleq \sqrt{\frac{a_i \bar{\gamma}}{2m + a_i \bar{\gamma}}} . \tag{22.86}$$

Figure 22.21(b) shows the average BEP of the various MPSK schemes for transmission over a flat Rayleigh channel, where $m = 1$. The BEP of MPSK over the AWGN channel given in (22.80) and that over a Nakagami channel given in (22.85) will be used in comparing the performance of adaptive PSK schemes.

The parameters of our nine-mode adaptive PSK scheme are summarised in Table 22.5 following the definitions of our generic model used for the adaptive modulation schemes developed in Section 22.7.1. The models of other adaptive PSK schemes employing a different number of modes can be readily obtained by increasing or reducing the number of columns in Table 22.5. Since the number of modes is $K = 9$, we have $K + 1 = 10$ mode-switching levels, which are hosted by the vector $\mathbf{s} = \{s_k \mid k = 0, 1, 2, \cdots, 9\}$. Let us assume $s_0 = 0$ and $s_9 = \infty$. In order to evaluate the performance of the nine-mode adaptive PSK scheme, we

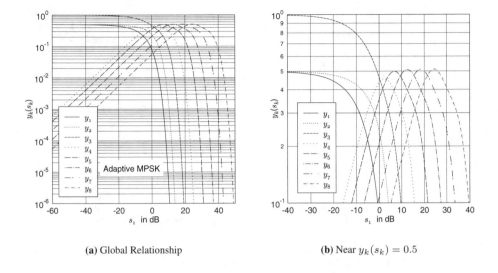

(a) Global Relationship   (b) Near $y_k(s_k) = 0.5$

Figure 22.22: 'Lagrangian-free' functions of (22.61) for a nine-mode adaptive PSK scheme. For a given value of s_1, there exist two solutions for s_k satisfying $y_k(s_k) = y_1(s_1)$. However, only the higher value of s_k satisfies the constraint of $s_{k-1} \le s_k$, $\forall\, k$.

have to obtain the optimum switching levels first. Let us evaluate the 'Lagrangian-free' functions defined in (22.61), using the parameters given in Table 22.5 and the BEP expressions given in (22.80). The 'Lagrangian-free' functions of our nine-mode adaptive PSK scheme are depicted in Figure 22.22. We can observe that there exist two solutions for s_k satisfying $y_k(s_k) = y_1(s_1)$ for a given value of s_1, which are given by the crossover points over the horizontal lines at the various coordinate values scaled on the vertical axis. However, only the higher value of s_k satisfies the constraint of $s_{k-1} \le s_k$, $\forall\, k$. The enlarged view near $y_k(s_k) = 0.5$ seen in Figure 22.22(b) reveals that $y_4(s_4)$ may have no solution of $y_4(s_4) = y_1(s_1)$, when $y_1(s_1) > 0.45$. One option is to use a constant value of $s_4 = 2.37$dB, where $y_4(s_4)$ reaches its peak value. The other option is to set $s_4 = s_3$, effectively eliminating 16PSK from the set of possible modulation modes. It was found that both policies result in the same performance up to four effective decimal digits in terms of the average BPS throughput.

Upon solving $y_k(s_k) = y_1(s_1)$, we arrive at the relationships sought between the first optimum switching level s_1 and the remaining optimum switching levels s_k. Figure 22.23(a) depicts these relationships. All the optimum switching levels, except for s_1 and s_2, approach their asymptotic limit monotonically, as s_1 decreases. A decreased value of s_1 corresponds to an increased value of the average SNR. Figure 22.23(b) illustrates the optimum switching levels of a seven-mode adaptive PSK scheme operating over a Rayleigh channel associated with $m = 1$ at the target BEP of $P_{th} = 10^{-2}$. These switching levels were obtained by solving (22.68). The optimum switching levels show a steady decrease in their values as the average SNR increases, until it reaches the avalanche SNR value of $\bar{\gamma} = 35$dB, beyond which always the highest-order PSK modulation mode, namely 64PSK, is activated.

Having highlighted the evaluation of the optimum switching levels for an adaptive PSK

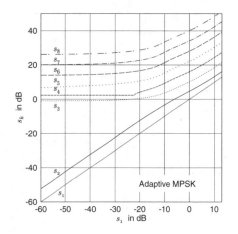

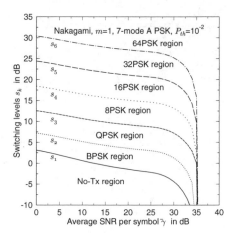

(a) Relationship between the optimum switching levels of a nine-mode PSK scheme

(b) Optimum switching levels of a seven-mode adaptive PSK scheme

Figure 22.23: Optimum switching levels. (a) Relationships between s_k and s_1 in a nine-mode adaptive PSK scheme. (b) Optimum switching levels for 7-mode adaptive PSK scheme operating over a Rayleigh channel at the target BEP of $P_{th} = 10^{-2}$

scheme, let us now consider the associated performance results. We are reminded that the average BEP of our optimised adaptive scheme remains constant at $P_{avg} = P_{th}$, provided that the average SNR is less than the avalanche SNR. Hence, the average BPS throughput and the relative SNR gain of our APSK scheme in comparison to the corresponding fixed-mode modem are our concern.

Let us now consider Figure 22.24, where the average BPS throughput of the various adaptive PSK schemes operating over a Rayleigh channel associated with $m = 1$ are plotted, which were designed for the target BEP of $P_{th} = 10^{-2}$ and $P_{th} = 10^{-3}$. The markers '$\otimes$' and '$\odot$' represent the required SNR of the various fixed-mode PSK schemes, while achieving the same target BER as the adaptive schemes, operating over an AWGN channel and a Rayleigh channel, respectively. It can be observed that introducing an additional constituent mode into an adaptive PSK scheme does not make any impact on the average BPS throughput, when the average SNR is relatively low. For example, when the average SNR $\bar{\gamma}$ is less than 10dB in Figure 22.24(a), employing more than four APSK modes for the adaptive scheme does not improve the average BPS throughput. In comparison to the various fixed-mode PSK modems, the adaptive modem achieved the SNR gains between 4dB and 8dB for the target BEP of $P_{th} = 10^{-2}$ and 10dB to 16dB for the target BEP of $P_{th} = 10^{-3}$ over a Rayleigh channel. Since no adaptive scheme operating over a fading channel can outperform the corresponding fixed-mode counterpart operating over an AWGN channel, it is interesting to investigate the performance differences between these two schemes. Figure 22.24 suggests that the required SNR of our adaptive PSK modem achieving 1BPS for transmission over a Rayleigh channel is approximately 1dB higher than that of fixed-mode BPSK operating over an AWGN channel. Furthermore, this impressive performance can be achieved by employing

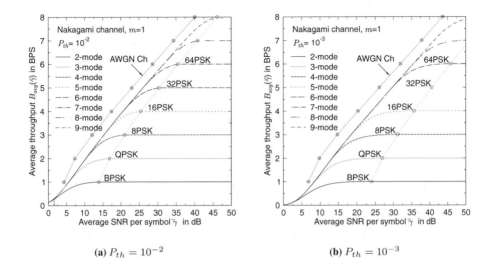

(a) $P_{th} = 10^{-2}$ **(b)** $P_{th} = 10^{-3}$

Figure 22.24: The average BPS throughput of various adaptive PSK schemes operating over a Rayleigh channel ($m = 1$) at the target BEP of (a) $P_{th} = 10^{-2}$ and (b) $P_{th} = 10^{-3}$. The markers '$\otimes$' and '$\odot$' represent the required SNR of the corresponding fixed-mode PSK scheme, while achieving the same target BEP as the adaptive schemes, operating over an AWGN channel and a Rayleigh channel, respectively

only three modes, namely No-Tx, BPSK and QPSK for the adaptive PSK modem. For other BPS throughput values, the corresponding SNR differences are in the range of 2dB to 3dB, while maintaining the BEP of $P_{th} = 10^{-2}$ and 4dB for the BEP of $P_{th} = 10^{-3}$.

We observed in Figure 22.24 that the average BPS throughput of the various adaptive PSK schemes is dependent on the target BEP. Hence, let us investigate the BPS performances of the adaptive modems for the various values of target BEPs using the results depicted in Figure 22.25. The average BPS throughputs of a nine-mode adaptive PSK scheme are represented as various types of lines without markers depending on the target average BERs, while those of the corresponding fixed PSK schemes are represented as various types of lines with markers according to the key legend shown in Figure 22.25. We can observe that the difference between the required SNRs of the adaptive schemes and fixed schemes increases, as the target BEP decreases. It is interesting to note that the average BPS curves of the adaptive PSK schemes seem to converge to a set of densely packed curves, as the target BEP decreases to values around $10^{-4} - 10^{-6}$. In other words, the incremental SNR required for achieving the next target BEP, which is an order of magnitude lower, decreases as the target BEP decreases. On the other hand, the incremental SNR for the same scenario of fixed modems seems to remain nearly constant at 10dB. Comparing Figure 22.25(a) and Figure 22.25(b), we find that this seemingly constant incremental SNR of the fixed-mode modems is reduced to about 5dB, as the fading becomes less severe, i.e. when the fading parameter becomes $m = 2$.

Let us now investigate the effects of the Nakagami fading parameter m on the average BPS throughput performance of various adaptive PSK schemes by observing Figure 22.26. The BPS throughput of the various fixed PSK schemes for transmission over an AWGN chan-

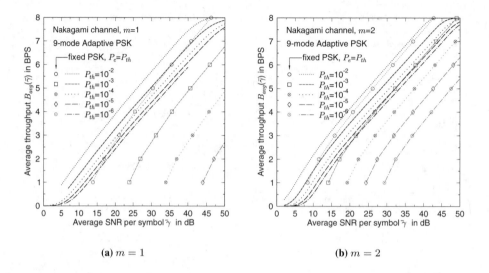

(a) $m = 1$ **(b)** $m = 2$

Figure 22.25: The average BPS throughput of a nine-mode adaptive PSK scheme operating over a Nakagami fading channel (a) $m = 1$ and (b) $m = 2$. The markers represent the SNR required for achieving the same BPS throughput and the same average BEP as the adaptive schemes

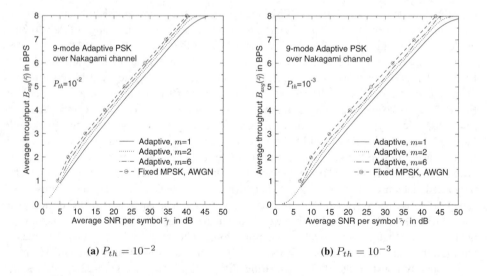

(a) $P_{th} = 10^{-2}$ **(b)** $P_{th} = 10^{-3}$

Figure 22.26: The effects of the Nakagami fading parameter m on the average BPS throughput of a nine-mode adaptive PSK scheme designed for the target BEP of (a) $P_{th} = 10^{-2}$ and (b) $P_{th} = 10^{-3}$. As m increases, the average throughput of the adaptive modem approaches the throughput of fixed PSK modems operating over an AWGN channel

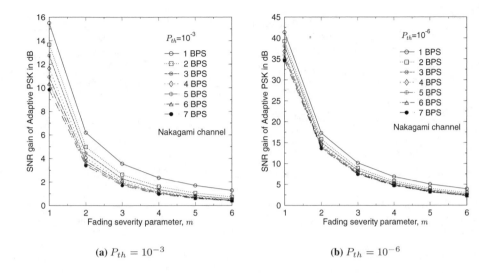

(a) $P_{th} = 10^{-3}$ (b) $P_{th} = 10^{-6}$

Figure 22.27: The SNR gain of adaptive PSK schemes in comparison to the corresponding fixed-mode
PSK schemes yielding the same BPS throughput for the target BEP of (a) $P_{th} = 10^{-3}$
and (b) $P_{th} = 10^{-6}$. The performance advantage of employing adaptive PSK schemes
decreases, as the fading becomes less severe

nel is depicted in Figure 22.26 as the ultimate performance limit achievable by the adaptive
schemes operating over Nakagami fading channels. For example, when the channel exhibits
Rayleigh fading, i.e. when the fading parameter becomes $m = 1$, the adaptive PSK schemes
show 3dB to 4dB SNR penalty compared to their fixed-mode counterparts operating over the
AWGN channel. Compared to fixed-mode BPSK, the adaptive scheme required only a 1dB
higher SNR. As the fading becomes less severe, the average BPS throughput of the adaptive
PSK schemes approaches that of fixed-mode PSK operating over the AWGN channel. For
the target BEP of $P_{th} = 10^{-3}$, the SNR gap between the BPS throughput curves becomes
higher. The adaptive PSK scheme operating over the Rayleigh channel required 4dB to 5dB
higher SNR for achieving the same throughput compared to the fixed PSK schemes operating
over the AWGN channel.

Figure 22.27 summarises the relative SNR gains of our adaptive PSK schemes over the
corresponding fixed PSK schemes. For the target BEP of $P_{th} = 10^{-3}$ the relative SNR
gain of the nine-mode adaptive scheme compared to BPSK changes from 15.5dB to 1.3dB,
as the Nakagami fading parameter changes from 1 to 6. Observing Figure 22.27(a) and
Figure 22.27(b) we conclude that the advantages of employing adaptive PSK schemes are
more pronounced when

1) the fading is more severe,

2) the target BEP is lower, and

3) the average BPS throughput is lower.

Having studied the range of APSK schemes, let us in the next section consider the family

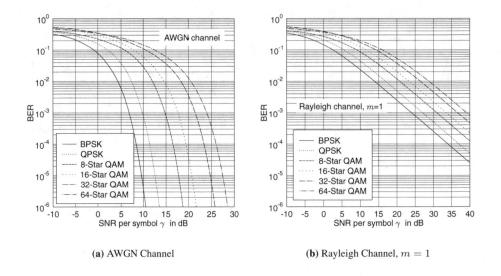

(a) AWGN Channel   **(b)** Rayleigh Channel, $m = 1$

Figure 22.28: The average BEP of various SQAM modulation schemes

of adaptive coherently detected Star-QAM schemes.

22.9.1.2   Adaptive Coherent Star QAM Schemes

In this section, we study the performance of adaptive coherent QAM schemes employing Type-I Star constellations [180]. Even though non-coherent Star QAM (SQAM) schemes are more popular owing to their robustness to fading without requiring pilot symbol assisted channel estimation and Automatic Gain Control (AGC) at the receiver, the results provided in this section can serve as benchmark results for non-coherent Star QAM schemes and the coherent Square QAM schemes.

The BEP of coherent Star QAM over an AWGN channel is derived in Appendix 31.1. It is shown that their BEP can be expressed as:

$$p_{SQAM}(\gamma) \simeq \sum_{i} A_i \, Q(\sqrt{a_i \gamma}), \qquad (22.87)$$

where A_i and a_i are given in Appendix 31.1 for 8-Star, 16-Star, 32-Star and 64-Star QAM. The SNR-dependent optimum ring ratios were also derived in Appendix 31.1 for these Star QAM modems. Figure 22.28(a) shows the BEP of BPSK, QPSK, 8-Star QAM, 16-Star QAM, 32-Star QAM and 64-Star QAM employing the optimum ring ratios over the AWGN channel. Comparing Figure 22.21(a) and Figure 22.28(a), we can observe that 16-Star QAM, 32-Star QAM and 64-Star QAM are more power-efficient than 16 PSK, 32 PSK and 64 PSK, respectively. However, the envelope power of the Star QAM signals is not constant, unlike that of the PSK signals. Following an approach similar to that used in (22.84) and (22.85), the average BEP of the various SQAM schemes over a flat Nakagami-m fading channel can

k	0	1	2	3	4	5	6
m_k	0	2	4	8	16	32	64
b_k	0	1	2	3	4	5	6
c_k	0	1	1	1	1	1	1
mode	No Tx	BPSK	QPSK	8-Star	16-Star	32-Star	64-Star

Table 22.6: Parameters of a seven-mode adaptive Star QAM scheme following the definitions developed in Section 22.7.1 for the generic adaptive modulation model

be expressed as:

$$P_{SQAM}(\bar{\gamma}) = \sum_i A_i \left[\tfrac{1}{2}(1 - \mu_i)\right]^m \sum_{j=0}^{m-1} \binom{m-1+j}{j} \left[\tfrac{1}{2}(1 + \mu_i)\right]^j, \qquad (22.88)$$

where μ_i is defined as:

$$\mu_i \triangleq \sqrt{\frac{a_i \bar{\gamma}}{2m + a_i \bar{\gamma}}}. \qquad (22.89)$$

Figure 22.28(b) shows the average BEP of various SQAM schemes for transmission over a flat Rayleigh channel, where $m = 1$. It can be observed that the 16-Star, 32-Star and 64-Star QAM schemes exhibit SNR advantages of around 3.5dB, 4dB, and 7dB compared to 16-PSK, 32-PSK and 64-PSK schemes at a BEP of 10^{-2}. The BEP of SQAM for transmission over the AWGN channel given in (22.87) and that over a Nakagami channel given in (22.88) will be used in comparing the performance of the various adaptive SQAM schemes.

The parameters of a seven-mode adaptive Star QAM scheme are summarised in Table 22.6 following the definitions of the generic model developed in Section 22.7.1 for adaptive modulation schemes. Since the number of modes is $K = 7$, we have $K + 1 = 8$ mode-switching levels hosted by the vector $\mathbf{s} = \{s_k \mid k = 0, 1, 2, \cdots, 7\}$. Let us assume that $s_0 = 0$ and $s_7 = \infty$. Then, we have to determine the optimum values for the remaining six switching levels using the technique developed in Section 22.8.4. The 'Lagrangian-free' functions corresponding to a seven-mode Star QAM scheme are depicted in Figure 22.29 and the relationships obtained for the switching levels are displayed in Figure 22.30(a). We can observe that as seen for APSK in Figure 22.22 there exist two solutions for s_6 satisfying $y_6(s_6) = y_1(s_1)$ for a given value of s_1, when $y_1 \leq 0.382$. However, only the higher value of s_k satisfies the constraint of $s_6 \geq s_5$. When $s_1 \leq 7.9$dB, the optimum value of s_6 should be set to s_5, in order to guarantee $s_6 \geq s_5$. Figure 22.30(b) illustrates the optimum switching levels of a seven-mode adaptive Star QAM scheme operating over a Rayleigh channel at the target BEP of $P_{th} = 10^{-2}$. These switching levels were obtained by solving (22.68). The optimum switching levels show a steady decrease in their values, as the average SNR increases, until they reach the avalanche SNR value of $\bar{\gamma} = 28.5$dB, beyond which always the highest-order modulation mode, namely 64-Star QAM, is activated.

Let us now investigate the associated performance results. We are reminded that the average BEP of our optimised adaptive scheme remains constant at $P_{avg} = P_{th}$, provided

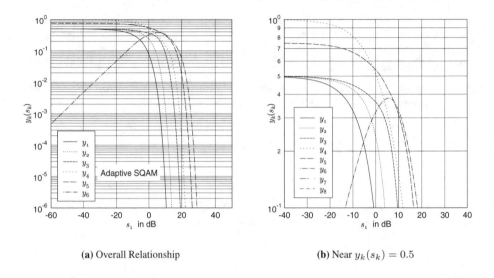

(a) Overall Relationship   **(b)** Near $y_k(s_k) = 0.5$

Figure 22.29: 'Lagrangian-free' functions of (22.61) for a seven-mode adaptive Star QAM scheme

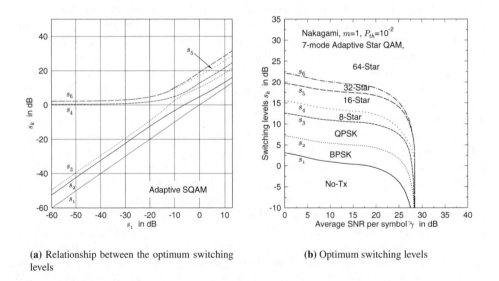

(a) Relationship between the optimum switching levels   **(b)** Optimum switching levels

Figure 22.30: Optimum switching levels of a seven-mode Adaptive Star QAM scheme. (a) Relationships between s_k and s_1. (b) Optimum switching levels of a seven-mode adaptive Star
QAM scheme operating over a Rayleigh channel at the target BEP of $P_{th} = 10^{-2}$

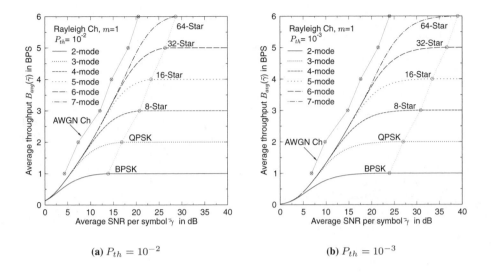

(a) $P_{th} = 10^{-2}$

(b) $P_{th} = 10^{-3}$

Figure 22.31: The average BPS throughput of the various adaptive Star QAM schemes operating over a Rayleigh fading channel associated with $m = 1$ at the target BEP of (a) $P_{th} = 10^{-2}$ and (b) $P_{th} = 10^{-3}$. The markers '$\otimes$' and '$\odot$' represent the required SNR of the corresponding fixed-mode Star QAM schemes, while achieving the same target BEP as the adaptive schemes, operating over an AWGN channel and a Rayleigh channel, respectively

that the average SNR is less than the avalanche SNR. Hence, the average BPS throughput and the SNR gain of our adaptive modem in comparison to the corresponding fixed-mode modems are our concern.

Let us first consider Figure 22.31, where the average BPS throughput of the various adaptive Star QAM schemes operating over a Rayleigh channel associated with $m = 1$ is shown at the target BEP of $P_{th} = 10^{-2}$ and $P_{th} = 10^{-3}$. The markers '$\otimes$' and '$\odot$' represent the required SNR of the corresponding fixed-mode Star QAM schemes, while achieving the same target BEP as the adaptive schemes, operating over an AWGN channel and a Rayleigh channel, respectively. Comparing Figure 22.24(a) and Figure 22.31(a), we find that the tangent of the average BPS curves of the adaptive Star QAM schemes is higher than that of adaptive PSK schemes. Explicitly, the tangent of the Star QAM schemes is around 0.3BPS/dB, whereas that of the APSK schemes was 0.18BPS/dB. This is due to the more power-efficient constellation arrangement of Star QAM in comparison to the single-ring constellations of the PSK modulations schemes. In comparison to the corresponding fixed-mode Star QAM modems, the adaptive modem achieved an SNR gain of 6dB to 8dB for the target BEP of $P_{th} = 10^{-2}$ and 12dB to 16dB for the target BEP of $P_{th} = 10^{-3}$ over a Rayleigh channel. Compared to the fixed-mode Star QAM schemes operating over an AWGN channel, our adaptive schemes approached their performance within about 3dB in terms of the required SNR value, while achieving the same target BEP of $P_{th} = 10^{-2}$ and $P_{th} = 10^{-3}$.

Since Figure 22.31 suggests that the relative SNR gain of the adaptive schemes is dependent on the target BER, let us investigate the effects of the target BEP in more detail.

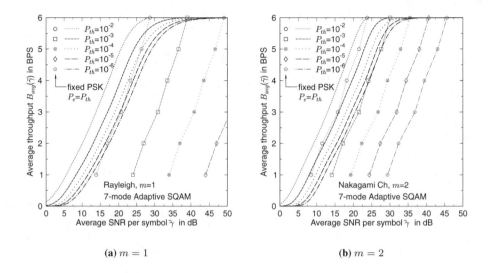

(a) $m = 1$ **(b)** $m = 2$

Figure 22.32: The average BPS throughput of a seven-mode adaptive Star QAM scheme operating over a Nakagami fading channel (a) $m = 1$ and (b) $m = 2$. The markers represent the SNR required by the fixed-mode schemes for achieving the same BPS throughput and the same average BER as the adaptive schemes

Figure 22.32 shows the BPS throughput of the various adaptive schemes at the target BEP of $P_{th} = 10^{-2}$ to $P_{th} = 10^{-6}$. The average BPS throughput of a seven-mode adaptive Star QAM scheme is represented with the aid of the various line types without markers, depending on the target average BERs, while those of the corresponding fixed-mode Star QAM schemes are represented as various types of lines having markers according to the legends shown in Figure 22.32. We can observe that the difference between the SNRs required for the adaptive schemes and fixed schemes increases, as the target BEP decreases. The fixed-mode Star QAM schemes require additional SNRs of 10dB and 6dB in order to achieve an order of magnitude lower BEP for the Nakagami fading parameters of $m = 1$ and $m = 2$, respectively. However, our adaptive schemes require additional SNRs of only 1dB to 3dB for achieving the same goal.

Let us now investigate the effects of the Nakagami fading parameter m on the average BPS throughput performance of the various adaptive Star QAM schemes by observing Figure 22.33. The BPS throughput of the fixed-mode Star QAM schemes for the transmission over an AWGN channel is depicted in Figure 22.33 as the ultimate performance limit achievable by the adaptive schemes operating over Nakagami fading channels. As the Nakagami fading parameter m increases from 1 to 2 and to 6, the SNR gap between the adaptive schemes operating over a Nakagami fading channel and the fixed-mode schemes decreases. When the average SNR is less than $\bar{\gamma} \leq 6$dB, the average BPS throughput of our adaptive schemes decreases, when the fading parameter m increases. The rationale of this phenomenon is that as the channel becomes more and more like an AWGN channel, the probability of activating the BPSK mode is reduced, resulting in more frequent activation of the No-Tx mode and hence the corresponding average BPS throughput inevitably decreases.

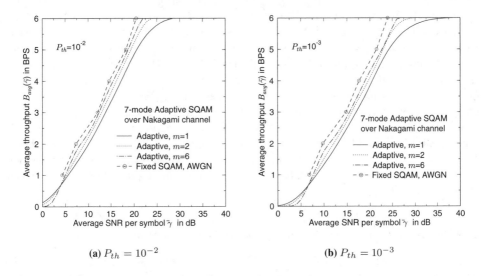

(a) $P_{th} = 10^{-2}$ **(b)** $P_{th} = 10^{-3}$

Figure 22.33: The effects of the Nakagami fading parameter m on the average BPS throughput of a seven-mode adaptive Star QAM scheme at the target BEP of (a) $P_{th} = 10^{-2}$ and (b) $P_{th} = 10^{-3}$. As m increases, the average throughput of the adaptive modem approaches the throughput of the fixed-mode Star QAM modems operating over an AWGN channel

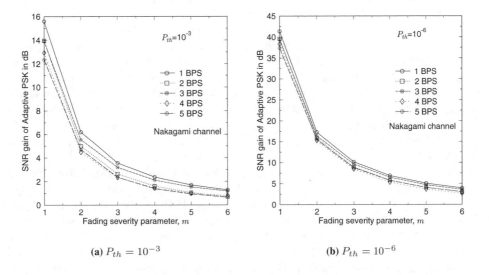

(a) $P_{th} = 10^{-3}$ **(b)** $P_{th} = 10^{-6}$

Figure 22.34: The SNR gain of the various adaptive Star QAM schemes in comparison to the fixed-mode Star QAM schemes yielding the same BPS throughput at the target BEP of (a) $P_{th} = 10^{-3}$ and (b) $P_{th} = 10^{-6}$. The advantage of the adaptive Star QAM schemes decreases, as the fading becomes less severe

The effects of the Nakagami fading factor m on the SNR gain of our adaptive Star QAM scheme can be observed in Figure 22.34. As expected, the relative SNR gain of the adaptive schemes at a throughput of 1 BPS is the highest among the BPS throughputs considered. However, the order observed in terms of the SNR gain of the adaptive schemes does not strictly follow the increasing BPS order at the target BEP of $P_{th} = 10^{-3}$ and $P_{th} = 10^{-6}$, as it did for the adaptive PSK schemes of Section 22.9.1.1. Even though the adaptive Star QAM schemes exhibit a higher throughput than the adaptive PSK schemes, the SNR gains compared to their fixed-mode counterparts are more or less the same, showing typically less than 1dB difference, except for the 5 BPS throughput scenario, where the adaptive QAM scheme gained up to 1.3dB more in terms of the required SNR than the adaptive PSK scheme.

Having studied the performance of a range of adaptive Star QAM schemes, in the next section we consider adaptive modulation schemes employing the family of square-shaped QAM constellations.

22.9.1.3   Adaptive Coherent Square QAM Modulation Schemes

Since coherent Square M-ary QAM (MQAM) is the most power-efficient M-ary modulation scheme [180] and the accurate channel estimation becomes possible with the advent of Pilot Symbol Assisted Modulation (PSAM) techniques [139, 140, 178], *Otsuki*, *Sampei* and *Morinaga* proposed the employment of coherent square QAM as the constituent modulation modes for an adaptive modulation scheme [179] instead of non-coherent Star QAM modulation [177]. In this section, we study the various aspects of this adaptive square QAM scheme employing the optimum switching levels of Section 22.8.4. The closed form BEP expressions of square QAM over an AWGN channel can be found in (22.2) and that over a Nakagami channel can be expressed using a similar form given in (22.88). The optimum switching levels of adaptive Square QAM were studied in Section 22.8.4 as an example.

The average BEP of our six-mode adaptive Square QAM scheme operating over a flat Rayleigh fading channel is depicted in Figure 22.35(a), which shows that the modem maintains the required constant target BER, until it reaches the BER curve of the specific fixed-mode modulation scheme employing the highest-order modulation mode, namely 256-QAM, and then it follows the BEP curve of the 256-QAM mode. The various grey lines in the figure represent the BEP of the fixed constituent modulation modes for transmission over a flat Rayleigh fading channel. An arbitrarily low target BEP could be maintained at the expense of a reduced throughput.

The average throughput is shown in Figure 22.35(b) together with the estimated channel capacity of the narrow-band Rayleigh channel [181, 182] and with the throughput of several variable-power, variable-rate modems reported in [184]. Specifically, Goldsmith and Chua [184] studied the performance of their variable-power variable-rate adaptive modems based on a BER bound of m-ary Square QAM, rather than using an exact BER expression. Since our adaptive Square QAM schemes do not vary the transmission power, our scheme can be regarded as a sub-optimal policy viewed for their respective [184]. However, the throughput performance of Figure 22.35(b) shows that the SNR degradation is within 2dB in the low-SNR region and within half a dB in the high-SNR region, in comparison to the ideal continuously variable-power adaptive QAM scheme employing a range of hypothetical continuously variable-BPS QAM modes [184], represented as the 'Goldsmith 1' scheme in the figure. *Goldsmith* and *Chua* [184] also reported the performance of a variable-power

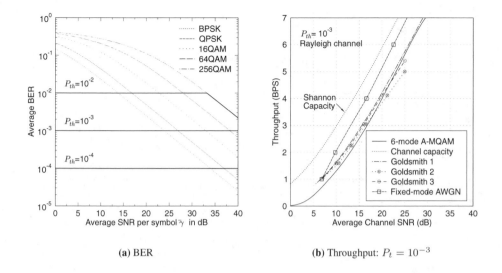

(a) BER   **(b)** Throughput: $P_t = 10^{-3}$

Figure 22.35: The average BEP and average throughput performance of a six-mode adaptive Square
QAM scheme operating over a flat Rayleigh channel ($m = 1$). (a) The constant target average BEP is maintained over the entire range of the average SNR values up to
the avalanche SNR. (b) The average BPS throughput of the equivalent constant-power
adaptive scheme is compared to *Goldsmith*'s schemes [184]. The 'Goldsmith 1' and
'Goldsmith 2' schemes represent a variable-power adaptive scheme employing hypothetical continuously variable-BPS QAM modulation modes and Square QAM modes,
respectively. The 'Goldsmith 3' scheme represents the simulation results associated with
a constant-power adaptive Square QAM reported in [184]

discrete-rate and a constant-power discrete-rate scheme, which we represented as the 'Goldsmith 2' and 'Goldsmith 3' scenarios in Figure 22.35(b), respectively. Since their results are
based on approximate BER formulas, the average BPS throughput performance of the 'Goldsmith 3' scheme is optimistic, when the average SNR γ is less than 17dB. Considering that
our scheme achieves the maximum possible throughput for the given average SNR value with
the aid of the globally optimised switching levels, the average throughput of the 'Goldsmith
3' scheme is expected to be lower than that of our scheme, as is the case when the average
SNR γ is higher than 17dB.

Figure 22.36(a) depicts the average BPS throughput of our various adaptive Square QAM
schemes operating over a Rayleigh channel associated with $m = 1$ at the target BEP of $P_{th} =
10^{-3}$. Figure 22.36(a) shows that even though the constituent modulation modes of our
adaptive schemes do not include 3, 5 and 7-BPS constellations, the average BPS throughput
steadily increases without undulations. Compared to the fixed-mode Square QAM schemes
operating over an AWGN channel, our adaptive schemes require additional SNRs of less
than 3.5dB, when the throughput is below 6.5 BPS. The comparison of the average BPS
throughputs of the adaptive schemes employing PSK, Star QAM and Square QAM modems,
as depicted in Figure 22.36(b), confirms the superiority of Square QAM over the other two
schemes in terms of the required average SNR for achieving the same throughput and the

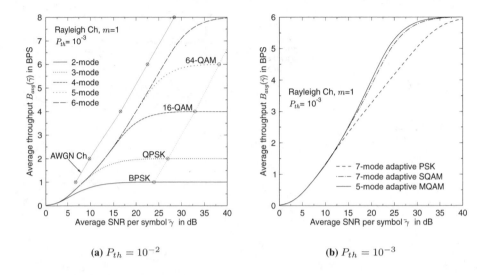

(a) $P_{th} = 10^{-2}$ **(b)** $P_{th} = 10^{-3}$

Figure 22.36: The average BPS throughput of various adaptive Square QAM schemes operating over a Rayleigh channel ($m = 1$) at the target BEP of $P_{th} = 10^{-3}$. (a) The markers '$\otimes$' and '$\odot$' represent the required SNR of the corresponding fixed-mode Square QAM schemes achieving the same target BEP as the adaptive schemes, operating over an AWGN channel and a Rayleigh channel, respectively. (b) The comparison of the various adaptive schemes employing PSK, Star QAM and Square QAM as the constituent modulation modes

same target average BEP. Since all these three schemes employ BPSK, QPSK as the second and the third constituent modulation modes, their throughput performance shows virtually no difference, when the average throughput is less than or equal to $B_{avg} = 2$ BPS.

Let us now investigate the effects of the Nakagami fading parameter m on the average BPS throughput performance of the adaptive Square QAM schemes observing Figure 22.37. The BPS throughput of the fixed-mode Square QAM schemes over an AWGN channel is depicted in Figure 22.37 as the ultimate performance limit achievable by the adaptive schemes operating over Nakagami fading channels. Similar observations can be made for the adaptive Square QAM scheme, as for the adaptive Star QAM arrangement characterised in Figure 22.33. A specific difference is, however, that the average BPS throughput recorded for the fading parameter of $m = 6$ exhibits an undulating curve. For example, an increased m value results in a limited improvement of the corresponding average BPS throughput near the throughput values of 2.5, 4.5 and 6.5 BPS. This is because our adaptive Square QAM schemes do not use 3-, 5- and 7-BPS constituent modems, unlike the adaptive PSK and adaptive Star QAM schemes. Figure 22.38 depicts the corresponding optimum mode-switching levels for the six-mode adaptive Square QAM scheme. The black lines represent the switching levels, when the Nakagami fading parameter is $m = 6$ and the grey lines when $m = 1$. In general, the lower the switching levels, the higher the average BPS throughput of the adaptive modems. When the Nakagami fading parameter is $m = 1$, the switching levels decrease monotonically, as the average SNR increases. However, when the fading severity parameter is

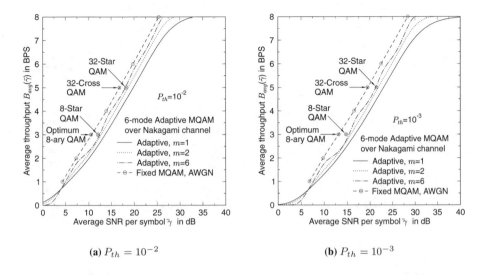

(a) $P_{th} = 10^{-2}$ **(b)** $P_{th} = 10^{-3}$

Figure 22.37: The effects of the Nakagami fading parameter m on the average BPS throughput of a seven-mode adaptive Square QAM scheme at the target BEP of (a) $P_{th} = 10^{-2}$ and (b) $P_{th} = 10^{-3}$. As m increases, the average throughput of the adaptive modem approaches the throughput of the corresponding fixed Square QAM modems operating over an AWGN channel

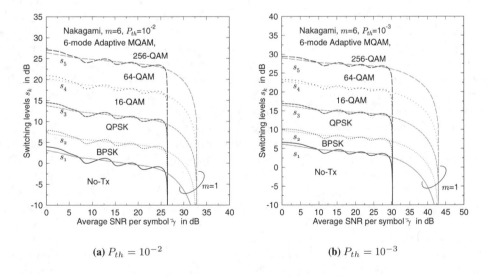

(a) $P_{th} = 10^{-2}$ **(b)** $P_{th} = 10^{-3}$

Figure 22.38: The switching levels of the six-mode adaptive Square QAM scheme operating over Nakagami fading channels at the target BER of (a) $P_{th} = 10^{-2}$ and (b) $P_{th} = 10^{-3}$. The bold lines are used for the fading parameter of $m = 6$ and the grey lines are for $m = 1$

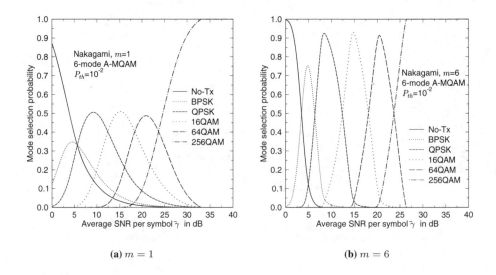

(a) $m = 1$ **(b)** $m = 6$

Figure 22.39: The mode selection probability of a six-mode adaptive Square QAM scheme operating over Nakagami fading channels at the target BEP of $P_{th} = 10^{-2}$. When the fading becomes less severe, the mode selection scheme becomes more 'selective' in comparison to that for $m = 1$

$m = 6$, the switching levels fluctuate, exhibiting several local minima around 8dB, 15dB and 21dB. In the extreme case of $m \to \infty$, i.e. when operating over an AWGN-like channel, the switching levels would be $s_1 = s_2 = 0$ and $s_k = \infty$ for other k values in the SNR range of 7.3dB $< \bar{\gamma} < $ 14dB, $s_1 = s_2 = s_3 = 0$ and $s_4 = s_5 = \infty$ when we have 14dB $< \bar{\gamma} < $ 20dB, $s_k = 0$ except for $s_5 = \infty$ when the SNR is in the range of 20dB $< \bar{\gamma} < $ 25dB and finally, all $s_k = 0$ for $\forall k$, when $\bar{\gamma} > $ 25dB, when considering the fixed-mode Square QAM performance achieved over an AWGN channel represented by markers '⊙' in Figure 22.37. Observing Figure 22.39, we find that our adaptive schemes become highly 'selective', when the Nakagami fading parameter becomes $m = 6$, exhibiting narrow triangular shapes. As m increases, the shapes will eventually converge to Kronecker delta functions.

A possible approach to reducing the undulating behaviour of the average BPS throughput curve is the introduction of a 3-BPS and a 5-BPS mode as additional constituent modem modes. The power-efficiency of 8-Star QAM and 32-Star QAM is insufficient for maintaining a linear growth of the average BPS throughput, as we can observe in Figure 22.37. Instead, the most power-efficient 8-ary QAM scheme (see page 279 of [388]) and the so-called 32-ary cross-shaped QAM scheme have a potential of reducing these undulation effects. However, since we observed in Section 22.9.1.1 and Section 22.9.1.2 that the relative SNR advantage of employing adaptive Square QAM rapidly reduces, when the Nakagami fading parameter increases, even though the additional 3-BPS and 5-BPS modes are also used, there seems to be no significant benefit in employing non-square shaped additional constellations.

Again, we can observe in Figure 22.37 that when the average SNR is less than $\bar{\gamma} \leq $ 6dB, the average BPS throughput of our adaptive Square QAM scheme decreases, as the Nakagami fading parameter m increases. As we discussed in Section 22.9.1.2, this is due to the less

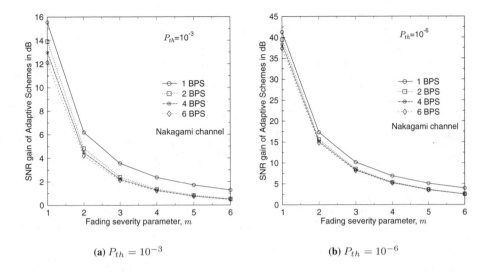

(a) $P_{th} = 10^{-3}$ (b) $P_{th} = 10^{-6}$

Figure 22.40: The SNR gain of the six-mode adaptive Square QAM scheme in comparison to the various fixed-mode Square QAM schemes yielding the same BPS throughput at the target BEP of (a) $P_{th} = 10^{-3}$ and (b) $P_{th} = 10^{-6}$. The performance advantage of the adaptive Square QAM schemes decreases, as the fading becomes less severe

frequent activation of the BPSK mode in comparison to the 'No-Tx' mode, as the channel variation is reduced.

The effects of the Nakagami fading factor m on the relative SNR gain of our adaptive Square QAM scheme can be observed in Figure 22.40. The less severe the fading, the smaller the relative SNR advantage of employing adaptive Square QAM in comparison to its fixed-mode counterparts. Except for the 1-BPS mode, the SNR gains become less than 0.5dB, when m is increased to 6 at the target BEP of $P_{th} = 10^{-3}$. The trend observed is the same at the target BEP of $P_{th} = 10^{-6}$, showing relatively higher gains in comparison to the $P_{th} = 10^{-3}$ scenario.

22.9.2 Performance over Narrow-band Rayleigh Channels Using Antenna Diversity

In the last section, we observed that the adaptive modulation schemes employing Square QAM modes exhibit the highest BPS throughput among the schemes investigated, when operating over Nakagami fading channels. Hence, in this section we study the performance of the adaptive Square QAM schemes employing antenna diversity operating over independent Rayleigh fading channels. The BEP expression of the fixed-mode coherent BPSK scheme can be found on Page 781 of [388] and those of coherent Square QAM can readily be extended using the equations in (22.2) and (22.3). Furthermore, the antenna diversity scheme operating over independent narrow-band Rayleigh fading channels can be viewed as a special case of the two-dimensional (2D) Rake receiver analysed in Appendices 31.2 and 31.3. The performance of antenna-diversity assisted adaptive Square QAM schemes can be readily

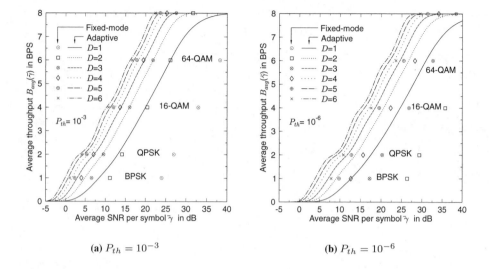

(a) $P_{th} = 10^{-3}$ 					(b) $P_{th} = 10^{-6}$

Figure 22.41: The average BPS throughput of the MRC-aided antenna-diversity assisted adaptive Square QAM scheme operating over independent Rayleigh fading channels at the target average BEP of (a) $P_{th} = 10^{-3}$ and (b) $P_{th} = 10^{-6}$. The markers represent the corresponding fixed-mode Square QAM performances

analysed using the technique developed in Section 22.8.4.

Figure 22.41 depicts the average BPS throughput performance of our adaptive schemes employing Maximal Ratio Combining (MRC) aided antenna diversity [381] (Chapters 5 and 6) operating over independent Rayleigh fading channels at the target average BEP of $P_{th} = 10^{-3}$ and $P_{th} = 10^{-6}$. The markers represent the performance of the corresponding fixed-mode Square QAM modems in the same scenario. The average SNRs required achieving the target BEP of the fixed-mode schemes and that of the adaptive schemes decrease, as the antenna diversity order increases. However, the differences between the required SNRs of the adaptive schemes and their fixed-mode counterparts also decrease, as the antenna diversity order increases. The SNRs of both schemes required for achieving the target BEPs of $P_{th} = 10^{-3}$ and $P_{th} = 10^{-6}$ are displayed in Figure 22.42, where we can observe that dual antenna diversity is sufficient for the fixed-mode schemes in order to obtain half of the achievable SNR gain of the six-antenna aided diversity scheme, whereas triple-antenna diversity is required for the adaptive schemes operating in the same scenario. The corresponding first switching levels s_1 are depicted in Figure 22.43 for different orders of antenna diversity up to an order of six. As the antenna diversity order increases, the avalanche SNR becomes lower and the switching-threshold undulation effects begin to appear. The required values of the first switching level s_1 are within a range of about 1dB and 0.5dB for the target BEPs of $P_{th} = 10^{-3}$ and $P_{th} = 10^{-6}$, respectively, before the avalanche SNR is reached. This suggests that the optimum mode-switching levels are more dependent on the target BEP than on the number of diversity antennas.

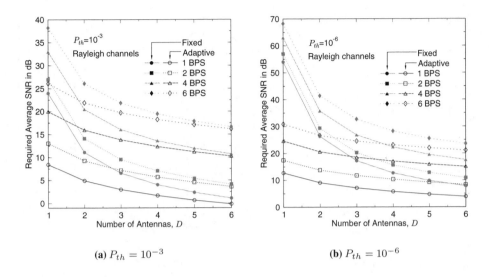

(a) $P_{th} = 10^{-3}$ **(b)** $P_{th} = 10^{-6}$

Figure 22.42: The SNR required for the MRC-aided antenna-diversity assisted adaptive Square QAM schemes and the corresponding fixed-mode modems operating over independent Rayleigh fading channels at the target average BEP of (a) $P_{th} = 10^{-3}$ and (b) $P_{th} = 10^{-6}$

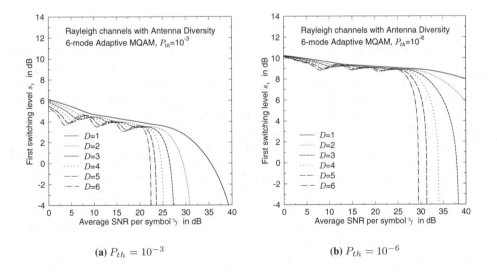

(a) $P_{th} = 10^{-3}$ **(b)** $P_{th} = 10^{-6}$

Figure 22.43: The first switching level s_1 of the MRC-aided antenna-diversity assisted adaptive Square QAM scheme operating over independent Rayleigh fading channels at the target average BEP of (a) $P_{th} = 10^{-3}$ and (b) $P_{th} = 10^{-6}$

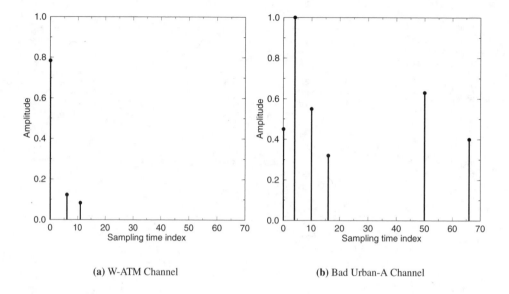

(a) W-ATM Channel   (b) Bad Urban-A Channel

Figure 22.44: Multi-path Intensity Profiles (MIPs) of the Wireless Asynchronous Transfer Mode (W-ATM) indoor channel of Figure 17.1 and that of the Bad-Urban Reduced-model A (BU-RA) channel [676]

22.9.3 Performance over Wideband Rayleigh Channels using Antenna Diversity

Wideband fading channels are characterised by their multi-path intensity profiles (MIP). In order to study the performance of the various adaptive modulation schemes, we employ two different MIP models in this section, namely the shortened Wireless Asynchronous Transfer Mode (W-ATM) channel of Figure 17.3 for an indoor scenario and a Bad-Urban Reduced-model A (BU-RA) channel [676] for a hilly urban outdoor scenario. Their MIPs are depicted in Figure 22.44.   The W-ATM channel exhibits short-range, low-delay multi-path components, while the BU-RA channel exhibits six higher-delay multi-path components. Again, let us assume that our receivers are equipped with MRC Rake receivers [677], employing a sufficiently higher number of Rake fingers, in order to capture all the multi-path components generated by our channel models. Furthermore, we employ antenna diversity [381] (Chapter 5) at the receivers. This combined diversity scheme is often referred to as a two-dimensional (2D) Rake receiver [678] (Page 263). The BEP of the 2D Rake receiver transmission over wide-band independent Rayleigh fading channels is analysed in Appendix 31.2. A closed-form expression for the mode-specific average BEP of a 2D-Rake assisted adaptive Square QAM scheme is also given in Appendix 31.3. Hence, the performance of our 2D-Rake assisted adaptive modulation scheme employing the optimum switching levels can be readily obtained.

   The average BPS throughputs of the 2D-Rake assisted adaptive schemes operating over the two different types of wideband channel scenarios are presented in Figure 22.45 at the target BEP of $P_{th} = 10^{-2}$. The throughput performance depicted corresponds to the

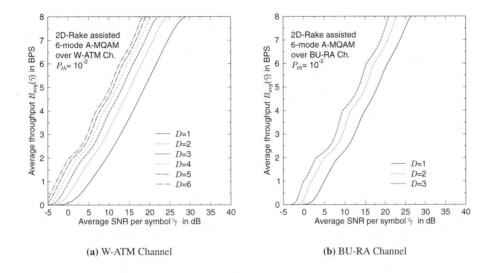

(a) W-ATM Channel   (b) BU-RA Channel

Figure 22.45: The effects of the number of diversity antennas D on the average BPS throughput of the 2D-Rake assisted six-mode adaptive Square QAM scheme operating over the wideband independent Rayleigh fading channels characterised in Figure 22.44 at the target BEP of $P_{th} = 10^{-2}$

upper-bound performance of Direct-Sequence Code Division Multiple Access (DS-CDMA) or Multi-Carrier CDMA employing Rake receivers and the MRC-aided diversity assisted scheme in the absence of Multiple Access Interference (MAI). We can observe that the BPS throughput curves undulate, when the number of antennas D increases. This effect is more pronounced for transmission over the BU-RA channel, since the BU-RA channel exhibits six multi-path components, increasing the available diversity potential of the system approximately by a factor of two in comparison to that of the W-ATM channel. The performance of our adaptive scheme employing more than three antennas for transmission over the BU-RA channel could not be obtained owing to numerical instability, since the associated curves become similar to a series of step-functions, which is not analytic in mathematical terms. A similar observation can be made in the context of Figure 22.46, where the target BEP is $P_{th} = 10^{-3}$. Comparing Figure 22.45 and Figure 22.46, we observe that the BPS throughput curves corresponding to $P_{th} = 10^{-3}$ are similar to shifted versions of those corresponding to $P_{th} = 10^{-2}$, which are shifted in the direction of increasing SNRs. On the other hand, the BPS throughput curves corresponding to $P_{th} = 10^{-3}$ undulate more dramatically. When the number of antennas is $D = 3$, the BPS throughput curves of the BU-RA channel exhibit a stair-case-like shape. The corresponding mode switching levels and mode selection probabilities are shown in Figure 22.47. Again, the switching levels heavily undulate. The mode-selection probability curve of BPSK has a triangular shape, increasing linearly, as the average SNR $\bar{\gamma}$ increases to 2.5dB and decreasing linearly again as $\bar{\gamma}$ increases from 2.5dB. On the other hand, the mode-selection probability curve of QPSK increases linearly and decreases exponentially, since no 3-BPS mode is used. This explains why the BPS throughput curves increase in a near-linear fashion in the SNR range of 0 to 5dB and in a stair-case fash-

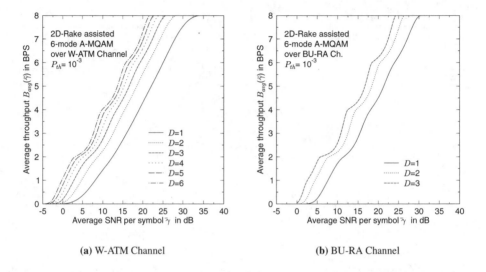

(a) W-ATM Channel   **(b)** BU-RA Channel

Figure 22.46: The effects of the number of diversity antennas D on the average BPS throughput of the 2D-Rake assisted six-mode adaptive Square QAM scheme operating over the wideband independently Rayleigh fading channels characterised in Figure 22.44 at the target BEP of $P_{th} = 10^{-3}$

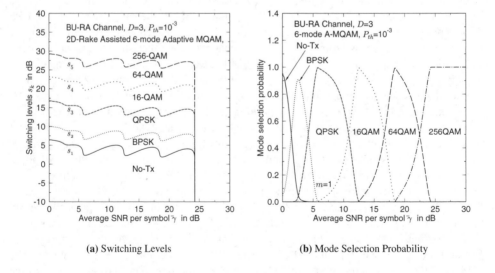

(a) Switching Levels   **(b)** Mode Selection Probability

Figure 22.47: The mode switching levels and mode selection probability of the 2D-Rake assisted six-mode adaptive Square QAM scheme using $D = 3$ antennas operating over the BU-RA channel characterised in Figure 22.44(b) at the target BEP of $P_{th} = 10^{-3}$

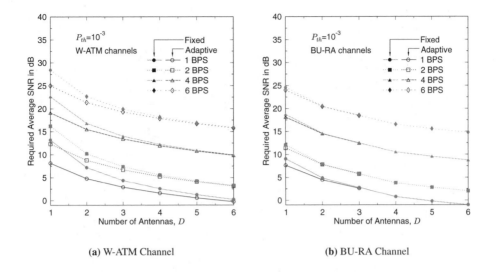

(a) W-ATM Channel   **(b)** BU-RA Channel

Figure 22.48: The average SNRs required for achieving the target BEP of $P_{th} = 10^{-3}$ by the 2D-Rake assisted adaptive schemes and by the fixed-mode schemes operating over (a) the W-ATM channel and (b) the BU-RA channel

ion beyond that point. We can conclude that the stair-case like shape in the upper SNR range of SNR is a consequence of the absence of the 3-BPS, 5-BPS and 7-BPS modulation modes in the set of constituent modulation modes employed. As we discussed in Section 22.9.1.3, this problem may be mitigated by introducing power-efficient 8 QAM, 32 QAM and 128 QAM modes.

The average SNRs required achieving the target BEP of $P_{th} = 10^{-3}$ by the 2D-Rake assisted adaptive scheme and the fixed-mode schemes operating over wide-band fading channels are depicted in Figure 22.48. Since the fixed-mode schemes employing Rake receivers are already enjoying the diversity benefit of multi-path fading channels, the SNR advantages of our adaptive schemes are less than 8dB and 2.6dB over the W-ATM channel and over the BU-RA channel, respectively, even when a single antenna is employed. This relatively small SNR gain in comparison to those observed over narrow-band fading channels in Figure 22.42 erodes as the number of antennas increases. For example, when the number of antennas is $D = 6$, the SNR gains of the adaptive schemes operating over the W-ATM channel of Figure 22.44(a) become virtually zero, where the combined channel becomes an AWGN-like channel. On the other hand, $D = 3$ number of antennas is sufficient for the BU-RA channel for exhibiting such a behaviour, since the underlying multi-path diversity provided by the six-path BU-RA channel is higher than that of the tree-path W-ATM channel.

22.9.4 Uncoded Adaptive Multi-Carrier Schemes

The performance of the various adaptive Square QAM schemes has also been studied in the context of multi-carrier systems [180, 643, 679] . The family of Orthogonal Frequency Division Multiplex (OFDM) [592] systems converts frequency selective Rayleigh channels

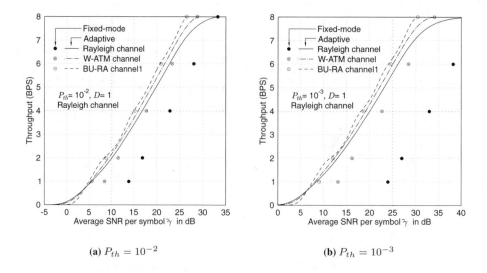

(a) $P_{th} = 10^{-2}$ (b) $P_{th} = 10^{-3}$

Figure 22.49: The average BPS throughput of the adaptive schemes and fixed-mod schemes transmission over a narrow-band Rayleigh channel, the W-ATM channel and BU-RA channel of Figure 22.44 at the target BEP of (a) $P_{th} = 10^{-2}$ and (b) $P_{th} = 10^{-3}$.

into frequency non-selective or flat Rayleigh channels for each sub-carrier, provided that the number of sub-carriers is sufficiently high. The power and bit allocation strategy of adaptive OFDM has attracted substantial research interests [180]. OFDM is particularly suitable for combined time-frequency domain processing [643]. Since each sub-carrier of an OFDM system experiences a flat Rayleigh channel, we can apply adaptive modulation for each sub-carrier independently from other sub-carriers. Although a practical scheme would group the sub-carriers into similar-quality sub-bands for the sake of reducing the associated modem mode signalling requirements. The performance of this AQAM assisted OFDM (A-OFDM) scheme is identical to that of the adaptive scheme operating over flat Rayleigh fading channels, characterised in Section 22.9.2.

MC-CDMA [252, 255] receiver can be regarded as a frequency domain Rake-receiver, where the multiple carriers play a similar role to that of the time-domain Rake fingers. Our simulation results showed that the single-user BEP performance of MC-CDMA employing multiple antennas is essentially identical to that of the time-domain Rake receiver using antenna diversity, provided that the spreading factor is higher than the number of resolvable multi-path components in the channel. Hence, the throughput of the Rake-receiver over the three-path W-ATM channel [180] and the six-path BU-RA channel [676] studied in Section 22.9.3 can be used for investigating the upper-bound performance of adaptive MC-CDMA schemes over these channels. Figure 22.49 compares the average BPS throughput performances of these schemes, where the throughput curves of the various adaptive schemes are represented as three different types of lines, depending on the underlying channel scenarios, while the fixed-mode schemes are represented as three different types of markers. The solid line corresponds to the performance of A-OFDM and the marker '•' corresponds to that of the fixed-mode OFDM. On the other hand, the dotted lines correspond to the BPS throughput performance of adaptive MC-CDMA operating over wideband channels and the markers '⊙' and '⊗' to those of the fixed-mode MC-CDMA schemes.

It can be observed that fixed-mode MC-CDMA has the potential to outperform A-OFDM, when the underlying channel provides sufficient diversity due to the high number of resolvable multi-path components. For example, the performance of fixed-mode MC-CDMA operating over the W-ATM channel of Figure 22.44(a) is slightly lower than that of A-OFDM for the BPS range of less than or equal to 6 BPS, owing to the insufficient diversity potential of the wide-band channel. On the other hand, fixed-mode MC-CDMA outperforms A-OFDM, when the channel is characterised by the BU-RA model of Figure 22.44(b). We have to consider several factors, in order to answer, whether fixed-mode MC-CDMA is better than A-OFDM. Firstly, fully loaded MC-CDMA, which can transmit the same number of symbols as OFDM, suffers from multi-code interference and our simulation results showed that the SNR degradation is about 2-4dB at the BEP of 10^{-3}, when the Minimum Mean Square Error Block Decision Feedback Equaliser (MMSE-BDFE) [680] based joint detector is used at the receiver. Considering these SNR degradations, the throughput of fixed-mode MC-CDMA using the MMSE-BDFE joint detection receiver falls just below that of the A-OFDM scheme, when the channel is characterised by the BU-RA model. On the other hand, the adaptive schemes may suffer from inaccurate channel estimation/prediction and modem mode signalling feedback delay [184]. Hence, the preference order of the various schemes may depend on the channel scenario encountered, on the interference effects and other practical issues, such as the aforementioned channel estimation accuracy, feedback delays, etc.

22.9.5 Concatenated Space-Time Block Coded and Turbo Coded Symbol-by-Symbol Adaptive OFDM and Multi-Carrier CDMA[3]

In the previous sections we studied the performance of uncoded adaptive schemes. Since a Forward Error Correction (FEC) code reduces the SNR required for achieving a given target BEP at the expense of a reduced BPS throughput, it is interesting to investigate the performance of adaptive schemes employing FEC techniques. These investigations will allow us to gauge, whether channel coding is capable of increasing the system's effective throughput, when aiming for a specific target BER. Another important question to be answered is whether there are any further potential performance advantages, when we combine adaptive modulation with space-time coding. We note in advance that our related investigations are included here with a view to drawing the reader's attention to the associated system design trade-offs, rather than providing an indepth comparative study of adaptive modulation and space-time coding. Hence for reasons of space here we will be unable to elaborate on the philosophy of space-time coding, we will simply refer to the associated literature for background reading [171].

A variety of FEC techniques have been used in the context of adaptive modulation schemes. In their pioneering work on adaptive modulation, Webb and Steele [177] used a set of binary BCH codes. Vucetic [682] employed various punctured convolutional codes in response to the time-variant channel status. On the other hand, various Trellis Coded Modulation (TCM) [494, 495] schemes were used in the context of adaptive modulation by Alamouti and Kallel [683], Goldsmith and Chua [684], as well as Hole, Holm and

---

[3]This section was based on collaborative research with the contents of [681].

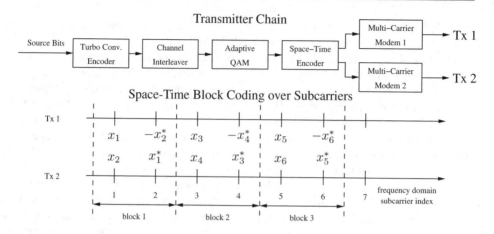

Figure 22.50: Transmitter structure and space-time block encoding scheme

Øien [209]. Keller, Liew and Hanzo studied the performance of Redundant Residue Number System (RRNS) codes in the context of adaptive multi-carrier modulation [685, 686]. Various turbo coded adaptive modulation schemes have been investigated also by Liew, Wong, Yee and Hanzo [203, 687, 688]. With the advent of space-time (ST) coding techniques [616, 618, 619], various concatenated coding schemes combining ST coding and FEC coding can be applied in adaptive modulation schemes. In this section, we investigate the performance of various concatenated space-time block-coded and turbo-coded adaptive OFDM and MC-CDMA schemes.

Figure 22.50 portrays the stylised transmitter structure of our system. The source bits are channel coded by a half-rate turbo convolutional encoder [689] using a constraint length of $K = 3$ as well as an interleaver having a memory of $L = 3072$ bits and interleaved by a random block interleaver. Then, the AQAM block selects a modulation mode from the set of no transmission, BPSK, QPSK, 16-QAM and 64-QAM depending on the instantaneous channel quality perceived by the receiver, according to the SNR-dependent optimum switching levels derived in Section 22.8.4. It is assumed that the perfectly estimated channel quality experienced by receiver A is fed back to transmitter B superimposed on the next burst transmitted to receiver B. The modulation mode switching levels of our AQAM scheme determine the average BEP as well as the average throughput.

The modulated symbol is now space-time encoded. As seen at the bottom of Figure 22.50, Alamouti's space-time block code [616] is applied across the frequency domain. A pair of the adjacent sub-carriers belonging to the same space-time encoding block is assumed to have the same channel quality. We employed a Wireless Asynchronous Transfer Mode (W-ATM) channel model of Figure 17.1 transmitting at a carrier frequency of 60GHz, at a sampling rate of 225MHz and employing 512 sub-carriers. Specifically, we used a three-path fading channel model, where the average SNR of each path is given by $\bar{\gamma}_1 = 0.79192\bar{\gamma}$, $\bar{\gamma}_2 = 0.12424\bar{\gamma}$ and $\bar{\gamma}_3 = 0.08384\bar{\gamma}$. The Multi-path Intensity Profile (MIP) of the W-ATM channel is illustrated in Figure 22.44(a) in Section 22.9.3. Each channel associated with a different antenna is assumed to exhibit independent fading.

The simulation results related to our uncoded adaptive modems are presented in Fig-

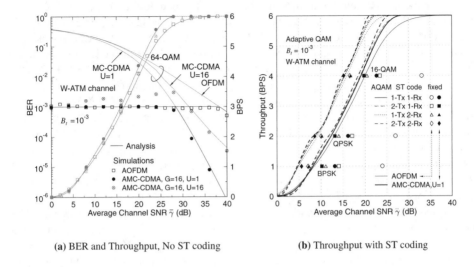

(a) BER and Throughput, No ST coding

(b) Throughput with ST coding

Figure 22.51: Performance of uncoded five-mode AOFDM and AMC-CDMA. The target BEP is $B_t = 10^{-3}$ when transmitting over the W-ATM channel model of Figurefig:10+420. (a) The constant average BEP is maintained for AOFDM and single user AMC-CDMA, while 'full-user' AMC-CDMA exhibits a slightly higher average BEP due to the residual MUI. (b) The SNR gain of the adaptive modems decreases, as ST coding increases the diversity order. The BPS curves appear in pairs, corresponding to AOFDM and AMC-CDMA - indicated by the thin and thick lines, respectively - for each of the four different ST code configurations. The markers represent the SNRs required by the fixed-mode OFDM and MC-CDMA schemes for maintaining the target BER of 10^{-3} in conjunction with the four ST-coded schemes considered

ure 22.51. Since we employed the optimum switching levels derived in Section 22.8.4, both our adaptive OFDM (AOFDM) and the adaptive single-user MC-CDMA (AMC-CDMA) modems maintain the constant target BER of 10^{-3} up to the 'avalanche' SNR value, and then follow the BER curve of the 64-QAM mode. However, 'full-user' AMC-CDMA, which is defined as an AMC-CDMA system supporting $U = 16$ users with the aid of a spreading factor of $G = 16$ and employing the MMSE-BDFE Joint Detection (JD) receiver [690], exhibits a slightly higher average BER than the target of $B_t = 10^{-3}$ due to the residual Multi-User Interference (MUI) of the imperfect joint detector. Since in Section 22.8.4 we derived the optimum switching levels based on a single-user system, the levels are no longer optimum, when residual MUI is present. The average throughputs of the various schemes expressed in terms of BPS steadily increase and at high SNRs reach the throughput of 64-QAM, namely 6 BPS. The throughput degradation of 'full-user' MC-CDMA imposed by the imperfect JD was within a fraction of a dB. Observe in Figure 22.51(a) that the analytical and simulation results are in good agreement, which we denoted by the lines and distinct symbols, respectively.

The effects of ST coding on the average BPS throughput are displayed in Figure 22.51(b). Specifically, the thick lines represent the average BPS throughput of our AMC-CDMA scheme, while the thin lines represent those of our AOFDM modem. The four pairs of hollow and filled markers associated with the four different ST-coded AOFDM and AMC-CDMA scenarios considered represent the BPS throughput versus SNR values associated with fixed-mode OFDM and fixed-mode MMSE-BDFE JD assisted MC-CDMA schemes. Specifically, observe for each of the 1, 2 and 4 BPS fixed-mode schemes that the right-most markers, namely the circles, correspond to the 1-Tx / 1-Rx scenario, the squares to the 2-Tx / 1-Rx scheme, the triangles to the 1-Tx / 2-Rx arrangement and the diamonds to the 2-Tx / 2-Rx scenarios. First of all, we can observe that the BPS throughput curves of OFDM and single-user MC-CDMA are close to each other, namely within 1 dB for most of the SNR range. This is surprising, considering that the fixed-mode MMSE-BDFE JD assisted MC-CDMA scheme was reported to exhibit around 10dB SNR gain at a BEP of 10^{-3} and 30dB gain at a BEP of 10^{-6} over OFDM [257]. This is confirmed in Figure 22.51(b) by observing that the SNR difference between the o and • markers is around 10dB, regardless whether the 4, 2 or 1 BPS scenario is concerned.

Let us now compare the SNR gains of the adaptive modems over the fixed modems. The SNR difference between the BPS curve of AOFDM and the fixed-mode OFDM represented by the symbol o at the same throughput is around 15dB. The corresponding SNR difference between the adaptive and fixed-mode 4, 2 or 1 BPS MC-CDMA modem is around 5dB. More explicitly, since in the context of the W-ATM channel model of Figure 17.1 fixed-mode MC-CDMA appears to exhibit a 10dB SNR gain over fixed-mode OFDM, the additional 5dB SNR gain of AMC-CDMA over its fixed-mode counterpart results in a total SNR gain of 15dB over fixed-mode OFDM. Hence ultimately the performance of AOFDM and AMC-CDMA becomes similar.

Let us now examine the effect of ST block coding. The SNR gain of the fixed-mode schemes due to the introduction of a 2-Tx / 1-Rx ST block code is represented as the SNR difference between the two right-most markers, namely circles and squares. These gains are nearly 10dB for fixed-mode OFDM, while they are only 3dB for fixed-mode MC-CDMA modems. However, the corresponding gains are less than 1dB for both adaptive modems, namely for AOFDM and AMC-CDMA. Since the transmitter power is halved due to using two Tx antennas in the ST codec, a 3dB channel SNR penalty was already applied to the curves in Figure 22.51(b). The introduction of a second receive antenna instead of a second transmit antenna eliminates this 3dB penalty, which results in a better performance for the 1-Tx/2-Rx scheme than for the 2-Tx/1-Rx arrangement. Finally, the 2-Tx / 2-Rx system gives around 3-4dB SNR gain in the context of fixed-mode OFDM and a 2-3dB SNR gain for fixed-mode MC-CDMA, in both cases over the 1-Tx / 2-Rx system. By contrast, the SNR gain of the 2-Tx / 2-Rx scheme over the 1-Tx / 2-Rx based adaptive modems was, again, less than 1dB in Figure 22.51(b). More importantly, for the 2-Tx / 2-Rx scenario the advantage of employing adaptive modulation erodes, since the fixed-mode MC-CDMA modem performs as well as the AMC-CDMA modem in this scenario. Moreover, the fixed-mode MC-CDMA modem still outperforms the fixed-mode OFDM modem by about 2dB. We conclude that since the diversity-order increases with the introduction of ST block codes, the channel quality variation becomes sufficiently small for the performance advantage of adaptive modems to erode. This is achieved at the price of a higher complexity due to employing two transmitters and two receivers in the ST coded system.

When channel coding is employed in the fixed-mode multi-carrier systems, it is expected that OFDM benefits more substantially from the frequency domain diversity than MC-CDMA, which benefited more than OFDM without channel coding. The simulation

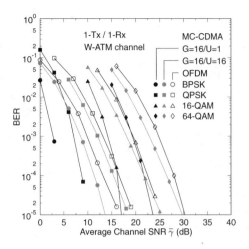

Figure 22.52: Performance of turbo convolutional coded fixed-mode OFDM and MC-CDMA for transmission over the W-ATM channel of Figure 17.1, indicating that JD MC-CDMA still outperforms OFDM. However, the SNR gain of JD MC-CDMA over OFDM is reduced to 1-2dB at a BEP of 10^{-4}

results depicted in Figure 22.52 show that the various turbo-coded fixed-mode MC-CDMA systems consistently outperform OFDM. However, the SNR differences between the turbo-coded BER curves of OFDM and MC-CDMA are reduced considerably.

The performance of the concatenated ST block coded and turbo convolutional coded adaptive modems is depicted in Figure 22.53. We applied the optimum set of switching levels designed in Section 22.8.4 for achieving an uncoded BEP of 3×10^{-2}. This uncoded target BEP was stipulated after observing that it is reduced by half-rate, $K = 3$ turbo convolutional coding to a BEP below 10^{-7}, when transmitting over AWGN channels. However, our simulation results yielded zero bit errors, when transmitting 10^9 bits, except for some SNRs, when employing only a single antenna.

Figure 22.53(a) shows the BEP of our turbo coded adaptive modems, when a single antenna is used. We observe in the figure that the BEP reaches its highest value around the 'avalanche' SNR point, where the adaptive modulation scheme consistently activates 64-QAM. The system is most vulnerable around this point. In order to interpret this phenomenon, let us briefly consider the associated interleaving aspects. For practical reasons we have used a fixed interleaver length of $L = 3072$ bits. When the instantaneous channel quality was high, the $L = 3072$ bits were spanning a shorter time-duration during their passage over the fading channel, since the effective BPS throughput was high. Hence the channel errors appeared more bursty than in the lower-throughput AQAM modes, which conveyed the $L = 3072$ bits over a longer time duration, hence dispersing the error bursts over a longer duration of time. The uniform dispersion of erroneous bits versus time enhances the error correction power of the turbo code. On the other hand, in the SNR region beyond the 'avalanche' SNR point seen in Figure 22.53(a) the system exhibited a lower uncoded BER, reducing the coded BEP even further. This observation suggests that further research ought to determine the set of switching thresholds directly for a coded adaptive system, rather than by simply

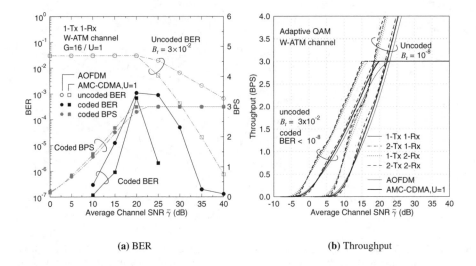

(a) BER   (b) Throughput

Figure 22.53: Performance of the concatenated ST block coded and turbo convolutional coded adaptive OFDM and MC-CDMA systems communicating over the W-ATM channel of Figure 17.1. The uncoded target BEP is 3×10^{-2}. The coded BEP was less than 10^{-8} for most of the SNR range, resulting in virtually error free transmission. (a) The coded BEP becomes higher near the 'avalanche' SNR point, when a single antenna was used. (b) The coded adaptive modems have SNR gains up to 7dB compared to their uncoded counterparts achieving a comparable average BER

estimating the uncoded BER, which is expected to result in near-error-free transmission.

We can also observe that the turbo coded BEP of AOFDM is higher than that of AMC-CDMA in the SNR range of 10-20dB, even though the uncoded BER is the same. This appears to be the effect of the limited exploitation of frequency domain diversity of coded OFDM, compared to MC-CDMA, which leads to a more bursty uncoded error distribution, hence degrading the turbo coded performance. The fact that ST block coding aided multiple antenna systems show virtually error free performance corroborates our argument.

Figure 22.53(b) compares the throughputs of the coded adaptive modems and the uncoded adaptive modems exhibiting a comparable average BER. The SNR gains due to channel coding were in the range of 0dB to 8dB, depending on the SNR region and on the scenarios employed. Each bundle of throughput curves corresponds to the scenarios of 1-Tx/1-Rx OFDM, 1-Tx/1-Rx MC-CDMA, 2-Tx/1-Rx OFDM, 2-Tx/1-Rx MC-CDMA, 1-Tx/2-Rx OFDM, 1-Tx/2-Rx MC-CDMA, 2-Tx/2-Rx OFDM and 2-Tx/2-Rx MC-CDMA starting from the far right curve, when viewed for throughput values higher than 0.5 BPS. The SNR difference between the throughput curves of the ST and turbo coded AOFDM and those of the corresponding AMC-CDMA schemes was reduced compared to the uncoded performance curves of Figure 22.51(b). The SNR gain owing to ST block coding assisted transmit diversity in the context of AOFDM and AMC-CDMA was within 1dB due to the halved transmitter power. Therefore, again, ST block coding appears to be less effective in conjunction with adaptive modems.

In conclusion, the performance of ST block coded constant-power adaptive multi-carrier modems employing optimum SNR-dependent modem mode switching levels were investigated in this section. The adaptive modems maintained the constant target BEP stipulated, while maximising the average throughput. As expected, it was found that ST block coding reduces the relative performance advantage of adaptive modulation, since it increases the diversity order and eventually reduces the channel quality variations. When turbo convolutional coding was concatenated to the ST block codes, near-error-free transmission was achieved at the expense of halving the average throughput. Compared to the uncoded system, the turbo coded system was capable of achieving a higher throughput in the low SNR region at the cost of a higher complexity. The study of the relationship between the uncoded BEP and the corresponding coded BEP showed that adaptive modems obtain higher coding gains than that of fixed modems. This was due to the fact that the adaptive modem avoids burst errors even in deep channel fades by reducing the number of bits per modulated symbol eventually to zero.

22.10 Summary

Following a brief introduction to several fading counter-measures, a general model was used to describe several adaptive modulation schemes employing various constituent modulation modes, such as PSK, Star QAM and Square QAM, as one of the attractive fading counter-measures. In Section 22.7.3.1, the closed form expressions were derived for the average BER, the average BPS throughput and the mode selection probability of the adaptive modulation schemes, which were shown to be dependent on the mode-switching levels as well as on the average SNR. After reviewing in Section 22.8.1, 22.8.2 and 22.8.3 the existing techniques devised for determining the mode-switching levels, in Section 22.8.4 the optimum switching levels achieving the highest possible BPS throughput while maintaining the average target BEP were developed based on the Lagrangian optimization method.

Then, in Section 22.9.1 the performance of uncoded adaptive PSK, Star QAM and Square QAM was characterised, when the underlying channel was a Nakagami fading channel. It was found that an adaptive scheme employing a k-BPS fixed-mode as the highest throughput constituent modulation mode was sufficient for attaining all the benefits of adaptive modulation, while achieving an average throughput of up to $(k-1)$ BPS. For example, a three-mode adaptive PSK scheme employing No-Tx, 1-BPS BPSK and 2-BPS QPSK modes attained the maximum possible average BPS throughput of 1 BPS and hence adding higher-throughput modes, such as 3-BPS 8-PSK to the three-mode adaptive PSK scheme resulting in a four-mode adaptive PSK scheme did not achieve a better performance across the 1 BPS throughput range. Instead, this four-mode adaptive PSK scheme extended the maximal achievable BPS throughput by any adaptive PSK scheme to 2 BPS, while asymptotically achieving a throughput of 3 BPS as the average SNR increases.

On the other hand, the relative SNR advantage of adaptive schemes in comparison to fixed-mode schemes increased as the target average BER became lower and decreased as the fading became less severe. More explicitly, less severe fading corresponds to an increased Nakagami fading parameter m, to an increased number of diversity antennas, or to an increased number of multi-path components encountered in wide-band fading channels. As the fading becomes less severe, the average BPS throughput curves of our adaptive Square QAM schemes exhibit undulations owing to the absence of 3-BPS, 5-BPS and 7-BPS square QAM modes.

The comparisons between fixed-mode MC-CDMA and adaptive OFDM (AOFDM) were made based on different channel models. In Section 22.9.4 it was found that fixed-mode MC-CDMA might outperform adaptive OFDM, when the underlying channel provides sufficient diversity. However, a definite conclusion could not be drawn since in practice MC-CDMA might suffer from MUI and AOFDM might suffer from imperfect channel quality estimation and feedback delays.

Concatenated space-time block coded and turbo convolutional-coded adaptive multi-carrier systems were investigated in Section 22.9.5. The coded schemes reduced the required average SNR by about 6dB-7dB at throughput of 1 BPS achieving near error-free transmission. It was also observed in Section 22.9.5 that increasing the number of transmit antennas in adaptive schemes was not very effective, achieving less than 1dB SNR gain, due to the fact that the transmit power per antenna had to be reduced in order to limit the total transmit power for the sake of fair comparison.

Part IV

Advanced QAM: Turbo-Equalised Adaptive TCM, TTCM, BICM, BICM-ID and Space-Time Coding Assisted OFDM and CDMA Systems

Chapter 23

Capacity and Cutoff Rate of Gaussian and Rayleigh Channels

23.1 Introduction

An important accomplishment of information theory is the determination of the channel capacity, C, which quantifies the maximum achievable transmission rate, C^*, of a system communicating over a bandlimited channel, while maintaining an arbitrarily low probability of error. Given the fact that the available bandwidth of all transmission media is limited, it is desirable to transmit information as bandwidth-efficiently, as possible. This implies transmitting as many bits per Hertz, as possible. In recent years the available wireless communications frequency bands have been auctioned by the American, British, German and other goverments to service provider companies at a high price and therefore it is of great commercial interest to exploit the available bandwidth as best as possible. Quantifying these information theoretic limits is the objective of this chapter. Given these limits, in the rest of the book we will aim for quantifying the various systems' ability to perform as close to the limits as possible. This issue was first discussed in a rudimentary fashion in the context of Figure 2.4 and here we will considerably deepen our approach.

The units of the channel capacity C and relative or normalised channel capacity $C^* = C/T$ are bit per symbol and bit per second, respectively, where T is the symbol period. The capacity of a Single-Input Single-Output (SISO) AWGN channel was quantified by Shannon in 1948 [550]. Since then, substantial research efforts have been invested in finding channel codes that would produce an arbitrarily low probability of error at a transmission rate close to C^*. Normalising the channel capacity with respect to the bandwidth occupied yields another useful parameter, namely the bandwidth efficiency η. A lower bound to the channel capacity referred to as the channel's cutoff rate is another useful parameter. The cutoff rate, R_0, has been also referred to as the "practically achievable capacity", since the complexity of a coded system becomes substantially higher, when communicating near R_0, in comparison to transmissions at rates below R_0 [493, 501, 691].

23.2 Channel Capacity

Let the input and output of the Discrete Memoryless Channel (DMC) [388] be X and Y, respectively, where X may assume one of K discrete-amplitude values, while Y can be one of J legitimate discrete-amplitude values. The assignment of $x = a_k$ and $y = b_j$ corresponds to encountering two specific *events*. Let the probability of encountering each event be denoted as:

$$p(k) = P(x = a_k),$$
(23.1)

$$p(j) = P(y = b_j),$$
(23.2)

while the conditional probability of receiving $y = b_j$, given that $x = a_k$ was transmitted be denoted as:

$$p(j|k) = P(y = b_j | x = a_k).$$
(23.3)

Mutual information is by definition a measure of "information about the event $x = a_k$ provided by the occurrence of the event $y = b_j$", which is defined as [692]:

$$I_{X;Y}(a_k; b_j) = \log_2\left[\frac{p(k|j)}{p(k)}\right] \text{ [bit]},$$
(23.4)

where the base of the logarithm is 2 and hence the units of mutual information are bits. The average mutual information, $I(X;Y)$, is the expectation of $I_{X;Y}(a_k; b_j)$ expressed in Equation 23.4, which yields:

$$I(X;Y) = \sum_{k=1}^{K}\sum_{j=1}^{J} p(k,j)\log_2\left[\frac{p(k|j)}{p(k)}\right] \text{ [bit/symbol]},$$
(23.5)

where the unit of bit/symbol is used for indicating the number of bits conveyed per transmitted symbol. By using the probability identities [672] of $p(x|y) = \frac{p(y|x)p(x)}{p(y)}$ and $p(x,y) = p(y|x)p(x)$, derived from Bayes' rule, the average mutual information is rewritten as:

$$I(X;Y) = \sum_{k=1}^{K}\sum_{j=1}^{J} p(j|k)p(k)\log_2\left[\frac{p(j|k)}{p(j)}\right] \text{ [bit/symbol]}.$$
(23.6)

The channel capacity is defined as the highest possible average mutual information obtained by finding the specific set of input symbol probability assignments, $\{p(k); k = 1, \ldots, K\}$, which maximises $I(X;Y)$. The DMC capacity, C_{DMC}, may be written as [692, p. 74]:

$$C_{\text{DMC}} = \max_{\{p(k); k=1,\ldots,K\}} \sum_{k=1}^{K}\sum_{j=1}^{J} p(j|k)p(k)\log_2\left[\frac{p(j|k)}{p(j)}\right] \text{ [bit/symbol]}.$$
(23.7)

Naturally, in practice we do not have control over the probability of the channel's input symbols and hence depending on the specific probabilities of the channel input symbols we may not be able to approach the capacity of the channel.

23.2.1   Vector Channel Model

It was argued in [502, pp. 348–351] that bandlimited signals having a finite energy may be described as vectors having a dimensionality of:

$$N \;=\; 2WT, \tag{23.8}$$

provided that the following assumptions are satisfied:

1) the waveform is constrained to an ideal lowpass or bandpass bandwidth, W; and

2) the waveform is limited to the time interval, $0 \leq t \leq T$.

Strictly speaking assumptions 1 and 2 cannot be fulfilled simultaneously, because according to the properties of the Fourier transform a finite-bandwidth signal has an infinite time-domain support and vice versa. Let us, however, briefly consider a full-response Minimum Shift Keying (MSK) [388] modulator, where the modulated signal's spectrum has a sinc-function shape. The first and highest spectral side-lobe of the sinc-shaped spectrum is about 20 dB lower than the main spectral lobe, as seen for example in Figure 13.15 of [369]. Hence we may argue that although the time-domain signalling pulse is time-limited, the corresponding spectrum has a low energy outside the main spectral lobe. This line of argument may be continued by considering for example partial-response Gaussian MSK (GMSK) signalling, which results in substantially lower spectral side-lobes than MSK, again as seen for example in Figure 13.15 of [369]. In fact, the time-domain Gaussian pulse is known to have the smoothest possible time-domain signalling pulse evolution, resulting in the most compact spectral-domain representation and a more-or-less bandlimited spectrum. Hence loosely speaking we may argue that although according to the Fourier transform the spectral-domain support of a finite-duration signal is infinite, for practical reasons the signal may be considered both time- and band-limited. Similar arguments are also valid in case of full-response Nyquist signalling based system such as those considered in this monograph.

   A set of functions is said to be orthonormal, if the functions are orthogonal to each other and they are also normalised to unit energy within a common interval of $T = T_2 - T_1$, yielding [693, p. 153]:

$$\int_{T_1}^{T_2} \phi_i(t)\phi_j(t)\, dt = \left\{ \begin{array}{ll} 1, & i = j, \\ 0, & i \neq j. \end{array} \right. \tag{23.9}$$

Orthonormal functions $\{\phi_n(t)\}$ can be generated using a variety of different methods, such as the Fourier series and Gram-Schmidt procedures [388, pp. 167-173]. However, the dimensionality of the signal space remains $N = 2WT$, as long as the signals are time- and band-limited in the sense as argued above. Specifically, a bandlimited continuous-time signalling waveform, $x(t)$, can be expressed as a linear combination of orthonormal functions,

$\{\phi_n(t)\}$, as:

$$x(t) = \sum_{n=1}^{N} x[n]\phi_n(t),\qquad\qquad (23.10)$$

which is sufficiently accurately defined by $N = 2WT$ number of coefficients, when $x(t)$ is sufficiently close to zero outside the interval T. Furthermore, the coefficients, $x[n]$, can be obtained from:

$$x[n] = \int_{t=0}^{T} x(t)\phi_n(t)\,dt\qquad\qquad (23.11)$$

for all n, where the integration is over the signalling period T. Let us now represent the channel's input $\mathbf{x} = (x[1],\ldots,x[N])$ and output $\mathbf{y} = (y[1],\ldots,y[N])$ as N-component, i.e. N-dimensional real-valued vectors. Note that $\mathbf{x}$ and $\mathbf{y}$ may be discrete-valued or continuous-valued, depending on the channel encountered.

A relative of the DMC is the Continuous-Input Continuous-Output Memoryless Channel (CCMC) [388], where the corresponding coefficients of $\mathbf{x}$ and $\mathbf{y}$ are continuous-valued as indicated below:

$$x[n] \in [-\infty, +\infty],\qquad\qquad (23.12)$$

$$y[n] \in [-\infty, +\infty],\; n = 1,\ldots,N.\qquad\qquad (23.13)$$

This model is applicable to a scenario employing an analogue modulation scheme, such as amplitude, phase or frequency modulation. The channel capacity of the DMC in Equation 23.7 can be extended for the CCMC as [692]:

$$C_{\text{CCMC}} = \max_{p(\mathbf{x})} \underbrace{\int_{-\infty}^{\infty} \cdots \int_{-\infty}^{\infty}}_{\text{2N-fold}} p(\mathbf{y}|\mathbf{x})p(\mathbf{x})\log_2\left[\frac{p(\mathbf{y}|\mathbf{x})}{p(\mathbf{y})}\right]\,d\mathbf{x}\,d\mathbf{y}\;[\text{bit/symbol}],\;(23.14)$$

where again, $\mathbf{x} = (x[1],\ldots,x[N])$ and $\mathbf{y} = (y[1],\ldots,y[N])$ are N-dimensional signals at the channel input and output, respectively.

Another relative of the DMC is the Discrete-Input Continuous-Output Memoryless Channel (DCMC) [388], where the channel input belongs to the discrete set of M-ary values:

$$\mathbf{x} \in \{\mathbf{x}_m : m = 1,\ldots,M\}.\qquad\qquad (23.15)$$

More explicitly, the channel input $\mathbf{x}_m = (x_m[1],\ldots,x_m[N])$ is a $\log_2(M)$-bit symbol having N discrete-valued coefficients. By contrast, the channel output $\mathbf{y}$ has continuous-valued coefficients:

$$y[n] \in [-\infty, +\infty],\; n = 1,\ldots,N.\qquad\qquad (23.16)$$

The channel capacity for the DCMC can also be derived from Equation 23.7 as [694]:

$$C_{\text{DCMC}} = \max_{p(\mathbf{x}_1)...p(\mathbf{x}_M)} \sum_{m=1}^{M} \underbrace{\int_{-\infty}^{\infty} \cdots \int_{-\infty}^{\infty}}_{\text{N-fold}} p(\mathbf{y}|\mathbf{x}_m)p(\mathbf{x}_m) \log_2 \left[\frac{p(\mathbf{y}|\mathbf{x}_m)}{p(\mathbf{y})} \right] d\mathbf{y} \text{ [bit/symbol]},$$

(23.17)

where again $\mathbf{x}_m = (x_m[1], \ldots, x_m[N])$ is the N-dimensional M-ary symbol at the channel's input while $\mathbf{y} = (y[1], \ldots, y[N])$ is the N-dimensional signal at the channel's output.

23.2.2   The Capacity of AWGN Channels

The Shannon bound of an AWGN channel is obtained by finding the capacity of a continuous-input continuous-output AWGN channel, where the modulated signal itself, $x(t)$, may be modelled by bandlimited Gaussian noise [1] at the channel input, which is contaminated by the AWGN channel noise $n(t)$. After bandlimiting, the samples of both noise sources are taken at the Nyquist rate. These samples are independent identical distributed (iid) Gaussian random variables with zero mean having a variance of σ^2 for $x(t)$ and $N_0/2$ for $n(t)$. The resultant sampled waveforms can be described as vectors of N discrete-time but continuous-valued samples, where $N = 2WT$ is the signal dimensionality defined in Equation 23.8. Upon exploiting that the Probability Density Functions (PDFs) of $p(\mathbf{x})$, $p(\mathbf{y})$ and $p(\mathbf{y}|\mathbf{x})$ are Gaussian, the Shannon bound can be derived from Equation 23.14 as [550, 695]:

$$\begin{aligned} C_{\text{CCMC}}^{\text{AWGN}} &= WT \log_2(1 + \gamma) \text{ [bit/symbol]}, \\ &= \frac{N}{2} \log_2(1 + \gamma) \text{ [bit/symbol]}, \end{aligned}$$

(23.18)

where γ is the Signal to Noise ratio (SNR). Note that when the channel input is a continuous-valued variable corresponding to an analogue modulation scheme, the capacity is only restricted either by the signalling energy and hence γ or by the bandwidth W [388]. Therefore we will refer to the capacity of the CCMC as the *unrestricted bound*.

Let us now consider the achievable capacity of DCMC, when transmitting the N-dimensional M-ary signals using Equation 23.17. Assuming equiprobable M-ary input symbols conveying $\log_2(M)$ bit/symbol information, we have:

$$p(\mathbf{x}_m) = \frac{1}{M}, \quad m = 1, \ldots, M.$$

(23.19)

The conditional probability of receiving $\mathbf{y}$ given that $\mathbf{x}$ was transmitted when communicating over an AWGN channel is determined by the PDF of the noise, yielding:

$$p(\mathbf{y}|\mathbf{x}_m) = \prod_{n=1}^{N} \frac{1}{\sqrt{\pi N_0}} \exp\left(\frac{-(y_n - x_{mn})^2}{N_0} \right),$$

(23.20)

where $N_0/2$ is the channel's noise variance. Note that $p(\mathbf{y}|\mathbf{x}_m)$ is also referred to as the chan-

---

[1]Naturally, the information to be transmitted is not an AWGN process. However, it was shown by Shannon [550] that it is beneficial to render the modulated signal input to the channel as 'AWGN-like' as possible.

nel's transition probability. By substituting Equations 23.19 and 23.20 into Equation 23.17, the capacity expression of the DCMC can be simplified to [695, 696]:

$$C_{DCMC}^{AWGN} = \log_2(M) - \frac{1}{M(\sqrt{\pi})^N} \cdot \qquad (23.21)$$

$$\sum_{m=1}^{M} \underbrace{\int_{-\infty}^{\infty} \cdots \int_{-\infty}^{\infty}}_{N\text{-fold}} \exp\left(-|\mathbf{t}|^2\right) \log_2\left[\sum_{i=1}^{M} \exp\left(-2\mathbf{t} \cdot \mathbf{d}_{mi} - |\mathbf{d}_{mi}|^2\right)\right] d\mathbf{t} \text{ [bit/symbol]},$$

where $\mathbf{d}_{mi} = (\mathbf{x}_m - \mathbf{x}_i)/\sqrt{N_0}$ and $\mathbf{t} = (t[1], \ldots, t[N])$ is an integration variable. The DCMC capacity given by Equation 23.22 can be determined using numerical integration. More specifically, the integration can be approximated using the Gauss-Hermite Quadrature method [294, 695]. Let us also represent the effect of the AWGN channel as an N-dimensional additive noise vector given by $\mathbf{n} = (n[1], \ldots, n[N])$. The average SNR can be determined from [495, 695] as:

$$\gamma = \frac{E[x_m^2(t)]}{E[n^2(t)]},$$

$$= \frac{\frac{1}{M}\sum_{m=1}^{M} |\mathbf{x}_m|^2}{\sum_{n=1}^{N} E[n^2[n]]},$$

$$= \frac{E_s}{NN_0/2}, \qquad (23.22)$$

where E_s is the average energy of the N-dimensional M-ary symbol $\mathbf{x}_m$ and $N\frac{N_0}{2}$ is the average energy of the N-dimensional AWGN $\mathbf{n}$. Hence, if E_s is normalised to unity, from Equation 23.22 we have $N_0 = 2/(N\gamma)$.

On the other hand, it was shown in [495] that the channel capacity of the DCMC for $N = 2$-dimensional M-ary signalling can also be obtained using:

$$C_{DCMC}^{AWGN} = \log_2(M) - \qquad (23.23)$$

$$\frac{1}{M}\sum_{m=1}^{M} E\left[\log_2\sum_{i=1}^{M} \exp\left(\frac{-|\mathbf{x}_m + \mathbf{n} - \mathbf{x}_i|^2 + |\mathbf{n}|^2}{N_0}\right)\right] \text{ [bit/symbol]},$$

where $\mathbf{n}$ is the complex AWGN having a variance of $N_0/2$ per dimension. The expectation E[.] in Equation 23.24 is taken over $\mathbf{n}$ and it can be determined using the Monte Carlo averaging method.

23.2.3 The Capacity of Uncorrelated Rayleigh Fading Channels

Let us define $\mathbf{h} = h_i + jh_q$ as the *complex* uncorrelated Rayleigh fading coefficient, where h_i and h_q are the in-phase and quadrature-phase coefficients, respectively. Specifically, h_i and h_q are zero mean iid Gaussian random variables, each having a variance of $\sigma_r^2 = 1/2$. Note that σ_r^2 is normalised to $1/2$ so that the average energy of $|\mathbf{h}|^2$ is unity. Furthermore, the coefficient $\chi_2^2 = |\mathbf{h}|^2 = h_i^2 + h_q^2$ of the Rayleigh fading channel is a chi-squared distributed random variable with two degrees of freedom. The corresponding PDF is given by [388]:

$$p(\chi_2^2) \quad = \quad \frac{1}{2\sigma_r^2} \exp\left(-\frac{\chi_2^2}{2\sigma_r^2}\right). \tag{23.24}$$

The capacity of continuous-input continuous-output uncorrelated (memoryless) Rayleigh fading channels can be evaluated based on the capacity formula of the Gaussian channel given in Equation 23.18 by simply weighting the SNR γ of the Gaussian channel by the probability of encountering the specific SNR determined by the Rayleigh fading magnitude χ_2^2, i.e. $\chi_2^2 \gamma$. Then the resultant capacity value must be averaged, either by integration or summation over the legitimate range of the SNR given by $\chi_2^2 \gamma$, yielding [180, 697]:

$$C_{\text{CCMC}}^{\text{RAY}} \quad = \quad \text{E}\left[\frac{N}{2}\log_2(1 + \chi_2^2 \gamma)\right] \text{ [bit/symbol]}, \tag{23.25}$$

where the expectation is taken over χ_2^2.

The capacity of the DCMC for $N = 2$-dimensional M-ary *complex* signals, such as the classic PSK [388], can be derived from Equation 23.24 as follows:

$$C_{\text{DCMC}}^{\text{RAY}} \quad = \quad \log_2(M) - \frac{1}{M}\sum_{m=1}^{M}\text{E}\left[\log_2\sum_{i=1}^{M}\exp(\Phi_i^m)\right] \text{ [bit/symbol]}, \tag{23.26}$$

where we have:

$$\Phi_i^m \quad = \quad \frac{-|\mathbf{h}(\mathbf{x}_m - \mathbf{x}_i) + \mathbf{n}|^2 + |\mathbf{n}|^2}{N_0},$$

$$= \quad \frac{-|\chi_2^2(\mathbf{x}_m - \mathbf{x}_i) + \mathbf{\Omega}|^2 + |\mathbf{\Omega}|^2}{\chi_2^2 N_0}. \tag{23.27}$$

More explicitly the capacity of DCMC depends on M, $\mathbf{x}_m$, $\mathbf{x}_i$, χ_2^2 and $\mathbf{\Omega} = \mathbf{h}^*\mathbf{n}$, which is the effective AWGN having a zero mean and a variance of $\chi_2^2 N_0/2$ per dimension. The expectation in Equation 23.26 is taken over the Rayleigh-faded magnitude χ_2^2 and the effective AWGN $\mathbf{\Omega}$. The expectation can be estimated using the Monte Carlo averaging method.

For the general case of M-ary *complex* signals having $N \geq 2$ dimensions, such as L-orthogonal PSK signalling [698, 699] of Section 23.5.5, we have:

$$\Phi_i^m \quad = \quad \sum_{n=1}^{N} \frac{-|\chi_2^2[n](x_m[n] - x_i[n]) + \Omega[n]|^2 + |\Omega[n]|^2}{\chi_2^2[n]N_0}, \tag{23.28}$$

where $\chi_2^2[j] = \chi_2^2[j+1]$ for $j \in \{1, 3, 5\ldots\}$ since a *complex* channel has two dimensions and $\Omega[n]$ is the dth dimension of the N-dimensional AWGN having a zero mean and a variance of $\chi_2^2[n]N_0/2$ per each of the N dimensions. In this case, the expectation in Equation 23.26 is taken over $\chi_2^2[n]$ and $\Omega[n]$. Note that the relationship between Equations 23.27 and 23.28 for $N = 2$ complex signals is that we have $\mathbf{x}_k = x_k[1] + jx_k[2]$ for $k \in \{1, \ldots, M\}$, $\chi_2^2 = \chi_2^2[1] = \chi_2^2[2]$ and $\Omega = \Omega[1] + j\Omega[2]$.

Note that when the channel is *real*, where only the in-phase coefficient h_i is considered, the uncorrelated Rayleigh fading coefficient is given by $h = h_i$. Explicitly, h_i is a zero mean iid Gaussian random variable having a variance of $\sigma_r^2 = 1$. We also have $\chi_1^2 = h^2$, which is a chi-squared distributed random variable with one degree of freedom. Hence, for the case of N-dimensional M-ary *real* signals, such as M-ary orthogonal signalling [388] having $N = M$ or for pulse amplitude modulation schemes [388] having $N = 1$, we have:

$$\Phi_i^m = \sum_{n=1}^{N} \frac{-|\chi_1^2[n](x_m[n] - x_i[n]) + \Omega[n]|^2 + |\Omega[n]|^2}{\chi_1^2[n]N_0}, \tag{23.29}$$

where $\Omega[n]$ is the dth dimension of the N-dimensional AWGN having a zero mean and a variance of $\chi_1^2[n]N_0/2$ per dimension. In this case, the expectation in Equation 23.26 is taken over $\chi_1^2[n]$ and $\Omega[n]$.

23.3 Channel Cutoff Rate

The cutoff rate R_0 of the channel is defined as a channel capacity related quantity so that for any $R < R_0$, it is possible to construct a channel code having a block length n and coding rate of at least R capable of maintaining an average error probability that obeys $P_e \leq 2^{-n(R_0-R)}$ [388]. As mentioned before, R_0 has also been referred to as the "practically achievable capacity" of channel coded systems, where communication at rates above R_0 is typically far more complex to implement than at rates below R_0 [493,501,691]. For example, as soon as the coding rate reaches R_0, the expected number of computation per nodes in the context of sequential decoding [501] tends to infinity. Although it is maintained that turbo codes are indeed capable of operating at rates above R_0, their decoding does become substantially more complex, as R exceeds R_0.

Apart from the above complexity-related context, R_0 is also used as an analytical bound limiting the bit error ratio performance of various classes of random codes designed for specific channels [365,700]. Furthermore, R_0 constitutes a lower bound of the channel capacity and it is more straightforward to compute compared to the channel capacity. In general, the cutoff rate associated with M-ary QAM/PSK signalling and a Rician fading channel in the presence of perfect channel magnitude and phase estimates is given by [701–703]:

$$R_0 = 2\log_2(M) - \log_2\left(\sum_{m=1}^{M}\sum_{i=1}^{M} C(\mathbf{x}_m, \mathbf{x}_i)\right) \text{ [bit/symbol]}, \tag{23.30}$$

where $C(\mathbf{x}_m, \mathbf{x}_i)$ is the Chernoff bound on the pairwise error probability expressed as [701, 703, 704]:

$$C(\mathbf{x}_m, \mathbf{x}_i) = \frac{1+K}{1+K+\frac{1}{4}|\mathbf{d}_{mi}|^2} \times \exp\left(-\frac{K\frac{1}{4}|\mathbf{d}_{mi}|^2}{1+K+\frac{1}{4}|\mathbf{d}_{mi}|^2}\right), \tag{23.31}$$

where we have $|\mathbf{d}_{mi}|^2 = |\mathbf{x}_m - \mathbf{x}_i|^2/N_0$ and K is the Rician factor. For an AWGN channel having a Rician factor of $K = \infty$ we have:

$$C(\mathbf{x}_m, \mathbf{x}_i) = \exp\left(-\frac{1}{4}|\mathbf{d}_{mi}|^2\right). \tag{23.32}$$

By contrast, for the other extreme scenario of encountering a Rayleigh fading channel, where $K = 0$, we have:

$$C(\mathbf{x}_m, \mathbf{x}_i) = \frac{1}{1 + \frac{1}{4}|\mathbf{d}_{mi}|^2}. \tag{23.33}$$

Note that we will apply Equations 23.30, 23.32 and 23.33 in Section 23.5 for the computation of the cutoff rate for a range of M-ary digital signalling sets, when communicating over AWGN and Rayleigh fading channels.

23.4   Bandwidth Efficiency

The capacity analysis of the CCMC and DCMC provided in Section 23.2 determines the maximum number of information bits conveyed per transmitted symbol, as a function of the SNR. The system's bandwidth efficiency may be expressed as the capacity C normalised by the product of the bandwidth W occupied and the symbol period T, given by:

$$\eta = \frac{C}{WT} = \frac{C}{N/2} \text{ [bit/s/Hz]}, \tag{23.34}$$

where the associated unit is bit/s/Hz. The bandwidth efficiency of the CCMC may be expressed as:

$$\begin{aligned}
\eta_{\text{CCMC}} &= \frac{C_{\text{CCMC}}}{WT} \\
&= \begin{cases} \log_2(1 + \gamma) \text{ [bit/s/Hz]}, & \text{AWGN Channel}, \\ \mathrm{E}\left[\log_2(1 + \chi_2^2\gamma)\right] \text{ [bit/s/Hz]}, & \text{Rayleigh channel}. \end{cases}
\end{aligned} \tag{23.35}$$

We will refer to the bandwidth efficiency of CCMC as the normalised unrestricted bound. The bandwidth efficiency curve may be plotted as a function of the bit energy to noise spectral density ratio (E_b/N_0), which can be determined from the SNR γ as:

$$\frac{E_b}{N_0} = \frac{\gamma}{\eta}. \tag{23.36}$$

Note that the bandwidth efficiency of the CCMC may be directly computed using Equations 23.18, 23.34 and 23.36, yielding:

$$\begin{aligned}
\eta &= \log_2(1 + \eta\frac{E_b}{N_0}), \\
\frac{E_b}{N_0} &= \frac{2^\eta - 1}{\eta}.
\end{aligned} \tag{23.37}$$

It is interesting to note that as the bandwidth efficiency η of the CCMC tends to zero, by using L'Hôpital's rule [672], we arrive at:

$$
\begin{aligned}
\lim_{\eta \to 0} \frac{E_b}{N_0} &= \lim_{\eta \to 0} \frac{2^\eta - 1}{\eta}, \\
&= \lim_{\eta \to 0} 2^\eta \ln(2), \\
&= \ln(2), \\
&\approx -1.59 \, [\mathrm{dB}].
\end{aligned}
\tag{23.38}
$$

More explicitly, Equation 23.38 suggests that as E_b/N_0 approaches -1.59 dB, the bandwidth efficiency of the CCMC approaches zero.

It is also useful to normalise the cutoff rate R_0 with respect to the product of the bandwidth W occupied and the symbol period T, yielding:

$$
R_{0,\eta} = \frac{R_0}{WT} = \frac{R_0}{N/2} \, [\mathrm{bit/s/Hz}],
\tag{23.39}
$$

where the associated unit is bit/s/Hz. The normalised cutoff rate of the channel $R_{0,\eta}$ may be used for direct comparison with η of Equation 23.34.

23.5 Channel Capacity and Cutoff Rate of M-ary Modulation

In this section we will quantify the capacity, cutoff rate, bandwidth efficiency and normalised cutoff rate of the AWGN and the uncorrelated Rayleigh fading channels for a range of M-ary signalling sets based on the DCMC model. The unrestricted bounds of the AWGN and uncorrelated Rayleigh fading channels based on the CCMC model are also plotted for comparisons. Explicitly, the capacity of the AWGN CCMC and DCMC is computed using Equation 23.18 and Equation 23.22, respectively. By contrast, the capacity of the uncorrelated Rayleigh fading CCMC may be computed using Equation 23.25, while that of the DCMC is quantified using Equations 23.26, 23.28 and 23.29. On the other hand, the cutoff rate of the channel is determined using Equation 23.30 as well as Equations 23.32 and 23.33 for the AWGN and Rayleigh fading channels, respectively. Finally, the bandwidth efficiency curves of the DCMC and CCMC are evaluated using Equations 23.34 and 23.35, while the normalised cutoff rate of the DCMC is computed using Equation 23.39.

23.5.1 Introduction

An M-ary modulator is a device that maps each of the discrete-time symbols belonging to a set of M alphabets into one of the M continuous-time analogue waveforms suitable for transmission over the physical channel. There are many types of modulation techniques, differing in the way they manipulate an electromagnetic signal. Such manipulations include changing the amplitude, frequency or phase angle of a sinusoidal signal, the polarisation of the electromagnetic radiation or the pulse position within a modulation interval.

There are several different signalling pulse shaping functions. Most often, pulse shaping is carried out in the frequency-domain by designing a Nyquist filter having a certain roll-off factor α, as shown in Figure 4.6. Alternatively, pulse shaping may be implemented in the time-domain, as seen in Figure 4.15. The choice of the pulse shaping function influences the spectrum of the transmitted signal. More specifically, rectangular signalling pulses give rise to an infinite bandwidth requirement. By contrast, the raised cosine time-domain pulse shaping principle of Figure 4.6 results in a more compact spectrum. On one hand, the half-cycle sinusoidal time-domain pulse shaping function was utilised in the Minimum Shift Keying (MSK) scheme of [388, pp. 197-199]. On the other hand, the Q^2PSK and Q^2AM schemes outlined in Chapter 13 employ both sinusoidal and cosinusoidal pulse shaping functions. For the sake of simplicity, we will employ the rectangular time-domain pulse shaping function for illustrating the implementation of M-ary modulation techniques. We define the signalling pulse duration as T_p and the modulated symbol duration as T_s. As seen in Figure 4.5 for example, the baseband equivalent filter response of a rectangular frequency-domain Nyquist filter associated with $\alpha = 0$ spans from $-f_N = -1/2T_p$ to $f_N = 1/2T_p$. Hence the bandwidth required is given by $W = 2f_N = 1/T_p$ and the number of signal dimensions is given by Equation 23.8 as $N = 2WT_s$, when obeying the assumptions made in Section 23.2.1.

As explained in Section 13.2, we note that the MSK scheme exploits only two out of the four possible signalling dimensions available. On the other hand, when the channel is strictly bandlimited, the Q^2PSK and Q^2AM signalling schemes outlined in Chapter 13 have an identical bandwidth efficiency to their classic QPSK and QAM counterparts, despite being more difficult to implement. Furthermore, the MSK, Q^2PSK and Q^2AM signalling schemes require two different carrier frequencies as we have shown briefly in Equations 13.3 and 13.8 of Chapter 13. Therefore, we do not study the family of MSK, Q^2PSK and Q^2AM signalling schemes in this chapter.

23.5.2 M-ary Phase Shift Keying

M-ary Phase Shift Keying (PSK) constitutes a signalling scheme, where the $\log_2(M)$-bit information to be transmitted is mapped to M number of phases of the transmitted carrier. The modulated signalling waveforms may be expressed as:

$$
\begin{aligned}
x_m(t) &= \sqrt{\frac{2E_s}{T}} \cos\left(w_0 t - \frac{2\pi m}{M}\right), \quad m = 1, \ldots, M, \ 0 \le t \le T, & (23.40) \\
&= x_m[1]\phi_1(t) + x_m[2]\phi_2(t), \quad m = 1, \ldots, M, \ 0 \le t \le T, & (23.41)
\end{aligned}
$$

where w_0 is the carrier frequency in radians per second and the orthonormal basis functions are given by:

$$
\phi_1(t) = \sqrt{\frac{2}{T}} \cos(w_0 t), \ 0 \le t \le T, \tag{23.42}
$$

$$
\phi_2(t) = \sqrt{\frac{2}{T}} \sin(w_0 t), \ 0 \le t \le T, \tag{23.43}
$$

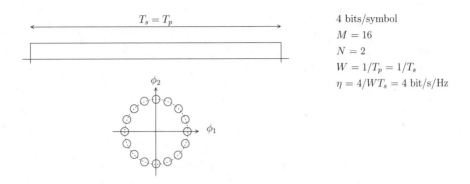

Figure 23.1: A PSK signalling example conveying 4 information bits per symbol

while the coefficients of the signalling vector, $\mathbf{x}_m = x_m[1] + jx_m[2] = (x_m[1], x_m[2])$, are given by:

$$x_m[1] = \sqrt{E_s} \cos(\frac{2\pi m}{M}), \quad 1 \leq m \leq M, \tag{23.44}$$

$$x_m[2] = \sqrt{E_s} \sin(\frac{2\pi m}{M}), \quad 1 \leq m \leq M. \tag{23.45}$$

Specifically, each phasor $\mathbf{x}_m$ of the PSK signalling set has an equal energy and its signalling phasor constellation is mapped to a circle of radius $\sqrt{E_s}$. Figure 23.1 depicts an example of PSK signalling having $M = 16$ for the sake of conveying 4 information bits per symbol. In the context of classic M-ary PSK signalling, the duration of the rectangular signalling pulse T_p equals the modulated symbol duration T_s, as illustrated in Figure 23.1. Therefore, the bandwidth required is $W = 1/T_p = 1/T_s$ Hz and the number of dimensions offered by the signal space is $N = 2WT_s = 2$. During a symbol duration T_s, a phasor $\mathbf{x}_m$ chosen from the M legitimate phasors in the constellation is transmitted. Since we have $WT_s = N/2 = 1$, the asymptotic value of the bandwidth efficiency is similar to that of the achievable capacity of the classic PSK signalling sets. Note that when we have $M = 2$, 2-PSK signalling utilises only 1 out of the 2 possible signalling dimensions.

We may conclude that PSK signalling does not constitute an efficient scheme, since it has to obey the *PSK limit* [705], which is significantly lower than the unrestricted bound of Equation 23.18. The PSK limit for AWGN channels is given by [705, pp. 276-279]:

$$C_{\text{PSK LIMIT}}^{\text{AWGN}} = \log_2 \sqrt{\frac{4\pi}{e} \frac{E_s}{N_0}} \quad [\text{bit/symbol}], \tag{23.46}$$

where we have $\frac{E_s}{N_0} = \gamma$ according to Equation 23.22, since the signalling dimensionality is $N = 2$. On other hand, the PSK limit valid for the Rayleigh fading channel may be derived from Equation 23.46 by weighting the SNR γ of the Gaussian channel by its probability of occurrence given by the Rayleigh-faded magnitude χ_2^2, which was defined in Section 23.2.3,

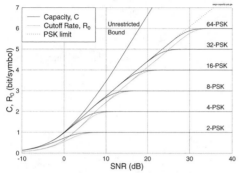

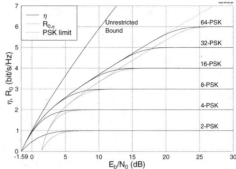

(a) The capacity C and cutoff rate R_0 are computed using Equations 23.18, 23.22, 23.30 and 23.32

(b) The bandwidth efficiency η and normalised cutoff rate $R_{0,\eta}$ are determined using Equations 23.34 and 23.39

Figure 23.2: M-ary PSK characteristics for $M = 2, 4, 8, 16, 32$ and 64 when communicating over an AWGN channel

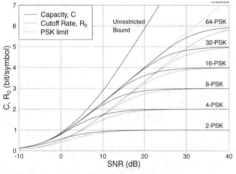

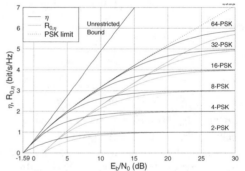

(a) The capacity C and cutoff rate R_0 are computed using Equations 23.25, 23.26, 23.28, 23.30 and 23.33

(b) The bandwidth efficiency η and normalised cutoff rate $R_{0,\eta}$ are determined using Equations 23.34 and 23.39

Figure 23.3: M-ary PSK characteristics for $M = 2, 4, 8, 16, 32$ and 64 when communicating over a Rayleigh fading channel

and then averaging it over χ_2^2 yielding:

$$C_{\text{PSK LIMIT}}^{\text{RAY}} = \mathrm{E}\left[\log_2\sqrt{\frac{4\pi}{e}\chi_2^2\frac{E_s}{N_0}}\right] \quad [\text{bit/symbol}], \tag{23.47}$$

where the expectation is evaluated with respect to χ_2^2. Note that the $\frac{E_b}{N_0}$ value of the PSK limit curve is given by $\frac{\gamma}{C_{\text{PSK LIMIT}}}$ and the normalised PSK limit is given by:

$$\eta_{\text{PSK LIMIT}} = \frac{C_{\text{PSK LIMIT}}}{N/2} \quad [\text{bit/s/Hz}]. \tag{23.48}$$

Figures 23.2 and 23.3 show the capacity, cutoff rate, bandwidth efficiency, normalised cutoff rate and PSK limit of the M-ary PSK signals, when communicating over AWGN and Rayleigh fading channels, respectively. As we can observe from Figure 23.2(a), at a capacity of $b = 3$ bit/symbol the SNR performance of the $2^{b+1} = 16$-PSK scheme is about 3 dB better than that of the $2^b = 8$-PSK scheme, when communicating over an AWGN channel. Even more significantly, when communicating over a Rayleigh fading channel, the SNR performance of the 16-PSK scheme is about 13 dB better than that of the 8-PSK scheme at a capacity of 3 bit/symbol, as is shown in Figure 23.3(a). However, at a capacity of $b = 3$ bit/symbol, the PSK schemes having $M > 2^{b+1}$ yield very little additional SNR gain in comparison to 16-PSK. More explicitly, all PSK signalling schemes having $M > 16$ perform virtually identically to 16-PSK at a capacity of $b = 3$ bit/symbol, when communicating over AWGN channels, as evidenced by Figure 23.2(a). When communicating over uncorrelated Rayleigh fading channels, an SNR gain of less than 0.5 dB is obtained by a PSK signalling schemes having $M > 16$ in comparison to 16-PSK at a capacity of $b = 3$ bit/symbol, as shown in Figure 23.3(a). Similar observations are also true for $b \in \{1, 2, \ldots, 6\}$, as it is evidenced by Figures 23.2 and 23.3. Therefore, in order to approach the achievable capacity of b bit/symbol, it is better to employ 2^{b+1}-PSK, rather than 2^b-PSK. At first sight this statement may seem inplausible, however, we will show in Chapter 24 that this is exactly the motivation of Ungerböck's Trellis Coded Modulation (TCM) scheme, where the modulation constellation size is doubled for the sake of accommodating an extra bit. This extra bit is used in TCM for error correction, potentially allowing us to operate without errors at the cost of a higher complexity but at a lower SNR, i.e. to approach the capacity limit more closely. As a further observation, by doubling M from 2^b to 2^{b+1} most of the total achievable gain can be obtained, when aiming for a capacity of $C = b$ bit/symbol and any further expansion of the modulation constellation is only likely to yield marginal SNR benefits.

By comparing Figures 23.2 and 23.3, we notice that the SNR or E_b/N_0 gap between the capacity and cutoff rate of the uncorrelated Rayleigh fading channel is wider than that observed for the AWGN channel. For example, at a capacity of 3 bit/symbol the SNR gap between the capacity curve and cutoff rate curve of 16-PSK communicating over AWGN channels and uncorrelated Rayleigh fading channels is about 1 dB (as shown in Figure 23.2(a)) and 4 dB (as shown in Figure 23.3(a)), respectively. This implies that it is harder to reach the capacity of the uncorrelated Rayleigh fading channel compared to that of the AWGN channel. Additionally, the SNR gap between the capacity of M-ary PSK and the unrestricted bound becomes larger for increasing values of M at a capacity of $(\log_2(M) - 1)$ bit/symbol. For example, when communicating over uncorrelated Rayleigh fading channels, the SNR gap between the capacity curve of 4-PSK and the unrestricted bound at 1 bit/symbol is only about 1 dB. By contrast, the SNR gap between the capacity curve of 64-PSK and the unrestricted bound is approximately 10 dB, as we can observe from Figure 23.3(a). This is a consequence of the convergence of the PSK curves to the ultimate PSK limit mentioned earlier.

23.5.3 M-ary Quadrature Amplitude Modulation

M-ary Quadrature Amplitude Modulation (QAM) may be viewed as a combination of two independent Pulse Amplitude Modulation (PAM) schemes. The modulated signalling waveforms may be expressed as in Equation 23.41 and the two orthonormal basis functions are similar to that of PSK, which are given by Equations 23.42 and 23.43. Specifically, the QAM signalling set maps each message block onto a rectangular phasor constellation based on the

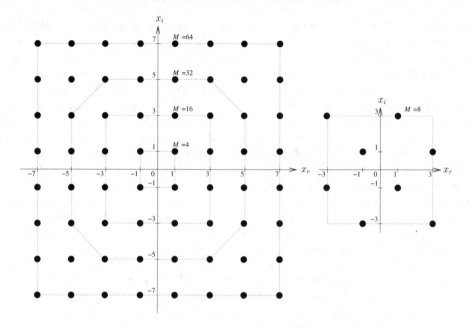

Figure 23.4: QAM constellations for $M = 4, 8, 16, 32$ and 64

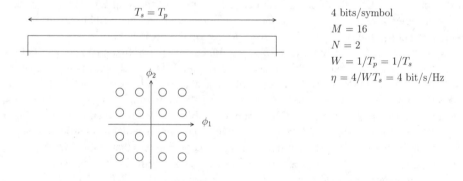

4 bits/symbol
$M = 16$
$N = 2$
$W = 1/T_p = 1/T_s$
$\eta = 4/WT_s = 4$ bit/s/Hz

Figure 23.5: A QAM signalling example conveying 4 information bits per symbol

coefficients of $\mathbf{x}_m$ as follows:

$$\mathbf{x}_m = x_m[1] + jx_m[2], \quad x_m[1] \in x_r, \quad x_m[2] \in x_i, \quad m = 1, \dots, M, \quad (23.49)$$

where x_r and x_i are the values of $\mathbf{x}_m$ mapped to the real and imaginary axis of the signal constellation. The QAM signalling set may also be viewed as a combined amplitude and phase modulation scheme. The signal space diagrams of QAM constellations used in this chapter are shown in Figure 23.4. Note that the 8-QAM constellation seen in Figure 23.4, which was originally proposed in [495], exhibits a higher minimum Euclidean distance compared to the rectangular 8-QAM of [388, p. 180], although its peak-to-mean envelope and its phase-jitter resilience defined in Section 4.1 are inferior.

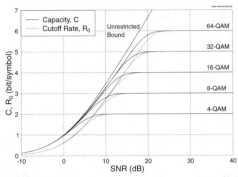

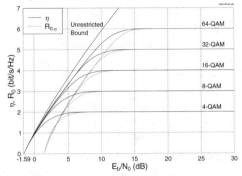

(a) The capacity C and cutoff rate R_0 are computed using Equations 23.18, 23.22, 23.30 and 23.32

(b) The bandwidth efficiency η and normalised cutoff rate $R_{0,\eta}$ are determined using Equations 23.34 and 23.39

Figure 23.6: M-ary QAM characteristics for M = 4, 8, 16, 32 and 64 when communicating over AWGN channel

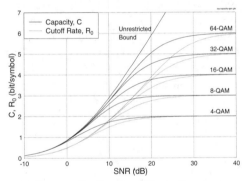

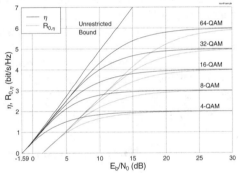

(a) The capacity C and cutoff rate R_0 are computed using Equations 23.25, 23.26, 23.28, 23.30 and 23.33

(b) The bandwidth efficiency η and normalised cutoff rate $R_{0,\eta}$ are determined using Equations 23.34 and 23.39

Figure 23.7: M-ary QAM characteristics for M = 4, 8, 16, 32 and 64 when communicating over a Rayleigh fading channel

Figure 23.5 characterises QAM signalling having $M = 16$ for the sake of conveying 4 information bits per symbol. Specifically, in the 16-QAM scheme, we have $x_r = -3, -1, 1, 3$ and $x_i = -3, -1, 1, 3$. In the classic M-ary QAM signalling scheme, the duration of the rectangular signalling pulse T_p equals the modulated symbol duration T_s, as depicted in Figure 23.5. Therefore, the bandwidth required is $W = 1/T_p = 1/T_s$ Hz and the number of dimensions of the signalling space is $N = 2WT_s = 2$. During a symbol duration T_s, a phasor $\mathbf{x}_m$ chosen from the M legitimate phasors of the constellation is transmitted. Since we have $WT_s = N/2 = 1$, the asymptotic value of the bandwidth efficiency is similar to that of the achievable capacity of the classic QAM signalling sets.

Figures 23.6 and 23.7 show the achievable capacity, cutoff rate, bandwidth efficiency and normalised cutoff rate of the family of M-ary QAM signals, when communicating over AWGN and Rayleigh fading channels, respectively. Similar to our finding in the context of the M-ary PSK results of Section 23.5.2, in order to achieve a capacity of b bit/symbol, it is better to employ 2^{b+1}-ary QAM, rather than the QAM scheme having $M = 2^b$ or $M > 2^{b+1}$. Explicitly, by doubling M from 2^b to 2^{b+1} most of the achievable capacity gain may be obtained when aiming for a capacity of $C = b$ bit/symbol. For example, as evidenced in Figure 23.7(a), the SNR required for 8-QAM, 16-QAM, 32-QAM and 64-QAM is about 28 dB, 12 dB, 11 dB and 11 dB, respectively, when communicating over uncorrelated Rayleigh fading channels at a capacity of $b = 3$ bit/symbol. It is also harder to approach the capacity of the Rayleigh fading channel compared to the AWGN channel in the context of QAM, as a consequence of having a wider SNR gap between the capacity and cutoff rate of Rayleigh fading channels compared to that of the AWGN channels. For example, at a capacity of 3 bit/symbol the SNR gap between the capacity curve and cutoff rate curve of 16-QAM when communicating over AWGN channels and uncorrelated Rayleigh fading channels is about 2 dB (as shown in Figure 23.6(a)) and 5 dB (as shown in Figure 23.7(a)), respectively. However, as seen by comparing Figures 23.3(a) and 23.7(a), the SNR performance difference between the unrestricted bound and the capacity of M-ary QAM is significantly smaller than that of the M-ary PSK scheme of Section 23.5.2. As we can see from Figures 23.3(a) and 23.7(a), the SNR requirement of 5 bit/symbol signalling at the unrestricted bound as well as when using 64-QAM and 64-PSK communicating over uncorrelated Rayleigh fading channels is approximately 17 dB, as well as 20 dB and 26 dB, respectively. This indicates that QAM is potentially more bandwidth efficient than PSK. Again, a practical manifestation of this statement will be detailed in the context of TCM in Chapter 24, where the expanded signalling constellation accommodates an error correction code, potentially allowing the expanded phasor constellation to approach the capacity more closely owing to its better error resilience, despite its reduced minimum distance amongst the constellation points.

23.5.4 M-ary Orthogonal Signalling

M-ary orthogonal signalling constitutes a transmission scheme, where $\log_2(M)$ number of bits are mapped to M orthogonal waveforms, as for example in the IS-95 CDMA standard, which is also known as cdmaOne [706]. The bandwidth requirement of M-ary orthogonal signalling is given by [388, p. 283]:

$$W = \frac{M}{2T_s},\tag{23.50}$$

where T_s is the symbol duration and hence this signalling scheme may also be interpreted as a collection of phasor points in the $N = 2WT_s = M$-dimensional phasor space, where only one phasor point is located on each of the M coordinate axes. The M-dimensional signalling vectors can be represented as [388, 695]:

$$
\begin{aligned}
\mathbf{x}_1 &= \sqrt{E_s}(1,0,\ldots,0) = \sqrt{E_s}\phi_1, \\
\mathbf{x}_2 &= \sqrt{E_s}(0,1,\ldots,0) = \sqrt{E_s}\phi_2, \\
&\vdots \\
\mathbf{x}_M &= \sqrt{E_s}(0,0,\ldots,1) = \sqrt{E_s}\phi_M,
\end{aligned}\tag{23.51}
$$

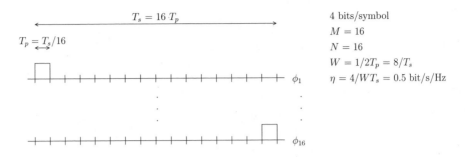

Figure 23.8: An orthogonal signalling example conveying 4 information bits per symbol

where each $\mathbf{x}_m$, $m \in \{1, \ldots, M\}$, in the M-dimensional space is located at a distance of $\sqrt{E_s}$ from the origin. The orthonormal basis function ϕ_m is M-dimensional:

$$\phi_m = (\phi_m[1], \phi_m[2], \ldots, \phi_m[M]), \tag{23.52}$$

which may be constructed from non-overlapping signalling pulses as follows:

$$\phi_m[i] = \begin{cases} 1, & i = m, \\ 0, & i \neq m. \end{cases} \tag{23.53}$$

Note that M-ary orthogonal signalling is more power-efficient and more error-resilient, but less bandwidth efficient compared to classic M-ary PSK and QAM [388, p. 284]. Figure 23.8 depicts an example of the M-ary orthogonal signalling scheme having $M = N = 16$ for the sake of conveying 4 information bits per symbol. In M-ary orthogonal signalling, the duration of the rectangular pulse is given by $T_p = T_s/M$, as seen in Figure 23.8. However, the bandwidth required is given by Equation 23.50 as $W = \frac{M}{2T_s} = 1/2T_p$ Hz, which is different from that of the QAM and PSK schemes. The number of dimensions of the signalling space is $N = 2WT_s = M$. During a symbol duration T_s, only one pulse duration is active, while the rest are inactive, when the orthonormal basis functions are constructed as non-overlapping pulses according to Equation 23.53. Since we have $WT_s = N/2 = M/2$, the asymptotic value of bandwidth efficiency is different from that of the capacity of the M-ary orthogonal signalling sets for values of $M > 2$.

Figures 23.9 and 23.10 show the capacity, cutoff rate, bandwidth efficiency and normalised cutoff rate of M-ary orthogonal signalling, when communicating over AWGN and Rayleigh fading channels, respectively. Note that the unrestricted bound of Equation 23.18 is dependent on $WT = N/2$, hence it is different for different dimensionality values N. However, the normalised unrestricted bound expressed in Equation 23.35 is independent of $WT = N/2$. As shown in Figures 23.9(a) and 23.10(a), the channel capacity curves reach the asymptotic performance of $\log_2(M)$ bit/symbol at low SNRs, when increasing the value of M. This trend is different from the channel capacity curves recorded for classic PSK and QAM. As depicted in Figures 23.9(b) and 23.10(b), unlike for classic PSK and QAM signals, the bandwidth efficiency of M-ary orthogonal signalling is further away from the unrestricted bound at low E_b/N_0 values as M decreases. This phenomenon is a consequence of having non-zero centre of gravity or mean in M-ary orthogonal signalling, which is given

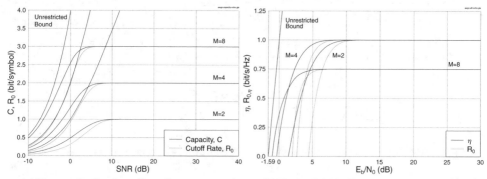

(a) The capacity C and cutoff rate R_0 are computed using Equations 23.18, 23.22, 23.30 and 23.32

(b) The bandwidth efficiency η and normalised cutoff rate $R_{0,\eta}$ are determined using Equations 23.34 and 23.39

Figure 23.9: M-ary orthogonal characteristics for $M = 2$, 4 and 8 when communicating over an AWGN channel

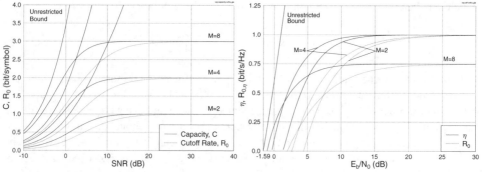

(a) The capacity C and cutoff rate R_0 are computed using Equations 23.25, 23.26, 23.29, 23.30 and 23.33

(b) The bandwidth efficiency η and normalised cutoff rate $R_{0,\eta}$ are determined using Equations 23.34 and 23.39

Figure 23.10: M-ary orthogonal characteristics for $M = 2$, 4 and 8 when communicating over a Rayleigh fading channel

by [707, p. 245]:

$$\bar{\mathbf{x}} = \frac{1}{M} \sum_{m=1}^{M} \mathbf{x}_m = \frac{\sqrt{E_s}}{M} \sum_{m=1}^{M} \phi_m, \tag{23.54}$$

and the energy of $\bar{\mathbf{x}}$ is:

$$|\bar{\mathbf{x}}|^2 = \frac{E_s}{M^2} \sum_{m=1}^{M} |\phi_m|^2 = \frac{E_s}{M}. \tag{23.55}$$

Therefore, the actual transmitted power used for conveying information is $E_s - |\bar{\mathbf{x}}|^2 =$

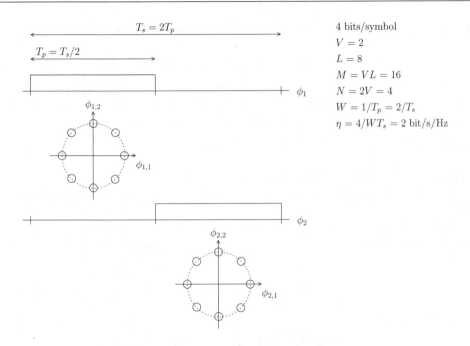

Figure 23.11: An L-orthogonal PSK signalling example conveying 4 bits per symbol

$E_s\left(\frac{M-1}{M}\right)$, which is lower than the total transmitted power of E_s. The corresponding E_b/N_0 value at 0 bit/s/Hz may be calculated as: $E_b/N_0 = -1.59 - 10\log_{10}\left(\frac{M-1}{M}\right)$ dB. Hence, as M increases, the bandwidth efficiency curves converge more closely to the Shannon bound. However, since we have $WT = M/2 > 1$ for $M > 2$, the asymptotic value of the bandwidth efficiency is a factor of $M/2$ lower than that of the capacity of the M-ary orthogonal signalling set. Therefore, as $M \to \infty$, we have $\eta = \frac{C}{M/2} \to 0$, which implies having a zero bandwidth efficiency. Similar to the related findings for classic PSK and QAM, the SNR gap between the capacity and cutoff rate of the uncorrelated Rayleigh fading channel is wider than that of the AWGN channel. For example, at a capacity of 2 bit/symbol the SNR gap between the capacity curve and cutoff rate curve of 8-orthogonal signalling communicating over AWGN channels and uncorrelated Rayleigh fading channels is about 2 dB (as shown in Figure 23.9(a)) and 5.5 dB (as shown in Figure 23.10(a)), respectively. This indicates that it is harder to reach the capacity of the uncorrelated Rayleigh fading channel compared to that of the AWGN channel.

23.5.5 L-Orthogonal PSK Signalling

L-orthogonal PSK signalling constitutes a hybrid form of M-ary orthogonal and classic PSK signalling [698, 699], comprised of V number of independent L-ary PSK subsets. Therefore, the total number of available signalling waveforms is $M = VL$ and hence the number of bits transmitted per signalling symbol is $\log_2(VL)$. The total number of dimensions is $N = 2V$.

The vector representation of L-orthogonal PSK signalling may be formulated as:

$$\mathbf{x}_m = \mathbf{x}_l^{LPSK}\phi_v, \quad m = 1, \dots, M, \quad l = m\%L, \quad v = \left(\frac{m-l}{V} + 1\right), \quad (23.56)$$

where $m\%L$ is the remainer of m/L and $\mathbf{x}_l^{LPSK}$ is the classic 2-dimensional L-ary PSK signal vector, which obeys the form of Equations 23.44 and 23.45, yielding:

$$\mathbf{x}_l^{LPSK} = \left(x_l^{LPSK}[1], x_l^{LPSK}[2]\right), \quad l = 1, \dots, L. \quad (23.57)$$

Furthermore, the orthonormal basis function $\phi_v = (\phi_v[1], \phi_v[2], \dots, \phi_v[V])$ is a vector of V elements, which may be constructed as a set of non-overlapping pulses defined in Equation 23.53.

Specifically, the vector of an L-orthogonal PSK signalling set having $L = 8$ and $V = 2$ may be formulated as:

$$
\begin{aligned}
\mathbf{x}_1 &= \mathbf{x}_1^{8PSK}(1,0) = \left(x_1^{8PSK}[1], x_1^{8PSK}[2], 0, 0\right), \\
\mathbf{x}_2 &= \mathbf{x}_2^{8PSK}(1,0) = \left(x_2^{8PSK}[1], x_2^{8PSK}[2], 0, 0\right), \\
&\ \ \vdots \\
\mathbf{x}_8 &= \mathbf{x}_8^{8PSK}(1,0) = \left(x_8^{8PSK}[1], x_8^{8PSK}[2], 0, 0\right), \\
\mathbf{x}_9 &= \mathbf{x}_1^{8PSK}(0,1) = \left(0, 0, x_1^{8PSK}[1], x_1^{8PSK}[2]\right), \\
\mathbf{x}_{10} &= \mathbf{x}_2^{8PSK}(0,1) = \left(0, 0, x_2^{8PSK}[1], x_2^{8PSK}[2]\right), \\
&\ \ \vdots \\
\mathbf{x}_{16} &= \mathbf{x}_8^{8PSK}(0,1) = \left(0, 0, x_8^{8PSK}[1], x_8^{8PSK}[2]\right), \quad (23.58)
\end{aligned}
$$

where the total number of legitimate waveforms is $M = VL = 16$ and hence the number of bits per symbol is $\log_2(M) = 4$. Explicitly, the L-orthogonal PSK signalling set having $L = 8$ and $V = 2$ is illustrated in Figure 23.11. As we can see from Figure 23.11, the duration of the rectangular pulse is given by $T_p = T_s/V$ and during a modulated symbol duration T_s, only one signalling pulse of duration T_p is active, while the rest of them are inactive when the orthonormal basis functions are constructed as non-overlapping pulses according to Equation 23.53. A phasor $\mathbf{x}_m$ chosen from the $L = 8$ legitimate phasors in the 8-PSK constellation is transmitted, when the corresponding pulse-slot is active. The bandwidth required is $W = 1/T_p = V/T_s$, hence the number of dimensions of the signal space is $N = 2WT_s = 2V$. Therefore, we have:

$$V = WT_s = \frac{N}{2}. \quad (23.59)$$

Since we have $WT_s = N/2 = V$, the asymptotic value of bandwidth efficiency is a factor of V lower than the capacity. When $V = 1$, we have $L = M$ and hence L-orthogonal PSK signalling becomes analogous to classic M-ary PSK signalling. Note that L-orthogonal PSK signalling requires only $V = N/2$ number of timeslots for conveying a symbol, whereas the M-ary orthogonal signalling waveforms of Section 23.5.4 require N number of timeslots for

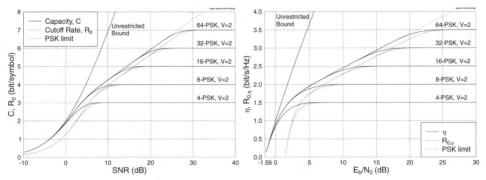

(a) The capacity C and cutoff rate R_0 are computed using Equations 23.18, 23.22, 23.30 and 23.32

(b) The bandwidth efficiency η and normalised cutoff rate $R_{0,\eta}$ are determined using Equations 23.34 and 23.39

Figure 23.12: L-orthogonal PSK characteristics for $V = 2$ and $L = 4, 8, 16, 32$ and 64 when communicating over AWGN channel

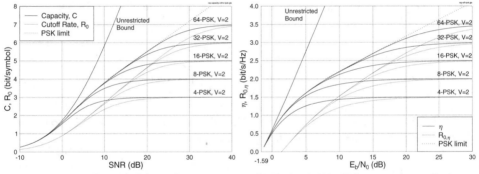

(a) The capacity C and cutoff rate R_0 are computed using Equations 23.25, 23.26, 23.28, 23.30 and 23.33

(b) The bandwidth efficiency η and normalised cutoff rate $R_{0,\eta}$ are determined using Equations 23.34 and 23.39

Figure 23.13: L-orthogonal PSK, $V = 2$ and $L = 4, 8, 16, 32$ and 64 when communicating over a Rayleigh fading channel

conveying a symbol. Hence, for a given number of dimensions N, the achievable transmission rate of L-orthogonal PSK signalling is a factor of two higher than that of M-ary orthogonal signalling. Furthermore, at a given dimensionality N there are $M = VL = NL/2$ number of waveforms in the context of L-orthogonal PSK signalling, which is a factor of $L/2$ times higher than that of classic M-ary orthogonal signalling. However, at a given value of M, L-orthogonal PSK signalling is $V = M/L$ times less bandwidth efficient than the classic M-ary PSK signalling scheme.

Figures 23.12 and 23.13 portray the capacity, cutoff rate, bandwidth efficiency and normalised cutoff rate of L-orthogonal PSK signalling schemes having $V = 2$, when communicating over AWGN and uncorrelated Rayleigh fading channels, respectively. Similar to classic PSK signalling, most of the achievable gain has already been attained by doubling the value of L and very little additional gain may be achieved, if L is further increased. More explicitly, as we can observe from Figure 23.13(a), the 16-orthogonal PSK having $V = 2$ is approximately 14 dBs better than the 8-orthogonal PSK having $V = 2$ at a capacity of 4 bit/symbol. It is also harder to approach the capacity of the uncoded Rayleigh fading channel compared to the AWGN channel, since there is a wider SNR gap between the capacity and cutoff rate of the Rayleigh fading channel compared to that of the AWGN channel.

Since we have $M = VL$ for the L-orthogonal PSK signalling, the capacity of L-orthogonal PSK signalling is $\log_2\left(\frac{VL}{L}\right) = \log_2(V)$ bit higher than that of the classic L-ary PSK subset. However, the bandwidth efficiency of L-orthogonal PSK signalling is $\frac{V\log_2(L)}{\log_2(VL)}$ times lower than that of its L-ary PSK subset. For example, 16-PSK signalling has a capacity of $C = 4$ bit/symbol and bandwidth efficiency of $\eta = 4$ bit/s/Hz. By contrast, the L-orthogonal PSK signalling employing $V = 2$ number of 16-PSK subsets has $C = 5$ and $\eta = 2.5$. Therefore, L-orthogonal PSK signalling having $V = 2$ and $L = 16$ is $5 - 4 = 1$ bit higher than 16-PSK in terms of capacity, but it is $4/2.5 = 1.6$ times lower than 16-PSK in terms of bandwidth efficiency, where $\log_2(V) = 1$ and $\frac{V\log_2(L)}{\log_2(VL)} = 1.6$.

As we can see from Figures 23.12 and 23.13, L-orthogonal PSK signalling also has to obey the PSK limit. In general, an L-orthogonal PSK signalling scheme employing V number of timeslots for transmitting an $N = 2V$ dimensional signal has an extra throughput of $\log_2(V)$ bits compared to classical L-level PSK signalling. Therefore, the ultimate limit of L-orthogonal PSK signalling, when communicating over AWGN channels, may be derived from Equation 23.46 as:

$$C_{L\text{-ORTHO PSK LIMIT}}^{\text{AWGN}} = \log_2\left(\sqrt{\frac{4\pi}{e}\frac{E_s}{N_0}}\right) + \log_2(V) \text{ [bit/symbol]}, \qquad (23.60)$$

where $\frac{E_s}{N_0} = V\gamma$ according to Equation 23.22, since we have $V = N/2$ and γ is the SNR. Therefore, Equation 23.60 can be further simplified to:

$$C_{L\text{-ORTHO PSK LIMIT}}^{\text{AWGN}} = \log_2\left(\sqrt{\frac{4\pi}{e}V\gamma}\right) + \log_2(V) \text{ [bit/symbol]},$$

$$= \log_2\left(\sqrt{\frac{4\pi}{e}\gamma}\right) + \frac{3}{2}\log_2(V) \text{ [bit/symbol]}, \qquad (23.61)$$

$$= C_{\text{PSK LIMIT}}^{\text{AWGN}} + \frac{3}{2}\log_2(V) \text{ [bit/symbol]}, \qquad (23.62)$$

where $C_{\text{PSK LIMIT}}^{\text{AWGN}}$ is the ultimate limit of PSK signalling ($V = 1$) given by Equation 23.46. Similar to Equation 23.62, the ultimate limit for L-orthogonal PSK signalling, when communicating over uncorrelated Rayleigh fading channels can be expressed as:

$$C_{L\text{-ORTHO PSK LIMIT}}^{\text{RAY}} = C_{\text{PSK LIMIT}}^{\text{RAY}} + \frac{3}{2}\log_2(V) \text{ [bit/symbol]}, \qquad (23.63)$$

where $C_{\text{PSK LIMIT}}^{\text{RAY}}$ was given by Equation 23.47. Hence, when V varies, the curve of L-orthogonal PSK limit is shifted by a constant of $\frac{3}{2}\log_2 V$, but the slope of the curve remains unchanged. The L-orthogonal PSK limit curves calculated for $V = 1, 2, 4$ and 8 are illustrated in Figures 23.14(a) and 23.15(a). In terms of the achievable bandwidth efficiency, according

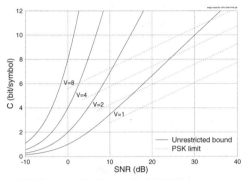

(a) The unrestricted bound and PSK limit are computed using Equations 23.18 and 23.62

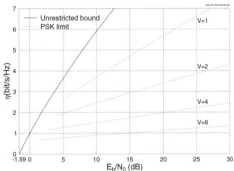

(b) The unrestricted bound and PSK limit normalised to WT are determined using Equations 23.34 and 23.64

Figure 23.14: L-orthogonal PSK characteristics for $V = 1, 2, 4,$ and 8 when communicating over an AWGN channel

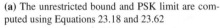

to Equations 23.34 and 23.59, we may express the normalised L-orthogonal PSK limit as:

$$\eta_{L\text{-ORTHO PSK LIMIT}} = \frac{C_{L\text{-ORTHO PSK LIMIT}}}{V} \text{ [bit/s/Hz]}. \qquad (23.64)$$

Therefore, the gradient of the normalised L-orthogonal PSK limit is reduced by a factor of V, as portrayed in Figures 23.14(b) and 23.15(b). From Figures 23.12, 23.13, 23.14 and 23.15, we can see that L-orthogonal PSK signalling becomes inefficient for a large value of L and V. For example, when communicating over uncorrelated Rayleigh fading channels, the SNR gap between the capacity curve of 4-orthogonal PSK and the unrestricted bound at 2 bit/symbol is only about 1.5 dB. By contrast, the SNR gap between the capacity curve of 64-orthogonal PSK and the unrestricted bound is approximately 13.5 dB, as we can observe from Figure 23.13(a), when we have $V = 2$. Furthermore, the L-orthogonal PSK limit curves is further away from the unrestricted bound for higher values of V, as we can observe from Figure 23.15(b). More specifically, at a bandwidth efficiency of $\eta = 2$ bit/s/Hz the E_b/N_0 gap between the unrestricted bound and L-orthogonal PSK limit curves plotted for $V = 2$ and $V = 4$ is about 4.5 dB and 19.5 dB, respectively, as evidenced by Figure 23.15(b).

23.5.6 L-Orthogonal QAM Signalling

The novel message of this section is that L-orthogonal signalling may also incorporate QAM subsets instead of the PSK subsets, which have to obey the ultimate PSK limit. Explicitly, L-orthogonal QAM signalling constitutes a hybrid form of M-ary orthogonal and QAM signalling. Similar to L-orthogonal PSK, it comprised of V independent L-ary QAM subsets. The total number of legitimate waveforms is $M = VL$ and the number of transmitted bits per symbol is $\log_2(VL)$. The total number of dimensions is $N = 2V$. The vector representation of L-orthogonal QAM signalling may be formulated as:

$$\mathbf{x}_m = \mathbf{x}_l^{LQAM} \phi_v, \quad m = 1, \dots, M, \quad l = m\%L, \quad v = \left(\frac{m-l}{V} + 1\right), \qquad (23.65)$$

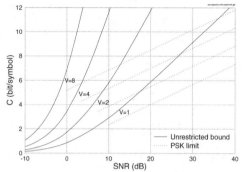 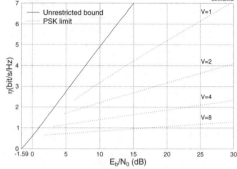

(a) The unrestricted bound and PSK limit are computed using Equations 23.25 and 23.63

(b) The unrestricted bound and PSK limit normalised to WT are determined using Equations 23.34 and 23.64

Figure 23.15: L-orthogonal PSK characteristics for $V = 1, 2, 4,$ and 8 when communicating over a Rayleigh fading channel

where $\mathbf{x}_l^{LQAM}$ is the classic 2-dimensional L-ary QAM signal vector given by Equation 23.49. The orthonormal basis function ϕ_v may be constructed as a set of non-overlapping pulses outlined in Equation 23.53 similar to that of the L-orthogonal PSK signalling characterised in Equation 23.56.

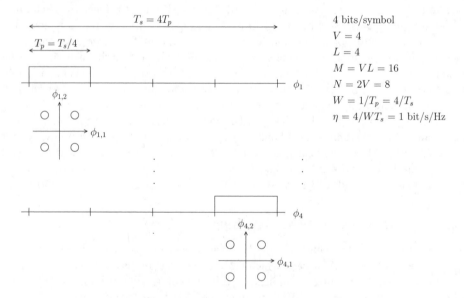

4 bits/symbol
$V = 4$
$L = 4$
$M = VL = 16$
$N = 2V = 8$
$W = 1/T_p = 4/T_s$
$\eta = 4/WT_s = 1 \text{ bit/s/Hz}$

Figure 23.16: An L-orthogonal QAM signalling example conveying 4 bits per symbol

Figure 23.16 depicts an example of the L-orthogonal QAM signalling scheme having $L = 4$ and $V = 4$, which conveys 4 information bits per symbol. In the L-orthogonal QAM signalling scheme, the duration of the rectangular signalling pulse is given by $T_p = T_s/V$, as was shown in Figure 23.5. Therefore, the bandwidth required is $W = 1/T_p = V/T_s$ Hz and the number of dimensions of the signalling space is $N = 2WT_s = 2V$. A phasor $\mathbf{x}_m$ chosen from the L legitimate phasors in the 4-QAM constellation is transmitted for a signalling pulse duration of T_p, followed by silence for the rest of the $(V-1)$ timeslots, since the orthonormal basis functions are constructed as non-overlapping pulses according to Equation 23.53. Since we have $WT_s = N/2 = V$, the asymptotic value of the bandwidth efficiency of L-orthogonal QAM signalling is a factor of V lower than the capacity. More explicitly, the vector of an L-orthogonal QAM signalling set having $L = 4$ and $V = 4$ may be expressed as:

$$\mathbf{x}_1 = \mathbf{x}_1^{4QAM}(1,0,0,0) = \left(x_1^{4QAM}[1], x_1^{4QAM}[2], 0,0,0,0,0,0 \right),$$

$$\mathbf{x}_2 = \mathbf{x}_2^{4QAM}(1,0,0,0) = \left(x_2^{4QAM}[1], x_2^{4QAM}[2], 0,0,0,0,0,0 \right),$$

$$\mathbf{x}_3 = \mathbf{x}_3^{4QAM}(1,0,0,0) = \left(x_3^{4QAM}[1], x_3^{4QAM}[2], 0,0,0,0,0,0 \right),$$

$$\mathbf{x}_4 = \mathbf{x}_4^{4QAM}(1,0,0,0) = \left(x_4^{4QAM}[1], x_4^{4QAM}[2], 0,0,0,0,0,0 \right),$$

$$\vdots$$

$$\mathbf{x}_{13} = \mathbf{x}_1^{4QAM}(0,0,0,1) = \left(0,0,0,0,0,0, x_1^{4QAM}[1], x_1^{4QAM}[2] \right),$$

$$\mathbf{x}_{14} = \mathbf{x}_2^{4QAM}(0,0,0,1) = \left(0,0,0,0,0,0, x_2^{4QAM}[1], x_2^{4QAM}[2] \right),$$

$$\mathbf{x}_{15} = \mathbf{x}_3^{4QAM}(0,0,0,1) = \left(0,0,0,0,0,0, x_3^{4QAM}[1], x_3^{4QAM}[2] \right),$$

$$\mathbf{x}_{16} = \mathbf{x}_4^{4QAM}(0,0,0,1) = \left(0,0,0,0,0,0, x_4^{4QAM}[1], x_4^{4QAM}[2] \right), \quad (23.66)$$

where the number of signalling waveforms is $M = VL = 16$ and the number of signalling dimensions is $N = 2V = 8$. Again, when we have $V = 1$, the L-orthogonal QAM scheme becomes a classic QAM signalling scheme. We can expect the L-orthogonal QAM scheme to exhibit a better performance, in terms of bandwidth efficiency versus E_b/N_0 and capacity versus SNR than that of L-orthogonal PSK, since the performance of QAM studied in Section 23.5.3 was shown to be better than that of PSK studied in Section 23.5.2.

Figures 23.17 and 23.18 characterise the capacity, cutoff rate, bandwidth efficiency and normalised cutoff rate of L-orthogonal QAM signalling schemes, when communicating over both AWGN and uncorrelated Rayleigh fading channels, respectively. Similar to classic QAM signalling, most of the achievable gain is already attained by doubling the modulation levels M from L to $2L$, when aiming for a capacity of $\log_2(VL)$ bit/symbol. More explicitly, as we can observe from Figure 23.18(a), the 16-orthogonal QAM having $V = 2$ is approximately 18 dBs better than the 8-orthogonal QAM having $V = 2$ at a capacity of 4 bit/symbol. Again, the SNR gap between the capacity and cutoff rate of the uncorrelated Rayleigh fading channel is wider than that of the AWGN channel. Specifically, the SNR gap between the capacity and cutoff rate of the uncorrelated Rayleigh fading channel is approximately 6 dBs at a capacity of 4 bit/symbol, in the context of 16-orthogonal QAM having $V = 2$ as shown in Figure 23.18(a). By contrast, as we can see from Figure 23.17(a), the corresponding SNR gap between the capacity and cutoff rate of AWGN is only about 2 dBs. Hence it is harder to approach the capacity of the uncorrelated Rayleigh fading channel compared to the AWGN channel for the L-orthogonal QAM signalling scheme. Similar

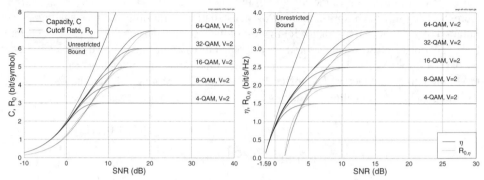

(a) The capacity C and cutoff rate R_0 are computed using Equations 23.18, 23.22, 23.30 and 23.32

(b) The bandwidth efficiency η and normalised cutoff rate $R_{0,\eta}$ are determined using Equations 23.34 and 23.39

Figure 23.17: L-orthogonal QAM, $V = 2$ and $L = 4, 8, 16, 32$ and 64 for AWGN channel

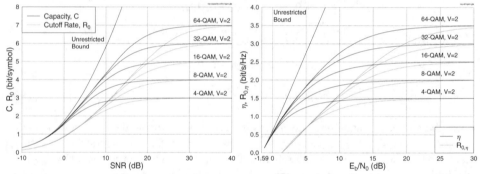

(a) The capacity C and cutoff rate R_0 are computed using Equations 23.25, 23.26, 23.28, 23.30 and 23.33

(b) The bandwidth efficiency η and normalised cutoff rate $R_{0,\eta}$ are determined using Equations 23.34 and 23.39

Figure 23.18: L-orthogonal QAM, $V = 2$ and $L = 4, 8, 16, 32$ and 64 for Rayleigh fading channel

to the L-orthogonal QAM signalling, the L-orthogonal QAM signalling is capable of achieving $\log_2(V)$ bits higher throughput than that of the classic L-ary QAM subset, at a cost of $\frac{V \log_2(L)}{\log_2(VL)}$ times lower bandwidth efficiency than that of the L-ary QAM subset. Although the SNR gap between L-orthogonal QAM signalling and the unrestricted bound becomes wider upon increasing L and V, yet this gap is significantly narrower than that of L-orthogonal PSK signalling. For example, the L-orthogonal QAM scheme outperforms the L-orthogonal PSK scheme by an SNR gain of 7 dB, when employing $L = 64$ and $V = 2$ at a capacity of 6 bit/symbol, as evidenced by Figures 23.13(a) and 23.18(a). Since we have $WT_s = V$, the asymptotic value of the bandwidth efficiency is V times lower than that of the capacity, as shown in Figures 23.17 and 23.18.

Note that both L-orthogonal PSK signalling and L-orthogonal QAM signalling schemes have twice the bandwidth efficiency compared to the M-ary orthogonal signalling scheme at a given number of modulation levels M. For instance, the $M = 8$-orthogonal signalling scheme can only achieve a throughput of 0.75 bit/s/Hz, as seen in Figure 23.10(b). However, both L-orthogonal PSK signalling and L-orthogonal QAM signalling having $M = VL = 2 \times 4 = 8$ may achieve a throughput of 1.5 bit/s/Hz, as seen in Figures 23.13(b) and 23.18(b). Furthermore, it was shown in [388] that orthogonal signalling schemes, such as M-ary orthogonal signalling is more error resilient than non-orthogonal signalling schemes, such as the classic QAM arrangement. Hence we can expect the L-orthogonal QAM signalling scheme to be more error resilient than the classic QAM signalling arrangement.

23.6 Summary

In this chapter, we have studied the capacity C, cutoff rate R_0, bandwidth efficiency η and the normalised cutoff rate $R_{0,\eta}$ of M-ary PSK, M-ary QAM, M-ary orthogonal signalling as well as the hybrid of PSK/QAM and orthogonal signalling schemes. The novel contributions of this chapter are:

- the introduction of Equation 23.28 for evaluating the performance of N-dimensional M-ary signalling schemes communicating over uncorrelated Rayleigh fading channels;

- the introduction of Equation 23.47, which quantifies the ultimate PSK limit for transmission over uncorrelated Rayleigh fading channels;

- the introduction of L-orthogonal QAM signalling, as an extension of L-orthogonal PSK signalling;

- the introduction of Equations 23.56 and 23.65 for representing L-orthogonal signalling employing PSK and QAM subsets, respectively, and

- the introduction of Equations 23.62 and 23.63 for quantifying the ultimate limits of L-orthogonal PSK signalling.

This study quantified the maximum achievable capacity for a range of M-ary digital signalling set for transmission over both AWGN and Rayleigh fading channels, in the quest for more error-resilient, power-efficient and bandwidth-efficient channel coding schemes.

As we discussed in Section 23.5, it is beneficial to double the number of modulation levels M from 2^b to 2^{b+1}, when aiming for a capacity of b bit/symbol in all the M-ary signalling schemes studied. In the forthcoming chapters, we will study a range of bandwidth efficient coded modulation schemes based on both M-ary PSK and QAM signalling, where the number of modulation levels is increased by introducing an extra parity bit in each of the original b-bit information symbol. Since the L-orthogonal QAM signalling scheme was shown in Section 23.5.6 to be more bandwidth efficient compared to the M-ary orthogonal signalling and L-orthogonal PSK signalling schemes, as well as being more error-resilient than the classic PSK and QAM schemes, future research might show the benefits of designing coded modulation schemes based on the family of L-orthogonal QAM signalling scheme.

Chapter 24

Coded Modulation Theory

24.1 Motivation

The objective of channel coding is to combat the effects of channel impairment and thereby aid the receiver in its decision-making process. The design of a good channel coding and modulation scheme depends on a range of contradictory factors, some of which are portrayed in Figure 24.1. Specifically, given a certain transmission channel, it is always feasible to design a coding and modulation system which is capable of further reducing the Bit Error Ratio (BER) and/or Frame Error Ratio (FER) achieved. The gain quantified in terms of the bit energy reduction at a certain BER/FER, achieved by the employment of channel coding with respect to the uncoded system is termed the coding gain. However, this implies further investments in terms of the required implementational complexity and coding/interleaving delay as well as reducing the effective throughput. Different solutions accrue, when designing a coding and modulation scheme, which aim for optimising different features. For example, in a power-limited scenario, the system's bandwidth can be extended for the sake of accommodating a low rate code. By contrast, the effective throughput of the system can be reduced for

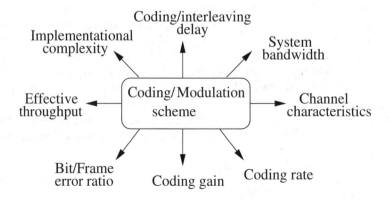

Figure 24.1: Factors affecting the design of channel coding and modulation scheme

the sake of absorbing more parity information. To elaborate further, in a bandwidth-limited and power-limited scenario a more complex, but a higher coding gain code can be employed. The system's effective throughput can be increased by increasing the coding rate at the cost of sacrificing the achievable transmission integrity. The coding and modulation scheme's design also depends on the channel's characteristics. More specifically, the associated bit and frame error statistics change, when the channel exhibits different statistical characteristics.

On the other hand, a joint channel coding and modulation scheme can be designed by employing high rate channel coding schemes in conjunction with multidimensional or high level modulation schemes. In this coded modulation scheme a coding gain may be achieved without bandwidth expansion. In this part of the book a variety of coded modulation assisted systems will be proposed and investigated in mobile wireless propagation environments.

24.2   A Historical Perspective on Coded Modulation

The history of channel coding or Forward Error Correction (FEC) coding dates back to Shannon's pioneering work [356] in 1948, in which he showed that it is possible to design a communication system with any desired small probability of error, whenever the rate of transmission is smaller than the capacity of the channel. While Shannon outlined the theory that explained the fundamental limits imposed on the efficiency of communications systems, he provided no insights into how to actually approach these limits. This motivated the search for codes that would produce arbitrarily small probability of error. Specifically, Hamming [357] and Golay [358] were the first to develop practical error control schemes. Convolutional codes [359] were later introduced by Elias in 1955, while Viterbi [360] invented a maximum likelihood sequence estimation algorithm in 1967 for efficiently decoding convolutional codes. In 1974, Bahl proposed the more complex Maximum A-Posteriori (MAP) algorithm, which is capable of achieving the minimum achievable BER.

The first successful application of channel coding was the employment of convolutional codes [359] in deep-space probes in the 1970s. However, for years to come, error control coding was considered to have limited applicability, apart from deep-space communications. Specifically, this is a power-limited scenario, which has no strict bandwidth limitation. By contrast mobile communications systems constitute a power- and bandwidth-limited scenario. In 1987, a bandwidth efficient Trellis Coded Modulation (TCM) [361] scheme employing symbol-based channel interleaving in conjunction with Set-Partitioning (SP) [362] assisted signal labelling was proposed by Ungerböck. Specifically, the TCM scheme, which is based on combining convolutional codes with multidimensional signal sets, constitutes a bandwidth efficient scheme that has been widely recognised as an efficient error control technique suitable for applications in mobile communications [363]. Another powerful coded modulation scheme utilising bit-based channel interleaving in conjunction with Gray signal labelling, which is referred to as Bit-Interleaved Coded Modulation (BICM), was proposed by Zehavi [364] as well as by Caire, Taricco and Biglieri [365]. Another breakthrough in the history of error control coding is the invention of turbo codes by Berrou, Glavieux and Thitimajshima [366] in 1993. Convolutional codes were used as the component codes and decoders based on the MAP algorithm were employed. The results proved that a performance close to the Shannon limit can be achieved in practice with the aid of binary codes. The attractive properties of turbo codes have attracted intensive research in this area [367–369].

As a result, turbo coding has reached a state of maturity within just a few years and was standardised in the recently ratified third-generation (3G) mobile radio systems [370].

However, turbo codes often have a low coding rate and hence require considerable bandwidth expansion. Therefore, one of the objectives of turbo coding research is the design of bandwidth-efficient turbo codes. In order to equip the family of binary turbo codes with a higher spectral efficiency, BICM-based Turbo Coded Modulation (TuCM) [371] was proposed in 1994. Specifically, TuCM uses a binary turbo encoder, which is linked to a signal mapper, after its output bits were suitably punctured and multiplexed for the sake of transmitting the desired number of information bits per transmitted symbol. In the TuCM scheme of [371] Gray-coding based signal labelling was utilised. For example, two 1/2-rate Recursive Systematic Convolutional (RSC) codes are used for generating a total of four turbo coded bits and this bit stream may be punctured for generating three bits, which are mapped to an 8PSK modulation scheme. By contrast, in separate coding and modulation scheme, any modulation schemes for example BPSK, may be used for transmitting the channel coded bits. Finally, without puncturing, 16QAM transmission would have to be used for maintaining the original transmission bandwidth. Turbo Trellis Coded Modulation (TTCM) [372] is a more recently proposed channel coding scheme that has a structure similar to that of the family of turbo codes, but employs TCM schemes as its component codes. The TTCM symbols are transmitted alternatively from the first and the second constituent TCM encoders and symbol-based interleavers are utilised for turbo interleaving and channel interleaving. It was shown in [372] that TTCM performs better than the TCM and TuCM schemes at a comparable complexity. In 1998, iterative joint decoding and demodulation assisted BICM referred to as BICM-ID was proposed in [373, 374], which uses SP-based signal labelling. The aim of BICM-ID is to increase the Euclidean distance of BICM and hence to exploit the full advantage of bit interleaving with the aid of soft-decision feedback based iterative decoding [374]. Many other bandwidth efficient schemes using turbo codes have been proposed in the literature [368], but we will focus our study on TCM, BICM, TTCM and BICM-ID schemes in the context of wireless channels in this part of the book.

The radio spectrum is a scarce resource. Therefore, one of the most important objectives in the design of digital cellular systems is the efficient exploitation of the available spectrum, in order to accommodate the ever-increasing traffic demands. Trellis-Coded Modulation (TCM) [495], which will be detailed in Section 24.3, was proposed originally for Gaussian channels, but it was further developed for applications in mobile communications [362, 708]. Turbo Trellis-Coded Modulation (TTCM) [709], which will be discussed in Section 24.5, is a more recent joint coding and modulation scheme that has a structure similar to that of the family of power-efficient binary turbo codes [366, 367], but employs TCM schemes as component codes. TTCM [709] requires approximately 0.5 dB lower Signal-to-Noise Ratio (SNR) at a Bit Error Ratio (BER) of 10^{-4} than binary turbo codes when communicating using 8PSK over Additive White Gaussian Noise (AWGN) channels. TCM and TTCM invoked Set Partitioning (SP) based signal labelling, as will be discussed in the context of Figure 24.8 in order to achieve a higher Euclidean distance between the unprotected bits of the constellation, as we will show during our further discourse. It was shown in [495] that parallel trellis transitions can be associated with the unprotected information bits; as we will discuss in Figure 24.3(b), this reduced the decoding complexity. Furthermore, in our TCM and TTCM oriented investigations random symbol interleavers, rather than bit interleavers, were utilised, since these schemes operate on the basis of symbol, rather than bit, decisions.

Another coded modulation scheme distinguishing itself by utilising bit-based interleaving in conjunction with Gray signal constellation labelling is referred to as Bit-Interleaved Coded Modulation (BICM) [364]. More explicitly, BICM combines conventional convolutional codes with several independent bit interleavers, in order to increase the achievable diversity order to the binary Hamming distance of a code for transmission over fading channels [364], as will be shown in Section 24.6.1. The number of parallel bit interleavers equals the number of coded bits in a symbol for the BICM scheme proposed in [364]. The performance of BICM is better than that of TCM over uncorrelated or perfectly interleaved narrowband Rayleigh fading channels, but worse than that of TCM in Gaussian channels owing to the reduced Euclidean distance of the bit-interleaved scheme [364], as will be demonstrated in Section 24.6.1. Recently iterative joint decoding and demodulation assisted BICM (BICM-ID) was proposed in an effort to further increase the achievable performance [212,373,710–713], which uses SP-based signal labelling. The approach of BICM-ID is to increase the Euclidean distance of BICM, as will be shown in Section 24.7, and hence to exploit the full advantage of bit interleaving with the aid of soft-decision feedback-based iterative decoding [374].

In this chapter we embark on a study of the above-mentioned TCM, TTCM, BICM and BICM-ID schemes in the context of Phase Shift Keying (PSK) and Quadrature Amplitude Modulation (QAM) schemes. Specifically, the code generator polynomials of 4-level QAM (4QAM) or Quadrature PSK (QPSK), 8-level PSK (8PSK), 16-level QAM (16QAM) and 64-level QAM (64QAM) will be given in Tables 24.1, 24.2, 24.3 and 24.4.

24.3   Trellis-Coded Modulation

The basic idea of TCM is that instead of sending a symbol formed by $\bar{m}$ information bits, for example two information bits for 4PSK, we introduce a parity bit, while maintaining the same effective throughput of 2 bits/symbol by doubling the number of constellation points in the original constellation to eight, i.e. by extending it to 8PSK. As a consequence, the redundant bit can be absorbed by the expansion of the signal constellation, instead of accepting a 50% increase in the signalling rate, i.e. bandwidth. A positive coding gain is achieved when the detrimental effect of decreasing the Euclidean distance of the neighbouring phasors is outweighed by the coding gain of the convolutional coding incorporated.

Ungerböck has written an excellent tutorial paper [361], which fully describes TCM, and which this section is based upon. TCM schemes employ redundant non-binary modulation in combination with a finite state Forward Error Control (FEC) encoder, which governs the selection of the coded signal sequences. Essentially the expansion of the original symbol set absorbs more bits per symbol than required by the data rate, and these extra bit(s) are used by a convolutional encoder which restricts the possible state transitions amongst consecutive phasors to certain legitimate constellations. In the receiver, the noisy signals are decoded by a trellis-based soft-decision maximum likelihood sequence decoder. This takes the incoming data stream and attempts to map it onto each of the legitimate phasor sequences allowed by the constraints imposed by the encoder. The best fitting symbol sequence having the minimum Euclidean distance from the received sequence is used as the most likely estimate of the transmitted sequence.

Simple four-state TCM schemes, where the four-state adjective refers to the number of possible states that the encoder can be in, are capable of improving the robustness of

8PSK-based TCM transmission against additive noise in terms of the required SNR by 3dB compared to conventional uncoded 4PSK modulation. With the aid of more complex TCM schemes the coding gain can reach 6 dB [361]. As opposed to traditional error correction schemes, these gains are obtained without bandwidth expansion, or without the reduction of the effective information rate. Again, this is because the FEC encoder's parity bits are absorbed by expanding the signal constellation in order to transmit a higher number of bits per symbol. The term 'trellis' is used, because these schemes can be described by a state transition diagram similar to the trellis diagrams of binary convolutional codes [714]. The difference is that in the TCM scheme the trellis branches are labelled with redundant non-binary modulation phasors, rather than with binary code symbols.

24.3.1   TCM Principle

We now illustrate the principle of TCM using the example of a four-state trellis code for 8PSK modulation, since this relatively simple case assists us in understanding the principles involved.

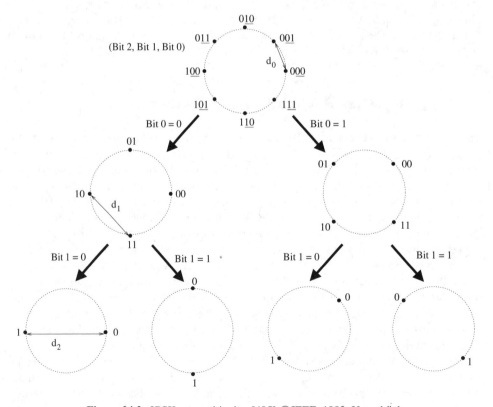

Figure 24.2: 8PSK set partitioning [495] ©IEEE, 1982, Ungerböck

The partitioned signal set proposed by Ungerböck [361, 495] is shown in Figure 24.2, where the binary phasor identifiers are now not Gray encoded. Observe in the figure that the Euclidean distance amongst constellation points is increased at every partitioning step.

The underlined last two bits, namely bit 0 and bit 1, are used for identifying one of the four partitioned sets, while bit 2 finally pinpoints a specific phasor in each partitioned set.

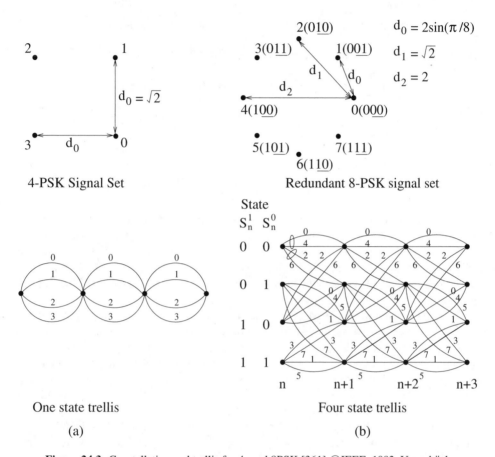

Figure 24.3: Constellation and trellis for 4- and 8PSK [361] ©IEEE, 1982, Ungerböck

The signal sets and state transition diagrams for (a) uncoded 4PSK modulation and (b) coded 8PSK modulation using four trellis states are given in Figure 24.3, while the corresponding four-state encoder-based modulator structure is shown in Figure 24.4. Observe that after differential encoding bit 2 is fed directly to the 8PSK signal mapper, while bit 1 is half-rate convolutionally encoded by a two-stage four-state linear circuit. The convolutional encoder adds the parity bit, bit 0, to the sequence, and again these two protected bits are used for identifying which constellation subset the bits will be assigned to, while the more widely spaced constellation points will be selected according to the unprotected bit 2.

The trellis diagram for 4PSK is a trivial one-state trellis, which portrays uncoded 4PSK from the viewpoint of TCM. Every connected path through the trellis represents a legitimate signal sequence where no redundancy-related transition constraints apply. In both systems, starting from any state, four transitions can occur, as required for encoding two bits/symbol. The four parallel transitions in the state trellis diagram of Figure 24.3(a) do not restrict the sequence of 4PSK symbols that can be transmitted, since there is no channel coding and

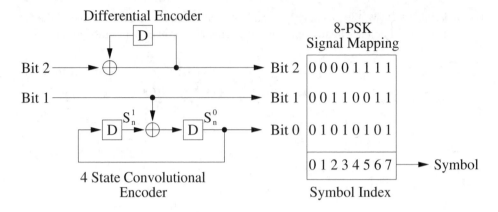

Figure 24.4: Encoder for the four-state 8PSK trellis [361] ©IEEE, 1982, Ungerböck

therefore all trellis paths are legitimate. Hence the optimum detector can only make nearest-phasor-based decisions for each individual symbol received. The smallest distance between the 4PSK phasors is $\sqrt{2}$, denoted as d_0, and this is termed the free distance of the uncoded 4PSK constellation. Each 4PSK symbol has two nearest neighbours at this distance. Each phasor is represented by a two-bit symbol and transitions from any state to any other state are legitimate.

The situation for 8PSK TCM is a little more complex. The trellis diagram of Figure 24.3(b) is constituted by four states according to the four possible states of the shift-register encoder of Figure 24.4, which we represent by the four vertically stacked bold nodes. Following the elapse of a symbol period a new two-bit input symbol arrives and the convolutional encoder's shift register is clocked. This event is characterised by a transition in the trellis from state S_n to state S_{n+1}, tracking one of the four possible paths corresponding to the four possible input symbols.

In the four-state trellis of Figure 24.3(b) associated with the 8PSK TCM scheme, the trellis transitions occur in pairs and the states corresponding to the bold nodes are represented by the shift-register states S_n^0 and S_n^1 in Figure 24.4. Owing to the limitations imposed by the convolutional encoder of Figure 24.4 on the legitimate set of consecutive symbols, only a limited set of state transitions associated with certain phasor sequence is possible. These limitations allow us to detect and to reject illegitimate symbol sequences, namely those which were not legitimately produced by the encoder, but rather produced by the error-prone channel. For example, when the shift register of Figure 24.4 is in state (0,0), only the transitions to the phasor points (0,2,4,6) are legitimate, while those to phasor points (1,3,5,7) are illegitimate. This is readily seen, because the linear encoder circuit of Figure 24.4 cannot produce a non-zero parity bit from the zero-valued input bits and hence the symbols (1,3,5,7) cannot be produced when the encoder is in the all-zero state. Observe in the 8PSK constellation of Figure 24.3(b) that the underlined bit 1 and bit 0 identify four twin-phasor subsets, where the phasors are opposite to each other in the constellation and hence have a high intra-subset separation. The unprotected bit 2 is then invoked for selecting the required phasor point within the subset. Since the redundant bit 0 constitutes also one of the shift-register state bits, namely S_n^0, from the initial states of $(S_n^1, S_n^0) = (0,0)$ or $(1,0)$ only the even-valued phasors (0,2,4,6) having $S_n^0 = 0$ can emerge, as also seen in Figure 24.3(b). Similarly, if we have

$(S_n^1, S_n^0) = (0,1)$ or $(1,1)$ associated with $S_n^0 = 1$ then the branches emerging from these lower two states of the trellis in Figure 24.3(b) can only be associated with the odd-valued phasors of (1,3,5,7).

There are other possible codes, which would result in, for example, four distinct transitions from each state to all possible successor states, but the one selected here proved to be the most effective [361]. Within the 8PSK constellation we have the following distances: $d_0 = 2\sin(\pi/8)$, $d_1 = \sqrt{2}$ and $d_2 = 2$. The 8PSK signals are assigned to the transitions in the four-state trellis in accordance with the following rules:

1) Parallel trellis transitions are associated with phasors having the maximum possible distance, namely (d_2), between them, which is characteristic of phasor points in the subsets (0,4), (1,5), (2,6) and (3,7). Since these parallel transitions belong to the same subset of Figure 24.3(b) and are controlled by the unprotected bit 2, symbols associated with them should be as far apart as possible.

2) All four-state transitions originating from, or merging into, any one of the states are labelled with phasors having a distance of *at least* $d_1 = \sqrt{2}$ between them. These are the phasors belonging to subsets (0,2,4,6) or (1,3,5,7).

3) All 8PSK signals are used in the trellis diagram with equal probability.

Observe that the assignment of bits to the 8PSK constellation of Figure 24.3(b) does not obey Gray coding and hence adjacent phasors can have arbitrary Hamming distances between them. The bit mapping and encoding process employed was rather designed for exploiting the high Euclidean distances between sets of points in the constellation. The underlined bit 1 and bit 0 of Figure 24.3(b) representing the convolutional codec's output are identical for all parallel branches of the trellis. For example, the branches labelled with phasors 0 and 4 between the identical consecutive states of (0,0) and (0,0) are associated with (bit 1) = 0 and (bit 0) = 0, while the uncoded bit 2 can be either '0' or '1', yielding the phasors 0 and 4, respectively. However, owing to appropriate code design this unprotected bit has the maximum protection distance, namely $d_2 = 2$, requiring the corruption of phasor 0 into phasor 4, in order to inflict a single bit error in the position of bit 2.

The effect of channel errors exhibits itself at the decoder by diverging from the trellis path encountered in the encoder. Let us consider the example of Figure 24.5, where the encoder generated the phasors 0-0-0 commencing from state (0,0), but owing to channel errors the decoder's trellis path was different from this, since the phasor sequence 2-1-2 was encountered. The so-called free distance of a TCM scheme can be computed as the lower one of two distances. Namely, the Euclidean distances between the phasors labelling the parallel branches in the trellis of Figure 24.3(b) associated with the uncoded bit(s), which is $d_2 = 2$ in our example, as well as the distances between trellis paths diverging and remerging after a number of consecutive trellis transitions, as seen in Figure 24.5 in the first and last of the four consecutive (0,0) states. The lower one of these two distances characterises the error resilience of the underlying TCM scheme, since the error event associated with it will be the one most frequently encountered owing to channel effects. Specifically, if the received phasors are at a Euclidean distance higher than half of the code's free distance from the transmitted phasor, an erroneous decision will be made. It is essential to ensure that by using an appropriate code design the number of decoded bit errors is minimised in the most likely

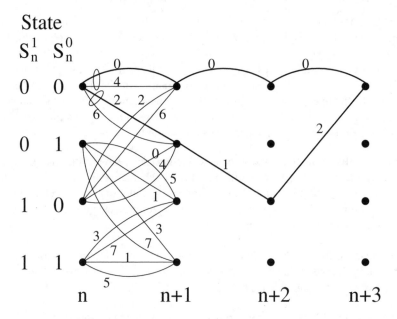

State

$S_n^1 \, S_n^0$

Figure 24.5: Diverging trellis paths for the computation of d_{free}. The parallel paths labelled by the symbols 0 and 4 are associated with the uncoded bits '0' and '1', respectively, as well as with the furthest phasors in the constellation of Figure 24.3(b)

error events, and this is akin to the philosophy of using Gray coding in a non-trellis-coded constellation.

The Euclidean distance between the phasors of Figure 24.3(b) associated with the parallel branches is $d_2 = 2$ in our example. The distance between the diverging trellis paths of Figure 24.3(b) labelled by the phasor sequences of 0-0-0 and 2-1-2 following the states $\{(0,0),(0,0),(0,0),(0,0)\}$ and $\{(0,0),(0,1),(1,0),(0,0)\}$ respectively, portrayed in Figure 24.5, is inferred from Figure 24.3(b) as d_1-d_0-d_1. By inspecting all the remerging paths of the trellis in Figure 24.3(b) we infer that this diverging path has the shortest accumulated Free Euclidean Distance (FED) that can be found, since all other diverging paths have higher accumulated FED from the error-free 0-0-0 path. Furthermore, this is the only path having the minimum free distance of $\sqrt{d_1^2 + d_0^2 + d_1^2}$. More specifically, the free distance of this TCM sequence is given by:

$$
\begin{aligned}
d_{free} &= min\{d_2; \sqrt{d_1^2 + d_0^2 + d_1^2}\} \\
&= min\{2; \sqrt{2 + (2.\sin\frac{\pi}{8})^2 + 2}\}.
\end{aligned}
\tag{24.1}
$$

Explicitly, since the term under the square root in Equation 24.1 is higher than $d_2 = 2$, the free distance of this TCM scheme is given ultimately by the Euclidean distance between the parallel trellis branches associated with the uncoded bit 2, i.e.:

$$
d_{free} = 2.
\tag{24.2}
$$

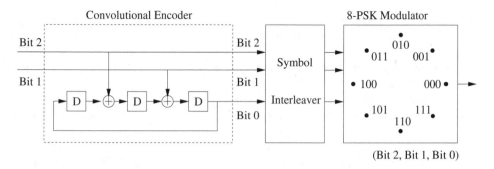

(Bit 2, Bit 1, Bit 0)

Figure 24.6: Ungerböck's RSC encoder and modulator forming the TCM encoder. The SP-based mapping of bits to the constellation points was highlighted in Figure 24.2.

The free distance of the uncoded 4PSK constellation of Figure 24.3(a) was $d_0 = \sqrt{2}$ and hence the employment of TCM has increased the minimum distance between the constellation points by a factor of $g = \frac{d_{free}^2}{d_0^2} = \frac{2^2}{(\sqrt{2})^2} = 2$, which corresponds to 3 dB. There is only one nearest-neighbour phasor at $d_{free} = 2$, corresponding to the π-rotated phasor in Figure 24.3(b). Consequently the phasor arrangement can be rotated by π, while retaining all of its properties, but other rotations are not admissible.

The number of erroneous decoded bits induced by the diverging path 2-1-2 is seen from the phasor constellation of Figure 24.3(b) to be 1-1-1, yielding a total of three bit errors. The more likely event of a bit 2 error, which is associated with a Euclidean distance of $d_2 = 2$, yields only a single bit error.

Soft-decision-based decoding can be accomplished in two steps. The first step is known as subset decoding, where within each phasor subset assigned to parallel transitions, i.e. to the uncoded bit(s), the phasor closest to the received channel output in terms of Euclidean distance is determined. Having resolved which of the parallel paths was more likely to have been encountered by the encoder, we can remove the parallel transitions, hence arriving at a conventional trellis. In the second step the Viterbi algorithm is used for finding the most likely signal path through the trellis with the minimum sum of squared Euclidean distances from the sequence of noisy channel outputs received. Only the signals already selected by the subset decoding are considered. For a description of the Viterbi algorithm the reader is referred to references [504, 715].

24.3.2 Optimum TCM Codes

Ungerböck's TCM encoder is a specific convolutional encoder selected from the family of Recursive Systematic Convolutional (RSC) codes [495], which attaches one parity bit to each information symbol. Only $\tilde{m}$ out of $\bar{m}$ information bits are RSC encoded and hence only $2^{\tilde{m}}$ branches will diverge from and merge into each trellis state. When not all information bits are RSC encoded, i.e. $\tilde{m} < \bar{m}$, $2^{\bar{m}-\tilde{m}}$ parallel transitions are associated with each of the $2^{\tilde{m}}$ branches. Therefore a total of $2^{\tilde{m}} \times 2^{\bar{m}-\tilde{m}} = 2^{\bar{m}}$ transitions occur at each trellis stage. The memory length ν of a code defines the number of shift-register stages in the encoder. Figure 24.6 shows the TCM encoder using an eight-state Ungerböck code [495], which has a high FED for the sake of attaining a high performance over AWGN channels.

It is a systematic encoder, which attaches an extra parity bit to the original 2-bit information word. The resulting 3-bit codewords generated by the 2-bit input binary sequence are then interleaved by a symbol interleaver in order to disperse the bursty symbol errors induced by the fading channel. Then, these 3-bit codewords are modulated onto one of the $2^3 = 8$ possible constellation points of an 8PSK modulator.

The connections between the information bits and the modulo-2 adders, as shown in Figure 24.6, are given by the generator polynomials. The coefficients of these polynomials are defined as:

$$H^j(D) := h_\nu^j.D^\nu + h_{\nu-1}^j.D^{\nu-1} + \ldots + h_1^j.D + h_0^j, \tag{24.3}$$

where D represents the delay due to one register stage. The coefficient h_i^j takes the value of '1', if there is a connection at a specific encoder stage or '0', if there is no connection. The polynomial $H^0(D)$ is the feedback generator polynomial and $H^j(D)$ for $j \geq 1$ is the generator polynomial associated with the jth information bit. Hence, the generator polynomial of the encoder in Figure 24.6 can be described in binary format as:

$$\begin{aligned} H^0(D) &= 1001 \\ H^1(D) &= 0010 \\ H^2(D) &= 0100, \end{aligned}$$

or equivalently in octal format as:

$$\begin{aligned} \mathbf{H(D)} &= \begin{bmatrix} H^0(D) & H^1(D) & H^2(D) \end{bmatrix} \\ &= \begin{bmatrix} 11 & 02 & 04 \end{bmatrix}. \end{aligned} \tag{24.4}$$

Ungerböck suggested [495] that all feedback polynomials should have coefficients $h_\nu^0 = h_0^0 = 1$. This guarantees the realisability of the encoders shown in Figures 24.4 and 24.6. Furthermore, all generator polynomials should also have coefficients $h_\nu^j = h_0^j = 0$ for $j > 0$. This ensures that at time n the input bits of the TCM encoder have no influence on the parity bit to be generated, nor on the input of the first binary storage element in the encoder. Therefore, whenever two paths diverge from or merge into a common state in the trellis, the parity bit must be the same for these transitions, whereas the other bits differ in at least one bit [495]. Phasors associated with diverging and merging transitions therefore have at least a distance of d_1 between them, as we can see from Figure 24.3(b). Table 24.1 summarises the generator polynomials of some TCM codes, which were obtained with the aid of an exhaustive computer search conducted by Ungerböck [361], where $\tilde{m}$ ($\leq \bar{m}$) indicates the number of information bits to be encoded, from the $\bar{m}$ information bits in a symbol.

24.3.3 TCM Code Design for Fading Channels

It was shown in Section 24.3.1 that the design of TCM for transmission over AWGN channels is motivated by the maximisation of the FED, d_{free}. By contrast, the design of TCM concerned for transmission over fading channels is motivated by minimising the length of the shortest error event path and the product of the branch distances along that particular path [708].

Code	State, ν	$\tilde{m}$	$H^0(D)$	$H^1(D)$	$H^2(D)$
4QAM	8, 3	1	13	06	-
4QAM	64, 6	1	117	26	-
8PSK	8, 3	2	11	02	04
8PSK	32, 5	2	45	16	34
8PSK	64, 6	2	103	30	66
8PSK	128, 7	2	277	54	122
8PSK	256, 8	2	435	72	130
16QAM	64, 6	2	101	16	64

Table 24.1: Ungerböck's TCM codes [361, 495, 716, 717], where ν denotes the code memory and octal format is used for representing the generator polynomial coefficients.

The average bit error probability of TCM using M-ary PSK (MPSK) [495] for transmission over Rician channels at high SNRs is given by [708]:

$$P_b \cong \frac{1}{B} C \left(\frac{(1 + \bar{K})e^{-\bar{K}}}{E_s/N_0} \right)^{\mathsf{L}} ; E_s/N_0 \gg \bar{K} \qquad (24.5)$$

where C is a constant that depends on the weight distribution of the code, which quantifies the number of trellises associated with all possible Hamming distances measured with respect to the all-zero path [370]. The variable B in Equation 24.5 is the number of binary input bits of the TCM encoder during each transmission interval, i.e. the information bits per symbol, while $\bar{K}$ is the Rician fading parameter [370] and E_s/N_0 is the channel's symbol energy to noise spectral density ratio. Furthermore, L is defined as the 'length' of the shortest error event path in [718] or as the Effective Code Length (ECL) in [701, 719] or as the code's diversity in [708]. Explicitly, L is expressed as the number of erroneously decoded TCM symbols asociated with the shortest error event path. Note that, in conventional TCM each trellis branch is labelled by one TCM symbol. Therefore, L can be expressed as the number of trellis branches having an erroneously decoded symbol, in the shortest error event path. Most of the time, L is equal to the number of trellis branches on this path. It is clear from Equation 24.5 that P_b varies inversely proportionally with $(E_s/N_0)^{\mathsf{L}}$ and this ratio can be increased by increasing the code's diversity [708]. More specifically, in [718], the authors pointed out that the shortest error event paths are not necessarily associated with the minimum accumulated FED error events. For example, let the all-zero path be the correct path. Then the code characterised by the trellis seen in Figure 24.7 exhibits a minimum squared FED of:

$$\begin{aligned} d_{free}^2 &= d_1^2 + d_0^2 + d_1^2 \\ &= 4.585, \end{aligned} \qquad (24.6)$$

from the 0-0-0 path associated with the transmission of three consecutive 0 symbols from the path labelled with the transmitted symbols of 6-7-6. However, this is not the shortest error event path, since its length is $\mathsf{L} = 3$, which is longer than the path labelled with transmitted symbols of 2-4, which has a length of $\mathsf{L} = 2$ and a FED of $d_{free}^2 = d_1^2 + d_2^2 = 6$. Hence, the 'length' of the shortest error event path is $\mathsf{L} = 2$ for this code, which, again, has a squared Euclidean distance of 6. In summary, the number of bit errors associated with the above

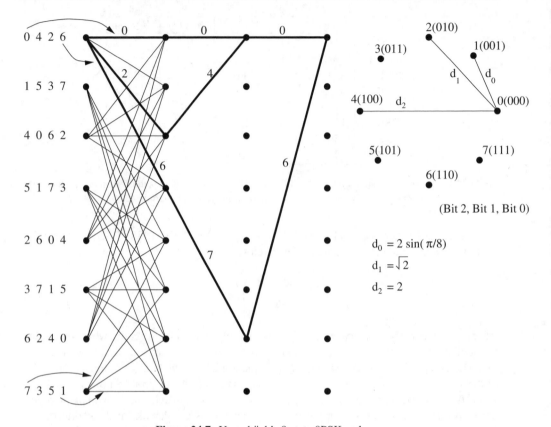

Figure 24.7: Ungerböck's 8-state 8PSK code

$L = 3$ and $L = 2$ shortest error event paths is seven and two, respectively, clearly favouring the $L = 2$ path, which had a higher accumulated FED of 6 than that of the 4.585 FED of the $L = 3$ path. Hence, it is worth noting that if the code was designed based on the minimum FED, it may not minimise the number of bit errors. Hence, as an alternative design approach, in Section 24.6 we will study BICM, which relies on the shortest error event path L or the bit-based Hamming distance of the code and hence minimises the BER.

The design of coded modulation schemes is affected by a variety of factors. A high squared FED is desired for AWGN channels, while a high ECL and a high minimum product distance are desired for fading channels [708]. In general, a code's diversity or ECL is quantified in terms of the length of the shortest error event path L, which may be increased, for example, by simple repetition coding, although at the cost of reducing the effective data rate proportionally. Alternatively, space-time-coded multiple transmitter/receiver structures can be used, which increase the scheme's cost and complexity. Finally, simple interleaving can be invoked, which induces latency. In our approach, symbol-based interleaving is employed in order to increase the code's diversity.

24.3.4   Set Partitioning

As we have seen in Figure 24.5, if higher-order modulation schemes, such as 16QAM or 64QAM, are used, parallel transitions may appear in the trellis diagram of the TCM scheme, when not all information bits are convolutional channel encoded or when the number of states in the convolutional encoder has to be kept low for complexity reasons. As noted before, in order to avoid encountering high error probabilities, the parallel transitions should be assigned to constellation points exhibiting a high Euclidean distance. Ungerböck solved this problem by introducing the set partitioning technique. Specifically, the signal set is split into a number of subsets, such that the minimum Euclidean distance of the signal points in the new subset is increased at every partitioning step.

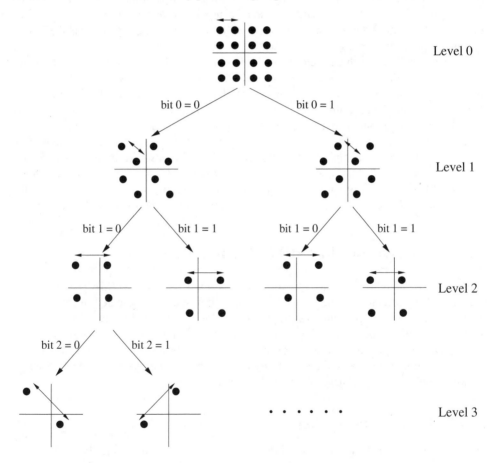

Figure 24.8: Set partitioning of a 16QAM signal constellation. The minimum Euclidean distance at a partition level is denoted by the line between the signal points [495] ©IEEE, 1982, Ungerböck

In order to elaborate a little further, Figure 24.8 illustrates the set partitioning of 16QAM. Here we used the $R = \frac{3}{4}$-rate code of Table 24.1. This is a relatively high-rate code, which would not be sufficiently powerful if we employed it for protecting all three original information bits. Moreover, if we protect for example two out of the three information bits, we can use a more potent $\frac{2}{3}$-rate code for the protection of the more vulnerable two information bits and leave the most error-resilient bit of the 4-bit constellation unprotected. This is justifiable, since we can observe in Figure 24.8 that the minimum Euclidean distance of the constellation points increases from Level 0 to Level 3 of the constellation partitioning tree. This indicates that the bits labelling or identifying the specific partitions have to be protected by the RSC code, since they label phasors that have a low Euclidean distance. By contrast, the intra-set distance at Level 3 is the highest, suggesting a low probability of corruption. Hence the corresponding bit, bit 3, can be left unprotected. The partitioning in Figure 24.8 can be continued, until there is only one phasor or constellation point left in each subset. The intra-subset distance increases as we traverse down the partition tree. The first partition level, *Level* 0, is labelled by the parity bit, and the next two levels by the coded bits. Finally, the uncoded bit labels the lowest level, *Level* 3, in the constellation, which has the largest minimum Euclidean distance.

Conventional TCM schemes are typically decoded/demodulated with the aid of the appropriately modified Viterbi Algorithm (VA) [720]. Furthermore, the VA is a maximum likelihood sequence estimation algorithm, which does not guarantee that the Symbol Error Ratio (SER) is minimised, although it achieves a performance near the minimum SER. By contrast, the symbol-based MAP algorithm [709] guarantees the minimum SER, albeit at the cost of a significantly increased complexity. Hence the symbol-based MAP algorithm has been used for the decoding of TCM sequences. We will, however, in Section 24.5, also consider Turbo TCM (TTCM), where instead of the VA-based sequence estimation, symbol-by-symbol-based soft information has to be exchanged between the TCM decoders of the TTCM scheme. Hence in the next section we will present the symbol-based MAP algorithm.

24.4   The Symbol-based MAP Algorithm

In this section, the non-binary or symbol-based Maximum-A-Posteriori (MAP) decoding algorithm will be presented. The binary MAP algorithm was first presented in [582], while the non-binary MAP algorithm was proposed in [709]. A reduced-complexity version of the MAP algorithm, operating in the logarithmic domain (log-domain) after transforming the operands and the operations to this domain will also be presented. In our forthcoming discourse we use $p(x)$ to denote the probability of the event x, and, given a symbol sequence y_k, we denote by y_a^b the sequence of symbols given by $y_a, y_{a+1}, \ldots, y_b$.

24.4.1   Problem Description

The problem that the MAP algorithm has to solve is presented in Figure 24.9. An information source produces a sequence of N information symbols u_k, $k = 1, 2, \ldots, N$. Each information symbol can assume M different values, i.e. $u_k \in \{0, 1, \ldots, M - 1\}$, where M is typically a power of two, so that each information symbol carries $\bar{m} = log_2 M$ information bits. We assume here that the symbols are to be transmitted over an AWGN channel. To this end, the $\bar{m}$-bit symbols are first fed into an encoder for generating a sequence of N channel symbols $x_k \in X$, where X denotes the set of complex values belonging to some phasor constellations such as an increased-order QAM or PSK constellation, having M possible val-

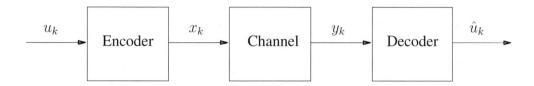

Figure 24.9: The transmission system

ues carrying $\mathsf{m} = log_2\mathsf{M}$ bits. Again, the channel symbols are transmitted over an AWGN channel and the received symbols are:

$$y_k = x_k + n_k, \tag{24.7}$$

where n_k represents the complex AWGN samples. The received symbols are fed to the decoder, which has the task of producing an estimate $\hat{u}_k$ of the $2^{\bar{\mathsf{m}}}$-ary information sequence, based on the 2^{m}-ary received sequence, where $\mathsf{m} > \bar{\mathsf{m}}$. If the goal of the decoder is that of minimising the number of symbol errors, where a symbol error occurs when $u_k \neq \hat{u}_k$, then the best decoder is the MAP decoder [582]. This decoder computes the A Posteriori Probability (APP) $A_{k,m}$ for every $2^{\bar{\mathsf{m}}}$-ary information symbol u_k that the information symbol value was m given the received sequence, i.e. computes $A_{k,m} = p(u_k = m|y_1^N)$, for $m = 0, 1, \ldots, M - 1$, $k = 1, 2, \ldots, N$. Then it decides that the information symbol was the one having the highest probability, i.e. $\hat{u}_k = m$ if $A_{k,m} \geq A_{k,i}$ for $i = 0 \ldots M - 1$. In order to realise a MAP decoder one has to devise a suitable algorithm for computing the APP.

In order to compute the APP, we must specify how the encoder operates. We consider a trellis encoder. The operation of a trellis encoder can be described by its trellis. The trellis seen in Figure 24.10, is constituted by $(N + 1) \cdot S$ nodes arranged in $(N + 1)$ columns of S nodes. There are M branches emerging from each node, which arrive at nodes in the immediately following column. The trellis structure repeats itself identically between each pair of columns.

It is possible to identify a set of paths originating from the nodes in the first column and terminating in a node of the last column. Each path will comprise exactly N branches. When employing a trellis-encoder, the input sequence unambiguously determines a single path in the trellis. This path is identified by labelling the M branches emerging from each node by the M possible values of the original information symbols, although only the labelling of the first branch at $m = 0$ and the last branch at $m = M - 1$ are shown in Figure 24.10 due to space limitations. Then, commencing from a specified node in the first column, we use the first input symbol, u_1, to decide which branch is to be chosen. If $u_1 = m$, we choose the branch labelled with m, and move to the corresponding node in the second column that this branch leads to. In this node we use the second information symbol, u_2, for selecting a further branch and so on. In this way the information sequence identifies a path in the trellis. In order to complete the encoding operation, we have to produce the symbols to be transmitted over the channel, namely $x_1, x_2, \ldots, x_N$ from the information symbols $u_1, u_2, \ldots, u_N$. To this end we add a second label to each branch, which is the corresponding phasor constellation point that is transmitted when the branch is encountered.

In a trellis it is convenient to attach a time index to each column, from 0 to N, and to

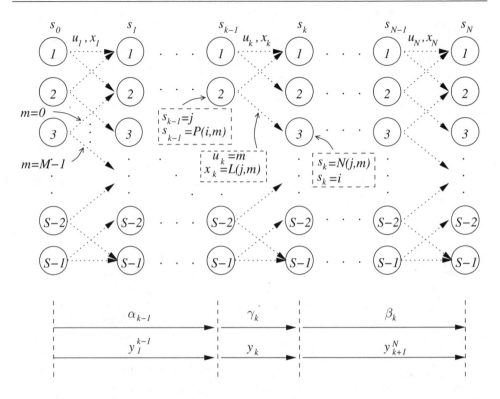

Figure 24.10: The non-binary trellis and its labelling, where there are M branches emerging from each node.

number the nodes in each column from 0 to $S - 1$. This allows us to introduce the concept of trellis states at time k. Specifically, during the encoding process, we say that the trellis is in state i at time k, and write $s_k = i$, if the path determined by the information sequence crosses the i-th node of the k-th column. The structure of a trellis encoder is specified by two functions. The first function is $N(j, m) \in \{0, 1, \ldots, S - 1\}$, which specifies the trellis' next state, namely $s_k = N(j, m)$, when the information symbol is $u_k = m$ and the previous state is $s_{k-1} = j$ as seen in Figure 24.10. In order to specify the symbol transmitted when this branch is encountered, we use the function $L(j, m) \in X$. To summarize, there is a branch leading from state $s_{k-1} = j$ to state $s_k = N(j, m)$, which is encountered if the input symbol is $u_k = m$, and the corresponding transmitted symbol is $L(j, m)$. It is useful to consider a third function, $P(i, m) \in \{0, 1, \ldots, S - 1\}$ specifying the previous state $s_{k-1} = P(i, m)$ of the trellis when the present state is $s_k = i$, and the last original information symbol is $u_k = m$ as seen in Figure 24.10. The aim of the MAP decoding algorithm is to find the path in the trellis that is associated with the most likely transmitted symbols, i.e. that of minimising the Symbol Error Ratio (SER). By contrast, the VA-based detection of TCM signals aims for identifying the most likely transmitted symbol sequence, which does not automatically guarantee attaining the minimum SER.

24.4.2   Detailed Description of the Symbol-based MAP Algorithm

Having described the problem to be solved by the MAP decoder and the encoder structure, we now seek an algorithm capable of computing the APP, i.e. $A_{k,m} = p(u_k = m | y_1^N)$. The easiest way of computing these probabilities is by determining the sum of a different set of probabilities, namely $p(u_k = m, s_k = i, s_{k-1} = j | y_1^N)$, where again, y_1^N denotes the symbol sequence $y_1, y_2, \ldots, y_N$. This is because we can devise a recursive way of computing the second set of probabilities, as we traverse through the trellis from state to state which reduces the detection complexity. Thus, we write:

$$A_{k,m} = p(u_k = m | y_1^N) = \sum_{i,j=0}^{S-1} p(u_k = m, s_k = i, s_{k-1} = j | y_1^N), \qquad (24.8)$$

where the summation implies adding all probabilities associated with the nodes j and i labelled by $u_k = m$ and the problem is now that of computing $p(u_k = m, s_k = i, s_{k-1} = j | y_1^N)$. As a preliminary consideration we note that this probability is zero, if the specific branch of the trellis emerging from state j and merging into state i is not labelled with the input symbol m. Hence, we can eliminate the corresponding terms of the summation. Thus, upon denoting the specific set of pairs i, j, by I_m for which a trellis branch labelled with m exists that traverses from state j to state i, we can rewrite Equation 24.8 as:

$$A_{k,m} = \sum_{i,j \in I_m} p(u_k = m, s_k = i, s_{k-1} = j | y_1^N). \qquad (24.9)$$

If $i, j \in I_m$, then we can compute the probabilities $p(u_k = m, s_k = i, s_{k-1} = j | y_1^N)$ as [582, 721]:

$$p(u_k = m, s_k = i, s_{k-1} = j | y_1^N) = \frac{1}{p(y_1^N)} \cdot \beta_k(i) \cdot \alpha_{k-1}(j) \cdot \gamma_k(j, m), \qquad (24.10)$$

where

$$\begin{aligned}
\beta_k(i) &= p(y_{k+1}^N | s_k = i) \\
\alpha_{k-1}(j) &= p(y_1^{k-1}, s_{k-1} = j) \\
\gamma_k(j, m) &= p(y_k, u_k = m | s_{k-1} = j).
\end{aligned} \qquad (24.11)$$

In order to simplify our discourse, we defer the proof of Equation 24.10 to Section 24.4.3, where we also show how the $\alpha_{k-1}(j)$ values and the $\beta_k(i)$ values can be efficiently computed using the $\gamma_k(j, m)$ values. In our forthcoming discourse we study the $\gamma_k(j, m)$ values and further simplify Equation 24.9.

The first simplification is to note that we do not necessarily need the exact $A_{k,m}$ values, but only their ratios. In fact, for a fixed k, the vector $A_{k,m}$, being a probability vector, must sum to unity. Thus, by normalising the sum in Equation 24.9 to unity, we can compute the exact value of $A_{k,m}$ from $\bar{A}_{k,m}$ with the aid of:

$$\bar{A}_{k,m} = C_k \cdot A_{k,m}. \qquad (24.12)$$

For this reason we will omit the common normalisation factor of $C_k = \frac{1}{p(y_1^N)}$ in Equation 24.10. Then, upon substituting Equation 24.10 into Equation 24.9 we have:

$$\bar{A}_{k,m} = \sum_{i,j \in I_m} \beta_k(i) \cdot \alpha_{k-1}(j) \cdot \gamma_k(j,m). \tag{24.13}$$

A second simplification is to note that in Equation 24.13 the value of i is uniquely specified by the pair j and m, since $i, j \in I_m$. Specifically, since i is the state reached after emerging from state j when the input symbol is m, we have $i = N(j,m)$ where $N(j,m)$ was defined at the end of Section 24.4.1. Thus, we can rewrite [1] Equation 24.13 as:

$$\bar{A}_{k,m} = \sum_{j=0}^{S-1} \beta_k(N(j,m)) \cdot \alpha_{k-1}(j) \cdot \gamma_k(j,m). \tag{24.14}$$

Before we proceed, it is worth presenting Bayes' rule, which is applied repeatedly throughout this section. This rule gives the joint probability of "a and b", $P(a,b)$, in terms of the conditional probability of "a given b", $P(a|b)$, as:

$$P(a,b) = P(a|b) \cdot P(b) = P(b|a) \cdot P(a). \tag{24.15}$$

Two useful consequences of Bayes' rule are:

$$P(a,b,c) = P(a|b,c) \cdot P(b,c) \tag{24.16}$$

and

$$
\begin{aligned}
P(a,b|c) &= \frac{P(a,b,c)}{P(c)} \\
&= \frac{P(a,b,c)}{P(b,c)} \cdot \frac{P(b,c)}{P(c)} \\
&= P(a|b,c).P(b|c).
\end{aligned} \tag{24.17}
$$

Let us now consider the term $\gamma_k(j,m) = p(y_k, u_k = m | s_{k-1} = j)$ of Equation 24.11, which can be rewritten using the relationship of Equation 24.17 as:

$$\gamma_k(j,m) = p(y_k, u_k = m | s_{k-1} = j) = p(y_k | u_k = m, s_{k-1} = j) \cdot p(u_k = m | s_{k-1} = j) \tag{24.18}$$

Let us now study the multiplicative terms at the right of Equation 24.18, where $p(y_k | u_k = m, s_{k-1} = j)$ is the probability that we receive y_k, when the branch emerging from state $s_{k-1} = j$ of Figure 24.10 labelled with the information symbol $u_k = m$ is encountered. When this branch is encountered, the symbol transmitted is $x_k = L(j,m)$, as seen in Figure 24.10. Thus, the probability of receiving the sample y_k, given that the previous state was

---

[1]Equivalently, we could note that in Equation 24.13, we have $j = P(i,m)$, since $i, j \in I_m$ and rewrite Equation 24.13 as:

$$\bar{A}_{k,m} = \sum_{i=0}^{S-1} \beta_k(i) \cdot \alpha_{k-1}(P(i,m)) \cdot \gamma_k(P(i,m),m).$$

$s_{k-1} = j$ and the transition symbol encountered was $u_k = m$ can be written as:

$$p(y_k|u_k = m, s_{k-1} = j) = p(y_k|x_k = L(j,m)) = \eta_k(j,m). \qquad (24.19)$$

By remembering that $y_k = x_k + n_k$, and that n_k is the complex AWGN, we can compute $\eta_k(j,m)$ as [722]:

$$\eta_k(j,m) = e^{\frac{-|y_k - L(j,m)|^2}{2\sigma^2}}, \qquad (24.20)$$

where $\sigma^2 = N_0/2$ is the noise's variance and N_0 is the noise's Power Spectral Density (PSD). In verbal terms, Equation 24.20 indicates that the probability expressed in Equation 24.19 is a function of the distance between the received noisy sample y_k and the transmitted noiseless sample $x_k = L(j,m)$. Observe in Equation 24.20 that we dropped the multiplicative factor of $\frac{1}{2\pi\sigma^2}$, since it constitutes another scaling factor, which can be viewed as comprised in the constant C_k associated with $\bar{A}_{k,m} = C_k A_{k,m}$. As to the second multiplicative term at the right-hand side of Equation 24.18, note that $p(u_k = m|s_{k-1} = j) = p(u_k = m)$, since the original information to be transmitted is independent of the previous trellis state. The probabilities:

$$\Pi_{k,m} = p(u_k = m) \qquad (24.21)$$

are the *a priori* probabilities of the information symbols. Typically the information symbols are independent and equiprobable, hence $\Pi_{k,m} = 1/M$. However, if we have some prior knowledge about the transmitted symbols, this can be used as their *a priori* probability. As we will see, a turbo decoder will have some *a priori* knowledge about the transmitted symbols after the first iteration. We now rewrite Equation 24.18 using Equation 24.19 and the *a priori* probabilities as:

$$\gamma_k(j,m) = \Pi_{k,m} \cdot \eta_k(j,m). \qquad (24.22)$$

Then, by substituting Equation 24.22 into Equation 24.14 and exchanging the order of summations we can portray the APPs in their final form, yielding:

$$\bar{A}_{k,m} = \Pi_{k,m} \cdot \sum_{j=0}^{S-1} \beta_k(N(j,m)) \cdot \alpha_{k-1}(j) \cdot \eta_k(j,m). \qquad (24.23)$$

24.4.3 Recursive Metric Update Formulae

In this section we will deduce Equation 24.10. Figure 24.10 visualises the intervals, namely α_{k-1}, γ_k and β_k in the trellis for a given k, as well as the symbols received in these intervals, namely $\mathbf{y}_1^{k-1}, \mathbf{y}_k$ and $\mathbf{y}_{k+1}^N$, where γ is the so-called branch transition metric, α is the so-called forward recursive variable and β is the so-called backward recursive variable. As the first step of decoding, we have to compute all the values of γ_k using Equation 24.22, which depend only on the current received symbol y_k, for $k = 1, \cdots, N$. Then, we can compute α_{k-1} and β_k based on these γ_k values with the aid of Equation 24.11.

Now, we commence our discourse by considering the additive terms in Equation 24.9, which we formulated with the aids of Bayes' rule in Equations 24.15 to 24.17 as:

$$p(u_k = m, s_k = i, s_{k-1} = j | y_1^N) = \frac{1}{p(y_1^N)} \cdot p(y_1^N, u_k = m, s_k = i, s_{k-1} = j), \quad (24.24)$$

and consider the term $p(u_k = m, s_k = i, s_{k-1} = j, y_1^N)$. We can write

$$p(u_k = m, s_k = i, s_{k-1} = j, y_1^N) = p(u_k = m, s_k = i, s_{k-1} = j, y_1^k, y_{k+1}^N)$$

$$= p(y_{k+1}^N | u_k = m, s_k = i, s_{k-1} = j, y_1^k) \cdot p(u_k = m, s_k = i, s_{k-1} = j, y_1^k). \quad (24.25)$$

Let us now simplify the first multiplicative term of Equation 24.25 by noting that if the current state s_k is known, the decoded output sequence probability is not affected by either the previous state s_{k-1}, the input symbol u_k or the previous received symbol sequence y_1^k. Thus, Equation 24.25 can be rewritten as

$$p(u_k = m, s_k = i, s_{k-1} = j, y_1^N) = p(y_{k+1}^N | s_k = i) \cdot p(u_k = m, s_k = i, s_{k-1} = j, y_1^k)$$

$$= p(y_{k+1}^N | s_k = i) \cdot p(y_1^{k-1} | u_k = m, s_k = i, s_{k-1} = j, y_k) \cdot p(y_k, u_k = m, s_k = i, s_{k-1} = j). \quad (24.26)$$

Again, we simplify the second multiplicative term of Equation 24.26 by noting that, if s_{k-1} is known, the received symbol sequence y_1^{k-1} is not affected by either s_k, u_k or y_k, hence we can rewrite Equation 24.26 as

$$p(u_k = m, s_k = i, s_{k-1} = j | y_1^N) = p(y_{k+1}^N | s_k = i) \quad (24.27)$$
$$\cdot \; p(y_1^{k-1} | s_{k-1} = j) \cdot p(y_k, u_k = m, s_k = i, s_{k-1} = j).$$

By multiplying and dividing the second and the third multiplicative term, respectively, with $p(s_{k-1} = j)$, we can rearrange Equation 24.28 to

$$p(u_k = m, s_k = i, s_{k-1} = j | y_1^N) = p(y_{k+1}^N | s_k = i) \quad (24.28)$$
$$\cdot \; p(y_1^{k-1} s_{k-1} = j) \cdot p(y_k, u_k = m, s_k = i | s_{k-1} = j).$$

Then, by introducing

$$\beta_k(i) = p(y_{k+1}^N | s_k = i) \quad (24.29)$$

and

$$\alpha_{k-1}(j) = p(y_1^{k-1}, s_{k-1} = j) \quad (24.30)$$

we have

$$p(u_k = m, s_k = i, s_{k-1} = j | y_1^N) = \beta_k(i) \cdot \alpha_{k-1}(j) \cdot p(y_k, u_k = m, s_k = i | s_{k-1} = j). \quad (24.31)$$

If $i, j \notin I_m$, the above probability is zero, since no branch exists leading from state j to state i, when the information symbol is m. Thus, we assume $i, j \in I_m$. In this case we can simplify the second multiplicative term of Equation 24.31 as

$$p(y_k, u_k = m, s_k = i | s_{k-1} = j) = p(y_k, u_k = m | s_{k-1} = j). \quad (24.32)$$

Upon defining
$$\gamma_k(j, m) = p(y_k, u_k = m | s_{k-1} = j) \qquad (24.33)$$
and upon substituting Equation 24.32 and Equation 24.33 in Equation 24.31 we obtain

$$p(u_k = m, s_k = i, s_{k-1} = j, y_1^N) = \beta_k(i) \cdot \alpha_{k-1}(j) \cdot \gamma_k(j, m), \qquad (24.34)$$

and upon substituting Equation 24.34 in Equation 24.24 we obtain Equation 24.10, QED.

24.4.3.1 Backward Recursive Computation of $\beta_k(i)$

Let us now highlight how the values $\beta_k(i)$ can be used, in order to 'backward' recursively compute $\beta_{k-1}(P(i, m) = j)$ from $\beta_k(i)$. With the aid of the definition in Equation 24.29 we have
$$\beta_{k-1}(j) = p(y_k^N | s_{k-1} = j) = p(y_k, y_{k+1}^N | s_{k-1} = j), \qquad (24.35)$$
which can be reformulated in terms of $p(y_k, y_{k+1}^N, s_k = i | s_{k-1} = j)$, by summing these probabilities for all the trellis states $i = 0 \ldots (S-1)$, which are reached from $s_{k-1} = j$, yielding:

$$\beta_{k-1}(j) = \sum_{i=0}^{S-1} p(y_k, y_{k+1}^N, s_k = i | s_{k-1} = j).$$

This can be reformatted using Equations 24.15-24.17 as:

$$\beta_{k-1}(j) = \sum_{i=0}^{S-1} p(y_{k+1}^N | y_k, s_k = i, s_{k-1} = j) \cdot p(y_k, s_k = i | s_{k-1} = j). \quad (24.36)$$

With reference to the trellis diagram of Figure 24.10 we note that the received symbol sequence y_{k+1}^N is not affected by y_k and s_{k-1}, if s_k is given. Thus, from Equations 24.36, 24.29 and 24.15-24.17 we obtain:

$$\beta_{k-1}(j) = \sum_{i=0}^{S-1} p(y_{k+1}^N | s_k = i) \cdot p(y_k, s_k = i | s_{k-1} = j) = \sum_{i=0}^{S-1} \beta_k(i) \cdot p(y_k, s_k = i | s_{k-1} = j).$$
$$(24.37)$$

Let us now consider the summation over the index range of $i = 0 \ldots (S-1)$, and note that for a fixed j the probability $p(y_k, s_k = i | s_{k-1} = j)$ will be non-zero only, if a branch exists that leads from state j to state i. Thus, there are only M specific values of i, which contribute to the summation, namely the values of $i = N(j, m)$ for some m. We can thus rewrite Equation 24.37 as

$$\beta_{k-1}(j) = \sum_{m=0}^{M-1} \beta_k(N(j, m)) \cdot p(y_k, s_k = N(j, m) | s_{k-1} = j), \qquad (24.38)$$

where for the second multiplicative term we have $p(y_k, s_k = N(j, m)|s_{k-1} = j) = p(y_k, u_k = m|s_{k-1} = j) = \gamma_k(j, m)$. Hence we can write

$$\beta_{k-1}(j) = \sum_{m=0}^{M-1} \beta_k(N(j, m))\gamma_k(j, m). \tag{24.39}$$

Equation 24.39 facilitates the 'backward' recursive calculation of the $\beta_k(N(j, m) = i)$ values, commencing from $\beta_N(N(j, m) = i)$. In order to determine this boundary value we note that by using Equation 24.39 for computing $\beta_{N-1}(j)$ we have

$$\beta_{N-1}(j) = p(y_N|s_{N-1} = j) = \sum_{m=0}^{M-1} \beta_N(N(j, m)) \cdot p(y_N, u_N = m|s_{N-1} = j) \tag{24.40}$$

and that in order to render the above expression true we have to choose

$$\beta_N(N(j, m)) = \beta_N(i) = 1. \tag{24.41}$$

24.4.3.2   Forward Recursive Computation of $\alpha_k(i)$

In this section we recursively derive the values $\alpha_k(N(j, m) = i)$ from $\alpha_{k-1}(P(i, m) = j)$. Upon exploiting Equation 24.30 we have

$$\alpha_k(i) = p(y_1^k, s_k = i) = p(y_k, y_1^{k-1}, s_k = i). \tag{24.42}$$

We can compute the right-hand side form of Equation 24.42 using the probability $p(y_k, y_1^{k-1}, s_{k-1} = j, s_k = i)$ by summing these probabilities for all the trellis states $j = 0 \dots (S - 1)$, from which the state $s_k = i$ is reached, as follows:

$$\alpha_k(i) = \sum_{j=0}^{S-1} p(y_k, y_1^{k-1}, s_{k-1} = j, s_k = i).$$

This can be reformatted using Equations 24.15-24.17 as:

$$\alpha_k(i) = \sum_{j=0}^{S-1} p(y_k, s_k = i|s_{k-1} = j, y_1^{k-1}) \cdot p(y_1^{k-1}, s_{k-1} = j). \tag{24.43}$$

With reference to the trellis diagram of Figure 24.10 we note that the received symbol sequence y_1^{k-1} has no effect on the first multiplicative term of Equation 24.43, if s_{k-1} is given. Thus, from Equation 24.43 we obtain

$$\alpha_k(i) = \sum_{j=0}^{S-1} p(y_k, s_k = i|s_{k-1} = j) \cdot p(y_1^{k-1}, s_{k-1} = j)$$

and with the aid of definition in Equation 24.30 we have:

$$\alpha_k(i) = \sum_{j=0}^{S-1} p(y_k, s_k = i | s_{k-1} = j) \cdot \alpha_{k-1}(j). \tag{24.44}$$

Let us now consider the summation over the index range of $j = 0 \ldots (S - 1)$ and note that for a fixed i, the probability of $p(y_k, s_k = i | s_{k-1} = j)$ will be non-zero only, if a branch exists from state j to state i. Thus, there are only M non-zero values of j, which contribute to the summation in Equation 24.44, namely the values $j = P(i, m)$ for a given m. We can thus rewrite Equation 24.44 as

$$\alpha_k(i) = \sum_{m=0}^{M-1} \alpha_{k-1}(P(i, m)) \cdot p(y_k, s_k = i | s_{k-1} = P(i, m)). \tag{24.45}$$

For the second multiplicative term of Equation 24.45 we have $p(y_k, s_k = i | s_{k-1} = P(i, m)) = p(y_k, u_k = m | s_{k-1} = P(i, m)) = \gamma_k(P(i, m), m)$, hence we can write

$$\alpha_k(i) = \sum_{m=0}^{M-1} \alpha_{k-1}(P(i, m)) \cdot \gamma_k(P(i, m), m). \tag{24.46}$$

Equation 24.46 allows the recursive calculation of the $\alpha_{k-1}(P(i, m) = j)$ values, commencing from $\alpha_0(j)$. In order to determine this boundary value we note that $\alpha_0(j) = p(s_0 = j)$, i.e. $\alpha_0(j)$ is the *a priori* probability of the first state j leading to state i. Conventionally, we commence the encoding from the first state, i.e. from state $j = 0$. In this case the boundary conditions are:

$$\alpha_0(j) = \begin{cases} 1 & \text{if } j = 0 \\ 0 & \text{if } j \neq 0 \end{cases} \tag{24.47}$$

Let us now consider how the above recursive computations can be carried out more efficiently in the logarithmic domain.

24.4.4 The MAP Algorithm in the Logarithmic-Domain

In this section we will describe the operation of the MAP algorithm in logarithmic domain (log-domain). In 1995, Robertson proposed the Log-Map algorithm [637], which dramatically reduces the complexity of the MAP algorithm, while attaining an identical performance to that of the MAP algorithm. The Max-Log-MAP algorithm constitutes a further substantial simplification, which performs, however, suboptimally compared to the Log-MAP algorithm. Specifically, in the log-domain multiplications correspond to additions, which are significantly less demanding in terms of computational complexity. A further simplification accrues by using the Jacobian logarithm [637] as follows:

$$\begin{aligned} g(\Phi_1, \Phi_2) &= ln(e^{\Phi_1} + e^{\Phi_2}) \\ &= max\{\Phi_1, \Phi_2\} + ln(1 + e^{-|\Phi_1 - \Phi_2|}) \\ &= max\{\Phi_1, \Phi_2\} + f_c(|\Phi_1 - \Phi_2|), \end{aligned} \tag{24.48}$$

where the summation of $e^{\Phi_1} + e^{\Phi_2}$ is replaced by selecting the maximum of the terms Φ_1 and Φ_2 and adding a correction term f_c that depends on the Euclidean distance of both terms. For the summation of more than two terms, i.e. for example for the summations seen in Equations 24.39 and 24.46, nesting of the $g(\Phi_1, \Phi_2)$ terms in Equation 24.48 can be carried out as follows:

$$ln(\sum_{i=1}^{I} e^{\Phi_i}) = g(\Phi_I, g(\Phi_{I-1}, \ldots g(\Phi_3, g(\Phi_2, \Phi_1)) \ldots)). \tag{24.49}$$

The correction term f_c in Equation 24.48 can be determined with the aid of three different methods:

- The Exact-Log-MAP algorithm, which is characterised by calculating the exact value of the correction term f_c as:

$$f_c = ln(1 + e^{-|\Phi_1 + \Phi_2|}). \tag{24.50}$$

  The corresponding performance is identical to that of the MAP algorithm.

- The Approx-Log-MAP algorithm invokes an approximation of the correction term f_c. Robertson [637] found that a look-up table containing eight values for f_c, ranging between 0 and 5, gives practically the same performance as the Exact-Log-MAP algorithm.

- The Max-Log-MAP algorithm, which retains only the maximum value in Equation 24.48, hence ignoring the correction term f_c. However, the Approx-Log-MAP algorithm is only marginally more complex, than the Max-Log-MAP algorithm, although it has a superior performance.

For these reasons, our simulations have been carried out by employing the Approx-Log-MAP algorithm. Explicitly, an addition operation is substituted with an addition, a subtraction, a table look-up and a maximum-search operation according to Equation 24.48, when the Approx-Log-MAP algorithm is employed.

24.4.5   Symbol-based MAP Algorithm Summary

Let us now summarize the operations of the symbol-based MAP algorithm using Figure 24.11. We assume that the *a priori* probabilities $\Pi_k(i)$ in Equation 24.21 were known. These are either all equal to $1/M$ or they are constituted by additional external information. The first step is to compute the set of probabilities $\eta_k(i, m)$ from Equation 24.20 as:

$$\eta_k(j, m) = e^{\frac{-|y_k - L(j,m)|^2}{2\sigma^2}}. \tag{24.51}$$

From these and the *a priori* probabilities, the $\gamma_k(i, m)$ values are computed according to Equation 24.22 as

$$\gamma_k(j, m) = \Pi_{k,m} \cdot \eta_k(j, m). \tag{24.52}$$

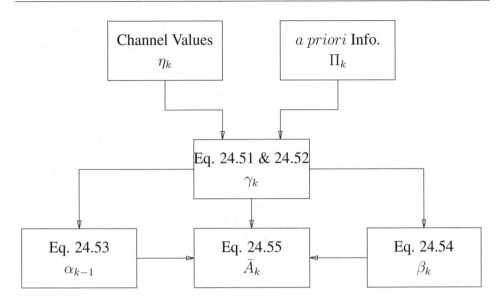

Figure 24.11: Summary of the symbol-based MAP algorithm operations

The above values are then used to recursively compute the values $\alpha_{k-1}(j)$ employing Equations 24.46 and 24.47 as

$$\alpha_k(i) = \sum_{m=0}^{M-1} \alpha_{k-1}(P(i,m)) \cdot \gamma_k(P(i,m),m), \qquad (24.53)$$

and the values $\beta_k(i)$ using Equations 24.39 and 24.41 as

$$\beta_{k-1}(j) = \sum_{m=0}^{M-1} \beta_k(N(j,m)) \cdot \gamma_k(j,m). \qquad (24.54)$$

Finally, the APP can be obtained using Equation 24.23

$$\bar{A}_{k,m} = \Pi_{k,m} \cdot \sum_{j=0}^{S-1} \beta_k(N(j,m)) \cdot \alpha_{k-1}(j) \cdot \eta_k(j,m). \qquad (24.55)$$

When considering the implementation of the MAP algorithm, one can opt for computing and storing the $\eta_k(j,m)$ values, and use these values together with the *a priori* probabilities for determining the values $\gamma_k(j,m)$ during decoding. In order to compute the probabilities $\eta_k(j,m)$ it is convenient to separately evaluate the exponential function of Equation 24.51 for every k and for every possible value of the transmitted symbol. As described in Section 24.4.1, a sequence of N information symbols was produced by the information source and each information symbol can assume M possible values, while the number of encoder states is S. There are $\mathsf{M} = 2 \cdot M$ possible transmitted symbols, since the size of the original signal constellation was doubled by the trellis encoder. Thus, $N \cdot 2 \cdot M$ evaluations of the exponential function of Equation 24.51 are needed. Using the online computation of the $\gamma_k(j, m)$ values, two multiplications are required for computing one additive term in each of Equation 24.53 and 24.54, and there are $N \cdot S$ terms to be computed, each requiring M terms to be summed. Hence $2 \cdot N \cdot M \cdot S$ multiplications and $N \cdot M \cdot S$ additions are required for computing the forward recursion α or the backward recursion β. Approximately three multiplications are required for computing each additive term in Equation 24.55, and there are $N \cdot M$ terms to be computed, each requiring S terms to be summed. Hence, the total implementational complexity entails $7 \cdot N \cdot M \cdot S$ multiplications, $3 \cdot N \cdot M \cdot S$ summations and $N \cdot 2 \cdot M$ exponential function evaluations, which is directly proportional to the length N of the transmitted sequence, to the number of code states S and to the number of different values M assumed by the input symbols.

The computational complexity can be reduced by implementing the algorithm in the log-domain, where the evaluation of the exponential function in Equation 24.51 is avoided. The multiplications and additions in Equations 24.52 to 24.55 are replaced by additions and Jacobian comparisons, respectively. Hence the total implementational complexity imposed is $7 \cdot N \cdot M \cdot S$ additions and $3 \cdot N \cdot M \cdot S$ Jacobian comparisons.

When implementing the MAP decoder presented here it is necessary to control the dynamic range of the likelihood terms computed in Equations 24.53 to 24.55. This is because these values tend to become lower and lower due to the multiplication of small values. The dynamic range can be controlled by normalising the sum of the $\alpha_k(i)$ and the $\beta_k(i)$ values to unity at every particular k symbol. The resulting symbol values will not be affected, since the normalisation only affects the scaling factors C_k in Equation 24.12. However, this problem can be avoided, when the MAP algorithm is implemented in the log-domain.

To conclude, let us note that the MAP decoder presented here is suitable for the decoding of finite-length, preferably short, sequences. When long sequences are transmitted, the employment of this decoder is impractical, since the associated memory requirements increase linearly with the sequence-length. In this case the MAP decoder has to be modified. A MAP decoder designed for long sequences was first presented in [723]. An efficient implementation, derived by adapting the algorithm of [582], was proposed by Piazzo in [724]. Having described the symbol-based MAP algorithm, let us now consider Turbo TCM (TTCM) and the way it invokes the MAP procedure.

24.5 Turbo Trellis-Coded Modulation

24.5.1 TTCM Encoder

It is worth describing the signal set dimensionality ($\bar{D}$) [725, 726] before we proceed. For a specific $2\bar{D}$ code, we have one $2\bar{D}$ symbol per codeword. For a general multidimensional code having a dimensionality of $D = 2 \cdot n$ where $n > 0$ is an integer, one $D\bar{D}$ codeword is comprised of n $2\bar{D}$ sub-codewords. The basic concept of the multidimensional signal mapping [725] is to assign more than one $2\bar{D}$ symbol to one codeword, in order to increase the spectral efficiency, which is defined as the number of information bits transmitted per channel symbol. For instance, a $2\bar{D}$ 8PSK TCM code seen in Table 24.2 maps $n = \frac{D}{2} = 1$ three-bit $2\bar{D}$ symbol to one $2\bar{D}$ codeword, where the number of information bits per $2\bar{D}$ codeword is $\bar{m} = 2$ yielding a spectral efficiency of $\bar{m}/n = 2$ information bits per symbol. However, a $4\bar{D}$ 8PSK TCM code seen in Table 24.2 maps $n = \frac{D}{2} = 2$ three-bit $2\bar{D}$ symbols to one six-bit $4\bar{D}$ codeword using the mapping rule of [725], where the number of information bits per $4\bar{D}$ codeword is $\bar{m} = 5$, yielding a spectral efficiency of $\bar{m}/n = 2.5$ information bits per symbol. However, during our further discourse we only consider $2\bar{D}$ signal sets.

Employing TTCM [709] avoids the obvious disadvantage of rate loss that one would incur when applying the principle of parallel concatenation to TCM without invoking puncturing. Specifically, this is achieved by puncturing the parity information in a particular manner, so that all information bits are sent only once, and the parity bits are provided alternatively by the two component TCM encoders. The TTCM encoder is shown in Figure 24.12, which comprises two identical TCM encoders linked by a symbol interleaver.

Let the memory of the interleaver be N symbols. The number of modulated symbols per block is $N.n$, where $n = \frac{D}{2}$ is an integer and D is the number of dimensions of the signal set. The number of information bits transmitted per block is $N.\bar{m}$, where $\bar{m}$ is the number of information bits per symbol. The encoder is clocked at a rate of $n.T$, where T is the symbol duration of each transmitted $2^{(\bar{m}+1)/n}$-ary $2\bar{D}$ symbol. At each step, $\bar{m}$ information bits are input to the TTCM encoder and n symbols each constituted by $\bar{m} + 1$ bits are transmitted, yielding a coding rate of $\frac{\bar{m}}{\bar{m}+1}$.

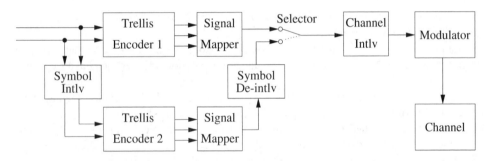

Figure 24.12: Schematic of the TTCM encoder. The selector enables the transmission of the information bits only once and selects alternative parity bits from the constituent encoders seen at the top and bottom [709] ©IEEE, 1998, Robertson and Wörz.

Each component TCM encoder consists of an Ungerböck encoder and a signal mapper. The first TCM encoder operates on the original input bit sequence, while the second TCM encoder manipulates the interleaved version of the input bit sequence. The signal mapper translates the codewords into complex symbols using the SP-based labelling method of Section 24.3.4. A complex symbol represents the amplitude and phase information passed to the modulator in the system seen in Figure 24.12. The complex output symbols of the signal mapper at the bottom of Figure 24.12 are symbol de-interleaved according to the inverse operation of the interleaver. Again, the interleaver and de-interleaver are symbol interleavers [727]. Owing to invoking the de-interleaver of Figure 24.12 at the output of the component encoder seen at the bottom, the TTCM codewords of both component encoders have identical information bits before the selector. Hence, the selector that alternatively selects the symbols of the upper and lower component encoders is effectively a puncturer that punctures the parity bits of the output symbols.

The output of the selector is then forwarded to the channel interleaver, which is, again, another symbol interleaver. The task of the channel interleaver is to effectively disperse the bursty symbol errors experienced during transmission over fading channels. This increases

Code	State, ν	$\tilde{m}$	$H^0(D)$	$H^1(D)$	$H^2(D)$	$H^3(D)$	$d^2_{free}/\triangle^2_0$
$2\bar{D}$, 8PSK	4, 2	2	07	02	04	-	
$2\bar{D}$, 8PSK	8, 3	2	11	02	04	-	3
$4\bar{D}$, 8PSK	8, 3	2	11	06	04	-	3
$2\bar{D}$, 8PSK	16, 4	2	23	02	10	-	3
$4\bar{D}$, 8PSK	16, 4	2	23	14	06	-	3
$2\bar{D}$, 16QAM	8, 3	3	11	02	04	10	2
$2\bar{D}$, 16QAM	16, 4	3	21	02	04	10	3
$2\bar{D}$, 64QAM	8, 3	2	11	04	02	-	3
$2\bar{D}$, 64QAM	16, 4	2	21	04	10	-	4

Table 24.2: 'Punctured' TCM codes exhibiting the best minimum distance for 8PSK, 16QAM and 64QAM, where octal format is used for specifying the generator polynomials [709] ©IEEE, 1998, Robertson and Wörz. The notation $\bar{D}$ denotes the dimensionality of the code, ν denotes the code memory, $\triangle^2_0$ denotes the squared Euclidean distance of the signal set itself and d^2_{free} denotes the squared FED of the TCM code.

the diversity order of the code [708, 718]. Finally, the output symbols are modulated and transmitted through the channel.

Table 24.2 shows the generator polynomials of some component TCM codes that can be employed in the TTCM scheme. These generator polynomials were obtained by Robertson and Wörz [709] using an exhaustive computer search of all polynomials and finding the one that maximises the minimal Euclidean distance, taking also into account the alternative selection of parity bits for the TTCM scheme. In Table 24.2, $\tilde{m}$ denotes the number of information bits to be encoded from the $\bar{m}$ number of information bits in a symbol, $\triangle^2_0$ denotes the squared Euclidean distance of the signal set itself, i.e. after TCM signal expansion, and d^2_{free} denotes the squared FED of the TCM constituent codes, as defined in Section 24.3.1. Since $d^2_{free}/\triangle^2_0 > 0$, the 'punctured' TCM codes constructed in Table 24.2 exhibit a positive coding gain in comparison to the uncoded but expanded signal set, although not necessarily in comparison to the uncoded and unexpanded original signal set. Nonetheless, the design target is to provide a coding gain also in comparison to the uncoded and unexpanded original signal set at least for the targeted operational SNR range of the system.

Considering the 8PSK example of Table 24.2, where $\triangle^2_0 = d^2_{8PSK}$ applies, we have $d^2_{free}/d^2_{8PSK} = 3$. However, when we compare the 'punctured' 8PSK TCM codes with the original uncoded QPSK, signal set we have $d^2_{free}/d^2_{QPSK} = d^2_{free}/2 = 0.878$ [709], which implies attaining a negative coding gain. However, when the iterative decoding scheme of TTCM is invoked, we attain a significant positive coding gain, as we will demonstrate in the following chapters.

24.5.2 TTCM Decoder

The concept of *a priori*, *a posteriori* and *extrinsic* information is illustrated in Figure 24.13. The associated concept is portrayed in more detail in Figure 24.14. The TTCM decoder structure of Figure 24.14(b) is similar to that of binary turbo codes shown in Figure 24.14(a), except that there is a difference in the nature of the information passed from one decoder to the other and in the treatment of the very first decoding step. Specifically, each

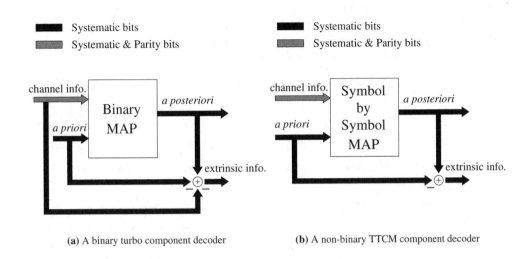

(a) A binary turbo component decoder   (b) A non-binary TTCM component decoder

Figure 24.13: Schematic of the component decoders for binary turbo codes and non-binary TTCM

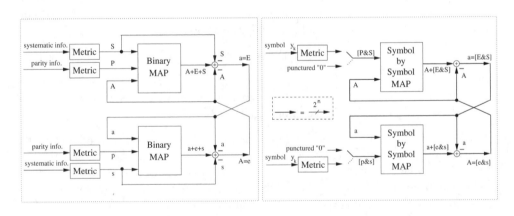

(a) Binary Turbo Decoder at step k   (b) TTCM Decoder at step k

Figure 24.14: Schematic of the decoders for binary turbo codes and TTCM. Note that the labels and arrows apply only to one specific information bit for the binary turbo decoder, or a group of $\bar{m}$ information bits for the TTCM decoder [709] ©IEEE, 1998, Robertson and Wörz. The interleavers/de-interleavers are not shown and the notations P, S, A and E denote the parity information, systematic information, *a priori* probabilities and *extrinsic* probabilities, respectively. Upper (lower) case letters represent the probabilities of the upper (lower) component decoder.

decoder alternately processes its corresponding encoder's channel-impaired output symbol, and then the other encoder's channel-impaired output symbol.

In a binary turbo coding scheme the component encoders' output can be split into three additive parts for each information bit u_k at step k, when operating in the logarithmic or LLR domain [637] as shown in Figure 24.14(a), which are:

1) the systematic component (S/s), i.e. the corresponding received systematic value for bit u_k;

2) the *a priori* or *intrinsic* component (A/a), i.e. the information provided by the other component decoder for bit u_k; and

3) the *extrinsic* information component related to bit u_k (E/e), which depends not on bit u_k itself but on the surrounding bits.

These components are impaired by independent noise and fading effects. In turbo codes, only the *extrinsic* component should be passed on to the other component decoder, so that the *intrinsic* information directly related to a bit is not reused in the other component decoder [366]. This measure is necessary in turbo codes for avoiding the prevention of achieving iterative gains, due to the dependence of the constituent decoders' information on each other.

However, in a symbol-based non-binary TTCM scheme the $\bar{m}$ systematic information bits and the parity bit are transmitted together in the same non-binary symbol. Hence, the systematic component of the non-binary symbol, namely the original information bits, cannot be separated from the *extrinsic* component, since the noise and/or fading that affects the parity component also affects the systematic component. Therefore, in this scenario the symbol-based information can be split into only two components:

1) the *a priori* component of a non-binary symbol (A/a), which is provided by the other component decoder, and

2) the inseparable *extrinsic* as well as systematic component of a non-binary symbol $([E\&S]/[e\&s])$, as can be seen from Figure 24.14(b).

Each decoder passes only the latter information to the next component decoder while the *a priori* information is removed at each component decoder's output, as seen in Figure 24.14(b), where, again, the *extrinsic* and systematic components are inseparable.

As described in Section 24.5.1, the number of modulated symbols per block is $N \cdot n$, with $n = \frac{D}{2}$, where D is the number of dimensions of the signal set. Hence for a $2\bar{D}$ signal set we have $n = 1$ and the number of modulated symbols per block is N. Therefore the symbol interleaver of length N will interleave a block of N complex symbols. Let us consider $2\bar{D}$ modulation having a coding rate of $\frac{m}{\bar{m}+1}$ for the following example.

The received symbols are input to the 'Metric' block of Figure 24.15, in order to generate a set of $M = 2^{\bar{m}+1}$ symbol probabilities for quantifying the likelihood that a certain symbol of the M-ary constellation was transmitted. The selector switches seen at the input of the 'Symbol by Symbol MAP' decoder select the current symbol's reliability metric, which is produced at the output of the 'Metric' block, if the current symbol was not punctured by the corresponding encoder. Otherwise puncturing will be applied where the probabilities of the various legitimate symbols at index k are set to 1 or to 0 in the log-domain. The upper (lower) case letters denote the set of probabilities of the upper (lower) component decoder, as

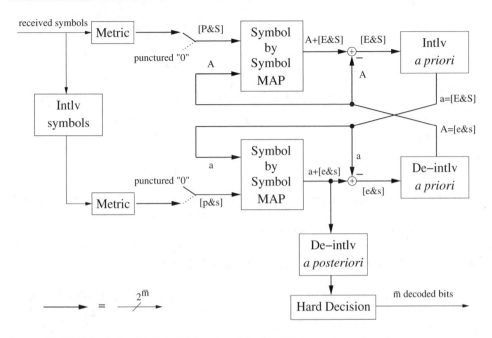

Figure 24.15: Schematic of the TTCM decoder. P, S, A and E denote the parity information, systematic information, *a priori* probabilities and *extrinsic* probabilities, respectively. Upper (lower) case letters represent the probabilities of the upper (lower) component decoder

shown in the figure. The 'Metric' block provides the decoder with the inseparable parity and systematic ($[P\&S]$ or $[p\&s]$) information, and the second input to the decoder is the *a priori* (A or a) information provided by the other component decoder. The MAP decoder then provides the *a posteriori* ($A + [E\&S]$ or $a + [e\&s]$) information at its output. Then A (or a) is subtracted from the *a posteriori* information, so that the same information is not used more than once in the other component decoder, since otherwise the component decoders' corresponding information would become dependent on each other, which would preclude the achievement of iteration gains. The resulting $[E\&S$ or $e\&s]$ information is symbol interleaved (or de-interleaved) in order to present the a (or A) input for the other component decoder in the required order. This decoding process will continue iteratively, in order to offer an improved version of the set of symbol reliabilities for the other component decoder. One iteration comprises the decoding of the received symbols by both the component decoders once. Finally, the *a posteriori* information of the lower component decoder will be de-interleaved in order to extract $\bar{m}$ decoded information bits per symbol. Hard decision implies selecting the specific symbol which exhibits the maximum *a posteriori* probability associated with the $\bar{m}$-bit information symbol from the $2^{\bar{m}}$ probability values. Having described the operation of the symbol-based TTCM technique, which does not protect all transmitted bits of the symbols, let us now consider bit-interleaved coded modulation as a design alternative.

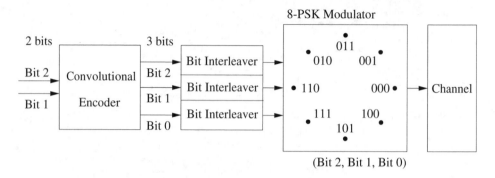

Figure 24.16: BICM encoder schematic employing independent bit interleavers and and protecting all transmitted bits. Instead of the SP-based labelling of TCM in Figure 24.2 here Gray labelling is employed [364] ©IEEE, 1992, Zehavi

24.6 Bit-Interleaved Coded Modulation

Bit-Interleaved Coded Modulation (BICM) was proposed by Zehavi [364] with the aim of increasing the diversity order of Ungerböck's TCM schemes which was quantified in Section 24.3.3. Again, the diversity order of a code is defined as the 'length' of the shortest error event path expressed in terms of the number of trellis stages encountered, before remerging with the all-zero path [718] or, equivalently, defined as the minimum Hamming distance of the code [365] where the diversity order of TCM using a symbol-based interleaver is the minimum number of different symbols between the erroneous path and the correct path along the shortest error event path. Hence, in a TCM scenario having parallel transitions, as shown in Figure 24.5, the code's diversity order is one, since the shortest error event path consists of one branch. This implies that parallel transitions should be avoided in TCM codes if possible, and if there were no parallel branches, any increase in diversity would be obtained by increasing the constraint length of the code. Unfortunately no TCM codes exist where the parallel transitions associated with the unprotected bits are avoided. In order to circumvent this problem, Zehavi's idea [364] was to render the code's diversity equal to the smallest number of different bits, rather than to that of the different channel symbols, by employing bit-based interleaving, as will be highlighted below.

24.6.1 BICM Principle

The BICM encoder is shown in Figure 24.16. In comparison to the TCM encoder of Figure 24.6, the differences are that BICM uses independent bit interleavers for all the bits of a symbol and non-systematic convolutional codes, rather than a single symbol-based interleaver and systematic RSC codes protecting some of the bits. The number of bit interleavers equals the number of bits assigned to the non-binary codeword. The purpose of bit interleaving is:

- to disperse the bursty errors induced by the correlated fading and to maximise the diversity order of the system;

- to render the bits associated with a given transmitted symbol uncorrelated or independent of each other.

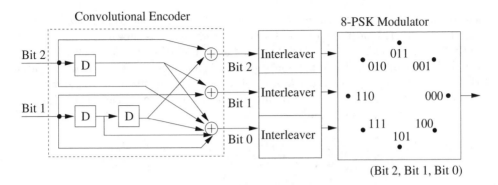

Figure 24.17: Paaske's non-systematic convolutional encoder, bit-based interleavers and modulator forming the BICM encoder [364,728], where none of the bits are unprotected and instead of the SP-based labelling as seen in Figure 24.2 here Gray labelling is employed

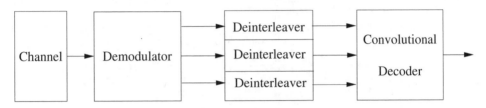

Figure 24.18: BICM decoder [364]

The interleaved bits are then grouped into non-binary symbols, where Gray-coded labelling is used for the sake of optimising the performance of the BICM scheme. The BICM encoder uses Paaske's non-systematic convolutional code proposed on page 331 of [728], which exhibits the highest possible free Hamming distance, hence attaining optimum performance over Rayleigh fading channels. Figure 24.17 shows Paaske's non-systematic eight-state code of rate-$2/3$, exhibiting a free bit-based Hamming distance of four. The BICM decoder implements the inverse process, as shown in Figure 24.18. In the demodulator module six bit metrics associated with the three bit positions, each having binary values of 0 and 1, are generated from each channel symbol. These bit metrics are de-interleaved by three independent bit de-interleavers, in order to form the estimated codewords. Then the convolutional decoder of Figure 24.18 is invoked for decoding these codewords, generating the best possible estimate of the original information bit sequence.

From Equation 24.5 we know that the average bit error probability of a coded modulation scheme using MPSK over Rayleigh fading channels at high SNRs is inversely proportional to $(E_s/N_0)^L$, where E_s/N_0 is the channel's symbol energy to noise spectral density ratio and L is the minimum Hamming distance or the code's diversity order. When bit-based interleavers are employed in BICM instead of the symbol-based interleaver employed in TCM, the minimum Hamming distance of BICM is quantified in terms of the number of different bits between the erroneous path in the shortest error event and the correct path. Since in BICM the bit-based minimum Hamming distance is maximised, BICM will give a lower bit error probability in Rayleigh fading channels than that of TCM that maximises the FED. Again, the design of BICM is aimed at providing maximum minimum Hamming distance, rather than providing maximum FED, as in TCM schemes. Moreover, we note that attaining a maximum FED is desired for transmission over Gaussian channels, as shown in Section 24.3.1. Hence, the performance of BICM is not as good as that of TCM in AWGN channels. The reduced FED of BICM is due to the 'random' modulation imposed by the 'random' bit interleavers [364], where the m-bit BICM symbol is randomised by the m number of bit interleavers. Again, $\bar{m}$ denotes the number of information bits, while m denotes the total number of bits in a 2^m-ary modulated symbol.

Rate	State, ν	$g^{(1)}$	$g^{(2)}$	$g^{(3)}$	$g^{(4)}$	d_{free}
1/2	8, 3	15	17	-	-	5
(4QAM)	16, 4	23	35	-	-	7
	64, 6	133	171	-	-	10
2/3	8, 3	4	2	6	-	4
(8PSK)		1	4	7	-	
	16, 4	7	1	4	-	5
		2	5	7	-	
	64, 6	64	30	64	-	7
		30	64	74	-	
3/4	8, 3	4	4	4	4	4
(16QAM)		0	6	2	4	
		0	2	5	5	
	32, 5	6	2	2	6	5
		1	6	0	7	
		0	2	5	5	
	64, 6	6	1	0	7	6
		3	4	1	6	
		2	3	7	4	

Table 24.3: Paaske's non-systematic convolutional codes, page 331 of [728], where ν denotes the code memory and d_{free} denotes the free Hamming distance. Octal format is used for representing the generator polynomial coefficients.

Rate	State, ν	$g^{(1)}$	$g^{(2)}$	puncturing matrix	d_{free}
5/6	8, 3	15	17	1 0 0 1 0	3
(64QAM)				0 1 1 1 1	
	64, 6	133	171	1 1 1 1 1	3
				1 0 0 0 0	

Table 24.4: Rate-Compatible Punctured Convolutional (RCPC) codes [729, 730], where ν denotes the code memory and d_{free} denotes the free Hamming distance. Octal format is used for representing the generator polynomial coefficients.

Table 24.3 summarises the parameters of a range of Paaske's non-systematic codes utilised in BICM. For a rate-k/n code there are k generator polynomials, each having n coefficients. For example, $\mathbf{g_i} = (g^0, g^1, \ldots, g^n)$, $i \leq k$, is the generator polynomial associated with the ith information bit. The generator matrix of the encoder seen in Figure 24.17

is:

$$\mathbf{G(D)} = \left[\begin{array}{ccc} 1 & D & 1+D \\ D^2 & 1 & 1+D+D^2 \end{array} \right], \tag{24.56}$$

while the equivalent polynomial expressed in octal form is given by:

$$\mathbf{g_1} = \left[\begin{array}{ccc} 4 & 2 & 6 \end{array} \right] \quad \mathbf{g_2} = \left[\begin{array}{ccc} 1 & 4 & 7 \end{array} \right]. \tag{24.57}$$

Observe in Table 24.3 that Paaske generated codes of rate-1/2, 2/3 and 3/4, but not 5/6. In order to study rate-5/6 BICM/64QAM, we created the required punctured code from the rate-1/2 code of Table 24.3. Table 24.4 summarises the parameters of the Rate-Compatible Punctured Convolutional (RCPC) codes that can be used in rate 5/6 BICM/64QAM schemes. Specifically, rate-1/2 codes were punctured according to the puncturing matrix of Table 24.4 in order to obtain the rate-5/6 codes, following the approach of [729, 730]. Let us now consider the operation of BICM with the aid of an example.

24.6.2 BICM Coding Example

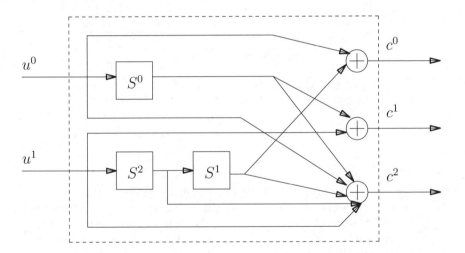

Figure 24.19: Paaske's non-systematic convolutional encoder [728].

Considering Paaske's eight-state convolutional code [728] in Figure 24.19 as an example, the BICM encoding process is illustrated here. The corresponding generator polynomial is shown in Equation 24.57. A two-bit information word, namely $u = (u^1, u^0)$, is encoded in each cycle in order to form a three-bit codeword, $c = (c^2, c^1, c^0)$. The encoder has three shift registers, namely S^0, S^1 and S^2, as shown in the figure. The three-bit binary contents of these registers represent eight states, as follows:

$$S = (S^2, S^1, S^0) \in \{000, 001, \ldots, 111\} = \{0, 1, \ldots, 7\}. \tag{24.58}$$

The input sequence, u, generates a new state S and a new codeword c at each encoding cycle. Table 24.5 illustrates the codewords generated and the associated state transitions. The encoding process can also be represented with the aid of the trellis diagram of Fig-

State $S = (S^2, S^1, S^0)$	Information Word $u = (u^1,\ u^0)$			
	$00 = 0$	$01 = 1$	$10 = 2$	$11 = 3$
$000 = 0$	$000 = 0$	$101 = 5$	$110 = 6$	$011 = 3$
$001 = 1$	$110 = 6$	$011 = 3$	$000 = 0$	$101 = 5$
$010 = 2$	$101 = 5$	$000 = 0$	$011 = 3$	$110 = 6$
$011 = 3$	$011 = 3$	$110 = 6$	$101 = 5$	$000 = 0$
$100 = 4$	$100 = 4$	$001 = 1$	$010 = 2$	$111 = 7$
$101 = 5$	$010 = 2$	$111 = 7$	$100 = 4$	$001 = 1$
$110 = 6$	$001 = 1$	$100 = 4$	$111 = 7$	$010 = 2$
$111 = 7$	$111 = 7$	$010 = 2$	$001 = 1$	$100 = 4$
Codeword $c = (c^2,\ c^1, c^0)$				
$000 = 0$	$000 = 0$	$001 = 1$	$100 = 4$	$101 = 5$
$001 = 1$	$000 = 0$	$001 = 1$	$100 = 4$	$101 = 5$
$010 = 2$	$000 = 0$	$001 = 1$	$100 = 4$	$101 = 5$
$011 = 3$	$000 = 0$	$001 = 1$	$100 = 4$	$101 = 5$
$100 = 4$	$010 = 2$	$011 = 3$	$110 = 6$	$111 = 7$
$101 = 5$	$010 = 2$	$011 = 3$	$110 = 6$	$111 = 7$
$110 = 6$	$010 = 2$	$011 = 3$	$110 = 6$	$111 = 7$
$111 = 7$	$010 = 2$	$011 = 3$	$110 = 6$	$111 = 7$
Next State $S = (S^2,\ S^1,\ S^0)$				

Table 24.5: The codeword generation and state transition table of the non-systematic convolutional encoder of Figure 24.19. The state transition diagram is seen in Figure 24.20

ure 24.20. Specifically, the top part of Table 24.20 contains the codewords $c = (c^2,\ c^1, c^0)$ as a function of the encoder state $S = (S^2, S^1, S^0)$ as well as that of the information word $u = (u^1,\ u^0)$, while the bottom section contains the next states, again as a function of S and u. For example, if the input is $u = (u^1,\ u^0) = (1,1) = 3$ when the shift register is in state $S = (S^2, S^1, S^0) = (1,1,0) = 6$, the shift register will change its state to state $S = (S^2, S^1, S^0) = (1,1,1) = 7$ and $c = (c^2,\ c^1, c^0) = (0,1,0) = 2$ will be the generated codeword. Hence, if the input binary sequence is $\{01\ 10\ 01\ 00\ 10\ 10 \rightarrow\}$ with the rightmost being the first input bit, the corresponding information words are $\{1\ 2\ 1\ 0\ 2\ 2 \rightarrow\}$. Before any decoding takes place, the shift register is initialised to zero. Therefore, as seen at the right of Figure 24.20, when the first information word of $u_1 = 2$ arrives, the state changes from $S^{-1} = 0$ to $S = 4$, generating the first codeword $c_1 = 6$ as seen in the bottom and top sections of Table 24.5, respectively. Then the second information word of $u_2 = 2$ changes the state from $S^{-1} = 4$ to $S = 6$, generating the second codeword of $c_2 = 2$. The process continues in a similiar manner according to the transition table, namely Table 24.5. The codewords generated as seen at the right of Figure 24.20 are $\{4\ 0\ 0\ 1\ 2\ 6 \rightarrow\}$, and the state transitions are $\{2 \leftarrow 4 \leftarrow 1 \leftarrow 2 \leftarrow 6 \leftarrow 4 \leftarrow 0\}$. Then the bits constituting the codeword sequence are interleaved by the three bit interleavers of Figure 24.17, before they are assigned to the corresponding 8PSK constellation points.

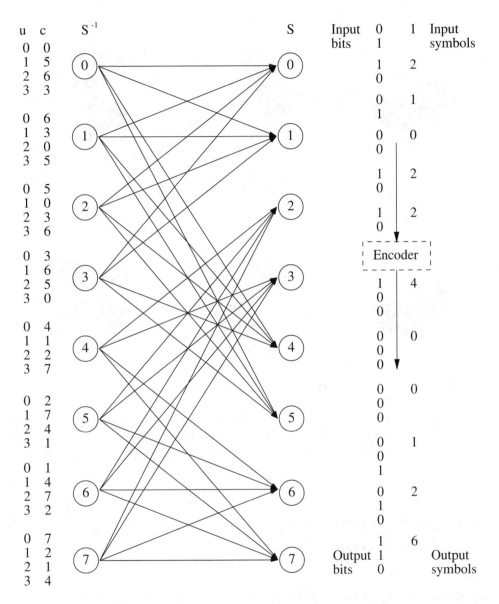

Figure 24.20: Trellis diagram for Paaske's eight-state convolutional code, where u indicates the information word, c indicates the codeword, S^{-1} indicates the previous state and S indicates the current state. As an example, the encoding of the input bit sequence of $\{011001001010 \rightarrow\}$ is shown at the right. The encoder schematic is portrayed in Figure 24.19, while the state transitions are summarised in Table 24.5

(Bit 2, Bit 1, Bit 0)

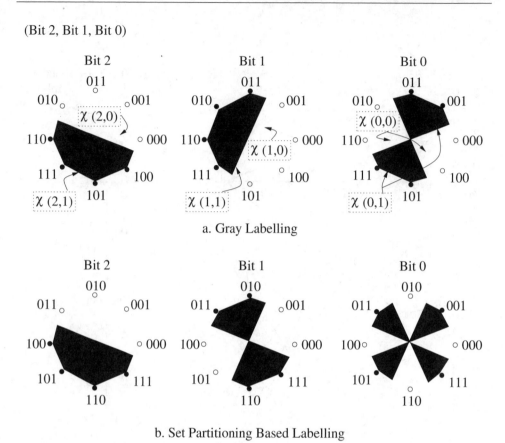

a. Gray Labelling

b. Set Partitioning Based Labelling

Figure 24.21: SP and Gray labelling methods for 8PSK and the corresponding subset partitioning for each bit, where $\chi(i,b)$ defined in Equation 24.64 refers to the subset of the modulation constellation for Bit i where Bit $i = b \in \{0,1\}$ [373] ©IEEE, 1999, Li and Ritcey

24.7   Bit-Interleaved Coded Modulation with Iterative Decoding

BICM using Iterative Decoding (BICM-ID) was proposed by Li [373, 710] for further improving the FED of Zehavi's BICM scheme, although BICM already improved the diversity order of Ungerböck's TCM scheme. This FED improvement can be achieved with the aid of combining SP-based constellation labelling, as in TCM, and by invoking soft-decision feedback from the decoder's output to the demodulator's input, in order to exchange soft-decision-based information between them. As we will see below, this is advantageous, since upon each iteration the channel decoder improves the reliability of the soft information passed to the demodulator.

(Bit 2, Bit 1, Bit 0)

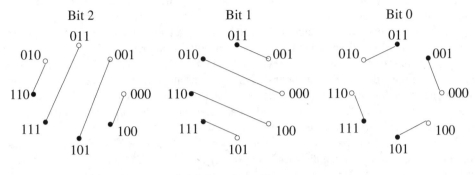

a. Gray Labelling

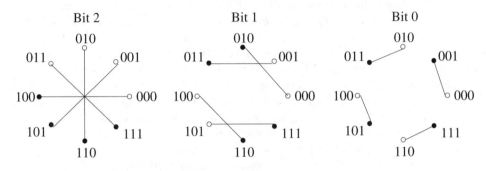

b. Set Partitioning Labelling

Figure 24.22: Iterative decoding translates the 8PSK scheme into three parallel binary sub-channels, each associated with a BPSK constellation selected from the four possible signal sets [373] ©IEEE, 1999, Li and Ritcey

24.7.1 Labelling Method

Let us now consider the mapping of the interleaved bits to the phasor constellation in this section. Figure 24.21 shows the process of subset partitioning for each of the three bit positions for both Gray labelling and in the context of SP labelling. The shaded regions shown inside the circle correspond to the subset $\chi(i, 1)$ defined in Equation 24.64, and the unshaded regions to $\chi(i, 0)$, $i = 0, 1, 2$, where i indicates the bit position in the three-bit BICM/8PSK symbol. These are also the decision regions for each bit, if hard-decision-based BICM demodulation is used for detecting each bit individually. The two labelling methods seen in Figure 24.21 have the same intersubset distances, although a different number of nearest neighbours. For example, $\chi(0, 1)$, which denotes the region where bit 0 equals to 1, is divided into two regions in the context of Gray labelling, as can be seen in Figure 24.21(a). By contrast, in the context of SP labelling seen in Figure 24.21(b), $\chi(0, 1)$ is divided into four regions. Clearly, Gray labelling has a lower number of nearest neighbours compared to SP-based labelling. The higher the number of nearest neighbours, the higher the chances for a bit to be decoded into the wrong region. Hence, Gray labelling is a more appropriate mapping during the first decoding iteration, and hence it was adopted by the non-iterative BICM scheme of Figure 24.18.

During the second decoding iteration in BICM-ID, given the feedback information representing Bit 1 and Bit 2 of the coded symbol, the constellation associated with Bit 0 is confined to a pair of constellation points, as shown on the right of Figure 24.22. Therefore, as far as Bit 0 is concerned, the 8PSK phasor constellation is translated into four binary constellations, where one of the four possible specific BPSK constellations is selected by the feedback Bit 1 and Bit 2. The same is true for the constellations associated with both Bit 1 and Bit 2, given the feedback information of the corresponding other two bits.

In order to optimise the second-pass decoding performance of BICM-ID, one must maximise the minimum Euclidean distance between any two points of all the $2^{m-1} = 4$ possible phasor pairs at the left (Bit 2), centre (Bit 1) and the right (Bit 0) of Figure 24.22. Clearly, SP-based labelling serves this aim better, when compared to Gray labelling, since the corresponding minimum Euclidean distance of SP-based labelling is higher than that of Gray labelling for both Bit 1 and Bit 2, as illustrated at the left and the centre of Figure 24.22. Although the first-pass performance is important, in order to prevent error precipitation due to erroneous feedback bits, the error propagation is effectively controlled by the soft feedback of the decoder. Therefore, BICM-ID assisted by soft decision feedback uses SP labelling.

Specifically, the desired high Euclidean distance for Bit 2 in Figure 24.22(b) is only attainable when Bit 1 and Bit 0 are correctly decoded and fed back to the SP-based demodulator. If the values to be fed back are not correctly decoded, the desired high Euclidean distance will not be achieved and error propagation will occur. On the other hand, an optimum convolutional code having a high binary Hamming distance is capable of providing a high reliability for the decoded bits. Therefore, an optimum convolutional code using appropriate signal labelling is capable of 'indirectly' translating the high binary Hamming distance between coded bits into a high Euclidean distance between the phasor pairs portrayed in Figure 24.22. In short, BICM-ID converts a 2^m-ary signalling scheme to m independent parallel binary schemes by the employment of m number of independent bit interleavers and involves an iterative decoding method. This simultaneously facilitates attaining a high diversity order with the advent of the bit interleavers, as well as achieving a high FED with the aid of the iterative decoding and SP-based labelling. Hence, BICM-ID effectively combines powerful binary codes with bandwidth-efficient modulation.

24.7.2   Interleaver Design

The interleaver design is important as regards the performance of BICM-ID. In [711], Li introduced certain constraints on the design of the interleaver, in order to maximise the minimum Euclidean distance between the two points in the 2^{m-1} possible specific BPSK constellations. However, we advocate a simpler approach, where the m number of interleavers used for the 2^m-ary modulation scheme are generated randomly and separately, without any interactions between them. The resultant minimum Euclidean distance is less than that of the scheme proposed in [711], but the error bursts inflicted by correlated fading are expected to be randomised effectively by the independent bit interleavers. This was expected to give a better performance over fading channels at the cost of a slight performance degradation over AWGN channels, when compared to Li's scheme [711]. However, as we will demonstrate in the context of our simulation results in Section 25.2.2, our independent random interleaver design and Li's design perform similarly.

Having described the labelling method and the interleaver design in the context of BICM-ID, let us now consider the operation of BICM-ID with the aid of an example.

24.7.3 BICM-ID Coding Example

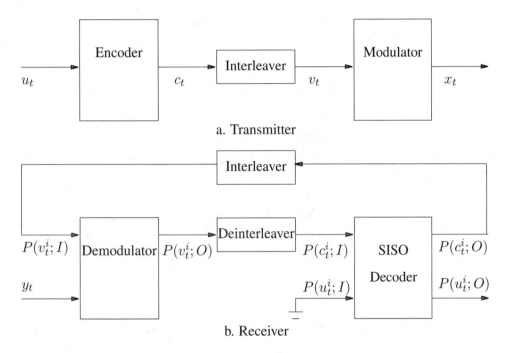

Figure 24.23: The transmitter and receiver modules of the BICM-ID scheme using soft-decision feedback [710] ©IEEE, 1998, Li

The BICM-ID scheme using soft-decision feedback is shown in Figure 24.23. The interleavers used are all bit-based, as in the BICM scheme of Figure 24.17, although for the sake of simplicity here only one interleaver is shown. A Soft-Input Soft-Output (SISO) [731] decoder is used in the receiver module and the decoder's output is fed back to the input of the demodulator. The SISO decoder of the BICM-ID scheme is actually a MAP decoder that computes the *a posteriori* probabilities for the non-systematically channel-coded bits and the original information bits.

For an (n, k) binary convolutional code the encoder's input symbol at time t is denoted by $u_t = [u_t^0, u_t^1, \ldots, u_t^{k-1}]$ and the coded output symbol by $c_t = [c_t^0, c_t^1, \ldots, c_t^{n-1}]$, where u_t^i or c_t^i is the ith bit in a symbol as defined in the context of Table 24.5 and Figure 24.20. The coded bits are interleaved by m independent bit interleavers, then m interleaved bits are grouped together in order to form a channel symbol $v_t = [v_t^0, v_t^1, \ldots, v^{m-1}]$ as seen in Figure 24.23(a), for transmission using 2^m-ary modulation. Let us consider 8PSK modulation, i.e. m = 3 as an example.

A signal labelling method μ maps the symbol v_t to a complex phasor according to $x_t =$

$\mu(v_t)$, $x_t \in \chi$, where the 8PSK signal set is defined as $\chi = \{\sqrt{E_s}\, e^{j2n\pi/8},\ n = 0,\ldots,7\}$ and E_s is the energy per transmitted symbol. In conjunction with a rate-2/3 code, the energy per information bit is $E_b = E_s/2$. For transmission over Rayleigh fading channels using coherent detection, the received discrete time signal is:

$$y_t = \rho_t x_t + n_t, \tag{24.59}$$

where ρ_t is the Rayleigh-distributed fading amplitude [370] having an expectation value of $E[\rho_t^2] = 1$, while n_t is the complex AWGN exhibiting a variance of $\sigma^2 = N_0/2$ where N_0 is the noise's PSD. For AWGN channels we have $\rho_t = 1$ and the Probability Density Function (PDF) of the non-faded but noise-contaminated received signal is expressed as [722]:

$$P(y_t | x_t, \rho_t) = \frac{1}{2\pi\sigma^2}\, e^{-\frac{1}{2}\left(\frac{n_t}{\sigma}\right)^2}, \tag{24.60}$$

where $\sigma^2 = N_0/2$ and the constant multiplicative factor of $\frac{1}{2\pi\sigma^2}$ does not influence the shape of the distribution and hence can be ignored when calculating the branch transition metric η, as described in Section 24.4.5. For AWGN channels, the conditional PDF of the received signal can be written as:

$$P(y_t | x_t) = e^{-\frac{|y_t - x_t|^2}{2\sigma^2}}. \tag{24.61}$$

Considering AWGN channels, the demodulator of Figure 24.23(b) takes y_t as its input for computing the confidence metrics of the bits using the maximum APP criterion [374]:

$$P(v_t^i = b | y_t) = \sum_{x_t \in \chi(i,b)} P(x_t | y_t), \tag{24.62}$$

where $i \in \{0,1,2\}$, $b \in \{0,1\}$ and $x_t = \mu(v_t)$. Furthermore, the signal after the demodulator of Figure 24.23 is described by the demapping of the bits $[\nabla^0(x_t), \nabla^1(x_t), \nabla^2(x_t)]$ where $\nabla^i(x_t) \in \{0,1\}$ is the value of the ith bit of the three-bit label assigned to x_t. With the aid of Bayes' rule in Equations 24.15 to 24.17 we obtain:

$$P(v_t^i = b | y_t) = \sum_{x_t \in \chi(i,b)} P(y_t | x_t) P(x_t), \tag{24.63}$$

where the subset $\chi(i,b)$ is described as:

$$\chi(i,b) = \{\mu([\nabla^0(x_t), \nabla^1(x_t), \nabla^2(x_t)]) \mid \nabla^j(x_t) \in \{0,1\}, j \neq i\}, \tag{24.64}$$

which contains all the phasors for which $\nabla^i(x_t) = b$ holds. For 8PSK, where $\mathsf{m} = 3$, the size of each such subset is $2^{\mathsf{m}-1} = 4$ as portrayed in Figure 24.21. This implies that only the *a priori* probabilities of $\mathsf{m} - 1 = 2$ bits out of the total of $\mathsf{m} = 3$ bits per channel symbol have to be considered, in order to compute the bit metric of a particular bit.

Now using the notation of Benedetto *et al.* [731], the *a priori* probabilities of an original uncoded information bit at time index t and bit index i, namely u_t^i being 0 and 1, are denoted by $P(u_t^i = 0; I)$ and $P(u_t^i = 1; I)$ respectively, while I refers to the *a priori* probabilities of the bit. This notation is simplified to $P(u_t^i; I)$, when no confusion arises, as shown in Figure 24.23. Similarly, $P(c_t^i; I)$ denotes the *a priori* probabilities of a legitimate coded bit at time index t and position index i. Finally, $P(u_t^i; O)$ and $P(c_t^i; O)$ denote the *extrinsic a posteriori* information of the original information bits and coded bits, respectively.

The *a priori* probability $P(x_t)$ in Equation 24.63 is unavailable during the first-pass decoding, hence an equal likelihood is assumed for all the 2^m legitimate symbols. This renders the *extrinsic a posteriori* bit probabilities, $P(v_t^i = b; O)$, equal to $P(v_t^i = b|y_t)$, when ignoring the common constant factors. Then, the SISO decoder of Figure 24.23(b) is used for generating the *extrinsic a posteriori* bit probabilities $P(u_t^i; O)$ of the information bits, as well as the *extrinsic a posteriori* bit probabilities $P(c_t^i; O)$ of the coded bits, from the de-interleaved probabilities $P(v_t^i = b; O)$, as seen in Figure 24.23(b). Since $P(u_t^i; I)$ is unavailable, it is not used in the entire decoding process.

During the second iteration $P(c_t^i; O)$ is interleaved and fed back to the input of the demodulator in the correct order in the form of $P(v_t^i; I)$, as seen in Figure 24.23(b). Assuming that the probabilities $P(v_t^0; I)$, $P(v_t^1; I)$ and $P(v_t^2; I)$ are independent by the employment of three independent bit interleavers, we have for each $x_t \in \chi$:

$$
\begin{aligned}
P(x_t) &= P(\mu([\nabla^0(x_t), \nabla^1(x_t), \nabla^2(x_t)])) \\
&= \prod_{j=0}^{2} P(v_t^j = \nabla^j(x_t); I),
\end{aligned}
\tag{24.65}
$$

where $\nabla^j(x_t) \in \{0, 1\}$ is the value of the jth bit of the three-bit label for x_t. Now that we have the *a priori* probability $P(x_t)$ of the transmitted symbol x_t, the *extrinsic a posteriori* bit probabilities for the second decoding iteration can be computed using Equations 24.63 and 24.65, yielding:

$$
\begin{aligned}
P(v_t^i = b; O) &= \frac{P(v_t^i = b|y_t)}{P(v_t^i = b; I)} \\
&= \sum_{x_t \in \chi(i,b)} \left(P(y_t|x_t) \prod_{j \neq i} P(v_t^j = \nabla^j(x_t); I) \right) \\
&\qquad i \in \{0, 1, 2\}, \quad b \in \{0, 1\}.
\end{aligned}
\tag{24.66}
$$

As seen from Equation 24.66, in order to recalculate the metric for a bit we only need the *a priori* probabilities of the other two bits in the same channel symbol. After interleaving in the feedback loop of Figure 24.23, the regenerated bit metrics are tentatively soft demodulated again and the process of passing information between the demodulator and decoder is continued. The final decoded output is the hard-decision-based *extrinsic* bit probability $P(u_t^i; O)$.

24.8 Summary

In this chapter we have studied the conceptual differences between four coded modulation schemes in terms of coding structure, signal labelling philosophy, interleaver type and decoding philosophy. The symbol-based non-binary MAP algorithm was also highlighted, when operating in the log-domain.

In the next chapter, we will proceed to study the performance of TCM, BICM, TTCM and BICM-ID over non-dispersive propagation environments.

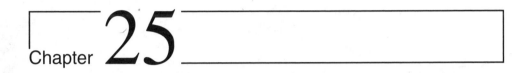

Chapter 25

Coded Modulation Performance in Non-dispersive Propagation Environments

25.1 Introduction

Having described the operating principles of the TCM, BICM, TTCM and BICM-ID schemes in Chapter 24, in Section 25.2 their performance will be initially evaluated for transmission over non-dispersive narrowband channels. By contrast, in Section 25.4 we will investigate their performance, when communicating over wideband fading channels in the context of a multi-carrier Orthogonal Frequency Division Multiplexing (OFDM) system [180]. However, we will assume that the number of subcarriers is sufficiently high for assuming that each OFDM subcarrier experiences narrowband, non-dispersive channel conditions.

25.2 Coded Modulation in Narrowband Channels

In this section, a comparative study of TCM, TTCM, BICM and BICM-ID schemes over both Gaussian and uncorrelated narrowband Rayleigh fading channels is presented in the context of 8PSK, 16QAM and 64QAM. We comparatively study the associated decoding complexity, the effects of the encoding block length and the achievable bandwidth efficiency. It will be shown that TTCM constitutes the best compromise scheme, followed by BICM-ID.

25.2.1 System Overview

The schematic of the coded modulation schemes under consideration is shown in Figure 25.1. The source generates random information bits, which are encoded by one of the TCM, TTCM or BICM encoders. The coded sequence is then appropriately interleaved and used for modulating the waveforms according to the symbol mapping rules. For a narrowband

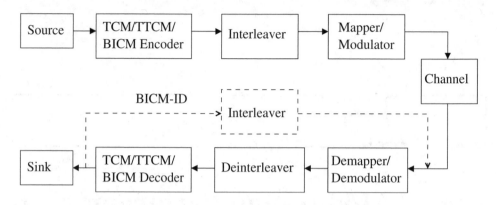

Figure 25.1: System overview of different coded modulation schemes.

Rayleigh fading channel in conjunction with coherent detection, the relationship between the transmitted discrete time signal x_t and the received discrete time signal y_t is given by:

$$y_t = \rho_t x_t + n_t, \tag{25.1}$$

where ρ_t is the Rayleigh-distributed fading amplitude having an expected value of $E[\rho_t^2] = 1$, while n_t is the complex AWGN having a variance of $\sigma^2 = N_0/2$ where N_0 is the noise's Power Spectral Density (PSD). For AWGN channels we have $\rho_t = 1$. The receiver consists of a coherent demodulator followed by a de-interleaver and one of the TCM, TTCM or BICM decoders. TTCM schemes consist of two component TCM encoders and two parallel decoders. In BICM-ID schemes the decoder output is appropriately interleaved and fed back to the demodulator input, as shown in Figure 25.1.

The log-domain branch metric required for the maximum likelihood decoding of TCM and TTCM over fading channels is given by the squared Euclidean distance between the faded transmitted symbol x_t and the noisy received symbol y_t, which is formulated as:

$$\pi_t = |y_t - \rho_t x_t|^2. \tag{25.2}$$

By contrast, the corresponding branch metric for BICM and BICM-ID is formed by summing the de-interleaved bit metrics λ of each coded bit v_t^i which quantifies the reliability of the corresponding symbol, yielding:

$$\pi_t = \sum_{i=0}^{\bar{m}} \lambda(v_t^i = b), \tag{25.3}$$

where i is the bit position of the coded bit in a constellation symbol, $\bar{m}$ is the number of information bits per symbol and $b \in (0, 1)$. The number of coded bits per symbol is $(\bar{m} + 1)$, since the coded modulation schemes add one parity bit to the $\bar{m}$ information bits by doubling the original constellation size, in order to maintain the same spectral efficiency of $\bar{m}$ bits/s/Hz.

Rate	State	$\tilde{m}$	H^0	H^1	H^2	H^3
2/3	8	2	11	02	04	-
(8PSK)	64 *	2	103	30	66	-
3/4	8	3	11	02	04	10
(16QAM)	64 *	3	101	16	64	-
5/6	8	2	11	02	04	-
(64QAM)	64 *	2	101	16	64	-

Table 25.1: 'Punctured' TCM codes with best minimum distance for PSK and QAM, ©Robertson and Wörz [709]. '*' indicates Ungerböck's TCM codes [495]. Two-dimensional $(2\bar{D})$ modulation is utilised. Octal format is used for representing the generator polynomials H^i and $\tilde{m}$ denotes the number of coded information bits out of the total $\bar{m}$ information bits in a modulated symbol.

The BICM bit metrics $\tilde{\lambda}$ before the de-interleaver are defined as [373]:

$$\tilde{\lambda}(v_t^i = b) = \sum_{x \in \chi(i,b)} |y_t - \rho_t x|^2, \qquad (25.4)$$

where $\chi(i, b)$ is the signal set, for which the bit i of the symbol has a binary value b.

To elaborate a little further, the coded modulation schemes that we comparatively studied are Ungerböck's TCM [495], Robertson's TTCM [709], Zehavi's BICM [364] and Li's BICM-ID [374]. Table 25.1 shows the generator polynomials of both the TCM and TTCM codes in octal format. These are Recursive Systematic Convolutional (RSC) codes that add one parity bit to the information bits. Hence, the coding rate of a $2^{\bar{m}+1}$-ary PSK or QAM signal is $R = \frac{\bar{m}}{\bar{m}+1}$. The number of decoding states associated with a code of memory ν is 2^ν. When the number of protected/coded information bits $\tilde{m}$ is less than the total number of original information bits $\bar{m}$, there are $(\bar{m} - \tilde{m})$ uncoded information bits and $2^{\bar{m}-\tilde{m}}$ parallel transitions in the trellis of the code. Parallel transitions assist in reducing the decoding complexity and the memory required, since the dimensionality of the corresponding trellis is smaller than that of a trellis having no parallel branches.

Table 25.2 shows the generator polynomials for the BICM and BICM-ID codes in octal format. These codes are non-systematic convolutional codes having a maximum free Hamming distance. Again, only one extra bit is added to the information bits. Hence, the achievable coding rate and the bandwidth efficiency are similar to that of TCM and TTCM for the $2^{\bar{m}+1}$-ary modulation schemes used. In order to reduce the required decoding memory, the BICM and BICM-ID schemes based on 64QAM were obtained by puncturing the rate-1/2 codes following the approach of [730], since for a non-punctured rate-5/6 code there are $2^{(\bar{m}=5)} = 32$ branches emerging from each trellis state for a block length of L_B, whereas for the punctured rate-1/2 code, there are only $2^{(\bar{m}=1)} = 2$ branches emerging from each trellis state for a block length of $\bar{m}L_B = 5L_B$. Therefore the required decoding memory is reduced by a factor of $\frac{2^{\bar{m}} \cdot L_B}{2^1 \cdot \bar{m} \cdot L_B} = 3.2$.

Soft-decision trellis decoding utilising the Log-Maximum A Posteriori (Log-MAP) algorithm [637] was invoked for the decoding of the coded modulation schemes. As discussed in Section 24.4.5, the Log-MAP algorithm is a numerically stable version of the MAP algorithm operating in the log-domain, in order to reduce its complexity and to mitigate the numerical

Rate	State	g^1	g^2	g^3	g^4	d_{free}
2/3	8	4	2	6	-	4
(8PSK)	($\nu=3$)	1	4	7	-	
	16	7	1	4	-	5
	($\nu=4$)	2	5	7	-	
	64	15	6	15	-	7
	($\nu=6$)	6	15	17	-	
3/4	8	4	4	4	4	4
(16QAM)	($\nu=3$)	0	6	2	4	
		0	2	5	5	
	32	6	2	2	6	5
	($\nu=5$)	1	6	0	7	
		0	2	5	5	

Rate	State	g^1	g^2	Puncturing	d_{free}
5/6	8	15	17	1 0 0 1 0	3
(64QAM)	($\nu=3$)			0 1 1 1 1	
	64	133	171	1 1 1 1 1	3
	($\nu=6$)			1 0 0 0 0	

Table 25.2: Top table shows the generator polynomials of Paaske's code, p. 331 of [728]. Bottom table shows those of the rate-compatible puncture convolutional codes [730]. ν is the code memory and d_{free} is the free Hamming distance. Octal format is used for the polynomial coefficients g^i, while '1' and '0' in the puncturing matrix indicate the position of the unpunctured and punctured coded bits, respectively.

problems associated with the MAP algorithm [582].

25.2.2 Simulation Results and Discussion

In this section we study the performance of TCM, TTCM, BICM and BICM-ID using computer simulations. The complexity of the coded modulation schemes is compared in terms of the number of decoding states, and the number of decoding iterations. For a TCM or BICM code of memory ν, the corresponding complexity is proportional to the number of decoding states namely to $S = 2^\nu$. Since TTCM schemes invoke two component TCM codes, a TTCM code with t iterations and using an S-state component code exhibits a complexity proportional to $2.t.S$ or $t.2^{\nu+1}$. As for BICM-ID schemes, only one decoder is used but the demodulator is invoked in each decoding iteration. However, the complexity of the demodulator is assumed to be insignificant compared to that of the channel decoder. Hence, a BICM-ID code with t iterations using an S-state code exhibits a complexity proportional to $t.S$ or $t.2^\nu$.

25.2.2.1 Performance over AWGN Channels

It is important to note that in terms of the total number of trellis states the decoding complexity of 64-state TCM and 8-state TTCM using two TCM decoders in conjunction with four iterations can be considered similar. The same comments are valid also for 16-state BICM-

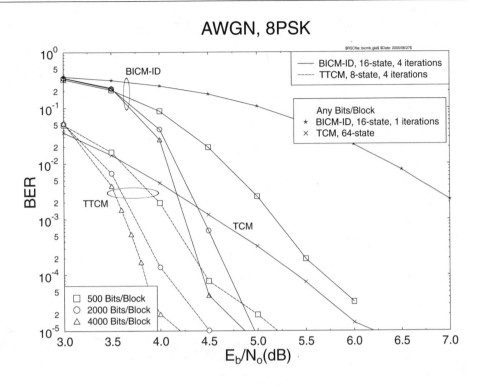

Figure 25.2: Effects of block length on the TCM, TTCM and BICM-ID performance in the context of an 8PSK scheme for transmissions over AWGN channels

ID using four iterations or for 8-state BICM-ID using eight iterations. In our forthcoming discourse we will always endeavour to compare schemes of similar decoding complexity, unless otherwise stated. Figure 25.2 illustrates the effects of interleaving block length on the TCM, TTCM and BICM-ID performance in an 8PSK scheme over AWGN channels. It is clear from the figure that a high interleaving block length is desired for the iterative TTCM and BICM-ID schemes. The block length does not affect the BICM-ID performance during the first pass, since it constitutes a BICM scheme using SP-based phasor labelling. However, if we consider four iterations, the performance improves, converging faster to the Error-Free-Feedback (EFF) bound[1] [373] for larger block lengths. At a BER of 10^{-4} a 500-bit block length was about 1 dB inferior in terms of the required SNR to the 2000-bit block length in the context of the BICM-ID scheme. A slight further SNR improvement was obtained for the 4000-bit block length. In other words, the advantage of BICM-ID over TCM for transmissions over AWGN channels is more significant for larger block lengths. The 8-state TTCM performance also improves, when using four iterations, as the block length is increased and, on the whole, TTCM is the best performer in this scenario.

Figure 25.3 shows the effects of the decoding complexity on the TCM, TTCM, BICM and BICM-ID schemes' performance in the context of an 8PSK scheme for transmissions over AWGN channels using a block length of 4000 information bits (2000 symbols). Again,

---

[1]The EFF bound is defined as the BER upper bound performance achieved for the idealised situation, when the decoded values fed back to the demodulator in Figure 25.1 are error free.

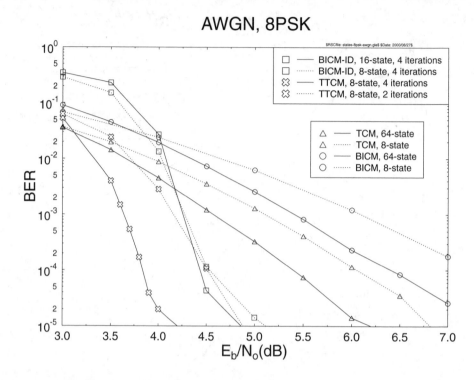

Figure 25.3: Effects of decoding complexity on the TCM, TTCM, BICM and BICM-ID schemes' performance in the context of an 8PSK scheme for transmissions over AWGN channels using a block length of 4000 information bits (2000 symbols)

the 64-state TCM, 64-state BICM, 8-state TTCM using four iterations and 16-state BICM-ID along with four iterations exhibit a similar decoding complexity. At a BER of 10^{-4}, TTCM requires about 0.6 dB lower SNR than BICM-ID, 1.6 dB less energy than TCM and 2.5 dB lower SNR than BICM. When the decoding complexity is reduced such that 8-state codes are used in the TCM, BICM and BICM-ID schemes, their corresponding performance becomes worse than that of the 64-state codes, as shown in Figure 25.3. In order to be able to compare the associated performance with that of 8-state BICM-ID using four iterations, 8-state TTCM along with two iterations is employed. Observe that due to the insufficient number of iterations, TTCM exhibits only marginal advantage over BICM-ID.

Figure 25.4 shows the performance of TCM, TTCM and BICM-ID invoking 16QAM for transmissions over AWGN channels using a block length of 6000 information bits (2000 symbols). Upon comparing 64-state TCM with 32-state BICM-ID using two iterations, we observed that BICM-ID outperforms TCM for E_b/N_0 values in excess of 6.8 dB. However, 8-state BICM-ID using an increased number of iterations, such as eight, outperforms the similar complexity 32-state BICM-ID scheme employing two iterations as well as 64-state TCM. An approximately 1.2 dB E_b/N_0 gain was obtained at a BER of 10^{-4} for 8-state BICM-ID using eight iterations over 64-state TCM at a similar decoding complexity. Comparing 8-state TTCM using two iterations and 8-state BICM-ID employing four iterations reveals that BICM-ID performs better for the E_b/N_0 range of 5.7 dB to 7 dB. When the number of

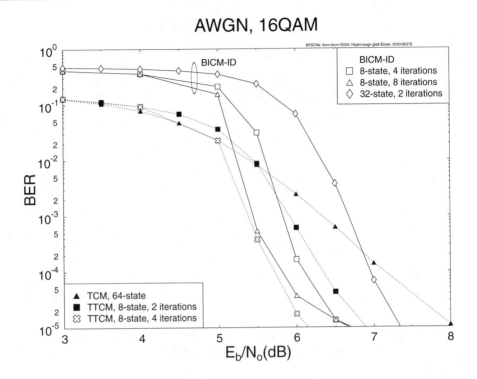

Figure 25.4: Performance comparison of TCM, TTCM and BICM-ID employing 16QAM for transmissions over AWGN channels using a block length of 6000 information bits (2000 symbols)

iterations is increased to four for TTCM and to eight for BICM-ID, TTCM exhibits a better performance, as seen in Figure 25.4.

Owing to the associated SP, the intra-subset distance of TCM and TTCM increases as we traverse down the partition tree of Figure 24.8, for example. It was shown in [709] that we only need to encode $\tilde{m} = 2$ out of $\bar{m} = 5$ information bits in the 64QAM/TTCM to attain target BERs around 10^{-5} in AWGN channels. Hence in this scenario there are $2^{\bar{m}-\tilde{m}} = 8$ parallel transitions due to the $\bar{m} - \tilde{m} = 3$ uncoded information bits in the trellis of 64QAM/TTCM. Figure 25.5 illustrates the performance of TCM, TTCM, BICM and BICM-ID using 64QAM over AWGN channels. When using a block length of 10000 information bits (2000 symbols), 8-state TTCM invoking four iterations is the best candidate, followed by the similar complexity 8-state BICM-ID scheme employing eight iterations. Again, TCM performs better than BICM in AWGN channels. When a block length of 1250 information bits (250 symbols) was used, both TTCM and BICM-ID experienced a performance degradation. It is also seen in Figure 25.5 that BICM-ID performs close to TTCM, when a longer block length is used.

25.2.2.2 Performance over Uncorrelated Narrowband Rayleigh Fading Channels

The uncorrelated Rayleigh fading channels implied using an infinite-length interleaver over narrowband Rayleigh fading channels. Figure 25.6 shows the performance of 64-state TCM, 64-state BICM, 8-state TTCM using four iterations and 16-state BICM-ID employing four

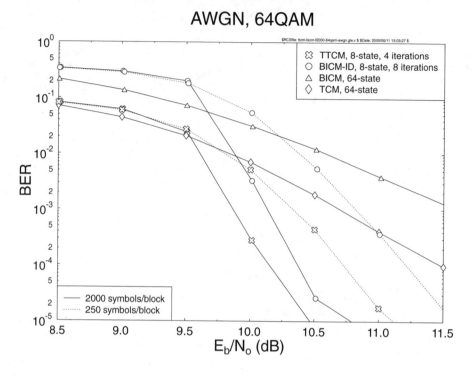

Figure 25.5: Performance comparison of TCM, TTCM, BICM and BICM-ID using 64QAM over AWGN channels

iterations in the context of an 8PSK scheme communicating over uncorrelated narrowband Rayleigh fading channels using a block length of 4000 information bits (2000 symbols). These four coded modulation schemes have a similar complexity. As can be seen from Figure 25.6, TTCM performs best, followed by BICM-ID, BICM and TCM. At a BER of 10^{-4}, TTCM performs about 0.7 dB better in terms of the required E_b/N_0 value than BICM-ID, 2.3 dB better than BICM and 4.5 dB better than TCM. The error floor of TTCM [709] was lower than the associated EFF bound of BICM-ID. However, the BERs of TTCM and BICM-ID were identical at $E_b/N_0 = 7$ dB.

Figure 25.7 compares the performance of TCM, TTCM and BICM-ID invoking 16QAM for communicating over uncorrelated narrowband Rayleigh fading channels using a block length of 6000 information bits (2000 symbols). Observe that 32-state BICM-ID using two iterations outperforms 64-state TCM for E_b/N_0 in excess of 9.6 dB. At the same complexity, 8-state BICM-ID invoking eight iterations outperforms 64-state TCM beyond $E_b/N_0 = 8.2$ dB. Similarly to 8PSK, the coding gain of BICM-ID over TCM in the context of 16QAM is more significant over narrowband Rayleigh fading channels compared to AWGN channels. Near E_b/N_0 of 11 dB the 8-state BICM-ID scheme approaches the EFF bound, hence 32-state BICM-ID using two iterations exhibits a better performance due to its lower EFF bound. Observe also that 8-state BICM-ID using four iterations outperforms 8-state TTCM employing two iterations in the range of $E_b/N_0 = 8.5$ dB to 12.1 dB. Increasing the number of iterations only marginally improves the performance of BICM-ID, but results in a significant gain for

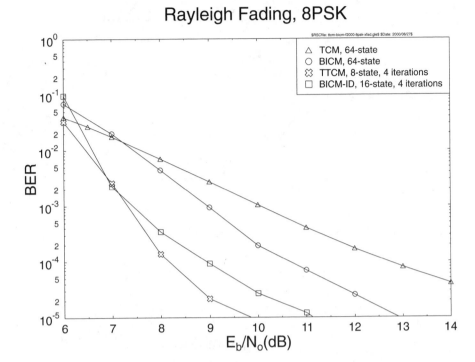

Figure 25.6: Performance comparison of TCM, TTCM, BICM and BICM-ID for 8PSK transmissions over uncorrelated Rayleigh fading channels using a block length of 4000 information bits (2000 symbols)

TTCM. The performance of 8-state TTCM using four iterations is better than that of 8-state BICM-ID along with eight iterations for E_b/N_0 values in excess of 9.6 dB.

Figure 25.8 illustrates the performance of TCM, TTCM, BICM and BICM-ID when invoking 64QAM for communicating over uncorrelated narrowband Rayleigh fading channels. Using a block length of 10000 information bits (2000 symbols), 64-state BICM performs better than 64-state TCM for E_b/N_0 values in excess of 15 dB. BICM-ID exhibits a lower error floor than TTCM in this scenario, since BICM-ID protects all the five information bits, while TTCM protects only two information bits of the six-bit 64QAM symbol. The three unprotected information bits of TCM and TTCM render these schemes less robust to the bursty error effects of the uncorrelated fading channel. If we use a TCM or TTCM code generator that encodes all the five information bits, a better performance is expected. Reducing the block length from 2000 symbols to 250 symbols resulted in a small performance degradation for TTCM, but yielded a significant degradation for BICM-ID.

25.2.2.3 Coding Gain versus Complexity and Interleaver Block Length

In this section, we will investigate the coding gain (G) of the coded modulation schemes utilising an 8PSK scheme versus the Decoding Complexity (DC) and the Interleaver Block Length (IL) at a BER of 10^{-4}. The coding gain G is measured by comparing to the uncoded

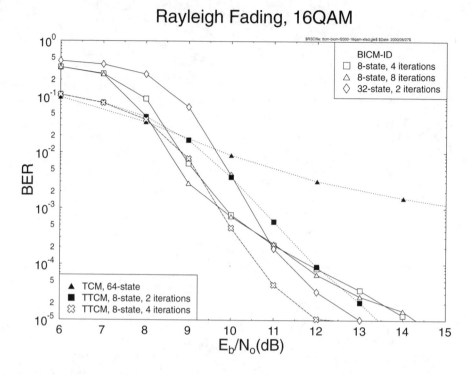

Figure 25.7: Performance comparison of TCM, TTCM and BICM-ID for 16QAM transmissions over uncorrelated narrowband Rayleigh fading channels transmitting 2000 symbols/block (6000 information bits/block)

4PSK scheme, which exhibits a BER of 10^{-4} at $E_b/N_0 = 8.35$ dB and $E_b/N_0 = 35$ dB for transmissions over AWGN channels and uncorrelated narrowband Rayleigh fading channels, respectively. Again, the DC is measured using the associated number of decoding states and the notations S and t represent the number of decoding states and the number of decoding iterations, respectively. Hence, the relative complexity of TCM, BICM, TTCM and BICM-ID is given by S, S, $2tS$ and tS, respectively. The IL is measured in terms of the number of information bits in the interleaver.

Figure 25.9 portrays the coding gain G versus DC plot of the coded modulation schemes for 8PSK transmissions over (a) AWGN channels and (b) uncorrelated narrowband Rayleigh fading channels, using an IL of 4000 information bits (2000 symbols). At a DC as low as 8, the non-iterative TCM scheme exhibits the highest coding gain G for transmissions over AWGN channels, as seen in Figure 25.9(a). By contrast, the BICM scheme exhibits the highest coding gain G for transmissions over uncorrelated narrowband Rayleigh fading channels, as seen in Figure 25.9(b). However, for a DC higher than 16, the iterative TTCM and BICM-ID schemes exhibit higher coding gains than their non-iterative counterparts for transmission over both channels.

For the iterative schemes different combinations of t and S may yield different performances at the same DC. For example, the coding gain G of BICM-ID in conjunction with $tS = 8 \times 8$ is better than that of $tS = 4 \times 16$ at DC=64 for transmissions over AWGN chan-

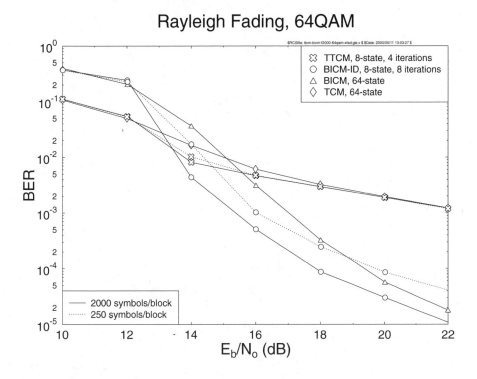

Figure 25.8: Performance comparison of TCM, TTCM, BICM and BICM-ID using 64QAM for transmissions over uncorrelated narrowband Rayleigh fading channels

nels, as seen in Figure 25.9(a), since BICM-ID invoking a constituent code associated with $S = 16$ has not reached its optimum performance at iteration $t = 4$. However, the coding gain G of BICM-ID in conjunction with $tS = 4 \times 16$ is better than that of $tS = 8 \times 8$ at DC = 64, when communicating over uncorrelated narrowband Rayleigh fading channels, as seen in Figure 25.9(b). This is because BICM-ID invoking a constituent code associated with $S = 8$ has reached its EFF bound at iteration $t = 4$, while BICM-ID invoking a constituent code associated with $S = 16$ has not reached its EFF bound, because the EFF bound for code associated with $S = 16$ is lower than that of a code associated with $S = 18$. In general, the coding gain G of TTCM is the highest for DC values in excess of 32 for transmissions over both channels.

Figure 25.10 portrays the coding gain G versus IL plot of the coded modulation schemes for 8PSK transmissions over (a) AWGN channels and (b) uncorrelated narrowband Rayleigh fading channels in conjunction with a DC of 64 both with and without code termination [363]. We can observe in Figure 25.10(a) that IL affects the performance of the schemes using no code termination, since the shorter the IL, the higher the probability of the decoding trellis terminating at a wrong state. For transmissions over AWGN channels and upon using code-terminated schemes, only the performance of the BICM-ID scheme is affected by the IL, since the performance of the scheme communicating over AWGN channels depends on the FED, while the high FED of BICM-ID depends on the reliability of the feedback values. Therefore, when the IL is short, BICM-ID suffers from a performance degradation. However, the other

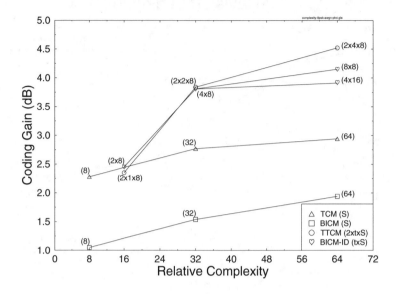

(a) AWGN channels

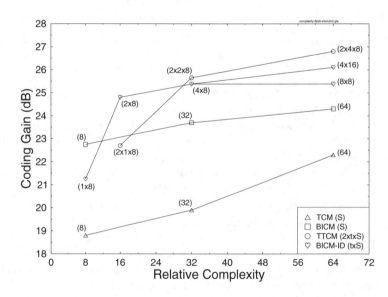

(b) uncorrelated narrowband Rayleigh fading channels

Figure 25.9: Coding gain at BER = 10^{-4} over the uncoded 4PSK scheme, against the decoding complexity of TCM, TTCM, BICM and BICM-ID for 8PSK transmissions over (a) AWGN channels and (b) uncorrelated narrowband Rayleigh fading channels, using an interleaver block length of 4000 information bits (2000 symbols). The notations S and t represent the number of decoding states and the number of decoding iterations, respectively

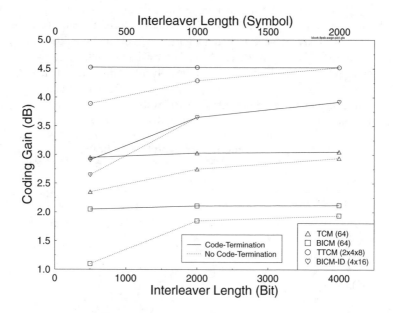

(a) AWGN channels

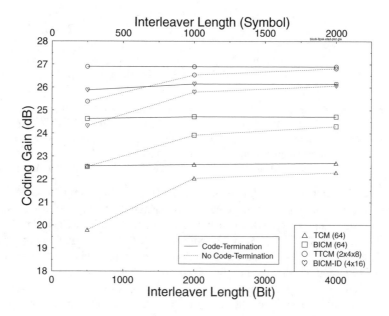

(b) uncorrelated narrowband Rayleigh fading channels ·

Figure 25.10: Coding gain at a BER of 10^{-4} over the uncoded 4PSK scheme, against the IL of TCM, TTCM, BICM and BICM-ID for 8PSK transmissions over (a) AWGN channels and (b) uncorrelated narrowband Rayleigh fading channels, invoking a DC of 64 applying code termination or no code termination.

schemes are not affected by the IL when communicating over AWGN channels, as seen in Figure 25.10(a), since there are no bursty channel errors to be dispersed by the interleaver and hence there is no advantage in utilising a long IL. To elaborate a little further, as seen in Figure 25.10 for transmissions over uncorrelated narrowband Rayleigh fading channels using code-terminated schemes, the IL does not significantly affect the performance of the schemes, since the error events are uncorrelated in the uncorrelated Rayleigh fading scenario. These results constitute the upper bound performance achievable when an infinitely long interleaver is utilised for rendering the error events uncorrelated.

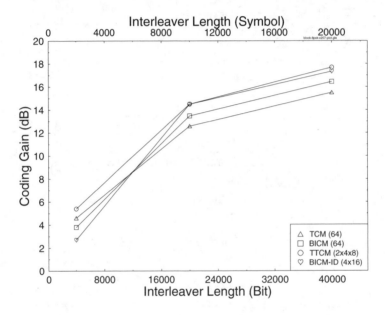

Figure 25.11: Coding gain at a BER of 10^{-4} over the uncoded 4PSK scheme, against the IL of TCM, TTCM, BICM and BICM-ID for 8PSK transmissions over correlated narrowband Rayleigh fading channels, invoking a DC of 64 applying code termination. The normalised Doppler frequency of the channel is 3.25×10^{-5}, which corresponds to a Baud rate of 2.6 MBaud, a carrier frequency of 1.9 GHz and a vehicular speed of 30 mph

Figure 25.11 portrays the coding gain G versus IL plot of the coded modulation schemes for 8PSK transmissions over correlated narrowband Rayleigh fading channels, in conjunction with a decoding complexity of 64, when applying code termination. The normalised Doppler frequency of the channel is 3.25×10^{-5}, which corresponds to a Baud rate of 2.6 MBaud, a carrier frequency of 1.9 GHz and a vehicular speed of 30 mph. This is a slow fading channel and hence the fading envelope is highly correlated. It is demonstrated by Figure 25.11 that the coding gain G of all coded modulation schemes improves as the IL increases. This is because the MAP decoder is unable to perform at its best when the channel errors occur in bursts. However, the performance improves when the error bursts are dispersed by the employment of a long interleaver. In general, TTCM is the best performer for a variety of IL values. However, BICM-ID is the worst performer for an IL of 4000 bits, while performing similarly to TTCM for long IL values.

On one hand, TCM performs better than BICM for short IL values, which follows the performance trends observed for transmissions over AWGN channels, as shown in Figure 25.10(a). This is because slowly fading channels are highly correlated and hence they behave as near-Gaussian channels, where TCM is at its best, since TCM was designed for Gaussian channels. By contrast, although BICM was designed for fading channels, when the channel-induced error bursts are inadequately dispersed owing to the employment of a short IL, the performance of BICM suffers. In other words, when communicating over slowly fading channels, extremely long interleavers may be necessary for over-bridging the associated long fades and for facilitating the dispersion of bursty transmission errors, which is a prerequisite for the efficient operation of channel codecs.

On the other hand, BICM performs better than TCM for long IL values, which is reminiscent of the performance trends observed when communicating over uncorrelated Rayleigh fading channels, as evidenced by Figure 25.10(b). This is justified, since the correlation of the fading channel is broken when a long IL is employed for dispersing the error bursts.

25.2.3 Conclusion

In conclusion, at a given complexity TCM performs better than BICM in AWGN channels, but worse in uncorrelated narrowband Rayleigh fading channels. However, BICM-ID using soft-decision feedback outperforms TCM and BICM for transmissions over both AWGN and uncorrelated narrowband Rayleigh fading channels at the same DC. TTCM has shown superior performance over the other coded modulation schemes studied, but exhibited a higher error floor for the 64QAM scheme due to the presence of uncoded information bits for transmissions over uncorrelated narrowband Rayleigh fading channels. Comparing the coding gain against the DC, the iterative decoding schemes of TTCM and BICM-ID are capable of providing a high coding gain even in conjunction with a constituent code exhibiting a short memory length, although only at the cost of a sufficiently high number of decoding iterations, which may imply a relatively high decoding complexity. Comparing the achievable coding gain against the IL, TTCM is the best performer for a variety of ILs, while the performance of BICM-ID is highly dependent on the IL for transmissions over both AWGN and Rayleigh fading channels.

25.3 Orthogonal Frequency Division Multiplexing

Owing to the Inter-Symbol Interference (ISI) inflicted by wideband channels, the employment of channel equalisers is essential in supporting the operation of the coded modulation schemes considered. However, the employment of channel equalisers will be discussed only at a later stage, namely, in Chapter 26. As an attractive design alternative, here a Discrete Fourier Transform (DFT) based multi-carrier OFDM scheme [180] will be utilised for removing the ISI. Specifically, as long as the OFDM subcarrier signalling rate is sufficiently low since the number of subcarriers is sufficiently high, the individual OFDM subchannels experience non-dispersive narrowband fading conditions.

25.3.1 Orthogonal Frequency Division Multiplexing Principle

In this section we briefly introduce Frequency Division Multiplexing (FDM), also referred to as Orthogonal Multiplexing (OMPX), as a means of dealing with the problems of frequency-selective fading encountered when transmitting over a high-rate wideband radio channel. The fundamental principle of orthogonal multiplexing originates from Chang [143], and over the years a number of researchers have investigated this technique [144, 146]. Despite its conceptual elegance, until recently its employment has been mostly limited to military applications because of the associated implementational difficulties. However, it has recently been adopted as the new European Digital Audio Broadcasting (DAB) standard; it is also a strong candidate for Digital Terrestrial Television Broadcast (DTTB) and for a range of other high-rate applications, such as 155 Mbit/s wireless Asynchronous Transfer Mode (ATM) local area networks. These wide-ranging applications underline its significance as an alternative technique to conventional channel equalisation in order to combat signal dispersion [147, 155, 337].

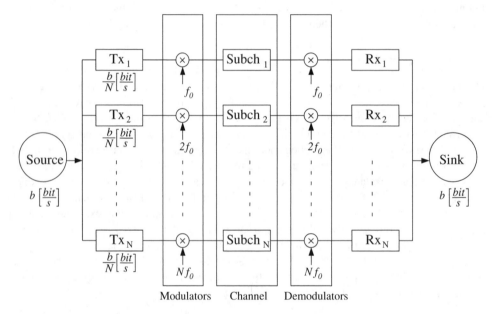

Figure 25.12: Simplified block diagram of the orthogonal parallel modem

In the FDM scheme of Figure 25.12 the serial data stream of a traffic channel is passed through a serial-to-parallel converter, which splits the data into a number of parallel subchannels. The data in each sub-channel are applied to a modulator, such that for N channels there are N modulators whose carrier frequencies are $f_0, f_1, \ldots, f_{N-1}$. The centre frequency difference between adjacent channels is Δf and the overall bandwidth W of the N modulated carriers is $N\Delta f$.

These N modulated carriers are then combined to give a FDM signal. We may view the serial-to-parallel converter as applying every Mth symbol to a modulator. This has the effect of interleaving the symbols entered into each modulator, hence symbols $S_0, S_N, S_{2N}, \ldots$ are applied to the modulator whose carrier frequency is f_0. At the receiver the received FDM signal is demultiplexed into N frequency bands, and the N modulated signals are demodulated. The baseband signals are then recombined using a parallel-to-serial converter.

The main advantage of the above FDM concept is that because the symbol period has been increased, the channel's delay spread is a significantly shorter fraction of a symbol period than in the serial system, potentially rendering the system less sensitive to ISI than the conventional N times higher-rate serial system. In other words, in the low-rate sub-channels the signal is no longer subject to frequency-selective fading, hence no channel equalisation is necessary.

A disadvantage of the FDM approach shown in Figure 25.12 is its increased complexity in comparison to the conventional system caused by employing N modulators and filters at the transmitter and N demodulators and filters at the receiver. It can be shown that this complexity can be reduced by employing the DFT, typically implemented with the aid of the Fast Fourier Transform (FFT) [180]. The sub-channel modems can use almost any modulation scheme, and 4- or 16-level QAM is an attractive choice in many situations.

The FFT-based QAM/FDM modem's schematic is portrayed in Figure 25.13. The bits provided by the source are serial/parallel converted in order to form the n-level Gray-coded symbols, N of which are collected in TX buffer 1, while the contents of TX buffer 2 are being transformed by the IFFT in order to form the time-domain modulated signal. The Digital-to-Analogue (D/A) converted, low-pass filtered modulated signal is then transmitted via the channel and its received samples are collected in RX buffer 1, while the contents of RX buffer 2 are being transformed to derive the demodulated signal. The twin buffers are alternately filled with data to allow for the finite FFT-based demodulation time. Before the data are Gray coded and passed to the data sink, they can be equalised by a low-complexity method, if there are some dispersions within the narrow sub-bands. For a deeper tutorial exposure the interested reader is referred to reference [180].

25.4 Coded Modulation Aided Orthogonal Frequency Division Multiplexing

25.4.1 Introduction

The coded modulation schemes of Chapter 24 are here amalgamated with an OFDM modem. Specifically, the coded symbols are mapped to the OFDM modulator at the transmitter, while the symbols output by the OFDM demodulator are channel decoded at the receiver.

When the channel is frequency selective and OFDM modulation is used, the received symbol is given by the product of the transmitted frequency-domain OFDM symbol and the frequency-domain transfer function of the channel. This direct relationship facilitates the employment of simple frequency-domain channel equalisation techniques. Essentially, if an estimate of the complex frequency-domain transfer function H_k is available at the receiver, channel equalisation can be performed by dividing each received value by the corresponding frequency-domain channel transfer function estimate. The channel's frequency-domain transfer function can be estimated with the aid of known frequency-domain pilot subcarriers inserted into the transmitted signal's spectrum [180]. These known pilots effectively sample the channel's frequency-domain transfer function according to the Nyquist frequency. These frequency-domain samples then allow us to recover the channel's transfer function between the frequency-domain pilots with the aid of interpolation. In addition to this simple form of channel equalisation, another advantage of the OFDM-based modulation is that it turns a channel exhibiting memory into a memoryless one, where the memory is the influence of the past transmitted symbols on the value of the present symbol.

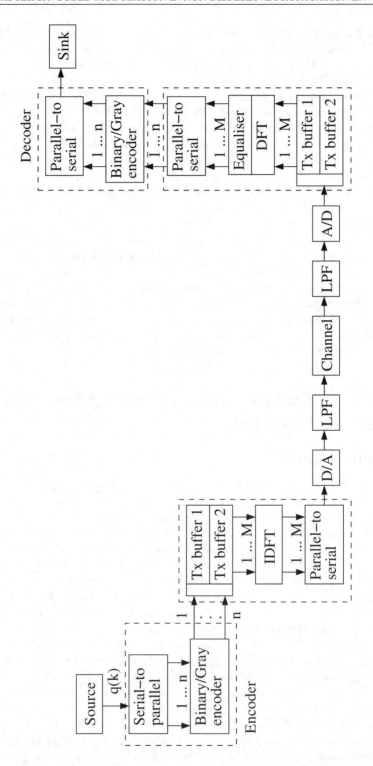

Figure 25.13: FFT-based OFDM modem schematic ©Webb, Hanzo, 1994, [180]

In the literature the performance of TCM and TTCM assisted OFDM modems has been investigated in [732] when communicating over various dispersive Rayleigh fading channels. In [733], a multilevel coding and BICM assisted OFDM modem employing noncoherent reception was investigated in dispersive Rayleigh fading environments. In this section, the performance of TCM, TTCM, BICM and BICM-ID schemes will be investigated using the Pan-European Digital Video Broadcasting (DVB) standard's OFDM scheme [180] for transmissions over the 12-path COST207 [676] Hilly Terrain (HT) Rayleigh fading channels, which characterise a rather hostile, strongly dispersive environment.

25.4.2 System Overview

The encoder produces a block of N_i channel symbols to be transmitted. These symbols are transmitted by the OFDM modulator. As the OFDM modulator transmits N_u modulated symbols per OFDM symbol, if $N_u = N_i$ then the whole block of N_u modulated symbols can be transmitted in a single OFDM symbol. We refer to this case as the single-symbol mapping based scenario. By contrast, if $N_u < N_i$, then more than one OFDM symbol is required for the transmission of the channel-coded block. We refer to this case as the multiple-symbol mapping based scenario. Both the single- and the multiple-symbol scenarios are interesting. The single-symbol scenario is more appealing from an implementation point of view, as it is significantly more simple. However, it is well known that the performance of a turbo-coded scheme improves upon increasing the IL. Since the number of subcarriers in an OFDM system is limited by several factors, such as for example the oscillator's frequency stability, by using the single-symbol solution we would be limited in terms of the TTCM block length as well. Thus the multiple-symbol solution also has to be considered in order to fully exploit the advantages of the TTCM scheme. Since the single-symbol based scheme is conceptually more simple, we will consider this scenario first. Its extension to the multiple-symbol scenario is straightforward.

In the single carrier system discussed in Section 26.3.1 the received signal is given by $y_k = x_k * h_k + n_k$, where $*$ denotes the convolution of the transmitted sequence x_k with the channel's impulse response h_k. An equaliser is used for removing the ISI before channel decoding, giving $\tilde{y}_k = \tilde{x}_k + \tilde{n}_k$. The associated branch metrics can be computed as:

$$
\begin{aligned}
P(\tilde{y}_k|\tilde{x}_k) &= e^{-\frac{|\tilde{n}|^2}{2\sigma^2}} \\
&= e^{-\frac{|\tilde{y}_k - \tilde{x}_k|^2}{2\sigma^2}}.
\end{aligned}
\tag{25.5}
$$

However, in a multi-carrier OFDM system, the received signal is given by $Y_k = X_k \cdot H_k + N_k$, which facilitates joint channel equalisation and channel decoding by computing the branch metrics as:

$$
P(y_k|x_k) = e^{-\frac{|Y_k - H_k \cdot X_k|^2}{2\sigma^2}},
\tag{25.6}
$$

where H_k is the channel's frequency-domain transfer function at the centre frequency of the kth subcarrier. Hence, as long as the the channel transfer function estimation is of sufficiently high quality, simple frequency-domain equalisation could be invoked during the decoding process. If iterative channel decoding is invoked, the channel transfer function estimation is expected to improve during the consecutive iterative steps, in a fashion known in the context of turbo equalisation [734]. Indeed, a performance as high as that in conjunction with perfect

OFDM Parameters	
Total number of subcarriers, N	2048 (2K mode)
Number of effective subcarriers, N_u	1705
OFDM symbol duration T_s	224 μs
Guard interval	$T_s/4 = 56\mu$s
Total symbol duration (inc. guard interval)	280 μs
Consecutive subcarrier spacing $1/T_s$	4464 Hz
DVB channel spacing	7.61 MHz
QPSK and QAM symbol period	7/64 μs
Baud rate	9.14 MBaud

Table 25.3: Parameters of the OFDM module [438]

channel estimation can be attained [735].

Let us now consider the effect of the channel interleaver. When the channel is frequency selective, it exhibits frequency-domain nulls, which may obliterate several OFDM subcarriers. Thus the quality of several consecutive received subcarrier symbols will be low. If the quality is inferior, the channel decoder is unable to correctly estimate the transmitted symbols. When, however, a channel interleaver is present, the received symbols are shuffled before channel decoding and hence these clusters of corrupted subcarriers are dispersed. Thus, after the channel interleaver we expect to have only isolated low-quality subcarriers surrounded by unimpaired ones. In this case the decoder is more likely to be able to recover the symbol transmitted on the corrupted subcarrier, using the redundancy added by the channel coding process, conveyed by the surrounding unimpaired subcarriers.

Finally, we consider the multiple-symbol scenario. The system requires only a minor modification. When more than one OFDM symbol is required for transmitting the block of channel-coded symbols, the OFDM demodulator has to store both the demodulated symbols and the channel transfer function estimates in order to form a whole channel-coded block. This block is then fed to the channel interleaver and then to the channel decoder, together with the channel transfer function estimate. Exactly the same MAP decoder as in the single-symbol case can be used for performing the joint channel equalisation and channel decoding. Similarly, the function of the channel interleaver is the same as in the single-symbol scenario.

25.4.3   Simulation Parameters

The DVB standard's OFDM scheme [180] was used for this study. The parameters of the OFDM DVB system are presented in Table 25.3. Since the OFDM modem has 2048 subcarriers, the subcarrier signalling rate is effectively 2000 times lower than the maximum DVB transmission rate of 20 Mbit/s, corresponding to about 10 kbit/s. At this sub-channel rate, the individual sub-channels can be considered nearly frequency-flat.

The channel model employed is the 12-path COST207 [676] HT type of impulse response, exhibiting a maximum relative path delay of 19.9 μs. The unfaded impulse response is depicted in Figure 25.14. The carrier frequency is 500 MHz and the sampling rate is 7/64 μs. Each of the channel paths was faded independently, obeying a Rayleigh fading distribution, according to a normalised Doppler frequency of 10^{-5} [370]. This corresponds to a

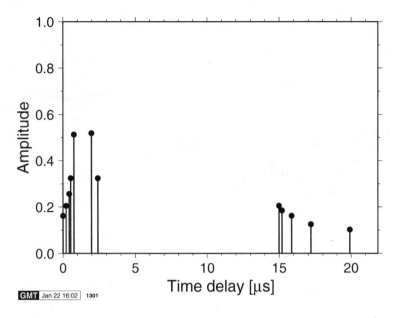

Figure 25.14: COST207 Hilly Terrain (HT) type of impulse response [676]

worst-case vehicular velocity of about 200 km/h.

25.4.4 Simulation Results and Discussion

In this section, the system performance of the OFDM-based coded modulation schemes is evaluated using QPSK, 8PSK and 16QAM. The coding rate of the coded modulation schemes changes according to the modem mode used. Hence the effective throughput for QPSK is 1 bit/subcarrier, for 8PSK it is 2 bits/subcarrier and finally for 16QAM it is 3 bits/subcarrier.

Figure 25.15 shows the performance of the integrated systems in conjunction with a channel interleaver of 5000 symbols using a QPSK modem. We can see from Figure 25.15 that at a BER of 10^{-4} BICM performs about 1.5 dB better than TCM. When the iterative schemes of TTCM and BICM-ID are used, a better performance is achieved. Specifically, in conjunction with eight iterations, TTCM and BICM-ID approach their best achievable performance. Clearly, in this scenario TTCM is superior to BICM-ID.

Figure 25.16 illustrates the performance of the systems considered, which maintain an effective throughput of 2 bits/subcarrier using a channel interleaver length of 5000 symbol duration. All schemes exhibit a similar complexity, however, TTCM using eight iterations is slightly more complex. Specifically, in order to be able to compare the associated complexities we assumed that the number of decoder trellis states determined the associated implementation/decoding complexity. For example, a memory-length $\nu = 3$ TTCM scheme had $2^{\nu} = 8$ trellises per decoding iteration and there are two decoders. Hence, after four iterations we encounter a total of $8 \cdot 2 \cdot 4 = 64$ trellis states. Therefore a 64-state, $\nu = 6$ TCM scheme has a similar complexity to a $\nu = 4$ TTCM arrangement using four iterations. Again, TTCM is the best scheme. At a BER of 10^{-4}, TTCM performs about 0.7 dB better than BICM-ID, 1.8 dB better than BICM and 3.0 dB better than TCM at a similar complexity. Another 0.6 dB

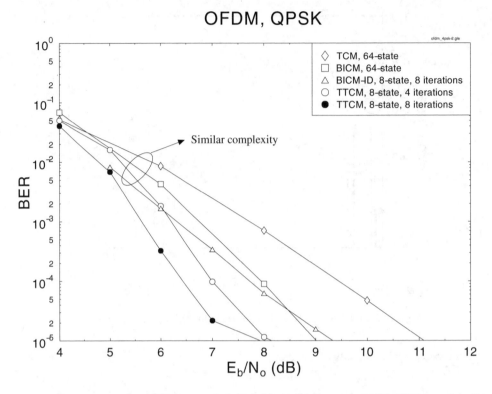

Figure 25.15: Comparison of TCM, TTCM, BICM and BICM-ID using QPSK-OFDM modem for transmissions over the Rayleigh fading COST207 HT channel of Figure 25.14 at a normalised Doppler frequency of 10^{-5}. The OFDM parameters are listed in Table 25.3, while the coded modulation parameters were summarised in Section 25.2.1.

gain was attained by TTCM, when the number of turbo iteration was increased from four to eight.

Figure 25.17 compares the performance of the systems maintaining a throughput of 3 bits/subcarrier using a channel interleaver of 5000 symbols. Again, the systems exhibited a similar complexity, except for TTCM employing eight iterations. There is one uncoded information bit in the 4-bit 16QAM symbol of the rate-3/4 TCM scheme having 64 states, as can be seen from its generator polynomial in Table 25.1. For this reason, this TCM scheme is only potent at lower SNRs, while exhibiting modest performance improvements in the higher SNR region, as demonstrated by Figure 25.17. The rest of the schemes do not have uncoded information bits in their 16QAM symbols. BICM-ID outperforms BICM for E_b/N_0 values in excess of about 1.2 dB, while it is inferior in comparison to TTCM by about 0.2 dB at a BER of 10^{-4}. Note that BICM-ID starts to exhibit an error floor at a BER around 10^{-4}, while TTCM starts to exhibit an error floor at a BER around 10^{-5}.

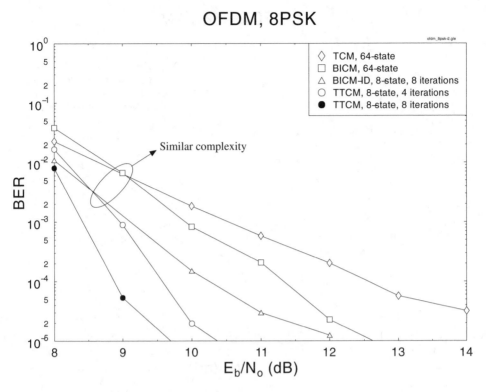

Figure 25.16: Comparison of TCM, TTCM, BICM and BICM-ID using 8PSK-OFDM modem for transmissions over the Rayleigh fading COST207 HT channel of Figure 25.14 at a normalised Doppler frequency of 10^{-5}. The OFDM parameters are listed in Table 25.3 while the coded modulation parameters were summarised in Section 25.2.1.

25.4.5 Conclusion

In this section OFDM was integrated with the coded modulation schemes of Chapter 24. The performance of TCM, TTCM, BICM and BICM-ID assisted OFDM was investigated for transmissions over the dispersive COST207 HT Rayleigh fading channel of Figure 25.14 using QPSK, 8PSK and 16QAM modulation modes. TTCM was found to be the best compromise scheme, followed by BICM-ID, BICM and TCM.

25.5 Summary

In Section 25.2 the performance of four coded modulation schemes was evaluated for transmissions over AWGN and narrowband fading channels. Explicitly, it was shown in Sections 25.2.2.1 and 25.2.2.2 that at a given complexity TCM performed better than BICM in AWGN channels, but performed worse than BICM in uncorrelated narrowband Rayleigh fading channels. However, BICM-ID outperformed TCM and BICM for transmissions over both AWGN and uncorrelated narrowband Rayleigh fading channels at the same decoding complexity. TTCM, however, has shown superior performance over the other coded modulation

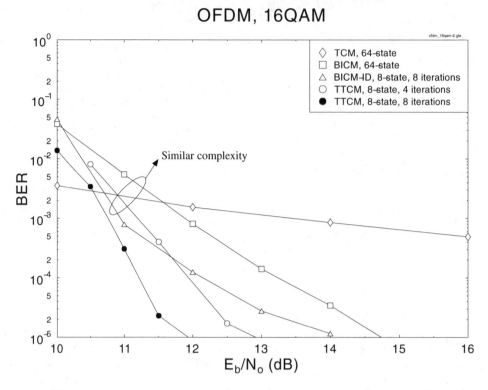

Figure 25.17: Comparison of TCM, TTCM, BICM and BICM-ID using 16QAM-OFDM modem for transmissions over the Rayleigh fading COST207 HT channel of Figure 25.14 at a normalised Doppler frequency of 10^{-5}. The OFDM parameters are listed in Table 25.3 while the coded modulation parameters were summarised in Section 25.2.1.

schemes studied. By contrast, in Section 25.4 we evaluated the performance of the coded modulation schemes when communicating over the wideband COST207 HT Rayleigh fading channels, by assuming that the number of subcarriers was sufficiently high for maintaining narrowband channel conditions in each OFDM subchannel. In the context of 8PSK-based OFDM, TTCM performs about 0.7 dB better than BICM-ID, 1.8 dB better than BICM and 3.0 dB better than TCM at a BER of 10^{-4}, as evidenced by Figure 25.16.

The TTCM assisted scheme was found to represent the most attractive trade-off in terms of its performance versus complexity balance, closely followed by the BICM-ID assisted scheme, both in the context of the conventional single carrier system of Section 25.2 as well as in the multi-carrier OFDM system of Section 25.4, when communicating over non-dispersive propagation environment.

In summary, the coding gain values exhibited by the coded modulation schemes studied in this chapter were tabulated in Tables 25.4 and 25.5. Specifically, we summarised the performance gain exhibited by the various coded modulation schemes of Section 25.2, when communicating over non-dispersive AWGN channels and uncorrelated Rayleigh fading channels in Table 25.4. The coding gains exhibited by the OFDM-based coded modulation schemes of Section 25.4 communicating over the COST207 HT Rayleigh fading channels were tabulated in Table 25.5.

Code	Modulation/ BPS	AWGN Channels				Flat Uncorrelated Rayleigh Fading Channels			
		E_b/N_o (dB)		Gain (dB)		E_b/N_o (dB)		Gain (dB)	
		BER				BER			
		10^{-3}	10^{-5}	10^{-3}	10^{-5}	10^{-3}	10^{-5}	10^{-3}	10^{-5}
Uncoded	BPSK/1	6.75	9.52	0.00	0.00	24.00	43.50	0.00	0.00
TCM	QPSK/1	2.81	4.42	3.94	5.10	6.29	9.88	17.71	33.62
TTCM	QPSK/1	**1.27**	**1.90**	5.48	7.62	**3.40**	**4.28**	20.60	39.22
BICM	QPSK/1	2.60	4.16	4.15	5.36	5.00	7.46	19.00	36.04
BICM-ID	QPSK/1	2.00	4.07	4.75	5.45	5.05	8.50	18.95	35.00
Uncoded	QPSK/2	6.75	9.52	0.00	0.00	24.00	43.50	0.00	0.00
TCM	8PSK/2	4.57	6.20	2.18	3.32	10.05	17.00	13.95	26.50
TTCM	8PSK/2	**3.64**	**4.20**	3.11	5.32	**7.32**	**9.84**	16.68	33.66
BICM	8PSK/2	5.41	7.46	1.34	2.06	8.95	12.90	15.05	30.60
BICM-ID	8PSK/2	4.25	4.86	2.50	4.66	7.43	11.12	16.57	32.38
Uncoded	8PSK/3	10.02	12.95	0.00	0.00	26.15	46.00	0.00	0.00
TCM	16QAM/3	6.32	8.01	3.70	4.94	15.54	34.00	10.61	12.00
TTCM	16QAM/3	**5.38**	**6.15**	4.64	6.80	**9.72**	**12.10**	16.43	33.90
BICM	16QAM/3	8.50	10.10	1.52	2.85	12.76	15.50	13.39	30.50
BICM-ID	16QAM/3	5.45	6.65	4.57	6.30	9.75	14.30	16.40	31.70
Uncoded	16QAM/4	10.51	13.37	–	–	26.78	47.00	–	–
TCM	64QAM/5	10.70	12.20	–	–	22.80	43.00	–	–
TTCM	64QAM/5	**9.86**	**10.46**	–	–	22.80	43.00	–	–
BICM	64QAM/5	11.60	13.50	–	–	**17.02**	23.50	–	–
BICM-ID	64QAM/5	10.14	10.78	–	–	15.38	**22.00**	–	–
Uncoded	64QAM/6	14.74	17.72	–	–	30.37	50.00	–	–

Table 25.4: Coding gain values of the various coded modulation schemes studied in this chapter, when communicating over non-dispersive AWGN channels and uncorrelated Rayleigh fading channels. All of the coded modulation schemes exhibited a similar decoding complexity in terms of the number of decoding states, which was equal to 64 states. An interleaver block length of 2000 symbols was employed. The performance of the best scheme is printed in bold

Code	Modulation/ BPS	OFDM, COST207 HT Rayleigh fading channels			
		E_b/N_o (dB)		Gain (dB)	
		BER			
		10^{-3}	10^{-5}	10^{-3}	10^{-5}
uncoded	BPSK/1	24.18	43.70	0.00	0.00
TCM	QPSK/1	7.72	10.07	16.46	33.63
TTCM	QPSK/1	**6.20**	**8.12**	17.98	35.58
BICM	QPSK/1	6.75	9.00	17.43	34.70
BICM-ID	QPSK/1	6.32	9.31	17.86	34.39
uncoded	QPSK/2	24.18	43.70	0.00	0.00
TCM	8PSK/2	10.53	15.68	13.65	28.02
TTCM	8PSK/2	**8.96**	**10.36**	15.22	33.34
BICM	8PSK/2	9.91	12.61	14.27	31.09
BICM-ID	8PSK/2	9.11	12.20	15.07	31.50
uncoded	8PSK/3	26.47	46.02	0.00	0.00
TCM	16QAM/3	11.19	33.90	15.28	12.12
TTCM	16QAM/3	13.42	**12.90**	13.05	33.12
BICM	16QAM/3	11.88	14.73	14.59	31.29
BICM-ID	16QAM/3	**10.94**	14.10	15.53	31.92
uncoded	16QAM/4	27.17	46.60	–	–
uncoded	64QAM/6	30.65	50.10	–	–

Table 25.5: Coding gain values of the various OFDM-based coded modulation schemes studied in this chapter, when communicating over COST207 HT Rayleigh fading channels. All of the coded modulation schemes exhibited a similar decoding complexity in terms of the number of decoding states, which was equal to 64 states. An interleaver block length of 5000 symbols was employed and the corresponding simulation parameters were shown in Table 25.3. The performance of the best scheme is printed in bold

Chapter 26

Coded Modulation Assisted Channel Equalised Systems

26.1 Introduction

Having studied the attainable performance of coded modulation when communicating over narrow-band fading channels in the previous chapter, in this chapter we will further investigate the performance of the TCM, BICM, TTCM and BICM-ID coded modulation schemes in the context of single carrier systems communicating over wideband fading channels. Channel equalisation was discussed in detail in the context of Chapter 7, hence in this chapter we will not delve into channel equalisation in much detail, we rather focus our attention on the related coded modulation and system design aspects. Further specific details concerning both Kalman filter and classification based Radial Basis Function (RBF) equalisers may be found in [736].

Owing to the Inter-Symbol Interference (ISI) inflicted by wideband channels, the employment of channel equalisers is essential. Hence the conventional Decision Feedback Equaliser (DFE) is introduced in Section 26.3.1, while in Section 26.4 we will evaluate the performance of a conventional DFE-aided wideband burst-by-burst adaptive coded modulation system. Specifically, in this adaptive coded modulation scheme [124] a higher-order modulation mode is employed, when the instantaneous estimated channel quality is high in order to increase the number of Bit Per Symbol (BPS) transmitted and, conversely, a more robust lower-order modulation mode is employed when the instantaneous channel quality is low, in order to improve the mean BER performance.

As another design alternative, the channel equalisation procedure can be formulated as a symbol classification problem with the advent of employing a Radial Basis Function (RBF)-based equaliser [736, 737]. The performance of this system can be further improved with the aid of *turbo equalisation* (TEQ) [369, 736]. More specifically, turbo equalisation is a joint channel equalisation and channel decoding scheme, where the equaliser is fed both by the channel outputs plus the soft decisions provided by the channel decoder. This process is invoked for a number of iterations. An overview of RBF-based channel equalisation is given in Section 26.5.1, while the TEQ scheme utilising a symbol-based MAP decoder is introduced in Section 26.6. A novel RBF-TEQ based CM scheme is presented in Section 26.7. Our discussions evolve further by introducing the reduced-complexity I/Q-TEQ philosophy in Section 26.8, while a reduced complexity RBF-I/Q-TEQ based CM scheme is proposed in Section 26.9.

Let us now commence our discourse by studying the nature of the ISI in the following section.

26.2 Intersymbol Interference

The mobile radio channels [370] can typically be characterised by band-limited linear filters. If the modulation bandwidth exceeds the coherence bandwidth of the radio channel, ISI occurs and the modulation pulses are spread or dispersed in the time domain. The ISI, inflicted by band-limited frequency, selective time-dispersive channels, distorts the transmitted signals. At the receiver, the linearly distorted signal has to be equalised in order to recover the information.

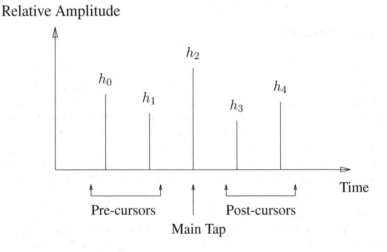

Figure 26.1: Channel Impulse Response (CIR) having pre-cursors, the main tap and post-cursors

The linearly distorted instantaneous signal received over the dispersive channel can be visualised as the superposition of the channel's response due to several information symbols in the past and in the future. Figure 26.1 shows the Channel's Impulse Response (CIR) exhibiting three distinct parts. The main tap h_2 possesses the highest relative amplitude. The taps before the main tap, namely h_0 and h_1, are referred to as pre-cursors, whereas those following the main tap, namely h_3 and h_4, are referred to as post-cursors.

The energy of the wanted signal is received mainly over the path described by the main channel tap. However, some of the received energy is contributed by the convolution of the pre-cursors with future symbols and the convolution of the post-cursor with past symbols, which are termed pre-cursor ISI and post-cursor ISI, respectively. Thus the received signal is constituted by the superposition of the wanted signal, pre-cursor ISI and post-cursor ISI.

26.3   Decision Feedback Equaliser

Channel equalisers that are utilised for compensating the effects of ISI can be classified struc-
turally as linear equalisers or DFEs. They can be distinguished also on the basis of the criterion used for optimising their coefficients. When applying the Minimum Mean Square Error
(MMSE) criterion, the equaliser is optimised such that the mean squared error between the
distorted signal and the actual transmitted signal is minimised. For time varying dispersive
channels, a range of adaptive algorithms can be invoked for updating the equaliser coefficients
and for tracking the channel variations [180].

26.3.1   Decision Feedback Equaliser Principle

The simple Zero Forcing Equaliser (ZFE) [738] forces all the impulse response contributions of the concatenated system constituted by the channel and the equaliser to zero at the
signalling instants nT for $n \neq 0$, where T is the signalling interval duration. The ZFE
provides gain in the frequency domain at frequencies where the channel's transfer function
experiences attenuation and vice versa. However, both the signal and the noise are enhanced
simultaneously and hence the ZFE is ineffective owing to the associated noise enhancement
effects. Furthermore, no finite-gain ZFE can be designed for channels that exhibit spectral
nulls [180, 739].

Linear MMSE equalisers [738] are designed for mitigating both the pre-cursor ISI and
the post-cursor ISI, as defined in Section 26.2. The MMSE equaliser is more intelligent than
the ZFE, since it jointly minimises the effects of both the ISI and noise. Although the linear
MMSE equaliser approaches the same performance as the ZFE at high SNRs, an MMSE
solution does exist for all channels, including those that exhibit spectral nulls.

The idea behind the DFE [180, 738, 739] is that once an information symbol has been
detected and decided upon, the ISI that these detected symbols inflicted on future symbols can
be estimated and the corresponding ISI can be cancelled before the detection of subsequent
symbols.

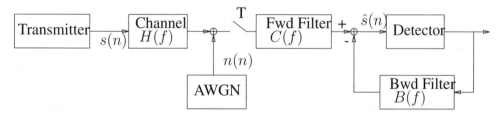

Figure 26.2: Schematic of the transmission system portraying the feedforward (Fwd) and backward
(Bwd) oriented filter of the DFE, where $C(f)$ and $B(f)$ are the corresponding frequency-
domain transfer functions, respectively.

The DFE employs a feedforward filter and a backward-oriented filter for combating the
effects of dispersive channels. Figure 26.2 shows the general block diagram of the transmission system employing a DFE. The forward-oriented filter partially eliminates the ISI
introduced by the dispersive channel. The feedback filter, in the absence of decision errors,
is fed with the ISI-free transmitted signal in order to further reduce the ISI.

The feedback filter, denoted as the Bwd Filter in Figure 26.2, receives the detected symbol. Its output is then subtracted from the estimates generated by the forward filter, denoted as the Fwd Filter, in order to produce the detector's input. Since the feedback filter uses the ISI-free signal as its input, the feedback loop mitigates the ISI without introducing enhanced noise into the system. The drawback of the DFE is that when wrong decisions are fed back into the feedback loop, error propagation is inflicted and the BER performance of the equaliser is degraded.

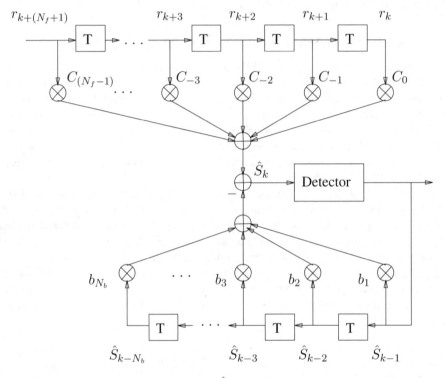

Figure 26.3: Structure of the DFE where r_k and $\hat{S}_k$ denote the received signal and detected symbol, respectively, while C_m, b_q represent the coefficient taps of the forward- and backward-oriented filters, respectively

The detailed DFE structure is shown in Figure 26.3. The feedforward filter is constituted by the coefficients or taps labelled as $C_0 \rightarrow C_{N_f-1}$, where N_f is the number of taps in the feedforward filter, as shown in the figure. The causal feedback filter is constituted by N_b feedback taps, denoted as $b_1 \rightarrow b_{N_b}$. Note that the feedforward filter contains only the present input signal r_k, and future input signals $r_{k+1} \ldots r_{k+(N_f+1)}$, which implies that no latency is inflicted. Therefore, the feedforward filter eliminates only the pre-cursor ISI, but not the post-cursor ISI. By contrast, the feedback filter mitigates the ISI caused by the past data symbols, i.e. post-cursor ISI. Since the feedforward filter only eliminates the pre-cursor ISI, the noise enhancement effects are less problematic in DFEs compared to the linear MMSE equaliser.

Here, the MMSE criterion [738] is used for deriving the optimum coefficients of the feedforward section of the DFE. The Mean Square Error (MSE) between the transmitted signal, s_k, and its estimate, $\hat{s}_k$, at the equaliser's output is formulated as:

$$MSE = E[|s_k - \hat{s}_k|^2], \tag{26.1}$$

where $E[|s_k - \hat{s}_k|^2]$ denotes the expected value of $|s_k - \hat{s}_k|^2$. In order to minimise the MSE, the orthogonality principle [740] is applied, stating that the residual error of the equaliser, $e_k = s_k - \hat{s}_k$, is orthogonal to the input signal of the equaliser, r_k, when the equaliser taps are optimal, yielding:

$$E[e_k r_{k+l}^*] = 0, \tag{26.2}$$

where the superscript * denotes conjugation. Following Cheung's approach [37, 180], the optimum coefficient of the feedforward section can be derived from the following set of N_f equations:

$$\sum_{m=0}^{N_f-1} C_m \left[\sum_{v=0}^{l} h_v^* h_{v+m-l} \sigma_S^2 + N_0 \delta_{m-l} \right] = h_l^* \sigma_S^2, \; l = 0 \ldots N_f - 1, \tag{26.3}$$

where σ_S^2 and $N_0/2$ are the signal and noise variance, respectively, while h^* denotes the complex conjugate of the CIR and δ is the delta function. By solving these N_f simultaneous equations, the equaliser coefficients, C_m, can be obtained. For the feedback filter the following set of N_b equations were used, in order to derive the optimum feedback coefficient, b_q [37]:

$$b_q = \sum_{m=0}^{N_f-1} C_m h_{m+q}, \; q = 1 \ldots N_b. \tag{26.4}$$

26.3.2 Equaliser Signal to Noise Ratio Loss

The equaliser's performance can be measured in terms of the equaliser's SNR loss, BER performance and MSE [739]. Here the SNR loss is considered, since this parameter will be used in next section.

The SNR loss of the equaliser was defined by Cheung [37] as:

$$SNR_{loss} = SNR_{input} - SNR_{output}, \tag{26.5}$$

where SNR_{input} is the SNR measured at the equaliser's input, given by:

$$SNR_{input} = \frac{\sigma_S^2}{2\sigma_N^2}, \tag{26.6}$$

with σ_S^2 being the average received signal power, assuming wide sense stationary conditions, and σ_N^2 is the variance of the AWGN.

The equaliser's output contains the wanted signal, the effective Gaussian noise, the residual ISI and the ISI caused by the past data symbols. In order to simplify the calculation of SNR_{output}, we assume that the SNR is high and hence we consider the low-BER range, where effectively correct bits are fed back to the DFE's feedback filter. Thus the post-cursor

ISI is completely eliminated from the equaliser's output. Hence SNR_{output} is given by [37]:

$$SNR_{output} = \frac{\text{Wanted Signal Power}}{\text{Residual ISI Power} + \text{Effective Noise Power}}, \qquad (26.7)$$

where the residual ISI is assumed to be an extra noise source that possessed a Gaussian distribution. Therefore, we have:

$$\text{Wanted Signal Power} = E\left[\left|s_k \sum_{m=0}^{N_f-1} C_m h_m\right|^2\right], \qquad (26.8)$$

$$\text{Effective Noise Power} = N_0 \sum_{i=0}^{N_f-1} |C_i|^2, \qquad (26.9)$$

and:

$$\text{Residual ISI Power} = \sum_{q=-(N_f-1)}^{-1} E\left[|f_q s_{k-q}|^2\right], \qquad (26.10)$$

where $f_q = \sum_{m=0}^{N_f-1} C_m h_{m+q}$ and the remaining notations are accrued from Figure 26.3. By substituting Equations 26.8, 26.9 and 26.10 into 26.7, the SNR_{output} can be written as [739]:

$$SNR_{output} = \frac{E\left[\left|s_k \sum_{m=0}^{N_f-1} C_m h_m\right|^2\right]}{\sum_{q=-(N_f-1)}^{-1} E\left[|f_q s_{k-q}|^2\right] + N_0 \sum_{i=0}^{N_f-1} |C_i|^2}. \qquad (26.11)$$

Following this rudimentary introduction to channel equalisation, we focus our attention on quantifying the performance of various wideband Coded Modulation (CM) schemes, referring the reader to [180] for an in-depth discourse on channel equalisation.

26.4 Decision Feedback Equaliser Aided Adaptive Coded Modulation

In this section, DFE-aided wideband Burst-by-Burst (BbB) adaptive TCM, TTCM, BICM and BICM-ID schemes are proposed and characterised in performance terms, when communicating over the COST207 Typical Urban (TU) wideband fading channel. These schemes are evaluated using a practical near-instantaneous modem mode switching regime, in the context of *System I* which invokes channel interleaving over one transmission burst and *System II* which invokes channel interleaving over four transmission bursts. Explicitly, *System I* exhibited a factor four lower delay in overall modem mode signalling, and hence it was capable of more prompt modem mode reconfiguration. By contrast, *System II* was less agile in terms of modem mode reconfiguration, but benefited from a longer interleaver delay. We will show in Section 26.4.4 that a substantially improved BPS and BER performance was achieved by *System II* in comparison to *System I*. We will also show that BbB adaptive TTCM is the best performer in comparison to other adaptive CM schemes, in the context of *System II* at a similar decoding complexity, when aiming for a target BER of below 0.01%.

26.4.1 Introduction

In general, fixed-mode transceivers fail to adequately counteract the time varying nature of the mobile radio channel and hence typically result in bursts of transmission errors. By contrast, in BbB adaptive schemes [124] a higher-order modulation mode is employed when the instantaneous estimated channel quality is high in order to increase the number of BPS transmitted and, conversely, a more robust lower-order modulation mode is employed when the instantaneous channel quality is low, in order to improve the mean BER performance. In the literature the performance of adaptive schemes was investigated mostly in the context of narrowband Rayleigh fading channels. Specifically, the performance of uncoded adaptive schemes and coded adaptive schemes has been investigated for transmissions over narrowband Rayleigh fading channels in [124, 192, 195, 219, 741, 742] and [193, 208, 211, 743–747], respectively. By contrast, a turbo-coded wideband adaptive scheme assisted by a DFE was investigated in [615]. In this section, a DFE-based coded modulation assisted wideband adaptive scheme is investigated.

In our practical approach the transmitter A obtains the channel quality estimate generated by receiver B upon receiving the transmission of transmitter B. In other words, the modem mode required by receiver B is superimposed on the transmission burst of transmitter B. Hence a delay of one transmission burst duration is incurred. In the literature, adaptive coding designed for time varying channels using outdated fading estimates has been investigated in [208].

Over wideband fading channels the DFE employed will eliminate most of the ISI. Consequently, the MSE at the output of the DFE can be calculated and used as the metric invoked for switching the modulation modes [192]. This ensures that the performance is optimised by employing equalisation and BbB adaptive CM jointly, in order to combat both the signal power fluctuations and the time variant ISI of the wideband channel.

In Section 26.4.2, the system's schematic is outlined. In Section 26.4.3, the performance of various fixed-mode CM schemes is evaluated, while Section 26.4.4 contains the detailed characterisation of the BbB adaptive CM schemes in the context of *System I* and *System II*. Finally, we will conclude with our findings in Section 26.4.5.

26.4.2 System Overview

The multipath channel model is characterised by its discretised symbol-spaced COST207 Typical Urban (TU) CIR [748], as shown in Figure 26.4. Each path is faded independently according to a Rayleigh distribution and the corresponding normalised Doppler frequency is 3.25×10^{-5}, the system's Baud rate is 2.6 MBaud, the carrier frequency is 1.9 GHz and the vehicular speed is 30 mph. The DFE incorporated 35 feed forward taps and 7 feedback taps and the transmission burst structure used is shown in Figure 26.5. When considering a Time Division Multiple Access (TDMA)/Time Division Duplex (TDD) system providing 16 slots per 4.615 ms TDMA frame, the transmission burst duration is 288 μs, as specified in the Pan-European FRAMES proposal [749].

The following assumptions are stipulated. First, we assume that the equaliser is capable of estimating the CIR perfectly with the aid of the equaliser training sequence of Figure 26.5. Second, the CIR is time-invariant for the duration of a transmission burst, but varies from burst to burst according to the normalised Doppler frequency, which corresponds to assuming

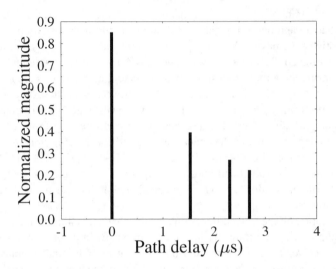

Figure 26.4: The impulse response of a COST207 Typical Urban (TU) channel [748]

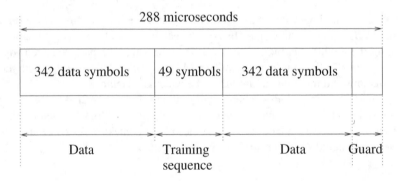

Figure 26.5: Transmission burst structure of the FMA1 non-spread data as specified in the FRAMES proposal [749]

that the CIR is slowly varying. We refer to this scenario as encountering burst-invariant fading. The error propagation imposed by erroneous decisions of the DFE will degrade the estimated performance, but the effect of error propagation is left for further study. Here we assumed that correct symbol decisions are input to the DFE's feedback filter in order to obtain the upper bound performance of the system. It is expected that the more vulnerable higher-throughput modem modes are more prone to error propagation effects, but their influence may be compensated by slightly increasing the switching thresholds. This allows us to maintain the target BER but reduces the throughput. At the receiver, the CIR is estimated, and is then used for calculating the DFE coefficients [180]. Subsequently, the DFE is used for equalising the ISI-corrupted received signal. In addition, both the CIR estimate and the DFE feedforward coefficients are utilised for computing the SNR at the output of the DFE. More specifically, by assuming that the residual ISI is near-Gaussian distributed and that the probability of decision feedback errors is negligible, the SNR at the output of the DFE, γ_{dfe}, is calculated as [37]:

$$\gamma_{dfe} = \frac{\text{Wanted Signal Power}}{\text{Residual ISI Power + Effective Noise Power}}.$$

$$= \frac{E\left[\left|s_k\sum_{m=0}^{N_f}C_mh_m\right|^2\right]}{\sum_{q=-(N_f-1)}^{-1}E\left[\left|\sum_{m=0}^{N_f-1}C_mh_{m+q}s_{k-q}\right|^2\right] + N_0\sum_{m=0}^{N_f}|C_m|^2}, (26.12)$$

where C_m and h_m denote the DFE's feedforward coefficients and the CIR, respectively. The transmitted signal is represented by s_k and N_0 denotes the noise spectral density. Finally, the number of DFE feedforward coefficients is denoted by N_f.

The equaliser's SNR, γ_{dfe}, in Equation 26.12, is then compared against a set of adaptive modem mode switching thresholds f_n, and subsequently the appropriate modulation mode is selected [187, 192]. The modem mode required by receiver B is then fed back to transmitter A. The modulation modes that are utilised in this scheme are 4QAM, 8PSK, 16QAM and 64QAM [180].

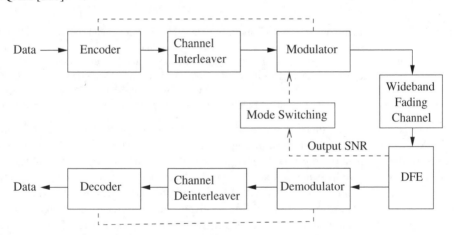

Figure 26.6: *System I* employing channel interleaving over one transmission burst. The equaliser's output SNR is used for selecting a suitable modulation mode, which is fed back to the transmitter on a burst-by-burst basis.

The simplified block diagram of the BbB adaptive CM *System I* is shown in Figure 26.6, where channel interleaving over one transmission burst is employed. Transmitter A extracts the modulation mode required by receiver B from the reverse-link transmission burst in order to adjust the adaptive CM mode suitable for the currently experienced instantaneous channel quality. This incurs one TDMA/TDD frame delay between estimating the actual channel condition at receiver B and the selected modulation mode of transmitter A. Better channel quality prediction can be achieved using the techniques proposed in [175]. We invoke four encoders, each adding one parity bit to each information symbol, yielding the coding rate of 1/2 in conjunction with the CM mode of 4QAM, 2/3 for 8PSK, 3/4 for 16QAM and 5/6 for

64QAM.

The design of CM schemes contrived for fading channels relies on the time and space diversity provided by the associated channel coder [708, 718]. Diversity may be achieved by repetition coding, which reduces the effective data rate, space-time coded multiple transmitter/receiver structures [626], which increases cost and complexity, or by simple interleaving, which induces latency. In [683] adaptive TCM schemes were designed for narrowband fading channels utilising repetition-based transmissions during deep fades along with ideal channel interleavers and assuming zero delay for the feedback of the channel quality information.

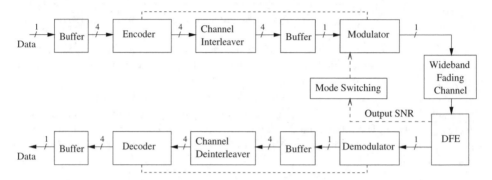

Figure 26.7: *System II* employing a channel interleaver length of four TDMA/TDD bursts. Data are entered into the input buffer on a burst-by-burst basis and the modulator modulates coded data read from the output buffer for transmission on a burst-by-burst basis. The encoder and channel interleaver as well as the decoder and channel de-interleaver operate on a four-burst basis. The equaliser's output SNR during the fourth burst is used for selecting a suitable modulation mode and fed back to the transmitter on the reverse-link burst.

Figure 26.7 shows the block diagram of *System II*, where symbol-based channel interleaving over four transmission bursts, or equivalently 4×684 symbols is utilised, in order to disperse the bursty symbol errors. Hence, the coded modulation module assembles four bursts using an identical modulation mode, so that they can be interleaved using the symbol-by-symbol random channel interleaver without the need of adding dummy bits. Then, these four-burst CM packets are transmitted to the receiver. Once the receiver has received the fourth burst, the equaliser's output SNR for this most recent burst is used for choosing a suitable modulation mode. The selected modulation mode is fed back to the transmitter on the reverse-link burst. Upon receiving the modulation mode required by receiver B (after one TDMA frame delay), the coded modulation module assembles four bursts of data from the input buffer for coding and interleaving, which are then stored in the output buffer ready for the next four bursts' transmission. Thus the first transmission burst exhibits one TDMA/TDD frame delay and the fourth transmission burst exhibits a delay of four frames which is the worst-case scenario.

26.4.3   Fixed-Mode Based Performance

Before characterising the proposed wideband BbB adaptive scheme, the BER performance of the fixed modem modes of 4QAM, 8PSK, 16QAM and 64QAM are studied in the context of *System I* where channel interleaving over one transmission burst is utilised and *System*

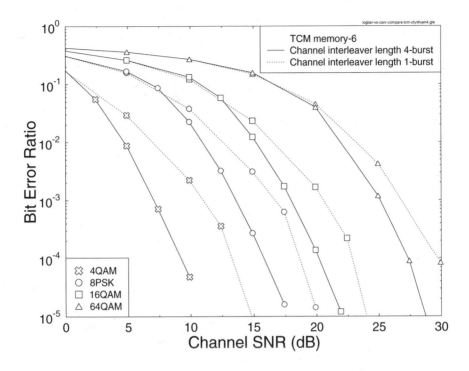

Figure 26.8: TCM performance of each individual modulation mode over the Rayleigh fading COST207 TU channel of Figure 26.4. A TCM code memory of 6 was used, since it had a similar decoding complexity to TTCM in conjunction with four iterations using a component TCM code memory of 3. The normalised Doppler frequency was 3.25×10^{-5}

II where channel interleaving over four transmission bursts is invoked. These results are shown in Figure 26.8 for TCM, and in Figure 26.9 for TTCM. The random TTCM symbol-interleaver memory was set to 684 symbols, corresponding to the number of data symbols in the transmission burst structure of Figure 26.5, where the resultant number of bits was the $BPS \times 684$. A channel interleaver of 4×684 symbols was utilised for *System II*, where the number of coded bits was $(BPS + 1) \times 4 \times 684$ bits, since one parity bit was added to each CM symbol.

As expected, in Figures 26.8 and 26.9 the BER performance of *System II* was superior compared to that of *System I*, although at the cost of an associated higher transmission delay. The SNR gain difference between *System II* and *System I* was about 5 dB in the TTCM/4QAM mode, but this difference reduced for higher-order modulation modes. Again, this gain was obtained at the cost of a four-burst channel interleaving delay. This SNR gain difference shows the importance of time diversity in CM schemes.

TTCM has been shown to be more efficient than TCM for transmissions over AWGN channels and narrowband fading channels [709, 732]. Here, we illustrate the advantage of TTCM in comparison to TCM over the dispersive or wideband Gaussian CIR of Figure 26.4, as seen in Figure 26.10. In conclusion, TTCM is superior to TCM in a variety of channels.

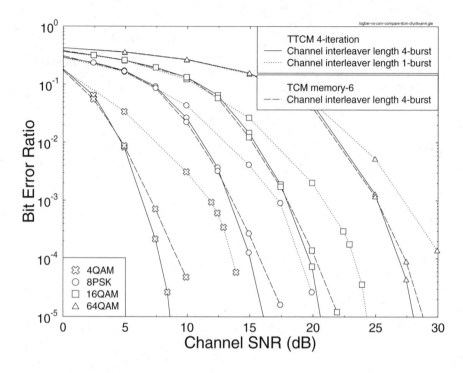

Figure 26.9: TTCM performance of each individual modulation mode over the Rayleigh fading COST207 TU channel of Figure 26.4. A component TCM code memory of 3 was used and the number of turbo iterations was four. The performance of the TCM code with memory 6 utilising a channel interleaver was also plotted for comparison. The normalised Doppler frequency was 3.25×10^{-5}

Figure 26.11 shows the fixed modem modes' performance for TCM, TTCM, BICM and BICM-ID in the context of *System II*. For the sake of a fair comparison of the decoding complexity, we used a TCM code memory of 6, TTCM code memory of 3 in conjunction with four turbo iterations, BICM code memory of 6 and a BICM-ID code memory of 3 in conjunction with eight decoding iterations. However, BICM-ID had a slightly higher decoding complexity, since the demodulator was invoked in each BICM-ID iteration, whereas in the BICM, TCM and TTCM schemes the demodulator was only visited once in each decoding process. As illustrated in the figure, the BICM scheme performed marginally better than the TCM scheme at a BER below 0.01%, except in the 64QAM mode. Hence, adaptive BICM is also expected to be better than adaptive TCM in the context of *System II*, when a target BER of less than 0.01% is desired. This is because when the channel interleaver depth is sufficiently high, the diversity gain of the BICM's bit interleaver is higher than that of the TCM's symbol interleaver [364, 365].

Let us now compare the performance of the BbB adaptive CM *system I* and *II*.

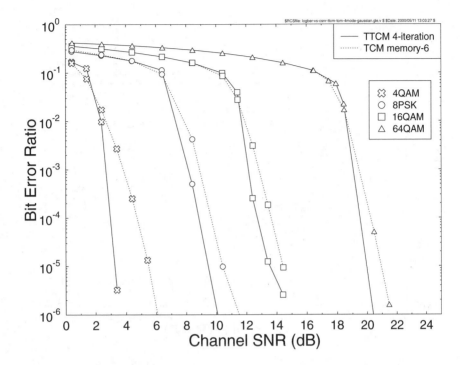

Figure 26.10: TTCM and TCM performance of each individual modulation mode for transmissions over the unfaded COST207 TU channel of Figure 26.4. The TTCM scheme used component TCM codes of memory 3 and the number of turbo iterations was four. The performance of the TCM scheme in conjunction with memory 6 was plotted for comparison with the similar-complexity TTCM scheme

26.4.4 System I and System II Performance

The modem mode switching mechanism of the adaptive schemes is characterised by a set of switching thresholds, the corresponding random TTCM symbol interleavers and the component codes, as follows:

$$
\text{Modulation Mode} = \begin{cases} 4QAM, I_0 = 684, R_0 = 1/2 & \text{if } \gamma_{DFE} \leq f_1 \\ 8PSK, I_1 = 1368, R_1 = 2/3 & \text{if } f_1 < \gamma_{DFE} \leq f_2 \\ 16QAM, I_2 = 2052, R_2 = 3/4 & \text{if } f_2 < \gamma_{DFE} \leq f_3 \\ 64QAM, I_3 = 3420, R_3 = 5/6 & \text{if } \gamma_{DFE} > f_3, \end{cases} \quad (26.13)
$$

where $f_n, n = 1 \ldots 3$, are the equaliser's output SNR thresholds, while I_n represents the random TTCM symbol interleaver size in terms of the number of bits, which is not used for the other CM schemes. The switching thresholds f_n were chosen experimentally, in order to maintain a BER of below 0.01%, and these thresholds are listed in Table 26.1.

Let us consider the adaptive TTCM scheme in order to investigate the performance of *System I* and *System II*. The performance of *System I* employing channel interleaving over

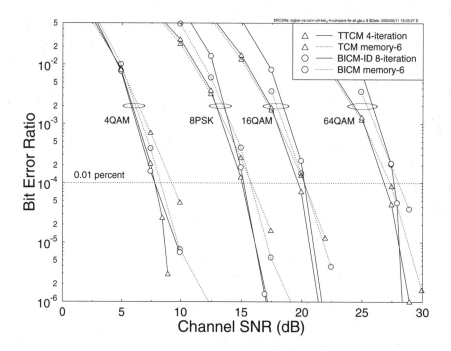

Figure 26.11: BER performance of the fixed modem modes of 4QAM, 8PSK, 16QAM and 64QAM
utilising TCM, TTCM, BICM and BICM-ID schemes in the context of *System II* for
transmissions over the COST207 TU channel of Figure 26.4. For the sake of maintaining
a similar decoding complexity, we used a TCM code memory of 6, TTCM code memory
of 3 in conjunction with four turbo iterations, BICM code memory of 6 and a BICM-ID
code memory of 3 in conjunction with eight decoding iterations. However, BICM-ID had
a slightly higher complexity than the other systems, since the demodulator module was
invoked eight times as compared to only once for its counterparts during each decoding
process. The normalised Doppler frequency was 3.25×10^{-5}

BER < 0.01 %		Switching Thresholds		
Adaptive System Type		f_1	f_2	f_3
TCM, Memory 3	System I	19.56	23.91	30.52
	System II	17.17	21.91	29.61
TCM, Memory 6	System I	19.56	23.88	30.07
	System II	17.14	21.45	29.52
TTCM, 4 iterations	System I	19.69	23.45	30.29
	System II	16.66	21.40	28.47
BICM, Memory 3	System I	19.94	24.06	31.39
BICM-ID, 8 iterations	System II	16.74	21.45	28.97

Table 26.1: The switching thresholds were set experimentally for transmissions over the COST207 TU
channel of Figure 26.4, in order to achieve a target BER of below 0.01%. *System I* utilises
a channel interleaver length of one TDMA/TDD burst, while *System II* uses a channel
interleaver length of four TDMA/TDD bursts.

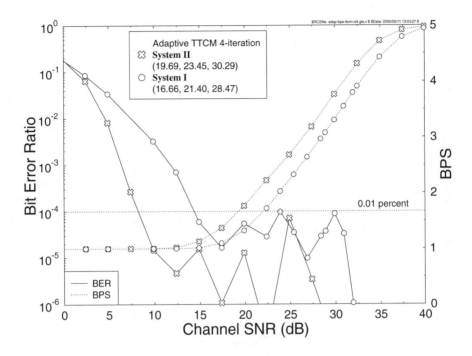

Figure 26.12: BER and BPS performance of adaptive TTCM for transmissions over the COST207 TU channel of Figure 26.4, using four turbo iterations in *System I* (with a channel interleaver length of one burst) and in *System II* (with a channel interleaver length of four bursts) for a target BER of less than 0.01%. The legends indicate the associated switching thresholds expressed in dB, as seen in the brackets. The normalised Doppler frequency was 3.25×10^{-5}

one transmission burst was found to be identical to that of the same scheme employing no channel interleaving. This is because in the context of the burst-invariant fading scenario the channel behaves like a dispersive Gaussian channel, encountering a specific fading envelope and phase trajectory across a transmission burst. The CIR is then faded at the end or at the commencement of each transmission burst. Hence the employment of a channel interleaver having a memory of one transmission burst would not influence the distribution of the channel errors experienced by the decoder. The BER and BPS performances of both adaptive TTCM systems using four iterations are shown in Figure 26.12, where we observed that the throughput of *System II* was superior to that of *System I*. Furthermore, the overall BER of *System II* was lower than that of *System I*. In order to investigate the switching dynamics of both systems, the mode switching together with the equaliser's output SNR was plotted versus time at an average channel SNR of 25 dB in Figures 26.13 and 26.14. Observe in Table 26.1 that the switching thresholds f_n of *System II* are lower than those of *System I*, since the fixed-mode-based results of *System II* in Figure 26.9 were better. Hence higher-order modulation modes were chosen more frequently than in *System I*, giving a better BPS throughput. From Figures 26.13 and 26.14, it is clear that *System I* was more flexible in terms of mode switching, while *System II* benefited from higher diversity gains due to the four-burst channel

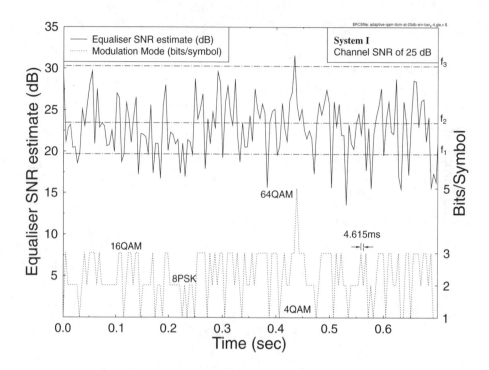

Figure 26.13: Channel SNR estimate and BPS versus time plot for adaptive TTCM for transmissions
over the COST207 TU channel of Figure 26.4, using four turbo iterations in *System I* at
an average channel SNR of 25 dB, where the modulation mode switching is based upon
the equaliser's output SNR, which is compared to the switching thresholds f_n defined in
Table 26.1. The duration of one TDMA/TDD frame is 4.615 ms. The TTCM mode can
be switched after one frame duration

interleaver. This diversity gain compensated for the loss of switching flexibility, ultimately
providing a better performance in terms of BER and BPS, as seen in Figure 26.12.

In our next endeavour, the adaptive CM schemes of *System I* and *System II* are compared.
Figure 26.15 shows the BER and BPS performance of *System I* for adaptive TTCM using
four iterations, adaptive TCM of memory 3 (which was the component code of our TTCM
scheme), adaptive TCM of memory 6 (which had a similar decoding complexity to our TTCM
scheme) and adaptive BICM of memory 3. As can be seen from the fixed-mode results of
Figures 26.8 and 26.9 in the previous section, TCM and TTCM performed similarly in terms
of their BER, when channel interleaving over one transmission burst was used for the slow
fading wideband COST207 TU channel of Figure 26.4. Hence, they exhibited a similar
performance in the context of the adaptive schemes of *System I*, as shown in Figure 26.15.
Even the TCM scheme of memory 3 associated with a lower complexity could give a similar
BER and BPS performance. This shows that the equaliser plays a dominant role in *System
I*, where the coded modulation schemes could not benefit from sufficient diversity due to the
lack of interleaving. Also shown in Figure 26.15 is that adaptive TCM exhibited a better BPS
throughput and BER performance than adaptive BICM, due to employing an insufficiently

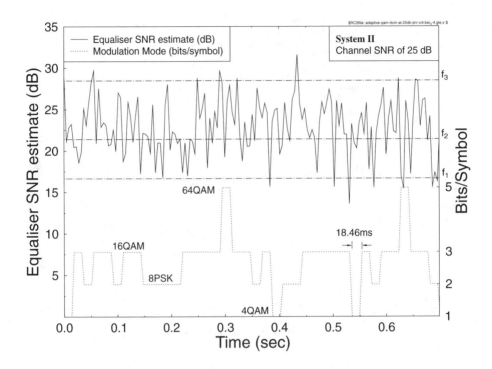

Figure 26.14: Channel SNR estimate and BPS versus time plot for adaptive TTCM for transmissions over the COST207 TU channel of Figure 26.4, using four turbo iterations in *System II* at an average channel SNR of 25 dB, where the modulation mode switching is based upon the equaliser's output SNR which is compared to the switching thresholds f_n defined in Table 26.1. The duration of one TDMA/TDD frame is 4.615 ms. The TTCM mode is maintained for four frame durations, i.e. for 18.46 ms

high channel interleaving depth for the BICM scheme, for transmissions over our slow fading wideband channels.

When channel interleaving over four transmission bursts is introduced in *System II*, the bursty symbol errors are dispersed. Figure 26.16 illustrates the BER and BPS performance of *System II* for adaptive TTCM using four iterations, adaptive TCM of memory 3, adaptive TCM of memory 6 and adaptive BICM-ID of memory 3 in conjuction with eight decoding iterations. The performance of all these schemes improved in the context of *System II*, as compared to the corresponding schemes in *System I*. The TCM scheme of memory 6 had a lower BER than TCM of memory 3, and also exhibited a small BPS improvement. As observed in Figure 26.11, we noticed that BICM-ID had the worst performance at low SNRs in each modulation modes compared to other CM schemes. However, it exhibited a steep slope and therefore at high SNRs it approached the performance of the TTCM scheme. As shown in Figure 26.16, the adaptive TTCM scheme exhibited a better BPS throughput than adaptive BICM-ID, since TTCM had a better performance in fixed modem modes at a BER of 0.01%. However, adaptive BICM-ID exhibited a lower BER performance than adaptive TTCM owing to the high steepness of the BER curve of BICM-ID in its fixed modem modes.

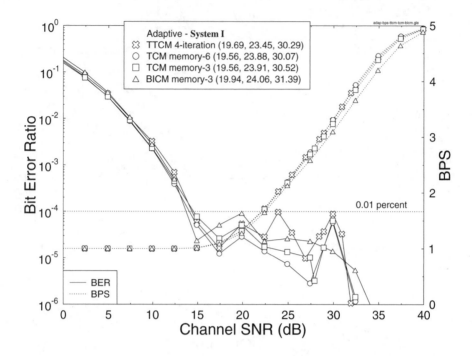

Figure 26.15: BER and BPS performance of adaptive TCM, TTCM and BICM in the context of *System I*, for transmissions over the Rayleigh fading COST207 TU channel of Figure 26.4. The switching mechanism was characterised by Equation 26.13. The switching thresholds were set experimentally, in order to achieve a BER of below 0.01%, as shown in Table 26.1. The normalised Doppler frequency was 3.25×10^{-5}

In summary, we have observed BER and BPS gains for the adaptive CM schemes of *System II* in comparison to the adaptive CM schemes of *System I*. Adaptive TTCM exhibited a superior BPS performance in comparison to other adaptive CM schemes in the context of *System II*.

26.4.5   Conclusions

In this section, BbB adaptive TCM, TTCM, BICM and BICM-ID were proposed for transmissions over wideband fading channels and they were characterised in performance terms when communicating over the COST207 TU fading channel. When channel interleaving over one transmission burst is invoked in *System I*, the performance of the system was dominant by the channel equaliser. Nonetheless, adaptive TCM still outperformed adaptive BICM in the context of *System I*, as shown in Figure 26.15. When observing the associated BPS curves, adaptive TTCM exhibited up to 2.5 dB SNR gain when the channel interleaver depth was increased from one to four transmission bursts, as evidenced in Figure 26.12. Upon comparing the BPS curves, adaptive TTCM also exhibited up to 0.7 dB SNR gain compared to adaptive TCM of the same complexity in the context of *System II* for a target BER of less than 0.01%, as shown in Figure 26.16. Finally, the BPS performance of adaptive BICM-ID was

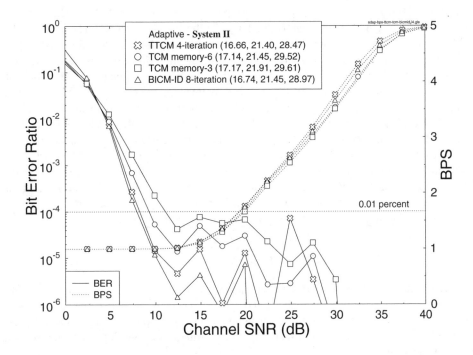

Figure 26.16: BER and BPS performance of adaptive TCM, TTCM and BICM-ID in the ccontext of *System II*, for transmissions over the Rayleigh fading COST207 TU channel of Figure 26.4. The switching mechanism was characterised by Equation 26.13. The switching thresholds were set experimentally, in order to achieve a BER of below 0.01%, as shown in Table 26.1. The normalised Doppler frequency was 3.25×10^{-5}.

only marginally worse than that of adaptive TTCM in the context of *System II*, as illustrated in Figure 26.16.

26.5 Radial Basis Function-Based Equalisation

In the forthcoming sections, we will further investigate the Coded Modulation (CM) schemes in the context of a Radial Basis Function (RBF) assisted turbo equalised system [170].

The RBF based equaliser [170, 737] studied constitutes a non-linear equalisation scheme, which formulates the channel equalisation procedure as a classification problem. More explicitly, in conventional equalisers of Section 26.3.1 the received signal is linearly filtered with the aid of the channel equaliser, which is aiming for mimicking the inverse of the CIR. By contrast, given the CIR, the RBF based equaliser determines all possible channel outputs engendered by the set of legitimate transmitted symbols and then classifies each received symbol into the nearest legitimate channel output, which allows us to infer the specific symbol transmitted. The application of non-linear RBF based equalisers has been studied in conjunction with channel codecs [170, 203, 736], space-time codecs [207] as well as turbo equalisation [204]. The BER performance of RBF-based Turbo Equalisation (RBF-TEQ) was presented in [204] in the context of Quadrature Amplitude Modulation (QAM), which was found similar to that of the Conventional Trellis-based Turbo Equalisation (CT-TEQ) [750]. The RBF-assisted schemes are, however, capable of maintaining a lower complexity than their conventional trellis-based counterparts, when communicating over both dispersive Gaussian and Rayleigh fading channels, while maintaining a similar performance. The complexity of the RBF-TEQ scheme can be further reduced by invoking the In-phase/Quadrature-phase Turbo Equalisation (I/Q-TEQ) technique, while maintaining a similar performance to that of the CT-TEQ [736]. Explicitly the philosophy of carrying out the equalisation of the in phase and quadrature phase components separately is appealing, since the dimensionality of the I and Q components is significantly lower than that of the complex constellation, which reduces the equaliser's complexity. However, this principle can only be invoked in conjunction with turbo equalisation where the associated gross simplification of considering the I and Q components in isolation and hence disregarding their channel-induced cross-coupling is compensated for by the turbo-equaliser's consecutive iterations [369].

26.5.1 RBF-Based Equaliser Principle

As already mentioned above, the characteristics of the transmitted sequence can be exploited by capitalising on the finite-duration CIR and by considering the channel equalisation process as a geometric classification problem. This approach was first proposed by Gibson, Siu and Cowan [751], who investigated utilising nonlinear structures offered by neural networks as channel equalisers [736].

After filtering the received signal with the aid of a filter mimicking the inverse of the CIR, a linear equaliser performs the classification into symbols in conjunction with a decision device, which is often a simple sign function. The decision boundary is constituted by the locus of all noisy channel outputs, where the output of the linear equaliser is zero. In general, the linear equaliser invokes a hyperplane as its decision boundary, and hence it implements a sub-optimum classification strategy. Gibson *et al.* [751] have shown examples of linearly non-separable channels, when the decision delay is zero and the channel is of non-minimum phase nature. The linear separability of the channel outputs depends on the equaliser order, m, on the decision delay τ and in situations where the channel characteristics are time varying, it may not be possible to specify appropriate values of m and τ, which will guarantee linear separability.

According to Chen, Gibson and Cowan [752], the above shortcomings of the linear equaliser are circumvented by a Bayesian approach [753] to obtaining an optimal equalisation solution. In general, the optimal Bayesian decision boundary is constituted by a hypersurface, rather than just a hyper-plane in the m-dimensional observation space and the realisation of this nonlinear boundary requires a nonlinear decision capability, which can be provided by neural networks. The so-called Radial Basis Function (RBF) network is ideal for channel equalisation applications, since it is capable of realising the optimal Bayesian equalisation solution [754]. Therefore, RBF equalisers can be derived directly from theoretical considerations related to optimal detection and all our prior knowledge concerning detection problems [753] can be exploited.

Briefly, the structure of the RBF network [755] consists of three different layers, as shown in Figure 26.17. The input layer is constituted by p source nodes. A set of $\mathcal{M}$ nonlinear activation functions $\varphi_i, i = 1, \ldots, \mathcal{M}$, constitutes the hidden second layer where $\mathcal{M}$ is the

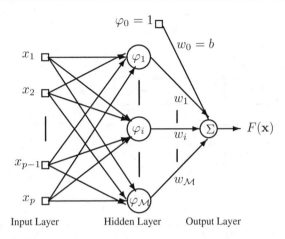

Figure 26.17: Architecture of a radial basis function based network

number of independent basis functions of the RBF network. The output of the RBF network is provided by the third layer, which is comprised of output nodes. Figure 26.17 shows only one output node. This construction is based on the fundamentals of neural network design [755]. As suggested by the terminology, the activation functions in the hidden layer take the form of radial basis functions [755], each centred around one of the $\mathcal{M}$ independent basis functions of the RBF network.

Radial functions are characterised by their responses that decrease or increase monotonically with distance from a central point, $\mathbf{c}$, i.e. as the Euclidean norm $\|\mathbf{x} - \mathbf{c}\|$ is increased, where $\mathbf{x} = [x_1\ x_2\ \ldots\ x_p]^T$ is the input vector of the RBF network. The central points in the vector $\mathbf{c}$ are often referred to as the RBF centres. Therefore, the radial basis functions take the form of:

$$\varphi_i(\mathbf{x}) = \varphi(\|\mathbf{x} - \mathbf{c}_i\|), \qquad i = 0, \ldots, \mathcal{M}. \tag{26.14}$$

This justifies the 'radial' terminology. Referring to Figure 26.17, the RBF network can be represented mathematically as follows:

$$F(\mathbf{x}) = \sum_{i=0}^{\mathcal{M}} w_i \varphi_i(\mathbf{x}), \tag{26.15}$$

where w_i is the ith weight of the RBF network. The additive bias b in Figure 26.17 is absorbed into the summation as w_0 by including an extra basis function φ_0, whose activation function is set to 1. For a detailed description of the RBF network design, the interested reader is referred to [736] for example.

The architecture of the RBF equaliser designed for an M-ary multilevel modem scenario is shown in Figure 26.18.[1] Observe in Figure 26.18 that each of the M possible symbols has a dedicated sub-RBF network. Note furthermore that the output of each sub-RBF network gives the corresponding conditional density function ζ_i and this output value can be

---

[1]Explicitly, M determines the number of possible symbols constituted by $m = \log_2(\mathsf{M})$ bits, while $\mathcal{M}$ is the number of RBF centres, which is determined by all possible combinations of the legitimate transmitted symbols and by the length of the CIR.

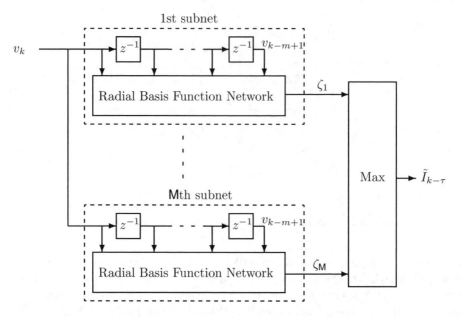

Figure 26.18: RBF equaliser for M-level modems, where $\{v_k\}$ is the channel output sequence, $\tilde{I}_{k-\tau}$ is the estimate of the transmitted signal $I_{k-\tau}$ and ζ_i is the conditional density function of the ith sub-RBF network

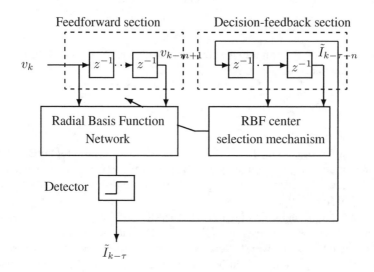

Figure 26.19: RBF equaliser with decision feedback

used for generating soft decision channel decoder inputs in conjunction with error correction techniques [369]. Figure 26.19 illustrates the RBF assisted Decision Feedback Equaliser (RBF-DFE) for the specific scenario of a binary modulation scheme. This is also equivalent to one sub-RBF network of M-level modems. Observe in the figure that in contrast to conventional DFEs, where the output of the feedback section is subtracted from that of the feedforward section for the sake of cancelling the ISI component imposed by the symbols that have already been detected, here the feedback section is employed for assisting the operation of the feedforward section where a subset of centres was selected for a particular decision based on the feedback symbol vector. More explicitly, some of the channel outputs can be eliminated from further classification-related comparisons, since the corresponding symbols have already been determined [736]. The structure of RBF-DFE is specified by the equaliser's decision delay τ, the feedforward order m and the feedback order n.

Determine the feedback state	
$n_{s,i}(m+2) - 2\mathsf{M} + n_{s,f}$	subtraction and addition
$n_{s,f}$	multiplication
$n_{s,f}$	division
$n_{s,i} - \mathsf{M} + 1$	max
$n_{s,i} - \mathsf{M}$	table look-up

Table 26.2: Reduced computational complexity per equalised output sample of an M-ary Jacobian RBF DFE based on scalar centres. The Jacobian RBF DFE has m inputs and $n_{s,i}$ hidden RBF nodes, which are derived from the $n_{s,f}$ number of scalar centres.

The Bayesian-based RBF equaliser has a high computational complexity imposed by the evaluation of the nonlinear exponential functions in Equation 26.19 and owing to the high number of additions/subtractions and multiplications/divisions required for the estimation of each symbol. However, the output of the RBF network can be generated in the logarithmic domain by using the Jacobian logarithm [637] described in Section 24.4.4, in order to avoid the computation of exponentials and to reduce the number of multiplications performed. The RBF equaliser using the Jacobian logarithm is referred to as the *Jacobian RBF equaliser*. The complexity of the Jacobian RBF equaliser can be further reduced by exploiting the fact that the elements of the vector of noiseless channel outputs constituting the channel states $\mathbf{r}_i, i = 1, \ldots, n_s$ correspond to the convolution of a sequence of $(\bar{L}+1)$ transmitted symbols and $(\bar{L}+1)$ CIR taps. These vector elements are referred to as the scalar channel states $r_l, l = 1, \ldots, n_{s,f} (= \mathsf{M}^{\bar{L}+1})$ and we could use the method advocated by Patra and Mulgrew [756] for reducing the computational load [736]. Following the procedure of Sections 8.10 and 10.2 in [736], it can be shown that the corresponding computional complexity per equalised output sample of an M-ary Jacobian RBF DFE is given by the values shown in Table 26.2. All systems presented in this chapter employed the reduced complexity M-ary Jacobian RBF-DFE of [736].

Having presented a brief overview of the RBF equaliser, we will proceed to introduce the turbo equalisation technique using the symbol-based MAP decoder of Section 24.4.

26.6   Turbo Equalisation Using Symbol-Based MAP Decoder

In the RBF DFE-based systems discussed in Section 26.5.1 channel equalisation and channel decoding ensued independently. However, it is possible to improve the receiver's performance, if the equaliser is fed by the channel outputs plus the soft decisions provided by the channel decoder, invoking a number of iterative processing steps. This novel receiver scheme was first proposed by Douillard *et al.* [734] for a convolutional coded Binary Phase Shift Keying (BPSK) system, using a similar principle to that of turbo codes and hence it was termed *turbo equalisation* [369]. This scheme is illustrated in Figure 26.21, which will be detailed during our forthcoming discourse. Gertsman and Lodge [757] extended this work and showed that the iterative process of turbo equalisation is capable of compensating for the performance degradation imposed by imperfect channel estimation. In [758] turbo equalisation was implemented in conjunction with turbo coding, rather than conventional convolutional coding, by Raphaeli and Zarai, which demonstrated an increased performance gain owing to turbo coding as well as with the advent of enhanced ISI mitigation achieved by turbo equalisation.

26.6.1   Principle of Turbo Equalisation Using Symbol-Based MAP Decoder

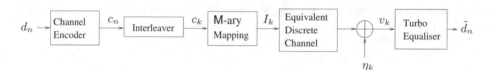

Figure 26.20: Serially concatenated coded M-ary system using the turbo equaliser, which performs the equalisation, demodulation and channel decoding iteratively

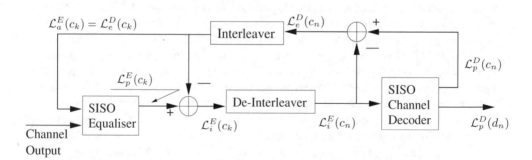

Figure 26.21: Iterative turbo equalisation schematic

The principles of bit-based iterative turbo decoding [689] were modified appropriately for employment of the symbol-based M-ary coded modulation system of Figure 26.20. As seen in the figure, a data symbol d_n is fed into the channel encoder in order to yield a channel encoded symbol c_n of $\mathsf{m} = \log_2(\mathsf{M})$ bits. Then the interleaved channel encoded symbol c_k is mapped to an M-ary symbol before transmission. In this scheme the channel is viewed as an 'inner encoder' of a serially concatenated arrangement, since it can be modelled with the aid of a tapped delay line similar to that of a convolutional encoder [734, 759], as demonstrated in Section 16.5 of [369]. At the receiver the equaliser and decoder employ a Soft-In Soft-Out (SISO) algorithm, such as the optimal Maximum *A Posteriori* (MAP) algorithm [369, 582] or the Log-MAP algorithm [369, 637]. The SISO equaliser processes the *a priori* information associated with the coded symbol c_k transmitted over the channel and – in conjunction with the channel output values v_k – computes the *a posteriori* information concerning the coded symbol. The soft values of the coded bits constituting the channel coded symbol c_k are normally quantified in the form of the log-likelihood ratio [734], however, here we will quantify them in the form of the symbol probabilities using the symbol-based MAP decoder of Section 24.4. Note that in the context of turbo equalisation the *a posteriori* information concerning all the *coded* bits is required, whereas in the context of conventional turbo channel decoding only the *a posteriori* information of the *information* bits are computed.

In our description of the turbo equaliser depicted in Figure 26.21 [2] we have used the notation $\mathcal{L}^E$ and $\mathcal{L}^D$ for denoting the Log-domain Probability (LP) values output by the SISO equaliser and SISO decoder, respectively. The subscripts e, i, a and p were used to represent the *extrinsic* LP, the combined channel and *extrinsic* LP, the *a priori* LP and the *a posteriori* LP, respectively. Referring to Figure 26.21, the SISO equaliser processes the channel outputs and the *a priori* information $\mathcal{L}_a^E(c_k)$ of the coded symbol, and generates the *a posteriori* LP values $\mathcal{L}_p^E(c_k)$ of the interleaved coded symbol c_k seen in Figure 26.20. Before passing the above *a posteriori* LPs generated by the SISO equaliser to the SISO decoder of Figure 26.21, the contribution of the decoder — which is represented in the form of the *a priori* information $\mathcal{L}_a^E(c_k)$ — accruing from the previous iteration must be removed, in order to yield the combined channel and *extrinsic* information $\mathcal{L}_i^E(c_k)$ seen in Figure 26.21. To expound a little further, the channel and *extrinsic* information are referred to as 'combined', since they are intrinsically bound and cannot be separated. However, note that at the initial iteration stage no *a priori* information is available yet. To elaborate further, the *a priori* information $\mathcal{L}_a^E(c_k)$ was removed at this stage, in order to prevent the decoder from processing its own output information, which would result in overwhelming the decoder's current reliability-estimation characterising the coded bits, i.e. the *extrinsic* information. The combined channel and *extrinsic* LP values are channel-deinterleaved – as seen in Figure 26.21 – in order to yield $\mathcal{L}_i^E(c_n)$, which is then passed to the SISO channel decoder. Subsequently, the channel decoder computes the *a posteriori* LP values $\mathcal{L}_p^D(c_n)$ of the coded symbol. The *a posteriori* LPs generated at the output of the channel decoder are constituted by the *extrinsic* LP $\mathcal{L}_e^D(c_n)$ and the channel-deinterleaved combined channel and *extrinsic* LP $\mathcal{L}_i^E(c_n)$ extracted from the equaliser's *a posteriori* LP $\mathcal{L}_p^E(c_k)$. The *extrinsic* part can be interpreted as the incremental information concerning the current symbol obtained through the decoding process from all the information available due to all surrounding symbols imposed by the code constraints, but excluding the information directly conveyed by the symbol. This information can be calculated by subtracting on a symbol-by-symbol basis the LP values $\mathcal{L}_i^E(c_n)$ at the input of the decoder from the *a posteriori* LP values $\mathcal{L}_p^D(c_n)$ at the channel decoder's

---

[2]The SISO channel decoder block may contain any of the coded modulation decoders such as the TCM, BICM, TTCM and BICM-ID decoders. The TTCM and BICM-ID decoders themselves have a number of inner iterations and the number of inner and outer iterations is adjusted such that all the different schemes encounter the same total number of trellis stages.

output, as seen also in Figure 26.21, yielding:

$$\mathcal{L}_e^D(c_n) = \mathcal{L}_p^D(c_n) - \mathcal{L}_i^E(c_n). \tag{26.16}$$

The *extrinsic* information $\mathcal{L}_e^D(c_n)$ of the coded symbol is then interleaved as shown in Figure 26.21, in order to yield $\mathcal{L}_e^D(c_k)$, which is fed back in the required symbol-order to the channel equaliser, where it is used as the *a priori* information $\mathcal{L}_a^E(c_k)$ in the next equalisation iteration. This constitutes the first iteration. Again, it is important that only the channel-interleaved *extrinsic* part – i.e. $\mathcal{L}_e^D(c_k)$ of $\mathcal{L}_p^D(c_n)$ – is fed back to the equaliser, since the interdependence between the *a priori* information $\mathcal{L}_a^E(c_k) = \mathcal{L}_e^D(c_k)$ used by the equaliser and the previous decisions of the equaliser should be minimised. This independence assists in obtaining the equaliser's reliability-estimation of the coded symbols for the current iteration, without being 'influenced' by its previous estimations. Ideally, the *a priori* information should be based on an independent estimation. As argued above, this is the reason that the *a priori* information $\mathcal{L}_a^E(c_k)$ is subtracted from the *a posteriori* LP value $\mathcal{L}_p^E(c_k)$ at the output of the channel equaliser in Figure 26.21, before passing the LP values to the channel decoder. In the final iteration, the *a posteriori* LPs $\mathcal{L}_p^D(d_n)$ of the information symbols are computed by the channel decoder.

Previous turbo equalisation research has implemented the SISO equaliser using the Soft-Output Viterbi Algorithm (SOVA) [734], the optimal MAP algorithm [750] and linear filters [760]. We will now introduce the RBF based equaliser as the SISO equaliser, which will be employed in the context of turbo equalisation using the symbol-based MAP algorithm.

26.7 RBF Assisted Turbo Equalisation of Coded Modulation Schemes

The RBF network based equaliser is capable of utilising the *a priori* information $\mathcal{L}_a^E(c_k)$ provided by the channel decoder of Figure 26.21, in order to improve its performance. This *a priori* information can be assigned namely to the weights of the RBF network [761]. In turn, the RBF equaliser provides the decoder with the *a posteriori* information $\mathcal{L}_p^E(c_k)$ concerning the coded symbol. We will now provide a brief overview of symbol-based coded modulation assisted, RBF aided turbo equalisation. Note that this procedure is different from the separate bit-based channel coding and modulation philosophy outlined in Section 11.2 of [736].

26.7.1 System Overview

The conditional Probability Density Function (PDF) of the ith symbol, $i = 1, \ldots, \mathrm{M}$, associated with the ith subnet of the M-ary RBF channel equaliser having a feedforward order of

m is given by [736]:

$$f^i_{RBF}(\mathbf{v}_k) = \sum_{j=1}^{n_{s,i}} w^i_j \varphi(|\mathbf{v}_k - \mathbf{c}^i_j|), \tag{26.17}$$

$$w^i_j = p^i_j (2\pi\sigma^2_N)^{-m/2}, \tag{26.18}$$

$$\varphi(x) = \exp\left(\frac{-x^2}{2\sigma^2_N}\right) \tag{26.19}$$

$$i = 1, \ldots, \mathsf{M}, \qquad j = 1, \ldots, n_{s,i}$$

where $\mathbf{c}^i_j$, w^i_j, $\varphi(\cdot)$ and $2\sigma^2_N$ are the RBF's centres, weights, activation function and width, respectively, and σ^2_N is the noise variance of the channel. The actual number of channel states $n_{s,i}$ is determined by the specific design of the algorithm invoked, but in general we aim for reducing the number of channel states from the optimum number of $\mathsf{M}^{m+\bar{L}-1}$, where m is the equaliser feedforward order and $\bar{L} + 1$ is the (CIR) duration [756, 762, 763], to a lower value for the sake of reducing the computational complexity.

The term $\mathbf{v}_k$ in Equation 26.17 is the received symbol sequence, as shown in Figure 26.20. Explicitly, $\mathbf{v}_k$ hosts the channel outputs observed by the mth order equaliser, which can be expressed in an m-dimensional vectorial form as:

$$\mathbf{v}_k = \begin{bmatrix} v_k & v_{k-1} & \cdots & v_{k-m+1} \end{bmatrix}^T. \tag{26.20}$$

The channel input states are hosted by the vector $\mathbf{s}_j$, which is also referred to as the channel input vector. Explicitly, this vector is given by the jth possible combination of the $(\bar{L} + m)$ number of transmitted symbols, namely by:

$$\mathbf{s}_j = \begin{bmatrix} s_{j1} & \cdots & s_{jp} & \cdots & s_{j(\bar{L}+m)} \end{bmatrix}^T. \tag{26.21}$$

In order to arrive at the Bayesian equalisation solution [369, 754] the RBF centres $\mathbf{c}^i_j$ are assigned the values of the channel output states $\mathbf{r}^i_j$. The channel output states $\mathbf{r}_j$ is the product of the CIR matrix $\mathbf{H}$ and the channel input states $\mathbf{s}_j$. $\mathbf{r}_j$ is also referred to as the channel output vector and it is represented as [369]:

$$\mathbf{r}_j = \mathbf{H}\mathbf{s}_j, \tag{26.22}$$

where the z-transform of the CIR $h(t)$ having a memory of $\bar{L}$ symbols is represented by $H(z) = \sum_{n=0}^{\bar{L}} h_n z^{-n}$ and $\mathbf{H}$ is an $m \times (m + \bar{L})$ matrix given by the CIR taps as follows:

$$\mathbf{H} = \begin{bmatrix} h_0 & h_1 & \cdots & h_{\bar{L}} & \cdots & 0 \\ 0 & h_0 & \cdots & h_{\bar{L}-1} & \cdots & 0 \\ \vdots & \vdots & & & & \vdots \\ 0 & 0 & h_0 & \cdots & h_{\bar{L}-1} & h_{\bar{L}} \end{bmatrix}. \tag{26.23}$$

The RBF weights w^i_j correspond to the *a priori* probability of the channel states $p^i_j = P(\mathbf{r}^i_j)$, $i = 1, \ldots, \mathsf{M}, j = 1, \ldots, n_{s,i}$, as shown in Equation 26.18. The probability p^i_j of the channel states $\mathbf{r}_j^i$, and therefore the weights of the RBF equaliser can be derived from the *a priori* information $\mathcal{L}_a^E(c_k)$ estimated by the symbol-based MAP channel decoder. Explicitly, $\mathcal{L}_a^E(c_k)$ is the interleaved version of the extrinsic information $\mathcal{L}_e^D(c_n)$ in Equation 26.16. More specifically, we derived the *a posteriori* LP value of the $\mathsf{m}(= \log_2 \mathsf{M})$-bit coded symbol as:

$$\mathcal{L}_p^D(c_n) = \begin{cases} \ln(\bar{A}_{n,d}), & \text{if } c_n = L(j,d) \text{ exist,} \\ \ln(0), & \text{otherwise,} \end{cases} \tag{26.24}$$

where $\bar{A}_{n,d}$ is the *a posteriori* probability of the information symbol d defined in Equation 24.55, while $ln(0) = -\infty$ can be substituted by a large negative value, $L(j,d)$ is the corresponding coded symbol when the information symbol is d and the previous trellis state is j. Explicitly, we have $d = \{0, \dots, 2^{\mathsf{m}-1} - 1\}$ when a rate-$(\mathsf{m} - 1)/\mathsf{m}$ channel codec is employed.

Based on Equation 26.22 – for a time-invariant CIR and assuming that the symbols in the sequence $\mathbf{s}_j$ are statistically independent of each other – the probability of the received channel output states $\mathbf{r}_j$ is given by:

$$\begin{aligned} P(\mathbf{r}_j) &= P(\mathbf{s}_j) \\ &= P(s_{j1} \cap \dots s_{jp} \cap \dots s_{j(\bar{L}+m)}) \\ &= \prod_{p=1}^{\bar{L}+m} P(s_{jp}) \qquad j = 1, \dots, n_{s,i}. \end{aligned} \tag{26.25}$$

The transmitted symbol vector component s_{jp} – i.e. the pth symbol in the vector of Equation 26.21– is given by $\mathsf{m} = \log_2 \mathsf{M}$ number of bits $b_{jp1}, b_{jp2}, \dots, b_{jpm}$, which constitute the coded symbol c_k. Explicitly, the transmitted symbol vector component s_{jp} is mapped to the coded symbol c_k. Therefore we have:

$$P(s_{jp}) = \exp\left(\mathcal{L}_a^E(c_k)\right), \qquad s_{jp} = c_k. \tag{26.26}$$

Note that the probability $P(\mathbf{r}_j)$ of the channel output states and therefore also the RBF weights defined in Equation 26.18 are time-variant, since the values of $\mathcal{L}_p^D(c_n)$ are time-variant. Hence, referring to Equations 26.25 and 26.26, the probability $P(\mathbf{r}_j)$ of the channel output states can be represented in terms of the symbol LP $\mathcal{L}_a^E(c_k)$ as follows:

$$\begin{aligned} P(\mathbf{r}_j) &= P(\mathbf{s}_j) \\ &= \prod_{\substack{p=1 \\ s_{jp}=c_k}}^{\bar{L}+m} \exp\left(\mathcal{L}_a^E(c_k)\right) \qquad j = 1, \dots, n_{s,i}. \end{aligned} \tag{26.27}$$

In summary, the computation of the PDF $f_{RBF}^i(\mathbf{v}_k)$ of the ith symbol in Equation 26.17, $i = 1, \dots, \mathsf{M}$, which is associated with the ith subnet of the M-ary RBF channel equaliser, requires the knowledge of the channel states' *a priori* probability p_j^i, when determining the RBF weights w_j^i, as shown in Equation 26.18. Explicitly, we have $p_j^i = P(\mathbf{r}_j^i)$ and note that for a specific subnet i we can suppress the index i for the sake of brevity. Finally, $P(\mathbf{r}_j)$ can be computed from Equation 26.27 using the *a priori* information $\mathcal{L}_a^E(c_k)$. Explicitly,

$\mathcal{L}_a^E(c_k)$ is the interleaved version of the extrinsic information $\mathcal{L}_e^D(c_n)$ of Equation 26.16, and the *a posteriori* information $\mathcal{L}_p^D(c_n)$ can be obtained from Equation 26.24. Therefore, we have demonstrated how the soft output $\mathcal{L}_a^E(c_k)$ provided by the symbol-based MAP channel decoder of Figure 26.21 can be utilised by the RBF equaliser.

On the other hand, the ith subnet of the M-ary RBF equaliser provides the *a posteriori* LP value of the ith coded symbol c_k^i according to:

$$\mathcal{L}_p^E(c_k^i) \;=\; \ln\left(\frac{f_{RBF}^i(\mathbf{v}_k)}{\sum_{all\ l} f_{RBF}^l(\mathbf{v}_k)}\right), \qquad l = 1, \ldots, \mathsf{M}. \qquad (26.28)$$

where $f_{RBF}^i(\mathbf{v}_k)$ was defined by Equation 26.17, while the term $\sum_{all\ l} f_{RBF}^l(\mathbf{v}_k)$ is a normalisation factor, ensuring that we have $\sum_{all\ i} \exp(\mathcal{L}_p^E(c_k^i)) = 1$ and the received sequence $\mathbf{v}_k$ is defined in Equation 26.20.

26.7.2 Simulation Results and Discussion

We employed the Jacobian RBF-DFE of [736, 764], which reduced the complexity of the RBF equaliser by utilising the Jacobian logarithmic function [637] and decision feedback assisted RBF-centre selection [736, 761] as well as a TEQ scheme using a symbol-based MAP channel decoder. The parameters used for the RBF-DFE scheme are: τ=2, m=3 and n=1. The transmitted $(\mathsf{m}-1)$-bit information symbols are encoded by the rate-$(\mathsf{m}-1)/\mathsf{m}$ CM encoder, interleaved and mapped to an M-ary modulated symbol where $\mathsf{M} = 2^{\mathsf{m}}$. We utilised 16QAM in order to obtain an effective transmission throughput of $\mathsf{m} - 1$=3 information Bits Per Symbol (BPS). All the 16QAM-based CM schemes employed exhibited a similar decoding complexity for the sake of a fair comparison. More specifically, a component TCM (or BICM) code memory of 3 was used for the TTCM (or BICM-ID) scheme. The number of iterations for TTCM (BICM-ID) was fixed to 4 (8) and hence the iterative scheme exhibited a similar decoding complexity to that of the TCM (BICM) code of memory 6 when quantified in terms of the number of coding states according to the explanations given in Section 25.2.2. Their corresponding generator polynomials can be found from Tables 24.1, 24.2, 24.3 and 24.4.

The transmission burst structure used in this system is the FMA1 non-spread data burst specified by the Pan-European FRAMES proposal [749], which is shown in Figure 26.22. When considering a Time Division Multiple Access (TDMA) system having 16 slots per $4.615ms$ TDMA frame, the transmission burst length is 288 μs, as shown in Figure 26.22. In our investigations, the transmission delay was limited to approximately $8 \times 4.615ms =$ $37ms$. This corresponds to a transmission delay of 8 TDMA frames and a channel interleaver depth of $8 \times 684 = 5472$ symbols can be employed.

A two-path, symbol-spaced CIR of equal tap weights was used, which can be expressed as $h(t) = 0.707 + 0.707z^{-1}$, where $\bar{L} = 1$ and the Rayleigh fading statistics obeyed a normalised Doppler frequency of 3.25×10^{-5}. The fading magnitude and phase were kept constant for the duration of a transmission burst, a condition which we refer to as employing transmission burst-invariant fading. The Least Mean Square (LMS) algorithm [388] was employed for estimating the CIR based on the training sequence of the transmission burst, as seen in Figure 26.22. Iterative CIR estimation was invoked, where the initial LMS CIR estimation step-size used was 0.05, which was reduced to 0.01 for the second and the subsequent

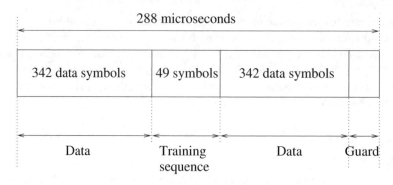

non-spread data burst

Figure 26.22: Transmission burst structure of the FMA1 non-spread speech burst of the FRAMES proposal [749]

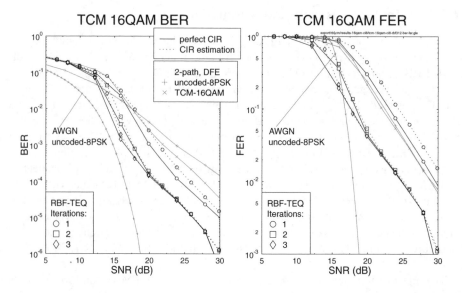

Figure 26.23: The BER and FER versus channel SNR performance of the **RBF-TEQ-TCM** 16QAM scheme, when communicating over the equally-weighted two-path Rayleigh fading CIR. The initial LMS CIR estimation step-size used was 0.05, which was reduced to 0.01 for the second and the subsequent iterations. Our simulation results using perfect CIR estimation are also shown for comparison. The normalised Doppler frequency was 3.25×10^{-5}

iterations. This LMS-aided CIR estimation was outlined in details in [736].

Figures 26.23 to 26.26 illustrate the BER and FER versus channel SNR performance of the RBF-TEQ scheme assisted by 16QAM-based TCM, TTCM, BICM and BICM-ID, when communicating over the equally-weighted two-path Rayleigh fading CIR. As we can see from the figures, the systems employing CIR estimations, rather than perfect CIR knowledge, exhibited some performance loss compared to the ideal systems employing perfect

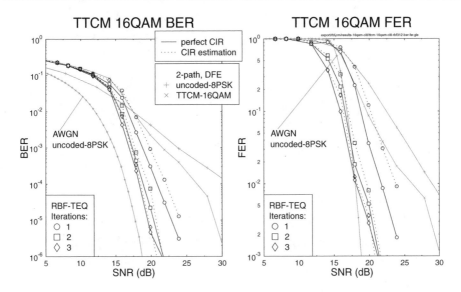

Figure 26.24: The BER and FER versus channel SNR performance of the **RBF-TEQ-TTCM** 16QAM scheme, when communicating over the equally-weighted two-path Rayleigh fading CIR. The initial LMS CIR estimation step-size used was 0.05, which was reduced to 0.01 for the second and the subsequent iterations. Our simulation results using perfect CIR estimation are also shown for comparison. The normalised Doppler frequency was 3.25×10^{-5}.

CIR estimation, but the associated losses reduced rapidly, when the number of TEQ iteration was increased. The BER and FER performance of the identical-throughput uncoded 8PSK scheme communicating over non-dispersive AWGN channels was used as a benchmarker for the 16QAM-based RBF-TEQ arrangement using various CM schemes, when communicating over dispersive Rayleigh fading channels. We found that the BER curves of the TTCM, BICM and BICM-ID assisted schemes are only about 2 dB away from the benchmarker at a BER of 10^{-5}. However, the TCM assisted scheme suffers from an error floor owing to the existence of unprotected bits in the TCM coded symbols. On the other hand, the FER performance of the TTCM, BICM and BICM-ID assisted RBF-TEQ schemes was found in Figures 26.24 to 26.26 to be better than that of the benchmarker at low SNR values. Furthermore, it was found from our simulations that the achievable performance gain remained only marginal, when more than three TEQ iterations were employed.

In order to compare the performance of the RBF-TEQ assisted CM scheme to that of the conventional DFE assisted CM scheme extensively studied in Section 26.4, in Figures 26.23 to 26.26 we have plotted the BER and FER performance of the conventional DFE assisted uncoded-8PSK and those of the CM-16QAM schemes, when communicating over the two-path Rayleigh fading channels considered, where perfect CIR knowledge was assumed. The conventional DFE's feedforward order m and feedback order n were set to seven and one, respectively, since we found from our simulations that further increasing the values of m and n yielded no further performance improvement when communicating over the two-path Rayleigh fading channels considered. Specifically, the conventional DFE exhibits a lower

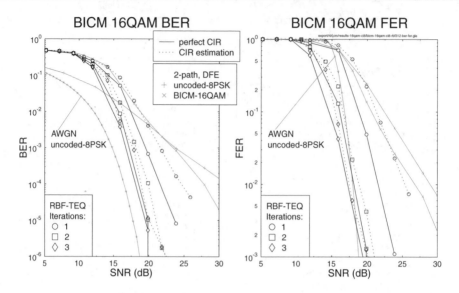

Figure 26.25: The BER and FER versus channel SNR performance of the **RBF-TEQ-BICM** 16QAM
scheme, when communicating over the equally-weighted two-path Rayleigh fading CIR.
The initial LMS CIR estimation step-size used was 0.05, which was reduced to 0.01
for the second and the subsequent iterations. Our simulation results using perfect CIR
estimation are also shown for comparison. The normalised Doppler frequency was
3.25×10^{-5}.

complexity than that of the RBF-DFE. However, the performance of the conventional DFE
scheme is lower than that of its RBF-DFE counterpart owing to experiencing an error floor
in the high SNR region [736]. From Figures 26.23 to 26.26 we noticed that the conventional
DFE assisted CM-16QAM schemes exhibited approximately 2 to 3 dB coding/SNR gain [3]
compared to the identical-throughput conventional DFE assisted uncoded-8PSK scheme at a
BER of 10^{-4}. However, the achievable coding gain of the various CM schemes was significantly increased, when the RBF-TEQ scheme was employed, although this was achieved at a
concomitantly higher complexity owing to employing an increased number of iterations. Explicitly, a coding gain in excess of 10 dB was achievable at a BER of 10^{-4} by the various CM
schemes against the identical-throughput conventional DFE assisted uncoded-8PSK scheme,
when the RBF-TEQ scheme invoked 3 iterations.

Figure 26.27 shows the BER and FER versus channel SNR performance of RBF-TEQ for
various CM aided 16QAM schemes, when communicating over the equally-weighted two-
path Rayleigh fading CIR and utilising iterative LMS-based CIR estimation. It is illustrated

---

[3]In our investigation we will be consistently plotting BER versus SNR as well as FER versus SNR curves. These
curves may be readily converted to BER versus E_b/N_0 and FER versus E_b/N_0 curves by shifting them according
to the associated coding rate. However, all curves would have to be shifted by the same amount and hence, the
achievable coding gain remains identical to the achievable SNR gain. This is because coding gain is defined on the
E_b/N_0 scale, while E_b represents the energy of the useful information bits. Therefore, in our further discourse
we will refers to SNR gain and coding gain interchangeably. More specifically, in order to convert the SNR scale
to E_b/N_0 scale, all curves require a shift of $10 \log(3) = 4.77$ dB to the left of the SNR scale when the effective
throughput is 3 BPS.

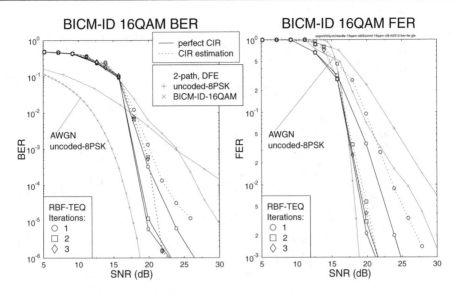

Figure 26.26: The BER and FER versus channel SNR performance of the **RBF-TEQ-BICM-ID** 16QAM scheme, when communicating over the equally-weighted two-path Rayleigh fading CIR. The initial LMS CIR estimation step-size used was 0.05, which was reduced to 0.01 for the second and the subsequent iterations. Our simulation results using perfect CIR estimation are also shown for comparison. The normalised Doppler frequency was 3.25×10^{-5}.

in Figure 26.27 that the RBF-TEQ-BICM scheme attained the highest TEQ gain compared to its counterparts. The RBF-TEQ-BICM scheme is also the best performer in terms of the achievable FER, but the RBF-TEQ-TTCM arrangement is the best performer in terms of the BER attained.

26.7.3   Conclusion

We found in Figure 26.27 that the BER performance of the TTCM, BICM and BICM-ID assisted RBF-TEQ schemes was only about 2 dB away from that of the identical-throughput uncoded 8PSK scheme communicating over AWGN channels. The achievable coding gain of the various CM schemes was significantly increased, when employing the RBF-TEQ scheme of Section 26.7 rather than the conventional DFE. However, the achievable performance gain remained only marginal, when more than three TEQ iterations were employed. The RBF-TEQ-BICM scheme is the best performer in terms of the achievable FER, but the RBF-TEQ-TTCM arrangement has the edge in terms of the BER attained.

Having studied the performance of the RBF-TEQ arrangement employing various CM schemes, let us now commence our discourse on employing CM schemes in the context of the reduced complexity In-phase/Quadrature-phase TEQ [736,765] system to be described in Section 26.8.

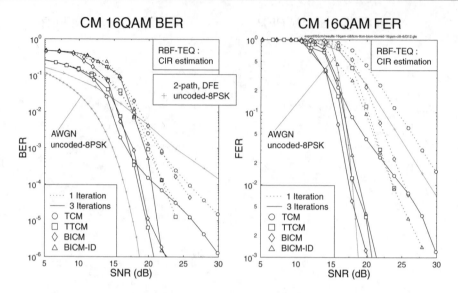

Figure 26.27: The BER and FER versus channel SNR performance of the **RBF-TEQ for various CM 16QAM scheme**, when communicating over the equally-weighted two-path Rayleigh fading CIR. The initial LMS CIR estimation step-size used was 0.05, which was reduced to 0.01 for the second and the subsequent iterations. The iteration gains of TEQ can be observed by comparing the performance of the first and third TEQ iteration of the systems. The normalised Doppler frequency was 3.25×10^{-5}.

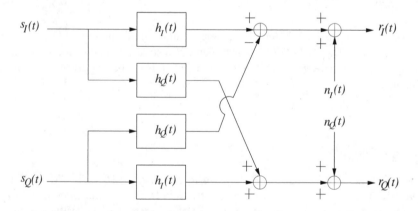

Figure 26.28: The complex channel model. After transmission over the complex channel $h(t)$, the received signal $r(t)$ becomes dependent on the in-phase component $s_I(t)$ and quadrature-phase component $s_Q(t)$ of the transmitted signal, as expressed in Equations 26.29 and 26.30

26.8 In-phase/Quadrature-phase Turbo Equalisation

In this section, we denote the modulated signal by $s(t)$, which is transmitted over the dispersive channel characterised by the CIR $h(t)$. The signal is also contaminated by the zero-mean Additive White Gaussian Noise (AWGN) $n(t)$ exhibiting a variance of $\sigma_N^2 = N_0/2$, where

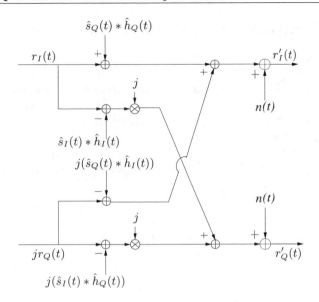

Figure 26.29: The process of removing the dependency of $r_I(t)$ and $r_Q(t)$ on the quadrature components of the transmitted signals, namely on $s_I(t)$ and $s_Q(t)$, to give $r'_I(t)$ and $r'_Q(t)$, respectively. In this figure, it is assumed that the CIR estimation is perfect, i.e. that we have $\hat{h}_I(t) = h_I(t)$ as well as $\hat{h}_Q(t) = h_Q(t)$ and that the transmitted signals are known, giving $\hat{s}_I(t) = s_I(t)$ and $\hat{s}_Q(t) = s_Q(t)$. In this case, perfect decoupling is achieved. However, in practice these estimates have to be generated at the receiver

N_0 is the single-sided noise power spectral density. The received signal $r(t)$ is then formulated as:

$$
\begin{aligned}
r(t) &= s(t) * h(t) + n(t) \\
&= [s_I(t) + js_Q(t)] * [h_I(t) + jh_Q(t)] \\
&+ n_I(t) + jn_Q(t) \\
&= r_I(t) + jr_Q(t),
\end{aligned}
\tag{26.29}
$$

where we have:

$$
\begin{aligned}
r_I(t) &= s_I(t) * h_I(t) - s_Q(t) * h_Q(t) + n_I(t) \\
r_Q(t) &= s_I(t) * h_Q(t) + s_Q(t) * h_I(t) + n_Q(t),
\end{aligned}
\tag{26.30}
$$

since the CIR $h(t)$ is complex-valued and therefore consists of the I component $h_I(t)$ and Q component $h_Q(t)$. On the same note, $s_I(t)$ and $s_Q(t)$ are the I and Q components of $s(t)$ in Figure 26.28, while $n_I(t)$ and $n_Q(t)$ denote the corresponding AWGN components. Both of the received I/Q signals, namely $r_I(t)$ and $r_Q(t)$ of Equation 26.30 become dependent on both $s_I(t)$ and $s_Q(t)$ due to the cross-coupling effect imposed by the complex channel. Hence a conventional channel equaliser, regardless of whether it is an iterative or non-iterative equaliser, would have to consider the effects of this cross-coupling.

26.8.1　In-phase/Quadrature-phase Turbo Equalisation Principle

Initially we can ignore the channel-induced cross-coupling of the received signal's quadrature components and then we compensate for this gross simplification with the aid of the turbo equaliser. More explicitly, this simplification would result in an unacceptable performance degradation in the context of conventional non-iterative channel equalisation. However, the employment of the iterative turbo equalisation technique allows us to compensate for the above simplification during the consecutive iterations. Therefore we can compute the I and Q components of the decoupled channel output $r'(t)$, as though they were dependent on $s_I(t)$ or $s_Q(t)$ only, as portrayed in Figure 26.29 in the context of the following equations:

$$
\begin{aligned}
r'_I(t) &= s_I(t) * h(t) + n_I(t) \\
&= s_I(t) * h_I(t) + j[s_I(t) * h_Q(t)] + n(t) \\
r'_Q(t) &= -s_Q(t) * h(t) + n_Q(t) \\
&= -(s_Q(t) * h_I(t) + j[s_Q(t) * h_Q(t)]) + n(t).
\end{aligned}
$$
$$(26.31)$$

More explicitly, the removal of the cross-coupling is facilitated by generating the estimates $\hat{s}_I(t)$ and $\hat{s}_Q(t)$ of the transmitted signal [760] with the aid of the reliability information generated by the channel decoder and then by cancelling the cross-coupling effects imposed by the channel, yielding $r'_I(t)$ and $r'_Q(t)$, respectively, as seen in Figure 26.29. In the ideal scenario, where perfect knowledge of both the CIR and that of the transmitted signal is available, it is plausible that the channel-induced cross-coupling between the quadrature components can be removed. However, when unreliable symbol estimates are generated owing to the channel-impaired low-confidence reliability values, errors are introduced in the decoupling operation. Nonetheless, we will show that the associated imperfect decoupling effects are compensated with the aid of the iterative turbo equalisation process in its consecutive iterations.

Following the above decoupling operation, the modified complex channel outputs, namely $r'_I(t)$ and $r'_Q(t)$ of Figure 26.29, respectively, can be viewed as the result of convolving both quadrature components independently with the complex CIR on each quadrature arm. Consequently, we can equalise $s_I(t)$ and $s_Q(t)$ independently, hence reducing the number of channel states and the associated complexity quite significantly. Again, note that in Equation 26.31 we have assumed that perfect signal regeneration and perfect decoupling are achieved at the receiver, in order to highlight the underlying principle of the reduced complexity equaliser.

26.9　RBF Assisted Reduced Complexity In-phase and Quadrature-phase Turbo Equalisation of Coded Modulation Schemes

In the RBF-I/Q-EQ scheme we utilised the principle of separate I/Q equalisation outlined as in Section 26.8, where two separate RBF equalisers was used for the in-phase and quadrature component of the transmitted symbols. The in-phase-RBF-EQ has RBF centres, which con-

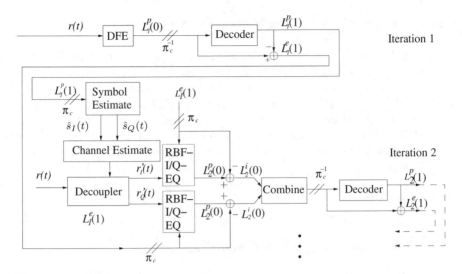

Figure 26.30: Schematic of the turbo equaliser employing a DFE and a SISO channel decoder in the first turbo equalisation iteration. In subsequent iterations, two RBF-I/Q-EQs and one SISO channel decoder are employed. The notation π_c represents a channel interleaver, while π_c^{-1} is used to denote a channel deinterleaver.

sist of the in-phase decoupled channel output $r'_I(t)$ of Equation 26.30 and vice versa for the quadrature-RBF-EQ. The number of possible channel output states is reduced, since the decoupled channel output $r'(t)$ is dependent on $\sqrt{M}$ number of possible in-phase or quadrature-phase transmitted symbols instead of the original M number of possible symbols.

26.9.1   System Overview

Figure 26.30 illustrates the schematic of the turbo equaliser utilising two reduced-complexity RBF-I/Q equalisers. The same notations employed in Section 26.6 is used in this section. The subscripts in Figure 26.30 are used for representing the iteration index, while the argument within the brackets ( ) indicates the index of the receiver stage, where the equalisers are denoted as receiver stage 0, while the channel decoder as receiver stage 1.

The conventional minimum mean square error DFE seen at the top left corner of Figure 26.30 is used for the first turbo equalisation iteration for providing soft decisions in the form of the LP $L_1^p(0)$ to the CM decoder. Invoking the DFE at the first iteration constitutes a low-complexity approach to providing an initial estimate of the transmitted symbols, as compared to the more complex RBF-I/Q-EQ. The symbol-based MAP channel decoder of Figure 26.30 generates the *a posteriori* LP $L_1^p(1)$ and from that, the *extrinsic* information of the encoded symbols $L_1^e(1)$ is extracted. In the next iteration, the *a posteriori* LP $L_1^p(1)$ is used for regenerating estimates of the I and Q components of the transmitted signal, namely $\hat{s}_I(t)$ and $\hat{s}_Q(t)$, as seen in the 'Symbol Estimate' block of Figure 26.30. The *a posteriori* information was transformed from the logarithmic domain to modulated symbols using the approach employed in [760]. The estimated transmitted quadrature components $\hat{s}_I(t)$ and $\hat{s}_Q(t)$ are then convolved with the estimate of the CIR $h(t)$. At the decoupler block of Figure 26.30, the resultant signal is used for removing the cross-coupling effect, as seen in

Equation 26.30, according to Equation 26.31 from both quadrature components of the trans-mitted signal, yielding $r'_I(t)$ and $r'_Q(t)$.

After the decoupling operation, $r'_I(t)$ and $r'_Q(t)$ are passed to the RBF-I/Q-EQ in the schematic of Figure 26.30. In addition to these received quadrature signals, the RBF-I/Q-EQ also processes the *a priori* information received, which is constituted by the *extrinsic* LPs $L_1^e(1)$ derived from the previous iteration, and generates the *a posteriori* information $L_2^p(0)$. Subsequently, the combined channel and *extrinsic* information $L_2^i(0)$ is extracted from both RBF-I/Q-EQs in Figure 26.30 and combined, before being passed to the Log-MAP channel decoder. As in the first turbo equalisation iteration, the *a posteriori* and *extrinsic* information of the encoded symbol, namely $L_2^p(1)$ and $L_2^e(1)$, respectively, are evaluated. Subsequent turbo equalisation iterations obey the same sequence of operations, until the iteration termi-nation criterion is met.

26.9.2   Simulation Results and Discussion

In this section, we will study the performance of a number of RBF-I/Q-TEQ schemes employ-ing various CM schemes. Similar simulation parameters to those outlined in Section 26.7.2 are used. The RBF-DFE based TEQ is specified by the equaliser's decision delay τ, the feed-forward order m and the feedback order n. The number of RBF nodes is $n_{s,i} = \bar{M}^{\bar{L}+m-n}$ and the number of scalar channel states of the Jacobian RBF equaliser is $n_{s,f} = \bar{M}^{\bar{L}+1}$, where we have $\bar{M} = M$ for the non-I/Q based full-complexity RBF-TEQ system, while $\bar{M} = \sqrt{M}$ for the I/Q based RBF-TEQ system. The estimated computational complexity of gener-ating the *a posteriori* LP for the Jacobian RBF equaliser is summarised in Table 26.2, where $n_{s,i}(m+2) - 2\bar{M} + n_{s,f}$ number of additions/subtractions and $2n_{s,f}$ multiplications/divisions are required. Here, we employed $\tau = 2$, $m = 3$ and $n = 1$ for the RBF-TEQ, as well as $m = 7$ and $n = 1$ for the conventional DFE. Therefore, the 'per-iteration' complexity of the full-RBF-TEQ expressed in terms of the number of additions/subtractions and multiplica-tions/divisions is about 20704 and 512, respectively, while that of the RBF-I/Q-TEQ is about 328 and 32, respectively.

Figures 26.31 to 26.34 illustrate the BER and FER versus channel SNR performance of the RBF-I/Q-TEQ scheme incorporating into 16QAM-based TCM, TTCM, BICM and BICM-ID, when communicating over the equally-weighted two-path Rayleigh fading CIR. As we can see from the figures, the systems employing LMS-based CIR estimation exhibited only marginal performance losses compared to the ideal systems employing perfect CIR es-timation. This is because the RBF-I/Q-TEQ scheme reduces the effect of error propagation, since the set of RBF centres to be selected using the DFE mechanism is reduced from $\mathcal{M}^n$ to $\mathcal{M}^{n/2}$ [736, 761]. Again the BER performance of the identical-throughput uncoded 8PSK scheme communicating over the non-dispersive AWGN channels was used as a benchmark for the 16QAM-based RBF-I/Q-TEQ arrangement using various CM schemes, which com-municated over the dispersive two path Rayleigh fading channels. It was found from our simulations that the achievable performance gain remained only marginal, when more than six TEQ iterations were employed.

The BER and FER performance of the conventional DFE assisted uncoded-8PSK and CM-16QAM schemes employing perfect CIR knowledge was also plotted in Figures 26.31 to 26.34 for comparison to the RBF-I/Q-TEQ assisted CM scheme when communicating

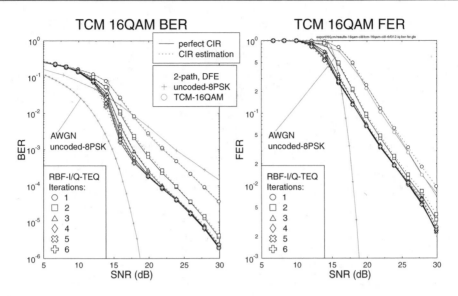

Figure 26.31: The BER and FER versus channel SNR performance of the **RBF-I/Q-TEQ-TCM**
16QAM scheme, when communicating over the equally-weighted two-path Rayleigh
fading CIR. The initial LMS CIR estimation step-size used was 0.05, which was reduced
to 0.01 for the second and the subsequent iterations. Our simulation results using perfect
CIR estimation are also shown for comparison. The normalised Doppler frequency was
3.25×10^{-5}.

over the two-path Rayleigh fading channels considered. Explicitly, the first iteration of the
RBF-I/Q-TEQ-CM scheme employed a conventional DFE rather than the RBF-DFE, hence
the corresponding performance is identical to that of the conventional DFE assisted CM-
16QAM schemes characterised in Figures 26.31 to 26.34. Specifically, the achievable coding
gain of the various 16QAM-based RBF-I/Q-TEQ assisted CM schemes against the identical-
throughput conventional DFE assisted uncoded-8PSK scheme increases, as the number of
iterations increases. Again, the achievable coding gain of the various RBF-I/Q-TEQ assisted
CM schemes is significantly higher than that of the conventional DFE assisted CM schemes,
albeit this is achieved at a higher complexity. Nonetheless, the complexity of the RBF-I/Q-
TEQ scheme still remains lower than that of the conventional trellis-based TEQ as argued
in [204, 736].

Figure 26.35 shows the BER and FER versus channel SNR performance of RBF-I/Q-TEQ
for various CM aided 16QAM schemes, when communicating over the equally-weighted
two-path Rayleigh fading CIR and utilising the iterative LMS-based CIR estimation of [736].
It is explicit in Figure 26.35 that RBF-I/Q-TEQ-BICM scheme obtained the highest TEQ
gains compared to its counterparts. The RBF-I/Q-TEQ-BICM scheme is also the best performer in terms of the achievable FER, but the RBF-I/Q-TEQ-TTCM scheme is the best
performer in terms of the BER attained. Let us now compare the performance of the RBF-
I/Q-TEQ-CM scheme to that of the RBF-TEQ-CM scheme in Figure 26.36. It is found from
Figure 26.36 that the performance of the RBF-I/Q-TEQ-CM scheme having six iterations is
similar to that of RBF-TEQ-CM having three iterations, except for the RBF-I/Q-TEQ-TCM

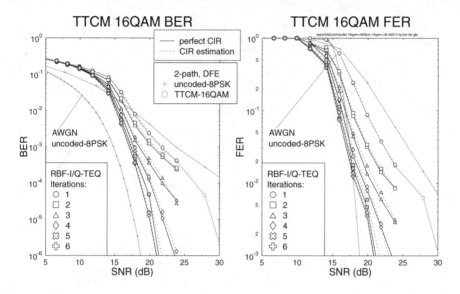

Figure 26.32: The BER and FER versus channel SNR performance of the **RBF-I/Q-TEQ-TTCM**
16QAM scheme, when communicating over the equally-weighted two-path Rayleigh
fading CIR. The initial LMS CIR estimation step-size used was 0.05, which was reduced
to 0.01 for the second and the subsequent iterations. Our simulation results using perfect
CIR estimation are also shown for comparison. The normalised Doppler frequency was
3.25×10^{-5}.

scheme, where the achievable FER performance is about one dB inferior in comparison to
that of the RBF-TEQ-TCM scheme.

Figure 26.37 illustrates the BER and FER versus channel SNR performance of the TTCM
assisted RBF-I/Q-TEQ and RBF-TEQ schemes on an iteration-by-iteration basis. In terms of
the attainable BER, the performance of the first three iterations of RBF-I/Q-TEQ-TTCM is
inferior to that of the first iteration of RBF-TEQ-TTCM for BER values below 10^{-4}. In terms
of the achievable FER, the performance of the first two iterations of RBF-I/Q-TEQ-TTCM
is inferior to that of the first iteration of RBF-TEQ-TTCM for FER values below 10^{-2}. This
is due to the employment of a conventional DFE during the first iteration of the RBF-I/Q-
TEQ-TTCM scheme, as well as owing to the imperfect I/Q decoupling effects, when unreliable symbol estimates are employed. However, more reliable symbol estimates become
available with the aid of the iterative TEQ scheme during the forthcoming iterations and the
performance of RBF-I/Q-TEQ-TTCM becomes comparable to that of the full-complexity
RBF-TEQ-TTCM arrangement eventually, where the performance of RBF-I/Q-TEQ-TTCM
having eight iterations is identical to that of RBF-TEQ-TTCM having four iterations for BER
values below 10^{-4} and FER values below 10^{-2}, as shown in Figure 26.37. Note that the complexity imposed by the conventional DFE during the first RBF-I/Q-TEQ iteration is insignificant compared to that of the remaining RBF based iterations. Hence, we should compare
the complexity of the RBF-DFE assisted scheme using seven iterations in the eight-iteration
aided RBF-I/Q-TEQ-TTCM scheme shown in Figure 26.37, to that of the four-iteration full
RBF-TEQ-TTCM scheme shown in Figure 26.37. Therefore, it can be shown that complexity

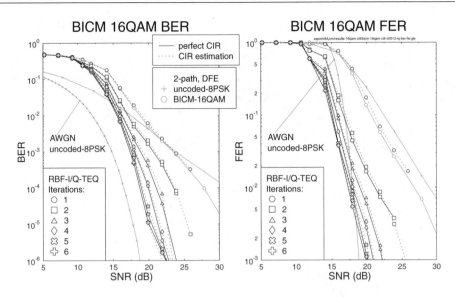

Figure 26.33: The BER and FER versus channel SNR performance of the **RBF-I/Q-TEQ-BICM 16QAM** scheme, when communicating over the equally-weighted two-path Rayleigh fading CIR. The initial LMS CIR estimation step-size used was 0.05, which was reduced to 0.01 for the second and the subsequent iterations. Our simulation results using perfect CIR estimation are also shown for comparison. The normalised Doppler frequency was 3.25×10^{-5}.

reduction factors of $\frac{4}{7} \cdot \frac{20704}{328} \approx 36$ and $\frac{4}{7} \cdot \frac{512}{32} \approx 9$ were obtained in terms of the required number of additions/subtractions and multiplications/divisions, respectively.

26.9.3 Conclusions

We found that the RBF-I/Q-TEQ scheme employing LMS-based CIR estimation exhibited only marginal performance losses compared to the ideal systems employing perfect CIR estimation. This is because the effect of error propagation was reduced significantly, when employing the RBF-I/Q-TEQ scheme, compared to that of the complex-valued RBF-I/Q-TEQ scheme. The achievable performance gain remained only marginal, when more than six TEQ iterations were employed by the CM assisted RBF-I/Q-TEQ scheme.

A significant complexity reduction was obtained by the RBF-I/Q-TEQ-CM scheme, when compared to the complex-valued RBF-TEQ-CM arrangement, while achieving virtually the same performance. Specifically, complexity reduction factors of 36 and 9 were achieved in terms of the required number of additions/subtractions and multiplications/divisions, respectively.

26.10 Summary

In this chapter the performance of four single-carrier coded modulation schemes employing channel equalisers was evaluated for transmissions over wideband channels. Specifically, the

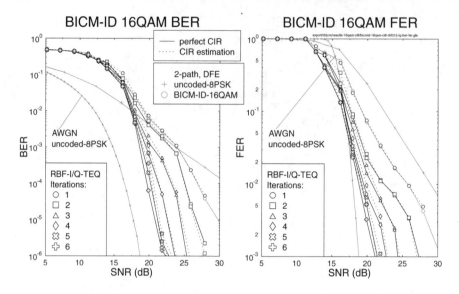

Figure 26.34: The BER and FER versus channel SNR performance of the **RBF-I/Q-TEQ-BICM-ID** 16QAM scheme, when communicating over the equally-weighted two-path Rayleigh fading CIR. The initial LMS CIR estimation step-size used was 0.05, which was reduced to 0.01 for the second and the subsequent iterations. Our simulation results using perfect CIR estimation are also shown for comparison. The normalised Doppler frequency was 3.25×10^{-5}.

performance of the BbB adaptive conventional DFE-assisted coded modulation scheme was investigated in Section 26.4, where an improved BER and BPS performance was attained in comparison to that of the fixed-mode based coded modulation schemes. It was found that the employment of a channel interleaver having a memory of one transmission burst was insufficient for randomising the bursty channel errors induced by the slowly-fading COST207 TU channels. Systems that employ a channel interleaver having a longer memory will attain a higher diversity gain, but suffer from a reduced flexibility in terms of modulation mode switching. However, the diversity gain achieved by employing a channel interleaver spanning over four transmission bursts compensated for the loss of switching flexibility, ultimately providing a better performance in terms of both the achievable BER and BPS, as seen in Figure 26.12. TTCM was found to be the best performer, followed by BICM-ID, TCM and BICM in the context of the conventional DFE-assisted adaptive coded modulation scheme.

In Section 26.7, RBF-based turbo equaliser [736] was amalgamated with the coded modulation schemes communicating over wideband fading channels. Specifically, both the proposed RBF-TEQ-CM and the RBF-I/Q-TEQ-CM schemes were investigated under dispersive Rayleigh fading channel conditions using 16QAM for maintaining an effective throughput of 3 BPS as discussed in Sections 26.7 and 26.9, respectively. In general, the BER performance of both the 16QAM-based RBF-TEQ-CM and RBF-I/Q-TEQ-CM schemes when communicating over wideband fading channels, was found to be only about 2 dB away from that of the identical-throughput uncoded 8PSK scheme communicating over AWGN channels. Our simulation results show significant complexity reductions for the RBF-I/Q-TEQ-

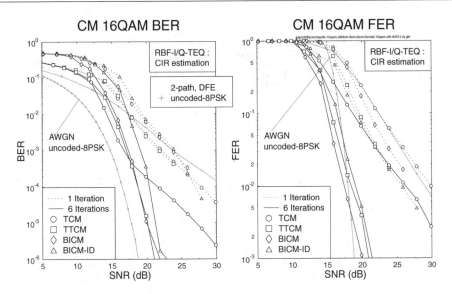

Figure 26.35: The BER and FER versus channel SNR performance of the **RBF-I/Q-TEQ for various CM** 16QAM scheme, when communicating over the equally-weighted two-path Rayleigh fading CIR. The initial LMS CIR estimation step-size used was 0.05, which was reduced to 0.01 for the second and the subsequent iterations. The iteration gains of TEQ can be observed by comparing the performance of the first and third TEQ iteration of the systems. The normalised Doppler frequency was 3.25×10^{-5}.

CM scheme when compared to complex-valued RBF-TEQ-CM, while achieving virtually the same performance. This was demonstrated in Figures 26.36 and 26.37. Amongst the four CM schemes, the best performer was TTCM followed by BICM, BICM-ID and TCM in terms of the achievable BER, as shown in Figure 26.27 for the RBF-TEQ scheme [736] and in Figure 26.35 for the RBF-I/Q-TEQ scheme. However, in terms of the FER attained the best performer was BICM, followed by TTCM, BICM-ID and TCM, as was demonstrated in Figures 26.27 and 26.35.

We also compared the performance of the RBF-TEQ-CM and RBF-I/Q-TEQ-CM schemes to that of the conventional DFE assisted CM scheme which was characterised in Section 26.4. More specifically, the conventional DFE assisted CM scheme was integrated into a burst-by-burst adaptive CM system in Section 26.4 based on the corresponding fixed mode's performance. Although the adaptive CM system was not investigated in the context of RBF-TEQ [736], nonetheless we can expect that the RBF-TEQ assisted adaptive CM system will maintain a significantly better performance compared to that of the conventional DFE assisted adaptive CM system of Section 26.4, since the coding gain of the fixed mode RBF-TEQ-CM and RBF-I/Q-TEQ-CM schemes is significantly higher than that of their conventional DFE based counterpart, as we demonstrated in Sections 26.7 and 26.9. Although the complexity of the RBF-TEQ is higher than that of the conventional DFE, the RBF-assisted schemes are capable of maintaining a lower complexity than that of their conventional trellis-based counterparts, when communicating over both dispersive Gaussian and Rayleigh fading channels, while maintaining a similar performance [204, 736].

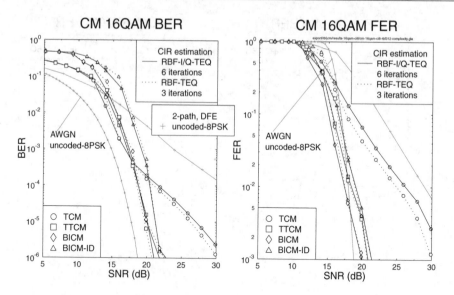

Figure 26.36: The BER and FER versus channel SNR performance of the **RBF-I/Q-TEQ** and **RBF-TEQ** for various CM 16QAM scheme, when communicating over the equally-weighted two-path Rayleigh fading CIR. The initial LMS CIR estimation step-size used was 0.05, which was reduced to 0.01 for the second and the subsequent iterations. The RBF-I/Q-TEQ-CM scheme employs six iterations and the RBF-TEQ-CM scheme employs three iterations. The normalised Doppler frequency was 3.25×10^{-5}.

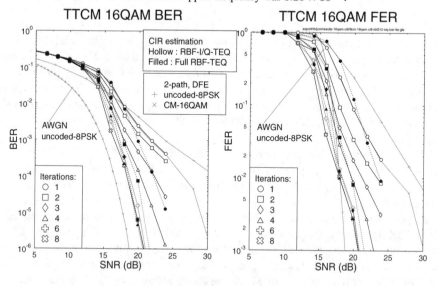

Figure 26.37: The BER and FER versus channel SNR performance of the **RBF-I/Q-TEQ-TTCM** and **RBF-TEQ-TTCM** 16QAM scheme, when communicating over the equally-weighted two-path Rayleigh fading CIR. The initial LMS CIR estimation step-size used was 0.05, which was reduced to 0.01 for the second and the subsequent iterations. The normalised Doppler frequency was 3.25×10^{-5}.

In summary, the coding gain exhibited by the coded modulation schemes studied in this chapter was summarised in Tables 26.3, 26.4 and 26.5. Specifically, in Table 26.3 we summarise the performance gain exhibited by the DFE-based coded modulation schemes of Section 26.4, when communicating over the COST207 TU Rayleigh fading channel. Furthermore, the performance gain of the OFDM-based coded modulation schemes of Section 25.4 employing a similar set of parameters and communicating also over the COST207 TU Rayleigh fading channel was incorporated in Table 26.3 for comparison. Specifically, the total number of OFDM subcarriers was 1024 (1K mode [438]) and the number of effective subcarriers was 684. From Table 26.3 we can see that the DFE-based coded modulation schemes of Section 26.4 perform better than the OFDM-based coded modulation schemes of Section 25.4, when targetting a BER of 10^{-3} and 10^{-5}. In Tables 26.4 and 26.5 the coding gain exhibited by the RBF-TEQ-CM and RBF-I/Q-TEQ-CM schemes communicating over 2-path Rayleigh fading channels was shown.

COST207 TU Rayleigh fading channels		DFE				OFDM			
		E_b/N_o (dB)		Gain (dB)		E_b/N_o (dB)		Gain (dB)	
Code	Modulation/	BER				BER			
	BPS	10^{-3}	10^{-5}	10^{-3}	10^{-5}	10^{-3}	10^{-5}	10^{-3}	10^{-5}
uncoded	BPSK/1	12.35	18.10	0.00	0.00	25.00	44.85	0.00	0.00
TCM	QPSK/1	7.02	11.20	5.33	6.90	11.76	16.00	13.24	28.85
TTCM	QPSK/1	6.35	**8.62**	6.00	9.48	**10.72**	**11.65**	14.28	33.20
BICM	QPSK/1	6.67	9.72	5.68	8.38	11.16	14.09	13.84	30.76
BICM-ID	QPSK/1	**6.21**	9.58	6.14	8.52	10.88	14.34	14.12	30.51
uncoded	QPSK/2	12.71	18.59	0.00	0.00	25.00	44.85	0.00	0.00
TCM	8PSK/2	10.57	14.74	2.14	3.85	14.96	19.84	10.04	25.01
TTCM	8PSK/2	**10.36**	**13.07**	2.35	5.52	14.18	**15.97**	10.82	28.88
BICM	8PSK/2	11.04	14.04	1.67	4.55	14.88	17.97	10.12	26.88
BICM-ID	8PSK/2	10.91	**13.07**	1.80	5.52	**14.05**	16.98	10.95	27.87
uncoded	8PSK/3	16.91	22.31	0.00	0.00	27.34	47.10	0.00	0.00
TCM	16QAM/3	13.18	17.05	3.73	5.03	18.20	36.10	9.14	11.00
TTCM	16QAM/3	**13.13**	**15.81**	3.78	6.50	16.36	**18.56**	10.98	28.54
BICM	16QAM/3	13.65	16.99	3.26	5.53	17.07	20.37	10.27	26.73
BICM-ID	16QAM/3	14.12	16.15	2.79	6.16	**16.32**	19.20	11.02	27.90
uncoded	16QAM/4	17.98	23.75	–	–	28.00	47.85	–	–
TCM	64QAM/5	**18.12**	21.80	–	–	24.48	43.62	–	–
TTCM	64QAM/5	**18.12**	**21.03**	–	–	24.28	43.50	–	–
BICM	64QAM/5	19.01	22.91	–	–	22.55	**26.66**	–	–
BICM-ID	64QAM/5	19.33	21.08	–	–	**21.30**	26.82	–	–
uncoded	64QAM/6	20.42	27.42	–	–	31.61	51.28	–	–

Table 26.3: Coding gain values of the various DFE-based coded modulation schemes studied in this chapter, when communicating over the COST207 TU Rayleigh fading channel. All of the coded modulation schemes exhibited a similar decoding complexity in terms of the number of decoding states, which was equal to 64 states. An interleaver block length of $4 \times 684 = 2736$ symbols was employed and the corresponding simulation parameters were shown in Section 26.4.2. Furthermore, the performance of the OFDM-based coded modulation schemes employing a similar set of parameters, when communicating over the COST207 TU Rayleigh fading channel was also considered here for comparison. Specifically, the total number of OFDM subcarriers was 1024 (1K mode [438]) and the number of effective subcarriers was 684. The performance of the best scheme is printed in bold.

RBF-TEQ, 2-path Rayleigh Fading Channels		First iteration performance				Third iteration performance			
Code	Modulation/ BPS	E_b/N_o (dB) BER		Gain (dB) BER		E_b/N_o (dB) BER		Gain (dB) BER	
		10^{-3}	10^{-5}	10^{-3}	10^{-5}	10^{-3}	10^{-5}	10^{-3}	10^{-5}
Uncoded/AWGN	8PSK/3	10.03	12.97	–	–	10.03	12.97	–	–
Uncoded/DFE	8PSK/3	18.88	31.63	0.00	0.00	18.88	31.63	0.00	0.00
TCM	16QAM/3	16.63	25.83	2.25	5.80	**11.98**	21.48	6.90	10.15
TTCM	16QAM/3	**15.03**	**19.23**	3.85	12.40	12.42	**14.88**	6.46	16.75
BICM	16QAM/3	16.87	23.43	2.01	8.20	12.95	15.23	5.93	16.40
BICM-ID	16QAM/3	15.35	21.43	3.53	10.20	14.64	16.47	4.24	15.16

Table 26.4: Coding gain values of the various RBF-TEQ assisted coded modulation schemes studied in this chapter, when communicating over 2-path Rayleigh fading channels. All of the coded modulation schemes exhibited a similar decoding complexity in terms of the number of decoding states, which was equal to 64 states. An interleaver block length of $8 \times 684 = 5472$ symbols was employed and the corresponding simulation parameters were summarised in Section 26.7.2. The performance of the best scheme is printed in bold

RBF-I/Q-TEQ, 2-path Rayleigh Fading Channels		First iteration performance				Sixth iteration performance			
Code	Modulation/ BPS	E_b/N_o (dB) BER		Gain (dB) BER		E_b/N_o (dB) BER		Gain (dB) BER	
		10^{-3}	10^{-5}	10^{-3}	10^{-5}	10^{-3}	10^{-5}	10^{-3}	10^{-5}
Uncoded/AWGN	8PSK/3	10.03	12.97	–	–	10.03	12.97	–	–
Uncoded/DFE	8PSK/3	18.88	31.63	0.00	0.00	18.88	31.63	0.00	0.00
TCM	16QAM/3	17.38	27.13	1.50	4.50	**11.97**	22.28	6.91	9.35
TTCM	16QAM/3	**17.00**	26.75	1.88	4.88	12.32	15.30	6.56	16.33
BICM	16QAM/3	18.88	26.73	0.00	4.90	12.59	**15.19**	6.29	16.44
BICM-ID	16QAM/3	19.20	**24.73**	-0.32	6.90	14.33	16.28	4.55	15.35

Table 26.5: Coding gain values of the various RBF-I/Q-TEQ assisted coded modulation schemes studied in this chapter, when communicating over 2-path Rayleigh fading channels. All of the coded modulation schemes exhibited a similar decoding complexity in terms of the number of decoding states, which was equal to 64 states. An interleaver block length of $8 \times 684 = 5472$ symbols was employed and the corresponding simulation parameters were summarised in Sections 26.7.2 and 26.9.2. The performance of the best scheme is printed in bold

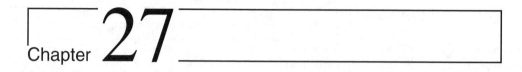

Chapter **27**

Coded Modulation Assisted
Code-Division Multiple Access

27.1 Introduction

In this chapter we embark on studying the performance of Coded Modulation (CM) assisted Direct Sequence (DS) Code-Division Multiple Access (CDMA) systems. Two sub-optimum Multi-User Detection (MUD) schemes will be utilised, namely the Minimum Mean Square Error-based Decision Feedback Equaliser (MMSE-DFE) MUD and Genetic Algorithm (GA) based MUD.

Joint Detection (JD) [690] receivers are derivatives of the family of well-known single-user channel equalisers, which were originally designed for equalising signals that have been impaired by Inter-Symbol Interference (ISI) due to traversing through multipath channels. The MMSE-DFE based JD (JD-MMSE-DFE) scheme constitutes a powerful approach to mitigating the effects of Multiple Access Interference (MAI) and ISI [766], while at the same time improving the system's performance by benefiting from the multipath diversity provided by dispersive channels. In the literature, TCM and BICM schemes were investigated in the context of a coherent DS-CDMA receiver using an MMSE receiver, when communicating over flat Rayleigh fading channels [767]. However, in this chapter a JD-MMSE-DFE arrangement will be employed for assisting the operation of the TCM, TTCM, BICM and BICM-ID schemes invoked for transmissions over dispersive Rayleigh fading channels in the context of a DS-CDMA system.

On the other hand, GAs have been used for efficiently solving combinatorial optimisation problems in numerous applications [768]. Recently, GA assisted MUD has been studied using Binary-Phase-Shift-Keying (BPSK) in the context of a CDMA system [769–771]. When compared to Verdu's optimum MUD scheme [334], the GA-MUD is capable of significantly reducing the computational complexity imposed.

The rest of this chapter is organised as follows. The CM assisted JD-MMSE-DFE based MUD is presented in Section 27.2, while an Adaptive CM assisted JD-MMSE-DFE based MUD is investigated in Section 27.3. Finally, the performance of CM assisted GA based

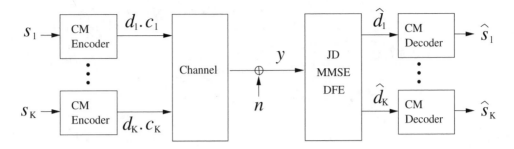

Figure 27.1: Block diagram of the concatenated CM and JD-MMSE-DFE scheme

MUD is evaluated in Section 27.4.

27.2   CM Assisted JD-MMSE-DFE-Based CDMA

The block diagram of the CM-assisted Joint Detection based CDMA (CM-JD-CDMA) system is shown in Figure 27.1. There are K users in the system, where each user is assigned a spreading code. At first, the $2^{\bar{m}}$-ary information symbol S_k of user k, is encoded by the CM encoder to an $2^{\bar{m}+1}$-ary signal, d_k, by adding a parity bit to the original information symbol of $\bar{m}$ information bits. Then, d_k is spread by the spreading code c_k of user k before transmission through the channel. In this uplink scenario each user transmits his/her signal through different channels using a single transmit antenna per user. At the Base Station (BS), we consider one receive antenna for all users. The JD-MMSE-DFE subsystem of the BS's receiver jointly detects all users' signals. The estimate of the signal $\hat{d}_k$ of user k, is then fed from JD-MMSE-DFE to the CM decoder for generating the decoded output $\hat{S}_k$.

27.2.1   The JD-MMSE-DFE Subsystem

The conventional detector used for DS-CDMA systems is the classic matched filter [334]. The matched filter is capable of maximising the SNR at the required sampling instant at its output for a given received waveform [180]. The Whitening Matched Filter (WMF) [680,690] is an extension of the conventional data estimation technique that uses a bank of matched filters, one for each user. JD schemes can be viewed as extensions to the WMF.

27.2.1.1   DS-CDMA System Model

Before highlighting the structure of the JD-MMSE-DFE subsystem, let us consider the structure of the system matrix $\mathbf{A}$ for the K-user CDMA system in Figure 27.2. A synchronous system was considered here for simplicity. However, an asynchronous system matrix can also be constructed with the knowledge of the users' delays, provided that the value of the delay can be exactly determined. The combined impulse response $\mathbf{b}_n^{(k)}$ of user k, constituted by the convolution of the spreading sequence $\mathbf{c}^{(k)}$ and the Channel Impulse Response (CIR)

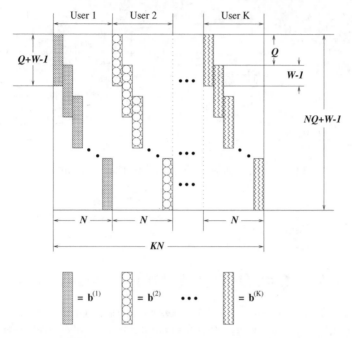

Figure 27.2: Example of the system matrix **A** for a K-user CDMA system, where $\mathbf{b}^{(1)}$, $\mathbf{b}^{(2)}$ and $\mathbf{b}^{(K)}$ are column vectors representing the combined impulse responses of user 1, 2 and K respectively in Equation 27.2. The notations are as follows : N denotes the number of coded symbols transmitted by each transmitter, Q represents the number of chips in each spreading sequence and W indicates the length of the wideband Channel Impulse Response (CIR).

$\mathbf{h}_n^{(k)}$ formulated as:

$$
\begin{aligned}
\mathbf{b}_n^{(k)} &= (b_n^{(k)}(1), b_n^{(k)}(2), \dots, \\
&\qquad b_n^{(k)}(l), \dots, b_n^{(k)}(Q+W-1))^T \\
&= \mathbf{c}^{(k)} * \mathbf{h}_n^{(k)}, \\
&\text{for } k = 1 \dots \mathsf{K}; \quad n = 1, \dots N,
\end{aligned}
$$

(27.1)

(27.2)

where K represents the total number of users, N denotes the number of coded symbols transmitted by each transmitter, Q represents the number of chips in each spreading sequence and W indicates the length of the wideband CIR. The system matrix of user k, $\mathbf{A}^{(k)}$ is represented by:

$$
[\mathbf{A}^{(k)}]_{in} = \begin{cases} b_n^{(k)}(l) & \text{for } i = (n-1)Q+l; \\ & n = 1, \dots, N; \\ & l = 1, \dots, Q+W-1; \\ 0 & \text{otherwise.} \end{cases}
$$

(27.3)

The overall system matrix can be constructed by appending the matrix $\mathbf{A}^{(k)}$ of each of the K users column-wise:

$$\mathbf{A} = (\mathbf{A}^{(1)}, \mathbf{A}^{(2)}, \ldots, \mathbf{A}^{(k)}, \ldots, \mathbf{A}^{(K)}). \tag{27.4}$$

Therefore, the discretised received composite signal can be represented in matrix form as:

$$
\begin{aligned}
\mathbf{y} &= \mathbf{A}\mathbf{d} + \mathbf{n}, \\
\mathbf{y} &= (y_1, y_2, \ldots, y_{NQ+W-1})^T,
\end{aligned} \tag{27.5}
$$

where $\mathbf{n} = (n_1, n_2, \ldots, n_{NQ+W-1})^T$, is the noise sequence having a variance of σ^2. The covariance matrix of the noise is given by:

$$\mathbf{R}_n = E[\mathbf{n}.\mathbf{n}^H] = \sigma^2 \mathbf{I}_{(NQ+W-1)}, \tag{27.6}$$

where $\mathbf{I}_{(NQ+W-1)}$ is the identity matrix having the dimension of $[NQ + W - 1] \times [NQ + W - 1]$. The composite signal vector $\mathbf{y}$ has $(NQ + W - 1)$ elements for a transmission burst of length N symbols.

The basic concept of joint detection is centred around processing the received composite signal vector, $\mathbf{y}$, in order to determine the transmitted data vector, $\mathbf{d}$ of the K number of users supported. This concept is encapsulated in the following equation:

$$\hat{\mathbf{y}} = \mathbf{S}\hat{\mathbf{d}} = \mathbf{M}\mathbf{y}, \tag{27.7}$$

where $\mathbf{S}$ is a square matrix having dimensions of $(KN \times KN)$ and the matrix $\mathbf{M}$ is a $[KN \times (NQ+W-1)]$ dimensional matrix. These two matrices determine the type of joint detection algorithm.

Having considered the system model describing the DS-CDMA system, let us now proceed to the description of the joint detection algorithm employed during our further studies.

27.2.1.2 Minimum Mean Square Error Decision Feedback Equaliser Based Joint Detection Algorithm

The principle behind MMSE estimation is the minimisation of the error between the data vector estimate, $\hat{\mathbf{d}}$, and the actual data vector, $\mathbf{d}$. Hence the MMSE algorithm jointly minimises the effects of both MAI, ISI and noise. Explicitly, the MMSE estimator minimises the simple quadratic form [680]:

$$Q(\hat{\mathbf{d}}) = E[(\mathbf{d} - \hat{\mathbf{d}})^H (\mathbf{d} - \hat{\mathbf{d}})]. \tag{27.8}$$

Upon invoking the well-known Orthogonality Principle [740], in order to minimise the Mean Squared Error (MSE), the error vector $\mathbf{e} = \mathbf{d} - \hat{\mathbf{d}}$ has to be set orthogonal by the MMSE equaliser to the estimator's input vector $\mathbf{y}$. This implies that:

$$E[(\mathbf{d} - \hat{\mathbf{d}})\mathbf{y}^H] = \mathbf{0}, \tag{27.9}$$

where $\mathbf{0}$ is a matrix with all of its elements being zero-valued. If we let $\hat{\mathbf{d}} = \mathbf{M}\mathbf{y}$, where $\mathbf{M}$

is a linear estimator, then:

$$
\begin{aligned}
E[(\mathbf{d} - \mathbf{My})\mathbf{y}^H] &= \mathbf{0} \\
E[(\mathbf{dy}^H - \mathbf{Myy}^H)] &= \mathbf{0} \\
E[(\mathbf{dy}^H)] - \mathbf{M}\,E[\mathbf{yy}^H] &= \mathbf{0} \\
\mathbf{R}_{dy} - \mathbf{MR}_y &= \mathbf{0} \\
\therefore \mathbf{M} &= \mathbf{R}_{dy}\mathbf{R}_y^{-1},
\end{aligned}
\tag{27.10}
$$

where $\mathbf{R}_{dy} = E[\mathbf{dy}^H]$ and $\mathbf{R}_y = E[\mathbf{yy}^H]$. For the special case of Equation 27.5, i.e. when $\mathbf{y} = \mathbf{Ad} + \mathbf{n}$, we have [772]:

$$
\begin{aligned}
\mathbf{R}_{dy} &= E[\mathbf{d}(\mathbf{Ad} + \mathbf{n})^H] \\
&= E[\mathbf{dd}^H\mathbf{A}^H + \mathbf{dn}^H],
\end{aligned}
\tag{27.11}
$$

and assuming that the transmitted data vector, $\mathbf{d}$, and the noise vector, $\mathbf{n}$, are uncorrelated with each other, i.e. when $E[\mathbf{dn}^H] = 0$, we arrive at:

$$
\begin{aligned}
\mathbf{R}_{dy} &= E[\mathbf{dd}^H\mathbf{A}^H] \\
&= \mathbf{R}_d\mathbf{A}^H,
\end{aligned}
\tag{27.12}
$$

where $\mathbf{R}_d = E[\mathbf{dd}^H]$. Furthermore, the covariance matrix, $\mathbf{R}_y$, of the received vector $\mathbf{y}$ in Equation 27.10 is given by:

$$
\begin{aligned}
\mathbf{R}_y &= E[(\mathbf{Ad} + \mathbf{n})(\mathbf{Ad} + \mathbf{n})^H] \\
&= E[\mathbf{Add}^H\mathbf{A}^H + \mathbf{nn}^H] \\
&= \mathbf{AR}_d\mathbf{A}^H + \mathbf{R}_n.
\end{aligned}
\tag{27.13}
$$

Substituting Equations 27.12 and 27.13 into Equation 27.10, we get:

$$
\begin{aligned}
\mathbf{M} &= \mathbf{R}_d\mathbf{A}^H(\mathbf{AR}_d\mathbf{A}^H + \mathbf{R}_n)^{-1} \tag{27.14} \\
&= (\mathbf{A}^H\mathbf{R}_n^{-1}\mathbf{A} + \mathbf{R}_d^{-1})^{-1}\mathbf{A}^H\mathbf{R}_n^{-1}. \tag{27.15}
\end{aligned}
$$

Finally, by substituting the MMSE detector expression of Equation 27.15 into the Equation $\hat{\mathbf{d}} = \mathbf{My}$, we arrive at:

$$
\hat{\mathbf{d}}_{\text{MMSE-BLE}} = (\mathbf{A}^H\mathbf{R}_n^{-1}\mathbf{A} + \mathbf{R}_d^{-1})^{-1}\mathbf{A}^H\mathbf{R}_n^{-1}\mathbf{y},
\tag{27.16}
$$

which is the data estimate generated by the MMSE Block Linear Equaliser (MMSE-BLE) or, in other words, the output of the feed-forward filter of the MMSE-DFE. On the other hand, the output of the WMF is given by [773]:

$$
\hat{\mathbf{d}}_{\text{WMF}} = \mathbf{A}^H\mathbf{R}_n^{-1}\mathbf{y}.
\tag{27.17}
$$

Observing Equations 27.17 and 27.16, the output of the MMSE-BLE is given by the output of the WMF in Equation 27.17, multiplied by the matrix $(\mathbf{A}^H\mathbf{R}_n^{-1}\mathbf{A} + \mathbf{R}_d^{-1})$. The

Cholesky decomposition [774] is performed on the matrix $(\mathbf{A}^H \mathbf{R}_n^{-1} \mathbf{A} + \mathbf{R}_d^{-1})$ [680] giving:

$$\mathbf{A}^H \mathbf{R}_n^{-1} \mathbf{A} + \mathbf{R}_d^{-1} = (\mathbf{DU})^H \mathbf{DU}, \qquad (27.18)$$

where $\mathbf{U}$ is an upper triangular matrix, where all the elements on its main diagonal have the value of one and $\mathbf{D}$ is a diagonal matrix having real-valued elements. With the aid of Equation 27.5, the output of the WMF from Equation 27.17 can be rearranged as:

$$
\begin{aligned}
\hat{\mathbf{d}}_{\text{WMF}} &= \mathbf{A}^H \mathbf{R}_n^{-1} \mathbf{A} \mathbf{d} + \mathbf{A}^H \mathbf{R}_n^{-1} \mathbf{n} \\
&= (\mathbf{A}^H \mathbf{R}_n^{-1} \mathbf{A} + \mathbf{R}_d^{-1}) \mathbf{d} - \mathbf{R}_d^{-1} \mathbf{d} + \mathbf{A}^H \mathbf{R}_n^{-1} \mathbf{n}.
\end{aligned}
\qquad (27.19)
$$

Applying the matrix $((\mathbf{DU})^H)^{-1}$ to both sides of Equation 27.19 and invoking Equation 27.18, yields:

$$
\begin{aligned}
((\mathbf{DU})^H)^{-1} \hat{\mathbf{d}}_{\text{WMF}} &= ((\mathbf{DU})^H)^{-1}(\mathbf{A}^H \mathbf{R}_n^{-1} \mathbf{A} + \mathbf{R}_d^{-1})\mathbf{d} - ((\mathbf{DU})^H)^{-1}\mathbf{R}_d^{-1}\mathbf{d} + \\
& \quad ((\mathbf{DU})^H)^{-1}\mathbf{A}^H \mathbf{R}_n^{-1}\mathbf{n} \\
&= \mathbf{DUd} - ((\mathbf{DU})^H)^{-1}\mathbf{R}_d^{-1}\mathbf{d} + \\
& \quad ((\mathbf{DU})^H)^{-1}\mathbf{A}^H \mathbf{R}_n^{-1}\mathbf{n}.
\end{aligned}
\qquad (27.20)
$$

Again, since $\mathbf{D}$ is a diagonal matrix, it can be treated as a scaling factor and removed by multiplying both sides of Equation 27.20 with its inverse, giving the resultant vector of $\hat{\mathbf{y}}$ as:

$$
\begin{aligned}
\hat{\mathbf{y}} &= (\mathbf{D})^{-1}((\mathbf{DU})^H)^{-1}\hat{\mathbf{d}}_{\text{WMF}} \\
&= (\mathbf{D})^{-1}\mathbf{DUd} - (\mathbf{D})^{-1}((\mathbf{DU})^H)^{-1}\mathbf{R}_d^{-1}\mathbf{d} + \\
& \quad (\mathbf{D})^{-1}((\mathbf{DU})^H)^{-1}\mathbf{A}^H \mathbf{R}_n^{-1}\mathbf{n} \\
&= \mathbf{Ud} - \mathbf{e} + \mathbf{z} \\
&= \mathbf{d} + (\mathbf{U} - \mathbf{I}_{KN})\mathbf{d} - \mathbf{e} + \mathbf{z},
\end{aligned}
\qquad
\begin{aligned}
(27.21) \\
\\
\\
\\
(27.22)
\end{aligned}
$$

where

$$\mathbf{e} = (\mathbf{D})^{-1}((\mathbf{DU})^H)^{-1}\mathbf{R}_d^{-1}\mathbf{d}, \qquad (27.23)$$

and

$$\mathbf{z} = (\mathbf{D})^{-1}((\mathbf{DU})^H)^{-1}\mathbf{A}^H \mathbf{R}_n^{-1}\mathbf{n}. \qquad (27.24)$$

Therefore, in the MMSE-DFE, the received vector, $\mathbf{y}$, is processed by a WMF, followed by the feed-forward filter that is represented by the matrix $(\mathbf{D})^{-1}((\mathbf{DU})^H)^{-1}$. The output of the feed-forward filter is given as [690]:

$$
\begin{aligned}
\hat{\mathbf{y}} &= (\mathbf{D})^{-1}((\mathbf{DU})^H)^{-1}\mathbf{A}^H \mathbf{R}_n^{-1}\mathbf{y} \qquad && (27.25) \\
&= \mathbf{My}, && (27.26)
\end{aligned}
$$

where

$$\mathbf{M} = (\mathbf{D})^{-1}((\mathbf{DU})^H)^{-1}\mathbf{A}^H \mathbf{R}_n^{-1}. \qquad (27.27)$$

The matrix $\mathbf{M}$ represents the combination of the WMF and feed-forward filter. Since $\mathbf{U}$ is an upper triangular matrix having ones on its main diagonal, $(\mathbf{U} - \mathbf{I}_{KN})$ is an upper triangular matrix having zeros on its main diagonal. Therefore, from Equation 27.22 the i-th element of the vector $\hat{\mathbf{y}}$ is given by:

$$\hat{y}_i = d_i + \sum_{j=i+1}^{J} [\mathbf{U} - \mathbf{I}_{KN}]_{i,j} \, d_j - e_i + z_i, \tag{27.28}$$

where $J = KN$. The summation is only from $j = i + 1$ to J, since $(\mathbf{U} - \mathbf{I}_{KN})$ is an upper triangular matrix having zeros on its main diagonal. For $i = J$, we have:

$$\hat{y}_J = d_J - e_J + z_J. \tag{27.29}$$

The i-th data estimate is $Q(\hat{y}_J)$, where $Q\{.\}$ is the thresholding or quantisation operator performed in a threshold detector. For $i = 1, \ldots, J - 1$, if the estimates of the data are obtained in decreasing order of i, then for each $\hat{y}_i$, the data estimates $\hat{d}_{i+1}, \ldots, \hat{d}_J$ have already been obtained. Therefore, the data estimate, $\hat{d}_i$ is given by:

$$\hat{d}_i = d_i + \sum_{j=i+1}^{J} [\mathbf{U} - \mathbf{I}_{KN}]_{i,j} \, (d_j - \hat{d}_j) - e_i + z_i. \tag{27.30}$$

If the data estimates, $\hat{d}_j$ are accurately estimated, i.e. if we have $\hat{d}_j = d_j$, Equation 27.30 becomes:

$$\hat{d}_i = d_i - e_i + z_i, \tag{27.31}$$

giving an MMSE estimate of the data symbol. For the MMSE-DFE, the feedback operator, $\mathbf{S}$ can be derived from Equation 27.30 as:

$$\mathbf{S} = \mathbf{U} - \mathbf{I}_{KN}, \tag{27.32}$$

where $\mathbf{S}$ is an upper diagonal matrix and all the elements on its main diagonal have values of zero. If the data symbols are estimated in descending order of j, the previously estimated data symbols of $\hat{d}_j$ are fed into the threshold detector for obtaining a hard decision value of $\hat{t}_j$. The hard decision value $\hat{t}_j$ is multiplied with the feedback operator $\mathbf{S}$ and then deducted from the feed-forward filter's output $\hat{y}_i$, yielding [690]:

$$
\begin{aligned}
\hat{d}_i(\text{MMSE-DFE}) &= \hat{y}_i - \sum_{j=i+1}^{J=KN} [\mathbf{U} - \mathbf{I}_{KN}]_{ij} \, \hat{t}_j \\
&= \hat{y}_i - \sum_{j=i+1}^{J=KN} ([\mathbf{U}]_{ij} - [\mathbf{I}_{KN}]_{ij}) \, \hat{t}_j \\
&= \hat{y}_i - \sum_{j=i+1}^{J=KN} [\mathbf{U}]_{ij} \, \hat{t}_j,
\end{aligned}
\tag{27.33}
$$

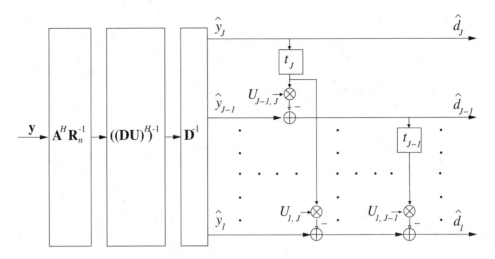

Figure 27.3: Structure of the MMSE-DFE-based receiver. The composite received vector, $\mathbf{y}$, is processed through the WMF, which is represented by the matrix $(\mathbf{A}^H \mathbf{R}_n^{-1})$ and this operation is expressed in Equation 27.17. The output of the WMF is then passed through the feed-forward filter which is constructed by the lower triangular matrix of $((\mathbf{DU})^H)^{-1}$ and the scaling factor matrix of $\mathbf{D}^{-1}$. These two operations were expressed in Equation 27.21. The output of the feed-forward filter is then processed, in order to obtain the data estimates, $\hat{\mathbf{d}}_j$, where the estimates are obtained in the order of $j = J, J-1, \ldots, 1$. As the data estimates, $\hat{\mathbf{d}}_j$, are obtained, they are fed back into the receiver, where they are multiplied by the elements in the upper triangular matrix, $[\mathbf{U}]_{ij} = U_{ij}$, as shown in Equation 27.33

where we have $[\mathbf{I}_{KN}]_{ij} = 0$ for $i \neq j$.

27.2.1.3 Algorithm Summary

The schematic of the MMSE-DFE is shown in Figure 27.3. The operations required for obtaining the JD-MMSE-DFE data estimates can be summarised as follows. First we construct the system matrix $\mathbf{A}$ in Equation 27.4. Then we obtain the output of the WMF as in Equation 27.17:

$$\hat{\mathbf{d}}_{\mathbf{WMF}} = \mathbf{A}^H \mathbf{R}_n^{-1} \mathbf{y},$$

where $\mathbf{A}^H$ is the conjugate transpose of $\mathbf{A}$ and $\mathbf{R}_n^{-1}$ is the inverse of the noise covariance matrix $\mathbf{R}_n$. Next, Cholesky decomposition [294] of the matrix $(\mathbf{A}^H \mathbf{R}_n^{-1} \mathbf{A} + \mathbf{R}_d^{-1})$ is performed as in Equation 27.18:

$$\mathbf{A}^H \mathbf{R}_n^{-1} \mathbf{A} + \mathbf{R}_d^{-1} = (\mathbf{DU})^H \mathbf{DU},$$

where $\mathbf{R}_d^{-1}$ is the inverse of the signal's covariance matrix $\mathbf{R}_d$, $\mathbf{D}$ is a diagonal matrix having real-valued elements and $\mathbf{U}$ is an upper triangular matrix, where all the elements on the main diagonal have the value of one. Consequently, the feed-forward filter output is obtained by

Parameter	Value
Doppler frequency	80 Hz
Spreading ratio, Q	16
Chip rate	2.167 MBaud
Normalised Doppler frequency	$80/(2.167\times10^6) = 3.69\times10^{-5}$ Hz
Frame burst structure	FRAMES Mode 1 Spread burst 1 [775] in Figure 27.4
No. of symbols per JD block, N	28
Modulation mode	QPSK, 8PSK, 16QAM
CM subsystem	TCM (ν=6)
	BICM (ν=6)
	TTCM (ν=3, 4 iterations)
	BICM-ID (ν=3, 8 iterations)

Table 27.1: Simulation parameters of the CM-JD-CDMA system

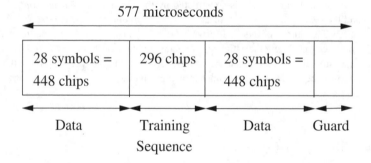

Figure 27.4: Transmission burst structure of the FMA1 spread speech/data burst of the FRAMES proposal [775]

solving Equation 27.25:

$$\hat{\mathbf{y}} = (\mathbf{D})^{-1}((\mathbf{DU})^H)^{-1} \, \mathbf{A}^H \mathbf{R}_n^{-1} \mathbf{y}$$
$$= (\mathbf{D})^{-1}((\mathbf{DU})^H)^{-1} \, \hat{\mathbf{d}}_{\mathbf{WMF}}.$$

Finally the feedback operation from Equation 27.33 is invoked for producing the final estimate of the coded symbol, yielding:

$$\hat{d}_i(\text{MMSE-DFE}) = \hat{y}_i - \sum_{j=i+1}^{J=KN} [\mathbf{U}]_{ij}\hat{t}_j.$$

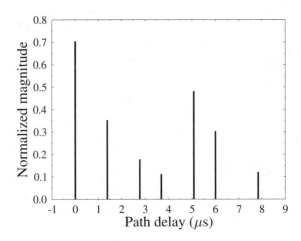

Figure 27.5: Normalised channel impulse response of the COST 207 [676] *seven path Bad Urban channel.*

27.2.2   Simulation Parameters

Let us now investigate the performance of the proposed schemes using the simulation parameters shown in Table 27.1, where 16-chip random spreading codes were utilised by each user. The transmission frame structure used is the FMA1 spread speech/data burst of the FRAMES proposal [775], which is shown in Figure 27.4. The channel model used is the COST 207 [676] seven path Bad Urban channel shown in Figure 27.5. Each path is faded according to independent Rayleigh fading statistics, as described by the parameters of Table 27.1. We assumed that the receiver perfectly knows the CIRs although in reality this has to be estimated with the aid of the training sequence of the transmission frame shown in Figure 27.4. The fading envelope was kept constant for the duration of the transmission burst of 577 μs and it was faded immediately before the next transmission burst. The CM subsystems shown in Table 27.1 exhibit a similar complexity, where their corresponding generator polynomials can be found from Tables 24.1, 24.2, 24.3 and 24.4.

In our performance evaluations, the uncoded QPSK system was compared to the CM assisted 8PSK system when aiming for an effective throughput of 2 information Bits Per Symbol (BPS). Similarly, the uncoded 8PSK system was compared to the CM assisted 16QAM system for a throughput of 3 BPS. Channel interleavers of length 112 symbols or 1120 symbols were utilised, which correspond to 2 transmission burst duration of 1.154 ms or 20 transmission burst durations of 11.54 ms, respectively.

27.2.3   Simulation Results and Discussion

Figure 27.6 shows the Bit Error Ratio (BER) versus Signal to Noise Ratio (SNR) performance of CM-JD-CDMA schemes for transmissions over the COST 207 [676] seven path

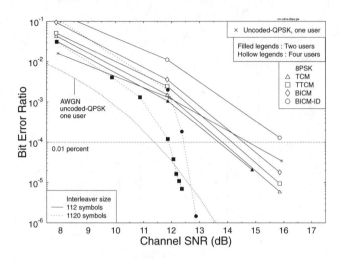

Figure 27.6: Bit Error Ratio (BER) versus Signal to Noise Ratio (SNR) performance of the various CM-JD-CDMA schemes for transmissions over the COST 207 [676] seven path Bad Urban channel of Figure 27.5 using the transmission burst structure of the FMA1 spread speech/data burst of the FRAMES proposal [775] shown in Figure 27.4 utilising the simulation parameters of Table 27.1 for a throughput of **2 BPS**.

Bad Urban channel shown in Figure 27.5 using the transmission burst structure of the FMA1 spread speech/data burst of the FRAMES proposal [775] shown in Figure 27.4, when utilising the simulation parameters of Table 27.1 and maintaining a throughput of **2 BPS**. Considering the performance of the TCM-JD-CDMA scheme using 8PSK modulation, the performance difference between the two- and four-user scenario is marginal indicated by the hollow and filled triangles in Figure 27.6 due to employing the powerful JD-MMSE-DFE scheme. Comparing the single-user uncoded-QPSK scheme represented by the cross, with the four-user CM-8PSK schemes utilising an interleaver length of 112 symbols at BER = 10^{-4}, performance gains can be observed for all CM-8PSK schemes over the uncoded-QPSK scheme, except for the BICM-ID-8PSK scheme. It can be also observed in Figure 27.6 that the TCM-8PSK scheme constitutes the best candidate, showing an SNR gain of 1 dB at BER = 10^{-4}. Due to the short interleaver length used, the TTCM and BICM-ID iterative decoding schemes were unable to perform efficiently. By contrast, the interleaver length does not dramatically affect the performance of the non-iterative TCM and BICM schemes [712]. Comparing the two-user TCM-8PSK scheme with the two-user TTCM-8PSK and BICM-ID-8PSK schemes utilising an interleaver length of 1120 symbols at BER = 10^{-4}, TTCM-8PSK and BICM-ID-8PSK show a 1.8 dB and 1.2 dB SNR gain over TCM-8PSK, respectively. For a throughput of 2BPS, TTCM-8PSK utilising an interleaver length of 1120 symbols is the best candidate, exhibiting an SNR gain of 2.9 dB over the uncoded-QPSK scheme at BER = 10^{-4}. The performance of the single-user uncoded QPSK system communicating over the non-dispersive AWGN channel is also plotted in Figure 27.6 as a benchmark. This benchmark is outperformed by both the TTCM-8PSK and BICM-ID-8PSK schemes at low BERs, when an interleaver length of 1120 symbols was utilised.

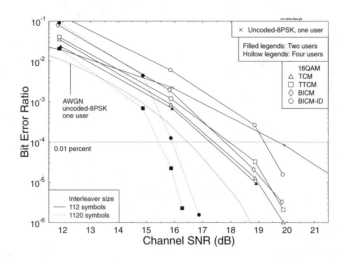

Figure 27.7: BER versus SNR performance of the various CM-JD-CDMA schemes for transmissions over the COST 207 [676] seven path Bad Urban channel of Figure 27.5 using the transmission burst structure of the FMA1 spread speech/data burst of the FRAMES proposal [775] shown in Figure 27.4 utilising the simulation parameters of Table 27.1 for a throughput of **3 BPS**.

Figure 27.7 shows the BER versus SNR performance of CM-JD-CDMA schemes for transmissions over the COST 207 [676] seven path Bad Urban channel shown in Figure 27.5 using the transmission burst structure of the FMA1 spread speech/data burst of the FRAMES proposal [775] shown in Figure 27.4 utilising the simulation parameters of Table 27.1. In this the effective throughput is **3 BPS**. Comparing the performance of the TCM-JD-CDMA scheme using 16QAM, the performance difference between the scheme supporting two- and four-user is again marginal due to invoking the powerful JD-MMSE-DFE scheme. Comparing the single-user uncoded-8PSK scheme with the four-user CM-16QAM schemes utilising an interleaver length of 112 symbols at BER = 10^{-4}, performance gains can be observed for all CM-16QAM schemes over the uncoded-8PSK scheme. The TCM-16QAM scheme constitutes the best candidate, showing an SNR gain of 2.3 dB. Comparing the two-user TCM-16QAM scheme with the two-user TTCM-16QAM and BICM-ID-16QAM schemes utilising an interleaver length of 1120 symbols at BER = 10^{-4}, TTCM-16QAM and BICM-ID-16QAM show 1.8 dB and 1.3 dB SNR gains over TCM-16QAM, respectively. For the effective throughput of 3BPS, TTCM-16QAM utilising an interleaver length of 1120 symbols is the best candidate showing an SNR gain of 4.2 dB over the uncoded 8PSK scheme at BER = 10^{-4}. The performance of the single-user uncoded 8PSK scheme communicating over the non-dispersive AWGN channel is also plotted in Figure 27.7 as a benchmark. The benchmark is outperformed by both the TTCM-16QAM and BICM-ID-16QAM schemes utilising an interleaver length of 1120 symbols.

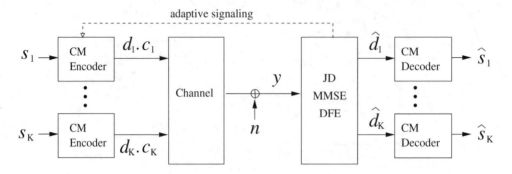

Figure 27.8: Block diagram of the BbB ACM-JD-CDMA scheme

27.2.4   Conclusion

In this section, TCM, TTCM, BICM and BICM-ID assisted JD-MMSE-DFE-based CDMA schemes were proposed and evaluated in performance terms over the COST 207 [676] seven path Bad Urban channel. For systems using a short interleaver length of 112 symbols, TCM was found to be the best candidate when providing a throughput of 2 and 3 BPS. However, for systems that can afford a longer delay due to utilising a long interleaver length, TTCM was found to be the best candidate for providing a throughput of 2 and 3 BPS.

27.3   Adaptive CM Assisted JD-MMSE-DFE Based CDMA

In this section, a Burst-by-Burst Adaptive Coded Modulation-Aided Joint Detection-Based CDMA (ACM-JD-CDMA) scheme is proposed for wireless communications and characterised in performance terms, when communicating over the UMTS Terrestrial Radio Access (UTRA) wideband vehicular Rayleigh fading channel. The block diagram of the ACM-JD-CDMA system is shown in Figure 27.8, which is similar to that shown in Figure 27.1 of Section 27.2 except for the adaptive signalling link used by the JD-MMSE-DFE for informing the CM-encoder regarding the choice of the suitable modulation mode.

More explicitly, the instantaneous channel quality is estimated by the JD-MMSE-DFE and the index or identifier of the highest-throughput modulation mode still capable of maintaining the required target BER or FER can then be communicated to the transmitter via explicit signalling in a closed-loop scheme. Conversely, in an open-loop scheme, by assuming reciprocity in the uplink and downlink channels of Wideband Time Division Duplex (TDD) CDMA systems, the modulation mode required for downlink transmission is chosen according to the channel quality estimate related to the uplink and vice versa. Channel reciprocity issues in TDD/CDMA systems have been investigated by Miya *et al.* [579], Kato *et al.* [580] and Jeong *et al.* [666].

In adaptive modulation schemes the issue of modulation mode signalling between the transmitter and receiver is of salient importance, especially in the context of Symbol-by-Symbol (SbS) adaptive systems [197], where the transmitter is capable of transmitting symbols in different modem modes, depending on the channel conditions. Naturally, the receiver has to synchronise with the transmitter in terms of the SbS-adapted mode sequence, in order to correctly demodulate the received symbols. However, in slow-fading pedestrian indoor channels in conjunction with a high transmission rate, the channel conditions do not change dramatically during a transmission burst. Hence a Burst-by-Burst (BbB) – rather than SbS – adaptive scheme can be employed, where the same modulation mode is used throughout one transmission burst. As a benefit, the mode signalling mechanism of the BbB adaptive scheme is simpler than that of the SbS adaptive scheme, since the BbB adaptive scheme signals one modulation mode per burst, while the SbS adaptive scheme has to signal and/or predict a sequence of modulation modes per burst [197].

To elaborate a little further, the design issues of modulation mode signalling between the transmitter and receiver were discussed for example in the contributions of Lau *et al.* [197, 776], Otsuki *et al.* [179] and Torrance *et al.* [196]. Given the above range of AQAM mode signalling solutions in the literature, in this section we refrained from considering this issue and have assumed perfect modulation mode signalling. More specifically we employed a closed-loop controlled BbB adaptive scheme, where the receiver requests the highest-throughput modulation mode capable of maintaining the target-FER from the remote transmitter for the next transmission burst according to the expected channel conditions. The associated signalling delay will result in a slight performance degradation, since during this time the channel-quality will change. A range of long-term channel-quality prediction techniques were proposed by Duel-Hallen *et al.* in [175]. The impact of signalling delay will be investigated in Section 27.3.4. A range of other important design issues of adaptive CDMA can be found in [168, 180, 197, 777–779].

27.3.1 Modem Mode Adaptation

Joint detection CDMA is suitable for combining with ACM, because the implementation of the joint detection algorithms does not require any knowledge of the modulation mode used. The system matrix required for joint detection is constructed by using only the Channel Impulse Response (CIR) estimates and the spreading sequences of all the users. Therefore, the joint detection receiver does not have to be reconfigured, when the modulation mode is switched and its complexity is essentially independent of the modulation mode used.

In joint detection systems the Signal to Interference and Noise Ratio (SINR) of each user at the output of the MMSE-DFE can be calculated by using the channel estimates and the spreading sequences of all the users. By assuming that the transmitted data symbols and the noise samples are uncorrelated, the expression for calculating the SINR, γ_o, of the n-th symbol transmitted by the k-th user was given by Klein *et al.* [680] as:

$$\gamma_o(j) = \frac{\text{Wanted Signal Power}}{\text{Res. MAI and ISI Power + Eff. Noise Power}}$$

$$= g_j^2 [\mathbf{D}]_{j,j}^2 - 1, \quad \text{for } j = n + N(k-1), \quad (27.34)$$

where SINR is the ratio of the wanted signal power to the residual MAI and ISI power plus the effective noise power. The number of users in the system is K and each user transmits N symbols per transmission burst. The matrix $\mathbf{D}$ is a diagonal matrix that is obtained with the aid of the Cholesky decomposition [774] of the matrix used for linear MMSE equalisation of the CDMA system [680]. The notation $[\mathbf{D}]_{j,j}$ represents the elements in the j-th row and j-th column of the matrix $\mathbf{D}$ and the value g_j is the amplitude of the j-th symbol.

After the output SINR is calculated, the best-matching modulation mode maximising the throughput, while maintaining the required target BER or FER, is chosen accordingly and communicated to the transmitter. Let us denote the choice of modulation modes by V_m, where the total number of modulation modes is $M_o = 4$ and $m = 1, 2, \ldots, M_o$. The modulation mode having the lowest number of constellation points is V_1 and the one associated with the highest is V_{M_o}. The rules used for switching the modulation modes are as follows:

$$
\begin{aligned}
\Gamma_o(k) \leq t_1 & \implies V_1 = 4QAM \\
t_1 < \Gamma_o(k) \leq t_2 & \implies V_2 = 8PSK \\
t_2 < \Gamma_o(k) \leq t_3 & \implies V_3 = 16QAM \\
t_3 \leq \Gamma_o(k) & \implies V_4 = 64QAM,
\end{aligned}
$$

where $\Gamma_o(k)$ is the SINR of the k-th user at the output of the MMSE-DFE, which was calculated by using Equation 27.34 and $\Gamma_o(k) = \frac{1}{N} \sum_{n=1}^{N} \gamma_o(j)$, $j = n + N(k-1)$. The values $(t_1, \cdots, t_{M_o-1})$ represent the switching thresholds used for activating the modulation modes, where we have $t_1 < t_2 < \cdots < t_{M_o-1}$.

The mode switching thresholds were adjusted for maintaining the target performance requirements, such as a fixed BER or FER. In 1996 Torrance et al. [185] proposed a set of mode switching levels s optimised for achieving the highest average BPS throughput while maintaining the target average BER. The method was based on defining a specific combined cost-function for transmission over narrowband Rayleigh fading channels, which incorporated both the BPS throughput as well as the target average BER of the system. Powell's optimisation was invoked for finding a set of mode switching thresholds, which were constant, regardless of the actual channel Signal to Noise Ratio (SNR) encountered, i.e. irrespective of the prevalent instantaneous channel conditions. However, in 2001 Choi et al. [186] have recognised that a higher BPS throughput can be achieved, if under high channel SNR conditions the activation of high-throughput AQAM modes is further encouraged by lowering the switching thresholds. More explicitly, a set of SNR-dependent mode switching levels was proposed [186], which keeps the average BER constant, while maximising the achievable throughput. We note that, the set of switching levels derived in [185, 187] is based on Powell's multidimensional optimisation technique [188] and hence the optimisation process may become trapped in a local minimum. This problem was overcome by Choi et al. upon deriving an optimum set of switching levels [186], when employing the Lagrangian multiplier technique. It was shown that this set of switching levels results in the global optimum in a sense that the corresponding AQAM scheme obtains the maximum possible average BPS throughput, while maintaining the target average BER. A further approach was proposed by Lau et al. in [197], where an eight-mode adaptive modulation scheme was proposed for DS-CDMA using an adaptive M-ary orthogonal modulator. More specifically, a rate-$1/9$ channel code and a 2^9-ary orthogonal modulator were employed in Mode-0, while a rate-$1/2$ channel code and a 2^2-ary orthogonal modulator were employed in Mode-7. The adaptive mode thresholds in this scheme were chosen in 3 dB steps, where the overall performance was controlled by a single parameter, which was the threshold value of mode 1.

However, the optimum switching thresholds depend on the accuracy of the instantaneous channel estimation. Specifically, the accuracy of the channel estimation depends on the channel quality prediction method employed and also on the signalling delay incurred, where the

Parameter	Value
Carrier Frequency	1.9 GHz
Vehicular Speed	30 mph
Doppler frequency	85 Hz
System Baud rate	3.84 MBd
Normalised Doppler frequency	$85/(3.84 \times 10^6) = 2.21 \times 10^{-5}$ Hz
Channel type	UMTS Vehicular Channel A
Number of paths in channel	6
Receiver type	JD-MMSE-DFE
No. of symbols per JD block	20
Data modulation	Adaptive Coded Modulation (4QAM, 8PSK, 16QAM, 64QAM)
CM subsystem	TCM (ν=6)
	BICM (ν=6)
	TTCM (ν=3, 4 iterations)
	BICM-ID (ν=3, 8 iterations)

Table 27.2: CM and channel parameters

longer the signalling delay the more 'outdated' the channel estimation. As mentioned above, efficient channel quality prediction techniques were proposed by Duel-Hallen *et al.* in [175]. On the other hand, the effects of outdated channel estimation were investigated in [208]. In this section we considered the practical scenario, where the channel estimation is outdated. In general, this deficiency could be caused by an inaccurate channel quality prediction or due to the associated signalling delay. More specifically, the modulation mode signalling in our system is delayed by one transmission frame duration and hence the channel estimation is outdated by $10ms$, which is the UTRA frame length, as shown in Table 27.3. The switching thresholds of the system studied here were set experimentally by finding the SINR values, where the constituent fixed-mode QAM schemes were capable of maintaining the required target FER, when communicating over the UTRA Rayleigh fading channel. On the other hand, the switching thresholds of the idealistic benchmark system used in conjunction with perfect channel estimation and zero signalling delay were set experimentally by finding the SINR values, where the constituent fixed-mode QAM schemes were capable of maintaining the required target FER, when communicating over the non-fading, non-dispersive Gaussian channel. In general, these two systems represent two extreme cases, where the former system provides the worst performance and the latter system the best possible performance.

27.3.2 Channel Model and System Parameters

Table 27.2 shows the modulation and channel parameters employed. The multipath channel model is characterised by its discretised chip-spaced UTRA vehicular channel A [780]. The corresponding CIR is shown in Figure 27.9, where each path is faded independently according to the Rayleigh distribution. The CM subsystems shown in Table 27.2 exhibit a similar complexity and their corresponding generator polynomials can be found from Tables 24.1, 24.2, 24.3 and 24.4. The transmission burst structure of the modified UTRA Burst 1 [168]

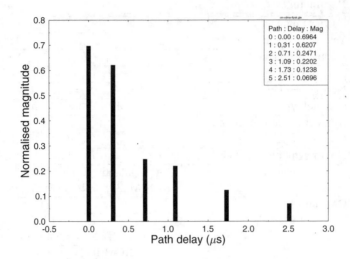

Figure 27.9: UTRA vehicular channel A [780]

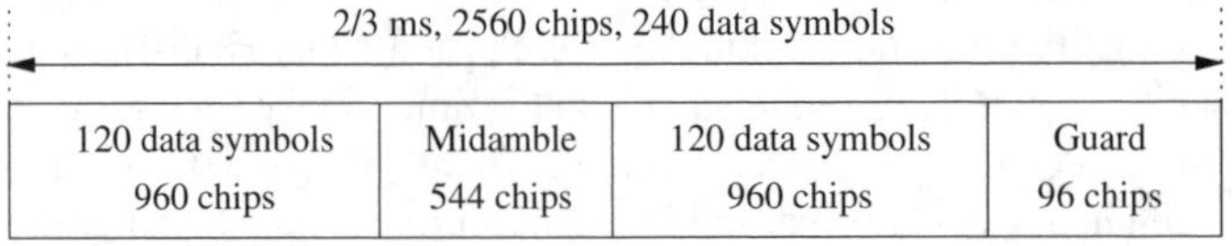

2/3 ms, 2560 chips, 240 data symbols			
120 data symbols	Midamble	120 data symbols	Guard
960 chips	544 chips	960 chips	96 chips

Figure 27.10: A modified UTRA Burst 1 [168] with a spreading factor of 8. The original UTRA burst has 244 data symbols

using a spreading factor of eight is shown in Figure 27.10. The number of data symbols per JD block is 20, hence the original 244-symbol long UTRA Burst 1 was slightly modified to host a burst of 240 data symbols, which is a multiple of 20. The remaining system parameters are shown in Table 27.3, where there are 15 time slots in one UTRA frame and we assign one slot for one group of CDMA users. More specifically, each group employed a similar system configuration but communicated with the base station employing different time slots.

In general, the total number of users supportable by the uplink CDMA system can be increased by using a higher spreading factor at the cost of a reduced throughput, since the system's chip rate was fixed at 3.84 Mcps as shown in Table 27.2. Another option for increasing the number of users supported is by assigning more uplink time slots for new groups of users. In our study, we investigate the proposed system using one time slot only. Hence the data symbol rate per slot per user is 24 KBd for a spreading factor of eight. Table 27.4 shows the operational-mode specific transceiver parameters for the TTCM-assisted ACM-JD-CDMA scheme. Specifically, the corresponding lowest bitrate is 23.4 Kbit/s in the 4QAM mode and the highest bitrate is 117 Kbit/s in the 64QAM mode.

Features	Value
Multiple access	CDMA, TDD
No. of Slots/Frame	15
Spreading factor	8
Frame length	10 ms
Slot length	2/3 ms
Data Symbols/Slot/User	240
No. of Slot/Group	1
User Data Symbol Rate (KBd)	240/10 = 24
System Data Symbol Rate (KBd)	24x15 = 360
Chips/Slot	2560
Chips/Frame	2560x15=38400
User Chip Rate (KBd)	2560/10 = 256
System Chip Rate (MBd)	38.4/10 = 3.84
System Bandwidth (MHz)	3.84 x 3/2 = 5.76
Eff. User Bandwidth (kHz)	5760/15 = 384

Table 27.3: Generic system features of the reconfigurable multi-mode transceiver, using the spread data burst 1 of UTRA proposal [168, 780] shown in Figure 27.10

Features	Multi-rate System			
Mode	4QAM	8PSK	16QAM	64QAM
Transmission Symbols	240			
Bit/Symbol	2	3	4	6
Transmission bits	480	720	960	1440
Code Termination Symbols	6			
Data Symbols	234			
Coding Rate	1/2	2/3	3/4	5/6
Information Bit/Symbol	1	2	3	5
Information Bits	234	468	708	1170

Table 27.4: Operational-mode specific transceiver parameters for TTCM

27.3.3   Performance of the Fixed Modem Modes

Let us investigate the BER and FER performance of the various CM schemes using the fixed modulation modes of 4QAM, 8PSK, 16QAM and 64QAM in Figures 27.11 and 27.12, respectively. In terms of BER, TCM assisted JD-CDMA scheme exhibits the best performance followed by TTCM, BICM and BICM-ID. However, in terms of FER, the TTCM assisted JD-CDMA scheme exhibits the best performance followed by TCM, BICM-ID and BICM. More specifically, TTCM exploits the time-diversity imposed by the turbo interleaver. However, a turbo interleaver length of 240 symbols is insufficient for TTCM to attain its optimum BER performance when communicating over slow Rayleigh fading channels due to the limited time-diversity experienced. On the other hand, TTCM is a powerful scheme employing iterative turbo decoding, which is capable of correcting all the bit errors in moderately corrupted

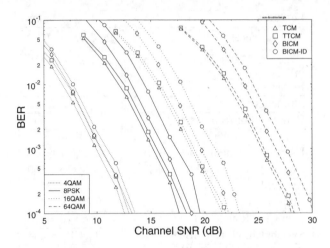

Figure 27.11: BER versus channel SNR performance of the four fixed modem modes, using the TCM, TTCM, BICM and BICM-ID coded modulation schemes, when supporting K = 2 users. The simulation parameters are listed in Table 27.2 and the simulations were conducted over the six-path UTRA vehicular channel A of Figure 27.9

frames, but which increases the BER in the severely corrupted frames due to the error precipitation effects encountered in the iterative turbo decoding process. Hence the BER of TTCM is far from its optimum, when the time-diversity of the system is limited. By contrast, its FER is always better than that of the other CM schemes at the same decoding complexity. BICM-ID is another scheme employing iterative decoding. However, it was found in [373, 712] that BICM-ID typically requires a higher order of time-diversity, and hence a longer channel interleaver than the TTCM scheme in order to achieve its optimum performance. Therefore, its BER becomes the worst, while its FER is better than that of BICM in our system. As for the non-iterative TCM and BICM schemes, TCM performs better than BICM in the slow Rayleigh fading channels studied.

Figure 27.12 also illustrates the FER performance of the uncoded fixed modulation modes of BPSK, 4QAM, 8PSK and 64QAM having a throughput of 1, 2, 3 and 6 BPS, respectively. For a throughput of 1 information BPS at 5% FER, TTCM-4QAM is about 5 dB more power efficient than the uncoded BPSK scheme while requiring the same bandwidth, as shown in Figure 27.12. Similarly, the coding gain of TTCM at throughputs of 2 and 3 BPS is about 3 dB and 4 dB respectively, at the same bandwidth requirement, while maintaining a FER of 5%. It is clear that the employment of CM schemes allow us to mitigate the effects of transmission errors effectively without any bandwidth expansion.

Based on these fixed-mode performance trends, we set the switching thresholds for our system experimentally by finding the corresponding average SINR values, where the constituent fixed-mode QAM schemes were capable of maintaining the required target FER, when communicating over the UTRA Rayleigh fading channels. More specifically, the thresholds required for maintaining a given target FER can be obtained by plotting the FER versus SINR curves for the constituent fixed-mode QAM schemes, when communicating over

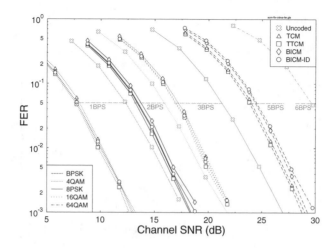

Figure 27.12: FER versus channel SNR performance of the four fixed modem modes, using the TCM, TTCM, BICM and BICM-ID coded modulation schemes, when supporting K = 2 users. The simulation parameters are listed in Table 27.2 and the simulations were conducted when communicating over the six-path UTRA vehicular channel A of Figure 27.9

the UTRA Rayleigh fading channel, where the intersections of the curves with a horizontal line at a specific target FER are the switching thresholds. The switching thresholds of the system benefiting from perfect channel estimation and a zero signalling delay were obtained by plotting the FER versus SINR curves for the constituent fixed-mode QAM schemes when communicating over non-fading, non-dispersive Gaussian channels.

27.3.4 Performance of the Adaptive Modes

In this section, we will investigate the performance of the TTCM-assisted ACM-JD-CDMA scheme, when communicating over the UTRA Rayleigh fading channels. When maintaining a target FER of 5% in conjunction with a mode signalling delay of $10ms$, the probability of each modulation mode being chosen versus the channel SNR for transmission in the TTCM-assisted ACM-JD-CDMA system is illustrated in Figure 27.13. As we can observe from the figure, at low values of the channel SNR, the 4QAM mode was chosen with the highest probability, thus predominantly having a throughput of 1 information BPS. However, as the channel quality improved the channel SNR increased, thus allowing the 8PSK, 16QAM and eventually 64QAM modes to be activated more often, which resulted in an increased average BPS performance.

The FER and BPS performance of the TTCM-assisted ACM-JD-CDMA scheme designed for a target FER of 5% and for supporting K = 2 and 4 users is shown in Figure 27.14. The FER was below the target FER of 5% and the BPS throughput improved, as the channel SNR increased. Since we employed switching thresholds which are constant over the SNR range, in the region of SNR = 10 dB the FER followed the trend of 4QAM. Similarly, for SNRs between 17 and 20 dB FER trend of 16QAM was predominantly obeyed. In both of these

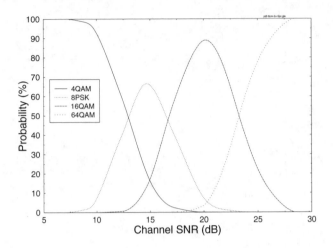

Figure 27.13: The probability of each modulation mode being chosen versus channel SNR for transmission in the TTCM-assisted ACM-JD-CDMA system when maintaining a target FER of 5% supporting K=2 users. The simulation parameters are listed in Table 27.2 and the simulations were conducted when communicating over the six-path UTRA vehicular channel A of Figure 27.9

SNR regions a significantly lower FER was achieved than the target FER. We note, however that it is possible to increase the FER to the value of the target FER, at these SNR regions for the sake of obtaining extra BPS gains by employing a set of switching thresholds, where the thresholds are varied as a function of the SNR [186], but this was set aside for further research.

From Figure 27.14 we also notice that the performance difference between the $K = 2$ and 4 scenario is only marginal with the advent of the powerful JD-MMSE-DFE scheme. Specifically, there was only about one dB SNR penalty, when the number of users increased from two to four for the 4QAM and 64QAM modes at both low and high SNRs, respectively. The FER of the system supporting $K = 4$ users was still below the target FER, when the switching thresholds derived for $K = 2$ users were employed. In general, the SNR penalty is less than one dB, when the number of users supported is increased from $K = 2$ to 4 users. Note that the delay spread of the chip-spaced UTRA vehicular channel A in Figure 27.9 is 2.51 μs corresponding to $2.51 \times 3.84 \approx 10$ chip duration for the 3.84 MBd Baud rate of our system, as seen in Table 27.2. Hence the delay spread is longer than the spreading code length ($Q = 8$ chips) used in our system and therefore the resultant ISI in the system is significantly higher than that of the system employing a higher spreading factor, such as $Q > 10$ chips. These findings illustrated the efficiency of the JD-MMSE-DFE scheme in combating the high ISI and MAI of the system. More importantly, the employment of the JD-MMSE-DFE scheme in our system generalised our results recorded for the $K = 2$ users scenario to that of a higher number of users, since the performance penalty associated with supporting more users was found marginal.

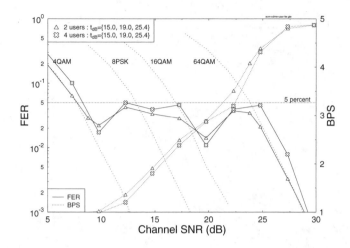

Figure 27.14: The average FER and BPS performance versus channel SNR of the TTCM-assisted ACM-JD-CDMA system supporting K = 2 and 4 users. The same set of switching thresholds t_{dB} was employed as shown in the legend. The FER curves for the same system but using fixed modulation modes of 4QAM, 8PSK, 16QAM and 64QAM for K = 2 users are also shown. The simulation parameters are listed in Table 27.2 and the simulations were conducted when communicating over the six-path UTRA vehicular channel A of Figure 27.9

27.3.5 Effects of Estimation Delay and Switching Thresholds

Let us now investigate the effect of mode signalling delay on the performance of the TTCM-assisted ACM-JD-CDMA scheme in Figure 27.15. The performance of the ideal scheme, where the channel quality estimation is perfect without any signalling delay is compared to that of the more practical scheme, where the channel quality estimation is imperfect and outdated by the delay of one frame duration of $10\mu s$. For a target FER of 5%, the ideal scheme exhibited a higher BPS throughput than the practical scheme. More specifically, at a target FER of 5%, about 2.5 dB SNR gain is achieved by the ideal scheme in the SNR region spanning from 8 dB to 27 dB. A channel quality signalling delay of one frame duration certainly represents the worst case scenario. In general, the shorter the signalling delay, the better the performance of the adaptive system. Hence the performance of the zero-delay and one-frame delay schemes represent the lower-bound and upper-bound performance, respectively, for practical adaptive systems.

Let us now investigate the performance of the practical one-frame delay TTCM-assisted ACM-JD-CDMA system using three switching threshold sets designed for maintaining target FER of 10%, 5% and 1% in Figure 27.16. As the values of the switching thresholds increases, the FER and/or the BPS throughput of the system reduces. The FER performance trends observed in Figure 27.14 also appear in Figure 27.16 for the various switching threshold sets. More explicitly, the FER curves seen in Figure 27.16 exhibit undulations, where the ACM-JD-CDMA FER curves tend to follow those of the predominant fixed modes at a given channel SNR, before gradually switching to the higher throughput modem mode. This is a

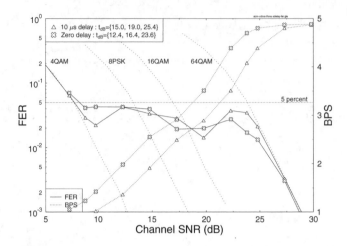

Figure 27.15: The average FER and BPS performance versus channel SNR of the TTCM-assisted
ACM-JD-CDMA system supporting $K = 2$ users. Two different sets of switching thresholds t_{dB} were employed as shown in the legend for systems having zero and $10\mu s$ mode
signalling delay. The FER curves for the same system but using fixed modulation modes
of 4QAM, 8PSK, 16QAM and 64QAM for $K = 2$ users are also shown. The simulation
parameters are listed in Table 27.2 and the simulations were conducted when communicating over the six-path UTRA vehicular channel A of Figure 27.9

consequence of the employment of SNR-independent switching thresholds in the one-frame
mode signalling delay system. These three schemes illustrate the associated trade-offs between the achievable BPS and FER performance of the system. Specifically, systems that are
capable of tolerating a higher FER exhibit the benefit of a higher BPS throughput.

27.3.6   Conclusion

In this section, TTCM-assisted burst-by-burst adaptive coded modulation aided joint detection based CDMA has been studied, when communicating over the UTRA wideband vehicular Rayleigh fading channel. The adaptive transceiver is capable of operating in four different
CM modes, namely using 4QAM, 8PSK, 16QAM and 64QAM. Various CM schemes were
used in our experiments. It was shown in Figure 27.12 that the CM schemes exhibited significant SNR gains over a wide range of BPS throughputs, while the JD-MMSE-DFE was
shown to be robust against both the MAI and ISI in Figure 27.14.

The advantage of the adaptive coded modulation scheme is that when invoking higher-
order modulation modes in case of encountering a higher instantaneous channel quality, the
coding rate approaches unity. For example, the coding rate of 4QAM was 1/2, while that of
64QAM was 5/6. As a result, this near-unity coding rate allows the system to maintain a high
effective throughput, while maintaining the required transmission integrity. More explicitly,
the proposed scheme guaranteed the same performance, as the lowest- and highest-order
fixed-mode modulation schemes in the low and high channel SNR range, respectively. As
an added benefit, between these extreme SNRs the effective bitrate increased smoothly, as

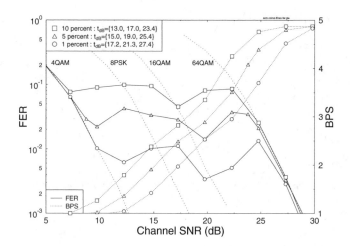

Figure 27.16: The average FER and BPS performance versus channel SNR of the TTCM-assisted ACM-JD-CDMA system supporting $K = 2$ users. Three different sets of switching thresholds t_{dB} were employed as shown in the legend for maintaining target FERs of 10%, 5% and 1%. The FER curves for the same system but using fixed modulation modes of 4QAM, 8PSK, 16QAM and 64QAM for $K = 2$ users are also shown. The simulation parameters are listed in Table 27.2 and the simulations were conducted when communicating over the six-path UTRA vehicular channel A of Figure 27.9

the channel SNR increased, while maintaining a near-constant FER. In this study, the lower and upper performance bounds of the TTCM-assisted ACM-JD-CDMA scheme have been provided in Figure 27.15 and the tradeoff between the achievable BPS and FER performance has been studied in Figure 27.16 by employing different mode switching threshold sets.

The proposed TTCM-assisted ACM-JD-CDMA scheme constitutes a promising practical communication system guaranteeing reliable transmission providing an effective bitrate ranging from 23.4 Kbit/s/slot to 117 Kbit/s/slot.

Next, we will proceed to study the performance of the CM assisted CDMA system using the GA Based MUD.

27.4 CM Assisted GA-Based Multiuser Detection Aided CDMA

In this section we propose a novel M-ary Coded Modulation assisted Genetic Algorithm based Multiuser Detection (CM-GA-MUD) scheme [172]. for synchronous CDMA systems. The performance of the proposed scheme was investigated using QPSK, 8PSK and 16QAM when communicating over AWGN and narrowband Rayleigh fading channels. When compared with the optimum MUD scheme, the GA-MUD subsystem is capable of reducing the computational complexity significantly. On the other hand, the CM subsystem is capable of obtaining considerable coding gains despite being fed with sub-optimal information provided

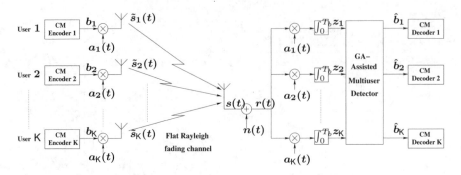

Figure 27.17: Block diagram of the K-user synchronous CDMA uplink model in a flat Rayleigh fading channel

by the GA-MUD output.

27.4.1   Introduction

The optimal CDMA MUD [334, 781] based on the Maximum-Likelihood (ML) detection rule performs an exhaustive search of all the possible combinations of the users' transmitted bit or symbol sequences and then selects the most likely combination as the detected bit or symbol sequence. Since an exhaustive search is conducted, the computational complexity of the detector increases exponentially with the number of users as well as with the number of levels in the modulation scheme employed. Since a CDMA system is required to support a large number of users, it is impractical to implement the optimum ML multiuser detector owing to its excessive complexity. This complexity constraint led to numerous so-called suboptimal multiuser detection [172, 782] proposals.

GAs have been used for efficiently solving combinatorial optimisation problems in many applications [768]. Recently, GA assisted MUD has been studied using BPSK modulation in the context of a CDMA system [769–771]. In an effort to increase the system's performance with the aid of channel coding, but without increasing the required bandwidth, in this section we will investigate the performance of the CM assisted Genetic Algorithm Based Multiuser Detection (CM-GA-MUD) using M-ary modulation modes.

27.4.2   System Overview

In our study, each user invokes a CM encoder, which provides a block of N modulated symbols before spreading. We consider a synchronous CDMA uplink as illustrated in Figure 27.17, where K users simultaneously transmit data packets of equal length using M-ary modulation to a single receiver. The transmitted signal of the kth user can be expressed in an equivalent lowpass representation as :

$$\tilde{s}_k(t) = \sum_{n=0}^{N-1} b_k^n a_k(t - nT_b), \quad \forall k = 1, \ldots, K \tag{27.35}$$

where $a_k(t)$ is the kth user's signature sequence, T_b is the data symbol duration, N is the number of data symbols transmitted in a packet and $b_k^n = \sqrt{\xi_k^n} e^{j\theta_k^n}$ represents the nth coded-modulated M-ary symbol of the kth user, where ξ_k^n and θ_k^n are the kth user's signal energy and phase of the nth transmitted symbol, respectively. More explicitly, b_k^n denotes a complex CM-coded symbol in the range of $0, 1, \ldots, M-1$, where M is the number of possible constellation points in the M-ary modulation, which is equals to $4, 8$ and 16 for QPSK, 8PSK and 16QAM. The superscript n can be omitted, since no dispersion-induced interference is inflicted by symbols outside a single symbol duration T_b in narrowband channel.

Each user's signal $\tilde{s}_k(t)$ is assumed to propagate over a narrowband slowly Rayleigh fading channel, as shown in Figure 27.17 and the fading envelope of each path is statistically independent for all users. The complex lowpass CIR for the link between the kth user's transmitter and the base station's receiver, as shown in Figure 27.17, can be written as :

$$h_k(t) = \alpha_k(t) e^{j\phi_k(t)} \delta(t), \quad \forall k = 1, \ldots, K \qquad (27.36)$$

where the amplitude $\alpha_k(t)$ is a Rayleigh distributed random variable and the phase $\phi_k(t)$ is uniformly distributed between $[0, 2\pi)$.

Hence, when the kth user's spread spectrum signal $\tilde{s}_k(t)$ given by Equation 27.35 propagates through a Rayleigh fading channel having an impulse response given by Equation 27.36, the resulting output signal $s_k(t)$ over a single symbol duration can be written as :

$$s_k(t) = \alpha_k b_k a_k(t) e^{j\phi_k}, \quad \forall k = 1, \ldots, K \qquad (27.37)$$

Upon combining Equation 27.37 for all K users, the received signal at the receiver, which is denoted by $r(t)$ in Figure 27.17, can be written as :

$$r(t) = s(t) + n(t), \qquad (27.38)$$

where $s(t) = \sum_{k=1}^{K} s_k(t)$ is the sum of the resultant output signals of all users and $n(t)$ is the zero-mean AWGN having independent real and imaginary components, each having a double-sided power spectral density of $\sigma^2 = N_0/2$.

The joint optimum decision rule for the M-ary modulated K-user CDMA system based on the synchronous system model can be derived from that of the BPSK-modulated system [334], which is expressed in vectorial notation as:

$$\Omega(b) = 2\Re \left[b^H C^* Z \right] - b^H C^* R C b, \qquad (27.39)$$

where

$$
\begin{aligned}
C &= \operatorname{diag}\left[\alpha_1 e^{j\phi_1}, \ldots, \alpha_K e^{j\phi_K} \right] \\
b &= [b_1, \ldots, b_K]^T, \\
Z &= ouput\ vector\ of\ the\ matched\ filters.
\end{aligned}
$$

More specifically, $(.)^H$ is the complex conjugate transpose of the matrix $(.)$ and $(.)^*$ is the complex conjugate of the matrix $(.)$. For BPSK modulation the term b^H in Equation 27.39 is replaced by b^T, which is the transpose of the matrix b, since only the real component is considered in the context of BPSK modulation.

The decision rule for the optimum CDMA multiuser detection scheme based on the maximum likelihood criterion is to choose the specific symbol combination b, which maximises the correlation metric of Equation 27.39, yielding:

$$\hat{b} = \arg\left\{\max_{b} \left[\Omega\left(b\right)\right]\right\}. \qquad (27.40)$$

Here, the optimum decision vector $\hat{b}$ represents the *hard decision* values for a specific K-symbol combination of the K users during a symbol period. Based on the *hard decision* vector component $\hat{b}_k$ of vector $\hat{b}$, we derived the log-likelihood channel metrics for the kth user's CM decoder for all the M possible M-ary modulated symbols as:

$$P_{k,m}(\hat{b}_k|b_{k,m}) = -\frac{|\hat{b}_k - b_{k,m}|^2}{2\sigma^2}, \qquad (27.41)$$

where $b_{k,m}$ is the mth phasor of user k in the constellation space and $m \in \{0,\dots,\mathsf{M}-1\}$. Note that it is possible to obtain the *soft decision* metrics for the optimum-MUD, although its employment imposes a higher complexity. Specifically, given $r(t)$ and all possible $\Omega\left(b\right)$ values, we derived the *soft decision* metrics as:

$$
\begin{aligned}
P_{k,m}(r(t)|b_{k,m}) &= ln\left\{\sum_{\substack{if\ b_{k,m}=b_k}}^{all\ possible\ \boldsymbol{b}} exp\left(-\frac{1}{2\sigma^2}\int_0^{T_b}|r(t)-s(t)|^2\,dt\right)\right\} \\
&= ln\left\{\sum_{\substack{if\ b_{k,m}=b_k}}^{all\ possible\ \boldsymbol{b}} exp\left(-\frac{1}{2\sigma^2}\int_0^{T_b}|r(t)|^2\,dt - \Omega\left(b\right)\right)\right\}.
\end{aligned}
$$
$$(27.42)$$

The maximisation of Equation 27.39 is a combinatorial optimisation problem. Specifically, Equation 27.39 has to be evaluated for each of the M^K possible combinations of the M-ary modulated symbols for the K users, in order to find the vector b that maximises the correlation metric of Equation 27.39. Explicitly, since there are M^K different possible vectors b, the optimum multiuser detection has a complexity that increases exponentially with the number of users K and the modulation mode employed M.

In this section, we aim at reducing the complexity of the optimum MUD, which performs M^K full search, by employing the sub-optimum GA-based MUD, which performs only a partial search. Hence, the sub-optimum decision vector $\hat{b}$ output by the GA-MUD is input to the CM decoder for generating the final estimate of the information. Note that the *hard decision* metric of Equation 27.41 is based on using only the optimum correlation metric of $\Omega\left(b\right)$, which is operated by the low-complexity GA-MUD. However, the *soft decision* metric of Equation 27.42 is based on evaluating all possible correlation metrics $\Omega\left(b\right)$ according to Equation 27.39, which requires a full search. Hence, the proposed reduced-complexity CM-GA-MUD scheme employed the lower-complexity *hard decision* metric.

27.4.3 The GA-Assisted Multiuser Detector Subsystem

For a detailed description of the GA-MUD, the interested readers are referred to the literature [369, 769]. A brief description of the GA-MUD is given below.

The flowchart depicting the structure of the genetic algorithm adopted for our GA-assisted multiuser detection technique is shown in Figure 27.18. Firstly, an initial population consisting of P number of so-called *individuals* is created, where P is known as the population size. Each individual represents a legitimate K-dimensional vector of M-ary modulated symbols constituting the solution of the given optimisation problem. In other words, an individual can be considered as a K-dimensional vector consisting of the M-ary modulated symbol's decision variables to be optimised.

In order to aid our GA-assisted search at the beginning, we employed the hard decisions offered by the matched filter outputs Z which were denoted as:

$$\hat{\boldsymbol{b}}_{MF} = \left[\hat{b}_{1,MF}, \hat{b}_{2,MF}, \ldots, \hat{b}_{K,MF} \right], \tag{27.43}$$

where $\hat{b}_{l,MF}$ for $l = 1, \ldots, K$ is given by:

$$\hat{b}_{l,MF} = arg \left\{ \min_{b} \left| z_l - \alpha_l e^{j\phi_l} b \right| \right\}. \tag{27.44}$$

In Equation 27.44 a multiplication by $\alpha_l e^{j\phi_l}$ is necessary for coherent detection, because the phase rotation introduced by the channel has to be taken into account. A different randomly 'mutated' version [769, 783] of the hard decision vector $\hat{\boldsymbol{b}}_{MF}$ of Equation 27.43 was assigned to each of the individuals in the initial population, where the same probability of mutation, namely p_m was adopted for all individuals. Note that we cannot assign the same hard decision vector $\hat{\boldsymbol{b}}_{MF}$ to all the individuals, since the process of incest prevention [784] is invoked, which will not allow identical individuals to mate.

The so-called fitness value [768] associated with each individual in the population is evaluated by substituting the candidate solution represented by the individual under consideration into the objective function, as indicated by the 'Evaluation' block of Figure 27.18. Individuals having the T number of highest fitness values are then placed in a so-called *mating pool* [768, 769] where $2 \leq T \leq P$. Using a kind of natural selection scheme [768] together with the genetically-inspired operators of *crossover* [783] and *mutation* [783], the individuals in the mating pool are then evolved to a new population. Our objective function, or synonymously, fitness value is defined by the correlation metric of Equation 27.39. Here, the legitimate solutions are the M^K possible combinations of the K-symbol vector b, where there are $log_2(M)$ bits in each of the M-ary symbols. Hence, each individual will take the form of a K-symbol vector corresponding to the K users' M-ary symbols during a single symbol interval. We will denote the pth individual here as $\tilde{\boldsymbol{b}}_p(y) = \left[\tilde{b}_{p,1}(y), \ldots, \tilde{b}_{p,K}(y) \right]$, where y denotes the yth generation. Our goal is to find the specific individual that corresponds to the highest fitness value in the sense of Equation 27.39. In order to ensure that the fitness values are positive for all combinations of b for the so-called *fitness-proportionate* selection scheme [768], we modify the correlation metric of Equation 27.39 according to [785]:

$$\exp\left\{ \Omega\left(\boldsymbol{b} \right) \right\} = \exp\left\{ 2\Re\left[\boldsymbol{b}^H \boldsymbol{C}^* \boldsymbol{Z} \right] - \boldsymbol{b}^H \boldsymbol{C}^* \boldsymbol{R} \boldsymbol{C} \boldsymbol{b} \right\}. \tag{27.45}$$

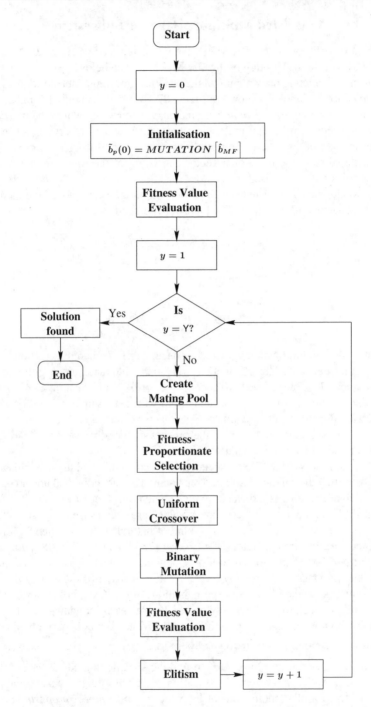

Figure 27.18: A flowchart depicting the structure of the genetic algorithm adopted for our GA-assisted
multiuser detection technique

The associated probability of fitness-proportionate selection p_i of the ith individual is defined as [768]:

$$p_i = \frac{f_i}{\sum_j^T f_j}, \tag{27.46}$$

where f_i is the fitness value associated with the ith individual. Once a pair of parents is selected, the *uniform crossover* [769, 783, 786] and *binary mutation* [783] operations are then applied to this pair of parents.

The *crossover* [783] operation is a process in which arbitrary decision variables are exchanged between a pair of selected parents, 'mimicking the biological recombination process between two single-chromosome organisms'. Hence, the crossover operation creates two new individuals, known as *offspring* in GA parlance [768], which have a high probability of having better fitness values than their parents. In order to generate P number of new offspring, P/2 number of crossover operations are required. A new pair of parents is selected from the mating pool for each crossover operation. The newly created offspring will form the basis of the new population. In a *uniform crossover* operation [786], a so-called *crossover mask* is invoked. The crossover mask is a vector consisting of randomly generated 1s and 0s of equal probability, having a length equal to that of the individuals. Bits or M-ary symbols are exchanged between the selected pair of parents at locations corresponding to a 1 in the crossover mask. While it was shown in [787] that the uniform crossover operation has a higher probability of destroying a schema, it is also capable of creating new schemata.

During the *mutation* operation [783], each decision variable in the offspring is perturbed, i.e. corrupted, with a probability of p_m, by either a predetermined or a random value. This allows new areas in the search space to be explored. The mutation probability of a decision variable is usually low, in the region of 0.1-0.01 [768]. This value is often reduced throughout the search, when the optimisation is likely to approach the final solution. In a *binary mutation* operation [783], there are only two possible values for each binary decision variable hosted by an individual. Hence, when mutation is invoked for a particular bit, the value of the bit is toggled to the other possible value. For example, a bit of logical '1' is changed to a logical '0' and vice versa.

In order to ensure that high-merit individuals are not lost from one generation to the next, the best or a few of the best individuals are copied into the forthcoming generation, replacing the worst offspring of the new population. This technique is known as *elitism* [783]. In our application, we will terminate the GA-assisted search at the Yth generation and the individual associated with the highest fitness value at this point will be the detected solution.

27.4.4 Simulation Parameters

The configuration of the GA employed in our system is shown in Table 27.5. Soft decision trellis decoding utilising the Log-MAP algorithm [637] was invoked for channel decoding. The similar complexity CM schemes utilised are: TCM using a code memory of six, BICM using a code memory of six, TTCM using a code memory of three in conjunction with four decoding iterations and finally, BICM-ID using a code memory of four employing four decoding iterations. Their corresponding code generator polynomials can be found from Tables 24.1, 24.2, 24.3 and 24.4.

Setup/Parameter	Method/Value
Individual initialisation method	Mutation of $\hat{b}_{MF}$ of Equation 27.43
Selection method	Fitness-proportionate
Crossover operation	Uniform crossover
Mutation operation	Standard binary mutation
Elitism	Yes
Incest Prevention	Yes
Population size, P	$K \cdot M$
Mating pool size, T	$T \leq P$ depends on the no. of non-identical individuals
Probability of mutation, p_m	0.1
Termination generation, Y	$\frac{1}{2} K \cdot M$

Table 27.5: The configuration of the GA employed in our system, where K is the number of CDMA users and M is the number of modulation levels

27.4.5   Simulation Results and Discussion

Our performance metric is the average BER evaluated over the course of several GA generations. The detection time of the GA is governed by the number of generations Y required, in order to obtain a reliable decision. The computational complexity of the GA, quantified in the context of the total number of objective function evaluations, is related to P × Y. Since our GA-assisted multiuser detector is based on optimising the modified correlation metric of Equation 27.45, the computational complexity is deemed to be acceptable, if there is a significant amount of reduction in comparison to the optimum multiuser detector, which requires M^K objective function evaluations, in order to reach the optimum decision.

The BER versus E_b/N_0, performance of the QPSK-based CM-GA-MUD schemes is shown in Figures 27.19 and 27.20 for transmissions over AWGN channel and uncorrelated Rayleigh fading channel, respectively. The simulation parameters were summarised in Table 27.5. A 'codeword length' of 1000 symbols and a spreading factor of 31 chips were employed. As determined by the 'codeword length', the turbo interleaver of TTCM and the internal bit interleavers of BICM and BICM-ID had a memory of 1000 symbol duration. The employment of an uncorrelated Rayleigh fading channel implies ideal channel interleaving, which has an infinitely long interleaver depth.

It is widely recognised that a QPSK signal consists of two orthogonal BPSK signals in a single-user scenario and that the associated BERs of BPSK and QPSK are identical in terms of E_b/N_0. Hence the single-user bounds for QPSK modulation shown in Figure 27.19 for AWGN channel and Figure 27.20 for uncorrelated Rayleigh fading channel, are identical to that of the BPSK modulation. However, the orthogonality of the in-phase and quadrature-phase BPSK signals is corrupted by the MAI when a QPSK signal is transmitted in a CDMA system. Hence the BER of QPSK signal is not identical to that of BPSK signals in the context of a MAI-limited CDMA environment. Therefore, the uncoded QPSK performance of a K = 10-user CDMA system is worse than that of the single-user bounds illustrated in Figures 27.19 and 27.20.

Note that the computational complexity of the GA-MUD is $\frac{M^K}{P \times Y} = 1310.72$ times lower

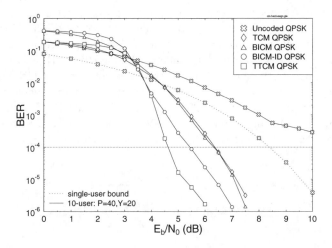

Figure 27.19: BER versus E_b/N_0 performance of the various CM-GA-MUD schemes for transmissions over the AWGN channel employing QPSK and utilising the simulation parameters of Table 27.5. A codeword length of 1000 symbols and a spreading factor of 31 chips were employed

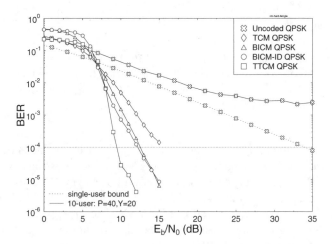

Figure 27.20: BER versus E_b/N_0 performance of the various CM-GA-MUD schemes for transmissions over the uncorrelated Rayleigh fading channel employing QPSK and utilising the simulation parameters of Table 27.5. A codeword length of 1000 symbols and a spreading factor of 31 chips were employed

than that of the optimum MUD, when supporting $K = 10$ *users employing QPSK modulation* [1]. The penalty for this complexity reduction is the BER error floor experienced by the GA-MUD schemes at high SNRs, as shown in Figures 27.19 and 27.20. However, this disadvantage is eliminated, when the CM schemes are utilised. In particular, the TTCM assisted GA-MUD constitutes the best candidate, followed by the BICM-ID assisted GA-MUD,. as evidenced in Figures 27.19 and 27.20 for transmissions over the AWGN and uncorrelated Rayleigh fading channels encountered. More specifically, for a throughput of 1 BPS and a target BER of 10^{-4}, the $K = 10$-user TTCM-GA-MUD assisted CDMA system is capable of providing SNR gains of about 4 and 25 dB in AWGN and perfectly interleaved narrowband Rayleigh fading channels, respectively, against the single-user bounds of the uncoded BPSK scheme.

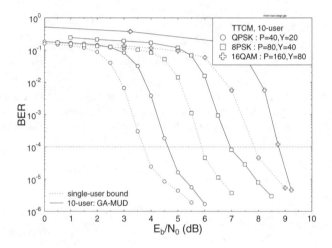

Figure 27.21: BER versus E_b/N_0 performance of the TTCM-GA-MUD scheme for transmissions over the AWGN channel employing QPSK, 8PSK and 16QAM and utilising the simulation parameters of Table 27.5. A codeword length of 1000 symbols and a spreading factor of 31 chips were employed

Next, let us study the performance of the TTCM-GA-MUD scheme in conjunction with QPSK, 8PSK and 16QAM in Figures 27.21 and 27.22 for transmissions over both AWGN and uncorrelated Rayleigh fading channels, respectively. The *hard decision*-based single-user performance bounds for TTCM are also plotted in Figures 27.21 and 27.22 as benchmarks. We found that the performance of the TTCM-GA-MUD scheme, which supports 10 users is comparable to that of the single-user TTCM benchmark, in the 1, 2 and 3 BPS effective throughput modes associated with QPSK, 8PSK and 16QAM, respectively. In the higher-

---

[1] We note that Verdu's full-search based optimum MUD exhibits an excessive complexity and hence it is not a practical benchmark. However, it exhibits a performance which is similar to the achievable single-user performance. Therefore our benchmark is the single-user performance which may be approach using a variety of different reduced-complexity MUD schemes. The GA-MUD's complexity may be best characterised by quantifying the number of objective function evaluation required for achieving near single-user performance. The closest relative of the GA-based MUD family is that of the M-algorithm [788,789] and T-algorithm [790,791], both of which would constitute an appropriate benchmark scheme.

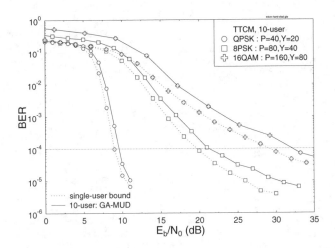

Figure 27.22: BER versus E_b/N_0 performance of the TTCM-GA-MUD scheme for transmissions over the uncorrelated Rayleigh fading channel employing QPSK, 8PSK and 16QAM and utilising the simulation parameters of Table 27.5. A codeword length of 1000 symbols and a spreading factor of 31 chips were employed

throughput modes, it was achievable by doubling the population size P and the number of generations Y of the TTCM-GA-MUD, every time when the BPS throughput was increased by one, as shown in the legends of Figures 27.21 and 27.22. More specifically, we found that for detecting K users each employing M-ary modulation, the rule of thumb quantifying the required complexity of the specific TTCM-GA-MUD scheme considered is given by $P \times Y \approx \frac{1}{2}K^2M^2$, where we require $P \approx KM$ and $Y \approx \frac{1}{2}KM$, which attains a performance similar to that of the optimum MUD. *Hence, the computational complexity reductions obtained by the TTCM-GA-MUD compared to that of the optimum MUD can be construed to be about* $F \approx \frac{M^K}{\frac{1}{2}K^2M^2} = 2M^{K-2}K^{-2}$. For the specific example of K = 10 users, the complexity reduction factors F for QPSK, 8PSK and 16QAM were 1.3×10^3, 6.7×10^5 and 8.6×10^7, respectively. Despite these huge complexity reduction gains, the BER penalty for TTCM-GA-MUD was only around 0.5 to 2 dB at a BER of 10^{-4} compared to the single-user benchmark, when communicating over the AWGN and uncorrelated Rayleigh fading channels employing QPSK, 8PSK and 16QAM, as evidenced by Figures 27.21 and 27.22.

In Figure 27.23, we show the BER versus E_b/N_0 performance of the TTCM-GA-MUD and that of the TTCM-optimum-MUD schemes, when communicating over AWGN channels using QPSK, 8PSK and 16QAM TTCM. It is shown in Figure 27.23 that the optimum-MUD exhibits an approximately 0.3 dB performance loss both in conjunction with *hard-* and *soft decisions*, when K increased from one to four. For K = 10 users the performance loss of TTCM-GA-MUD was about 1 and 3 dB, as shown in Figure 27.23, compared to the TTCM-assisted single-user schemes employing *hard-* and *soft-decisions*, respectively, regardless whether QPSK, 8PSK or 16QAM modulation schemes were used. The TTCM-GA-MUD aided QPSK, 8PSK and 16QAM systems supporting K = 10 users achieved an SNR-reduction of 3.8, 1.4 and 2.9 dB at the effective throughputs of 1, 2 and 3 BPS in

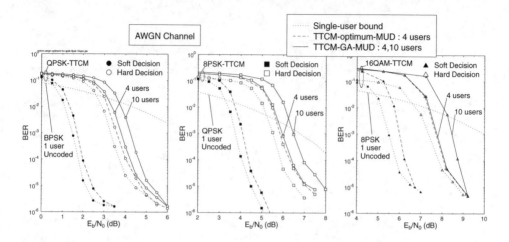

Figure 27.23: BER versus E_b/N_0 performance of the TTCM-GA-MUD and TTCM-optimum-MUD schemes, when communicating over non-dispersive AWGN channels; left: effective throughput of 1 BPS; middle: effective throughput of 2 BPS; right: effective throughput of 3 BPS

comparison to the single-user bounds of the uncoded schemes of BPSK, QPSK and 8PSK, respectively, without extending the required bandwidth.

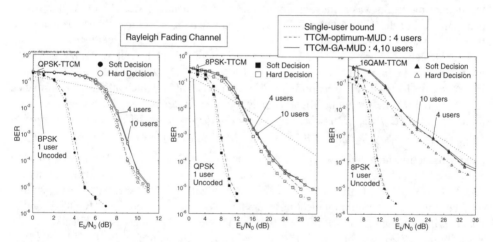

Figure 27.24: BER versus E_b/N_0 performance of the TTCM-GA-MUD and TTCM-optimum-MUD schemes, when communicating over non-dispersive Rayleigh fading channels; left: effective throughput of 1 BPS; middle: effective throughput of 2 BPS; right: effective throughput of 3 BPS

Similar performances trend can also be observed in Figure 27.24 for the TTCM-GA-MUD and TTCM-optimum-MUD schemes, when communicating over uncorrelated Rayleigh fading channels. Despite the huge complexity reduction and the support of K = 10 users, the TTCM-GA-MUD aided QPSK, 8PSK and 16QAM systems were still capable of achieving an SNR-reduction of 25, 13 and 4 dB at the effective throughputs of 1, 2 and 3 BPS in comparison to the single-user bounds of the uncoded schemes of BPSK, QPSK and 8PSK, respectively, within the same bandwidth.

27.4.6   Conclusion

In this section, TCM, TTCM, BICM and BICM-ID assisted GA-based MUD schemes were proposed and evaluated in performance terms when communicating over the AWGN and narrowband Rayleigh fading channels encountered. It was shown that the GA-MUD is capable of significantly reducing the computational complexity of the optimum-MUD, but experiences a GA-induced error floor at high SNRs due to invoking an insufficiently large population size and a low number of generations as demonstrated in Figures 27.19 and 27.20. However, with the advent of the bandwidth-efficient CM schemes proposed, this problem is eliminated. When comparing the four CM schemes at the same decoding complexity, TTCM was found to be the best candidate for assisting the operation of the GA-MUD system.

When higher throughput M-ary modulation schemes were investigated with the advent of TTCM-GA-MUD arrangements, we found that the complexity of TTCM-GA-MUD was dramatically lower than that of the optimum-MUD, with the penalty of only 0.5 to 2 dB SNR loss compared to that of the TTCM single-user bound as evidenced in Figures 27.21 and 27.22. Furthermore, the reduced-complexity TTCM-GA-MUD schemes are capable of attaining further coding gains with respect to the uncoded single-user modulation schemes while maintaining a similar throughput and bandwidth, even when supporting a high number of users, as shown in Figures 27.23 and 27.24. We note furthermore that the BER curves of TTCM-GA-MUD in Figures 27.23 and 27.24 also show a tendency to have a residual BER, which is the characteristic of all turbo coding-based arrangements.

27.5   Summary

In this chapter the CM schemes introduced in Chapter 24 were investigated in the context of CDMA environments with the aid of JD-MMSE-DFE based MUD scheme in Sections 27.2 and 27.3, and with the aid of GA-MUD in Section 27.4.

It was found in Section 27.2 that the TCM-assisted JD-MMSE-DFE based MUD scheme was the best candidate for employment in systems using a short channel interleaver, while the TTCM-assisted JD-MMSE-DFE based MUD was the best candidate for systems using a long channel interleaver. The burst-by-burst adaptive TTCM-assisted JD-MMSE-DFE based MUD was investigated in Section 27.3, where the complexity of the joint detection algorithms remained constant upon changing the CM modes. The proposed adaptive scheme constitutes a promising practical communication system, guaranteeing reliable transmission while providing an effective bitrate ranging from 23.4 Kbit/s/slot to 117 Kbit/s/slot for transmission over the UTRA environment of Figure 27.9 and Table 27.2.

A reduced complexity M-ary CM-GA-MUD scheme was proposed and investigated in Section 27.4. It was shown in Section 27.4 that the CM-GA-MUD scheme is capable of attaining further coding gains with respect to the uncoded single-user modulation schemes at a similar throughput and in the same bandwidth, when supporting a high number of CDMA users. At the same time, the CM-GA-MUD scheme also attains a significant complexity reduction with respect to the optimum-MUD.

In summary, the coding gain values exhibited by the coded modulation schemes studied in this chapter were collated in Tables 27.6 and 27.7. Specifically, we summarised the performance gains exhibited by the JD-DFE assisted coded modulation schemes of Section 27.3 communicating over the UTRA Rayleigh fading channels in Tables 27.6. By contrast, the coding gains exhibited by the GA-MUD-CM schemes of Section 27.4 communicating over non-dispersive AWGN channels and uncorrelated Rayleigh fading channels were summarised in Table 27.7.

		JD-DFE, UTRA Rayleigh fading channels							
		E_b/N_o (dB)		Gain (dB)		E_b/N_o (dB)		Gain (dB)	
Code	Modulation/	BER				FER			
	BPS	10^{-3}	10^{-5}	10^{-3}	10^{-5}	10^{-1}	10^{-2}	10^{-1}	10^{-2}
uncoded	BPSK/1	11.42	17.40	0.00	0.00	11.00	14.79	0.00	0.00
TCM	QPSK/1	**10.00**	**14.19**	1.42	3.21	6.79	10.17	4.21	4.62
TTCM	QPSK/1	10.41	**14.19**	1.01	3.21	**6.36**	**9.77**	4.64	5.02
BICM	QPSK/1	10.41	14.80	1.01	2.60	6.48	9.96	4.52	4.83
BICM-ID	QPSK/1	10.98	14.34	0.44	3.06	6.48	10.06	4.52	4.73
uncoded	QPSK/2	11.52	17.39	0.00	0.00	11.74	15.31	0.00	0.00
TCM	8PSK/2	**12.19**	**16.42**	-0.67	0.97	9.19	12.58	2.55	2.73
TTCM	8PSK/2	12.54	16.52	-1.02	0.87	**9.03**	**12.37**	2.71	2.94
BICM	8PSK/2	13.25	17.13	-1.73	0.26	9.57	12.91	2.17	2.40
BICM-ID	8PSK/2	14.38	17.84	-2.86	-0.45	9.27	12.64	2.47	2.67
uncoded	8PSK/3	14.49	20.63	0.00	0.00	15.27	19.01	0.00	0.00
TCM	16QAM/3	**13.96**	**18.38**	0.53	2.25	11.23	14.63	4.04	4.38
TTCM	16QAM/3	14.23	18.54	0.26	2.09	**11.00**	**14.25**	4.27	4.76
BICM	16QAM/3	14.89	18.93	-0.40	1.70	11.23	14.50	4.04	4.51
BICM-ID	16QAM/3	16.21	19.07	-1.72	1.56	11.06	14.38	4.21	4.63
uncoded	16QAM/4	15.11	21.23	–	–	15.98	19.58	–	–
TCM	64QAM/5	**18.31**	**22.78**	–	–	15.79	19.15	–	–
TTCM	64QAM/5	18.53	23.09	–	–	**15.61**	**18.96**	–	–
BICM	64QAM/5	19.65	23.21	–	–	16.26	19.57	–	–
BICM-ID	64QAM/5	20.62	24.01	–	–	16.39	19.79	–	–
uncoded	64QAM/6	19.25	25.20	–	–	20.59	24.22	–	–

Table 27.6: Coding gain values attained by the various JD-DFE assisted coded modulation schemes studied in this chapter, when communicating over the UTRA Rayleigh fading channels. All of the coded modulation schemes exhibited a similar decoding complexity in terms of the number of decoding states, which was equal to 64 states. An interleaver block length of 240 symbols was employed and the corresponding simulation parameters were summarised in Section 27.3.2. The performance of the best scheme is printed in bold

CM-GA-MUD		AWGN Channels				Flat Uncorrelated Rayleigh Fading Channels			
		E_b/N_o (dB)		Gain (dB)		E_b/N_o (dB)		Gain (dB)	
Code	Modulation/	BER				BER			
	BPS	10^{-3}	10^{-5}	10^{-3}	10^{-5}	10^{-3}	10^{-5}	10^{-3}	10^{-5}
Uncoded	BPSK/1	6.75	9.52	0.00	0.00	24.00	43.50	0.00	0.00
TCM	QPSK/1	5.45	7.15	1.30	2.37	12.12	21.00	11.88	22.50
TTCM	QPSK/1	**4.22**	**5.22**	2.53	4.30	**8.76**	**11.00**	15.24	32.50
BICM	QPSK/1	5.35	7.06	1.40	2.46	10.40	14.62	13.60	28.88
BICM-ID	QPSK/1	4.57	6.34	2.18	3.18	9.68	14.80	14.32	28.70
Uncoded	QPSK/2	6.75	9.52	0.00	0.00	24.00	43.50	0.00	0.00
TTCM	8PSK/2	**6.48**	**7.90**	0.27	1.62	**17.15**	**30.10**	6.85	13.40
Uncoded	8PSK/3	10.02	12.95	0.00	0.00	26.15	46.00	0.00	0.00
TTCM	16QAM/3	**8.45**	**9.11**	1.57	3.84	**24.00**	**45.00**	2.15	1.00

Table 27.7: Coding gain values of the various of GA-MUD assisted coded modulation schemes studied in this chapter, when communicating over non-dispersive AWGN channels and uncorrelated Rayleigh fading channels. All of the coded modulation schemes exhibited a similar decoding complexity in terms of the number of decoding states, which was equal to 64 states. An interleaver block length of 1000 symbols was employed and the corresponding simulation parameters were summarised in Section 27.4.4. The performance of the best scheme is printed in bold.

Chapter **28**

Coded Modulation and Space Time Block Coding Aided CDMA

28.1 Introduction

In this chapter, we will further develop the Coded Modulation (CM) schemes studied in Chapter 24 by amalgamating them with Space Time Block Coding (STBC) [616, 617] and investigate the achievable performance improvements in the light of the complexity investments required.

In numerous practical situations the wireless channels are neither highly time selective nor significantly frequency selective. This motivated numerous researchers to investigate space diversity techniques with the aim of improving the system's performance. Classic receiver diversity [792] has been widely used at the base stations of both the GSM and IS-136 systems. As an additional performance enhancement, recently the family of transmit diversity techniques [369, 793] was studied extensively for employment at the base station, since it is more practical to have multiple transmit antennas at the base station than at the mobile station. Furthermore, upgrading the base station instead of the mobile stations has the potential of benefiting numerous mobile stations in a cost-efficient manner. Space-Time Trellis Coding (STTC) pioneered by Tarokh *et. al.* [619] incorporates jointly designed channel coding, modulation, transmit diversity and optional receiver diversity [369]. In an attempt to reduce the associated decoding complexity, Alamouti proposed STBC employing two transmit antennas. Alamouti's scheme was later generalised to an arbitrary number of transmit antennas [617].

Furthermore, in this chapter we will argue that further diversity gain may also be attained by exploiting the independent nature of the fading experienced by the In-phase (I) and Quadrature-phase (Q) components of the complex-valued transmitted signal with the aid of two independent IQ interleavers. Alternatively, the multipath components exhibited by the dispersive Rayleigh fading channel constitute a further often used source of useful diversity potential.

This chapter is organised as follows. The proposed STBC aided IQ-interleaved CM (IQ-

CM) scheme will be studied in Section 28.2, while the novel concept of a STBC assisted double-spread Rake receiver assisted CDMA scheme will be studied in Section 28.3. Finally, an amalgamated system based on the combination of the above two schemes with be investigated in Section 28.4 and our conclusions are offered in Section 28.5.

28.2 Space-Time Block Coded IQ-Interleaved Coded Modulation

28.2.1 Introduction

In this section we will present the philosophy of a set of STBC-assisted IQ-interleaved TCM and TTCM schemes designed for transmission over both non-dispersive AWGN and for flat Rayleigh fading channels. Specifically, we will demonstrate that the proposed schemes are capable of quadrupling the achievable diversity order of the conventional symbol-interleaved TCM and TTCM schemes, when two transmit antennas are employed. The increased diversity order of the proposed schemes provides significant additional coding gains, when communicating over non-dispersive Rayleigh fading channels, which is achieved without compromising the coding gain attainable over Gaussian channels. This is a significant achievement, since the design criteria of schemes destined for those different channels are also different. Hence in general a scheme optimised for one of these channels is typically suboptimum, when communicating over the other. The BICM as well as BICM-ID schemes introduced in Sections 24.6 and 24.7, respectively, are also incorporated into the proposed system and their performance is compared to that of TCM and TTCM schemes highlighted in Sections 24.3 and 24.5, respectively.

To elaborate a little further, it was shown in [708] that the maximisation of the minimum Hamming distance measured in terms of the number of different symbols between any two transmitted symbol sequences is the key design criterion of TCM schemes contrived for flat Rayleigh fading channels, in particular when communicating at high Signal-to-Noise Ratios (SNR). In an effort to increase the achievable time diversity, a multidimensional TCM scheme utilising a symbol interleaver and two encoders was proposed by S. Al-Semari and T. Fuja in [794], where the individual encoders generate the I and Q components of the complex transmitted signal, respectively. Another efficient TCM scheme using constellation rotation was proposed by B. D. Jelicic and S. Roy in [795], which utilised two separate channel interleavers for interleaving the I and Q components of the complex-valued transmitted signals, but assumed the absence of I/Q *cross-coupling*, which can only be realistically expected when communicating over fading channels exhibiting a real-valued Channel Impulse Response (CIR).

In order to improve the performance of the existing state-of-the-art systems, in this section we proposed a novel system, which amalgamates STBC [616] with IQ-interleaved TCM, TTCM, BICM and BICM-ID schemes using no constellation rotation.

28.2.2 System Overview

The block diagram of the STBC based IQ-interleaved (STBC-IQ) TCM/TTCM system is shown in Figure 28.1. Specifically, we employ two transmitters and one receiver in the STBC

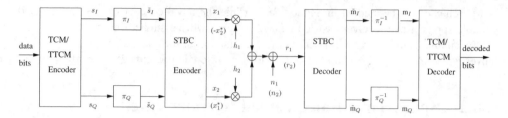

Figure 28.1: Block diagram of the STBC-based IQ-interleaved system. The notations π and π^{-1} denote the interleaver and deinterleaver, while the variables in the round brackets () denote the STBC signals transmitted during the second symbol period.

scheme. Furthermore, we invoke two independent IQ interleavers in the TCM/TTCM arrangement, which are denoted as π_I and π_Q in the block diagram of Figure 28.1. We denote the IQ-interleaved modulated signal by $\tilde{s} = \tilde{s}_I + j\tilde{s}_Q$, which is transmitted over the flat Rayleigh fading channel having a complex fading coefficient of $h = \alpha e^{j\theta}$ with the aid of two STBC transmitters [369]. As seen in Figure 28.1, during the first symbol period of the STBC transmission the signals $x_1 = \tilde{s}_1$ and $x_2 = \tilde{s}_2$ are transmitted, while during the second symbol period, the signals $-x_2^*$ and x_1^* are emitted from the transmit antennas 1 and 2, respectively, as in Alamouti's G_2 code [616]. Specifically, G_2 is defined as [616, 617]:

$$ G_2 \;=\; \left(\begin{array}{cc} x_1 & x_2 \\ -x_2^* & x_1^* \end{array} \right), \tag{28.1}$$

where two symbols, x_1 and x_2, were transmitted using two transmit antennas over the duration of two symbol periods. Hence, a coding rate of unity was achieved. We assume that both the fading envelope and the phase of the CIR taps remains constant across the two STBC time slots. The signal is also contaminated by the zero-mean AWGN n, exhibiting a variance of $\sigma^2 = N_0/2$, where N_0 is the single-sided noise power spectral density. It can readily be seen from Figure 28.2 that the two signals received during the two consecutive symbol periods of the STBC scheme can be represented in a matrix form as $\mathbf{r} = \mathbf{A} \cdot \mathbf{x} + \mathbf{n}$:

$$ \left(\begin{array}{c} r_1 \\ r_2^* \end{array} \right) = \left(\begin{array}{cc} h_1 & h_2 \\ h_2^* & -h_1^* \end{array} \right) \left(\begin{array}{c} x_1 \\ x_2 \end{array} \right) + \left(\begin{array}{c} n_1 \\ n_2^* \end{array} \right), \tag{28.2}$$

where $\mathbf{A}$ is termed the system matrix and x^* denotes the complex conjugate of the transmitted symbol x. At the combiner/decoupler block of Figure 28.2, the received vector $\mathbf{r}$ is multiplied with the conjugate transpose of $\mathbf{A}$, namely with $\mathbf{A}^H$, yielding:

$$ \begin{aligned} \hat{\mathbf{x}} &= \mathbf{A}^H \cdot \mathbf{r} \\ \left(\begin{array}{c} \hat{x}_1 \\ \hat{x}_2 \end{array} \right) &= (\alpha_1^2 + \alpha_2^2) \left(\begin{array}{c} x_1 \\ x_2 \end{array} \right) + \left(\begin{array}{c} \hat{n}_1 \\ \hat{n}_2 \end{array} \right), \end{aligned} \tag{28.3}$$

where the terms $\hat{n}_1 = h_1^* n_1 + h_2 n_2^*$ and $\hat{n}_2 = h_2^* n_1 + h_1 n_2^*$ represent the resultant noise.

Note that as usual, owing to the complex-valued nature of the CIR, the I (or Q) component of the received signal r_i, namely $r_{i,I}$ (or $r_{i,Q}$) where we have $i \in \{1, 2\}$, is dependent on both the I and Q components of x_1 and x_2, namely on $x_{1,I}$, $x_{1,Q}$, $x_{2,I}$ and $x_{2,Q}$. This phenomenon

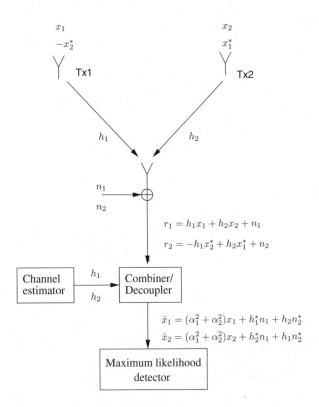

Figure 28.2: Baseband representation of the STBC G_2 code employing two transmitters and one receiver.

is often referred to as IQ crosscoupling. More explicitly, we have:

$$r_{1,I} = h_{1,I}x_{1,I} - h_{1,Q}x_{1,Q} + h_{2,I}x_{2,I} - h_{2,Q}x_{2,Q} + n_{1,I}, \qquad (28.4)$$

owing to the cross-coupling effect imposed by the complex-valued CIR. It is however desirable to decouple them, so that we can compute the I (or Q) branch metrics m_I (or m_Q) in Figure 28.1 for a particular transmitted symbol x_i independently, as a function of only $x_{1,I}$ and $x_{2,I}$ (or $x_{1,Q}$ and $x_{2,Q}$). Surprisingly, this may be achieved without carrying out any explicit decoupling operation for the STBC based IQ-interleaved TCM/TTCM scheme. More specifically, this is achieved because the signals already have been decoupled during the STBC decoding operation at the combiner/decoupler block of Figure 28.2, when the received vector $\mathbf{r}$ is multiplied with $\mathbf{A}^H$, yielding $\hat{\mathbf{x}} = \mathbf{A}^H \cdot \mathbf{r}$ in Equation 28.3. Explicitly, the transmitted signal estimate of $\hat{x}_1 = (\alpha_1^2 + \alpha_2^2)x_1 + \hat{n}_1$ is the decoupled version of r_1, where $\hat{x}_{1,I}$ (or $\hat{x}_{1,Q}$) is independent of $x_{1,Q}$ and $x_{2,Q}$ (or $x_{1,I}$ and $x_{2,I}$).

Before we proceed further to the computation of the branch metrics for the STBC based IQ-TCM/TTCM scheme, let us first derive the branch metrics for the single-transmitter based IQ-TCM/TTCM scheme. For a single-transmitter scheme emitting the signal $x = x_I + jx_Q$, which is conveyed over narrowband fading channels exhibiting a complex-valued CIR of $h = \alpha e^{j\theta}$, the corresponding received signal is $r = hx + n$ and the I/Q-decoupled signal can be computed from:

$$\hat{x} = h^*r,$$
$$= \alpha^2 x + h^* n. \tag{28.5}$$

The corresponding symbol-based branch metrics can be calculated from:

$$\tilde{m}(x|\hat{x}, h) = -\frac{|r - hx|^2}{2\sigma^2}, \tag{28.6}$$

$$= -\frac{|h^*r - h^*hx|^2}{2\sigma^2 |h^*|^2}, \tag{28.7}$$

$$= -\frac{|\hat{x} - \alpha^2 x|^2}{2\sigma^2 \alpha^2}, \tag{28.8}$$

$$= -\frac{|\hat{x} - Dx|^2}{2\sigma^2 D}, \tag{28.9}$$

where we have $D = \alpha^2$, which can be physically interpreted as a fading envelope-related variable or a diversity quantity. Since $\hat{x}$, which was computed from Equation 28.5, is the decoupled version of the received signal r, we can compute the corresponding I and Q branch metrics based on Equation 28.9 as:

$$\tilde{m}(x|\hat{x}, h) = -\frac{|\hat{x}_I - D_I x_I|^2}{2\sigma^2 D_I} - \frac{|\hat{x}_Q - D_Q x_Q|^2}{2\sigma^2 D_Q}$$
$$= \tilde{m}_I(x_I|\hat{x}_I, D_I) + \tilde{m}_Q(x_Q|\hat{x}_Q, D_Q), \tag{28.10}$$

where $\tilde{m}_I$ and $\tilde{m}_Q$ denote the I and Q branch metrics, respectively, before the IQ interleavers as shown in Figure 28.1. Furthermore, we have $D_I = D_Q = D = \alpha^2$.

Let us now compute the I and Q branch metrics for the STBC scheme by first studying the similarity between $\hat{x} = h^*r$ of Equation 28.5 and $\hat{x} = \mathbf{A}^H \cdot \mathbf{r}$ of Equation 28.3. Clearly, $\hat{x}$ and $\hat{x}$ are the decoupled versions of their corresponding received signals, hence the branch metrics for the STBC scheme can be computed similarly to that of the single-transmitter scheme. Although, in the STBC scheme we cannot employ a relationship similar to Equation 28.6, we can directly utilise a relationship similar to Equation 28.9. Specifically, the symbol-based branch metrics for $x_i, i \in \{1, 2\}$ of the STBC scheme can be calculated from:

$$\tilde{m}(x_i|\hat{x}_i, D) = -\frac{|\hat{x}_i - Dx_i|^2}{2\sigma^2 D}, \tag{28.11}$$

where we have $D = (\alpha_1^2 + \alpha_2^2)$ for the STBC G_2 scheme. Hence, the associated IQ branch metrics of the STBC coded signal $x_i = x_{i,I} + jx_{i,Q}$ can be derived from $\hat{x}_i = \hat{x}_{i,I} + j\hat{x}_{i,Q}$

as:

$$
\begin{aligned}
\tilde{\mathrm{m}}(x_i|\hat{x}_i, D) &= -\frac{\left|\hat{x}_{i,I} - D_I x_{i,I}\right|^2}{2\sigma^2 D_I} - \frac{\left|\hat{x}_{i,Q} - D_Q x_{i,Q}\right|^2}{2\sigma^2 D_Q} \\
&= \tilde{\mathrm{m}}_I(x_{i,I}|\hat{x}_{i,I}, D_I) + \tilde{\mathrm{m}}_Q(x_{i,Q}|\hat{x}_{i,Q}, D_Q),
\end{aligned} \tag{28.12}
$$

where again, we have $D_I = D_Q = D = (\alpha_1^2 + \alpha_2^2)$. The effect of the associated second-order transmit diversity attained may be observed in the context of the term $(\alpha_1^2 + \alpha_2^2)$ in Equation 28.3.

Note that $\tilde{\mathrm{m}}_I$ and $\tilde{\mathrm{m}}_Q$ share the same fading channel-envelope related D value for the same transmitted signal of $x(=\tilde{s})$ before the IQ interleavers, as seen in Figure 28.1. However, owing to rearranging the transmitted IQ components by the IQ deinterleavers of π_I^{-1} and π_Q^{-1} seen in Figure 28.1, m_I and m_Q will be associated with different fading envelope values of $D_I \neq D_Q$ after the IQ deinterleavers. The branch metrics of the TCM/TTCM-coded complex-valued signal s can be computed by summing the I and Q branch metrics as follows:

$$
\mathrm{m}(s) = \mathrm{m}_I(x_I = s_I) + \mathrm{m}_Q(x_Q = s_Q). \tag{28.13}
$$

Since there are two independent IQ components for a complex-valued TCM/TTCM symbol and since they are independently interleaved by the interleavers π_I and π_Q in Figure 28.1, m_I and m_Q are exposed to independent fading and hence they have the potential of providing independent diversity for a particular symbol. More explicitly, *since we have $D_I \neq D_Q$ for a particular coded symbol s, the IQ-interleaved TCM/TTCM scheme may be expected to double the achievable diversity order compared to its symbol-interleaved counterpart.* Therefore the achievable Hamming distance between the transmitted symbol sequences of the proposed IQ-interleaved TCM/TTCM scheme is based on the sum of the number of different I and Q components between the different transmitted messages, rather than on the number of different complex-valued symbols, which was the case in the context of conventional symbol-interleaved TCM/TTCM.

We have also amalgamated the proposed STBC-IQ scheme of Figure 28.1 with the BICM and BICM-ID schemes of Sections 24.6 and 24.7. More specifically, in addition to their internal bit-interleavers [369], two extra random interleavers were invoked for separately interleaving the I and Q components of their bit-interleaved complex symbol $s = s_I + j s_Q$ for yielding $\tilde{s} = \tilde{s}_I + j\tilde{s}_Q$, as was illustrated in Figure 28.1.

28.2.3 Simulation Results and Discussion

We evaluated the performance of the proposed schemes using 16QAM in the context of the non-iterative 64-state TCM and BICM schemes, as well as in conjunction with the iterative 8-state TTCM arrangement using four decoding iterations and along with the 8-state BICM-ID arrangement using eight decoding iterations. As in Section 25.2.2, the rationale of using 64 and 8 states, respectively, was that the TCM/BICM and TTCM/BICM-ID schemes considered here exhibit a similar decoding complexity expressed in terms of the total number of trellis states. Explicitly, since there are two 8-state TTCM decoders, which are invoked in four iterations, we encounter a total of $2 \times 8 \times 4 = 64$ TTCM trellis states. By contrast, only a single 8-state BICM-ID decoder is required, which is invoked in eight iterations, involving

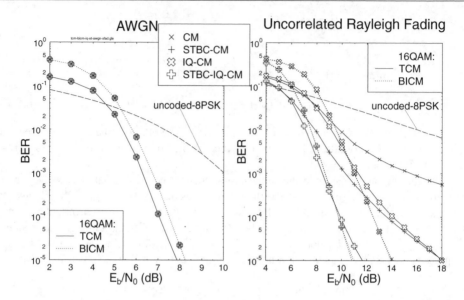

Figure 28.3: BER versus E_b/N_0 performance of the 16QAM-based **TCM** and **BICM** schemes, when communicating over AWGN and uncorrelated flat Rayleigh fading channels. The legends are described in Footnote 1. An interleaved codeword length of 1000 symbols was used and the performance of the uncoded 8PSK scheme is also plotted for benchmarking the CM schemes, all having an effective throughput of **3 BPS**.

a total of $8 \times 8 = 64$ BICM-ID trellis states. The effective throughput was 3 Bits Per Symbol (BPS) for all the 16QAM based CM schemes, since a coding rate of 3/4 was used.

In Figure 28.3, we portray the Bit Error Ratio (BER) versus signal to noise ratio per bit, namely E_b/N_0, performance of the 16QAM-based TCM and BICM schemes, when communicating over AWGN as well as over uncorrelated flat Rayleigh fading channels.[1] An interleaved codeword length of 1000 symbols was used and the BER performance curve of the uncoded 8PSK scheme is also plotted for benchmarking the schemes, all having an effective throughput of 3 BPS. As illustrated in Figure 28.3, all the TCM/BICM schemes associated with the conventional CM, STBC-CM, IQ-CM and STBC-IQ-CM arrangements exhibit a similar performance in AWGN channels. This is because no space diversity or time diversity is attainable over Gaussian channels despite using multiple transmitters and interleaving. On the other hand, the TCM scheme performs approximately 0.5 dB better, than the BICM scheme, when communicating over AWGN channels, since it has a higher Euclidean distance than that of BICM, which is the decisive criterion in the context of AWGN channels.

By contrast, when communicating over uncorrelated flat Rayleigh fading channels, the BER curve of IQ-TCM merged with that of STBC-TCM in the high-SNR region of Figure 28.3, since they both exhibit twice the diversity potential compared to conventional TCM. As seen in Figure 28.3, with the advent of STBC-IQ-TCM a further 6.4 dB gain can be obtained at a BER of 10^{-5} compared to the IQ-TCM and STBC-TCM schemes. By contrast,

---

[1] CM: the conventional CM scheme; STBC-CM: the STBC based conventional CM scheme; IQ-CM: the proposed IQ-interleaved CM scheme using no STBC; STBC-IQ-CM: the proposed STBC-based IQ-interleaved CM scheme.

the BICM scheme exhibits only transmit diversity gain but no IQ diversity gain, as we can observe in Figure 28.3. This is because when further randomising the output bit-sequence of a random interleaver, we simply arrive at a different but similarly randomised bit-sequence. In the context of 16QAM BICM having four bits per symbol, four bit-interleavers were employed for interleaving the four modulated bits in parallel. On the other hand, each of the I and Q interleavers will be further randomising two BICM output bit-sequences in parallel. Therefore, no interleaving gain is attainable, when employing extra I and Q interleavers in the context of 16QAM BICM. However, this will not be true in the context of 8PSK BICM, as we will argue in Section 28.4.3. Upon returning to Figure 28.3, since in BICM the bit-based minimum Hamming distance of the codewords is maximised [369], which is the decisive criterion in the context of Rayleigh fading channels, BICM will benefit from a lower bit error probability in Rayleigh fading channels than that of TCM, because TCM maximises the free Euclidean distance of bits within the modulated signal constellation. Note that the performance of the conventional TCM scheme is significantly lower than that of conventional BICM owing to the existence of unprotected bits in the 16QAM TCM symbol. However, the achievable coding gain of the STBC-IQ-TCM scheme is only about 0.8 dB less than that of the STBC-IQ-BICM scheme at a BER of 10^{-5}.

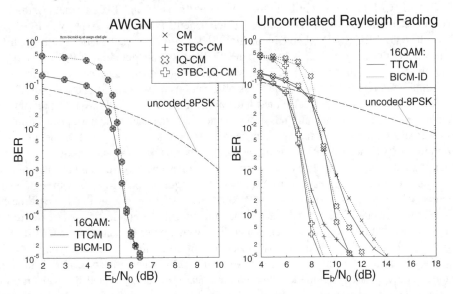

Figure 28.4: BER versus E_b/N_0 performance of the 16QAM-based **TTCM** and **BICM-ID** schemes, when communicating over AWGN and uncorrelated flat Rayleigh fading channels. The legends are described in Footnote 1. An interleaved codeword length of 1000 symbols was used and the performance of the uncoded 8PSK scheme is also plotted for benchmarking the CM schemes, all having an effective throughput of **3 BPS**.

Let us now study in Figure 28.4 the BER versus E_b/N_0 performance of the 16QAM-based iterative TTCM and BICM-ID schemes, when communicating over AWGN and uncorrelated flat Rayleigh fading channels, again, in the context of the conventional CM, STBC-CM, IQ-CM and STBC-IQ-CM arrangements defined in Footnote 1. An interleaved codeword length of 1000 symbols was used and the performance of the uncoded 8PSK scheme was also plotted as a benchmark. As portrayed in Figure 28.4, the TTCM scheme exhibits a better performance than BICM-ID in the low SNR region, when communicating over AWGN channels, although their BER curves converge beyond $E_b/N_0 = 5.8$ dB. Again, no space and no time diversity gain was achieved, when communicating over Gaussian channels.

From Figure 28.4 we infer that similarly to the performance trends observed for the TCM schemes of Figure 28.3, the BER curve of IQ-TTCM merged with that of STBC-TTCM in the high-SNR region, when communicating over the uncorrelated flat Rayleigh fading channels, since they both exhibit twice the diversity potential compared to conventional TTCM. However, the achievable diversity/coding gain of TTCM was found to be lower than that of TCM owing to the fact that TTCM has already achieved part of its total attainable diversity gain with the aid of its iterative turbo decoding procedure, assisting in achieving time diversity. Nonetheless, at a BER of 10^{-5} the performance of STBC-IQ TTCM is about 5.1 dB better than that of the conventional TTCM scheme. Note from Figure 28.4 that the BICM-ID scheme is capable of exploiting the IQ diversity owing to employing iterative detection, when communicating over uncorrelated flat Rayleigh fading channels. Although IQ-BICM-ID exhibits a performance, which is about 1 dB lower than that of IQ-TTCM at BER=10^{-5}, nonetheless the coding gain of STBC-IQ-BICM-ID is only marginally lower than that of STBC-IQ-TTCM. However, unlike in the context of the TCM and TTCM schemes, the BER performance of the IQ-BICM-ID scheme – which benefits from IQ diversity – was lower than that of the STBC-BICM-ID scheme in the high-SNR region, where the latter exhibits a transmit diversity of order two. This is because the IQ diversity gain potential of IQ-BICM-ID is a benefit that is turned into reality by the iterative decoding, rather than accruing automatically from employing IQ interleaving alone. This observations may be confirmed in Figure 28.4.

Note that encountering uncorrelated flat Rayleigh fading channels would imply that the channel interleaving has an infinitely long memory or that the vehicular speed is infinite. However, practical Rayleigh fading channels exhibit correlated fading and the degree of the correlation experienced depends on the associated normalised Doppler frequency. Let us now investigate the performance of the proposed schemes under correlated flat Rayleigh fading channel conditions having a normalised Doppler frequency of 3.25×10^{-5} in Figure 28.5, where an interleaved codeword length of 10000 symbols was used. Note that the BER performance of the uncoded 8PSK benchmark is the same, when communicating over uncorrelated and correlated flat Rayleigh Fading channels. However, the performance of the CM schemes degrades, when the fading exhibits a high degree of correlation. As portrayed at the left of Figure 28.5, the performance of the conventional TCM scheme becomes lower than that of the uncoded 8PSK benchmark, when communicating over correlated flat Rayleigh fading channels. However, with the advent of IQ interleaving, the performance of the IQ-TCM scheme improved significantly and it becomes better than that of BICM (or IQ-BICM) under these conditions. On the other hand, STBC-IQ-TCM performs better than STBC-BICM (or STBC-IQ-BICM), although STBC-TCM performs worse than STBC-BICM, when the slowly fading channel exhibits a normalised Doppler frequency of 3.25×10^{-5}, as evidenced in Figure 28.5. However, IQ-TCM is outperformed by STBC-TCM, when communicating over correlated flat Rayleigh fading channels. This is because the two STBC transmitter antennas were arranged sufficiently far apart, so that their transmitted signals experience independent channel fading, whereas the IQ-interleaved signals suffer from correlated channel fading, as a result of the limited-memory 10000-symbol interleaver.

The performance of the TTCM and BICM-ID schemes communicating over correlated

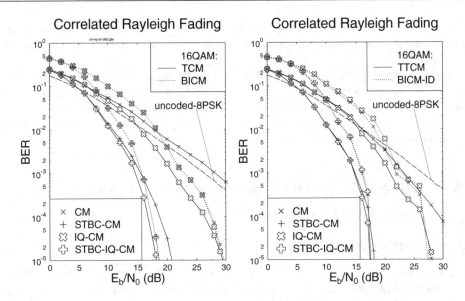

Figure 28.5: BER versus E_b/N_0 performance of the 16QAM-based **TCM, BICM, TTCM** and **BICM-ID** schemes, when communicating over correlated flat Rayleigh fading channels having a normalised Doppler frequency of 3.25×10^{-5}. The legends are described in Footnote 1. An interleaved codeword length of 10000 symbols was used and the performance of the uncoded 8PSK scheme is also plotted for benchmarking the CM schemes, all having an effective throughput of **3 BPS**.

flat Rayleigh fading channels was also shown at the right of Figure 28.5. Specifically, IQ-BICM-ID (or STBC-IQ-BICM-ID) shows no advantage over its conventional BICM-ID (or STBC-BICM-ID) counterpart, when communicating over slowly fading channels. By contrast, IQ-TTCM still outperforms conventional TTCM by approximately 2 dB under these conditions. However, the performance of the STBC-IQ-TTCM arrangement is only marginally better than that of its STBC-TTCM counterpart.

28.2.4 Conclusion

In this section we proposed a set of novel STBC-IQ aided CM schemes for transmissions over both AWGN and Rayleigh fading channels. The STBC-IQ-TCM, STBC-IQ-TTCM and STBC-IQ-BICM-ID schemes are capable of providing significant diversity gains over their conventional counterparts. Specifically, in case of the uncorrelated flat Rayleigh fading channel, coding gains of 26.1 dB, 28.2 dB, 26.9 dB and 28.1 dB were achieved over the identical-throughput uncoded 8PSK benchmark at a BER of 10^{-4} by the STBC-IQ-TCM, STBC-IQ-TTCM, STBC-IQ-BICM and STBC-IQ-BICM-ID schemes, respectively. All schemes achieved an effective throughput of 3 BPS without bandwidth expansion. From Figures 28.3, 28.4 and 28.5 STBC-IQ-TTCM was found to be the best scheme, when communicating over AWGN as well as over uncorrelated and correlated flat Rayleigh fading channels.

For systems requiring the reduced complexity of a single-transmitter scheme, the IQ-interleaved TCM/TTCM scheme is still capable of doubling the achievable diversity potential of conventional symbol-interleaved TCM/TTCM with the aid of a single transmit antenna, although the IQ diversity attainable decreased, when the fading channel exhibited a higher correlation.

Having studied the proposed STBC-IQ aided CM schemes for transmission over both AWGN and non-dispersive Rayleigh fading channels, we will now invoke Direct Sequence (DS) spreading and study a STBC-based CDMA system designed for transmission over dispersive Rayleigh fading channels in the following section. Then in Section 28.4 we will amalgamate it with STBC-IQ CM schemes.

28.3 Space-Time Block Coding-Assisted Double-Spread Rake Receiver-Based CDMA

28.3.1 Introduction

In this section we will present a novel STBC-assisted Double-Spread Rake Receiver (DoS-RR) based CDMA scheme designed for downlink (base to mobile) transmissions over dispersive Rayleigh fading channels, where all the transmitted user signals arriving at a particular mobile station experience identical propagation conditions. The design philosophy of this system is that of achieving the maximum possible diversity gain, when communicating over dispersive fading channels. This is achieved by systematically incorporating performance enhancements, which is capable of providing transmit diversity, time diversity and multipath diversity. Specifically, the proposed STBC-assisted DoS-RR (STBC-DoS-RR) scheme invokes Turbo Convolutional (TC) [366] channel coding and STBC for achieving both time and space diversity gains in addition to the path diversity gain provided by the dispersive Rayleigh fading environments. We will demonstrate that the double-spreading mechanism of the STBC-DoS-RR scheme is capable of detecting the wanted user's channel-impaired wideband signals with the aid of a low-complexity Rake receiver, without resorting to employing a complex interference cancellation scheme. The performance of the proposed scheme was investigated using Quadrature-Phase-Shift-Keying (QPSK), when communicating over dispersive channels. It will be shown that the wideband scheme advocated is capable of achieving E_b/N_0 gains up to 35 dB in comparison to an uncoded single-transmitter scheme communicating over narrowband channels.

Space-Time Coding (STC) schemes [369, 616, 617, 619] were originally proposed for transmissions over narrowband fading channels. When encountering wideband channels, Multi-Carrier Code Division Multiple Access (MC-CDMA) and Orthogonal Frequency Division Modulation (OFDM) [180] can be utilised for converting the wideband channel to numerous narrowband channels. However, a detrimental effect of employing STTCs or STBCs is that not only the desired signal, but also the Multiple Access Interference (MAI) and the Co-Channel Interference (CCI) are enhanced due to the multiple antenna-based transmission diversity scheme and hence the effective average SNR and SIR may not be substantially improved. Nonetheless, the fading-induced fluctuation of the SNR and SIR are improved, which results in overall performance improvements. These are the most important limiting factors of employing space-time coding in CDMA systems [796]. By contrast, Multi-User Detection (MUD) and Interference Cancellation (IC) constitute two efficient ways of combating the MAI and CCI, although their complexity may become excessive, especially, when the

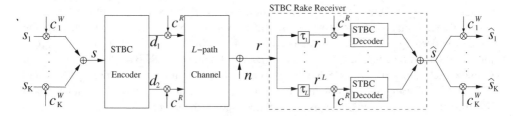

Figure 28.6: Block diagram of the space-time block coding assisted double-spread Rake receiver based downlink CDMA scheme

number of users supported is high. On the other hand, a low-complexity Rake receiver can be used for obtaining path diversity in CDMA systems, provided that the maximum propagation path delay is shorter than a fraction of the original symbol period, and hence the ISI can be neglected but still sufficiently high for encountering a number of resolvable multipath components after Direct Sequence (DS) spreading. Furthermore, the multipath intensity profile typically has an exponentially decaying shape in realistic channel conditions, where the longer the path delay, the lower the magnitude of the path. Hence the ISI induced by a long-delay path is typically insignificant and can be ignored. It is worth noting that in [797], an adaptive Space-Time Spreading (STS) assisted DS-CDMA scheme was proposed, which is capable of maintaining an arbitrary diversity order and transmission integrity by adaptively controlling the amount of STS. Based on these arguments the employment of a simple Rake receiver is often more practical than that of a complex IC scheme, especially in a downlink scenario imposing identical propagation conditions on all users.

In our proposed system, a double-spreading assisted Rake receiver is used. Specifically, in the first spreading operation Walsh code-based unique user signatures are used for distinguishing the users and hence for mitigating the effects of MAI, while in the second spreading step a random code is employed for DS spreading and hence for attaining path diversity. Additionally, STBC is invoked for achieving transmit diversity, while a power efficient TC channel code [366] is used for obtaining time diversity.

28.3.2   System Description

The block diagram of the STBC-DoS-RR based CDMA scheme proposed for downlink (base to mobile) transmissions is shown in Figure 28.6. The information destined for user-i, where we have $i \in \{1 \dots K\}$, is first encoded by the TC channel encoder to yield S_i. Then, S_i is spread by the Walsh code c_i^W, where each user is assigned a unique Walsh code. The sum S of the signals of all users is then passed to the STBC encoder on a chip-by-chip basis. Here Alamouti's G_2 code [616] associated with two transmit antennas is used for STBC. The STBC encoder yields two chips, namely d_1 and d_2, one for each transmit antenna. Both chips are then DS spread by the same random code for the sake of attaining multipath diversity, namely c^R, before transmission through the L-path dispersive Rayleigh fading channels associated with the two transmission antennas. A single STBC-based Rake receiver is utilised at the mobile station. In the Rake receiver we have L number of delayed replicas of the transmitted signal, resulting in the received signals $r^1 \dots r^L$, where r^i is synchronised to the ith resolvable path having a path delay of τ_i. Each of the signals $r^1 \dots r^L$ will be DS despread by the random code c^R. Due to the employment of the G_2 STBC encoder, each of the L number of DS despread signals benefits from a transmit diversity order of two. Consequently, the STBC decoder is invoked L times, namely once for each of the L number of DS despread signals for obtaining a path diversity of order L. The L number of STBC decoded signals are summed, yielding $\hat{S}$ of Figure 28.6, which benefits from a total diversity order of $2L$. Then, the signal $\hat{S}_i$ of user-i can be obtained by despreading $\hat{S}$ with the aid of the unique Walsh signature sequence c_i^W. Finally, $\hat{S}_i$ is channel decoded by the TC channel decoder in order to yield the estimate of the information transmitted by user-i.

28.3.2.1   Double-Spreading Mechanism

The family of orthogonal Walsh codes is well known for its attractive zero cross-correlation property [798] expressed as:

$$\sum_{l=1}^{Q} c_i^W[l] \cdot c_j^W[l] \quad = \quad 0; \text{ for } i \neq j, \tag{28.14}$$

where $i, j \in \{1 \dots \mathsf{K}\}$ and $c_i[l] \in \{+1, -1\}$. There are Q chips in a code and the maximum number of Walsh codes of length Q is Q, which allows us to support a maximum number of $\mathsf{K} = Q$ users. Therefore, it is possible to support Q users employing Walsh codes of length Q, without encountering any MAI, provided that the codes' orthogonality is not destroyed by the channel. However, the off-peak auto-correlation of Walsh codes is relatively high [798], which is quantified as:

$$\sum_{l=1}^{Q} c_i^W[m] \cdot c_i^W[l] \quad \neq \quad 0; \text{ for } m, i \in \{1 \dots Q\}. \tag{28.15}$$

Hence, the good cross-correlation properties of orthogonal Walsh codes may be destroyed by the multipath interference encountered in dispersive channels. In contrast to orthogonal Walsh codes, random codes exhibit high cross-correlation, but low off-peak auto-correlation values [798]. Therefore non-orthogonal random codes can be processed by a Rake receiver for obtaining path diversity as a benefit of their low auto-correlation value. However, without additional Walsh-spreading they are only applicable for single user scenarios, since the associated high cross-correlation will result in excessive MAI, i.e. poor 'user-separation'.

To elaborate a little further, the double-spreading mechanism introduced here is different from the conventional method of combining orthogonal Walsh codes with non-orthogonal random or Pseudo-Noise (PN) scrambling codes. The mechanism of both methods is illustrated in Figure 28.7. Specifically, in the context of the conventional method the chip duration of the Walsh code is identical to that of the random code. For the transmission of a B-symbol burst using orthogonal Walsh codes having a spreading factor of Q, there are Q chips per information symbol, resulting in $B \times Q$ chips per transmission block. After each of the information symbols has been DS spread by the Q-chip orthogonal Walsh code, a non-orthogonal random code of length $B \times Q$ chips is multiplied by the $B \times Q$-chip transmission block on a chip-by-chip basis in order to decrease the auto-correlation of the orthogonal Walsh codes. Although the resultant spread signal exhibits a reduced auto-correlation, the cross-correlation of the various users' signals is no longer zero.

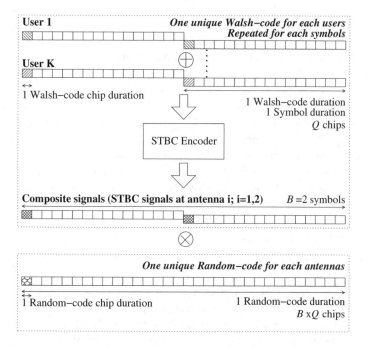

(a) Conventional-spreading mechanism

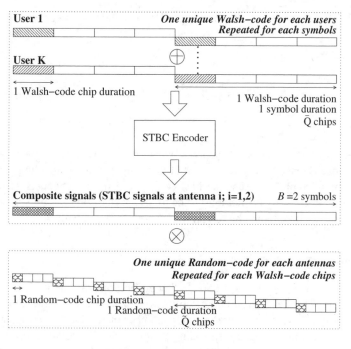

(b) Double-spreading mechanism

Figure 28.7: Mechanism of conventional-spreading and double-spreading employing Walsh codes and Random codes

By contrast, the proposed double-spreading mechanism invokes $\bar{Q}$-chip orthogonal Walsh codes for supporting $K = \bar{Q}$ users with the aid of DS spreading of each symbol to $\bar{Q}$ chips. Then each of the $\bar{Q}$ chips is further spread by a $\tilde{Q}$-chip non-orthogonal random code for attaining multipath diversity, where there are a total of Q chips per information symbol and we have $\bar{Q} \times \tilde{Q} = Q$. The first spreading operation spreads each information symbol of user-i to $\bar{Q}$ chips, using the Walsh code c_i^W. Then the $\tilde{Q}$-chip random code further spreads each of the $\bar{Q}$ number of resultant chips to $\tilde{Q}$ chips. Hence there are a total of $\bar{Q} \times \tilde{Q} = Q$ chips per information symbol also according to the conventional method highlighted in the previous paragraph, although the spreading mechanism used is different. In other words, here the same $\tilde{Q}$-chip random code repeats itself every $\bar{Q}$ chips, namely for each of the Walsh-code chips. Note that the lth chip of the $\bar{Q}$-chip Walsh code of all users is spread by the same random code, which is conveyed via the same channel and experiences an identical multipath interference of:

$$I[l] = I_k[l] = I_i[l], \tag{28.16}$$

where $k, i \in \{1 \ldots \bar{Q}\}$ and $I_k[l]$ is the interference imposed on the lth chip of the Walsh code of user-k, $c_k^W[l]$. If a slow fading channel is encountered, the Channel Impulse Response (CIR) of the multipath components of the channel can be assumed to be time-invariant for the duration of $Q = \bar{Q} \times \tilde{Q}$ chips. Since each chip of the $\bar{Q}$-chip Walsh code is further spread by the same random code as seen in Figure 28.7(b), each of the $\bar{Q}$ chips, again, experiences identical multipath interference in the slow fading channels considered, yielding:

$$I = I[l] = I[m]; \quad l, m \in \{1 \ldots \bar{Q}\}. \tag{28.17}$$

With the aid of Equations 28.14, 28.16 and 28.17 the resultant cross-correlation of the Q-chip codes used by the double-spreading mechanism of Figure 28.7(b) can be shown to be zero, which is expressed as:

$$
\begin{aligned}
\sum_{l=1}^{\bar{Q}} (c_i^W[l] \cdot I_i[l]) \cdot (c_k^W[l] \cdot I_k[l]) &= I^2 \cdot \sum_{l=1}^{\bar{Q}} c_i^W[l] \cdot c_k^W[l], \\
&= 0; \text{ for } i \neq k. \tag{28.18}
\end{aligned}
$$

This implies encountering no MAI, hence requiring no IC. Therefore attaining near-single-user performance is feasible with the advent of this simple double-spreading method without the employment of complex IC schemes in slow-fading dispersive channels, provided that the factor $\bar{Q}$ lower user capacity is tolerable. If the user-capacity is at premium, a factor $\bar{Q}$ higher user-capacity can be maintained by tolerating the high complexity of the IC.

28.3.2.2   Space-Time Block Coded Rake Receiver

The main idea behind Rake receivers [677] is that of capturing the energy of several multipath components of the channels in a spread spectrum system, provided that the different multipath delays are multiples of the chip duration. The number of resolvable multipath components can be increased by increasing the spreading factor used. The Rake receiver consists of a number of matched filters, often referred to as Rake fingers. The maximum number of Rake fingers required is equal to the number of the resolvable paths. These Rake receiver fingers are synchronised to each path and the corresponding delayed replicas of the transmitted signals are combined coherently or noncoherently. Assuming a perfect knowledge of the CIR and multipath delays, the Rake receiver is equivalent to a Maximal Ratio Combining (MRC) scheme having a diversity order L [799]. In a conventional single-transmitter scenario the ith delayed replica of the transmitted signal, r^i, is first DS despread and then multiplied with the complex conjugate of the corresponding CIR tap. However, when using transmit diversity, each of the DS despread signals is conveyed to one of the STBC decoders shown in Figure 28.6. In other words, there are L number of STBC decoders, each corresponding to one resolvable path of the dispersive channel.

In the STBC scheme using Alamouti's G_2 code [616], the encoding requires two time slots and two transmit antennas for transmitting two symbols, namely x_1 and x_2. Specifically, G_2 was defined in Equation 28.1 [616, 617], but we repeat it here for the sake of the reader's convenience:

$$ G_2 = \begin{pmatrix} x_1 & x_2 \\ -x_2^* & x_1^* \end{pmatrix}, $$

where, x^* denotes the complex conjugate of the signal x. In time slot 1, signals x_1 and x_2 are transmitted, while in time slot 2, signals $-x_2^*$ and x_1^* are emitted from the transmit antennas 1 and 2, respectively. We assume that the fading amplitude is constant across the two STBC time slots, where each time slot has a duration of $\tilde{Q}$ chips, or one chip of c^W. Hence we can write:

$$ h_p^i(t) = h_p^i(t+T) = h_p^i = \alpha_p^i e^{j\theta_p^i}; \quad p \in \{1, 2\}, \tag{28.19} $$

where T is the time slot duration, while h_p^i is the CIR associated with path i and transmit antenna p. The received signals after despreading are:

$$ \begin{aligned} r_1^i = r^i(t) &= h_1^i \cdot x_1 + h_2^i \cdot x_2 + n_1^i, \\ r_2^i = r^i(t+T) &= -h_1^i \cdot x_2^* + h_2^i \cdot x_1^* + n_2^i, \end{aligned} \tag{28.20} $$

where r_t^i is the received signal at instant t for path i, while n_t^i is the random variable representing the sum of the receiver noise at instant t and the multipath interference imposed by path i after despreading. Equivalently, the two signals received over path i can be represented in matrix form as:

$$ \begin{aligned} \begin{pmatrix} r_1^i \\ (r_2^i)^* \end{pmatrix} &= \begin{pmatrix} h_1^i & h_2^i \\ (h_2^i)^* & -(h_1^i)^* \end{pmatrix} \begin{pmatrix} x_1 \\ x_2 \end{pmatrix} + \begin{pmatrix} n_1^i \\ (n_2^i)^* \end{pmatrix} \\ \mathbf{r}^i &= \mathbf{A}^i \cdot \mathbf{x} + \mathbf{n}^i, \end{aligned} \tag{28.21} $$

where $\mathbf{A}^i$ is termed the system matrix associated with path i. Since the transmitted signal vector $\mathbf{x}$ can be factorised out from the STBC received signal vector $\mathbf{r}^i$, as shown in Equation 28.21, due to the orthogonality of the G_2 code, transmit diversity can be obtained by multiplying $\mathbf{r}^i$ with the conjugate transpose of $\mathbf{A}^i$, namely with $\mathbf{A}_i^H$ during the STBC decoding process. With the aid of the Rake receiver we can sum all the L number of STBC decoded

Parameter	Value
Doppler frequency (Hz)	80
Vehicular velocity (mph)	30
First Spreading ratio, Q	4
Second Spreading ratio, $\tilde{Q}$	8
Total Spreading ratio, Q	32
Chip rate (MBaud)	2.167
Modulation mode	QPSK
STBC code	G_2 [617]
Number of transmitters	2
Number of receivers	1
TC coding rate	1/2
TC constraint length	3
TC decoding iterations	8
TC interleaver length (bit)	1000
Channel A power profile	Equal/Constant
Channel B power profile	Exponential decay
Number of resolvable paths (chip-spaced)	1 to 5

Table 28.1: Parameters of the STBC-DoS-RR CDMA scheme for downlink transmissions

outputs for generating the estimate of:

$$\hat{\mathbf{x}} = \sum_{i=1}^{L}(\mathbf{A_i^H} \cdot \mathbf{r^i}),$$

$$\begin{pmatrix} \hat{x}_1 \\ \hat{x}_2 \end{pmatrix} = \left(\sum_{i=1}^{L} ((\alpha_1^i)^2 + (\alpha_2^i)^2) \right) \begin{pmatrix} x_1 \\ x_2 \end{pmatrix} + \mathbf{N}, \qquad (28.22)$$

where $\mathbf{N}$ is the resultant noise plus interference for all the L paths, while the diversity of order $2L$ can be observed in the context of the term $\sum_{i=1}^{L} ((\alpha_1^i)^2 + (\alpha_2^i)^2)$.

Having exploited both the space and multipath diversity, now the TC decoder is invoked for attaining further time diversity. Based on Equation 28.22 the soft inputs of the TC decoder can readily be computed. Finally, the estimates of the original information symbols are obtained at the output of the TC decoder. Here, two identical 1/2-rate Recursive Systematic Convolutional (RSC) codes having a constraint length of three were utilised by the turbo encoder, while soft decision trellis decoding utilising the binary Log-Maximum A Posteriori (Log-MAP) algorithm [637] was invoked for turbo decoding. For detailed discussions on TC channel coding, the interested reader is referred to [366].

28.3.2.3 Channel Model and System Parameter Design

In this section the performance of the proposed scheme will be evaluated using the simulation parameters shown in Table 28.1. We denote the conventional single-transmitter scenario as the 'G_1-coded' scheme, which is invoked for comparison with the two-transmitter based G_2

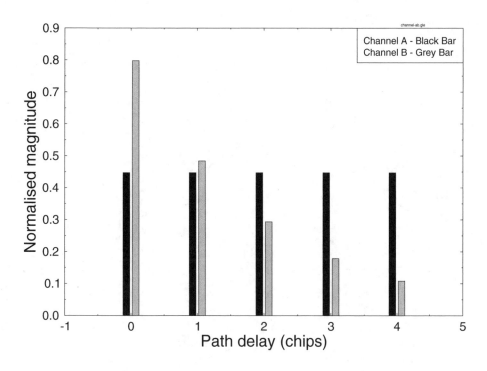

Figure 28.8: The normalised magnitude versus chip delay profile for the 5-path Channel A and Channel B

code. We assumed that the receiver determined the CIRs perfectly, while the fading envelope and phase was maintained at a constant level for the duration of two STBC timeslots or for $2Q=64$ chip duration. Two different channels were used in our investigation. Channel A exhibits a CIR having equal-power chip-spaced taps, while the CIR of Channel B exhibits an exponentially decaying power for each of the multipath components. Each path was faded according to independent Rayleigh fading statistics, as described by the parameters of Table 28.1. Figure 28.8 shows the normalised magnitude versus chip delay profile for the 5-path CIR of Channel A and Channel B, where the total power of the paths was normalised to unity. The delay of each path is expressed in terms of chip durations, where the chip duration is equal to the reciprocal of the system's chip rate. We will consider the presence of one to five resolvable paths for Channel A and Channel B.

28.3.3 Simulation Results and Discussion

Let us now compare the performance of the proposed double-spreading scheme and that of the conventional spreading scheme. Figure 28.9 shows the Bit Error Ratio (BER) versus signal to noise ratio per information bit (E_b/N_0) performance of both the double-spreading and that of the conventional spreading schemes in a multi-user scenario for transmissions over Channel A, when using only two resolvable paths, utilising the simulation parameters of Table 28.1. The separation of the two CIR taps was one chip duration.

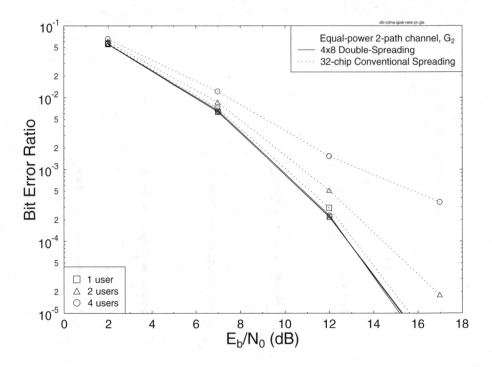

Figure 28.9: Bit Error Ratio (BER) versus signal to noise ratio per information bit (E_b/N_0) performance of both the double-spreading and conventional spreading schemes assisted by a
Rake receiver in a multi-user scenario for transmissions over Channel A exhibiting two
resolvable paths, utilising the simulation parameters of Table 28.1. The separation of the
two CIR taps was one chip interval.

We observe in Figure 28.9 that conventional spreading experienced performance degradations, when the number of users increased, since the orthogonality of the 32-chip Walsh
codes used by the conventional spreading scheme was destroyed by the multipath interference
imposed by the two-path channel. By contrast, the double-spreading scheme experienced no
performance degradations, when supporting up to $\bar{Q} = 4$ users, since the multipath interference can be rejected, as suggested by Equation 28.18. Hence the orthogonality of the 4-chip
Walsh codes was preserved by the double-spreading scheme.

Although the conventional spreading method is also capable of attaining multipath diversity with the aid of the signal spreading imposed by the random scrambling code, the multipath components will inflict severe interference, rather than yielding beneficial path diversity,
when the number of users is high. This is illustrated in Figure 28.10, where the performance
of the conventional spreading scheme degrades, as the channel exhibits an increasing number
of multipath components, when supporting four users. By contrast, the performance of the
double-spreading scheme improved, as the channel exhibited a higher number of multipath
components, although the path diversity gains eventually saturated, when a high number of
paths was encountered.

Having investigated the performance gains achieved using the double-spreading scheme,

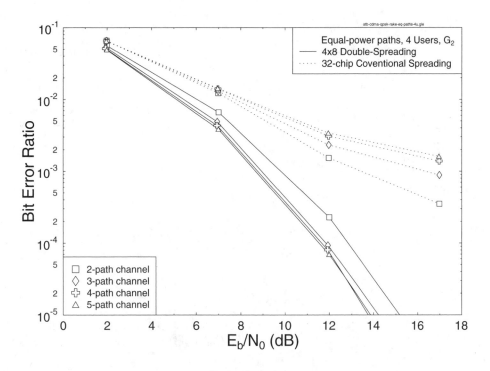

Figure 28.10: BER versus E_b/N_0 performance of both the double-spreading and conventional spreading schemes assisted by a Rake receiver in a multipath scenario for transmissions over Channel A, utilising the simulation parameters of Table 28.1.

let us now apply the double-spreading scheme in conjunction with the G_2 STBC arrangement utilising two transmitters, as well as along with the single transmitter scenario, which we have denoted as the 'G_1-coded' scheme in Section 28.3.2.3. Explicitly, the G_2 code exhibits twice the diversity gain in comparison to its 'G_1-coded' counterpart. For example, the performance of the G_2 code in single-path Rayleigh channel is identical to that of the G_1 code in the two-path Channel A, as seen from Figure 28.11. However, when targeting a total diversity order of four, the performance of the G_2 code used for transmission over the chip-spaced two-path Channel A is slightly better than that of the G_1 code in the chip-spaced four-path Channel A. This is because the equal power paths in Channel A result in a higher multipath interference, when the channel exhibits a longer delay spread, where the delay spreads of the four-path and two-path channels constitute 50% and 25% of the total duration of the 8-chip random code, respectively. Hence, when aiming for attaining a certain diversity gain, the multipath interference associated with G_2 is less severe than that in conjunction with G_1, when communicating over dispersive channels having longer delay spreads.

When transmitting over Channel B, where the power of the paths decays exponentially with time, the multipath interference will be less severe in a long-delay-spread environment, although the achievable multipath diversity gain will be decreased. These effects can be studied in Figure 28.12, where the performance of the G_2 code recorded in a single-path scenario is seen to be better than that of the G_1 code employed for transmission over the two-

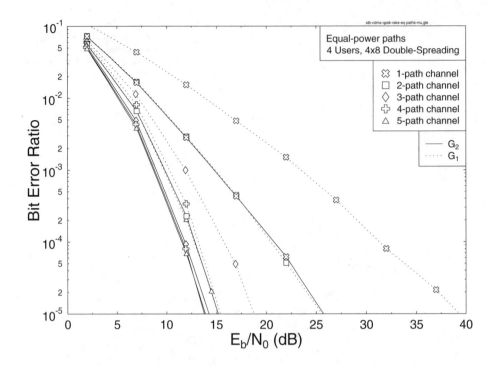

Figure 28.11: BER versus E_b/N_0 performance of the STBC-DoS-RR scheme without TC channel code using the G_2 code of Equation 28.1 and the G_1 code defined in Section 28.3.2.3, for transmissions over Channel A, utilising the simulation parameters of Table 28.1, when supporting four users.

path Channel B. Similarly, the performance of the G_2 code used for transmission over the two-path Channel B is significantly better than that of the G_1 code in the four-path Channel B. Hence, we conclude that the diversity gain of G_2 is higher than that of G_1 in a realistic dispersive channel having negative exponentially decaying CIR, which exhibits a reduced power for the longer delay multipath components.

The employment of power efficient TC channel coding further enhances the achievable coding gain of the STBC-DoS-RR scheme. As we can see from Figures 28.13 and 28.14, a high coding gain was obtained by the TC-coded systems compared to the uncoded systems characterised in Figures 28.11 and 28.12. We can observe that at a BER of 10^{-5}, the TC-coded STBC-DoS-RR schemes utilising the G_2 code required only $E_b/N_0 = 4.25$ dB, when communicating over the 5-path Channel A, as shown in Figure 28.13. By contrast, the uncoded STBC-DoS-RR scheme utilising G_1 required an E_b/N_0 of 39.3 dB in a narrowband channel, as shown in Figure 28.11, resulting in a total diversity plus coding gain of 35.05 dB. However, we observed from Figure 28.13 that most of this total diversity plus coding gain was already achieved in the 3-path Channel A. Furthermore, we can see from Figure 28.14 that the total gain achieved in the 5-path Channel B for the TC-coded STBC-DoS-RR scheme utilising G_2 was 33.2 dB. By observing Figures 28.13 and 28.14, we notice that the relative performance gain attained is the highest, when the diversity is on the order of two, either util-

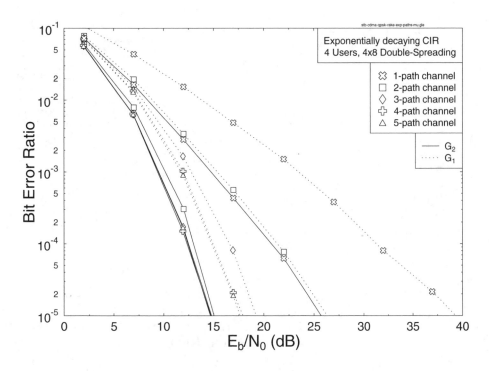

Figure 28.12: BER versus E_b/N_0 performance of STBC-DoS-RR scheme without TC channel code using the G_2 code of Equation 28.1 and the G_1 code defined in Section 28.3.2.3, for transmissions over Channel B, utilising the simulation parameters of Table 28.1, when supporting four users.

ising G_2 in a narrowband channel for attaining transmit diversity, or utilising G_1 in a 2-path channel for achieving multipath diversity.

The coding plus diversity gain versus the number of resolvable paths at a BER of 10^{-5} for the STBC-DoS-RR scheme against the benchmarker of the uncoded STBC-DoS-RR scheme utilising G_1 in a narrowband channel are summarised in Figure 28.15. The benchmarker required an E_b/N_0 of 39.3 dB for a BER of 10^{-5} as shown in Figure 28.11. It is clear from the figure that the additional path diversity gain attainable is only marginal, when having more than three resolvable paths for the TC or STBC assisted schemes.

28.3.4 Conclusions

In this section, a novel space-time coding-assisted double-spreading aided Rake receiver-based CDMA scheme was proposed and characterised in performance terms, which was contrived for downlink transmissions over dispersive Rayleigh fading channels. The double-spreading mechanism is capable of attaining a near-single-user performance, when using a low-complexity Rake-receiver, rather than a complex IC scheme. Additionally, it is capable of yielding a performance equivalent to that of maximal ratio combining of order L in L-path channels, even when utilising only a single Rake receiver. Explicitly, the transmit diversity

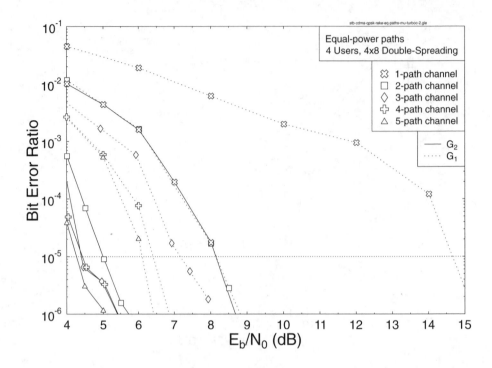

Figure 28.13: BER versus E_b/N_0 performance of the STBC-DoS-RR scheme in conjunction with TC channel coding using the G_2 and G_1 codes for transmissions over Channel A, utilising the simulation parameters of Table 28.1, when supporting four users.

and path diversity constitute two independent sources of diversity gain, capable of acting in unison.

It was shown in Figures 28.12 and 28.15 that the employment of transmit diversity is more advantageous than that of multipath diversity in practical dispersive Rayleigh fading channels exhibiting an exponentially decaying CIR. When targeting a certain diversity gain, the multipath interference sensitivity of STBC is lower than that of a single-transmitter based scheme, especially when communicating over wideband channels having a long delay spread, as investigated in Figures 28.11 and 28.15. With the advent of TC channel coding the proposed scheme required a low bit energy for attaining a low BER. *In conclusion, near-single-user performance can be achieved even with the aid of a single Rake receiver, i.e. without complex IC schemes, at the cost of supporting a factor of $\bar{Q}$ reduced number of users.*

Having introduced the STBC-IQ aided CM scheme in Section 28.2 and the STBC-DoS-RR scheme in Section 28.3, we will now amalgamate these two schemes for the sake of jointly exploiting their benefits and investigate the achievable performance in the next section.

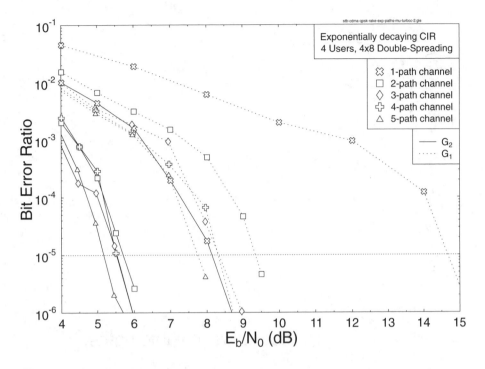

Figure 28.14: BER versus E_b/N_0 performance of the STBC-DoS-RR scheme in conjunction with TC channel coding using the G_2 and G_1 codes for transmissions over Channel B, utilising the simulation parameters of Table 28.1, when supporting four users.

28.4 STBC IQ-Interleaved Coded Modulation Assisted Double-Spread Rake Receiver-Based CDMA

28.4.1 Introduction

In Section 28.2 we argue that STBC-based IQ-interleaved CM schemes are capable of providing significant diversity gains over their conventional counterparts, when communicating over flat Rayleigh fading channels using a 16QAM modulation scheme. It is important to emphasise that this can be achieved without compromising the coding gain attainable over Gaussian channels. It was found that the achievable transmit diversity was maintained, although the IQ diversity was decreased, when the fading channel exhibited correlation.

Furthermore in Section 28.3, we demonstrated that the STBC-based scheme designed for transmission over non-dispersive fading channels can also be invoked for transmission over dispersive fading channels for the sake of benefiting from multipath diversity. Specifically, when a double-spreading aided Rake receiver is employed, near-single-user performance may be achieved without the aid of complex IC schemes, when transmitting from the base station to the mobile station over dispersive fading channels using QPSK modulation.

In this section, the STBC-IQ-CM scheme proposed in Section 28.2 will be amalgamated with the DoS-RR scheme proposed in Section 28.3 for downlink transmission over 1-path

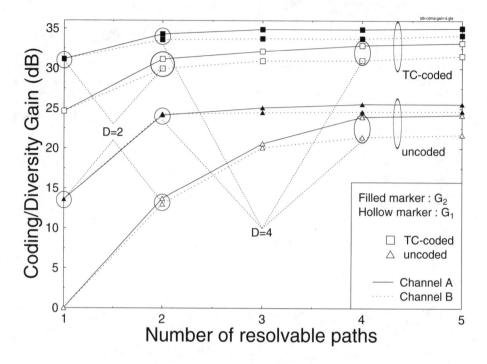

Figure 28.15: Coding plus diversity gain versus the number of resolvable paths at a BER of 10^{-5} for the STBC-DoS-RR scheme. The notation D denotes the diversity order and the uncoded STBC-DoS-RR scheme utilising the G_1 code, when communicating over a narrowband channel was used as the benchmarker exhibiting zero coding/diversity gain. The simulation parameters of Table 28.1 were utilised and the number of users was four.

and 2-path Rayleigh fading channels, when supporting four CDMA users, each using 8PSK modulation.

28.4.2   System Description

The block diagram of the amalgamated system can readily be constructed by concatenating the schematics of Figures 28.1 and 28.6, which is shown in Figure 28.16. More specifically, the channel coded signal S_1 of user one, which is shown in Figure 28.16, is generated from the signals $\tilde{s}_I$ and $\tilde{s}_Q$ of Figures 28.1 and 28.16, where we have $S_1 = \tilde{s}_I + j\tilde{s}_q$. At the decoder, the I and Q components of the signal $\hat{S}_1$, which are shown in Figures 28.6 and 28.16, will be used for computing the signals $\tilde{m}_I$ and $\tilde{m}_Q$ according to Equation 28.12, which are seen in both Figure 28.1 and Figure 28.16.

The CM schemes employed are the non-iterative 64-state TCM and BICM schemes, the iterative 8-state TTCM arrangement using four decoding iterations and the 8-state BICM-ID arrangement invoking eight decoding iterations. Furthermore, we will also employ Turbo Coded Modulation (TuCM) [371] for this study. More specifically, TuCM is an extension of TC codes, where the output bits of a binary TC encoder are suitably punctured and multi-

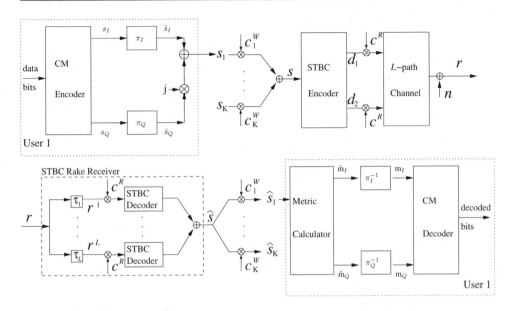

Figure 28.16: Block diagram of the STBC-IQ-CM assisted DoS-RR scheme. The notations π and π^{-1} denote the interleaver and deinterleaver.

plexed before the Gray-labelled signal mapper, for achieving the desired number of information bits per transmitted symbol. For example, two 1/2-rate RSC codes are used for generating a total of four turbo coded bits and this bit stream may be punctured for generating three bits, which are mapped to an 8PSK modulation scheme. By contrast, in separate coding and modulation scheme, any modulation schemes for example BPSK, may be used for transmitting the channel coded bits. Finally, without puncturing, 16QAM transmission would have to be used for maintaining the original transmission bandwidth. Explicitly, a TuCM arrangement employing two 1/2-rate and 4-state RSC component codes using eight decoding iterations will be employed. All the CM schemes exhibit a similar decoding complexity expressed in terms of the total number of trellis states, as detailed in Section 28.2.3. The effective throughput of all the 8PSK-based CM schemes is 2 BPS and hence the performance of the uncoded QPSK scheme communicating over flat Rayleigh fading channels will be used as a benchmarker. A codeword length of 1000 symbols was used.

28.4.3   Simulation Results and Discussion

Let us now characterise the performance of the STBC-IQ-CM assisted DoS-RR scheme using the channel parameters and CDMA parameters shown in Table 28.1. The 2-path channel exhibits two equal-power, chip-spaced CIR taps. Specifically, four arrangements will be studied in conjunction with each of the CM schemes, which are denoted in the legends of Figures 28.17 to 28.20 as CM: the conventional CM scheme; STBC-CM: the STBC-based conventional CM scheme; IQ-CM: the proposed IQ-interleaved CM scheme using no STBC; and STBC-IQ-CM: the proposed STBC-based IQ-interleaved CM scheme. A codeword length of 1000 symbols was used and the performance of the uncoded QPSK(8PSK) scheme communicating over AWGN and flat Rayleigh fading channels is also plotted for benchmarking the

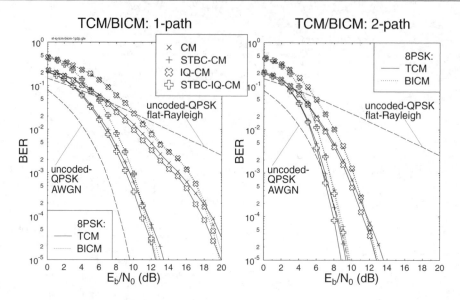

Figure 28.17: BER versus E_b/N_0 performance of the 8PSK-based non-iterative **TCM** and **BICM** schemes, when communicating over correlated Rayleigh fading channels exhibiting either one or two resolvable chip-spaced, equal weight paths. The legends are described in Footnote 1. A codeword length of 1000 symbols was used and the performance of the uncoded QPSK scheme communicating over AWGN and flat Rayleigh fading channels is also plotted for benchmarking the CM schemes having an effective throughput of **2 BPS**.

CM schemes having an effective throughput of **2(3) BPS**. Note that the BER performance of the uncoded QPSK/8PSK benchmarker is the same, when communicating over both uncorrelated and correlated flat Rayleigh Fading channels.

In Figure 28.17, we portray the BER versus E_b/N_0 performance of the 8PSK-based non-iterative TCM and BICM schemes, when communicating over correlated flat Rayleigh fading channels exhibiting either one or two resolvable paths. As we can infer from Figure 28.17, the TCM schemes perform slightly better than the corresponding BICM arrangements across most of the SNR region, although the BICM arrangements marginally outperform the non-IQ-interleaved TCM arrangements in the high-SNR region. However, the channel-properties become more Gaussian-like, when the diversity gain of the system is high. To elaborate a little further, the main contributors of diversity gain in the system are the transmit-diversity and path-diversity, when communicating over slowly fading channels. Therefore, the STBC-TCM arrangement outperforms its STBC-BICM counterpart, when communicating over the 2-path channels, since the TCM schemes originally contrived for AWGN channels outperform the BICM schemes designed for Rayleigh fading channels, in the context of Gaussian-like channels. On the other hand the IQ-interleaved TCM arrangements perform similar to or better than their BICM counterparts, owing to the extra IQ diversity gain achieved as evidenced by Figure 28.17. Note that the performance of the STBC-CM schemes communicating over single-path channels, is similar to that of the corresponding CM schemes communicating over 2-path channels, owing to the independent/uncorrelated nature of the fading

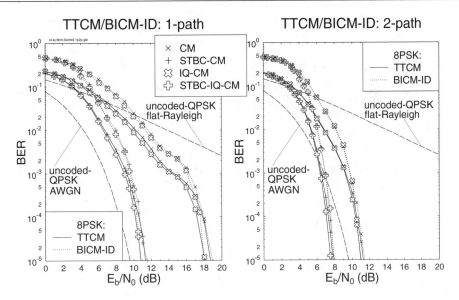

Figure 28.18: BER versus E_b/N_0 performance of the 8PSK-based iterative **TTCM** and **BICM-ID** schemes, when communicating over correlated Rayleigh fading channels exhibiting either one or two resolvable chip-spaced, equal weight paths. The legends are described in Footnote 1. A codeword length of 1000 symbols was used and the performance of the uncoded QPSK scheme communicating over AWGN and flat Rayleigh fading channels is also plotted for benchmarking the CM schemes having an effective throughput of **2 BPS**.

experienced by the signals arriving from two different transmit antennas or received via two independent paths.

Note also that in the context of 16QAM studied in Figure 28.3, there was no IQ-interleaving gain for the 16QAM IQ-BICM scheme, since the I (or Q) interleaver will only further randomise two BICM output bit-sequences in parallel, resulting in two similarly randomised bit-sequences. By contrast, in the context of 8PSK, the bits representing the I and Q components cannot be bit-interleaved, while the 8PSK-based BICM scheme seen in Figure 24.16 employed three independent random parallel bit-interleavers. As a result, the I and Q random interleavers of the 8PSK scheme randomise the quadrature components, rather than randomising the BICM bit-sequences. Therefore this scheme contributes another source of interleaving and hence potentially improves the achievable diversity. As it was evidenced in Figure 28.17, the IQ-interleaved 8PSK-based BICM arrangements benefit from an additional IQ diversity gain in comparison to the BICM arrangements, although this gain is only marginal due to communicating over the slowly fading channels of Table 28.1.

Let us now investigate the achievable performance of the iterative TTCM and BICM-ID arrangements in Figure 28.18. Again, the main contributors of diversity gain in the system are the transmit-diversity and path-diversity, when communicating over the slowly fading channels of Table 28.1. Nonetheless, up to 0.8 dB and 1.0 dB gains may be provided by IQ-TTCM and IQ-BICM-ID, respectively, compared to conventional TTCM and BICM-ID, when communicating over the slowly fading single-path channels of Table 28.1. On the other

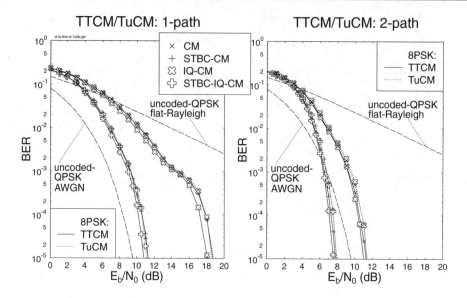

Figure 28.19: BER versus E_b/N_0 performance of the 8PSK-based non-iterative **TCM** and **TuCM** schemes, when communicating over correlated Rayleigh fading channels exhibiting either one or two resolvable chip-spaced, equal weight paths. The legends are described in Footnote 1. A codeword length of 1000 symbols was used and the performance of the uncoded QPSK scheme communicating over AWGN and flat Rayleigh fading channels is also plotted for benchmarking the CM schemes having an effective throughput of **2 BPS**.

hand, the TTCM arrangements perform better than or equal to the BICM-ID arrangements. It is also shown in Figures 28.17 and 28.18 that the TTCM scheme is the best performer in the set of TCM, BICM and BICM-ID schemes.

Below we will continue our discourse by comparing the performance of the TTCM and TuCM arrangements in Figure 28.19. Explicitly, the TuCM scheme employs only one bit-interleaver for supporting the operation of the 3-bit-based 8PSK modulation, unlike the 8PSK BICM scheme of Figure 24.16 where the modulated bits were randomised in parallel using three bit-interleavers. The TuCM arrangement exhibits no further IQ-diversity gain, when employing two extra IQ interleavers, as we can observe in Figure 28.19. More explicitly, as portrayed in Figure 28.19, the TuCM arrangements – both with or without IQ-interleaving – perform worse than or similarly to the TTCM arrangements refraining from using IQ-interleaving. This implies that the IQ-interleaved BICM-ID scheme is expected to outperform the TuCM scheme for transmission over the slowly fading channels of Table 28.1.

Based on Figures 28.17 to 28.19, we concluded that that TTCM is the best performer amongst the various CM schemes. Let us now compare the performance of the 8PSK and 16QAM-based TTCM schemes, when communicating over correlated Rayleigh fading channels exhibiting either one or two resolvable chip-spaced, equal-weight paths using Figure 28.20. Explicitly, a further coding gain of 7 to 8 dB can be attained by the 8PSK-based STBC-(IQ)-TTCM scheme, when communicating over channels exhibiting a single resolvable path, or by (IQ-)TTCM when communicating over channels exhibiting two resolvable

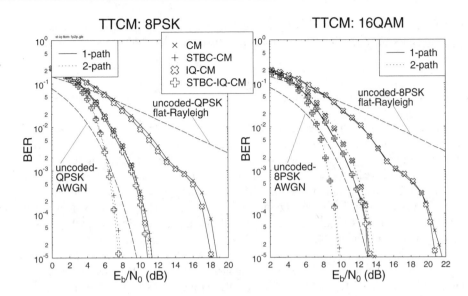

Figure 28.20: BER versus E_b/N_0 performance of the 8PSK and 16QAM-based iterative **TTCM** schemes, when communicating over correlated Rayleigh fading channels exhibiting either one or two resolvable chip-spaced, equal weight paths. The legends are described in Footnote 1. A codeword length of 1000 symbols was used and the performance of the uncoded QPSK(8PSK) scheme communicating over AWGN and flat Rayleigh fading channels is also plotted for benchmarking the CM schemes having an effective throughput of **2(3) BPS**.

paths, with respect to (IQ-)TTCM when communicating over channels exhibiting one resolvable path, as shown at the left of Figure 28.20. This is owing to doubling the diversity order with the aid of transmit-diversity or multipath-diversity. However, as evidenced by Figure 28.20, the achievable further coding gain with respect to the performance of the TTCM schemes exhibiting a diversity order of two was reduced to about 3.5 to 4 dB, when the diversity order was quadrupled. Similar trends can also be observed for the 16QAM-based TTCM arrangements, as shown on the right of Figure 28.20, when providing an effective throughput of 3 BPS.

28.4.4   Conclusion

In conclusion, in these section we have demonstrated that the proposed STBC-IQ-CM assisted DoS-RR scheme is capable of exploiting transmit-diversity, multipath-diversity and IQ-diversity and hence provides significant coding/diversity gains over the identical-throughput uncoded schemes without bandwidth expansion, while supporting multiple downlink CDMA users. Near single-user performance was also achieved without using complex IC schemes. It was shown that the STBC-IQ-CM scheme designed for square QAM modulation in Section 28.2 can also be employed in the context of 8PSK modulation. Furthermore, we have shown in Section 28.3 that the DoS-RR scheme of Figure 28.6 employing QPSK modulation can also be extended to 8PSK and 16QAM for achieving a higher effective throughput without experiencing any BER error floor, although only a simple Rake receiver is employed for supporting multiple downlink CDMA users in dispersive Rayleigh fading channels.

28.5   Summary

In this chapter we have proposed two schemes for achieving diversity gain. Specifically, in Section 28.2 a combination of transmit-diversity and IQ-diversity was exploited for achieving a substantial diversity gain using high-order modulation schemes, while in Section 28.3, we have exploited transmit-diversity and multipath-diversity using a simple Rake receiver benefiting from the proposed double spreading in the context of low-level modulation scheme. Finally, the two designs were amalgamated for yielding a new scheme which is capable of exploiting transmit-diversity, IQ-diversity and multipath-diversity by employing the CM schemes studied in Chapter 24.

Specifically, coding gains of 26.1 dB, 28.2 dB, 26.9 dB and 28.1 dB were achieved over the identical-throughput uncoded 8PSK benchmarker at a BER of 10^{-4} by the 16QAM based STBC-IQ-TCM, STBC-IQ-TTCM, STBC-IQ-BICM and STBC-IQ-BICM-ID schemes, respectively, when communicating over the uncorrelated flat Rayleigh fading channel, as evidenced in Figures 28.3 and 28.4. It was shown in Figures 28.3 and 28.4 that the IQ-TCM and IQ-TTCM schemes were capable of doubling the achievable diversity potential of the conventional symbol-interleaved TCM and TTCM schemes. Explicitly, similar performances were achieved by both IQ-TCM and STBC-TCM, as well as by IQ-TTCM and STBC-TTCM. However, it was shown in Figure 28.5 that the IQ diversity attainable decreased, when the fading channel envelope exhibited correlation.

It was illustrated in Section 28.3.3 that with the advent of the double-spreading based Rake receiver, near-single-user performance can be achieved without employing complex interference cancellation schemes, although this complexity reduction is achieved at the cost of supporting a factor of $\tilde{Q}$ reduced number of users. Specifically, transmit diversity and path diversity constitute two independent sources of diversity gain in the context of the STBC-DoS-RR CDMA downlink. It was shown in Figures 28.12 and 28.15 that the employment of transmit diversity is more advantageous than that of the multipath diversity in practical dispersive Rayleigh fading channels exhibiting an exponentially decaying CIR.

The proposed STBC-IQ-CM assisted DoS-RR scheme was capable of exploiting transmit-diversity, multipath-diversity and IQ-diversity, hence providing a significant coding/diversity gain over the similar throughput uncoded scheme, which is achieved without bandwidth expansion, while supporting a number of CDMA users. It was shown in Section 28.4.3 that the main contributors of diversity gain in the system were the transmit-diversity and path-diversity, when communicating over the slowly fading channels of Table 28.1. Based on Figures 28.17 to 28.19, the STBC-IQ-TTCM assisted DoS-RR scheme was found to be the best performer amongst the various CM schemes studied.

In summary, the coding gain values exhibited by the various coded modulation schemes studied in this chapter were tabulated in Tables 28.2 and 28.3. Specifically, in Tables 28.2 we summarised the performance gain exhibited by the STBC-assisted and/or IQ-interleaved coded modulation schemes of Section 28.2 communicating over non-dispersive uncorrelated as well as correlated correlated Rayleigh fading channels. The coding gains exhibited by the STBC-assisted and/or IQ-interleaved DoS-RR-based coded modulation schemes of Section 28.4 communicating over both 1-path and 2-path Rayleigh fading channels were summarised in Table 28.3.

STBC-IQ-CM		Flat Uncorrelated Rayleigh Fading Channels				Flat Correlated Rayleigh Fading Channels			
		E_b/N_o (dB)		Gain (dB)		E_b/N_o (dB)		Gain (dB)	
Code	Modulation/	BER				BER			
	BPS	10^{-3}	10^{-5}	10^{-3}	10^{-5}	10^{-3}	10^{-5}	10^{-3}	10^{-5}
Uncoded	8PSK/3	26.15	46.00	0.00	0.00	26.15	46.00	0.00	0.00
TCM	16QAM/3	15.54	34.00	10.61	12.00	28.15	45.00	-2.00	1.00
IQ-TCM	16QAM/3	11.32	18.10	14.83	27.90	21.00	29.45	5.15	16.55
STBC-TCM	16QAM/3	10.29	18.10	15.86	27.90	15.47	20.75	10.68	25.25
STBC-IQ-TCM	16QAM/3	8.47	11.65	17.68	34.35	**14.58**	18.11	11.57	27.89
TTCM	16QAM/3	9.88	13.86	16.27	32.14	23.15	32.85	3.00	13.15
IQ-TTCM	16QAM/3	9.31	11.22	16.84	34.78	20.80	28.20	5.35	17.80
STBC-TTCM	16QAM/3	7.66	11.30	18.49	34.70	14.75	18.07	11.40	27.93
STBC-IQ-TTCM	16QAM/3	**7.30**	**8.72**	18.85	37.28	14.65	**17.36**	11.50	28.64
BICM	16QAM/3	11.00	14.00	15.15	32.00	22.90	29.66	3.25	16.34
IQ-BICM	16QAM/3	11.00	13.96	15.15	32.04	22.90	29.45	3.25	16.55
STBC-BICM	16QAM/3	8.66	10.90	17.49	35.10	15.66	18.60	10.49	27.40
STBC-IQ-BICM	16QAM/3	8.66	10.85	17.49	35.15	15.66	18.55	10.49	27.45
BICM-ID	16QAM/3	9.78	14.03	16.37	31.97	21.85	27.45	4.30	18.55
IQ-BICM-ID	16QAM/3	9.36	12.20	16.79	33.80	22.88	28.01	3.27	17.99
STBC-BICM-ID	16QAM/3	7.44	9.65	18.71	36.35	16.00	17.50	10.15	28.50
STBC-IQ-BICM-ID	16QAM/3	7.40	8.95	18.75	37.05	16.12	17.45	10.03	28.55

Table 28.2: Coding gain values of the STBC-assisted and/or IQ-interleaved coded modulation schemes of Section 28.2 when communicating over non-dispersive uncorrelated and correlated Rayleigh fading channels. The coded modulation schemes exhibited a similar decoding complexity in terms of the number of decoding states, which was equal to 64 states. A turbo interleaver block length of 1000 symbols was employed when communicating over uncorrelated Rayleigh fading channels and an interleaver block length of 10000 symbols was employed when communicating over correlated Rayleigh fading channels. The correlated Rayleigh fading channels exhibit a normalised Doppler frequency of 3.25×10^{-5}. The corresponding simulation parameters were summarised in Section 28.2.3 The performance of the best scheme is printed in bold

STBC-IQ-CM DoS-RR CDMA		1-path Correlated Rayleigh Fading Channels				2-path Correlated Rayleigh Fading Channels			
		E_b/N_o (dB)		Gain (dB)		E_b/N_o (dB)		Gain (dB)	
Code	Modulation/	BER				BER			
	BPS	10^{-3}	10^{-5}	10^{-3}	10^{-5}	10^{-3}	10^{-5}	10^{-3}	10^{-5}
Uncoded (1-path)	QPSK/2	24.00	43.50	0.00	0.00	24.00	43.50	0.00	0.00
TCM	8PSK/2	15.13	21.49	8.87	22.01	9.67	13.60	14.33	29.90
IQ-TCM	8PSK/2	13.55	19.96	10.45	23.54	9.38	12.79	14.62	30.71
STBC-TCM	8PSK/2	9.54	13.44	14.46	30.06	7.00	9.35	17.00	34.15
STBC-IQ-TCM	8PSK/2	9.12	12.48	14.88	31.02	6.74	8.81	17.26	34.69
TTCM	8PSK/2	14.62	18.70	9.38	24.80	9.25	11.27	14.75	32.23
IQ-TTCM	8PSK/2	14.35	18.06	9.65	25.44	9.18	10.99	14.82	32.51
STBC-TTCM	8PSK/2	9.22	11.31	14.78	32.19	6.51	7.79	17.49	35.71
STBC-IQ-TTCM	8PSK/2	**9.00**	**10.80**	15.00	32.70	**6.40**	**7.60**	17.60	35.90
BICM	8PSK/2	16.28	20.23	7.72	23.27	10.15	12.81	13.85	30.69
IQ-BICM	8PSK/2	16.28	19.75	7.72	23.75	10.15	12.95	13.85	30.55
STBC-BICM	8PSK/2	9.96	12.96	14.04	30.54	7.53	9.82	16.47	33.68
STBC-IQ-BICM	8PSK/2	9.96	12.61	14.04	30.89	7.46	9.71	16.54	33.79
BICM-ID	8PSK/2	16.29	18.94	7.71	24.56	9.72	11.42	14.28	32.08
IQ-BICM-ID	8PSK/2	16.12	18.00	7.88	25.50	9.77	10.95	14.23	32.55
STBC-BICM-ID	8PSK/2	9.72	11.50	14.28	32.00	6.98	7.95	17.02	35.55
STBC-IQ-BICM-ID	8PSK/2	9.57	**10.80**	14.43	32.70	6.95	7.90	17.05	35.60
TuCM	8PSK/2	14.70	18.60	9.30	24.90	11.39	11.39	12.61	32.11
IQ-TuCM	8PSK/2	14.70	18.75	9.30	24.75	9.55	11.35	14.45	32.15
STBC-TuCM	8PSK/2	9.28	11.22	14.72	32.28	6.75	7.94	17.25	35.56
STBC-IQ-TuCM	8PSK/2	9.28	11.11	14.72	32.39	6.75	7.92	17.25	35.58
Uncoded (1-path)	8PSK/3	26.15	46.00	0.00	0.00	26.15	46.00	0.00	0.00
TTCM	16QAM/3	16.30	21.56	9.85	24.44	11.62	13.77	14.53	32.23
IQ-TTCM	16QAM/3	16.30	28.88	9.85	17.12	11.62	13.32	14.53	32.68
STBC-TTCM	16QAM/3	11.00	13.24	15.15	32.76	8.52	9.86	17.63	36.14
STBC-IQ-TTCM	16QAM/3	11.00	13.10	15.15	32.90	8.52	9.69	17.63	36.31

Table 28.3: Coding gain values of the STBC-assisted and/or IQ-interleaved DoS-RR-based coded modulation schemes of Section 28.4 when communicating over both 1-path and 2-path Rayleigh fading channels. The coded modulation schemes exhibited a similar decoding complexity in terms of the number of decoding states, which was equal to 64 states. An interleaver block length of 1000 symbols was employed and the corresponding simulation parameters were summarised in Section 28.3.2.3. The performance of the best scheme is printed in bold

Chapter 29

Comparative Study of Various Coded Modulation Schemes

29.1 Comparative Study

In this chapter a brief comparative study of the main findings of our coded modulation related investigations will be presented. These discussions will be followed by a range of ideas on a number of possible future research topics and directions.

Part IV of the book investigated the application of coded modulation schemes for transmission over wireless fading channels. Specifically, four coded modulation schemes, namely TCM, TTCM, BICM and BICM-ID, were evaluated in the context of various transceivers.

In Chapter 24, we commenced our discussions by outlining the principles of TCM, TTCM, BICM and BICM-ID. The conceptual differences amongst these coded modulation schemes were outlined in terms of their coding structure, signal labelling philosophy, interleaver type and decoding philosophy. The TCM and TTCM schemes employed a symbol-based interleaver, while the BICM and BICM-ID schemes employed several parallel bit-based interleavers. As for symbol labelling, the BICM scheme employed Gray-coded signal labelling, while the remaining three schemes employed SP-based signal labelling. More specifically, the conventional TCM scheme is designed for maximising the minimum Euclidean distance between the unprotected bits of the constellation. However, this design criterion is only suitable for AWGN channels. By contrast, maximising the free Hamming distance is the main design criterion for fading channels. Therefore, the TCM schemes designed for fading channels avoid having parallel trellis transitions, in order to maximise the associated free Hamming distance between the codewords. By contrast, the conventional BICM scheme is designed for communicating over fading channels, where the bit-based free Hamming distance is maximised. However, the associated minimum Euclidean distance of the various bits of the phasor constellation is considerably lower than that of the TCM scheme. On the other hand, the BICM-ID scheme was designed with a good performance over both AWGN and fading channels. Specifically, the corresponding minimum Euclidean distance of the various bits of the phasor constellation is further increased with the advent of iterative demodulation and BICM decoding, without sacrificing the high free Hamming distance between the codewords obtained by the non-iterative BICM scheme. Another coded modulation scheme designed for attaining a good performance over both AWGN and fading channels is the TTCM scheme, where the turbo coding principle is invoked by employing TCM schemes as the constituent codes. Explicitly, the free Hamming distance between the codewords is maximised with the advent of the turbo interleaver and two constituent encoders. The symbol-based MAP algorithm was also detailed in this chapter, where a MAP algorithm operating in the logarithmic domain was invoked by the coded modulation schemes throughout Part IV of the book.

In Chapter 25, a comparative study of the TCM, TTCM, BICM and BICM-ID schemes was carried out, while communicating over both AWGN and uncorrelated narrowband Rayleigh fading channels. Explicitly, it was shown in Sections 25.2.2.1 and 25.2.2.2 that at a given complexity TCM performed better than BICM in AWGN channels, but worse than BICM in uncorrelated narrowband Rayleigh fading channels, in the context of 8PSK, 16QAM and 64QAM. However, BICM-ID outperformed both TCM and BICM for transmissions over both AWGN and uncorrelated narrowband Rayleigh fading channels at the same decoding complexity. Finally, TTCM showed a superior performance in comparison to the other three coded modulation schemes studied. When employing 64QAM, the trends changed. Specifically, TCM and TTCM were outperformed by BICM and BICM-ID in the context of 64QAM scheme, when communicating over uncorrelated narrowband Rayleigh fading channels. This was deemed to be a consequence of the presence of several uncoded information bits in the TCM/TTCM coded 64QAM symbol. When comparing the achievable coding gain plotted against the interleaver length in Section 25.2.2.3, TTCM was the best performer for a variety of interleaver lengths, while the performance of BICM-ID was better than TCM and BICM only when a long interleaver length was employed. In Section 25.4, multi-carrier OFDM was integrated with the various coded modulation schemes, where we assumed that the individual OFDM subcarriers experienced narrowband fading, when communicating over the wideband COST207 HT Rayleigh fading channels. The TTCM scheme was found to constitute the best compromise, followed by BICM-ID, BICM and TCM in the context of the OFDM scheme. In summary, the TTCM-assisted scheme was found to represent the most attractive trade-off in terms of its achievable performance and complexity, closely followed by the BICM-ID-assisted scheme, which was our conclusion both in the context of the conventional single-carrier system of Section 25.2 as well as the multi-carrier OFDM system of Section 25.4, in the non-dispersive propagation environments. Explicitly, the coding gain values of the coded modulation schemes studied in this chapter were summarised in Tables 25.4 and 25.5.

Based on Tables 25.4 and 25.5, the BPS throughput versus E_b/N_0 performance of the various coded modulation schemes communicating over AWGN channels as well as communicating over OFDM environment was plotted in Figure 29.1. As we can observe in Figure 29.1(a), the performance curve of the TTCM scheme employing a code memory of three and four iterations is about 2 to 3 dB away from the Shannon limit when targeting a BER of 10^{-5} in the context of AWGN channels. When aiming for BER = 10^{-3}, these curves are expected to be closer to the Shannon limit, as we can infer from Table 25.4. Figure 29.1(b) portrays the performance of the OFDM-based coded modulation schemes of Section 25.4 when communicating over the COST207 HT Rayleigh fading channel. When the number of subcarriers is sufficiently high, each OFDM subcarrier experiences narrowband, non-dispersive

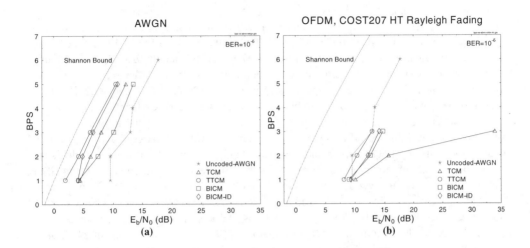

Figure 29.1: The BPS throughput versus E_b/N_0 performance of the coded modulation schemes of Chapter 24 when communicating over (a) AWGN channels and (b) OFDM environment. The corresponding results were extracted from Tables 25.4 and 25.5, respectively, at BER $= 10^{-5}$. The corresponding simulation parameters were summarised in Section 25.2.2 and Table 25.3, respectively.

channel conditions. Therefore, the performance trend of the OFDM-based coded modulation schemes communicating over dispersive Rayleigh fading channels is similar to that of the conventional coded modulation schemes communicating over flat uncorrelated Rayleigh fading channels, as we can observe in Tables 25.4 and 25.5. Note that the performance of the uncoded schemes communicating over flat Rayleigh fading channels was very poor, as it is shown in Table 25.4. Therefore, the OFDM-based uncoded schemes communicating over the COST207 HT Rayleigh fading channel was also very poor, as shown in Table 25.5. However, the performance can be further improved with the advent of employing Walsh-Hadamard spreading [736]. Specifically, when spreading the information carried by each subchannels to all subcarriers with the aid of orthogonal spreading code, the corruption of a few chips owing to low quality subcarriers is unlikely to result in a corrupted subcarrier symbol. Hence, this spreading operation typically improves the achievable OFDM performance.

In Chapter 26, the performance of the various coded modulation schemes was evaluated for transmission over wideband fading channels, by incorporating single-carrier channel equalised arrangements. Specifically, the DFE-aided adaptive coded modulation schemes were characterised in performance terms, when communicating over the COST207 TU fading channels in Section 26.4, where the transceivers' performance was optimised by employing channel equalisation and adaptive coded modulation jointly. A delay of one transmission burst duration was incurred in the channel quality estimation process. The MSE at the output of the DFE was used as the metric invoked for switching amongst the various coded modulation modes, in order to combat both the signal power fluctuations as well as the effects of the time-variant ISI imposed by the wideband channel. We found in Sections 26.4.4 that the diversity gain obtained by the employment of a longer channel interleaver spanning over four transmission bursts in the adaptive coded modulation was more than sufficient for compensating for the loss of modem mode switching flexibility imposed by the four-burst duration mode-switching mechanism, ultimately providing a better performance in terms of the achievable BER and BPS performance. The TTCM scheme was found to be the best performer followed by BICM-ID, TCM and BICM in the context of the DFE-assisted adaptive coded modulation schemes. Furthermore, in Sections 26.7 and 26.9, an RBF-based turbo equaliser was also incorporated into the coded modulation schemes designed for transmission over wideband fading channels. Specifically, various RBF-TEQ-CM and RBF-I/Q-TEQ-CM schemes were investigated for transmission over wideband Rayleigh fading channels in Sections 26.7 and 26.9, respectively. Our simulation results provided in Sections 26.7.2 and 26.9.2 showed that the BER performance of both the 16QAM-based RBF-TEQ-CM and that of the RBF-I/Q-TEQ-CM schemes communicating over wideband fading channels was only about 2 dB inferior in comparison to that of the identical-throughput uncoded 8PSK scheme communicating over benign AWGN channels. As illustrated in Figure 26.36, a significant complexity reduction was achieved by the RBF-I/Q-TEQ-CM scheme, when compared to the complex-valued RBF-TEQ-CM scheme, while achieving virtually the same performance. In particular, complexity reduction factors of 36 and 9 were attained in terms of the required number of additions/subtractions and multiplications/divisions, respectively, by the RBF-I/Q-TEQ-TTCM scheme, when compared to the RBF-TEQ-TTCM arrangement. As illustrated in Figures 26.27 and 26.35, the best performer was TTCM followed by BICM, BICM-ID and TCM in terms of the achievable BER. However, in terms of the FER attained the best performance was achieved by BICM, followed by TTCM, BICM-ID and TCM in the context of both RBF-TEQ and RBF-I/Q-TEQ. We have also compared the performance of the RBF-TEQ-CM and RBF-I/Q-TEQ-CM schemes to that of the conventional DFE-assisted CM scheme in Sections 26.7.2 and 26.9.2. Explicitly, a significant performance gain was achieved by the RBF-TEQ-CM and RBF-I/Q-TEQ-CM schemes in comparison to the conventional DFE-assisted coded modulation scheme in the context of 16QAM. The coding gain values of the channel equalised coded modulation schemes characterised in this chapter were summarised in Tables 26.3 and 26.4.

Based on Tables 26.3 and 26.4, the BPS throughput versus E_b/N_0 performance of the DFE and RBF-TEQ-assisted coded modulation schemes was illustrated in Figure 29.2 when communicating over dispersive Rayleigh fading channels. Note that the signals at the output of the DFE exhibit Gaussian-like statistics, since the anti-fading effects of the path-diversity were exploited by the channel equaliser. Therefore, the performance of the DFE-assisted uncoded schemes is significantly better than that of the OFDM-assisted uncoded schemes, as shown in Figure 29.2(a) as well as in Tables 25.5 and 26.3. For the sake of a fair comparison, the performance of the OFDM-based TTCM scheme employing a similar set of parameters for communicating over the COST207 TU Rayleigh fading channel was also plotted in Figure 29.2(a). Specifically, the total number of OFDM subcarriers was 1024 (1K mode) and the number of effective subcarriers was 684. As portrayed in Figure 29.2(a), the performance of OFDM-TTCM is about 3 to 4 dB lower than that of the DFE-TTCM scheme in the context of 4QAM, 8PSK and 16QAM. However, the performance curve of OFDM-TTCM-64QAM exhibits a BER error floor, similar to that of TTCM-64QAM, when communicating over flat uncorrelated Rayleigh fading channels, as we have seen in Figure 25.8. Again, this phenomenon is owing to the presence of several uncoded information bits in the TTCM-coded 64QAM symbol. Furthermore, DFE-TTCM benefits from the dispersion-induced multipath diversity provided by the channel while OFDM-TTCM experiences flat fading in each of the subcarriers and hence no multipath diversity gain is achieved. Next, DFE-CM scheme is com-

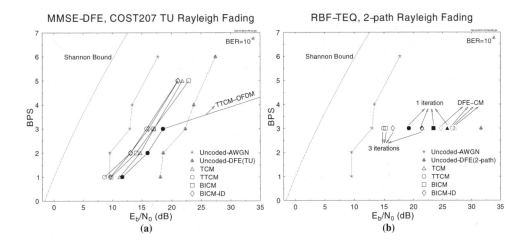

Figure 29.2: The BPS throughput versus E_b/N_0 performance of (a) the DFE-assisted coded modulation schemes of Section 26.4 and of (b) the RBF-TEQ-assisted coded modulation schemes of Section 26.7, when communicating over dispersive Rayleigh fading channels. The corresponding results were extracted from Tables 26.3 and 26.4, respectively, at BER $= 10^{-5}$. The corresponding simulation parameters were summarised in Sections 26.4.2 and 26.7.2, respectively

pared to the RBF-TEQ-CM arrangement in Figure 29.2(b) in the context of 2-path Rayleigh fading channels. Explicitly, the performance of RBF-TEQ-CM is significantly better than that of the DFE-CM scheme, which is achieved at the expense of a higher complexity owing to the employment of the iterative TEQ.

In Chapter 27, the performance of coded modulation assisted DS-CDMA systems was evaluated. Explicitly, coded modulation assisted JD-MMSE-DFE based CDMA schemes were proposed and evaluated in performance terms, when communicating over the COST 207 seven-path Bad Urban channels in Section 27.2. The powerful JD-MMSE-DFE-aided scheme was capable of mitigating the effects of MAI and ISI effectively. We found in Section 27.2.3 that CDMA/TCM was the best candidate, when a short interleaver length was employed, whereas CDMA/TTCM was found to be the best choice for systems that can afford a longer delay owing to utilising a high interleaver length, in the context of CDMA-based slow fading wideband channels. The performance of the JD-MMSE-DFE-aided coded modulation schemes has also been evaluated when communicating over the UTRA wideband vehicular Rayleigh fading channels in Section 27.3. It was observed in Section 27.3.3 that the CDMA/TTCM scheme was the best performer in terms of the attainable FER. Therefore, a JD-MMSE-DFE based adaptive CDMA/TTCM scheme was studied in the context of communicating over the UTRA [370] wideband vehicular Rayleigh fading channels. Explicitly, the output SINR of the JD-MMSE-DFE was utilised as the metric invoked for switching the operational modes of the TTCM modem. The effect of modem mode signalling delay was studied in Figure 27.15, where an idealistic scheme having zero signalling delay was compared to a practical scheme having $10\mu s$ signalling delay. It was found that at a target FER of 5%, an approximately 2.5 dB SNR gain was exhibited by the ideal scheme in the SNR region spanning from 8 dB to 27 dB. In Figure 27.16, the systems that were designed for maintain-

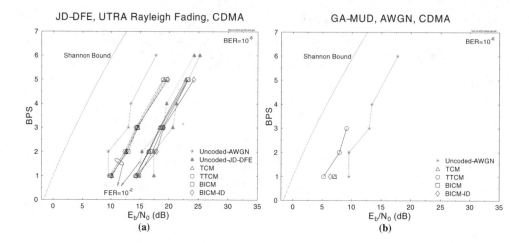

Figure 29.3: The BPS throughput versus E_b/N_0 performance according to Tables 27.6 and 27.7, respectively, at a BER of 10^{-5} for (a) JD-DFE-assisted coded modulation schemes and (b) GA-MUD-assisted coded modulation schemes, when communicating in CDMA environments. The corresponding simulation parameters were summarised in Sections 27.3.2 and 27.4.4, respectively

ing a higher FER were shown to exhibit the benefit of a higher BPS throughput, which was achieved by invoking a different switching threshold set designed for maintaining a higher FER. The proposed adaptive TTCM-assisted CDMA scheme constitutes a promising practical communication system, which is capable of guaranteeing reliable transmission, while providing an effective bitrate ranging from $23.4 \ Kbit/s$ to $117 \ Kbit/s$, in the context of the UTRA environment.

As an alternative design, in Section 27.4 an M-ary coded modulation assisted GA-based MUD scheme was proposed and characterised. It was shown in Figures 27.19 and 27.20 that the uncoded GA-MUD was capable of significantly reducing the computational complexity of the optimum-MUD, but experienced an error floor at high SNRs due to invoking an insufficiently large GA population size and a low number of GA generations. However, this problem was eliminated with the aid of CM schemes. Explicitly, the CM-GA-MUD scheme attained a significant complexity reduction with respect to the optimum-MUD and the TTCM scheme was found to be the best candidate amongst the four coded modulation schemes invokes for assisting the operation of the GA-MUD system. Furthermore, it was shown in Figures 27.23 and 27.24 that the CM-GA-MUD scheme was capable of attaining further coding gains with respect to the uncoded single-user schemes, while maintaining a similar throughput and bandwidth, even when supporting a high number of CDMA users. Tables 27.6 and 27.7 summarise the performance gains attained by the coded modulation schemes of Chapter 24 communicating over the CDMA-based fading environment, when employing a JD-MMSE-DFE and a GA-based MUD, respectively.

Based on Tables 27.6 and 27.7, the BPS throughput versus E_b/N_0 performance both the JD-DFE and GA-MUD-assisted coded modulation schemes communicating in CDMA environment was illustrated in Figure 29.3[1]. As we can observe in Figure 29.3(a), the coding gain

---

[1]Note in Figure 29.3(a) that we have also plotted the achievable BPS performance of the various coded modula-

achieved at a BER of 10^{-5} by the JD-DFE-assisted coded modulation schemes against the JD-DFE-assisted uncoded scheme is modest, when communicating over the UTRA Rayleigh fading channel. This performance limitation is a consequence of the employment of a short channel interleaver having a duration of 240 symbols, when communicating over slow Rayleigh fading channels. However, in terms of the achievable FER the coding gain of the JD-DFE-assisted coded modulation schemes is significantly higher. As we can observe from Figure 29.3(a), the BPS performance of the JD-DFE-assisted TTCM scheme is significantly better than that of the JD-DFE-assisted uncoded scheme, when targeting a FER of 10^{-2}. In Figure 29.3(b), the performance of the GA-MUD-assisted coded modulation schemes of Section 27.4 was portrayed when communicating over the AWGN channels. Specifically, the BER performance curve of GA-MUD-assisted TTCM is about 5 to 6 dB away from the Shannon bound.

Finally, in Chapter 28, we studied a CM-assisted STBC scheme. Explicitly, in Section 28.2 we proposed a set of novel STBC based IQ-interleaved CM schemes designed for transmission over both AWGN and Rayleigh fading channels. Specifically, the associated minimum Hamming distance of the encoded sequences of the CM schemes can be doubled with the advent of employing two separate I and Q interleavers, while the minimum Euclidean distance was not changed. The STBC scheme provided both space and time diversity in conjunction with the employment of multiple transmit antennas. It was observed in Figures 28.3 and 28.4 that coding gains of 26.1 dB, 28.2 dB, 26.9 dB and 28.1 dB were achieved for comparison to the identical-throughput uncoded 8PSK benchmarker at a BER of 10^{-4} by the 16QAM-based STBC-IQ-TCM, STBC-IQ-TTCM, STBC-IQ-BICM and STBC-IQ-BICM-ID schemes, respectively, when communicating over narrowband fading channels. It was found from Figures 28.3 and 28.4 that the IQ-TCM and IQ-TTCM schemes were capable of doubling the achievable diversity potential of the conventional symbol-interleaved TCM and TTCM schemes without requiring STBC. Explicitly, a similar performance was observed for both IQ-TCM and STBC-TCM as well as by IQ-TTCM and STBC-TTCM. However, it was shown in Figure 28.5 that the IQ diversity attainable decreased, when the fading channel exhibited a higher correlation.

On the other hand, a novel STBC-assisted double-spread Rake receiver-based CDMA scheme was proposed and characterised in performance terms for downlink transmissions, when communicating over wideband Rayleigh fading channels, as was shown in Section 28.3. More specifically, it was illustrated in Section 28.3.3 that the double-spreading mechanism was capable of attaining a near-single-user performance with the aid of a low-complexity Rake receiver, i.e. without employing a complex interference cancellation scheme, at the cost of supporting a reduced number of users. Additionally, the double-spread Rake receiver was capable of yielding a performance equivalent to that of maximal ratio combining of order L when transmitting over L-path channels. Specifically, transmit diversity and path diversity constitute two independent sources of diversity gain in the context of the proposed STBC-assisted double-spreading aided Rake receiver-based CDMA scheme. It was shown in Figures 28.12 and 28.15 that second-order transmit diversity was more advantageous than path diversity having the same order in the context of practical wideband Rayleigh fading channels exhibiting an exponentially decaying CIR. This is plausible, since the low-energy CIR taps contribute only modestly towards improving the achievable path diversity gain.

---

tion schemes, when aiming for FER $= 10^{-2}$ instead of BER $= 10^{-5}$. The BER corresponding to FER $= 10^{-2}$ can be inferred from Figures 27.11 and 27.12 for the various coded modulation schemes.

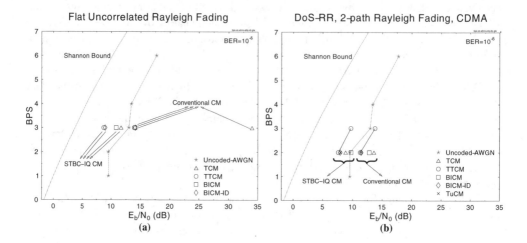

Figure 29.4: The BPS throughput versus E_b/N_0 performance of the STBC-IQ coded modulation scheme of Section 28.2 when communicating over (a) flat uncorrelated Rayleigh fading channels channels and (b) 2-path Rayleigh fading channels. The results were extracted from Tables 28.2 and 28.3, respectively, at BER = 10^{-5}. The corresponding simulation parameters were summarised in Sections 28.2.3 and 28.3.2.3, respectively

When targeting a certain diversity gain, the multipath interference sensitivity of STBC was lower than that of a single-transmitter based scheme, when communicating over wideband channels having a long delay spread, as evidenced by Figures 28.11 and 28.15.

Lastly, the STBC-IQ-aided CM scheme was amalgamated with a DoS-RR scheme in Section 28.4. Explicitly, the proposed STBC-IQ-CM-assisted DoS-RR scheme was capable of exploiting transmit-diversity, multipath-diversity as well as IQ-diversity and hence provided a significant coding/diversity gain over the uncoded scheme, which was achieved without bandwidth expansion, while supporting a number of CDMA users. It was shown in Section 28.4.3 that the main contributors of diversity gain in the system were the transmit-diversity and path-diversity, when communicating over slowly fading channels. We have also invoked TuCM in the proposed scheme, however, the TTCM-assisted scheme once again proved to be the best performer amongst all the schemes studied. Tables 28.2 and 28.3 summarise the diversity/coding gain attained by the coded modulation schemes with the advent of employing IQ-interleaving and multiple transmit antennas of Section 28.2 as well as a double-spread Rake receiver of Section 28.4 in the CDMA environment.

Based on Tables 28.2 and 28.3, the BPS throughput versus E_b/N_0 performance of the STBC-IQ coded modulation schemes of Section 28.2 was plotted in Figure 29.4. These curves were generated for BER = 10^{-5}. As we observed in Figure 29.4(a), the performance of the STBC-IQ-assisted coded modulation schemes of Section 28.2 was found to be better than that of the uncoded schemes communicating over AWGN channels, when communicating over flat uncorrelated Rayleigh fading channels. When comparing the performance of STBC-IQ-CM to that of the conventional CM schemes in Figure 29.4(a), an approximately 5 to 6 dB gain was attained by the TTCM and BICM-ID schemes. In the context of CDMA, the performance of the STBC-IQ-assisted DoS-RR based coded modulation arrangement of Section 28.4 was illustrated in Figure 29.4(b) when communicating over 2-path Rayleigh fading channels. Explicitly, the performance of the STBC-IQ-assisted DoS-RR based coded modulation schemes communicating over 2-path Rayleigh fading channels in a downlink CDMA system was found to be still better than that of the single-user uncoded schemes communicating over AWGN channels.

29.2 Suggestions for Further Research

As demonstrated in Chapter 26, the performance of the RBF-TEQ-aided CM schemes is very promising. Therefore, the investigation of coded modulation techniques employing a reduced-complexity MAP-based TEQ [734] schemes as well as TEQ-based iterative MUD [800] schemes may be beneficial. Another area of promising research is the employment of a TEQ-based scheme in adaptive coded modulation for yielding a better performance in terms of the achievable BER and BPS, which is attained at the cost of a higher complexity. On the other hand, the coded modulation assisted GA-MUD scheme investigated in Section 27.4 may be combined both with space-time coding and multi-carrier CDMA.

As we have proposed in Chapter 28, the IQ-interleaved coded modulation schemes, especially TCM and TTCM, exhibit a better performance than that of the conventional coded modulation schemes with the advent of employing two extra I and Q interleavers, when communicating over fading channels. Therefore, investigations based on IQ-interleaved coded modulation constitute a promising topic for future research. Explicitly, investigations invoking OFDM and multi-carrier CDMA [801] schemes as well as the Multi-Input Multi-Output (MIMO) equaliser [802] in the context of IQ-interleaved coded modulation, are of immediate interest.

Finally, the concept of coded modulation can also be invoked for application in other coding schemes for the sake of achieving a coding gain without bandwidth expansion. Specifically, research into Low Density Parity Check code [803] based block coded modulation is also of interest for transmission over wireless channels. In addition, research on jointly designing IQ-TTCM in conjunction with space-time trellis coding for MIMO schemes and multi-carrier CDMA schemes appears also promising.

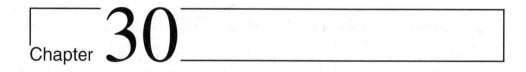

Chapter **30**

QAM-Based Terrestrial and Satellite Video Broadcasting to Mobile and Stationary Receivers

C.S. Lee, L. Hanzo, T. Keller, S. Vlahoyiannatos

30.1 DVB-T for Mobile Receivers[1]

30.1.1 Background and Motivation

Following the standardisation of the pan-European digital video broadcasting (DVB) systems, we have begun to witness the arrival of digital television services in the home. However, for a high proportion of business and leisure travellers it is desirable to have access to DVB services while on the move. Although it is feasible to access these services with the aid of dedicated DVB receivers, these receivers may also find their way into the laptop computers of the near future. These intelligent laptops may also become the portable DVB receivers of wireless in-home networks.

In recent years three DVB standards have emerged in Europe for terrestrial [340], cable-based [804] and satellite-oriented [805] delivery of DVB signals. The more hostile propagation environment of the terrestrial system requires concatenated Reed–Solomon (RS) [370, 806] and rate-compatible punctured convolutional coding (RCPCC) [370, 806] combined with orthogonal frequency division multiplexing (OFDM) based modulation, which was detailed throughout the previous chapters. By contrast, the more benign cable and

---

[1] This section is based on C. S. Lee, T. Keller and L. Hanzo: OFDM-based turbo-coded hierarchical and non-hierarchical terrestrial mobile digital video broadcasting, in *IEEE Trans. Broadcasting*, March 2000, ©2000 IEEE. Personal use of this material is permitted. However, permission to reprint/republish this material for advertising or promotional purposes or for creating new collective works for resale or redistribution to servers or lists, or to refuse any copyrighted component of this work in other works must be obtained from the IEEE.

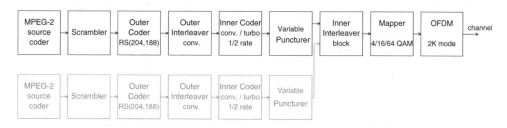

Figure 30.1: Schematic of the DVB terrestrial transmitter functions

satellite based media facilitate the employment of multilevel modems using up to 256-level quadrature amplitude modulation (QAM). These schemes are capable of delivering high definition video at bit rates of up to 20 Mbit/s in stationary broadcast-mode distributive wireless scenarios.

Recently, there have been a range of DVB system performance studies in the literature [807–810]. Against this background, in this chapter we have proposed turbo coding improvements to the terrestrial DVB system [340] and investigated its performance under hostile mobile channel conditions. We have also studied various video bitstream partitioning and channel coding schemes both in the so-called hierarchical and non-hierarchical transceiver modes to be discussed during our further discourse and compared their performance.

The rest of this section is organised as follows. A brief overview of the enhanced turbo-coded and standard DVB terrestrial scheme is presented in Section 30.1.2, while the channel model is described in Section 30.1.3. Following this, the performance of the improved DVB terrestrial system employing the so-called non-hierarchical and hierarchical format [340] is examined in a mobile environment in Sections 30.1.4 and 30.1.6.

30.1.2   DVB Terrestrial Scheme

The block diagram of the DVB terrestrial (DVB-T) transmitter [340] is shown in Figure 30.1, which consists of an MPEG-2 video encoder, channel coding modules and an OFDM modem [811]. The bitstream generated by the MPEG-2 encoder is packetized into frames of 188-byte long. The video data in each packet is then randomized by the scrambler of Figure 30.1. The specific details concerning the scrambler have not been included in this chapter, since these may be obtained from the DVB-T standard [340].

Due to the poor error resilience of the MPEG-2 video codec, powerful concatenated channel coding is employed. The concatenated channel codec of Figure 30.1 comprises a shortened Reed–Solomon (RS) outer code and an inner convolutional encoder. The 188-byte MPEG-2 video packet is extended by the Reed–Solomon encoder [370, 806] with parity information to facilitate error recovery in order to form a 204-byte packet. The Reed–Solomon decoder can then correct up to 8 erroneous bytes for each 204-byte packet. Following this, the RS-coded packet is interleaved by a convolutional interleaver and further protected by a half-rate inner convolutional encoder using a constraint length of 7 [370, 806].

Furthermore, the overall code rate of the concatenated coding scheme can be adapted by variable puncturing which supports code rates of 1/2 (no puncturing) as well as 2/3, 3/4, 5/6 and 7/8. The parameters of the convolutional encoder are summarised in Table 30.1.

Convolutional Coder Parameters	
Code rate	$1/2$
Constraint length,	7
n	2
k	1
Generator polynomials (octal format)	171, 133

Table 30.1: Parameters of the $CC(n, k, K)$ convolutional inner encoder of the DVB-T modem

If only one of the two branches of the transmitter in Figure 30.1 is utilised, the DVB-T modem is said to be operating in its non-hierarchical mode. In this mode, the modem can have a choice of QPSK, 16-QAM or 64-QAM modulation constellations, all of which were considered in detail throughout our previous chapters.

A second video bitstream can also be multiplexed with the first one by the inner inter-leaver, when the DVB modem is in its so-called hierarchical mode [340]. The choice of modulation constellations in this mode is between 16-QAM and 64-QAM. We shall be employing this transmission mode, when the standard data partitioning scheme of [340] is used to split the incoming MPEG-2 video bitstream into two video bit-protection classes with one class having a higher grade of protection or priority than the other one. Recall from Chapter 5 that 16-QAM exhibited two different-integrity subchannels, while 64-QAM had three different-BER subchannels. Hence the higher priority video bits will be mapped to the MSBs of the modulation constellation points and the lower priority video bits to the LSBs of the QAM-constellation. For 16-QAM and 64-QAM, the two MSBs of each 4-bit or 6-bit QAM symbol will contain the more important video data. The lower priority video bits will then be mapped to the lower significance 2 bits and 4 bits of 16-QAM and 64-QAM, respectively, where the BER differences amongst the different subchannels were around a factor 2-3 in Chapter 5 over AWGN channels.

These QPSK, 16-QAM or 64-QAM symbols are then distributed over the OFDM carriers, as discussed in the previous chapters. The parameters of the OFDM system are presented in Table 30.2.

OFDM Parameters	
Total number of subcarriers	2048 (2K mode)
Number of effective subcarriers	1705
OFDM symbol duration T_s	224 μs
Guard interval	$T_s/4 = 56\mu$s
Total symbol duration (inc. guard interval)	280 μs
Consecutive subcarrier spacing $1/T_s$	4464 Hz
DVB channel spacing	7.61 MHz
QPSK and QAM symbol period	7/64 μs

Table 30.2: Parameters of the OFDM module used in the DVB-T modem [340]

Besides implementing the standard DVB-T system as a benchmark, we have improved the system by replacing the convolutional coder with a turbo codec [366,689]. The turbo codec's parameters used in our investigations are displayed in Table 30.3. The block diagram of the turbo encoder is shown in Figure 30.2. The turbo encoder is constructed of two component encoders. Each component encoder is a half-rate convolutional encoder, whose parameters are listed in Table 30.3. The two component encoders are used to encode the same input bits, although the input bits of the second component encoder are interleaved before encoding. The output bits of the two component codes are punctured and multiplexed, in order to form a single output bitstream. The component encoder used here is known as a half-rate recursive systematic convolutional encoder (RSC) [812]. It generates one parity bit and one systematic output bit for every input bit. In order to provide an overall coding rate of $R = 1/2$, half the output bits from the two encoders must be punctured. The puncturing arrangement used in our work is to transmit all the systematic bits from the first encoder and every other parity bit from both encoders [813]. We note here that one iteration of the turbo decoder involves two so-called Logarithmic Maximum A-Posteriori (log-MAP) [637] decoding operations, which we repeated for the 8 iterations. Hence, the total turbo decoding complexity is about 16 times higher than a constraint length $K = 3$ constituent convolutional decoding. Therefore the turbo decoder exhibits a similar complexity to the $K = 7$ convolutional decoder.

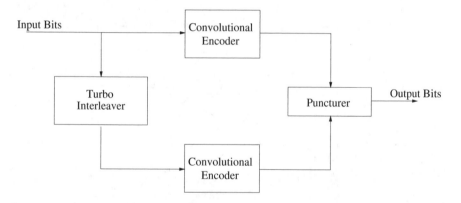

Figure 30.2: Block diagram of turbo encoder

In this section, we have given an overview of the standard and enhanced DVB-T system, which we have used in our experiments. Readers interested in further details of the DVB-T system are referred to the DVB-T standard [340]. The performance of the standard DVB-T system and the turbo-coded system is characterised in Section 30.1.4 and 30.1.6 for non-hierarchical and hierarchical transmissions, respectively. Let us now briefly consider the multipath channel model used in our investigations.

30.1.3 Terrestrial Broadcast Channel Model

The channel model employed in this study was the 12-path COST 207 [676] hilly terrain (HT) type impulse response, with a maximal relative path delay of 19.9 μs. This channel was selected in order to provide a worst-case propagation scenario for the DVB-T system

Turbo Coder Parameters	
Turbo code rate	1/2
Input block length	17 952 bits
Interleaver type	Random
Number of turbo decoder iterations	8
Turbo Encoder Component Code Parameters	
Component code encoder type	Recursive Systematic Convolutional (RSC)
Component code decoder type	log-MAP [637]
Constraint length	3
n	2
k	1
Generator polynomials (octal format)	7, 5

Table 30.3: Parameters of the inner turbo encoder used to replace the DVB-T system's $K = 7$ convolutional encoder of Table 30.1 (RSC: recursive systematic code)

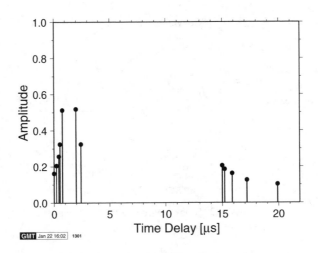

Figure 30.3: COST 207 hilly terrain (HT) type impulse response

employed in our study.

In the system characterised here, we have used a carrier frequency of 500 MHz and a sampling rate of 7/64 μs. Each of the channel paths was faded independently obeying a Rayleigh fading distribution, according to a normalised Doppler frequency of 10^{-5} [370]. This corresponds to a worst-case vehicular velocity of about 200 km/h. The unfaded impulse response is depicted in Figure 30.3. For the sake of completeness we note that the standard COST 207 channel model was defined in order to facilitate the comparison of different GSM implementations [370] under identical conditions. The associated bit rate was

271 kbit/s, while in our investigations the bit rate of DVB-quality transmissions can be as high as 20 Mbit/s, where there is a higher number of resolvable multipath components within the dispersion-range considered. However, the performance of various wireless tranceivers is well understood by the research community over this standard COST 207 channel and hence its employment is beneficial in benchmarking terms. Furthermore, since the OFDM modem has 2048 subcarriers, the subcarrier signalling rate is effectively 2000-times lower than our maximum DVB-rate of 20 Mbit/s, corresponding to 10 kbit/s. At this sub-channel rate, the individual sub-channel can be considered nearly frequency-flat. In summary, in conjunction with the 200 km/h vehicular speed the investigated channel conditions constitute a pessimistic scenario.

Let us now consider the performance of the so-called non-hierarchical DVB-T scheme in the next section. By contrast, the hierarchical system's performance will be characterised in the next but one section. To be more explicit concerning the hierarchical versus non-hierarchical modes suffice to say here that on the basis of the different sensitivity of the transmitted video bits the hierarachical mode allows us to implement unequal error protection schemes, as was detailed for example in [814].

In a low complexity implementation, two protection classes may be envisaged. The higher priority class would contain all the important and error sensitive MPEG2 video synchronisation and header information and some of the more important low frequency variable-length coded Discrete Cosine Transform (DCT) coefficients [815]. The lower priority class would then contain the remaining less important, higher frequency variable-length coded DCT coefficients. Again, the performance of this partitioning process will be characterised in the context of the hierarchical DVB [340] transmission scheme [814] in Section 30.1.6. Let us, however, first consider the architecture of the investigated DVB system in the next section.

30.1.4 Non-Hierarchical OFDM DVB System Performance

In this section we shall elaborate on our findings when the convolutional code used in the standard non-hierarchical DVB scheme [340] is replaced by a turbo code. We will invoke a range of standard-compliant schemes as benchmarks. The 704×576-pixel HDTV-resolution "Football" video sequence was used in our experiments. The MPEG-2 decoder employs a simple error concealment algorithm to fill in missing portions of the reconstructed image in the event of decoding errors. The concealment algorithm will select the specific portion of the previous reconstructed image, which corresponds to the missing portion of the current image, in order to conceal the errors.

In Figure 30.4(a) and (b) the bit error rate (BER) performance of the various modem modes in conjunction with our diverse channel coding schemes is portrayed over stationary, narrowband additive white Gaussian noise channels (AWGN), where the turbo codec exhibits a significantly steeper BER reduction in comparison to the convolutionally coded arrangements.

Specifically, comparing the performance of the various turbo and convolutional codes for QPSK and 64-QAM at a BER of 10^{-4}, the turbo code exhibited an additional coding gain of about 2.24 dB and 3.7 dB respectively, when using half-rate codes in Figure 30.4(a) and (b). Below we will now quantify the associated video performance in terms of the so-called Peak

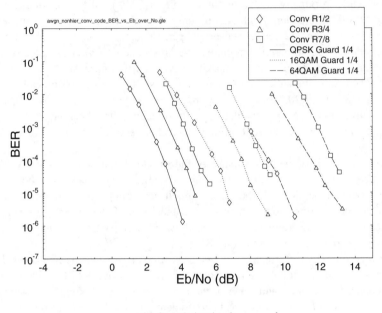

(a) Convolutional code

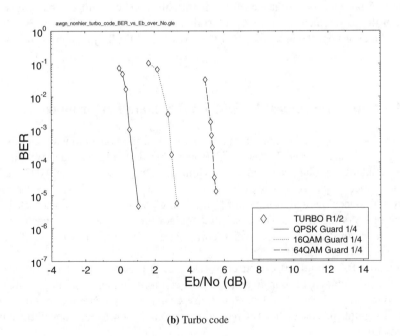

(b) Turbo code

Figure 30.4: BER after (a) convolutional decoding and (b) turbo decoding for the DVB-T scheme over stationary non-dispersive **AWGN** channels for **non-hierarchical transmission** [814], ©IEEE, 2000, Lee, Keller, Hanzo

Signal-to-Noise Ratio (PSNR), which is defined as follows:

$$\text{PSNR} = 10 \, log_{10} \frac{\sum_{n=0}^{N} \sum_{m=0}^{M} 255^2}{\sum_{n=0}^{N} \sum_{m=0}^{M} \Delta^2}, \tag{30.1}$$

where Δ is the difference between the uncoded pixel value and the reconstructed pixel value. The variables M and N refer to the dimension of the image. The maximum possible 8-bit represented pixel luminance value of 255 was used in Equation 30.1 in order to mitigate the PSNR's dependence on the video material used. The average PSNR is then the mean of the PSNR values computed for all the images constituting the video sequence.

The associated PSNR versus channel Signal-to-Noise Ratio (SNR) graphs in Figure 30.7 demonstrate that approximately 2 dB and 3.5 dB lower channel SNRs are required in conjunction with the rate 1/2 turbo codec for QPSK and 64-QAM, respectively, than for convolutional coding, in order to maintain high reconstructed video quality. The term unimpaired as used in Figure 30.7 and Figure 30.8 refers to the condition, where the PSNR of the MPEG-2 decoder's reconstructed image at the receiver is the same as the PSNR of the same image generated by the local decoder of the MPEG-2 video encoder, corresponding to the absence of channel – but not MPEG-2 coding – impairments.

Comparing the BER performance of the 1/2 rate convolutional decoder in Figure 30.5(a) and the so-called log-MAP [637] turbo decoder using eight iterations in Figure 30.5(b) for QPSK modulation over the worst-case fading mobile channel of Figure 30.3, we observe that at a BER of about 10^{-4} the turbo code provided an additional coding gain of 6 dB in comparison to the convolutional code. By contrast, for 64-QAM using similar codes, a 5 dB coding gain was observed at this BER.

Similar observations were also made with respect to the average PSNR versus channel signal–to–noise ratio (SNR) plots of Figure 30.8. For the QPSK modulation mode and a 1/2 coding rate, the turbo code required an approximately 5.5 dB lower channel SNR for maintaining near-unimpaired video quality than the convolutional code.

Comparing Figure 30.5(a) and Figure 30.6(a), we note that the Reed–Solomon decoder becomes effective in lowering the bit error probability of the transmitted data further below the BER threshold of 10^{-4}. From these figures we also observe that the rate-3/4 convolutional code is unsuitable for transmission over the highly dispersive hilly terrain channel used in this experiment, when 64-QAM is employed. When the rate-7/8 convolutional code is used, both the 16-QAM and 64-QAM schemes perform poorly. As for the QPSK modulation scheme, a convolutional code rate as high as 7/8 can still provide a satisfactory performance after Reed–Solomon decoding.

In conclusion, Tables 30.4 and 30.5 summarise the system performance in terms of the channel SNR (CSNR) required for maintaining less than 2 dB PSNR video degradation. It was observed that, at this PSNR degradation, decoding errors were still perceptually unnoticeable to the viewer due to the 30 frame/s refresh rate, although the typical still frame shown in Figure 30.9 in this scenario exhibits some degradation. It is important to underline once again that the $K = 3$ turbo code and the $K = 7$ convolutional code exhibited comparable complexities. The higher performance of the turbo codec facilitates, for example, the employment of turbo-coded 16-QAM at a similar channel SNR, where convolutional-coded QPSK can be invoked. This in turn allows us to double the bit rate within the same bandwidth and hence improve the video quality. In the next section, we first outline the philosophy of the

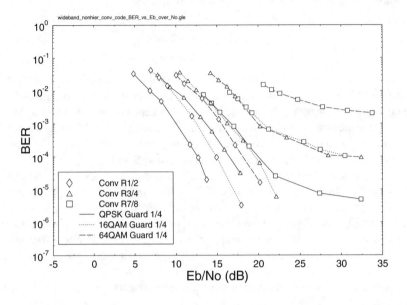

(a) Convolutional code

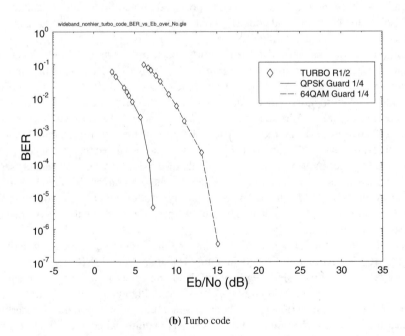

(b) Turbo code

Figure 30.5: BER after (a) convolutional decoding and (b) turbo decoding for the DVB-T scheme over the **wideband fading channel** of Figure 30.3 for **non-hierarchical transmission** [814], ©IEEE, 2000, Lee, Keller, Hanzo

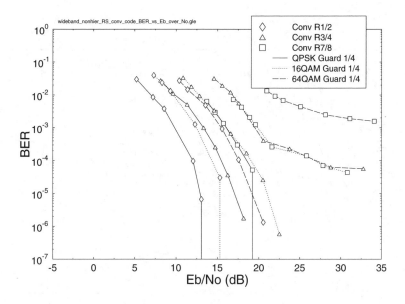

(a) RS and onvolutional code

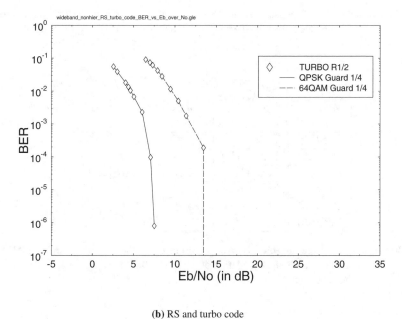

(b) RS and turbo code

Figure 30.6: BER after (a) RS and convolutional decoding and (b) RS and turbo decoding for the DVB-T scheme over the **wideband fading channel** of Figure 30.3 for **non-hierarchical transmission** [814], ©IEEE, 2000, Lee, Keller, Hanzo

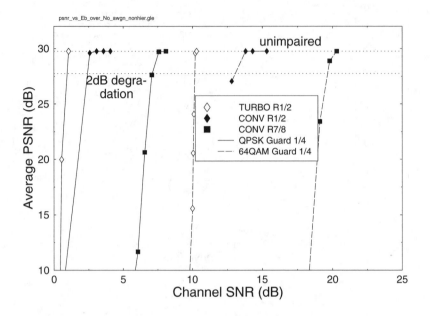

Figure 30.7: Average PSNR versus channel SNR of the DVB scheme [340] over non-dispersive **AWGN** channels for **non-hierarchical transmission** [814], ©IEEE, 2000, Lee, Keller, Hanzo

Mod.	Code	CSNR (dB)	E_b/N_0	BER
QPSK	Turbo (1/2)	1.02	1.02	6×10^{-6}
64-QAM	Turbo (1/2)	9.94	5.17	2×10^{-3}
QPSK	Conv (1/2)	2.16	2.16	1.1×10^{-3}
64-QAM	Conv (1/2)	12.84	8.07	6×10^{-4}
QPSK	Conv (7/8)	6.99	4.56	2×10^{-4}
64-QAM	Conv (7/8)	19.43	12.23	3×10^{-4}

Table 30.4: Summary of the **non-hierarchical** performance results over non-dispersive **AWGN** channels tolerating a PSNR degradation of 2 dB. The BER measure refers to BER after Viterbi or turbo decoding

MPEG2 video data partitioning scheme, assisting in invoking unequal video bit protection. Then in the next but one section we shall present the results of our investigations employing the DVB-T system [340] in a hierarchical transmission scenario.

psnr_vs_Eb_over_No_wideband_nonhier.gle

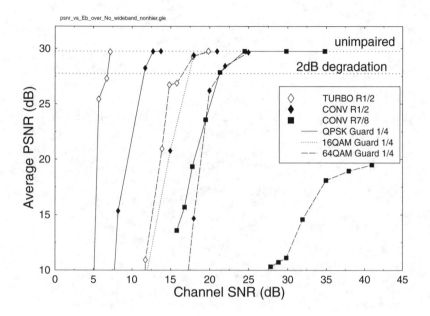

Figure 30.8: Average PSNR versus channel SNR of the DVB scheme [340] over the **wideband fading channel** of Figure 30.3 for **non-hierarchical transmission** [814], ©IEEE, 2000, Lee, Keller, Hanzo

Mod.	Code	CSNR (dB)	E_b/N_0	BER
QPSK	Turbo (1/2)	6.63	6.63	2.5×10^{-4}
64-QAM	Turbo (1/2)	15.82	11.05	2×10^{-3}
QPSK	Conv (1/2)	10.82	10.82	6×10^{-4}
64-QAM	Conv (1/2)	20.92	16.15	7×10^{-4}
QPSK	Conv (7/8)	20.92	18.49	3×10^{-4}

Table 30.5: Summary of the **non-hierchical** performance results over **wideband fading channels** tolerating a PSNR degradation of 2 dB. The BER measure refers to BER after Viterbi or turbo decoding

30.1.5 Video Data Partitioning Scheme

The philosophy of the hierarchical transmission mode is that the natural BER difference of a factor 2 to 3 found in Chapter 5 for the 16-QAM modem is exploited for providing unequal error protection for the FEC-coded video streams B3 and B4 of Figure 30.10. If the sensitivity of the video bits requires a different BER ratio between the B3 and B4 streams, the choice of the FEC codes protecting the video streams B1 and B2 of Figure 30.10 can be appropriately

Figure 30.9: Frame 79 of the "Football" sequence, which illustrates the visual effects of minor decoding errors at a BER of 2.10^{-4} after convolutional decoding. The PSNR degradation observed is approximately 2 dB. The sequence was coded using a rate 7/8 convolutional code and transmitted emplying QPSK modulation [814], ©IEEE, 2000, Lee, Keller, Hanzo

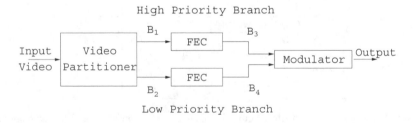

Figure 30.10: Video partitioning for the DVB-T system operating in hierarchical mode

adjusted to equal out or to augment these differences. The FEC module of Figure 30.10 represents the concatenated coding system, consisting of a Reed–Solomon codec [370] and a convolutional codec [370]. The modulator can invoke both 16-QAM and 64-QAM. We shall now use an example to illustrate the choice of the various partitioning ratios summarised in Table 30.6.

We shall assume that 64-QAM is selected and the high and low priority video partitions employ rate 1/2 and 3/4 convolutional codes, respectively. This scenario is portrayed in the

Modulation	Conv. Code Rate (High Priority)	Conv. Code Rate (Low Priority)	Ratio (High - B1 : Low - B2)
16-QAM	1/2	1/2	1 : 1
	1/2	2/3	3 : 4
	1/2	3/4	2 : 3
	1/2	5/6	3 : 5
	1/2	7/8	4 : 7
	2/3	1/2	4 : 3
64-QAM	1/2	1/2	1 : 2
	1/2	2/3	3 : 8
	1/2	**3/4**	**1 : 3**
	1/2	5/6	3 : 10
	1/2	7/8	2 : 7
	2/3	1/2	2 : 3

Table 30.6: The bit rate partitioning ratios based on the modulation mode and code rates selected for the DVB-T hierarchical mode. The line in bold corresponds to our worked example

third line of the 64-QAM section of Table 30.6. We do not have to take the Reed–Solomon code rate into account, since both partitions invoke the same Reed–Solomon codec. Based on these facts and upon referring to Figure 30.10, the input bit rates B_3 and B_4 of the modulator must be in the ratio 1:2, since the two MSBs of the 64-QAM constellation are assigned to the high priority video partition and the remaining four bits to the low priority video partition.

At the same time, the ratio of B_3 to B_4 is related to the ratio of B_1 to B_2 with the FEC redundancy taken into account, requiring

$$
\begin{aligned}
\frac{B_3}{B_4} &= \frac{2 \times B_1}{\frac{4}{3} \times B_2} \stackrel{64-QAM}{=} \frac{1}{2} \\
&= \frac{3}{2} \cdot \frac{B_1}{B_2} \stackrel{64-QAM}{=} \frac{1}{2} \\
\frac{B_1}{B_2} &= \frac{1}{2} \times \frac{2}{3} \\
&= \frac{1}{3}.
\end{aligned}
\tag{30.2}
$$

If, for example, the input video bit rate to the data partitioner module is 1 Mbit/s, the output bit rate of the high and low priority partition would be $B_1 = 250$ kbit/s and $B_2 = 750$ kbit/s, respectively, according to the ratio indicated by Equation 30.2.

Let us consider here the 16-QAM modem and refer to the equally split rate 1/2 convolutional coded high and low priority scenario as Scheme 1. Furthermore, the 16-QAM rate 1/3 convolutional coded high priority data and rate 2/3 convolutional coded low priority data based scenario is referred to here as Scheme 2. Finally, the 16-QAM rate 2/3 convolutional coded high priority data and rate 1/3 coded low priority data based partitioning scheme is termed as Scheme 3. We then programmed the partitioning scheme of Figure 30.10 for maintaining the required splitting ratio B_1/B_2, as seen in Table 30.7. This was achieved by continuously adjusting the so-called Priority Break Point (PBP) governing the corresponding split of the bitstream into the priority or protection classes using the corresponding al-

	Modulation	Conv. Code Rate (High Prior. – B1)	Conv. Code Rate (Low Prior. – B2)	Ratio (High : Low) (B1 : B2)
Scheme 1	16-QAM	1/2	1/2	1 : 1
Scheme 2	16-QAM	1/3	2/3	1 : 2
Scheme 3	16-QAM	2/3	1/3	2 : 1

Table 30.7: Summary of the three schemes employed in our investigations into the performance of the data partitioning scheme. The FEC-coded high priority video bitstream B3, as shown in Figure 30.10, was mapped to the high priority 16-QAM sub-channel, while the low priority B4-stream to the low priority 16-QAM sub-channel

forithms of [814]. The 704 × 576-pixel "Football" high definition television (HDTV) video sequence was used in these investigations. Having outlined the operation of the data partitioning scheme, let us now characterise its performance in the context of the overall system in the next section.

30.1.6   Hierarchical OFDM DVB System Performance

Below we will invoke the above DVB-T hierarchical scheme in a mobile broadcasting scenario. We shall also demonstrate the improvements which turbo codes offer when replacing the convolutional code in the standard scheme. Hence the convolutional codec in both the high and low priority partitions was replaced by the turbo codec. We have also investigated replacing only the high priority convolutional codec with the turbo codec, pairing the 1/2 rate turbo codec in the high priority partition with the convolutional codec in the low priority partition. Again, the "Football" sequence was used in these experiments. Partitioning was carried out using the schematic of Figure 30.10 as well as the detailed partitioning algorithms of [814]. The FEC-coded high priority video partition B3 of Figure 30.10 was mapped to the higher integrity 16-QAM or 64-QAM sub-channel. By contrast, the low priority partition B4 of Figure 30.10 was directed to the lower integrity 16-QAM or 64-QAM sub-channel. Finally, no specific mapping was required for QPSK, since it exhibits no sub-channels. We note, however, that further design trade-offs become feasible, when reversing the above mapping rules. Indeed, this is necessary for example in conjunction with Scheme 2 of Table 30.7, since the high number of bits in the low priority partition render it more sensitive than the high priority partition. Again, according to Chapter 5 the 16-QAM sub-channels exhibit a factor of 2 to 3 BER difference, which improves the robustness of the 'reverse-mapped' Scheme 2 of Table 30.7.

Referring to Figure 30.11 and comparing the performance of the 1/2 rate convolutional code and turbo code at a BER of 10^{-4} for the low priority partition, the turbo code, employing 8 iterations, exhibited a coding gain of about 6.6 dB and 5.97 dB for 16-QAM and 64-QAM, respectively. When the number of turbo decoding iterations was reduced to 4, the coding gains offered by the turbo code over that of the convolutional code were 6.23 dB and 5.7 dB for 16-QAM and 64-QAM respectively. We observed that by reducing the number of iterations to 4 halved the associated complexity, but the turbo code exhibited a coding loss of only about 0.37 dB and 0.27 dB in comparison to the 8-iteration scenario for 16-QAM and

64-QAM, respectively. Hence the computational complexity of the turbo codec can be halved by sacrificing only a small amount of coding gain. The substantial coding gain provided by turbo coding is also reflected in the PSNR versus channel SNR graphs of Figure 30.13. In order to achieve transmission with very low probability of error, Figure 30.13 demonstrated that approximately 5.72 dB and 4.56 dB higher channel SNRs are required by the standard scheme compared to the scheme employing turbo coding, when using 4 iterations in both partitions. We have only shown the performance of turbo coding for the low priority partition in Figures 30.11(b) and 30.12(b), since the turbo or convolutional-coded high priority partition was received with very low probability of error after Reed–Solomon decoding for the range of SNRs used.

We also observed that the rate 3/4 and rate 7/8 convolutional codes in the low priority partition were unable to provide sufficient protection to the transmitted video bits, as it becomes evident from Figures 30.11(a) and 30.12(a). In these high coding rate scenarios, due to the presence of residual errors even after the Reed–Solomon decoder, the decoded video exhibited some decoding errors, which is evidenced by the flattening of the PSNR versus channel SNR curves in Figure 30.13(a), before reaching the error-free PSNR.

A specific problem when using the data partitioning scheme in conjunction with the high priority partition being protected by the rate 1/2 code and the low priority partition protected by the rate 3/4 and rate 7/8 codes was that when the low priority partition data was corrupted, the error-free high priority data available was insufficient for concealing the errors, as discussed in Section 30.1.5. We have also experimented with the combination of rate 2/3 convolutional coding and rate 1/2 convolutional coding, in order to protect the high and low priority data, respectively. From Figure 30.13(a) we observed that the performance of this 2/3 rate and 1/2 rate combination approached that of the rate 1/2 convolutional code in both partitions. This was expected, since now more data can be inserted into the high priority partition. Hence, in the event of decoding errors in the low priority data, we had more error-free high priority data that could be used to reconstruct the received image.

Our last combination investigated involved using rate 1/2 turbo coding and convolutional coding for the high and low priority partitions, respectively. Comparing Figures 30.14 and 30.13(a), the channel SNRs required for achieving unimpaired video transmission were similar in both cases. This was expected, since the turbo-convolutional combination's video performance is dependent on the convolutional code's performance in the low priority partition.

Finally, comparing Figures 30.13 and 30.8, we found that the unimpaired PSNR condition was achieved at similar channel SNRs for the hierarchical and non-hierarchical schemes, suggesting that the data partitioning scheme had not provided sufficient performance improvements in the context of the mobile DVB scheme to justify its added complexity. Again, this was a consequence of relegating a high proportion of video bits to the low integrity partition.

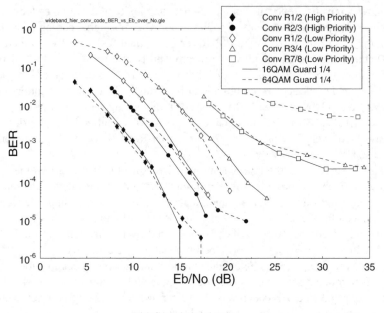

(a) Convolutional code

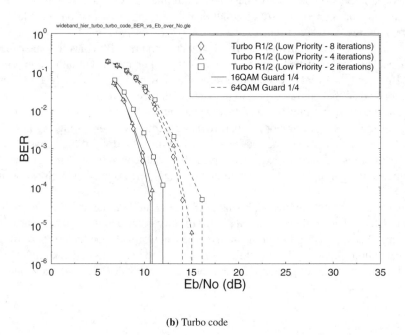

(b) Turbo code

Figure 30.11: BER after (a) convolutional decoding and (b) turbo decoding for the **DVB-T hierarchical scheme** over the **wideband fading channel** of Figure 30.3 using the schematic of Figure 30.10. In (b), the BER of the turbo or convolutional-coded high priority partition is not shown [814], ©IEEE, 2000, Lee, Keller, Hanzo

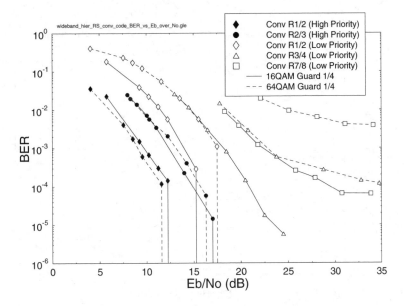

(a) RS and convolutional code

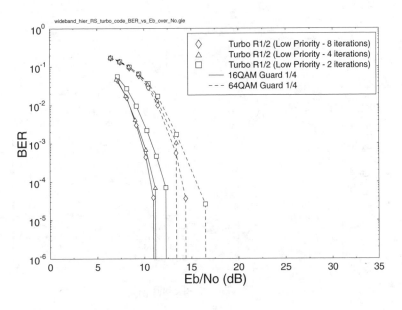

(b) RS and turbo code

Figure 30.12: BER after (a) RS and convolutional decoding and (b) RS and turbo decoding for the **DVB-T hierarchical scheme** over the **wideband fading channel** of Figure 30.3 using the schematic of Figure 30.10. In (b), the BER of the turbo or convolutional-coded high priority partition is not shown [814], ©IEEE, 2000, Lee, Keller, Hanzo

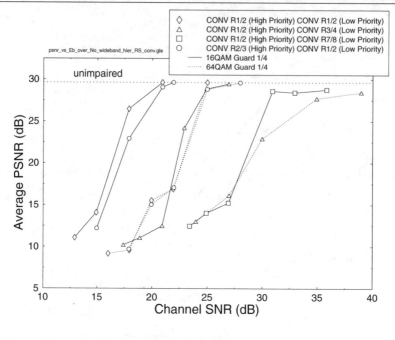

(a) Standard scheme

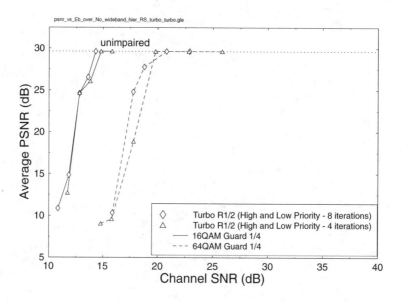

(b) Turbo code in both partitions

Figure 30.13: Average PSNR versus channel SNR for (a) standard DVB scheme [340] and (b) system with turbo coding employed in both partitions, for transmission over the **wideband fading channel** of Figure 30.3 for **hierarchical transmission** using the schematic of Figure 30.10 [814], ©IEEE, 2000, Lee, Keller, Hanzo

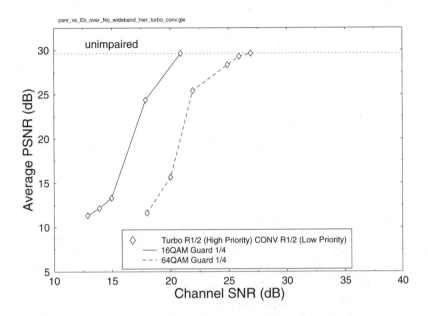

psnr_vs_Eb_over_No_wideband_hier_turbo_conv.gle

Figure 30.14: Average PSNR versus channel SNR of the DVB scheme, employing turbo coding in the high priority partition and convolutional coding in the low priority partition, over the **wideband fading channel** of Figure 30.3 for **hierarchical transmission** using the schematic of Figure 30.10 [814], ©IEEE, 2000, Lee, Keller, Hanzo

30.2 Satellite-based Video Broadcasting [2]

30.2.1 Background and Motivation

As mentioned at the beginning of this chapter, in recent years three harmonised DVB standards have emerged in Europe for terrestrial [340], cable-based [804] and satellite-oriented [805] delivery of DVB signals. The dispersive wireless propagation environment of the terrestrial system characterised in the previous section requires concatenated Reed–Solomon (RS) [370, 806] and rate compatible punctured convolutional coding (RCPCC) [370, 806] combined with OFDM based modulation. The satellite-based system employs the same concatenated channel coding arrangement, as the terrestrial scheme, while the cable-based system refrains from using concatenated channel coding, opting for RS coding only. The performance of both of the latter schemes can be improved upon invoking blind-equalised multi-level modems, although the associated mild dispersion or linear distortion does not

[2]This section is based on C.S. Lee, S. Vlahoyiannatos and L. Hanzo: Satellite based turbo-coded, blind-equalised 4-QAM and 16-QAM digital video broadcasting, *IEEE Trans. Broadcasting*, March 2000 [491] ©2000 IEEE. Personal use of this material is permitted. However, permission to reprint/republish this material for advertising or promotional purposes or for creating new collective works for resale or redistribution to servers or lists, or to refuse any copyrighted component of this work in other works must be obtained from the IEEE.

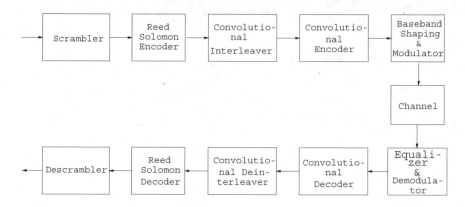

Figure 30.15: Schematic of the DVB satellite system

necessarily require channel equalisation. However, since we propose invoking turbo-coded 4-bit/symbol 16-QAM in order to improve the system's performance at the cost of increased complexity, in this chapter we additionally invoked blind channel equalisers. This is further justified by the associated high video transmission rates, where the channel's dispersion may become a more dominant performance limitation.

Finally, as in the DVB-T system of the previous section, the video codec used in all three systems is the Motion Pictures Expert Group's MPEG-2 codec. These standardisation activities were followed by a variety of system performance studies in the open literature [816–819]. Against this background, similarly to the enhanced DVB-T system of the previous section, we suggested turbo coding improvements to the satellite-based DVB system [805]. Here we present the results of our performance studies for the proposed system under dispersive channel conditions in conjunction with a variety of blind channel equalisation algorithms. The transmitted power requirements of the standard system employing convolutional codecs can be reduced upon invoking more complex, but more powerful turbo codecs. Alternatively, the standard quaternary or 2-bit/symbol system's bit error rate (BER) versus signal-to-noise ratio (SNR) performance can almost be matched by a turbo-coded 4-bit/symbol 16-QAM scheme, while doubling the achievable bit rate within the same bandwidth and hence improving the associated video quality. This is achieved at the cost of an increased system complexity.

The remainder of this section is organised as follows. A succinct overview of the turbo-coded and standard DVB satellite scheme is presented in Section 30.2.2, while our channel model is described in Section 30.2.3. A brief summary of the blind equaliser algorithms employed is presented in Section 30.2.4. Following this the performance of the improved DVB satellite system is examined for transmission over a dispersive two-path channel in Section 30.2.5, before our conclusions and future work areas are presented in Section 30.3.

30.2.2   DVB Satellite Scheme

The block diagram of the DVB satellite (DVB-S) system [805] is shown in Figure 30.15, which is composed of a MPEG-2 video encoder (not shown in the diagram), channel coding modules and a quadrature phase shift keying (QPSK) modem. The bitstream generated by the MPEG-2 encoder is packetized into frames of 188-byte long. The video data in each packet is then randomized by the scrambler. Similarly to the DVB-T scheme, the details concerning the scrambler have not been included in this chapter, since these may be found in the DVB-S standard [805].

Due to the poor error resilience of the MPEG-2 video codec, powerful concatenated channel coding is employed. The concatenated channel codec comprises a shortened Reed–Solomon (RS) outer code and an inner convolutional encoder. The 188-byte MPEG-2 video packet is extended by the Reed–Solomon encoder [370, 806] with parity information to facilitate error recovery to form a 204-byte packet. The Reed–Solomon decoder can then correct up to 8 erroneous bytes for each 204-byte packet. Following this, the RS-coded packet is interleaved by a convolutional interleaver and further protected by a half-rate inner convolutional encoder with a constraint length of 7 [370, 806].

Furthermore, the overall code rate of the concatenated coding scheme can be adapted by variable puncturing, not shown in the figure, which supports code rates of $1/2$ (no puncturing) as well as $2/3$, $3/4$, $5/6$ and $7/8$. The parameters of the convolutional encoder are summarised in Table 30.8.

Convolutional Coder Parameters	
Code rate	$1/2$
Constraint length	7
n	2
k	1
Generator polynomials (octal format)	171, 133

Table 30.8: Parameters of the $CC(n, k, K)$ convolutional inner encoder of the DVB-S modem

In addition to implementing the standard DVB-S system as a benchmark, we have improved the system's performance with the aid of a turbo codec [366, 689]. The block diagram of the turbo encoder is shown in Figure 30.16. The turbo encoder is constructed of two component encoders. Each component encoder is a half-rate convolutional encoder, whose parameters are listed in Table 30.9. The two component encoders are used to encode the same input bits, although the input bits of the second component encoder are interleaved before encoding. The output bits of the two component codes are punctured and multiplexed, in order to form a single output bitstream. The component encoder used here is known as a half-rate recursive systematic convolutional encoder (RSC) [812]. It generates one parity bit and one systematic output bit for every input bit. In order to provide an overall coding rate of one half, half the output bits from the two encoders must be punctured. The puncturing arrangement used in our work is to transmit all the systematic bits from the first encoder and every other parity bit from both encoders.

Readers interested in further details of the DVB-S system are referred to the DVB-S standard [805]. The performance of the standard DVB-S system and the performance of the turbo-coded system are characterised in Section 30.2.5. Let us now briefly consider the multipath channel model used in our investigations.

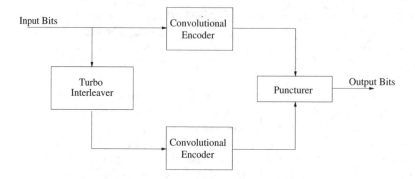

Figure 30.16: Block diagram of turbo encoder

Turbo Coder Parameters	
Turbo code rate	1/2
Input block length	17 952 bits
Interleaver type	Random
Number of turbo decoder iterations	8
Turbo Encoder Component Code Parameters	
Component code encoder type	Convolutional Encoder (RSC)
Component code decoder type	log-MAP [637]
Constraint length	3
n	2
k	1
Generator polynomials (octal format)	7, 5

Table 30.9: Parameters of the inner turbo encoder used to replace the DVB-S system's convolutional coder (RSC: recursive systematic code)

30.2.3   Satellite Channel Model

The DVB-S system was designed to operate in the 12 GHz frequency band (K-band). Within this frequency band, tropospheric effects such as the transformation of electromagnetic energy into thermal energy due to induction of currents in rain and ice crystals lead to signal attenuations [820, 821]. In the past 20 years, various researchers have concentrated their efforts on attempting to model the satellite channel, typically within a land mobile satellite channel scenario. However, the majority of the work conducted, for example, by Vogel and his colleagues [822–825] concentrated on modelling the statistical properties of a narrowband satellite channel in lower frequency bands, such as the 870 MHz UHF band and the 1.5 GHz L-band.

However, our high bit rate DVB satellite system requires a high bandwidth, hence the video bitstream is exposed to dispersive wideband propagation conditions. Recently, Saunders *et. al.* [826, 827] have proposed the employment of multipath channel models to study

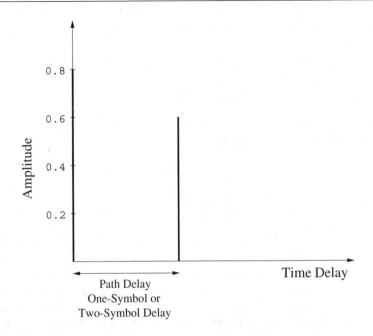

Figure 30.17: Two-path satellite channel model with either a one-symbol or two-symbol delay

the satellite channel, although their study was concentrated on the L-band and S-band only.

Due to the dearth of reported work on wideband satellite channel modelling in the K-band, we have adopted a simpler approach. The channel model employed in this study was the two-path (nT)-symbol spaced impulse response, where T is the symbol duration. In our studies we used $n = 1$ and $n = 2$ (Figure 30.17). This corresponds to a stationary dispersive transmission channel. Our channel model assumed that the receiver had a direct line-of-sight with the satellite as well as a second path caused by a single reflector probably from a nearby building or due to ground reflection. The ground reflection may be strong, if the satellite receiver dish is only tilted at a low angle.

Based on these channel models, we studied the ability of a range of blind equaliser algorithms to converge under various path delay conditions. In the next section we provide a brief overview of the various blind equalisers employed in our experiments, noting that the readers who are mainly interested in the system's performance may proceed directly to our performance analysis section, Section 30.2.5. Recall, furthermore, that an in-depth discourse on blind channel equalisation was provided in Chapter 7 and the next section offers only a rudimentary overview of the topic.

30.2.4 Blind Equalisers

In this section the blind equalisers used in the system are presented. The following blind equalisers have been studied:

1) The modified constant modulus algorithm (MCMA) [457].

2) The Benveniste–Goursat algorithm (B-G) [51].

3) The stop-and-go algorithm (S-a-G) [57].

4) The per-survivor processing (PSP) algorithm [60].

We will now briefly introduce these algorithms.
    First we define the variables that we will use:

$$\mathbf{y}(n) = [y(n + N_1), \ldots, y(0), \ldots, y(n - N_2)]^T \tag{30.3}$$

$$\mathbf{c}^{(n)} = [c_{-N_1}, \ldots, c_0, \ldots, c_{N_2}]^T \tag{30.4}$$

$$z(n) = \left(\mathbf{c}^{(n)}\right)^T \mathbf{y}(n) = \mathbf{y}^T(n)\mathbf{c}^{(n)} \tag{30.5}$$

where $\mathbf{y}(n)$ is the received symbol vector at time n, containing the $N_1 + N_2 + 1$ most recent received symbols, while N_1, N_2 are the number of equaliser feedback and feedforward taps, respectively. Furthermore, $\mathbf{c}^{(n)}$ is the equaliser tap vector, consisting of the equaliser tap values and $z(n)$ is the equalised symbol at time n, given by the convolution of the received signal with the equaliser's impulse response, while $()^T$ stands for matrix transpose. Note that the variables of Equations 30.3–30.5 assume complex values, when multilevel modulation is employed.

The **modified CMA** (MCMA) is an improved version of Godard's well-known *constant modulus algorithm (CMA)* [52]. The philosophy of the CMA is based on forcing the magnitude of the equalised signal to a constant value. In mathematical terms the CMA is based on minimising the cost function:

$$J^{(CMA)} = E\left[\left(|z(n)|^2 - R_2\right)^2\right], \tag{30.6}$$

where R_2 is a suitably chosen constant and $E[]$ stands for the expectation. Similarly to the CMA, the MCMA, which was proposed by Wesolowsky [457], forces the real and imaginary parts of the complex signal to the constant values of $R_{2,R}$ and $R_{2,I}$ respectively, according to the equaliser tap update equation of [457]:

$$\mathbf{c}^{(n+1)} = \mathbf{c}^{(n)} - \lambda \cdot \mathbf{y}^*(n) \cdot \{Re[z(n)] \cdot ((Re[z(n)])^2 - R_{2,R})$$
$$+ jIm[z(n)] \cdot ((Im[z(n)])^2 - R_{2,I})\}, \tag{30.7}$$

where λ is the step size parameter and the $R_{2,R}$, $R_{2,I}$ constant parameters of the algorithm are defined as:

$$R_{2,R} = \frac{E\left[(Re[a(n)])^4\right]}{E\left[(Re[a(n)])^2\right]} \tag{30.8}$$

$$R_{2,I} = \frac{E\left[(Im[a(n)])^4\right]}{E\left[(Im[a(n)])^2\right]}, \tag{30.9}$$

where $a(n)$ is the transmitted signal at time instant n.

The **Benveniste–Goursat** (B-G) algorithm [51] is an amalgam of the Sato's algorithm [49] and the decision directed (DD) algorithm of Chapter 7. Strictly speaking, the decision directed algorithm is not a blind equalisation technique, since its convergence is highly dependent on the channel.

This algorithm estimates the error between the equalised signal and the detected signal as:

$$\epsilon^{DD}(n) = z(n) - \hat{z}(n), \tag{30.10}$$

where $\hat{z}(n)$ is the receiver's estimate of the transmitted signal at time instant n. Similarly to the DD algorithm's error term, the Sato error [49] is defined as:

$$\epsilon^{Sato}(n) = z(n) - \gamma \cdot csgn(z(n)), \tag{30.11}$$

where γ is a constant parameter of the Sato algorithm, defined as:

$$\gamma = \frac{E\left[(Re\,[a(n)])^2\right]}{E\left[|Re\,[a(n)]|\right]} = \frac{E\left[(Im\,[a(n)])^2\right]}{E\left[|Im\,[a(n)]|\right]} \tag{30.12}$$

and $csgn(x) = sign(Re\{x\}) + jsign(Im\{x\})$ is the complex sign function. The B-G algorithm combines the above two error terms into one:

$$\epsilon^G(n) = k_1 \cdot \epsilon^{DD}(n) + k_2 \cdot |\epsilon^{DD}(n)| \cdot \epsilon^{Sato}(n), \tag{30.13}$$

where the two error terms are suitably weighted by the constant parameters k_1 and k_2 in Equation (30.13). Using this error term, the B-G equaliser updates the equaliser coefficients according to the following equaliser tap update equations [51]:

$$\mathbf{c}^{(n+1)} = \mathbf{c}^{(n)} - \lambda \cdot \mathbf{y}^*(n) \cdot \epsilon^G(n). \tag{30.14}$$

In our investigations, the weights were chosen as $k_1 = 1$, $k_2 = 5$, so that the Sato error was weighted more heavily, than the DD error.

Again, the **stop-and-go** (S-a-G) algorithm [57] is a variant of the decision directed algorithm of Chapter 7, where at each equaliser coefficient adjustment iteration, the update is enabled or disabled depending on whether the update is likely to be correct. The update equations of this algorithm are given by [57]

$$\mathbf{c}^{(n+1)} = \mathbf{c}^{(n)} - \lambda \cdot \mathbf{y}^*(n) \cdot [f_{n,R} \cdot Re\{\epsilon^{DD}(n)\} + jf_{n,I} \cdot Im\{\epsilon^{DD}(n)\}] \tag{30.15}$$

where $*$ stands for the complex conjugate, $\epsilon^{DD}(n)$ is the decision directed error as in Equation (30.10) and the binary functions $f_{n,R}$, $f_{n,I}$ enable or disable the update of the equaliser according to the following rule. If the sign of the Sato error (the real or the imaginary part independently) is the same as the sign of the decision directed error, then the update takes place, otherwise it does not.

In mathematical terms, this is equivalent to [57]:

$$f_{n,R} = \begin{cases} 1 & \text{if } sgn(Re[\epsilon^{DD}(n)]) = sgn(Re[\epsilon^{Sato}(n)]) \\ 0 & \text{if } sgn(Re[\epsilon^{DD}(n)] \neq sgn(Re[\epsilon^{Sato}(n)]) \end{cases} \tag{30.16}$$

$$f_{n,I} = \begin{cases} 1 & \text{if } sgn(Im\{\epsilon^{DD}(n)\}) = sgn(Im\{\epsilon^{Sato}(n)\}) \\ 0 & \text{if } sgn(Im\{\epsilon^{DD}(n)\}) \neq sgn(Im\{\epsilon^{Sato}(n)\}). \end{cases} \tag{30.17}$$

For a blind equaliser, this condition provides us with a measure of the probability of the coefficient update being correct.

The **PSP algorithm** [60] is based on employing convolutional coding and hence it is a trellis-based sequence estimation technique, in which the channel is not known *a priori*. Hence an iterative channel estimation technique is employed, in order to estimate the channel jointly with the modulation symbol. In this sense, an initial channel is used and the estimate is updated at each new symbol's arrival.

In our case the update was based on the *least mean squares (LMS)* estimates, according to the following channel-tap update equations [60]:

$$\hat{\mathbf{h}}^{(n+1)} = \hat{\mathbf{h}}^{(n)} + \lambda \cdot \hat{\mathbf{a}}^*(n) \cdot \left(y(n) - \hat{\mathbf{a}}^T(n)\hat{\mathbf{h}}^{(n)} \right), \tag{30.18}$$

where $\hat{\mathbf{h}}^{(n)} = (\hat{h}_{-L_1}^{(n)}, \dots, \hat{h}_o^{(n)}, \dots, \hat{h}_{L_2}^{(n)})^T$ is the estimated (for one surviving path) channel tap vector at time instant n, $\hat{\mathbf{a}}(n) = (\hat{a}(n+L_1), \dots, \hat{a}(0), \dots, \hat{a}(n-L_2))^T$ is the associated estimated transmitted symbol vector and $y(n)$ is the actually received symbol at time instant n.

Each of the surviving paths in the trellis carries not only its own signal estimation, but also its own channel estimation. Moreover, convolutional decoding can take place jointly with this channel and data estimation procedure, leading to improved bit error rate (BER) performance. Summary of the various equalisers' parameters is given in Table 30.10.

	Step size λ	No. of equaliser taps	Initial tap vector
Benveniste–Goursat	5×10^{-4}	10	$(1.2, 0, \cdots, 0)$
Modified CMA	5×10^{-4}	10	$(1.2, 0, \cdots, 0)$
Stop-and-go	5×10^{-4}	10	$(1.2, 0, \cdots, 0)$
PSP (1 sym delay)	10^{-2}	2	$(1.2, 0)$
PSP (2 sym delay)	10^{-2}	3	$(1.2, 0, 0)$

Table 30.10: Summary of the equaliser parameters used in the simulations.   The tap vector $(1.2, 0, \cdots, 0)$ indicates that the first equaliser coefficient is initialised to the value 1.2, while the others are initialised to 0

Having described the components of our enhanced DVB-S system, let us now consider the overall system's performance.

30.2.5   Performance of the DVB Satellite System

In this section, the performance of the DVB-S system was evaluated by means of simulations. Two modulation types were used, namely the standard QPSK and the enhanced 16-QAM schemes. The channel model of Figure 30.17 was employed. The first channel model had a one-symbol second-path delay, while in the second one the path-delay corresponded to the period of two symbols. The average BER versus SNR per bit performance was evaluated after the equalisation and demodulation process, as well as after Viterbi [370] or turbo decoding [689]. The SNR per bit or E_b/N_o is defined as follows:

$$\text{SNR per bit} = 10 \, log_{10} \frac{\bar{S}}{\bar{N}} + \delta \tag{30.19}$$

where $\bar{S}$ is the average received signal power, $\bar{N}$ is the average received noise power and δ, which is dependent on the type of modulation scheme used and channel code rate (R), is defined as follows:

$$\delta = 10 \, log_{10} \frac{1}{\text{R} \times \text{Bits per modulation symbol}}. \tag{30.20}$$

Our results are further divided into two subsections for ease of discussion. First, we will present the system performance over the one-symbol delay two-path channel in Section 30.2.5.1. Next, the system performance over the two-symbol delay two-path channel is presented in Section 30.2.5.2. Finally, a summary of the system performance is provided in Section 30.2.5.3.

30.2.5.1 Transmission over the Symbol-Spaced Two-Path Channel

The linear equalisers' performance was quantified and compared using QPSK modulation over the one-symbol delay two-path channel model of Figure 30.18. Since all the equalisers have similar BER performance, only the modified CMA results are shown in the figure.

The equalised performance over the one symbol-spaced channel was inferior to that over the non-dispersive AWGN channel. However, as expected, it was better than without any equalisation. Another observation for Figure 30.18 was that the different punctured channel coding rates appeared to give slightly different bit error rates after equalisation. This was because the linear blind equalisers required uncorrelated input bits in order to converge. However, the input bits were not entirely random when convolutional coding was used. The consequences of violating the zero-correlation constraint are not generally known. Nevertheless, two potential problems were apparent. Firstly, the equaliser may diverge from the desired equaliser equilibrium [54]. Secondly, the performance of the equaliser is expected to degrade, owing to the violation of the randomness requirement, which is imposed on the input bits in order to ensure that the blind equalisers will converge.

Since the channel used in our investigations was static, the first problem was not encountered. Instead, the second problem was what we actually observed. Figure 30.19 quantifies the equalisers' performance degradation due to convolutional coding. We can observe a 0.1 dB SNR degradation, when the convolutional codec creates correlation among the bits for this specific case.

The average BER curves after Viterbi or turbo decoding are shown in Figure 30.20(a). In this figure, the average BER over the non-dispersive AWGN channel after turbo decoding constitutes the best case performance, while the average BER of the one-symbol delay two-path MCMA-equalised rate 7/8 convolutionally coded scenario exhibits the worst case performance. Again, in this figure only the modified CMA was featured for simplicity. The performance of the remaining equalisers was characterised in Figure 30.20(b). Clearly, the

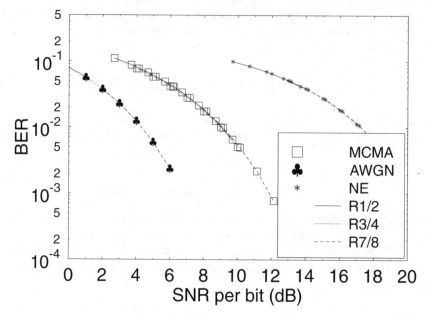

(a) After equalisation and demodulation

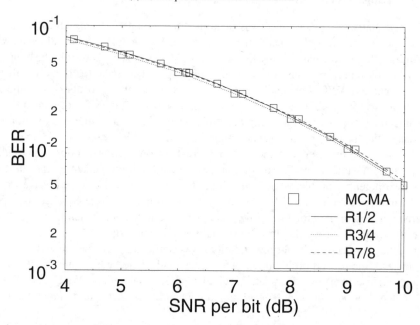

(b) Same as (a) but enlarged in order to show performance difference of the blind equaliser, when different convolutional code rates are used.

Figure 30.18: Average BER versus SNR per bit performance after equalisation and demodulation employing **QPSK** modulation and **one-symbol delay channel** (**NE** = non-equalised; **MCMA** = modified constant modulus algorithm) [491], ©IEEE, 2000, Lee, Vlahoyiannatos, Hanzo

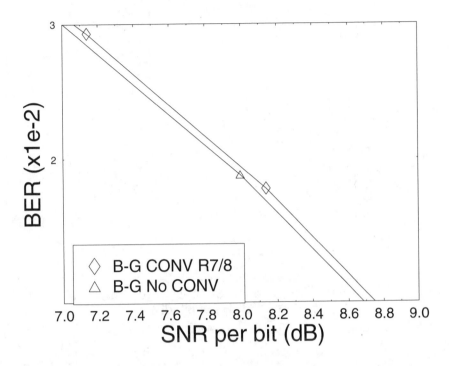

Figure 30.19: Average BER versus SNR per bit performance after equalisation and demodulation employing **QPSK** modulation and the **one-symbol delay two-path channel** of Figure 30.17, for the Benveniste–Goursat algorithm, where the input bits are random (No CONV) or correlated (CONV 7/8) as a result of convolutional coding having a coding rate of 7/8 [491], ©IEEE, 2000, Lee, Vlahoyiannatos, Hanzo

performance of all the linear equalisers investigated was similar.

It is observed in Figure 30.20(a) that the combination of the modified CMA blind equaliser with turbo decoding exhibited the best SNR performance over the one-symbol delay two-path channel. The only comparable alternative was the PSP algorithm. Although the performance of the PSP algorithm was better at low SNRs, the associated curves cross over and the PSP algorithm's performance became inferior below the average BER of 10^{-3}. Although not shown in Figure 30.20, the Reed–Solomon decoder, which was concatenated to either the convolutional or the turbo decoder, became effective, when the average BER of its input was below approximately 10^{-4}. In this case, the PSP algorithm performed by at least 1 dB worse in the area of interest, which is at an average BER of 10^{-4}.

A final observation in the context of Figure 30.20(a) is that when convolutional decoding was used, the associated E_b/N_o performance of the rate 1/2 convolutional coded scheme appeared slightly inferior to that of the rate 3/4 and the rate 7/8 scenarios beyond certain E_b/N_o values. This was deemed to be a consequence of the fact that the 1/2 rate encoder introduced more correlation into the bitstream than its higher rate counterparts and this degraded the

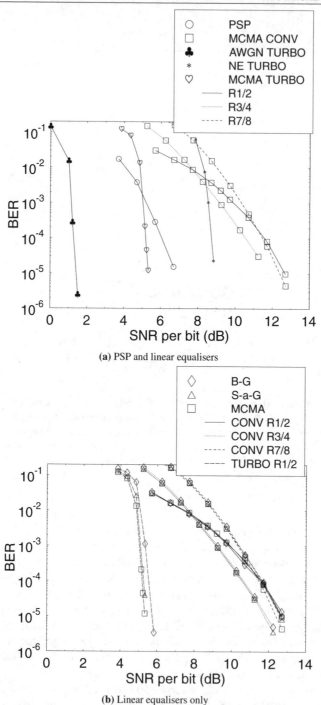

(a) PSP and linear equalisers

(b) Linear equalisers only

Figure 30.20: Average BER versus SNR per bit performance after convolutional or turbo decoding for **QPSK** modulation and **one-symbol delay channel** (**NE** = non-equalised; **B-G** = Benveniste–Goursat; **S-a-G** = stop-and-go; **MCMA** = modified constant modulus algorithm; **PSP** = per-survivor processing) [491], ©IEEE, 2000, Lee, Vlahoyiannatos, Hanzo

performance of the blind channel equalisers, which performed best, when fed with random bits.

Having considered the QPSK case, we shall now concentrate on the enhanced system, which employed 16-QAM under the same channel and equaliser conditions. In Figure 30.21 and Figure 30.22, the performance of the DVB system employing 16-QAM is presented. Again, for simplicity, only the modified CMA results are given. In this case the ranking order of the different coding rates followed our expectations more closely in the sense that the lowest coding rate of 1/2 was the best performer, followed by rate 3/4 codec, in turn followed by the least powerful rate 7/8 codec.

The stop-and-go algorithm has been excluded from these results, since it does not converge for high SNR values. This happens because the equalisation procedure is only activated when there is a high probability of correct decision directed equaliser update. In our case, the equaliser is initialised far from its convergence point and hence the decision directed updates are unlikely to be correct. In the absence of noise this leads to the update algorithm being permanently de-activated. If noise is present though, then some random perturbations from the point of the equaliser's initialization can activate the stop-and-go algorithm and can lead to convergence. We made this observation at medium SNR values in our simulation study. For high SNR values, the algorithm did not converge.

It is also interesting to compare the performance of the system for the QPSK and 16-QAM schemes. When the one-symbol delay two-path channel model of Figure 30.17 was considered, the system was capable of supporting the use of 16-QAM with the provision of an additional SNR per bit of approximately 4–5 dB. This observation was made by comparing the performance of the DVB system when employing the modified CMA and the half-rate convolutional or turbo code in Figure 30.20 and Figure 30.22 at a BER of 10^{-4}. Although the original DVB satellite system only employs QPSK modulation, our simulations had shown that 16-QAM can be employed equally well for the range of blind equalisers that we have used in our work. This allowed us to double the video bit rate and hence to substantially improve the video quality. *The comparison of Figures 30.20 and 30.22 also reveals that the extra SNR requirement of approximately 4–5 dB of 16-QAM over QPSK can be eliminated by employing turbo coding at the cost of a higher implementational complexity. This allowed us to accommodate a doubled bit rate within a given bandwidth, which improved the video quality.*

30.2.5.2   Transmission over the Two-Symbol-Delay Two-Path Channel

In Figure 30.23 (only for the Benveniste-Goursat algorithm for simplicity) and Figure 30.24 the corresponding BER results for the two-symbol delay two-path channel of Figure 30.17 are given for QPSK. The associated trends are similar to those in Figures 30.18 and 30.20, although some differences can be observed, as listed below:

- The "cross-over point" is where the performance of the PSP algorithm becomes inferior to the performance of the modified CMA in conjunction with turbo decoding. This point is now at 10^{-4}, which is in the range where the RS decoder guarantees an extremely low probability of error.

- The rate 1/2 convolutional decoding is now the best performer, when convolutional decoding is concerned, while the rate 3/4 scheme exhibits the worst performance.

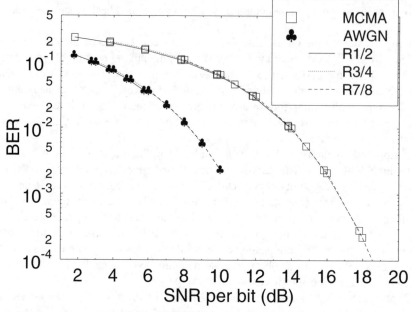

(a) After equalisation and demodulation

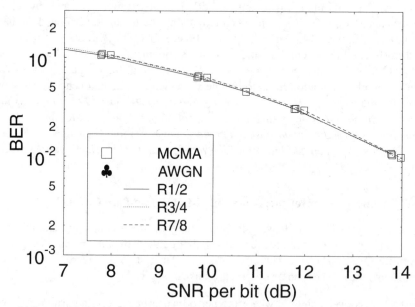

(b) Same as (a) but enlarged in order to show performance difference of the blind equaliser, when different convolutional code rates are used.

Figure 30.21: Average BER versus SNR per bit after equalisation and demodulation for **16-QAM** over the **one-symbol delay two-path channel** of Figure 30.17 (**MCMA** = modified constant modulus algorithm) [491], ©IEEE, 2000, Lee, Vlahoyiannatos, Hanzo

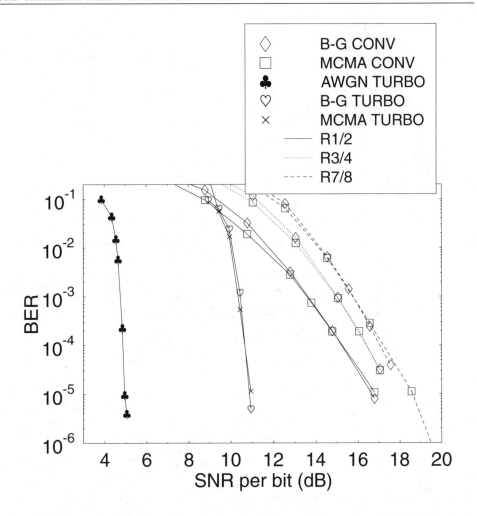

Figure 30.22: Average BER versus SNR per bit after Viterbi or turbo decoding for **16-QAM** over the
one-symbol delay two-path channel of Figure 30.17 (**B-G** = Benveniste–Goursat; **S-a-
G** = stop-and-go; **MCMA** = modified constant modulus algorithm; **PSP** = per-survivor
processing) [491], ©IEEE, 2000, Lee, Vlahoyiannatos, Hanzo

Finally, in Figure 30.25, the associated 16-QAM results are presented. Notice that the
stop-and-go algorithm was again excluded from the results. Furthermore, we observe a high
performance difference between the Benveniste–Goursat algorithm and the modified CMA.
In the previous cases we did not observe such a significant difference. The difference in this
case is that the channel exhibits an increased delay spread. This illustrated the capability of
the equalisers to cope with more widespread multipaths, while keeping the equaliser order
constant at 10. The Benveniste–Goursat equaliser was more efficient than the modified CMA
in this case.

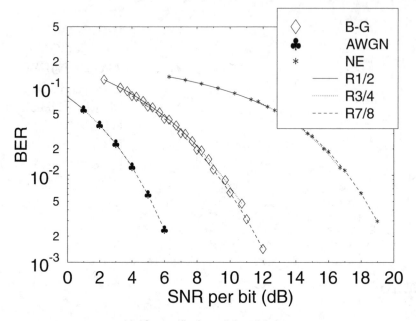

(a) After equalisation and demodulation

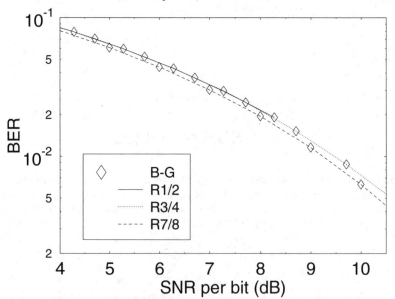

(b) Same as (a) but enlarged in order to show performance difference of the blind equaliser, when different convolutional code rates are used.

Figure 30.23: Average BER versus SNR per bit performance after equalisation and demodulation for **QPSK** modulation over the **two-symbol delay two-path channel** of Figure 30.17 (**B-G** = Benveniste–Goursat) [491], ©IEEE, 2000, Lee, Vlahoyiannatos, Hanzo

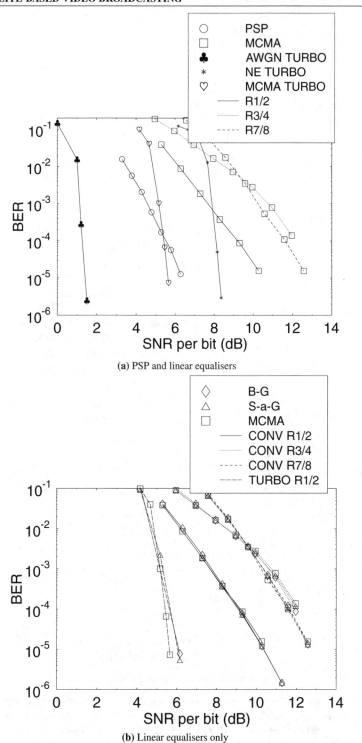

(a) PSP and linear equalisers

(b) Linear equalisers only

Figure 30.24: Average BER versus SNR per bit performance after convolutional or turbo decoding for **QPSK** modulation over the **two-symbol delay two-path channel** of Figure 30.17 (**B-G** = Benveniste–Goursat; **S-a-G** = stop-and-go; **MCMA** = modified constant modulus algorithm; **PSP** = per-survivor processing) [491], ©IEEE, 2000, Lee, Vlahoyiannatos, Hanzo

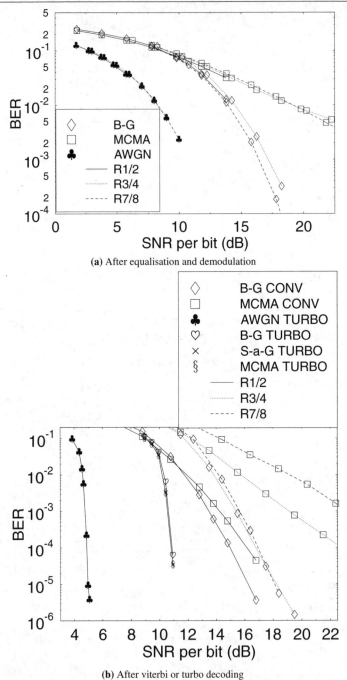

(a) After equalisation and demodulation

(b) After viterbi or turbo decoding

Figure 30.25: Average BER versus SNR per bit performance (a) after equalisation and demodulation and (b) after Viterbi or turbo decoding for *16-QAM* over the **two-symbol delay two-path channel** of Figure 30.17 (**B-G** = Benveniste–Goursat; **S-a-G** = stop-and-go; **MCMA** = modified constant modulus algorithm; **PSP** = per-survivor processing) [491], ©IEEE, 2000, Lee, Vlahoyiannatos, Hanzo

It is interesting to note that in this case the performance of the different coding rates was again in the expected order: the rate 1/2 scheme is the best, followed by the rate 3/4 scheme and then the rate 7/8 scheme.

If we compare the performance of the system employing QPSK and 16-QAM over the two-symbol delay two-path channel of Figure 30.17, we again observe that **16-QAM can be incorporated into the DVB system, if an extra 5 dB of SNR per bit is affordable in power budget terms. Here, only the B-G algorithm is worthwhile considering out of the three linear equalisers of Table 30.10.** This observation was made by comparing the performance of the DVB system when employing the Benveniste–Goursat equaliser and the half-rate convolutional coder in Figure 30.24 and Figure 30.25.

30.2.5.3 Performance Summary of the DVB-S System

Table 30.11 provides an approximation of the convergence speed of each blind equalisation algorithm of Table 30.10. It is clear that PSP exhibited the fastest convergence, followed by the Benveniste–Goursat algorithm. In our simulations the convergence was quantified by observing the slope of the BER curve, and finding when this curve reached the associated residual BER, implying that the BER has reached its steady–state value. Figure 30.26 gives an illustrative example of the equaliser's convergence for 16-QAM. The stop-and-go algorithm converges significantly slower than the other algorithms, which can also be seen from Table 30.11. This happens because, during start-up, the algorithm is deactivated most of the time, an effect which becomes more severe with increasing QAM order.

	B-G	MCMA	S-a-G	PSP
QPSK 1 sym	73	161	143	0.139
QPSK 2 sym	73	143	77	0.139
16-QAM 1 sym	411	645	1393	
16-QAM 2 sym	359	411	1320	

Table 30.11: Equaliser convergence speed (in miliseconds) measured in the simulations, given as an estimate of time required for convergence when 1/2 rate puncturing is used (x sym = x-symbol delay two-path channel and x can take either the value 1 or 2)

Figure 30.27 portrays the corresponding reconstructed video quality in terms of the average PSNR versus channel SNR (CSNR) for the one-symbol delay and two-symbol delay two-path channel models of Figure 30.17. Table 30.12 and Table 30.13 provide a summary of the DVB satellite system's performance tolerating a PSNR degradation of 2 dB, which was deemed to be nearly imperceptible in terms of subjective video degradations. The average BER values quoted in the tables refer to the average BER achieved after Viterbi or turbo decoding. The channel SNR is quoted in association with the 2 dB average video PSNR degradation, since the viewer will begin to perceive video degradations due to erroneous decoding of the received video around this threshold.

Tables 30.14 and 30.15 provide a summary of the SNR per bit required for the various system configurations. The BER threshold of 10^{-4} was selected here, since at this average BER after Viterbi or turbo decoding, the RS decoder becomes effective, guaranteeing near error-free performance. This also translates into near unimpaired reconstructed video quality.

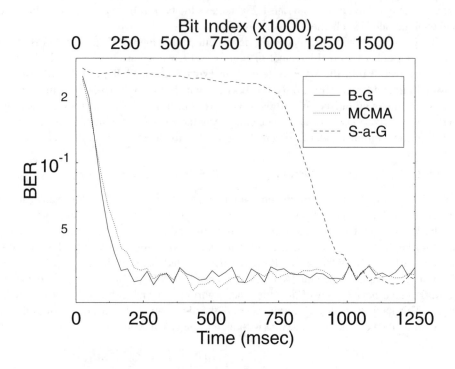

Figure 30.26: Learning curves for 16-QAM, one-symbol delay two-path channel at SNR = 18 dB [491], ©IEEE, 2000, Lee, Vlahoyiannatos, Hanzo

Mod.	Equaliser	Code	CSNR (dB)	E_b/N_0
QPSK	PSP	R=1/2	5.3	5.3
QPSK	MCMA	Turbo (1/2)	5.2	5.2
16-QAM	MCMA	Turbo (1/2)	13.6	10.6
QPSK	MCMA	Conv (1/2)	9.1	9.1
16-QAM	MCMA	Conv (1/2)	17.2	14.2
QPSK	MCMA	Conv (3/4)	11.5	9.7
16-QAM	MCMA	Conv (3/4)	20.2	15.4
QPSK	B-G	Conv (7/8)	13.2	10.8
16-QAM	B-G	Conv (7/8)	21.6	16.2

Table 30.12: Summary of performance results over the dispersive one-symbol delay two-path AWGN channel of Figure 30.17 tolerating a PSNR degradation of 2 dB

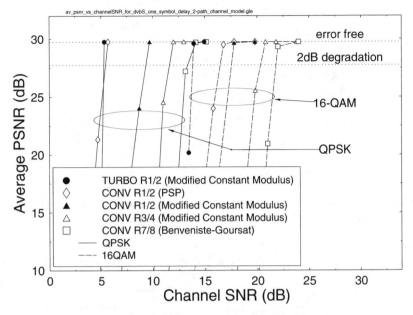

(a) One-symbol delay two-path channel model

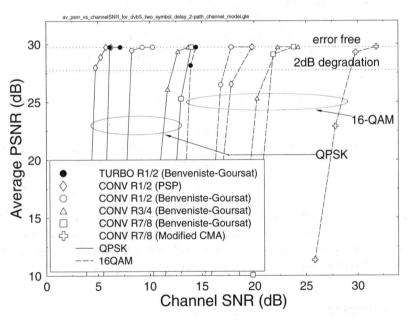

(b) Two-symbol delay two-path channel model

Figure 30.27: Average PSNR versus channel SNR for (a) the one-symbol delay two-path channel model and (b) the two-symbol delay two-path channel model of Figure 30.17 at a video bit rate of 2.5 Mbit/s using the "Football" video sequence [491], ©IEEE, 2000, Lee, Vlahoyiannatos, Hanzo

Mod.	Equaliser	Code	CSNR (dB)	E_b/N_0
QPSK	PSP	R=1/2	4.7	4.7
QPSK	B-G	Turbo (1/2)	5.9	5.9
16-QAM	B-G	Turbo (1/2)	13.7	10.7
QPSK	B-G	Conv (1/2)	8.0	8.0
16-QAM	B-G	Conv (1/2)	17.0	14.0
QPSK	B-G	Conv (3/4)	12.1	10.3
16-QAM	B-G	Conv (3/4)	21.1	16.3
QPSK	B-G	Conv (7/8)	13.4	11.0
16-QAM	MCMA	Conv (7/8)	29.2	23.8

Table 30.13: Summary of performance results over the dispersive two-symbol delay two-path AWGN channel of Figure 30.17 tolerating a PSNR degradation of 2 dB

Mod.	Equaliser	Code	E_b/N_0
QPSK	PSP	R=1/2	6.1
QPSK	MCMA	Turbo (1/2)	5.2
16-QAM	MCMA	Turbo (1/2)	10.7
QPSK	MCMA	Conv (1/2)	11.6
16-QAM	MCMA	Conv (1/2)	15.3
QPSK	MCMA	Conv (3/4)	10.5
16-QAM	MCMA	Conv (3/4)	16.4
QPSK	B-G	Conv (7/8)	11.8
16-QAM	B-G	Conv (7/8)	17.2

Table 30.14: Summary of system performance results over the dispersive one-symbol delay two-path AWGN channel of Figure 30.17 tolerating an average BER of 10^{-4}, which was evaluated after Viterbi or turbo decoding but before RS decoding

Finally, in Table 30.16 the QAM symbol rate or baud rate is given for different puncturing rates and for different modulation schemes, based on the requirement of supporting a video bit rate of 2.5 Mbit/sec. We observe that the baud rate is between 0.779 and 2.73 MBd, depending on the coding rate and the number of bits per modulation symbol.

30.3 Summary

In this chapter we commenced our discourse by investigating the performance of a turbo-coded terrestrial DVB system in a mobile environment in Section 30.1. A range of system performance results were presented based on the standard DVB-T scheme as well as on an improved turbo-coded scheme. The convolutional code specified in the standard system was replaced by turbo coding, which resulted in a substantial coding gain of around 5 dB. It

Mod.	Equaliser	Code	E_b/N_0
QPSK	PSP	R=1/2	5.6
QPSK	B-G	Turbo (1/2)	5.7
16-QAM	B-G	Turbo (1/2)	10.7
QPSK	B-G	Conv (1/2)	9.2
16-QAM	B-G	Conv (1/2)	15.0
QPSK	B-G	Conv (3/4)	12.0
16-QAM	B-G	Conv (3/4)	16.8
QPSK	B-G	Conv (7/8)	11.7
16-QAM	MCMA	Conv (7/8)	26.0

Table 30.15: Summary of system performance results over the dispersive two-symbol delay two-path AWGN channel of Figure 30.17 tolerating an average BER of 10^{-4}, which was evaluated after Viterbi or turbo decoding but before RS decoding

Punctured Rate	4-QAM Baud Rate (MBd)	16-QAM Baud Rate (MBd)
1/2	2.73	1.37
3/4	1.82	0.909
7/8	1.56	0.779

Table 30.16: The channel bit rate for the three different punctured coding rates and for the two modulations

is important to underline once again that the $K = 3$ turbo code and the $K = 7$ convolutional code exhibited comparable complexities. The higher performance of the turbo codec facilitates, for example, the employment of turbo-coded 16-QAM at a similar SNR, where convolutional-coded QPSK can be invoked. This in turn allows us to double the video bit rate within the same bandwidth and hence to improve the video quality. We have also applied data partitioning to the MPEG-2 video stream to gauge its efficiency in increasing the error resilience of the video codec. However, from these experiments we found that the data partitioning scheme did not provide substantial improvements compared to the non-partitioned video transmitted over the non-hierarchical DVB-T system. Our future work will be focused on extending this DVB-T system study to incorporate various types of channel models, as well as on investigating the effects of different Doppler frequencies on the system. Further work will also be dedicated to trellis coded modulation (TCM) and turbo trellis coded modulation (TTCM) based OFDM. The impact of employing various types of turbo interleavers on the system performance is also of interest. A range of further wireless video communications issues are addressed in [168, 828]. Let us now consider a variety of satellite-based turbo-coded blind-equalised multilevel modulation-assisted video broadcasting schemes.

The above terrestrial DVB system studies were continued by investigating the performance of a turbo-coded DVB system in a satellite-based broadcast environment in Section 30.2. A range of system performance results was presented based on the standard DVB-S scheme, as well as on a turbo-coded scheme in conjunction with blind-equalised QPSK/16-QAM. *The convolutional code specified in the standard system was replaced by turbo coding, which resulted in a substantial coding gain of approximately 4–5 dB. We have also shown that 16-QAM can be utilised instead of QPSK if an extra 5 dB SNR per bit gain is added to the link budget. This extra transmitted power requirement can be eliminated upon invoking the more complex turbo codec, which requires lower transmitted power for attaining the same performance as the standard convolutional codecs.*

Our future work will be focused on extending the DVB satellite system to support mobile users in the reception of satellite broadcast signals. The use of blind turbo equalisers will also be investigated in comparison to conventional blind equalisers. Further work will also be dedicated to trellis coded modulation (TCM) and turbo trellis coded modulation (TTCM) based orthogonal frequency division multiplexed (OFDM) and single-carrier equalised modems. The impact on the system performance by employing various types of turbo interleavers and turbo coded is also of interest. A range of further wireless video communications issues are addressed in [168].

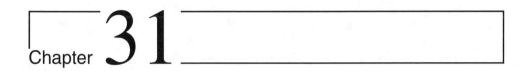

Chapter 31

Appendix

31.1 BER Analysis of Type-I Star-QAM

The Star Quadrature Amplitude Modulation (SQAM) technique of Chapter 10, also known as Amplitude-modulated Phase Shift Keying (APSK), employs circular constellations, rather than the rectangular constellation or Square QAM, as in Chapter 3. Although Square QAM has the maximum possible minimum Euclidean distance amongst its phasors given a constant average symbol power, in some situations Star QAM may be preferred due to its relatively simple detector and for its low Peak-to-Average Power Ratio (PAPR) compared to the Square QAM scheme of Chapter 3. Since differentially detected non-coherent Star QAM signals are robust against fading effects, many researchers analysed its Bit Error Ratio (BER) performance for transmission over AWGN channels [404], Rayleigh fading channels [142, 404] as well as Rician fading channels [829]. The effects diversity reception on its BER were also studied when communicating over Rayleigh fading channels [830, 831]. The BER of coherent 16 Star QAM was also analysed for transmission over AWGN channels [404] as well as when communicating over *Nakagami-m* fading channels [832]. However, the BER of Star-QAM schemes other than 16-level Star QAM, such as 8, 32 and 64-level Star QAM, has not been studied.

31.1.1 Coherent Detection

The BER of coherent star QAM schemes employing the Type-I constellation of Figure 4.1 employed for communicating over AWGN channels may be analysed using the signal-space method [404, 673, 832, 833]. The phasor constellations of the various Type-I Star QAM schemes are illustrated in Figure 31.1 and 31.6. Let us first consider 8-level or 3 bit/symbol Star QAM. In Figure 31.1(a), a is the radius of the inner ring, while $a\beta$ is the radius of the outer ring. The ring ratio is given by $a\beta/a = \beta$. The three bits, namely b_1, b_2 and b_3, are assigned as shown in Figure 31.1(a), representing Gray coding for each ring using the bits $b_1 b_2$. The third bit, namely b_3, indicates which ring of the constellation is encountered. The

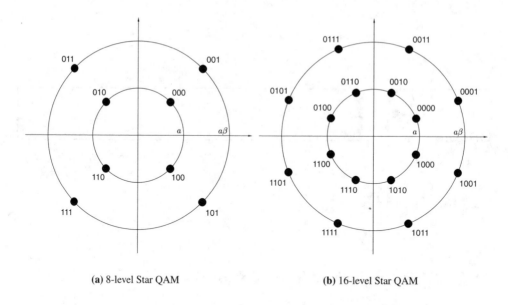

(a) 8-level Star QAM   **(b)** 16-level Star QAM

Figure 31.1: Type-I constellations of Star QAM using two constellation rings

average symbol power is given as:

$$E_s = \frac{4a^2 + 4a^2\beta^2}{8} = \frac{1}{2}a^2(1+\beta^2) \, . \tag{31.1}$$

In order to normalise the constellations so that the average symbol power becomes unity, a should be given as:

$$a = \sqrt{\frac{2}{1+\beta^2}} \, . \tag{31.2}$$

In terms of the signal space, the modulation scheme with respect to b_3 is an Amplitude Shift Keying (ASK) scheme. The decision rule related to bit b_3 is specified in Figure 31.2(a). The BER of bit b_3 can be expressed as:

$$P_{b_3} = \frac{1}{2}\left[Q\left(\frac{a(\beta-1)}{2}\sqrt{2\gamma} \right) + Q\left(\frac{a(\beta+3)}{2}\sqrt{2\gamma} \right) \right]$$
$$+ \frac{1}{2}\left[Q\left(\frac{a(\beta-1)}{2}\sqrt{2\gamma} \right) - Q\left(a(\beta+1)\sqrt{2\gamma} \right) \right] \tag{31.3}$$

$$\simeq Q\left(\frac{a(\beta-1)}{2}\sqrt{2\gamma} \right) \tag{31.4}$$

$$= Q\left(\sqrt{\frac{(\beta-1)^2}{1+\beta^2}\gamma} \right) , \tag{31.5}$$

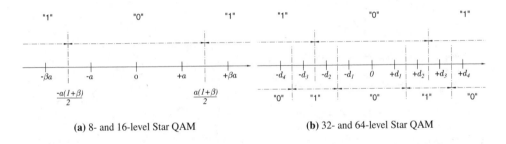

(a) 8- and 16-level Star QAM

(b) 32- and 64-level Star QAM

Figure 31.2: Magnitude-bit decision regions for various Type-I Star QAM constellations

where the Gaussian Q-function is defined as $Q(x) = \frac{1}{\sqrt{2\pi}} \int_x^\infty e^{-y^2/2} dy$ and γ is the SNR per symbol. Since bits b_1 and b_2 corresponds to Gray coded QPSK signals, their BER can be expressed as:

$$P_{b_1} = P_{b_2} \simeq \frac{1}{2} Q \left(\frac{a}{\sqrt{2}} \sqrt{2\gamma} \right) + \frac{1}{2} Q \left(\frac{a\beta}{\sqrt{2}} \sqrt{2\gamma} \right) \tag{31.6}$$

$$= \frac{1}{2} Q \left(\sqrt{\frac{2}{1+\beta^2}\gamma} \right) + \frac{1}{2} Q \left(\sqrt{\frac{2\beta^2}{1+\beta^2}\gamma} \right) . \tag{31.7}$$

Hence, the average BER of an 8-level Star QAM scheme communicating over an AWGN channel can be expressed as:

$$P_8 = \frac{1}{3} P_{b_1} + \frac{1}{3} P_{b_2} + \frac{1}{3} P_{b_3} \tag{31.8}$$

$$\simeq \frac{1}{3} \left[Q \left(\sqrt{\frac{2}{1+\beta^2}\gamma} \right) + Q \left(\sqrt{\frac{2\beta^2}{1+\beta^2}\gamma} \right) + Q \left(\sqrt{\frac{(\beta-1)^2}{1+\beta^2}\gamma} \right) \right] . \tag{31.9}$$

The BER of (31.9) is plotted in Figure 31.3(a) as a function of the ring ratio β for various values of the SNR per symbol γ. We can observe that the BER of 8-level Star QAM reaches its minimum, when the ring ratio is $\beta \simeq 2.4$. This is not surprising, considering that the ring ratio should be $\beta = 1 + \sqrt{2}$ in order to make the Euclidean distances between an inner ring constellation point and its three adjacent constellation points the same. However, the optimum ring ratio β_{opt}, where the BER reaches its minimum is SNR dependent. The optimum ring ratio versus the SNR per symbol is plotted in Figure 31.3(b). It can be observed that when the SNR is lower than 8dB, the optimum ring ratio increases sharply. Since the corresponding BER improvement was however less than 0.1dB even at SNRs near 0dB, the fixed ring ratio of $\beta = 1 + \sqrt{2}$ can be used for all SNR values. Figure 31.4(a) compares the BER of 8-level Star QAM and 8-PSK. Observe that 8-level Star QAM exhibits an approximately 1dB SNR performance gain, when the SNR is below 2dB, but above this SNR the SNR gain becomes marginal..

Let us now consider 16-level Star QAM. The corresponding phasor constellation is given in Figure 31.1(b). Since the average symbol power is the same as that of 8-level Star QAM, Equation 31.2 can be used for determining a. The BER analysis for the fourth bit, namely

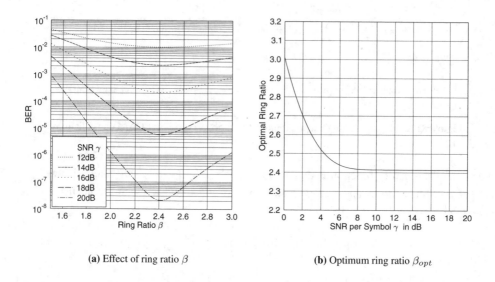

(a) Effect of ring ratio β (b) Optimum ring ratio β_{opt}

Figure 31.3: BER of Gray-mapped 8-level Star QAM for transmission over AWGN channels

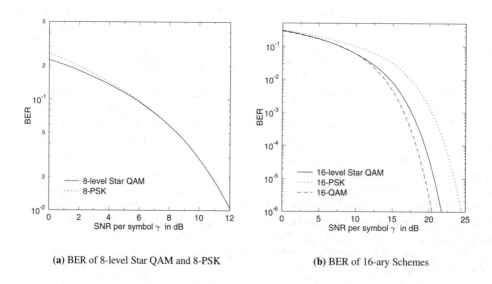

(a) BER of 8-level Star QAM and 8-PSK   (b) BER of 16-ary Schemes

Figure 31.4: BER comparison of 8-ary and 16-ary modulation schemes for transmission over AWGN
channels

for b_4 is exactly the same as that of 8-level Star QAM and the corresponding value of P_{b_4} is given in (31.5). Since the first three bits, namely b_1, b_2 and b_3, are 8-PSK modulated, their BER can be expressed as:

$$P_{b_1} = P_{b_2} = P_{b_3} = \frac{1}{2} P_{8PSK}(a^2\gamma) + \frac{1}{2} P_{8PSK}(a^2\beta^2\gamma) \,. \tag{31.10}$$

Lu, Letaief, Chuang and Liou found an accurate approximation of the BER of Gray-coded MPSK, which is given by [673]:

$$P_{MPSK} \simeq \frac{2}{\log_2 M} \sum_{i=1}^{2} Q\left(\sqrt{2\sin^2\left(\frac{2i-1}{M}\pi\right)\gamma}\right) \,, \tag{31.11}$$

where γ is the SNR per symbol. Hence, the BER of (31.10) can be expressed as:

$$P_{b_1} = P_{b_2} = P_{b_3} \tag{31.12}$$

$$\simeq \frac{1}{3}\left[Q\left(\sqrt{\frac{4\sin^2(\pi/8)}{1+\beta^2}\gamma}\right) + Q\left(\sqrt{\frac{4\sin^2(3\pi/8)}{1+\beta^2}\gamma}\,,\right)\right]$$

$$+ \frac{1}{3}\left[Q\left(\sqrt{\frac{4\beta^2\sin^2(\pi/8)}{1+\beta^2}\gamma}\right) + Q\left(\sqrt{\frac{4\beta^2\sin^2(3\pi/8)}{1+\beta^2}\gamma}\right)\right] \,. \tag{31.13}$$

Now, the average BER of a 16-level Star QAM scheme for transmission over AWGN channel can be expressed as:

$$P_{16} = \frac{1}{4}P_{b_1} + \frac{1}{4}P_{b_2} + \frac{1}{4}P_{b_3} + \frac{1}{4}P_{b_4} \tag{31.14}$$

$$\simeq \frac{1}{4}\left[Q\left(\sqrt{\frac{4\sin^2(\pi/8)}{1+\beta^2}\gamma}\right) + Q\left(\sqrt{\frac{4\sin^2(3\pi/8)}{1+\beta^2}\gamma}\,,\right) + Q\left(\sqrt{\frac{4\beta^2\sin^2(\pi/8)}{1+\beta^2}\gamma}\right)\right.$$

$$\left. + Q\left(\sqrt{\frac{4\beta^2\sin^2(3\pi/8)}{1+\beta^2}\gamma}\right) + Q\left(\sqrt{\frac{(\beta-1)^2}{1+\beta^2}\gamma}\right)\right] \,. \tag{31.15}$$

The BER of (31.15) is plotted in Figure 31.5(a) as a function of the ring ratio β for the various values of the SNR per symbol γ. We can observe that the ring ratio of $\beta \simeq 1.8$ minimises the BER of 16-level Star QAM, when communicating over AWGN channels [404]. This is also expected, since the ring ratio should be $\beta = 1 + 2\cos(3\pi/8) = 1.7654$ in order to render the Euclidean distances between an inner-ring constellation point and its three adjacent constellation points the same. The actual optimum ring ratio β_{opt}, where the BER reaches its minimum is plotted in Figure 31.5(b). As for 8-level Star QAM, even though the optimum ratio is SNR dependent, the difference between the BER corresponding to the optimum ring ratio and that corresponding to the constant ring ratio of $\beta = 1 + 2\cos(3\pi/8)$ is negligible. Figure 31.4(b) compares the BER of 16-level Star QAM, 16-PSK and 16-level Square QAM. We found that the BER performance of 16-level Star QAM is inferior to that of 16-level

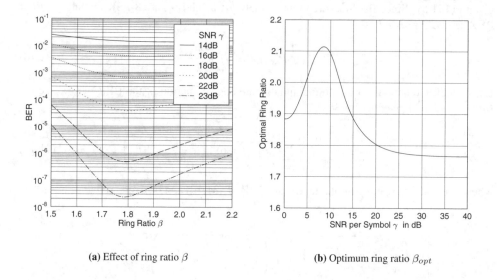

(a) Effect of ring ratio β (b) Optimum ring ratio β_{opt}

Figure 31.5: BER of Gray mapped 16-level Star QAM for transmission over AWGN channels

Square QAM. Viewing the corresponding performance from the perspective of the required SNR per symbol, the former requires an approxmately 1.3dB high SNR for maintaining the BER of 10^{-6}. By contrast, it requires a 2.7dB lower symbol-SNR, than 16-PSK.

Having considered the family of twin-ring constellations, let us focus our attention on two four-ring constellations. The Type-I constellations of 32-level and 64-level Star QAM scheme are depicted in Figure 31.6. The last two bits of a symbol are Gray coded in the 'radial direction' and they are four-level ASK modulated. Let us assume that the Gray coding scheme for the bits "$b_4 b_5$" of 32-level Star QAM in the 'radial direction' is given as "00", "01", "11" and "10", when viewing it from the inner most ring to the outer rings. The decision regions of these bits were illustrated in Figure 31.2(b). The remaining bits are also Gray coded along each of the four rings and PSK modulated. Let us denote the radius of each ring as d_1, $d_2 = d_1 \beta_1$, $d_3 = d_1 \beta_2$ and $d_4 = d_1 \beta_3$, where β_1, β_2 and β_3 are the corresponding ring ratios of each ring. Since the average power per symbol E_s is given as:

$$E_s = \frac{d_1^2 + d_2^2 + d_3^2 + d_4^2}{4} = \frac{d_1^2}{4}\left(1 + \beta_1^2 + \beta_2^2 + \beta_3^2\right), \qquad (31.16)$$

the value of d_1 required for normalising the average power to unity can be expressed as:

$$d_1 = \sqrt{\frac{4}{1 + \beta_1^2 + \beta_2^2 + \beta_3^2}}. \qquad (31.17)$$

Inspecting Figure 31.2(b), the BER of the fourth bit of 32-level Star QAM can be formulated

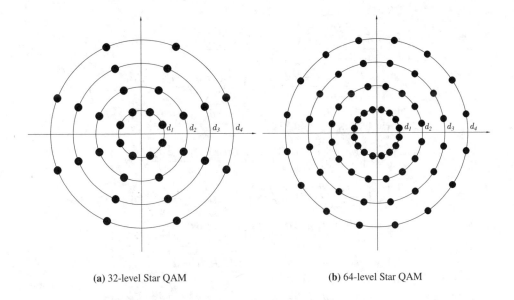

(a) 32-level Star QAM   (b) 64-level Star QAM

Figure 31.6: Type-I constellations of Star QAM schemes having four rings

as:

$$
P_{b_4} = \frac{1}{4}\left[Q\left(\frac{d_2 + d_3 - 2d_1}{2}\sqrt{2\gamma}\right) + Q\left(\frac{d_2 + d_3 + 2d_1}{2}\sqrt{2\gamma}\right)\right]
$$
$$
+ \frac{1}{4}\left[Q\left(\frac{d_3 - d_2}{2}\sqrt{2\gamma}\right) + Q\left(\frac{d_3 + 3d_2}{2}\sqrt{2\gamma}\right)\right]
$$
$$
+ \frac{1}{4}\left[Q\left(\frac{d_3 - d_2}{2}\sqrt{2\gamma}\right) - Q\left(\frac{3d_3 + d_2}{2}\sqrt{2\gamma}\right)\right]
$$
$$
+ \frac{1}{4}\left[Q\left(\frac{2d_4 - d_2 - d_3}{2}\sqrt{2\gamma}\right) - Q\left(\frac{2d_4 + d_2 + d_3}{2}\sqrt{2\gamma}\right)\right] \tag{31.18}
$$
$$
\simeq \frac{1}{4}\left[Q\left(\sqrt{\frac{2(\beta_1 + \beta_2 - 2)^2}{1 + \beta_1^2 + \beta_2^2 + \beta_3^2}\gamma}\right) + 2\,Q\left(\sqrt{\frac{2(\beta_2 - \beta_1)^2}{1 + \beta_1^2 + \beta_2^2 + \beta_3^2}\gamma}\right)\right.
$$
$$
\left. + Q\left(\sqrt{\frac{2(2\beta_3 - \beta_1 - \beta_2)^2}{1 + \beta_1^2 + \beta_2^2 + \beta_3^2}\gamma}\right)\right]. \tag{31.19}
$$

The decision regions depicted in the lower part of Figure 31.2(b) are valid for the fifth bit, namely b_5. The BER of b_5 can be expressed as:

$$
\begin{aligned}
P_{b_5} = \frac{1}{4} &\left[Q\left(\frac{d_2 - d_1}{2} \sqrt{2\gamma} \right) + Q\left(\frac{d_2 + 3\,d_1}{2} \sqrt{2\gamma} \right) \right. \\
&\left. - Q\left(\frac{d_3 + d_4 - 2\,d_1}{2} \sqrt{2\gamma} \right) - Q\left(\frac{d_3 + d_4 + 2\,d_1}{2} \sqrt{2\gamma} \right) \right] \\
+ \frac{1}{4} &\left[Q\left(\frac{d_2 - d_1}{2} \sqrt{2\gamma} \right) + Q\left(\frac{d_3 + d_4 - 2\,d_2}{2} \sqrt{2\gamma} \right) \right. \\
&\left. - Q\left(\frac{d_1 + 3\,d_2}{2} \sqrt{2\gamma} \right) + Q\left(\frac{d_3 + d_4 + 2\,d_2}{2} \sqrt{2\gamma} \right) \right] \\
+ \frac{1}{4} &\left[Q\left(\frac{2\,d_3 - d_1 - d_2}{2} \sqrt{2\gamma} \right) + Q\left(\frac{d_4 - d_3}{2} \sqrt{2\gamma} \right) \right. \\
&\left. - Q\left(\frac{d_1 + d_2 + 2\,d_3}{2} \sqrt{2\gamma} \right) + Q\left(\frac{3\,d_3 + d_4}{2} \sqrt{2\gamma} \right) \right] \\
+ \frac{1}{4} &\left[Q\left(\frac{d_4 - d_3}{2} \sqrt{2\gamma} \right) - Q\left(\frac{2\,d_4 - d_1 - d_2}{2} \sqrt{2\gamma} \right) \right. \\
&\left. + Q\left(\frac{d_1 + d_2 + 2\,d_4}{2} \sqrt{2\gamma} \right) - Q\left(\frac{d_3 + 3\,d_4}{2} \sqrt{2\gamma} \right) \right] .
\end{aligned}
\tag{31.20}
$$

The expression of the BER P_{b_5} can be accurately approximated as ;

$$
\begin{aligned}
P_{b_5} \simeq \frac{1}{4} &\left[2\,Q\left(\sqrt{\frac{2(\beta_1 - 1)^2}{1 + \beta_1^2 + \beta_2^2 + \beta_3^2}}\,\gamma \right) + Q\left(\sqrt{\frac{2(\beta_2 + \beta_3 - 2\,\beta_1)^2}{1 + \beta_1^2 + \beta_2^2 + \beta_3^2}}\,\gamma \right) \right. \\
&\left. + Q\left(\sqrt{\frac{2(2\,\beta_2 - 1 - \beta_1)^2}{1 + \beta_1^2 + \beta_2^2 + \beta_3^2}}\,\gamma \right) + 2\,Q\left(\sqrt{\frac{2(\beta_3 - \beta_2)^2}{1 + \beta_1^2 + \beta_2^2 + \beta_3^2}}\,\gamma \right) \right] .
\end{aligned}
\tag{31.21}
$$

Let us now find the BER of the PSK modulated bits b_1, b_2 and b_3. Since they are 8-PSK modulated, the BER can be expressed using the results of [673] as:

$$P_{b_1} = P_{b_2} = P_{b_3} = \frac{1}{4}P_{8PSK}(d_1^2\gamma) + \frac{1}{4}P_{8PSK}(d_2^2\gamma) + \frac{1}{4}P_{8PSK}(d_3^2\gamma) + \frac{1}{4}P_{8PSK}(d_4^2\gamma)$$

(31.22)

$$\simeq \frac{1}{6}\left[Q\left(\sqrt{\frac{8\sin^2(\pi/8)}{1+\beta_1^2+\beta_2^2+\beta_3^2}\gamma} \right) + Q\left(\sqrt{\frac{8\sin^2(3\pi/8)}{1+\beta_1^2+\beta_2^2+\beta_3^2}\gamma} \right) \right]$$

$$+ \frac{1}{6}\left[Q\left(\sqrt{\frac{8\,\beta_1^2\sin^2(\pi/8)}{1+\beta_1^2+\beta_2^2+\beta_3^2}\gamma} \right) + Q\left(\sqrt{\frac{8\,\beta_1^2\sin^2(3\pi/8)}{1+\beta_1^2+\beta_2^2+\beta_3^2}\gamma} \right) \right]$$

$$+ \frac{1}{6}\left[Q\left(\sqrt{\frac{8\,\beta_2^2\sin^2(\pi/8)}{1+\beta_1^2+\beta_2^2+\beta_3^2}\gamma} \right) + Q\left(\sqrt{\frac{8\,\beta_2^2\sin^2(3\pi/8)}{1+\beta_1^2+\beta_2^2+\beta_3^2}\gamma} \right) \right]$$

$$+ \frac{1}{6}\left[Q\left(\sqrt{\frac{8\,\beta_3^2\sin^2(\pi/8)}{1+\beta_1^2+\beta_2^2+\beta_3^2}\gamma} \right) + Q\left(\sqrt{\frac{8\,\beta_3^2\sin^2(3\pi/8)}{1+\beta_1^2+\beta_2^2+\beta_3^2}\gamma} \right) \right]. \quad (31.23)$$

Hence, the BER of 32-level Star QAM can be expressed as:

$$P_{32} = \frac{1}{5}(3P_{b_1} + P_{b_4} + P_{b_5}),$$

(31.24)

where P_{b_1}, P_{b_4} and P_{b_5} are given in Equations (31.23), (31.19) and (31.21), respectively. The optimum ring ratios of 32-level Star QAM are depicted in Figure 31.7(a). The optimum ring ratios converge to $\beta_1 = 1.77$, $\beta_2 = 2.541$ and $\beta_3 = 3.318$. Note that the first optimum ring ratio is the same as the optimum ring ratio of 16-level Star QAM and the corresponding distances between the second, the third and the fourth rings are approximately equal, as one would expect in an effort to maintain an identical distance amongst the constellations points. Figure 31.7(b) compares the BER of 32-level Star QAM and 32-PSK. We found that the SNR gain of 32-level Star QAM over 32-PSK is 4.6dB at a BER of 10^{-6}.

The BER of 64-level Star QAM can be obtained using the same procedure employed for determing the BER of 32-level Star QAM, considering that now the bits b_1, b_2, b_3 and b_4 are 16-PSK modulated on each ring. The BER of the last two bits, P_{b_5} and P_{b_6} are the same as those given in (31.19) and (31.21), respectively. On the other hand, the BER of the 16-PSK

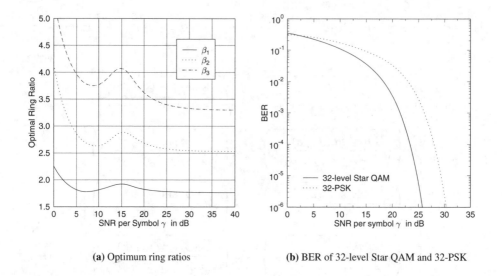

(a) Optimum ring ratios   (b) BER of 32-level Star QAM and 32-PSK

Figure 31.7: BER of 32 Star QAM over AWGN channel

modulated bits of 64-level Star QAM can be expressed as:

$$
\begin{aligned}
P_{b_1} =& P_{b_2} = P_{b_3} = P_{b_4} \\
=& \frac{1}{4} P_{16PSK}(d_1^2 \gamma) + \frac{1}{4} P_{16PSK}(d_2^2 \gamma) + \frac{1}{4} P_{16PSK}(d_3^2 \gamma) + \frac{1}{4} P_{16PSK}(d_4^2 \gamma) \quad (31.25) \\
\simeq& \frac{1}{8} \left[Q\left(\sqrt{\frac{8 \sin^2(\pi/16)}{1 + \beta_1^2 + \beta_2^2 + \beta_3^2} \gamma} \right) + Q\left(\sqrt{\frac{8 \sin^2(3\pi/16)}{1 + \beta_1^2 + \beta_2^2 + \beta_3^2} \gamma} \right) \right] \\
&+ \frac{1}{8} \left[Q\left(\sqrt{\frac{8 \beta_1^2 \sin^2(\pi/16)}{1 + \beta_1^2 + \beta_2^2 + \beta_3^2} \gamma} \right) + Q\left(\sqrt{\frac{8 \beta_1^2 \sin^2(3\pi/16)}{1 + \beta_1^2 + \beta_2^2 + \beta_3^2} \gamma} \right) \right] \\
&+ \frac{1}{8} \left[Q\left(\sqrt{\frac{8 \beta_2^2 \sin^2(\pi/16)}{1 + \beta_1^2 + \beta_2^2 + \beta_3^2} \gamma} \right) + Q\left(\sqrt{\frac{8 \beta_2^2 \sin^2(3\pi/16)}{1 + \beta_1^2 + \beta_2^2 + \beta_3^2} \gamma} \right) \right] \\
&+ \frac{1}{8} \left[Q\left(\sqrt{\frac{8 \beta_3^2 \sin^2(\pi/16)}{1 + \beta_1^2 + \beta_2^2 + \beta_3^2} \gamma} \right) + Q\left(\sqrt{\frac{8 \beta_3^2 \sin^2(3\pi/16)}{1 + \beta_1^2 + \beta_2^2 + \beta_3^2} \gamma} \right) \right] . \quad (31.26)
\end{aligned}
$$

Hence, the average BER of 64-level Star QAM can be expressed as:

$$
P_{64} = \frac{1}{6}(4P_{b_1} + P_{b_5} + P_{b_6}), \quad (31.27)
$$

where P_{b_1} is given in (31.26), where P_{b_5} and P_{b_6} are given in (31.19) and (31.21), respec-

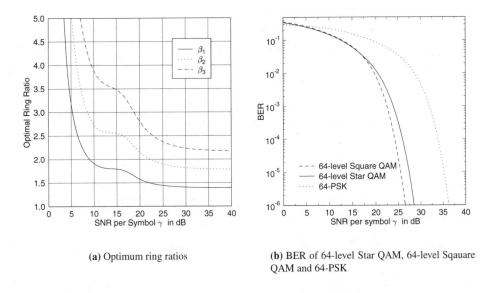

(a) Optimum ring ratios

(b) BER of 64-level Star QAM, 64-level Sqauare QAM and 64-PSK

Figure 31.8: BER of 64-level Star QAM for transmission over AWGN channels

tively. The optimum ring ratios of 64-level Star QAM are depicted in Figure 31.8(a). The optimum ring ratios converge to $\beta_1 = 1.4$, $\beta_2 = 1.81$ and $\beta_3 = 2.23$. It was observed that the SNR difference between the optimised BER and that employing the asymptotic ring ratio is at most 1dB in the SNR range of 5dB to 15dB. Figure 31.8(b) compares the BER of 64-level Star QAM, 64-level Square QAM and 64-PSK. We found that the SNR gain of 64-level Star QAM over 64-PSK is 7.7dB at the BER of 10^{-6}. The 64-level Square QAM arrangement is the most power-efficient scheme, which exhibits 2dB SNR gain over 64-level Star QAM at the BER of 10^{-6}.

31.2 Two-Dimensional Rake Receiver

31.2.1 System Model

The schematic of our Rake-receiver and D-antenna diversity assisted adaptive Square QAM (AQAM) system is illustrated in Figure 31.9. A band-limited equivalent low-pass m-ary QAM signal $s(t)$, having a spectrum of $S(f) = 0$ for $|f| > 1/2\,W$, is transmitted over time variant frequency selective fading channels and received by a set of D RAKE-receivers. Each Rake-receiver [388, 677] combines all the resolvable multi-path components using Maximal Ratio Combining (MRC). The combined signals of the D-antenna assisted Rake-receivers are summed and demodulated using the estimated channel quality information. The estimated signal-to-noise ratio is fed back to the transmitter and it is used for deciding upon the most appropriate m-ary square QAM modulation mode to be used during the next transmission burst. We assume that the channel quality is estimated perfectly and it is available at the

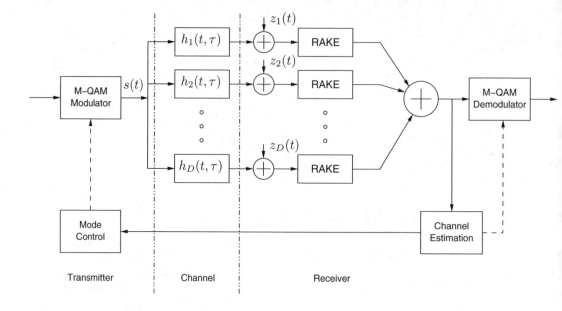

Figure 31.9: Equivalent low-pass model of a D-th order antenna diversity based RAKE-receiver assisted AQAM system

transmitter immediately. The effects of channel estimation error and feedback delay on the performance of AQAM were studied for example by Goldsmith and Chua [184].

The low-pass equivalent impulse response of the channel between the transmitter and the d-th antenna may be represented as [388]:

$$h_d(t, \tau) = \sum_{n=1}^{N} h_{d,n}(t)\, \delta\left(\tau - \frac{n}{W}\right),$$ (31.28)

where $\{h_{d,n}(t)\}$ is a set of independent complex valued stationary random Gaussian processes. The maximum number of resolvable multi-path components N is given by $\lfloor T_m W \rfloor + 1$, where T_m is the multi-path delay spread of the channel [388]. Hence, the low-pass equivalent received signal $r_d(t)$ at the d-th antenna can be formulated as:

$$r_d(t) = \sum_{n=1}^{N} h_{d,n}(t) s\left(t - \frac{n}{W}\right) + z_d(t),$$ (31.29)

where $z_d(t)$ is a zero mean Gaussian random process having a two-sided power spectral density of $N_o/2$. Let us assume that the fading is sufficiently slow or $(\Delta t)_c \ll T$, where $(\Delta t)_c$ is the channel's coherence time [370] and T is the signalling period. Then, $h_{d,n}(t)$ can be simplified to $h_{d,n}(t) = \alpha_{d,n} e^{j\phi_{d,n}}$ for the duration of signalling period T, where the fading magnitude $\alpha_{d,n}$ is assumed to be Rayleigh distributed and the phase $\phi_{d,n}$ is assumed to be uniformly distributed.

31.2.2 BER Analysis of Fixed-mode Square QAM

An ideal RAKE receiver [677] combines all the signal powers scattered over N paths in an optimal manner, so that the instantaneous Signal-to-Noise Ratio (SNR) per symbol at the RAKE receiver's output can be maximised [388]. The noise at the RAKE receiver's output is known to be Gaussian [388]. The SNR, γ_d, at the d-th ideal RAKE receiver's output is given as [388]:

$$\gamma_d = \sum_{n=1}^{N} \gamma_{d,n} \, , \tag{31.30}$$

where $\gamma_{d,n} = E/N_o\,\alpha_{d,n}^2$ and $\{\alpha_{d,n}\}$ is assumed to be normalised, such that $\sum_{n=1}^{N} \alpha_{d,n}^2$ becomes unity. Since we assumed that each multi-path component has an independent Rayleigh distribution, the characteristic function of γ_d can be represented as [388] (Page 802):

$$\psi_{\gamma_d}(j\upsilon) = \prod_{n=1}^{N} \frac{1}{1 - j\upsilon\bar{\gamma}_{d,n}} \, , \tag{31.31}$$

where $\gamma_{d,n} = E/N_o\mathrm{E}[\alpha_{d,n}^2]$. Let us assume furthermore that each of the D diversity channels has the same multi-path intensity profile (MIP), although in practical systems each antenna may experience a different MIP. Under this assumption, $\bar{\gamma}_{d,n}$ in (31.31) can be written as $\bar{\gamma}_n$. The total SNR per symbol, γ, at the output of the demodulator depicted in Figure 31.9 is given as:

$$\gamma = \sum_{d=1}^{D} \gamma_d \, , \tag{31.32}$$

while the characteristic function of the SNR per symbol γ, under the assumption of independent identical diversity channels, can be formulated as:

$$\psi_{\gamma}(j\upsilon) = \prod_{n=1}^{N} \frac{1}{(1 - j\upsilon\bar{\gamma}_n)^D} \, . \tag{31.33}$$

Applying the technique of Partial Fraction Expansion (PFE) [672], $\psi_{\gamma}(j\upsilon)$ can be expressed as:

$$\psi_{\gamma}(j\upsilon) = \sum_{d=1}^{D} \sum_{n=1}^{N} \Lambda_{D-d+1,n} \frac{1}{(1 - j\upsilon\bar{\gamma}_n)^d} \, . \tag{31.34}$$

Let us now determine the constant coefficients $\Lambda_{d,n}$. Equating (31.33) with (31.34) and substituting $j\upsilon = -p$, we have

$$\prod_{i=1}^{N} \frac{1}{(1 + p\bar{\gamma}_i)^D} = \sum_{d=1}^{D} \sum_{i=1}^{N} \Lambda_{D-d+1,i} \frac{1}{(1 + p\bar{\gamma}_i)^d} \, . \tag{31.35}$$

Multiplying by $(1 + p\bar{\gamma}_n)^D$ at both sides, (31.35) becomes:

$$\prod_{\substack{i=1 \\ i \neq n}}^{N} \frac{1}{(1 + p\bar{\gamma}_i)^D} = \sum_{d=1}^{D} \sum_{\substack{i=1 \\ i \neq n}}^{N} \Lambda_{D-d+1,i} \frac{1}{(1 + p\bar{\gamma}_i)^d} + \sum_{d=1}^{D} \Lambda_{d,n} (1 + p\bar{\gamma}_n)^{d-1} . \qquad (31.36)$$

Setting the $(d-1)$th derivatives with respect to p and substituting $p = -1/\bar{\gamma}_n$, we have:

$$\frac{d^{d-1}}{dp^{d-1}} \left[\prod_{\substack{i=1 \\ i \neq n}}^{N} (p\bar{\gamma}_i + 1)^{-D} \right]_{p = -1/\bar{\gamma}_n} = (d-1)! \, \bar{\gamma}_n^{(d-1)} \Lambda_{d,n} . \qquad (31.37)$$

Hence, $\Lambda_{d,n}$ is given as:

$$\Lambda_{d,n} \triangleq \frac{1}{(d-1)! \, \bar{\gamma}_n^{(d-1)}} \varphi_{d,n} (-1/\bar{\gamma}_n) , \qquad (31.38)$$

where $\varphi_{d,n}(x)$ is defined as:

$$\varphi_{d,n}(x) \triangleq \frac{d^{d-1}}{dp^{d-1}} \left[\prod_{\substack{i=1 \\ i \neq n}}^{N} (p\bar{\gamma}_i + 1)^{-D} \right]_{p = x} . \qquad (31.39)$$

Upon setting the derivatives directly, $\varphi_{d,n}(-1/\bar{\gamma}_n)$ of (31.39) can be represented recursively as:

$$\begin{aligned} \varphi_{1,n}(-1/\bar{\gamma}_n) &= \pi_n^D \\ \varphi_{d,n}(-1/\bar{\gamma}_n) &= \sum_{i=1}^{d-1} \left[C_{d,i} \, D \, \varphi_{d-i,n}(-1/\bar{\gamma}_n) \sum_{\substack{j=1 \\ j \neq n}}^{N} \left(\frac{\bar{\gamma}_n \bar{\gamma}_j}{\bar{\gamma}_n - \bar{\gamma}_j} \right)^i \right] , \end{aligned} \qquad (31.40)$$

where π_n is defined as:

$$\pi_n \triangleq \prod_{\substack{i=1 \\ i \neq n}}^{N} \frac{\bar{\gamma}_n}{\bar{\gamma}_n - \bar{\gamma}_i} \qquad (31.41)$$

and the doubly indexed coefficient $C_{d,i}$ of (31.40) can also be expressed recursively as:

$$\begin{aligned} C_{d,1} &= -1 & \text{for all } d \\ C_{d,d} &= 0 & \text{for } d > 1 \\ C_{d,i} &= -(i-1)\, C_{d-1,i-1} + C_{d-1,i} & \text{for } d > i . \end{aligned} \qquad (31.42)$$

The PDF of γ, $f_{\bar{\gamma}}(\gamma)$, can be found by applying the inverse Fourier transform to $\psi_\gamma(j\upsilon)$

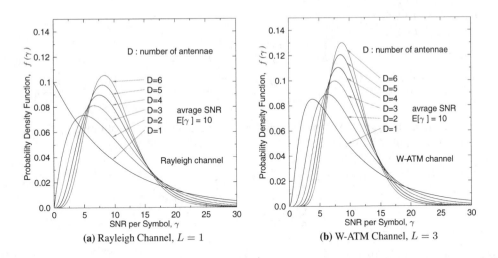

Figure 31.10: PDF of γ given in (31.43) for an average SNR per symbol of $E[\gamma] = 10$dB

in (31.34), which is given on Page 781 of Section (14-4-13)] in [388]:

$$f_{\bar{\gamma}}(\gamma) = \sum_{d=1}^{D} \sum_{n=1}^{N} \Lambda_{D-d+1,n} \frac{1}{(d-1)! \, \bar{\gamma}_n^d} \gamma^{d-1} e^{-\gamma/\bar{\gamma}_n} . \tag{31.43}$$

Figure 31.10 shows the PDF of the SNR per symbol over both a narrow-band Rayleigh channel and the dispersive Wireless Asynchronous Transfer Mode (W-ATM) channel of [180]. Specifically, the W-ATM channel is a 3-path indoor channel, where the average SNR for each path is given as $\bar{\gamma}_1 = 0.79192\bar{\gamma}$, $\bar{\gamma}_2 = 0.12424\bar{\gamma}$ and $\bar{\gamma}_3 = 0.08384\bar{\gamma}$.

Since we now have the PDF $f_{\bar{\gamma}}(\gamma)$ of the channel SNR, let us calculate the average BEP of m-ary square QAM employing Gray mapping. The average BEP P_e can be expressed as [180, 388]:

$$P_e = \int_0^{\infty} p_m(\gamma) f(\gamma) d\gamma , \tag{31.44}$$

where $p_m(\gamma)$ is the BER of m-ary square QAM employing Gray mapping over Gaussian channels [180]:

$$p_m(\gamma) = \sum_i A_i Q(\sqrt{a_i \gamma}) , \tag{31.45}$$

where $Q(x)$ is the Gaussian Q-function defined as $Q(x) \triangleq \frac{1}{\sqrt{2\pi}} \int_x^{\infty} e^{-t^2/2} dt$ and $\{A_i, a_i\}$ is a set of modulation mode dependent constants. For the modulation modes associated with

$m = 2, 4, 16$ and 64, the sets $\{A_i, a_i\}$ are given as [180, 667]:

$$
\begin{array}{lll}
m = 2, & \text{BPSK} & \{(1,2)\} \\
m = 4, & \text{QPSK} & \{(1,1)\} \\
m = 16, & \text{16-QAM} & \left\{ \left(\frac{3}{4}, \frac{1}{5}\right), \left(\frac{2}{4}, \frac{3^2}{5}\right), \left(-\frac{1}{4}, \frac{5^2}{5}\right) \right\} \\
m = 64, & \text{64-QAM} & \left\{ \left(\frac{7}{12}, \frac{1}{21}\right), \left(\frac{6}{12}, \frac{3^2}{21}\right), \left(-\frac{1}{12}, \frac{5^2}{21}\right), \left(\frac{1}{12}, \frac{9^2}{21}\right), \left(-\frac{1}{12}, \frac{13^2}{21}\right) \right\}.
\end{array}
$$

$$\tag{31.46}$$

The average BEP of m-ary QAM in our scenario can be calculated by substituting $p_m(\gamma)$ of (31.45) and $f_{\bar{\gamma}}(\gamma)$ of (31.43) into (31.44):

$$
P_{e,m}(\bar{\gamma}) = \int_0^\infty \sum_i A_i Q(\sqrt{a_i \gamma}) f_{\bar{\gamma}}(\gamma) d\gamma \tag{31.47}
$$

$$
= \sum_i A_i P_e(\bar{\gamma}; a_i), \tag{31.48}
$$

where each constituent BEP $P_e(\bar{\gamma}; a_i)$ is defined as:

$$
P_e(\bar{\gamma}; a_i) = \int_0^\infty Q(\sqrt{a_i \gamma}) \, f_{\bar{\gamma}}(\gamma) \, d\gamma. \tag{31.49}
$$

Using the similarity of $f_{\bar{\gamma}}(\gamma)$ in (31.43) and the PDF of the SNR of a D-antenna diversity-assisted Maximal Ratio Combining (MRC) system transmitting over flat Rayleigh channels [388] (Page 781), the closed form solution for the component BEP $P_e(\bar{\gamma}; a_i)$ can be expressed as:

$$
P_e(\bar{\gamma}; a_i) = \sum_{d=1}^{D} \sum_{n=1}^{N} \frac{1}{\sqrt{2\pi}} \int_0^\infty \int_{\sqrt{2\gamma}}^\infty e^{-x^2/2} \Lambda_{D-d+1,n} \frac{1}{(d-1)! \bar{\gamma}_n^d} \gamma^{d-1} e^{-\gamma/\bar{\gamma}_n} \, dx \, d\gamma
$$

$$\tag{31.50}$$

$$
= \sum_{d=1}^{D} \sum_{n=1}^{N} \left[\Lambda_{D-d+1,n} \left\{ \tfrac{1}{2}(1 - \mu_n) \right\}^d \sum_{i=0}^{d-1} \binom{d-1+i}{i} \left\{ \tfrac{1}{2}(1 + \mu_n) \right\}^i \right],
$$

$$\tag{31.51}$$

where $\mu_n \triangleq \sqrt{\frac{a_i \bar{\gamma}_n}{2 + a_i \bar{\gamma}_n}}$ and the average SNR per symbol is $\bar{\gamma} = D \sum_{n=1}^{N} \bar{\gamma}_n$. Substituting $P_e(\bar{\gamma}; a_i)$ of (31.51) into (31.48), the average BEP of an m-ary QAM Rake receiver using antenna diversity can be expressed in a closed form.

Let us consider the performance of BPSK by setting $D=1$ or $N=1$. When the number of antennae is one, i.e. $D=1$, $P_{e,2}(\bar{\gamma})$ is reduced to

$$P_{e,BPSK} = \sum_{n=1}^{N} \Lambda_{1,n} \left\{ \tfrac{1}{2}(1 - \mu_n) \right\} \tag{31.52}$$

$$= \sum_{n=1}^{N} \pi_n \left\{ \tfrac{1}{2}(1 - \mu_n) \right\} , \tag{31.53}$$

which is identical to the result given on Page 802 of Section (14-5-28) in [388]. On the other hand, when the channels exhibit flat fading, i.e. $L=1$, our system is reduced to a D-antenna diversity-based MRC system transmitting over D number of flat Rayleigh channels. In this case, $\Lambda_{D-d+1,n}$ of (31.38) becomes zero for all values of d, except for $\Lambda_{1,1}=1$ when $d=D$, and the average BPSK BEP in this scenario becomes:

$$P_{e,BPSK} = \sum_{d=1}^{D} \left[\Lambda_{D-d+1,1} \left\{ \tfrac{1}{2}(1 - \mu_1) \right\}^d \sum_{i=0}^{d-1} \binom{d-1+i}{i} \left\{ \tfrac{1}{2}(1 + \mu_1) \right\}^i \right] \tag{31.54}$$

$$= \left\{ \tfrac{1}{2}(1 - \mu_1) \right\}^D \sum_{i=0}^{D-1} \binom{D-1+i}{i} \left\{ \tfrac{1}{2}(1 + \mu_1) \right\}^i , \tag{31.55}$$

which is also given on Page 781 of Section (14-4-15) in [388].

31.3 Mode Specific Average BEP of an Adaptive Modulation Scheme

A closed form solution for the 'mode-specific average BER' of a Maximal Ratio Combining (MRC) receiver using Dth order antenna diversity over independent Rayleigh channels is derived, where the 'mode-specific average BER' refers to the BER of the adaptive modulation scheme, while activating one of its specific constituent modem modes. The PDF $f_{\bar{\gamma}}(\gamma)$ of the channel SNR γ was formulated in Section (14-4-13)] of [388] as:

$$f_{\bar{\gamma}}(\gamma) = \frac{1}{(D-1)! \, \bar{\gamma}^D} \gamma^{D-1} e^{-\gamma/\bar{\gamma}} , \quad \gamma \geq 0 , \tag{31.56}$$

where $\bar{\gamma}$ is the average channel SNR. Since the PDF of the instantaneous channel SNR γ over a Nakagami fading channel is given as:

$$f_{\bar{\gamma}}(\gamma) = \left(\frac{m}{\bar{\gamma}} \right)^m \frac{\gamma^{m-1}}{\Gamma(m)} e^{-m\gamma/\bar{\gamma}} , \quad \gamma \geq 0 , \tag{31.57}$$

the following results can also be applied to a Nakagami fading channel with a simple change of variable given as $D = m$ and $\bar{\gamma} = \bar{\gamma} / m$.

The mode-specific average BEP is defined as

$$P_r(\alpha, \beta; \bar{\gamma}, D, a) \triangleq \int_\alpha^\beta Q(\sqrt{a\gamma}) \, f_{\bar{\gamma}}(\gamma) \, d\gamma \tag{31.58}$$

$$= \int_\alpha^\beta Q(\sqrt{a\gamma}) \, \frac{1}{(D-1)! \, \bar{\gamma}^D} \, \gamma^{D-1} \, e^{-\gamma/\bar{\gamma}} \, d\gamma \,, \tag{31.59}$$

where $Q(x) \triangleq \frac{1}{\sqrt{2\pi}} \int_x^\infty e^{-t^2/2} \, dt$. Applying integration-by-part or $\int u \, dv = uv - \int v \, du$ and noting that

$$u = Q(\sqrt{a\gamma})$$

$$du = -\frac{\sqrt{a}}{2\sqrt{2\pi\gamma}} \, e^{-a\gamma/2}$$

$$dv = \frac{1}{(D-1)! \, \bar{\gamma}^D} \, \gamma^{D-1} \, e^{-\gamma/\bar{\gamma}}$$

$$v = -e^{-\gamma/\bar{\gamma}} \sum_{d=0}^{D-1} (\gamma/\bar{\gamma})^d \frac{1}{d!}$$

(31.58) becomes

$$P_r(\alpha, \beta; \bar{\gamma}, D, a) = \left[e^{-\gamma/\bar{\gamma}} \, Q(\sqrt{a\gamma}) \sum_{d=0}^{D-1} (\gamma/\bar{\gamma})^d \frac{1}{d!} \right]_\beta^\alpha - \sum_{d=0}^{D-1} I_d(\alpha, \beta) \,, \tag{31.60}$$

where

$$I_d(\alpha, \beta) = \int_\alpha^\beta \frac{\sqrt{a}}{2\sqrt{2\pi}} \frac{1}{d!} \, (\gamma/\bar{\gamma})^d \frac{1}{\sqrt{\gamma}} \, e^{-a\gamma/(2\mu^2)} \, d\gamma \tag{31.61}$$

and $\mu = \sqrt{\frac{a\bar{\gamma}}{a\bar{\gamma}+2}}$. Let us consider I_d for the case of $d = 0$:

$$I_0(\alpha, \beta) = \int_\alpha^\beta \frac{\sqrt{a}}{2\sqrt{2\pi}} \frac{1}{\sqrt{\gamma}} \, e^{-a\gamma/(2\mu^2)} \, d\gamma \,. \tag{31.62}$$

Upon introducing the variable $t^2 = a\gamma/\mu^2$ and exploiting that $d\gamma = 2\mu\sqrt{\gamma/a} \, ds$, we have:

$$I_0(\alpha, \beta) = [\mu \, Q(\sqrt{a\gamma}/\mu)]_\beta^\alpha \,. \tag{31.63}$$

Applying integration-by-part once to (31.61) yields

$$I_d(\alpha, \beta) = \left[\frac{\mu^2}{\sqrt{2a\pi}} \frac{1}{d!} \, (\gamma/\bar{\gamma})^d \frac{1}{\sqrt{\gamma}} \, e^{-a\gamma/(2\mu^2)} \right]_\beta^\alpha + \frac{2d-1}{a\bar{\gamma}d} \, \mu^2 \, I_{d-1} \,, \tag{31.64}$$

which is a recursive form with the initial value given in (31.63). For this recursive form of

(31.64), a non-recursive form of I_d can be expressed as:

$$
I_d(\alpha, \beta) = \left[\frac{\mu^2}{\sqrt{2a\pi}} \frac{\Gamma(d+\frac{1}{2})}{\bar{\gamma}^d \, \Gamma(d+1)} \sum_{i=1}^{d} \left(\frac{2\mu^2}{a} \right)^{d-i} \frac{\gamma^{i-\frac{1}{2}}}{\Gamma(i+\frac{1}{2})} e^{-a\gamma/(2\mu^2)} \right]_\beta^\alpha
$$
$$
+ \left[\left(\frac{2\mu^2}{a\bar{\gamma}} \right)^d \frac{1}{\sqrt{\pi}} \frac{\Gamma(d+\frac{1}{2})}{\Gamma(d+1)} \mu \, Q(\sqrt{a\gamma}/\mu) \right]_\beta^\alpha. \tag{31.65}
$$

By substituting $I_d(\alpha, \beta)$ of (31.65) into (31.60), the regional BER $P_r(\bar{\gamma}; a, \alpha, \beta)$ can be represented in a closed form.

Glossary

16QAM	16-level Quadrature Amplitude Modulation
3G	Third generation
4PSK	4-level Phase Shift Keying
4QAM	4-level Quadrature Amplitude Modulation
64QAM	64-level Quadrature Amplitude Modulation
8-DPSK	8-Phase Differential Phase Shift Keying
8PSK	8-level Phase Shift Keying
ACF	autocorrelation function
ADC	Analog–to–Digital Converter
ADM	adaptive delta modulation
ADPCM	Adaptive Differential Pulse Coded Modulation.
AGC	Automatic Gain Control
AM-PM	amplitude modulation and phase modulation
AOFDM	Adaptive Orthogonal Frequency Division Multiplexing
APP	A Posteriori Probability
ARQ	Automatic Repeat Request, Automatic request for retransmission of corrupted data
ATM	Asynchronous Transfer Mode
AWGN	Additive White Gaussian Noise

BbB	Burst-by-Burst
BCH	Bose-Chaudhuri-Hocquenghem, A class of forward error correcting codes (FEC)
BCM	block code modulation
BER	Bit error rate, the fraction of the bits received incorrectly
BICM	Bit Interleaved Coded Modulation
BICM-ID	Bit-Interleaved Coded Modulation with Iterative decoding
BPF	Bandpass Filter
BPS	Bits Per Symbol
BPSK	Binary Phase Shift Keying
BS	A common abbreviation for Base Station
CCI	Co-Channel Interference
CCITT	Now ITU, standardisation group
CD	Code Division, a multiplexing technique where signals are coded and then combined, in such a way that they can be separated using the assigned user signature codes at a later stage.
CDMA	Code Division Multiple Access
CIR	Carrier to Interference Ratio, same as SIR.
CISI	controlled inter-symbol interference
CM	Coded Modulation
CM-GA-MUD	Coded Modulation assisted Genetic Algorithm based Multiuser Detection
CM-JD-CDMA	Coded Modulation-assisted Joint Detection-based CDMA
CRC	Cyclic Redundancy Checksum
CT-TEQ	Conventional Trellis-based Turbo Equalisation
D/A	Digital to Analogue
DAB	Digital Audio Broadcasting

DC	Direct Current, normally used in electronic circuits to describe a power source that has a constant voltage, as opposed to AC power in which the voltage is a sine-wave. It is also used to describe things which are constant, and hence have no frequency component.
DECT	A Pan-European digital cordless telephone standard.
DFE	Decision Feedback Equalizer
DFT	Discrete Fourier Transform
DoS-RR	Double-Spreading aided Rake Receiver
DS	Direct Sequence
DTTB	Digital Terrestrial Television Broadcast
DTX	discontinuous transmission
DVB	Digital Video Broadcasting
ECL	The Effective Code Length or the "length" of the shortest error event path.
EFF	Error Free Feedback
EQ	Equaliser
E_b/N_0	Ratio of bit energy to noise power spectral density.
FD	Frequency Division, a multiplexing technique, where different frequencies are used for each communications link.
FDM	Frequency Division Multiplexing
FEC	Forward Error Correction
FED	Free Euclidean distance
FER	Frame error rate
FFT	Fast Fourier Transform
FSK	Frequency Shift Keying
G	Coding Gain
GA	Genetic Algorithm
GF	Galois field

GMSK	Gaussian Mean Shift Keying, a modulation scheme used by the Pan-European GSM standard by virtue of its spectral compactness.
GSM	A Pan-European digital mobile radio standard, operating at 900MHz.
HT	Hilly Terrain, channel impulse response of a hilly terrain environment.
I	The In-phase component of a complex quantity.
I/Q-TEQ	In-phase/Quadrature-phase Turbo Equalisation
IC	Interference Cancellation
ICI	Inter–Channel Interference
IF	Intermediate Frequency
IFFT	Inverse Fast Fourier Transform
IL	interleaver block length
IMD	Intermodulation Distortion
IQ-CM	IQ-interleaved Coded Modulation
ISI	Inter Symbol Interference, Inter Subcarrier Interference
JD	Joint Detection
JD-MMSE-DFE	Joint Detection scheme employing MMSE-DFE
LAR	Logarithmic area ratio
LMS	Least Mean Square, a stochastic gradient algorithm used in adapting the equalizer's coefficients in a non-stationary environment
log-domain	logarithmic-domain
LOS	Line–Of–Sight
LP	Logarithmic-domain Probability
LPF	low pass filter
LS	Least Square, a category of adaptive algorithms which uses recursive least squres methods in adapting the equalizer or channel estimators in a non-stationary environment
LSB	least significant bit

LSF	Least Squares Fitting
LTP	long term predictor
MAI	Multiple Access Interference
MAP	Maximum–A–Posteriori
MC-CDMA	Multi-Carrier Code Division Multiple Access
MDI	multi-dimensional interference
MIMO	Multi-Input Multi-Output
ML	Maximum Likelihood
MMSE	Minimum Mean Square Error
MMSE-BLE	Minimum Mean Square Error based Block Linear Equaliser
MMSE-DFE	Minimum Mean Square Error based Decision Feedback Equaliser
MPSK	M-ary Phase Shift Keying
MRC	Mixed Radix Conversion
MS	A common abbreviation for Mobile Station
MSE	Mean Square Error, a criterion used to optimised the coefficients of the equalizer such that the ISI and the noise contained in the received signal is jointly minimised.
MUD	Multi-User Detection
NLA	non-linear amplification
NLF	non-linear filtering
OFDM	Orthogonal Frequency Division Multiplexing
OMPX	Orthogonal Multiplexing
OOB	out of band
OQAM	offset quadrature amplitude modulation
OQPSK	offset quadrature phase shift keying
OSWE	one-symbol window equaliser
PAM	pulse amplitude modulation
PCM	pulse code modulation

PCN	Personal Communications Network
PD	phase detector
PDF	Probability Density Function
PLL	phase locked loop
PLMR	Public Land Mobile Radio
PN	Pseudo-Noise
PR	PseudoRandom
PSAM	Pilot symbol assisted modulation, a technique where known symbols (pilots) are transmitted regularly. The effect of channel fading on all symbols can then be estimated by interpolating between the pilots
PSD	Power Spectral Density
PSK	Phase Shift Keying
PSTN	Public switched telephone network
Q	The Quadrature-phase component of a complex quantity.
QAM	Quadrature Amplitude Modulation
QMF	Quadrature Mirror Filtering
QOS	Quality of Service
QPSK	Quaternary Phase Shift Keying
RBF	Radial Basis Function
RBF-DFE	RBF assisted Decision Feedback Equaliser
RBF-TEQ	Radial Basis Function based Turbo Equalisation
RCPC	Rate-Compatible Puncture Convolutional
RF	radio frequency
RLS	Recursive Least Squares, an adaptive filtering technique where a recursive method is used to adapt the filter tap weights such that the square of the error between the filter output and the desired response is minimized
RPE	regular pulse excited

RPE-LTP	Regular pulse excited codec with long term predictor
RRNS	Redundant Residual Number System
RS	Reed Solomon Codes
RSC	Recursive Systematic Convolutional
RSSI	Received Signal Strength Indicator, commonly used as an indicator of channel quality in a mobile radio network.
SbS	Symbol-by-Symbol
SER	Symbol Error Ratio
SINR	Signal to Interference plus Noise ratio, same as signal to noise ratio (SNR), when there is no interference.
SIR	Signal to Interference ratio
SISO	Soft-Input-Soft-Output
SNR	Signal to Noise Ratio, noise energy compared to the signal energy
SOVA	Soft-Output Viterbi Algorithm
SP	Set Partitioning
STB	Space-Time Block
STBC	Space-Time Block Coding
STBC-DoS-RR	Space-Time Block Coding-assisted Double-Spread Rake Receiver
STBC-IQ	Space-Time Block Coding based IQ-interleaved
STC	Space-Time Coding
STP	Short term predictor
STS	Space-Time Spreading
STT	Space-Time Trellis
STTC	Space-Time Trellis Coding
TC	Trellis Coded
TCM	trellis code modulation

TDD	Time-Division Duplex, a technique where the forward and reverse links are multiplexed in time.
TDMA	Time Division Multiple Access
TEQ	Turbo Equalisation
TTCM	Turbo Trellis Coded Modulation
TTIB	transparent tone in band
TU	Typical Urban, channel impulse response of an urban environment.
TuCM	Turbo Coded Modulation
TWT	travelling wave tube
UHF	ultra high frequency
UMTS	Universal Mobile Telecommunications System, a future Pan-European third generation mobile radio standard.
UTRA	UMTS Terrestrial Radio Access
VA	Viterbi Algorithm
VCO	voltage controlled oscillator
VE	Viterbi equalizer
WATM	Wireless Asynchronous Transfer Mode (ATM)
WMF	Whitening Matched Filter
WN	white noise
ZF	Zero Forcing, a criterion used to optimised the coefficients of the equalizer such that the ISI contained in the received signal is totally eliminated.
ZFE	Zero Forcing Equalizer.

Bibliography

[1] S. Stapleton and F. Costescu, "An adaptive predistorter for a power amplifier based on adjacent channel emissions," *IEEE Transactions on Vehicular Technology*, vol. 41, pp. 49–57, February 1992.

[2] S. S. L.D. Quach, "A post-distortion receiver for mobile communications," *IEEE Transactions on Vehicular Technology*, vol. 42, pp. 604–616, November 1993.

[3] S. Haykin, *Adaptive Filter Theory*. Englewood Cliffs, NJ, USA: Prentice-Hall, 1996.

[4] C. Cahn, "Performance of digital phase modulation communication systems," *IRE Transactions on Communications*, vol. CS-7, pp. 3–6, May 1959.

[5] C. Cahn, "Combined digital phase and amplitude modulation communication system," *IRE Transactions on Communications*, vol. CS-8, pp. 150–155, September 1960.

[6] J. Hancock and R. Lucky, "Performance of combined amplitude and phase modulated communications system," *IRE Transactions on Communicationss*, vol. CS-8, pp. 232–237, December 1960.

[7] C. Campopiano and B. Glazer, "A coherent digital amplitude and phase modulation scheme," *IRE Transactions on Communications Systems*, vol. CS-10, pp. 90–95, 1962.

[8] R. Lucky and J. Hancock, "On the optimum performance of m-ary systems having two degrees of freedom," *IRE Transactions on Communications*, vol. CS-10, pp. 185–192, June 1962.

[9] R. Lucky, J. Salz, and E. Weldon, *Principles of Data Communication*. New York, USA: McGraw-Hill, 1968.

[10] J. Salz, J. Sheenhan, and D. Paris, "Data transmission by combined AM and PM," *Bell Systems Technical Journal*, vol. 50, pp. 2399–2419, September 1971.

[11] E. Ho and Y. Yeh, "Error probability of a multilevel digital system with intersymbol interference and Gaussian noise," *Bell Systems Technical Journal*, vol. 50, pp. 1017–1023, March 1971.

[12] G. Foschini, R. Gitlin, and S. Weinstein, "Optimization of two-dimensional signal constellations in the presence of Gaussian noise," *IEEE Transactions on Communications*, vol. COM-22, pp. 28–38, January 1974.

[13] C. Thomas, M. Weidner, and S. Durrani, "Digital amplitude-phase keying with m-ary alphabets," *IEEE Transactions on Communications*, vol. COM-22, pp. 168–180, February 1974.

[14] M. Simon and J. Smith, "Carrier synchronization and detection of QASK signal sets," *IEEE Transactions on Communications*, vol. COM-22, pp. 98–106, February 1974.

[15] M. Simon and J. Smith, "Offset quadrature communications with decision feedback carrier synchronization," *IEEE Transactions on Communications*, vol. COM-22, pp. 1576–1584, October 1974.

[16] J. Smith, "Odd-bit quadrature amplitude-shift keying," *IEEE Transactions on Communications*, vol. COM-23, pp. 385–389, March 1975.

[17] K. Miyauchi, S. Seki, and H. Ishio, "New techniques for generating and detecting multilevel signal formats," *IEEE Transactions on Communications*, vol. COM-24, pp. 263–267, February 1976.

[18] W. Weber, "Differential encoding for multiple amplitude and phase shift keying systems," *IEEE Transactions on Communications*, vol. COM-26, pp. 385–391, March 1978.

[19] P. Dupuis, M. Joindot, A. Leclert, and D. Soufflet, "16 QAM modulation for high capacity digital radio system," *IEEE Transactions on Communications*, vol. COM-27, pp. 1771–1781, December 1979.

[20] I. Horikawa, T. Murase, and Y. Saito, "Design and performance of a 200mbit/s 16 QAM digital radio system," *IEEE Transactions on Communications*, vol. COM-27, pp. 1953–1958, December 1979.

[21] V. Prabhu, "The detection efficiency of 16-ary QAM," *Bell Systems Technical Journal*, vol. 59, pp. 639–656, April 1980.

[22] D. Morais and K. Feher, "NLA-QAM: A method for generating high power QAM signals through non-linear amplification," *IEEE Transactions on Communications*, vol. COM-30, pp. 517–522, March 1982.

[23] T. Hill and K. Feher, "A performance study of NLA 64-state QAM," *IEEE Transactions on Communications*, vol. COM-31, pp. 821–826, June 1983.

[24] D. Tufts, "Nyquist's problem - the joint optimisation of the transmitter and receiver in pulse amplitude modulation," *Proceedings of the IEEE*, vol. 53, pp. 248–260, March 1965.

[25] J. Smith, "The joint optimization of transmitted signal and receiving filter for data transmission filters," *Bell Systems Technical Journal*, vol. 44, pp. 2363–2392, December 1965.

[26] E. Hänsler, "Some properties of transmission systems with minimum mean square error," *IEEE Transactions on Communications Technology (Corresp)*, vol. COM-19, pp. 576–579, August 1971.

[27] T. Ericson, "Structure of optimum receiving filters in data transmission systems," *IEEE Transactions on Information Theory (Corresp)*, vol. IT-17, pp. 352–353, May 1971.

[28] G. Forney Jr, "Maximum likelihood sequence estimation of digital sequences in the presence of intersymbol interference," *IEEE Transactions on Information Theory*, vol. IT-18, pp. 363–378, May 1972.

[29] M. Austin, "Decision feedback equalization for fading dispersive channels," Tech. Rep. 461, M.I.T Research Lab. Electron, August 1971.

[30] P. Monsen, "Feedback equalization for fading dispersive channels," *IEEE Transactions on Information Theory*, vol. IT-17, pp. 1144–1153, January 1971.

[31] J. Salz, "Optimum mean square decision feedback equalization," *Bell Systems Techncial Journal*, vol. 52, pp. 1341–1373, October 1973.

[32] D. Falconer and G. Foschini, "Theory of mmse qam system employing decision feedback equalization," *Bell Systems Technical Journal*, vol. 52, pp. 1821–1849, November 1973.

[33] R. Price, "Non-linearly feedback equalized pam versus capacity for noisy filter channels," in *Rec. Int. Conf. Communication*, pp. 12–17, 1972.

[34] R. Lucky, "A survey of the communication theory literature : 1968–1973," *IEEE Transactions on Information Theory*, vol. IT-19, pp. 725–739, July 1973.

[35] C. Belfiore and J. Park Jr, "Decision feedback equalization," *Proceedings of the IEEE*, vol. 67, pp. 1143–1156, August 1979.

[36] S. Qureshi, "Adaptive equalization," in *Advanced Digital Communications Systems and Signal Processing Techniques* (K.Feher, ed.), pp. 640–713, Englewood Cliffs NJ, USA: Prentice-Hall, 1987.

[37] J.C. Cheung, *Adaptive Equalisers for Wideband TDMA Mobile Radio*. PhD thesis, Department of Electronics and Computer Science, University of Southampton, UK, 1991.

[38] J. Cheung and R. Steele, "Soft-decision feedback equalizer for continuous-phase modulated signals in wide-band mobile radio channels," *IEEE Transactions on Communications*, vol. 42, pp. 1628–1638, February/March/April 1994.

[39] J. Wu, A. Aghvami, and J. Pearson, "A reduced state soft decision feedback viterbi equaliser for mobile radio communications," in *Proceedings of IEEE International Symposium on Personal, Indoor and Mobile Radio Communications*, (Stockholm, Sweden), pp. 234–242, June 1994.

[40] J. Wu and A. Aghvami, "A new adaptive equalizer with channel estimator for mobile radio communications," *IEEE Transactions on Vehicular Technology*, vol. 45, pp. 467–474, August 1996.

[41] Y. Gu and T. Le-Ngoc, "Adaptive combined DFE/MLSE techniques for ISI channels," *IEEE Transactions on Communications*, vol. 44, pp. 847–857, July 1996.

[42]   D. Duttweiler, J. Mazo, and D. Messerschmitt, "An upper bound on the error probability on decision feedback equalization," *IEEE Transactions on Information Theory*, vol. IT-20, pp. 490–497, July 1974.

[43]   J. Smee and N. Beaulieu, "Error-rate evaluating of linear equalization and decision feedback equalization with error rate performance," *IEEE Transactions on Communications*, vol. 46, pp. 656–665, May 1998.

[44]   S. Altekar and N. Beaulieu, "Upper bounds to the error probability of decision feedback equalization," *IEEE Transactions on Communications*, vol. 39, pp. 145–157, January 1993.

[45]   M. Tomlinson, "New automatic equalizer employing modulo arithmetic," *IEE Electronics Letters*, vol. 7, pp. 138–139, March 1971.

[46]   H. Harashima and H. Miyakawa, "Matched transmission technique for channels with intersymbol interference," *IEEE Transactions on Communications*, vol. COM-20, pp. 774–780, August 1972.

[47]   M. Russell and J. Bergmans, "A technique to reduce error propagation in M-ary decision feedback equalization," *IEEE Transactions on Communications*, vol. 43, pp. 2878–2881, December 1995.

[48]   M. Chiani, "Introducing erasures in decision feedback equalization to reduce error propagation," *IEEE Transactions on Communications*, vol. 45, pp. 757–760, July 1997.

[49]   Y. Sato, "A method of self–recovering equalization for multilevel amplitude–modulation systems," *IEEE Transactions on Communications*, vol. COM–23, pp. 679–682, June 1975.

[50]   A. Benveniste, M. Goursat, and G. Ruget, "Robust identification of a nonminimum phase system: Blind adjustment of a linear equalizer in data communications," *IEEE Transactions on Automatic Control*, vol. 25, pp. 385–399,, June 1980.

[51]   M. Goursat and A. Benveniste, "Blind equalizers," *IEEE Transactions on Communications*, vol. COM–28, pp. 871–883, August 1984.

[52]   D. Godard, "Self-recovering equalization and carrier tracking in two-dimensional data communication systems," *IEEE Transactions on Communications*, vol. COM–28, pp. 1867–1875, November 1980.

[53]   G. Foschini, "Equalizing without altering or deleting data," *AT&T Technical Journal*, vol. 64, pp. 1885–1911, October 1985.

[54]   Z. Ding, R. Kennedy, B. Anderson, and R. Johnson, "Ill-convergence of Godard blind equalizers in data communications systems," *IEEE Transactions on Communications*, vol. COM–39, pp. 1313–1327, September 1991.

[55]   S. Bellini, "Bussgang techniques for blind equalisation," in *Proceedings of the IEEE Global Telecommunications Conference*, (Houston, TX, USA), pp. 1634–1640, December 1986.

[56]   J. Bussgang, "Cross–correlation functions of amplitude-distorted Gaussian signals," *MIT Research Laboratory Technical Report*, no. 216, 1952.

[57]   G. Picchi and G. Prati, "Blind equalization and carrier recovery using a "stop–and–go" decision–directed algorithm," *IEEE Transactions on Communications*, vol. COM–35, pp. 877–887, September 1987.

[58]   N. Seshadri, "Joint data and channel estimation using blind Trellis search techniques," *IEEE Transactions on Communications*, vol. 42, pp. 1000–1011, February–April 1994.

[59]   D. Forney, "Maximum-likelihood sequence estimation of digital sequences in the presence of intersymbol interference," *IEEE Transactions on Information Theory*, vol. 18, pp. 363–378, May 1972.

[60]   A. Polydoros, R. Raheli, and C. Tzou, "Per–survivor processing: a general approach to MLSE in uncertain environments," *IEEE Transactions on Communications*, vol. COM–43, pp. 354–364, February–April 1995.

[61]   A. Polydoros and K. Chugg, "MLSE for an unknown channel - Part I: Optimality considerations," *IEEE Transactions on Communications*, vol. 44, pp. 836–846, July 1996.

[62]   K. Chugg and A. Polydoros, "MLSE for an unknown channel - Part II: Tracking performance," *IEEE Transactions on Communications*, vol. 44, pp. 949–958, August 1996.

[63]   C. Antón-Haro, J. Fonolossa, and J. Fonolossa, "Blind channel estimation and data detection using hidden Markov models," *IEEE Transactions on Signal Processing*, vol. 45, pp. 241–247, January 1997.

[64]   H. Cirpan and M. Tsatsanis, "Blind receivers for nonlinearly modulated signals in multipath," *IEEE Transactions on Signal Processing*, vol. 47, pp. 583–586, February 1999.

[65] L. Favalli, A. Mecocci, and P. Savazzi, "Blind MLSE equalizer with fuzzy metric calculation for mobile radio environments," *Electronics Letters*, vol. 33, pp. 1841–1842, October 1997.

[66] K. Chugg, "Acquisition performance of blind sequence detectors using per-survivor processing," in *Proceedings of the 1997 47th IEEE Vehicular Technology Conference*, (Phoenix, USA), pp. 539–543, May 1997.

[67] K. Chugg, "Blind acquisition characteristics of PSP-based sequence detectors," *International Journal on Selected Areas in Communications*, vol. 16, pp. 1518–1529, October 1998.

[68] E. Baccarelli and R. Cusani, "Combined channel estimation and data detection using soft statistics for frequency selective fast-fading digital links," *IEEE Transactions on Communications*, vol. 46, pp. 424–427, April 1998.

[69] S. Chen and Y. Wu, "Maximum likelihood joint channel and data estimation using genetic algorithms," *IEEE Transactions on Signal Processing*, vol. 46, pp. 1469–1473, May 1998.

[70] L. Tong, G. Xu, and T. Kailath, "A new approach to blind identification and equalization of multipath channels," in *Proceedings of the 25th Asilomar Conference*, (Pacific Grove, Canada), pp. 856–860, 4–6 November 1991.

[71] E. Mulines, J. Cardoso, and S. Mayrargue, "Subspace methods for the blind identification of multichannel fir filters," *IEEE Transactions on Signal Processing*, vol. 43, pp. 516–525, February 1995.

[72] M. Tsatsanis and G. Giannakis, "Transmitter induced cyclostationarity for blind channel equalization," *IEEE Transactions on Signal Processing*, vol. 45, pp. 1785–1794, July 1997.

[73] A. Chevreuil, F. Desbouvries, A. Gorokhov, P. Loubaton, and C. Vignat, "Blind equalization in the presence of jammers and unknown noise: Solutions based on second-order cyclostationary statistics," *IEEE Transactions on Signal Processing*, vol. 46, pp. 259–263, January 1998.

[74] A. Chevreuil and P. Loubaton, "Blind second–order identification of FIR channels: Forced cyclostationarity and structured subspace method," *IEEE Signal Processing Letters*, vol. 4, pp. 204–206, July 1997.

[75] M. Tsatsanis and G. Giannakis, "Subspace methods for blind estimation of time-varying FIR channels," *IEEE Transactions on Signal Processing*, vol. 45, pp. 3084–3093, December 1997.

[76] Z. Ding, "Matrix outer-product decomposition method for blind multiple channel identification," *IEEE Transactions on Signal Processing*, vol. 45, pp. 3053–3061, December 1997.

[77] G. Giannakis and E. Serpedin, "Blind identification of ARMA channels with periodically modulated inputs," *IEEE Transactions on Signal Processing*, vol. 46, pp. 3099–3104, November 1998.

[78] G. Giannakis, "Filterbanks for blind channel identification and equalization," *IEEE Signal Processing Letters*, vol. 4, pp. 184–187, June 1997.

[79] R. Heath Jr. and G. Giannakis, "Exploiting input cyclostationarity for blind channel identification in OFDM systems," *IEEE Transactions on Signal Processing*, vol. 47, pp. 848–856, March 1999.

[80] H. Wong and J. Chambers, "Two-stage interference immune blind equaliser which exploits cyclostationary statistics," *Electronics Letters*, vol. 32, pp. 1763–1764, September 1996.

[81] H. Liu, G. Xu, L. Tong, and T. Kailath, "Recent developments in blind channel equalization: From cyclostationarity to subspace," *Signal Processing*, vol. 50, pp. 83–99, April 1996.

[82] Y. Hua, H. Yang, and W. Qiu, "Source correlation compensation for blind channel identification based on second order statistics," *IEEE Signal Processing Letters*, vol. 1, pp. 119–120, August 1994.

[83] Z. Ding, "Characteristics of band-limited channels unidentifiable from second-order cyclostationary statistics," *IEEE Signal Processing Letters*, vol. 3, pp. 150–152, May 1996.

[84] J. Xavier, V. Barroso, and J. Moura, "Closed-form blind channel identification and source separation in SDMA systems through correlative coding," *International Journal on Selected Areas in Communications*, vol. 16, pp. 1506–1517, October 1998.

[85] X. Wang and H. Poor, "Blind equalization and multiuser detection in dispersive CDMA channels," *IEEE Transactions on Communications*, vol. 46, pp. 91–103, January 1998.

[86] X. Wang and H. Poor, "Blind joint equalization and multiuser detection for DS-CDMA in uknowon correlated noise," *IEEE Transactions on Circuits and Systems II: Analog and Digital Signal Processing*, vol. 46, pp. 886–895, July 1999.

[87] J. Zhu, Z. Ding, and X.-R. Cao, "Column–anchored zeroforcing blind equalization for multiuser wireless FIR channels," *International Journal on Selected Areas in Communications*, vol. 17, pp. 411–423, March 1999.

[88] H. Zeng and L. Tong, "Blind channel–estimation using the second–order statistics algorithms," *IEEE Transactions on Signal Processing*, vol. 45, pp. 1919–1930, August 1997.

[89] D. Hatzinakos and C. Nikias, "Blind equalization using a tricepstrum based algorithm," *IEEE Transactions on Communications*, vol. 39, pp. 669–682, May 1991.

[90] D. Boss, K. Kammeyer, and T. Petermann, "Is blind channel estimation feasible in mobile communication systems ?; a study based on GSM," *International Journal on Selected Areas in Communications*, vol. 16, pp. 1479–1492, October 1998.

[91] T. Endres, S. Halford, C. Johnson, and G. Giannakis, "Blind adaptive channel equalization using fractionally-spaced receivers: A comparison study," in *Proceedings of the Conference on Information Sciences and Systems*, (Princeton, USA), 20–22 March 1996.

[92] C. Johnson Jr. and B. Anderson, "Godard blind equalizer error surface characteristics: White, zero–mean binary source case," *International Journal of Adaptive Control and Signal Processing*, vol. 9, pp. 301–324, July–August 1995.

[93] L. Tong and S. Perreau, "Analysis of a nonparametric blind equalizer for discrete-valued signals," *Proceedings of the IEEE*, vol. 86, pp. 1951–1968, March 1996.

[94] J. Proakis, *Digital Communications*. New York, USA: McGraw-Hill, 3rd ed., 1995.

[95] A. Nandi, *Blind Estimation using Higher-Order Statistics*. Dordrecht: Kluwer Academic Publishers, 1999.

[96] C. Becchetti, A. Cocco, and G. Jacovitti, "Performance comparison of second order based blind equalizers in data communication channels," in *Proceedings of the 1997 13th International Conference on Digital Signal Processing, DSP. Part 1 (of 2)*, vol. 1, (Santorini, Greece), pp. 147–150, 2–4 July 1997.

[97] M. Kristensson and B. Ottersten, "Asymptotic comparison of two blind channel identification algorithms," in *Proceedings of the 1997 1st IEEE Signal Processing Workshop on Signal Processing Advances in Wireless Communications, SPAWC'97*, pp. 361–364, 16–18 April 1997.

[98] J. Altuna and B. Mulgrew, "Comparison of cyclostationary blind equalization algorithms in the mobile radio environment," *International Journal of Adaptive Control and Signal Processing*, vol. 12, pp. 267–282, May 1998.

[99] K. Skowratananont and J. Chambers, "Comparison of blind channel estimation and equalisation techniques for a fading environment," in *Proceedings of the 1998 6th IEE Conference on Telecommunications*, no. 451, (Edinburgh, UK), pp. 27–31, 21 March – 2 April 1998.

[100] J. Shynk, P. Gooch, G. Krishnamurthy, and C. Chan, "Comparative performance study of several blind equalization algorithms," in *Proceedings of SPIE — The International Society for Optical Engineering*, vol. 1565, (San Diego, CA, USA), pp. 102–117, 22–24 July 1991.

[101] T. Schirtzinger, X. Li, and W. Jenkins, "Comparison of three algorithms for blind equalization based on the constant modulus error criterion," in *Proceedings of the 1995 International Conference on Acoustics, Speech, and Signal Processing*, vol. Part 2 (of 5), (Detroit, USA), pp. 1049–1052, 9–12 May 1995.

[102] T. Endres, S. Halford, C. Johnson Jr., and G. Giannakis, "Simulated comparisons of blind equalization algorithms for cold start-up applications," *International Journal of Adaptive Control and Signal Processing*, vol. 12, pp. 283–301, May 1998.

[103] K. Feher, ed., *Digital Communications – Satellite/Earth Station Engineering*. Englewood Cliffs, NJ, USA: Prentice-Hall, 1983.

[104] K.-T. Wu and K. Feher, "256-QAM modem performance in distorted channels," *IEEE Transactions on Communications*, vol. COM-33, pp. 487–491, May 1985.

[105] P. Mathiopoulos and K. Feher, "Performance evaluation of a 512-QAM system in distorted channels," *Proceedings Pt F*, vol. 133, pp. 199–204, April 1986.

[106] M. Borgne, "Comparison of high level modulation schemes for high capacity digital radio systems," *IEEE Transactions on Communications*, vol. COM-33, pp. 442–449, May 1985.

[107] M. Shafi and D. Moore, "Further results on adaptive equalizer improvements for 16 QAM and 64 QAM digital radio," *IEEE Transactions on Communications*, vol. COM-34, pp. pp59–66, January 1986.

[108] Y. Saito and Y. Nakamura, "256 QAM modem for high capacity digital radio system," *IEEE Transactions on Communications*, vol. COM-34, pp. 799–805, August 1986.

[109] A. Rustako, L. Greenstein, R. Roman, and A. Saleh, "Using times four carrier recovery in M-QAM digital radio receivers," *IEEE Journal on Selected Areas of Communications*, pp. 524–533, April 1987.

[110] C.-E. Sundberg, W. Wong, and R. Steele, "Logarithmic PCM weighted QAM transmission over Gaussian and Rayleigh fading channels," *IEE Proceedings Pt. F*, vol. 134, pp. 557–570, October 1987.

[111] R. Steele, C.-E. Sundberg, and W. Wong, "Transmission of log-PCM via QAM over Gaussian and Rayleigh fading channels," *IEE Proceedings*, vol. 134, Pt. F, pp. 539–556, October 1987.

[112] L. Hanzo, R. Steele, and P. Fortune, "A subband coding, BCH coding and 16-QAM system for mobile radio speech communication," *IEEE Transactions on Vehicular Technology*, vol. 39, pp. 327–340, November 1990.

[113] H. Sari and S. Moridi, "New phase and frequency detectors for carrier recovery in PSK and QAM systems," *IEEE Transactions on Communications*, vol. COM-36, pp. 1035–1043, September 1988.

[114] J.-I. Chuang, "The effects of time-delay spread on QAM with non-linearly switched filters in a portable radio communications channel," *IEEE Transactions on Communications*, vol. 38, pp. 9–13, February 1989.

[115] J. McGeehan and A. Bateman, "Phase-locked transparent tone in band (TTIB): A new spectrum configuration particularly suited to the transmission of data over SSB mobile radio networks," *IEEE Transactions on Communications*, vol. COM-32, no. 1, pp. 81–87, 1984.

[116] J. Matthews, "Cochannel performance of 16-level QAM with phase locked TTIB/FFSR processing," *IEE colloquium on multi-level modulation*, March 1990.

[117] P. Huish and G. Richman, "Increasing the capacity and quality of digital microwave radio," *IEE colloquium on multi-level modulation*, March 1990.

[118] W. Webb and R. Steele, "16-level circular QAM transmissions over a Rayleigh fading channel," *IEE colloquium on multi-level modulation*, March 1990.

[119] E. Issman and W. Webb, "Carrier recovery for 16-level QAM in mobile radio," *IEE colloquium on multi-level modulation*, March 1990.

[120] W. Peterson and E. Weldon Jr., *Error Correcting Codes*. Cambridge, MA, USA: MIT. Press, 2nd ed., August 1972. ISBN: 0262160390.

[121] W. Webb and R. Steele, "Equaliser techniques for QAM transmissions over dispersive mobile radio channels," *IEE Proceedings, Pt. I*, vol. 138, pp. 566–576, December 1991.

[122] W. Webb, "QAM, the modulation scheme for future mobile radio communications?," *IEE Electronics & Communications Journal*, vol. 4, pp. 1167–176, August 1992.

[123] W. Webb, "Modulation methods for PCNs," *IEEE Communications Magazine*, vol. 30, pp. 90–95, December 1992.

[124] R. Steele and W. Webb, "Variable rate QAM for data transmission over Rayleigh fading channels," in *Proceeedings of Wireless '91*, (Calgary, Alberta), pp. 1–14, IEEE, 1991.

[125] K. Feher, "Modems for emerging digital cellular mobile systems," *IEEE Transactions on Vehicular Technology*, vol. 40, pp. 355–365, May 1991.

[126] M. Iida and K. Sakniwa, "Frequency selective compensation technology of digital 16-QAM for microcellular mobile radio communication systems," in *Proceedings of IEEE VTC '92*, (Denver, CO, USA), pp. 662–665, IEEE, 10–13 May 1992.

[127] R. Castle and J. McGeehan, "A multilevel differential modem for narrowband fading channels," in *Proceedings of IEEE VTC '92*, (Denver, CO, USA), pp. 104–109, IEEE, 10–13 May 1992.

[128] D. Purle, A. Nix, M. Beach, and J. McGeehan, "A preliminary performance evaluation of a linear frequency hopped modem," in *Proceedings of IEEE VTC '92*, (Denver, CO, USA), pp. 120–124, IEEE, 10–13 May 1992.

[129] Y. Kamio and S. Sampei, "Performance of reduced complexity DFE using bidirectional equalizing in land mobile communications," in *Proceedings of IEEE VTC '92*, (Denver, CO, USA), pp. 372–376, IEEE, 10–13 May 1992.

[130] S. S. T. Nagayasu and Y. Kamio, "Performance of 16-QAM with decision feedback equalizer using interpolation for land mobile communications," in *Proceedings of IEEE VTC '92*, (Denver, CO, USA), pp. 384–387, IEEE, 10–13 May 1992.

[131] E. Malkamaki, "Binary and multilevel offset QAM, spectrum efficient modulation schemes for personal communications," in *Proceedings of IEEE VTC '92*, (Denver, CO, USA), pp. 325–328, IEEE, 10–13 May 1992.

[132] Z. Wan and K. Feher, "Improved efficiency CDMA by constant envelope SQAM," in *Proceedings of IEEE VTC '92*, (Denver, CO, USA), pp. 51–55, IEEE, 10–13 May 1992.

[133] H. Sasaoka, "Block coded 16-QAM/TDMA cellular radio system using cyclical slow frequency hopping," in *Proceedings of IEEE VTC '92*, (Denver, CO, USA), pp. 405–408, IEEE, 10–13 May 1992.

[134] P. Kenington, R. Wilkinson, and J. Marvill, "Broadband linear amplifier design for a PCN base-station," in *Proceedings of IEEE Vehicular Technology Conference (VTC'91)*, (St. Louis, MO, USA), pp. 155–160, IEEE, 19–22 May 1991.

[135] R. Wilkinson et al., "Linear transmitter design for MSAT terminals," in *Proceedings of 2nd International Mobile Satellite Conference*, June 1990.

[136] S. Stapleton, G. Kandola, and J. Cavers, "Simulation and analysis of an adaptive predistorter utilizing a complex spectral convolution," *IEEE Transactions on Vehicular Technology*, vol. 41, pp. 387–394, November 1992.

[137] A. Wright and W. Durtler, "Experimental performance of an adaptive digital linearized power amplifier," *IEEE Transactions on Vehicular Technology*, vol. 41, pp. 395–400, November 1992.

[138] M. Faulkner and T. Mattson, "Spectral sensitivity of power amplifiers to quadrature modulator misalignment," *IEEE Transactions on Vehicular Technology*, vol. 41, pp. 516–525, November 1992.

[139] J. Cavers, "An analysis of pilot symbol assisted modulation for Rayleigh fading channels," *IEEE Transactions on Vehicular Technology*, vol. 40, pp. 686–693, November 1991.

[140] S. Sampei and T. Sunaga, "Rayleigh fading compensation for QAM in land mobile radio communications," *IEEE Transactions on Vehicular Technology*, vol. 42, pp. 137–147, May 1993.

[141] T. Sunaga and S. Sampei, "Performance of multi-level QAM with post-detection maximal ratio combining space diversity for digital land-mobile radio communications," *IEEE Transactions on Vehicular Technology*, vol. 42, pp. 294–301, August 1993.

[142] F. Adachi and M. Sawahashi, "Performance analysis of various 16 level modulation schemes under RRayleigh fading," *Electronics Letters*, vol. 28, pp. 1579–1581, November 1992.

[143] R. W. Chang, "Synthesis of band-limited orthogonal signals for multichannel data transmission," *Bell Systems Technical Journal*, vol. 46, pp. 1775–1796, December 1966.

[144] M.S. Zimmermann and A.L. Kirsch, "The AN/GSC-10/KATHRYN/Variable Rate Data Modem for HF Radio," *IEEE Transactions on Communication Technology*, vol. CCM–15, pp. 197–205, April 1967.

[145] S. B. Weinstein and P. M. Ebert, "Data transmission by frequency division multiplexing using the discrete fourier transform," *IEEE Transactions on Communication Technology*, vol. COM–19, pp. 628–634, October 1971.

[146] L.J. Cimini, "Analysis and simulation of a digital mobile channel using orthogonal frequency division multiplexing," *IEEE Transactions on Communications*, vol. 33, pp. 665–675, July 1985.

[147] M. Alard and R. Lassalle, "Principles of modulation and channel coding for digital broadcasting for mobile receivers," *EBU Review, Technical No. 224*, pp. 47–69, August 1987.

[148] *Proceedings of 1st International Symposium,DAB*, (Montreux, Switzerland), June 1992.

[149] A. Peled and A. Ruiz, "Frequency domain data transmission using reduced computational complexity algorithms," in *Proceedings of International Conference on Acoustics, Speech, and Signal Processing, ICASSP'80*, vol. 3, (Denver, CO, USA), pp. 964–967, IEEE, 9–11 April 1980.

[150] B. Hirosaki, "An orthogonally multiplexed QAM system using the discrete fourier transform," *IEEE Transactions on Communications*, vol. COM-29, pp. 983–989, July 1981.

[151] H. Kolb, "Untersuchungen über ein digitales mehrfrequenzverfahren zur datenübertragung," in *Ausgewählte Arbeiten über Nachrichtensysteme*, no. 50, Universität Erlangen-Nürnberg, 1982.

[152] H. Schüssler, "Ein digitales Mehrfrequenzverfahren zur Datenübertragung," in *Professoren-Konferenz, Stand und Entwicklungsaussichten der Daten und Telekommunikation*, (Darmstadt, Germany), pp. 179–196, 1983.

[153] K. Preuss, "Ein Parallelverfahren zur schnellen Datenübertragung Im Ortsnetz," in *Ausgewählte Arbeiten über Nachrichtensysteme*, no. 56, Universität Erlangen-Nürnberg, 1984.

[154] R. Rückriem, "Realisierung und messtechnische Untersuchung an einem digitalen Parallelverfahren zur Datenübertragung im Fernsprechkanal," in *Ausgewählte Arbeiten über Nachrichtensysteme*, no. 59, Universität Erlangen-Nürnberg, 1985.

[155] I. Kalet, "The multitone channel," *IEEE Transactions on Communications*, vol. 37, pp. 119–124, February 1989.

[156] B. Hirosaki, "An analysis of automatic equalizers for orthogonally multiplexed QAM systems," *IEEE Transactions on Communications*, vol. COM-28, pp. 73–83, January 1980.

[157] L. Hanzo, R. Salami, R. Steele, and P. Fortune, "Transmission of digitally encoded speech at 1.2 Kbaud for PCN," *IEE Proceedings, Part I*, vol. 139, pp. 437–447, August 1992.

[158] P. Fortune, L. Hanzo, and R. Steele, "On the computation of 16-QAM and 64-QAM performance in Rayleigh-fading channels," *IEICE Transactions on Communications*, vol. E75-B, pp. 466–475, June 1992.

[159] R. Stedman, H. Gharavi, L. Hanzo, and R. Steele, "Transmission of subband-coded images via mobile channels," *IEEE Transactions on Circuits and Systems for Video Technology*, vol. 3, pp. 15–27, February 1993.

[160] X. Lin, L. Hanzo, R. Steele, and W. Webb, "A subband-multipulse digital audio broadcasting scheme for mobile receivers," *IEEE Transactions on Broadcasting*, vol. 39, pp. 373–382, December 1993.

[161] W. Webb, R. Steele, J. Cheung, and L. Hanzo, "A packet reservation multiple access assisted cordless telecommunications scheme," *IEEE Transactions on Vehicular Technology*, vol. 43, pp. 234–245, May 1994.

[162] L. Hanzo, W. Webb, R. Salami, and R. Steele, "On QAM speech transmission schemes for microcellular mobile PCNs," *European Transactions on Communications*, pp. 495–510, September/October 1993.

[163] L. Hanzo, J. Streit, R. Salami, and W. Webb, "A low-rate multi-level voice/video transceiver for personal communications," *Wireless Personal Communications*, vol. 2, no. 3, pp. 217–234, 1995.

[164] L. Hanzo, R. Stedman, R. Steele, and J. Cheung, "A mobile speech/video/data transceiver scheme," in *Proceedings of IEEE VTC '94*, (Stockholm, Sweden), pp. 452–456, IEEE, 8–10 June 1994.

[165] L. Hanzo, X. Lin, R. Steele, and W. Webb, "A mobile hi-fi digital audio broadcasting scheme," in *Proceedings of IEEE VTC '94*, (Stockholm, Sweden), pp. 1035–1039, IEEE, 8–10 June 1994.

[166] J. Woodard and L. Hanzo, "A dual-rate algebraic CELP-based speech transceiver," in *Proceedings of IEEE VTC '94*, vol. 3, (Stockholm, Sweden), pp. 1690–1694, IEEE, 8–10 June 1994.

[167] J. Streit and L. Hanzo, "A fractal video communicator," in *Proceedings of IEEE VTC '94*, (Stockholm, Sweden), pp. 1030–1034, IEEE, 8–10 June 1994.

[168] L. Hanzo and P. Cherriman and J. Streit, *Wireless Video Communications: From Second to Third Generation Systems, WLANs and Beyond*. IEEE Press, 2001. (For detailed contents please refer to http://www-mobile.ecs.soton.ac.uk.).

[169] L. Hanzo, F. Somerville, and J. Woodard, *Voice Compression and Communications: Principles and Applications for Fixed and Wireless Channels*. IEEE Press. 2001 (For detailed contents, please refer to http://www-mobile.ecs.soton.ac.uk.).

[170] L. Hanzo, C. Wong, and M. Yee, *Adaptive Wireless Transceivers*. John Wiley, IEEE Press, 2002. (For detailed contents, please refer to http://www-mobile.ecs.soton.ac.uk.).

[171] L. Hanzo, T. Liew, and B. Yeap, *Turbo Coding, Turbo Equalisation and Space-Time Coding*. John Wiley, IEEE Press, 2002. (For detailed contents, please refer to http://www-mobile.ecs.soton.ac.uk.).

[172] L. Hanzo, L. L. Yang, E. L. Kuan, and K. Yen, *Single- and Multi-Carrier CDMA*. John Wiley and IEEE Press, 2003.

[173] L. Hanzo and M. Münster and B-J. Choi and T. Keller, *OFDM versus MC-CDMA for Broadband Multi-User Communications, WLANs and Broadcasting*. John Wiley and IEEE press, 2003.

[174] J. F. Hayes, "Adaptive feedback communications," *IEEE Transactions on Communication Technology*, vol. 16, no. 1, pp. 29–34, 1968.

[175] A. Duel-Hallen and S. Hu and H. Hallen, "Long range prediction of fading signals," *IEEE Signal Processing Magazine*, vol. 17, pp. 62–75, May 2000.

[176] J. K. Cavers, "Variable rate transmission for Rayleigh fading channels," *IEEE Transactions on Communications Technology*, vol. COM-20, pp. 15–22, February 1972.

[177] W. T. Webb and R. Steele, "Variable rate QAM for mobile radio," *IEEE Transactions on Communications*, vol. 43, no. 7, pp. 2223–2230, 1995.

[178] M. Moher and J. Lodge, "TCMP—a modulation and coding strategy for rician fading channels," *IEEE Journal on Selected Areas in Communications*, vol. 7, pp. 1347–1355, December 1989.

[179] S. Otsuki, S. Sampei, and N. Morinaga, "Square QAM adaptive modulation/TDMA/TDD systems using modulation level estimation with Walsh function," *Electronics Letters*, vol. 31, pp. 169–171, February 1995.

[180] L. Hanzo, W. Webb, and T. Keller, *Single- and Multi-carrier Quadrature Amplitude Modulation*. New York, USA: IEEE Press-John Wiley, April 2000.

[181] W. Lee, "Estimate of channel capacity in Rayleigh fading environment," *IEEE Transactions on Vehicular Technology*, vol. 39, pp. 187–189, August 1990.

[182] A. Goldsmith and P. Varaiya, "Capacity of fading channels with channel side information," *IEEE Transactions on Information Theory*, vol. 43, pp. 1986–1992, November 1997.

[183] M. S. Alouini and A. J. Goldsmith, "Capacity of Rayleigh fading channels under different adaptive transmission and diversity-combining technique," *IEEE Transactions on Vehicular Technology*, vol. 48, pp. 1165–1181, July 1999.

[184] A. Goldsmith and S. Chua, "Variable rate variable power MQAM for fading channels," *IEEE Transactions on Communications*, vol. 45, pp. 1218–1230, October 1997.

[185] J. Torrance and L. Hanzo, "Optimisation of switching levels for adaptive modulation in a slow Rayleigh fading channel," *Electronics Letters*, vol. 32, pp. 1167–1169, 20 June 1996.

[186] B. J. Choi and L. Hanzo, "Optimum mode-switching levels for adaptive modulation systems," in *Proceedings of IEEE GLOBECOM 2001*, 2001.

[187] B. J. Choi, M. Münster, L. L. Yang, and L. Hanzo, "Performance of Rake receiver assisted adaptive-modulation based CDMA over frequency selective slow Rayleigh fading channel," *Electronics Letters*, vol. 37, pp. 247–249, February 2001.

[188] W. H. Press, S. A. Teukolsky, W. T. Vetterling, and B. P. Flannery, *Numerical Recipes in C*. Cambridge University Press, 1992.

[189] C. Tang, "An intelligent learning scheme for adaptive modulation," *Proceedings of the IEEE Vehicular Technology Conference*, pp. 718–719, Oct 2001.

[190] J. Torrance and L. Hanzo, "Upper bound performance of adaptive modulation in a slow Rayleigh fading channel," *Electronics Letters*, vol. 32, pp. 718–719, 11 April 1996.

[191] C. Wong and L. Hanzo, "Upper-bound of a wideband burst-by-burst adaptive modem," in *Proceeding of VTC'99 (Spring)*, (Houston, TX, USA), pp. 1851–1855, IEEE, 16–20 May 1999.

[192] C. Wong and L. Hanzo, "Upper-bound performance of a wideband burst-by-burst adaptive modem," *IEEE Transactions on Communications*, vol. 48, pp. 367–369, March 2000.

[193] H. Matsuoka and S. Sampei and N. Morinaga and Y. Kamio, "Adaptive Modulation System with Variable Coding Rate Concatenated Code for High Quality Multi-Media Communications Systems," in *Proceedings of IEEE VTC'96*, vol. 1, (Atlanta, GA, USA), pp. 487–491, IEEE, 28 April–1 May 1996.

[194] A. J. Goldsmith and S. G. Chua, "Adaptive coded modulation for fading channels," in *Proceedings of IEEE International Conference on Communications*, vol. 3, (Montreal, Canada), pp. 1488–1492, 8–12 June 1997.

[195] A.J. Goldsmith and S. Chua, "Variable-rate variable-power MQAM for fading channels," *IEEE Transactions on Communications*, vol. 45, pp. 1218–1230, October 1997.

[196] J. Torrance and L. Hanzo, "Demodulation level selection in adaptive modulation," *Electronics Letters*, vol. 32, pp. 1751–1752, 12 September 1996.

[197] V. Lau and S. Maric, "Variable rate adaptive modulation for DS-CDMA," *IEEE Transactions on Communications*, vol. 47, pp. 577–589, April 1999.

[198] S. Sampei, N. Morinaga, and Y. Kamio, "Adaptive modulation/TDMA with a BDDFE for 2 mbit/s multi-media wireless communication systems," in *Proceedings of IEEE Vehicular Technology Conference (VTC'95)*, vol. 1, (Chicago, USA), pp. 311–315, IEEE, 15–28 July 1995.

[199] J. Torrance and L. Hanzo, "Latency considerations for adaptive modulation in a slow Rayleigh fading channel," in *Proceedings of IEEE VTC'97*, vol. 2, (Phoenix, AZ, USA), pp. 1204–1209, IEEE, 4–7 May 1997.

[200] J. Torrance and L. Hanzo, "Statistical multiplexing for mitigating latency in adaptive modems," in *Proceedings of IEEE International Symposium on Personal, Indoor and Mobile Radio Communications, PIMRC'97*, (Marina Congress Centre, Helsinki, Finland), pp. 938–942, IEEE, 1–4 September 1997.

[201] T. Ue, S. Sampei, and N. Morinaga, "Symbol rate controlled adaptive modulation/TDMA/TDD for wireless personal communication systems," *IEICE Transactions on Communications*, vol. E78-B, pp. 1117–1124, August 1995.

[202] M. Yee and L. Hanzo, "Radial Basis Function decision feedback equaliser assisted burst-by-burst adaptive modulation," in *Proceedings of IEEE Global Telecommunications Conference (GLOBECOM)*, (Rio de Janeiro, Brazil), 5–9 December 1999.

[203] M. Yee, T. Liew, and L. Hanzo, "Radial basis function decision feedback equalisation assisted block turbo burst-by-burst adaptive modems," in *Proceedings of VTC '99 Fall*, (Amsterdam, Holland), pp. 1600–1604, 19-22 September 1999.

[204] M. S. Yee, B. L. Yeap, and L. Hanzo, "Radial basis function assisted turbo equalisation," in *Proceedings of IEEE Vehicular Technology Conference*, (Tokyo, Japan), pp. 640–644, IEEE, 15-18 May 2000.

[205] M. S. Yee and T. H. Liew and L. Hanzo, "Burst-by-burst adaptive turbo-coded radial basis function-assisted decision feedback equalization," *IEEE Transactions on Communications*, pp. 1935–1945, Nov. 2001.

[206] M. S. Yee and B. L. Yeap and L. Hanzo, "RBF-based decision feedback aided turbo equalisation of convolutional and space-time trellis coded systems," *IEE Electronics Letters*, pp. 1298–1299, October 2001.

[207] M. S. Yee, B. L. Yeap, and L. Hanzo, "Turbo equalisation of convolutional coded and concatenated space time trellis coded systems using radial basis function aided equalizers," in *Proceedings of Vehicular Technology Conference*, (Atlantic City, USA), pp. 882–886, Oct 7-11 2001.

[208] D. Göckel, "Adaptive coding for fading channels using outdated fading estimates," *IEEE Transactions on Communications*, vol. 47, pp. 844–855, June 1999.

[209] K. J. Hole, H. Holm, and G. E. Oien, "Adaptive multidimensional coded modulation over flat fading channels," *IEEE Journal on Selected Areas in Communications*, vol. 18, pp. 1153–1158, July 2000.

[210] D. Pearce, A. Burr, and T. Tozer, "Comparison of counter-measures against slow Rayleigh fading for TDMA systems," in *IEE Colloquium on Advanced TDMA Techniques and Applications*, (London, UK), pp. 9/1–9/6, IEE, 28 October 1996. digest 1996/234.

[211] V.K.N. Lau and M.D. Macleod, "Variable rate adaptive trellis coded QAM for high bandwidth efficiency applications in Rayleigh fading channels," in *Proceedings of IEEE Vehicular Technology Conference (VTC'98)*, (Ottawa, Canada), pp. 348–352, IEEE, 18–21 May 1998.

[212] S. X. Ng, C. H. Wong and L. Hanzo, "Burst-by-burst adaptive decision feedback equalized tcm, ttcm, bicm and bicm-id," in *International Conference on Communications (ICC)*, (Helsinki, Finland), pp. 3031–3035, June 2001.

[213] T. Suzuki, S. Sampei, and N. Morinaga, "Space and path diversity combining technique for 10 Mbits/s adaptive modulation/TDMA in wireless communications systems," in *Proceedings of IEEE VTC'96*, (Atlanta, GA, USA), pp. 1003–1007, IEEE, 28 April–1 May 1996.

[214] K. Arimochi, S. Sampei, and N. Morinaga, "Adaptive modulation system with discrete power control and predistortion-type non-linear compensation for high spectral efficient and high power efficient wireless communication systems," in *Proceedings of the IEEE International Symposium on Personal, Indoor and Mobile Radio Communications (PIMRC)*, (Helsinki, Finland), pp. 472–477, 1–4 September 1997.

[215] T. Ikeda, S. Sampei, and N. Morinaga, "TDMA-based adaptive modulation with dynamic channel assignment (AMDCA) for high capacity multi-media microcellular systems," in *Proceedings of IEEE Vehicular Technology Conference*, (Phoenix, USA), pp. 1479–1483, May 1997.

[216] T. Ue, S. Sampei, and N. Morinaga, "Adaptive modulation packet radio communication system using NP-CSMA/TDD scheme," in *Proceedings of IEEE VTC'96*, (Atlanta, GA, USA), pp. 416–421, IEEE, 28 April–1 May 1996.

[217] M. Naijoh, S. Sampei, N. Morinaga, and Y. Kamio, "ARQ schemes with adaptive modulation/TDMA/TDD systems for wireless multimedia communication systems," in *Proceedings of the IEEE International Symposium on Personal, Indoor and Mobile Radio Communications (PIMRC)*, (Helsinki, Finland), pp. 709–713, 1–4 September 1997.

[218] S. Sampei, T. Ue, N. Morinaga, and K. Hamguchi, "Laboratory experimental results of an adaptive modulation TDMA/TDD for wireless multimedia communication systems," in *Proceedings of IEEE International Symposium on Personal, Indoor and Mobile Radio Communications, PIMRC'97*, (Marina Congress Centre, Helsinki, Finland), pp. 467–471, IEEE, 1–4 September 1997.

[219] J.M. Torrance and L. Hanzo, "Latency and networking aspects of adaptive modems over slow indoors Rayleigh fading channels," *IEEE Transactions on Vehicular Technology*, vol. 48, no. 4, pp. 1237–1251, 1998.

[220] J. Torrance, L. Hanzo, and T. Keller, "Interference aspects of adaptive modems over slow Rayleigh fading channels," *IEEE Transactions on Vehicular Technology*, vol. 48, pp. 1527–1545, September 1999.

[221] A. Czylwik, "Adaptive OFDM for wideband radio channels," in *Proceeding of IEEE Global Telecommunications Conference, Globecom 96*, (London, UK), pp. 713–718, IEEE, 18–22 November 1996.

[222] P. Chow, J. Cioffi, and J. Bingham, "A practical discrete multitone transceiver loading algorithm for data transmission over spectrally shaped channels," *IEEE Transactions on Communications*, vol. 48, pp. 772–775, 1995.

[223] P. Bello, "Selective fading limitations of the KATHRYN modem and some system design considerations," *IEEE Trabsactions on Communications Technology*, vol. COM–13, pp. 320–333, September 1965.

[224] E. Powers and M. Zimmermann, "A digital implementation of a multichannel data modem," in *Proceedings of the IEEE International Conference on Communications*, (Philadelphia, USA), 1968.

[225] R. Chang and R. Gibby, "A theoretical study of performance of an orthogonal multiplexing data transmission scheme," *IEEE Transactions on Communication Technology*, vol. COM–16, pp. 529–540, August 1968.

[226] B. R. Saltzberg, "Performance of an efficient parallel data transmission system," *IEEE Transactions on Communication Technology*, pp. 805–813, December 1967.

[227] K. Fazel and G. Fettweis, eds., *Multi-Carrier Spread-Spectrum*. Dordrecht: Kluwer, 1997. ISBN 0-7923-9973-0.

[228] F. Classen and H. Meyr, "Synchronisation algorithms for an OFDM system for mobile communications," in *Codierung für Quelle, Kanal und Übertragung*, no. 130 in ITG Fachbericht, (Berlin), pp. 105–113, VDE–Verlag, 1994.

[229] F. Classen and H. Meyr, "Frequency synchronisation algorithms for OFDM systems suitable for communication over frequency selective fading channels," in *Proceedings of IEEE VTC '94*, (Stockholm, Sweden), pp. 1655–1659, IEEE, 8–10 June 1994.

[230] S. Shepherd, P. van Eetvelt, C. Wyatt-Millington, and S. Barton, "Simple coding scheme to reduce peak factor in QPSK multicarrier modulation," *Electronics Letters*, vol. 31, pp. 1131–1132, July 1995.

[231] A. E. Jones, T. A. Wilkinson, and S. K. Barton, "Block coding scheme for reduction of peak to mean envelope power ratio of multicarrier transmission schemes," *Electronics Letters*, vol. 30, pp. 2098–2099, December 1994.

[232] D. Wulich, "Reduction of peak to mean ratio of multicarrier modulation by cyclic coding," *Electronics Letters*, vol. 32, pp. 432–433, 1996.

[233] S. Müller and J. Huber, "Vergleich von OFDM–Verfahren mit reduzierter Spitzenleistung," in *2. OFDM–Fachgespräch in Braunschweig*, 1997.

[234] M. Pauli and H.-P. Kuchenbecker, "Neue Aspekte zur Reduzierung der durch Nichtlinearitäten hervorgerufenen Außerbandstrahlung eines OFDM–Signals," in *2. OFDM–Fachgespräch in Braunschweig*, 1997.

[235] T. May and H. Rohling, "Reduktion von Nachbarkanalstörungen in OFDM–Funkübertragungssystemen," in *2. OFDM–Fachgespräch in Braunschweig*, 1997.

[236] D. Wulich, "Peak factor in orthogonal multicarrier modulation with variable levels," *Electronics Letters*, vol. 32, no. 20, pp. 1859–1861, 1996.

[237] H. Schmidt and K. Kammeyer, "Adaptive Subträgerselektion zur Reduktion des Crest faktors bei OFDM," in *3. OFDM Fachgespräch in Braunschweig*, 1998.

[238] R. Dinis and A. Gusmao, "Performance evaluation of OFDM transmission with conventional and 2-branch combining power amplification schemes," in *Proceeding of IEEE Global Telecommunications Conference, Globecom 96*, (London, UK), pp. 734–739, IEEE, 18–22 November 1996.

[239] R. Dinis, P. Montezuma, and A. Gusmao, "Performance trade-offs with quasi-linearly amplified OFDM through a 2-branch combining technique," in *Proceedings of IEEE VTC'96*, (Atlanta, GA, USA), pp. 899–903, IEEE, 28 April–1 May 1996.

[240] R. Dinis, A. Gusmao, and J. Fernandes, "Adaptive transmission techniques for the mobile broadband system," in *Proceeding of ACTS Mobile Communication Summit '97*, (Aalborg, Denmark), pp. 757–762, ACTS, 7–10 October 1997.

[241] B. Daneshrad, L. Cimini Jr., and M. Carloni, "Clustered-OFDM transmitter implementation," in *Proceedings of IEEE International Symposium on Personal, Indoor, and Mobile Radio Communications (PIMRC'96)*, (Taipei, Taiwan), pp. 1064–1068, IEEE, 15–18 October 1996.

[242] M. Okada, H. Nishijima, and S. Komaki, "A maximum likelihood decision based nonlinear distortion compensator for multi-carrier modulated signals," *IEICE Transactions on Communications*, vol. E81B, no. 4, pp. 737–744, 1998.

[243] R. Dinis and A. Gusmao, "Performance evaluation of a multicarrier modulation technique allowing strongly nonlinear amplification," in *Proceedings of ICC 1998*, pp. 791–796, IEEE, 1998.

[244] T. Pollet, M. van Bladel, and M. Moeneclaey, "BER sensitivity of OFDM systems to carrier frequency offset and Wiener phase noise," *IEEE Transactions on Communications*, vol. 43, pp. 191–193, February/March/April 1995.

[245] H. Nikookar and R. Prasad, "On the sensitivity of multicarrier transmission over multipath channels to phase noise and frequency offset," in *Proceedings of IEEE International Symposium on Personal, Indoor, and Mobile Radio Communications (PIMRC'96)*, (Taipei, Taiwan), pp. 68–72, IEEE, 15–18 October 1996.

[246] W. Warner and C. Leung, "OFDM/FM frame synchronization for mobile radio data communication," *IEEE Transactions on Vehicular Technology*, vol. 42, pp. 302–313, August 1993.

[247] H. Sari, G. Karam, and I. Jeanclaude, "Transmission techniques for digital terrestrial TV broadcasting," *IEEE Communications Magazine*, pp. 100–109, February 1995.

[248] P. Moose, "A technique for orthogonal frequency division multiplexing frequency offset correction," *IEEE Transactions on Communications*, vol. 42, pp. 2908–2914, October 1994.

[249] K. Brüninghaus and H. Rohling, "Verfahren zur Rahmensynchronisation in einem OFDM-System," in *3. OFDM Fachgespräch in Braunschweig*, 1998.

[250] F. Daffara and O. Adami, "A new frequency detector for orthogonal multicarrier transmission techniques," in *Proceedings of IEEE Vehicular Technology Conference (VTC'95)*, (Chicago, USA), pp. 804–809, IEEE, 15–28 July 1995.

[251] M. Sandell, J.-J. van den Beek, and P. Börjesson, "Timing and frequency synchronisation in OFDM systems using the cyclic prefix," in *Proceedings of International Symposium on Synchronisation*, (Essen, Germany), pp. 16–19, 14–15 December 1995.

[252] N. Yee, J.-P. Linnartz, and G. Fettweis, "Multicarrier CDMA in indoor wireless radio networks," in *PIMRC'93*, pp. 109–113, 1993.

[253] A. Chouly, A. Brajal, and S. Jourdan, "Orthogonal multicarrier techniques applied to direct sequence spread spectrum CDMA systems," in *Proceedings of the IEEE Global Telecommunications Conference 1993*, (Houston, TX, USA), pp. 1723–1728, 29 November – 2 December 1993.

[254] G. Fettweis, A. Bahai, and K. Anvari, "On multi-carrier code division multiple access (MC-CDMA) modem design," in *Proceedings of IEEE VTC '94*, (Stockholm, Sweden), pp. 1670–1674, IEEE, 8–10 June 1994.

[255] K. Fazel and L. Papke, "On the performance of convolutionally-coded CDMA/OFDM for mobile communication system," in *PIMRC'93*, pp. 468–472, 1993.

[256] R. Prasad and S. Hara, "Overview of multicarrier CDMA," *IEEE Communications Magazine*, pp. 126–133, December 1997.

[257] B.-J. Choi, E.-L. Kuan, and L. Hanzo, "Crest–factor study of MC-CDMA and OFDM," in *Proceeding of VTC'99 (Fall)*, vol. 1, (Amsterdam, Netherlands), pp. 233–237, IEEE, 19–22 September 1999.

[258] Y. Li and N. Sollenberger, "Interference suppression in OFDM systems using adaptive antenna arrays," in *Proceedings of Globecom'98*, (Sydney, Australia), pp. 213–218, IEEE, 8–12 November 1998.

[259] Y. Li and N. Sollenberger, "Adaptive antenna arrays for OFDM systems with cochannel interference," *IEEE Transactions on Communications*, vol. 47, pp. 217–229, February 1999.

[260] Y. Li, L. Cimini, and N. Sollenberger, "Robust channel estimation for OFDM systems with rapid dispersive fading channels," *IEEE Transactions on Communications*, vol. 46, pp. 902–915, April 1998.

[261] C. Kim, S. Choi, and Y. Cho, "Adaptive beamforming for an OFDM sytem," in *Proceeding of VTC'99 (Spring)*, (Houston, TX, USA), IEEE, 16–20 May 1999.

[262] L. Lin, L. Cimini Jr., and J.-I. Chuang, "Turbo codes for OFDM with antenna diversity," in *Proceeding of VTC'99 (Spring)*, (Houston, TX, USA), IEEE, 16–20 May 1999.

[263] M. Münster, T. Keller, and L. Hanzo, "Co–channel interference suppression assisted adaptive OFDM in interference limited environments," in *Proceeding of VTC'99 (Fall)*, vol. 1, (Amsterdam, Netherlands), pp. 284–288, IEEE, 19–22 September 1999.

[264] J. Blogh and L. Hanzo, *3G Systems and Intelligent Networking*. John Wiley and IEEE Press, 2002. (For detailed contents, please refer to http://www-mobile.ecs.soton.ac.uk.).

[265] P. Höher, "TCM on frequency-selective land-mobile fading channels," in *International Workshop on Digital Communications*, (Tirrenia, Italy), pp. 317–328, September 1991.

[266] J. Chow, J. Cioffi, and J. Bingham, "Equalizer training algorithms for multicarrier modulation systems.," in *International Conference on Communications*, (Geneva, Switzerland), pp. 761–765, IEEE, May 1993.

[267] S. Wilson, R. E. Khayata, and J. Cioffi, "16QAM modulation with orthogonal frequency division multiplexing in a Rayleigh-fading environment," in *Vehicular Technology Conference*, vol. 3, (Stockholm, Sweden), pp. 1660–1664, IEEE, June 1994.

[268] J.-J. van den Beek, O. Edfors, M. Sandell, S. Wilson, and P. Börjesson, "On channel estimation in OFDM systems," in *Proceedings of Vehicular Technology Conference*, vol. 2, (Chicago, IL USA), pp. 815–819, IEEE, July 1995.

[269] O. Edfors, M. Sandell, J. van den Beek, S. K. Wilson, and P. Börjesson, "OFDM channel estimation by singular value decomposition," in *Proceedings of Vehicular Technology Conference*, vol. 2, (Atlanta, GA USA), pp. 923–927, IEEE, April 28 - May 1 1996.

[270] P. Frenger and A. Svensson, "A decision directed coherent detector for ofdm," in *Proceedings of Vehicular Technology Conference*, vol. 3, (Atlanta, GA USA), pp. 1584–1588, IEEE, Apr 28 - May 1 1996.

[271] V. Mignone and A. Morello, "CD3-OFDM: A novel demodulation scheme for fixed and mobile receivers," *IEEE Transactions on Communications*, vol. 44, pp. 1144–1151, September 1996.

[272] F. Tufvesson and T. Maseng, "Pilot assisted channel estimation for ofdm in mobile cellular systems," in *Proceedings of Vehicular Technology Conference*, vol. 3, (Phoenix, Arizona), pp. 1639–1643, IEEE, May 4-7 1997.

[273] P. Höher, S. Kaiser, and P. Robertson, "Two-dimensional pilot-symbol-aided channel estimation by Wiener filtering," in *International Conference on Acoustics, Speech and Signal Processing*, (Munich, Germany), pp. 1845–1848, IEEE, April 1997.

[274] P. Höher, S. Kaiser, and P. Robertson, "Pilot–symbol–aided channel estimation in time and frequency," in *Proceedings of Global Telecommunications Conference: The Mini–Conf.*, (Phoenix, AZ), pp. 90–96, IEEE, November 1997.

[275] O. Edfors, M. Sandell, J.-J. van den Beek, S. Wilson, and P. Börjesson, "OFDM channel estimation by singular value decomposition," *IEEE Transactions on Communications*, vol. 46, pp. 931–939, April 1998.

[276] F. Tufvesson, M. Faulkner, and T. Maseng, "Pre-compensation for rayleigh fading channels in time division duplex ofdm systems," in *Proceedings of 6th International Workshop on Intelligent Signal Processing and Communications Systems*, (Melbourne, Australia), pp. 57–33, IEEE, November 5-6 1998.

[277] M. Itami, M. Kuwabara, M. Yamashita, H. Ohta, and K. Itoh, "Equalization of orthogonal frequency division multiplexed signal by pilot symbol assisted multipath estimation," in *Proceedings of Global Telecommunications Conference*, vol. 1, (Sydney, Australia), pp. 225–230, IEEE, November 8-12 1998.

[278] E. Al-Susa and R. Ormondroyd, "A Predictor-based decision feedback channel estimation method for COFDM with high resilience to rapid time-variations," in *Proceedings of Vehicular Technology Conference*, vol. 1, (Amsterdam, Netherlands), pp. 273–278, IEEE, September 19-22 1999.

[279] B. Yang, K. Letaief, R. Cheng, and Z. Cao, "Robust and improved channel estimation for OFDM systems in frequency selective fading channels," in *Proceedings of Global Telecommunications Conference*, vol. 5, (Rio de Janeiro, Brazil), pp. 2499–2503, IEEE, December 5-9 1999.

[280] Y. Li, "Pilot-symbol-aided channel estimation for OFDM in wireless systems," *IEEE Transactions on Vehicular Technology*, vol. 49, pp. 1207–1215, July 2000.

[281] B. Yang, K. Letaief, R. Cheng, and Z. Cao, "Channel estimation for OFDM transmission in multipath fading channels based on parametric channel modeling," *IEEE Transactions on Communications*, vol. 49, pp. 467–479, March 2001.

[282] S. Zhou and G. B. Giannakis, "Finite-alphabet based channel estimation for OFDM and related multicarrier systems," *IEEE Transactions on Communications*, vol. 49, pp. 1402–1414, August 2001.

[283] X. Wang and K. Liu, "OFDM channel estimation based on time-frequency polynomial model of fading multipath channel," in *Proceedings of Vehicular Technology Conference*, vol. 1, (Atlantic City, NJ USA), pp. 460–464, IEEE, 7-11 October 2001.

[284] B. Yang, Z. Cao, and K. Letaief, "Analysis of low-complexity windowed DFT-based MMSE channel estimator for OFDM systems," *IEEE Transactions on Communications*, vol. 49, pp. 1977–1987, November 2001.

[285] B. Lu and X. Wang, "Bayesian blind turbo receiver for coded OFDM systems with frequency offset and frequency-selective fading," *IEEE Journal on Selected Areas in Communications*, vol. 19, pp. 2516–2527, December 2001.

[286] Y. Li and N. Sollenberger, "Clustered OFDM with channel estimation for high rate wireless data," *IEEE Transactions on Communications*, vol. 49, pp. 2071–2076, December 2001.

[287] M. Morelli and U. Mengali, "A comparison of pilot-aided channel estimation methods for OFDM systems," *IEEE Transactions on Signal Processing*, vol. 49, pp. 3065–3073, December 2001.

[288] M.-X. Chang and Y. Su, "Model-based channel estimation for OFDM signals in Rayleigh fading," *IEEE Transactions on Communications*, vol. 50, pp. 540–544, April 2002.

[289] M. Necker and G. Stüber, "Totally blind channel estimation for OFDM over fast varying mobile channels," in *Proceedings of International Conference on Communications*, (New York, NY USA), IEEE, April 28 - May 2 2002.

[290] B. Yang, Z. Cao, and K. Letaief, "Low complexity channel estimator based on windowed DFT and scalar Wiener filter for OFDM systems," in *Proceedings of International Conference on Communications*, vol. 6, (Helsinki, Finland), pp. 1643–1647, IEEE, June 11-14 2001.

[291] J. Deller, J. Proakis, and J. Hansen, *Discrete-Time Processing of Speech Signals*. Basingstoke: Macmillan Publishing Company, 1993.

[292] A. Duel-Hallen, S. Hu, and H. Hallen, "Long range prediction of fading signals," *IEEE Signal Processing Magazine*, vol. 17, pp. 62–75, May 2000.

[293] F. Tufvesson, *Design of Wireless Communication Systems - Issues on Synchronization, Channel Estimation and Multi-Carrier Systems*. Department of Applied Electronics, Lund University, Sweden, 2000.

[294] W.H. Press and S.A. Teukolsky and W.T. Vetterling and B.P. Flannery, *Numerical Recipes in C*. Cambridge: Cambridge University Press, 1992.

[295] T. Moon and W. Stirling, *Mathematical Methods and Algorithms for Signal Processing*. Englewood Cliffs, NJ, USA: Prentice Hall, 2000.

[296] Y. Li, N. Seshadri, and S. Ariyavisitakul, "Channel estimation for OFDM systems with transmitter diversity in mobile wireless channels," *IEEE Journal on Selected Areas in Communications*, vol. 17, pp. 461–471, March 1999.

[297] W. Jeon, K. Paik, and Y. Cho, "An efficient channel estimation technique for OFDM systems with transmitter diversity," in *Proceedings of International Symposium on Personal, Indoor and Mobile Radio Communications*, vol. 2, (Hilton London Metropole Hotel, London, UK), pp. 1246–1250, IEEE, 18-21 September 2000.

[298] Y. Li, "Optimum training sequences for OFDM systems with multiple transmit antennas," in *Proc. of Global Telecommunications Conference*, vol. 3, (San Francisco, United States), pp. 1478–1482, IEEE, 27 November – 1 December 2000.

[299] A. Mody and G. Stüber, "Parameter estimation for OFDM with transmit receive diversity," in *Proceedings of Vehicular Technology Conference*, vol. 2, (Rhodes, Greece), pp. 820–824, IEEE, 6-9 May 2001.

[300] Y. Gong and K. Letaief, "Low rank channel estimation for space-time coded wideband OFDM systems," in *Proceedings of Vehicular Technology Conference*, vol. 2, (Atlantic City Convention Center, Atlantic City, NJ USA), pp. 772–776, IEEE, 7-11 October 2001.

[301] W. Jeon, K. Paik, and Y. Cho, "Two-dimensional MMSE channel estimation for OFDM systems with transmitter diversity," in *Proceedings of Vehicular Technology Conference*, vol. 3, (Atlantic City Convention Center, Atlantic City, NJ USA), pp. 1682–1685, IEEE, 7-11 October 2001.

[302] F. Vook and T. Thomas, "MMSE multi-user channel estimation for broadband wireless communications," in *Proceedings of Global Telecommunications Conference*, vol. 1, (San Antonio, Texas, USA), pp. 470–474, IEEE, 25-29 November 2001.

[303] Y. Xie and C. Georghiades, "An EM-based channel estimation algorithm for OFDM with transmitter diversity," in *Proceedings of Global Telecommunications Conference*, vol. 2, (San Antonio, Texas, USA), pp. 871–875, IEEE, 25-29 November 2001.

[304] Y. Li, "Simplified channel estimation for OFDM systems with multiple transmit antennas," *IEEE Transactions on Wireless Communications*, vol. 1, pp. 67–75, January 2002.

[305] H. Bölcskei, R. Heath, and A. Paulraj, "Blind channel identification and equalization in OFDM-based multi-antenna systems," *IEEE Transactions on Signal Processing*, vol. 50, pp. 96–109, January 2002.

[306] H. Minn, D. Kim, and V. Bhargava, "A reduced complexity channel estimation for OFDM systems with transmit diversity in mobile wireless channels," *IEEE Transactions on Wireless Communications*, vol. 50, pp. 799–807, May 2002.

[307] S. Slimane, "Channel estimation for HIPERLAN/2 with transmitter diversity," in *International Conference on Communications*, (New York, NY USA), IEEE, 28 April – 2 May 2002.

[308] C. Komninakis, C. Fragouli, A. Sayed, and R. Wesel, "Multi-input multi-output fading channel tracking and equalization using Kalman estimation," *IEEE Transactions on Signal Processing*, vol. 50, pp. 1065–1076, May 2002.

[309] G. Foschini, "Layered space-time architecture for wireless communication in a fading environment when using multi-element antennas," *Bell Labs Technical Journal*, vol. Autumn, pp. 41–59, 1996.

[310] F. Vook and K. Baum, "Adaptive antennas for OFDM," in *Proceedings of IEEE Vehicular Technology Conference (VTC'98)*, vol. 2, (Ottawa, Canada), pp. 608–610, IEEE, 18–21 May 1998.

[311] X. Wang and H. Poor, "Robust adaptive array for wireless communications," *IEEE Transactions on Communications*, vol. 16, pp. 1352–1366, October 1998.

[312] K.-K. Wong, R.-K. Cheng, K. Letaief, and R. Murch, "Adaptive antennas at the mobile and base station in an OFDM/TDMA system," in *Proceedings of Global Telecommunications Conference*, vol. 1, (Sydney, Australia), pp. 183–190, IEEE, 8-12 November 1998.

[313] Y. Li and N. Sollenberger, "Interference suppression in OFDM systems using adaptive antenna arrays," in *Proceedings of Global Telecommunications Conference*, vol. 1, (Sydney, Australia), pp. 213–218, IEEE, 8-12 November 1998.

[314] G. Golden, G. Foschini, R. Valenzuela, and P. Wolniansky, "Detection algorithms and initial laboratory results using V-BLAST space-time communication architecture," *IEE Electronics Letters*, vol. 35, pp. 14–16, January 1999.

[315] Y. Li and N. Sollenberger, "Adaptive antenna arrays for OFDM systems with cochannel interference," *IEEE Transactions on Communications*, vol. 47, pp. 217–229, February 1999.

[316] P. Vandenameele, L. Van der Perre, M. Engels, and H. Man, "A novel class of uplink OFDM/SDMA algorithms for WLAN," in *Proceedings of Global Telecommunications Conference — Globecom'99*, vol. 1, (Rio de Janeiro, Brazil), pp. 6–10, IEEE, 5–9 December 1999.

[317] M. Speth, A. Senst, and H. Meyr, "Low complexity space-frequency MLSE for multi-user COFDM," in *Proceedings of Global Telecommunications Conference — Globecom'99*, vol. 1, (Rio de Janeiro, Brazil), pp. 2395–2399, IEEE, 5–9 December 1999.

[318] C. H. Sweatman, J. Thompson, B. Mulgrew, and P. Grant, "A comparison of detection algorithms including BLAST for wireless communication using multiple antennas," in *Proceedings of International Symposium on Personal, Indoor and Mobile Radio Communications*, vol. 1, (Hilton London Metropole Hotel, London, UK), pp. 698–703, IEEE, 18-21 September 2000.

[319] R. van Nee, A. van Zelst, and G. Awater, "Maximum likelihood decoding in a space-division multiplexing system," in *Proceedings of Vehicular Technology Conference*, vol. 1, (Tokyo, Japan), pp. 6–10, IEEE, 15-18 May 2000.

[320] G. Awater, A. van Zelst, and R. van Nee, "Reduced complexity space division multiplexing receivers," in *Proceedings of Vehicular Technology Conference*, vol. 1, (Tokyo, Japan), pp. 11–15, IEEE, 15-18 May 2000.

[321] A. van Zelst, R. van Nee, and G. Awater, "Space division multiplexing (SDM) for OFDM systems," in *Proceedings of Vehicular Technology Conference*, vol. 2, (Tokyo, Japan), pp. 1070–1074, IEEE, 15-18 May 2000.

[322] P. Vandenameele, L. V. D. Perre, M. Engels, B. Gyselinckx, and H. D. Man, "A combined OFDM/SDMA approach," *IEEE Journal on Selected Areas in Communications*, vol. 18, pp. 2312–2321, November 2000.

[323] X. Li, H. Huang, A. Lozano, and G. Foschini, "Reduced-complexity detection algorithms for systems using multi-element arrays," in *Proc. of Global Telecommunications Conference*, vol. 2, (San Francisco, United States), pp. 1072–1076, IEEE, 27 November – 1 December 2000.

[324] C. Degen, C. Walke, A. Lecomte, and B. Rembold, "Adaptive MIMO techniques for the UTRA-TDD mode," in *Proceedings of Vehicular Technology Conference*, vol. 1, (Rhodes, Greece), pp. 108–112, IEEE, 6-9 May 2001.

[325] X. Zhu and R. Murch, "Multi-input multi-output maximum likelihood detection for a wireless system," in *Proceedings of Vehicular Technology Conference*, vol. 1, (Rhodes, Greece), pp. 137–141, IEEE, 6-9 May 2001.

[326] J. Li, K. Letaief, R. Cheng, and Z. Cao, "Joint adaptive power control and detection in OFDM/SDMA wireless LANs," in *Proceedings of Vehicular Technology Conference*, vol. 1, (Rhodes, Greece), pp. 746–750, IEEE, 6-9 May 2001.

[327] F. Rashid-Farrokhi, K. Liu, and L. Tassiulas, "Transmit beamforming and power control for cellular wireless systems," *IEEE Journal on Selected Areas in Communications*, vol. 16, pp. 1437–1450, October 1998.

[328] A. van Zelst, R. van Nee, and G. Awater, "Turbo-BLAST and its performance," in *Proceedings of Vehicular Technology Conference*, vol. 2, (Rhodes, Greece), pp. 1282–1286, IEEE, 6-9 May 2001.

[329] A. Benjebbour, H. Murata, and S. Yoshida, "Performance of iterative successive detection algorithm with space-time transmission," in *Proceedings of Vehicular Technology Conference*, vol. 2, (Rhodes, Greece), pp. 1287–1291, IEEE, 6-9 May 2001.

[330] M. Sellathurai and S. Haykin, "A simplified diagonal BLAST architecture with iterative parallel-interference cancellation receivers," in *Proceedings of International Conference on Communications*, vol. 10, (Helsinki, Finland), pp. 3067–3071, IEEE, 11-14 June 2001.

[331] A. Bhargave, R. Figueiredo, and T. Eltoft, "A detection algorithm for the V-BLAST system," in *Proceedings of Global Telecommunications Conference*, vol. 1, (San Antonio, Texas, USA), pp. 494–498, IEEE, 25-29 November 2001.

[332] S. Thoen, L. Deneire, L. Van der Perre, and M. Engels, "Constrained least squares detector for OFDM/SDMA-based wireless networks," in *Proceedings of Global Telecommunications Conference*, vol. 2, (San Antonio, Texas, USA), pp. 866–870, IEEE, 25-29 November 2001.

[333] Y. Li and Z.-Q. Luo, "Parallel detection for V-BLAST system," in *Proceedings of International Conference on Communications*, (New York, NY USA), IEEE, 28 April – 2 May 2002.

[334] S. Verdú, *Multiuser Detection*. Cambridge: Cambridge University Press, 1998.

[335] J. Litva and T.-Y. Lo, *Digital Beamforming in Wireless Communications*. London: Artech House Publishers, 1996.

[336] P. Vandenameele, L. Van der Perre, M. Engels, B. Gyselinckx, and H. Man, "A novel class of uplink OFDM/SDMA algorithms: A statistical performance analysis," in *Proceedings of Vehicular Technology Conference*, vol. 1, (Amsterdam, Netherlands), pp. 324–328, IEEE, 19–22 September 1999.

[337] F. Mueller-Roemer, "Directions in audio broadcasting," *Journal Audio Engineering Society*, vol. 41, pp. 158–173, March 1993.

[338] G. Plenge, "DAB — a new radio broadcasting system — state of development and ways for its introduction," *Rundfunktech. Mitt.*, vol. 35, no. 2, 1991.

[339] ETSI, *Digital Audio Broadcasting (DAB)*, 2nd ed., May 1997. ETS 300 401.

[340] ETSI, *Digital Video Broadcasting (DVB); Framing Structure, Channel Coding and Modulation for Digital Terrestrial Television*, August 1997. EN 300 744 V1.1.2.

[341] P. Chow, J. Tu, and J. Cioffi, "A discrete multitone transceiver system for HDSL applications," *IEEE Journal on Selected Areas in Communications*, vol. 9, pp. 895–908, August 1991.

[342] P. Chow, J. Tu, and J. Cioffi, "Performance evaluation of a multichannel transceiver system for ADSL and VHDSL services," *IEEE Journal on Selected Areas in Communications*, vol. 9, pp. 909–919, August 1991.

[343] K. Sistanizadeh, P. Chow, and J. Cioffi, "Multi-tone transmission for asymmetric digital subscriber lines (ADSL)," in *Proceedings of ICC'93*, pp. 756–760, IEEE, 1993.

[344] ANSI, *ANSI/T1E1.4/94-007, Asymmetric Digital Subscriber Line (ADSL) Metallic Interface.*, August 1997.

[345] A. Burr and P. Brown, "Application of OFDM to powerline telecommunications," in *3rd International Symposium On Power-Line Communications*, (Lancaster, UK), 30 March – 1 April 1999.

[346] M. Deinzer and M. Stoger, "Integrated PLC-modem based on OFDM," in *3rd International Symposium On Power-Line Communications*, (Lancaster, UK), 30 March – 1 April 1999.

[347] R. Prasad and H. Harada, "A novel OFDM based wireless ATM system for future broadband multimedia communications," in *Proceeding of ACTS Mobile Communication Summit '97*, (Aalborg, Denmark), pp. 757–762, ACTS, 7–10 October 1997.

[348] C. Ciotti and J. Borowski, "The AC006 MEDIAN project — overview and state–of–the–art," in *Proc. ACTS Summit '96*, (Granada, Spain), pp. 362–367, 27–29 November 1996.

[349] J. Borowski, S. Zeisberg, J. Hübner, K. Koora, E. Bogenfeld, and B. Kull, "Performance of OFDM and comparable single carrier system in MEDIAN demonstrator 60GHz channel," in *Proceeding of ACTS Mobile Communication Summit '97*, (Aalborg, Denmark), pp. 653–658, ACTS, 7–10 October 1997.

[350] M. D. Benedetto, P. Mandarini, and L. Piazzo, "Effects of a mismatch in the in–phase and in–quadrature paths, and of phase noise, in QDCPSK-OFDM modems," in *Proceeding of ACTS Mobile Communication Summit '97*, (Aalborg, Denmark), pp. 769–774, ACTS, 7–10 October 1997.

[351] T. Rautio, M. Pietikainen, J. Niemi, J. Rautio, K. Rautiola, and A. Mammela, "Architecture and implementation of the 150 Mbit/s OFDM modem (invited paper)," in *IEEE Benelux Joint Chapter on Communications and Vehicular Technology, 6th Symposium on Vehicular Technology and Communications*, (Helsinki, Finland), p. 11, 12–13 October 1998.

[352] J. Ala-Laurila and G. Awater, "The magic WAND — wireless ATM network demonstrator system," in *Proceeding of ACTS Mobile Communication Summit '97*, (Aalborg, Denmark), pp. 356–362, ACTS, 7–10 October 1997.

[353] J. Aldis, E. Busking, T. Kleijne, R. Kopmeiners, R. van Nee, R. Mann-Pelz, and T. Mark, "Magic into reality, building the WAND modem," in *Proceeding of ACTS Mobile Communication Summit '97*, (Aalborg, Denmark), pp. 775–780, ACTS, 7–10 October 1997.

[354] E. Hallmann and H. Rohling, "OFDM-Vorschläge für UMTS," in *3. OFDM Fachgespräch in Braunschweig*, 1998.

[355] "Universal mobile telecommunications system (UMTS); UMTS terrestrial radio access (UTRA); concept evaluation," tech. rep., ETSI, 1997. TR 101 146.

[356] C. E. Shannon, "A mathematical theory of communication," *Bell System Technical Journal*, pp. 379–427, 1948.

[357] R. Hamming, "Error detecting and error correcting codes," *Bell System Technical Journal*, vol. 29, pp. 147–160, 1950.

[358] M. Golay, "Notes on digital coding," *Proceedings of the IEEE*, vol. 37, p. 657, 1949.

[359] P. Elias, "Coding for noisy channels," *IRE Conv. Rec. pt.4*, pp. 37–47, 1955.

[360] A. Viterbi, "Error bounds for convolutional codes and an asymptotically optimum decoding algorithm," *IEEE Transactions on Information Theory*, vol. IT-13, pp. 260–269, April 1967.

[361] G. Ungerböck, "Trellis-coded modulation with redundant signal sets. Part 1 and 2," *IEEE Communications Magazine*, vol. 25, pp. 5–21, February 1987.

[362] D. Divsalar and M. K. Simon, "The design of trellis coded MPSK for fading channel: Set partitioning for optimum code design," *IEEE Transactions on Communications*, vol. 36, pp. 1013–1021, September 1988.

[363] C. Schlegel, *Trellis Coding*. The Institute of Electrical and Electronics Engineers, Inc., New York: IEEE Press, 1997.

[364] E. Zehavi, "8-PSK trellis codes for a Rayleigh fading channel," *IEEE Transactions on Communications*, vol. 40, pp. 873–883, May 1992.

[365] G. Caire and G. Taricco and E. Biglieri, "Bit-Interleaved Coded Modulation," *IEEE Transactions on Information Theory*, vol. 44, pp. 927–946, May 1998.

[366] C. Berrou and A. Glavieux and P. Thitimajshima, "Near Shannon limit error-correcting coding and decoding: Turbo codes," in *Proceedings of the International Conference on Communications*, (Geneva, Switzerland), pp. 1064–1070, May 1993.

[367] *Proceedings of the International Symposium on Turbo Codes & Related Topics*, (Brest, France), 3–5 September 1997.

[368] D. J. Costello, A. Banerjee, T. E. Fuja and P. C. Massey, "Some reflections on the design of bandwidth efficient turbo codes," in *Proceedings of 4th ITG Conference on Source and Channel Coding*, no. 170 in ITG Fachbericht, (Berlin), pp. 357–363, VDE–Verlag, 28–30 January 2002.

[369] L. Hanzo, T.H. Liew and B.L. Yeap, *Turbo Coding, Turbo Equalisation and Space Time Coding for Transmission over Wireless channels*. New York, USA: John Wiley IEEE Press, 2002.

[370] R. Steele and L. Hanzo, eds., *Mobile Radio Communications: Second and Third Generation Cellular and WATM Systems*. New York, USA: IEEE Press - John Wiley & Sons, 2nd ed., 1999.

[371] S. L. Goff, A. Glavieux, and C. Berrou, "Turbo-codes and high spectral efficiency modulation," in *Proceedings of IEEE International Conference on Communications*, pp. 645–649, 1994.

[372] P. Robertson and T. Wörz, "Bandwidth-efficient turbo trellis-coded modulation using punctured component codes," *IEEE Journal on Selected Areas in Communications*, vol. 16, pp. 206–218, Feb 1998.

[373] X. Li and J.A. Ritcey, "Trellis-coded modulation with bit interleaving and iterative decoding," *IEEE Journal on Selected Areas in Communications*, vol. 17, April 1999.

[374] X. Li and J.A. Ritcey, "Bit-interleaved coded modulation with iterative decoding using soft feedback," *IEE Electronics Letters*, vol. 34, pp. 942–943, May 1998.

[375] J. Winters, "Smart antennas for wireless systems," *IEEE Personal Communications*, vol. 5, pp. 23–27, February 1998.

[376] R. Derryberry, S. Gray, D. Ionescu, G. Mandyam, and B. Raghothaman, "Transmit diversity in 3g cdma systems," *IEEE Communications Magazine*, vol. 40, pp. 68–75, April 2002.

[377] A. Molisch, M. Win, and J. Winters, "Space-time-frequency (stf) coding for mimo-ofdm systems," *IEEE Communications Letters*, vol. 6, pp. 370–372, September 2002.

[378] A. Molisch, M. Steinbauer, M. Toeltsch, E. Bonek, and R. Thoma, "Capacity of mimo systems based on measured wireless channels," *IEEE Journal on Selected Areas in Communications*, vol. 20, pp. 561–569, April 2002.

[379] D. Gesbert, M. Shafi, D.-S. Shiu, P. Smith, and A. Naguib, "From theory to practice: an overview of mimo space-time coded wireless systems," *IEEE Journal on Selected Areas in Communications*, vol. 21, pp. 281–302, April 2003.

[380] M. Shafi, D. Gesbert, D.-S. Shiu, P. Smith, and W. Tranter, "Guest editorial: Mimo systems and applications," *IEEE Journal on Selected Areas in Communications*, vol. 21, pp. 277–280, April 2003.

[381] W. Jakes Jr., ed., *Microwave Mobile Communications*. New York, USA: John Wiley & Sons, 1974.

[382] W. Lee, *Mobile Cellular Communications*. New York, USA: McGraw-Hill, 1989.

[383] R. Steele and L. Hanzo, eds., *Mobile Radio Communications*. Piscataway, NJ, USA: IEEE Press, 1999.

[384] D. Parsons, *The Mobile Radio Propagation Channel*. London: Pentech Press, 1992.

[385] D. Greenwood and L. Hanzo, "Characterisation of mobile radio channels," in Steele and Hanzo [383], ch. 2, pp. 92–185.

[386] R. Steele and V. Prabhu, "Mobile radio cellular structures for high user density and large data rates," *Proceedings of the IEE*, pp. 396–404, August 1985. Pt F.

[387] R. Steele, "The cellular environment of lightweight hand-held portables," *IEEE Communications Magazine*, pp. 20–29, July 1989.

[388] J. G. Proakis, *Digital Communications*. New York: Mc-Graw Hill International Editions, 3rd ed., 1995.

[389] K. Bullington, "Radio propagation at frequencies above 30 Mc/s," *Proceedings IRE 35*, pp. 1122–1136, 1947.

[390] R. Edwards and J. Durkin, "Computer prediction of service area for VHF mobile radio networks," *Proc of IRE*, vol. 116, no. 9, pp. 1493–1500, 1969.

[391] W. Webb, "Sizing up the microcell for mobile radio communications," *IEE Electronics and Communications Journal*, vol. 5, pp. 133–140, June 1993.

[392] M. Hata, "Empirical formula for propagation loss in land mobile radio," *IEEE Transactions on Vehicular Technology*, vol. 29, pp. 317–325, August 1980.

[393] Y. Okumura, E. Ohmori, T. Kawano, and K. Fukuda, "Field strength and its variability in VHF and UHF land mobile service," *Review of the Electrical Communication Laboratory*, vol. 16, pp. 825–873, September–October 1968.

[394] E. Green, "Radio link design for microcellular systems," *British Telecom Technology Journal*, vol. 8, pp. 85–96, January 1990.

[395] G. O. A. Rustako, N. Amitay and R. Roman, "Propagation measurements at microwave frequencies for microcellular mobile and personal communications," *Proceedings of 39th IEEE VTC*, pp. 316–320, 1989.

[396] J. Kiebler, "The design and planning of feeder links to broadcasting satellites," *IEEE Journal on Selected Areas of Communications*, vol. SAC-3, pp. 181–185, January 1985.

[397] C. Loo, "A statistical model for a land mobile radio satellite link," *IEEE Transactions on Vehicular Technology*, vol. VT-34, pp. 122–127, August 1985.

[398] C. Loo, "Digital transmission through a land mobile satellite channel," *IEEE Transactions on Communications*, vol. 38, pp. 693–697, May 1990.

[399] E. Lutz, D. Cygan, M. Dippold, F. Dolainsky, and W. Papke, "The land mobile satellite communications channel - recording, statistics and channel model," *IEEE Transactions on Vehicular Technology*, vol. 40, pp. 375–386, May 1991.

[400] J. Hagenauer, F. Dolainsky, E. Lutz, W. Papke, and R. Schweikert, "The maritime satellite communication channel – channel model, performance of modulation and coding," *IEEE Journal on Selected Areas in Communications*, vol. 5, pp. 701–713, May 1987.

[401] C. Loo, "Measurements and models of a land mobile satellite channel and their application to MSK signals," *IEEE Transactions on Vehicular Technology*, vol. VT-35, pp. 114–121, August 1987.

[402] H. Nyquist, "Certain factors affecting telegraph speed," *Bell System Technical Journal*, p. 617, April 1928.

[403] H. Raemer, *Statistical Communication Theory and Applications*. Englewood Cliffs, NJ, USA: Prentice-Hall, 1969.

[404] Y. Chow, A. Nix, and J. McGeehan, "Analysis of 16-apsk modulation in awgn and rayleigh fading channel," *Electronics Letters*, vol. 28, pp. 1608–1610, November 1992.

[405] N. Kingsbury, "Transmit and receive filters for QPSK signals to optimise the performance on linear and hard limited channels," *IEE Proceedings*, vol. 133, pp. 345–355, July 1986. Pt.F.

[406] B. Sklar, *Digital Communications–Fundamentals and Applications*. Englewood Cliffs, NJ, USA: Prentice-Hall, 1988.

[407] M. Schwartz, *Information Transmission, Modulation and Noise*. New York: McGraw-Hill, 1990.

[408] K. Feher, ed., *Advanced Digital Communications: Systems and Signal Processing*. Englewood Cliffs, NJ, USA: Prentice-Hall, 1987.

[409] A. Saleh and D. Cox, "Improving the power-added efficiency of FET amplifiers operating with varying envelope signals," *IEEE Transactions on Microwave Theory Technology*, vol. MTT-31, pp. 51–56, January 1983.

[410] D. Green, "Characterisation and compensation of nonlinearities in microwave transmitters," *IEEE Transactions on Microwave Theory Technology.*, vol. MTT-30, pp. 213–217, 1982.

[411] F. Casadevall, "The LINC transmitter," *RF Design*, pp. 41–48, February 1990.

[412] Y. Akaiwa and Y. Nagata, "Highly efficient digital mobile communications with a linear modulation method," *IEEE Journal on Selected Areas in Communications*, vol. SAC-5, pp. 890–895, June 1987.

[413] D. H. A. Bateman and R. Wilkinson, "Linear transceiver architectures," in *Proceedings of IEEE Vehicular Technology Conference*, pp. 478–484, 1988.

[414] A. Wright and W. Duntler, "Experimental performance of an adaptive digital linearised power amplifier," *IEEE Transactions on Vehicular Technology*, vol. 41, pp. 395–400, November 1992.

[415] S. Stapleton and L. Quach, "Reduction of adjacent channel interference using postdistortion," in *Proceedings of IEEE VTC '92*, (Denver, CO, USA), pp. 915–918, IEEE, 10–13 May 1992.

[416] J. Namiki, "An automatically controlled predistorter for multilevel quadrature amplitude modulation," *IEEE Transactions on Communications*, vol. COM-31, pp. 707–712, May 1983.

[417] T. Nojima and T. Konno, "Cuber predistortion linearizer for relay equipment in the 800 MHz band land mobile telephone system," *IEEE Transactions on Vehicular Technology*, vol. VT-34, pp. 169–177, November 1985.

[418] P. M. M. Nannicini and F. Oggioni, "Temperature controlled predistortion circuits for 64 QAM microwave power amplifiers," *IEEE Microwave Theory Tech. Dig.*, pp. 99–102, 1985.

[419] Y. Nagata, "Linear amplification technique for digital mobile communications," in *Proceedings of IEEE Vehicular Technology Conference (VTC'89)*, (San Francisco, CA, USA), pp. 159–164, IEEE, 1–3 May 1989.

[420] A. Saleh and J. Salz, "Adaptive linearization of power amplifiers in digital radio systems," *Bell Systems Technical Journal*, vol. 62, pp. 1019–1033, April 1983.

[421] B. Bunday, *Basic Optimisation Methods*. London: Edward Arnold, 1984.

[422] S. Stapleton and F. Costesu, "An adaptive pre-distortion system," in *Proceedings of IEEE VTC '92*, (Denver, CO, USA), pp. 690–693, IEEE, 10–13 May 1992.

[423] M. K. Simon, S. M. Hinedi, and W. C. Lindsey, *Communication Techniques – Signal Design and Detection*. Prentice Hall, 1995.

[424] L. Franks, "Carrier and bit synchronization — a tutorial review," *IEEE Transactions on Communications*, vol. COM-28, pp. 1107–1121, August 1980.

[425] R. Ziemer and R. Peterson, *Digital Communications and Spread Spectrum System*. New York, USA: Macmillan Publishing Company, 1985.

[426] L. Franks, "Synchronisation subsystems: Analysis and design," in Feher [103], ch. 7.

[427] A. Carlson, *Communication Systems*. New York, USA: McGraw-Hill, 1975.

[428] I. Wassell, *Digital mobile radio communication*. PhD thesis, University of Southampton, UK, 1991.

[429] R. Cupo. and R. Gitlin, "Adaptive carrier recovery systems for digital data communications receivers," *IEEE Journal on Selected Areas of Communications*, vol. 7, pp. 1328–1339, December 1989.

[430] W. Lindsey and M. Simon, "Carrier synchronisation and detection of polyphase signals," *IEEE Transactions on Communications*, pp. 441–454, June 1972.

[431] J. Smith, *Modern Communications Circuits*. New York, USA: McGraw Hill, 1986.

[432] M. Woodbury, "Inverting modified matrices," tech. rep., Statistical Research Group, Princeton University, Princeton, NJ, USA. Mem.Rep. 42.

[433] B. Picinbono, "Adaptive signal processing for detection and communication," in *Communication Systems and Random Process Theory* (J. Skwirzinsky, ed.), Alphen aan den Rijn, The Netherlands: Sijthof and Noordhoff, 1978.

[434] K. Murota and K. Hirade, "GMSK modulation for digital mobile radio telephony," *IEEE Transactions on Communications*, vol. 29, pp. 1044–1050, July 1981.

[435] L. Lopes, "GSM radio link simulation," tech. rep., University Research in Mobile Radio, 1990. IEE Colloquium.

[436] J. Anderson, T. Aulin, and C. Sundberg, *Digital Phase Modulation*. New York: Plenum Press, 1986.

[437] ETSI, *Digital Video Broadcasting (DVB): Framing Structure, Channel Coding and Modulation for 11/12 GHz Satellite Services*, August 1997. ETS 300 421.

[438] ETSI, *Digital Video Broadcasting (DVB): Framing Structure, Channel Coding and Modulation for Digital Terrestrial Television*, August 1997. ETS 300 744.

[439] ETSI, *Digital Video Broadcasting (DVB): Framing Structure, Channel Coding and Modulation for Cable Systems*, December 1997. ETS 300 429.

[440] L. Hanzo, W. Webb, and T. Keller, *Single- and Multi-carrier Quadrature Amplitude Modulation*. New York: John Wiley-IEEE Press, April 2000.

[441] L. Hanzo and W. Webb, *Modern Quadrature Amplitude Modulation – Principles and Applications for Fixed and Wireless Channels*. New York, USA: IEEE Press – John Wiley, 1994.

[442] S. Haykin, *Blind Deconvolution*. Englewood Cliffs, NJ: Prentice-Hall, 1st ed., 1994.

[443] D. Lainiotis, S. Katsikas, and S. Likothanassis, "Optimal seismic deconvolution," *Signal Processing*, vol. 15, pp. 375–404, December 1988.

[444] D. Huang and F. Gustafsson, "Sufficient output conditions for identifiability in blind equalization," *IEEE Transactions on Communications*, vol. 47, pp. 191–194, February 1999.

[445] L. Tong, G. Xu, and T. Kailath, "Blind identification and equalization based on second-order statistics: a time domain approach," *IEEE Transactions on Information Theory*, vol. 40, pp. 380–389, December 1994.

[446] S.-C. Pei and M.-F. Shih, "Fractionally spaced blind equalization using polyperiodic linear filtering," *IEEE Transactions on Communications*, vol. 46, pp. 16–19, January 1998.

[447] K. Dogancay and R. Kennedy, "Least squares approach to blind channel equalization," *Signal Processing*, vol. 58, pp. 63–78, April 1997.

[448] T. Endres, C. Johnson, and M. Green, "Robustness to fractionally-spaced equalizer length using the constant modulus criterion," *IEEE Transactions on Signal Processing*, vol. 47, pp. 544–548, February 1999.

[449] J. LeBlanc, I. Fijalkow, and C. Johnson Jr., "CMA fractionally spaced equalizers: Stationary points and stability under iid and temporally correlated sources," *International Journal of Adaptive Control and Signal Processing*, vol. 12, pp. 135–155, March 1998.

[450] M. Magarini, A. Spalvieri, and G. Tartara, "Asymptotic analysis of stabilisation technique for the blind fractionally spaced equaliser," *Electronics Letters*, vol. 32, pp. 1947–1948, October 1996.

[451] C. Papadias and D. Slock, "Fractionally spaced equalization of linear polyphase channels and related blind techniques based on multichannel linear prediction," *IEEE Transactions on Signal Processing*, vol. 47, pp. 641–654, March 1999.

[452] V. Yang and D. Jones, "A vector constant modulus algorithm for shaped constellation equalization," *IEEE Signal Processing Letters*, vol. 5, pp. 89–91, April 1998.

[453] O. Shalvi and E. Weinstein, "New criteria for blind deconvolution of nonminimum phase systems (channels)," *IEEE Transactions on Information Theory*, vol. 36, pp. 312–321, March 1990.

[454] O. Shalvi and E. Weinstein, "Super-exponential methods for blind deconvolution," *IEEE Transactions on Information Theory*, vol. 39, pp. 504–519, March 1993.

[455] H. Chiang and C. Nikias, "Adaptive deconvolution and identification of nonminimum phase FIR systems based on cumulants," *IEEE Transactions on Automatic Control*, vol. 35, pp. 36–47, January 1990.

[456] D. Boss and K.-D. Kammeyer, "Blind GSM channel estimation," in *Proceedings of the 1997 47th IEEE Vehicular Technology Conference*, (Phoenix, USA), pp. 1044–1048, 4–7 May 1997.

[457] K. Wesolowsky, "Analysis and properties of the modified constant modulus algorithm for blind equalization," *European Transactions on Telecommunication*, vol. 3, pp. 225–230, May–June 1992.

[458] J. Choi, I. Song, and R. Park, "Some convergence properties of Godard's quartic algorithm," *Signal Processing*, vol. 56, pp. 313–320, February 1997.

[459] Z. Ding, R. Johnson, and R. Kennedy, "On the (non)existence of undesirable equilibria of Godard blind equalizers," *IEEE Transactions on Signal Processing*, vol. 40, pp. 2425–2432, October 1992.

[460] Y. Li, K. Liu, and Z. Ding, "Length -and cost- dependent local minima of unconstrained blind channel equalizers," *IEEE Transactions on Signal Processing*, vol. 44, pp. 2726–2735, November 1996.

[461] Z. Ding, R. Kennedy, B. Anderson, and R. Johnson Jr., "Local convergence of the Sato blind equalizer and generalizations under practical constraints," *IEEE Transactions on Information Theory*, vol. 39, pp. 129–144, January 1993.

[462] Z. Ding and R. Kennedy, "On the whereabouts of local minima for blind adaptive equalizers," *IEEE Transactions on Circuits and Systems II: Analog and Digital Signal Processing*, vol. 39, pp. 119–123, February 1992.

[463] Z. Ding and R. Johnson Jr., "On the nonvanishing stability of undesirable equilibria for FTR Godard blind equalizers," *IEEE Transactions on Signal Processing*, vol. 41, pp. 1940–1944, May 1993.

[464] Y. Li and Z. Ding, "Convergence analysis of finite length blind adaptive equalizers," *IEEE Transactions on Signal Processing*, vol. 43, pp. 2120–2129, September 1995.

[465] H. Zeng, L. Tong, and C. Johnson, "Relationships between the constant modulus and Wiener receivers," *IEEE Transactions on Information Theory*, vol. 44, pp. 1523–1539, July 1998.

[466] P. Regalia and M. Mboup, "Undermodeled equalization: A characterization of stationary points for a family of blind criteria," *IEEE Transactions on Signal Processing*, vol. 47, pp. 760–770, March 1999.

[467] M. Gu and L. Tong, "Geometrical characterizations of constant modulus receivers," *IEEE Transactions on Signal Processing*, vol. 47, pp. 2745–2756, October 1999.

[468] Y. Li and K. Liu, "Static and dynamic convergence behaviour of adaptive blind equalizers," *IEEE Transactions on Signal Processing*, vol. 44, pp. 2736–2745, November 1996.

[469] V. Weerackody, S. Kassam, and K. Laker, "Conver gence analysis of an algorithm for blind equalization," *IEEE Transactions on Communications*, vol. 39, pp. 856–865, June 1991.

[470] W. Lee and K. Cheun, "Convergence analysis of the stop-and-go blind equalization algorithm," *IEEE Transactions on Communications*, vol. 47, pp. 177–180, February 1999.

[471] Y. Li and Z. Ding, "Global convergence of fractionally spaced Godard (CMA) adaptive equalizers," *IEEE Transactions on Signal Processing*, vol. 44, pp. 818–826, April 1996.

[472] Z. Ding, "On convergence analysis of fractionally spaced adaptive blind equalizers," *IEEE Transactions on Signal Processing*, vol. 45, pp. 650–657, March 1997.

[473] J. J. Shynk and C. K. Chan, "Performance surfaces of the constant modulus algorithm based on a conditional Gaussian model," *IEEE Transactions on Signal Processing*, vol. 41, pp. 1965–1969, May 1993.

[474] S. Douglas, A. Cichocki, and S. Amari, "Fast-convergence filtered regressor algorithms for blind equalisation," *Electronics Letters*, vol. 32, pp. 2114–2115, November 1996.

[475] C. Papadias and D. Slock, "Normalized sliding window constant modulus (CM) and decision-directed algorithms: a link between blind equalization and classical adaptive filtering," *IEEE Transactions on Signal Processing*, vol. 45, pp. 231–235, January 1997.

[476] J. Anderson and S. Mohan, "Sequential coding algorithms: A survey and cost analysis," *IEEE Transactions on Communications*, vol. 32, pp. 1689–1696, February 1984.

[477] Z. Xie, C. Rushforth, R. Short, and T. Moon, "Joint signal detection and parameter estimation in multiuser communications," *IEEE Transactions on Communications*, vol. 41, pp. 1208–1216, August 1993.

[478] A. Papoulis, *Probability, Random Variables, and Stochastic Processes*. New York, USA: McGraw-Hill, 2nd ed., 1984.

[479] S. Haykin, *Communications Systems*. New York, USA: John Wiley and Sons, 2nd ed., 1994.

[480] B. Noble and J. Daniel, *Applied Linear Algebra*. Englewood Cliffs, NJ, USA: Prentice-Hall, 3rd ed., 1986.

[481] D. Hatzinakos, "Blind equalization based on prediction and polycepstra principles," *IEEE Transactions on Communications*, vol. 43, pp. 178–181, February–April 1995.

[482] D. Hatzinakos, "Blind equalization using decision feedback prediction and tricepstrum principles," *Signal Processing*, vol. 36, pp. 261–276, April 1994.

[483] A. Bessios and C. Nikias, "POTEA: the power cepstrum and tricoherence equalization algorithm," *IEEE Transactions on Communications*, vol. 43, pp. 2667–2671, November 1995.

[484] A. Petropulu and C. Nikias, "Blind deconvolution of coloured signals based on higher–order cepstra and data fusion," *IEE Proceedings, Part F: Radar and Signal Processing*, vol. 140, pp. 356–361, December 1993.

[485] G. Kechriotis, E. Zervas, and E. Manolakos, "Using recurrent neural networks for adaptive communication channel equalization," *IEEE Transactions on Neural Networks*, vol. 5, pp. 267–278, March 1994.

[486] S. Amari and A. Cichocki, "Adaptive blind signal processing - neural network approaches," *Proceedings of the IEEE*, vol. 86, pp. 2026–2048, October 1998.

[487] C. You and D. Hong, "Nonlinear blind equalization scheme using complex-valued multilayer feedforward neural networks," *IEEE Transactions on Neural Networks*, vol. 9, pp. 1442–1455, November 1998.

[488] Y. Fang and T. Chow, "Blind equalization of a noisy channel by linear neural network," *IEEE Transactions on Neural Networks*, vol. 10, no. 4, pp. 918–924, 1999.

[489] S. Choi. and A. Cichocki, "Cascade neural networks for multichannel blind deconvolution," *Electronics Letters*, vol. 34, pp. 1186–1187, June 1998.

[490] S. Mo and B. Shafai, "Blind equalization using higher order cumulants and neural network," *IEEE Transactions on Signal Processing*, vol. 42, pp. 3209–3217, November 1994.

[491] L. H. C.S. Lee, S. Vlahoyiannatos, "Satellite based turbo-coded, blind-equalised 4-QAM and 16-QAM digital video broadcasting," *IEEE Transactions on Broadcasting*, vol. 46, pp. 23–34, March 2000.

[492] G. Forney Jr, R. Gallager, G. Lang, F. Longstaff, and S. Qureshi, "Efficient modulation for band-limited channels," *IEEE Journal on Selected Areas in Communications*, vol. 2, pp. 632–647, September 1984.

[493] J. Massey, "Coding and modulation in digital communications," in *Proceedings of International Zurich Seminar on Digital Communications 1994*, (Zurich, Switzerland), March 1974.

[494] H. Imai and S. Hirakawa, "A new multi-level coding method using error correcting codes," *IEEE Transactions on Information Theory*, vol. 23, pp. 371–377, May 1977.

[495] G. Ungerböck, "Channel coding with multilevel/phase signals," *IEEE Transactions on Information Theory*, vol. IT-28, pp. 55–67, January 1982.

[496] G. Ungerböck, "Trellis-coded modulation with redundant signal sets part 1: Introduction," *IEEE Communications Magazine*, vol. 25, pp. 5–11, February 1987.

[497] E. Biglieri and M. Luise, "Coded modulation and bandwidth-efficient transmission," in *Proceedings of the Fifth Tirrenia Internatioanal Workshop*, (Netherlands), 8–12 September 1991.

[498] "Special issue on coded modulation," *IEEE Communications Magazine*, vol. 29, December 1991.

[499] E. Biglieri, D. Divsalar, P. McLane, and M. Simon, *Introduction to Trellis Coded Modulation with Applications*. New York, USA: Macmillan Publishing Co., 1991.

[500] C. E. Shannon, *Mathematical Theory of Communication*. Chicago: University of Illinois Press, 1963.

[501] J. Wozencraft and R. Kennedy, "Modulation and demodulation for probabilistic coding," *IEEE Transactions on Information Theory*, vol. IT-12, pp. 291–297, 1966.

[502] J. Wozencraft and I. Jacobs, *Principles of Communications Engineering*. New York, USA: John Wiley, 1965.

[503] R. Blahut, *Theory and Practice of Error Control Codes*. Reading, MA, USA: Addison-Wesley, 1983.

[504] E. Berlekamp, *Algebraic Coding Theory*. New York, USA: McGraw-Hill, 1968.

[505] W. Peterson, *Error Correcting Codes*. Cambridge, MA, USA: MIT Press, 1st ed., 1961.

[506] A. Michelson and A. Levesque, *Error Control Techniques for Digital Communication*. New York, USA: John Wiley and Sons, 1985.

[507] K. Wong and L. Hanzo, "Channel coding," in Steele and Hanzo [383], ch. 4, pp. 347–488.

[508] *International Consultative Committee for Telephone and Telegraph Recommendations*. Geneva. V.29 - V.33.

[509] L. Wei, "Rotationally-invariant convolutional channel coding with expanded signal space, part I and II," *IEEE Transactions on Selected Areas in Comms*, vol. SAC-2, pp. 659–686, September 1984.

[510] K. Shanmugam, *Digital and Analog Communications Systems*. New York, USA: John Wiley and Sons, 1979.

[511] W. Lee, *Mobile Communications Engineering*. New York, USA: McGraw-Hill, 1982.

[512] I. Gradshteyn and I. Ryzhik, *Table of Integrals, Series and Products*. New York, USA: Academic Press, 1980.

[513] D. Yoon, D. Chang, N. Kim, and H. Woo, "Linear diversity analysis for M-ary square quadrature amplitude modulation over Nakagami fading channels," *ETRI Journal*, vol. 25, pp. 231–237, August 2003.

[514] P. Vitthaladevuni and M. Alouini, "A recursive algorithm for the exact BER computation of generalized hierarchical QAM constellations," *IEEE Transactions on Information Theroy*, vol. 49, pp. 297–307, January 2003.

[515] D. Y. K.K. Cho, "On the general BER expression of one- and two-dimensional amplitude modulations," *IEEE Transactions on Communications*, vol. 50, pp. 1074–1080, July 2002.

[516] P. Vitthaladevuni and M. Alouini, "BER computation of 4/M-QAM hierarchical constellations," *IEEE Transactions on Broadcasting*, vol. 47, pp. 228–239, September 2001.

[517] G. Saulnier and W. Raffety, "Pilot-aided modulation for narrowband satellite communications," in *Proceedings of Mobile Satellite Conference*, pp. 329–336, 1988.

[518] A. Bateman and J. McGeehan, "Feedforward transparent tone in band for rapid fading protection in multipath fading," in *IEE International Conference on Communications*, vol. 68, pp. 9–13, 1986.

[519] A. Bateman and J. McGeehan, "The use of transparent tone in band for coherent data schemes," in *IEEE International Conference on Communications*, (Boston, MA, USA), 1983.

[520] A. Bateman, G. Lightfoot, A. Lymer, and J. McGeehan, "Speech and data transmissions over a 942MHz TAB and TTIB single sideband mobile radio system," *IEEE Transactions on Vehicular Technology*, vol. VT-34, pp. 13–21, February 1985.

[521] A. Bateman and J. McGeehan, "Data transmissions over UHF fading mobile radio channels," *Proceedings of IEE*, vol. 131, no. Pt.F, pp. 364–374, 1984.

[522] J. McGeehan and A. Bateman, "A simple simultaneous carrier and bit synchronisation system for narrowband data transmissions," *Proceedings of IEE*, vol. 132, no. Pt.F, pp. 69–72, 1985.

[523] J. McGeehan and A. Bateman, "Theoretical and experimental investigation of feedforward signal regeneration," *IEEE Transactions on Vehicular Technology*, vol. VT-32, pp. 106–120, 1983.

[524] A. Bateman, "Feedforward transparent tone in band: Its implementation and applications," *IEEE Transactions on Vehicular Technology*, vol. 39, pp. 235–243, August 1990.

[525] M. Simon, "Dual pilot tone calibration technique," *IEEE Transactions on Vehicular Technology*, vol. VT-35, pp. 63–70, May 1986.

[526] M. Fitz, "A dual-tone reference digital demodulator for mobile communications," *IEEE Transactions on Vehicular Technology*, vol. VT-42, pp. 156–166, May 1993.

[527] S. Gamnathan and K. Feher, "Pilot tone aided QPRS systems for digital audio broadcasting," *IEEE Transactions on Broadcasting*, vol. 38, pp. 1–6, March 1992.

[528] F. Davarrin, "Mobile digital communications via tone calibration," *IEEE Transactions on Vehicular Technology*, vol. VT-36, pp. 55–62, May 1987.

[529] J. Cavers, "The performance of phase locked transparent tone in band with symmetric phase detection," *IEEE Transactions on Communications*, vol. 39, pp. 1389–1399, September 1991.

[530] J. Cavers, "Performance of tone calibration with frequency offset and imperfect pilot filter," *IEEE Transactions on Vehicular Technology*, vol. 40, pp. 426–434, May 1991.

[531] P. Martin and A. Bateman, "Practical results for a modem using linear mobile radio channels," in *Proceedings of IEEE Vehicular Technology Conference (VTC'91)*, (St. Louis, MO, USA), pp. 386–392, IEEE, 19–22 May 1991.

[532] D. Esteban and C. Galand, "Application of quadrature mirror filters to split band voice coding scheme," in *Proceedings of International Conference on Acoustics, Speech, and Signal Processing, ICASSP'77*, (Hartford, CT, USA), pp. 191–195, IEEE, 9–11 May 1977.

[533] J. Johnston, "A filter family designed for use in quadrature mirror filter banks," in *Proceedings of International Conference on Acoustics, Speech, and Signal Processing, ICASSP'80*, (Denver, CO, USA), pp. 291–294, IEEE, 9–11 April 1980.

[534] J. Lodge and M. Moher, "Time diversity for mobile satellite channels using trellis coded modulations," in *IEEE Global Telecommunications Conference*, (Tokyo, Japan), 1987.

[535] S. Sampei and T. Sunaga, "Rayleigh fading compensation method for 16-QAM in digital land mobile radio channels," in *Proceedings of IEEE Vehicular Technology Conference (VTC'89)*, (San Francisco, CA, USA), pp. 640–646, IEEE, 1–3 May 1989.

[536] J. Cavers, "Pilot symbol assisted modulation in fading and delay spread," in *Proceedings of IEEE VTC '93*, (Secaucus, NJ, USA), pp. 13–16, IEEE, 18–20 May 1993.

[537] J. Seymour and M. Fitz, "Improved carrier synchronisation techniques for mobile communications," in *Proceedings of IEEE VTC '93*, (Secaucus, NJ, USA), pp. 901–904, IEEE, 18–20 May 1993.

[538] AT&T Information Services, *A Trellis Coded Modulation Scheme that Includes Differential Encoding for 9600 Bit/Sec Full-Duplex, Two-Wire Modems*, August 1983. CCITT SG XVII.

[539] R. Salami, L. Hanzo, R. Steele, K. Wong, and I. Wassell, "Speech coding," in Steele and Hanzo [383], ch. 3, pp. 186–346.

[540] K. Larsen, "Short convolutional codes with maximal free distance for rate 1/2, 1/3 and 1/4," *IEEE Transactions on Information Theory*, vol. IT-19, pp. 371–372, May 1973.

[541] K. Wong, L. Hanzo, and R. Steele, "Channel coding for satellite mobile channels," *International Journal on Satellite Communications*, vol. 7, pp. 143–163, July–September 1989.

[542] P. Ho, J. Cavers, and J. Varaldi, "The effect of constellation density on trellis coded modulation in fading channels," in *Proceedings of IEEE VTC '92*, (Denver, CO, USA), pp. 463–467, IEEE, 10–13 May 1992.

[543] S. Fechtel and H. Meyr, "Combined equalisation, decoding and antenna diversity combining for mobile personal digital radiotransmission using feedforward synchronisation," in *Proceedings of IEEE VTC '93*, (Secaucus, NJ, USA), IEEE, 18–20 May 1993.

[544] R. Bultitude and G. Bedal, "Propagation characteristics on microcellular urban mobile radio channels at 910MHz," *IEEE Journal on Selected Areas in Communications*, vol. 7, pp. 31–39, January 1989.

[545] R. Bultitude, S. Mahmoud, and W. Sullivan, "A comparison of indoor radio propagation characteristics at 910MHz and 1.75GHz," *IEEE Journal on Selected Areas in Communications*, vol. 7, pp. 20–30, January 1989.

[546] H. Harmuth, *Transmission of Information by Orthogonal Time Functions*. Berlin: Springer Verlag, 1969.

[547] H. Harmuth, "On the transmission of information by orthogonal time functions," *AIEE*, July 1960.

[548] H. Harmuth, "Die Orthogonalteilung als Verallgemeinerung der Zeit- und Frequenzteilung," *AEÜ*, vol. 18, pp. 43–50, 1964.

[549] D. Saha and T. Birdsall, "Quadrature-quadrature phase shift keying," *IEEE Transactions on Communications*, vol. 37, pp. 437–448, May 1989.

[550] C. E. Shannon, "A mathematical theory of communication," *Bell System Technical Journal*, vol. 27, pp. 379–423 and 623–656, June and October 1948.

[551] H. Landau and H. Pollak, "Prolate spheroidal wave functions...," *Bell Systems Technical Journal*, vol. 41, pp. 1295–1336, July 1962.

[552] W. Lee, "Spectrum efficiency in cellular," *IEEE Transactions on Vehicular Technology*, vol. 38, pp. 69–75, May 1989.

[553] H. Kolb private communications.

[554] J. Lindner private communications.

[555] D. Schnidman, "A generalized Nyquist criterion and an optimum linear receiver for a pulse modulation system," *Bell Systems Technical Journal*, pp. 2163–2177, November 1967.

[556] W. V. Etten, "An optimum linear receiver for multiple channel digital transmission systems," *IEEE Transactions on Communications*, vol. COM-23, pp. 828–834, August 1975.

[557] A. Kaye and D. George, "Transmission of multiplexed PAM signals over multiple channel and diversity systems," *IEEE Transactions on Communications Technology*, vol. COM-18, pp. 520–525, October 1970.

[558] M. Aaron and D. Tufts, "Intersymbol interference and error probability," *IEEE Transactions on Information Theory*, vol. IT-12, pp. 26–34, January 1966.

[559] D. Tufts, "Nyquist's problem: The joint optimization of transmitter and receiver in pulse amplitude modulation," *Proceedings of IEEE*, vol. 53, pp. 248–259, March 1965.

[560] H. Schüssler, *Digitale Systeme zur Signalverarbeitung*. Berlin, Heidelberg, and New York: Springer Verlag, 1974.

[561] R. O'Neill and L. Lopes, "Performance of amplitude limited multitone signals," in *Proceedings of IEEE VTC '94*, (Stockholm, Sweden), IEEE, 8–10 June 1994.

[562] X. Li and L. Cimini, "Effects of clipping and filtering on the performance of OFDM," in *Proceedings of IEEE VTC'97*, (Phoenix, AZ, USA), pp. 1634–1638, IEEE, 4–7 May 1997.

[563] A. Garcia and M. Calvo, "Phase noise and sub-carrier spacing effects on the performance of an OFDM communications system," *IEEE Communications Letters*, vol. 2, pp. 11–13, January 1998.

[564] W. Robins, *Phase Noise in Signal Sources*, vol. 9 of *IEE Telecommunication series*. Peter Peregrinus Ltd., 1982.

[565] C. Tellambura, Y. Guo, and S. Barton, "Equaliser performance for HIPERLAN in indoor channels," *Wireless Personal Communications*, vol. 3, no. 4, pp. 397–410, 1996.

[566] T. Ojanperä, M. Gudmundson, P. Jung, J. Sköld, R. Pirhonen, G. Kramer, and A. Toskala, "FRAMES: - hybrid multiple access technology," in *Proceedings of IEEE ISSSTA'96*, (Mainz, Germany), pp. 334–338, IEEE, September 1996.

[567] M. Failli, "Digital land mobile radio communications COST 207," tech. rep., European Commission, 1989.

[568] J. Torrance and L. Hanzo, "Comparative study of pilot symbol assisted modem schemes," in *Proceedings of IEE Conference on Radio Receivers and Associated Systems (RRAS'95)*, pp. 36–41, September 1995.

[569] K. Fazel, S. Kaiser, P. Robertson, and M. Ruf, "A concept of digital terrestrial television broadcasting," *Wireless Personal Communications*, vol. 2, pp. 9–27, 1995.

[570] J. Kuronen, V.-P. Kaasila, and A. Mammela, "An all-digital symbol tracking algorithm in an OFDM system by using the cyclic prefix," in *Proc. ACTS Summit '96*, (Granada, Spain), pp. 340–345, 27–29 November 1996.

[571] M. Kiviranta and A. Mammela, "Coarse frame synchronization structures in OFDM," in *Proc. ACTS Summit '96*, (Granada, Spain), pp. 464–470, 27–29 November 1996.

[572] Z. Li and A. Mammela, "An all digital frequency synchronization scheme for OFDM systems," in *Proceedings of the IEEE International Symposium on Personal, Indoor and Mobile Radio Communications (PIMRC)*, (Helsinki, Finland), pp. 327–331, 1–4 September 1997.

[573] J. Bingham, "Method and apparatus for correcting for clock and carrier frequency offset, and phase jitter in multicarrier modems." U.S. Patent No. 5206886, 27 April 1993.

[574] T. de Couasnon, R. Monnier, and J. Rault, "OFDM for digital TV broadcasting," *Signal Processing*, vol. 39, pp. 1–32, 1994.

[575] P. Mandarini and A. Falaschi, "SYNC proposals." MEDIAN Design Note, January 1996.

[576] T. Keller and L. Hanzo, "Orthogonal frequency division multiplex synchronisation techniques for wireless local area networks," in *Proceedings of IEEE International Symposium on Personal, Indoor, and Mobile Radio Communications (PIMRC'96)*, vol. 3, (Taipei, Taiwan), pp. 963–967, IEEE, 15–18 October 1996.

[577] S.-G. Chua and A. Goldsmith, "Variable-rate variable-power mQAM for fading channels," in *Proceedings of IEEE VTC'96*, (Atlanta, GA, USA), pp. 815–819, IEEE, 28 April–1 May 1996.

[578] J. Torrance, *Adaptive Full Response Digital Modulation for Wireless Communications Systems*. PhD thesis, Department of Electronics and Computer Science, University of Southampton, UK, 1997.

[579] K. Miya, O. Kato, K. Homma, T. Kitade, M. Hayashi, and T. Ue, "Wideband CDMA systems in TDD-mode operation for IMT-2000," *IEICE Transactions on Communications*, vol. E81-B, pp. 1317–1326, July 1998.

[580] O. Kato, K. Miya, K. Homma, T. Kitade, M. Hayashi, and M. Watanabe, "Experimental performance results of coherent wideband DS-CDMA with TDD scheme," *IEICE Transactions on Communications.*, vol. E81-B, pp. 1337–1344, July 1998.

[581] T. Keller and L. Hanzo, "Blind-detection assisted sub-band adaptive turbo-coded OFDM schemes," in *Proceeding of VTC'99 (Spring)*, (Houston, TX, USA), pp. 489–493, IEEE, 16–20 May 1999.

[582] L.R. Bahl and J. Cocke and F. Jelinek and J. Raviv, "Optimal decoding of linear codes for minimising symbol error rate," *IEEE Transactions on Information Theory*, vol. 20, pp. 284–287, March 1974.

[583] T. Keller, M. Münster, and L. Hanzo, "A turbo-coded burst-by-burst adaptive wideband speech transceiver," *IEEE Journal on Selected Areas of Communications*, vol. 18, pp. 2363–2372, November 2000.

[584] T. Keller, J. Woodard, and L. Hanzo, "Turbo-coded parallel modem techniques for personal communications," in *Proceedings of IEEE VTC'97*, (Phoenix, AZ, USA), pp. 2158–2162, IEEE, 4–7 May 1997.

[585] T. Keller and L. Hanzo, "Adaptive orthogonal frequency division multiplexing schemes," in *Proceeding of ACTS Mobile Communication Summit '98*, (Rhodes, Greece), pp. 794–799, ACTS, 8–11 June 1998.

[586] C. E. Shannon, "Communication in the presence of noise," *Proceedings of the I.R.E.*, vol. 37, pp. 10–22, January 1949.

[587] L. Piazzo, "A fast algorithm for near-optimum power and bit allocation in OFDM systems," *IEE Electronics Letters*, vol. 35, pp. 2173 – 2174, 9 Decemebr 1990.

[588] T. Willink and P. Wittke, "Optimization and performance evaluation of multicarrier transmission," *IEEE Transactions on Information Theory*, vol. 43, pp. 426–440, March 1997.

[589] R. Fischer and J. Huber, "A new loading algorithm for discrete multitone transmission," in *Proceeding of IEEE Global Telecommunications Conference, Globecom 96*, (London, UK), pp. 713–718, IEEE, 18–22 November 1996.

[590] S. Lai, R. Cheng, K. Letaief, and R. Murch, "Adaptive trellis coded MQAM and power optimization for OFDM transmission," in *Proceedings of VTC'99 (Spring)*, (Houston, TX, USA), IEEE, 16–20 May 1999.

[591] D. Hughes-Hartogs, "Ensemble modem structure for imperfect transmission media." U.S Patents Nos. 4,679,227 (July 1988) 4,731,816 (March 1988) and 4,833,796 (May 1989).

[592] J. Bingham, "Multicarrier modulation for data transmission: an idea whose time has come," *IEEE Communications Magazine*, pp. 5–14, May 1990.

[593] L. Godara, "Applications of antenna arrays to mobile communications, part II: Beam-forming and direction-of-arrival considerations," *Proceedings of the IEEE*, vol. 85, pp. 1193–1245, August 1997.

[594] Y. Li, "Pilot-symbol-aided channel estimation for OFDM in wireless systems," in *Proceedings of VTC'99 (Spring)*, (Houston, TX, USA), IEEE, 16–20 May 1999.

[595] N. Szabo and R. Tanaka, *Residue Arithmetic and Its Applications to Computer Technology*. New York, USA: McGraw-Hill, 1967.

[596] R. Watson and C. Hastings, "Self-checked computation using residue arithmetic," *Proceedings of the IEEE*, vol. 54, pp. 1920–1931, December 1966.

[597] R. Pyndiah, "Iterative decoding of product codes: Block turbo codes," in ISTC'97 [367], pp. 71–79.

[598] P. Adde, R. Pyndiah, O. Raoul, and J.-R. Inisan, "Block turbo decoder design," in Copied [367], pp. 166–169.

[599] W. Jenkins and B. Leon, "The use of residue number system in the design of finite impulse response filters," *IEEE Transactions on Circuits Systems*, vol. CAS-24, pp. 191–201, April 1977.

[600] M. Soderstrand, "A high-speed, low-cost, recursive digital filter using residue number arithmetic," *Proceedings of IEEE*, vol. 65, pp. 1065–1067, July 1977.

[601] M. Soderstrand and E. Fields, "Multipliers for residue number arithmetic digital filters," *Electronics Letters*, vol. 13, pp. 164–166, March 1977.

[602] M. Soderstrand, W. Jenkins, and G. Jullien, *Residue Number System Arithmetic: Modern Applications in Digital Signal Processing*. New York, USA: IEEE Press, 1986.

[603] E. Claudio, G. Orlandi, and F. Piazza, "A systolic redundant residue arithmetic error correction circuit," *IEEE Transactions on Computers*, vol. 42, pp. 427–432, April 1993.

[604] H. Krishna, K.-Y. Lin, and J.-D. Sun, "A coding theory approach to error control in redundant residue number systems - Part I: theory and single error correction," *IEEE Transactions on Circuits Systems*, vol. 39, pp. 8–17, January 1992.

[605] J.-D. Sun and H. Krishna, "A coding theory approach to error control in redundant residue number systems — Part II: multiple error detection and correction," *IEEE Transactions on Circuits Systems*, vol. 39, pp. 18–34, January 1992.

[606] T. Liew, L.-L. Yang, and L. Hanzo, "Soft-decision redundant residue number system based error correction coding," in *Proceeding of VTC'99 (Fall)*, (Amsterdam, Netherlands), pp. 2974–2978, IEEE, 19–22 September 1999.

[607] L.-L. Yang and L. Hanzo, "Residue number system arithmetic assisted M-ary modulation," *IEEE Communications Letters*, vol. 3, pp. 28–30, February 1999.

[608] L.-L. Yang and L. Hanzo, "Performance of residue number system based DS-CDMA over multipath fading channels using orthogonal sequences," *ETT*, vol. 9, pp. 525–536, November–December 1998.

[609] H. Krishna and J.-D. Sun, "On theory and fast algorithms for error correction in residue number system product codes," *IEEE Transactions on Comput.*, vol. 42, pp. 840–852, July 1993.

[610] D. Chase, "A class of algorithms for decoding block codes with channel measurement information," *IEEE Transactions on Information Theory*, vol. IT-18, pp. 170–182, January 1972.

[611] J. Hagenauer, E. Offer, and L. Papke, "Iterative decoding of binary block and convolutional codes," *IEEE Transactions on Information Theory*, vol. 42, pp. 429–445, March 1996.

[612] H. Nickl, J. Hagenauer, and F. Burkett, "Approaching Shannon's capacity limit by 0.27 dB using simple Hamming codes," *IEEE Communications Letters*, vol. 1, pp. 130–132, September 1997.

[613] T. Liew, C. Wong, and L. Hanzo, "Block turbo coded burst-by-burst adaptive modems," in *Proceedings of Microcoll'99, Budapest, Hungary*, pp. 59–62, 21–24 March 1999.

[614] B. Yeap, T. Liew, J. Hamorsky, and L. Hanzo, "Comparative study of turbo equalisers using convolutional codes and block-based turbo-codes for GMSK modulation," in *Proceedings of VTC 1999 Fall*, (Amsterdam, Holland), pp. 2974–2978, 19-22 September 1999.

[615] C.H. Wong, T. H. Liew and L. Hanzo, "Burst-by-burst turbo coded wideband adaptive modulation with blind modem mode detection," *Proceedings of 4th ACTS Mobile Communications Summit 1999, Sorrento, Italy*, pp. 303–308, 8–11 June 1999.

[616] S. M. Alamouti, "A simple transmit diversity technique for wireless communications," *IEEE Journal on Selected Areas in Communications*, vol. 16, pp. 1451–1458, October 1998.

[617] H. J. V. Tarokh and A. Calderbank, "Space-time block codes from orthogonal designs," *IEEE Transactions on Information Theory*, vol. 45, pp. 1456–1467, May 1999.

[618] V. Tarokh, H. Jafarkhani, and A. R. Calderbank, "Space-time block coding for wireless communications: Performance results," *IEEE Journal on Selected Areas in Communications*, vol. 17, pp. 451–460, March 1999.

[619] V. Tarokh, N. Seshadri, and A. R. Calderbank, "Space-time codes for high data rate wireless communication: Performance criterion and code construction," *IEEE Transactions on Information Theory*, vol. 44, pp. 744–765, March 1998.

[620] N. Seshadri, V. Tarokh, and A. Calderbank, "Space-time codes for high data rate wireless communications: Code construction," in *Proceedings of IEEE Vehicular Technology Conference '97*, (Phoenix, Arizona), pp. 637–641, 1997.

[621] V. Tarokh and N. Seshadri and A. Calderbank, "Space-time codes for high data rate wireless communications: Performance criterion and code construction," in *Proc IEEE International Conference on Communications '97*, (Montreal, Canada), pp. 299–303, 1997.

[622] N. S. V. Tarokh, A. Naguib and A. Calderbank, "Space-time codes for high data rate wireless communications: Mismatch analysis," in *Proc IEEE International Conference on Communications '97*, (Montreal, Canada), pp. 309–313, 1997.

[623] A. F. Naguib, V. Tarokh, N. Seshadri, and A. R. Calderbank, "A space-time coding modem for high-data-rate wireless communications," *IEEE Journal on Selected Areas in Communications*, vol. 16, pp. 1459–1478, October 1998.

[624] V. Tarokh, A. Naguib, N. Seshadri, and A. R. Calderbank, "Space-time codes for high data rate wireless communication: Performance criteria in the presence of channel estimation errors, mobility, and multiple paths," *IEEE Transactions on Communications*, vol. 47, pp. 199–207, February 1999.

[625] R. Horn and C. Johnson, *Matrix Analysis*. New York: Cambridge University Press, 1988.

[626] A. Naguib, N. Seshdri, and A. Calderbank, "Increasing data rate over wireless channels: Space-time coding for high data rate wireless communications," *IEEE Signal Processing Magazine*, vol. 17, pp. 76–92, May 2000.

[627] G. Bauch, A. Naguib, and N. Seshadri, "MAP equalization of space-time coded signals over frequency selective channels," in *Proceedings of Wireless Communications and Networking Conference*, (New Orleans, USA), September 1999.

[628] G. Bauch and N. Al-Dhahir, "Reduced-complexity turbo equalization with multiple transmit and receive antennas over multipath fading channels," in *Proceedings of Information Sciences and Systems*, (Princeton, USA), pp. WP3 13–18, March 2000.

[629] D. Agrawal, V. Tarokh, A. Naguib, and N. Seshadri, "Space-time coded OFDM for high data-rate wireless communication over wideband channels," in *Proceedings of IEEE Vehicular Technology Conference*, (Ottawa, Canada), pp. 2232–2236, May 1998.

[630] Y. Li, N. Seshadri, and S. Ariyavisitakul, "Channel estimation for OFDM systems with transmitter diversity in mobile wireless channels," *IEEE Journal on Selected Areas in Communications*, vol. 17, pp. 461–471, March 1999.

[631] Y. Li, J. Chuang, and N. Sollenberger, "Transmitter diversity for OFDM systems and its impact on high-rate data wireless networks," *IEEE Journal on Selected Areas in Communications*, vol. 17, pp. 1233–1243, July 1999.

[632] W. Choi and J. Cioffi, "Space-time block codes over frequency selective fading channels," in *Proceedings of VTC 1999 Fall*, (Amsterdam, Holland), pp. 2541–2545, 19-22 September 1999.

[633] Z. Liu, G. Giannakis, A. Scaglione, and S. Barbarossa, "Block precoding and transmit-antenna diversity for decoding and equalization of unknown multipath channels," in *Proc 33rd Asilomar Conference Signals, Systems and Computers*, (Pacific Grove, Canada), pp. 1557–1561, 1-4 November 1999.

[634] Z. Liu and G. Giannakis, "Space-time coding with transmit antennas for multiple access regardless of frequency-selective multipath," in *Proc 1st Sensor Array and Multichannel SP Workshop*, (Boston, USA), 15-17 March 2000.

[635] T. Liew, J. Pliquett, B. Yeap, L.-L. Yang, and L. Hanzo, "Comparative study of space time block codes and various concatenated turbo coding schemes," in *PIMRC 2000*, (London, UK), pp. 741–745, 18-21 September 2000.

[636] T. Liew, J. Pliquett, B. Yeap, L.-L. Yang, and L. Hanzo, "Concatenated space time block codes and TCM, turbo TCM, convolutional as well as turbo codes," in *GLOBECOM 2000*, (San Francisco, USA), 27 Nov -1 Dec 2000.

[637] P. Robertson, E. Villebrun, and P. Höher, "A comparison of optimal and sub-optimal MAP decoding algorithms operating in the log domain," in *Proceedings of the International Conference on Communications*, (Seattle, United States), pp. 1009–1013, June 1995.

[638] G. Bauch, "Concatenation of space-time block codes and Turbo-TCM," in *Proceedings of IEEE International Conference on Communications*, (Vancouver, Canada), pp. 1202–1206, June 1999.

[639] G. Forney, "The Viterbi algorithm," *Proceedings of the IEEE*, vol. 61, pp. 268–278, March 1973.

[640] W. Webb and R. Steele, "Variable rate QAM for mobile radio," *IEEE Transactions on Communications*, vol. 43, pp. 2223–2230, July 1995.

[641] J. Torrance and L. Hanzo, "Performance upper bound of adaptive QAM in slow Rayleigh-fading environments," in *Proceedings of IEEE ICCS'96/ISPACS'96*, (Singapore), pp. 1653–1657, IEEE, 25–29 November 1996.

[642] H. Matsuako, S. Sampei, N. Morinaga, and Y. Kamio, "Adaptive modulation systems with variable coding rate concatenated code for high quality multi-media communication systems," in *Proceedings of IEEE Vehicular Technology Conference*, (Atlanta, USA), pp. 487–491, April 1996.

[643] T. Keller and L. Hanzo, "Adaptive multicarrier modulation: A convenient framework for time-frequency processing in wireless communications," *Proceedings of the IEEE*, vol. 88, pp. 611–642, May 2000.

[644] J. Torrance and L. Hanzo, "On the upper bound performance of adaptive QAM in a slow Rayleigh fading," *IEE Electronics Letters*, pp. 169–171, April 1996.

[645] Ö. F. Açikel and W. E. Ryan, "Punctured turbo-codes for BPSK/QPSK channels," *IEEE Transactions on Communications*, vol. 47, pp. 1315–1323, September 1999.

[646] L. Hanzo, "Bandwidth-efficient wireless multimedia communications," *Proceedings of the IEEE*, vol. 86, pp. 1342–1382, July 1998.

[647] S. Nanda, K. Balachandran, and S. Kumar, "Adaptation techniques in wireless packet data services," *IEEE Communications Magazine*, vol. 38, pp. 54–64, January 2000.

[648] T. Liew and L. Hanzo, "Space-time block coded adaptive modulation aided OFDM," in *Proceedings of GLOBECOM'2001*, (San Antonio, USA), pp. 136–140, IEEE, 26-29 November 2001.

[649] B. J. Choi and L. Hanzo, "Optimum mode-switching-assisted constant-power single- and multicarrier adaptive modulation," *IEEE Transactions on Vehicular Technology*, vol. 52, pp. 536–560, May 2003.

[650] T. Ottosson and A. Svensson, "On schemes for multirate support in DS-CDMA systems," *Wireless Personal Communications (Kluwer)*, vol. 6, pp. 265–287, March 1998.

[651] S. Spangenberg, D. Cruickshank, S. McLaughlin, G. Povey, and P. Grant, "Advanced multiuser detection techniques for downlink CDMA, version 2.0," tech. rep., Virtual Centre of Excellence in Mobile and Personal Communications Ltd (Mobile VCE), July 1999.

[652] S. Ramakrishna and J. Holtzman, "A comparison between single code and multiple code transmission schemes in a CDMA system," in *Proceedings of IEEE Vehicular Technology Conference (VTC'98)*, (Ottawa, Canada), pp. 791–795, IEEE, 18–21 May 1998.

[653] F. Adachi, K. Ohno, A. Higashi, T. Dohi, and Y. Okumura, "Coherent multicode DS-CDMA mobile radio access," *IEICE Transactions on Communications*, vol. E79-B, pp. 1316–1324, September 1996.

[654] T. Dohi, Y. Okumura, A. Higashi, K. Ohno, and F. Adachi, "Experiments on coherent multicode DS-CDMA," *IEICE Transactions on Communications*, vol. E79-B, pp. 1326–1332, September 1996.

[655] H. Schotten, H. Elders-Boll, and A. Busboom, "Adaptive multi-rate multi-code CDMA systems," in *Proceedings of the IEEE Vehicular Technology Conference (VTC)*, (Ottawa, Canada), pp. 782–785, 18–21 May 1998.

[656] M. Saquib and R. Yates, "Decorrelating detectors for a dual rate synchronous DS/CDMA channel," *Wireless Personal Communications (Kluwer)*, vol. 9, pp. 197–216, May 1999.

[657] A.-L. Johansson and A. Svensson, "Successive interference cancellation schemes in multi-rateDS/CDMA systems," in *Wireless Information Networks (Baltzer)*, pp. 265–279, 1996.

[658] A. Johansson and A. Svensson, "Multistage interference cancellation in multirate DS/CDMA on a mobile radio channel," in *Proceedings of the IEEE Vehicular Technology Conference (VTC)*, (Atlanta, GA, USA), pp. 666–670, 28 April–1 May 1996.

[659] M. Juntti, "Multiuser detector performance comparisons inmultirate CDMA systems," in *Proceedings of the IEEE Vehicular Technology Conference (VTC)*, (Ottawa, Canada), pp. 36–40, 18–21 May 1998.

[660] S. Kim, "Adaptive rate and power DS/CDMA communications in fading channels," *IEEE Communications Letters*, vol. 3, pp. 85–87, April 1999.

[661] S. Abeta, S. Sampei, and N. Morinaga, "Channel activation with adaptive coding rate and processing gain control for cellular DS/CDMA systems," in *Proceedings of IEEE VTC'96*, (Atlanta, GA, USA), pp. 1115–1119, IEEE, 28 April–1 May 1996.

[662] M. Hashimoto, S. Sampei, and N. Morinaga, "Forward and reverse link capacity enhancement of DS/CDMA cellular system using channel activation and soft power control techniques," in *Proceedings of the IEEE International Symposium on Personal, Indoor and Mobile Radio Communications (PIMRC)*, (Helsinki, Finland), pp. 246–250, 1–4 September 1997.

[663] S. Tateesh, S. Atungsiri, and A. Kondoz, "Link adaptive multi-rate coding verification system for CDMA mobile communications," in *Proceedings of the IEEE Global Telecommunications Conference (GLOBECOM)*, (London, UK), pp. 1969–1973, 18–22 November 1996.

[664] Y. Okumura and F. Adachi, "Variable-rate data transmission with blind rate detection for coherent DS-CDMA mobile radio," *IEICE Transactions on Communications*, vol. E81B, pp. 1365–1373, July 1998.

[665] J. Blogh, P. Cherriman, and L. Hanzo, "Adaptive beamforming assisted dynamic channel allocation," in *Proceeding of VTC'99 (Spring)*, (Houston, TX, USA), pp. 199–203, IEEE, 16–20 May 1999.

[666] I. Jeong and M. Nakagawa, "A novel transmission diversity system in TDD-CDMA," *IEICE Transactions on Communications*, vol. E81-B, pp. 1409–1416, July 1998.

[667] M. S. Alouini, X. Tand, and A. J. Goldsmith, "An adaptive modulation scheme for simultaneous voice and data transmission over fading channels," *IEEE Journal on Selected Areas in Communications*, vol. 17, pp. 837–850, May 1999.

[668] D. Yoon, K. Cho, and J. Lee, "Bit error probability of M-ary quadrature amplitude modulation," in *Proc. IEEE VTC 2000-Fall*, vol. 5, pp. 2422–2427, IEEE, September 2000.

[669] E. L. Kuan, C. H. Wong, and L. Hanzo, "Burst-by-burst adaptive joint-detection CDMA," in *Proc. of IEEE VTC'99 Fall*, vol. 2, (Amsterdam, Netherland), pp. 1628–1632, September 1999.

[670] M. Nakagami, "The M-distribution - A general formula of intensity distribution of rapid fading," in *Statistical Methods in Radio Wave Propagation* (W. C. Hoffman, ed.), pp. 3–36, Pergamon Press, 1960.

[671] I. S. Gradshteyn and I. M. Ryzhik, *Table of Integrals, Series and Products*. New York, USA: Academic Press, 1980.

[672] E. Kreyszig, *Advanced Engineering Mathematics*. John Wiley & Sons, Inc., 7th ed., 1993.

[673] J. Lu, K. B. Letaief, C. I. J. Chuang, and M. L. Lio, "M-PSK and M-QAM BER computation using signal-space concepts," *IEEE Transactions on Communications*, vol. 47, no. 2, pp. 181–184, 1999.

[674] T. Keller and L. Hanzo, "Adaptive modulation technique for duplex OFDM transmission," *IEEE Transactions on Vehicular Technology*, vol. 49, pp. 1893–1906, September 2000.

[675] G. S. G. Beveridge and R. S. Schechter, *Optimization: Theory and Practice*. McGraw-Hill, 1970.

[676] "COST 207: Digital land mobile radio communications, final report," tech. rep., Luxembourg, 1989.

[677] R. Price and E. Green Jr., "A communication technique for multipath channels," *Proceedings of the IRE*, vol. 46, pp. 555–570, March 1958.

[678] M. K. Simon and M. S. Alouini, *Digital Communication over Fading Channels: A Unified Approach to Performance Analysis*. John Wiley & Sons, Inc., 2000.

[679] C. Y. Wong, R. S. Cheng, K. B. Letaief, and R. D. Murch, "Multiuser OFDM with adaptive subcarrier, bit, and power allocation," *IEEE Journal on Selected Areas in Communications*, vol. 17, pp. 1747–1758, October 1999.

[680] A. Klein, G. Kaleh, and P. Baier, "Zero forcing and minimum mean square error equalization for multiuser detection in code division multiple access channels," *IEEE Transactions on Vehicular Technology*, vol. 45, pp. 276–287, May 1996.

[681] B. J. Choi, T. H. Liew, and L. Hanzo, "Concatenated space-time block coded and turbo coded symbol-by-symbol adaptive OFDM and multi-carrier CDMA systems," in *Proceedings of IEEE VTC 2001-Spring*, p. P.528, IEEE, May 2001.

[682] B. Vucetic, "An adaptive coding scheme for time-varying channels," *IEEE Transactions on Communications*, vol. 39, no. 5, pp. 653–663, 1991.

[683] S. M. Alamouti and S. Kallel, "Adaptive trellis-coded multiple-phased-shift keying Rayleigh fading channels," *IEEE Transactions on Communications*, vol. 42, pp. 2305–2341, June 1994.

[684] S. Chua and A. Goldsmith, "Adaptive coded modulation for fading channels," *IEEE Transactions on Communications*, vol. 46, pp. 595–602, May 1998.

[685] T. Keller, T. Liew, and L. Hanzo, "Adaptive rate RRNS coded OFDM transmission for mobile communication channels," in *Proceedings of VTC 2000 Spring*, (Tokyo, Japan), pp. 230–234, 15-18 May 2000.

[686] T. Keller, T. H. Liew, and L. Hanzo, "Adaptive redundant residue number system coded multicarrier modulation," *IEEE Journal on Selected Areas in Communications*, vol. 18, pp. 1292–1301, November 2000.

[687] T. Liew, C. Wong, and L. Hanzo, "Block turbo coded burst-by-burst adaptive modems," in *Proceedings of Microcoll'99*, (Budapest, Hungary), pp. 59–62, 21-24 March 1999.

[688] C. Wong, T. Liew, and L. Hanzo, "Turbo coded burst by burst adaptive wideband modulation with blind modem mode detection," in *ACTS Mobile Communications Summit*, (Sorrento, Italy), pp. 303–308, 8-11 June 1999.

[689] C. Berrou and A. Glavieux, "Near optimum error correcting coding and decoding: Turbo codes," *IEEE Transactions on Communications*, vol. 44, pp. 1261–1271, October 1996.

[690] P. Jung and J. Blanz, "Joint detection with coherent receiver antenna diversity in CDMA mobile radio systems," *IEEE Transactions on Vehicular Technology*, vol. 44, pp. 76–88, February 1995.

[691] J. Wozencraft and B. Reiffen, *Sequential Decoding*. Cambridge, MA, USA: MIT Press, 1961.

[692] R. Gallager, *Information Theory and Reliable Communication*. John Wiley and Sons, 1968.

[693] S. G. Wilson, *Digital Modulation and Coding*. Englewood Cliffs, NJ, USA: Prentice-Hall International Editions, 1996.

[694] M. Campanella and G. Mamola, "On the channel capacity for constant envelope signals with effective bandwidth constraint," *IEEE Transactions on Communications*, vol. 38, pp. 1164–1172, August 1990.

[695] P. E. McIllree, "Channel capacity calculations for M-ary n-dimensional signal sets," M.Eng thesis, The University of South Australia, 1995.

[696] P. E. McIllree, "Calculation of channel capacity for M-ary digital modulation signal sets," in *IEEE Singapore International Conference on Information Engineering*, (Singapore), pp. 639–643, September 1993.

[697] G. J. Foschini and M. J. Gans, "On limits of wireless communications in a fading environment when using multiple antennas," *Wireless Personal Communications*, vol. 6, pp. 311–335, March 1998.

[698] I. S. Reed and R. A. Scholtz, "N-orthogonal phase modulated codes," *IEEE Transactions on Information Theory*, vol. 12, pp. 388–395, July 1966.

[699] W. C. Lindsey, M. K. Simon, "L-orthogonal signal transmission and detection," *IEEE Transactions on Communications*, vol. COM-20, pp. 953–960, October 1972.

[700] A. Viterbi and J. Omura, *Principles of Digital Communication and Coding*. New York, USA: McGraw-Hill, 1979.

[701] C. Schlegel and D. J. Costello, "Bandwidth Efficient Coding for Fading Channels: Code Construction and Performance Analysis," *IEEE Journal on Selected Areas in Communications*, vol. 7, pp. 1356–1368, December 1989.

[702] S. Al-Semari and T. Fuja, "Performance analysis of coherent TCM systems with diversity reception in slow Rayleigh fading," *IEEE Transactions on Vehicular Technology*, vol. 48, pp. 198–212, January 1999.

[703] J. Ventura-Traveset, G. Caire, E. Biglieri and G. Taricco, "Impact of diversity reception on fading channels with coded modulation–part i: Coherent detection," *IEEE Transactions on Communications*, vol. 45, pp. 563–572, May 1997.

[704] D. Divsalar and M.K. Simon, "Trellis coded modulation for 4800-9600 bits/s transmission over a fading mobile satellite channel," *IEEE Journal on Selected Areas in Communications*, vol. 5, pp. 162–175, February 1987.

[705] R. E. Blahut, *Principles and Practice of Information Theory*. Reading, MA, USA: Addison-Wesley, 1987.

[706] L. Hanzo and L-L. Yang, E. L. Kuan and K. Yen, *Single- and Multi-Carrier CDMA*. New York, USA: John Wiley, IEEE Press, 2003.

[707] M. Kanefsky, *Communication Techniques for Digital and Analog Signals*. New York, USA: John Wiley, 1987.

[708] D. Divsalar and M. K. Simon, "The design of trellis coded MPSK for fading channel: Performance criteria," *IEEE Transactions on Communications*, vol. 36, pp. 1004–1012, September 1988.

[709] P. Robertson, T. Wörz, "Bandwidth-efficient turbo trellis-coded modulation using punctured component codes," *IEEE Journal on Selected Areas in Communications*, vol. 16, pp. 206–218, February 1998.

[710] X. Li and J.A. Ritcey, "Bit-interleaved coded modulation with iterative decoding," *IEEE Communications Letters*, vol. 1, November 1997.

[711] X. Li and J.A. Ritcey, "Bit-interleaved coded modulation with iterative decoding — Approaching turbo-TCM performance without code concatenation," in *Proceedings of CISS 1998*, (Princeton University, USA), March 1998.

[712] S. X. Ng, T. H. Liew, L-L. Yang and L. Hanzo, "Comparative study of TCM, TTCM, BICM and BICM-ID schemes," in *IEEE Vehicular Technology Conference*, (Rhodes, Greece), pp. 2450–2454, May 2001.

[757] M. Gertsman and J. Lodge, "Symbol-by-symbol MAP demodulation of CPM and PSK signals on Rayleigh flat-fading channels," *IEEE Transactions on Communications*, vol. 45, pp. 788–799, July 1997.

[758] D. Raphaeli and Y. Zarai, "Combined turbo equalization and turbo decoding," *IEEE Communications Letters*, vol. 2, pp. 107–109, April 1998.

[759] A. Knickenberg, B. L. Yeap, J. Hamorsky, M. Breiling, and L. Hanzo, "Non-iterative joint channel equalisation and channel decoding," in *Proceedings of Globecom'99*, (Rio de Janeiro, Brazil), pp. 442–446, 5-9 December 1999.

[760] A. Glavieux, C. Laot, and J. Labat, "Turbo equalization over a frequency selective channel," in *Proceedings of the International Symposium on Turbo Codes*, (Brest, France), pp. 96–102, 3-5 September 1997.

[761] M. Yee and L. Hanzo, "Multi-level radial basis function network based equalisers for Rayleigh channel," in *Proceeding of VTC'99 (Spring)*, (Houston, TX, USA), pp. 707–711, IEEE, 16–20 May 1999.

[762] S. Chen, B. Mulgrew, and S. McLaughlin, "Adaptive Bayesian equalizer with decision feedback," *IEEE Transactions on Signal Processing*, vol. 41, pp. 2918–2927, September 1993.

[763] E.-S. Chng, H. Yang, and W. Skarbek, "Reduced complexity implementation of Bayesian equaliser using local RBF network for channel equalisation problem," *Electronics Letters*, vol. 32, pp. 17–19, January 1996.

[764] M. S. Yee, T. H. Liew and L. Hanzo, "Burst-by-burst adaptive turbo-coded radial basis function-assisted decision feedback equalization," *IEEE Transactions on Communications*, vol. 49, pp. 1935–1945, November 2001.

[765] B. L. Yeap, C. H. Wong, and L. Hanzo, "Reduced complexity in-phase/quadrature-phase turbo equalisation with iterative channel estimation," in *IEEE International Communications Conference 2001*, (Helsinki, Finland), pp. 1395–1399, 11-15 June 2001.

[766] E.L. Kuan and C.H. Wong and L. Hanzo, "Comparative study of joint-detection and interference cancellation based burst-by-burst adaptive CDMA schemes," in *Proceedings of the IEEE Vehicular Technology Conference (VTC Fall)*, (Amsterdam, The Netherlands), pp. 653–657, 19–22 September 1999.

[767] J. R. Foerster and L. B. Milstein, "Coded modulation for a coherent DS-CDMA system employing an mmse receiver in a fading channel," *IEEE Transactions on Communications*, vol. 48, pp. 1909–1918, November 2000.

[768] D. E. Goldberg, *Genetic Algorithms in Search, Optimization, and Machine Learning*. Reading, Massachusetts: Addison-Wesley, 1989.

[769] K. Yen and L. Hanzo, "Hybrid genetic algorithm based multi-user detection schemes for synchronous CDMA systems," in *submitted to the IEEE Vehicular Technology Conference (VTC)*, (Tokyo, Japan), 2000.

[770] K. Yen and L. Hanzo, "Genetic algorithm assisted joint multiuser symbol detection and fading channel estimation for synchronous CDMA systems," *IEEE Journal on Selected Areas in Communications*, vol. 19, pp. 985–998, June 2001.

[771] S. Abedi and R. Tafazolli, "Genetically modified multiuser detection for code division multiple access systems," *IEEE Journal on Selected Areas in Communications*, vol. 20, pp. 463–473, February 2002.

[772] A. Whalen, *Detection of Signals in Noise*. New York, USA: Academic Press, 1971.

[773] E.A. Lee and D.G. Messerschmitt, *Digital Communication*. Dordrecht: Kluwer Academic Publishers, 1988.

[774] G. Golub and C. van Loan, *Matrix Computations*. North Oxford Academic, 1983.

[775] T. Ojanperä, A. Klein, and P.-O. Anderson, "FRAMES multiple access for UMTS," *IEE Colloquium (Digest)*, pp. 7/1–7/8, May 1997.

[776] V. K. N. Lau and M. Macleod, "Variable-rate adaptive trellis coded QAM for flat-fading channels," *IEEE Transactions on Communications*, vol. 49, pp. 1550–1560, September 2001.

[777] T.S. Lee and T.C. Tsai, "A partially adaptive CDMA interference canceller for multipath channels," *IEEE Vehicular Technology Conference*, vol. 2, pp. 917–921, May 2000.

[778] S. Kazi and L. Lucke, "A convolutionally-coded adaptive CDMA receiver architecture," *Signals, Systems and Computers. Thirty-Second Asilomar Conference*, vol. 2, pp. 1199–1203, 1998.

[779] S.W. Lei and V.K.N. Lau, "Adaptive interleaving for OFDM in TDD system," *IEE Proceedings on Communications*, vol. 148, no. 2, pp. 77–80, 2001.

[780] Special Mobile Group of ETSI, "UMTS: Selection procedures for the choice of radio transmission technologies of the UMTS," tech. rep., European Telecommunications Standard Institute (ETSI), France, 1998.

[781] S. Verdú, "Minimum probability of error for asynchronous Gaussian multiple-access channel," *IEEE Transactions on Communications*, vol. 32, pp. 85–96, January 1986.

[782] S. Moshavi, "Multi-user detection for DS-CDMA communications," *IEEE Communications Magazine*, vol. 34, pp. 124–136, October 1996.

[783] M. Mitchell, *An Introduction to Genetic Algorithms*. Cambridge, Massachusetts: MIT Press, 1996.

[784] L. J. Eshelman and J. D. Schaffer, "Preventing premature convergence in genetic algorithms by preventing incest," in *Proceedings of the Fourth International Conference on Genetic Algorithms* (R. K. Belew and L. B. Booker, eds.), (California, USA), pp. 115–122, Morgan Kaufmann, 1991.

[785] M. J. Juntti, T. Schlösser, and J. O. Lilleberg, "Genetic algorithms for multiuser detection in synchronous CDMA," in *IEEE International Symposium on Information Theory – ISIT'97*, (Ulm, Germany), p. 492, 1997.

[786] G. Syswerda, "Uniform crossover in genetic algorithms," in *Proceedings of the Third International Conference on Genetic Algorithms* (J. D. Schaffer, ed.), (California, USA), pp. 2–9, Morgan Kaufmann, 1989.

[787] W. Spears and K. De Jong, *Foundations of Genetic Algorithms*, ch. An Analysis of Multi-Point Crossover, pp. 301–315. California, USA: G. Rawlins, ed., Morgan Kaufmann, 1991.

[788] J. Anderson and S. Mohan, "Sequential coding algorithms: a survey and cost analysis," *IEEE Transactions on Communications*, vol. 32, pp. 169–176, February 1984.

[789] T. Hashimoto, "A list-type reduced-constraint generalization of the Viterbi algorithm," *IEEE Transactions on Information Theory*, vol. 33, pp. 866–876, November 1987.

[790] S. J. Simmons, "Breadth-first trellis decoding with adaptive effort," *IEEE Transactions on Communications*, vol. 38, pp. 3–12, January 1990.

[791] L. Rasmussen, T. Lim, and T. Aulin, "Breadth-first maximum likelihood detection in multiuser CDMA," *IEEE Transactions on Communications*, vol. 45, pp. 1176–1178, October 1997.

[792] P. Balaban, J. Salz, "Optimum diversity combining and equalization in digital data transmission with application to cellular mobile radio – Part I: Theoretical considerations," *IEEE Transactions on Communications*, vol. 40(5), pp. 885–894, 1992.

[793] A. Wittneben, "Base station modulation diversity for digital SIMULCAST," in *Proceedings of IEEE Vehicular Technology Conference*, pp. 505–511, May 1993.

[794] S. Al-Semari and T. Fuja, "I-Q TCM: Reliable communication over the Rayleigh fading channel close to the cuttoff rate," *IEEE Transactions on Information Theory*, vol. 43, pp. 250–262, January 1997.

[795] B. D. Jelicic and S. Roy, "Design of trellis coded QAM for flat fading and AWGN channels," *IEEE Transactions on Vehicular Technology*, vol. 44, pp. 192–201, February 1994.

[796] G. Klang, A. F. Naguib, "Transmit diversity based on space-time block codes in frequency selective Rayleigh fading DS-CDMA systems," *IEEE Vehicular Technology Conference*, pp. 264–268, Spring 2000.

[797] L.-L. Yang and L. Hanzo, "Performance of wideband CDMA using adaptive space-time spreading over multipath Nakagami fading channels," *IEEE Vehicular Technology Conference*, pp. 615–619, May 2002.

[798] L. Miller and J. Lee, *CDMA Systems Engineering Handbook*. London, UK: Artech House, 1998.

[799] J. Mar and H. Chen, "Performance analysis of cellular CDMA networks over frequency-selective fading channel," *IEEE Transactions on Vehicular Technology*, vol. 47, pp. 1234–1244, November 1998.

[800] M. C. Reed, C. B. Schlegel, P. D. Alexander, and J. A. Asenstorfer, "Iterative multiuser detection for CDMA with FEC: Near single user performance," *IEEE Transactions on Communication*, pp. 1693–1699, December 1998.

[801] R. Prasad and S. Hara, "Overview of multi-carrier CDMA," in *Proceedings of the IEEE International Symposium on Spread Spectrum Techniques and Applications (ISSSTA)*, (Mainz, Germany), pp. 107–114, 22–25 September 1996.

[802] C. Tidestav, A. Ahlén and M. Sternad, "Realiazable MIMO decision feedback equalizer: structure and design," *IEEE Transactions on Signal Processing*, vol. 49, pp. 121–133, January 2001.

[803] R. Gallager, "Low-density parity-check codes," *IEEE Transactions on Information Theory*, pp. 21–28, 1962.

[804] ETSI, *Digital Video Broadcasting (DVB): Framing Structure, Channel Coding and Modulation for Cable Systems*, December 1997. EN 300 429 V1.2.1.

[805] ETSI, *Digital Video Broadcasting (DVB): Framing Structure, Channel Coding and Modulation for 11/12 GHz Satellite Services*, August 1997. EN 300 421 V1.1.2.

[806] A. Michelson and A. Levesque, *Error Control Techniques for Digital Communication*. New York, USA: Wiley-Interscience, 1985.

[807] S. O'Leary and D. Priestly, "Mobile broadcasting of DVB-T signals," *IEEE Transactions on Broadcasting*, vol. 44, pp. 346–352, September 1998.

[808] W.-C. Lee, H.-M. Park, K.-J. Kang, and K.-B. Kim, "Performance analysis of Viterbi decoder using channel state information in COFDM system," *IEEE Transactions on Broadcasting*, vol. 44, pp. 488–496, December 1998.

[809] S. O'Leary, "Hierarchical transmission and COFDM systems," *IEEE Transactions on Broadcasting*, vol. 43, pp. 166–174, June 1997.

[810] L. Thibault and M. Le, "Performance evaluation of COFDM for digital audio broadcasting Part I: parametric study," *IEEE Transactions on Broadcasting*, vol. 43, pp. 64–75, March 1997.

[811] P. Shelswell, "The COFDM modulation system: the heart of digital audio broadcasting," *Electronics & Communication Engineering Journal*, vol. 7, pp. 127–136, June 1995.

[812] S. Wicker, *Error Control Systems for Digital Communication and Storage*. Englewood Cliffs, NJ, USA: Prentice-Hall, 1994.

[813] A. Barbulescu and S. Pietrobon, "Interleaver design for turbo codes," *IEE Electronics Letters*, pp. 2107–2108, December 1994.

[814] C. Lee, T. Keller, and L. Hanzo, "Turbo-coded hierarchical and non-hierarchical mobile digital video broadcasting," *IEEE Transaction on Broadcasting*, March 2000.

[815] B. Haskell, A. Puri, and A. Netravali, *Digital Video: An Introduction to MPEG-2*. Digital Multimedia Standards Series, London, UK: Chapman and Hall, 1997.

[816] G. Reali, G. Baruffa, S. Cacopardi, and F. Frescura, "Enhancing satellite broadcasting services using multiresolution modulations," *IEEE Transactions on Broadcasting*, vol. 44, pp. 497–506, December 1998.

[817] Y. Hsu, Y. Chen, C. Huang, and M. Sun, "MPEG-2 spatial scalable coding and transport stream error concealment for satellite TV broadcasting using Ka-band," *IEEE Transactions on Broadcasting*, vol. 44, pp. 77–86, March 1998.

[818] L. Atzori, F. D. Natale, M. D. Gregario, and D. Giusto, "Multimedia information broadcasting using digital TV channels," *IEEE Transactions on Broadcasting*, vol. 43, pp. 383–392, December 1997.

[819] W. Sohn, O. Kwon, and J. Chae, "Digital DBS system design and implementation for TV and data broadcasting using Koreasat," *IEEE Transactions on Broadcasting*, vol. 44, pp. 316–323, September 1998.

[820] J. Griffiths, *Radio Wave Propagation and Antennas : An Introduction*. Englewood Cliffs, NJ, USA: Prentice-Hall, 1987.

[821] M. Karaliopoulos and F.-N. Pavlidou, "Modelling the land mobile satellite channel: a review," *Electronics and Communication Engineering Journal*, vol. 11, pp. 235–248, October 1999.

[822] J. Goldhirsh and W. Vogel, "Mobile satellite system fade statistics for shadowing and multipath from roadside trees at UHF and L-band," *IEEE Transactions on Antennas and Propagation*, vol. 37, pp. 489–498, April 1989.

[823] W. Vogel and J. Goldhirsh, "Multipath fading at L band for low elevation angle, land mobile satellite scenarios," *IEEE Journal on Selected Areas in Communications*, vol. 13, pp. 197–204, February 1995.

[824] W. Vogel and G. Torrence, "Propagation measurements for satellite radio reception inside buildings," *IEEE Transactions on Antennas and Propagation*, vol. 41, pp. 954–961, July 1993.

[825] W. Vogel and U. Hong, "Measurement and modelling of land mobile satellite propagation at UHF and L-band," *IEEE Transactions on Antennas and Propagation*, vol. 36, pp. 707–719, May 1988.

[826] S. Saunders, C. Tzaras, and B. Evans, "Physical statistical propagation model for mobile satellite channel," tech. rep., European Commission, 1998.

[827] S. Saunders, *Antennas and Propagation for Wireless Communication Systems Concept and Design*. New York, USA: John Wiley and Sons, 1999.

[828] H. Gharavi and L. Hanzo, eds., *Proceedings of the IEEE*, vol. 87, October 1999.

[829] F. Adachi, "Error rate analysis of differentially encoded and detected 16APSK under Rician fading," *IEEE Transactions on Vehicular Technology*, vol. 45, pp. 1–12, February 1996.

[830] Y. C. Chow, A. R. Nix, and J. P. McGeehan, "Diversity improvement for 16-DAPSK in Rayleigh fading channel," *Electronics Letters*, vol. 29, pp. 387–389, February 1993.

[831] Y. C. Chow, A. R. Nix, and J. P. McGeehan, "Error analysis for circular 16-DAPSK in frquency-selective Rayleigh fading channels with diversity reception," *Electronics Letters*, vol. 30, pp. 2006–2007, November 1994.

[832] C. M. Lo and W. H. Lam, "Performance analysis of bandwidth efficient coherent modulation schemes with L-fold MRC and SC in Nakagami-m fading channels," in *Proceedings of IEEE PIMRC 2000*, vol. 1, pp. 572–576, September 2000.

[833] S. Benedetto, E. Biglierri, and V. Castellani, *Digital Transmission Theory*. Englewood Cliffs, NJ: Prentice-Hall, 1987.

Index

Author Index